30°E
60°E
90°E
120°E
150°E
180°
ARCTIC OCEAN
Barents Sea
FINLAND
SWEDEN
Helsinki
St. Petersburg
RUSSIA
Aldan
Sea of Okhotsk
Bering Sea
60°N
Moscow
Novosibirsk
Minsk
Kiev
Budapest
KAZAKHSTAN
Ulan Bator
MONGOLIA
Harbin
GEORGIA
Urumqi
Shenyang
Istanbul
Tashkent
KYRGYZSTAN
Beijing
N. KOREA
Sea of Japan
JAPAN
Rome
Ankara
ARMENIA
AZERBAIJAN
UZBEKISTAN
CHINA
Tianjin
Seoul
Athens
TURKEY
TURKMENISTAN
TAJIKISTAN
Tai'an
Qingdao
S. KOREA
Tokyo
Yokohama
Osaka
Mediterranean Sea
CYPRUS
SYRIA
LEBANON
ISRAEL
Baghdad
Tehran
Mashhad
Kabul
Islamabad
Xian
Zaozhuang
IRAQ
IRAN
AFGHANISTAN
Lhasa
Chengdu
Wuhan
Shanghai
Tripoli
Alexandria
Cairo
JORDAN
Lahore
BHUTAN
Chongqing
Ningbo
East China Sea
30°N
LIBYA
EGYPT
KUWAIT
BAHRAIN
PAKISTAN
New Delhi
NEPAL
Guangzhou (Canton)
Taipei
SAUDI ARABIA
UAE
QATAR
Karachi
INDIA
BANGLADESH
Kolkata (Calcutta)
Dhaka
TAIWAN
Ahmadabad
MYANMAR (BURMA)
Hong Kong (Xianggang)
Hanoi
NORTH PACIFIC OCEAN
Red Sea
OMAN
LAOS
Mumbai (Bombay)
Hyderabad
Bay of Bengal
Khartoum
ERITREA
YEMEN
VIETNAM
CHAD
Arabian Sea
THAILAND
PHILIPPINES
SUDAN
South China Sea
Bangkok
CAMBODIA
Manila
Philippine Sea
N'Djamena
Bangalore
Chennai (Madras)
DJIBOUTI
Addis Ababa
SRI LANKA
Ho Chi Minh City
CENTRAL AFRICAN REPUBLIC
PALAU
FEDERATED STATES OF MICRONESIA
MARSHALL ISLANDS
CAMEROON
ETHIOPIA
BRUNEI
Bangui
SOMALIA
Kuala Lumpur
MALAYSIA
CONGO
Mogadishu
MALDIVES
SINGAPORE
NAURU
KIRIBATI
0°
UGANDA
KENYA
RWANDA
Nairobi
DEM. REP. OF CONGO
BURUNDI
Kinshasa
TANZANIA
Dar es Salaam
SEYCHELLES
Jakarta
Surabaya
INDONESIA
PAPUA NEW GUINEA
SOLOMON ISLANDS
Bandung
Port Moresby
TUVALU
Luanda
TIMOR-LESTE
ANGOLA
MALAWI
COMOROS
Timor Sea
Darwin
SAMOA
ZAMBIA
INDIAN OCEAN
Coral Sea
Harare
MOZAMBIQUE
Mozambique Channel
VANUATU
FIJI
NAMIBIA
ZIMBABWE
MADAGASCAR
Antananarivo
TONGA
MAURITIUS
BOTSWANA
Gaborone
Maputo
AUSTRALIA
Pretoria
SWAZILAND
30°S
SOUTH AFRICA
LESOTHO
Perth
Great Australian Bight
Canberra
Sydney
Melbourne
Tasman Sea
Auckland
Wellington
NEW ZEALAND
60°S
ANTARCTICA
FINLAND
NORWAY
Helsinki
Oslo
Stockholm
SWEDEN
ESTONIA
St. Petersburg
LATVIA
DENMARK
Copenhagen
LITHUANIA
Minsk
UNITED KINGDOM
IRELAND
NETH.
Berlin
Warsaw
BELARUS
London
GERMANY
POLAND
BELGIUM
CZECH.
UKRAINE
Kiev
LUX.
SLOVAKIA
Paris
Vienna
AUS.
SWITZ.
HUNGARY
MOL.
FRANCE
SL.
CROATIA
ROMANIA
B&H
SERB.
BULGARIA
SPAIN
ITALY
MONT.
KO.
MAC.
Istanbul
PORTUGAL
Rome
ALB.
Madrid
GREECE
Ankara
TURKEY
Lisbon
Athens
Algiers
ALGERIA
TUNISIA

WORLD REGIONAL GEOGRAPHY CONCEPTS

THIRD EDITION

LYDIA MIHELIČ PULSIPHER
Geography Professor Emeritus,
University of Tennessee

ALEX A. PULSIPHER
Geographer and Independent Scholar

with the assistance of
CONRAD "MAC" GOODWIN
Anthropologist/Archaeologist and
Independent Scholar

W. H. Freeman and Company
A Macmillan Education Company
New York

**To Allan G. Pulsipher, who has been an excellent father and grandfather.
And to ideas that are ahead of their time. May we learn to catch up.**

Publisher: Steven Rigolosi
Developmental Editor: Debra Ginsberg
Senior Project Editor: Vivien Weiss
Marketing Manager: Taryn Burns
Marketing Assistant: Samantha Zimbler
Cover and Text Designer: Blake Logan
Art Manager: Matthew McAdams
Assistant and Supplements Editor: Stephanie Ellis
Maps: University of Tennessee, Cartographic Services Laboratory,
    Will Fontanez, Director; Maps.com
Photo Editors: Jennifer MacMillan, Nicholas Ciani
Photo Researcher: Alex Pulsipher
Production Manager: Susan Wein
Composition: Aptara®, Inc.
Printing and Binding: RR Donnelley
Front cover and title page (main image): Sami's Photography/Flickr Select/Getty Images
Front cover and title page (insets, top to bottom): John Lukuwi/AFP/Getty Images;
    Tang Ming Tung/Flickr RM/Getty Images; M. Lourdes Siracuza Cappi/Flickr Open/
    Getty Images; Khaled Desouki/AFP/Getty Images

Library of Congress Control Number: 2014942091
ISBN-13: 978-1-4641-1071-9
ISBN-10: 1-4641-1071-9

© 2015, 2012, 2009 by W. H. Freeman and Company. All rights reserved.
Printed in the United States of America
Second printing
W. H. Freeman and Company
41 Madison Avenue
New York, NY 10010
Houndmills, Basingstoke RG21 6XS, England

www.whfreeman.com/geography

ABOUT THE AUTHORS

Lydia Mihelič Pulsipher is a cultural-historical geographer who studies the landscapes of ordinary people through the lenses of archaeology, geography, and ethnography. She has contributed to several geography-related exhibits at the Smithsonian Museum of Natural History in Washington, DC, including "Seeds of Change," which featured the research she and Conrad Goodwin did in the eastern Caribbean. Lydia Pulsipher has ongoing research projects in the eastern Caribbean (historical archaeology) and in central Europe, where she is interested in various aspects of the post-Communist transition. Her graduate students have studied human ecology in the Caribbean and border issues and matters of national identity and exclusion in several central European countries. She has taught cultural, gender, European, North American, and Mesoamerican geography at the University of Tennessee at Knoxville since 1980; through her research, she has given many students their first experience in fieldwork abroad. Previously she taught at Hunter College and Dartmouth College. She received her BA from Macalester College, her MA from Tulane University, and her PhD from Southern Illinois University. For relaxation, she works in her gardens, makes jam, and bakes rhubarb pies.

Alex A. Pulsipher is an independent scholar in Knoxville, Tennessee, who has conducted research on vulnerability to climate change; sustainable communities; and the diffusion of green technologies in the United States. In the early 1990s, while a student at Wesleyan University in Connecticut, Alex spent time in South Asia working for a sustainable development research center. He then completed his BA at Wesleyan, writing his undergraduate thesis on the history of Hindu nationalism. Beginning in 1995, Alex contributed to the research and writing of the first edition of *World Regional Geography* with Lydia Pulsipher. In 1999 and 2000, he traveled to South America, Southeast Asia, and South Asia, where he collected information for the second edition of the text and for the Web site. In 2000 and 2001, he wrote and designed maps for the second edition. He participated in the writing of the fourth edition and in writing, restructuring the content of, and creating photo essays and maps for the fifth and sixth editions. Alex worked extensively on the first, second, and third editions of *World Regional Geography Concepts*, reorganizing content, writing, and researching photos. He has a master's degree in geography from Clark University in Worcester, Massachusetts.

Conrad McCall "Mac" Goodwin has assisted in the writing of *World Regional Geography* and the *Concepts* version in many ways. Mac, Lydia's husband, is an anthropologist and historical archaeologist with a BA in anthropology from the University of California, Santa Barbara; an MA in historical archaeology from the College of William and Mary; and a PhD in archaeology from Boston University. He specializes in sites created during the European colonial era in North America, the Caribbean, and the Pacific. He has particular expertise in the archaeology of agricultural systems, gardens, domestic landscapes, and urban spaces. In addition to work in archaeology and on the textbook, for the past 10 years he has been conducting research on wines and winemaking in Slovenia, and delivering papers on these topics at professional geography meetings. For relaxation, Mac works in his organic garden, builds stone walls (including a pizza oven), and is a slow-food chef.

BRIEF CONTENTS

	Preface	xiii
CHAPTER 1	Geography: An Exploration of Connections	1
CHAPTER 2	North America	58
CHAPTER 3	Middle and South America	108
CHAPTER 4	Europe	150
CHAPTER 5	Russia and the Post-Soviet States	192
CHAPTER 6	North Africa and Southwest Asia	230
CHAPTER 7	Sub-Saharan Africa	274
CHAPTER 8	South Asia	316
CHAPTER 9	East Asia	360
CHAPTER 10	Southeast Asia	404
CHAPTER 11	Oceania: Australia, New Zealand, and the Pacific	446
EPILOGUE	Antarctica	481
	Glossary	G-1
	Text Sources and Credits	TC-1
	Index	I-1

CONTENTS

CHAPTER 1
Geography: An Exploration of Connections — 1

- Where Is It? Why Is It There? Why Does It Matter? — 2
- What Is Geography? — 3
  - Physical and Human Geography 3
- Geographers' Visual Tools — 4
  - Understanding Maps 4 • Geographic Information Science (GISc) 7

THE DETECTIVE WORK OF PHOTO INTERPRETATION — 8

- The Region as a Concept — 9
  - Regions 9
- Thematic Concepts and Geographic Insights in This Book — 12

ENVIRONMENT — 12

- Human Impact on the Biosphere — 12
- Physical Geography — 13
  - Landforms: The Sculpting of the Earth 13 • Plate Tectonics 13 • Landscape Processes 13 • Climate 14 • Climate Regions 14 • Temperature and Air Pressure 16 • Precipitation 17
- Global Climate Change — 21
  - Drivers of Global Climate Change 21 • Climate-Change Impacts 23 • Vulnerability to Climate Change 23 • Responding to Climate Change 24
- Water — 26
  - Calculating Water Use per Capita 26 • Who Owns Water? Who Gets Access to It? 27 • Water Quality 27 • Water and Urbanization 28
- Food — 28
  - Agriculture: Early Human Impacts on the Physical Environment 28 • Agriculture and Its Consequences 29 • Modern Food Production and Food Security 29
- Globalization — 31
  - What Is the Global Economy? 31 • Workers in the Global Economy 32 • The Debate over Globalization and Free Trade 33
- Development — 34
  - Measuring Economic Development 35 • Geographic Patterns of Human Well-Being 36 • Sustainable Development and Political Ecology 36

POWER AND POLITICS — 38

- The Expansion of Political Freedoms — 38
  - What Factors Encourage the Expansion of Political Freedoms? 39
- Geopolitics — 39
  - International Cooperation 40

URBANIZATION — 42

- Why Are Cities Growing? — 42
  - Patterns of Urban Growth 43

POPULATION — 43

- Global Patterns of Population Growth — 43
  - Local Variations in Population Density and Growth 44
- Age and Sex Structures — 47
- Population Growth Rates and Wealth — 47
- Gender — 49
  - Gender Roles 49 • Gender Issues 50
- Sociocultural Issues — 51
  - Ethnicity and Culture: Slippery Concepts 51 • Values 52 • Religion and Belief Systems 52 • Language 53 • Race 53

PHOTO ESSAYS
- 1.3 Understanding Maps — 5
- 1.4 The Detective Work of Photo Interpretation — 8
- 1.7 Human Impacts on the Biosphere — 14
- 1.11 Climate Regions of the World — 18
- 1.15 Vulnerability to Climate Change — 24
- 1.22 Power and Politics — 40
- 1.24 Urbanization — 44
- 1.30 Major Religions of the World — 54

CHAPTER 2
North America — 58

- The North American Region — 60

PHYSICAL GEOGRAPHY AND ENVIRONMENTAL ISSUES — 61

- Landforms — 61
- Climate — 62
- Environment — 64
  - Climate Change and Air Pollution 64 • Water Resource Depletion, Pollution, and Marketization 64 • Loss of Habitat for Plants and Animals 68 • Oil Drilling 68 • Logging 68 • Coal Mining and Use 70 • Urbanization and Habitat Loss 70

HUMAN GEOGRAPHY — 70

- Human Patterns over Time — 70
  - The Peopling of North America 71 • The European Transformation 71 • Expansion West of the Mississippi and Great Lakes 73 • European Settlement and Native Americans 74 • The Changing Regional Composition of North America 75

viii Contents

Globalization and Development — 75
The Decline in Manufacturing Employment 75 • Growth of the Service Sector 76 • Globalization and Free Trade 76 • Repercussions of the Global Economic Downturn Beginning in 2007 78 • Interdependencies 79 • Women in the Economy 80 • North America's Changing Food-Production Systems 80 • Changing Transportation Networks and the North American Economy 83

Political Issues — 83
The Expansion of Political Freedoms in the United States and Canada 83 • Debt and Politics in the United States 84 • The United States and Canada Abroad 86 • Challenges to the United States' Global Power 86

Relationships Between Canada and the United States — 87
Asymmetries 87 • Similarities 87 • Democratic Systems of Government: Shared Ideals, Different Trajectories 87 • The Social Safety Net: Canadian and U.S. Approaches 89 • Gender in National Politics 90

Urbanization — 90

Population and Gender — 94
Gender and Fertility 95 • Aging in North America 95 • Population Distribution 96

Sociocultural Issues — 98
Immigration and Diversity 98 • Race and Ethnicity in North America 101 • Religion 103 • The American Family 103

PHOTO ESSAYS
2.4	Climates of North America	63
2.5	Vulnerability to Climate Change in North America	65
2.8	Human Impacts on the Biosphere in North America	69
2.9	Visual History of North America	72
2.17	Power and Politics in North America	84
2.21	Urbanization in North America	92

CHAPTER 3
Middle and South America — 108

The Middle and South American Region — 110

PHYSICAL GEOGRAPHY AND ENVIRONMENTAL ISSUES — 112

Physical Patterns — 112
Landforms 112 • Climate 114

Environmental Issues — 117
Tropical Forests, Climate Change, and Globalization 117 • Environmental Protection and Economic Development 118 • The Water Crisis 118

HUMAN GEOGRAPHY — 122

Human Patterns over Time — 122
The Peopling of Middle and South America 122 • European Conquest 123 • A Global Exchange of Crops and Animals 124 • The Legacy of Underdevelopment 124

Globalization and Development — 126

Economic Inequality and Income Disparity — 127
Phases of Economic Development 127 • The Current Era of Foreign Direct Investment 131 • The Informal Economy 131 • Regional Trade and Trade Agreements 132 • Food Production and Development 132

Power and Politics — 134
The Drug Trade and Conflict 135 • Foreign Involvement in the Region's Politics 136 • Recent Revolutionary Movements 138

Urbanization — 139
Migration and Urbanization 139

Population and Gender — 143
Slowing Population Growth 143 • Population Distribution 143

Sociocultural Issues — 144
Cultural Diversity 145 • The Family and Gender Roles 146 • Religion in Contemporary Life 146

PHOTO ESSAYS
3.5	Climates of Middle and South America	115
3.8	Human Impacts on the Biosphere in Middle and South America	119
3.10	Vulnerability to Climate Change in Middle and South America	121
3.11	Visual History of Middle and South America	122
3.19	Power and Politics in Middle and South America	137
3.21	Urbanization in Middle and South America	140

CHAPTER 4
Europe — 150

The European Region — 152

PHYSICAL GEOGRAPHY AND ENVIRONMENTAL ISSUES — 155

Physical Patterns — 155
Landforms 155 • Vegetation and Climate 155

Environmental Issues — 156
European Leadership in Response to Global Climate Change 156 • Europe's Impact on the Biosphere 159

HUMAN GEOGRAPHY — 164

Human Patterns over Time — 164
Sources of European Culture 164 • The Inequalities of Feudalism 165 • The Role of Urbanization in the Transformation of Europe 165 • European Colonialism: The Founding and Acceleration of Globalization 166 • Urban Revolutions in Industry and Politics 166 • Urbanization and Politics 168 • Two World Wars and Their Aftermath 169 • The Birth of the European Union 169

Globalization and Development — 170
Europe's Growing Service Economies 170 • The Euro and Debt Crises 172 • Food Production and the European Union 173

Contents ix

Power and Politics 175
The Politics of EU Expansion 175 • EU Governing Institutions 178 • NATO and the Rise of the European Union as a Global Peacemaker 178 • Social Welfare Systems and Their Outcomes 178

Urbanization 180
Population and Gender 180
Sociocultural Issues 184
Immigration and Migration: Needs and Fears 184 • Changing Gender Roles 186

PHOTO ESSAYS
4.4	Climates of Europe	157
4.5	Vulnerability to Climate Change in Europe	158
4.7	Human Impacts on the Biosphere in Europe	161
4.11	Visual History of Europe	166
4.18	Power and Politics in Europe	176
4.20	Urbanization in Europe	181

CHAPTER 5
Russia and the Post-Soviet States 192

The Russia and Post-Soviet States Region 194

PHYSICAL GEOGRAPHY AND ENVIRONMENTAL ISSUES 197
Physical Patterns 197
Landforms 197 • Climate and Vegetation 198

Environmental Issues 198
Urban and Industrial Pollution 201 • Nuclear Pollution 201 • The Globalization of Resource Extraction and Environmental Degradation 201 • Climate Change 204

HUMAN GEOGRAPHY 204
Human Patterns over Time 205
The Rise of the Russian Empire 205 • The Communist Revolution and Its Aftermath 206 • World War II and the Cold War 208

Globalization and Development 210
Oil and Gas Development: Fueling Globalization 210 • Food Production in the Post-Soviet Era 212

Power and Politics 214
Urbanization 218
Population and Gender 220
Gender and Life Expectancy 221 • Population Distribution 223

Sociocultural Issues 224
Gender: Challenges and Opportunities in the Post-Soviet Era 224 • Religious Revival in the Post-Soviet Era 226

PHOTO ESSAYS
5.5	Climates of Russia and the Post-Soviet States	199
5.6	Human Impacts on the Biosphere in Russia and the Post-Soviet States	200
5.8	Vulnerability to Climate Change in Russia and the Post-Soviet States	203
5.10	Visual History of Russia and the Post-Soviet States	206
5.16	Power and Politics in Russia and the Post-Soviet States	215
5.18	Urbanization in Russia and the Post-Soviet States	219

CHAPTER 6
North Africa and Southwest Asia 230

The North Africa and Southwest Asia Region 232

PHYSICAL GEOGRAPHY AND ENVIRONMENTAL ISSUES 235
Physical Patterns 235
Climate 235 • Landforms and Vegetation 235

Environmental Issues 237
An Ancient Heritage of Water Conservation 237 • Could There Be New Sources of Water? 239 • Water and Food Production 239 • Vulnerability to Climate Change 239

HUMAN GEOGRAPHY 243
Human Patterns over Time 243
Agriculture and the Development of Civilization 243 • Agriculture and Gender Roles 244 • The Coming of Monotheism: Judaism, Christianity, and Islam 244 • The Spread of Islam 246 • Western Domination, State Formation, and Antidemocratic Practices 247

Globalization and Development 249
Fossil Fuel Exports 249 • Economic Diversification and Growth 252

Power and Politics 254
The Arab Spring 254 • The Role of the Press, Media, and Internet in Political Change 256 • Democratization and Women 258

Three Worrisome Geopolitical Situations in the Region 258
Situation 1: Fifty Years of Trouble Between Iraq and the United States 258 • Situation 2: The State of Israel and the "Question of Palestine" 259 • Situation 3: Failure of the Arab Spring in Syria 262

Urbanization 263
Population and Gender 265
Changing Population Distribution 267

Sociocultural Issues 267
Families and Gender 267 • Gender Roles and Gendered Spaces 268 • The Rights of Women 269 • The Lives of Children 271

Contents

PHOTO ESSAYS

6.5	Climates of North Africa and Southwest Asia	236
6.7	Human Impacts on the Biosphere in North Africa and Southwest Asia	238
6.10	Vulnerability to Climate Change in North Africa and Southwest Asia	241
6.16	Visual History of North Africa and Southwest Asia	248
6.21	Power and Politics in North Africa and Southwest Asia	256
6.24	Urbanization in North Africa and Southwest Asia	264

CHAPTER 7
Sub-Saharan Africa   274

The Sub-Saharan Region   276

PHYSICAL GEOGRAPHY AND ENVIRONMENTAL ISSUES   278

Physical Patterns   278
Landforms 278 • Climate and Vegetation 279

Environmental Issues   279
Deforestation and Climate Change 279 • Agricultural Systems, Food, Water, and Vulnerability to Climate Change 282 • Wildlife and Climate Change 286

HUMAN GEOGRAPHY   288

Human Patterns over Time   288
The Peopling of Africa and Beyond 288 • Early Agriculture, Industry, and Trade in Africa 288 • The Scramble to Colonize Africa 290 • Power and Politics in the Aftermath of Independence 291

Globalization and Development   293
Successive Eras of Globalization 293 • Regional and Local Economic Development 296

Power and Politics   299
Gender, Power, and Politics 302

Urbanization   303

Population and Gender   305

Sociocultural Issues   309
Gender Issues 310 • Religion 311

PHOTO ESSAYS

7.5	Climates of Sub-Saharan Africa	280
7.6	Human Impacts on the Biosphere in Sub-Saharan Africa	281
7.8	Vulnerability to Climate Change in Sub-Saharan Africa	285
7.11	Visual History of Sub-Saharan Africa	288
7.21	Power and Politics in Sub-Saharan Africa	301
7.22	Urbanization in Sub-Saharan Africa	304

CHAPTER 8
South Asia   316

The South Asian Region   318

PHYSICAL GEOGRAPHY AND ENVIRONMENTAL ISSUES   320

Physical Patterns   320
Landforms 320 • Climate and Vegetation 321

Environmental Issues   322
South Asia's Vulnerability to Climate Change 324 • Responses to Water Issues Related to Global Climate Change 327 • Deforestation 329 • Industrial Air Pollution 329

HUMAN GEOGRAPHY   330

Human Patterns over Time   330
The Indus Valley Civilization 331 • A Series of Invasions 331 • Language and Ethnicity 331 • Religious Traditions 331 • Globalization and the Legacies of British Colonial Rule 334

Globalization and Development   338
Economic Trends 338 • Economic Reforms 338 • Offshore Outsourcing 340 • Free Trade Within South Asia 340 • Food Production and the Green Revolution 340 • Microcredit: A South Asian Innovation for the Poor 342

Power and Politics   343
Religious Nationalism 344 • The Growing Influence of Women and Young Voters 344 • Regional Conflicts 344

Urbanization   348
Mumbai 348

Population Patterns   351

Sociocultural Issues   354
The Texture of Village Life 354 • Social Patterns in the Status of Women 354 • Gender, Politics, and Power 355

PHOTO ESSAYS

8.5	Climates of South Asia	323
8.7	Vulnerability to Climate Change in South Asia	325
8.8	Human Impacts on the Biosphere in South Asia	328
8.10	Visual History of South Asia	332
8.20	Power and Politics in South Asia	345
8.21	Urbanization in South Asia	349

CHAPTER 9
East Asia   360

The East Asia Region   362

PHYSICAL GEOGRAPHY AND ENVIRONMENTAL ISSUES   364

Physical Patterns   364
Landforms 365 • Climate 365

Environmental Issues 367
Food Security and Sustainability 369 • Three Gorges Dam: The Power of Water 372 • Air Pollution: Choking on Success 374

HUMAN GEOGRAPHY 375
Human Patterns over Time 375
Bureaucracy and Imperial China 375 • Confucianism Molds East Asia's Cultural Attitudes 376 • Why Did China Not Colonize an Overseas Empire? 377 • European and Japanese Imperialism 377 • China's Turbulent Twentieth Century 377 • Japan Becomes a World Leader 380 • Chinese and Japanese Influences on Korea, Taiwan, and Mongolia 381
Globalization and Development 381
The Japanese Miracle 382 • Mainland Economies: Communists in Command 382 • Economic Reforms in China 384
Power and Politics 384
Pressures for Political Change in China 385 • Japan's Recent Political Shifts 389 • Political Tensions Between East Asian Countries 389
Urbanization 390
Transportation Improvements 390 • Hong Kong's Special Role 393
Population and Gender 394
Responding to an Aging Population 394 • The Legacies of China's One-Child Policy 395 • Population Distribution 396
Sociocultural Issues 397
East Asia's Most Influential Cultural Export: The Overseas Chinese 401

PHOTO ESSAYS
9.4	Climates of East Asia	366
9.5	Vulnerability to Climate Change in East Asia	368
9.9	Human Impacts on the Biosphere in East Asia	373
9.12	Visual History of East Asia	378
9.18	Power and Politics in East Asia	387
9.20	Urbanization in East Asia	391

CHAPTER 10
Southeast Asia 404

The Southeast Asia Region 406

PHYSICAL GEOGRAPHY AND ENVIRONMENTAL ISSUES 409
Physical Patterns 409
Landforms 409 • Climate and Vegetation 410
Environmental Issues 411
Deforestation 411 • Climate Change and Deforestation 411 • Climate Change and Food Production 415 • Climate Change and Water 417 • Responses to Climate Change 419

HUMAN GEOGRAPHY 419
Human Patterns over Time 419
Diverse Cultural Influences 419 • European Colonization 421 • Struggles for Independence 421
Globalization and Development 422
Economic Crisis and Recovery: The Perils of Globalization 424 • Regional Trade and ASEAN 425
Power and Politics 428
Southeast Asia's Authoritarian Tendencies 428 • Militarism and China 430 • Can the Expansion of Political Freedoms Help Bring Peace to Indonesia? 430
Urbanization 431
Emigration Related to Globalization 432
Population and Gender 435
Sociocultural Issues 438
Cultural and Religious Pluralism 438 • Gender Patterns in Southeast Asia 441 • Globalization and Gender: The Sex Industry 442

PHOTO ESSAYS
10.7	Climates of Southeast Asia	412
10.9	Human Impacts on the Biosphere in Southeast Asia	414
10.11	Vulnerability to Climate Change in Southeast Asia	418
10.13	Visual History of Southeast Asia	422
10.18	Power and Politics in Southeast Asia	429
10.19	Urbanization in Southeast Asia	433

CHAPTER 11
Oceania: Australia, New Zealand, and the Pacific 446

The Oceania Region 448

PHYSICAL GEOGRAPHY AND ENVIRONMENTAL ISSUES 450
Physical Patterns 450
Continent Formation 450 • Island Formation 451 • Climate 451 • Fauna and Flora 454
Environmental Issues 455
Global Climate Change 455 • Invasive Species and Food Production 457 • Globalization and the Environment in the Pacific Islands 457

HUMAN GEOGRAPHY 461
Human Patterns over Time 461
The Peopling of Oceania 461 • Arrival of the Europeans 463 • The Colonization of Australia and New Zealand 463 • Oceania's Shifting Global Relationships 463

Globalization and Development — 464
Globalization, Development, and Oceania's New Asian Orientation 464 • The Stresses of Asia's Economic Development "Miracle" on Australia and New Zealand 465 • The Advantages and Stresses of Tourism 467 • The Future: Diverse Global Orientations? 468

Power and Politics — 468

Urbanization — 470

Population and Gender — 472
Population Numbers and Distribution 472

Sociocultural Issues — 473
Ethnic Roots Reexamined 473 • Forging Unity in Oceania 476

PHOTO ESSAYS
11.5	Climates of Oceania	452
11.8	Vulnerability to Climate Change in Oceania	456
11.9	Human Impacts on the Biosphere in Oceania	458
11.14	Visual History of Oceania	464
11.17	Power and Politics in Oceania	469
11.18	Urbanization in Oceania	471

EPILOGUE: ANTARCTICA — 481
Glossary — G-1
Text Sources and Credits — TC-1
Index — I-1

PREFACE

In this text, we portray the rich diversity of human life across the world and humanize geographic issues by representing the daily lives of women, men, and children in the various regions of the globe. Our goal is to make global patterns of trade and consumption meaningful for students by showing how these patterns affect environments and people at the local level. This third edition of *World Regional Geography Concepts* strives to reach this goal with improvements to make this text as current, instructive, and visually appealing as possible. At just 481 pages, the text is designed to allow instructors to cover all world regions in a single semester.

NEW TO THE THIRD EDITION

Themes

Teaching world regional geography is never easy. Many instructors have found that focusing their courses on a few key ideas makes their teaching more effective and helps students retain information. With that goal in mind, we have identified five themes that provide a few basic hooks on which students can hang their growing knowledge of the world and each of its regions. These themes are listed here in the order in which they are covered in every chapter:

- Environment: How do issues of water scarcity, water pollution, and water management affect people and environments in a particular region? How do food production systems impact environments and societies in a region? What are the indications that climate change is underway? How might global climate change and changes in food production systems affect water resources? How are places, people, and ecosystems in a particular region vulnerable to the shifts that climate change may bring? Which human activities contribute significant amounts of greenhouse gases?

- Globalization and Development: How has a particular region been impacted by globalization, historically and currently? How are lives changing as flows of people, ideas, products, and resources become more global? How do shifts in economic, social, and other dimensions of development affect human well-being? What paths have been charted by the so-called developed world, and how are they relevant, or irrelevant, to the rest of the world? What new "homegrown" solutions are emerging from the so-called less developed countries?

- Power and Politics: What are the main differences in the ways that power is wielded in societies? Where are authoritarian modes of governance dominant? Where have political freedoms expanded the most? What kind of changes is the expansion of political freedoms bringing to different world regions? How are changes in the geopolitical order affecting current world events?

- Urbanization: Which forces are driving urbanization in a particular region? How have cities responded to growth? How are regions affected by the changes that accompany urbanization—for example, the growth of slums and changes in access to jobs, education, and health care?

- Population and Gender: What are the major forces driving population growth or decline in a region? How have changes in gender roles influenced population growth or decline? How are changes in life expectancy, family size, and the age of the population influencing population change?

Geographic Insights

For each chapter, the five themes form the basis of five learning objectives that we call **Geographic Insights**. These insights are stated at the beginning of each chapter and discussed at the relevant point in the text. To improve consistency, the Geographic Insights are now completely comparable across chapters. Geographic Insights are reviewed in "Things to Remember" sections found throughout the chapter, as well as in new questions posed in the "Geographic Insights Review and Self-Test" sections at the end of each chapter.

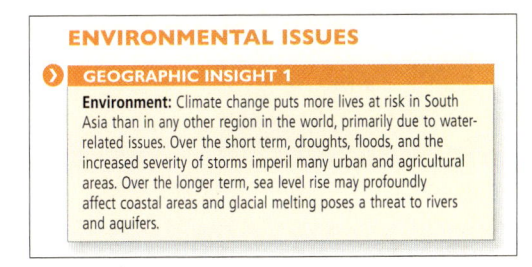

> **ENVIRONMENTAL ISSUES**
>
> **› GEOGRAPHIC INSIGHT 1**
>
> **Environment:** Climate change puts more lives at risk in South Asia than in any other region in the world, primarily due to water-related issues. Over the short term, droughts, floods, and the increased severity of storms imperil many urban and agricultural areas. Over the longer term, sea level rise may profoundly affect coastal areas and glacial melting poses a threat to rivers and aquifers.

Restructured Chapters

Each chapter has been restructured so that the discussions to which the Geographic Insights refer now occur in the same order in each chapter. We implemented this change to make it easier for teachers and students to navigate the complex topics of world regional geography. The number of themes has also been condensed, from nine in the previous edition to five in this edition. Discussions of climate change, food, and water now come under the heading of "Environmental Issues." Globalization and development are now discussed in one section. The theme of "democratization" has been broadened to include geopolitics and other political issues and is now called "Power and Politics." Some discussions of population issues have been linked with discussions of gender.

On the Bright Side

In light of the often overwhelming and at times depressing nature of the information presented in any world regional geography course, each chapter now has a series of new **On the Bright Side** commentaries that explore some of the more hopeful patterns and opportunities emerging within each region.

xiv Preface

Local Lives Photo Feature

Three **Local Lives** photo features in each region chapter add further human interest by showing regional customs related to foodways, people and animals, and festivals. Each photo has an extensive caption designed to pique students' curiosity.

FIGURE 8.27 LOCAL LIVES
Festivals in South Asia

A A village festival in Pakistan features *kabaddi*, a popular South Asian sport in which teams take turns sending a "raider" across a field center line. That person must tag, or in some cases wrestle to the ground, members of the other team and then return to his or her own side without taking a breath. Kabaddi has been played at the Indian National Games since 1939 and at the Asian Games since 1991. [Amir Mukhtar/FlickrVision/Getty Images]

B Celebrants in Kolkata, India, during Holi—a festival celebrating the end of winter and beginning of spring. Holi evolved from temple worship practices involving the application of color to statues. In a riotous and celebratory atmosphere, people of different ages, genders, castes, and economic backgrounds temporarily disregard their differences and hurl the colors of the coming spring at each other. [Sanjay Kanojia/AFP/Getty Images]

C Pilgrims during the 2010 Kumbh Mela bathe in the Ganga River at Haridwar. During this event, which is held every 3 years, Hindus purify themselves by bathing in the sacred waters of the river. In 2013, the 45-day-long event attracted over 100 million participants. [Pedro Ugarte/AFP/Getty Images]

New Photos

Every photo has been updated for currency in this edition. An ongoing aim of this text has been to awaken students to the circumstances of people around the world, and photos are a powerful way to accomplish this objective. This edition continues our tradition of promoting careful attention to photos by including in Chapter 1 a short lesson on photo interpretation. Students are encouraged to use these skills as they look at every photo in the text, and instructors are encouraged to use the photos as lecture themes and to help generate analytical class discussions.

Gonzalo Azumendi/age fotostock/Getty Images

Images by Ni-ree/Flickr/Getty Images

Kim Westerskov/Stone/Getty Images

Each photo was chosen to complement a concept or situation described in the text. All photos are numbered and referenced in the text, making it easier for students to integrate the text with the visuals as they read. Moreover, the photos—like all of the book's graphics, including the maps—have been given significant space and prominence in the page layout. The result is a visually engaging, dynamic, instructive, and up-to-date text.

At the beginning of each regional chapter, a series of photos surrounding the regional map introduces the reader to landscapes within the region. Subsequent figures that are **Photo Essays** illustrate particular themes. For example, each region has a photo essay about urbanization, including a map of national urbanization rates and large cities as well as photos that illustrate various aspects of urban life in the region.

FIGURE 2.5 PHOTO ESSAY: Vulnerability to Climate Change in North America

North America's wealth and its well-developed emergency response systems make it very resilient and reduce its overall vulnerability to climate change. However, certain regions are highly exposed to temperature increases, drought, hurricanes, and sea level rise.

A The location of Shishmaref, Alaska, on the Arctic Sea leaves it exposed to coastal erosion, which may be increasing in the area because of warmer air and sea temperatures.

B Florida's low elevations expose it to sea level rise and flooding during and after hurricanes.

The stark contrast in vulnerability between the United States and Mexico results from the countries' very different sensitivity and resilience to water scarcity. As temperatures rise, this already dry borderland area will have less water. The United States has a much better water infrastructure, reducing its sensitivity to drought, and its emergency response systems make it more resilient to water shortages than Mexico is. With such drastic differences, the U.S.–Mexico border could become an even more contentious zone as the climate changes.

C Higher temperatures raise the rate at which plants lose water, increasing the need for irrigation. Farms that get their water from shrinking aquifers are thus less able to increase irrigation, and so are highly sensitive to the rising temperatures that climate change is bringing.

D Hurricanes gain strength with warmer temperatures. Many low-lying coastal cities are very exposed and sensitive, though their resilience varies. Shown here is a fire that destroyed several homes in New Orleans when the flooding caused by Hurricane Katrina made it impossible for fire trucks to reach the fire.

To help instructors make use of all these new photo features in their teaching, the many photo essays and photo figures are accompanied by **Thinking Geographically** questions. The questions are found at the end of each chapter, and the answers can be found on this book's Web site, where they form the basis of computer-graded exercises that can be assigned and automatically graded and entered into the instructor's electronic gradebook.

Up-to-Date Content

Because the world is constantly changing, it is essential that a world regional geography text be as current as possible. Some major content areas of the book that have been updated include:

- Climate change and its environmental, political, and economic implications

- The global economic recession that began in 2008 and its effect on migrants, labor outsourcing, and job security in importing and exporting countries
- The role of terrorism in the realignment of power globally and locally
- The growing disparity in wealth in North America
- Domestic and global implications of the U.S. political, economic, and military stances
- Conflict over oil pollution in the Amazon
- Immigration and the ways it is changing countries economically and culturally
- Changing gender roles, particularly in developing countries
- Recent economic crises in the European Union and the consequences for the original EU members, new and potential member states, and the global community
- Growing tensions between Russia, its neighbors, the European Union, and the United States
- Political revolutions and conflict (the Arab Spring) in North Africa and Southwest Asia (often referred to as the Middle East)
- The new influence of Arabic media outlets, such as Al Jazeera
- Maps reflecting the new country of South Sudan (some data remain based on Sudan as a whole because South Sudan has not yet begun reporting statistics)
- Growing tensions over water in South Asia
- Disputes in East Asia over the Senkaku Islands
- Civil unrest in Thailand
- The consequences of global climate change on Pacific Island nations

Consistent Base Maps

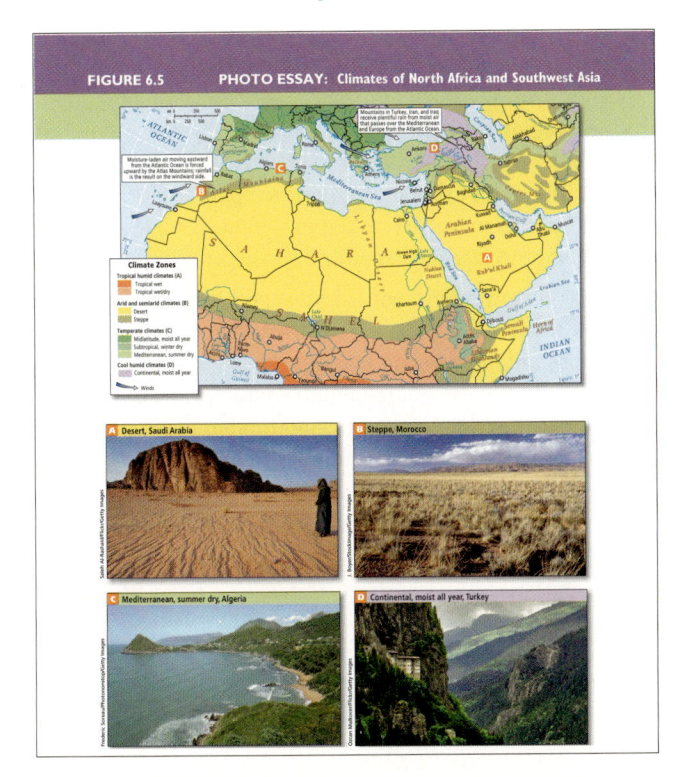

This edition focuses on improving further what has often been cited as a principal strength of this text: high-quality, relevant, and consistent maps. To help students make conceptual connections and to compare regions, every chapter contains the following:

- Regional map with landscape photos at the beginning of each chapter
- Political map
- Climate map with photos of different climate zones
- Map of the human impacts on the biosphere, with photo essay
- Map of the region's vulnerability to climate change, with photo essay
- Map of regional trends in power and politics, with photo essay
- Urbanization map, with photo essay
- Map of population density

Videos

More than 400 videos clips (an average of about 35 per chapter) are available with the third edition. Most videos are 2 to 6 minutes long and cover key issues discussed in the text. They can help instructors gain further expertise or can be used to generate class discussion. Each video is keyed to the text with an icon at the point in the discussion where it is most relevant. These videos, along with a related multiple-choice quiz, can be accessed at Geography LaunchPad (for more details, please see below). Questions can be automatically graded and entered into an electronic gradebook.

Visual Histories

These visual timelines for each region use images to illustrate key points in the region's history.

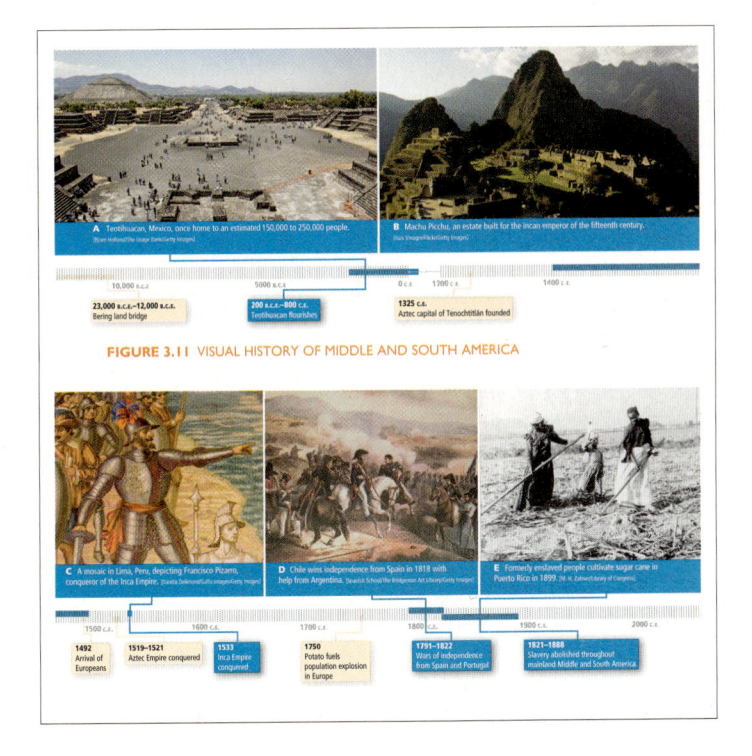

Things to Remember

At the close of every main section, a few concise statements review the important points in the section. The statements emphasize some key themes while encouraging students to think through the ways in which the material illustrates these points. They also review the Geographic Insights that begin each chapter.

Geographic Insights Review and Self-Test

At the end of each chapter, a series of questions, many tied to the chapter's Geographic Insights, encourage students to more broadly analyze the chapter content. These questions could be used for assignments, group projects, or class discussion.

Marginal Glossary of Key Terms

Terms important to the chapter content are boldfaced on first usage and defined on the page on which they first appear. Each term is listed at the end of the chapter, along with the number of the page where the term is defined. The key terms are also listed alphabetically and defined in the glossary at the end of the book.

THE ENDURING VISION: GLOBAL AND LOCAL PERSPECTIVES

The Global View

In addition to the new features and enhancements to the text, we retain the hallmark features that have made the first two editions of this text successful for instructors and students. For the third edition, we continue to emphasize global trends and the connections between regions that are changing lives throughout the world. The following linkages are explored in every chapter, as appropriate:

- The **multifaceted economic linkages** among world regions. These include (1) the effects of colonialism; (2) trade; (3) the role in the world economy of transnational corporations such as Walmart, Norilsk Nickel, Nike, and Apple; (4) the influence of regional trade organizations such as ASEAN and NAFTA; and (5) the changing roles of the World Bank and the International Monetary Fund as the negative consequences of structural adjustment programs become better understood. These issues are explored primarily in the Globalization and Development section of each chapter.

- **Migration.** Migrants are changing economic and social relationships in virtually every part of the globe. The societies they leave are changed radically by the migrants' absence, just as the host societies are altered by their presence. The text explores the local and global effects of foreign workers in places such as Japan, Europe, Africa, the Americas, and Southwest Asia, as well as the increasing number of refugees resulting from conflicts around the world. Also discussed are long-standing migrant groups, including the Overseas Chinese and the Indian diasporas. These topics are explored primarily in Population and Gender as well as the Sociocultural Issues sections of each chapter.

- **Gender issues** are covered in every chapter with the aim of addressing more completely the lives of ordinary people. Gender is intimately connected to other patterns, including internal and global migration, and these connections and other region-wide gender patterns are illustrated in a variety of maps and photos and in vignettes that illustrate gender roles as played out in the lives of individuals. The lives of children, especially with regard to their roles in families, are also covered, often in concert with the treatment of gender issues. These issues are explored primarily in Population and Gender as well as the Sociocultural Issues sections of each chapter.

The Local Level

We pay special attention in this book to the local scale—a town, a village, a household, an individual. Our hope is, first, that stories of individual people and families will make geography interesting and real to students; and second, that seeing the effects of abstract processes and trends on ordinary lives will dramatize the effects of these developments for students. Reviewers have mentioned that students particularly appreciate the personal vignettes, which are often stories of real people (with names disguised). For each region, we examine the following local responses:

- **Local lives:** We use photo essays to focus on particular regional customs and traditions as they relate to foodways, festivals, and the relationship of animals to the people of a region.

- **Cultural change:** We look closely at changes in the family, gender roles, and social organization in response to urbanization, modernization, and the global economy.

- **Impacts on well-being:** Ideas of what constitutes "well-being" differ from culture to culture, yet broadly speaking, people everywhere try to provide a healthful life for themselves in a community of their choosing. Their success in doing so is affected by local conditions, global forces, and their own ingenuity.

- **Issues of identity:** Paradoxically, as the world becomes more tightly knit through global communications and media, ethnic and regional identities often become stronger. The text examines how modern developments such as the Internet and related technologies are used to reinforce particular cultural identities, often bringing educated emigrants back to help with reforms or to facilitate rapid responses in crises.

- **Local attitudes toward globalization:** People often have ambivalent reactions to global forces: they are repelled by the seeming power of these forces, fearing effects on their own lives and livelihoods and on local traditional cultural values, but they are also attracted by the economic opportunities that may emerge from greater global integration. The text looks at how the people of a region react to cultural and economic globalization.

ALSO AVAILABLE

To better serve the different needs of diverse faculty and curricula, two other versions of this textbook are available.

World Regional Geography (with Subregions), Sixth Edition (1-4641-1070-0)

The sixth edition continues to employ a consistent structure for each chapter. Each chapter beyond the first is divided into three parts: **The Geographic Setting, Current Geographic Issues,** and **Subregions.**

The subregion coverage provides a descriptive characterization of particular countries and places within the region that expands on coverage in the main part of the chapter. For example, the sub-Saharan Africa chapter considers the West, Central, East, and Southern Africa subregions, providing additional insights into differences in well-being and into social and economic issues across the African continent.

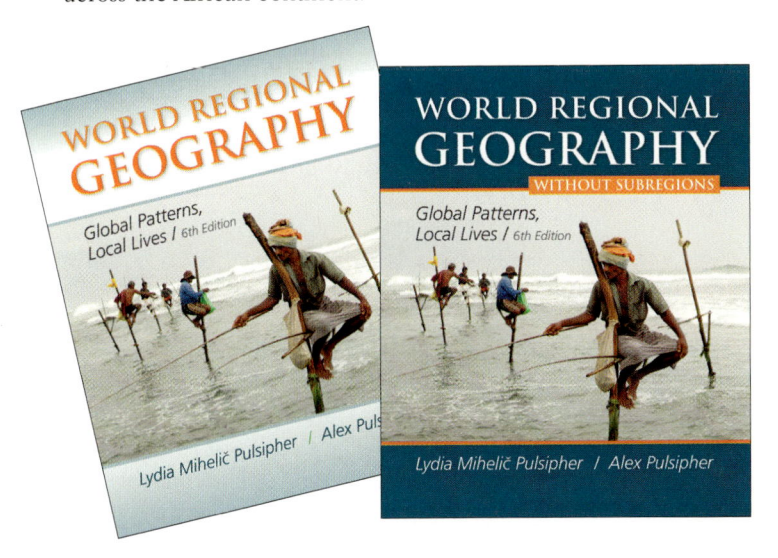

World Regional Geography Without Subregions, Sixth Edition (1-4641-1069-7)

The briefer version provides essentially the same main text coverage as the version described above, omitting only the subregional sections. This version contains all the pedagogy found in the main version.

FOR THE INSTRUCTOR: A WEALTH OF RESOURCES ONLINE AT GEOGRAPHY LAUNCHPAD

Macmillan Education LaunchPad

www.macmillanhighered.com/launchpad/pulsipherconcepts3e

The authors have taught world regional geography many times and understand the need for quick, accessible aids to instruction. Many of the new features were designed to streamline the job of organizing the content of each class session, with the goal of increasing student involvement through interactive discussions. Ease of instruction and active student involvement were the principal motivations behind the book's key features—the themes and geographic insights, the photo essays, the content maps that facilitate region-to-region comparisons, the photo features on local lives and regional customs, the "On the Bright Side" commentaries that highlight exciting and emerging reasons for optimism, and the wide selection of videos.

All of the following are available on Geography LaunchPad, a complete course management system that features full gradebook and reporting capacities. For a demo of Geography LaunchPad, please contact your W. H. Freeman/Macmillan Education sales consultant.

- All text images in PowerPoint and JPEG formats with enlarged labels for better projection quality.
- PowerPoint lecture outlines by Nicole C. James. The main themes of each chapter are outlined and enhanced with images from the book, providing a pedagogically sound foundation on which to build personalized lecture presentations.
- Instructor's Resource Manual by Jennifer Rogalsky, State University of New York, Geneseo, and Helen Ruth Aspaas, Virginia Commonwealth University, contains suggested lecture outlines, points to ponder for class discussion, and ideas for exercises and class projects. It is offered as chapter-by-chapter Word files to facilitate editing and printing.
- Test Bank by Rebecca Johns, University of South Florida, expanded from the original test bank created by Jason Dittmer, University College London, and Andy Walter, West Georgia University. The Test Bank is designed to match the pedagogical intent of the text and offers more than 2500 test questions (multiple choice, short answer, matching, true/false, and essay) in a Word format that makes it easy to edit, add, and resequence questions. A computerized test bank (powered by Diploma) with the same content is also available.
- Clicker questions by Rebecca Johns, University of South Florida. Prepared in Word, clicker questions allow instructors to jump-start discussions, illuminate important points, and promote better conceptual understanding during lectures.
- An integrated gradebook that records students' performance on online and video quizzes.

Course Management

All instructor and student resources are also available via Black-Board, Canvas, Angel, Moodle, Sakai, and Desire2Learn. W. H. Freeman offers a course cartridge that populates your site with content tied directly to the book. For access to a specific course cartridge, please contact your W. H. Freeman/Macmillan Education sales consultant.

W. H. Freeman *World Regional Geography* DVD

This DVD, available free to adopters of the third edition, gives students and instructors access to the fascinating personal stories of people from all over the world, building on the book's purpose of putting a face on geography. The DVD contains 35 projection-quality video clips from 3 to 7 minutes in length. An instructor's video manual is also included on the DVD.

FOR THE STUDENT: *WORLD REGIONAL GEOGRAPHY* ONLINE WITH GEOGRAPHY LAUNCHPAD

Macmillan Education LaunchPad

www.macmillanhighered.com/launchpad/pulsipherconcepts3e

A wealth of resources to support the textbook are available online on Geography LaunchPad:

- **Chapter quizzes:** These multiple-choice quizzes help students assess their mastery of each chapter.
- **Thinking Geographically questions:** These questions relate to select photos found throughout the book. The question sets form the basis of computer-graded exercises that can be assigned and automatically graded and entered into the instructor's online gradebook.
- **Thinking Critically About Geography:** These activities allow students to explore a set of current issues, such as deforestation, human rights, or free trade, and see how geography helps clarify our understanding of them. Linked Web sites are matched with a series of questions or with brief activities that help students think about the ways in which they themselves are connected to the places and people they read about in the text.
- **Blank outline maps:** Printable maps of the world and of each region are available for note taking, exam review, or both, as well as for preparing assigned exercises.
- **Flashcards:** Matching exercises teach vocabulary and definitions.
- **Audio pronunciation guide:** This spoken guide helps students learn to pronounce place names, regional terms, and names of historical figures.
- **World recipes and cuisines:** From *International Home Cooking*, the United Nations International School cookbook, these recipes provide students with the opportunity to explore foods from around the world.

Geography LaunchPad also offers users a set of unique resources not available anywhere else:

- **Map Builder software and Map Builder exercises:** The Map Builder program allows students to create layered thematic maps on their own, while Map Builder exercises offer a specific activity for each chapter in the third edition.
- **Map learning exercises:** Students can use these interactive activities to identify and locate countries, cities, and the major geographic features of each region. Instructors can assign these map exercises as homework or as quizzes; student results report directly to the instructor's online gradebook.
- An **e-Book** of *World Regional Geography Concepts*, complete and customizable. Students can quickly search the text and personalize it just as they would the printed version—complete with highlighting, bookmarking, and note-taking features.
- **A Guide to Using Google Earth** for the novice, plus step-by-step **Google Earth** exercises for each chapter.
- Selected articles from *Focus on Geography* magazine (one for each chapter in the textbook) and accompanying quizzes for each article.
- **Physical geography videos** for instructors who want to cover physical geography topics in more detail.
- Online **news feeds** for highly respected magazines such as the *Economist*.
- An **online atlas** prepared specifically for *World Regional Geography Concepts*.

Macmillan Education LearningCurve

Learning Curve is an integral part of Geography LaunchPad. It is an intuitive, fun, and highly effective formative assessment tool that is based on extensive educational research. Students can use Learning Curve to test their knowledge in a low-stakes environment that helps them improve their mastery of key concepts and prepare for lectures and exams. This adaptive quizzing engine moves students from basic knowledge through critical thinking and synthesis skills as they master content at each level. For a demo, visit www.learningcurveworks.com.

Also Available: Rand McNally's *Atlas of World Geography,* 176 pages

This atlas, available at a greatly reduced price when bundled with the textbook, contains:

- Fifty-two physical, political, and thematic maps of the world and continents; 49 regional, physical, political, and thematic maps; and dozens of metro-area inset maps
- Geographic facts and comparisons, covering topics such as population, climate, and weather
- A section on common geographic questions, a glossary of terms, and a comprehensive 25-page index

ACKNOWLEDGMENTS

The authors wish to acknowledge the many geographers whose insights and suggestions have informed this book.

World Regional Geography Concepts, Third Edition

Ola Ahlqvist
Ohio State University

Wayne Brew
Montgomery County Community College

Elizabeth Dudley-Murphy
University of Utah

Chad Garick
Jones County Junior College

Kari Jensen
Hofstra University

Timothy Kelleher
Florida State University

Mathias Le Bossé
Kutztown University of Pennsylvania

Fuyuan Liang
Western Illinois University

Chris Post
Kent State University, Stark

Amy Rock
Ohio University

Sarah Smiley
Kent State University, Salem

World Regional Geography, Sixth Edition

Victoria Alapo
Metropolitan Community College

Jeff Arnold
Southwestern Illinois College

Shaunna Barnhart
Pennsylvania State University

Dean Butzow
Lincoln Land Community College

Philip Chaney
Auburn University

Christine Hansell
Skyline College

Heidi LaMoreaux
Santa Rosa Junior College

Kent Mathewson
Louisiana State University

Julie Mura
Florida State University

Michael Noll
Valdosta State University

Tim Oakes
University of Colorado, Boulder

Kefa M. Otiso
Bowling Green State University

Sam Sweitz
Michigan Technological University

Jeff Ueland
Bemidji State University

Ben Wolfe
Metropolitan Community College, Blue River

World Regional Geography Concepts, Second Edition

Heike C. Alberts
University of Wisconsin, Oshkosh

John All
Western Kentucky University

Robert G. Atkinson
Tarleton State University

Christopher A. Badurek
Appalachian State University

Bradley H. Baltensperger
Michigan Technological University

Denis A. Bekaert
Middle Tennessee State University

Mikhail Blinnikov
St. Cloud State University

Mark Bonta
Delta State University

Patricia Boudinot
George Mason University

Lara M. P. Bryant
Keene State College

Craig S. Campbell
Youngstown State University

Bruce E. Davis
Eastern Kentucky University

L. Scott Deaner
Owens Community College

James V. Ebrecht
Georgia Perimeter College

Kenneth W. Engelbrecht
Metropolitan State College of Denver

Natalia Fath
Towson University

Alison E. Feeney
Shippensburg University

John H. Fohn II
Missouri State University, West Plains

Stephen Franklin
Coconino Community College

Joy Fritschle
West Chester University

Matthew Gerike
University of Missouri

Carol L. Hanchette
University of Louisville

Ellen R. Hansen
Emporia State University

Katie Haselwood-Weichelt
University of Kansas

Kari Jensen
Hofstra University

Cub Kahn
Oregon State University

Curtis A. Keim
Moravian College

Jeannine Koshear
Fresno City College

Hsiang-te Kung
University of Memphis

Chris Laingen
Eastern Illinois University

Leonard E. Lancette
Mercer University

Heidi Lannon
Santa Fe College

James Leonard
Marshall University

John Lindberg
Scott Community College

Max Lu
Kansas State University

Peter G. Odour
North Dakota State University

Kefa M. Otiso
Bowling Green State University

Lynn M. Patterson
Kennesaw State University

James Penn
Grand Valley State University

Paul E. Phillips
Fort Hays State University

Gabriel Popescu
Indiana University, South Bend

Jennifer Rahn
Samford University

Amanda Rees
Columbus State University

Robert F. Ritchie IV
Liberty University

Ginger L. Schmid
Minnesota State University, Mankato

Cynthia L. Sorrensen
Texas Tech University

Jennifer Speights-Binet
Samford University

Sam Sweitz
Michigan Technological University

Michael W. Tripp
Vancouver Island University

Julie L. Urbanik
University of Missouri, Kansas City

Acknowledgments

Irina Vakulenko
University of Texas, Dallas

Jean Vincent
Santa Fe College

Linda Wang
University of South Carolina, Aiken

Kelly Watson
Florida State University

Laura A. Zeeman
Red Rocks Community College

Sandra Zupan
University of Kentucky

World Regional Geography Concepts, First Edition

Gillian Acheson
Southern Illinois University, Edwardsville

Tanya Allison
Montgomery College

Keshav Bhattarai
Indiana University, Bloomington

Leonhard Blesius
San Francisco State University

Jeffrey Brauer
Keystone College

Donald Buckwalter
Indiana University of Pennsylvania

Craig Campbell
Youngstown State University

John Comer
Oklahoma State University

Kevin Curtin
George Mason University

Ron Davidson
California State University, Northridge

Tina Delahunty
Texas Tech University

Dean Fairbanks
California State University, Chico

Allison Feeney
Shippensburg University of Pennsylvania

Eric Fournier
Samford University

Qian Guo
San Francisco State University

Carole Huber
University of Colorado, Colorado Springs

Paul Hudak
University of North Texas

Christine Jocoy
California State University, Long Beach

Ron Kalafsky
University of Tennessee, Knoxville

David Keefe
University of the Pacific

Mary Klein
Saddleback College

Max Lu
Kansas State University

Donald Lyons
University of North Texas

Barbara McDade
University of Florida

Victor Mote
University of Houston

Darrell Norris
State University of New York, Geneseo

Gabriel Popescu
Indiana University, South Bend

Claudia Radel
Utah State University

Donald Rallis
University of Mary Washington

Pamela Riddick
University of Memphis

Jennifer Rogalsky
State University of New York, Geneseo

Tobie Saad
University of Toledo

Charles Schmitz
Towson University

Sindi Sheers
George Mason University

Ira Sheskin
University of Miami

Dmitri Siderov
California State University, Long Beach

Steven Silvern
Salem State College

Ray Sumner
Long Beach City College

Stan Toops
Miami University

Karen Trifonoff
Bloomsburg University of Pennsylvania

Jim Tyner
Kent State University

Michael Walegur
University of Delaware

Scott Walker
Northwest Vista College

Mark Welford
Georgia Southern University

World Regional Geography, Fifth Edition

Gillian Acheson
Southern Illinois University, Edwardsville

Greg Atkinson
Tarleton State University

Robert Begg
Indiana University of Pennsylvania

Richard Benfield
Central Connecticut State University

Fred Brumbaugh
University of Houston, Downtown

Deborah Corcoran
Missouri State University

Kevin Curtin
George Mason University

Lincoln DeBunce
Blue Mountain Community College

Scott Dobler
Western Kentucky University

Catherine Doenges
University of Connecticut, Stamford

Jean Eichhorst
University of Nebraska, Kearney

Brian Farmer
Amarillo College

Eveily Freeman
Ohio State University

Hari Garbharran
Middle Tennessee State University

Abe Goldman
University of Florida

Angela Gray
University of Wisconsin, Oshkosh

Ellen Hansen
Emporia State University

Nick Hill
Greenville Technical College

Johanna Hume
Alvin Community College

Edward Jackiewicz
California State University, Northridge

Rebecca Johns
University of South Florida, St. Petersburg

Suzanna Klaf
Ohio State University

Jeannine Koshear
Fresno City College

Brennan Kraxberger
Christopher Newport University

Heidi Lannon
Santa Fe College

Angelia Mance
Florida Community College, Jacksonville

Meredith Marsh
Lindenwood University

Linda Murphy
Blinn Community College

Monica Nyamwange
William Paterson University

Adam Pine
University of Minnesota, Duluth

Amanda Rees
Columbus State University

Benjamin Richason
St. Cloud State University

Amy Rock
Kent State University

Acknowledgments

Betty Shimshak
Towson University

Michael Siola
Chicago State University

Steve Smith
Missouri Southern State University

Jennifer Speights-Binet
Samford University

Emily Sturgess Cleek
Drury University

Gregory Taff
University of Memphis

Catherine Veninga
College of Charleston

Mark Welford
Georgia Southern University

Donald Williams
Western New England College

Peggy Robinson Wright
Arkansas State University, Jonesboro

World Regional Geography, Fourth Edition

Robert Acker
University of California, Berkeley

Joy Adams
Humboldt State University

John All
Western Kentucky University

Jeff Allender
University of Central Arkansas

David L. Anderson
Louisiana State University, Shreveport

Donna Arkowski
Pikes Peak Community College

Jeff Arnold
Southwestern Illinois College

Richard W. Benfield
Central Connecticut University

Sarah A. Blue
Northern Illinois University

Patricia Boudinot
George Mason University

Michael R. Busby
Murray State College

Norman Carter
California State University, Long Beach

Gabe Cherem
Eastern Michigan University

Brian L. Crawford
West Liberty State College

Phil Crossley
Western State College of Colorado

Gary Cummisk
Dickinson State University

Kevin M. Curtin
University of Texas, Dallas

Kenneth Dagel
Missouri Western State University

Jason Dittmer
University College London

Rupert Dobbin
University of West Georgia

James Doerner
University of Northern Colorado

Ralph Feese
Elmhurst College

Richard Grant
University of Miami

Ellen R. Hansen
Emporia State University

Holly Hapke
Eastern Carolina University

Mark L. Healy
Harper College

David Harms Holt
Miami University

Douglas A. Hurt
University of Central Oklahoma

Edward L. Jackiewicz
California State University, Northridge

Marti L. Klein
Saddleback College

Debra D. Kreitzer
Western Kentucky University

Jeff Lash
University of Houston, Clear Lake

Unna Lassiter
California State University, Long Beach

Max Lu
Kansas State University

Donald Lyons
University of North Texas

Shari L. MacLachlan
Palm Beach Community College

Chris Mayda
Eastern Michigan University

Armando V. Mendoza
Cypress College

Katherine Nashleanas
University of Nebraska, Lincoln

Joseph A. Naumann
University of Missouri, St. Louis

Jerry Nelson
Casper College

Michael G. Noll
Valdosta State University

Virginia Ochoa-Winemiller
Auburn University

Karl Offen
University of Oklahoma

Eileen O'Halloran
Foothill College

Ken Orvis
University of Tennessee

Manju Parikh
College of Saint Benedict and St. John's University

Mark W. Patterson
Kennesaw State University

Paul E. Phillips
Fort Hays State University

Rosann T. Poltrone
Arapahoe Community College, Littleton, Colorado

Waverly Ray
MiraCosta College

Jennifer Rogalsky
State University of New York, Geneseo

Gil Schmidt
University of Northern Colorado

Yda Schreuder
University of Delaware

Tim Schultz
Green River Community College, Auburn, Washington

Sinclair A. Sheers
George Mason University

D. James Siebert
North Harris Montgomery Community College, Kingwood

Dean Sinclair
Northwestern State University

Bonnie R. Sines
University of Northern Iowa

Vanessa Slinger-Friedman
Kennesaw State University

Andrew Sluyter
Louisiana State University

Kris Runberg Smith
Lindenwood University

Herschel Stern
MiraCosta College

William R. Strong
University of North Alabama

Ray Sumner
Long Beach City College

Rozemarijn Tarhule-Lips
University of Oklahoma

Alice L. Tym
University of Tennessee, Chattanooga

James A. Tyner
Kent State University

Robert Ulack
University of Kentucky

Jialing Wang
Slippery Rock University of Pennsylvania

Linda Q. Wang
University of South Carolina, Aiken

Keith Yearman
College of DuPage

Laura A. Zeeman
Red Rocks Community College

Acknowledgments

World Regional Geography, Third Edition

Kathryn Alftine
California State University, Monterey Bay

Donna Arkowski
Pikes Peak Community College

Tim Bailey
Pittsburg State University

Brad Baltensperger
Michigan Technological University

Michele Barnaby
Pittsburg State University

Daniel Bedford
Weber State University

Richard Benfield
Central Connecticut State University

Sarah Brooks
University of Illinois, Chicago

Jeffrey Bury
University of Colorado, Boulder

Michael Busby
Murray State University

Norman Carter
California State University, Long Beach

Gary Cummisk
Dickinson State University

Cyrus Dawsey
Auburn University

Elizabeth Dunn
University of Colorado, Boulder

Margaret Foraker
Salisbury University

Robert Goodrich
University of Idaho

Steve Graves
California State University, Northridge

Ellen Hansen
Emporia State University

Sophia Harmes
Towson University

Mary Hayden
Pikes Peak Community College

R. D. K. Herman
Towson University

Samantha Kadar
California State University, Northridge

James Keese
California Polytechnic State University

Phil Klein
University of Northern Colorado

Debra D. Kreitzer
Western Kentucky University

Soren Larsen
Georgia Southern University

Unna Lassiter
California State University, Long Beach

David Lee
Florida Atlantic University

Anthony Paul Mannion
Kansas State University

Leah Manos
Northwest Missouri State University

Susan Martin
Michigan Technological University

Luke Marzen
Auburn University

Chris Mayda
Eastern Michigan University

Michael Modica
San Jacinto College

Heather Nicol
State University of West Georgia

Ken Orvis
University of Tennessee

Thomas Paradis
Northern Arizona University

Amanda Rees
University of Wyoming

Arlene Rengert
West Chester University of Pennsylvania

B. F. Richason
St. Cloud State University

Deborah Salazar
Texas Tech University

Steven Schnell
Kutztown University

Kathleen Schroeder
Appalachian State University

Roger Selya
University of Cincinnati

Dean Sinclair
Northwestern State University

Garrett Smith
Kennesaw State University

Jeffrey Smith
Kansas State University

Dean Stone
Scott Community College

Selima Sultana
Auburn University

Ray Sumner
Long Beach City College

Christopher Sutton
Western Illinois University

Harry Trendell
Kennesaw State University

Karen Trifonoff
Bloomsburg University

David Truly
Central Connecticut State University

Kelly Victor
Eastern Michigan University

Mark Welford
Georgia Southern University

Wendy Wolford
University of North Carolina, Chapel Hill

Laura Zeeman
Red Rocks Community College

World Regional Geography, Second Edition

Helen Ruth Aspaas
Virginia Commonwealth University

Cynthia F. Atkins
Hopkinsville Community College

Timothy Bailey
Pittsburg State University

Robert Maxwell Beavers
University of Northern Colorado

James E. Bell
University of Colorado, Boulder

Richard W. Benfield
Central Connecticut State University

John T. Bowen Jr.
University of Wisconsin, Oshkosh

Stanley Brunn
University of Kentucky

Donald W. Buckwalter
Indiana University of Pennsylvania

Gary Cummisk
Dickinson State University

Roman Cybriwsky
Temple University

Cary W. de Wit
University of Alaska, Fairbanks

Ramesh Dhussa
Drake University

David M. Diggs
University of Northern Colorado

Jane H. Ehemann
Shippensburg University

Kim Elmore
University of North Carolina, Chapel Hill

Thomas Fogarty
University of Northern Iowa

James F. Fryman
University of Northern Iowa

Heidi Glaesel
Elon College

Ellen R. Hansen
Emporia State University

John E. Harmon
Central Connecticut State University

Michael Harrison
University of Southern Mississippi

Douglas Heffington
Middle Tennessee State University

Robert Hoffpauir
California State University, Northridge

Acknowledgments

Catherine Hooey
Pittsburg State University

Doc Horsley
Southern Illinois University, Carbondale

David J. Keeling
Western Kentucky University

James Keese
California Polytechnic State University

Debra D. Kreitzer
Western Kentucky University

Jim LeBeau
Southern Illinois University, Carbondale

Howell C. Lloyd
Miami University of Ohio

Judith L. Meyer
Southwest Missouri State University

Judith C. Mimbs
University of Tennessee, Chattanooga

Monica Nyamwange
William Paterson University

Thomas Paradis
Northern Arizona University

Firooza Pavri
Emporia State University

Timothy C. Pitts
Edinboro University of Pennsylvania

William Preston
California Polytechnic State University

Gordon M. Riedesel
Syracuse University

Joella Robinson
Houston Community College

Steven M. Schnell
Northwest Missouri State University

Kathleen Schroeder
Appalachian State University

Dean Sinclair
Northwestern State University

Robert A. Sirk
Austin Peay State University

William D. Solecki
Montclair State University

Wei Song
University of Wisconsin, Parkside

William Reese Strong
University of North Alabama

Selima Sultana
Auburn University

Suzanne Traub-Metlay
Front Range Community College

David J. Truly
Central Connecticut State University

Alice L. Tym
University of Tennessee, Chattanooga

World Regional Geography, First Edition

Helen Ruth Aspaas
Virginia Commonwealth University

Brad Bays
Oklahoma State University

Stanley Brunn
University of Kentucky

Altha Cravey
University of North Carolina, Chapel Hill

David Daniels
Central Missouri State University

Dydia DeLyser
Louisiana State University

James Doerner
University of Northern Colorado

Bryan Dorsey
Weber State University

Lorraine Dowler
Pennsylvania State University

Hari Garbharran
Middle Tennessee State University

Baher Ghosheh
Edinboro University of Pennsylvania

Janet Halpin
Chicago State University

Peter Halvorson
University of Connecticut

Michael Handley
Emporia State University

Robert Hoffpauir
California State University, Northridge

Glenn G. Hyman
International Center for Tropical Agriculture

David Keeling
Western Kentucky University

Thomas Klak
Miami University of Ohio

Darrell Kruger
Northeast Louisiana University

David Lanegran
Macalester College

David Lee
Florida Atlantic University

Calvin Masilela
West Virginia University

Janice Monk
University of Arizona

Heidi Nast
DePaul University

Katherine Nashleanas
University of Nebraska, Lincoln

Tim Oakes
University of Colorado, Boulder

Darren Purcell
Florida State University

Susan Roberts
University of Kentucky

Dennis Satterlee
Northeast Louisiana University

Kathleen Schroeder
Appalachian State University

Dona Stewart
Georgia State University

Ingolf Vogeler
University of Wisconsin, Eau Claire

Susan Walcott
Georgia State University

These world regional geography textbooks have been a family project many years in the making. Lydia Pulsipher came to the discipline of geography at the age of 5, when her immigrant father, Joe Mihelič, hung a world map over the breakfast table in their home in Coal City, Illinois, where he was pastor of the New Hope Presbyterian Church, and quizzed her on the location of such places as Istanbul. They soon moved to the Mississippi Valley of eastern Iowa, where Joe, then a professor at the Presbyterian theological seminary in Dubuque, continued his geography lessons on the passing landscapes whenever Lydia accompanied him on Sunday trips to small country churches.

Lydia's sons, Anthony and Alex, got their first doses of geography in the bedtime stories she told them. For plots and settings, she drew on the Caribbean colonial documents she had been reading for her dissertation. They first traveled abroad and learned about the hard labor of field geography when, at age 12 and 8, they were expected to help with the archaeological and ethnographic research conducted by Lydia and her colleagues on the eastern Caribbean island of Montserrat. It was Lydia's brother John Mihelič who first suggested that Lydia, Alex, and Mac write a book like this one, after he too came to appreciate geography. John has been a loyal cheerleader during the process, as

Acknowledgments

have family and friends in Knoxville, Montserrat, California, Slovenia, and beyond.

Graduate students and faculty colleagues in the geography department at the University of Tennessee have been generous in their support, serving as helpful impromptu sounding boards for ideas. Ken Orvis, especially, has advised us on the physical geography sections of all editions. Yingkui (Philip) Li provided information on glaciers and climate change; Russell Kirby wrote one of the vignettes, based on his research in Vietnam; Toby Applegate, Alex Pulsipher (in his capacity as an instructor), Michelle Brym, and Sara Beth Keough have helped us understand how to better assist instructors; and Ron Kalafsky, Tom Bell, Margaret Gripshover, and Micheline Van Riemsdijk have chatted with the authors many times on specific and broad issues related to this textbook.

Maps for this edition were conceived by Mac Goodwin and Alex Pulsipher and produced by Will Fontanez and the University of Tennessee cartography shop staff and by Maps.com under the direction of Mike Powers. Alex Pulsipher created and produced the photo essays and chose all the photos used in the book.

Liz Widdicombe and Sara Tenney at W. H. Freeman were the first to facilitate the idea that together we could develop a new direction for *World Regional Geography*, one that included the latest thinking in geography written in an accessible style and well illustrated with attractive, relevant maps and photos. In accomplishing this goal, we are especially indebted to our first developmental editor, Susan Moran, and to the W. H. Freeman staff for all they have done in the first years and since to ensure that this book is beautifully designed and well presented to the public.

We would also like to gratefully acknowledge the efforts of the following people at W. H. Freeman: Steven Rigolosi, publisher for this third edition, who has been extraordinarily supportive and resourceful; Debra Ginsberg, developmental editor, who has remained calm and congenial under great pressure; Vivien Weiss, senior project editor; Tom Digiano, marketing manager; Anna Paganelli, copyeditor; Blake Logan, design manager; Matt McAdams, art manager; Susan Wein, production manager; and Stephanie Ellis, assistant editor.

Given our ambitious new photo program, we are especially grateful for Blake Logan's brilliant work and responsiveness as designer for the third edition, as well as for Hilary Newman and Jennifer MacMillan's guidance and direction as our photo editors for the third edition. We are also grateful to the supplements authors, who have created what we think are unusually useful, up-to-date, and labor-saving materials for instructors who use our book.

WORLD REGIONAL GEOGRAPHY CONCEPTS

World Map

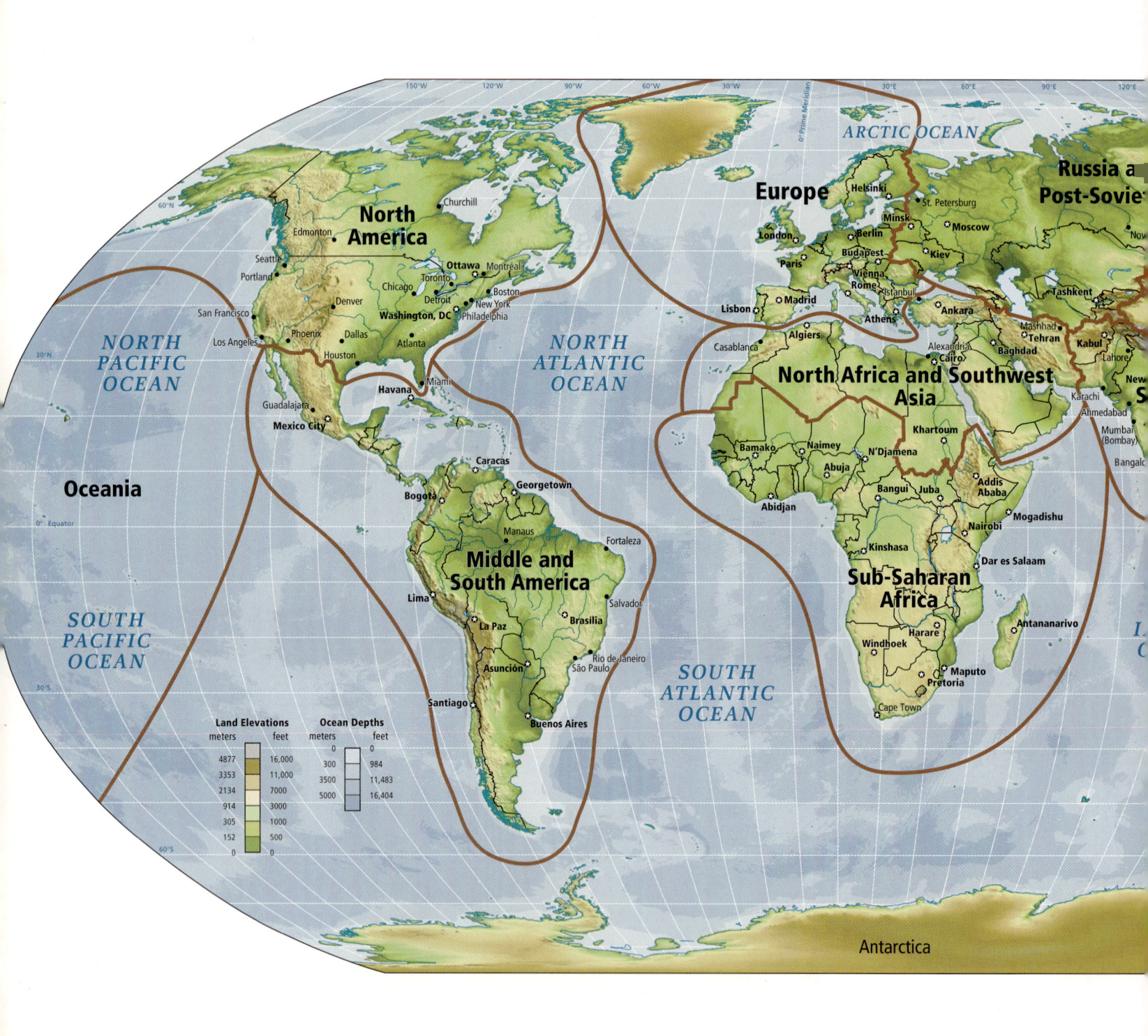

Regions:
- North America
- Middle and South America
- Europe
- Russia and Post-Soviet States
- North Africa and Southwest Asia
- Sub-Saharan Africa
- Oceania
- Antarctica

Oceans:
- Arctic Ocean
- North Pacific Ocean
- South Pacific Ocean
- North Atlantic Ocean
- South Atlantic Ocean

Cities labeled:

North America: Churchill, Edmonton, Seattle, Portland, San Francisco, Los Angeles, Phoenix, Denver, Dallas, Houston, Chicago, Detroit, Atlanta, Ottawa, Montréal, Toronto, Boston, New York, Philadelphia, Washington DC, Miami, Havana, Guadalajara, Mexico City

Middle and South America: Caracas, Bogotá, Georgetown, Manaus, Fortaleza, Lima, La Paz, Brasília, Salvador, Rio de Janeiro, São Paulo, Asunción, Santiago, Buenos Aires

Europe: Helsinki, St. Petersburg, London, Paris, Lisbon, Madrid, Berlin, Minsk, Moscow, Kiev, Budapest, Vienna, Rome, Athens, Istanbul

North Africa and Southwest Asia: Casablanca, Algiers, Alexandria, Cairo, Ankara, Baghdad, Tehran, Mashhad, Kabul, Lahore, Karachi, Ahmedabad, Khartoum, Bamako, Naimey, Abuja, N'Djamena

Sub-Saharan Africa: Abidjan, Bangui, Juba, Addis Ababa, Mogadishu, Nairobi, Kinshasa, Dar es Salaam, Harare, Antananarivo, Windhoek, Maputo, Pretoria, Cape Town

Asia: Tashkent, Mumbai (Bombay), Bangalore

Land Elevations

meters	feet
4877	16,000
3353	11,000
2134	7000
914	3000
305	1000
152	500
0	0

Ocean Depths

meters	feet
0	0
300	984
3500	11,483
5000	16,404

chapter 1
GEOGRAPHY: AN EXPLORATION OF CONNECTIONS

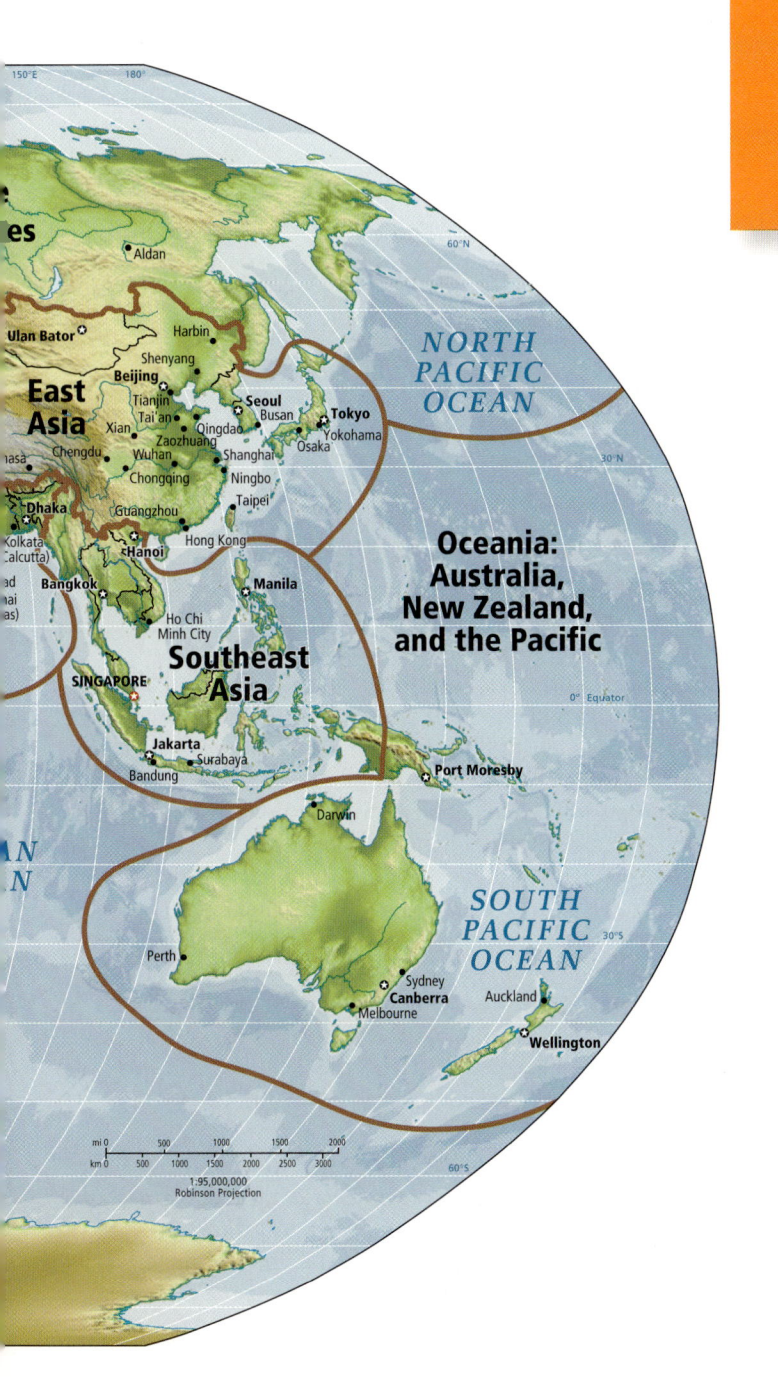

FIGURE 1.1 Regions of the world.

GEOGRAPHIC INSIGHTS

After you read this chapter, you will be able to discuss the following geographic insights as they relate to the five thematic concepts:

1. Environment: Humans are altering the planet at an unprecedented rate, causing sometimes drastic effects on ecosystems and climate. Multiple environmental factors often interact to influence the vulnerability of a location to the impacts of climate change. These vulnerabilities have a spatial pattern.

2. Globalization and Development: Global flows of information, goods, and people are transforming patterns of economic development. Local self-sufficiency is giving way to global interdependence as people and places are increasingly becoming connected, sometimes across vast distances.

3. Power and Politics: There are major differences across the world in the ways that power is wielded in societies. Modes of governing that are more authoritarian are based on the power of the state or community leaders. Modes that are based on notions of political freedom and democracy give the general public greater power over themselves and more of a role in deciding how policies are developed and governments are run. There are also many other ways of managing political power.

4. Urbanization: The development of urban manufacturing and service economies has pulled people into cities. Meanwhile, the mechanization of food production has drastically reduced the need for agricultural labor, thus pushing people out of rural areas.

5. Population and Gender: Population growth is slowing for a number of reasons, among them the increasing numbers of women who are delaying childbearing as they pursue educational and work opportunities outside the home.

WHERE IS IT? WHY IS IT THERE? WHY DOES IT MATTER?

Where are you? You may be in a house or a library or sitting under a tree on a fine fall afternoon. You are probably in a community (perhaps a college or university), and you are in a country (perhaps the United States) and a region of the world (perhaps North America, Southeast Asia, or the Pacific). Why are you where you are? Some answers are immediate, such as "I have an assignment to read." Other explanations are more complex, such as your belief in the value of an education, your career plans, and your or someone's willingness to sacrifice to pay your tuition. Even past social movements that opened up higher education to more than a fortunate few may help explain why you are where you are.

The questions *where* and *why* are central to geography. Think about a time when you had to find the site of a party on a Saturday night, the location of the best grocery store, or the fastest and safest route home. You were interested in location, spatial relationships, and connections between the environment and people. Those are among the interests of geographers.

Geographers seek to understand why different places have different sights, sounds, smells, and arrangements of features. They study what has contributed to the look and feel of a place, to the standard of living and customs of the people, and to the way people in one place relate to people in other places. Furthermore, geographers often think on several scales, from the local to the global. For example, when choosing the best location for a new grocery store, a geographer might consider the physical characteristics of potential sites, the socioeconomic circumstances of the neighborhood, and traffic patterns locally and in the city at large, as well as the store's location relative to the main population concentrations for the whole city. She would probably also consider national or even international transportation routes, possibly to determine cost-efficient connections to suppliers.

To make it easier to understand a geographer's many interests, try this exercise. Draw a map of your most familiar childhood landscape. Relax, and recall the objects and experiences that were most important to you there. If the place was your neighborhood, you might start by drawing and labeling your home. Then fill in other places you encountered regularly, such as your backyard, your best friend's home, or your school. Figure 1.2 shows the childhood landscape remembered by Julia Stump in Franklin, Tennessee.

Consider how your map reveals the ways in which your life was structured by space. What is the scale of your map? That is, how much space did you decide to illustrate on the map? The amount of space your map covers may represent the degree of freedom you had as a child, or how aware you were of the world around you. Were there places you were not supposed to go? Does your map reveal, perhaps subtly, such emotions as fear, pleasure, or longing? Does it indicate your sex, your ethnicity, or the makeup of your family? Did you use symbols to show certain features? In making your map and analyzing it, you have engaged in several aspects of geography:

- Landscape observation
- Descriptions of the Earth's surface and consideration of the natural environment

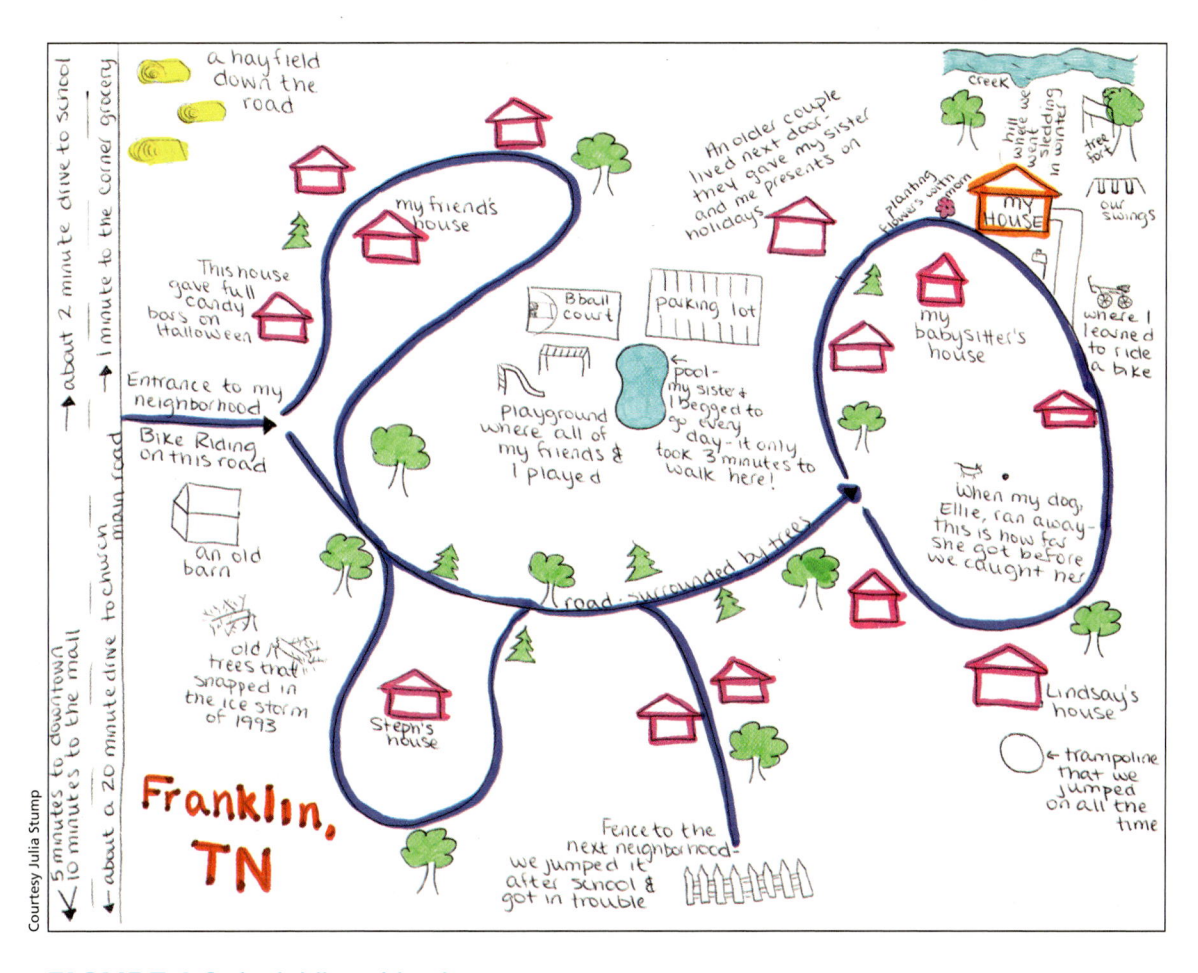

FIGURE 1.2 A childhood landscape map. Julia Stump drew this map of her childhood landscape in Franklin, Tennessee, as an exercise in Dr. Pulsipher's world geography class.

- Spatial analysis (the study of how people, objects, or ideas are related to one another across space)
- The use of different scales of analysis (your map probably shows the spatial features of your childhood at a detailed *local scale*)
- Cartography (the making of maps)

As you progress through this book and this course, you will acquire geographic information and skills that will help you achieve your goals, whatever they are. If you want to travel or work outside your hometown or simply understand local events within the context of world events, knowing how to practice geography will make your task easier and more engaging.

WHAT IS GEOGRAPHY?
Physical and Human Geography

The primary concern of both physical and human geography is the study of the Earth's surface and the interactive physical and human processes that shape the surface.

Geography is the study of our planet's surface and the processes that shape it. Yet this definition does not begin to convey the fascinating interactions of human and environmental forces that have given the Earth its diverse landscapes and ways of life.

Geography as an academic discipline is unique in that it links the physical sciences—such as geology, physics, chemistry, biology, and botany—with the social sciences—such as anthropology, sociology, history, economics, and political science. **Physical geography** generally focuses on how the Earth's physical processes work independently of humans, but physical geographers have become increasingly interested in how physical processes may affect humans and how humans affect these processes in return. **Human geography** is the study of the various aspects of human life that create the distinctive landscapes and regions of the world. Physical and human geography are often tightly linked. For example, geographers might try to understand:

- How and why people came to occupy a particular place
- How people use the physical aspects of that place (climate, landforms, and resources) and then modify them to suit their particular needs
- How people may create environmental problems
- How people interact with other places, far and near

physical geography the study of the Earth's physical processes: how they work and interact, how they affect humans, and how they are affected by humans

human geography the study of patterns and processes that have shaped human understanding, use, and alteration of the Earth's surface

4 CHAPTER 1 Geography: An Exploration of Connections

Geographers usually specialize in one or more fields of study, or subdisciplines. Some of these particular types of geography are mentioned over the course of the book. Despite their individual specialties, geographers often cooperate in studying **spatial interactions** between people and places and the **spatial distribution** of relevant phenomena. For example, in the face of increasing global warming, climatologists, cultural geographers, and economic geographers work together to understand the spatial distribution of carbon dioxide emissions, as well as the cultural and economic practices that might be changed to limit such emissions. This could take the form of redesigning urban areas so that people can live closer to where they work or encouraging food production in locations closer to where the food will be consumed.

Many geographers specialize in a particular region of the world, or even in one small part of a region. Regional geography is the analysis of the geographic characteristics of a particular place, the size and scale of which can vary radically. The study of a region can reveal connections among physical features and ways of life, as well as connections to other places. These links are key to understanding the present and the past, and are essential in planning for the future. This book follows a *world regional* approach, focusing on general knowledge about specific regions of the world. We will see just what geographers mean by *region* a little later in this chapter.

GEOGRAPHERS' VISUAL TOOLS

Among geographers' most important tools are maps, which they use to record, analyze, and explain spatial relationships, as you did on your childhood landscape map. Geographers who specialize in depicting geographic information on maps are called **cartographers**.

Understanding Maps

A map is a visual representation of space used to record, display, analyze, and explain spatial relationships. Figure 1.3 explains the various features of maps.

Legend and Scale The first thing to check on a map is the **legend**, which is usually a small box somewhere on the map that provides basic information about how to read the map, such as the meaning of the symbols and colors used (see parts A–C and the Legend box in Figure 1.3). Sometimes the scale of the map is also given in the legend.

In cartography, *scale* has a slightly different meaning than it does in general geographic analysis. **Scale** on a map refers to the relationship between the size of things on the map and the actual size they have on the surface of the Earth. It is usually represented by a scale bar (see Figure 1.3D–G) but is also sometimes represented by a ratio (for example 1:8000) or a fraction (1/8000), which indicates what one unit of measure on the map equals in the same units on the ground. For example, 1:8000 in. means that 1 inch on the map represents 8000 inches (about an eighth of a mile) on the surface of the Earth.

A scale of 1/800 is considered larger than a scale of 1/8000 because the features on a 1/800 scale map are larger and can be shown in greater detail. The larger the scale of the map, the smaller the area it covers. A larger-scale map shows things larger; a smaller-scale map shows more things—with each thing smaller, less visible. You can remember this with the following statement: "Things look larger on a larger-scale map."

In Figure 1.3, different *scales of imagery* are demonstrated using maps, photographs, and satellite images. Read the captions carefully to understand the scale being depicted in each image. Throughout this book, you will encounter different kinds of maps at different scales. Some will show physical features, such as landforms or climate patterns at the regional or global scale. Others will show aspects of human activities at these same regional or global scales—for example, the routes taken by drug traders. Yet other maps will show patterns of settlement or cultural features at the scale of countries or regions, or cities or even local neighborhoods (see Figure 1.3D–G).

It is important to keep the two types of scales used in geography—*map scale* and *scale of analysis*—distinct, because they have opposite meanings! In spatial analysis of a region such as Southwest Asia, scale refers to the spatial extent of the area that is being discussed. Thus a large-scale analysis means a large area is being explored. But in cartography, a large-scale map is one that shows a given area blown up so that fine detail is visible, while a small-scale map shows a larger area in much less detail. In this book, when we talk about scale we are referring to its meaning in spatial analysis (larger scale = larger area), unless we specifically indicate that we are talking about scale as used in cartography (larger scale = smaller area). In the Scale box in Figure 1.3, the largest-scale map is that on the left (D); the smallest is on the right (G).

Longitude and Latitude Most maps contain lines of latitude and longitude, which enable a person to establish a position on the map relative to other points on the globe. Lines of **longitude** (also called *meridians*) run from pole to pole; lines of **latitude** (also called *parallels*) run around the Earth parallel to the equator (see Figure 1.3H).

Both latitude and longitude lines describe circles, so there are 360° (the symbol ° refers to degrees) in each circle of latitude and 180° in each pole-to-pole semicircle of longitude. Each degree spans 60 minutes (minutes are designated with the symbol ′), and each minute has 60 seconds (which are designated with the symbol ″). Keep in mind that these are measures of relative linear space on a circle, not measures of time. They do not even represent real distance because the circles of latitude get successively smaller to the north and south of the equator until they become virtual dots at the poles.

---

spatial interaction the flow of goods, people, services, or information across space and among places

spatial distribution the arrangement of a phenomenon across the Earth's surface

cartographers geographers who specialize in depicting geographic information on maps

legend a small box somewhere on a map that provides basic information about how to read the map, such as the meaning of the symbols and colors used

scale (of a map) the proportion that relates the dimensions of the map to the dimensions of the area it represents; also, variable-sized units of geographical analysis from the local scale to the regional scale to the global scale

longitude the distance in degrees east and west of Greenwich, England; lines of longitude, also called meridians, run from pole to pole (the line of longitude at Greenwich is 0° and is known as the prime meridian)

latitude the distance in degrees north or south of the equator; lines of latitude run parallel to the equator, and are also called parallels

Geographers' Visual Tools   **5**

FIGURE 1.3 Understanding Maps

The Legend

Being able to read a map legend is crucial to understanding the maps in this book. The colors in the legend convey information about different areas on the map. In the population density map below, the lowest density (0–3 persons per square mile), is colored light tan. A part of North America with this density is shown in the map inset to the right of the legend. On the far right is a picture of this area. Two other densities (27–260 and more than 2600 people per square mile) are also shown in this manner.

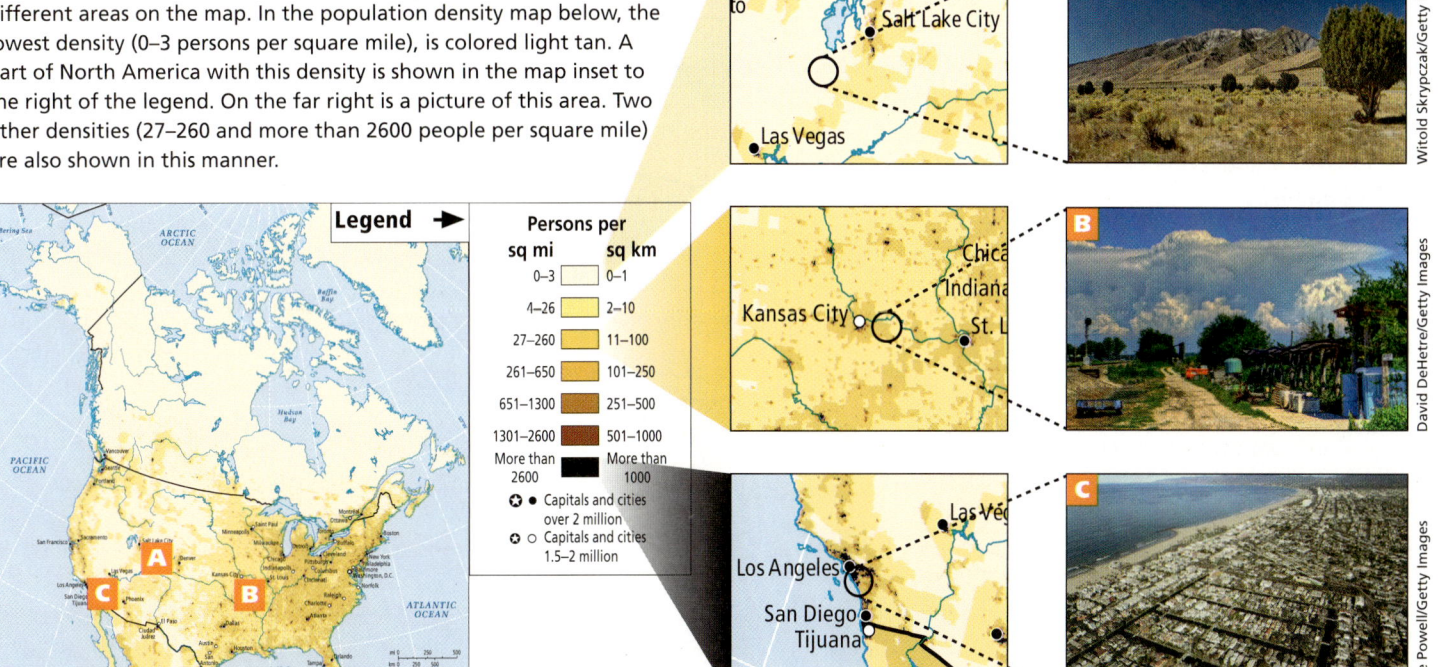

Legend

Persons per
sq mi	sq km
0–3	0–1
4–26	2–10
27–260	11–100
261–650	101–250
651–1300	251–500
1301–2600	501–1000
More than 2600	More than 1000

⊙ ● Capitals and cities over 2 million
⊙ ○ Capitals and cities 1.5–2 million

Witold Skrypczak/Getty Images
David DeHetre/Getty Images
Mike Powell/Getty Images

Scale

Maps often display information at different spatial scales, which means that lengths, areas, distances, and sizes can appear dramatically different on otherwise similar maps. This book often combines maps at several different scales with photographs taken by people at Earth's surface and photographs taken by satellites or astronauts in space. All of these visual tools convey information at a spatial scale. Here are some of the map scales you might encounter in this book and elsewhere. The scale is visible below each image.

USGS D   0 400 800 ft
USGS E   0 1600 3200 ft
USGS F   0 2600 5200 ft
USGS G   0 5200 10,400 ft

Representations of Scale

0   4000   8000 ft

This scale bar means that the length of the entire box represents 8000 feet on the ground.

mi 0 1000 2000 3000 4000 5000
km 0 1000 2000 3000 4000 5000 6000 7000 8000

This scale bar works like the one on the left, but shows lengths both in miles and kilometers.

1:8000

This means that 1 unit of measure (e.g., an inch or a fingerwidth) equals 8000 similar units of measure on the ground.

(continued on next page)

6 CHAPTER 1 Geography: An Exploration of Connections

FIGURE 1.3 Understanding Maps (continued)

Latitude and Longitude

H Lines of longitude and latitude form a global scale grid that can be used to designate the location of any place on the planet. [Source consulted: *The New Comparative World Atlas* (Maplewood, NJ: Hammond, 1997), p. 6]

The distance between lines of longitude decreases toward the poles.

Lines of latitude decrease in length as they they approach the poles.

Lines of latitude and longitude intersect at right angles.

Lines of latitude are parallel to each other.

The prime meridian is at 0° longitude and passes through Greenwich, England.

The equator is at 0° latitude.

The equator divides the globe into Northern and Southern hemispheres.

The half of the globe's surface west of the prime meridian is called the Western Hemisphere; the half to the east is called the Eastern Hemisphere.

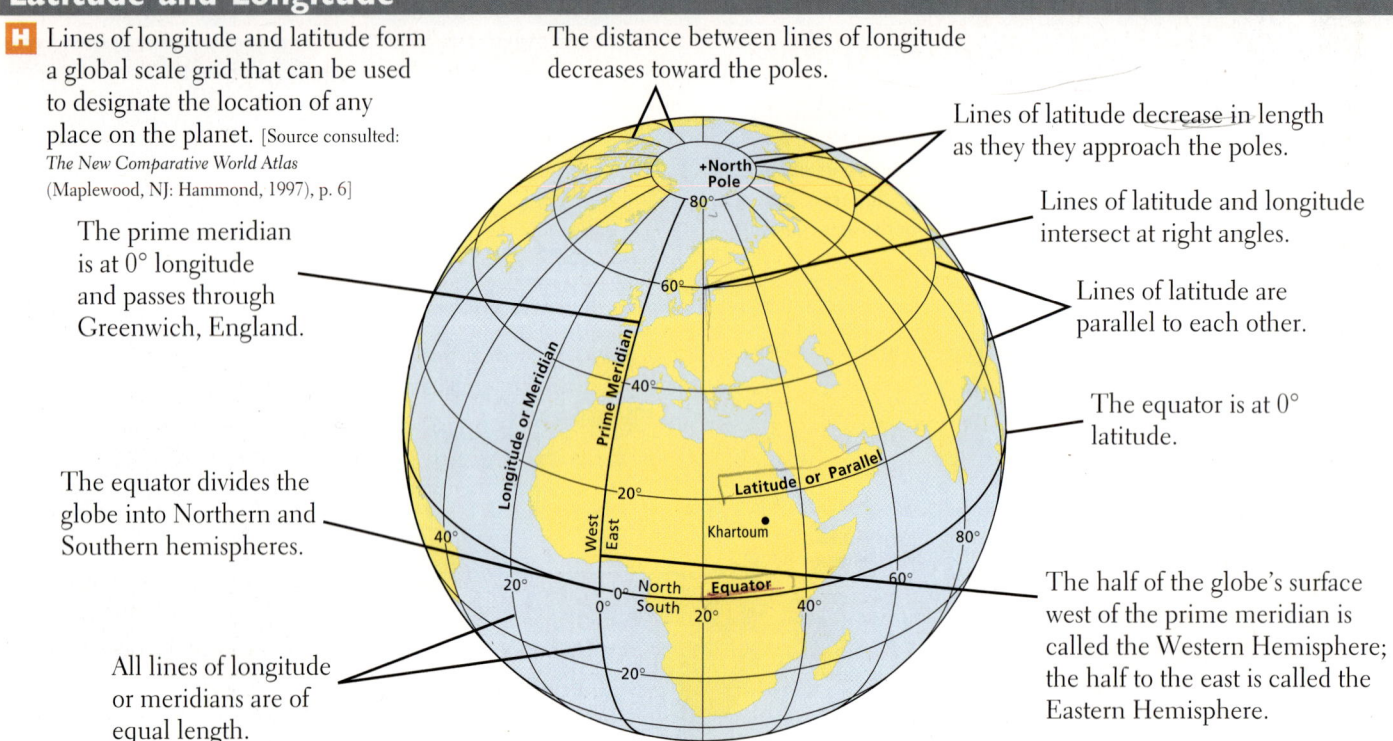

All lines of longitude or meridians are of equal length.

Projections

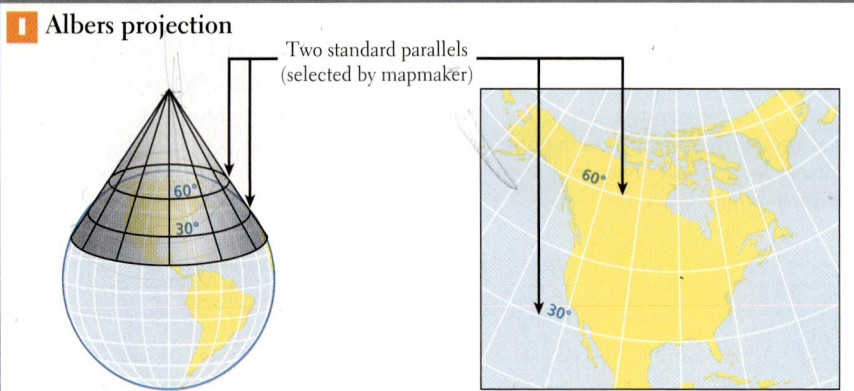

I Albers projection

Two standard parallels (selected by mapmaker)

Pros: Minimal distortion near two parallels (lines of latitude).

Cons: Areas farther away from these lines have distortion.

J Mercator projection

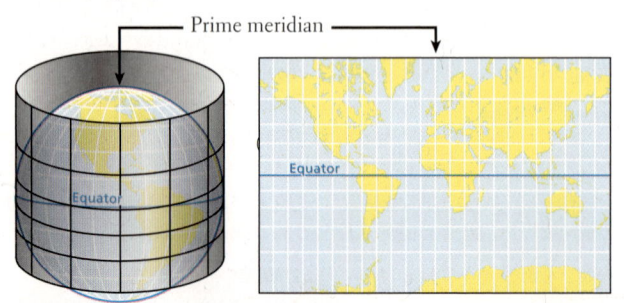

Prime meridian

Pros: A straight line between two points on this map gives an accurate compass direction between them. Minimal distortion within 15° of the equator.

Cons: Extreme distortion near the poles, especially above 60° latitude.

K Robinson projection

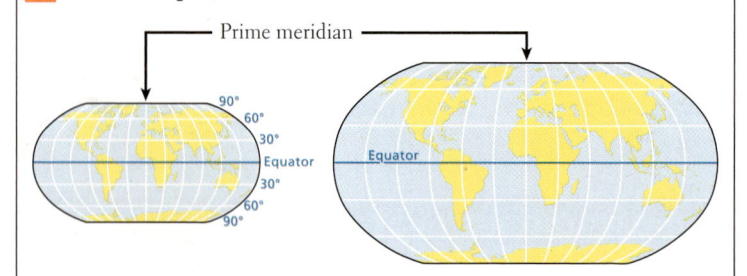

Prime meridian

Pros: Uninterrupted view of land and ocean. Less distortion in high latitudes than in the Mercator projection.

Cons: The shapes of landmasses are slightly distorted due to the curvature of the longitude lines.

The globe is also divided into hemispheres. The Northern and Southern hemispheres are on either side of the equator. The Western and Eastern hemispheres are defined as follows. The prime meridian, 0° longitude, runs from the North Pole through Greenwich, England, to the South Pole. The half of the globe's surface west of the prime meridian is called the Western Hemisphere; the half to the east is called the Eastern Hemisphere. The longitude lines both east and west of the prime meridian are labeled from 1° to 180° by their direction and distance in degrees from the prime meridian. For example, 20 degrees east longitude would be written as 20° E. The longitude line at 180° runs through the Pacific Ocean and is used roughly as the international date line; the calendar day officially begins when midnight falls at this line.

The equator divides the globe into the Northern and Southern hemispheres. Latitude is measured from 0° at the equator to 90° at the North or South Poles.

Lines of longitude and latitude form a grid that can be used to designate the location of a place. In Figure 1.3H, notice the dot that marks the location of Khartoum below the 20th parallel in eastern Africa. The position of Khartoum is 15° 35′ 17″ N latitude by 32° 32′ 3″ E longitude.

Map Projections Printed maps must solve the problem of showing the spherical Earth on a flat piece of paper. Imagine drawing a map of the Earth on an orange, peeling the orange, and then trying to flatten out the orange-peel map and transfer it exactly to a flat piece of paper. The various ways of showing the spherical surface of the Earth on flat paper are called **map projections**. All projections create some distortion. For maps of small parts of the Earth's surface, the distortion is minimal. Developing a projection for the whole surface of the Earth that minimizes distortion is much more challenging.

> **map projections** the various ways of showing the spherical Earth on a flat surface
>
> **Geographic Information Science (GISc)** the body of science that supports multiple spatial analysis technologies and keeps them at the cutting edge

For large midlatitude regions of the Earth that are mainly east/west in extent (North America, Europe, China, Russia), an *Albers projection* is often used. As you can see in Figure 1.3I, this is a conic, or cone-shaped, projection. The cartographer chooses two standard parallels (lines of latitude) on which to orient the map, and these parallels have no distortion. Areas along and between these parallels display minimal distortion. Areas farther to the north or south of the chosen parallels have more distortion. Although all areas on the map are proportional to areas on the ground, distortion of actual shape is inherent in the projection because, as previously discussed, parts of the globe are being projected onto flat paper.

The *Mercator projection* (see Figure 1.3J) has long been used by the general public, but geographers rarely use this projection because of its gross distortion near the poles. To make his flat map, the Flemish cartographer Gerhardus Mercator (1512–1594) stretched out the poles, depicting them as lines equal in length to the equator! As a result, for example, Greenland appears about as large as Africa, even though it is only about one-fourteenth Africa's size. Nevertheless, the Mercator projection is still useful for navigation because it portrays the shapes of landmasses more or less accurately, and because a straight line between two points on this map gives the compass direction between them. However, actual distance measurements are distorted.

The *Robinson projection* (see Figure 1.3K) shows the longitude lines bending toward the poles to give an impression of the Earth's curvature, and it has the advantage of showing an uninterrupted view of land and ocean; however, as a result, the shapes of landmasses are slightly distorted. In this book we often use the Robinson projection for world maps.

Maps are not unbiased. Most currently popular world map projections reflect the European origins of modern cartography. For example, Europe or North America is often placed near the center of the map, where distortion is minimal; other population centers, such as East Asia, are placed at the highly distorted periphery. For a less-biased study of the modern world, we need world maps that center on different parts of the globe. Another source of bias in maps is the convention that north is always at the top of the map. Some cartographers think that this can lead to a subconscious assumption that the Northern Hemisphere is somehow superior to the Southern Hemisphere.

Geographic Information Science (GISc)

The acronym **GISc** is now widespread and usually refers to **Geographic Information Science**, the body of science that supports spatial analysis technologies. GISc is multidisciplinary, using techniques from cartography (mapmaking), geodesy (measuring the Earth's surface), and photogrammetry (the science of making reliable measurements, especially by using aerial photography). Other sciences, such as cognitive psychology and spatial statistics (geomatics or geoinformatics), are increasingly being used to give greater depth and breadth to three-dimensional spatial analysis. GISc, then, can be used in medicine to analyze the human body, in engineering to analyze mechanical devices, in architecture to analyze buildings, in archaeology to analyze sites above and below ground, and in geography to analyze the Earth's surface and the space above and below the Earth's surface.

GISc is a burgeoning field in geography, with wide practical applications in government and business and in efforts to assess and improve human and environmental conditions. GIS (without the *c*) is an older term that refers to geographic information *systems* and is applied to the computerized analytical systems that are the tools of this newest of spatial sciences.

The now widespread use of GISc, particularly by governments and corporations, has dramatically increased the amount of information that is collected and stored, and changed the way it is analyzed and distributed. These changes create many new opportunities for solving problems, for example, by increasing the ability of local governments to plan future urban growth. However, these technologies also raise serious ethical questions. What rights do people have over the storage, analysis, and distribution of information about their location and movements, which can now be gathered from their cell phones? Should this information reside in the public domain? Should individuals have the right to have their location-based information suppressed from public view? Should a government or corporation have the right to sell information to anyone, without special permission, about where people spend their time and how frequently they go to particular places? Progress on these societal questions has not kept pace with the technological advances in GISc.

THE DETECTIVE WORK OF PHOTO INTERPRETATION

Most geographers use photographs to help them understand or explain a geographic issue or depict the character of a place. Interpreting a photo to extract its geographical information can sometimes be like detective work. Below are some points to keep in mind as you look at the pictures throughout this book. Try them out first with the photo in Figure 1.4.

(A) Landforms: Notice the lay of the land and the landform features. Is there any indication of how the landforms and humans have influenced each other? Is environmental stress visible?

(B) Vegetation: Notice whether the vegetation indicates a wet or dry, or warm or cold environment. Can you recognize specific species? Does the vegetation appear to be natural or influenced by human use?

(C) Material culture: Are there buildings, tools, clothing, foods, plantings, or vehicles that give clues about the cultural background, wealth, values, or aesthetics of the people who live where the picture was taken?

(D) What do the people in the photo suggest about the situation pictured?

(E) Can you see evidence of the global economy, such as goods that probably were not produced locally?

(F) Location: From your observations, can you tell where the picture was taken or narrow down the possible locations?

You can use this system to analyze any of the photos in this book and anywhere else. Practice by analyzing the photos in this book before you read their captions. Here is an example of how you could do this with Figure 1.4:

(A) Landforms:

1. The flat horizon suggests a plain or a river delta. *Environmental stress is visible in several places.*

2. This oily liquid doesn't look natural. Could it be crude oil? *What would have caused the landscape transformation? Maybe an oil spill?*

(B) Vegetation:

3. This looks like fairly rich vegetation. These could be palm trees or other types of plant life found in tropical climates. *Must be fairly wet and warm, possibly tropical.*

(C) Material culture:

There is not much that is obviously material culture here, just a single person. The whole area might be abandoned.

(D) People:

4. The clothing on this person doesn't look like he made it. It looks mass produced. *This suggests that he has access to goods produced some distance away, maybe in a nearby city. Or he could buy things in a market where imported goods are sold.*

(E) Global economy: See (D).

(F) Location: This could be somewhere tropical where there could have been an oil spill. Hint: Use this book! Look at Figure 6.19 on page 253 to see the member countries of OPEC (the Organization of the Petroleum Exporting Countries). The combination of the possible oil spill and the vegetation suggests that the photo could be of Venezuela, Ecuador, Nigeria, Angola, or Indonesia. Suggestion: Read chapters 3, 7, and 10!

FIGURE 1.4 Oil development and the environment. A man walks through swampy land in 2010. An international development company began extracting products from this area 50 years ago. A recent UN report stated that the area now needs a massive cleanup, which could take up to 30 years and cost more than a billion dollars, making it one of the biggest such efforts in the world. The area has had approximately 300 incidences of oil-related pollution each year since the 1970s, causing an unknown number of deaths.

THE REGION AS A CONCEPT

Regions

The concept of *region* is useful to geographers because it allows them to break up the world into manageable units in order to analyze and compare spatial relationships. Nonetheless, regions do not have rigid definitions and their boundaries are fluid.

A **region** is a unit of the Earth's surface that contains distinct patterns of physical features and/or distinct patterns of human development. It could be a desert region, a region that produces rice, or a region experiencing ethnic violence. Geographers rarely use the same set of attributes to describe any two regions. For example, the region of the southern United States might be defined by its distinctive vegetation, architecture, music, foods, and historical experience. Meanwhile Siberia, in eastern Russia, could be defined primarily by its climate, vegetation, remoteness, and sparse settlement.

region a unit of the Earth's surface that contains distinct patterns of physical features and/or distinct patterns of human development

Another issue in defining regions is that they may shift over time. The people and the land they occupy may change so drastically in character that they can no longer be thought of as belonging to a certain region, and become more closely aligned with another, perhaps adjacent, region. Examples of this are countries in Central Europe, such as Poland and Hungary, which, for more than 40 years, were closely aligned with Russia and the Soviet Union, a vast region that stretched across northern Eurasia to the Pacific (Figure 1.5A). Poland and Hungary's borders with western Europe were highly militarized and shut to travelers. In 2004, following the demise of the Soviet Union in the early 1990s and the drastic political and economic changes that then came about, Poland and Hungary became members of the European Union (EU; see Figure 1.5B). Their western borders are now open, while their eastern borders are now more heavily guarded in order to keep unwelcome immigrants and other influences out of the European Union. But through all this

FIGURE 1.5 Changing country alliances and relationships in Europe before 1989 and in 2013.
(A) Pre-1989 alignment of countries in Europe and the Soviet Union.

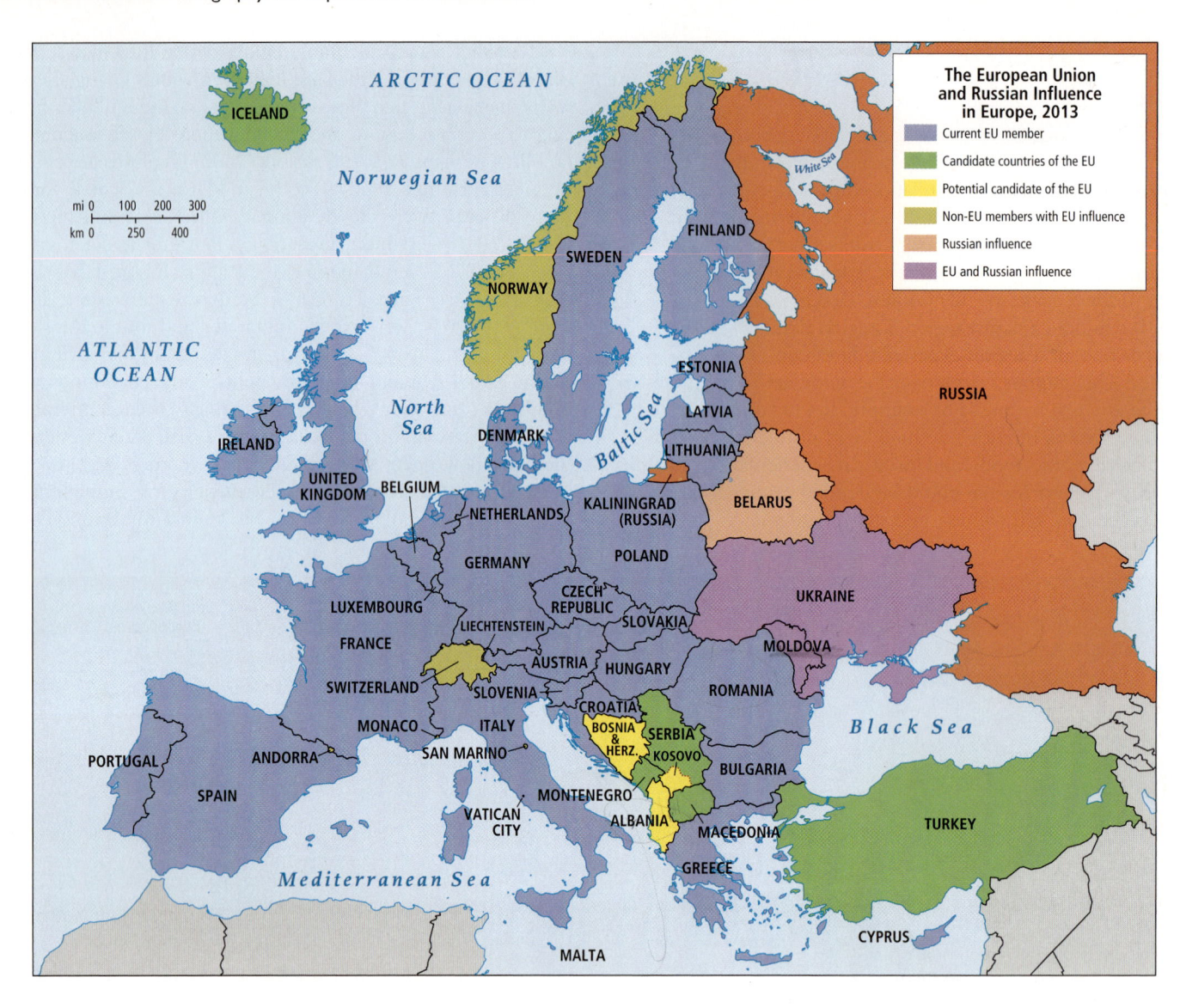

FIGURE 1.5 (B) Post-2013 alignments of the European Union and of Russia and the post-Soviet states.

change—on the ground, in the border regions—people share cultural features (language, religion, historical connections) even as they may define each other as having very different regional allegiances. For this reason, we say regional borders can be *fuzzy*, meaning that they are hard to determine precisely.

A recurring problem in world regional textbooks is the changing nature of regional boundaries and the fact that on the ground they are not clear lines but linear zones of fuzziness. For example, when we first designed this textbook in the mid-1990s, the changes to Europe were just beginning; its eastern limits were under revision as the Soviet Union disintegrated. There were also hints that as the Soviet Union disappeared, countries of Central Asia, long within the Soviet sphere—places like Kazakhstan, Turkmenistan, Kyrgyzstan, Uzbekistan—should perhaps have been defined as constituting a new region of their own. Some suggested that these four, plus Turkey, Iraq, Iran, Saudi Arabia, the United Arab Emirates, Afghanistan, Pakistan, and possibly even western China should become a new post-Soviet world region of Central Asia (Figure 1.6). The suggestion was that this region would be defined by what was thought to be a common religious heritage (Islam), plus long-standing cultural and historical ties and a difficult environment marked by water scarcity but also by rich oil and gas resources.

In fact, grouping these countries together in a single region would misrepresent the current situation as well as the past. A world region known as Central Asia may emerge eventually (and there are groups of interested parties discussing that possibility right now), but such a region will be slow to take shape and will be based on criteria very different from common religion, historical experiences, and environmental features. First of all, these supposed uniting features are, on closer inspection, quite diverse. There are many different versions of Islam practiced from the Mediterranean to western China and from southern Russia to the Hindu Kush. Also, to the extent that there are common historical experiences, they are actually linked more to European and Russian colonial exploitation than to an ancient and deeply uniting Central Asian cultural heritage. Finally, while environments generally defined by water scarcity are common to all the countries listed, oil and gas resources are not uniformly distributed at all. It could be that a Central Asian identity will eventually develop, perhaps centered on the leadership of Turkey or Dubai, which is trying hard to define itself

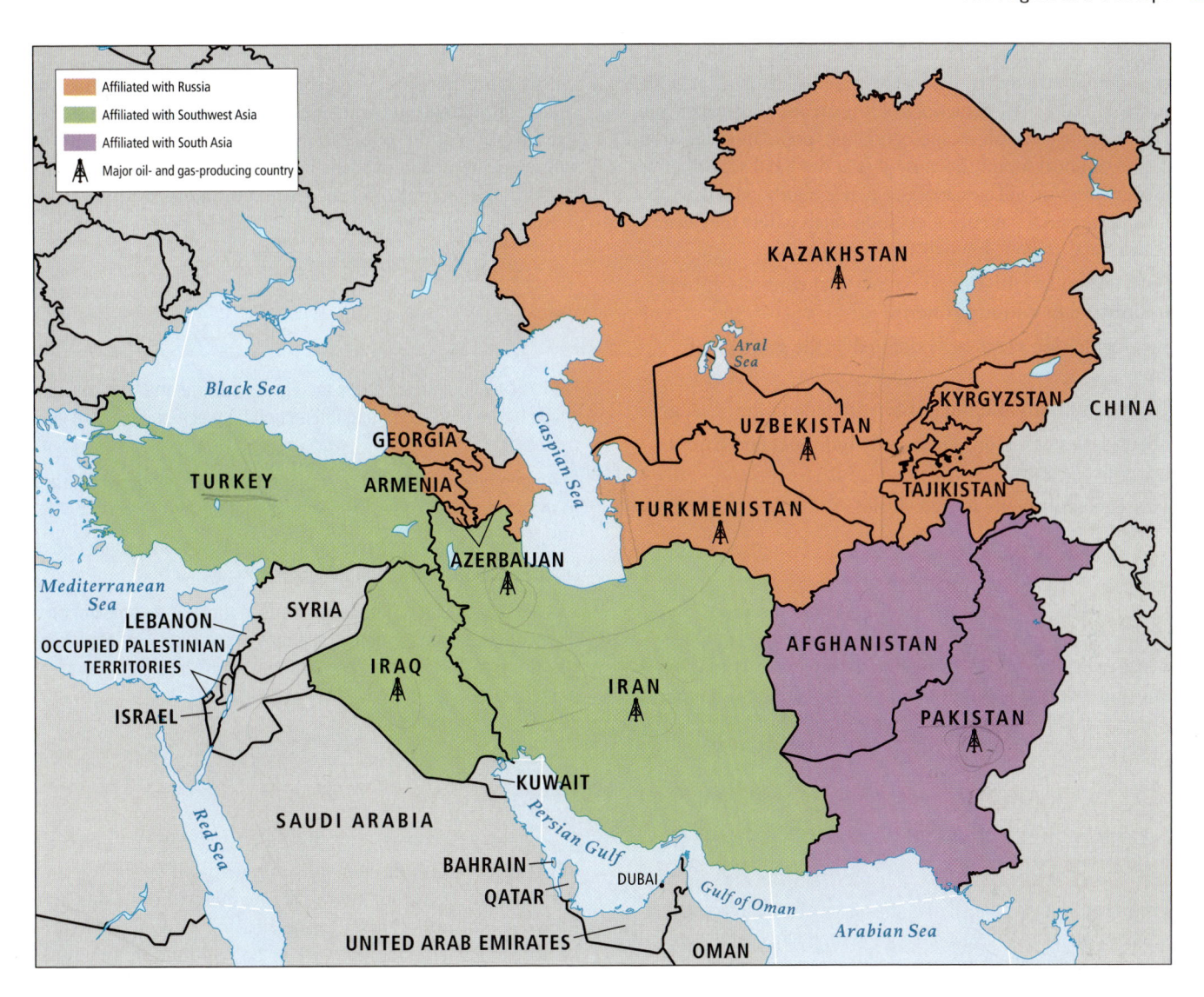

FIGURE 1.6 Hypothetical map of a possible Central Asian world region. A hypothetical region called Central Asia could consist of Kazakhstan, Turkmenistan, Kyrgyzstan, Tajikistan, and Uzbekistan, plus Turkey, Iraq, Iran, the Caucasus (Georgia, Armenia, and Azerbaijan), and also Afghanistan and Pakistan. For cultural reasons, it might even include parts of western China, but we will not consider that here. The Caucasus countries have had a long association with Central Asia; economically, they might find this a better association than one with Europe, which is halfheartedly courting them for their oil. Just which region Saudi Arabia, Jordan, Syria, Yemen, and Israel, would fall into is debatable—perhaps a region called the Eastern Mediterranean and North Africa. The Emirates, despite their location on the Arabian Peninsula, appear to be angling for a leading economic role in such a new Central Asian region.

as the affluent capital of such a region; but thus far, the region has not coalesced. Therefore, in this book, the countries listed above are to be found in the three different regions shown in Figure 1.6.

If regions are so difficult to define and describe, why do geographers use them? To discuss the whole world at once would be impossible, so geographers try to find a reasonable way to divide the world into manageable parts. There is nothing sacred about the criteria or the boundaries for the world regions we use. They are just practical aids to learning. In defining each of the world regions for this book, we have considered such factors as physical features, political boundaries, cultural characteristics, history, how the places now define themselves, and what the future may hold. We are constantly reevaluating regional boundaries and, in this edition, have made some changes. For example, the troubled new country of South Sudan, once part of North Africa and Southwest Asia, is now considered part of sub-Saharan Africa, to which it is more culturally aligned (see Chapter 7, page 278).

This book organizes the material into three *regional scales of analysis*: the **global scale**, the **world regional scale**, and the **local scale**. The term *scale of analysis* refers to the relative size of the area under discussion. At the global scale, explored in this chapter, the entire world is treated as a single area—a unity that is more and more relevant as our planet operates as a global system. We use the term *world region* for the largest divisions of the globe, such as East Asia and

global scale the level of geography that encompasses the entire world as a single unified area

local scale the level of geography that describes the space where an individual lives or works; a city, town, or rural area

North America (see Figure 1.1). We have defined ten world regions, each of which is covered in a separate chapter. In each regional chapter, we consider the interactions of human and physical geography in relation to cultural, social, economic, population, environmental, and political topics. Because it is at the local scale that we live our daily lives—in villages, towns, city neighborhoods—we show how global or regional patterns affect individuals where they live.

In summary, regions have the following traits:

- A region is a unit of the Earth's surface that contains distinct environmental or cultural patterns.
- No two regions are necessarily defined by the same set of attributes.
- Regional definitions and the territory included often change.
- The boundaries of regions are usually indistinct and hard to agree upon.
- Regions can vary greatly in size (scale).

THEMATIC CONCEPTS AND GEOGRAPHIC INSIGHTS IN THIS BOOK

Within the world regional framework, this book is also organized around five thematic concepts of special significance in the modern world. These concepts are the focus of the Geographic Insights of every chapter in this book: the environment, globalization and development, power and politics, urbanization, and population and gender. The sections that follow explain each of these five thematic concepts, how they tie in with the main concerns of geographers, and how they are related to each other.

THINGS TO REMEMBER

- Physical geographers study the Earth's surface and the processes that shape it. There is a continual interaction between physical and human processes.
- Among geographers' most important tools are maps, which they use to record, analyze, and explain spatial relationships.
- The careful analysis of photographs can lead to important understandings about places and geographic issues.
- A region is a unit of the Earth's surface that has a combination of distinct physical and/or human features; the complex of features can vary from region to region, and regional boundaries are rarely clear or precise.

ENVIRONMENT

> **GEOGRAPHIC INSIGHT 1**
>
> **Environment:** Humans are altering the planet at an unprecedented rate, causing sometimes drastic effects on ecosystems and climate. Multiple environmental factors often interact to influence the vulnerability of a location to the impacts of climate change. These vulnerabilities have a spatial pattern.

HUMAN IMPACT ON THE BIOSPHERE

From the beginning of human life, we have overused resources in seeking to improve our own living conditions, sometimes with disastrous consequences. What is new is the scale of human impacts on the planet, which can now be found virtually everywhere. In fact, geoscientists have recently identified a new geologic epoch, called the *Anthropocene*, which is defined as the time during which humans have had an overwhelming impact on Earth's biosphere. Just when the Anthropocene began is under debate.

As people have grown more aware of their environmental impacts, they have put forth numerous proposals to limit damage to the **biosphere**, defined here as the entirety of the Earth's integrated physical systems, with humans and their impacts included as part of nature. Societies have become so transformed by the intensive use of Earth's resources that reversing this level of use is enormously difficult. For example, how possible would it be for you and your entire family to live for even just one day without using any fossil fuels for transportation or home heating or cooling? Would you be able to get to school or work or be comfortable in your home? Intensive per capita resource consumption is now so deeply ingrained, especially in wealthy countries, that with just 20 percent of the world's population, the rich countries consume more than 80 percent of the available world resources.

Human consumption of natural resources is now being examined through the concept of the **ecological footprint**. This is a method of estimating the amount of biologically productive land and sea area needed to sustain a human at the average current standard of living for a given population (country). It is particularly useful for drawing comparisons. For example, about 4.5 acres is the worldwide average of the biologically productive area needed to support one person—this would be one individual's ecological footprint. Because of the lifestyle in the United States, one person's ecological footprint averages about 24 acres; it is about 18 acres in Canada, and just 4 acres in China. You can calculate your own footprint using the Global Footprint Network's calculator, at http://tinyurl.com/6d2wyl4. A similar concept more closely related to global warming is the *carbon footprint*, which measures the greenhouse gas emissions a person's activities produce. To calculate your family's carbon footprint, use Carbon Footprint's calculator, at http://tinyurl.com/2el2j5.

biosphere the entirety of the Earth's integrated physical spheres, with humans and other impacts included as part of nature

ecological footprint the amount of biologically productive land and sea area needed to sustain a person at the current average standard of living for a given population

Because the biosphere is a global ecological system that integrates all living things and their relationships, it is important to raise awareness that actions in widely separated parts of Earth have a cumulative effect on the whole. Figure 1.7 shows a global map of the relative intensity of human biosphere impacts. The map includes photo insets that show particular trouble spots in South America (see Figure 1.7E, F), Europe (see Figure 1.7A), South Asia (see Figure 1.7B), and Southeast Asia (see Figure 1.7C, D). However, to fully appreciate human impacts on the biosphere, some understanding of the underlying physical processes that shape the biosphere is needed.

PHYSICAL GEOGRAPHY

As discussed earlier in this chapter, physical geography is concerned with the processes that shape the Earth's landforms, climate, and vegetation. In this sense, physical geographers are similar to scientists from other disciplines who focus on these phenomena, though as geographers they often look at problems and depict the results of their analysis spatially. In this book, physical geography provides a backdrop for the many aspects of human geography we discuss. Physical geography is a fascinating, large, and growing field of study worth exploring in greater detail. What follows are just the basics of landforms and climate.

Landforms: The Sculpting of the Earth

The processes that create the world's varied **landforms**—mountain ranges, continents, and the deep ocean floor—are some of the most powerful and slow-moving forces on the planet. Originating deep beneath the Earth's surface, these *internal processes* can move entire continents, often taking hundreds of millions of years to do their work. However, it is *external processes* that form many of the Earth's landscape features, such as a beautiful waterfall or a rolling plain. These more rapid and delicate processes take place on the surface of the Earth. Geomorphologists study the processes that constantly shape and reshape the Earth's surface.

Plate Tectonics

Two key ideas related to internal processes in physical geography are the *Pangaea hypothesis* and *plate tectonics*. The geophysicist Alfred Wegener first suggested the Pangaea hypothesis in 1912. This hypothesis proposes that all the continents were once joined in a single vast continent called Pangaea (meaning "all lands"), which then fragmented over time into the continents we know today (Figure 1.8 on page 16). As one piece of evidence for his theory, Wegener pointed to the neat fit between the west coast of Africa and the east coast of South America.

For decades, most scientists rejected Wegener's hypothesis. We now know, however, that the Earth's continents have been assembled into supercontinents a number of times, only to break apart again. All of this activity is made possible by plate tectonics,

> **landforms** physical features of the Earth's surface, such as mountain ranges, river valleys, basins, and cliffs
>
> **plate tectonics** the scientific theory that the Earth's surface is composed of large plates that float on top of an underlying layer of molten rock; the movement and interaction of the plates create many of the large features of the Earth's surface, particularly mountains
>
> **Ring of Fire** the tectonic plate junctures around the edges of the Pacific Ocean; characterized by volcanoes and earthquakes
>
> **weathering** the physical or chemical decomposition of rocks by sun, rain, snow, ice, and the effects of life-forms
>
> **erosion** the process by which fragmented rock and soil are moved over a distance, primarily by wind and water

a process of continental motion discovered in the 1960s, long after Wegener's time.

The study of **plate tectonics** has shown that the Earth's surface is composed of large plates that float on top of an underlying layer of molten rock. The plates are of two types. Oceanic plates are dense and relatively thin, and they form the floor beneath the oceans. Continental plates are thicker and less dense. Much of their surface rises above the oceans, forming continents. These massive plates drift slowly, driven by the circulation of the underlying molten rock flowing from hot regions deep inside the Earth to cooler surface regions and back. The creeping movement of tectonic plates fragmented and separated Pangaea into pieces that are the continents we know today (see Figure 1.8E).

Plate movements influence the shapes of major landforms, such as continental shorelines and mountain ranges. Huge mountains have piled up on the leading edges of the continents as the plates carrying them collide with other plates, folding and warping in the process. Plate tectonics accounts for the long, linear mountain ranges that extend from Alaska to Chile in the Western Hemisphere and from Southeast Asia to the European Alps in the Eastern Hemisphere. The highest mountain range in the world, the Himalayas of South Asia, was created when what is now India, situated at the northern end of the Indian-Australian Plate, ground into Eurasia. The only continent that lacks these long, linear mountain ranges is Africa. Often called the "plateau continent," Africa is believed to have been at the center of Pangaea and to have moved relatively little since the breakup. However, as Figure 1.8E shows, parts of eastern Africa—the Somali Subplate and the Arabian Plate—continue to separate from the continent (the African Plate).

Humans encounter tectonic forces most directly as earthquakes and volcanoes. Plates slipping past each other create the catastrophic shaking of the landscape we know as an earthquake. Plates collide and one slips under the other in a process called *subduction*. Volcanoes arise at zones of subduction or sometimes in the middle of a plate, where gases and molten rock (called magma) can rise to the Earth's surface through fissures and holes in the plate. Volcanoes and earthquakes are particularly common around the edges of the Pacific Ocean, an area known as the **Ring of Fire** (Figure 1.9 on page 17).

Landscape Processes

The landforms created by plate tectonics have been further shaped by external processes, which are more familiar to us because we can observe them daily. One such process is **weathering**. Rock, exposed to the onslaught of sun, wind, rain, snow, ice, freezing and thawing, and the effects of life-forms (such as plant roots), fractures and decomposes into tiny pieces. These particles then become subject to another external process, **erosion**. During erosion, wind and water carry away rock particles and any associated decayed organic matter and deposit them in new locations. The deposition of eroded material can raise and flatten the land around a river,

FIGURE 1.7 PHOTO ESSAY: Human Impacts on the Biosphere

Humans have had enormous impacts on the biosphere. The map and insets show varying levels of these impacts on the biosphere. The map is derived from a synthesis of hundreds of studies. High-impact areas often have roads, railways, agriculture, or other intensive land uses. Low- to medium-impact areas have biodiversity loss and other disturbances related to human activity. [Sources consulted: United Nations Environment Programme, "Human impact, year 1700 (approximately)," and "Human impact, year 2002," (New York: United Nations Development Program), 2002, 2003, 2004, 2005, 2006, at http://www.grida.no/graphicslib/detail/human-impactyear-1700-approximately_6963# and http://www.grida.no/graphicslib/detail/human-impact-year-2002_157a]

where periodic flooding spreads huge quantities of silt. As small valleys between hills are filled in by silt, a **floodplain** is created. Where rivers meet the sea, floodplains often fan out roughly in the shape of a triangle, creating a **delta**. External processes tend to smooth out the dramatic mountains and valleys created by internal processes.

Human activity often contributes to external landscape processes. By altering the vegetative cover, agriculture and forestry expose the Earth's surface to sunlight, wind, and rain. These agents in turn increase weathering and erosion. Flooding becomes more common because the removal of vegetation limits the ability of the Earth's surface to absorb rainwater. As erosion increases, rivers may fill with silt, and deltas may extend into the oceans.

Climate

The processes associated with climate are generally more rapid than those that shape landforms. **Weather**, the short-term and spatially limited expression of climate, can change in a matter of minutes. **Climate** is the long-term balance of temperature and moisture that keeps weather patterns fairly consistent from year to year. By this definition, the last major global climate change took place about 15,000 years ago, when the glaciers of the last ice age began to melt.

Solar energy is the engine of climate. The Earth's atmosphere, oceans, and land surfaces absorb huge amounts of solar energy, and the differences in the amounts they absorb account for part of the variations in climate we observe. The most intense direct solar energy strikes the Earth more or less head-on in a broad band stretching about 30° north and south of the equator. The fact that Earth's axis sits at a 23° angle as it orbits the sun—an angle that does not change—means that where the band of greatest solar intensity strikes the Earth varies in regular sequence over the course of a year, creating seasons. Just how this yearly seasonal pattern works is illustrated by Figure 1.10 on page 17. The highest average temperatures on the Earth's surface are within this band of the strongest solar intensity. Moving away from the equator, solar energy strikes the Earth's surface less directly—at more of an obtuse (wide) angle. This wide angle reduces the heating effect of the sun's rays and results in lower average annual temperatures.

Climate Regions

Geographers have several systems for classifying the world's climates. The systems are based on the patterns of temperature and precipitation just described. This book uses a modification of the widely known *Köppen classification system*, which divides the world into several types of climate regions, labeled A through E on the climate map in Figure 1.11 on pages 18–19. As you look at the regions on this map, examine the photos, and read the accompanying climate

---

floodplain the flat land along a river where sediment is deposited during flooding

delta the triangular-shaped plain of sediment that forms where a river meets the sea

weather the short-term and spatially limited expression of climate that can change in a matter of minutes

climate the long-term balance of temperature and precipitation that characteristically prevails in a particular region

---

Human Impact, 2002

Land cover
- Forests
- Grasslands
- Deserts
- Tundra
- Ice

Human impact on land
- High impact
- Medium–high impact
- Low–medium impact

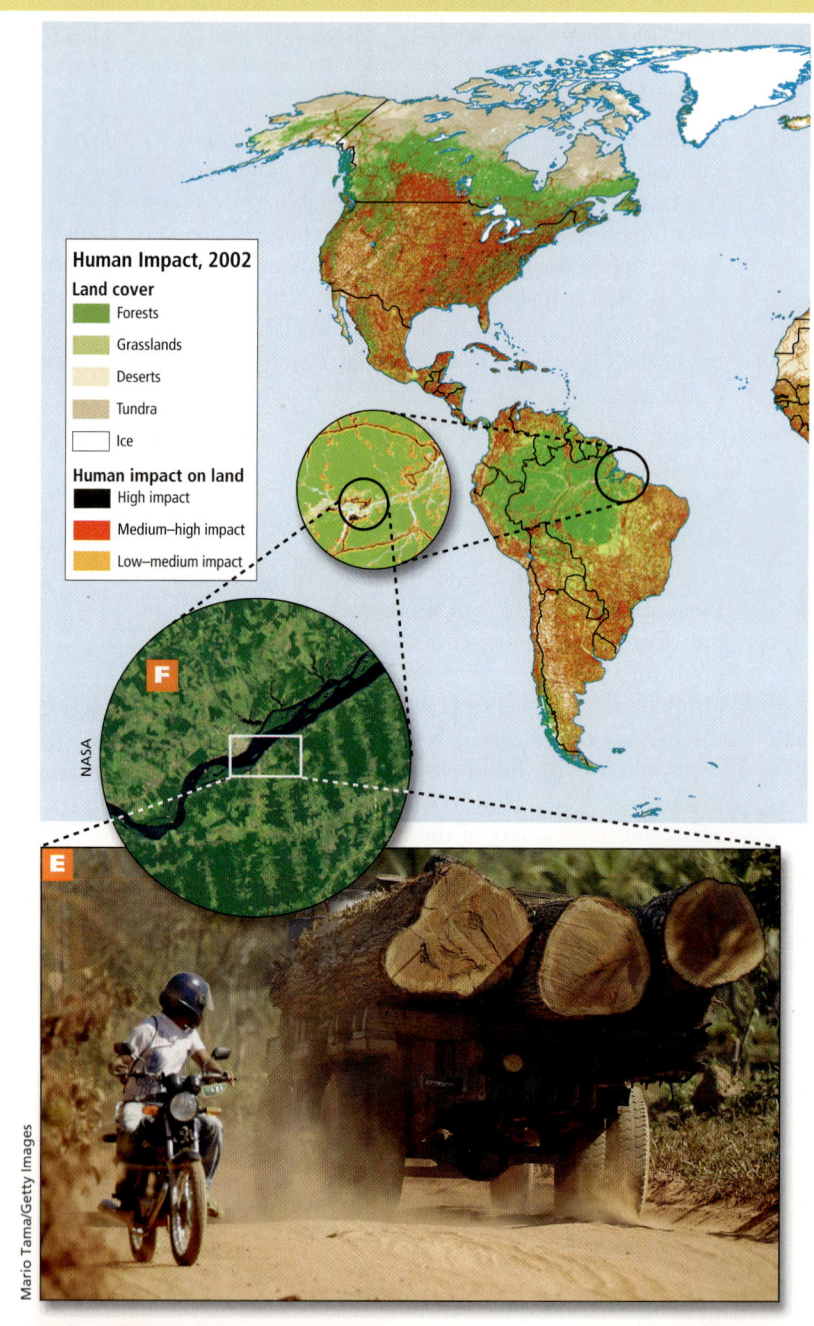

E and **F** **Development and deforestation.** In the Brazilian Amazon, deforestation often occurs in regularized spatial patterns, such as the "fishbone" pattern (see satellite image inset **F**). This pattern results from regulations that determine the location of roads used for settlement and logging. Whole logs are brought by road to rivers, where they are put on barges and taken to a port for export.

A Development and mining. Perhaps no other human activity has as striking an impact on the landscape as mining. This open-pit coal mine is located in one of the most industrialized and densely inhabited parts of Germany.

B War and political conflict. War can have a devastating effect on the environment. Women outside Kabul, Afghanistan, must now carry water by hand from distant sources that have not been polluted or damaged by war.

C and D Food and deforestation. Farmers practicing shifting cultivation (see Chapter 3, page 122) plant "hill rice" in Burma. Shifting cultivation is an ancient technique that can be sustained indefinitely, given sufficient land and fallow periods long enough for forest to regrow (20 years or more). Today, more and more forest is being turned over to short-fallow cultivation—3 to 6 years of cultivation—resulting in a loss of habitat and biodiversity (see inset **D**).

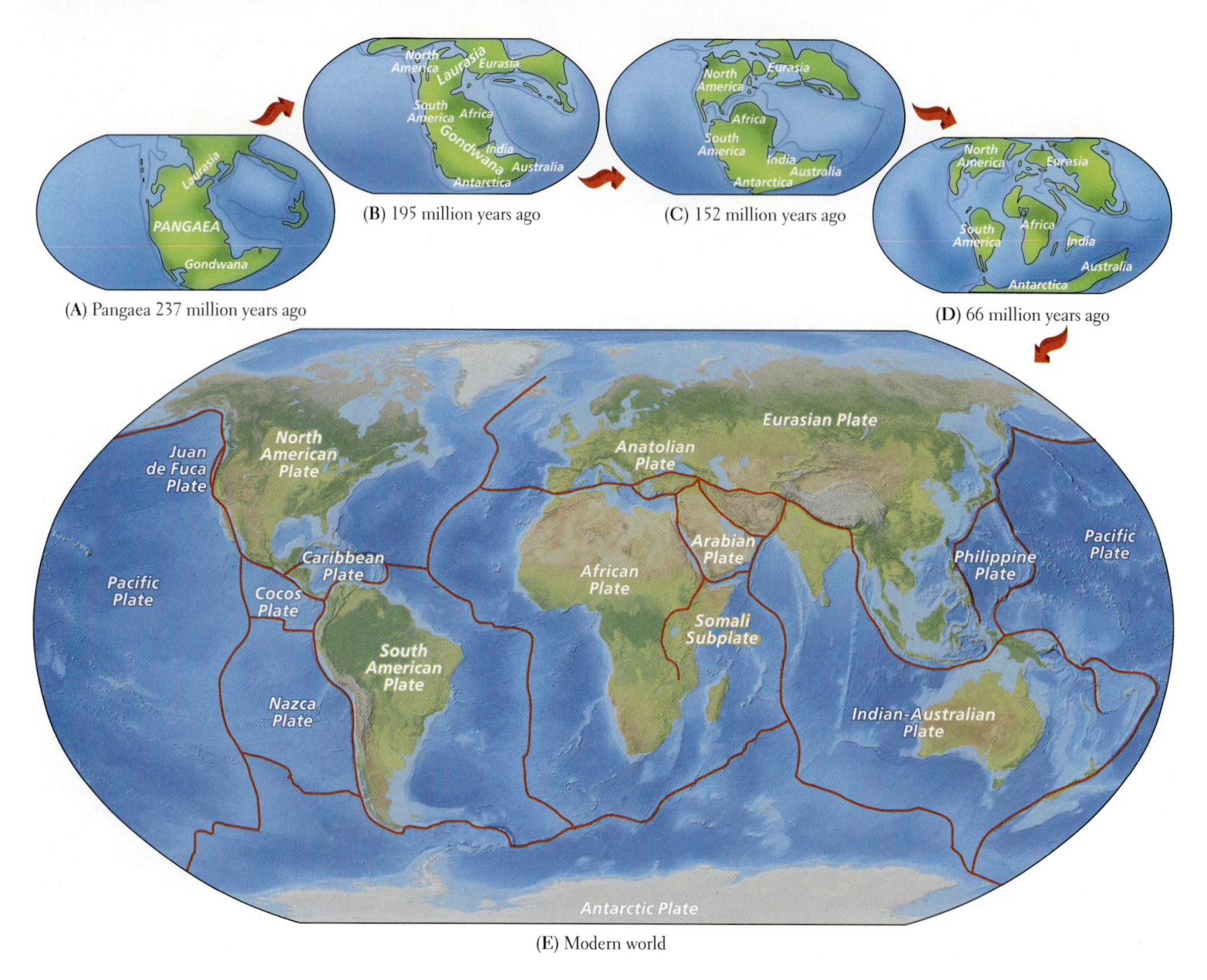

16 CHAPTER 1 Geography: An Exploration of Connections

FIGURE 1.8 The breakup of Pangaea. The modern world map **(E)** depicts the current boundaries of the major tectonic plates. Pangaea is only the latest of several global configurations that have coalesced and then fragmented over the last billion years. [Source consulted: Frank Press, Raymond Siever, John Grotzinger, and Thomas H. Jordan, *Understanding Earth*, 4th ed. (New York: W. H. Freeman, 2004), pp. 42–43]

descriptions, you will notice the importance of climate to vegetation. Each chapter includes a climate map about the region being discussed; when reading these maps, refer to the written descriptions in Figure 1.11 as necessary. Keep in mind that the sharp boundaries shown on climate maps are in reality much more gradual transitions.

Temperature and Air Pressure

The daily wind and weather patterns are largely a result of variations in solar energy absorption that create complex patterns of air temperature and *air pressure*. To understand air pressure, think of air as existing in a particular unit of space—for example, a column of air above a square foot of the Earth's surface. Air pressure is the amount of force (due to the pull of gravity) exerted by that column on that square foot of surface. Air pressure and temperature are related: The gas molecules in warm air are relatively far apart and are associated with low air pressure. In cool air, the gas molecules are relatively close together (dense) and are associated with high air pressure.

As the sun warms a unit of cool air, the molecules move farther apart. The air becomes less dense and exerts less pressure. Air tends to move from areas of higher pressure to areas of lower pressure, creating wind. If you have been to the beach on a hot day, you may have noticed a cool breeze blowing in off the water. This happens because land heats up (and cools down) faster than water, so on a hot day, the air over the land warms, rises, and becomes less dense than the air over the water. This causes the cooler, denser air to flow inland. At night the breeze often reverses direction, blowing from the now cooling land onto the now relatively warmer water.

Physical Geography **17**

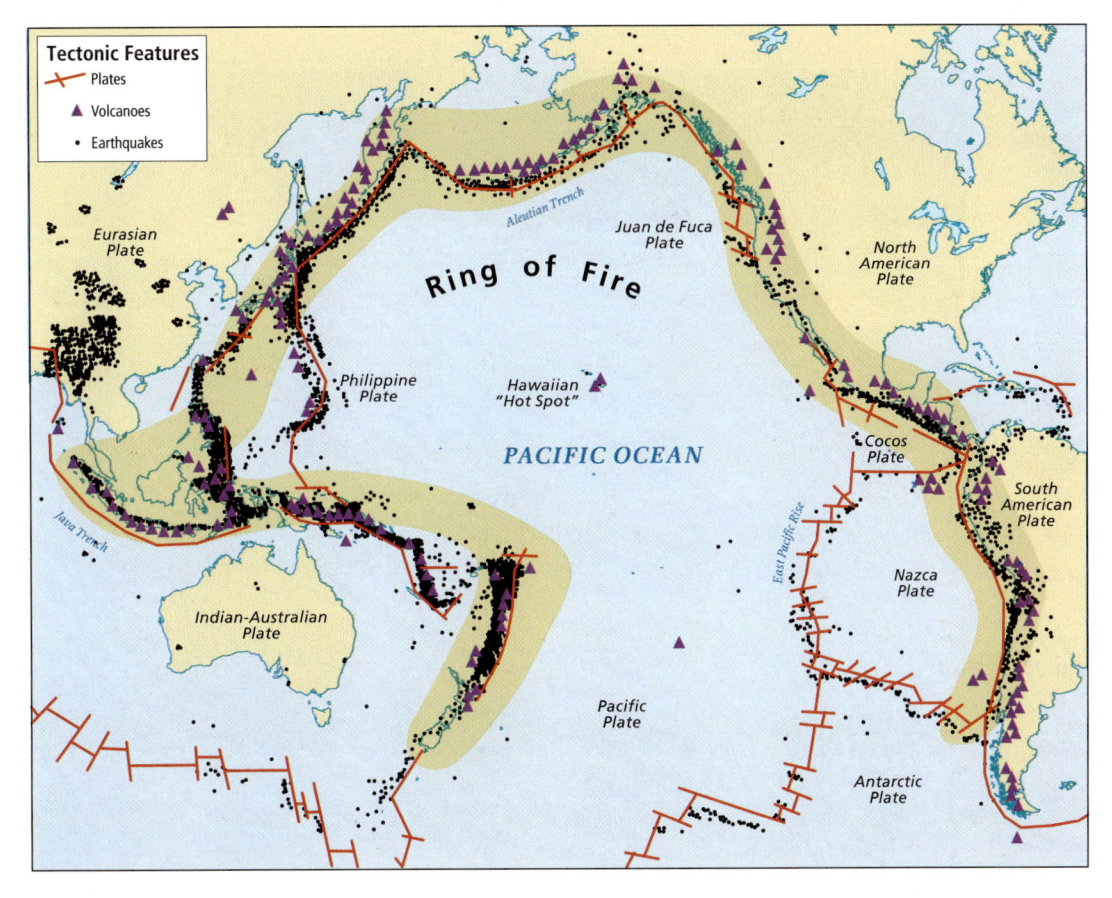

FIGURE 1.9 Ring of Fire. Volcanic formations encircling the Pacific Basin form the Ring of Fire, a zone of frequent earthquakes and volcanic eruptions. [Sources consulted: United States Geological Survey, *Active Volcanoes and Plate Tectonics*, "Hot Spots" and the "Ring of Fire," at http:/vulcan.wr.usgs.gov/Glossary/PlateTectonics/Maps/map_plate_tectonics_world.html; and Frank Press, Raymond Siever, John Grotzinger, and Thomas H. Jordan, *Understanding Earth*, 4th ed. (New York: W. H. Freeman, 2004), p. 27]

ranges, the air cools, and the moisture condenses to produce rainfall (Figure 1.12). This process, known as **orographic rainfall**, is most common in coastal areas where wind blows moist air from above the ocean onto the land and up the side of a coastal mountain range. Most of the moisture falls as rain as the cooling air rises along the coastal side of the range. On the inland side, the descending air warms and ceases to drop its moisture. The drier side of a mountain range is said to be in the *rain shadow*. Rain shadows may extend for hundreds of miles across the interiors of continents, as they do on the Mexican Plateau, or east of California's Pacific coastal ranges, or north of the Himalayas of Eurasia.

A central aspect of Earth's climate is the *rain belt* that exists in equatorial areas. Near the equator, moisture-laden tropical air is heated by the strong sunlight and rises to the point where it releases its moisture as rain. Neighboring nonequatorial areas also receive some of this moisture when seasonally shifting winds move the rain belt north and south of the equator. The huge downpours of the Asian summer monsoon are an example.

These air movements have a continuous and important influence on global weather patterns and are closely associated with land and water masses. Because continents heat up and cool off much more rapidly than the oceans that surround them, the wind tends to blow from the ocean to the land during summer and from the land to the ocean during winter. It is almost as if the continents were breathing once a year, inhaling in summer and exhaling in winter.

precipitation dew, rain, sleet, and snow
orographic rainfall rainfall produced when a moving moist air mass encounters a mountain range, rises, cools, and releases condensed moisture that falls as rain

Precipitation

Perhaps the most tangible way we experience changes in air temperature and pressure is through rain or snow. **Precipitation** (dew, rain, sleet, and snow) occurs primarily because warm air holds more moisture than cool air. When this warmer moist air rises to a higher altitude, its temperature drops, reducing its ability to hold moisture. The moisture condenses into drops to form clouds and may eventually fall as rain or snow.

Several conditions that encourage moisture-laden air to rise influence the patterns of precipitation around the globe. When moisture-bearing air is forced to rise as it passes over mountain

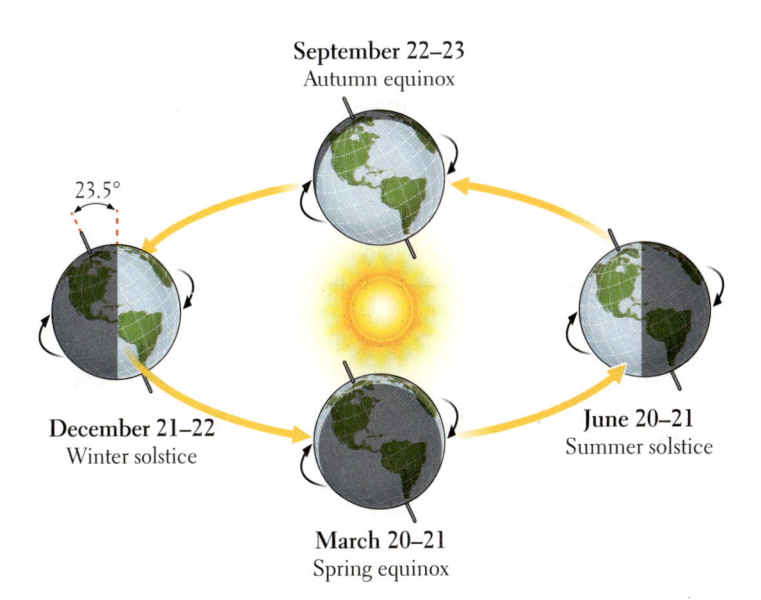

FIGURE 1.10 Diagram of the angle of Earth's orientation to the sun.

FIGURE 1.11   PHOTO ESSAY: Climate Regions of the World

A **Tropical humid climates.** In *tropical wet climates,* rain falls predictably every afternoon and usually just before dawn. The *tropical wet/dry climate,* also called a *tropical savanna,* has a wider range of temperatures and a wider range of rainfall fluctuation than the tropical wet climate.

B **Arid and semiarid climates.** *Deserts* generally receive very little rainfall (2 inches or less per year). Most of that rainfall comes in downpours that are extremely rare and unpredictable. *Steppes* have climates similar to those of deserts, but that are more moderate. They usually receive about 10 inches more rain per year than deserts and are covered with grass or scrub.

C **Temperate climates.** *Midlatitude temperate climates* are moist all year and have short, mild winters and long, hot summers. *Subtropical climates* differ from midlatitude climates in that subtropical winters are dry. *Mediterranean climates* have moderate temperatures but are dry in summer and wet in winter.

D **Cool humid climates.** Stretching across the broad interiors of Eurasia and North America are *continental climates,* which either have dry winters (northeastern Eurasia) or are moist all year (North America and north-central Eurasia). Summers in cool humid climates are short but can have very warm days.

E **Coldest climates.** *Arctic* and *high-altitude climates* are by far the coldest and are also among the driest. Although moisture is present, there is little evaporation because of the low temperatures. The Arctic climate is often called *tundra,* after the low-lying vegetation that covers the ground. The high-altitude version of this climate, which can be far from the Arctic, is more widespread and subject to greater daily fluctuations in temperature. High-altitude microclimates, such as those in the Andes and the Himalayas, can vary tremendously, depending on factors such as available moisture, orientation to the sun, and vegetation cover. As one ascends in altitude, the changes in climate loosely mimic those found as one moves from lower to higher latitudes. These changes are known as *temperature-altitude zones* (see Figure 3.6 on page 116).

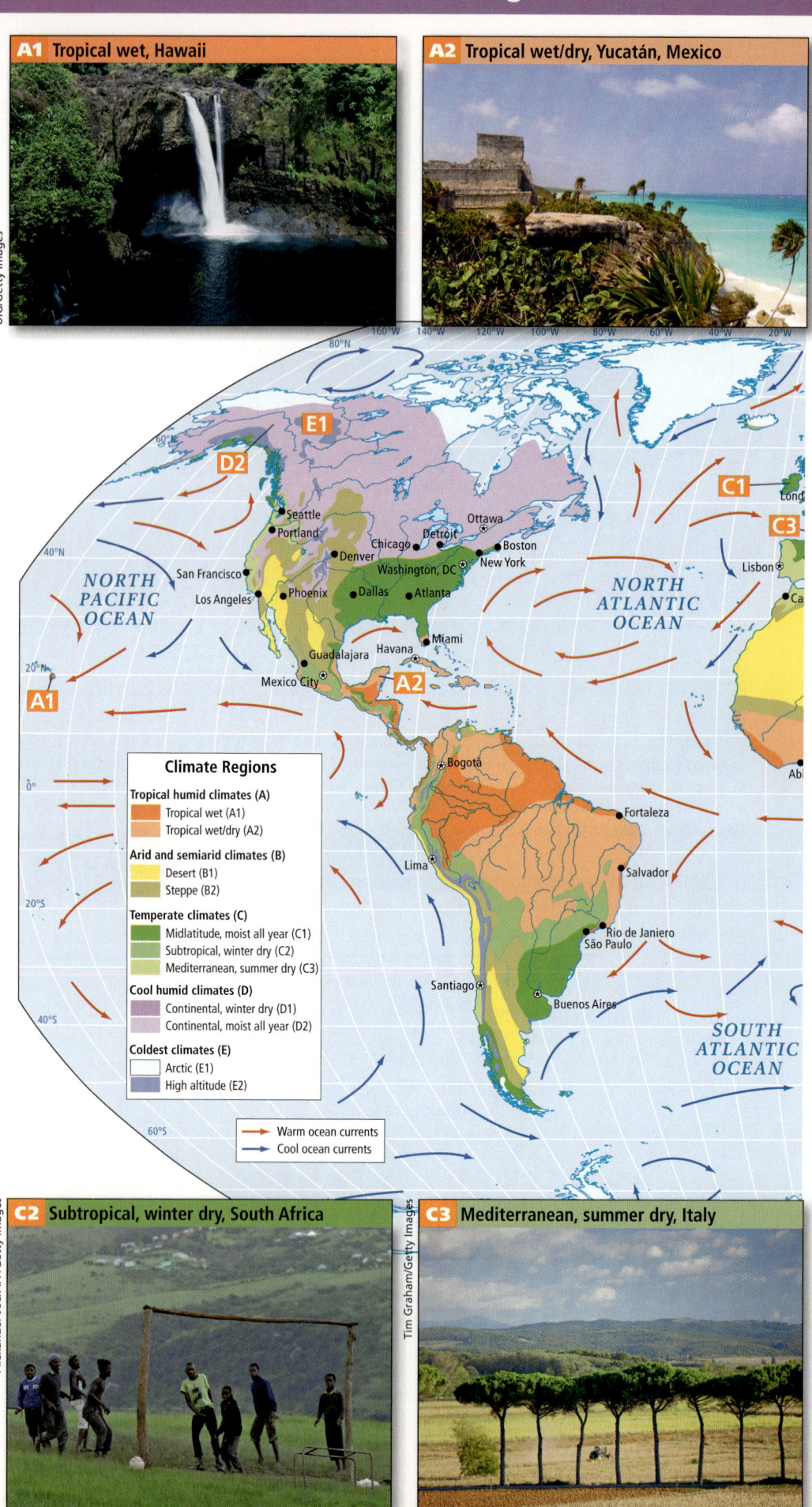

A1 Tropical wet, Hawaii

A2 Tropical wet/dry, Yucatán, Mexico

Climate Regions

- **Tropical humid climates (A)**
  - Tropical wet (A1)
  - Tropical wet/dry (A2)
- **Arid and semiarid climates (B)**
  - Desert (B1)
  - Steppe (B2)
- **Temperate climates (C)**
  - Midlatitude, moist all year (C1)
  - Subtropical, winter dry (C2)
  - Mediterranean, summer dry (C3)
- **Cool humid climates (D)**
  - Continental, winter dry (D1)
  - Continental, moist all year (D2)
- **Coldest climates (E)**
  - Arctic (E1)
  - High altitude (E2)

→ Warm ocean currents
→ Cool ocean currents

C2 Subtropical, winter dry, South Africa

C3 Mediterranean, summer dry, Italy

18

B1 Desert, Namibia

B2 Steppe, Mongolia

C1 Midlatitude, moist all year, United Kingdom

E2 High altitude, Tibet

D1 Continental, winter dry, Russia

D2 Continental, moist all year, Alaska

E1 Arctic tundra, Canada

FIGURE 1.12 Orographic rainfall (and rain shadow diagram).

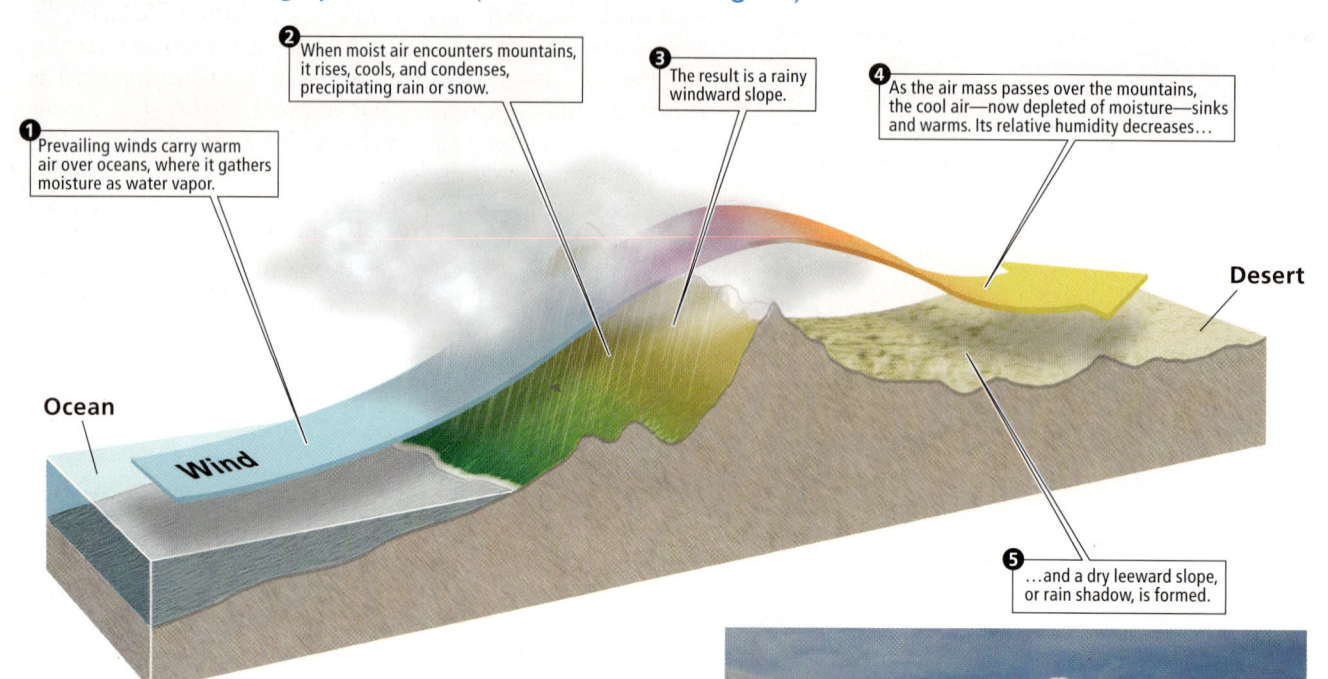

1 Prevailing winds carry warm air over oceans, where it gathers moisture as water vapor.

2 When moist air encounters mountains, it rises, cools, and condenses, precipitating rain or snow.

3 The result is a rainy windward slope.

4 As the air mass passes over the mountains, the cool air—now depleted of moisture—sinks and warms. Its relative humidity decreases…

5 …and a dry leeward slope, or rain shadow, is formed.

(A) A rain shadow diagram. [Source consulted: Frank Press, Raymond Siever, John Grotzinger, and Thomas H. Jordan, *Understanding Earth*, 4th ed. (New York: W. H. Freeman, 2004), p. 281]

(B) Owens Valley, California, sits in the rain shadow created by the Sierra Nevada mountain range, just visible on the left. Clouds that pass over here have already lost most of their moisture on the slopes of the Sierra Nevada.

In the summer **monsoon** season, the Eurasian continental landmass heats up, causing the overlying air to expand, become less dense, and rise. The somewhat cooler, yet moist, air of the Indian Ocean is drawn inland. The effect is so powerful that the equatorial rain belt is sucked onto the land (see Figure 8.4 on page 321). This results in tremendous, sometimes catastrophic, summer rains throughout virtually all of South and Southeast Asia and much of coastal and interior East Asia. The reverse happens in the winter as similar forces pull the equatorial rain belt south during the Southern Hemisphere's summer.

Much of the moisture that falls on North America and Eurasia is *frontal precipitation* caused by the interaction of large air masses of different temperatures and densities. These masses develop when air stays over a particular area long enough to take on the temperature of the land or sea beneath it. Often when we listen to a weather forecast, we hear about warm fronts or cold fronts. A *front* is the zone where warm and cold air masses come into contact, and it is always named after the air mass whose leading edge is moving into an area. At a front, the warm air tends to rise over the cold air, carrying warm clouds to a higher, cooler altitude. Rain or snow may follow. Much of the rain that falls along the outer edges of a hurricane is the result of frontal precipitation.

monsoon a wind pattern in which in summer months, warm, wet air coming from the ocean brings copious rainfall, and in winter, cool, dry air moves from the continental interior toward the ocean

THINGS TO REMEMBER

> **GEOGRAPHIC INSIGHT 1**
> - **Environment** Humans are altering the planet at an unprecedented rate, causing sometimes drastic effects on ecosystems and climate. Multiple environmental factors often interact to influence the vulnerability of a location to the impacts of climate change. These vulnerabilities have a spatial pattern.

- Landforms are shaped by internal processes, such as plate tectonics, and external processes, such as weathering.

- Climate is the long-term balance of temperature and precipitation that keeps weather patterns fairly consistent from year to year. Weather is the short-term and spatially limited expression of climate that can change in minutes.

- Seasons occur because the Earth's axis is positioned at a consistent angle of 23° as it orbits the sun over the course of a year.

- Variations in air pressure are caused by heating and cooling and account for phenomena such as wind and precipitation patterns.

- Precipitation occurs primarily because warm air holds more moisture than cool air. When this warmer moist air rises to a higher altitude, its temperature drops, reducing its ability to hold moisture, resulting in rain or snow. ■

GLOBAL CLIMATE CHANGE

Planet Earth is continually undergoing **climate change**, a slow shifting of climate patterns caused by the general cooling or warming of the atmosphere. The current trend of **global warming**—which refers to the observed warming of the Earth's climate as atmospheric levels of greenhouse gases increase—is extraordinary because it is happening more quickly than climate changes in the past and appears to be linked primarily to human agency. **Greenhouse gases** (**GHG**; carbon dioxide, methane, water vapor, and other gases) are essential to keeping the Earth's incoming and outgoing radiation balanced in such a way that the Earth's surface is maintained in a temperature range hospitable to life. Similar to the glass panes of a greenhouse, these gases allow solar radiation to pass through the atmosphere and strike the Earth's surface; the gases also allow much of this radiation to bounce back into space as surface radiation. But some radiation is re-reflected back to the Earth, keeping its surface (like the interior of a greenhouse) warmer than it would be if incoming and outgoing radiation were in balance. Figure 1.13 shows this system of incoming and outgoing radiation and the role of greenhouse gases in trapping some of the outgoing radiation and sending it back to warm the Earth. Now, as greenhouse gases are being released at accelerating rates by the burning of fossil fuels and by the effects of deforestation (see Figure 1.7A, C, E), the evidence indicates that the Earth is warming at a very fast rate.

Most scientists now agree that there is an urgent need to reduce greenhouse gas emissions to avoid catastrophic climate change in coming years. Climatologists, biogeographers, and other scientists are documenting long-term global warming and cooling trends by examining evidence in tree rings, fossilized pollen and marine creatures, and glacial ice. These data indicate that the twentieth century was the warmest century in 600 years, and that the current decade is the hottest on record. Evidence is mounting that these are not normal fluctuations. Very long-term climate-change patterns indicate that we should be heading into a cooling pattern, but instead it is estimated that, at present rates of warming, by 2100 average global temperatures could rise between 2.5°F and 10°F (about 2°C to 5°C).

However, a key problem is that those most responsible for global warming (the world's wealthiest and most industrialized countries) have the least incentive to reduce emissions because they are the least vulnerable to the changes global warming causes in the physical environment. Meanwhile, those most vulnerable to these changes (poor countries with low human development and with large slums in low-lying coastal wetlands) are the least responsible for the growth in greenhouse gases, and have the least power in the global geopolitical sphere; hence they have the least ability to affect the level of emissions (Figure 1.14). **330. U.S. GOVERNMENT SCIENTISTS CALL FOR URGENT ACTION ON GLOBAL WARMING**

> **climate change** a slow shifting of climate patterns due to the general cooling or warming of the atmosphere
>
> **global warming** the warming of the Earth's climate as atmospheric levels of greenhouse gases increase
>
> **greenhouse gases (GHG)** gases, such as carbon dioxide and methane, released into the atmosphere by human activities, which become harmful when released in excessive amounts

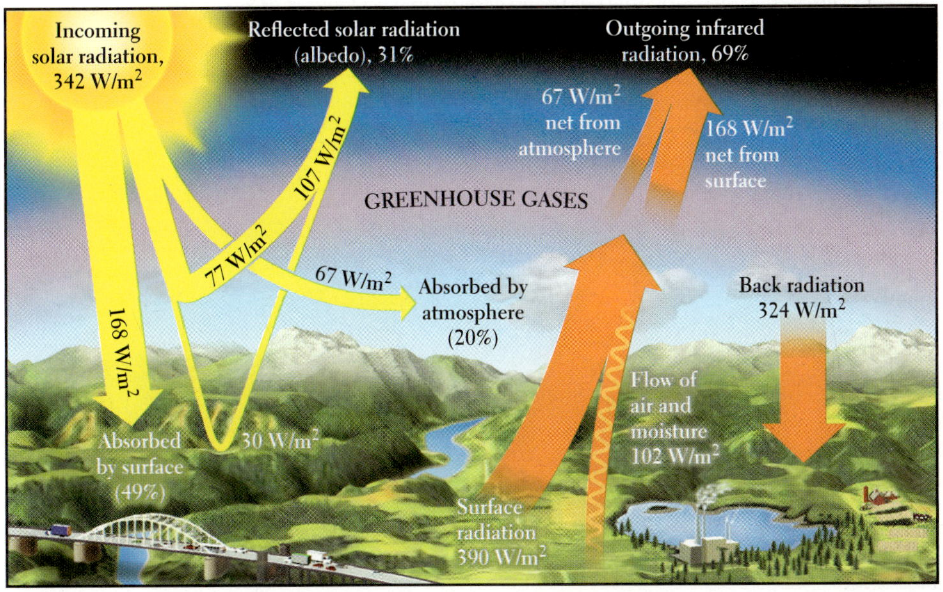

Drivers of Global Climate Change

Greenhouse gases exist naturally in the atmosphere. It is their heat-trapping ability that makes the Earth warm enough for life to exist. Increase their levels, as humans are doing now, and the Earth becomes warmer still.

Electricity generation, vehicles, industrial processes, and the heating of homes and businesses all burn large amounts of CO_2-producing fossil fuels such as coal, natural gas, and oil. Even the large-scale raising of grazing animals contributes methane through the animals' flatulence. Unusually large quantities of greenhouse gases from these sources are accumulating in the Earth's atmosphere, and their presence has already led to significant warming of the planet's climate.

Widespread deforestation worsens the situation. Living forests take in CO_2 from the atmosphere, release the oxygen, and store the carbon in their biomass. As more trees are cut down and their wood is used for fuel, more carbon enters the atmosphere, less is taken

FIGURE 1.13 The balance of incoming and outgoing radiation and the greenhouse effect. To maintain an even temperature, Earth has to balance energy coming in with energy going out. Energy coming in is mostly sunshine, and energy going out is mostly radiant heat. Here the sunshine, or incoming solar radiation, is shown in yellow: some reflects right back into outer space, a little gets absorbed in the air, and about half warms the ground. The numbers represent averages—obviously there is usually more sunshine at noon than at midnight! Heat, mainly infrared radiation, is shown in orange: quite a lot bounces and flows around near the surface in various forms. Clouds, dust, smoke, water vapor, and certain other gases tend to keep it there. But what finally reaches outer space almost exactly balances the amount of sunshine absorbed. These days, scientists find that extra greenhouse gases released by humans are causing Earth to retain extra energy—outgoing infrared radiation seems to average nearly 1 watt per square meter (W/m^2) less than incoming solar radiation, so average temperatures on Earth are rising.

FIGURE 1.14 Greenhouse gas emissions around the world in 2010, total per country and per capita. [Source consulted: UN Department of Economic and Social Affairs, Statistics Division, *Environmental Indicators: Greenhouse Gas Emissions*, 2009, at http://unstats.un.org/unsd/environment/air_greenhouse_emissions.htm]

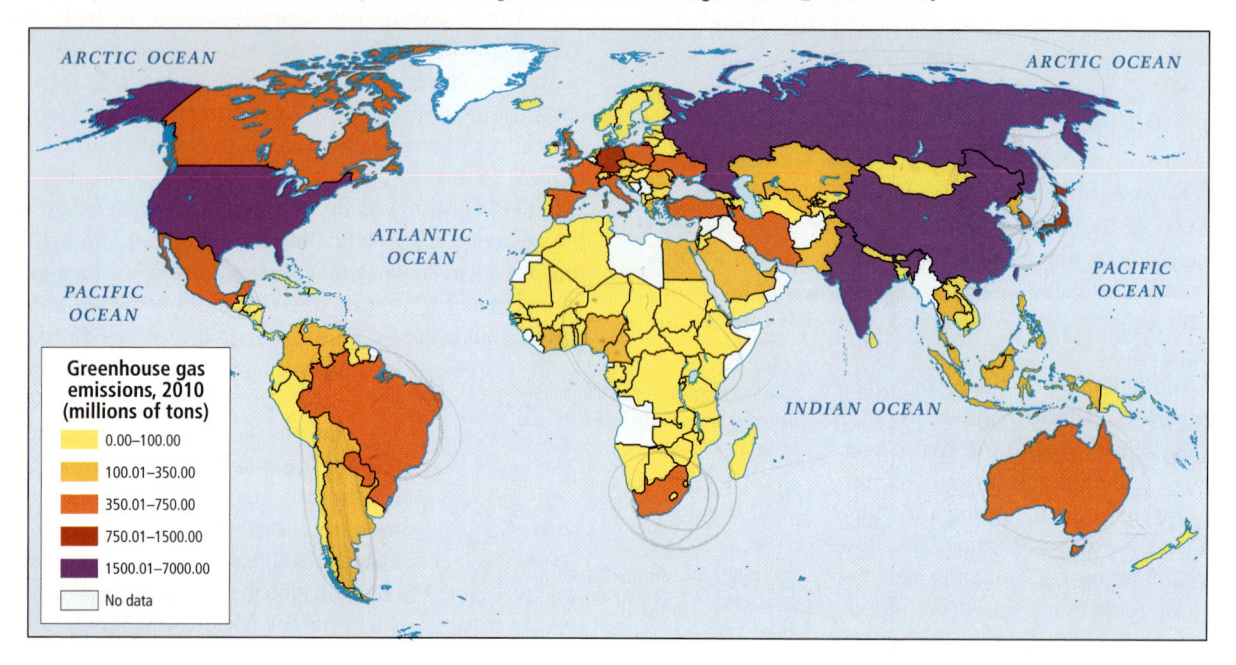

(A) Total emissions by millions of tons, 2010. China now has the highest amount of greenhouse gas (GHG) emissions (23.3 percent), and its rate has been rising sharply since 2005; the United States has the second most (18.11 percent), having had a small reduction since 2005; India increased emissions and now has the third most (5.78 percent), followed by Russia (5.67 percent) and Japan (4.01 percent). These top five countries contribute 56.9 percent of the world's greenhouse gas emissions.

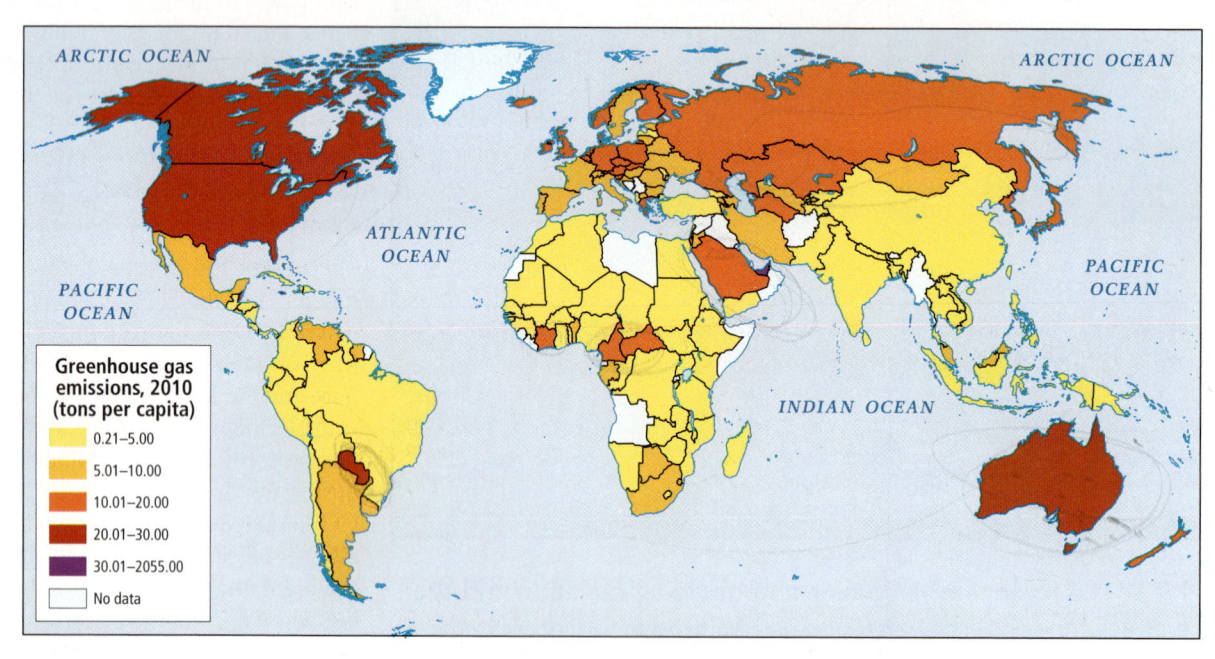

(B) Tons of emissions per capita. Jamaica leads, with 47.58 tons of GHG per capita, the United Arab Emirates are second (40.10 tons of GHG per capita), followed by Bahrain (34.9 tons per capita) and Paraguay (29.9 tons per capita). The United States is eighth, at 22.22 tons per capita. China is far down the list, at 3.39 tons per capita.

out, and less is stored. The loss of trees and other forest organisms produces as much as 30 percent of the buildup of CO_2 in the atmosphere. The use of fossil fuels accounts for the remaining 70 percent.

In percentages, the largest producers of total greenhouse gas emissions in 2010 were the industrialized countries and large, rapidly developing countries. The caption of Figure 1.14 lists the countries that are the most responsible for GHG emissions, total (A) and per capita (B). Note that the United States is among the leaders in both categories. For the period of 1859 to 1995, developed countries produced roughly 80 percent of the greenhouse gases from all types of industrial, home, and transportation sources, and developing countries produced 20 percent. But by 2007, the developing countries were catching up, accounting for nearly 30 percent of total emissions. As developing countries industrialize over the next century and continue to cut down their forests, they will release more and more greenhouse gases every year. If current patterns hold, greenhouse gas contributions by the developing countries will exceed those of the developed world by 2040.

Climate-Change Impacts

While it is not clear exactly what the impacts of rising global temperatures will be, it is clear that they will not be uniform across the globe. One prediction is that the glaciers and polar ice caps will melt, causing a corresponding rise in sea level. In fact, this phenomenon has been observable for several years. Satellite imagery analyzed by scientists at the National Aeronautics and Space Administration (NASA) shows that between 1979 and 2005—just 26 years—the polar ice caps shrank by about 23 percent. The amount of polar ice cap shrinkage wavers from year to year, with the ice caps regaining ice to some extent during winter months, but the overall trend in recent years is 12 percent shrinkage per decade. The polar ice caps normally reflect solar heat back into the atmosphere, but as the ice melts, the dark, open ocean absorbs solar heat. This warming of the oceans not only hastens ice cap melting, it is changing ocean circulation, the engine that drives weather and climate globally.

The melting of the ice caps also has several other effects. Already, trillions of gallons of meltwater have been released into the oceans. If this trend continues, at least 60 million people in coastal areas and on low-lying islands could be displaced by rising sea levels. Another issue is the melting of high mountain glaciers, which are a major source of water for many of the world's large rivers, such as the Ganga, the Indus, the Brahmaputra, the Huang He (Yellow), and the Chang Jiang (Yangtze). Similar melting effects on major rivers are expected in South America. Scientists have monitored mountain glaciers across the globe for more than 30 years; while some are growing, the majority are melting rapidly. Over the short term, melting mountain glaciers could result in the flooding of many rivers, but eventually river flows will decrease as mountain glaciers shrink or disappear entirely.

Over time, higher temperatures will shift northward in the Northern Hemisphere and southward in the Southern Hemisphere, bringing warmer climate zones to these regions. Such climate shifts could lead to the displacement of large numbers of people because the zones where specific crops can grow are likely to change. Animal and plant species that cannot adapt rapidly to the changes will disappear. Higher temperatures also will lead to stronger tornados and hurricanes because these storms are powered by warm, rising air (Figure 1.15). One example is Hurricane Sandy, the unusually large and powerful hurricane that struck the Atlantic Coast of North America in the autumn of 2012. In some areas, drought and water scarcity may also become more common since higher temperatures increase water evaporation rates from soils, vegetation, and bodies of water. Another effect of global warming is likely to be a shift in ocean currents. The result would be more chaotic and severe or dry weather, especially for places where climates are strongly influenced by ocean currents, such as western Europe (see "Vegetation and Climate" in Chapter 4, on pages 155–156).

Vulnerability to Climate Change

The vulnerability a place has to climate change can be thought of as the amount of risk its human or natural systems have of being damaged by such impacts as sea level rise, drought, flooding, or increased

---

A CASE STUDY OF VULNERABILITY

Mumbai, India

Three concepts are important in understanding a place's vulnerability to climate change: *exposure, sensitivity,* and *resilience*. Here we explore them in the context of the vulnerability to water-related climate-change impacts in Mumbai, India.

Exposure refers to the extent to which a place is exposed to climate-change impacts. A low-lying coastal city like Mumbai, India (see Figure 8.1 on page 316), is highly exposed to sea level rise. *Sensitivity* refers to how sensitive a place is to those impacts. Many of Mumbai's inhabitants, for example, are very sensitive to sea level rise because they are extremely poor and can only afford to live in low-lying slums that have no sanitation and therefore have polluted waterways nearby. If these waterways were to flood, they would spread epidemics of waterborne illness that could kill millions of people in the city and neighboring areas. *Resilience* refers to a place's ability to "bounce back" from the disturbances that climate-change impacts create. Mumbai's resilience to sea level rise is bolstered because despite its widespread poverty, it is the wealthiest city in South Asia. This wealth enables it to afford relief and recovery systems that could help it deal with sea level rise over the short and long term.

Over the short term, Mumbai benefits from more and better hospitals and emergency response teams than any other city in South Asia. Over the long term, Mumbai's well-trained municipal planning staff can create and execute plans to help sensitive populations, like people living in slums, adapt to sea level rise. This could be achieved, for example, through planned relocation to higher ground or by building sea walls and dikes that could keep sea waters out of low-lying slum areas. Of course, Mumbai's overall vulnerability is more complicated than these examples suggest because the city faces many more climate-change impacts than just sea level rise. However, these examples help us understand vulnerability to climate change as a combination of exposure, sensitivity, and resilience.

One effort at understanding the global pattern of vulnerability to climate change can be seen in Figure 1.15, which features a map of vulnerability to climate change and photos of the types of problems that are already being seen (see Figure 1.15A, C–E). One pattern is that places with low levels of human development tend to be more vulnerable to climate change. For example, many of the qualities that make Mumbai more sensitive to sea level rise are less present in urban areas in highly developed countries. New York City has virtually none of the large, unplanned lowland slums found in Mumbai. In addition, numerous world-class hospitals, emergency response teams, and large and well-trained municipal planning staffs boost New York's resilience. Despite all of New York City's wealth and resources, Hurricane Sandy showed that the metropolitan area is still very vulnerable to a large storm. ∎

storm intensity. Many of these vulnerabilities are water related; others are not. Scientists who study climate change agree that while we can take measures to minimize temperature increases, we can't stop them entirely, much less reverse those that have already occurred. We are going to have to live with and adapt to the impacts of climate change for quite some time. The first step in doing this is to understand how and where humans and ecosystems are especially vulnerable to climate change.

Responding to Climate Change

In 1997, an agreement known as the **Kyoto Protocol** was adopted. The protocol called for scheduled reductions in CO_2 emissions by the industrialized countries of North America, Europe, East Asia, and Oceania. The agreement also encouraged, though it did not require, developing countries to curtail their emissions. One hundred eighty-three countries had signed the agreement by 2009 (Canada withdrew in 2011). The only developed country that had not signed was the United States, which was then and still remains one of the world's largest per capita producers of CO_2.

In December 2009 in Copenhagen, 181 countries worked on a plan to halt the rising concentrations of CO_2 in the atmosphere by 2020. However, wealthy nations did not offer to curb emissions sufficiently enough to make a difference. The only progress that came out of Copenhagen was an agreement to help developing countries create clean-energy economies and otherwise adapt to climate change; but adequate funds were not allocated even for this.

Kyoto Protocol an amendment to a United Nations treaty on global warming, the Protocol is an international agreement, adopted in 1997 and in force in 2005, that sets binding targets for industrialized countries for reducing emissions of greenhouse gases

In December 2011, the UN Framework Convention on Climate Change (UNFCCC) convened, this time in Durban, South Africa, again with the goal of controlling emissions. After grueling negotiations, all 190 countries, including the top three emitters—China, the United States, and India—agreed to a plan to cut emissions significantly by 2020. But was this sufficient progress? According to climate scientists, 2020 is too late to begin major changes. Rather, to avoid a temperature rise of more than 2°C, carbon emissions would need to be in decline well before 2020. To achieve this, the agreements would need to be far more binding. Current research into CO_2 emissions suggests that no significant reductions have been made.

THINGS TO REMEMBER

- Human activities, such as burning of fossil fuels and deforestation, create large amounts of carbon dioxide, methane, and other greenhouse gases that trap heat in the atmosphere, causing global warming.

- The vulnerability a place has to climate change can be thought of as the amount of risk its human or natural systems have of being damaged by such impacts as sea level rise, drought, flooding, and increased storm intensity.

FIGURE 1.15 PHOTO ESSAY: Vulnerability to Climate Change

The map shows the overall vulnerability places around the world have to climate change, based on a combination of human and environmental factors. Areas in darkest brown are vulnerable to floods, hurricanes, droughts, sea level rise, or other impacts related to climate change. When a place is exposed to an impact that it is sensitive to and has little resilience to, it becomes vulnerable. For example, many places are exposed to drought, but generally speaking, those places with the poorest populations are the most sensitive. However, sensitivity to drought can be compensated for if adequate relief and recovery systems, such as emergency water and food distribution systems, are in place. These systems lend an area a level of resilience that can reduce its overall vulnerability to climate change. A place's vulnerability can be thought of as a combination of its sensitivity, exposure, and resilience in the face of multiple climate impacts.

E Climate change and hurricanes. Climatologists predict that hurricanes will increase in intensity as the planet warms. Indeed, there has been an increase in powerful storms in recent decades. Poverty (high sensitivity) and inadequate recovery systems (low resilience) make much of Central America particularly vulnerable to many hurricane-related impacts. Shown here is flood damage along the Aguan River in Honduras that was caused by Hurricane Mitch in 1999. More than 9000 people died in the storm, making Mitch the second most deadly hurricane in history.

A **United States: High resilience, low vulnerability.** Effective and well-funded recovery and relief systems give the United States high resilience to climate impacts. This contributes to generally low vulnerability. Shown here are ambulances in New York City responding to Hurricane Sandy in 2012.

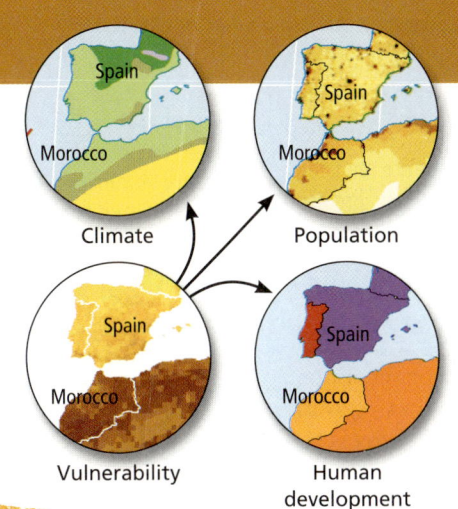

B **Spain and Morocco: The multiple dimensions of vulnerability.** A wide variety of information is used to make the global map of vulnerability shown below. For example, the contrast in vulnerability between Spain and Morocco relates to (among other things) differences in climate, population density and distribution, and human development (which is itself based on many factors).

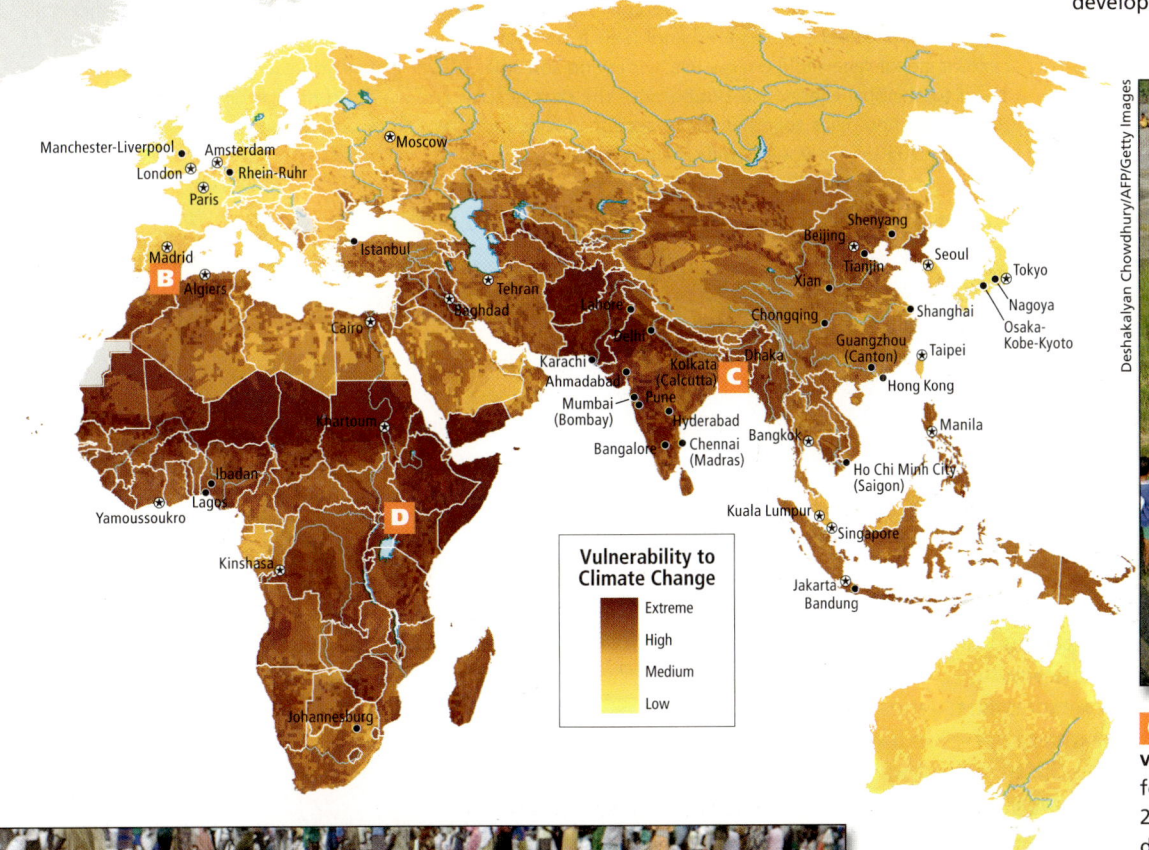

Climate • Population • Vulnerability • Human development

Vulnerability to Climate Change
- Extreme
- High
- Medium
- Low

C **India: Moderate resilience, high vulnerability.** Rural Indians line up for food and water after Cyclone Aila in 2009. Advances in government-led disaster recovery have increased India's resilience, and many areas no longer have the extreme vulnerability levels that neighboring Pakistan and Afghanistan share. Nevertheless, much of India remains in a state of high vulnerability, with high exposure and sensitivity to sea level rise, flooding, hurricanes (cyclones), drought, and other disturbances that climate change can create or intensify.

D **Northern Uganda and South Sudan: High to extreme vulnerability.** The overall situation in these areas is somewhat similar to that of India (C) but for different reasons. Because there is better access to water, sensitivity to drought is somewhat lower. However, armed conflict reduces resilience in much of Uganda, as many people have been forced to live in refugee camps and are dependent on food aid donated by foreigners. Shown here are refugees in northern Uganda, picking up bits of donated grain that has been dropped.

25

26   CHAPTER 1   Geography: An Exploration of Connections

- There has been little progress in making global agreements to control CO_2 emissions.
- While renewable resources are relatively underutilized today, many analysts predict a rapid increase in their use in coming decades. ∎

WATER

Water is emerging as the major resource issue of the twenty-first century. Demand for clean water skyrockets as people move out of poverty. A huge increase in water consumption occurs with the modernization of the production of foods, the manufacture of goods, the creation of energy, and the development of services (like cleaning or even entertainment). Modernization also inevitably creates *water pollution*. With clean, fresh water becoming scarce in so many parts of the world, water disputes are proliferating. This is especially true where rivers cross international borders and upstream users use more than their perceived fair share or pollute water for downstream users. Controversy also surrounds the sale of clean water. When water becomes a commodity rather than a free good, as it was before modern times, water prices can increase so much that the poor lose access, which then affects health and sanitation.

> **virtual water** the water used to produce a product, such as an apple or a pair of shoes
>
> **water footprint** all the water a person consumes, including both virtual water and the water they consume directly

ON THE BRIGHT SIDE

Renewable Energy

Renewable energy sources are those that can be replenished in a relatively short amount of time—for example, sunlight, wind, waves, or heat from deep inside the Earth. While these sources are relatively underutilized today, many analysts predict a rapid increase in their use in coming decades. This is because the costs of solar and wind power are declining, while the costs of fossil fuels are generally rising. For example, the cost of electricity generated by solar energy has fallen by 99% over the past 25 years and is now cheaper than electricity generated by fossil fuels in Germany, Japan, Spain, Italy, parts of India and China, as well as southern California and Hawaii. Fossil fuels are relatively expensive in all of these places due to a variety of factors, but by the end of the decade solar power is expected to be cheaper than fossil fuels in most of the developed world.

While the use of renewable energy is expanding quickly, it will take several decades for this growth to translate into significant reductions in the use of fossil fuels. About 80 percent of the energy used throughout the world today comes from fossil fuels (38 percent from petroleum, 26 percent form natural gas, and 16 percent from coal); 8 percent is from nuclear power; and 12 percent is from renewable sources. Hydroelectric energy, ethanol fuels from crops, and wood burning make up the vast majority of current global renewable energy use, with wind and solar power generation each accounting for less than 1% of total global energy use. But because the cost of solar and wind technologies continues to decrease, the two technologies together could account for as much as 34 percent of the energy used around the world by 2030. This would significantly reduce greenhouse gas emissions.

example, to produce 1 ton of corn in the United States requires 489 m^3 of virtual water, on average, whereas in India the same amount of corn requires 1935 m^3 of virtual water; in Mexico, 1744 m^3; and in the Netherlands, just 408 m^3. In the case of corn, water can be lost to *evapotranspiration* in the field, to the evaporation of standing irrigation water, and to evaporation as water flows to and from the field. An additional component of virtual water is that the water that becomes polluted in the production process is also lost to further use.

Calculating Water Use per Capita

Humans require an average of 5 to 13 gallons (20 to 50 liters) of clean water per day for basic domestic needs: drinking, cooking, and bathing/cleaning. Per capita domestic water consumption tends to increase as incomes rise; the average person in a wealthy country consumes as much as 20 times the amount of water, per capita, as the average person in a very poor country. However, domestic water consumption is only a fraction of a person's actual water consumption. **Virtual water** is the volume of water required to produce, process, and deliver a good or service that a person consumes. To grow an apple and ship it from the orchard to the consumer, for instance, requires many liters of water. When we add an individual's domestic water consumption to her virtual water consumption, we have that person's total **water footprint**. The more one consumes, the larger one's virtual water footprint. Table 1.1 shows the amounts of water used to produce some commonly consumed products. (As you look at Table 1.1 and read further, note that there are 1000 liters, or 263 gallons, in a cubic meter (m^3).)

Like domestic consumption, personal water footprints vary widely according to physical geography, standards of living, and rates of consumption (Figure 1.16). Moreover, the amount of virtual water used to produce 1 ton of a specific product varies widely from country to country because of climate conditions as well as agricultural and industrial technology and efficiency. For

TABLE 1.1	The global average virtual water content of everyday products*
Product[†]	**Virtual water content (in liters)**
1 potato	25
1 cup tea	35
1 kilogram of bread	1608
1 apple	125
1 glass of beer	75
1 glass of wine	120
1 egg	135
1 cup of coffee	140
1 glass of orange juice	170
1 pound of chicken meat	2000
1 hamburger	2400
1 pound of cheese	2500
1 pair of bovine leather shoes	8000

*Virtual water is the volume of water used to produce a product.
[†] To see the virtual water content of additional products, go to www.waterfootprint.org/?page=files/productgallery. [Source consulted: Arjen Y. Hoekstra and Ashok K. Chapagain, *Globalization of Water—Sharing the Planet's Freshwater Resources* (Malden, MA: Blackwell, 2008), p. 15, Table 2.2]

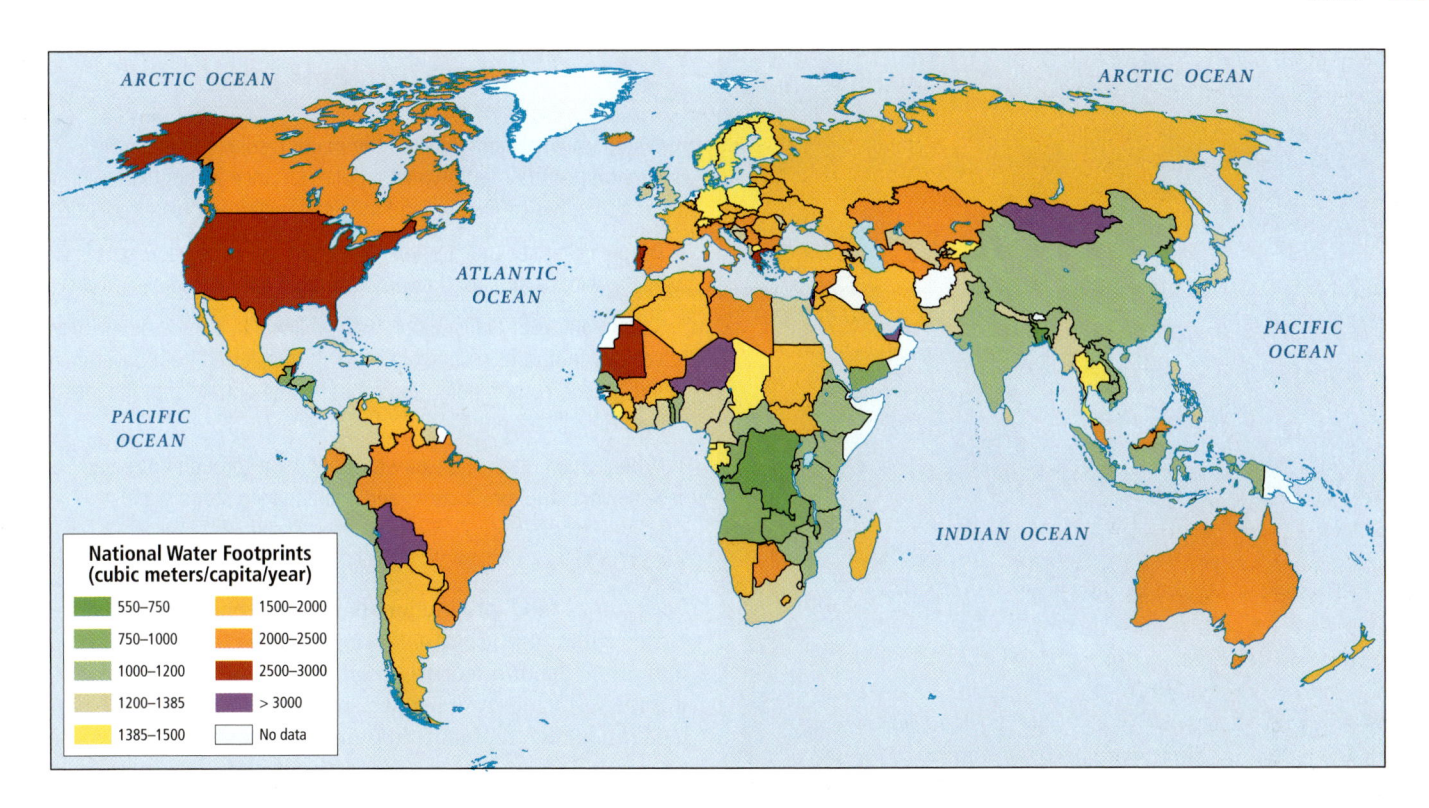

FIGURE 1.16 Map of national water footprints, 2005. Average national water footprint per capita (in cubic meters per capita per year). The color green indicates that the nation's water footprint is equal to or smaller than the global average. Countries in red and purple have a water footprint beyond the global average. (The latest year for which data are available is 2005.) [Sources consulted: *United Nations World Water Development Report 2: Water—A Shared Responsibility*. Published jointly in 2006 by the UN Educational, Scientific and Cultural Organization (UNESCO), Paris; and Berghahn Books, New York, pp. 391–392]

There are several Web sites designed to help people calculate their individual water footprint. Try Water Footprint's http://www.waterfootprint.org/?page=cal/waterfootprintcalculator_indv or *National Geographic*'s at http://environment.nationalgeographic.com/environment/freshwater/change-the-course/water-footprint-calculator/.

Who Owns Water? Who Gets Access to It?

Though many people consider water a human right that should not cost anything to access, water has become the third most valuable commodity after oil and electricity. Water in wells and running in streams and rivers is increasingly being *privatized*. This means that its ownership is being transferred from governments—which can be held accountable for protecting the rights of all citizens to access water—to individuals, corporations, and other private entities that manage the water primarily for profit. Governments often privatize water under the rationale that private enterprise will make needed investments that will boost the efficiency of water distribution systems and the overall quality of the water supply. Regardless of whether or not these potential gains are actually realized, privatization usually brings higher water costs to consumers. This can become quite controversial. For example, in 1999 in the city of Cochabamba, Bolivia, water costs rose beyond what the urban poor could afford following the sale of the city public water agency to a group of multinational corporations led by Bechtel of San Francisco, California. The result was a nationwide series of riots that led finally to the abandonment of privatization.

Water Quality

About one-sixth of the world's population does not have access to clean drinking water, and dirty water kills more than 6 million people each year. In an average year, more people die this way than in all of the world's armed conflicts. In the poorer parts of cities in the developing world, many people draw water with a pail from a communal spigot, sometimes from shallow wells, or simply from holes dug in the ground (Figure 1.17). Usually this water should be boiled before use, even for bathing. These water quality and access problems help explain why so many people are chronically ill and why 24,000 children under the age of 5 die every day from waterborne diseases.

In Europe and the United States, demand for higher water quality has resulted in a $100 billion bottled water industry. However, there are few standards of quality for bottled water; and in addition to generating mountains of plastic bottle waste, the bottled water industry often acquires its water from sources (springs, ponds, deep wells) that are publicly owned. Consumers thus pay for the same high-quality water they could consume for free as a public commodity. **245. CLEAN WATER PROJECT IMPROVES LIVES IN SENEGAL**

FIGURE 1.17 Access to water. A young girl gathers water from a shallow open well in a slum in Mumbai, India. Mumbai has vast, low-lying slums that have no sanitation and whose sewage pollutes nearby waterways. If these waterways were to flood, they would spread deadly epidemics of waterborne illnesses via open wells such as the one shown here.

> **THINKING GEOGRAPHICALLY** Why might this child be endangering her health by drinking this water?

THINGS TO REMEMBER

- Water scarcity is emerging as the major resource issue of the twenty-first century, with water supplies strained by population growth, skyrocketing per capita demand for clean water, and water pollution.
- People never see much of the water that they use because it is virtual water—water that is used to produce what they consume.
- Stream and river water, along with well water, is increasingly being privatized; that is, its ownership is being transferred from governments to individuals, corporations, and other private entities that manage the water primarily for profit.
- Urbanization often leads to water pollution through untreated sewage, industrial production, and mismanaged storm water.

FOOD

Over time, food production has undergone many changes. It started with hunting and gathering and over the millennia evolved into more labor-intensive, small-scale, subsistence agriculture. For the vast majority of people, subsistence remained the mode of food production for many thousands of years until a series of innovations and ideas that changed the way goods were manufactured. The **Industrial Revolution**, from about 1750 to 1850, opened the door to the development of what has become modern and mechanized commercial agriculture, involving the intensive use of machinery, fuel, and chemicals. Modern processes of food production, distribution, and consumption have greatly increased the supply and, to some extent, the security of food systems. However, this has come with a cost: it has created environmental pollution that may strain future food production and food security.

Agriculture: Early Human Impacts on the Physical Environment

Agriculture includes animal husbandry, or the raising of animals, as well as the cultivation of plants. The ability to produce food, as opposed to being dependent on hunting and gathering, led to a host of long-term changes. The human population began to grow more quickly, rates of natural resource use increased, permanent settlements eventually developed into towns and cities, and ultimately, human relationships became more formalized. Some would say these are the steps that led to civilization.

Very early humans hunted animals and gathered plants and plant products (seeds, fruits, roots, and fibers) for their food, shelter, and clothing. To successfully use these wild resources, humans developed an extensive folk knowledge of the needs of the plants and animals they favored. The transition from hunting in the wild to tending animals in pens and pastures and from gathering wild plant products to sowing seeds and tending plants in gardens, orchards, and fields probably took place gradually over thousands of years.

Where and when did plant cultivation and animal husbandry first develop? Genetic studies suggest that between 8000 and

Water and Urbanization

Urban development patterns dramatically affect the management of water. In most urban slum areas in poor countries, and even in places like the United States and Canada, crucial water management technologies such as sewage treatment systems can be entirely absent. Germ-laden human waste from toilets and kitchens may be deposited in urban gutters that drain into creeks, rivers, and bays. And yet, retrofitting wastewater collection and treatment systems in cities already housing several million inhabitants, is often deemed prohibitively costly, especially in poor countries.

Independent of sewage, water inevitably becomes polluted in cities as parking lots and rooftops replace areas that were once covered with natural vegetation. Rainwater quickly runs off these hard surfaces, collects in low places, and becomes stagnant instead of being absorbed into the ground. In urban slums, flooding can spread polluted water over wide areas, carrying it into homes and into local fresh water sources, as well as to places where children play (see Figure 1.17). Diseases such as malaria and cholera, carried in this water, can spread rapidly as a result. Fortunately, new technologies and urban planning methods are being developed that can help cities avoid these problems, but they are not yet in widespread use.

> **Industrial Revolution** a series of innovations and ideas that occurred broadly between 1750 and 1850, which changed the way goods were manufactured
>
> **agriculture** the practice of producing food through animal husbandry, or the raising of animals, and the cultivation of plants

20,000 years ago, people in many different places around the world independently learned to develop plants and animals for food through selective breeding, a process known as **domestication**.

Why did agriculture and animal husbandry develop in the first place? Certainly the desire for more secure food resources played a role, but the opportunity to trade may have been just as important. It is probably not a coincidence that many of the known locations of agricultural innovation lie along early trade routes—for example, along the Silk Road that runs through Central Asia from the eastern Mediterranean to China. In such locales, people would have had access to new information and new plants and animals brought by traders.

Agriculture and Its Consequences

Agriculture made possible the amassing of surplus stores of food for lean times, and allowed some people to specialize in activities other than food procurement. It also may have led to several developments now regarded as problems: rapid population growth, concentrated settlements where diseases could easily spread, environmental degradation, and paradoxically, malnutrition or even famine.

Through the study of human remains, archaeologists have learned that it was not uncommon for the nutritional quality of human diets to decline as people stopped eating diverse wild food species and began to eat primarily one or two species of domesticated plants and animals. Evidence of nutritional stress (shorter stature, malnourished bones and teeth) has been found repeatedly in human skeletons excavated in sites around the world where agriculture was practiced.

Whereas agriculture could support more people on a given piece of land than hunting and gathering, as populations expanded and as more land was turned over to agriculture, natural habitats were destroyed, reducing opportunities for hunting and gathering. Furthermore, the storage of food surpluses not only made it possible to trade food, but also made it possible for people to live together in larger communities, which marked the beginning of urban societies and, coincidentally, facilitated the spread of disease. Moreover, land clearing increased vulnerability to drought and other natural disasters that could wipe out an entire harvest. Thus, as ever-larger populations depended solely on cultivated food crops, episodic famine may have become more common and affected more people.

> **domestication** the process of developing plants and animals through selective breeding to live with and be of use to humans
>
> **food security** the ability of a state to consistently supply a sufficient amount of basic food to the entire population
>
> **green revolution** increases in food production brought about through the use of new seeds, fertilizers, mechanized equipment, irrigation, pesticides, and herbicides

Modern Food Production and Food Security

For most of human history, people lived in subsistence economies. Over the past five centuries of increasing global interaction and trade, though, people have become ever more removed from their sources of food. Today, occupational specialization means that food is increasingly mass-produced. Far fewer people work in agriculture than in the past, and now most people work for cash to buy food and other necessities.

A side effect of this dependence on money is that the **food security**—the ability of a state to consistently supply a sufficient amount of basic food to the entire population—of individuals and families can be threatened by economic disruptions, even in distant places. As countries become more involved with the global economy, they may import more food or their own food production may become vulnerable to price swings. For example, a crisis in food security began to develop in 2007 when the world price of corn spiked. Speculators in alternative energy, thinking that corn would be an ideal raw material with which to make ethanol—a substitute for gasoline—invested heavily in this commodity, creating a shortage. As a result, global corn prices rose beyond the reach of those who depended on corn as a food. Then, between the sharp price rise in oil in 2008 and the recession of 2007–2009, the global cost of basic foods rose 17 percent. When oil prices rise, all foods produced and transported with machines get more expensive.

There are other contributing causes to food insecurity. When the global recession—partially caused by rising oil prices—eliminated jobs in many world locations, migrant workers could no longer send remittances to their families, who then no longer had money with which to buy food. These episodes called into question the sustainability of current food production and acquisition systems. In developing countries, household economies were so ruined that parents sold important assets; went without food, to the detriment of their long-term health; and stopped sending children to school. UN statistics show real reversals of progress in human well-being in 2007–2008. Figure 1.18 identifies countries in which undernourishment is an ongoing problem and periodic food insecurity is especially intense.

Another way that modernized agriculture impacts food security is through its reliance on machines, chemical fertilizers, and pesticides. Paradoxically, when the shift to this kind of agriculture was introduced into developing countries like India and Brazil in the 1970s, it was called the **green revolution**. In fact, the green revolution is not "green" in the modern sense of being environmentally savvy. When successfully implemented, the results of green revolution agriculture were at first spectacular: soaring production levels along with high profits for those farmers who could afford the additional investment. In the early years of this movement, it seemed to scientists and developers that the world was literally getting greener. But often, poorer farmers couldn't afford the machinery and chemicals. Also, since greatly increased production leads to lower crop prices on the market, these poorer farmers lost money. To survive, they often were forced to sell their land and move to crowded cities. Here they joined masses of urban poor whose access to food was precarious.

Green revolution agriculture can also impact food security by damaging the environment. As rains wash fertilizers and pesticides into streams, rivers, and lakes, these bodies of water become polluted. Over time, the pollution destroys fish and other aquatic animals vital to food security. Hormones fed to farm animals to hasten growth may enter the human food chain. Soil degradation can also increase as green revolution techniques (such as mechanical plowing, tilling, and harvesting) leave soils exposed to rains that wash away natural nutrients and the soil itself. Indeed, many of the most agriculturally productive parts of

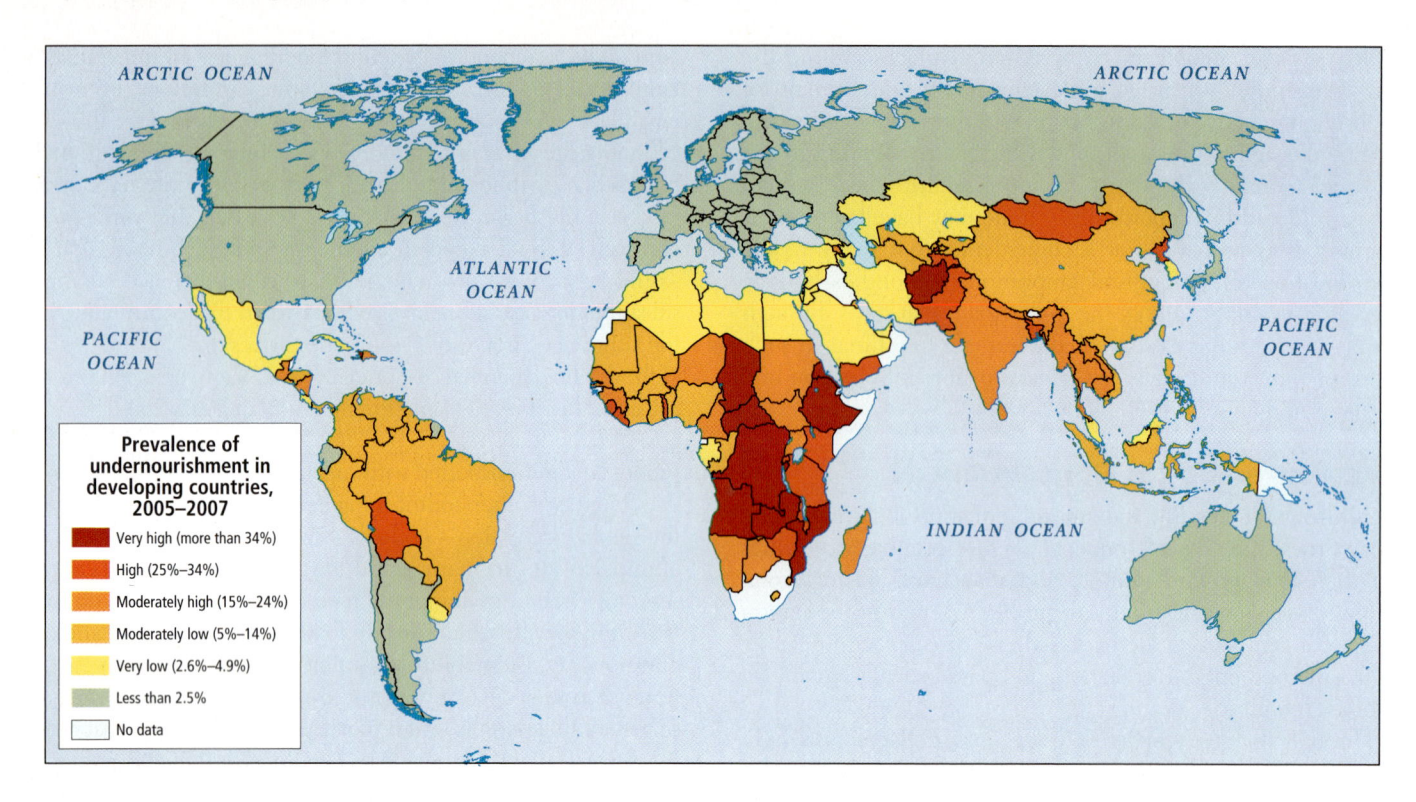

FIGURE 1.18 Global map of undernourishment, 2007. This map is based on 2007 data (the only data presently available), before the 2008 global recession put many people back into a state of hunger. What shows here is that the proportion of people suffering from undernourishment—the lack of adequate nutrition to meet their daily needs—had declined in the developing world, most notably in India, over the past several years. However, hundreds of millions of people remain affected by chronic hunger. As you can see from the map, those who are the most affected are people in much of sub-Saharan Africa; parts of South Asia; Mongolia and North Korea in East Asia; Bolivia in South America; and Haiti in the Caribbean. [Source consulted: Food and Agriculture Organization of the United Nations, FAO Statistics Division, Rome 2009: Map 14, Year 2003–2005, at http://www.fao.org/economic/ess/ess-publications/ess-yearbook/fao-statistical-yearbook-2007-2008/ghuman-welfare/en/]

North America, Europe, and Asia have already suffered moderate to serious loss of soil through erosion. Globally, soil erosion and other problems related to food production affect about 7 million square miles (2000 million hectares), putting at risk the livelihoods of a billion people.

At least in the short term, green revolution agriculture raised the maximum number of people that could be supported on a given piece of land, or its **carrying capacity**. However, it is unclear how sustainable these green revolution gains in food production are. In the 25 years between 1965 and 1990, total global food production rose between 70 and 135 percent (varying from region to region). In response, populations also rose quickly during this period. These successes in improving agricultural production and carrying capacities led the general public to assume that technological advances would perpetuate these increases; indeed, it is now estimated that to feed the population projected for 2050, global food output must increase by another 70 percent. Yet scientists from many disciplines estimate that within the next 50 years, environmental problems such as water scarcity and global climate change will limit, halt, or even reverse increases in food production. How can these discrepancies between expectations and realities be resolved?

carrying capacity the maximum number of people that a given territory can support sustainably with food, water, and other essential resources

genetic modification (GM) in agriculture, the practice of splicing together the genes from widely divergent species to achieve particular desirable characteristics

The technological advances that could make current agricultural systems more productive are increasingly controversial. In North America, **genetic modification (GM)**, the practice of splicing together the genes from widely divergent species to achieve particular characteristics, is being used to boost productivity. However, outside of North America, many worry about the side effects of such agricultural manipulation. Europeans have tried (unsuccessfully) to keep GM food products entirely out of Europe, fearing that they could lead to unforeseen ecological consequences or catastrophic crop failures. They point out that the main advance in GM agriculture has been the production of seeds that can tolerate high levels of environmentally damaging herbicides, such as Roundup. The use of GM crops thus could lead to more, not less, environmental degradation. Even if GM crops prove safe, in developing countries where farmers' budgets are tiny, genetically modified seeds are much more expensive than traditional seeds. They must be purchased anew each year because GM plants do not produce viable seeds, as do plants from traditional seeds.

As a result of the uncertainties of GM crops and the potential negative side effects of new agricultural technologies, many are returning to the much older idea of **sustainable agriculture**—farming that meets human needs without poisoning the environment or using up water and soil resources. Often these systems avoid chemical inputs entirely, as in the case of popular methods of *organic agriculture*. However, while these systems can be productive, they are less so than conventional green revolution systems and often require significantly more human labor, resulting in higher food prices. More dependence on sustainable and organic systems could therefore lead to food insecurity for some poor people, especially in cities.

According to the UN Food and Agriculture Organization, one-fifth of humanity subsists on a diet too low in total calories and vital nutrients to sustain adequate health and normal physical and mental development (see Figure 1.18). As we will see in later chapters, this massive hunger problem is partly due to political instability, corruption, and inadequate distribution systems. When food is scarce for whatever reason, it tends to go to those who have the money to pay for it. **3. DEFORESTATION: WORLDWIDE CONCERNS**

THINGS TO REMEMBER

- Genetic studies suggest that between 8000 and 20,000 years ago, people in many different places around the globe independently learned to use selective breeding to develop plants and animals for food.
- While modern processes of food production and distribution have greatly increased the supply of food, environmental damage and market disruptions could strain future food production and compromise food security.
- Many farmers are unable to afford the chemicals and machinery required for commercial agriculture or new, genetically modified seeds. Because their production is low, they cannot compete on price and may be forced to give up farming, often migrating to cities.
- Sustainable agriculture is farming that meets human needs without harming the environment or depleting water and soil resources. ■

GLOBALIZATION

GEOGRAPHIC INSIGHT 2

Globalization and Development: Global flows of information, goods, and people are transforming patterns of economic development. Local self-sufficiency is giving way to global interdependence as people and places are increasingly becoming connected, sometimes across vast distances.

The term **globalization** refers to the worldwide changes brought about by many types of flows that reach well beyond economics. For example, a person's attitudes and values may be modified because of being in contact with people and ideas from foreign cultures. This contact can be face to face, as a result of increased rates of migration, or virtual, via information technology, the reach of which is expanding along with the global economy. Globalization is the most complex and far-reaching of the thematic concepts described in this book.

Throughout the world, globalization is transforming patterns of economic development as local self-sufficiency is giving way to global interdependence and international trade. While for a time there was anticipation that globalization would lead to more prosperity for all, the economic recession that began in 2007 cast a spotlight on the unpredictable effects of global interdependence. In that year, economic disruptions in the United States and Europe resulted in powerful ripple effects that reached around the world. Foreclosures in the U.S. housing market meant that European banks that had invested in U.S. mortgages faltered and some failed. The governments of some indebted countries, such as Greece and Ireland, could not make loan payments, deepening the economic disruptions in Europe. Consumers in the United States and Europe shopped and vacationed less, and as a result, many workers across the world lost their jobs or saw their incomes decline.

What Is the Global Economy?

The **global economy** is the worldwide system in which goods, services, and labor are exchanged. Most of us participate in the global economy every day. For example, books like this can be made from trees cut down in Southeast Asia or Siberia and shipped to a paper mill in Oregon. Many books are now printed in Asia because labor costs are lower there. Globalization is not new. At least 2500 years ago, silk and other goods were traded along the Central Asian Silk Road that connected Greece and then Rome in the Mediterranean with distant China, an expanse of more than 4000 miles (6437 km).

> **sustainable agriculture** farming that meets human needs without poisoning the environment or using up water and soil resources
>
> **globalization** the growth of worldwide linkages and the changes these linkages are bringing about
>
> **global economy** the worldwide system in which goods, services, and labor are exchanged

European colonization was an early expansion of globalization. Starting in about 1500 C.E., European countries began extracting resources from distant parts of the world that they had conquered. The colonizers organized systems to process those resources into higher-value goods to be traded wherever there was a market. Sugarcane, for example, was grown on Caribbean and Brazilian plantations with slave labor from Africa (**Figure 1.19**) and made locally into crude sugar, molasses, and rum. It was then shipped to Europe and North America, where it was further refined and sold at considerable profit. The global economy grew as each region produced goods for export, rather than just for local consumption. At the same time, regions also became increasingly dependent on imported food, clothing, machinery, energy, and knowledge.

The new wealth derived from the colonies and the ready access to global resources led to Europe's Industrial Revolution. No longer was one woman producing the cotton or wool for clothes or fabric by spinning thread, weaving the thread into cloth, and sewing a garment. Instead, these separate tasks were spread out among many workers, often in distant places, with some people specializing in producing the fiber and others in spinning, weaving, or sewing. These innovations in efficiency were followed by labor-saving improvements such as mechanized reaping, spinning, weaving, and sewing.

32 CHAPTER 1 Geography: An Exploration of Connections

FIGURE 1.19 European use of colonial resources. Among the first global economic institutions were Caribbean plantations like the one shown here in a nineteenth-century painting of St. Croix Island. In the eighteenth century, thousands of sugar plantations in the British West Indies, subsidized by the labor of slaves, provided vast sums of money for England and helped fund the Industrial Revolution.

This larger-scale mechanized production accelerated globalization, as it created a demand for raw materials and a need for markets in which to sell finished goods. European colonizers managed to integrate the production and consumption of their colonial possessions in the Americas, Africa, and Asia. For example, in the British Caribbean colonies, hundreds of thousands of enslaved Africans wore rough garments made of cheap cloth woven in England from cotton grown in British India. The sugar that slaves produced on British-owned plantations with iron equipment from British foundries was transported to European markets in ships made in the British Isles of trees and resources from the North American colonies and other parts of the world.

Until the early twentieth century, much of the activity of the global economy took place within the colonial empires of a few European nations (Britain, France, the Netherlands, Germany, Belgium, Spain, and Portugal). By the 1960s, global economic and political changes brought an end to these empires, and now almost all colonial territories are independent countries. Nevertheless, the global economy persists in the form of banks and **multinational corporations**—such as Shell, Chevron, IBM, Walmart, Coca-Cola, Bechtel, Apple, British Petroleum, Toyota, Google, and Cisco—that operate across international borders. These corporations extract resources (including intellectual properties) from many places, make products in factories located where they can take advantage of cheap labor and transportation facilities, and market their products wherever they can make the most profit. Their global influence, wealth, and importance to local economies enable the multinational companies to influence the economic and political affairs of the countries in which they operate.

> **multinational corporation** a business organization that operates extraction, production, and/or distribution facilities in multiple countries

Workers in the Global Economy

VIGNETTE

Sixty-year-old Olivia lives near Soufrière on St. Lucia, an island in the Caribbean (Figure 1.20A). She, her daughter Anna, and her three grandchildren live in a wooden house surrounded by a leafy green garden dotted with fruit trees. Anna has a tiny shop

FIGURE 1.20 Workers in the global economy.

(A) Soufrière, St. Lucia.

(B) Setiya, along with other migrant workers, boards a bus that will take him to the construction site in Malacca, Malaysia. Later, during the recession that began in 2008, Setiya was expelled from the country, along with 500,000 other foreign workers.

(C) The trailer at the back of Tanya's lot, where her daughter lives.

at the side of the house, from which she sells various small everyday items and preserves that she and her mother make from the garden fruits.

On days when the cruise ships dock, Olivia heads to the market shed on the beach with a basket of homegrown goods. She calls out to the passengers as they near the shore, offering her spices and snacks for sale. In a good week she makes U.S.$70. Her daughter makes about U.S.$100 per week in the shop and is constantly looking for other ways to earn a few dollars.

Olivia and Anna support their family of five on about U.S.$170 a week (U.S.$8840 per year). From this income they take care of their bills and other purchases, including school fees for the granddaughter who will go to high school in the capital next year and perhaps college if she succeeds. Their livelihood puts them at or above the standard of living of most of their neighbors.

In Malacca, Malaysia, 30-year-old Setiya, an illegal immigrant from Tegal, Indonesia, is boarding a bus that will take him to a job site where he is helping build a new tourist hotel (see Figure 1.20B). Like a million other Indonesians attracted by the booming economy, he snuck into Malaysia, risking arrest, because in Malaysia average wages are four times higher than at home.

This is Setiya's second trip to Malaysia. His first trip was to do work legally on a Malaysian oil palm plantation. Upon arrival, however, Setiya found that he would have to work for 3 months just to pay off his boat fare from Indonesia. Not one to give in easily, he quietly went to another city and found a construction job earning U.S.$10 a day (about U.S.$2600 a year), which allowed him to send money home to his family in Indonesia.

In 2008, Malaysia announced it was expelling 500,000 foreign workers; Setiya was one of them. The country was suffering from growing unemployment due to the global recession, and its leaders wanted to save more jobs for local people by deporting foreign workers. When the recession is alleviated, young fathers from Indonesia (like Setiya) may once again risk trips on leaky boats to illegally enter Malaysia and Singapore in order to support their families.

Fifty-year-old Tanya works at a fast-food restaurant outside of Charleston, South Carolina, making less than U.S.$7.50 an hour. She had been earning U.S.$8 an hour sewing shirts at a textile plant until it closed and moved to Indonesia. Her husband is a delivery truck driver for a snack-food company.

Between them, Tanya and her husband make $30,000 a year, but from this income they must cover all their expenses, including their mortgage, gasoline, and car payments. In addition, they help their daughter, Rayna, who quit school after eleventh grade and married a man who is now out of work. They and their baby live at the back of the lot in an old mobile home (see Figure 1.20C).

With Tanya's now-lower wage (almost $1000 less a year), there will not be enough money to pay the college tuition for her son, who is in high school. He had hoped to become an engineer, and would have been the first in the family to go to college. For now, he is working at the local gas station.

These people, living worlds apart, are all part of the global economy. Workers around the world are paid startlingly different rates for jobs that require about the same skill level. Varying costs of living and varying local standards of wealth make a difference in how people live and how they perceive their own situation. Though Tanya's family has the highest income by far, they live in poverty compared to their neighbors, and their hopes for the future are dim. Olivia's family, on the other hand, are not well off, but they do not think of themselves as poor because they have what they need, others around them live in similar circumstances, and their children seem to have a future. They can subsist on local resources, and the tourist trade promises continued cash income. But their subsistence depends on circumstances beyond their control; in an instant, the cruise-line companies can choose another port of call. Setiya, by far the poorest, seems trapped by his status as an illegal worker, which robs him of many of his rights. Still, the higher pay that he can earn in Malaysia offers him a possible way out of poverty. [Source: From Lydia Pulsipher's and Alex Pulsipher's field notes. For detailed source information, see Text Sources and Credits.]

The Debate over Globalization and Free Trade

The term **free trade** refers to the unrestricted international exchange of goods, services, and capital. Free trade is an ideal that has not been achieved and probably never will be. Currently, all governments impose some restrictions on trade to protect their own national economies from foreign competition, although such restrictions are far fewer than in the 1980s. Restrictions take two main forms: *tariffs* and *import quotas*. Tariffs are taxes imposed on imported goods that increase the cost of those goods to the consumer, thus giving price advantages to competing, locally made goods. Import quotas set limits on the amount of a given good that may be imported over a set period of time, curtailing supply and keeping prices high, again to protect local producers.

> **free trade** the unrestricted international exchange of goods, services, and capital

These and other forms of trade protection are subjects of contention. Proponents of free trade argue that the removal of all tariffs and quotas encourages efficiency, lowers prices, and gives consumers more choices. Companies can sell to larger markets and take advantage of mass-production systems that lower costs further. As a result, businesses can grow faster, thereby providing people with jobs and opportunities to raise their standard of living. These pro–free trade arguments have been quite successful, and in recent decades, restrictions on trade imposed by individual countries have been greatly reduced. Several *regional trade blocs* have been formed; these are associations of neighboring countries that agree to lower trade barriers for one another. The main ones are the North American Free Trade Agreement (NAFTA), the European Union (EU), the Southern Common Market in South America (Mercosur), and the Association of Southeast Asian Nations (ASEAN).

One of the main global institutions that supports the ideal of free trade is the **World Trade Organization (WTO)**, whose stated mission is to lower trade barriers and to establish ground rules for international trade. Related institutions, the *World Bank* and the *International Monetary Fund* (IMF), both make loans to countries that need money to pay for economic development or to avoid financial crises. Before approving a loan, the World Bank or the IMF may require a borrowing country to reduce and eventually remove tariffs and import quotas. These requirements are part of larger "belt-tightening" measures that the IMF imposes on countries seeking loans, such as the requirement to close or privatize government enterprises and to reduce government services including education and health-care programs that benefit the poor.

Those opposed to free trade argue that it leads to a less regulated global economy that can be chaotic, resulting in rapid cycles of growth and decline that increase the disparity between rich and poor worldwide. Labor unions point out that as corporations relocate factories and services to poorer countries where wages are lower, jobs are lost in richer countries, creating poverty. In the poorer countries, multinational corporations often work with governments to prevent workers from organizing labor unions that could bargain for **living wages**, minimum wages high enough to support a healthy life. Environmentalists note that in newly industrializing countries, which often lack effective environmental protection laws, multinational corporations tend to use highly polluting and unsafe production methods to lower costs. Many fear that a "race to the bottom" in wages, working conditions, government services, and environmental quality is underway as countries compete for profits and potential investors. Multinational corporations that have recently agreed to address worker abuses include Nike, Walmart, and Apple.

Fair trade, proposed as an alternative to free trade, is intended to provide a fair price to producers and to uphold environmental and safety standards in the workplace. Economic relationships surrounding trade are rearranged in order to provide better prices for producers from developing countries. For example, "fair trade" coffee and chocolate are now sold widely in North America and Europe. Prices are somewhat higher for consumers, but the extreme profits of middlemen are eliminated. As a result, growers of coffee and cocoa beans who produce for fair trade companies can receive living wages and work under better conditions.

In evaluating free trade, globalization, and fair trade, consider how many of the things you own or consume were produced in the global economy—your computer, clothes, furniture, appliances, car, and foods. These products are cheaper for you to buy, and your standard of living is higher as a result of lower production costs as well as competition among many global producers. However, you or someone you know may have lost a job because a company moved to another location where labor and resources are cheaper. Underpaid workers (even children) working under harsh conditions that possibly generate high levels of pollution may have made those cheap products. If workers can't earn living wages, often some family members end up migrating, perhaps without the proper papers, to earn a better wage. Given all these factors, consider the advantages and drawbacks of both free trade and fair trade.

> **World Trade Organization (WTO)** a global institution made up of member countries whose stated mission is to lower trade barriers and to establish ground rules for international trade
>
> **living wages** minimum wages high enough to support a healthy life
>
> **fair trade** trade that values equity throughout the international trade system; now proposed as an alternative to free trade

ON THE BRIGHT SIDE

Reforms at the IMF and World Bank

In response to the now widely recognized failures of policies that rely on the power of markets to guide development, the IMF and the World Bank have made some changes. The standard set of loan repayment measures, known as *structural adjustment programs* (SAPs), have been replaced with *Poverty Reduction Strategy Papers,* or PRSPs. Instead of having a uniform policy aimed at loan repayment imposed on them, each country in need works with World Bank and IMF personnel to design a broad-based plan for both economic growth *and* poverty reduction. PRSPs still involve market-based solutions and are aimed toward reducing the role of government in the economy. These programs remain highly bureaucratic, but they do focus also on poverty reduction rather than just on development via structural adjustment. They promote the maintenance of education, health, social services, and broader participation in civil society. PRSPs also include the possibility that all or some of a country's debt be "forgiven" (written off by the IMF and the World Bank). This alleviates one of the worst problems of SAPs, which was that all the belt-tightening measures, instead of making it easier to pay off debt, sometimes sent a country into an economic decline that made loan repayment impossible.

THINGS TO REMEMBER

› GEOGRAPHIC INSIGHT 2 • **Globalization and Development** Global flows of information, goods, and people are transforming patterns of economic development. Local self-sufficiency is giving way to global interdependence as people and places are increasingly becoming connected, sometimes across vast distances.

• Workers around the world are paid startlingly different rates for jobs that require about the same skill level.

• Under true free trade, economic transactions are conducted without interference or regulation in an open marketplace. Under fair trade, an alternative to free trade, consumers are asked to pay a fair price to producers, and producers are expected to pay living wages and uphold environmental and safety standards in the workplace. ■

DEVELOPMENT

The economy is the forum in which people make their living, and resources are what they use to do so. *Extractive resources* are resources that must be mined from the Earth's surface (mineral ores) or grown from its soil (timber and plants). There are also *human resources*, such as skills and brainpower, which are used to transform extractive resources into useful products (such as refrigerators or bread) or bodies of knowledge (such as books or computer software). Economic activities are often divided into three *sectors of the economy*: the **primary sector** is based on **extraction** (mining, forestry, and agriculture); the **secondary sector** is **industrial production** (processing, manufacturing, and construction); and the **tertiary sector** is **services** (sales, entertainment, and financial). Of late, a fourth, or **quaternary sector**, has been added to cover intellectual pursuits such as education, research, and IT (information technology) development. Generally speaking, as people in a society shift from extractive activities, such as farming and mining, to industrial and service activities, their material standards of living rise—a process typically known as **development**.

The development process has several facets, one of which is a shift from economies based on extractive resources to those based on human resources. In many parts of the world, especially in poorer societies (often referred to as "underdeveloped" or "developing"), there are now shifts away from labor-intensive and low-wage, often agricultural, economies toward higher-wage but still labor-intensive manufacturing and service economies (including Internet-based economies). Meanwhile, the richest countries (often referred to as "developed") are lessening their dependence on labor-intensive manufacturing and shifting toward more highly skilled mechanized production or knowledge-based service and technology (quaternary) industries. As these changes take place, societies must provide adequate education, health care, and other social services to help their people contribute to economic development.

Measuring Economic Development

The most long-standing measure of development has been **gross domestic product (GDP) per capita**. GDP is simply an economic measure that refers to the total market value of all goods and services produced in a country in a given year. A closely related index that is now used more often by international agencies is *gross national income* (GNI), the total value of income in a country. When GDP or GNI is divided by the number of people in the country, the result is per capita GDP or GNI. This book now uses primarily the GNI per capita statistics.

Using GNI per capita as a measure of how well people are living has several disadvantages. First is the matter of wealth distribution. Because GNI per capita is an average, it can hide the fact that a country has a few fabulously rich people and a great mass of abjectly poor people. For example, a GNI per capita of U.S.$50,000 would be meaningless if a few lived on millions per year and most lived on less than $10,000 per year.

Second, the purchasing power of currency varies widely around the globe. A GNI of U.S.$18,000 per capita in Barbados might represent a middle-class standard of living, whereas that same amount in New York City could not buy even basic food and shelter. Because of these purchasing power variations, in this book GNI (and occasionally GDP) per capita figures have been adjusted for **purchasing power parity (PPP)**. PPP, usually indicated in parentheses after GDP or GNI, is the amount that the local currency equivalent of U.S. dollars will purchase in a given country. For example, according to the *Economist*, on January 14, 2012, a Big Mac at McDonald's in the United States cost U.S.$4.20. In the **Euro zone** (those countries in the European Union that use the Euro currency), the very same Big Mac cost the equivalent of U.S.$4.43, while in India it cost U.S.$1.62. Of course, for the consumer in India, where annual per capita GNI (PPP) is $3280, this would be a rather expensive meal. On the other hand, at $4.20, the Big Mac would be an economy meal in the United States, where the GNI per capita (PPP) is $45,640, or in the Euro area, where $34,000 is the average GNI (PPP) per capita (see "Big Mac Index," at http://www.economist.com/node/21542808).

A third disadvantage of using GDP (PPP) or GNI (PPP) per capita is that both measure only what goes on in the **formal economy**—all the activities that are officially recorded as part of a country's production. Many goods and services are produced outside formal markets, in the **informal economy**. Here, work is often bartered for food, housing, or services, or for cash payments made "off the books"—payments that are not reported to the government as taxable income. It is estimated that one-third or more of the world's work takes place in the informal economy. Examples of workers in this category include anyone who contributes to her/his own or someone else's well-being through bartered or off-the-books services such as housework, gardening, herding, animal care, or elder and child care. *Remittances*, or pay sent home by migrants, become part of the formal economy of the receiving society if they are sent through banks or similar financial institutions—because records are kept and taxes levied. If they are transmitted off the books (perhaps illegally), such as via mail or in cash, they become part of the informal economy.

There is a gender aspect to informal economies. Researchers studying all types of societies and cultures have shown that, on average, women perform about 60 percent of all the work done, and that much of this work is unpaid and in the informal

---

primary sector an economic sector of the economy that is based on extraction (see also *extraction*)

extraction mining, forestry, and agriculture

secondary sector an economic sector of the economy that is based on industrial production (see also *industrial production*)

industrial production processing, manufacturing, and construction

tertiary sector an economic sector of the economy that is based on services (see also *services*)

services sales, entertainment, and financial services

quaternary sector a sector of the economy that is based on intellectual pursuits such as education, research, and IT (information technology) development

development a term usually used to describe economic changes such as the greater productivity of agriculture and industry that lead to better standards of living or simply to increased mass consumption

gross domestic product (GDP) per capita the total market value of all goods and services produced within a particular country's borders and within a given year, divided by the number of people in the country

purchasing power parity (PPP) the amount that the local currency equivalent of U.S.$1 will purchase in a given country

Euro zone those countries in the European Union that use the Euro currency

formal economy all aspects of the economy that take place in official channels

informal economy all aspects of the economy that take place outside official channels

economy. Yet only the work women are paid for in the formal economy appears in the statistics, so economic figures per capita ignore much of the work women do. Statistics also neglect the contributions of millions of men and children who work in the informal economy as subsistence farmers, traders, service people, or seasonal laborers.

A fourth disadvantage of GDP (PPP) and GNI (PPP) per capita is that neither takes into consideration whether these levels of income are achieved at the expense of environmental sustainability, human well-being, or human rights.

Geographic Patterns of Human Well-Being

Some development experts, such as the Nobel Prize–winning economist Amartya Sen, advocate a broader definition of development that includes measures of **human well-being**. This term generally means a healthy and socially rewarding standard of living in an environment that is safe and sustainable. The following section explores the three measures of human well-being that are used in this book.

Global GNI per capita (PPP) is mapped in **Figure 1.21A**. Comparisons between regions and countries are possible, but as discussed above, GNI per capita figures ignore all aspects of development other than economic ones. For example, there is no way to tell from GNI per capita figures how quickly a country is consuming its natural resources, or how well it is educating its young, maintaining its environment, or seeking gender and racial equality. Therefore, along with the traditional GNI per capita figure, geographers increasingly use several other measures of development.

The second measure used in this book is the **United Nations Human Development Index (HDI)**, which calculates a country's level of well-being with a formula of factors that considers income adjusted to PPP, data on life expectancy at birth (an indicator of overall health care), and data on educational attainment (see Figure 1.21B).

The third measure of well-being used here, the **United Nations Gender Equality Index (GEI) rank**, is a composite measure reflecting the degree to which there is equality in achievements between women and men in three dimensions: reproductive health, empowerment, and the labor market (see Figure 1.21C). A high rank indicates that the genders are tending toward equality. Ranks are from most equal (1) to least equal (146).

Together, these three measures reveal some of the subtleties and nuances of well-being and make comparisons between countries somewhat more valid. Because the more sensitive indices (HDI and GEI) are also more complex than the purely economic GNI (PPP) per capita, they are still being refined by the United Nations.

> **human well-being** various measures of the extent to which people are able to obtain a healthy and socially rewarding standard of living in an environment that is safe and sustainable
>
> **United Nations Human Development Index (HDI)** an index that calculates a country's level of well-being, based on a formula of factors that considers income adjusted to PPP, data on life expectancy at birth, and data on educational attainment
>
> **United Nations Gender Equality Index rank (GEI)** a composite measure reflecting the degree to which there is inequality in achievements between women and men in three dimensions: reproductive health, empowerment, and the labor market. A high rank indicates that the genders are tending toward equality
>
> **sustainable development** the effort to improve current standards of living in ways that will not jeopardize those of future generations
>
> **political ecologists** geographers who study the interactions among development, politics, human well-being, and the environment

A geographer looking at these maps might make the following observations:

- The map of GNI per capita figures (see Figure 1.21A) shows a wide range of difference across the globe, with very obvious concentrations of high and low GNI per capita. The most populous parts of the world—China and India—have medium-low and low GNI, respectively, and sub-Saharan Africa has the lowest.
- The HDI rank map (see Figure 1.21B) shows a similar pattern, but look closely. Middle and South America, Southeast Asia, China, Italy, and Spain rank a bit higher on HDI than they do on GNI; several countries in southern Africa, as well as Iran, rank lower on HDI than on GNI, thus illustrating the disconnect between GNI per capita and human well-being.
- The map of GEI rank (see Figure 1.21C) shows some strange anomalies, with certain high-income countries, such as the United States (medium high) and Saudi Arabia (very low), ranking far lower on this index than on HDI. Meanwhile, China (high), which has a reputation for gender discrimination and a medium-low GNI, ranks higher than the United States on the GEI scale, in the same category with New Zealand and Ireland.

Sustainable Development and Political Ecology

The United Nations (UN) defines **sustainable development** as the effort to improve current living standards in ways that will not jeopardize those of future generations. Sustainability has only recently gained widespread recognition as an important goal—well after the developed parts of the world had already achieved high standards of living based on the mass consumption of resources that was accompanied by mass pollution of environments. However, sustainability is particularly important for the vast majority of the Earth's people who do not yet have an acceptable level of well-being. Without sustainable development strategies, efforts to improve living standards for those who need it most will increasingly be foiled by degraded or scarce resources.

Geographers who study the interactions among development, politics, human well-being, and the environment are called **political ecologists**. They are known for asking the "Development for whom?" question, meaning, "Who is actually benefiting from so-called development projects?" Political ecologists examine how the power relationships in a society affect the ways in which development proceeds. For instance, in a Southeast Asian country, the clearing of forests to grow oil palm trees might at first seem to benefit many people. It would create some jobs, earn profits for the growers, and raise tax revenues for

FIGURE 1.21 Global maps of human well-being. [Sources consulted: *Human Development Report 2011 Statistical Annex*, Tables 1 and 4, United Nations Development Programme, at http://www.undp.org/content/dam/undp/library/corporate/HDR/2011%20Global%20HDR/English/HDR_2011_EN_Tables.pdf]

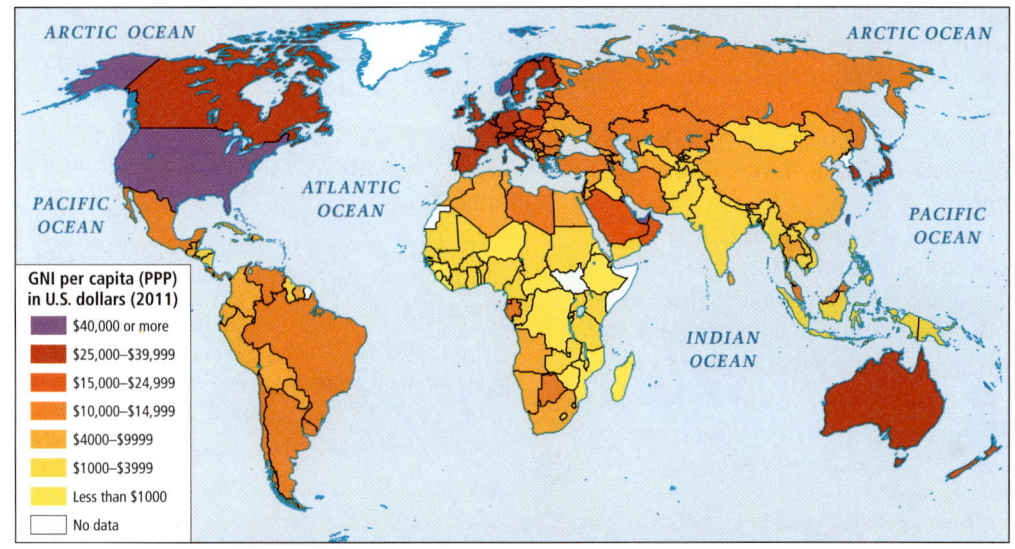

(A) Gross national income (GNI) per capita, adjusted for purchasing power parity (PPP).

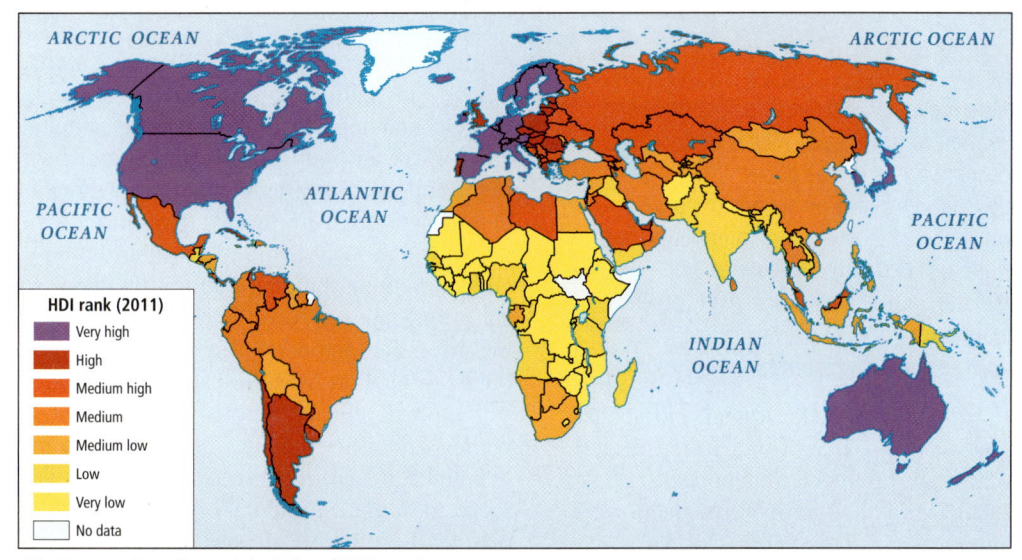

(B) Human Development Index (HDI) rank.

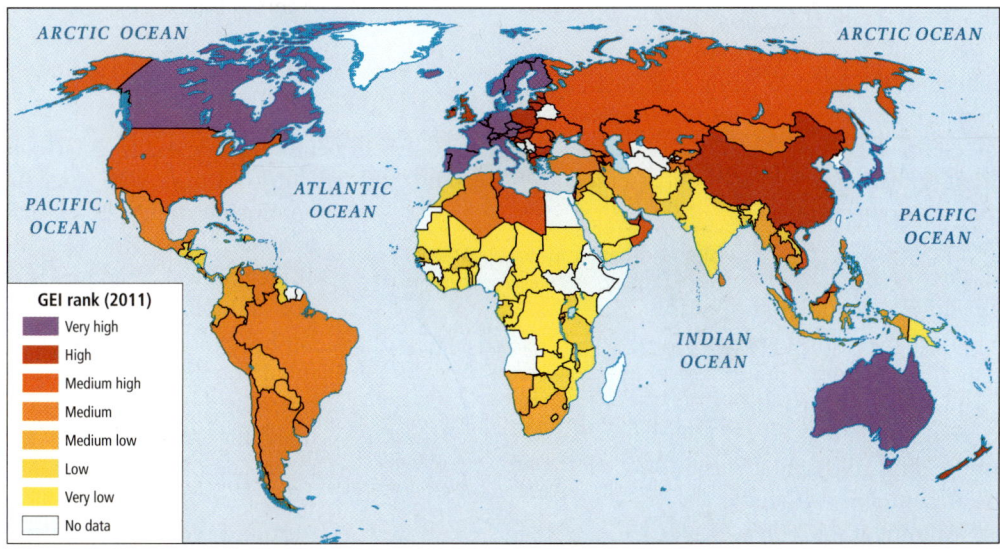

(C) Gender Equality Index (GEI) rank.

the government through the sale of palm oil, an important and widely used edible oil and industrial lubricant. However, these gains must be balanced against the loss of highly biodiverse tropical forest ecosystems and the human cultures that depend on them. Not only are forest dwellers losing their lands and means of livelihood to palm oil agribusiness, valuable knowledge that could be used to develop more sustainable uses of forest ecosystems is being lost as forest dwellers are forced to migrate to crowded cities, where their woodland skills are useless and therefore soon forgotten.

Political ecologists are raising awareness that development should be measured by the improvements brought to overall human well-being and long-term environmental quality, not just by the income created. By these standards, converting forests to oil palm plantations might appear less attractive, since only a few will benefit at the cost of widespread and often irreversible ecological and social disruption.

THINGS TO REMEMBER

- The term *development* has until recently referred to the rise in material standards of living that usually accompanies the shift from extractive economic activities, such as farming and mining, to industrial and service economic activities.
- Measures of development are being redefined to mean improvements in overall average well-being and progress in overall environmental sustainability.
- For development to happen, social services, such as education and health care, are necessary to enable people to contribute to economic growth.
- It is estimated that one-third or more of the world's work takes place in the informal economy, where work is often bartered for food, housing, or services, or for cash payments made "off the books" that are not reported to the government as taxable income.

POWER AND POLITICS

> **GEOGRAPHIC INSIGHT 3**
>
> **Power and Politics:** There are major differences across the world in the ways that power is wielded in societies. Modes of governing that are more authoritarian are based on the power of the state or community leaders. Modes that are based on notions of political freedom and democracy give the general public greater power over themselves and more of a role in deciding how policies are developed and governments are run. There are also many other ways of managing political power.

For geographers studying globalization, the social and spatial distribution of political power is an area of increasing interest (Figure 1.22). Recent years have produced what appears to be a trend toward political systems guided by competitive elections, a process often called **democratization**. This trend is of particular interest because it runs counter to **authoritarianism**, a form of government that subordinates individual freedom to the power of the state or elite regional and local leaders. In democratic systems of government, beyond the right to participate in free elections, average individuals have many other **political freedoms**, such as the following: freedom of speech (the right to express oneself in public and through the media), freedom of assembly (the right to gather together in groups to pursue common interests), freedom of movement (the right to travel, live, and work in any part of a state that a person is a citizen of), freedom from unreasonable searches and seizures (the right to privacy and protection from searches and seizures of individuals and their property by anyone not possessing a warrant granted by a court of law), freedom of the press (the right to communicate through any media without interference from the government or other entities), and freedom of religion (the right to practice or not practice any faith or spiritual path).

Geographers do not necessarily conclude that democracy and respect for political freedoms is the "best" political arrangement. Indeed, many geographers are critical of the imposition of democracy, often by foreign governments or organizations, in places where local people have not chosen democratic systems and where long-standing cultural traditions support other political arrangements. Nevertheless, few would deny that the shift toward more democratic systems of government and greater political freedom over the past century and into recent times is extremely significant, if not always peaceful. In this regard, geographers and other scholars are particularly interested in the role of political freedoms at the local level, in addition to the roles that social movements, international organizations, and a free media play in the exercise of power and politics. **23. PROMOTING DEMOCRACY: A CONTROVERSIAL THEME**

THE EXPANSION OF POLITICAL FREEDOMS

The twentieth century saw a steady expansion of some political freedoms throughout the world, with more and more countries holding elections of their leaders, at least at the national level. However the status of many other political freedoms is more complicated. For example, in the United States, recent revelations by former employees of federal government intelligence agencies have led many to question how well protected some political freedoms are in a country generally considered to be one of the more democratic and "politically free" places on Earth. The revelations surround secret mass surveillance programs in

---

democratization the transition toward political systems that are guided by competitive elections

authoritarianism a political system that subordinates individual freedom to the power of the state or of elite regional and local leaders

political freedoms the rights and capacities that support individual and collective liberty and public participation in political decision making

which the phone- and Internet-based communications of more than a billion people, including all U.S. citizens, are collected, stored, and analyzed by the federal government and various corporations that it hires. These activities have been criticized as undermining many political freedoms, including the freedom from unreasonable search and seizure, and, through intimidation, freedom of the press.

The Arab Spring movements, which commenced in early 2011 in a number of countries in North Africa and Southwest Asia, further highlight the complexity of the expansion of political freedoms. These movements showed that massive public demonstrations can achieve amazing turnovers of power, as was the case in Tunisia, Egypt, and Libya, but the outcomes are more complex and ambiguous than a straightforward expansion of political freedoms. While the demonstrations are an expression of certain political freedoms (freedom of assembly, freedom of speech) and have resulted in political reforms and elections in some cases, they have also provoked extremely repressive responses from authoritarian states and similarly forceful tactics from groups that oppose the state. The result can be widespread violence and constraints on political freedoms.

In Egypt, following demonstrations that precipitated the end of a notoriously corrupt authoritarian government that had ruled for more than 30 years, elections were held in 2012. An Islamist political party with authoritarian leanings, the Muslim Brotherhood, won an easy majority, causing concern that voters had quashed broader democratic reforms and endangered political freedoms. After a year of increasingly authoritarian rule, the Muslim Brotherhood was removed from power by a combination of protests and a military coup d'état (an overthrow of a government by the military). Supporters of the Muslim Brotherhood held violent protests in reaction, and the military responded with a crackdown in which thousands of demonstrators were killed and an unknown number imprisoned.

The case of Syria (see Figure 1.22B) is more extreme. Syrians held massive Arab Spring demonstrations in 2011 in an attempt to oust a regime that had ruled for more than 40 years. The military and police violently repressed the demonstrations, sparking a civil war in which more than 120,000 people have died and 5 million more have been displaced from their homes.

What Factors Encourage the Expansion of Political Freedoms?

Here are some of the most widely agreed-upon factors that support the expansion of political freedoms:

- **Peace:** Peace is essential to creating an environment in which people can, among other things, vote in "free and fair" elections, speak and gather freely, and use print and electronic media to voice their concerns.
- **Broad prosperity:** As a broader segment of the population gains access to more than the bare essentials of life, there is often a shift toward greater political freedom. Whether general prosperity must be in place before this occurs is still widely debated, as is the question of whether prosperity necessarily leads to any expansion of political freedoms.
- **Education:** Better-educated people tend to want a stronger voice in how they are governed. Although democracy has spread to countries with relatively undereducated populations, leaders in such places sometimes become more authoritarian once elected.
- **Civil society:** **Civil society** is made up of the social groups and traditions that function independently of the state and its institutions to foster a sense of unity and an informed common purpose among the general population. Civil society institutions can include the media, nongovernmental organizations (NGOs; discussed below and in Figure 1.23), political parties, universities, unions, and in some cases, religious organizations (see Figure 1.22C).

GEOPOLITICS

As the map in Figure 1.22 shows, high levels of political freedom are not enjoyed throughout the world. A possible explanation for this is that the expansion of political freedoms is sometimes at odds with **geopolitics**, the strategies that countries use to ensure that their own interests are served in relations with other countries. Geopolitics was perhaps most obvious during the *Cold War era*, the period from 1946 to the early 1990s when the United States and its allies in western Europe faced off against the Union of Soviet Socialist Republics (USSR) and its allies in eastern Europe and Central Asia. Ideologically, the United States promoted a version of free market **capitalism**—an economic system based on the private ownership of the means of production and distribution of goods, driven by the profit motive and characterized by a competitive marketplace. By contrast, the USSR and its allies favored what was called **communism**, but what was actually a state-controlled economy—a socialized system of public services and a centralized government in which citizens participated only indirectly through the Communist Party.

The Cold War became a race to attract the loyalties of unallied countries and to arm them. Sometimes the result was that authoritarian rulers were embraced as allies by one side or the other. Eventually, the Cold War influenced the internal and external policies of virtually every country in the world, often causing complex local issues to be oversimplified into a contest of democracy and capitalism versus communism.

In the post–Cold War period of the 1990s, geopolitics shifted. The Soviet Union dissolved, creating many independent states, nearly all of which began to implement some democratic and free market reforms. Globally, countries jockeyed for position in what looked like it might become a new era of trade and amicable

---

civil society the social groups and traditions that function independently of the state and its institutions to foster a sense of unity and an informed common purpose among the general population

geopolitics the strategies that countries use to ensure that their own interests are served in relations with other countries

capitalism an economic system based on the private ownership of the means of production and distribution of goods, driven by the profit motive and characterized by a competitive marketplace

communism an ideology, based largely on the writings of the German revolutionary Karl Marx, that calls on workers to unite to overthrow capitalism and establish an egalitarian society in which workers share what they produce; as practiced, communism was actually a socialized system of public services and a centralized government and economy in which citizens participated only indirectly through Communist Party representatives

prosperity rather than war. But throughout the 1990s, while the developed countries were in a period of unprecedented prosperity, many unresolved political conflicts emerged in southeastern Europe, Central and South America, Africa, Southwest Asia, and South Asia. Too often these disputes erupted into bloodshed and the systematic attempt to remove (through **ethnic cleansing**) or exterminate (through **genocide**) all members of a particular ethnic or religious group.

The new geopolitical era ushered in by the terrorist attacks on the United States on September 11, 2001, is still evolving. Because of the size and the global power of the United States, the attacks and the U.S. reactions to them affected virtually every international relationship, public and private. The ensuing adjustments, which will continue for years, are directly or indirectly affecting the daily lives of billions of people around the world.

International Cooperation

In addition to expanding political freedoms there is also a trend toward greater international cooperation. The prime example of this today is the **United Nations (UN)**, an assembly of 193 member states. The member states sponsor programs and agencies that focus on, among other things, economic development, general health and well-being, democratization, peacekeeping, and humanitarian aid. However, the UN rarely challenges a country's *sovereignty*, its right to conduct its internal affairs as it sees fit without interference from outside. Consequently, the UN often can enforce its rulings only through economic sanctions. While the UN does play a role in peacekeeping in troubled areas, there are no true UN military forces. Rather, there are troops from member states that wear UN designations on their uniforms and take orders from temporary UN commanders.

Nongovernmental organizations (NGOs) are an increasingly important embodiment of international cooperation. In such associations, individuals, often from widely differing backgrounds and locations, agree on political, economic, social, or environmental goals. For example, some NGOs, such as the World Wildlife Fund, work to protect the environment. Others, such as Doctors Without Borders, provide medical care to those who need it most, especially in conflict zones. The Red Cross and Red Crescent provide emergency relief after disasters, as do Oxfam, Catholic Charities, and Gift of the Givers (Figure 1.23).

---

ethnic cleansing the deliberate removal of an ethnic group from a particular area by forced migration

genocide the deliberate destruction of an ethnic, racial, or political group

United Nations (UN) an assembly of 193 member states that sponsors programs and agencies that focus on economic development, general health and well-being, democratization, peacekeeping assistance in "hot spots" around the world, humanitarian aid, and scientific research

nongovernmental organizations (NGOs) associations outside the formal institutions of government in which individuals, often from widely differing backgrounds and locations, share views and activism on political, social, economic, or environmental issues

---

FIGURE 1.22   PHOTO ESSAY: Power and Politics

The map and accompanying photo essay show two related trends in political power. Generally speaking, countries with fewer political freedoms and lower levels of democratization also have the most violent conflict. Countries are colored on the map according to their score on a "democracy index" created by the *Economist* magazine, which uses a combination of statistical indicators to capture elements crucial to the process of democratization. These elements include the ability of a country to hold peaceful elections that are accepted as fair and legitimate, people's ability to participate in elections and other democratic processes, the strength of civil liberties (such as a free media and the right to hold political gatherings and peaceful protests), and the ability of governments to enact the will of their citizens and be free of corruption. Also displayed on the map are major conflicts initiated or in progress since 1990 that have resulted in at least 10,000 casualties. Aspects of the connections between democratization and armed conflict are explored in the photo captions.

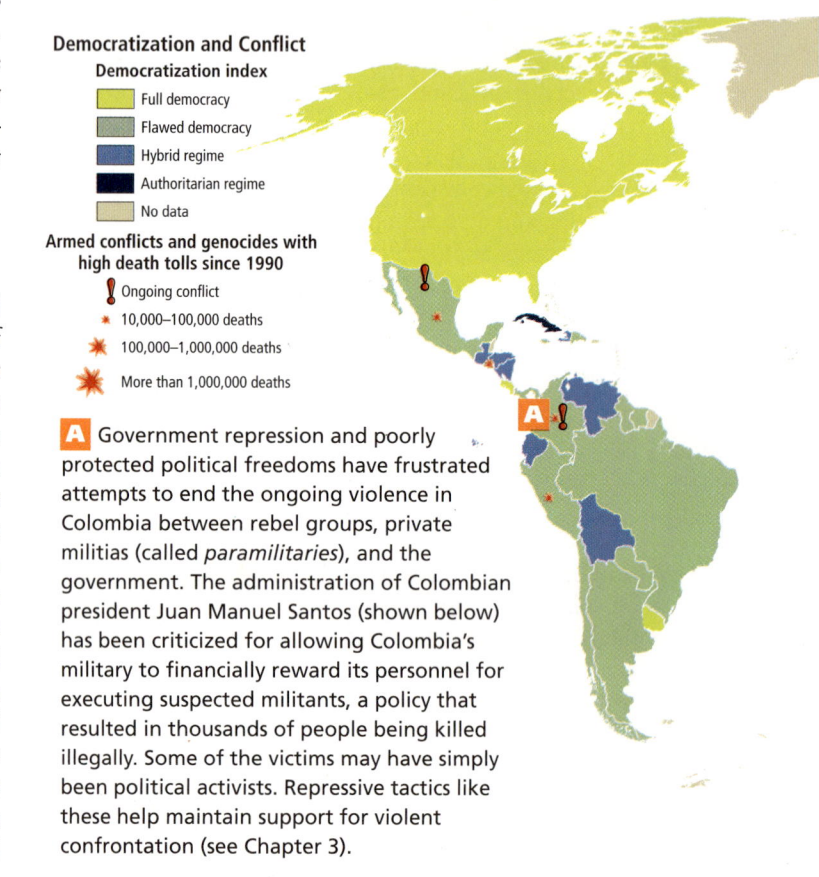

Democratization and Conflict

Democratization index
- Full democracy
- Flawed democracy
- Hybrid regime
- Authoritarian regime
- No data

Armed conflicts and genocides with high death tolls since 1990
- Ongoing conflict
- 10,000–100,000 deaths
- 100,000–1,000,000 deaths
- More than 1,000,000 deaths

A Government repression and poorly protected political freedoms have frustrated attempts to end the ongoing violence in Colombia between rebel groups, private militias (called *paramilitaries*), and the government. The administration of Colombian president Juan Manuel Santos (shown below) has been criticized for allowing Colombia's military to financially reward its personnel for executing suspected militants, a policy that resulted in thousands of people being killed illegally. Some of the victims may have simply been political activists. Repressive tactics like these help maintain support for violent confrontation (see Chapter 3).

40

B A Syrian woman holds an artillery shell fired by the Syrian military on civilians during the country's civil war. Sparked by demonstrations to gain more political freedom in 2011, part of the larger wave of protests known as the Arab Spring, Syria's civil war developed when the government responded with a harsh military crackdown on the civilian population. The resulting civil war has led to nearly 100,000 deaths, more than a million international refugees, and 2.5 to 3 million displaced people within Syria (see Chapter 6).

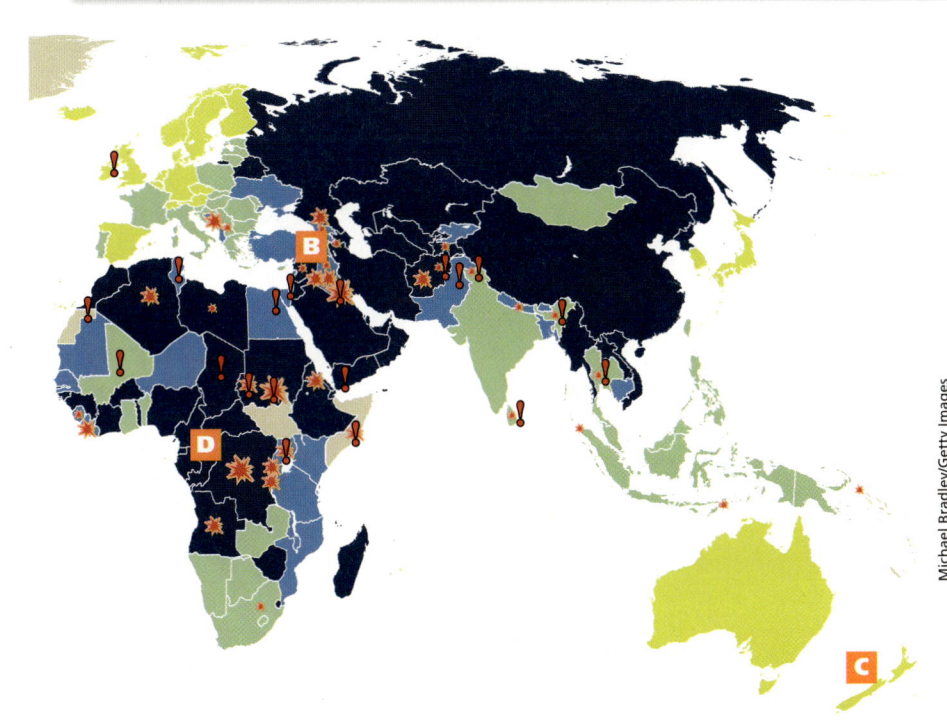

C A member of the Service and Food Workers Union in New Zealand pickets and solicits support from passing motorists for better wages and working conditions for the union's largely immigrant and female members (see Chapter 11). New Zealand has well-protected political freedoms and is among the world's most democratized nations.

D Political campaigns often become compromised during times of war or conflict. Shown here are supporters of the political opposition in Kinshasa, Congo, being dispersed by police with tear gas during elections in 2011. Accusations that the vote counting was being rigged added to tensions related to the country's ongoing civil war. The war has resulted in more than 5.4 million deaths so far, mostly due to disease and malnutrition among people displaced by the violence (see Chapter 7).

41

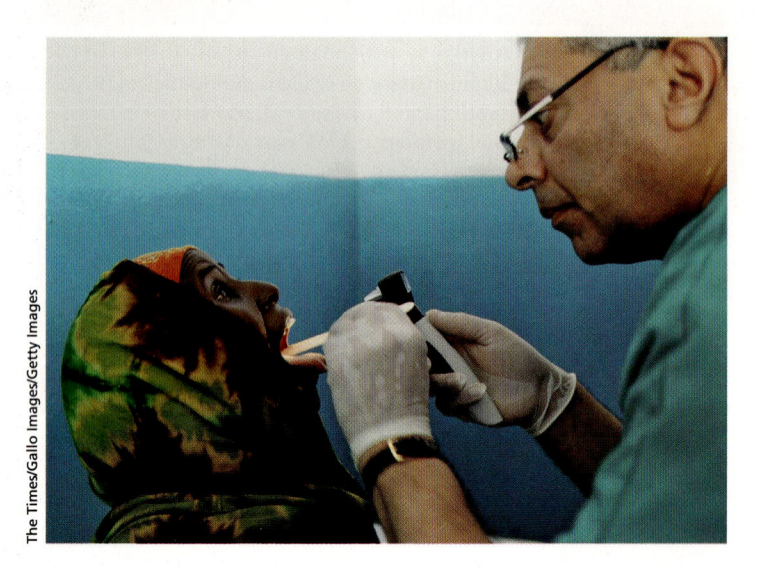

FIGURE 1.23 Nongovernmental organizations in action. A doctor examines a patient as part of an NGO-based effort to provide medical assistance and food aid to famine-stricken parts of Somalia. This doctor volunteers with the South Africa–based Gift of the Givers, the largest NGO started and staffed completely by Africans.

NGOs can be an important component of civil society, yet there is some concern that the power of huge international NGOs might undermine democratic processes, especially in small countries. Some critics feel that NGO officials are a powerful, do-gooder elite that does not interact sufficiently well with local people. A frequent target of such criticism is Oxfam International (a networked group of 17 NGOs), the world leader in emergency famine relief. Oxfam was a major provider of relief after the 2004 Indian Ocean tsunami, the 2010 Haiti earthquake, and the 2013 cyclone in the Philippines. It has now expanded to cover long-term efforts to reduce poverty and injustice, which Oxfam sees as the root causes of famine. This more politically active role has brought Oxfam into conflict with governments and officials capable of hampering the NGO's ability to achieve many of its goals at the local level. **22. NGOS PLAY LARGER ROLE IN WORLD AFFAIRS**

THINGS TO REMEMBER

> **GEOGRAPHIC INSIGHT 3**

- **Power and Politics** There are major differences across the world in the ways that power is wielded in societies. Modes of governing that are more authoritarian are based on the power of the state or community leaders. Modes that are based on notions of political freedom and democracy give the general public greater power over themselves and more of a role in deciding how policies are developed and governments are run. There are also many other ways of managing political power.

- There have been expansions over recent decades of some political freedoms throughout the world, with more and more countries holding elections of their leaders, at least at the national level. However, the status of many other political freedoms is more complicated.

- Some of the most widely agreed-upon factors that support the expansion of political freedoms are peace, broad prosperity, education, and civil society.

- The expansion of political freedoms is sometimes at odds with geopolitics, the strategies that countries use to ensure that their own interests are served in relations with other countries.

- NGOs can be an important component of civil society, yet there is some concern that the power of huge international NGOs might undermine democratic processes, especially in small countries.

URBANIZATION

> **GEOGRAPHIC INSIGHT 4**

Urbanization: The development of urban manufacturing and service economies has pulled people into cities. Meanwhile, the mechanization of food production has drastically reduced the need for agricultural labor, thus pushing people out of rural areas.

WHY ARE CITIES GROWING?

The mechanization of food production has drastically reduced the need for rural labor while also increasing the food supply to the point where large non-farming populations can be supported. This has tended to push people out of rural areas and into cities, where the development of manufacturing and service economies has created many jobs. This process is called the **push/pull phenomenon of urbanization**. Numerous cities, especially in poorer parts of the world, have been unprepared for the massive inflow of rural migrants, many of whom now live in polluted **slum** areas plagued by natural and human-made hazards, poor housing, and inadequate access to

In 1700, fewer than 7 million people, or just 10 percent of the world's total population, lived in cities, and only 5 cities had populations of several hundred thousand people or more. The world we live in today has been transformed by **urbanization**, the process whereby cities, towns, and suburbs grow as populations shift from rural to urban livelihoods. A little over half of the world's population now lives in cities, and there are more than 400 cities of more than 1 million people and 28 cities of more than 10 million people.

urbanization the process whereby cities, towns, and suburbs grow as populations shift from rural to urban livelihoods

push/pull phenomenon of urbanization conditions, such as political instability or economic changes, that encourage (push) people to leave rural areas, and urban factors, such as job opportunities, that encourage (pull) people to move to the urban area

slum densely populated area characterized by crowding, run-down housing, and inadequate access to food, clean water, education, and social services

Global Patterns of Population Growth   43

food, clean water, education, and social services (Figure 1.24C). Often a substantial portion of the migrants' cash income goes to support their still-rural families.

Patterns of Urban Growth

The most rapidly growing cities are in developing countries in Asia, Africa, and Middle and South America. The settlement pattern of these cities bears witness to their rapid and often unplanned growth, fueled in part by the steady arrival of masses of poor rural people looking for work (see Figure 1.24B). Cities like Mumbai (in India), Cairo (in Egypt), Nairobi (in Kenya), and Rio de Janeiro (in Brazil) sprawl out from a small affluent core, often the oldest part, where there are upscale businesses, fine old buildings, banks, shopping centers, and residences for wealthy people. Surrounding these elite landscapes are sprawling mixed commercial, industrial, and middle-class residential areas, interspersed with pockets of extremely dense slums. Also known as *barrios*, *favelas*, *hutments*, *shantytowns*, *ghettos*, and *tent villages*, these settlements provide housing for the poorest of the poor, who provide low-wage labor for the city. Housing is often self-built out of any materials the residents can find: cardboard, corrugated metal, masonry, scraps of wood and plastic. There are usually no building codes, no toilets with sewer connections, and there is little access to clean water. Electricity is often obtained from illegal and dangerous connections to nearby power lines. Schools are few and overcrowded, and transportation is provided only by informal, nonscheduled van-based services.

The UN estimates that currently more than a billion people live in urban slums, with that number to increase to 2 billion by 2030. Life in these areas can be insecure and chaotic as criminal gangs often assert control through violence and looting—all actions that are especially likely during periods of economic recession and political instability. **331. U.N. HABITAT AGENCY SAYS HALF THE WORLD POPULATION LIVES IN CITIES**

ON THE BRIGHT SIDE

Urban Migrant Success Stories

Slums are only part of the story of urbanization today. Those who are financially able to come to urban areas for education and complete their studies tend to find employment in modern industries and business services. They constitute the new middle class and leave their imprint on urban landscapes via the high-rise apartments they occupy and the shops and entertainment facilities they frequent (see Figure 1.24A and the figure map). Cities such as Mumbai in India, São Paulo in Brazil, Cape Town in South Africa, and Shanghai in China are now home to this more educated group of new urban residents, many of whom may have started life on farms and in villages.

In the past, most migrants to cities were young males, but increasingly they are young females. Cities offer women more opportunities to work outside the home than do rural areas. And urban jobs usually pay much more than rural jobs. Cities also provide women with better access to education, better health care, and more personal freedom.

THINGS TO REMEMBER

> **GEOGRAPHIC INSIGHT 4**
• **Urbanization** The development of urban manufacturing and service economies has pulled people into cities. Meanwhile, the mechanization of food production has drastically reduced the need for agricultural labor, thus pushing people out of rural areas.

• Today about half of the world's population live in cities; there are over 400 cities with more than 1 million people and 28 cities of more than 10 million people.

• For some, urbanization means improved living standards, while for others it means being forced into slums with inadequate food, water, and social services.

POPULATION

> **GEOGRAPHIC INSIGHT 5**

Population and Gender: Population growth is slowing for a number of reasons, among them the increasing numbers of women who are delaying childbearing as they pursue educational and work opportunities outside the home.

Over the last several hundred years, the human population has boomed, but growth rates are now slowing in most societies and, in a few, have even begun to decline (a process called *negative growth*). Much of this pattern has to do with changes in economic development, urbanization, and gender roles that have reduced incentives for large families.

GLOBAL PATTERNS OF POPULATION GROWTH

It took between 1 million and 2 million years (at least 40,000 generations) for humans to evolve and to reach a population of 2 billion, which happened around 1945. Then, remarkably, in just 66 years—by October of 2011—the world's population more than tripled to 7 billion (Figure 1.25). What happened to make the population grow so quickly in such a short time?

The explanation lies in changing relationships between humans and the environment. For most of human history, fluctuating food availability, natural hazards, and disease kept human death rates high, especially for infants. Out of many pregnancies, a couple might raise only one or two children. Also, pandemics such as the Black Death in Europe and Asia in the 1300s killed millions of people in a short time.

An astonishing upsurge in human population began about 1500, at a time when the technological, industrial, and scientific revolutions were beginning in some parts of the world. Human life expectancy increased dramatically, and more and more people lived long enough to reproduce successfully, often many times over. The result was an exponential pattern of growth (see Figure 1.25). This pattern is often called a *J curve* because the ever-shorter periods between doubling and redoubling of the population cause an abrupt upward swing in the growth line when depicted on a graph.

Today, the human population is growing in most regions of the world, more rapidly in some places than in others. The reason for this growth is that currently a very large group of young people has reached the age of reproduction and more will join them shortly. Nevertheless, the *rate* of global population growth is slowing. Since 1993 it has dropped from 1.7 percent per year to roughly 1.1 percent per year, which is where it stands today. If present slower growth trends continue, the world population may level off at between 7.8 billion and 10 billion before 2050. However, this projection is contingent on couples in less-developed countries having the education and economic security to choose to have smaller families and the ability to practice birth control using the latest information and technology.

In a few countries (especially Japan in Asia; Ukraine plus Romania, Bulgaria, and Hungary in Central Europe), the population is actually declining and rapidly aging, due primarily to low birth and death rates. This situation could prove problematic, as those who are elderly and dependent become more numerous, posing a financial burden on a declining number of working-age people. In addition, HIV-AIDS is affecting population patterns to varying extents in all world regions. In Africa, the epidemic is severe; as a result, several African countries have sharply lowered life expectancies among young and middle-aged adults. 🎥 **11. WORLD POPULATION TO BE CONCENTRATED IN DEVELOPING NATIONS, AS TOTAL EXPECTED TO REACH 9 BILLION BY 2050**

Local Variations in Population Density and Growth

If the more than 7 billion people on Earth today were evenly distributed across the land surface, they would produce an *average population density* of about 121 people per square mile (47 per square kilometer). But people are not evenly distributed (Figure 1.26). Nearly 90 percent of all people live north of the equator, and most of them live between 20° N and 60° N latitude. Even within that limited territory, people are concentrated on about 20 percent of the available land. They live mainly in zones that have climates warm and wet enough to support agriculture: along rivers, in lowland regions, or fairly close to the sea. In general, people are located where resources are available.

Usually, the variable that is most important for understanding population growth in a region is the rate of natural increase (often called the *growth rate*). The **rate of natural increase (RNI)** is the relationship in a given population between the number of people being born (the **birth rate**) and the number dying (the **death rate**), without regard to the effects of **migration** (the movement of people from one place or country to another).

---

rate of natural increase (RNI) the rate of population growth measured as the excess of births over deaths per 1000 individuals per year without regard for the effects of migration

birth rate the number of births per 1000 people in a given population, per unit of time (usually per year)

death rate the ratio of total deaths to total population in a specified community, usually expressed in numbers per 1000 or in percentages

migration the movement of people from one place or country to another, often for safety or economic reasons

---

FIGURE 1.24 PHOTO ESSAY: Urbanization

In the map, the color of the country indicates the percentage of the population living in urban areas. The blue circles represent the populations of the world's largest urban areas in 2013. [Sources consulted: *2011 World Population Data Sheet*, Population Reference Bureau, at http://www.prb.org/pdf11/2011populationdata-sheet_eng.pdf and *World Gazetteer*, at http://world-gazetteer.com/wg.php?x=&men=gcis&lng=en&des=wg&srt=npan&col=abcdefghinoq&msz=1500&pt=a&va=&srt=pnan]

A Cities have always been centers of innovation, entertainment, and culture, in large part because they draw both money and talented people. This resident of Shanghai is engaging in a current fad: parachuting off one of the new skyscrapers that now dominate the city's skyline.

Population Living in Urban Areas
- 83%–100%
- 65%–82%
- 47%–64%
- 29%–46%
- 11%–28%
- No data

Population of Metropolitan Areas, 2013
- 30 million
- 20 million
- 10 million

Note: Symbols on map are sized proportionally to metro area population

① **Global rank** (population 2013)

44

B A recent migrant to Dhaka pulls a cart loaded with goods. Many of the world's fastest-growing cities are attracting more people than they can support with decent jobs, housing, and infrastructure.

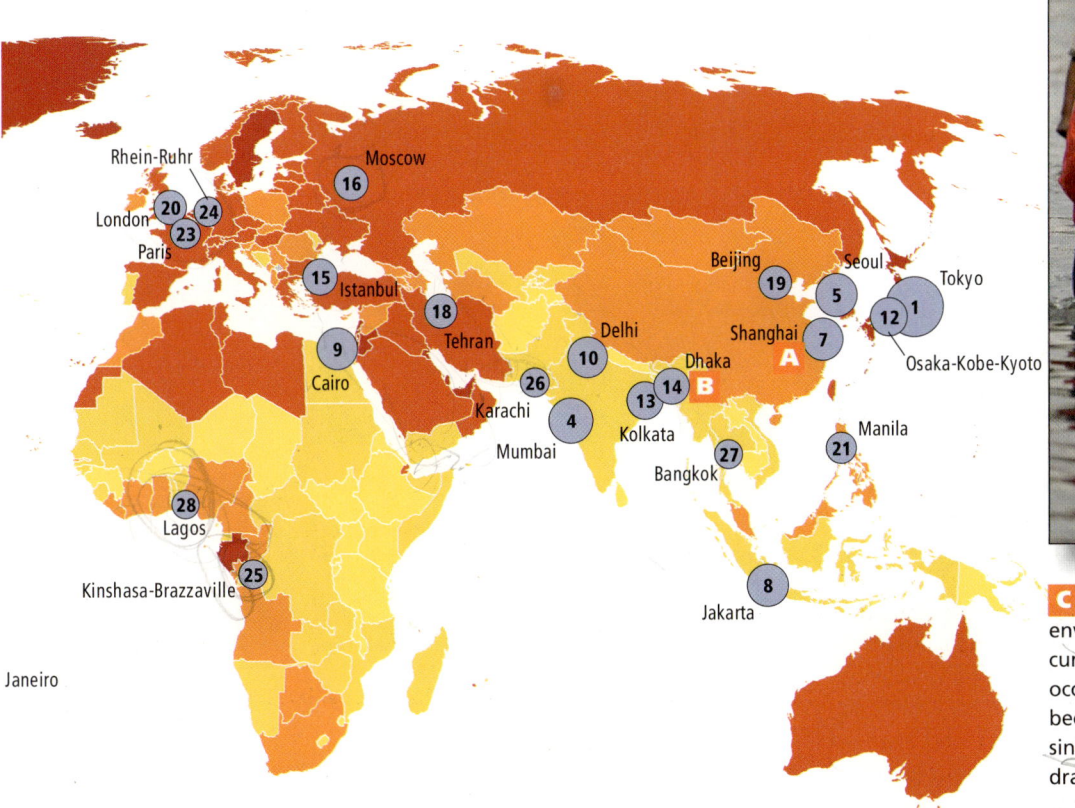

C Some cities struggle with major environmental problems. Mexico City, currently the world's second-largest city, occasionally suffers from severe flooding because of its location on an old, now-sinking lake bed and its antiquated drainage and sewage infrastructure.

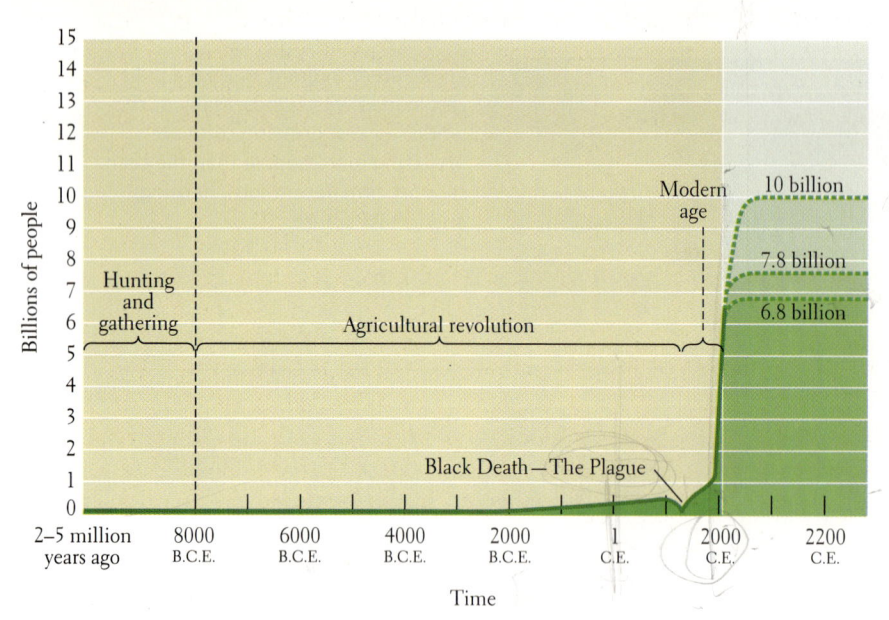

46 CHAPTER 1   Geography: An Exploration of Connections

FIGURE 1.25 Exponential population growth: The J curve. The curve's J shape is a result of successive doublings of the population. It starts out nearly flat, but as doubling time shortens, the curve bends ever more sharply upward. Note that B.C.E. (before the common era) is equivalent to B.C. (before Christ); C.E. (common era) is equivalent to A.D. (anno Domini). [Source consulted: G. Tyler Miller, Jr., *Living in the Environment*, 8th ed. (Belmont, CA: Wadsworth, 1994), p. 4]

The rate of natural increase is expressed as a percentage per year. For example, in 2011, the annual birth rate in Austria (in Europe) was 9 per 1000 people, and the death rate was 9 per 1000 people. Therefore, the annual rate of natural increase was 0 per 1000 (9 − 9 = 0), or 0 percent.

For comparison, consider Jordan (in Southwest Asia). In 2011, Jordan's birth rate was 31 per 1000, and the death rate was 4 per 1000. Thus the annual rate of natural increase was 27 per 1000 (31 − 4 = 27), or 2.7 percent per year. At this rate, Jordan's population will double in just 29 years.

total fertility rate (TFR) the average number of children that women in a particular population are likely to have at the present rate of natural increase

Total fertility rate (TFR), another term used to indicate trends in population, is the average number of children a woman in a particular population is likely to have during her reproductive years (15–49). The TFR in 2011 for Austrian women was 1.4; for Jordanian women, it was 3.8. As education rates for

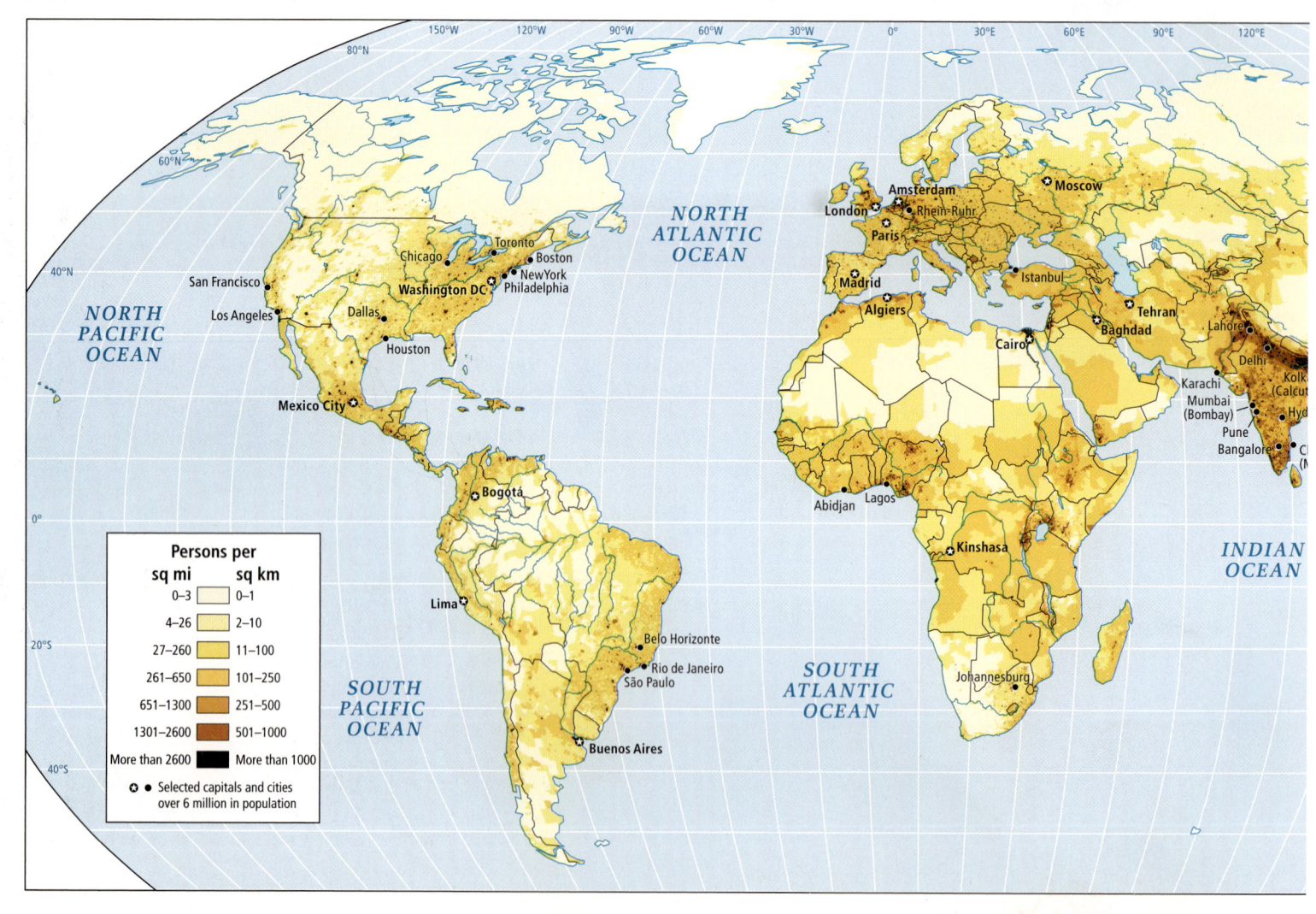

FIGURE 1.26 World population density.

women increase, and as they postpone childbearing into their late 20s, total fertility rates tend to decline.

Another powerful contributor to population growth is **immigration** (in-migration). In Europe, for example, the rate of natural increase is quite low, but the region's economic power attracts immigrants from throughout the world. Austria was attracting immigrants at the rate of 0.03 percent of its population in 2011, while Jordan was losing population to **emigration** (out-migration) at the rate of 0.04 percent of the population per year. In 2010, international migrants accounted for about 85 percent of the European Union's population growth; they were important additions to the labor force in an era of declining births. All across the world, people are on the move, trying to improve their circumstances; often they are fleeing war, natural disasters, or economic recessions; understandably, developed countries are favored destinations.

> **immigration** in-migration to a place or country (see also *migration*)
>
> **emigration** out-migration from a place or country (see also *migration*)
>
> **population pyramid** a graph that depicts the age and sex structures of a political unit, usually a country

AGE AND SEX STRUCTURES

The age and sex structures of a country's population reflect past and present social conditions and can help predict future population trends. The age distribution, or age structure, of a population is the proportion of the total population in each age group. The sex structure is the proportion of males and females in each age group.

The **population pyramid** is a graph that depicts age and sex structures. Consider the population pyramids for Austria and Jordan (Figure 1.27). Notice that Jordan's is a true pyramid, with a wide bottom, and that the largest groups are in the age categories 0 through 19.

By contrast, Austria's pyramid has an irregular vertical shape that tapers in toward the bottom. The base, which began to narrow about 40 years ago, indicates that there are now fewer people in the youngest age categories than in the teen, young adulthood, or middle-age groups; those over 70 greatly outnumber the youngest (ages 0 to 4). This age distribution is due to two trends: many Austrians now live to an advanced age, and in the last several decades, Austrian couples have chosen to have only one child, or none. If these trends continue, Austrians will need to support and care for large numbers of elderly people, and those responsible will be an ever-declining group of working-age people.

Population pyramids also reveal *sex imbalance* within populations. Look closely at the right (female) and left (male) halves of the pyramids in Figure 1.27. In several age categories, the sexes are not evenly balanced on both sides of the line. In the Austria pyramid, there are more women than men near the top (especially in the age categories of 70 and older). In Jordan, there are more males than females near the bottom (especially for ages 0 to 24).

Demographic research on the reasons behind statistical sex imbalance is relatively new, and many different explanations are proposed. In Austria, the predominance of elderly women reflects the fact that women live about 5 years longer than men in countries with long life expectancies (a trend that is still poorly understood). But the sex imbalance in younger populations has different explanations. The normal ratio worldwide is for about 95 females to be born for every 100 males. Because baby boys are somewhat weaker than girls on average, the ratio normally evens out naturally within the first 5 years. However, in many places the ratio is as low as 80 females to 100 males and continues throughout the life cycle. The widespread cultural preference for boys over girls (discussed in later chapters) becomes more observable as couples choose to have fewer children. Some fetuses, if identified as female, are purposely aborted; also, girls and women are sometimes fed less well and receive less health care than males, especially in poverty-stricken areas. Because of this, females are more likely to die, especially in early childhood. The gender imbalance of more males than females can be especially troubling for young adults, because the absence of females will mean that young men will find it hard to find a marriage mate.

POPULATION GROWTH RATES AND WEALTH

Although there is a wide range of variation, regions with slow population growth rates usually tend to be affluent, and regions with fast growth rates tend to have widespread poverty. The reasons for this difference are complicated; again, Austria and Jordan are useful examples.

FIGURE 1.27 Population pyramids for Austria and Jordan, 2011. [Sources consulted: "Population Pyramid of Germany," "Population Pyramid of Sweden," and "Population Pyramid of the European Union," *International Data Base*, U.S. Census Bureau, 2012, at http://www.census.gov/population/international/data/idb/informationGateway.php]

(A) Population pyramid for Austria.

(B) Population pyramid for Jordan.

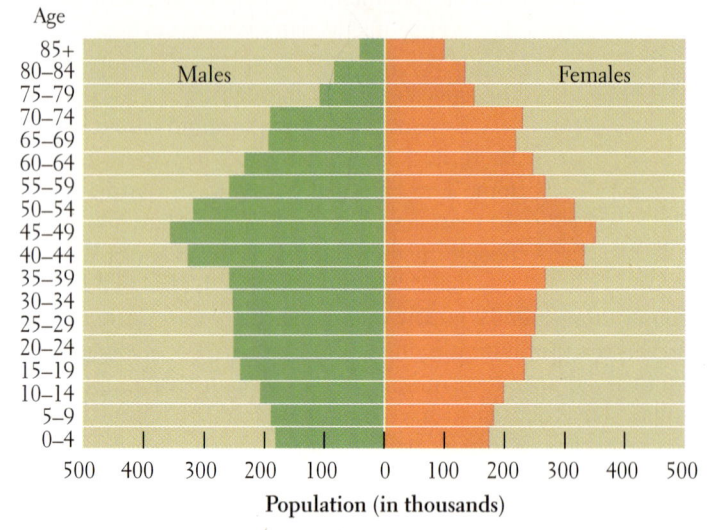

In 2013, Austria had an annual **gross national income (GNI) per capita** of $43,220 (PPP). This figure represents the total production of goods and services in a country in a given year, divided by the mid-year population; Austria's is among the highest in the world (the comparable figure for the United States is $50,610). Austria has a very low infant mortality rate of 3.7 per 1000. Its highly educated population is 100 percent literate, employed largely in technologically sophisticated industries and services. Large amounts of time, effort, and money are required to educate a child to compete in this economy. Many Austrian couples choose to have only one or two children, probably because of the cost of raising a child and because it is highly likely that children will survive to adulthood and thus will be available as companions for aging parents. Furthermore, the social welfare system in Austria ensures that citizens will not be impoverished in old age.

By contrast, Jordan has a GNI (PPP) per capita of $5730, in addition to a high infant mortality rate of 23 per 1000. Much everyday work is still done by hand, so each new child is a potential contributor to the family income at a young age. There is little social welfare for elderly citizens and a much greater risk of children not surviving into adulthood. Having more children helps ensure that someone will be there to provide care for aging parents and grandparents.

However, circumstances in Jordan have changed rapidly over the last 25 years. In 1985, Jordan's per capita GDP was $993, infant mortality was 77 per 1000, and women had an average of 8 children. Now Jordanian women have an average of only 3.8 children, with the need to educate those children increasingly recognized. Geographers would say that Jordan is going through a **demographic transition**, meaning that a period of high birth and death rates is giving way to a period of much lower birth and death rates (Figure 1.28). In the middle phases, however, when death rates decline more rapidly than birth rates (Jordan's death rate is just 4 per 1000 but its birth rate is 31 per 1000), population numbers can increase rapidly as attitudes about optimal family size slowly adjust. Populations may eventually stabilize, but because of the lag in the lowering of birth rates, population numbers may be significantly higher than they were at the beginning of the transition.

The demographic transition begins with the shift from subsistence to cash economies and from rural to urban ways of life. In a **subsistence economy**, a family, usually in a rural setting, produces most of its own food, clothing, and shelter, so there is little need for cash. Many children are needed to help perform the work that supports the family and birth rates are high. Most needed skills are learned around the home, farm, and surrounding lands, so expensive educations at technical schools and universities are not needed. Today, subsistence economies are disappearing as people seek cash with which to buy food and goods such as television sets and bicycles. In a **cash economy**, which tends to be urban but may be rural, skilled workers, well-trained specialists, and even farm laborers are paid in money. Each child needs years of education to qualify for a good cash-paying job and does not contribute to the family budget while in school. Having many children, therefore, is a drain on the family's resources. Perhaps most important, the higher development levels of cash economies mean that those populations are more likely to have better health care, increasing the likelihood that each child will survive to adulthood.

> **gross national income (GNI) per capita** the total production of goods and services in a country in a given year, divided by the mid-year population
>
> **demographic transition** the change from high birth and death rates to low birth and death rates that usually accompanies a cluster of other changes, such as change from a subsistence to a cash economy, increased education rates, and urbanization
>
> **subsistence economy** an economy in which families produce most of their own food, clothing, and shelter
>
> **cash economy** an economic system that tends to be urban but may be rural, in which skilled workers, well-trained specialists, and even farm laborers are paid in money

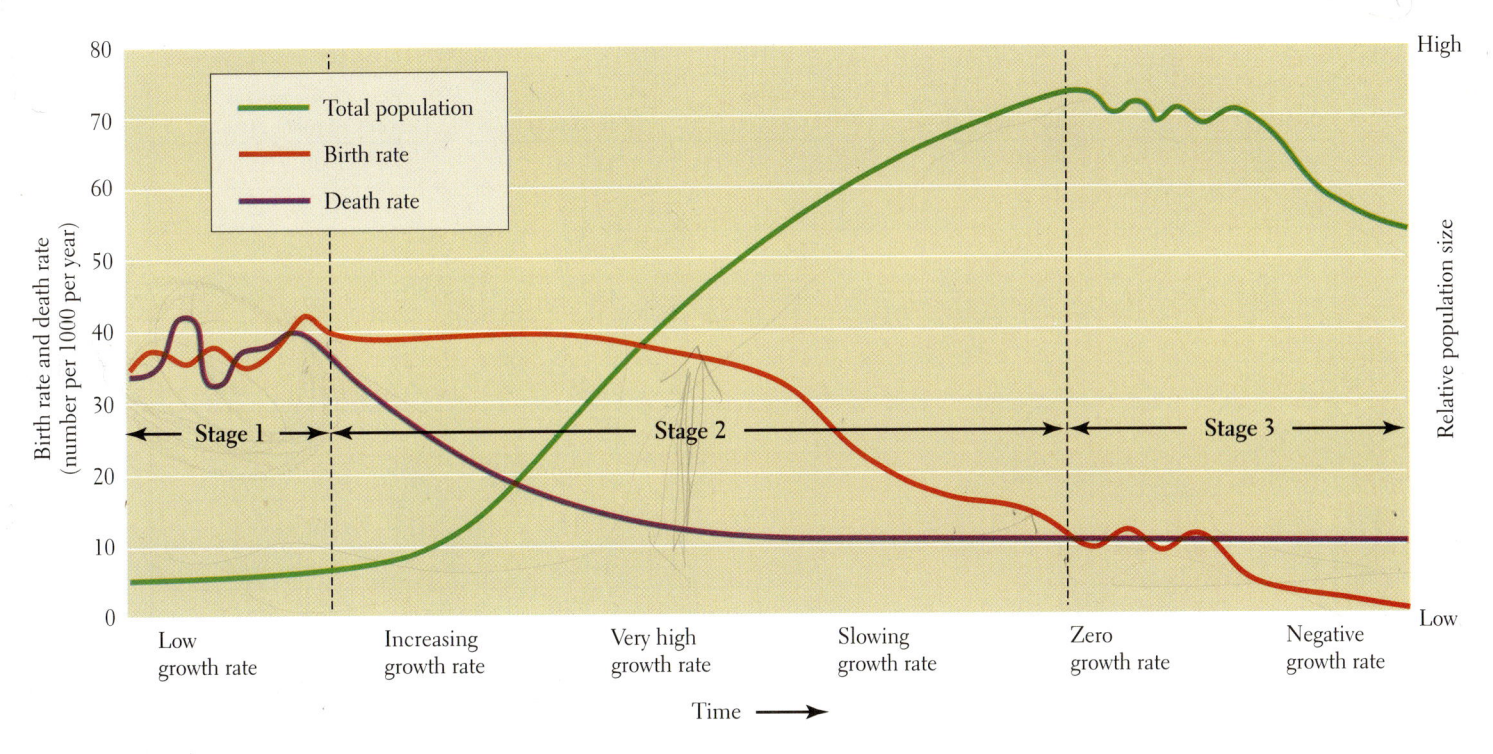

FIGURE 1.28 Demographic transition diagram. In traditional societies (Stage 1), both birth rates and death rates are usually high (left vertical axis), and population numbers (right vertical axis) remain low and stable. With advances in food production, education, and health care (Stage 2), death rates usually drop rapidly, but strong cultural values regarding reproduction remain, often for generations, so birth rates drop much more slowly, with the result that for decades or longer, the population continues to grow significantly. When changed social and economic circumstances enable most children to survive to adulthood and it is no longer necessary to produce a cadre of family labor (Stage 3), population growth rates slow and may eventually drift into negative growth. At this point, demographers say that the society has gone through the demographic transition. [Source consulted: G. Tyler Miller, Jr., *Living in the Environment*, 8th ed. (Belmont, CA.: Wadsworth, 1994), p. 218]

GENDER

Note the difference between the terms *gender* and *sex*. **Gender** indicates how a particular social group defines the differences between the sexes. **Sex** refers to the biological category of male or female but does not indicate how males or females may behave or identify themselves. Gender definitions and accepted behavior for the sexes can vary greatly from one social group to another. Here we consider gender.

For women, the historical and modern global gender picture is puzzlingly negative. In nearly every culture, in every region of the world, and for a great deal of recorded history, women have had (and still have) an inferior status. Exceptions are rare, although the intensity of this second-class designation varies considerably. On average, females have less access to education, medical care, and even food. They start work at a younger age and work longer hours than males. Around the world, people of both sexes still routinely accept the idea that males are more productive and intelligent than females. The puzzling question of how and why women became subordinate to men has not yet been well explored because, oddly enough, few thought the question significant until the last 50 years.

In nearly all cultures, families prefer boys over girls because, as adults, boys will have greater earning capacity (Table 1.2) and more power in society. This preference for boys has some unexpected side effects. For example, the preference may lead to fewer girls being born; eventually there will be a shortage of marriageable women, leaving many men without the hope of forming a family. Currently, there is concern in several Asian societies that the scarcity of young women could lead to antisocial behavior on the part of discouraged young men. **6. WOMEN STILL LAG BEHIND MEN IN TOP BOARDROOM JOBS**

Gender Roles

Geographers have begun to pay more attention to **gender roles**—the socially assigned roles for males and females—in different culture groups. In virtually all parts of the world, and for at least tens of thousands of years, the biological fact of maleness and femaleness has been translated into specific roles for each sex. The activities assigned to men and to women can vary greatly from culture to culture and from era to era, but they remain central to the ways societies function. Indeed, increasing attention to gender roles by geographers has been driven largely by interest in the shifts that occur when traditional ways are transformed by modernization.

There are some striking consistencies regarding traditional gender roles over time and around

---

gender the ways a particular social group defines the differences between the sexes

sex the biological category of male or female; does not indicate how males or females may behave or identify themselves

gender roles the socially assigned roles for males and females

TABLE 1.2	Comparisons of male and female income in selected countries, 2011			
Country	HDI rank	Female income (PPP U.S.$)	Male income (PPP U.S.$)	Female income as percent of male income
Austria	19	22,528	55,934	40
Barbados	47	15,119	23,507	64
Botswana	118	8823	17,952	49
Canada	6	30,005	45,763	66
Japan	12	20,572	44,892	46
Jordan	95	2456	8581	29
Kuwait	63	24,531	65,010	38
Poland	39	13,886	24,292	57
Russia	66	15,191	23,284	65
Saudi Arabia	56	7157	36,727	19
Sweden	10	32,990	41,830	79
United Kingdom	28	28,354	42,217	67
United States	4	35,346	56,918	62

Sources: Income data from "Estimated earned income," *The Global Gender Gap Report 2011* (Geneva: World Economic Forum), 44, table D3, http://reports.weforum.org/global-gender-gap-2011/.
HDI data from *United Nations Human Development Report 2011*, 128, table 1, http://hdr.undp.org/en/content/human-development-report-2011.

the globe. Men are expected to fulfill public roles, while women fulfill private roles. Certainly there are exceptions in every culture, and customs are changing, especially in wealthier countries. But generally, men work outside the home in positions such as executives, animal herders, hunters, farmers, warriors, or government leaders. Women keep house, bear and rear children, care for the elderly, grow and preserve food, and prepare the meals, among many other tasks. In nearly all cultures, women are defined as dependent on men—their fathers, husbands, brothers, or adult sons—even when the women may produce most of the family sustenance.

Gender Issues

Because their activities are focused on the home, women tend to marry early. One quarter of the girls in developing countries are mothers before they are 18. This is crucial in that pregnancy is the leading cause of death among girls aged 15 to 19 worldwide, primarily because immature female bodies are not ready for the stress of pregnancy and birth. Globally, babies born to women under 18 have a 60 percent higher chance of dying in infancy than do those born to women over 18.

Typically, women also have less access to education than men (around the world, 70 percent of youth who leave school early are girls). They are less likely to have access to information and paid employment, and so have less access to wealth and political power. When they do work outside the home (as is the case increasingly in every world region), women tend to fill lower-paid positions, such as laborers, service workers, or lower-level professionals. And even when they work outside the home, most women retain their household duties, so they work a *double day*.

Gender and sex categories can be confusing in matters such as the extent to which physical differences between males and females may affect their social roles. On average, males are stronger and taller, but females have comparable endurance and pain tolerance. Women's physical capabilities are somewhat limited during pregnancy and nursing but from the age of about 45, women are no longer subject to the limits of pregnancy, and most contribute in some significant way to the well-being of their adult children and grandchildren, an important social role. A growing number of biologists suggest that the evolutionary advantage of menopause in midlife is that it gives women the time, energy, and freedom to help succeeding generations thrive. This notion—sometimes labeled the *grandmother hypothesis*—seems to have worldwide validity (**Figure 1.29**). Certainly, grandfathers can play nurturing roles, but women tend to live 5 years longer than men.

Perhaps more than for any other culturally defined human characteristic, significant agreement exists that gender is important, but just how gender roles are defined varies greatly across places and over time.

Traditional notions of gender roles are now being challenged everywhere. In many countries, including conservative Muslim countries, females are acquiring education at higher rates than males. Although it will take females a while to catch up, eventually education should make women competitive with men for jobs and roles in public life as policy makers and government officials, not just as voters. Unless discrimination persists, women should also begin to earn pay equal to that of men (see Table 1.2).

Research data suggest that there is a ripple effect that benefits the whole group when developing countries pay attention to the needs of girls:

- Girls in developing countries who get 7 or more years of education marry 4 years later than average and have 2.2 fewer children.
- An extra year of secondary schooling over the average for her locale boosts a girl's lifetime income by 15 to 25 percent.
- The children of educated mothers are healthier and more likely to finish secondary school.
- When women and girls earn income, 90 percent of their earnings are invested in the family, compared to just 40 percent of males' earnings.

Considering only women's perspectives on gender, however, misses half the story. Men are also affected by strict gender expectations, often negatively. For most of human history, young men have borne a disproportionate share of burdensome physical tasks and dangerous undertakings. Until recently, mostly young men left home to migrate to distant, low-paying jobs. Overwhelmingly, it has been young men (the majority of soldiers) who die in wars or suffer physical and psychological injuries from combat.

An emerging body of research suggests that sexual violence is widespread and inextricably linked to gender roles. A recent survey of 10,000 men in six Asian countries indicated that one in four participants had raped a woman, usually a romantic partner. Seventy percent of the men who had committed rape said they felt entitled to do so. This and similar studies done in other countries suggest that rape is a widespread global phenomenon deeply rooted in power relations between genders, and that preventative measures are most effective during childhood and adolescence.

This book will return repeatedly to the question of gender disparities because they play such a central role in so many global problems and in creating a more just future for all people.

FIGURE 1.29 The grandmother hypothesis. In every culture and community worldwide, grandmothers contribute to the care and education of their grandchildren. This quality has played an essential role in human evolution. A grandmother in Xian, China, helps her toddler granddaughter learn to walk.

> **THINKING GEOGRAPHICALLY** What are five things grandmothers in every world region are likely to worry about regarding their grandchildren?

THINGS TO REMEMBER

> **GEOGRAPHIC INSIGHT 5**
- **Population and Gender** Population growth is slowing for a number of reasons, among them the increasing numbers of women who are delaying childbearing as they pursue educational and work opportunities outside the home.
- Over the last several hundred years, global population growth has been rapid. Although growth will continue for many years, rates are now slowing in most places and in a few places growth has reversed and the population is even shrinking.
- The circumstances that lead to lower population growth rates also lead to the overall aging of populations; these two phenomena are found nearly everywhere in the world today, but to varying degrees.
- Gender—the sexual category of a person—is both a biological and a cultural phenomenon. Gender indicates how a particular group defines the social differences between the sexes. Sex is the biological category of male or female.
- There are some global consistencies in gender disparities: typically, males have public roles and females have private roles; in every country, the average woman earns less than the average man.

SOCIOCULTURAL ISSUES

Many geographers are interested in the economic, social, and cultural practices of a people, and in the spatial patterns these factors create. Cultural geography focuses on culture as a complex of important distinguishing characteristics of human societies. **Culture** comprises everything people use to live that is not directly part of biological inheritance. Culture is represented by the ideas, materials, methods, and social arrangements that people have invented and passed on to subsequent generations, such as methods of producing food and shelter. Culture includes language, music, tools and technology, clothing, gender roles, belief systems, and moral codes, such as those prescribed in Confucianism, Islam, and Christianity.

culture all the ideas, materials, and institutions that people have invented to use to live on Earth that are not directly part of our biological inheritance

ethnic group a group of people who share a common ancestry and sense of common history, a set of beliefs, a way of life, a technology, and usually a common geographic location of origin

Ethnicity and Culture: Slippery Concepts

A group of people who share a location, a set of beliefs, a way of life, a technology, and usually a common ancestry and sense of common history form an **ethnic group**. The term *culture group* is often used interchangeably with ethnic group. Both of the concepts of culture and ethnicity are imprecise, especially as they are popularly used. For instance, as part of the modern globalization process, migrating people often move well beyond their customary cultural or ethnic boundaries to cities or even distant countries. In these new places they take on many new ways of life and beliefs—their culture actually changes, yet they still may identify with their cultural or ethnic origins.

The Kurds in Southwest Asia are an example of the tenacity of this ethnic or cultural group identity. Long before the U.S. war in Iraq, the Kurds were asserting their right to create their own country in the territory where they have lived as nomadic herders since before the founding of Islam. Syria, Iraq, Iran, and Turkey now claim parts of the traditional Kurdish area. Many Kurds are now educated urban dwellers, living and working in modern settings in Turkey, Iraq, Iran, or even London and New York, yet they actively support the cause of establishing a Kurdish homeland. Although urban Kurds think of themselves as ethnic Kurds and are so regarded in the larger society, they do not follow the traditional Kurdish way of life. We could argue that these urban Kurds have a new identity within the Kurdish culture or ethnic group. Or they may be in a *transcultural* position, moving from one culture to another.

Another problem with the imprecision of the concept of culture is that it is often applied to a very large group that shares only the most general of characteristics. For example, one often hears the terms American culture, African American culture, or Asian culture. In each case, the group referred to is far too large to share more than a few broad characteristics.

It might fairly be said, for example, that U.S. culture is characterized by beliefs that promote individual rights, autonomy, and individual responsibility. But when we look at specifics, just what constitutes the rights and responsibilities of the individual are quite debatable. In fact, American culture encompasses many subcultures that share some of the core set of beliefs but disagree over parts of the core and over a host of other matters. The same is true, in varying degrees, for all other regions of the world. When many culture groups live in close association, the society may be called **multicultural**.

multicultural society a society in which many culture groups live in close association

Values

Occasionally you will hear someone say, "After all is said and done, people are all alike," or "People ultimately all want the same thing." It is a heartwarming sentiment, but an oversimplification. True, we all want food, shelter, health, love, and acceptance; but culturally, people are not all alike, and that is one of the qualities that makes the study of geography interesting. We would be wise not to expect or even to want other people to be like us. Cultural diversity has helped humans to be successful and adaptable animals. The various cultures serve as a bank of possible strategies for responding to the social and physical challenges faced by the human species. The reasons for differences in behavior from one culture to the next are usually complex, but they are often related to differences in values. Consider the following vignette that contrasts the values and *norms* (accepted patterns of behavior based on values) held by modern and urban individualistic cultures with those held by rural, community-oriented cultures.

VIGNETTE One recent rainy afternoon, a beautiful 40-something Asian woman walked alone down a fashionable street in Honolulu, Hawaii. She wore high-heeled sandals, a flared skirt that showed off her long legs, and a cropped blouse that allowed a glimpse of her slim waistline. She carried a laptop case and a large fashionable handbag. Her long, shiny black hair was tied back. Everyone noticed and admired her because she exemplified an ideal Honolulu businesswoman: beautiful, self-assured, and rich enough to keep herself well-dressed.

In the village of this woman's grandmother—whether it be in Japan, Korea, Taiwan, or rural Hawaii—the dress that exposed her body to open assessment and admiration by strangers of both sexes would signal that she lacked modesty. The fact that she walked alone down a public street—unaccompanied by her father, husband, or female relatives—might even indicate that she was not a respectable woman. That she at the advanced age of 40 was investing in her own good looks might be assessed as pathetic self-absorption. Thus a particular behavior may be admired when judged by one set of values and norms but may be considered questionable or even disreputable when judged by another. [*Source: From Lydia Pulsipher's field notes. For detailed source information, see Text Sources and Credits.*] ■

If culture groups have different sets of values and standards, does that mean that there are no overarching human values or standards? This question increasingly worries geographers, who try to be sensitive both to the particularities of place and to larger issues of human rights. Those who lean too far toward appreciating difference could end up tacitly accepting inhumane behavior, such as the oppression of minorities or violence against women. Acceptance of difference does not preclude judgments of extreme customs or points of view. Nonetheless, although it is important to take a stand against cruelty of all sorts, deciding when and where to take that stand is rarely easy.

Religion and Belief Systems

The religions of the world are formal and informal institutions that embody value systems. Most have roots deep in history, and many include a spiritual belief in a higher power (such as God, Yahweh, or Allah) as the underpinning for their value systems. Today, religions often focus on reinterpreting age-old values for the modern world. Some formal religious institutions—such as Islam, Buddhism, and Christianity—proselytize; that is, they try to extend their influence by converting others. Others, such as Judaism and Hinduism, accept converts only reluctantly. Informal religions, often called *belief systems*, have no formal central doctrine and no firm policy on who may or may not be a practitioner. The fact that many people across the world combine informal religious beliefs with their more formal religious practices, of whatever persuasion, accounts for the very rich array of personal beliefs found in the world today.

Religious beliefs are often reflected in the landscape. For example, settlement patterns often demonstrate the central role of religion in community life: village buildings may be grouped around a mosque, a temple, a synagogue, or a church, and the same can be said for urban neighborhoods. In some places, religious rivalry is a major feature of the landscape. Certain spaces may be clearly delineated for the use of one group or another, as in Northern Ireland's Protestant and Catholic neighborhoods.

Religion has often been used to wield power. For example, during the era of European colonization, religion (Christianity) was used as a way to impose a change of attitude on conquered people.

Figure 1.30 shows the distribution of the major religious traditions on Earth today; it demonstrates some of the religious consequences of colonization. Note, for instance, the distribution of Roman Catholicism in the parts of the Americas, Africa, and Southeast Asia, all places colonized by European Catholic countries.

Religion can also spread through trade contacts. In the seventh and eighth centuries, Islamic people used a combination of trade and political power (and less often, actual conquest) to extend their influence across North Africa, throughout Central Asia, and eventually into South and Southeast Asia (see the Figure 6.15 map on page 247).

Language

Language is one of the most important criteria used in delineating cultural regions. The modern global pattern of languages reflects the complexities of human interaction and isolation over several hundred thousand years. Between 2500 and 3500 languages are spoken on Earth today, some by only a few dozen people in isolated places. Many languages have several *dialects*—regional variations in grammar, pronunciation, and vocabulary.

The geographic pattern of languages has continually shifted over time as people have interacted through trade and migration. The pattern changed most dramatically around 1500, when the languages of European colonists began to replace the languages of the people they conquered. For this reason, English, Spanish, Portuguese, or French are spoken in large patches of the Americas, Africa, Asia, and Oceania. Today, with increasing trade and instantaneous global communication, a few languages have become dominant. English is now the most important language of international trade, but Arabic, Spanish, Chinese, Hindi, and French are also widely used (Figure 1.31). At the same time, other languages are becoming extinct because children no longer learn them within families. 🎥 **250. NEW DOCUMENTARY FILM TRACKS LANGUAGES**

Race

Like ideas about gender roles, ideas about race affect human relationships everywhere on Earth. However, while race is of enormous social significance across the world, biologists tell us that from a scientific standpoint, race is a meaningless concept! The characteristics we popularly identify as race markers—skin color, hair texture, and face and body shape—have no significance as biological categories. All people now alive in the world are members of one species, *Homo sapiens sapiens*. For any supposed racial trait, such as skin color or hair texture or facial features,

ON THE BRIGHT SIDE

Altruism

Recognizing all the ills that have emerged from racism and similar prejudices, we need not infer that human history has been marked primarily by conflict and exploitation or that these conditions are inevitable. Actually, humans have probably been so successful as a species because of a strong inclination toward altruism, the willingness to sacrifice one's own well-being for the sake of others. It is probably our capacity for altruism that causes us such deep distress over the relatively infrequent occurrences of inhumane behavior.

race a social or political construct that is based on apparent characteristics such as skin color, hair texture, and face and body shape, but that is of no biological significance

---

there are wide variations within human groups. Meanwhile, many invisible biological characteristics, such as blood type and DNA patterns, cut across skin color distributions and other so-called *racial attributes* and are shared across what are commonly viewed as different races. In fact, over the last several thousand years there has been such massive gene flow among moving human populations that no modern group presents a discrete set of biological characteristics. Although any two of us may look quite different, from the biological point of view we are all simply *Homo sapiens sapiens* and are closely related.

Some of the easily visible features of particular human groups evolved to help them adapt to environmental conditions. For example, biologists have shown that people with darker skin (containing a high proportion of protective melanin pigment) evolved in regions close to the equator, where sunlight is most intense (Figure 1.32). All humans need the nutrient vitamin D, and sunlight striking the skin helps the body absorb vitamin D. Too much of the vitamin, however, can result in improper kidney functioning. Dark skin absorbs less vitamin D than light skin and thus would be a protective adaptation in equatorial zones. In higher latitudes, where the sun's rays are more dispersed, light skin facilitates the sufficient absorption of vitamin D; darker-skinned people at these higher latitudes may need to supplement vitamin D to be sure they get enough, since D deficiencies can result in several health risks. Meanwhile, light-skinned people with little protective melanin in their skin—if they live in equatorial or high-intensity sunlit zones (parts of Australia, for example)—will need to protect against too much vitamin D, serious sunburn, and skin cancer. Similar correlations have been observed between skin color, sunlight, and another essential vitamin, folate, which if deficient can result in birth defects.

Over time, race has acquired enormous social and political significance as humans from different parts of the world have encountered each other in situations of unequal power. *Racism*—the belief that genetic factors, usually visually apparent ones such as skin color, are a primary determinant of human abilities and even cultural traits—has often been invoked to justify the enslavement of particular groups, or confiscation of their land and resources. Race and its implications in North America will be covered in Chapter 2, and the topic will be discussed in several other world regions as well (see also a TED—Technology, Entertainment, and Design—talk by anthropologist Nina Jablonski, at http://tinyurl.com/l7wg7m).

👉 THINGS TO REMEMBER

- Cultural geographers seek to understand human variability on Earth through a variety of lenses: culture and ethnicity, religion, language, and race.

FIGURE 1.30   PHOTO ESSAY: Major Religions of the World

The small symbols on the map indicate a localized concentration of a particular religion within an area where another religion is predominant.

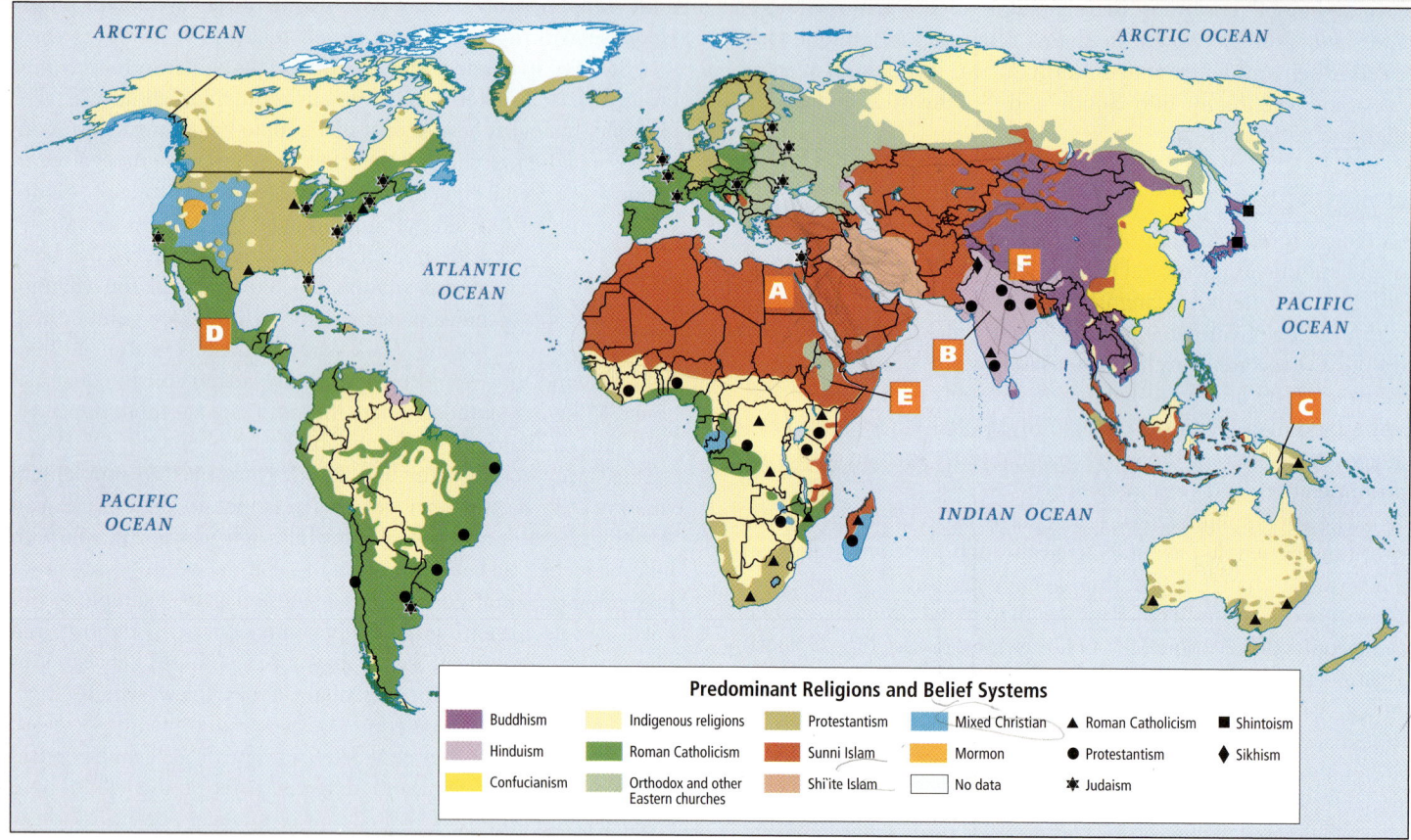

A Islam, Egypt

B Hinduism, India

C Indigenous religion, West Papua, Indonesia

D Indigenous religion, Mexico

E Christianity, Ethiopia

F Buddhism, Tibet

Predominant Religions and Belief Systems

- Buddhism
- Hinduism
- Confucianism
- Indigenous religions
- Roman Catholicism
- Orthodox and other Eastern churches
- Protestantism
- Sunni Islam
- Shi'ite Islam
- Mixed Christian
- Mormon
- No data
- ▲ Roman Catholicism
- ● Protestantism
- ✶ Judaism
- ■ Shintoism
- ◆ Sikhism

Sociocultural Issues **55**

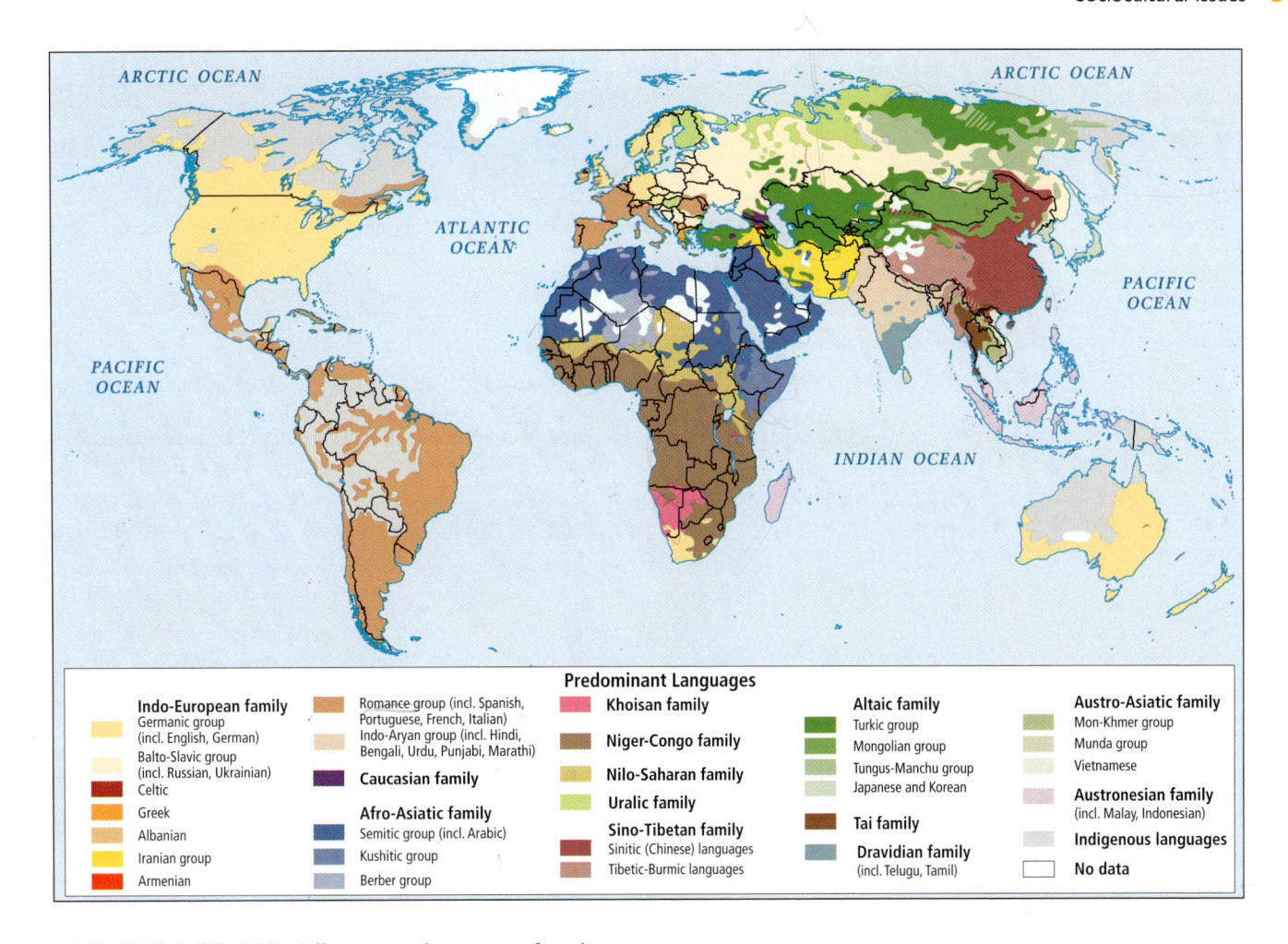

FIGURE 1.31 World's major language families. Distinct languages (Spanish and Portuguese, for example) are part of a larger group (Romance), which in turn is part of a language family (Indo-European).

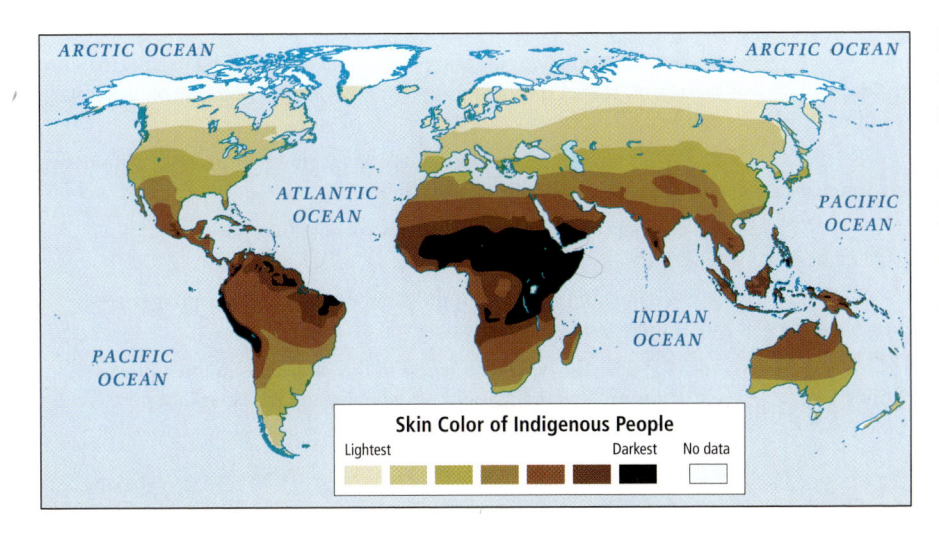

FIGURE 1.32 Skin color map for indigenous people as predicted from multiple environmental factors. Skin plays a twin role with respect to the sun: protection from excessive UV radiation and absorption of enough sunlight to trigger the production of vitamin D. [UNEP/GRID-Arendal at http://www.grida.no/graphicslib/detail/skin-colour-mapindigenous-people_8b88, Emmanuelle Bournay, cartographer. Data source consulted: G. Chaplin, "Geographic Distribution of Environmental Factors Influencing Human Skin Coloration," *American Journal of Physical Anthropology*, 125, 292–302, 2004; map updated in 2007]

- A group of people who share a location, a set of beliefs, a way of life, a technology, and usually a common ancestry and sense of common history form an ethnic group.

- Religion has often been used to wield power. For example, during the era of European colonization, religion (Christianity) was used to impose a change of attitude on conquered people.

- Today, with increasing trade and instantaneous global communication, a few languages have become dominant. English is now the most important language of international trade, but Arabic, Spanish, Chinese, Hindi, and French are also widely used. At the same time, other languages are becoming extinct because children no longer learn them within families.

- Race is biologically meaningless, yet it has acquired enormous social and political significance.

GEOGRAPHIC INSIGHTS
Review and Self-Test

1. Environment: Humans are altering the planet at an unprecedented rate, causing sometimes drastic effects on ecosystems and climate. Multiple environmental factors often interact to influence the vulnerability of a location to the impacts of climate change. These vulnerabilities have a spatial pattern.

- How is the place where you live vulnerable to climate change?
- Assess your exposure, sensitivity, and resilience to water-related climate-change impacts.

2. Globalization and Development: Global flows of information, goods, and people are transforming patterns of economic development. Local self-sufficiency is giving way to global interdependence as people and places are increasingly becoming connected, sometimes across vast distances.

- How is globalization evident in your life—from the clothes you wear, to your favorite foods, to your career plans?
- To what extent does your circle of friends show the effects of globalization?

3. Power and Politics: There are major differences across the world in the ways that power is wielded in societies. Modes of governing that are more authoritarian are based on the power of the state or community leaders. Modes that are based on notions of political freedom and democracy give the general public greater power over themselves and more of a role in deciding how policies are developed and governments are run. There are also many other ways of managing political power.

- Give several examples from recent world events of shifts from authoritarian modes of government to more democratic systems in which individuals have a greater say in the running of their governments.
- Are peace, broad prosperity, education, and civil society features of life in your home country? How do the political freedoms you enjoy shape your daily life?

4. Urbanization: The development of urban manufacturing and service economies has pulled people into cities. Meanwhile, the mechanization of food production has drastically reduced the need for agricultural labor, thus pushing people out of rural areas.

- How far back in your family history would you need to go to find ancestors who lived in a rural area and grew almost all their own food?
- Assuming you do not live in a rural area, how has urban life affected the diet and level of physical activity of you and/or your family?

5. Population and Gender: Population growth is slowing for a number of reasons, among them the increasing numbers of women who are delaying childbearing as they pursue educational and work opportunities outside the home.

- How is the shift toward more equality between the genders influencing population growth rates, patterns of economic development, and politics?
- How do your plans for a career and a family compare to those of your grandparents and great-grandparents?
- Ask your parents who the first woman in your family was who had a career. How, if at all, did this impact the number of children she had?

CRITICAL THINKING QUESTIONS

1. Some people argue that it is acceptable for people in the United States to consume at high levels because their consumerism keeps the world economy going. What are the weaknesses in this idea?

2. What are the causes of the huge global increases in migration, legal and illegal, that have taken place over the last 25 years?

3. What would happen in the global marketplace if all people earned a living wage?

4. As people live longer and decide not to raise large families, what are the impacts on society?

5. Given the threats posed by global warming, what are the most important steps to take now?

6. As you read about differing ways of life, values, and perspectives on the world, reflect on the appropriateness of force as a way of resolving conflicts.

7. What are some possible careers that would address one or more of the issues raised in particular chapters?

8. Reflect on the reasons why some people have much and others have little.

9. What would be some of the disadvantages and advantages of abolishing gender roles in any given culture?

10. Consider the ways that access to the Internet enhances prospects for better understanding between people, and the ways that it contributes to discord.

THINKING GEOGRAPHICALLY

Now that you have read about the connections between geographical regions, you should be able to answer the following questions about the photos in this chapter. To answer these questions online, go to Geography LaunchPad.

Human Impacts on the Biosphere (page 14)

A In which sector of the economy is mining?

B What form of pollution does this photo show most directly?

C, D What is the evidence that shifting cultivation may be contributing to deforestation in the region depicted?

E, F How is logging in Brazil linked to rising CO_2 levels and the global economy? Does your answer to (E) connect to you?

Vulnerability to Climate Change (page 24)

A Does this photo relate most to short-term or long-term resilience?

B Of the four thumbnail maps in this graphic, which depicts the information that best explains why Spain and Portugal are so much less vulnerable to climate change than is Morocco?

C What sign of an orderly response to disaster is visible in this picture?

D How can you tell that food supplies are low in this refugee camp?

Power and Politics (page 40)

A Why has Juan Manuel Santos's relationship with the Colombian military come under criticism?

B What about this photo suggests that the woman is a civilian, and not a member of a formal military unit?

C, D Of the factors mentioned in the text as necessary for democracy to flourish, which are obviously present in (C) and missing in (D)?

Urbanization (page 44)

A To what group of urban migrants does this skydiving young man probably belong?

B In what kind of neighborhood of Dhaka would you guess this man lives?

C This photo exemplifies what problem commonly faced by rapidly growing cities?

CHAPTER KEY TERMS

agriculture 28
authoritarianism 38
biosphere 12
birth rate 44
capitalism 39
carrying capacity 30
cartographers 4
cash economy 48
civil society 39
climate change 21
climate 14
communism 39
culture 51
death rate 44
delta 14
democratization 38
demographic transition 48
development 35
domestication 29
ecological footprint 12
emigration 47
erosion 13
ethnic group 51
ethnic cleansing 40
Euro zone 35
extraction 35
fair trade 34
floodplain 14
food security 29
formal economy 35
free trade 33
gender 49
gender roles 49
genetic modification (GM) 30
genocide 40

Geographic Information Science (GISc) 7
geopolitics 39
global economy 31
global scale 11
global warming 21
globalization 31
green revolution 29
greenhouse gases (GHG) 21
gross domestic product (GDP) per capita 35
gross national income (GNI) per capita 48
human geography 3
human well-being 36
immigration 47
industrial production 35
Industrial Revolution 28
informal economy 35
Kyoto Protocol 24
landforms 13
latitude 4
legend 4
living wages 34
local scale 11
longitude 4
map projections 7
migration 44
monsoon 20
multicultural society 52
multinational corporation 32
nongovernmental organizations (NGOs) 40
orographic rainfall 17
physical geography 3
plate tectonics 13
political ecologists 36
political freedoms 38

population pyramid 47
precipitation 17
primary sector 35
purchasing power parity (PPP) 35
push/pull phenomenon of urbanization 42
quaternary sector 35
race 53
rate of natural increase (RNI) 44
region 9
Ring of Fire 13
scale (of a map) 4
secondary sector 35
services 35
sex 49
slum 42
spatial distribution 4
spatial interaction 4
subsistence economy 48
sustainable agriculture 31
sustainable development 36
tertiary sector 35
total fertility rate (TFR) 46
United Nations (UN) 40
United Nations Gender Equality Index (GEI) rank 36
United Nations Human Development Index (HDI) 36
urbanization 42
virtual water 26
water footprint 26
weather 14
weathering 13
World Trade Organization (WTO) 34

Canadian Rockies, Alberta

D Canadian Rockies, Alberta

E Central lowlands, Louisiana

Land Elevations

meters	feet
4877	16,000
3353	11,000
2134	7000
914	3000
305	1000
152	500
0	0

mi 0   100   200   300   400   500
km 0   200   400   600   800

1:24,000,000
Azimuthal Equidistant Projection

chapter 2

NORTH AMERICA

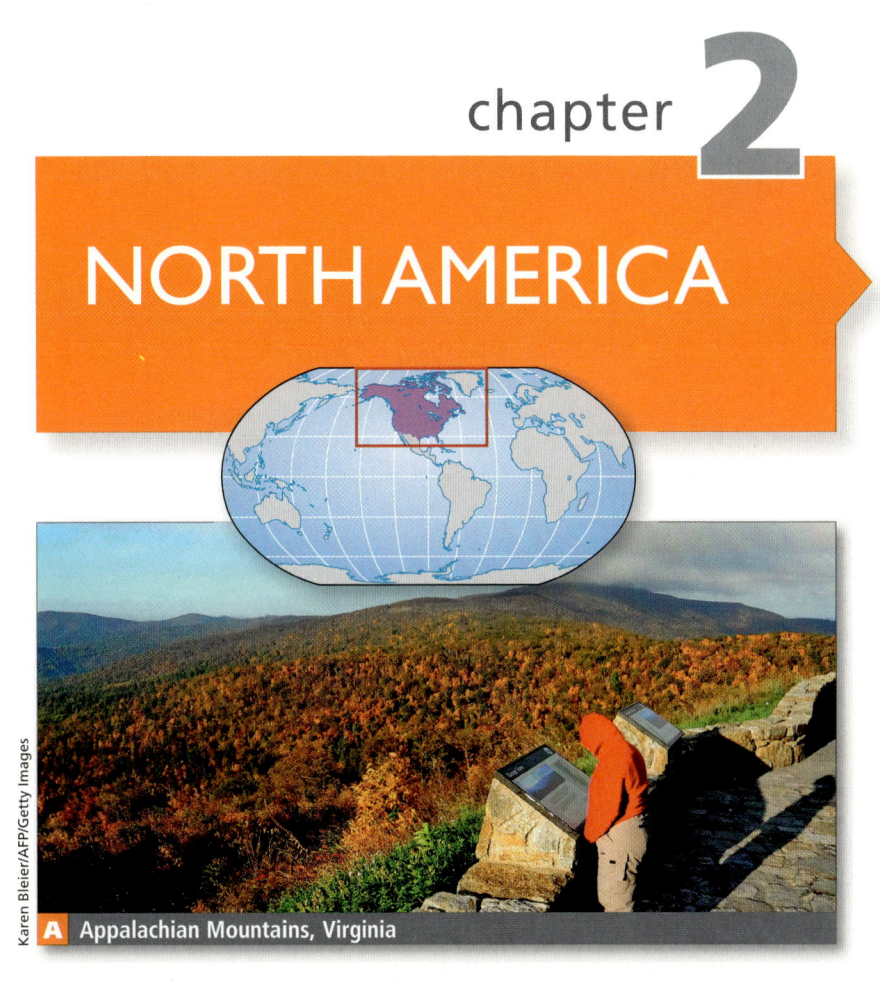

A Appalachian Mountains, Virginia

B Central lowlands, Minnesota

C Coastal lowlands, Charleston, South Carolina

FIGURE 2.1 Regional map of North America.

GEOGRAPHIC INSIGHTS

After you read this chapter, you will be able to discuss the following issues as they relate to the five thematic concepts:

1. Environment: North America's intensive use of resources has an enormous impact on the environment. Although home to only 5 percent of the world's population, North America produces 26 percent of the greenhouse gases released globally that are related to human activity. North American lifestyles have major environmental impacts, including the depletion and pollution of water resources and fisheries and the destruction of huge amounts of habitat for wild plants and animals.

2. Globalization and Development: Globalization has transformed economic development in North America, reorienting employment toward knowledge-intensive jobs that require education and training. Most of the manufacturing jobs upon which the region's middle class was built have either been moved abroad to take advantage of cheaper labor or have been replaced by technology. North America's demand for imported goods and its export of manufacturing jobs helps make it a major engine of globalization.

3. Power and Politics: North America has relatively high levels of political freedom, though in recent decades many of its residents have become disillusioned with the political process for a variety of reasons. While Canada plays a relatively modest political role abroad, the United States has enormous influence on the global political order, although its status as the world's predominant "superpower" is increasingly being challenged.

4. Urbanization: A dramatic change in the spatial patterns of cities and suburbs has profoundly affected life in this very urbanized region. Since World War II, North America's urban populations have increased by about 150 percent, but the amount of land they occupy has increased by almost 300 percent. This is primarily because of suburbanization and urban sprawl, which are companion processes to urbanization.

5. Population and Gender: Women's participation in North America's economy is beginning to rival that of men, contributing to more than two centuries of declining fertility rates as women delay childbearing to pursue education and careers. Declining childbirth rates play a major role in the aging of North American populations, which may slow economic growth.

The North American Region

North America (Figure 2.1) is one of the largest, wealthiest, and most politically powerful regions in the world. It encompasses many environments and a complex array of local cultures and economic activities that interact with each other across wide distances. Many of the trends discussed throughout this book are well advanced in North America.

The five thematic concepts in this book are explored as they arise in the discussion of regional issues, with interactions between two or more themes featured. Vignettes, like the one that follows about Javier Aguilar, illustrate one or more of the themes as they are experienced in individual lives.

VIGNETTE

Javier Aguilar, a 39-year-old father of three, has worked as an agricultural laborer in California's Central Valley for 20 years. He and hundreds of thousands like him tend the fields of crops (Figure 2.2) that feed the nation, especially during the winter months. Aguilar used to make $8.00 an hour, but now he is unemployed and standing in a church-sponsored food line. "If I don't work, [we] don't live. And here all the work is gone," he says, grimly.

In Mendota, also in California's Central Valley, young Latino men wait on street corners to catch a van to the fields. None come. By March of 2009, the unemployment rate in Mendota was 41 percent and rising, and Mayor Robert Silva said his community was dying on the vine. He saw the trouble spreading. Many small businesses were closing. Silva worried about the drug use, alcohol abuse, family violence, and malnutrition that can accompany severe unemployment in any community.

This level of unemployment in the heart of the nation's biggest producer of fruit, nuts, and vegetables was partly due to drought and partly to a global economic recession. The drought is related to natural dry cycles as well as to global climate change. Water is increasingly scarce in the Central Valley, and access to irrigation water has been cut to force conservation. Meanwhile, the global economic recession, which began in 2007 and continued into 2014, reduced overall demand for California's fresh fruits and vegetables as families turned to cheaper foods or those grown closer to home. All of these stresses have forced farmers to remove from production as much as 1 million of the 4.7 million acres of land once cultivated and irrigated in the Central Valley. This may eventually result in a loss of as many as 80,000 jobs

FIGURE 2.2 Lettuce harvest in California. Agricultural laborers pick lettuce in California's Central Valley.

> **THINKING GEOGRAPHICALLY** What factors might cause lettuce production to decline during a recession?

and as much as $2.2 billion in California agriculture and related industries. And even though by early 2014 agricultural production in Mendota had partially been revived, unemployment remained over 12 percent, nearly twice the national average.

The confluence of troubles in California's Central Valley has been particularly devastating for low-wage agricultural workers like Javier Aguilar because they have so little to fall back on in terms of savings, education, or skills. One potential silver lining, however, is that economic hardship often encourages unskilled workers like Javier to take advantage of free, government-sponsored adult education programs, and his children may also have more opportunity than he did to graduate from high school. For the time being, however, Javier's life is hard. [Source: New York Times; PBS Newshour. For detailed source information, see Text Sources and Credits.]

GLOBAL PATTERNS, LOCAL LIVES

What Makes North America a Region?

This world region is relatively easy to define. It consists of Canada and the United States, two countries that are linked because of their geographic proximity, similar history, and many common cultural, economic, and political features. The nature of the relationship between Canada and the United States is discussed in numerous places throughout the chapter.

Terms in This Chapter

The term *North America* is used to refer to both countries. Even though it is common on both sides of the border to call the people of Canada "Canadians" and people in the United States "Americans," this text uses the term *United States*, or *U.S.*, rather than *America*, for the United States. Other terms relate to the growing cultural diversity in this region. The text uses the term

> **Latino** a term used to refer to all Spanish-speaking people from Middle and South America, although their ancestors may have been European, African, Asian, or Native American

Latino to refer to all Spanish-speaking people from Middle and South America, although their ancestors may have been European, African, Asian, or Native American.

THINGS TO REMEMBER

- The global economic recession, which began in 2007 and continued into 2014, reduced overall demand for California's fresh fruits and vegetables as families turned to cheaper foods or those grown closer to home.

- The recession combined with drought forced farmers to remove from production as much as 1 million of the 4.7 million acres of land once cultivated and irrigated in the Central Valley. This may eventually result in a loss of as many as 80,000 jobs and as much as $2.2 billion in California agriculture and related industries.

PHYSICAL GEOGRAPHY AND ENVIRONMENTAL ISSUES

The continent of North America has almost every type of climate and a wide variety of landforms. Huge expanses of mountain peaks, ridges, and valleys meet expansive plains, long, winding rivers, myriad lakes, and extraordinarily lengthy coastlines. Here the focus is on a few of the most significant landforms.

LANDFORMS

A wide mass of mountains and basins, known as the Rocky Mountain zone, dominates western North America (see Figure 2.1D). It stretches down from the Bering Strait in the far north, through Alaska, and into Mexico. This zone formed about 200 million years ago when, as part of the breakup of the supercontinent Pangaea (see Figure 1.8 on page 16), the Pacific Plate pushed against the North American Plate, thrusting up mountains. These plates still rub against each other, causing earthquakes along the Pacific coast of North America.

The much older and hence more eroded Appalachian Mountains stretch along the eastern edge of North America from New Brunswick and Maine to Georgia. This range resulted from very ancient collisions between the North American Plate and the African Plate.

Between these two mountain ranges lies the huge central lowland of undulating plains that stretches from the Arctic to the Gulf of Mexico. This landform was created by the deposition of deep layers of material eroded from the mountains and carried to this central

North American region by wind and rain and by the rivers flowing east and west into what is now the Mississippi drainage basin.

During periodic ice ages over the last 2 million years, glaciers have covered the northern portion of North America. In the most recent ice age (between 10,000 and 25,000 years ago), the glaciers, sometimes as much as 2 miles (about 3 kilometers) thick, moved south from the Arctic, picking up rocks and soil and scouring depressions in the land surface. When the glaciers later melted, these depressions filled with water, forming the Great Lakes. Thousands of smaller lakes, ponds, and wetlands that stretch from Minnesota and Manitoba to the Atlantic were formed in the same way (see Figure 2.1B). Melting glaciers also dumped huge quantities of soil throughout the central United States. This soil, often many meters deep, provides the basis for large-scale agriculture but remains susceptible to wind and water erosion.

East of the Appalachians, the Atlantic coastal lowland stretches from New Brunswick to Florida. It then sweeps west to the southern reaches of the central lowland along the Gulf of Mexico. In Louisiana and Mississippi, much of this lowland is filled in by the Mississippi River delta—a low, flat, swampy transition zone between land and sea. The delta was formed by massive loads of silt deposited during floods over the past 150-plus million years by the Mississippi, North America's largest river system. The delta deposit originally began at what is now the junction of the Mississippi and Ohio rivers at Cairo, Illinois; slowly, as ever more sediment was deposited, the delta advanced 1000 miles (1600 kilometers) into the Gulf of Mexico.

Over the centuries, human activities such as deforestation, deep plowing, and heavy grazing have led to erosion and added to the silt load of the rivers. The construction of levees during the last 300 years along riverbanks has drastically reduced flooding. Because of this flood control, much of the silt that used to be spread widely across the lowlands during floods is being carried to the southern part of the Mississippi delta—a low, flat zone characterized by swamps, lagoons, and sandbars. The intrusion of silt is destroying wetlands and, as the silt load extends further out into deep waters, the extra weight is causing the delta to sink into the Gulf of Mexico, a process called *subsidence* (Figure 2.3).

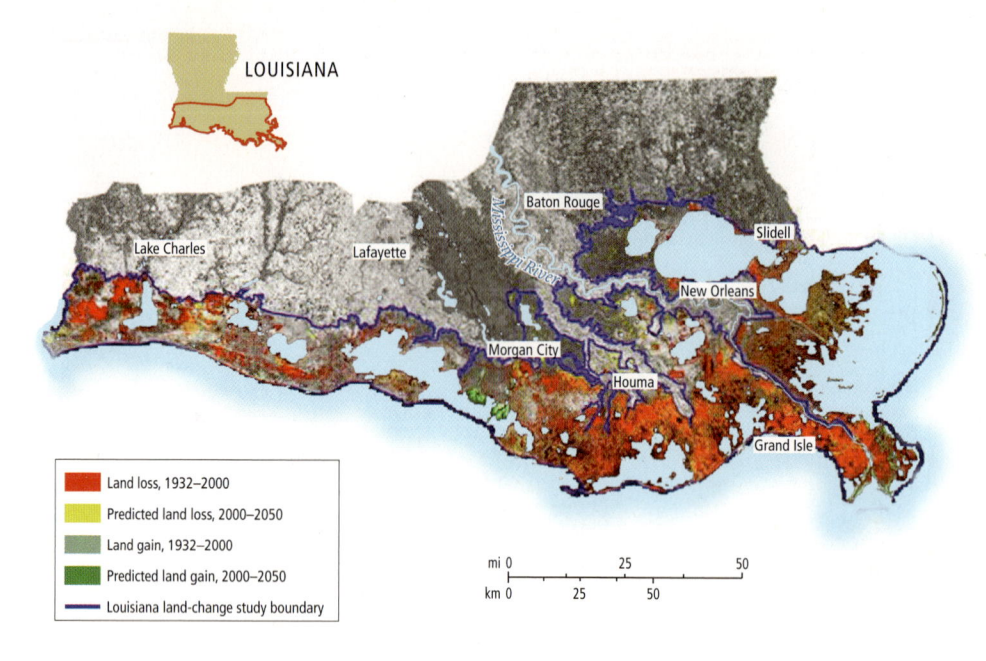

FIGURE 2.3 Wetland loss in Louisiana. The Louisiana coastline and the lower Mississippi River basin are vital to the nation's interests. They are the end point for the vast Mississippi drainage basin and provide coastal wildlife habitats, recreational opportunities, and transportation lanes that connect the vast interior of the country to the ocean and to offshore oil and gas. Most important, the wetlands provide a buffer against damage from hurricanes. Unfortunately, Louisiana has lost one-quarter of its total wetlands over the last century, largely because of human activity. The remaining 3.67 million acres constitute 14 percent of the total wetland area in the lower 48 states. [Source consulted: USGS/National Wetlands Research Center]

> **aquifers** ancient natural underground reservoirs of water

CLIMATE

The landforms across this continental expanse influence the movement and interaction of air masses and contribute to its enormous climate variety (Figure 2.4). Along the southern west coast of North America, the climate is generally mild (Mediterranean)—dry and warm in summer, cool and moist in winter. North of San Francisco, the coast receives moderate to heavy rainfall. East of the Pacific coastal mountains, climates are much drier because as the moist air sinks into the warmer interior lowlands, it tends to hold its moisture. This interior region becomes increasingly arid moving eastward across the Great Basin (see Figure 2.4C) and Rocky Mountains. Many dams and reservoirs for irrigation projects have been built to make agriculture and urbanization possible. Because of the low level of rainfall, however, efforts to extract water for agriculture and urban settlements are exceeding the capacity of ancient underground water basins (**aquifers**) to replenish themselves.

On the eastern side of the Rocky Mountains, the main source of moisture is the Gulf of Mexico. When the continent is warming in the spring and summer, the air masses above it rise, sucking in warm, moist, buoyant air from the Gulf. This air interacts with cooler, drier, heavier air masses moving into the central lowland from the north (see Figure 2.4A) and west, often creating violent thunderstorms and tornadoes. Generally, central North America is wettest in the eastern (see Figure 2.4B) and southern parts and driest in the north and west (see Figure 2.4C). Along the Atlantic coast, moisture is supplied by warm, wet air above the Gulf Stream—a warm ocean current that flows north from the eastern Caribbean and Florida and follows the coastline of the eastern United States and Canada before crossing the north Atlantic Ocean.

The large size of the North American continent creates wide temperature variations. Because land heats up and cools off more rapidly than water, temperatures in the interior of the continent are hotter in the summer and colder in the winter than in coastal areas, where temperatures are moderated by the oceans.

FIGURE 2.4   PHOTO ESSAY: Climates of North America

Climate Zones

Tropical humid climates (A)
- Tropical wet/dry

Arid and semiarid climates (B)
- Steppe
- Desert

Temperate climates (C)
- Midlatitude, moist all year
- Subtropical, winter dry
- Mediterranean, summer dry

Cool humid climates (D)
- Continental, moist all year

Coldest climates (E)
- Arctic
- High altitude

→ Winds
→ Ocean currents

Wet Pacific air blows into coastal mountains, bringing rain and moderate temperatures.

Continental effect makes winters colder and summers hotter.

Gulf Stream current brings warm tropical water, warming the air over the Atlantic coast.

Moist Gulf of Mexico air brings rain and moderate temperatures.

A Arctic, Alaska

B Temperate, midlatitude, Maryland

C Desert, Utah

THINGS TO REMEMBER

- North America consists of the United States and Canada.
- North America has two main mountain ranges, the Rockies and the Appalachians, separated by expansive plains through which run long, winding rivers.
- The size and variety of landforms influence the movement and interaction of air masses, creating enormous climatic variation. Because land heats up and cools off more rapidly than water, temperatures in the interior of the continent are higher in the summer and colder in the winter than in coastal areas. ■

ENVIRONMENT

> **GEOGRAPHIC INSIGHT 1**
>
> **Environment:** North America's intensive use of resources has an enormous impact on the environment. Although home to only 5 percent of the world's population, North America produces 26 percent of the greenhouse gases released globally that are related to human activity. North American lifestyles have major environmental impacts, including the depletion and pollution of water resources and fisheries and the destruction of huge amounts of habitat for wild plants and animals.

North America's wide range of resources, plus its seemingly limitless stretches of forest and grasslands, diverted attention for many years from the environmental impacts of settlement and development. However, it is becoming impossible to ignore the many environmental consequences of the North American lifestyle. This section focuses on a few of those consequences: climate change and air pollution, depletion and pollution of water resources and fisheries, and habitat loss.

> **smog** a combination of industrial emissions, car exhaust, and water vapor that frequently hovers as a yellow-brown haze over many North American cities, causing a variety of health problems.
>
> **acid rain** precipitation that has formed through the interaction of rainwater or moisture in the air with sulfur dioxide and nitrogen oxides emitted during the burning of fossil fuels, making it acidic

Climate Change and Air Pollution

On a per capita basis, North Americans contribute among the largest amounts of greenhouse gases to the Earth's atmosphere. Only a very few countries have higher per capita emissions. This is largely a result of North America's high consumption of fossil fuels, which is related to several factors. One of these is North America's dominant pattern of urbanization, characterized by vast and still-growing suburbs where people depend on their automobiles for almost all of their transportation needs. Freestanding dwellings and businesses spread out across the land also require more energy to heat and cool than do the densely packed, high-rise buildings typical of cities in most other world regions. North American industrial and agricultural production also depends very heavily on fossil fuels.

Canada's government was one of the first to commit to reducing the consumption of fossil fuels. Until recently, the United States resisted such moves, fearing damage to its economy. Both countries are now exploring alternative sources of energy, such as solar, wind, geothermal, and nuclear power. So far, neither country has been able to reduce its levels of greenhouse gas emissions, or even the rate at which these emissions are growing. However, both Canada and the United States possess the technological capabilities needed to lead the world in shifting over to cleaner sources of energy.

31. ENERGY REPORT
32. GREEN BUILDING

Vulnerability to Climate Change Both Canada and the United States are vulnerable to the effects of climate change. Dense population centers on the Gulf of Mexico and on the Atlantic coast are very exposed to hurricanes, which may become more violent as oceans warm (Figure 2.5B, D). Sea level rise and coastal erosion, caused by the thermal expansion of the oceans, are already affecting many coastal areas along the Arctic coast of North America (see Figure 2.5A). Here, at least 26 coastal villages are being forced to relocate inland, at an estimated cost of $130 million per village. Meanwhile, many arid farming zones will dry further as higher temperatures reduce soil moisture, making irrigation crucial (see Figure 2.6C). In all of these areas, resilience to climate change is bolstered over the short term by excellent emergency response and recovery systems. Long-term resilience is boosted by careful planning as well as by North America's large and diverse economy, which can provide alternative livelihoods to people who face significant exposure to climate impacts. **30. GLACIER NATIONAL PARK**

Air Pollution In addition to climate change, most greenhouse gases contribute to various forms of air pollution, such as smog and acid rain. **Smog** is a combination of industrial emissions, car exhaust, and water vapor that frequently hovers as a yellow-brown haze over cities, including many in North America, and causes a variety of health problems. These same emissions also result in **acid rain**, which is created when pollutants dissolve in falling precipitation and make the rain acidic. Acid rain can kill trees and, when concentrated in lakes and streams, poison fish and wildlife.

The United States, with its large population and extensive range of industries, is responsible for the vast majority of acid rain in North America. Because of continental weather and wind patterns, however, the area most affected by acid rain encompasses a wide swath on both sides of the eastern U.S.–Canada border (Figure 2.6). The eastern half of the continent, which includes the entire Eastern Seaboard from the Gulf Coast to Newfoundland, also has significant impacts from acid rain.

Water Resource Depletion, Pollution, and Marketization

People who live in the humid eastern part of North America find it difficult to believe that water is becoming scarce even there. Consider the case of Ipswich, Massachusetts, where the watershed is drying up as a result of overuse. There, innovators are saving precious water through conservation strategies in their homes and businesses. Elsewhere, as populations and per capita water usage grow, conflicts over water are becoming more and more common. **34. RAINWATER CASH**

FIGURE 2.5   PHOTO ESSAY: Vulnerability to Climate Change in North America

North America's wealth and its well-developed emergency response systems make it very resilient and reduce its overall vulnerability to climate change. However, certain regions are highly exposed to temperature increases, drought, hurricanes, and sea level rise.

A The location of Shishmaref, Alaska, on the Arctic Sea leaves it exposed to coastal erosion, which may be increasing in the area because of warmer air and sea temperatures.

Vulnerability to Climate Change
- Extreme
- High
- Medium
- Low

B Florida's low elevations expose it to sea level rise and flooding during and after hurricanes.

The stark contrast in vulnerability between the United States and Mexico results from the countries' very different sensitivity and resilience to water scarcity. As temperatures rise, this already dry borderland area will have less water. The United States has a much better water infrastructure, reducing its sensitivity to drought, and its emergency response systems make it more resilient to water shortages than Mexico is. With such drastic differences, the U.S.–Mexico border could become an even more contentious zone as the climate changes.

C Higher temperatures raise the rate at which plants lose water, increasing the need for irrigation. Farms that get their water from shrinking aquifers are thus less able to increase irrigation, and so are highly sensitive to the rising temperatures that climate change is bringing.

D Hurricanes gain strength with warmer temperatures. Many low-lying coastal cities are very exposed and sensitive, though their resilience varies. Shown here is a fire that destroyed several homes in New Orleans when the flooding caused by Hurricane Katrina made it impossible for fire trucks to reach the fire.

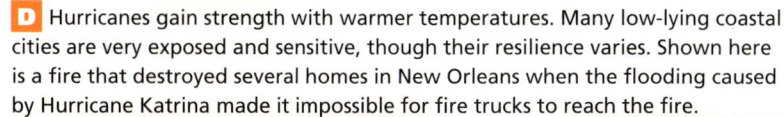

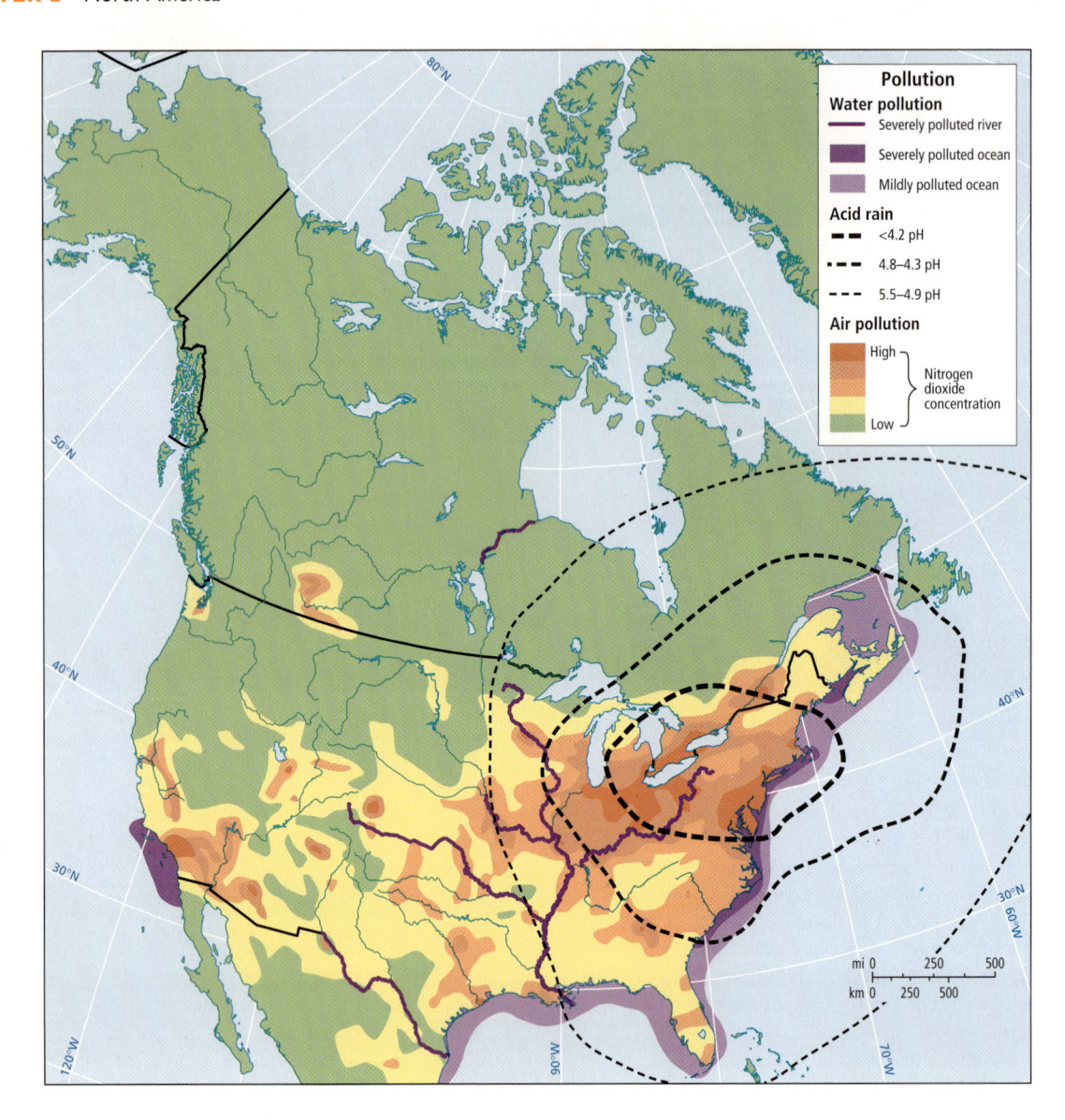

FIGURE 2.6 Air and water pollution in North America. This map shows two aspects of air pollution, as well as polluted rivers and coastal areas. Red and yellow indicate concentrations of nitrogen dioxide (NO_2), a toxic gas that comes primarily from the combustion of fossil fuels by motor vehicles and power plants. This gas interacts with rain to produce nitric acid, a major component of acid rain, as well as toxic organic nitrates that contribute to urban smog. The map also shows polluted coastlines (including all of the coastline from Texas to New Brunswick) as well as severely polluted rivers, which include much of the Mississippi River and its tributaries.

Water Depletion In North America, water becomes increasingly precious the farther west one goes. The Great Lakes Agreement (formally known as the Great Lakes–St. Lawrence River Basin Sustainable Water Resources Agreement) is an agreement between the eight states and two Canadian provinces that border this largest of the Earth's freshwater bodies (see the map in Figure 2.1) to manage the Great Lakes waters more wisely than they have been in the past. (See a clip from NPR's *Science Friday* of the *Inland Seas* video at http://www.sciencefriday.com/videos/watch/10113.)

Well before people were aware of the ramifications of making drastic changes to ecosystems, the city of Chicago, in order to clean up sewage it had been dumping into Lake Michigan, gained permission to reverse the flow of the Chicago River. The river now flows into the Mississippi, ultimately transferring Chicago's wastewater to the Gulf of Mexico. This diversion, which moves water at 3200 cubic feet per second from the Great Lakes, has opened the doors to requests from cities, not even on the shores of the Great Lakes, to gain access to the lake water. Cities in places as far away as Alabama have looked at the lakes as a potential water source for their growing needs.

On the North American Great Plains, rainfall varies considerably from year to year. To make farming more secure and predictable, taxpayers across the continent have subsidized the building of pumps and stock tanks for farm animals, and aqueducts and

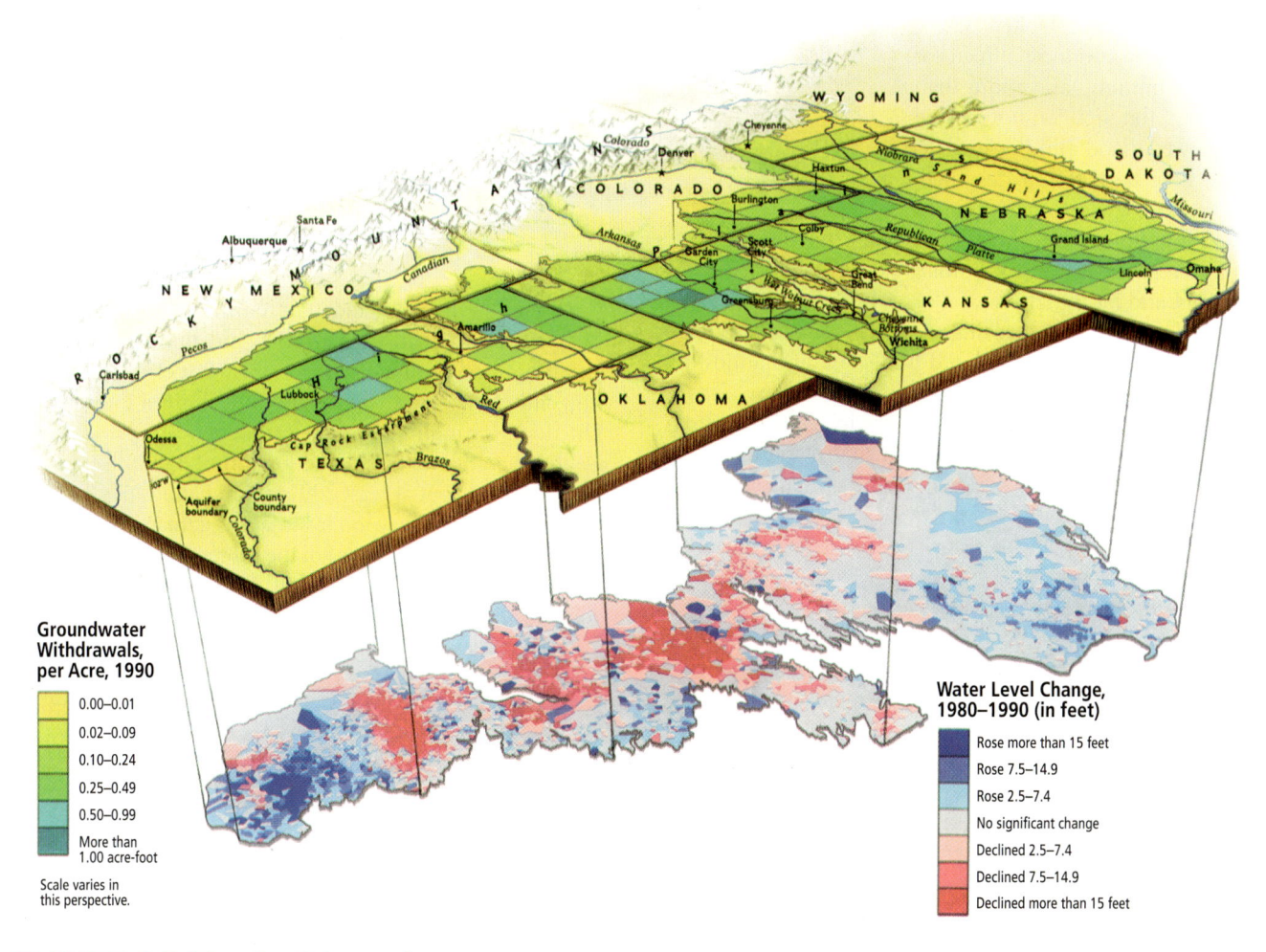

FIGURE 2.7 The Ogallala aquifer. Between the 1940s and the 1980s, the aquifer lost an average of 10 feet (3 meters) of water overall, and more than 100 feet (30 meters) of water in some parts of Texas. During the 1980s, though, there was a period of abundant rain and snow, which meant that water levels in the aquifer did not decline as much. However, in the Ogallala area, the climate fluctuates from moderately moist to very dry, and the dry periods are lengthening. A drought began in mid-1992 and has returned every few years, causing large agribusiness firms to pump Ogallala water to supplement scarce precipitation. Since 1992, water levels in the aquifer declined an average of 1.35 feet per year and now exceed replenishment rates many times over. [Sources consulted: *National Geographic* (March 1993): 84–85, with supplemental information from High Plains Underground Water Conservation District 1, Lubbock, Texas, at http://www.hpwd.com; Erin O'Brian, Biological and Agricultural Engineering, National Science Foundation Research Experience for Undergraduates, Kansas State University, 2001]

reservoirs for crop irrigation. However, more and more irrigation is being drawn from fossil water that has been stored over the millennia in aquifers. The *Ogallala aquifer* (Figure 2.7) underlying the Great Plains is the largest in North America. In parts of the Ogallala, water is being pumped out at rates that exceed natural replenishment by 10 to 40 times.

As mentioned in the opening vignette, fruit and vegetable crops in California are routinely irrigated with water from surrounding states. Such irrigation, involving expensive and massive engineering projects, accounts for some of the water that goes into the *virtual water footprint* of U.S. consumers, as discussed in Chapter 1 (page 26; see also Table 1.1 on page 26). This water also supplies the cities of Southern California, which are built on land that was once desert. Water is pumped from hundreds of miles away and over mountain ranges. California uses more energy to move water than some states use for all purposes. Moreover, irrigation in Southern California, especially that which is drawn from the Colorado River, deprives Mexico of this much-needed resource. The mouth of the Colorado (which is in Mexico) was once navigable. Because of massive diversions, it is now dry and sandy; only a mere trickle of water gets to Mexico.

Citizens in western North America are now recognizing that the use of scarce water for irrigating agriculture, raising livestock, and keeping lawns and golf courses green in desert environments is unsustainable. Conflicts over transporting water from wet regions to dry ones, or from sparsely inhabited to urban areas, are ongoing and have halted some new water projects. However, government subsidies have kept water artificially cheap, and in the past, new water supplies have always been found and harnessed, creating little incentive to change.

Water Pollution In the United States, 40 percent of rivers are too polluted for fishing and swimming, and more than 90 percent of the *riparian areas* (the interface between land and flowing surface water) have been lost or degraded. Pollution in the rivers of North America comes mainly as storm-water runoff from agricultural areas, urban and suburban developments, and industrial sites.

ON THE BRIGHT SIDE

Green Living at a Regional and Global Scale
North Americans are trying to figure out what they can do in their daily lives to ameliorate looming environmental crises. Solutions, such as *greener living*—which involves recycling, driving less, growing a food garden, and improving home energy efficiency—can collectively make a big difference. One major ongoing trend is the increased use of renewable energy, especially solar and wind power, which has grown by 50 percent and 33 percent (respectively) per year over the last decade.

In the 1970s, scientists studying coastal areas began noticing *dead zones* where water is so polluted that it supports almost no life. Dead zones occur near the mouths of major river systems that have been polluted by fertilizers and pesticides washed from farms and lawns when it rains. A large dead zone is in the Gulf of Mexico near the mouth of the Mississippi, and similar zones have been found in all U.S. coastal areas. Even Canada, where much lower population density means that rivers are generally cleaner, has dead zones on its western coast.

A recently discovered type of water pollution involves pharmaceuticals, such as antibiotics, that are excreted by humans and are not removed during water purification processes. These chemicals then make their way into rivers and lakes, where they enter the food system in drinking water or through fish.

35. DRUGS AND WATER SUPPLY

Water Marketization North Americans are used to paying for water, but the cost has usually been just high enough to cover extraction, purification, and delivery in pipes. Now, threats of polluted drinking water are beginning to change the way water is viewed. For example, in the last few years the public, in response to aggressive advertising, has been buying bottled water even when tap water is perfectly safe. A number of North American communities with abundant fresh water have agreed to sell water to beverage companies for bottling without understanding that massive water withdrawals from local aquifers can cause geologic subsidence, loss of aquatic habitats, and the depletion and pollution of natural wells and springs. Eventually, as these communities realize that their own low-cost access to water is threatened, they find that expensive litigation against the water bottling companies is often their only recourse.

Loss of Habitat for Plants and Animals

Before the European colonization of North America, which began soon after 1500, the environmental impact of humans in the region was relatively low. Though North America was by no means a pristine paradise when Europeans arrived, millions of acres of forests and grasslands that had served as habitats for native plants and animals were cleared to make way for European-style farms, cities, and industries (Figure 2.8). This was particularly true in the area that became the United States.

Oil Drilling

In many coastal and interior areas of North America, oil extraction is a large and potentially environmentally devastating industry. This often-overlooked reality was made clear in the spring and summer of 2010, when U.S. waters in the Gulf of Mexico became the site of the largest accidental marine oil spill in world history. In April of 2010, an explosion aboard the Deepwater Horizon, an offshore oil-drilling rig run by British Petroleum (BP), caused it to sink. One result was a massive leak from the rig's wellhead. Due to its location more than a mile beneath the sea surface, the wellhead could not be capped for almost 4 months, during which time it spewed out at least 200 million gallons of oil. While some of this oil made its way to the surface—damaging shorelines in all the Gulf Coast states—most of the oil remains beneath the surface because of BP's use of chemical dispersants. There is mounting concern that this oil poses an ongoing threat to the Gulf's many fish, shrimp, and other aquatic species. Moreover, the dispersants used by BP are a health concern because of their toxicity, which has already led to their being banned in Europe. Controversy surrounded the U.S. government's reopening of many Gulf fisheries in July of 2010, and independent studies did not concur with more optimistic government reports regarding the levels of both oil and dispersants in Gulf seafood. Clearly, the effects of the Deepwater Horizon spill will be felt for years to come.

In other places too, oil extraction has a dramatic effect on the environment. Along the northern coast of Alaska, the Trans-Alaska Pipeline runs southward for 800 miles to the Port of Valdez (see Figure 2.8E). Often running above ground to avoid shifting as the earth freezes and thaws, the pipeline poses a constant risk of rupture, which could potentially result in devastating oil spills. The pipeline also interferes with migrations of caribou and other animals that Alaska's indigenous people have depended on for food in the past. Protests about threats to the environment from oil extraction in Alaska tend to be quieted by the yearly rebate of several thousand dollars from oil revenues received by each Alaskan.

With one-tenth the population of the United States, Canada has the largest proven oil reserves in the world after Saudi Arabia, sufficient to meet its needs plus provide export capacity. It is the largest foreign supplier of oil to the United States, but more than 90 percent of this oil is in hard-to-access oil sands in western Canada. The costs of extracting this oil are high in terms of environmental impacts, energy expended, and water resources used. Transport of the extracted oil across the North American continent through the Keystone Pipeline system (part of which is already in use and part of which has yet to be built) also poses environmental and aesthetic threats.

Logging

Though widespread forest clearing for agriculture is now rare, logging is common throughout North America, where it remains especially important along the northern Pacific coast and in the southeastern United States. Logging in these areas provides

FIGURE 2.8 PHOTO ESSAY: Human Impacts on the Biosphere in North America

While parts of North America are relatively unaffected by humans, much of the region has had low-to-medium impacts from people, and the parts where most people live are very impacted.

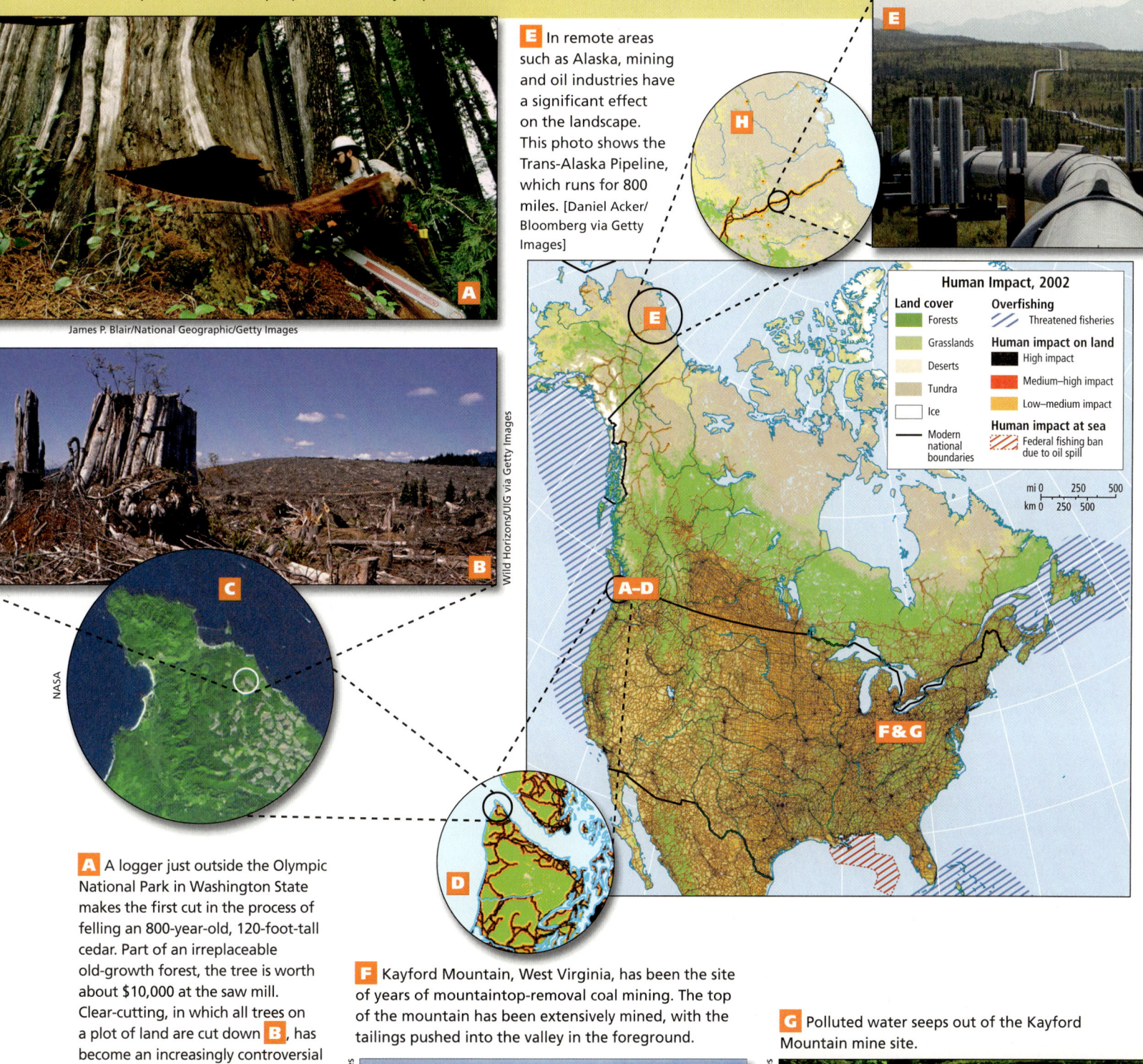

E In remote areas such as Alaska, mining and oil industries have a significant effect on the landscape. This photo shows the Trans-Alaska Pipeline, which runs for 800 miles. [Daniel Acker/Bloomberg via Getty Images]

Human Impact, 2002

Land cover
- Forests
- Grasslands
- Deserts
- Tundra
- Ice
- Modern national boundaries

Overfishing
- Threatened fisheries

Human impact on land
- High impact
- Medium–high impact
- Low–medium impact

Human impact at sea
- Federal fishing ban due to oil spill

A A logger just outside the Olympic National Park in Washington State makes the first cut in the process of felling an 800-year-old, 120-foot-tall cedar. Part of an irreplaceable old-growth forest, the tree is worth about $10,000 at the saw mill. Clear-cutting, in which all trees on a plot of land are cut down **B**, has become an increasingly controversial method of logging. Since 1971, more than 30 percent of the forests on Washington State's Olympic Peninsula **C** and **D** have been clear-cut. The clear-cutting of old-growth forests that have never been cut is a practice that has been declining in recent years, but it still continues in some places.

F Kayford Mountain, West Virginia, has been the site of years of mountaintop-removal coal mining. The top of the mountain has been extensively mined, with the tailings pushed into the valley in the foreground.

G Polluted water seeps out of the Kayford Mountain mine site.

most of the construction lumber and much of the paper used in Canada, the United States, and, increasingly, in parts of Asia.

Although the logging industry provides jobs and an exportable commodity, it has been depleting the continent's forests. Environmentalists have focused on the damage created by the logging industry, especially on **clear-cutting**, the dominant logging method used throughout North America. In this method, all trees on a given plot of land are cut down, regardless of age, health, or species (see Figure 2.8A–D). Clear-cutting destroys wild animal and plant habitats, thereby reducing species diversity. It also leaves forest soils uncovered and highly susceptible to erosion. Concerns about the environmental impacts of logging are heightened because of the dominance of service-sector jobs in the major logging states and provinces. For example, even in many remote areas of the Pacific Northwest, where logging was once the backbone of the economy, residents now depend on tourism and other occupations that rely on the beauty of intact forest ecosystems.

> **clear-cutting** a method of logging that involves cutting down all trees on a given plot of land, regardless of age, health, or species
>
> **urban sprawl** the encroachment of suburbs on agricultural land

Coal Mining and Use

In many remote interior areas of North America, coal mining is a major industry that is damaging to the environment. Strip mining, in which vast quantities of earth and rock are removed in order to extract underlying coal, can result in visual wastelands and in huge piles of mining waste called *tailings* that pollute waterways and threaten communities that depend on well water. Particularly damaging is a form of strip mining known as *mountaintop removal*, in which the whole top of a mountain is leveled and the tailings are pushed into surrounding valleys, resulting in the pollution of entire watersheds (see Figure 2.8F, G).

When coal is burned, ash is produced, and it is difficult to find a safe place to store the huge volumes of toxic ash that result from coal-fired power generation. Left to dry out, the material becomes airborne and contributes to air pollution, so it is stored wet. In December 2008, a large earthen dike of wet coal ash burst after a heavy rainstorm, spilling 1.5 billion gallons of toxic sludge (the largest industrial spill in U.S. history and roughly five times the amount of waste spilled in the Gulf of Mexico by BP's Deepwater Horizon) over 300 acres of beautiful lakeshore and forestland in rural Tennessee. The sludge, containing heavy metals and harmful chemicals, ruined the ecology of the immediate area and polluted the air and water for hundreds of square miles. Cleanup will continue until at least 2015 at a cost of well over a billion dollars.

Urbanization and Habitat Loss

An important aspect of urbanization is **urban sprawl** (see the "Farmland and Urban Sprawl" section on page 91). For several decades, middle- and upper-income urbanites throughout this region have sought lower-density suburban neighborhoods. Farms, forests, and other "undeveloped" land have given way to expansive, low-density urban and suburban residential developments where pavement, golf courses, office complexes, and shopping centers cover the landscape. In the process, natural habitats are being degraded even more intensely than they were by farming. The loss of farmland and natural habitat in the urban fringe affects recreational land and the ability to produce local, affordable food for urban populations.

As North American native plants and animals have been forced into ever smaller territories, many have died out entirely and been replaced by nonnative species (European and African grasses and the domestic cat, for example) brought in by humans either purposely or inadvertently (see Figure 2.23 on page 94). Estimates vary, but at least 4000 nonnative species have invaded North America. An example is the Asian snakehead fish, which is rapidly invading the Potomac River, where it eats baby bass and fiercely competes with native fish for food. **29. SNAKEHEAD REPORT**

THINGS TO REMEMBER

> **GEOGRAPHIC INSIGHT 1**
>
> - **Environment** North America's intensive use of resources has an enormous impact on the environment. Although home to only 5 percent of the world's population, North America produces 26 percent of the greenhouse gases released globally that are related to human activity. North American lifestyles have major environmental impacts, including the depletion and pollution of water resources and fisheries and the destruction of huge amounts of habitat for wild plants and animals.
>
> - While many parts of North America are exposed to multiple climate-change impacts, North America is also a very resilient region. Its overall vulnerability to climate change is generally low compared to that of other world regions.
>
> - Water pollution in North America is now a major problem, especially in the United States, where 40 percent of the rivers are too polluted for fishing or swimming.
>
> - Though widespread forest clearing for agriculture is now rare, logging is common throughout North America and remains especially important along the northern Pacific coast and in the southeastern United States.
>
> - In many remote interior areas of North America, coal mining is a major industry that is damaging to the environment.

HUMAN GEOGRAPHY

HUMAN PATTERNS OVER TIME

In prehistoric times, humans came from Eurasia via Alaska, dispersing to the south and east. Beginning in the 1600s, waves of European immigrants and enslaved Africans spread over the continent, primarily from east to west. Today, immigrants are coming mostly from all of Asia and from Middle and South America, arriving mainly in the Southwest and West, where immigrant populations are at their most concentrated. In addition, internal migration is still a defining characteristic of life for most North Americans, who are among the world's most mobile people. The average North American moves nearly 12 times in a lifetime.

The Peopling of North America

Recent evidence suggests that humans first came to North America from northeastern Asia at least 25,000 years ago and perhaps earlier, most arriving during an ice age. At that time, the global climate was cooler, polar ice caps were thicker, and sea levels were lower. The Bering land bridge, a huge, low landmass more than 1000 miles (1600 kilometers) wide, connected Siberia to Alaska. Bands of hunters crossed by foot or small boats into Alaska and traveled down the west coast of North America.

The Original Settling of North America By 15,000 years ago, humans had reached nearly to the tip of South America and had moved deep into that continent. By 10,000 years ago, global temperatures began to rise. As the ice caps melted, sea levels rose and the Bering land bridge was submerged beneath the sea.

Over thousands of years, the people settling in the Americas domesticated plants, created paths and roads, cleared forests, built permanent shelters, and sometimes created elaborate social systems. About 3000 years ago, corn was introduced from Mexico (into what is now the southwestern U.S. desert), as were other Mexican domesticated crops, particularly squash and beans. Such food crops are thought to have been closely linked to settled life and to North America's prehistoric population growth.

These foods provided surpluses that allowed some community members to engage in activities other than agriculture, hunting, and gathering, making possible large, city-like regional settlements. For example, by 1000 years ago, the urban settlement of Cahokia, including suburban settlements (in what is now central Illinois, across the Mississippi from St. Louis), covered 5 square miles (12 square kilometers) and was home to an estimated 30,000 people (Figure 2.9A). Here people could specialize in crafts, trade, or other activities beyond the production of basic necessities.

The Arrival of the Europeans North America was completely transformed by the sweeping occupation of the continent by Europeans. In the sixteenth century, Italian, Portuguese, and English explorers came ashore along the Eastern Seaboard of North America, and the Spanish explorer Hernando De Soto made his way from Florida deep into the heartland of the continent in the 1540s. In the early seventeenth century, the British established colonies along the Atlantic coast in what is now Virginia (1607) and Massachusetts (1620). The Dutch explored the Atlantic Seaboard looking for trading opportunities, and the French explored the northern interior of the continent, entering via the St. Lawrence River. Assisted by enslaved Africans, colonists and settlers from northern Europe built villages, towns, port cities, and plantations along the eastern coast over the next two centuries. By the mid-1800s, they had occupied most Native American lands into the central part of the continent.

Disease, Technology, and Native Americans The rapid expansion of European settlement was facilitated by the vulnerability of Native American populations to European diseases. Having been isolated from the rest of the world for many years, Native Americans had no immunity to diseases such as measles and smallpox. Transmitted by Europeans and Africans who had built up immunity to them, these diseases killed up to 90 percent of Native Americans within the first 100 years of contact. It is now thought that diseases spread by early expeditions, such as De Soto's into Florida, Georgia, Tennessee, and Arkansas, so decimated populations in the North American interior that fields and villages were abandoned, the forest grew back, and later explorers erroneously assumed the land had never been occupied.

Technologically advanced European weapons, trained dogs, and horses also took a large toll. Often the Native Americans had only bows and arrows. Some Native Americans in the Southwest acquired horses from the Spanish and learned to use them in warfare against the Europeans, but their other technologies could not compete. Numbers reveal the devastating effect of European settlement on Native American populations. Roughly 18 million Native Americans lived in North America in 1492. By 1542, after just a few Spanish expeditions, only half that number survived. By 1907, slightly more than 400,000, or a mere 2 percent, remained.

The European Transformation

European settlement erased many of the landscapes familiar to Native Americans and imposed new ones that fit the varied physical and cultural desires of the new occupants.

The Southern Settlements European settlement of eastern North America began with the Spanish in Florida in the mid-1500s and the establishment of the British colony of Jamestown in Virginia in 1607. By the late 1600s, large plantations in the colonies of Virginia, the Carolinas, and Georgia were cultivating crops such as tobacco and cotton, which became valuable exports.

To secure a large, stable labor force, Europeans brought enslaved Africans into North America beginning in 1619. Within 50 years, enslaved Africans were the dominant labor force on some of the larger Southern plantations (see Figure 2.9B). By the start of the Civil War in 1861, enslaved people made up about one-third of the population in the Southern states and were often a majority in the plantation regions. Working and living conditions on these plantations were often hazardous and brutal, with execution, beatings, and rape regular occurrences. Enslaved people were usually denied education in hopes that this would prevent them from escaping, and gatherings of any kind were often banned in order to make slave rebellions more difficult to organize. Even so, there were more than 250 documented slave rebellions in North America. The largest concentrations of African Americans in North America are still in the southeastern states (Figure 2.10).

The plantation system consolidated wealth in the hands of a small class of landowners who made up just 12 percent of Southerners in 1860. Planter elites kept taxes low and invested money from their exported crops in Europe or the more prosperous

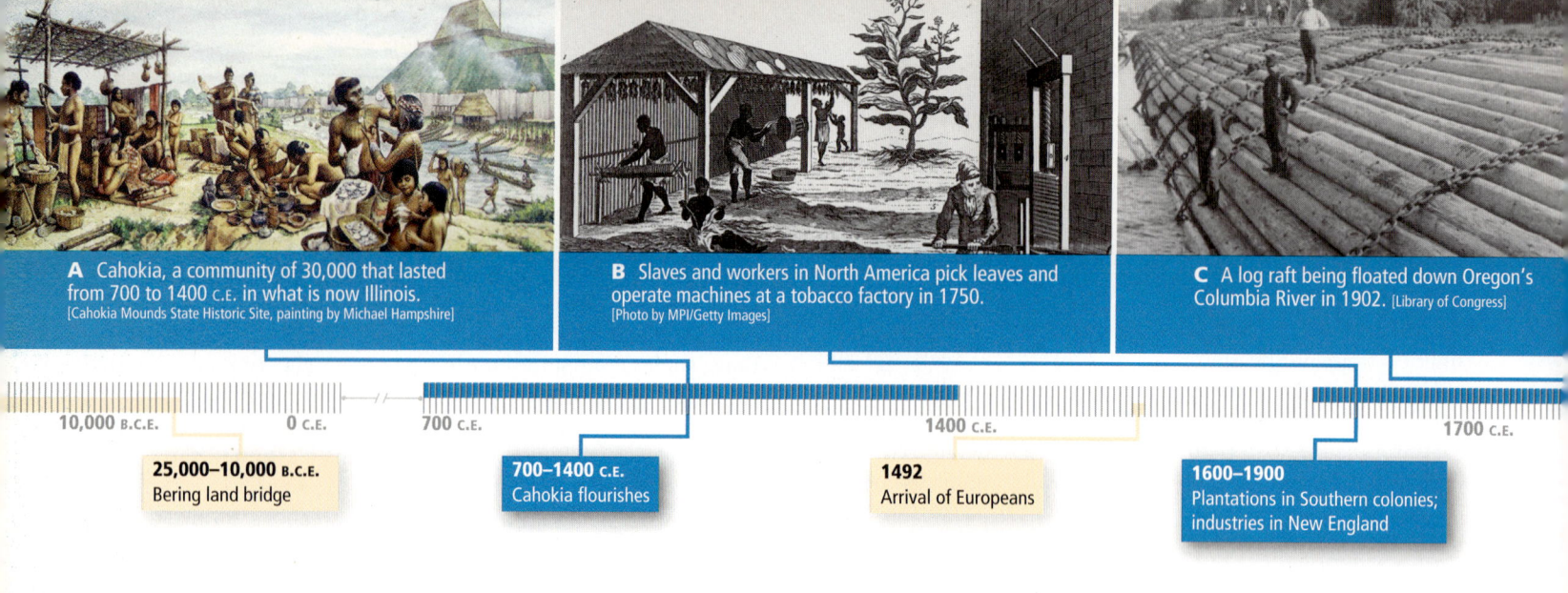

A Cahokia, a community of 30,000 that lasted from 700 to 1400 C.E. in what is now Illinois. [Cahokia Mounds State Historic Site, painting by Michael Hampshire]

B Slaves and workers in North America pick leaves and operate machines at a tobacco factory in 1750. [Photo by MPI/Getty Images]

C A log raft being floated down Oregon's Columbia River in 1902. [Library of Congress]

10,000 B.C.E. — 0 C.E. — 700 C.E. — 1400 C.E. — 1700 C.E.

25,000–10,000 B.C.E. Bering land bridge

700–1400 C.E. Cahokia flourishes

1492 Arrival of Europeans

1600–1900 Plantations in Southern colonies; industries in New England

FIGURE 2.9 VISUAL HISTORY OF NORTH AMERICA

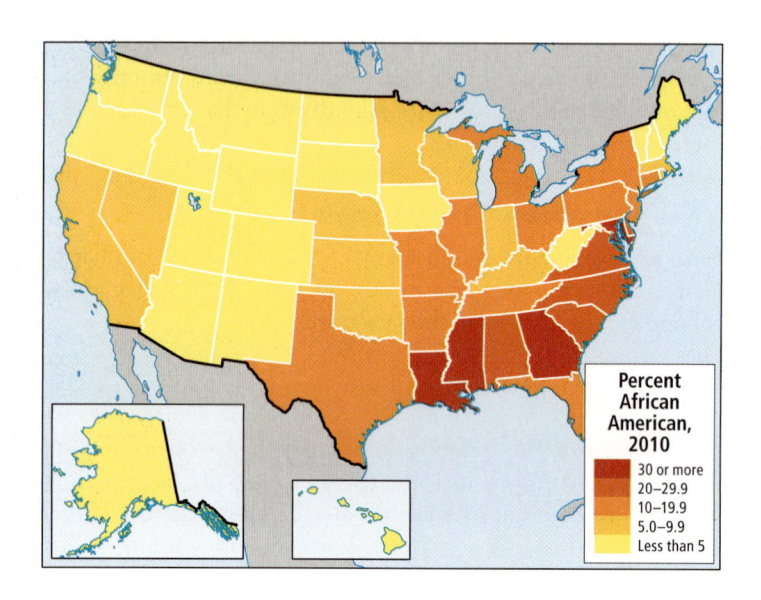

FIGURE 2.10 Percent of African American population in each state, 2010. The percent refers only to those persons who selected "Black, African Am., Negro" as their only race in the 2010 census. It does not include those who selected more than one race that included black. [Source consulted: "Black Population from Year 1790–2010," BlackDemographics.com, at http://www.blackdemographics.com/population/html]

Percent African American, 2010:
- 30 or more
- 20–29.9
- 10–19.9
- 5.0–9.9
- Less than 5

northern colonies, instead of in **infrastructure** at home. As a result, the road, rail, communication networks, and other facilities necessary for further economic growth in the South were rarely built.

More than half of Southerners were poor white farmers. Both they and the general slave population lived simply and their meager consumption did not provide much demand for goods. Because of this, there were few market towns and almost no industries. Plantations tended to import from Europe and the northern United States whatever they couldn't produce for themselves. So instead of generating a *multiplier effect*, enabling

infrastructure road, rail, and communication networks and other facilities necessary for economic activity

local enterprises like small shops, garment making, small restaurants and bars, manufacturing, and transportation and repair services to expand as a result of "spinoff" from the main industries they serve, economic growth was much slower in the South.

The inability of the federal government to reconcile the different needs of the agricultural export–oriented Southern economy and the industrializing Northern economies was one of the causes of the Civil War (1861–1865), perhaps equal to the abolition movement to free enslaved Africans and their descendants. After the war, while the victorious North returned to promoting its own industrial development, the plantation economy declined, and the South sank deeply into poverty. The South remained economically and socially underdeveloped well into the 1970s.

The Northern Settlements Throughout the seventeenth century, relatively poor subsistence farming communities dominated the colonies of New England and southeastern Canada. There were no plantations and few slaves, and not many cash crops were exported. What exports there were consisted of raw materials like timber, animal pelts, and fish from the Grand Banks off Newfoundland and the coast of Maine. Generally, farmers lived in interdependent communities that prized education, ingenuity, self-sufficiency, and thrift.

By the late 1600s, New England was implementing ideas and technology from Europe that led to the first industries. By the 1700s, diverse industries were supplying markets in North America and the Caribbean with metal products, pottery, glass, and textiles. By the early 1800s, southern New England, especially the region around Boston, became the center of manufacturing in North America. It drew largely on young male and female immigrant labor from French Canada and Europe.

The Mid-Atlantic Economic Core The colonies of New York, New Jersey, Pennsylvania, and Maryland eventually surpassed New England and southeastern Canada in population and in wealth.

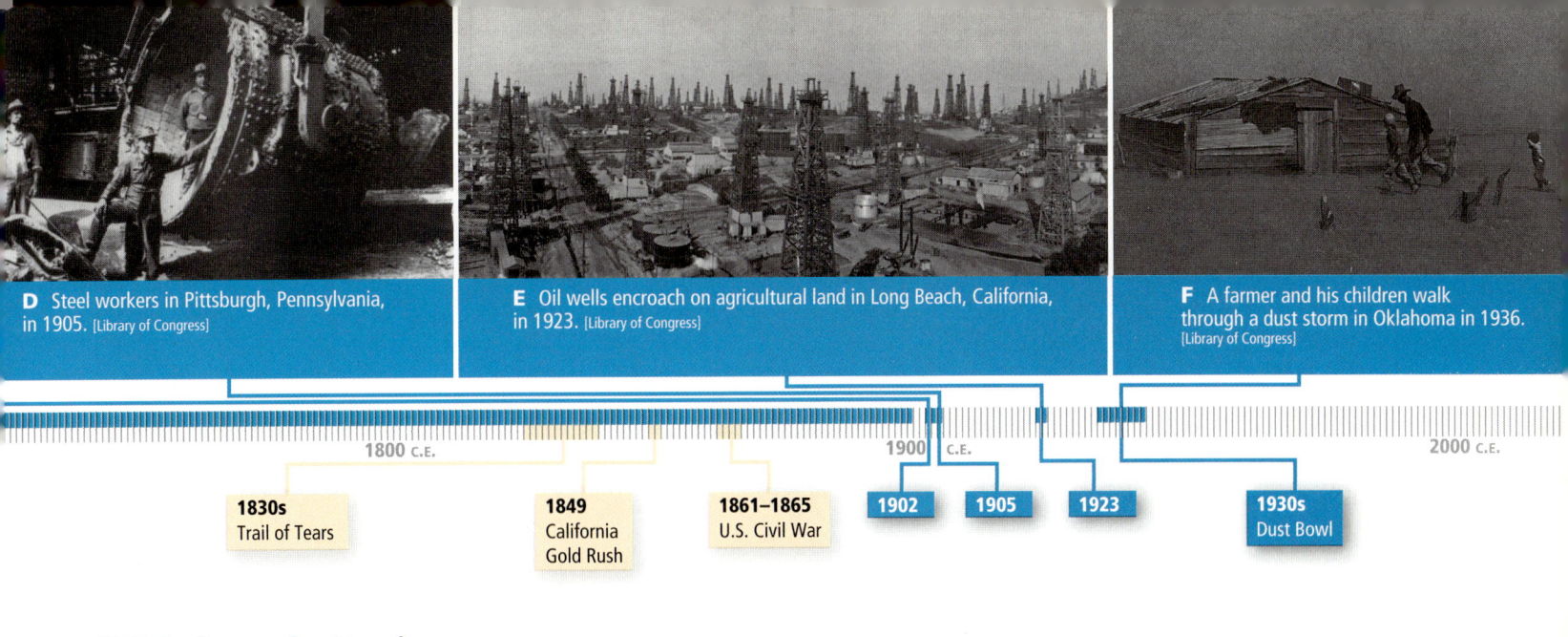

D Steel workers in Pittsburgh, Pennsylvania, in 1905. [Library of Congress]

E Oil wells encroach on agricultural land in Long Beach, California, in 1923. [Library of Congress]

F A farmer and his children walk through a dust storm in Oklahoma in 1936. [Library of Congress]

1830s	1849	1861–1865	1902	1905	1923	1930s
Trail of Tears	California Gold Rush	U.S. Civil War				Dust Bowl

NOTE: Timeline range is not to scale.

This mid-Atlantic region benefited from more fertile soils, a slightly warmer climate, multiple deepwater harbors, and better access to the resources of the interior. By the end of the Revolutionary War in 1783, the mid-Atlantic region was on its way to becoming the **economic core**, or the dominant economic region, of North America. Port cities such as New York, Philadelphia, and Baltimore prospered as the intermediaries for trade between Europe and the vast American continental interior.

economic core the dominant economic region within a larger region

In the early nineteenth century, both agriculture and manufacturing grew and diversified, drawing immigrants from much of northwestern Europe. As farmers became more successful, they bought mechanized equipment, appliances, and consumer goods made in nearby cities. By the mid-nineteenth century, the economy of the core was increasingly based on the steel industry, which spread westward to Pittsburgh and the Great Lakes industrial cities of Cleveland, Detroit, and Chicago. The steel industry relied on the mining of coal and iron ore deposits throughout the region and beyond. Steel became the basis for mechanization, and the region was soon producing heavy farm and railroad equipment, including steam engines (see Figure 2.9D).

By the early twentieth century, the economic core stretched from the Atlantic to St. Louis on the Mississippi (including many small industrial cities along the river, plus Chicago and Milwaukee), and from Ottawa to Washington, DC. It dominated North America economically and politically well into the middle of the twentieth century. Most other areas produced food and raw materials for the core's markets and depended on the core's factories for manufactured goods.

Expansion West of the Mississippi and Great Lakes

The east-to-west trend of settlement continued as land in the densely settled eastern parts of the continent became too expensive for new immigrants. By the 1840s, immigrant farmers from central and northern Europe, as well as European descendants born in eastern North America, were pushing their way beyond the Great Lakes and across the Mississippi River, north and west into the Great Plains of Canada and the United States (Figure 2.11).

The Great Plains Much of the land west of the Great Lakes and the Mississippi River was dry grassland or prairie. The soil usually proved very productive in wet years, and the area became known as North America's *breadbasket*. But the naturally arid character of this land eventually created an ecological disaster for Great Plains farmers. In the 1930s, after 10 especially dry years, a series of devastating dust storms blew away topsoil by the ton. This hardship was made worse by the widespread economic depression of the 1930s. Many Great Plains farm families packed up what they could and left what became known as the Dust Bowl (see Figure 2.9F), heading west to California and other states on the Pacific coast.

The Mountain West and Pacific Coast Some Europeans, alerted to the possibilities farther west, skipped over the Great Plains entirely. By the 1840s, they were coming to the valleys of the Rocky Mountains, to the Great Basin, and to the well-watered and fertile coastal zones of what was then known as the Oregon Territory and California. News of the discovery of gold in California in 1849 created the *Gold Rush*, which drew thousands of people with the prospect of getting rich quickly. The vast majority of gold seekers were unsuccessful, however, and by 1852 they had to look for employment elsewhere. Farther north, logging eventually became a major industry (see Figure 2.9C).

The extension of railroads across the continent in the nineteenth century facilitated the transportation of manufactured goods to the West as well as raw materials and eventually fresh produce to the East. Today, the coastal areas of this region, often called the Pacific Northwest, have thriving, diverse, high-tech economies and growing populations. Perhaps in response to their history, residents of the Pacific Northwest are on the forefront of so many efforts to reduce human impacts on the environment that the region has been nicknamed "Ecotopia."

FIGURE 2.11 Nineteenth-century transportation. [Sources consulted: James A. Henretta, W. Elliot Brownlee, David Brody, and Susan Ware, *America's History*, 2nd ed. (New York: Worth, 1993), pp. 400–401; James L. Roark, Michael P. Johnson, Patricia Cline Cohen, Sarah Stage, Alan Lawson, and Susan M. Hartmann, *The American Promise: A History of the United States*, 3rd ed. (Boston: Bedford/St. Martin's, 2005), p. 601]

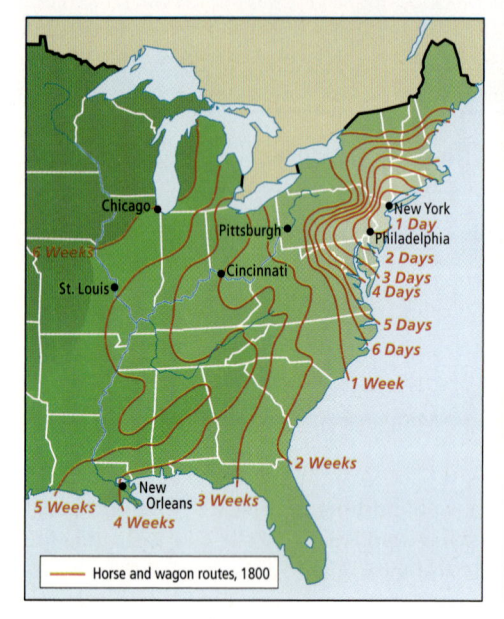

(A) Travel times from New York City, 1800. It took a day to travel by wagon from New York City to Philadelphia and a week to go to Pittsburgh.

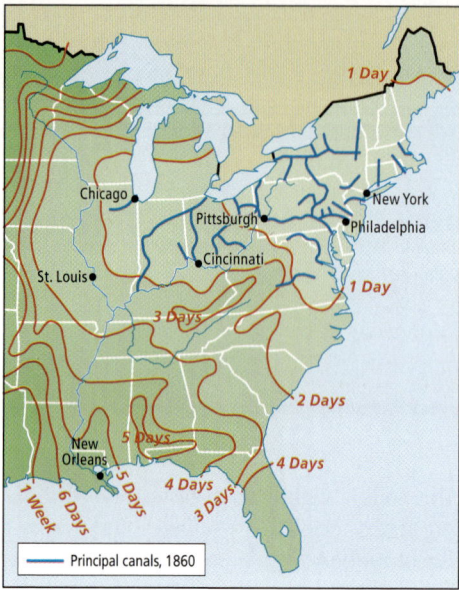

(B) Travel times from New York City, 1860. The travel time from New York to Philadelphia was now only 2 or 3 hours and to Pittsburgh less than a day because people could go part of the way via canals (dark blue). Via the canals, the Great Lakes, and rivers, they could easily reach principal cities along the Mississippi River, and travel was less expensive and onerous than it had been.

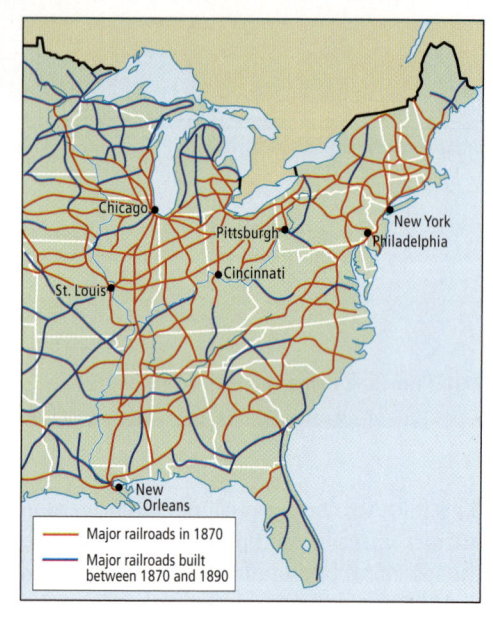

(C) Railroad expansion by 1890. With the building of railroads, which began in the decade before the Civil War, the mobility of people and goods increased dramatically. By 1890, railroads crossed the continent, though the network was most dense in the eastern half.

The Southwest People from the Spanish colony of Mexico first colonized the Southwest in the late 1500s. Their settlements were sparse. As immigrants from the United States expanded into the region, drawn by the cattle-raising industry, Mexico found it progressively more difficult to maintain control, and by 1850, nearly the entire Southwest was under U.S. control.

By the twentieth century, a vibrant agricultural economy had developed in central and southern California, supported by massive government-sponsored water-movement and irrigation projects. The mild Mediterranean climate made it possible to grow vegetables almost year-round. With the advent of refrigerated railroad cars, fresh California vegetables could be sent to the major population centers of the East. Southern California's economy rapidly diversified to include oil (see Figure 2.9E), entertainment, and a variety of engineering- and technology-based industries.

European Settlement and Native Americans

As settlement relentlessly expanded west, Native Americans (called First Nations people in Canada) who had survived early encounters with Europeans and were living in the eastern part of the continent occupied land that European newcomers wished to use. During the 1800s, almost all the surviving Native Americans were killed in innumerable skirmishes with European newcomers, absorbed into the societies of the Europeans (through intermarriage and acculturation), or forcibly relocated west to relatively small reservations with few resources. The largest relocation, in the 1830s, involved the Choctaw, Seminole, Creek, Chickasaw, and Cherokee of the southeastern states. These people had already adopted many European methods of farming, building, education, government, and religion. Nevertheless, they were rounded up by the U.S. Army and marched to Oklahoma, along a route that became known as the Trail of Tears because of the more than 4000 Native Americans who died along the way.

As Europeans occupied the Great Plains and prairies, many of the reservations were further shrunk or relocated onto even less desirable land. Today, reservations cover just over 2 percent of the land area of the United States.

In Canada the picture is somewhat different. Reservations now cover 20 percent of Canada, mostly because of the creation of the Nunavut Territory (now known simply as Nunavut) in 1999 in the far north and the ceding of Northwest Territory land to the Tłı̨chǫ First Nation (also known as the Dogrib) in 2003 (see the

Figure 2.1 map). These Canadian First Nations stand out as having won the right to legal control of their lands. In contrast to the United States, it had been unusual for native groups in Canada to have legal control of their territories.

After centuries of mistreatment, many Native American and First Nations people still live in poverty and, as in all communities under severe stress, rates of alcohol and drug addiction and violence are high. However, in recent decades some tribes have found avenues to greater affluence on the reservations and territories by establishing manufacturing industries; developing fossil fuel, uranium, and other mineral deposits under their lands; or opening gambling casinos. One measure of this economic resurgence is population growth. Expanding from a low of 400,000 in 1907, the Native American population, at less than 2 percent of the total population, stood at approximately 6 million in 2010 in the United States. In Canada, First Nations people number over 1.2 million, or 3.8 percent of the population.

The Changing Regional Composition of North America

The regions of European-led settlement still remain in North America, but they are now less distinctive. The economic core region is less dominant in industry, which has spread to other parts of the continent. Some regions that were once dependent on agriculture, logging, or mineral extraction now have high-tech industries as well. The West Coast, in particular, has blossomed with a high-tech economy and a rapidly growing population that includes many immigrants from Asia and Middle and South America. The West Coast also benefits from North America's trade with Asia, which now surpasses trade with Europe in volume and value.

THINGS TO REMEMBER

- Recent evidence suggests that humans first came to North America from northeastern Asia at least 25,000 years ago and perhaps earlier, most arriving during an ice age.
- By the late 1600s, New England was implementing ideas and technology from Europe and developing some of the first industries in North America.
- By the early twentieth century, North America's economic core was well established. Stretching from the Atlantic Ocean to St. Louis, Chicago, and Milwaukee, and from Ottawa to Washington, DC, it dominated North America economically and politically well into the middle of the twentieth century.
- The push to settle the Great Plains, the Mountain West, and the Pacific Coast attracted many immigrants, from both Mexico and Europe, interested in farming, mining, and cattle raising.
- In 1492, roughly 18 million Native American and First Nations people lived in North America. By 1542, after only a few Spanish expeditions, there were half that many. By 1907, only about 2 percent of the original population remained; however, by 2010, the Native American and First Nations populations had partially rebounded to approximately 7 million.

GLOBALIZATION AND DEVELOPMENT

GEOGRAPHIC INSIGHT 2

Globalization and Development: Globalization has transformed economic development in North America, reorienting employment toward knowledge-intensive jobs that require education and training. Most of the manufacturing jobs upon which the region's middle class was built have either been moved abroad to take advantage of cheaper labor or have been replaced by technology. North America's demand for imported goods and its export of manufacturing jobs helps make it a major engine of globalization.

46. U.S. GEOGRAPHY REPORT

Like their political systems, the economic systems of Canada and the United States have much in common. Both countries evolved from societies based mainly on family farms. Both then had an era of industrialization followed by a transformation to a primarily service-based economy, and both have important technology sectors and an economic influence that reaches worldwide.

The Decline in Manufacturing Employment

By the 1960s, the geography of manufacturing was changing. In the old economic core, higher pay and benefits and better working conditions won by labor unions led to increased production costs. This threatened the high profits demanded by the owners, managers, and shareholders of manufacturing corporations. A number of companies began moving their factories to the southeastern United States where wages were lower and corporate profits higher because of the absence of labor unions.

44. U.S. LABOR TRANSITION REPORT

In 1994, the **North American Free Trade Agreement (NAFTA)**—a free trade agreement that added Mexico to the 1989 economic arrangement between the United States and Canada—was passed. Many manufacturing industries, such as clothing, electronic assembly, and auto parts manufacturing, began moving farther south to Mexico or overseas. In these locales, labor was vastly cheaper. Further, employers saved on production costs because laws mandating environmental protection as well as safe and healthy workplaces were absent or less strictly enforced.

> **North American Free Trade Agreement (NAFTA)** a free trade agreement made in 1994 that added Mexico to the 1989 economic arrangement between the United States and Canada

Another factor in the decline of manufacturing employment has been automation. The steel industry provides an illustration. In 1980, huge steel plants, most of them in the economic core, employed more than 500,000 workers. At that time, it took about 10 person-hours and cost about $1000 to produce 1 ton of steel. Spurred by more efficient foreign competitors, the North American steel industry applied new technology to lower production costs, improve efficiency, and increase production. By 2006, steel was being produced at the rate of 0.44 person-hours per ton and at a cost of about $165 per ton. As a result, the steel industry in the United States reorganized, producing much of the steel in small, highly efficient mini-mills distributed throughout the United States. In total, the steel industry now employs fewer than half the workers it did in 1980. Throughout North America, this efficiency trend has resulted in far fewer people producing more of a given product at a far lower cost than was the case 30 years ago. Remarkably, even as employment in manufacturing has declined over the last three decades, the actual amount of industrial production has steadily increased.

Growth of the Service Sector

The economic base of North America is now a broad *tertiary sector* in which people are engaged in various services such as transportation, utilities, wholesale and retail trade, health, leisure, maintenance, finance, government, information, and education.

As of 2014, in both Canada and the United States about 80 percent of jobs and a similar percentage of the GNI were in the tertiary, or service, sector. There are high-paying jobs in all the service categories, but low-paying jobs are far more common. The largest private employer in the United States is the discount retail chain Walmart (1.4 million employees), where the average wage is $12 an hour, or $24,000 a year, full time. This is just barely above the poverty level for a family of four in the United States. Because many Walmart employees are part time, a large percentage of them do not receive benefits, including health care. Walmart creates mostly retail jobs because, for the most part, its wares are manufactured abroad.

The Knowledge Economy An important subcategory of the service sector involves the creation, processing, and communication of information—what is often called the *knowledge economy*, or the *quaternary sector*. The knowledge economy includes workers who manage information, such as those employed in finance, journalism, higher education, research and development, and many aspects of health care. It also includes the *information technology* or *IT* sector, which deals with computer software and hardware and the management of digital data.

Industries that rely on the use of computers and the Internet to process and transport information are freer to locate where they wish than were the manufacturing industries of the old economic core, which depended on locally available material resources such as steel and coal. These newer industries are more dependent on skilled managers, communicators, thinkers, and technicians, and are often located near major universities and research institutions.

Crucial to the knowledge economy is the Internet, which was first widely available in North America and has emerged as an economic force more rapidly there than in any other region in the world. With only 5 percent of the world's population, North America accounted for 11.4 percent of the world's Internet users in 2014. Roughly 78 percent of the population of the United States uses the Internet, as does 85 percent of the Canadian population, compared to 73 percent of the European Union and 39 percent of the world as a whole. The total economic impact of the Internet in North America is hard to assess, but retail Internet sales increase every year. Indeed, although overall purchases were down during the recession of 2008 to 2011, online purchases in the United States and Canada steadily increased.

The Internet has entered many aspects of life in North America. In the political sphere, social networking (Facebook, Twitter, YouTube, and others) now plays a large role in recruiting volunteers and eliciting cash contributions, especially during recent U.S. election cycles. Beginning in 2009, social networking through Facebook, Twitter, and YouTube became an integral part of public communication, often referred to on daily TV news programming. It is now a major source of information for North Americans about international events, such as the Arab Spring demonstrations around the Mediterranean.

The growth of Internet-based activity in North America makes access to the Internet increasingly crucial. Unfortunately, a **digital divide**—a discrepancy in access to information technology between small, rural, and poor areas and large, wealthy cities—has developed, because about a quarter of the North American population is not yet able to afford computers and Internet connections.

Globalization and Free Trade

The United States and Canada are major engines of globalization that impact the world economy through the size and technological sophistication of their economies. Together, they are almost as large as the economy of the entire European Union. North America's advantageous position in the global economy is also a reflection of its geopolitical influence—its ability to mold the pro-globalization free trade policies that suit the major corporations and the governments of North America.

Free trade has not always been emphasized the way it is now. Before its rise to prosperity and global dominance, trade barriers were important aids to North American development. For example, when it became independent of Britain in 1776, the new U.S. government imposed tariffs and quotas on imports and gave subsidies to domestic producers. This protected fledgling domestic industries and commercial agriculture, allowing its economic core region to flourish.

Now, because both are wealthy and globally competitive exporters, Canada and the United States see tariffs and quotas in other countries as obstacles to North America's economic expansion abroad. Thus, they usually advocate heavily for trade barriers to be reduced worldwide. Critics of these free trade policies point out a number of inconsistencies in the current North American position on free trade. First, North America once needed tariffs and quotas to protect its firms, much like many currently poorer countries still need to do. Furthermore, contrary to their own free trade precepts, both the United States and Canada still give significant subsidies to their farmers. These subsidies make it possible for North American farmers to sell their crops on the world market at such low prices that farmers elsewhere are hurt or even driven out of business (see the vignette on page 60). For example, many Mexican farmers have lost their small farms because of competition from large U.S. corporate farms, which receive subsidies from the U.S. government. Corporate farms in the United States can now sell their produce in Mexico or even relocate there under NAFTA agreements. The critics add that beyond agriculture, in North America, the benefits of

> **digital divide** the discrepancy in access to information technology between small, rural, and poor areas and large, wealthy cities that contain major governmental research laboratories and universities

Globalization and Development 77

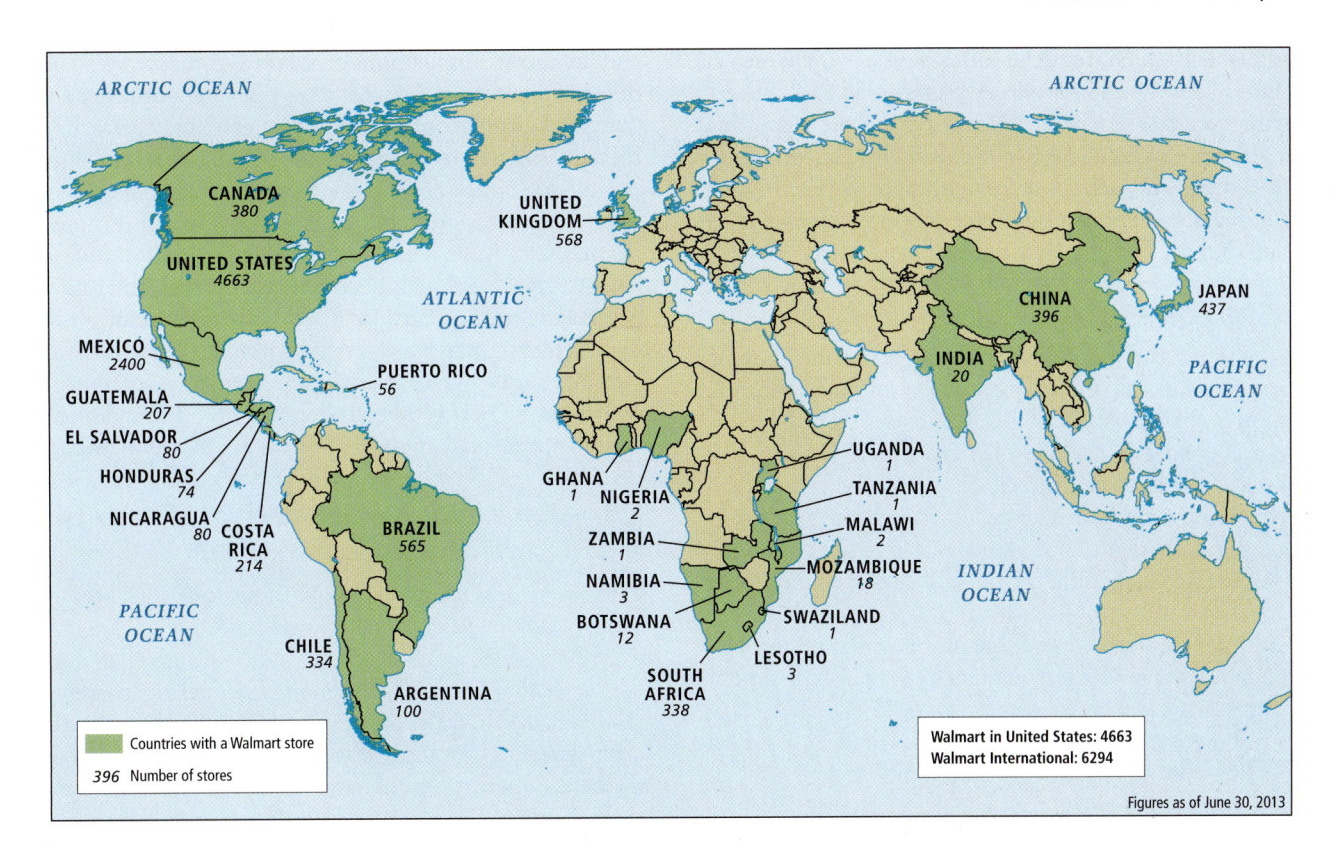

FIGURE 2.12 Walmart on the global scale. As of June 2013, Walmart had 4663 store operations in the United States and 6294 in 27 other countries. Walmart draws its products from more than 70 countries, deals with over 61,000 U.S. businesses, and is instrumental in generating over 3 million U.S. jobs. Walmart itself employs more than 1.3 million workers in the United States and 900,000 workers in other countries. [Source consulted: "Wal-Mart Stores, Inc. Data Sheet—Worldwide Unit Details January 2012," Walmart, at http://news.walmart.com/news-archive/2012/02/22/wal-mart-stores-inc-data-sheet-worldwide-unit-details-january-2012]

free trade go mostly to large manufacturers and businesses and their managers, while many workers end up losing their jobs to cheaper labor overseas, or see their incomes stagnate.

NAFTA Trade between the United States and Canada has been relatively unrestricted for many years. The process of reducing trade barriers began formally with the Canada–U.S. Free Trade Agreement of 1989. Mexico was included with the creation of NAFTA in 1994. The major long-term goal of NAFTA is to increase the amount of trade between Canada, the United States, and Mexico. Today, it is the world's largest trading bloc in terms of the GDP of its member states.

The extent of NAFTA's impacts are hard to assess because it is difficult to tell whether the many observable changes have been caused by the agreement itself or by other changes in regional and global economies. However, a few things are clear. NAFTA has increased trade, and many companies are making higher profits because they now have larger markets. Since 1990, exports among the three countries have increased in value by more than 300 percent. By value, NAFTA's exports to the world economy have increased by about 300 percent for the United States and Canada and by 600 percent for Mexico. Some U.S. companies, such as Walmart, expanded aggressively into Mexico after NAFTA was passed. Mexico now has more Walmarts (2400 retail stores) than any country except the United States, which has 4663 (Figure 2.12). Canada has 380 Walmarts.

NAFTA seems to have worsened the perennial tendency of the United States to spend more money on imports than it earns from exports. This imbalance is called a **trade deficit**. Before NAFTA, the United States had much smaller trade deficits with Mexico and Canada. After the agreement was signed, these deficits rose dramatically, especially with Mexico. For example, between 1994 and 2009, the value of U.S. exports to Mexico increased by about 153 percent, while the value of imports increased 265 percent.

NAFTA has also resulted in a net loss of about 1 million jobs in the United States. Increased imports from Mexico and Canada have displaced about 2 million U.S. jobs, while increased exports to these countries have created only about 1 million jobs. Some new NAFTA-related jobs do pay up to 18 percent more than the average North American wage. However, those jobs are usually in different locations than the ones that were lost and the people who take them tend to be younger and more highly skilled than those who lost jobs. Former factory workers often end up with short-term contract jobs or low-skill jobs that pay the minimum wage and carry no benefits.

As the drawbacks and benefits of NAFTA are being assessed, talk of extending it to the entire Western Hemisphere has stalled.

trade deficit the extent to which the money earned by exports is exceeded by the money spent on imports

Such an agreement, which would be called the Free Trade Area of the Americas (FTAA), would have its own drawbacks and benefits. A number of countries, including Brazil, Bolivia, Ecuador, and Venezuela, are wary of being overwhelmed by the U.S. economy. Even in the absence of such an agreement, trade between North America and Middle and South America is growing faster than trade with Asia and Europe. This emerging trade is discussed in Chapter 3.

The Asian Link to Globalization Another way in which the North American economy is becoming globalized is through the lowering of trade barriers with Asia. One huge category of trade with Asia is the seemingly endless variety of goods imported from China—everything from underwear to the chemicals used to make prescription drugs. China's lower wages make its goods cheaper than similar products imported from Mexico, despite Mexico's proximity and membership in NAFTA. Indeed, many factories that first relocated to Mexico from the southern United States have now moved to China to take advantage of its enormous supply of cheap labor. Despite adjustments in the Chinese economy that may raise prices, trade with China promises to remain quite robust for some time.

U.S. and Canadian companies also want to take advantage of China's vast domestic markets. For example, the U.S. fast-food chain KFC now has more than 2100 locations in 450 cities in China, and its business is growing rapidly there. The U.S. government has a powerful incentive to encourage such overseas expansion by companies like KFC that are headquartered in the United States and whose profits are taxable by the federal government.

Asian investment in North America is also growing. For example, Japanese and Korean automotive companies have located plants in North America to be near their most important pool of car buyers—commuting North Americans. They often establish their plants in the rural mid-South of the United States or in southern Canada, close to arteries of the Interstate Highway System. Here the Asian companies have found a ready, inexpensive labor force. These workers can access high-quality housing in rural settings within a commute of 20 miles (32 kilometers) or so from secure automotive jobs that pay reasonably well and include health and retirement benefit packages.

Japanese and Korean carmakers succeeded in North America even as U.S. auto manufacturers such as General Motors were struggling. This was primarily because more advanced Japanese automated production systems—requiring fewer but better-educated workers—produce higher-quality cars, which sell better both in North America and around the world. Some analysts predict that foreign carmakers will eventually take over the entire North American market, while others predict that the old American car companies, several of which are in the process of restructuring, will maintain and even expand market share by turning out better-built and more fuel-efficient cars.

IT Jobs Face New Competition from Developing Countries By the early 2000s, globalization was resulting in the *offshore outsourcing* of information technology (IT) jobs. A range of jobs—from software programming to telephone-based, customer-support services—shifted to lower-cost areas outside North America. By the middle of 2003, an estimated 500,000 IT jobs had been outsourced, and another 3.3 million are forecasted to follow by 2020. During the recent recession, IT jobs were still being created at a faster rate than they were being eliminated by economic contraction, yet the overall trend pointed to fewer and fewer of the world's total IT jobs staying in North America. New IT centers are now located in India, China, Southeast Asia, the Baltic states in North Europe, Central Europe, and Russia. In these areas, large pools of highly trained, English-speaking young people work for wages that are 20 to 40 percent of their American counterparts' pay. Some argue that rather than depleting jobs, outsourcing will actually help create jobs in North America by saving corporations money, which will then be reinvested in new ventures. The viability of this pro-outsourcing argument remains unproven. **45. U.S. COMPETITIVENESS**

Repercussions of the Global Economic Downturn Beginning in 2007

The severe worldwide economic downturn that began in 2007 came on the heels of a long upward trajectory of global economic expansion. A booming housing industry and related growth in the banks that finance home mortgages fueled this growth in the United States. In 2007, the housing industry collapsed as it became clear that much of the growth in previous years had been based on banks allowing millions of buyers to purchase homes with mortgages that were well beyond their means. The growth in the construction industry, based on erroneous assumptions, came to a sudden halt. When too many homebuyers could no longer afford their mortgage payments, the banks that had lent them money and/or the institutions to which the banks had sold these mortgages started to fail. The bank failures produced worldwide ripple effects because so many foreign banks were involved in the U.S. housing market. Between September 2008 and March 2009, the U.S. stock market fell by nearly half, wiping out the savings and pensions of millions of Americans. Similar plunges followed in foreign stock markets, ultimately resulting in a worldwide economic downturn because businesses could no longer find money to fund expansion.

As the recession intensified, job losses in the United States caused a sharp drop in consumption, which further affected world markets, including those in Canada. Thanks to its strong regulatory controls, Canada did not have bank failures. However, because so much of the Canadian economy is linked to exports and imports from the United States (see Figure 2.13), Canada underwent a massive slowdown. During three quarters in 2009, its GNI declined by 3.3 percent and its exports fell by 16 percent. Canada also lost many jobs but its recovery was quicker than that of the United States, possibly because household consumption was buttressed by Canada's stronger social safety net (see page 89).

Efforts to deal with the difficulties that caused the recession, as well as the long-term problems made worse by it, have yet to meet with much success. The financial industry has used well-funded lobbying to counter attempts to better regulate the banks that, through their reckless lending practices, caused the recession.

The recession worsened the trend toward larger wealth disparities that has been underway since the 1970s. Disparities are most extreme in the United States, where in 2014 the wealthiest 1 percent of households owned 40 percent of the country's total wealth, and the next wealthiest 19 percent owned roughly another

Globalization and Development   79

53 percent, leaving the remaining 80 percent of the population with only 7 percent of the wealth. Most North Americans are not in favor of this kind of wealth distribution, but there are many political barriers that make it difficult to reduce such inequality (see page 89).

Interdependencies

Canada and the United States are perhaps most intimately connected by their long-standing economic relationship. The two countries engage in mutual tourism, direct investment, migration, and most of all, trade. The equivalent of nearly U.S.$1.5 billion is traded daily between the two countries. Canada is a larger market for U.S. goods than are all 27 countries in the European Union. By 2005, that trade relationship had evolved into a two-way flow of U.S.$1 trillion annually (Figure 2.13). Each country is the other's largest trading partner. In 2010, fifty percent of Canada's imports came from the United States and 75 percent of its exports went to the United States. The United States, in turn, sells 19 percent

Tourism: Between 2006 and 2011, the number of tourists from each country visiting the other fluctuated between 8 and 20 million each way, depending on the economic conditions and security measures.

Trade: The value of goods and services traded between the two countries in 2011 was about $597.4 billion, down from $600 billion in 2008. In 2011, as in the past, Canada exported more to the United States than the United States exported to Canada. In 2011, Canada's trade surplus with the United States was $35.6 billion, but in 2008, the surplus was nearly twice as high. These changes are largely the result of the global recession that began in 2008, which hit the United States harder than it hit Canada.

Direct Investment: In 2010, Canadian investments in the United States were $306 billion, and U.S. investments in Canada were $206 billion. Typically, Canadian companies invest more heavily in the United States because of the attractively large market, roughly 10 times the size of Canada's.

Pollutants: Airborne pollutants, which cause acid rain (among other negative effects), cross the border in both directions, but Canada receives on average about three times more air pollutants from the United States than the reverse. This is primarily the result of high-volume truck traffic and the prevailing winds. The two countries have agreements to reduce the pollution that crosses the border, but Canada has apparently been more successful than the United States in these efforts.

Immigration: For many years there has been a pattern of immigration asymmetry, with more Canadians coming to the United States than the reverse. On average over the last decade, about 24,000 Canadians have annually come to the United States, and about one-third as many people from the United States have moved to Canada. In 2010, the numbers were substantially reduced, probably as a result of the recession and U.S. restrictions on immigration.

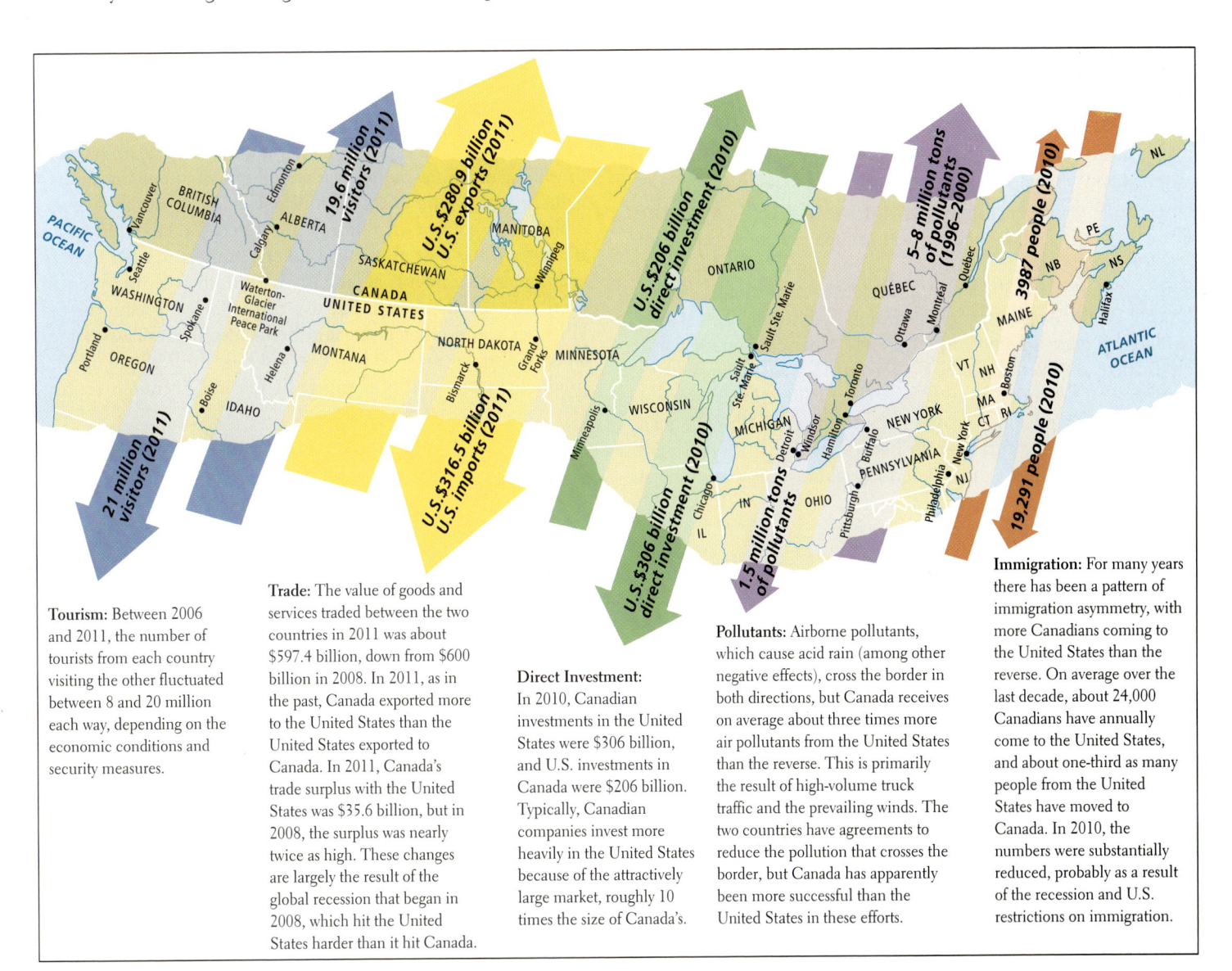

FIGURE 2.13 Transfers of tourists, goods, investment, pollution, and immigrants between the United States and Canada. Canada and the United States have the world's largest trading relationship. The flows of goods, money, and people across the long Canada–U.S. border are essential to both countries. However, because of its relatively small population and economy, Canada is more reliant on the United States than the United States is on Canada. All amounts shown are in U.S. dollars. [Sources consulted: *National Geographic*, February 1990: 106–107, and augmented with data from "Table 1: International Trips to Canada," Statistics Canada, at http://www.statcan.gc.ca/pub/66-001-p/2011012/t001-eng.htm; *International Visitation to the United States: A Statistical Summary of U.S. Visitation (2011)*, U.S. Department of Commerce, Office of Travel and Tourism Industries, at http://travel.trade.gov/outreachpages/download_data_table/2011_Visitation_Report.pdf; "Foreign Trade: Trade in Goods with Canada," U.S. Census Bureau, at http://www.census.gov/foreign-trade/balance/c1220.html; "U.S. Relations With Canada," U.S. Department of State, August 23, 2013, at http://www.state.gov/r/pa/ei/bgn/2089.htm; "Immigration Overview: Permanent and Temporary Residents: Canada—Permanent Residents by Category and Source Area," Citizenship and Immigration Canada, at http://www.cic.gc.ca/english/resources/statistics/facts2010/permanent/08.asp]

of its exports to Canada and buys 14 percent of its imports from Canada.

Notice, however, that there is asymmetry even in the realm of interdependencies: Canada's smaller economy is much more dependent on the United States than the reverse. Nonetheless, as many as 1 million U.S. jobs are dependent on the relationship with Canada.

Women in the Economy

While women have made steady gains in achieving equal pay and overall participation in the labor force, there are still important ways in which they lag behind their male counterparts. On average, U.S. and Canadian female workers earn about 80 cents for every dollar that male workers earn for doing the same job (Figure 2.14). For example, a female architect earns approximately 80 percent of what a male architect earns for performing comparable work. This situation is actually an improvement over previous decades. During World War II, when large numbers of women first started working in male-dominated jobs, North American female workers earned, on average, only 57 percent of what male workers earned. The advances made by this older generation of women and the ones that followed have transformed North American workplaces. For the first time in history, women now represent more than half of the North American labor force, though most still work for male managers.

North America's Changing Food-Production Systems

Agriculture remains the spatially dominant feature of North American landscapes, yet less than 1 percent of North Americans are engaged directly in agriculture. North America benefits from an abundant supply of food; the region produces food for foreign as well as domestic consumers (Figure 2.15). At one time, exports of agricultural products were the backbone of the North American economy. However, because of growth in other sectors, agriculture now accounts for less than 1.2 percent of the United States' GDP and less than 2 percent of Canada's. Moreover, because both countries are so involved in the global economy, an increasing amount of food in both countries is imported.

43. FOOD GLOBALIZATION

The shift to mechanized agriculture in North America brought about sweeping changes in employment and farm management. In 1790, agriculture employed 90 percent of the American workforce; in 1890, it employed 50 percent. Until 1910, thousands of very productive family-owned farms, located over much of the United States and southern Canada, provided for most domestic consumption and the majority of all exports. Today, the vast majority of these family farms have been replaced by farms owned by corporations.

Family Farms Give Way to Agribusiness Family farms began to mechanize and use chemical fertilizers in the late nineteenth century. Mechanical corn-seed planters and steam-powered threshing machines, along with methods of supplying farmers with large amounts of plant nutrients such as nitrogen, phosphate, and potassium, reduced the need for labor on farms. Because of the cost of this machinery, farmers needed to make ever-larger investments in land in order for their farms to remain profitable. By the 1940s, the use of pesticides (chemicals that kill insects and other pests) and herbicides (chemicals that kill weeds) also became widespread, adding further to both the productivity and cost of farming. By the mid-twentieth century,

---

ON THE BRIGHT SIDE

Women in Business and Education

Throughout North America, the number of women entrepreneurs is on the rise, and women start nearly half of all new businesses. While women-owned businesses tend to be small and less financially secure than those in which men have dominant control, credit opportunities for businesswomen are improving as women go into upper-level jobs in banking. Change is also coming to the private sector, where a 2012 study showed that male executives now prefer to hire qualified women because they are particularly ambitious and willing to gain advanced qualifications.

In secondary and higher education, North American women have equaled or exceeded the level of men in most categories. In 2008 in the United States, 33 percent of women aged 25 and over held an undergraduate degree, compared to just 26 percent of men. This imbalance is likely to increase because in 2010, U.S. women between the ages of 25 and 29 received 7 percent more undergraduate degrees than men did.

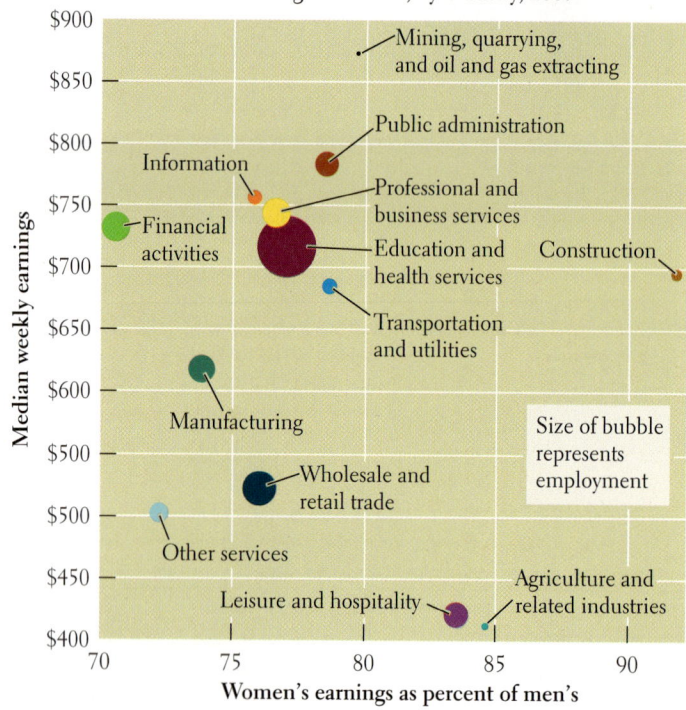

FIGURE 2.14 Women's earnings and employment by industry, 2009. On average, women earned less than 80 percent of men's median income in 2009 if they worked full time in a wage or salary job: $630 a week compared to men's $819. [Source consulted: "Women's Earnings and Employment by Industry, 2009," U.S. Department of Labor, Bureau of Labor Statistics, February 16, 2011, at http://www.bls.gov/opub/ted/2011/ted_20110216.htm]

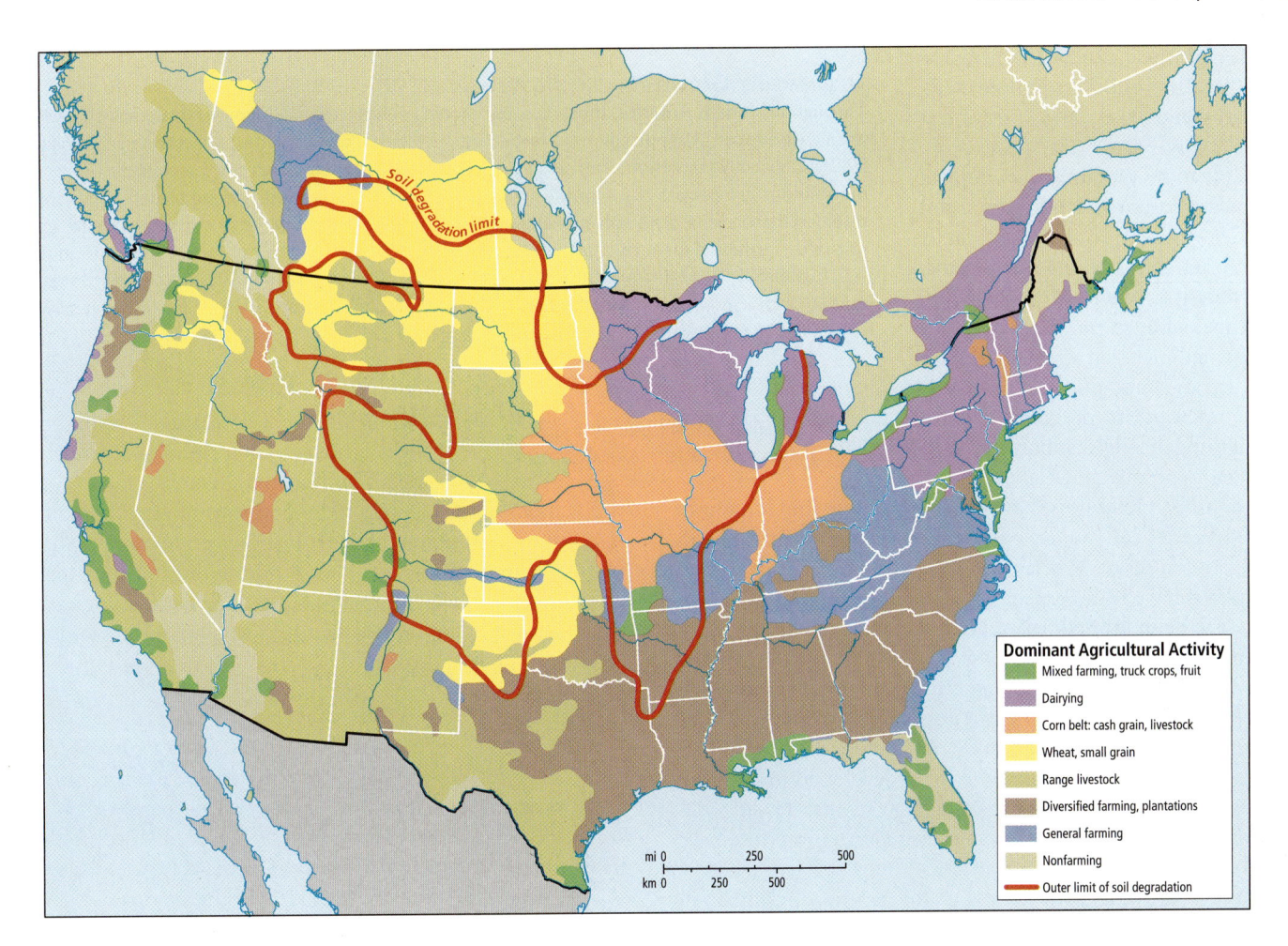

FIGURE 2.15 Agriculture in North America. Throughout much of North America, some type of agriculture is possible. The major exceptions are the northern parts of Canada and Alaska and the dry mountain and basin region (the continental interior) that lies between the Great Plains and the Pacific coastal zone. However, in some marginal areas, such as Southern California, southern Arizona, and the Utah Valley, irrigation is needed for cultivation. (Hawaii is not included here because it is covered in Oceania, Chapter 11.) [Source consulted: Arthur Getis and Judith Getis, eds., *The United States and Canada: The Land and the People* (Dubuque, IA: William C. Brown, 1995), p. 165]

the number of farms began to decline rapidly, as only wealthier farmers could invest in these green revolution methods (see Chapter 1, page 29). Some farmers prospered (see the vignette on page 60), while many with fewer resources sold their land, hoping to make a profit sufficient for retirement. Indebtedness and bankruptcies have become increasingly common for both large and small farmers.

A major part of the transformation of North American agriculture has been the growth of large **agribusiness** corporations that sell machinery, seeds, and chemicals. These corporations may also produce and process crops themselves on land they own, or they may contract to purchase crops from independent farms. The financial resources of agribusiness corporations has facilitated the transition to green revolution methods, enabling large investments in the research and development of new products. The corporations can also provide loans and cash to individual farmers, often as a part of contracts that leave farmers with little actual control over which crops are grown and what methods are used.

agribusiness the business of farming conducted by large-scale operations that purchase, produce, finance, package, and distribute agricultural products

While green revolution agriculture provides a wide variety of food at low prices for North Americans, the shift to these production methods has depressed local economies and created social problems in many rural areas. Communities in places such as the Great Plains of both Canada and the United States were once made up of farming families with similar middle-class incomes, social standing, and commitment to the region. Today, farm communities are increasingly composed of a few wealthy farmer-managers amid a majority of poor, often migrant Latino or Asian laborers who work on large farms and in food-processing plants for wages that are too low to provide a decent standard of living. These workers also struggle to be accepted into the communities where they live.

Food Production and Sustainability Can green revolution agriculture like that practiced by successful North American corn farmers persist over time? Many modern strategies to increase yields, including the use of chemical fertilizers, pesticides, and herbicides, can have negative impacts. These methods

82 CHAPTER 2 North America

can threaten the health of farm workers and nearby residents, pollute nearby streams and lakes, and even affect distant coastal areas (see page 68). Some irrigation methods can deplete scarce water resources and reduce soil fertility over time. In addition, many North American farming areas have lost as much as one-third of their topsoil because of deep plowing and other farming methods that create soil erosion. A major emerging health issue is that many crops are designed to be processed into high-sugar, high-carbohydrate foods that contribute to obesity and diabetes.

Recently, researchers have been studying the long-term impacts of genetically modified (GMO) crops on human health and the environment. While most scientific studies suggest that GMO plants have little impact on human health, many of these studies have been funded or even conducted by the agribusiness corporations that create GMO crops, making it possible that the studies might be biased. The biggest documented impact that GMO plants have had on the environment is in the way they have influenced the use of herbicides and pesticides. While some varieties of GMO crops have been developed to be more resistant to insects and other pests, and so require fewer pesticides, others can tolerate and may even require more intensive use of herbicides.

There is growing concern about the raising of animals in "factory farms" in which cows, pigs, or poultry are raised in crowded conditions and fed chemicals to make them gain weight. As with other forms of chemically intensive agriculture, the most well-documented threats to human health are to farmworkers and nearby residents who are exposed daily to many hazardous chemicals. The vast quantities of animal waste these farms produce can also severely pollute the air and nearby streams.

Critics of these food production systems often argue that the government subsidies currently being given to farmers should be used as a tool for reform. They advocate directing subsidies away from factory farms and large-scale, chemically intensive crop production and toward small farmers who are willing to use methods that have fewer impacts.

> **organically grown** products produced without chemical fertilizers and pesticides

ON THE BRIGHT SIDE

Organically Grown

Throughout North America, there is a burgeoning revival of small family farms that supply **organically grown** (produced without chemical fertilizers, herbicides, or pesticides) vegetables and fruits and grass-fed meat directly to consumers. Farmers' markets have popped up across the country (Figure 2.16), and more and more people are now buying locally grown organic foods, paying higher prices than those in traditional grocery stores. This movement is gaining such favor that large corporate farms are beginning to see the potential for high profits in sustainable (that is, organic), if not locally grown, food production.

Sustainable food production may cause a decrease in harmful impacts on the environment as farms shift to organic methods and consumers learn the advantages of paying more for higher-quality, toxin-free vegetables and fruits.

VIGNETTE

The flip side of this story of the decline in North America of the family farm way of life is represented by the burgeoning prosperity of farmers like 37-year-old Twitter user Brandon Hunnicutt, who runs his family's 3600-acre (1457-hectare) farm above the vast Ogallala Aquifer in Nebraska. He uses the aquifer water to irrigate in dry times and his computer to garner the latest information on nearly every aspect of his operation—from the diagnosis of a crop disease to the right moment for marketing his crops. In the midst of the global recession, he was selling corn at twice the price he received a few years before.

How can this be? In 2007, the rising price of petroleum products spurred the U.S. government to require that ethanol made from corn be added to gasoline in order to lower dependency on foreign energy sources. Now more than one-third of the corn produced in the United States is converted to ethanol in order to fuel cars and trucks. By 2008, the demand for corn had radically increased, as had its price. As the recession deepened, the value of the U.S. dollar was lowered in order to spur exports. This gave U.S. corn producers an advantage in the world market. Meanwhile, as the economies of Asian countries continued to grow, Asia began buying more and more corn-based food products, especially sweeteners. Bad weather in Russia and Ukraine, which also produce grains, further increased the demand for corn. Corn prices rose globally, to the great advantage of large U.S. producers like Brandon Hunnicutt but to the great anguish of the poor across the world, who had formerly relied on cheap corn as a mainstay in their diets. As shown in the discussion of food security in Chapter 1 (page 29), the vicissitudes of the global economy can bring very different outcomes to different parts of the world. Financial security for farmers like the Hunnicutts can mean *food insecurity* for the poor in Africa, Asia, and Middle and South America. [Source: NPR staff, All Things Considered. For detailed source information, see Text Sources and Credits.] ■

FIGURE 2.16 A Place of the Heart. A Tennessee wife-and-husband team sell produce from their farm, A Place of the Heart, at a local farmers' market. They also deliver a basketful of vegetables each week to a set of customers who pay a seasonal fee of $700 for the produce.

Changing Transportation Networks and the North American Economy

An extensive network of road and air transportation that enables the high-speed delivery of people and goods is central to the productivity of North America's economy. The development of the road system began with the first European settlement in the 1500s in Florida, but until the development of the first railways in the 1820s, most goods moved on rivers, canals, and coastal seas. By the 1870s, the railroads had formed a nationwide network that dominated transportation until inexpensive, mass-produced automobiles were developed in the 1920s. Beginning in the 1950s, the growth of automobile- and truck-based transportation was helped by the Interstate Highway System—a 45,000-mile (72,000-kilometer) network of high-speed, multilane roads that still dominates transportation today. Because this network is connected to the vast system of local roads, it can be used to deliver manufactured products faster and with more flexibility than can be done using the rail system. The highways have thus made it possible to disperse industry and related services into suburban and rural locales across the country, where labor, land, and living costs are lower.

After World War II, air transportation also enabled economic growth in North America. The primary niche of air transportation is business travel, because face-to-face contact remains essential to American business culture despite the growth of telecommunications and the Internet. Because many industries are widely distributed in numerous medium-size cities, air service is organized as a *hub-and-spoke network*. Hubs are strategically located airports, such as those in Atlanta, Chicago, Dallas, and Los Angeles. These airports serve as collection and transfer points for passengers and cargo continuing on to smaller cities and towns. Most airports are also located near major highways, which provide an essential link for high-speed travel and cargo shipping.

THINGS TO REMEMBER

> **GEOGRAPHIC INSIGHT 2**
> - **Globalization and Development** Globalization has transformed economic development in North America, reorienting employment toward knowledge-intensive jobs that require education and training. Most of the manufacturing jobs upon which the region's middle class was built have either been moved abroad to take advantage of cheaper labor or have been replaced by technology. North America's demand for imported goods and its export of manufacturing jobs helps make it a major engine of globalization.

- North America's advantageous position in the global economy is a reflection of its size, technological sophistication, and geopolitical influence—all of which enable it to mold the pro-globalization free trade policies that suit the major corporations of North America.

- The major long-term goal of NAFTA is to increase the amount of trade between Canada, the United States, and Mexico. NAFTA is currently the world's largest trading bloc in terms of the GDP of its member states.

- Flows of trade and investment between North America and Asia have increased dramatically in recent decades.

- By the early 2000s, globalization was resulting in the offshore outsourcing of hundreds of thousands of information technology (IT) jobs.

- North American farms have become highly mechanized, chemically intensive operations that need few workers but require huge amounts of land to be profitable.

- In the twentieth century, the mass production of inexpensive automobiles and trucks, as well as the Interstate Highway System, fundamentally changed how people and goods move across the continent.

POLITICAL ISSUES

> **GEOGRAPHIC INSIGHT 3**
>
> **Power and Politics:** North America has relatively high levels of political freedom, though in recent decades many of its residents have become disillusioned with the political process for a variety of reasons. While Canada plays a relatively modest political role abroad, the United States has enormous influence on the global political order, although its status as the world's predominant "superpower" is increasingly being challenged.

The Expansion of Political Freedoms in the United States and Canada

North Americans' relatively high levels of political freedom are part of a long-term trend toward more openness in the political process. Voting rights have steadily expanded over the history of both countries. They have been extended from including only white male property owners (U.S., mid-1700s; Canada, 1758) to including all adult white males (U.S., 1856; Canada 1898), to including men of African descent (U.S., 1870; Canada, 1837), women (U.S., 1920; Canada, 1918), and indigenous peoples (U.S., 1924; Canada 1960).

The ways in which candidates for elected office are selected has also opened up in both countries. In the past, a few political insiders selected the candidates that they wanted to run for a particular office. Starting in the 1920s in the United States, candidates for office began to be selected via primaries, which are elections that determine who a political party will nominate to run as their candidate for a particular office. For example, in 2008, Barack Obama was chosen to be the Democratic Party's candidate for the office of president of the United States only after he won a series of primaries held in states throughout the country in which he and a number of other candidates ran. After he won the primary, Obama went on to run in the general election, which he also won. Primaries have only recently been adopted in Canada, in part because they are a much more expensive way to choose candidates.

Political Disillusionment While the overall trajectory has been toward more openness in the political process, giving individuals more influence over their governance, there has been a general disillusionment with politics in both the United States and Canada. Evidence of this can be seen in lower *voter turnout*—the percentage of people who decide to cast a vote in an election. Voter turnout has been declining since the 1960s, with the United States having a generally lower voter turnout than Canada: 57.5 percent of voting-age U.S. citizens voted in the presidential election of 2012, and 61.1 percent of voting-age Canadians voted in 2011.

One possible reason for the lower turnout in the United States is the frustration created by the role of money in politics. A potential candidate for a major office in the United States must now spend millions or, in the case of the office of the U.S. president, more than a billion dollars, to win an election. This cost is a major barrier to people who want to enter into politics, as only those able to raise large amounts of money have any chance to win. Expensive elections also mean that successful candidates have to spend more of their time raising money in order to be reelected and less time doing their job representing the people who elected them. As a result, politicians tend to focus on wealthy people and corporations who can make the donations that will help them get reelected. This trend in politics runs counter to the ideal that a person's income should not determine their ability to influence their government. Canada's somewhat higher voter turnout rates may be the result of Canadian laws that regulate the amount of money candidates and their political parties can spend on elections.

Other explanations of the disillusionment with the political process focus on the role of television, which gained popularity in the 1960s, at about the same time that voter turnout started to decline. More time spent watching television is correlated with lower levels of community involvement and less participation in the political process.

Debt and Politics in the United States

There is much worry and political maneuvering in the United States surrounding the national debt. The debt consists of the money the United States borrows by issuing treasury securities to cover the expenses it has that exceed income from taxes and other revenues. There are two components to the U.S. debt: debt held by the public, including that held by private investors, the Federal Reserve System, and foreign, state, and local governments ($12.3 trillion as of December 2013); and debt held in accounts administered by the federal government, such as that borrowed from the Social Security Trust Fund ($4.9 trillion as of December 2013).

To whom does the United States owe its debt? About 70 percent of the total debt is owed to U.S. citizens who own treasury securities, and to various federal government entities, like the Social Security Trust Fund. About 29 percent is owed to foreign governments (8 percent to China, 5 percent to Japan, 2 percent to the United Kingdom, and smaller amounts to Brazil, Taiwan, and Hong Kong).

Proposed solutions highlight the differences between the two major political parties in the United States. Most Republicans believe that the debt is dangerous and needs to be reduced immediately by drastically reducing government spending, especially on social services such as the Affordable Health Care Act (often called "Obamacare"; see page 89) and various kinds of aid to the poor. Many Democrats hold that some government spending, especially military spending and subsidies to large corporations, should be cut.

FIGURE 2.17 PHOTO ESSAY: Power and Politics in North America

Political freedoms are well protected and democratization is at a relatively high level in North America. The United States is often thought of as using its power to promote political freedoms and democratization on a global scale. There is some truth to this. However, strategic and economic interests often play a greater role in shaping U.S. actions abroad. For example, some U.S. officials who planned the Iraq War point out that the desire to control Iraq's oil resources and those of its neighbors influenced U.S. actions there more than did the promotion of democracy. If the main U.S. interest abroad were the promotion of democracy, then sub-Saharan Africa would be a major focus of U.S. foreign aid and military installations, especially given the many violent conflicts that plague this region. As the map below shows, sub-Saharan Africa receives relatively little aid and has few U.S. military bases. Nevertheless, the United States has increased its presence there in recent years.

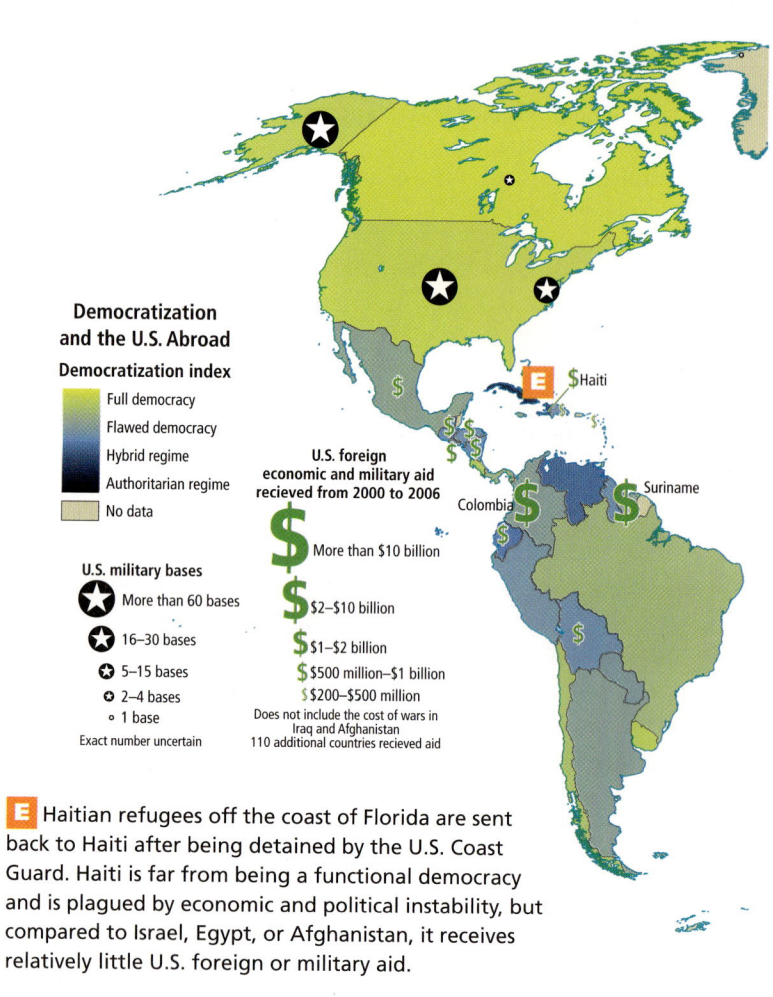

Democratization and the U.S. Abroad

Democratization index:
- Full democracy
- Flawed democracy
- Hybrid regime
- Authoritarian regime
- No data

U.S. military bases:
- More than 60 bases
- 16–30 bases
- 5–15 bases
- 2–4 bases
- 1 base
- Exact number uncertain

U.S. foreign economic and military aid recieved from 2000 to 2006:
- More than $10 billion
- $2–$10 billion
- $1–$2 billion
- $500 million–$1 billion
- $200–$500 million

Does not include the cost of wars in Iraq and Afghanistan
110 additional countries recieved aid

E Haitian refugees off the coast of Florida are sent back to Haiti after being detained by the U.S. Coast Guard. Haiti is far from being a functional democracy and is plagued by economic and political instability, but compared to Israel, Egypt, or Afghanistan, it receives relatively little U.S. foreign or military aid.

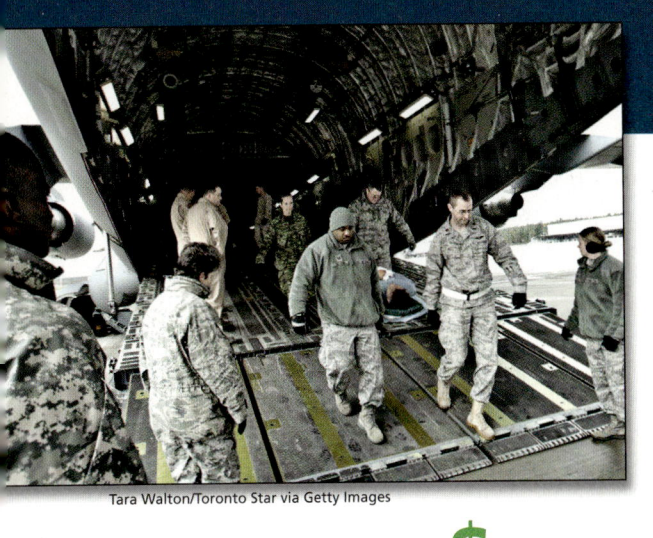

A A Canadian soldier wounded in Afghanistan is lifted off a plane at Ramstein Air Base, one of 260 U.S. bases in Germany. The large U.S. military presence in Europe is a legacy of World War II and the Cold War era that followed. The purpose of the military presence was to discourage potential aggression from the (now-defunct) Soviet Union against U.S. allies in Western Europe.

Tara Walton/Toronto Star via Getty Images

B Afghan women line up to vote in the country's first legitimate elections in decades. A large part of U.S. foreign aid in Afghanistan is used to promote democratic practices. Afghanistan became a major recipient of U.S. aid and military intervention only after the attacks of September 11, 2001 (masterminded by the Al Qaeda terrorist network, based in part in Afghanistan). Before 9/11, U.S. foreign aid to Afghanistan was much lower and U.S. military bases there were nonexistent.

Courtesy USAID

C A U.S. military base in Japan, one of 130 in the country, is surrounded by dense urban development. These bases are first and foremost a projection of U.S. power designed to counter any future aggression by China, North Korea, the former Soviet Union, South Korea, or by Japan itself.

Yoshikazu Tsuno/AFP/Getty Images

D A Kenyan police officer investigates a burning road block set up in the aftermath of disputed presidential elections in 2013. Democracy is fragile throughout much of sub-Saharan Africa, and elections are often plagued by violence. While U.S. financial support for democracy in sub-Saharan Africa is growing, U.S. foreign aid to the area is relatively small and historically the United States has done little to promote democracy in the region. The United States has few strategic or economic interests, and few military bases, in the region.

Till Muellenmeister/AFP/Getty Images

But they also argue that new revenues should be raised by taxing the very rich. These differences between the two major political parties show up repeatedly in the United States, as the government's ability to raise revenue via taxation and the ways it spends this revenue are central to almost every major political issue.

The United States and Canada Abroad

The United States and Canada have dramatically different roles in the global geopolitical order. The United States is recognized as the most powerful country in the world, with the world's largest economy and a military budget that year to year is larger than those of the next largest 12 or 13 countries combined. While the promotion of democracy and political freedoms is an official goal of U.S. foreign policy, in practice, the United States tends to focus its activities abroad more on its own economic and strategic military interests, which are often linked. This can be seen, for example, in the recent U.S. war in Iraq, which numerous planners of the war within the Bush Administration now acknowledge was motivated more by a desire to control access to Iraq's lucrative oil reserves than to bring democracy to Iraq (Figure 2.17).

U.S. economic and strategic interests are reflected in the global distribution of its military bases and its spending on aid to foreign governments. Three main concentrations of bases and spending, where the United States has for years had strong strategic and economic interests, can be seen on the map in Figure 2.17 in (1) Europe; (2) North Africa and Southwest Asia (especially Egypt, Israel, Iraq, Jordan), and Afghanistan and Pakistan in South Asia; and (3) island and peninsular East Asia. The first and third concentrations of bases and spending (those in Europe and East Asia) relate to strategic and economic interests dating from World War II that have remained relevant due to the Cold War (see Chapter 1, page 39), the subsequent collapse of the Soviet Union (see Chapter 5, pages 208–209), the continued large volume of trade between the United States and Europe, and the rapid growth of U.S. trade with East Asia since the opening up of China (see Chapter 9, page 384; see also Figure 2.17C). The second concentration relates directly to U.S. interests in Iraq, Iran, Afghanistan, Pakistan, Israel, and Egypt, and the oil and mineral resources of South and Southwest Asia in general.

The map in Figure 2.17 also shows that there are many places where the United States does not have many bases and where spending on foreign aid is at low or moderate levels. These are generally places where the United States has fewer strategic and economic interests. In part because of the poverty and political instability of sub-Saharan Africa, for instance, the United States has had few such interests in the region (see Figure 2.17D). (The same could be said of Haiti in the Caribbean; see Figure 2.17E). If U.S. policies to support democracy abroad were a strong factor in its allocation of assistance, one might expect that the focus of U.S. spending on military assistance and foreign aid would be in the parts of sub-Saharan Africa and elsewhere that have low levels of democratization. And yet the United States has few bases and spends little on military assistance and foreign aid in these parts of the world.

- 39. CONSEQUENCES REPORT
- 40. SECURITY AND PERSONAL LIBERTY REPORT
- 41. U.S. IMAGE REPORT

Canada's role on the world stage is dramatically different than that of the United States, with a more "live and let live" approach. While trade is a similarly strong motivator for Canada's foreign policies and foreign aid projects, Canada has far fewer strategic military interests abroad. Arguably, Canada's greatest geopolitical impact comes from its ability to influence the United States. As the largest trading partner of the United States, and one with considerable energy resources, Canada has helped to reduce U.S. dependence on foreign oil. This relationship was highlighted by the 9/11 attacks and the subsequent Iraq War. At that time (early 2000s), about 25 percent of the oil imported into the United States came from the Organization of the Petroleum Exporting Countries (OPEC), which was mostly made up of the countries along the Persian Gulf, where the terrorists had originated. Recent data show that 46 percent of oil imported into the United States comes from countries that are not part of OPEC. Of this 46 percent, Canada supplies 27.9 percent, Mexico supplies 10.8 percent, and 20 other countries supply the remaining 7.3 percent (Figure 2.18A).

Recent plans to extract fuel from the oil sands of western Canada relate to both Canadian and U.S. strategic interests in reducing dependence on foreign oil. The oil sands could potentially double the amount of oil that Canada exports to the United States and remove much of Canada's need to import oil. The main reason that Canada imports any oil at all is that most of its oil fields are in the west, far from Canadian population centers and closer to U.S. facilities that can refine the oil. Most of this oil stays in the United States, though some is sold back to the more heavily populated eastern provinces. The rest of the oil imported into Canada comes mostly from OPEC sources. This dependence makes many people in both Canada and the United States uncomfortable, especially given the often antagonistic relationships both countries have with the Persian Gulf states. Thus, despite the many environmental costs of developing Canada's oil sands, both the United States and Canada have strong strategic and economic interests in doing so.

Challenges to the United States' Global Power

A number of challenges to U.S. global power have emerged in recent decades. The wars in Iraq and Afghanistan exposed the limits of the United States' ability to bring about change in less powerful countries. Meanwhile, a new alliance is challenging the United States and its allies in Europe.

Immediately after the attacks on New York City and Washington, DC, on September 11, 2001 (9/11), the international community extended warm sympathy to the United States and generally supported then-president George W. Bush in his launching of the *War on Terror*, defined as a defense of the American way of life and of democratic principles. The first target was Afghanistan, which was then thought to be host to Osama bin Laden and the elusive Al Qaeda network that openly claimed credit for masterminding the 9/11 attacks. The aim was to capture bin Laden—accomplished in 2011, under President Barack Obama, when bin Laden was killed by U.S. Army Special Forces in Abbottabad, Pakistan—and remake Afghanistan into a stable ally, which still had not been accomplished by 2014.

Although NATO (North Atlantic Treaty Organization) forces (including Canadian troops) joined U.S. forces in Afghanistan, the war proved difficult to resolve because of heavy resistance from tribal leaders within the country and from militant Muslim insurgents in adjacent Pakistan. Some of the loss of momentum in Afghanistan can be attributed to the fact that in the spring of 2003, President Bush brought the War on Terror to Iraq, and U.S. strategic attention and troop support were diverted.

The wars in Iraq and Afghanistan eventually ended with few concrete gains, despite heavy losses of civilian life. U.S. combat troops withdrew from Iraq in 2011 and from Afghanistan in 2014, and both countries remain plagued by violence. In the case of Iraq, most of the lucrative contracts to develop the country's oil resources went to companies in countries that opposed the war, such as China. Meanwhile, international public opinion polls showed that the U.S. image abroad was badly damaged by what was seen as an unjust war aimed mainly at controlling the resources of foreign countries (see Figure 2.17A, B, and map).

Since 2001, an alliance of four countries has emerged as a symbol of growing resistance to the global power of the United States and its allies. Together, these countries—Brazil, Russia, India, and China (known collectively as BRIC)—account for over a quarter of the world's land area and more than 40 percent of its population, and by 2027 they will together form a larger group of economies than the United States and its major allies (Canada, France, Germany, Italy, Japan, and the United Kingdom), which are collectively known as the G7. Meeting at yearly summits, the BRIC countries are committed to establishing a "more multi-polar world order" in which the United States and its allies are less dominant. The BRIC countries are the most prominent of several groups of rapidly developing countries that are publicly challenging the current global geopolitical order.

RELATIONSHIPS BETWEEN CANADA AND THE UNITED STATES

Citizens of Canada and the United States share many characteristics and concerns. Indeed, in the minds of many people—especially those in the United States—the two countries are one. Yet that is hardly the case. Three key factors characterize the interaction between Canada and the United States: *asymmetries*, *similarities*, and *interdependencies*.

Asymmetries

Asymmetry means "lack of balance." Although the United States and Canada occupy about the same amount of space (Figure 2.19), much of Canada's territory is cold and sparsely inhabited. The U.S. population is about ten times the Canadian population. While Canada's economy is one of the largest and most productive in the world, producing U.S.$1.4 trillion (PPP) in goods and services in 2011, it is dwarfed by the U.S. economy, which is more than ten times larger, at $15.04 trillion (PPP) in 2011.

In international affairs, Canada quietly supports civil society efforts abroad, while the United States is an economic, military, and political superpower preoccupied with maintaining its role as a world leader. In framing foreign affairs policy, the United States regards Canada's position only as an afterthought, in part because the country is such a secure ally. But managing its relationship with the United States is a top foreign policy priority for Canada. As former Canadian Prime Minister Pierre Trudeau once told the U.S. Congress, "Living next to you is in some ways like sleeping with an elephant: No matter how friendly and even-tempered the beast, one is affected by every twitch and grunt."

Similarities

Notwithstanding the asymmetries, the United States and Canada have much in common. Both are former British colonies and have retained English as the dominant language. Both also experienced settlement and exploration by the French. From their common British colonial experience, they developed comparable democratic political traditions. Both are federations (of states or provinces), and both are representative democracies, with similar legal systems.

Not the least of the features they share is a 4200-mile (6720-kilometer) border, which until 2009 contained the longest sections of unfortified political boundary in the world. For years, the Canadian border had just 1000 U.S. guards, while the Mexican border, which is half as long, had nearly 10,000 agents. In 2009, the Obama administration decided to equalize surveillance of the two borders for national security reasons. Where some rural residents once passed in and out of Canada and the United States unobserved many times in the course of a routine day, now there are drone aircraft with night-vision cameras and cloud-piercing radar scanning the landscape for smugglers, illegal immigrants, and terrorists. Canadians and Americans, who formerly thought of their common border as nonexistent, now endure intrusive technology, barricades, and crossing delays. Tourism and commerce are also being negatively affected by these efforts to keep both countries safer. **42. U.S. BORDER SECURITY REPORT**

Canada and the United States share many other landscape similarities. Their cities and suburbs look much the same. The billboards that line their highways and freeways advertise the same brand names. Shopping malls and satellite business districts have followed suburbia into the countryside, encouraging comparable types of mass consumption and urban sprawl. The two countries also share similar patterns of ethnic diversity that developed in nearly identical stages of immigration from abroad.

Democratic Systems of Government: Shared Ideals, Different Trajectories

Canada and the United States have similar democratic systems of government, but there are differences in the way power is divided between the federal government and provincial or state governments. There are also differences in the way the division of power has changed since each country became independent.

Both countries have a federal government in which a union of states (United States) or provinces (Canada) recognizes the sovereignty of a central authority, while states/provinces and local governments retain many governing powers. In both Canada and the United States, the federal government has an elected executive branch, elected legislatures, and an appointed judiciary. In Canada, the executive branch is more closely bound to follow the will of the legislature. At the same time, the Canadian federal

88 CHAPTER 2 North America

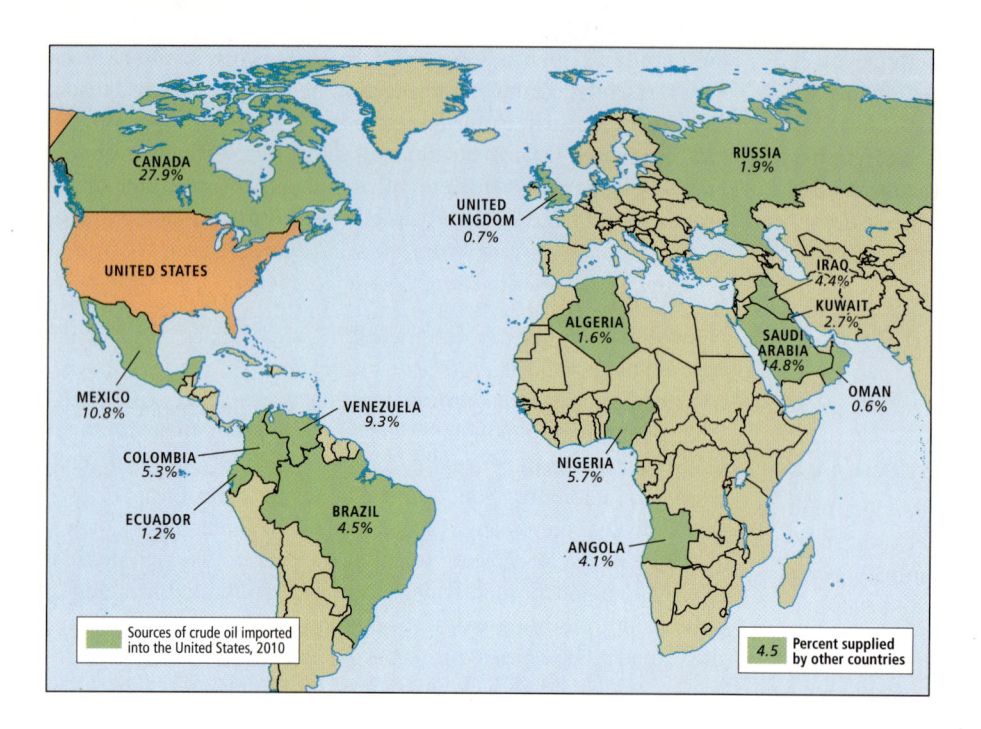

FIGURE 2.18 (A) Sources of average daily crude oil imports into the United States, 2012. The United States is dependent on crude oil from many locations around the world and imports nearly 60 percent of the crude it uses. This map shows the average percentage of daily crude (millions of barrels) imported into the United States from the top 15 exporting nations in September 2012. The United States produced the remaining 40 percent from its own sources. Canada's exports into the United States rose from 21 percent in September 2009 to nearly 28 percent by September 2010. [Sources consulted: "U.S. Imports by Country of Origin," U.S. Energy Information Administration, at http://www.eia.gov/dnav/pet/pet_move_impcus_a2_nus_epc0_im0_mbblpd_m.htm; "Supply and Disposition," U.S. Energy Information Administration, at http://www.eia.gov/dnav/pet/pet_sum_snd_d_nus_mbbl_m_cur.htm]

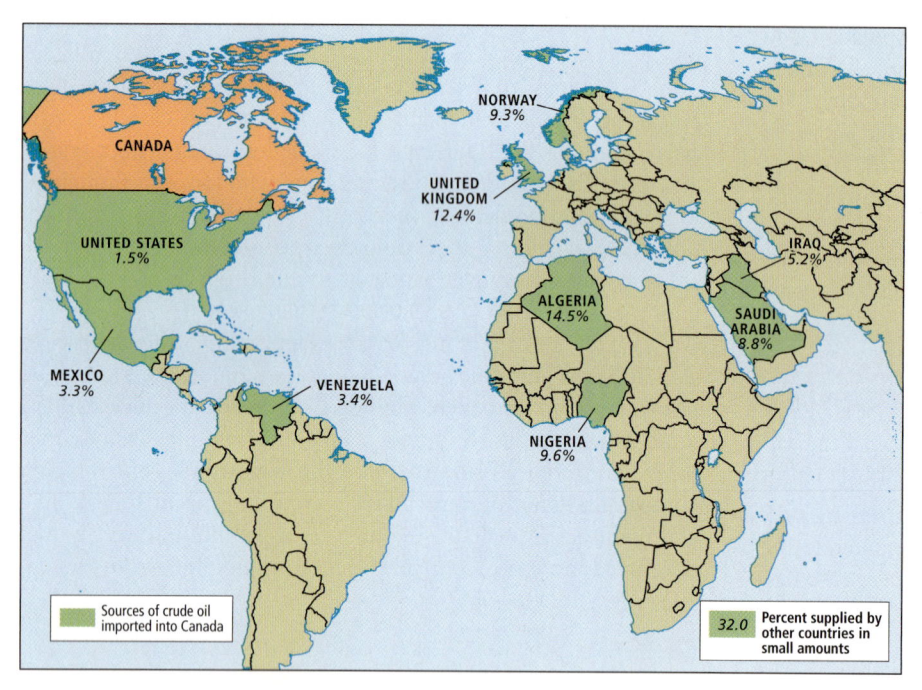

FIGURE 2.18 (B) Sources of average daily crude oil imports into Canada, 2010. In 2010, Canada produced 56.1 percent of the crude oil it used from its own resources, much of it from western Canada. More than half of the crude oil from western Canada was shipped to the United States; western Canada retained most of the rest. However, Canada also imported 43.9 percent of its crude oil, most of which went to its eastern provinces. [Sources consulted: *Statistical Handbook*, Canadian Association of Petroleum Producers, at http://www.capp.ca/library/statistics/handbook/Pages/default.aspx; *Statistical Handbook for Canada's Upstream Petroleum Industry*, Canadian Association of Petroleum Producers, November 2011, at http://www.capp.ca/GetDoc.aspx?DocId=184463&DT=NTV]

government has more and stronger powers (at least constitutionally) than does the U.S. federal government.

Over the years, both the Canadian and U.S. federal governments have moved away from the original intentions of their constitutions. Canada's initially strong federal government has become somewhat weaker, largely in response to demands by provinces, such as the French-speaking province of Québec (see the map in Figure 2.1), for more autonomy over local affairs.

The more limited federal government that the United States had at its outset has expanded its powers. The U.S. federal government's original source of power was its mandate to regulate trade between states. Over time, this mandate has been interpreted ever more broadly. Now the U.S. federal government has a powerful effect on life even at the local level, primarily through its ability to dispense federal tax monies via such means as grants for school systems, federally assisted housing, military bases, drug and food regulation and enforcement, urban renewal, and the rebuilding of interstate highways. Money for these programs is withheld if state and local governments do not conform to federal standards. This practice has made some poorer states dependent on the federal government. However, it has also encouraged some state and local governments to enact more enlightened laws than they might have done otherwise. For example, in the 1960s the federal government promoted civil rights for African American citizens by requiring states to end racial segregation in schools in order to receive federal support for their school systems.

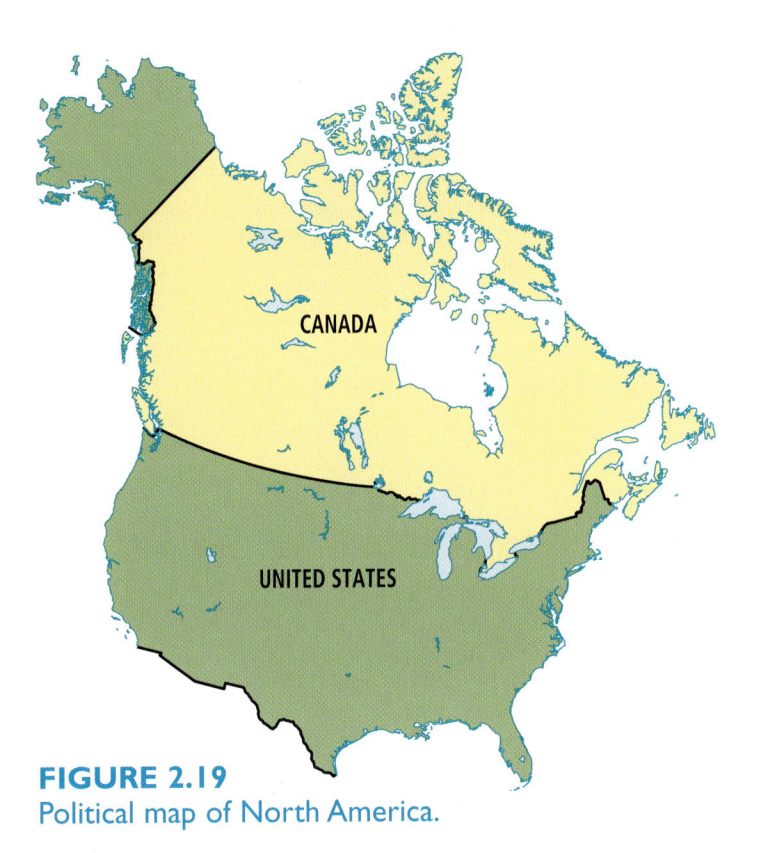

Relationships Between Canada and the United States   89

FIGURE 2.19
Political map of North America.

Two Health-Care Systems The contrasts between health-care systems in the two countries are indeed striking. Reformed in the 1970s, the Canadian system is heavily subsidized and covers 100 percent of the population. In the United States, health care has been largely private, relatively expensive, and tied to employment, which made changing jobs difficult for those with preexisting conditions. Also, many small businesses did not provide health insurance coverage for employees, who had to pay for it themselves or go without. A recent reform, dubbed "Obamacare" after President Obama, who pushed for the change, will cover most of the 47 million people who previously had no coverage; and government subsidies for care for the elderly, disabled, children, veterans, and the poor will continue.

The financial viability and medical outcomes of Obamacare are not yet known, but the most recent statistics show why change was sought. In 2012, the United States spent more per capita on health care than any other industrialized country—$8915 per capita, for a total of 17.2 percent of its GDP per capita. Canada spent $5948 per capita, or just 11.4 percent of its GDP per capita (GNI figures are not available)—yet Canada had better health outcomes, outranking

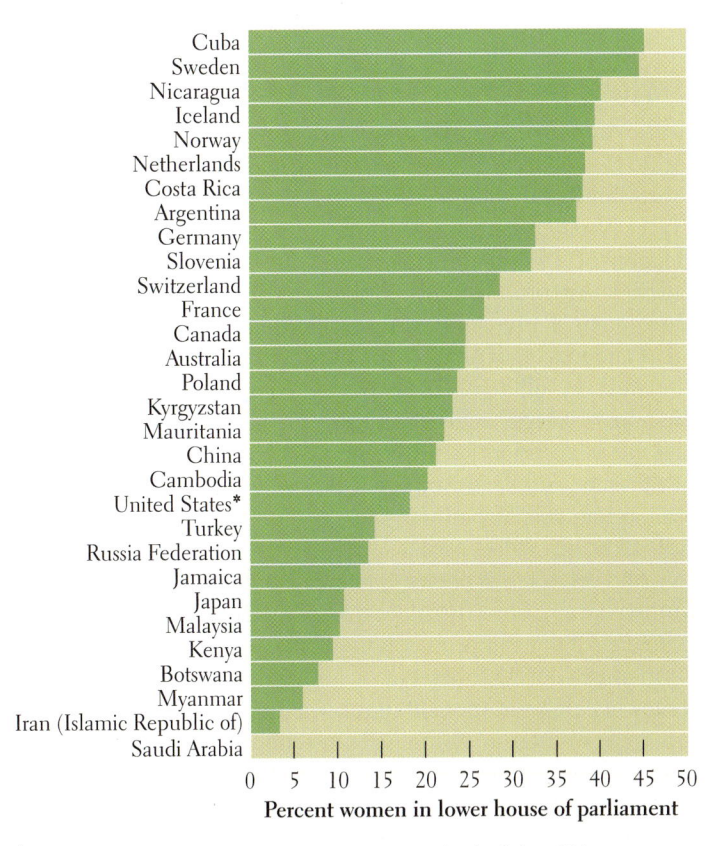

The Social Safety Net: Canadian and U.S. Approaches

The Canadian and U.S. governments have responded differently to workers who have been displaced by economic change. Ultimately, these differences derive from prevailing political positions and widely held notions each country has about what the government's role in society should be. In Canada there is broad political support for a robust **social safety net**, the services provided by the government—such as welfare, unemployment benefits, and health care—that prevent people from falling into extreme poverty. In the United States there is much less support for these programs and a great deal of contention over nearly all efforts to strengthen the U.S. social safety net.

social safety net the services provided by the government—such as welfare, unemployment benefits, and health care—that prevent people from falling into extreme poverty

For many decades, Canada has spent more per capita than the United States on social programs. These programs, especially unemployment benefits, have generally made the financial lives of working Canadians more secure. These policies also reflect the workings of Canada's democracy; voters have supported tax hikes to fund several major expansions of the social safety net during the twentieth century.

Compared to Canada, the United States provides much less of a social safety net. For years the prevailing political argument against the expansion of the U.S. social safety net has been that the system offers lower taxes for businesses, corporations, and the wealthy, which in turn invest their surpluses in expansions or in new businesses, thus creating jobs and new taxpayers. These benefits of low taxes are assumed to *trickle down* to those most in need. The fit of this position with the reality of what actually happens has often been challenged, especially during times of economic recession.

*Percentage for the entire U.S. Congress is used, which has 100 senators (20 are women) and 435 congresspeople (77 are women).

FIGURE 2.20 Percentage of women in the lower house of parliament (legislature) in selected countries, October 2012. [Sources consulted: Revised December 26, 2012, according to the results of 2012 U.S. election, at "Record Number of Women Will Serve in Congress; New Hampshire Elects Women to All Top Posts," CAWP, November 7, 2012, at http://www.cawp.rutgers.edu/press_room/news/documents/PressRelease_11-07-12.pdf; for the world results, "World Classification," Women in International Parliaments, Inter-Parliamentary Union, at http://www.ipu.org/wmn-e/classif.htm]

TABLE 2.1 Health-related indexes for Canada and the United States

Country	Health care cost as a percentage of GDP	Percentage of population with no insurance	Deaths per 1000†	Infant mortality per 1000 live births in 2009*	Maternal mortality per 100,000 live births in 2008‡	Life expectancy at birth (years) in 2011*	Annual health expenditures per capita (PPP U.S.$) in 2009
Canada	10.9	0	7	6	7	78.5	$5452§
United States	16.2	15.5	8	8	17	81	$8086**

Sources: *United Nations Development Program, *Human Development Report 2011* (New York: United Nations Development Programme), Tables 1, 9, and 10, at http://hdr.undp.org/en/reports/global/hdr2011/. Retrieved January 2012.
†Population Reference Bureau, *2011 World Population Data Sheet*, at http://www.prb.org/Publications/Datasheets/2011/world-population-data-sheet/data-sheet.aspx. Retrieved January 2012.
‡Medical News Today, "Maternal Mortality Rises in the USA, Canada and Denmark and Falls in China, Egypt, Ecuador and Bolivia," April 13, 2010, at http://www.medicalnewstoday.com/articles/185154.php.
§News Medical, "Health Care Spending in Canada Expected to Reach $183.1 Billion in 2009: CIHI," November 19, 2009, at http://www.news-medical.net/news/20091119/Health-care-spending-in-Canada-expected-to-reach-241831-billion-in-2009-CIHI.aspx.
**Centers for Medicare & Medicaid Services, "National Health Expenditure Data," at https://www.cms.gov/NationalHealthExpendData/25_NHE_Fact_Sheet.asp#TopOfPage. Retrieved July 2012.

the United States on most indicators of overall health, such as infant mortality, maternal mortality, and life expectancy (Table 2.1).

48. SICKO REPORT

Gender in National Politics

North America has some powerful political contradictions with regard to gender. While women voters are a potent political force, successful female politicians are not as numerous as one might expect. Women cast the deciding votes in the U.S. presidential election of 1996, voting overwhelmingly for Bill Clinton. In the 2006 interim elections, 55 percent of women voted for the Democratic Party candidates, registering strong anti-war sentiment and making it possible for Democrats to gain many seats in Congress. In 2008 and 2012, Barack Obama won 55 percent of the female vote. Deciding factors in the election included issues that are often priorities for women, such as reproductive rights, health care, family leave, equal pay, day care, and ending the wars in Iraq and Afghanistan.

49. WOMEN VOTING REPORT

And yet in terms of electing female political leaders to office, women have not had the success warranted by their powerful role as voters. Of the 535 people in the U.S. Congress as of 2014, only 99 members, or 18.5 percent, were women. Gender equity was somewhat closer at the state level, where 24.1 percent of legislators were women as of the 2012 elections. In Canada, women have had a little more success. As of 2014, Canadian women constituted 25 percent of the House of Commons in Parliament and held 24 percent of provincial legislative seats. Neither country compares well to the world at large in that, as Figure 2.20 shows, over the last decade other countries have added substantial female representation to legislatures.

Things are somewhat more equitable for Canadian women at the executive level. In 2011, Canada elected four women as provincial governors and another as executive of the Nunavut Territory. (There are ten provinces and three territories.) After the 2012 election, the United States had 7 out of a possible 50 female governors. Canada briefly had a female prime minister in 1993. While the United States has never had a female president, in 2008, Hillary Clinton was the first woman to have a serious chance at becoming the U.S. president when she competed with Barack Obama to become the Democratic Party candidate.

THINGS TO REMEMBER

> **GEOGRAPHIC INSIGHT 3**
• **Power and Politics** North America has relatively high levels of political freedom, though in recent decades many of its residents have become disillusioned with the political process for a variety of reasons. While Canada plays a relatively modest political role abroad, the United States has enormous influence on the global political order, although its status as the world's predominant "superpower" is increasingly being challenged.

• While an oft-stated ideal of U.S. official foreign policy (it is less of one for Canada) is to promote democracy abroad, the actual global distribution of U.S. spending on foreign aid and military assistance indicates that U.S. strategic and economic interests take precedence over promoting democracy.

• Canada continues to have a more robust social safety net, including health care, than does the United States.

• While women voters are a potent political force in the United States and Canada, the number of elected female politicians has not been commensurate with the percentage of women in the general population of each country. ■

URBANIZATION

> **GEOGRAPHIC INSIGHT 4**

Urbanization: A dramatic change in the spatial patterns of cities and suburbs has profoundly affected life in this very urbanized region. Since World War II, North America's urban populations have increased by about 150 percent, but the amount of land they occupy has increased by almost 300 percent. This is primarily because of suburbanization and urban sprawl, which are companion processes to urbanization.

Today, close to 80 percent of North Americans live in **metropolitan areas**—cities of 50,000 or more, plus their surrounding suburbs and towns. Most of these people live in car-dependent suburbs built since World War II, where urban life bears little resemblance to that in the central cities of the past.

In the nineteenth and early twentieth centuries, cities in Canada and the United States consisted of dense inner cores and less-dense urban peripheries that graded quickly into farmland. Starting in the early 1900s, central cities began losing population and investment, while urban peripheries—the **suburbs**—began growing (Figure 2.21). Workers were drawn by the opportunity to raise their families in single-family homes in secure and pleasant surroundings on lots large enough for vegetable gardens and recreation. Most continued to work in the city, traveling to and from on streetcars.

After World War II, suburban growth dramatically accelerated as cars became affordable to more people (Figure 2.22). In the United States, suburban growth was also encouraged by the federally funded Interstate Highway System, a network of free high-speed roads that passed through most cities, providing easy access to surrounding land along the interstate. Canada, with a less extensive interstate highway system and slightly fewer people who could afford cars, had less suburban growth.

With the growing popularity of the family car, interest in publicly funded, non-car-based forms of transportation for these spreading urban areas dwindled. This process was hastened by a consortium of automobile manufacturers, tire companies, and oil companies that purchased, dismantled, and replaced many urban streetcars and light-rail trains with bus systems.

As North American suburbs grew and spread out, nearby cities often coalesced into a single urban mass. The term **megalopolis** was originally coined to describe the 500-mile (800-kilometer) band of urbanization stretching from Boston through New York City, Philadelphia, and Baltimore, to south of Washington, DC. Other megalopolis formations in North America include the San Francisco Bay Area, Los Angeles and its environs, the region around Chicago, and the stretch of urban development from Seattle–Tacoma to Vancouver, British Columbia.

This pattern of *urban sprawl* requires residents to drive automobiles to complete most daily activities such as grocery shopping and commuting to work. Two major side effects of the dependence on vehicles that comes with urban sprawl are air pollution and emission of the greenhouse gases that contribute to climate change. Another important environmental consequence is the habitat loss that results when suburban development expands into farmland, forests, grasslands, and deserts.

Farmland and Urban Sprawl Urban sprawl drives farmers from land that is located close to urban areas, because farmland on the urban fringe is very attractive to real estate developers. The land is cheap compared to urban land, and it is easy to build roads and houses on since farms are generally flat and already cleared. As farmland is turned into suburban housing, property taxes go up and surrounding farmers who can no longer afford to keep their land sell it to housing developers. In North America each year, 2 million acres of agricultural and forest lands make way for urban sprawl.

Advocates of farmland preservation argue that beyond food and fiber, farms also provide economic diversity, soul-soothing scenery, and habitat for some wildlife. For example, the town of Pittsford, New York (a suburb of Rochester), decided that farms were a positive influence on the community. Mark Greene's 400-acre, 200-year-old farm lay at the edge of town. As the population grew, the chances of the farm remaining in business for another generation looked dim. Land prices and property taxes rose, and the Greene family could not meet their tax payments. Residents in the new suburban homes sprouting up on what had been neighboring farms pushed local officials to halt normal farm practices, such as noisy nighttime harvesting or planting, spreading smelly manure, and importing bees to pollinate fruit trees. Pittsford, however, decided to stand by the farmers by issuing $410 million in bonds so that it could pay Greene and six other farmers for promises that they would not sell their 1200 acres to developers, but would instead continue to farm them.

Smart Growth and Livability The term *smart growth* has been coined for a range of policies aimed at stopping sprawl by making existing urban areas more "livable." *Livability* relates to factors that lead to a higher quality of life in urban settings, such as safety, good schools, affordable housing, quality health care, numerous and well-maintained parks that offer recreational opportunities for people and their pets (Figure 2.23), and well-developed public transportation systems. Some indexes of livability for major cities worldwide are put out every year by *the Economist* magazine and the Mercer Quality of Living Survey.

In smart growth planning, environmental benefits are envisioned as well. The focus on life lived locally, where mass transit and walking replace the family car and where gardening for food is a community entertainment, will result in lower energy use, air

> **metropolitan areas** cities of 50,000 or more and their surrounding suburbs and towns
>
> **suburbs** populated areas along the peripheries of cities
>
> **megalopolis** an area formed when several cities expand so that their edges meet and coalesce

VIGNETTE

For Canada's *First Nations people* (the preferred term in Canada for Native Americans), the chance to indulge in the extravagance of urban sprawl is viewed positively. The Tsawwassen First Nation is one of 630 native groups that are negotiating with the Canadian government to gain full control over their ancestral territories. The land they recently received by a treaty lies 20 miles south of Vancouver, near the Strait of St. George. Eager to finally enjoy some of the fruits of development, they are planning a 2-million-square-foot "big box" and indoor shopping mall in the midst of what has been rural farmland. The surrounding communities, which had previously been able to shield their village-like settlements against urban sprawl, feel that there are enough shopping opportunities already. The Tsawwassen see the mall as an overdue chance for the prosperity that their neighbors already enjoy. ∎

FIGURE 2.21   PHOTO ESSAY: Urbanization in North America

North America is very urbanized, with 79 percent of the population living in cities. Some of the wealthiest cities in the world are located here, though only a few have high levels of "livability." Many cities, especially those in the United States, are characterized by sprawling development patterns that make people dependent on their automobiles. [Sources consulted: *2011 World Population Data Sheet*, Population Reference Bureau, at http://www.prb.org/pdf11/2011population-data-sheet_eng.pdf; *World Gazeteer*, at http://world-gazeteer.com/wg.php?x=&men=gcis&lng=en&des=wg&srt=npan&col=abcdefghinoq&msz=1500&pt=a&va=&srt=pnan]

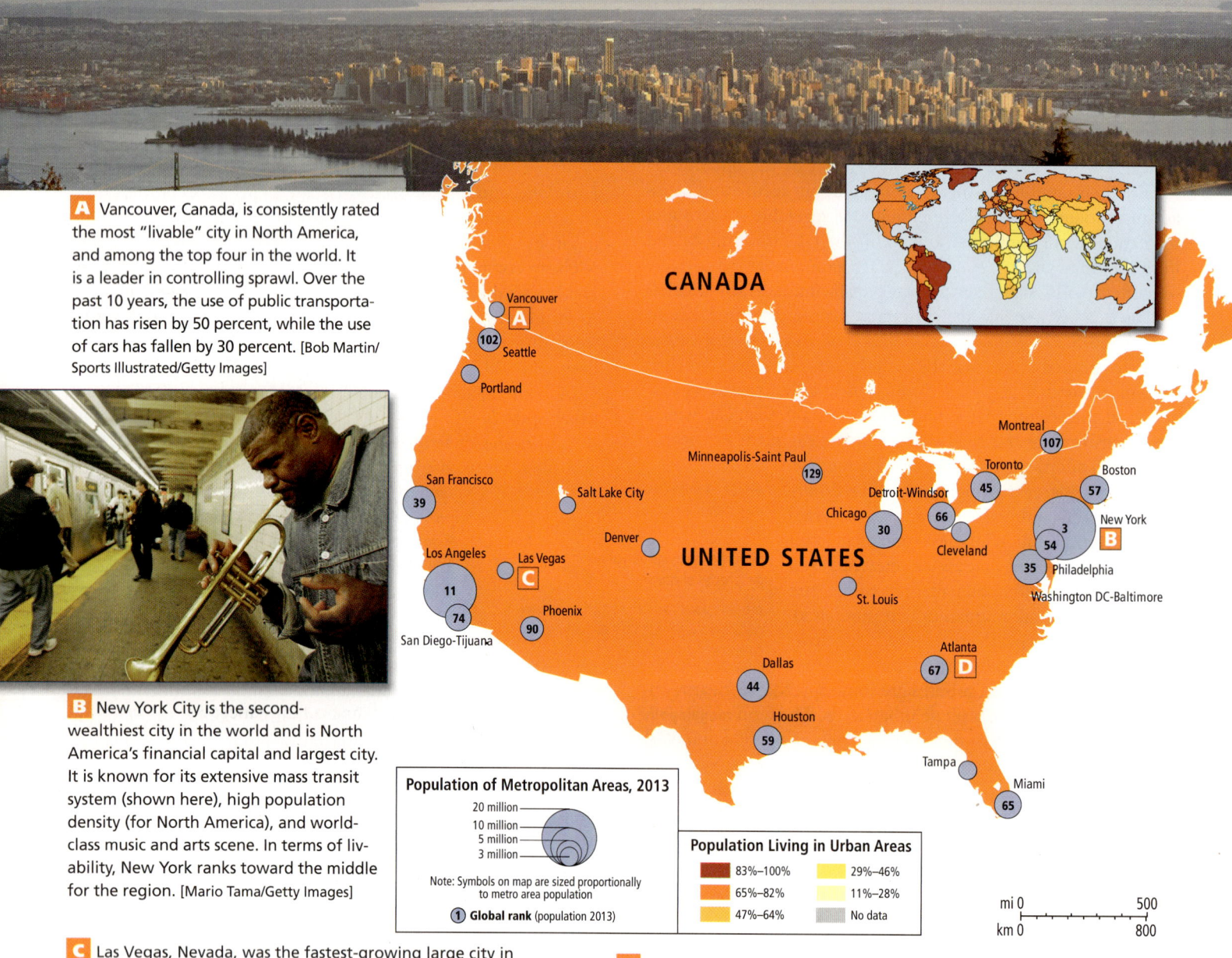

A Vancouver, Canada, is consistently rated the most "livable" city in North America, and among the top four in the world. It is a leader in controlling sprawl. Over the past 10 years, the use of public transportation has risen by 50 percent, while the use of cars has fallen by 30 percent. [Bob Martin/Sports Illustrated/Getty Images]

B New York City is the second-wealthiest city in the world and is North America's financial capital and largest city. It is known for its extensive mass transit system (shown here), high population density (for North America), and world-class music and arts scene. In terms of livability, New York ranks toward the middle for the region. [Mario Tama/Getty Images]

C Las Vegas, Nevada, was the fastest-growing large city in North America between 2000 and 2007, with 31.8 percent growth. However, the recession that began in 2007 led to population decline by 2010 as jobs in tourism shrank. For many, leaving was complicated by the inability to sell homes that had lost value. Las Vegas generally ranks toward the lower end of livability for North America.

D Morning traffic in Atlanta, Georgia, which added almost a million people between 2000 and 2007, more than any other city in North America. Atlanta is among the least dense and most sprawling cities in the region and depends almost entirely on cars for transportation. Atlanta is generally in the lower middle rankings of livability for North America.

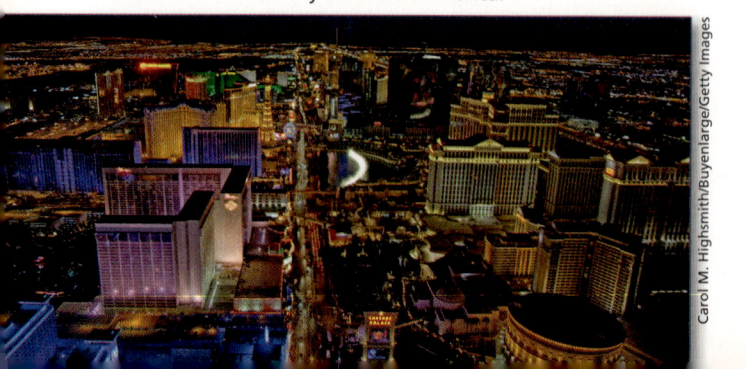

Population of Metropolitan Areas, 2013
- 20 million
- 10 million
- 5 million
- 3 million

Note: Symbols on map are sized proportionally to metro area population

① **Global rank** (population 2013)

Population Living in Urban Areas
- 83%–100%
- 65%–82%
- 47%–64%
- 29%–46%
- 11%–28%
- No data

Cities (global rank): Vancouver 102, Seattle, Portland, San Francisco 39, Los Angeles 11, San Diego-Tijuana 74, Las Vegas, Salt Lake City, Denver, Phoenix 90, Minneapolis-Saint Paul 129, Dallas 44, Houston 59, Chicago 30, St. Louis, Detroit-Windsor 66, Cleveland, Toronto 45, Montreal 107, Boston 57, New York 3, Philadelphia 54, Washington DC-Baltimore 35, Atlanta 67, Tampa, Miami 65

pollution, and CO_2 emissions. While smart growth and the shift toward livability are just starting to take hold throughout North America, there are a few cities, such as Vancouver, British Columbia (see Figure 2.21A), and Portland, Oregon, where these principles are much further along in implementation.

Some older cities are enjoying a renaissance as people move back to them in search of greater livability. Places like New York City are arguably the furthest along in crucial aspects of smart growth, given their high density and widespread use of mass transit (see Figure 2.21B). However, the high cost of living in these places can reduce their livability.

It may be a while before smart growth and livability make much of an impact in some of North America's fastest-growing large cities, such as Dallas, Atlanta, and Phoenix, where in recent years vast, car-dependent suburbs have spread out over enormous areas. Not surprisingly, none of these cities ranks high on livability indexes. Car dependence and rapid growth have brought Atlanta massive traffic jams and some of the worst air quality in the United States (see Figure 2.21D). Livability in the Las Vegas metro area is reduced by poor air and water quality as well as high crime rates (see Figure 2.21C).

Inner-city decay is another impact of sprawl on livability. This is especially true in the United States, where many inner cities are dotted with large tracts of abandoned former industrial land and neighborhoods debilitated by persistent poverty and loss of jobs. Old industrial sites that once held factories or rail yards are called **brownfields**. Because they are often contaminated with chemicals and covered with obsolete structures, blacktop, and concrete, they can be very expensive to redevelop for other uses. Also left behind in the inner cities are the least-skilled and least-educated citizens, many of whom were drawn in generations ago by the promise of jobs that have since moved out to the suburbs or overseas. Often the majority population in these inner cities is a mixture of African Americans, Asians, Latinos, and new immigrants. Some are relatively affluent and are leading efforts to renew old city centers. Others are in great need of the very services—health care, schools, and social support (including churches, synagogues, and mosques)—that have moved to the suburbs.

The new emphasis on smart growth and livability has also reinforced the **gentrification** of old, urban residential districts. As affluent people invest substantial sums of money in renovating old houses and apartments, poor inner-city residents can be displaced in the process. The effect of gentrification on the displaced poor appears to be somewhat less harsh in Canada than in the United States, primarily because Canada's stronger social safety net better ensures social services, housing, and help with housing maintenance. Some U.S. cities

> **brownfields** old industrial sites whose degraded conditions pose obstacles to redevelopment
>
> **gentrification** the renovation of old urban districts by affluent investment, a process that often displaces poorer residents

ON THE BRIGHT SIDE

The Greening of Detroit

Detroit, once the center of American auto industries, began to decline in the 1970s after a series of poorly made gas-guzzling automobiles tarnished the reputation of the car makers that had once provided the basis of the economy for the "Motor City." Faced with stiff competition from car makers in Japan, companies like Ford and General Motors cut thousands of jobs and the city's population shrank.

Forty years later, the depressing decay of a now much smaller Detroit is beginning to sprout with greenery and flowers as inner-city dwellers discover the joys of gardening on urban plots once occupied by houses that have since burned down. These green oases draw neighbors together and reconnect people with natural cycles. Children learn that food is something one can grow, not just buy.

FIGURE 2.22 Urban sprawl in Phoenix, Arizona.

(A) Phoenix grew rapidly between 1950 and 2000, resulting in a demand for housing. The photo is of Sun City, a new, car-oriented suburb designed for senior citizens, located about 30 miles from Phoenix's center.

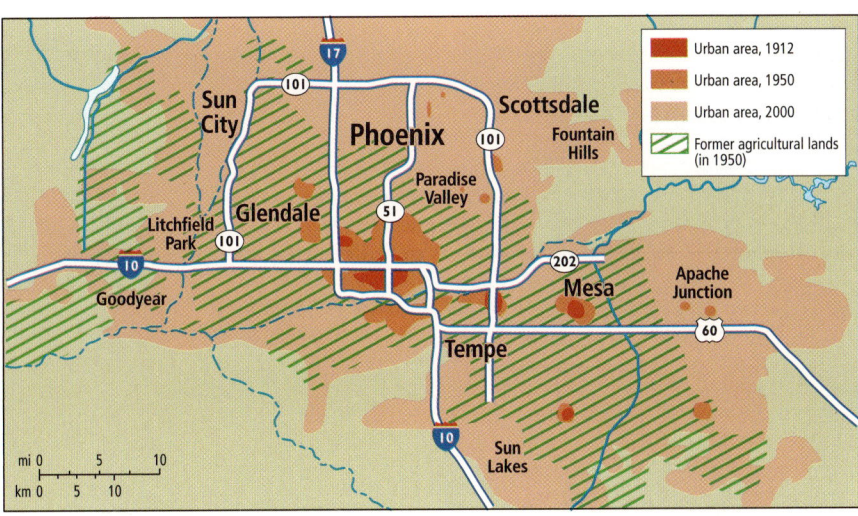

(B) This map shows the original urban settlement and how it has grown from 1912 to the present. A preference for single-family homes means that the city is sprawling into the surrounding desert rather than expanding vertically into high-rise apartments.

FIGURE 2.23 LOCAL LIVES
People and Animals in North America

A Labrador retrievers are the most popular breed of dog in North America. Fishers on the island of Newfoundland, part of Canada's province of Newfoundland and Labrador, bred these dogs to retrieve fishing nets, which accounts for the breed's webbed paws and love of water. [Education Images/UIG via Getty Images]

B The ancestors of the Maine coon cat, one of the largest of the popular domestic cat breeds, were developed to control rodent populations aboard ships trading along the shore of Maine and other parts of the Atlantic coast of North America. [Kasia Wandycz/Paris Match via Getty Images]

C Alaskan Malamutes pull a sled, which is the job that they were bred for by the Kuuvangmiut (formerly known as the Mahlemut) Inuit tribe of northwestern Alaska. Their large size and heavy build makes them ideal for pulling heavily loaded sleds. [Yuri Yuriev/AFP/Getty Images]

FIGURE 2.24 New Urbanism. Celebration, Florida, built in the 1990s, was designed to resemble older cities such as Charleston, South Carolina; New Orleans, Louisiana; Alexandria, Virginia; and Vicksburg, Mississippi.

THINGS TO REMEMBER

> **GEOGRAPHIC INSIGHT 4**

- **Urbanization** A dramatic change in the spatial patterns of cities and suburbs has profoundly affected life in this very urbanized region. Since World War II, North America's urban populations have increased by about 150 percent, but the amount of land they occupy has increased by almost 300 percent. This is primarily because of suburbanization and urban sprawl, which are companion processes to urbanization.

- After World War II, suburban growth dramatically accelerated as cars became affordable to more people.

- Urban sprawl drives farmers from land that is located close to urban areas, because farmland on the urban fringe is very attractive to real estate developers.

- The term *smart growth* has been coined for a range of policies aimed at stopping sprawl by making existing urban areas more "livable."

- One of the major impacts of urban sprawl, especially in the United States, is inner-city decay. Many inner cities are dotted with large tracts of abandoned former industrial land and have neighborhoods that are debilitated by persistent poverty and job loss. ■

POPULATION AND GENDER

> **GEOGRAPHIC INSIGHT 5**

Population and Gender: Women's participation in North America's economy is beginning to rival that of men, contributing to more than two centuries of declining fertility rates as women delay childbearing to pursue education and careers. Declining childbirth rates play a major role in the aging of North American populations, which may slow economic growth.

(such as Knoxville, Tennessee; Portland, Oregon; and Charlotte, North Carolina) have initiated *New Urbanism* projects to rehouse inner-city residents in pleasant, newly built, walkable urban neighborhoods with conveniently located services (Figure 2.24).

Gender and Fertility

North America is well along the demographic transition, with the number of children the average woman will have in her lifetime (also known as the *fertility rate*) declining since the early 1800s. This has happened in response to a number of factors: increasing economic development, improved health care, urbanization, and more women participating in the workforce.

All of these factors are connected in how they influence fertility. For example, social scientists have documented that increasing economic development is associated with decreased fertility. It is also usually associated with more women participating in the workforce, higher rates of urbanization, and improved health care. Similarly, urbanization is also associated with more women in the workforce, as families that move to the city are more likely to need the cash income that an adult woman who works outside the home can provide. This in turn is associated with lower fertility because women who have careers tend to delay childbearing. With regard to health care, improvements in the quality of health care and better access to medical professionals and medicines tend to lower the fertility rate as families choose to have fewer children because more will survive into adulthood. This choice is made easier with better access to birth control. All of these factors have come together to influence both gender and fertility in North America.

By the early 1800s, small numbers of North American women were starting to work in urban factories, where they had better access to health care. As more women chose this path, the fertility rate fell steadily from a high of 7 births per woman in 1800 to 3 births per woman in the 1920s. After the rapid rise in fertility following World War II, there was a sharp fertility decline as North American women worked more and gained access to birth control via "the pill," which became widespread in the 1960s. Since a low in the 1970s, North American fertility rates have remained more or less flat, at a level just below replacement level, contributing to an overall aging of populations.

Aging in North America

During the twentieth century, the number of older North Americans grew rapidly. In 1900, one in 25 individuals was over the age of 65; by 2010, the number was 1 in 8. By 2050, when most of the current readers of this book will be over 50, it is likely that 1 in 5 North Americans will be elderly.

What Causes a Population to Age? This aging of the region's population relates both to longer life expectancy, which increases the number of older people, and to a declining fertility rate, which decreases the number of younger people (Figure 2.25). While life expectancy increased steadily throughout the twentieth century, fertility rates declined significantly after the 1960s, when more women chose to obtain more education and pursue careers, resulting in their having fewer children.

The number of North Americans over the age of 65 is already high by global standards and will increase dramatically over the next 20 years. This will result from the combination of the marked jump in birth rate that took place after World War II during the *baby boom* of 1947 to 1964, and the ensuing drop in birth rate that occurred after 1965, as women had fewer children. The baby boomers constitute the largest age group in North America.

Dilemmas of Aging Aging populations in developed countries like Canada and the United States present us with unfamiliar and as-yet unresolved dilemmas. On the one hand, it is widely agreed

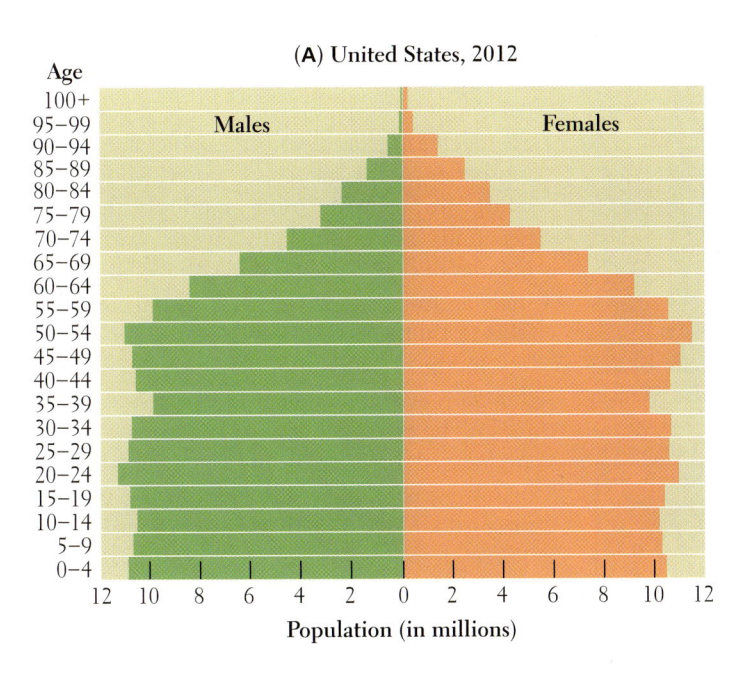

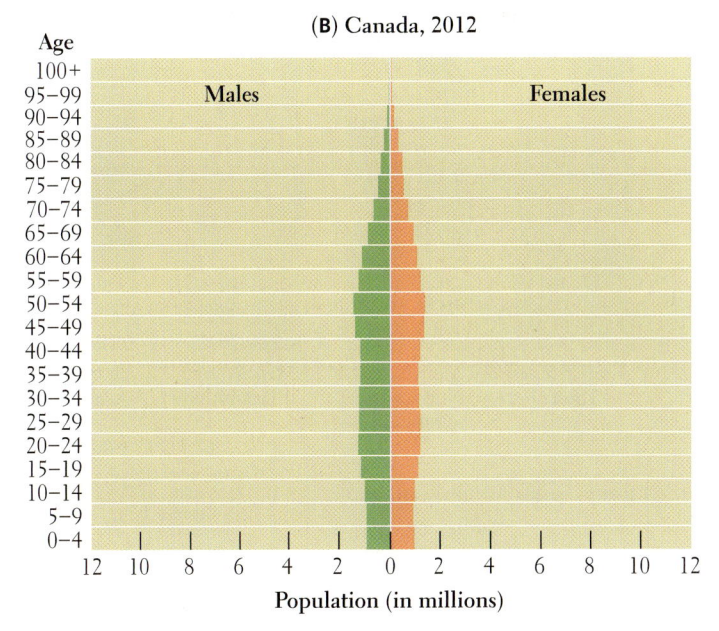

FIGURE 2.25 Population pyramids for the United States and Canada, 2012. The "baby boomers," born between 1947 and 1964, constitute the largest age group in North America, as indicated by the wider middle portion of these population pyramids. [Source consulted: International Data Base, U.S. Census Bureau, at http://www.census.gov/population/international/data/idb/infromationGateway.php]

that globally, population growth should be reduced to lessen the environmental impact of human life on Earth, especially in those societies (such as in North America) that consume the most. On the other hand, slower population growth means that there will be fewer working-age people to keep the economy going and to provide the financial and physical help the increasing number of elderly people will require.

Given that the population is aging, should retirement ages be extended from 65 to perhaps 75? Might this deprive young people of access to jobs? These worries have greatly preoccupied Europeans, for whom retirement ages can be as low as 58 and whose populations are aging faster than those of North America. Economists who have studied this issue say that it clearly makes sense to extend the retirement age. First, this means those over 65 will remain self-supporting longer; second, the talents and experience of these people will remain in circulation; third, statistics across the developed countries attest to the fact that high elderly employment rates are associated with high youth *employment*, not unemployment. Working people of any age spend their incomes and create jobs by so doing.

Eventually most elderly people will need special attention of some sort. To avoid the social isolation that so many elderly are already encountering, it will be necessary to develop affordable alternative living arrangements. Figure 2.26 shows how the elderly of various ethnic groups in the United States are living now. Notice how elderly women in all groups are the ones most likely to live alone. Although North Americans of all ages are choosing to live alone more than they did in the past, this can be problematic for the elderly as physical frailty increases and incomes shrink. Day care and cohousing for the elderly, where residents look after each other with the aid of a small staff, are two affordable strategies to provide care and companionship.

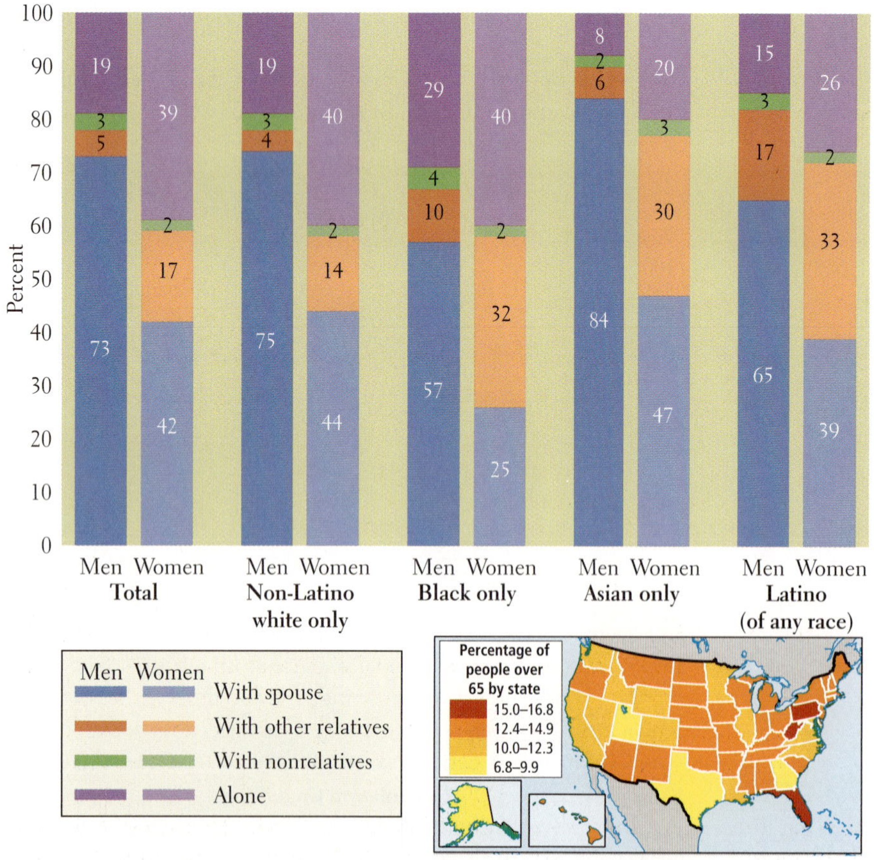

Population Distribution

The population map of North America (Figure 2.27) shows the uneven distribution of the more than 352 million people who live here. Canadians make up just under one-tenth (35.3 million) of North America's population. They live primarily in southeastern Canada, close to the border with the United States. The population of the United States is over 317 million, with many people moving from the old economic core into other regions of the country that are now growing much faster. Just how this trend will play out, however, is uncertain. The U.S. Census Bureau predicts that the Northeast and Middle West will grow more slowly than the South and West. Canada's national statistics agency predicts a similar pattern of western and southward movement of population.

The Geography of Population Change in North America

Every year, almost one-fifth of the U.S. population and two-fifths of Canada's population relocate. Some people are changing jobs and moving to cities; others are attending school, retiring to a warmer climate or a smaller city or town; and others are merely moving across town or to the suburbs or the countryside. Still other people are arriving from outside the region as immigrants.

In many farm towns and rural areas in the *Middle West*, or *Midwest* (the large central farming region of North America), populations are shrinking. As family farms are consolidated under corporate ownership, labor needs are decreasing and young people are choosing better-paying careers in cities. Midwestern cities are growing only modestly but are becoming more ethnically diverse, with rising populations of Latinos and Asians in places such as Indianapolis, St. Louis, and Chicago.

52. LATINO POLITICAL POWER

In the western mountainous interior, traditionally settlement has been light (see Figure 2.27). The principal reasons for this low density are rugged topography and lack of rain, and in northern or high-altitude zones, a growing season that is too short to sustain agriculture. There are some population clusters in irrigated agricultural areas, such as in the Utah Valley, near rich mineral deposits or resort areas. The gambling economy and frenetic construction activity generated by real estate speculation account for several knots of dense population at the southern end of the region. Until the recession beginning in 2007, Las Vegas,

FIGURE 2.26 Living arrangements of U.S. people age 65 and over by sex, race, and ethnicity, 2007.

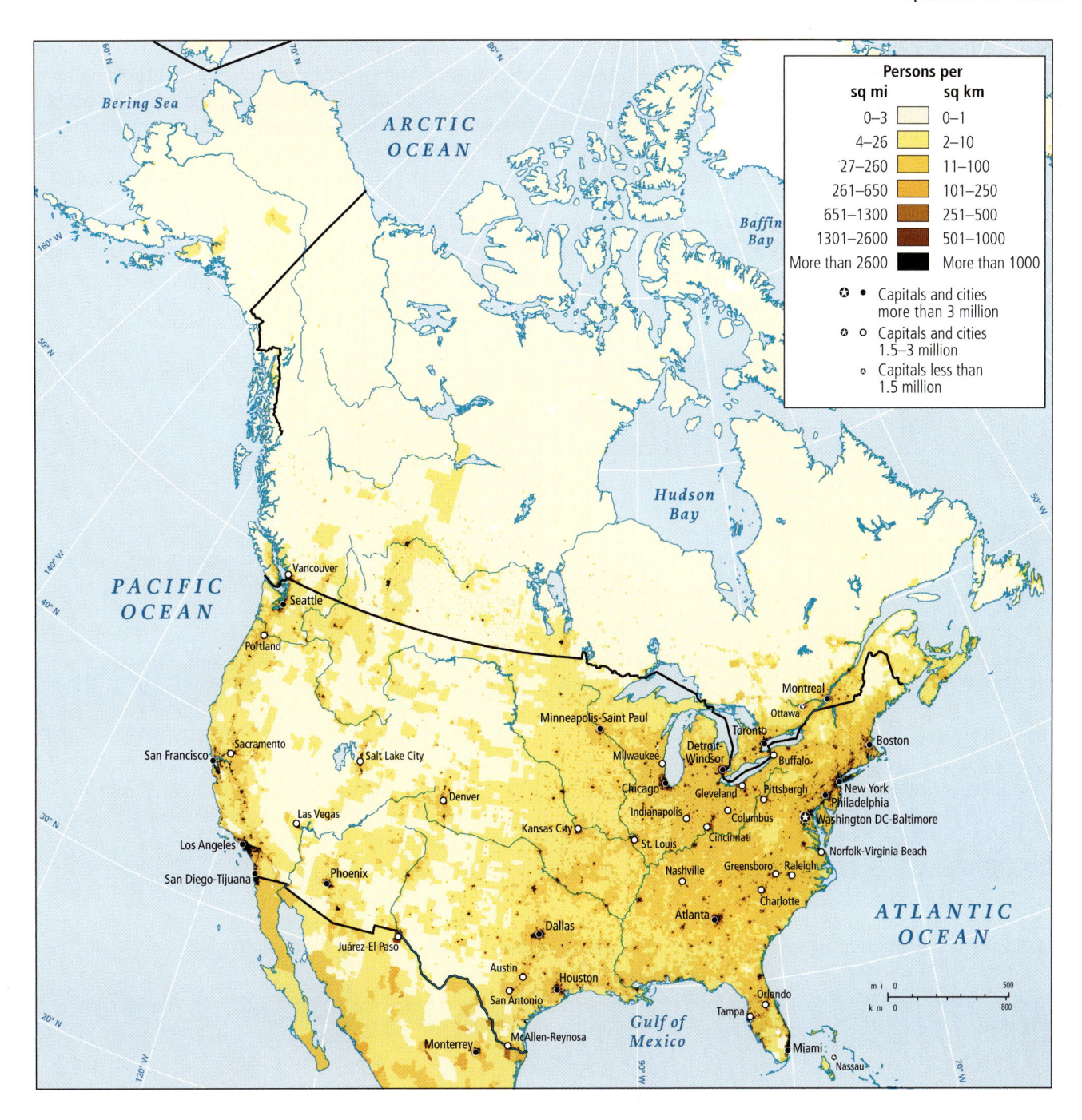

Population and Gender 97

FIGURE 2.27 Population density in North America.

Nevada, was the fastest-growing large city in the United States (see Figure 2.21C). But by mid-2009, approximately 67,000 homes in Las Vegas were in foreclosure, tourist arrivals were sharply down, and hotel construction projects were abruptly halted, with crane jibs left dangling in the air. In 2013, the situation had improved only slightly, with Nevada still having the second-highest rate of homes in foreclosure in the entire United States (Florida was first).

Along the Pacific coast, a band of growing population centers stretches north from San Diego to Vancouver (see Figure 2.21A) and includes Los Angeles, San Francisco, Portland, and Seattle. These are all port cities engaged in trade around the **Pacific Rim** (all the countries that border the Pacific Ocean). Over the past several decades, these North American cities have become centers of technological innovation.

Pacific Rim a term that refers to all the countries that border the Pacific Ocean

The rate of natural increase in North America (0.5 percent per year) is low, less than half the rate of the rest of the Americas (1.2 percent). Still, North Americans are adding to their numbers fast enough through births and immigration that the population could reach more than 422 million by 2050.

THINGS TO REMEMBER

> **GEOGRAPHIC INSIGHT 5**

- **Population and Gender** Women's participation in North America's economy is beginning to rival that of men, contributing to more than two centuries of declining fertility rates as women delay childbearing to pursue education and careers. Declining childbirth rates play a major role in the aging of North American populations, which may slow economic growth.

- North America's population is rapidly aging, and by 2050, one in five people will be over the age of 65. There will be fewer young people to work, pay taxes, and take care of the elderly.

- The population of North America is changing in distribution and becoming more diverse.

- North Americans are highly mobile. Every year, for a variety of reasons, almost one-fifth of the U.S. population and two-fifths of the Canadian population move to a different location.

SOCIOCULTURAL ISSUES

North America is increasingly diverse, with new immigrants from across the globe adding to a dominant culture that has European roots.

Immigration and Diversity

Immigration has played a central role in populating both the United States and Canada. Most people in North America descend from European immigrants, but there are many who have roots in Africa, Asia, Middle and South America, and Oceania. New waves of migration from Middle and South America and parts of Asia promise to make North America a region where people of non-European or mixed descent will become the majority (Figure 2.28). Most major North American cities are already characterized by ethnic diversity, and in some, recent immigration has led to near majorities of foreign-born residents. Houston, Texas, for example, was the most ethnically diverse metropolitan area in the United States in 2010, with 40 percent of the population Caucasian, 36.7 percent Latino, 16.8 percent African American, and 6.5 percent Asian. 50. IMMIGRATION AND POPULATION REPORT

In the United States, the spatial pattern of immigration is also changing. For decades, immigrants settled mainly in coastal or border states such as New York, Florida, Texas, or California. However, since about 1990, immigrants have been increasingly settling in interior states such as Illinois, Colorado, Nevada, and Utah (see Figure 2.28). The influence of immigrants can also be seen in many aspects of life in North America, including the popularity of ethnic cuisines (Figure 2.29).

> **push factors** factors that get people to consider the drastic move of leaving family and friends and a familiar place to strike out into the unknown, with what are usually unknown resources

Why Do People Decide to Immigrate to North America?
The decision to leave one's homeland is usually not an easy one. There are almost always **push factors** that cause people to consider the drastic move of leaving family and friends and familiar places to strike out into the unknown, with what are usually limited resources. These push factors can be civil or political unrest, or some kind of discrimination, but most often the lack of economic opportunity and the curbs on upward mobility that cause people to overcome the natural resistance to leaving home. Recently, the forces of globalization have often been the root causes of these push factors. For example, when NAFTA

FIGURE 2.28 Percent of total foreign-born people within each state in (A) 2000 and (B) 2010. [Sources consulted: "Percent of People Who Are Foreign Born—United States—Places by State; for Puerto Rico Universe: Total Population, 2010 American Community Survey 1-Year Estimates," American FactFinder, U.S. Census Bureau, at http://factfinder2.census.gov/faces/tableservices/jsf/pages/productview.xhtml?pid=ACS_10_1YR_GCT0501.US13PR&prodType=table]

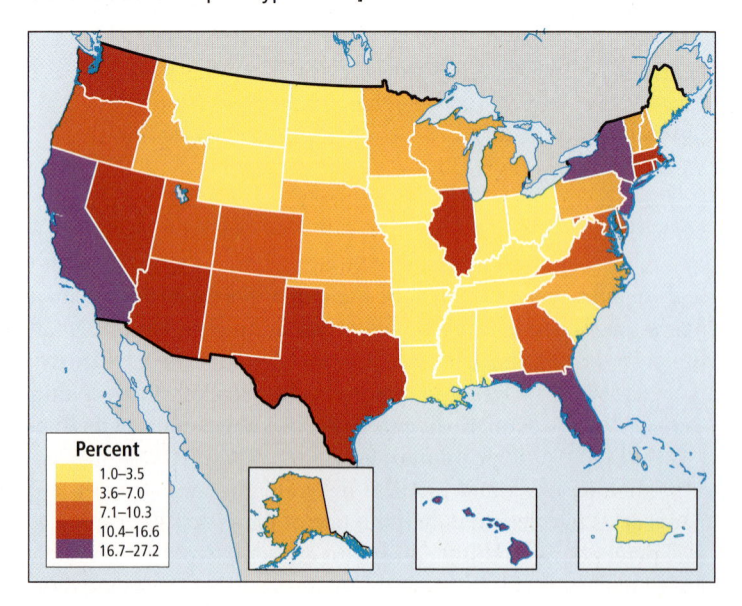

(A) Percentage of foreign-born of total population within each state in 2000.

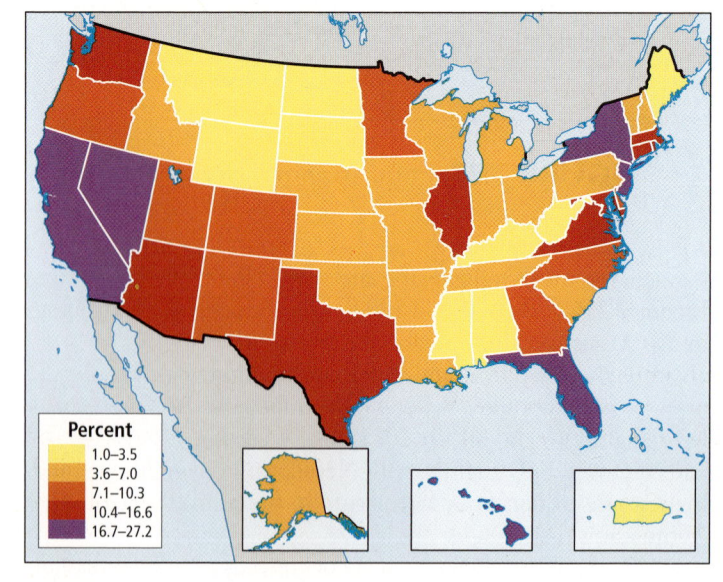

(B) Percentage of foreign-born of total population within each state in 2010.

Sociocultural Issues **99**

FIGURE 2.29 LOCAL LIVES
Foodways in North America

A Pork ribs are smoked and slow cooked in the "pit" at Spoon's Barbecue in Charlotte, North Carolina. Originally, Native Americans would first dig a pit or trench and fill it with burning coals or hot rocks. They would place meat on top and either slow cook it in the open at a low temperature for several hours or cover the entire pit and let it cook for a day or more. Most barbeque is now cooked in a metal smoker, still called a *pit*, which is often attached to a trailer so that it can be brought to special events. [Scott Olson/Getty Images]

B Poutine, a French Canadian dish, now hugely popular throughout Canada, consists of French fries served with cheese curds (a tasty by-product of cheese making), covered in gravy. One story holds that a take-out customer of Fernand LaChance's restaurant in Warwick, Québec, in the 1950s, asked for the combination, to which the owner responded "Ça va faire une maudite poutine!" ("That's going to make a cursed mess!"), thus giving the dish its name. The gravy was later added to keep the fries warm. [Christinne Muschi/Toronto Star via Getty Images]

C New England–style clam chowder served in a sourdough bread bowl is a classic San Francisco dish that springs from two American traditions. New England clam chowder was developed by fishers and consists of clams and other seafood, potatoes, and seasonings, cooked in a broth to which hard "sea biscuits" were mixed to thicken the chowder. In San Francisco, sourdough bread came via pioneer families and was later popularized by French bakers. It has a particularly sour, tangy flavor that may be the result of the climate of the San Francisco Bay Area. [Andrew McKinney/Dorling Kindersley/Getty Images]

encouraged large U.S. agribusinesses to relocate to Mexico to grow corn on a massive scale, small Mexican corn farmers were dislodged from land they had traditionally worked; others found they could not compete on price and had to give up farming. Yet all these farmers had families to feed and educate, so often fathers and their brothers and sons migrated north in search of work. In Texas, California, or Tennessee, they worked in construction or as field laborers, lived frugally, and sent most of their earnings home. Some entered the United States legally, but many—because of the complexities and costs of getting papers to immigrate—entered illegally. Legal U.S. immigration visas are available in far fewer numbers than both the potential immigrants and the jobs available for them to fill.

Do New Immigrants Cost U.S. Taxpayers Too Much Money? Many North Americans are concerned that immigrants burden schools, hospitals, and government services. And yet numerous studies have shown that, over the long run, immigrants contribute more to the U.S. economy than they cost. Legal immigrants have passed an exhaustive screening process that assures they will be self-supporting. As a result, most start to work and pay taxes within a week or two of their arrival in the country. Those immigrants, who draw on taxpayer-funded services such as welfare, tend to be legal refugees fleeing a major crisis in their homeland; they are dependent only in the first few years after they arrive. More than one-third of immigrant families are firmly within the middle class, with incomes of $45,000 or more. Even undocumented (illegal) immigrants play important roles as payers of payroll taxes, sales taxes, and indirect property taxes through rent. Because they fear deportation, illegal immigrants are also the least likely to take advantage of taxpayer-funded social services.

On average, immigrants are healthier and live longer than native U.S. residents, according to a 2004 study by the National Institutes of Health (NIH). They therefore represent less drain on the health care and social service systems than do native residents. The NIH attributes this difference to a stronger work ethic, a healthier lifestyle that includes more daily physical activity, and the more nutritious eating patterns of new residents compared to those of U.S. society at large. Unfortunately, these healthy practices tend to diminish the longer immigrants are in the country, and the more healthy status does not carry over to immigrants' children, who are nearly as likely to suffer from obesity as native-born children.

Do Immigrants Take Jobs Away from U.S. Citizens? The least educated, least skilled American workers are the most likely to end up competing with immigrants for jobs. In a local area, a large pool of immigrant labor can drive down wages in fields like roofing, landscaping, and general construction. Immigrants with little education now fill many of the very lowest-paid service, construction, and agricultural jobs.

It is often argued that U.S. citizens have rejected these jobs because of their low pay, which results in immigrants being needed to fill the jobs. Others say that these jobs might pay more and thus be more attractive to U.S. citizens if there were not a large pool of immigrants ready to do the work for less pay. Research has failed to clarify the issue. Some studies show that immigrants have driven down wages by 7.4 percent for U.S. natives without a high school diploma. However, other studies show no drop in wages at all. Employers across the country speak of the superior work ethic of immigrants, especially in the construction industry, which makes them happy to pay these workers decent wages.

Professionals in the United States occasionally compete with highly trained immigrants for jobs, but such competition is usually in occupations where there is a scarcity of native-born people who are trained to fill these positions (Figure 2.30).

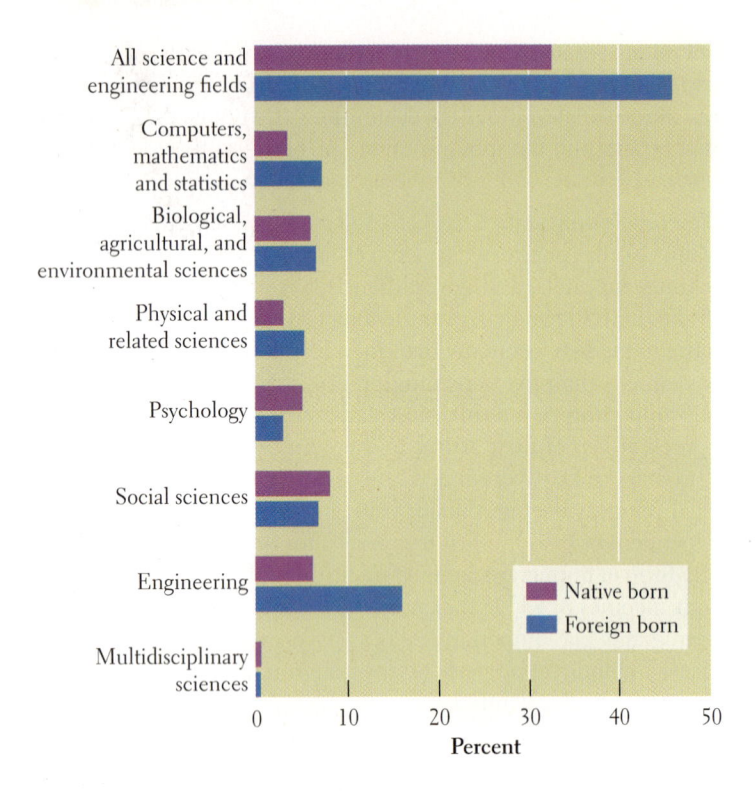

FIGURE 2.30 Nativity and field of study among those 25 and older with a science or engineering degree, 2010. [Source consulted: Christine Gambino and Thomas Gryn, "The Foreign Born with Science and Engineering Degrees: 2010," *American Community Survey Briefs*, Figure 2, p. 3, U.S. Census Bureau, November 2011, at http://www.census.gov/prod/2011pubs/acsbr10-06.pdf]

The computer engineering industry, for example, regularly recruits abroad in such places as India, where there is a surplus of highly trained workers. In this case, it is unclear whether these skilled immigrants are driving down wages. Until the recession caused jobs to be eliminated, it was quite clear that there were not enough sufficiently trained Americans to fill the available positions. In fact, India's growth in IT jobs and related services has been so impressive that many who came to North America to work in these jobs have elected to go back to India to work in the burgeoning IT industry in India, a development that has not been welcomed by their American employers. **253. SKILLED FOREIGN WORKERS IN U.S. MAY HAVE TO LEAVE**

Are Too Many Immigrants Being Admitted to the United States? Many people are concerned that immigrants are coming in such large numbers that they will strain resources here. There is some validity to this. Immigrants and their children accounted for 78 percent of the U.S. population growth in the 1990s. At the current rate of entry, by the year 2050, the U.S. population would reach 422.6 million (down 20 million from estimates in 2008), with immigration accounting for the majority of the increase. But one of the problems with making these projections is that no one really knows how much undocumented (illegal) immigration there is. Most research indicates that it has reached unprecedented levels over the past 30 years, possibly exceeding legal immigration rates since the mid-1990s. While estimates of the illegal immigrant population currently in the United States range from 7 to 20 million, reports in 2012 indicated that migration—legal and illegal—was down to less than half that of previous years.

Undocumented immigrants tend to lack skills, and they are not screened for criminal background, as are all legal immigrants. However, research also shows that undocumented immigrants are not only less likely to partake of social services, but also less likely to participate in criminal behavior than the general population, with only tiny percentages of them having committed offenses. Further analysis of the 2010 census will help researchers formulate factually based conclusions about the effects of illegal immigration.

47. AMERICAN CENSUS REPORT
51. IMMIGRATION LABOR SHORTAGE REPORT

Research shows that the decision to migrate without legal documentation is a very difficult one, undertaken because of severely limited economic, educational, and social opportunities at home. Few Americans are aware that children of all ages get caught up in these migration patterns, and their migratory experiences remain largely unrecorded. Despite tougher border security and declining adult immigration nationally and internationally, the numbers of unaccompanied children stopped at the border continually increase, illustrating yet one more of the complexities of immigration.

VIGNETTE

Every youth who arrives at the border detention shelter in Arizona writes a life story in a creative writing class. A 14-year-old indigenous girl from Honduras started hers: "At 10 years old, my papa started to tell me the good things and the bad things." The next day the teacher asked what this meant. After sitting in silence for 30 minutes, the girl wrote: "While my father is a good man, he did not always do good things. Some of these things created a bad situation for me." Afraid to tell her mom, she called her brothers to help her escape.

Three years before, at the peak of the global recession, her then-teenage brothers, unable to find work in their rural area, decided to take the month-long journey to the United States by jumping trains through Mexico and then walking for days across the mountainous and arid terrain of Texas. Once out of Texas, they found migrant farm work in Southern states. The brothers were working in Florida when they spoke to their sister on the phone.

Within days, the brothers sent their sister all of their savings—$8000—and arranged for a *coyote* to accompany the girl to the United States. As an unaccompanied and attractive female, she was more vulnerable to exploitation, abuse, and sex trafficking. Like most detained in the border shelter, she is unwilling to discuss what happened on her journey, describing it as "the necessary suffering to become someone." Luckily, she made it without falling victim to drug trafficking and fatal gang violence along the border.

In the 2 months it took her to arrive, she lost contact with her brothers. Lacking family contacts and money, her deportation is likely. Her biggest desire is "to work hard and give the money back to my brothers." [Source: Material for this vignette was pieced together by Lydia Pulsipher in April 2012 from personal correspondence with Elizabeth Kennedy and Stuart Aitken, geographers who participate in the ISYS Unaccompanied Minors project.]

FIGURE 2.31 LOCAL LIVES
Festivals in North America

A Mardi Gras Indian Chief Golden Comanche leads a procession through a neighborhood in New Orleans, Louisiana. Mardi Gras, which means "Fat Tuesday," marks the last day before the start of Lent, a 40-day period of fasting and abstinence in Christianity that precedes Easter. Mardi Gras Indians originated as a tribute to Native American communities that had harbored runaway African American slaves. [Mario Tama/Getty Images]

B A temporary sculpture at Burning Man, a weeklong "art event and transitory community based on radical self-expression and self-reliance." Begun in 1986 as a bonfire on the beach in San Francisco, Burning Man is guided by principles of inclusive participation, anticommercialism, and the goal of not leaving any physical trace of the festival after it happens each year. [David McNew/Newsmakers/Getty Images]

C Bull riding is an event at the Calgary Stampede, a 10-day rodeo, exhibition, and festival that was started in 1886 to help draw settlers from the east. Billed as the "Greatest Outdoor Show on Earth," the Stampede draws more than a million people each year. The highlight of today's Stampede is its rodeo, which offers more prize money than any other. [Photo by Todd Korol/Sports Illustrated/Getty Images]

Race and Ethnicity in North America

Despite strong scientific evidence to the contrary, people across the world still perceive skin color and other visible anatomical features to be significant markers of intelligence and ability. This racialized view of skin color is partially a remnant of European colonialism, which was based on assigning some to a perpetually low status while giving others privilege. As discussed previously (see Chapter 1, page 53), the science of biology tells us there is no scientific validity to such assumptions and practices. The same is true for **ethnicity**, which is the cultural counterpart to race, in that people may ascribe overwhelming (and unwarranted) significance to cultural characteristics such as religion, family structure, or gender customs. Thus, race and ethnicity are very important sociocultural factors not because they *have* to be, but because people *make* them so.

Extending Equal Opportunity Numerous surveys show that a large majority of Americans of all backgrounds favor equal opportunities for minority groups. Nonetheless, in both the United States and Canada, many middle-class African Americans, Native Americans, and Latinos (and to a lesser extent, Asian Americans) report experiencing both overt and covert discrimination that affects them economically as well as socially and psychologically. And indeed, even a cursory examination of statistics on access to health care, education, and financial services shows that, on average, Americans have quite unequal experiences based on their racial and ethnic characteristics.

Diversity is increasing across North America as is the acceptance of diversity. Take, for example, the enthusiasm with which various cultural festivals are attended by Americans regardless of their ethnic heritage (Figure 2.31). Still, the issue of race remains important in both the United States and Canada, where whiteness is considered the norm and African Americans, Native Americans and First Nations people, and Latinos often find themselves seen as exceptions to the norm. Prejudice has clearly hampered the ability of these three groups to reach social and economic equality with Americans of other ethnic backgrounds. In the United States—and to a considerably lesser degree in Canada—despite the removal of legal barriers to equality, these groups (with the notable exception of Asians) still have higher poverty rates and thus lower life expectancies, higher infant mortality rates, lower levels of academic achievement, and more unemployment than other groups.

ethnicity the quality of belonging to a particular culture group

Figure 2.32 shows the changes in the ethnic composition of the North American population from 1950 to 2010, and the projected changes for 2050. In 2001, Latinos overtook African Americans as the largest minority group in the United States. Because of a higher birth rate and a high immigration rate, the Latino population increased by 58 percent in the 1990s, to 14.7 percent of the total population by 2010. By 2010, Asian Americans made up just 5 percent of the U.S. population, but their numbers increased by 80 percent between 1990 and 2010. Of the foreign-born population in Canada in 1981, Asians made up 14.1 percent and those of European birth, 66.7 percent. Asians appear to be the fastest-growing nonnative-born group in both countries. It is estimated that if present trends continue, Asians will comprise 55 percent of North Americans born abroad by 2031.

Income Discrepancies Over the past few decades, many non–Euro-Americans and those of mixed heritage have joined the middle class, achieving success in the highest ranks of government and business. In particular, African Americans have completed

U.S. Population by Race and Ethnicity, 1950, 2010, and 2050 (projected)

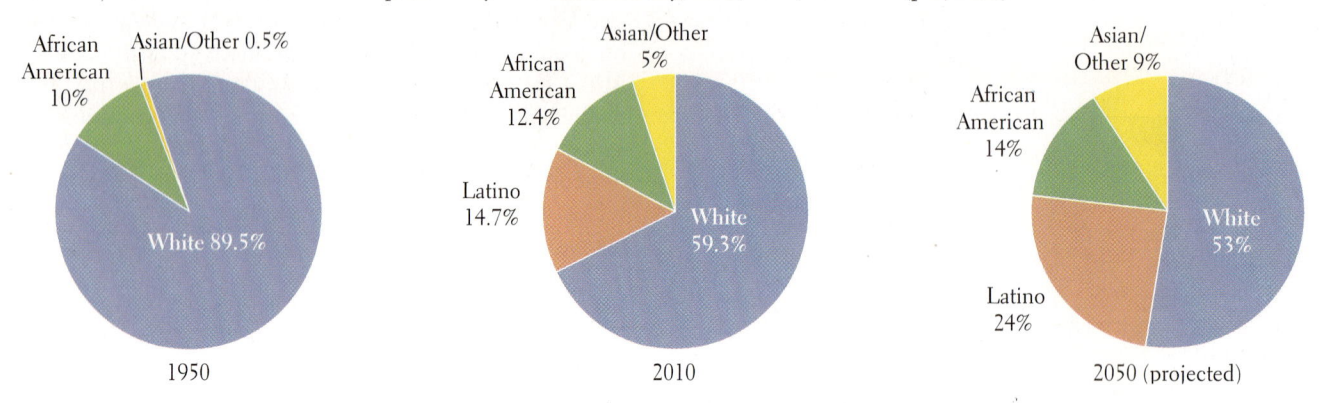

1950:
- White 89.5%
- African American 10%
- Asian/Other 0.5%

2010:
- White 59.3%
- Latino 14.7%
- African American 12.4%
- Asian/Other 5%

2050 (projected):
- White 53%
- Latino 24%
- African American 14%
- Asian/Other 9%

FIGURE 2.32 The changing ethnic composition in the United States and Canada, 1950, 2010, and 2050 (projected). As the percentage of ethnic minorities increases in the United States, the percentage of whites decreases. By 2050, if present reproductive trends continue, whites will constitute only slightly more than half of the population. [Sources consulted: Jorge del Pinal and Audrey Singer, "Generations of Diversity: Latinos in the United States," *Population Bulletin* 52 (October 1997): 14; U.S. Census Bureau, "Race by Sex, for the United States, Urban and Rural, 1950, and for the United States, 1850 to 1940," Census of Population, 1950, Volume 2, Part 1, United States Summary, Table 36, 1953, at http://www2.census.gov/prod2/decennial/documents/21983999v2p1ch3.pdf; http://2010.census.gov/news/releases/operations/cb11-cn125.html; http://www12.statcan.gc.ca/census-recensement/2006/dp-pd/tbt/Rp-eng.cfm?LANG=E&APATH=3&DETAIL=0&DIM=0&FL=A&FREE=0&GC=0&GID=0&GK=0&GRP=1&IPD=92342&PRID=0&PTYPE=88971,97154&S=0&SHOWALL=0&SUB=0&Temporal=2006&THEME=80&VID=0&VNAMEE=&NAMEF=]

advanced degrees in large numbers, and more than one-third now live in the suburbs. Yet, as overall groups, African Americans, Latinos, and Native Americans remain the country's poorest and least educated people (Figure 2.33).

The Culture of Poverty Is anything other than prejudice holding back some in these ethnic minority groups? Some social scientists suggest that persistently disadvantaged Americans of all ethnic backgrounds may suffer from a *culture of poverty*, meaning that poverty itself forces coping strategies that are counterproductive to social advancement. Plans to get higher education, for example, must be abandoned because of lack of funds. This and low social status have bred the perception among the poor that there is no hope for those trapped in the culture of poverty and therefore no point in trying to succeed. This perception is perpetuated and even exaggerated by a daily existence on the fringes of American mainstream life. The reality is that these poorest people rarely encounter examples of success, and they have few opportunities to complete a collegiate education, be hired into jobs with livable wages, or find a decent dwelling.

One component in the culture of poverty is the growing numbers of low-income, single-parent families. In 2005, only 23 percent of Euro-American children and 17 percent of Asian American children lived in single-parent families, while 65 percent of African Americans, 49 percent of Native Americans, and 36 percent of Latinos did. The reasons for these stark differences are often systemic. For example, many U.S. states prohibit or severely limit the amount of income support allowed to intact families who fall on hard times. In these states, it is easier to qualify for welfare if the family is missing one parent. Fathers who lose their jobs may simply absent themselves so the mothers can qualify for assistance. In these situations, children usually stay with their mothers, and fathers often are no longer active in their support and upbringing. This perpetuates the idea that males are not active, participating family members. The assistance packages always fall far short of meeting actual needs, and the enormous responsibilities of both child rearing and breadwinning are often left in the hands of undereducated young mothers who, being in need themselves, are unable to help their children advance.

Another particularly geographic aspect of the *culture of poverty* is that it is part of a larger problem of economic and

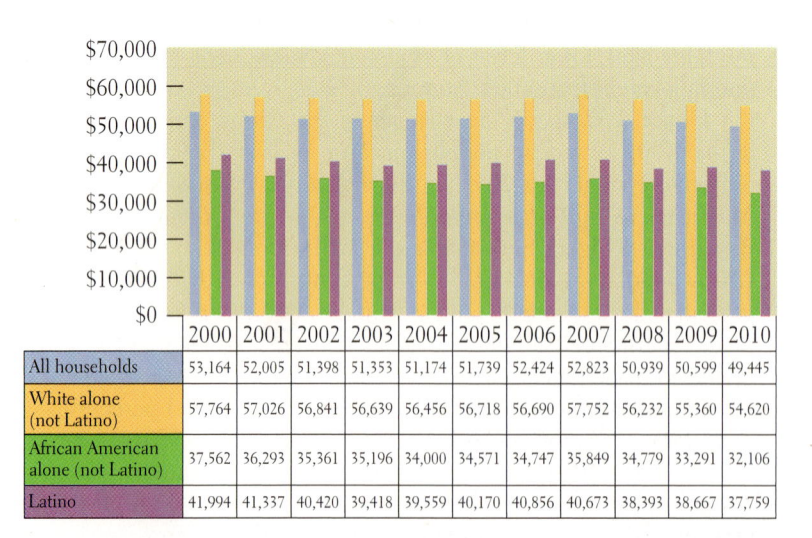

	2000	2001	2002	2003	2004	2005	2006	2007	2008	2009	2010
All households	53,164	52,005	51,398	51,353	51,174	51,739	52,424	52,823	50,939	50,599	49,445
White alone (not Latino)	57,764	57,026	56,841	56,639	56,456	56,718	56,690	57,752	56,232	55,360	54,620
African American alone (not Latino)	37,562	36,293	35,361	35,196	34,000	34,571	34,747	35,849	34,779	33,291	32,106
Latino	41,994	41,337	40,420	39,418	39,559	40,170	40,856	40,673	38,393	38,667	37,759

FIGURE 2.33 Median household income by race and ethnicity, in U.S.$, 2000–2010. [Source consulted: "What Happened to Incomes in 2010? Analysis of Income Estimates for 2010," Diagram 1, National Urban League Policy Institute, September 15, 2011, at http://www.nul.org/sites/default/files/Income_2010_CPS.pdf]

social spatial segregation based on class. In both the United States and Canada, the increasingly prosperous middle class, of whatever race or ethnicity, has moved to the suburbs—even to gated residential enclaves. Their children seldom encounter children from poorer families. Those who are successful have difficulty developing empathy for the poor because they do not know any poor people; in turn, the very poor rarely have the chance to associate with models of economic success.

Religion

Because so many early immigrants to North America were Christian in their home countries, Christianity is currently the predominant religious affiliation in North America. In surveys taken in 2012, seventy-three percent of those living in the United States identified themselves as Christian, as did 67 percent of those in Canada. In the United States, most Christians are Protestants; in Canada, most are Catholic. Despite the numerical dominance of Christianity, about 15 percent of people in both countries reported no religious affiliation. This group includes atheists, agnostics, and the majority who simply claim no religion by choice. Minority religions are growing in both countries. Virtually every medium-sized city across the region has at least one synagogue, mosque, and Buddhist or Hindu temple. In some localities, adherents of Judaism, Islam, Hinduism, or Buddhism are numerous enough to constitute a prominent cultural influence (see, for example, California and Ontario in Figure 2.34).

nuclear family a family consisting of a married father and mother and their children

ON THE BRIGHT SIDE

More Discussion of Race and Racism

In 2008 and again in 2012, the United States made history by electing its first African American president. Some commentators mistakenly heralded this as the end of racism and of the need to even talk about race. As if to prove them wrong, an avalanche of racially charged stories entered the news cycle of the mainstream national media. For example, in 2012 national news media focused on the shooting of Trayvon Martin, an unarmed black teenager, by George Zimmerman, a neighborhood watch coordinator in the gated community near Orlando, Florida, where Martin was living temporarily. Coverage of these stories produced an outpouring of discussion, protests, and other actions aimed at raising awareness of race in the United States.

Lost in much of the discussion was the fact that shootings like this, even though they are quite common, have historically rarely made even the local news, much less the national news. Currently, most major U.S. national news outlets feature an entire section on "race and ethnicity," at least in their online offerings, which was entirely absent just a few years ago. It may be a long time before any person's or organization's claims to be "beyond race" ring true in the United States, but at least the subject is receiving more attention and being more openly discussed.

The Geography of Christian Subgroups There are many versions of Christianity in North America, and their geographic distributions are closely linked to the settlement patterns of the immigrants who brought them here (see Figure 2.34). Roman Catholicism dominates in regions where Latino, French, Irish, and Italian people have settled—in southern Louisiana, the Southwest, and the far Northeast in the United States, and in Québec and other parts of Canada. Lutheranism is dominant where Scandinavian people have settled, primarily in Minnesota and the eastern Dakotas. Mormons dominate in Utah.

Baptists, particularly Southern Baptists and other evangelical Christians, are prominent in the "Bible Belt," which stretches across the Southeast from Texas and Oklahoma to the Eastern Seaboard. Evangelical Christianity is such an important part of community life in the South that frequently the first question newcomers to the region are asked is what church they attend.

The Relationship of Religion and Politics Just how interactive religion and politics should be in North American life has long been a controversial issue in the United States, more so than in Canada. This is true in large part because the framers of the U.S. Constitution, in an effort to ensure religious freedom, supported the idea that church and state should remain separate. In the past three decades, however, many conservative Christians have successfully pushed for a closer integration of religion and public life. Their political goals include banning abortion, promoting prayer in the public schools, teaching the biblical version of creation instead of teaching evolution, and preventing gays and lesbians from participating openly in the public sphere and marrying each other.

New immigrants have brought their own faiths and belief systems, and they are contributing to the debate about religion and public life. Muslims, as yet a tiny minority, have had trouble gaining the same rights as other religious minorities, such as Buddhists or Jews. Some immigrants leave their traditional faith and adopt another. Such is the case with the 15 percent of Latino immigrants to the United States who have left the Roman Catholic faith, either while still in their home country or after immigrating, and are now evangelical Christians. In Canada, Catholicism is growing.

National surveys have consistently indicated that a substantial majority of North Americans favor the separation of church and state and support personal choice in belief and behavior.

The American Family

The family is often seen as endangered by North America's fast-changing culture. A century ago, most North Americans lived in large extended families of several generations. Families pooled their incomes and shared chores. Aunts, uncles, cousins, siblings, and grandparents were almost as likely to provide daily care for a child as were the mother and father. Though it has long been present in society, the **nuclear family**, consisting of a married father and mother and their children, became especially widespread in the post-1900 industrial age.

The Nuclear Family Becomes a Shaky Norm Beginning after World War I, and especially after World War II, many young people left their large kin groups on the farm and migrated to distant cities, where they established new nuclear families. Soon suburbia, with its many similar single-family homes, seemed to provide the perfect domestic space for the emerging nuclear family.

104 CHAPTER 2 North America

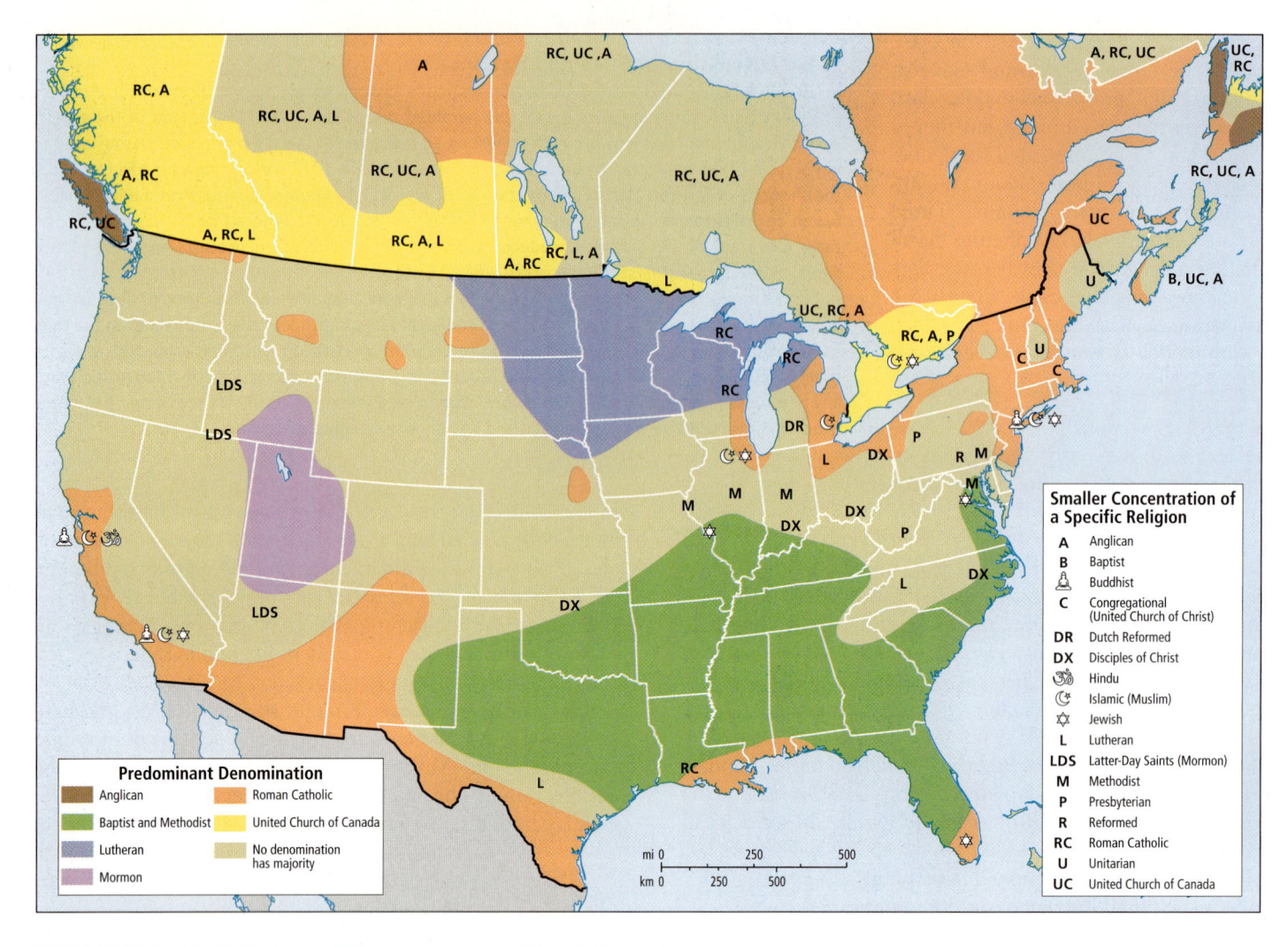

FIGURE 2.34 Religious affiliations across North America. [Source consulted: Jerome Fellmann, Arthur Getis, and Judith Getis, *Human Geography* (Dubuque, IA: Brown & Benchmark, 1997), p. 164]

This compact family type suited industry and business because it had no firm ties to other relatives and so was portable. Many North Americans born since 1950 moved as many as ten times before reaching adulthood. The grandparents, aunts, and uncles who were left behind missed helping raise the younger generation and had no one to look after them in old age. This separation of older from younger generations is one reason nursing homes for the elderly proliferated.

In the 1970s, the nuclear family encountered challenges from many areas, including changing gender roles. Suburban sprawl meant onerous commutes to jobs for men and long, lonely days at home for women. Women began to want their own careers, and rising consumption patterns made their incomes increasingly useful to family economies. By the 1980s, seventy percent of working-age women were in the paid workforce, compared with 30 percent of their mothers' generation.

Once employed, however, a woman could not easily move to a new job location if she had an upwardly mobile husband. Also, working women could not manage all of the family's housework and child care, as well as a job. Some married men began to handle part of the household management and child care, but the demand for commercial child care grew sharply. With family no longer around to strengthen the marital bond and help with child care, and with the new possibility women in unhappy marriages had of being able to financially support themselves, divorce rates rose into the mid-1970s, while birth rates fell. Those who married after 1975, however, have had a slowly declining rate of divorce. This may be related to the growing number of couples with college educations, as this group is less likely to divorce.

The Current Diversity of American Family Types There is no longer a typical American household, only an increasing diversity of household forms and ways of family life (**Figure 2.35**). In 1960, the nuclear family—households consisting of heterosexually married couples with children—comprised 74.3 percent of U.S. households. This number declined to 56 percent by 1990 and to 49.3 percent by 2010. Family households headed by a single person (female or male) rose from 10.7 percent in 1960, to 15 percent in 1990, to 18 percent in 2010. Nonfamily households (unrelated by blood or marriage) rose from 15 percent of the total in 1960, to 29 percent in 1990, to nearly 34 percent in 2010.

Household Composition, 1990 versus 2010

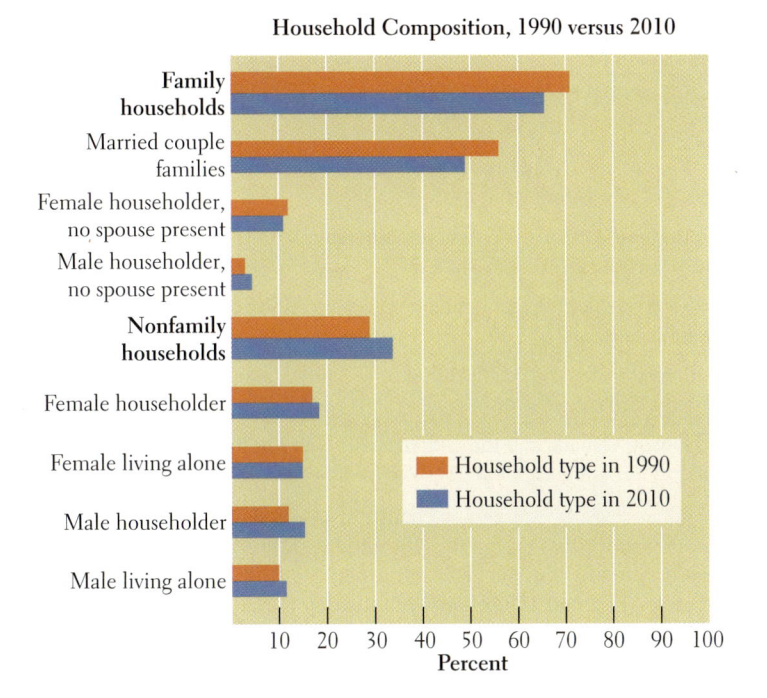

[Bar chart showing percentages for Family households, Married couple families, Female householder no spouse present, Male householder no spouse present, Nonfamily households, Female householder, Female living alone, Male householder, Male living alone — comparing Household type in 1990 vs Household type in 2010]

FIGURE 2.35 U.S. households by type, 1990–2010.
[Source consulted: "Selected Population Profile in the United States: 2007–2009 American Community Survey 3-Year Estimates," American FactFinder, U.S. Census Bureau, at http://factfinder2.census.gov/faces/tableservices/jsf/pages/productview.xhtml?pid=ACS_09_3YR_S0201&prodType=table]

While most of these families function well, a few of these new forms may not provide adequately for the welfare of children. In 2006, more than 29 percent of U.S. children lived in single-parent households. Although most single parents are committed to rearing their children well, the responsibilities can be overwhelming.

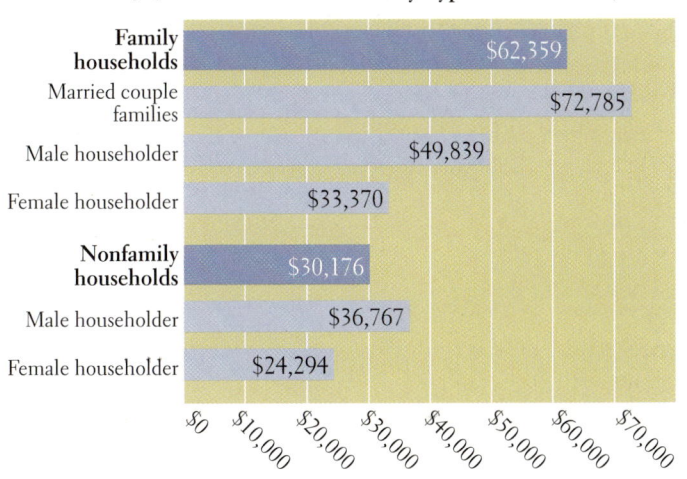

Children and Poverty in North America Single-parent families tend to be hampered by economic hardship and lack of education (Figure 2.36). Young women head the vast majority of single-parent households, and their incomes, on average, are less than two-thirds those of single male heads of household. Low income and low levels of education are closely linked, as Figure 2.36B shows. One result of the pattern of single-parent households having lower education levels and incomes is that children in the United States are disproportionately poor. In 2010 in the United States, 22 percent of children aged 0 to 17 (16.4 million) lived in poverty, whereas only 11.4 percent of adults did. In Canada, 14.2 percent of children were poor. By comparison, in Sweden, just 2.4 percent of children lived in poverty; in Ireland, 12.4 percent; and in Poland, 12.7 percent.

THINGS TO REMEMBER

- North America has some of the world's wealthiest cities, most of which are in the eastern United States. Nearly 80 percent of North Americans now live in metropolitan areas.

- A steady increase in migration from Middle and South America and parts of Asia promises to make North America a region where most people are of non-European descent.

- Statistics on access to health care, education, and financial services by race and ethnicity show that, on average, Americans have quite uneven experiences based on their racial and ethnic characteristics.

- There is no longer a typical American household, only an increasing diversity of household forms.

- A large percentage of children in the United States live in poverty, compared to other developed countries.

(A) Median U.S. Income by Type of Household, 2007

- Family households: $62,359
- Married couple families: $72,785
- Male householder: $49,839
- Female householder: $33,370
- Nonfamily households: $30,176
- Male householder: $36,767
- Female householder: $24,294

(B) Mean U.S. Income by Education, 2007

- Less than 9th grade: $21,484
- High school graduate (includes equivalency): $31,286
- Some college, no degree: $33,009
- Associate degree: $39,746
- Bachelor's degree: $57,181
- Master's degree: $70,186
- Professional degree: $120,978
- Doctorate degree: $95,565

FIGURE 2.36 Median U.S. income by (A) type of household and (B) level of education, 2007.
[Sources consulted: **(A)** U.S. Census Bureau, "Money Income of Households—Distribution by Income Level and Selected Characteristics: 2007," Income, Expenditures, Poverty, & Wealth: Household Income, Table 676, 2010, at http://www.census.gov/compendia/statab/cats/income_expenditures_poverty_wealth/household_income.html; **(B)** U.S. Census Bureau, "Mean Earnings by Highest Degree Earned: 2007," Education: Educational Attainment, Table 227, 2010, at http://www.census.gov/compendia/statab/cats/education/educational_attainment.html]

GEOGRAPHIC INSIGHTS

North America: Review and Self-Test

1. Environment: North America's intensive use of resources has an enormous impact on the environment. Although home to only 5 percent of the world's population, North America produces 26 percent of the greenhouse gases released globally that are related to human activity. North American lifestyles have major environmental impacts, including the depletion and pollution of water resources and fisheries and the destruction of huge amounts of habitat for wild plants and animals.

- What aspects of North American lifestyles have increased emissions of greenhouse gases? What efforts have been made to control these emissions?
- How does urban sprawl influence the average North American's consumption of gasoline?
- How might the livable city movement and smart growth affect greenhouse gas emissions? What are some responses to reducing greenhouse gases that you would be willing to take part in yourself?
- What activities in North America are major contributors to water pollution and water scarcity, and what are the resulting impacts for simple water use and beyond?
- What effect do green revolution agriculture techniques have on aquatic ecosystems?

2. Globalization and Development: Globalization has transformed economic development in North America, reorienting employment toward knowledge-intensive jobs that require education and training. Most of the manufacturing jobs upon which the region's middle class was built have either been moved abroad to take advantage of cheaper labor or have been replaced by technology. North America's demand for imported goods and its export of manufacturing jobs helps make it a major engine of globalization.

- How has globalization transformed the North American job market, including in the steel industry?
- How did fair trade principles lead to NAFTA goals, and what has been the effect of NAFTA on manufacturing jobs and immigration in North America?
- What are some exceptions to U.S. support of free trade principles?

3. Power and Politics: North America has relatively high levels of political freedom, though in recent decades many of its residents have become disillusioned with the political process for a variety of reasons. While Canada plays a relatively modest political role abroad, the United States has enormous influence on the global political order, although its status as the world's predominant "superpower" is increasingly being challenged.

- What role does the urge to spread democracy play in the foreign policy of North America?
- How does Canada's approach to foreign relations differ from that of the United States?
- What role did the ideal of democracy play in justifying the Iraq and Afghanistan wars?
- To what extent is the allocation of U.S. foreign aid linked to U.S. strategic military interests?
- How do Canada and the United States differ in their support for government programs that help low-income people?

4. Urbanization: A dramatic change in the spatial patterns of cities and suburbs has profoundly affected life in this very urbanized region. Since World War II, North America's urban populations have increased by about 150 percent, but the amount of land they occupy has increased by almost 300 percent. This is primarily because of suburbanization and urban sprawl, which are companion processes to urbanization.

- What role has car ownership played in suburban growth?
- How does urban sprawl impact farmers?
- What is *smart growth*?

5. Population and Gender: Women's participation in North America's economy is beginning to rival that of men, contributing to more than two centuries of declining fertility rates as women delay childbearing to pursue education and careers. Declining childbirth rates play a major role in the aging of North American populations, which may slow economic growth.

- How do education and work opportunities for contemporary women affect the North American population—and in what ways does this affect you?
- What are factors that have created an aging North American population, and how do those factors connect with immigration and household types in Canada and the United States?
- What are three ways in which you personally are likely to be affected by the aging of North America's population?
- What is the connection between single-parent households and the fact that 22 percent of U.S. children (14 percent of Canadian) live in poverty?

CRITICAL THINKING QUESTIONS

1. Discuss the ways in which North American culture is adapted to the high rate of spatial (geographic) mobility engaged in by Americans.

2. North American family types are changing. Explain the general patterns and discuss whether or not the nuclear family is or should be the sought-after norm.

3. The influence of globalization is now felt in small, even isolated, places in North America. Pick an example from the text or from your own experience and explain at least four ways in which this place is now connected to the global economy or global political patterns.

4. Some say that North Americans profit from having undocumented workers produce goods and services. Explain why this may be the case and discuss how the situation should be changed or left as is.

5. How do you think life in North America would differ if the United States was no longer the world's "superpower"? How might U.S. attitudes

toward the rest of the world be modified? Will these modifications will be useful or destructive in the long run?

6. When you compare the economies of the United States and Canada and the ways they are related, what are the most important factors to mention?

7. When U.S. farmers or producers get subsidies from the federal government, what is intended? Why do farmers and other food producers in poor countries say that this practice hurts them?

8. The United States and Canada have quite different approaches in their treatment of citizens who experience difficulties in life. Explain those different approaches and the apparent philosophies behind them.

⟩ THINKING GEOGRAPHICALLY

Now that you have read about North America, you should be able to answer the following questions about the photos in this chapter. To answer these questions online, go to Geography LaunchPad.

Vulnerability to Climate Change (page 65)

A Describe the coastal erosion illustrated by this photo.

B What elements of this community's emergency response system may be compromised by this flooding?

C How might irrigation reduce vulnerability to climate change but also result in more greenhouse gas emissions?

D Why are stronger hurricanes more likely as the climate warms?

Human Impacts on the Biosphere (page 69)

A Why might it make economic and environmental sense to keep old-growth trees, such as the one shown here, standing?

E What are some threats to the environment posed by the Trans-Alaska Pipeline?

F What kind of mining is mountaintop removal?

G What clues are there that this water may be toxic?

Visual History of North America (page 72)

A How did food crops such as corn, beans, and squash influence the development of settlements like Cahokia?

B Why did the plantation system inhibit the establishment of roads, communication networks, and small enterprises that could have boosted other forms of economic activity in the Southern colonies?

C What features of the economy of the Pacific Northwest today might surprise these men?

D How did the steel industry stimulate development in the "economic core" of North America?

E What is the evidence of economic diversification in this photo?

F What were the causes of the Dust Bowl?

Power and Politics (page 84)

A What economic interests does the United States have in Europe?

B What strategic and economic interests make it unlikely that the United States will withdraw completely from Afghanistan and Iraq in the near future?

D What has kept U.S. economic interests in Africa at a low level relative to those in other regions?

E What has kept U.S. economic interests in Haiti at a low level relative to those in other countries?

Urbanization (page 92)

A Is Vancouver's growth pattern typical for North America?

B What factors reduce livability in New York?

C What factors reduce livability in Las Vegas?

D What factors reduce livability in Atlanta?

CHAPTER KEY TERMS

acid rain 64
agribusiness 81
aquifers 62
brownfields 93
clear-cutting 70
digital divide 76
economic core 73
ethnicity 101
gentrification 93
infrastructure 72
Latino 61
megalopolis 91
metropolitan areas 91
North American Free Trade Agreement (NAFTA) 75
nuclear family 103
organically grown 82
Pacific Rim 97
push factors 98
smog 64
social safety net 89
suburbs 91
trade deficit 77
urban sprawl 70

Map of Latin America

A Sierra Madre, Mexico
B Andes, Chile
C Tierra del Fuego, Argentina

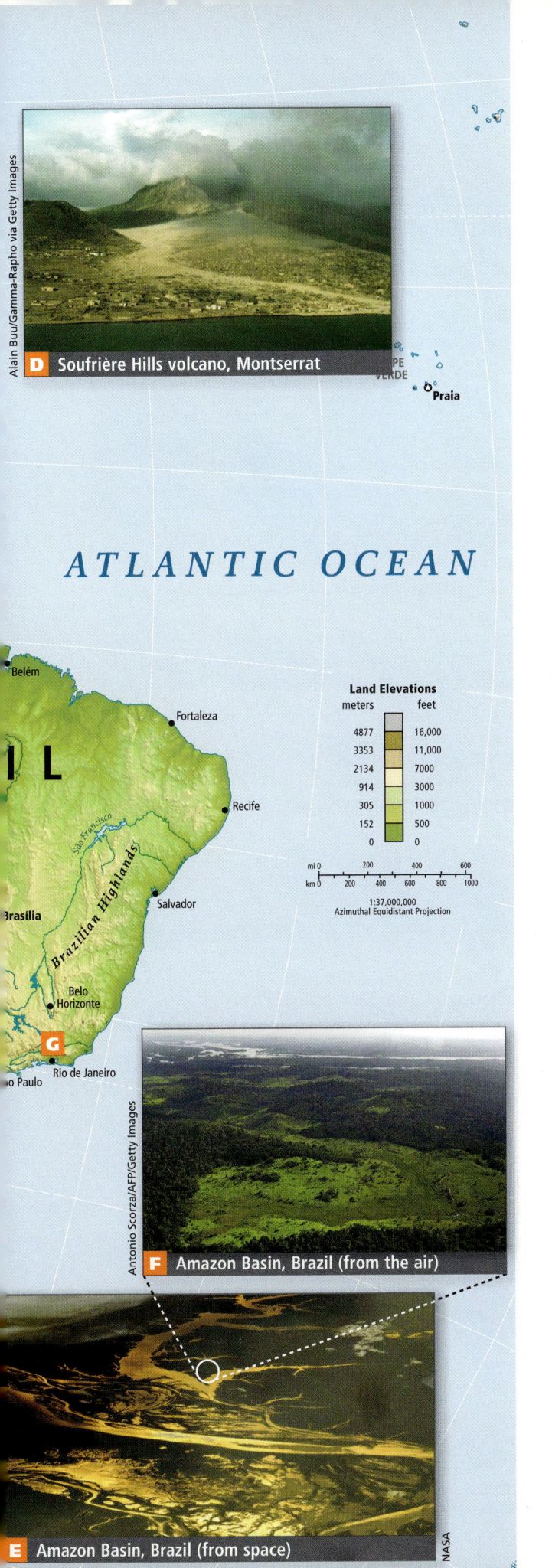

chapter 3
MIDDLE AND SOUTH AMERICA

D Soufrière Hills volcano, Montserrat

E Amazon Basin, Brazil (from space)

F Amazon Basin, Brazil (from the air)

G Rio de Janeiro, Brazil

H Yucatán lowlands, Mexico (with Mayan temple)

I Pampas, Argentina

ATLANTIC OCEAN

Land Elevations

meters	feet
4877	16,000
3353	11,000
2134	7000
914	3000
305	1000
152	500
0	0

1:37,000,000
Azimuthal Equidistant Projection

FIGURE 3.1 Regional map of Middle and South America.

109

110 CHAPTER 3 Middle and South America

GEOGRAPHIC INSIGHTS

After you read this chapter, you will be able to discuss the following issues as they relate to the five thematic concepts:

1. Environment: Deforestation in this region contributes significantly to global climate change through the removal of trees, which as living plants naturally absorb carbon dioxide, a greenhouse gas. Once the trees are cut or burned, they release large amounts of carbon dioxide into the atmosphere. Additionally, despite the region's overall abundant water resources, some areas are experiencing a water crisis related to climate change, inadequate water infrastructure, and intensified use of water.

2. Globalization and Development: The integration of this region with the global economy has left it with the widest gap between rich and poor in the world. Long-term failures to address poverty, economic instability, and the flow of resources and money out of the region have resulted in many conflicts and inspired numerous efforts at reform. Nevertheless, in recent years several countries in this region have emerged as global economic leaders.

3. Power and Politics: After decades of rule by elites and the military, punctuated by disruptive foreign military interventions, political freedoms in Middle and South America are expanding. Almost all countries now have multiparty political systems and democratically elected governments. However, the international illegal drug trade continues to be a source of violence and corruption in the region.

4. Urbanization: Since the early 1970s, Middle and South America has experienced rapid urban growth as rural people migrate to cities and towns. A lack of planning to accommodate the massive rush to the cities has created densely occupied urban landscapes that often lack adequate support services and infrastructure.

5. Population and Gender: During the early twentieth century, the combination of cultural and economic factors and improvements in health care created a population explosion. By the late twentieth century, improved living conditions, expanded access to education and medical care, urbanization, and changing gender roles were all working together to reduce population growth.

The Middle and South American Region

Middle and South America (Figure 3.1), like North America, is a region of great physical and cultural diversity; but here the cultural mix is richer and there are wider disparities of wealth than in any other world region. In spite of the many similarities between the two regions, modern history in Middle and South America has been very different from that of North America.

The five thematic concepts in this book are explored as they arise in the discussion of regional issues. Vignettes—like the one that follows about the Secoya—illustrate one or more of the themes as they are experienced in individual lives.

GLOBAL PATTERNS, LOCAL LIVES

VIGNETTE

The boat trip down the Aguarico River in Ecuador took me into a world of magnificent trees, river canoes, and houses built high up on stilts to avoid floods. I was there to visit the Secoya, a group of 350 indigenous people locked in negotiations with the U.S. oil company Occidental Petroleum over Occidental's plans to drill for oil on Secoya lands. Oil revenues supply 40 percent of the Ecuadorian government's budget and are essential to paying off its national debt. The government had threatened to use military force to compel the Secoya to allow drilling.

The Secoya wanted to protect themselves from pollution and cultural disruption. As Colon Piaguaje, chief of the Secoya, put it to me, "A slow death will occur. Water will be poorer. Trees will be cut. We will lose our culture and our language, alcoholism will increase, as will marriages to outsiders, and eventually we will disperse to other areas." Given all the impending changes, Chief Piaguaje asked Occidental to use the highest environmental standards in the industry. He also asked the company to establish a fund to pay for the educational and health needs of the Secoya people.

Like the Secoya, indigenous peoples around the world are facing environmental and cultural disruption arising from economic development efforts. Chief Piaguaje based his predictions for the future on what has happened in other parts of the Ecuadorian Amazon that have already had several decades of oil development.

The U.S. company Texaco was the first major oil developer to establish operations in Ecuador. From 1964 to 1992, its pipelines and waste ponds leaked almost 17 million gallons of oil into the Amazon Basin, enough to fill about 1900 fully loaded oil tanker trucks, or 35 Olympic-size swimming pools. Although Texaco sold its operations to the government and left Ecuador in 1992, its oil wastes continue to leak into the environment from hundreds of open pits (Figure 3.2).

In 1993, some 30,000 people sued Texaco in New York State, where the company (now owned by and called Chevron) is headquartered, for damages from the pollution. Those suing were both indigenous people and settlers who had established farms along Texaco/Chevron's service roads. Several epidemiological studies concluded that contamination from oil has contributed to higher rates of childhood leukemia, cancer, and spontaneous abortions among people who live near the pollution created by Texaco/Chevron. Oil extraction has had many other negative effects on the environment. Air and water pollution have increased the rates of illness. The wildlife that the Secoya used to depend on, such as tapirs, has disappeared almost entirely because of overhunting by new settlers from the highlands who are working in the oil industry.

In 2002, the Ecuadorian suit against Chevron was dismissed by the U.S. Court of Appeals, which argued that it had no jurisdiction in Ecuador. The case was refiled in Ecuador in 2003, and 8 years later an Ecuadorian court ruled against Chevron, assessing damages of U.S.$18 billion. Chevron appealed to the Ecuadorian Supreme Court, which cut this amount in half. However, because Chevron no longer has any assets in Ecuador, the plaintiffs are preparing to file another lawsuit to collect the U.S.$9 billion in a country where Chevron does have assets, such as Canada. *[Sources: Alex Pulsipher's field notes; Amazon Watch, 2006; Oxfam America, 2005; National Public Radio. For detailed source information, see Text Sources and Credits.]* ■

FIGURE 3.2 Pollution from oil development in Ecuador. A local resident samples one of the several hundred open waste pits that Texaco left behind in the Ecuadorian Amazon. Wildlife and livestock trying to drink from these pits are often poisoned or drowned. After heavy rains, the pits overflow, polluting nearby streams and wells.

The rich resources of Middle and South America have attracted outsiders since the first voyage of Christopher Columbus in 1492. Europe's encounter with this region marked a major expansion of the global economy. However, during most of the period of expansion, Middle and South America occupied a disadvantaged position in global trade, supplying cheap raw materials that aided the Industrial Revolution in Europe and then North America but reaping few of the profits. These extractive industries did little to advance economic development within the region, as most profits went to foreign investors, and the negative environmental effects were largely ignored. In recent years, countries such as Ecuador, Mexico, Bolivia, Brazil, and Venezuela have worked to control their own resources, develop local manufacturing and service-based industries, keep profits at home, and limit environmental pollution. Meanwhile, trade blocs within the region are creating conditions in which these countries can prosper from trade with each other.

The Ecuadorians' efforts to secure a damage settlement against a powerful multinational corporation is indicative of changing attitudes in this region and others (Figure 3.3) toward outside developers. Governments are now somewhat more wary when they lure investors to develop new extractive, manufacturing, and service industries. Local people are more aware that they must be vigilant to ensure that development serves their interests.

What Makes Middle and South America a Region?
Physically, the Middle and South America region consists of Mexico, which geologically is part of the North American continent; the **isthmus** (land bridge) of Central America; and the continent of South America. For the last 500 years, the region of Middle and South America has been defined by a colonial past very different from that of Canada and the United States. Most of the countries in Middle and South America were at one time colonies of Spain. The exceptions are Brazil, which was a colony of Portugal, and a few small countries that were possessions of Britain, France, the Netherlands, Germany, or Denmark (see Figure 3.12 on page 124).

Today, Middle and South America is a region of contrasts and disparities. Culturally, this region has large **indigenous** populations that have contributed to every aspect of life, blending with and changing the European, African, and Asian cultures introduced by the colonists. Social stratification based on class, race, and gender is notable; economically, the gap between rich and poor is the widest of any world region. Politically, the region's more than three dozen countries exhibit a range of governing ideologies, from the socialism of Cuba to the capitalism of Chile. Yet despite these contrasts and disparities, there are significant commonalities across the region, such as the Spanish language, Catholicism, and increasing development trajectories connected to the global economy.

isthmus a narrow strip of land that joins two larger land areas

indigenous native to a particular place or region

Middle America in this book, a region that includes Mexico, Central America, and the islands of the Caribbean

South America the continent south of Central America

Terms in This Chapter
In this book, **Middle America** refers to Mexico, Central America (the narrow ribbon of land, or isthmus, that extends south of Mexico to South America), and the islands of the Caribbean (Figure 3.4). **South America** refers to the continent south of Central America. The term *Latin America* is not used in this book because it describes the region only in terms of the Roman (Latin-speaking) origins of the former colonial powers of Spain and Portugal. It ignores the region's large indigenous groups, its African, Asian, and Northern European populations, as well as the many mixed cultures, often called *mestizo* cultures, which have emerged. In this chapter, we use the term *indigenous groups* or *peoples* rather than *Native Americans* to refer to the native inhabitants of the region.

112 CHAPTER 3 Middle and South America

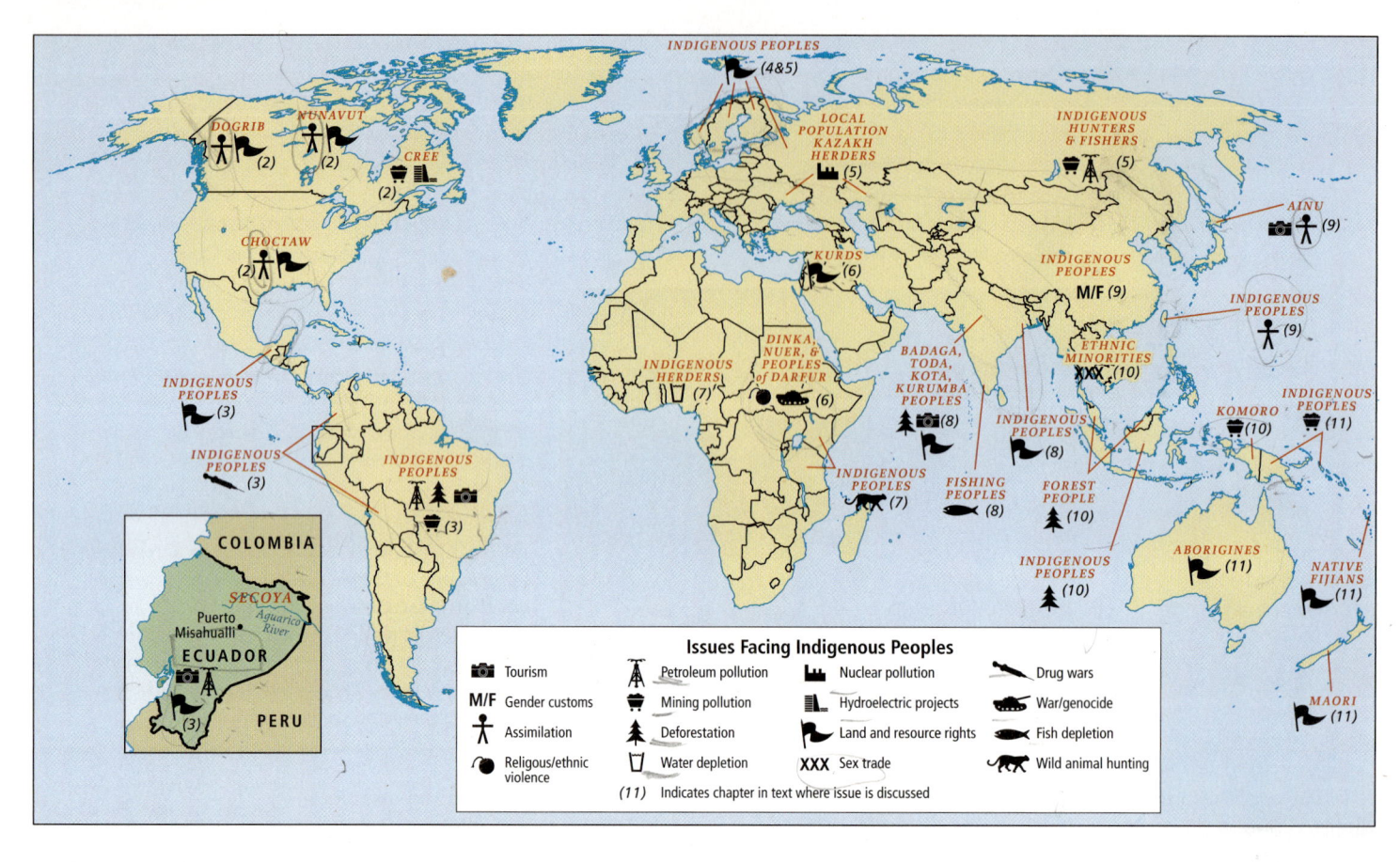

FIGURE 3.3 Indigenous peoples and environmental issues in this text. Issues relating to indigenous peoples are mentioned in many places in this book. In all cases, the issues are in some way related to interactions with the outside world and usually to uses of indigenous peoples' resources by the outside world.

THINGS TO REMEMBER

- The region of Middle and South America, heavily impacted by European colonialism, is making a shift away from raw materials–based industries to more profitable manufacturing and service-based industries.

- As people in this region gain more control over their own resources, some are searching for more sustainable ways to develop economically.

PHYSICAL GEOGRAPHY AND ENVIRONMENTAL ISSUES

PHYSICAL PATTERNS

Middle and South America extend south from the midlatitudes of the Northern Hemisphere across the equator through the Southern Hemisphere, nearly to Antarctica (see Figure 3.1 on page 108). This vast north-south expanse combines with variations in altitude to create the wide range of climates in the region. Tectonic forces have shaped the primary landforms of this huge territory to form an overall pattern of highlands to the west and lowlands to the east.

Landforms

There are a wide variety of landforms in Middle and South America, and this variety accounts for the many different climatic zones in the region. But for ease in learning, landforms are here divided into just two categories: highlands and lowlands.

Highlands A nearly continuous chain of mountains stretches along the western edge of the American continents for more than 10,000 miles (16,000 kilometers) from Alaska in the north to Tierra del Fuego at the southern tip of South America (see Figure 3.1C). The middle part of this long mountain chain is known as the Sierra Madre in Mexico (see Figure 3.1A), by various names in Central America, and as the Andes in South America (see Figure 3.1B, C). It was formed by a lengthy **subduction zone**, which runs thousands of miles along the western coast of the continents (see Figure 1.8 on page 16). Here, two oceanic plates—the

subduction zone a zone where one tectonic plate slides under another

FIGURE 3.4 Political map of Middle and South America.

Cocos Plate and the Nazca Plate—plunge beneath three continental plates—the North American Plate, the Caribbean Plate, and the South American Plate.

In a process that continues today, the leading edge of the overriding plates crumples to create mountain chains. In addition, molten rock from beneath the Earth's crust ascends to the surface through fissures in the overriding plate to form volcanoes. Such volcanoes are the backbone of the highlands that run through Middle America and the Andes of South America (see Figure 3.1). Although these volcanic and earthquake-prone highlands have been a major barrier to transportation, communication, and settlement, people now live close to quiescent volcanoes and in earthquake-prone zones, which can pose deadly hazards (for example, the 2010 earthquake in Peru).

The chain of high and low mountainous islands in the eastern Caribbean is also volcanic in origin, created as the Atlantic Plate thrusts under the eastern edge of the Caribbean Plate. It is not unusual for volcanoes to erupt in this active tectonic zone. On the island of Montserrat, for example, people have been living with an active, and sometimes deadly, volcano for more than a decade (see Figure 3.1D). Eruptions have taken the form of violent blasts of superheated rock, ash, and gas (known as *pyroclastic flows*) that move down the volcano's slopes at speeds upward of 450 miles (700 kilometers) per hour. The unusually strong earthquake in Haiti in January 2010 was also the result of plate tectonics.

Lowlands Vast lowlands extend over most of the land to the east of the western mountains. In Mexico, east of the Sierra Madre, a coastal plain borders the Gulf of Mexico (see Figure 3.1H). Farther south, in Central America, wide aprons of sloping land descend to the Caribbean coast. In South America, a huge wedge of lowlands, widest in the north, stretches from the Andes east to the Atlantic Ocean. These South American lowlands are interrupted in the northeast and the southeast by two modest highland zones: the Guiana Highlands and the Brazilian Highlands (see Figure 3.1G). Elsewhere in the lowlands, grasslands cover huge, flat expanses, including the llanos of Venezuela, Colombia, and Brazil, and the pampas of Argentina (see Figure 3.1I).

The largest feature of the South American lowlands is the Amazon Basin, drained by the Amazon River and its tributaries (see Figure 3.1E, F). This basin lies within Brazil and the neighboring countries to its west. The Earth's largest remaining expanse of tropical rain forest gives the Amazon Basin global significance as a reservoir of **biodiversity**. Hundreds of thousands of plant and animal species live here.

The basin's water resources are also astounding. Twenty percent of the Earth's flowing surface waters exist here, running in rivers so deep that ocean liners can steam 2300 miles (3700 kilometers) upriver from the Atlantic Ocean all the way to Iquitos, jokingly referred to as Peru's "Atlantic seaport." The vast Amazon River system starts as streams high in the Andes. These streams eventually unite to become rivers that flow eastward toward the Atlantic. Once they reach the flat land of the Amazon Plain, their velocity slows abruptly; fine soil particles, or **silt**, then sink to the riverbed. When the rivers flood, silt and organic material transported by the floodwaters renew the soil of the surrounding areas, nourishing millions of acres of tropical forest. Not all of the Amazon Basin is rain forest, however. Variations in weather and soil types, as well as human activity, have created grasslands and seasonally dry deciduous tropical forests in some areas.

> **biodiversity** the variety of life forms to be found in a given area
>
> **silt** fine soil particles
>
> **temperature-altitude zones** regions of the same latitude that vary in climate according to altitude
>
> **trade winds** winds that blow from the northeast and the southeast toward the equator

Climate

From the jungles of the Caribbean and the Amazon to the high, glacier-capped peaks of the Andes to the parched moonscape of the Atacama Desert and the frigid fjords of Tierra del Fuego, the range of climates in Middle and South America is enormous (Figure 3.5). Climates are essentially the result of interactions between temperature and moisture. In this region, the wide range of temperatures reflects both the great distance the landmass spans on either side of the equator and the tremendous variations in altitude across the region's landmass (the highest point in the Americas is Aconcagua in Argentina, at 22,841 feet [6962 meters]). Patterns of precipitation are affected both by the local shape of the land and by global patterns of wind and ocean currents that bring moisture in varying amounts.

Temperature-Altitude Zones Four main **temperature-altitude zones**, shown in Figure 3.6, are commonly recognized in the region. As altitude increases, the temperature of the air decreases by about 1°F per 300 feet (1°C per 165 meters) of elevation. Thus temperatures are highest in the lowlands, which are known in Spanish as the *tierra caliente*, or "hot land." The *tierra caliente* extends up to about 3000 feet (1000 meters), and in some parts of the region these lowlands cover wide expanses. Where moisture is adequate, tropical rain forests thrive, as does a wide range of tropical crops, notably bananas, sugarcane, cacao, and pineapples. Many coastal areas of the *tierra caliente*, such as northeastern Brazil, have become zones of plantation agriculture that support populations of considerable size.

Between 3000 and 6500 feet (1000 to 2000 meters) is the cooler *tierra templada* ("temperate land"). The year-round, spring-like climate of this zone drew large numbers of indigenous people in the distant past and, more recently, has drawn Europeans. Here, crops such as roses, corn, beans, squash, various green vegetables, wheat, and coffee are grown.

Between 6500 and 12,000 feet (2000 to 3600 meters) is the *tierra fria* ("cool land"). A variety of crops, such as wheat, fruit trees, potatoes, and cool-weather vegetables—cabbage and broccoli, for example—do very well at this altitude. Many animals, including dogs, llamas, sheep, and guinea pigs (Figure 3.7 on page 117), are raised for pets, food, and fiber. Several modern population centers are in this zone, including Mexico City, Mexico, and Quito, Ecuador.

Above 12,000 feet (3600 meters) is the *tierra helada* ("frozen land"). In the highest reaches of this zone, vegetation is almost absent, and mountaintops emerge from under snow and glaciers. A remarkable feature of such tropical mountain zones is that in a single day of strenuous hiking, one can encounter many of the climate types found on Earth.

Precipitation The pattern of precipitation throughout the region is influenced by the interaction of global wind patterns with mountains and ocean currents (see Figure 1.11 on pages 18–19). The **trade winds** sweep off the Atlantic, bringing heavy seasonal rains to places roughly 23° north and south of the equator (see the Figure 3.5 map). Winds from the Pacific bring seasonal rains to the west coast of Central America, but mountains block those rains from reaching the Caribbean side, which receives heavy rainfall from the northeast trade winds.

The Andes are a major influence on precipitation in South America. They block the rains borne by the trade winds off the Atlantic into the Amazon Basin and farther south, creating a rain shadow on the western side of the Andes in northern Chile and southwestern Peru (see Figure 3.5). Southern Chile is in the path of eastward-trending winds that sweep off of the Southern Ocean, bringing steady, cold rains that support forests similar to those of the Pacific Northwest in North America. The Andes block this flow of wet, cool air and divert it to the north. They thereby create another extensive rain shadow on the eastern side of the mountains along the southeastern coast of Argentina (Patagonia).

Adjacent oceans and their currents also influence the pattern of precipitation. Along the west coasts of Peru and Chile, the cold surface waters of the Peru Current bring cold air that cannot carry much moisture. The combined effects of the Peru Current and the central Andes rain shadow have created what is possibly the world's driest desert, the Atacama of northern Chile (see Figure 3.5B).

El Niño One aspect of the Peru Current that is only partly understood is its tendency to change direction every few years (on an irregular cycle, possibly linked to sunspot activity). When this

FIGURE 3.5   **PHOTO ESSAY: Climates of Middle and South America**

Northeast trade winds bring heavy seasonal rains.

Seasonal winds bring rains.

Peru Current brings cold surface waters. Air above is very dry.
El Niño brings warm water instead of cold every few years.

Rain shadow: The Andes block winds off the Atlantic.

Southeast trade winds bring rains.

Globe-encircling eastward-blowing winds bring steady cold rains.

Rain shadow: The Andes block rains coming from the west.

Climate Zones

Tropical humid climates (A)
- Tropical wet
- Tropical wet/dry

Arid and semiarid climates (B)
- Desert
- Steppe

Temperate climates (C)
- Midlatitude, moist all year
- Subtropical, winter dry
- Mediterranean, summer dry

Cool humid climates (D)
- Continental, winter dry
- Continental, moist all year

Coldest climates (E)
- Arctic
- High altitude

→ Winds
→ Ocean currents

A Tropical wet, Belém, Brazil
Holger Leue/Lonely Planet Images/Getty Images 

B Desert, Atacama, Chile 
Veronique Durruty/Gamma-Rapho via Getty Images

C Continental, moist all year, Tierra del Fuego 
DEA/A. Garozzo/De Agostini/Getty Images

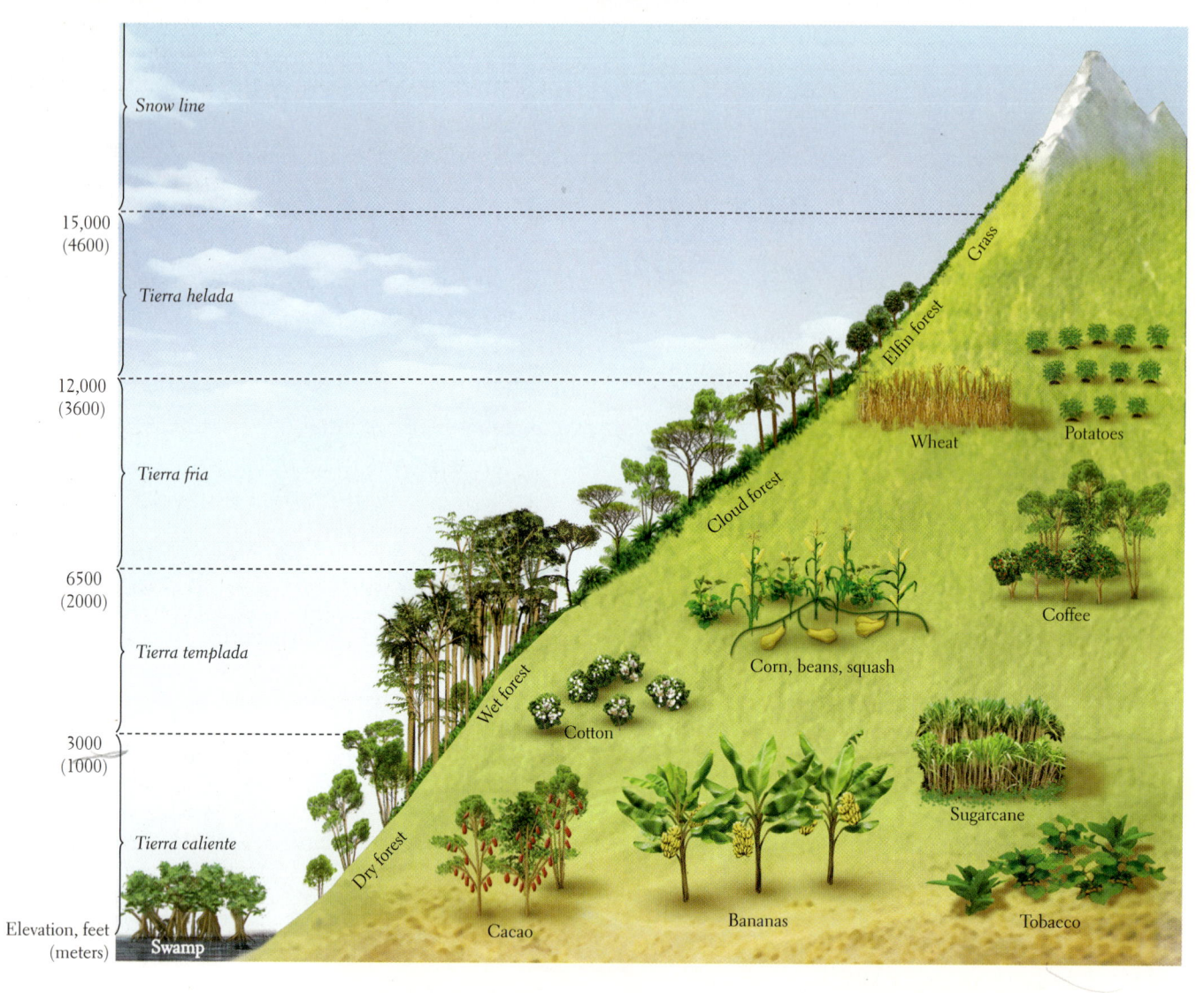

FIGURE 3.6 Temperature-altitude zones of Middle and South America. Temperatures tend to decrease as altitude increases, resulting in changes in the natural vegetation on mountainsides, as shown here. The same is true for crops, some of which are suited to lower, warmer elevations and some to higher, cooler ones.

happens, warm water flows eastward from the western Pacific, bringing warm water and torrential rains, instead of cold water and dry weather, to parts of the west coast of South America. The phenomenon was named **El Niño**, or "the Christ Child," by Peruvian fishermen, who noticed that when it does occur, it reaches its peak around Christmastime.

El Niño also has global effects, bringing cold air and drought to normally warm and humid western Oceania and unpredictable weather patterns to Mexico and the southwestern United States. The El Niño phenomenon in the western Pacific is discussed further in Chapter 11, where Figure 11.6 (page 453) illustrates its trans-Pacific effects.

El Niño periodic climate-altering changes, especially in the circulation of the Pacific Ocean, now understood to operate on a global scale

Hurricanes In this region, many coastal areas are threatened by powerful storms that can create extensive damage and loss of life. These form annually, primarily in the Atlantic Ocean north of the equator and close to Africa. A tropical storm begins as a group of thunderstorms. A few hurricanes also form in the southeastern Pacific and can affect the western coasts of Middle America before turning west toward Hawaii. When enough warming wet air comes together, the individual storms organize themselves into a swirling spiral of wind that moves across the Earth's surface. The highest wind speeds are found at the edge of the eye, or center, of the storm. Once wind speeds reach 75 miles (121 kilometers) per hour, such a storm is officially called a *hurricane*. Hurricanes usually last about 1 week; because they draw their energy from warm surface waters, they slow down and eventually dissipate as they move over cooler water or land. As the populations of coastal areas grow, more people are being exposed to the effects of hurricanes. Some scientists also think that climate change is leading to an increase in the number and intensity of hurricanes (see Figure 3.10B on page 121).

FIGURE 3.7 LOCAL LIVES
People and Animals in Middle and South America

A The Mexican hairless dog, or *xolo*, was bred some 3000 years ago as a hunting dog and companion, and for food. The xolo was considered sacred by the Aztecs, Maya, and other indigenous groups, all of whom believed the dogs helped their masters' souls pass safely through the dangers of the underworld realm of Mictlan. [Raphael Gaillarde/Gamma-Rapho via Getty Images]

B *Cuy*, or guinea pigs, have been raised throughout the Andes region for at least 5000 years. Primarily a source of meat, they are also considered spiritual mediums by traditional Andean healers who use them to diagnose illnesses in people. [Education Images/UIG via Getty Images]

C A llama being led by a girl in traditional Quechua dress in Cusco, Peru. Domesticated by pre-Incan peoples thousands of years ago, llamas were raised for their fur, their meat, and their labor as pack animals, and later they started to be used as guard animals for sheep, which were introduced by the Spanish. [Michele Burgess/Photolibrary/Getty Images]

THINGS TO REMEMBER

- Middle and South America extend south from the midlatitudes of the Northern Hemisphere across the equator through the Southern Hemisphere, nearly to Antarctica.

- A nearly continuous chain of mountains stretches along the western edge of the American continents for more than 10,000 miles (16,000 kilometers), from Alaska in the north to Tierra del Fuego at the southern tip of South America.

- The rain forests of the Amazon Basin are planetary treasures of biodiversity that also play a key role in regulating the Earth's climate.

- There are four temperature-altitude zones in the region that influence where and how people live and the crops they can grow.

- Significant environmental hazards include earthquakes, volcanic eruptions, and hurricanes.

ENVIRONMENTAL ISSUES

> ### GEOGRAPHIC INSIGHT 1
>
> **Environment:** Deforestation in this region contributes significantly to global climate change through the removal of trees, which as living plants naturally absorb carbon dioxide, a greenhouse gas. Once the trees are cut or burned, they release large amounts of carbon dioxide into the atmosphere. Additionally, despite the region's overall abundant water resources, some areas are experiencing a water crisis related to climate change, inadequate water infrastructure, and intensified use of water.

Environments in Middle and South America have long inspired concern about the use and misuse of the Earth's resources. Millennia before Europeans arrived in this region, human settlements in the Americas had major environmental impacts. However, today's impacts are particularly severe because both population density and per capita consumption have increased so dramatically. Moreover, local environments now supply global demands.

Tropical Forests, Climate Change, and Globalization

A period of rapid deforestation that has had global repercussions began in the 1970s in Brazil with the construction of the Trans-Amazon Highway. Migrant farmers followed the new road into the rain forest. They began clearing terrain to grow crops, prompting a few observers to warn of an impending crisis of deforestation. Initially, concern focused on the loss of plant and animal species, as the rain forests of the Amazon Basin are some of the most biodiverse on the planet. (Note that the Amazon includes Colombia, Peru, Bolivia, and Brazil; see Figure 3.1 on page 108.) However, climate change now dominates concerns about deforestation in the Amazon and the rest of this region. As explained in Chapter 1 (see pages 21–23), forests release oxygen and absorb carbon dioxide (CO_2), the greenhouse gas most responsible for global warming. The loss of large forests, such as those in the Amazon Basin, contributes to global warming by releasing CO_2 once locked in the woody bodies of trees. This CO_2 release happens as trees are burned to make way for crops and roads. Also, when there are fewer trees, less CO_2 can be absorbed from the atmosphere. Together, all of the Earth's tropical rain forests absorb about 18 percent of the CO_2 added to the atmosphere yearly by human activity. Because 50 percent of the Earth's remaining tropical rain forests are in South America, keeping these forests intact is crucial to controlling climate change.

Middle and South American rain forests are being diminished by multiple human impacts (Figure 3.8). One of the biggest impacts is caused by the clearing of land to raise cattle and grow crops such as soybeans (for animal feed), sugarcane (for ethanol), and African oil palm (for cooking oil) (see Figures 3.8C–E). Brazil now ranks as the world's fourth-largest emitter of greenhouse gases, after the United States, China, and Indonesia. In the Amazon, Middle America, and elsewhere in the region, forests are cleared to create pastures for beef cattle, many of which are destined for the U.S. fast-food industry. If deforestation continues in Middle America at the present rates, the natural forest cover will be entirely gone in 20 years.

Hardwood logging and the extraction of underlying minerals, including oil, gas, and precious stones, also contribute to deforestation. Investment capital in logging industries is increasingly coming from Asian multinational companies that have turned to the Amazon forests after having logged as much as 50 percent of the tropical forests in Southeast Asia. The access roads that have been built to support these activities help accelerate deforestation by opening new forest areas to migrants. The governments of Peru, Ecuador, and Brazil encourage impoverished urban people to occupy cheap land along the newly built roads (see Figure 3.8A), and provide chainsaws to the settlers to help remove the trees. However, the settlers have had a difficult time learning to cultivate the poor soils of the Amazon. After a few years of farming, they often abandon the land, now eroded and depleted of nutrients, and move on to new plots. Ranchers sometimes then buy the worn-out land from these failed small farmers to use as cattle pastures (see Figure 3.8D).

While Middle and South America contribute significantly to global warming by adding CO_2 to the Earth's atmosphere, many people in this region are particularly vulnerable to the increasing threats of climate change. Figure 3.10 on page 121 illustrates several such cases: shortages of clean water brought on by intensifying droughts and glacial melting; vulnerability to rising sea levels; and the effects of increasingly violent storms.

There is now an international effort to combat climate change by preserving the world's remaining forests—especially the crucial tropical rain forests found in this region. One of the cheapest ways to reduce greenhouse gas emissions is to integrate sustainable yet profitable agroforestry (tree cropping) into existing forests. When well funded and carefully executed, such efforts have the potential to save forests throughout the region. However, implementing complex agroforestry programs requires cooperation between diverse groups (see page 415 in Chapter 10).

Environmental Protection and Economic Development

In the past, governments in the region argued that economic development was so desperately needed that environmental regulations were an unaffordable luxury. Now, though, these governments are beginning to embrace both economic development and environmental protection. One example of this new approach is ecotourism.

> **ecotourism** nature-oriented vacations, often taken in endangered and remote areas, usually by travelers from affluent nations

Ecotourism Many countries are now trying to earn money from the beauty of still-intact natural environments through **ecotourism**—in which nature-oriented vacations, often taken in endangered areas, are offered to travelers usually hailing from affluent cities or foreign nations. Ecotourists travel to these places so they can see and appreciate ecosystems and wildlife that do not exist where they live (Figure 3.9; see also "On the Bright Side" on page 120).

Ecotourism has its downsides, however. Mismanaged, it can be similar to other kinds of tourism that damage the environment and return little to the surrounding community. While the profits of ecotourism can potentially be used to benefit local communities and environments, the profit margins may be small.

VIGNETTE

Puerto Misahualli, a small river boomtown in the Ecuadorian Amazon, is currently enjoying significant economic growth. Its prosperity is due to the many European, North American, and other foreign travelers who come for experiences that will bring them closer to the now-legendary rain forests of the Amazon.

The array of ecotourism offerings can be perplexing. One indigenous man offers to be a visitor's guide for as long as desired, traveling by boat and on foot, camping out in "untouched forest teeming with wildlife." His guarantee that they will eat monkeys and birds does not seem to promise the nonintrusive, sustainable experience the visitor might be seeking. At a well-known "eco-lodge," visitors are offered a plush room with a river view, a chlorinated swimming pool, and a fancy restaurant serving "international cuisine." All of this is on a private, 740-acre nature reserve separated from the surrounding community by a wall topped with broken glass. It seems more like a fortified resort than an eco-lodge.

By contrast, the solar-powered Yachana Lodge has simple rooms and local cuisine. Its knowledgeable resident naturalist is a veteran of many campaigns to preserve Ecuador's wilderness. Profits from the lodge fund a local clinic, a high school, and various programs that teach sustainable agricultural methods that protect the fragile Amazon soils while increasing farmers' earnings from surplus produce. The nonprofit group running the Yachana Lodge—the Foundation for Integrated Education and Development—earns just barely enough to sustain the clinic, the school, and the agricultural programs. *[Source: Alex Pulsipher's field notes in Ecuador.]* ■

The Water Crisis

Although Middle and South America receive more rainfall than any other world region and have three of the world's six largest rivers (in volume), parts of the region are experiencing water crises. Most of the factors causing the water crises are induced by humans, who in turn are now exposed to a variety of

FIGURE 3.8   PHOTO ESSAY: Human Impacts on the Biosphere in Middle and South America

While there are a wide variety of human impacts on the environments of this region, here we focus on the processes of land cover and land use change that lead to the conversion of forests to grazing land or farmland. The forces guiding this process are complex, driven by poor people's need for livelihoods, governments' desire to assert control over lightly populated areas, and the demands for wood, meat, and food that arise in distant urban centers and the global market. Together, these forces have led to a rapid loss of forest cover throughout much of this region. Brazil loses more forest cover each year than does any other country on the planet.

A A newly built road in Suriname.

B A family on a river raft in the Peruvian Amazon.

C African oil palm is planted on land recently cleared for agriculture.

D Cattle on land that has just been burned in Rondônia, Brazil.

E Soy fields recently cleared of forest in Mato Grosso, Brazil.

Human Impact, 2002

Land cover:
- Forests
- Grasslands
- Deserts
- Tundra
- Ice
- National boundaries

Overfishing:
- Threatened fisheries

Human impact on land:
- High impact
- Medium–high impact
- Low–medium impact

Acid rain:
- 4.8–4.3 pH
- 5.5–4.9 pH

water-related stresses. For example, the air pollution that hangs over so many cities in Middle and South America contributes to global warming, which in turn accelerates glacial melting in the Andes. The glaciers feed rivers that are the main source of water for millions of people (see Figure 3.10D). Should the glaciers actually disappear, many rivers will run much lower for parts of the year, straining communities and industries that depend on them.

Lax environmental policies have allowed industries and cities to pollute both air and water with few restraints. Waterways along the Mexican border with the United States are polluted by numerous factories set up to take advantage of NAFTA-related trade with the United States and Canada (see the discussion of maquiladoras on page 131). Only a small percentage of the often highly toxic water discharges are treated and disposed of properly. Even in Mexico City, the region's largest and wealthiest city, it is estimated that as much as 90 percent of urban wastewater goes untreated.

Rapid urbanization, combined with corruption, inadequate investment, and misguided policies, has left many people without access to clean water or sanitation. In some of the largest cities, the water infrastructure is so inadequate that as much as 50 percent of the fresh water is lost because of leaky pipes (see Figure 3.10A). Often as much as 80 percent of the population has no access to decent sanitation. Toilets are sometimes entirely lacking, leaving people to relieve themselves on the city streets. This poses a major health hazard, and as local water resources become too polluted to be drinkable, many cities must bring in potable water from distant areas.

Behind these immediate problems lie systemic policy failures and inadequate planning, such as that in Cochabamba, Bolivia (discussed in Chapter 1 on page 27). There, efforts to improve the city's inadequate water supply system focused on "marketizing" the water system. The water supply, long thought of as a public resource, was sold to a group of foreign corporations led by Bechtel of San Francisco, California. The hope was that Bechtel, in return for profits, would make investments in infrastructure that Cochabamba's notoriously corrupt water utility would not make. Unfortunately, Bechtel, unfamiliar with the actual living conditions of the majority of its citizens, immediately increased water prices to levels that few urban residents could afford, while doing little to improve water supply or delivery systems. At one point Bechtel even charged urban residents for water taken from their own wells and rainwater harvested off their roofs! Popular protests forced Bechtel to abandon Cochabamba's water utility, which remains plagued by corruption and an inadequate infrastructure.

ON THE BRIGHT SIDE

Alternatives to Deforestation

Ecotourism is now the most rapidly growing segment of the global tourism and travel industry, which by some measures is the world's largest industry, accounting for $3.5 trillion in annual expenditures. Many Middle and South American nations also have spectacular national parks that can provide a basis for ecotourism. As an alternative to deforestation, development based on ecotourism has the potential to preserve this region's biodiversity, reduce emissions of greenhouse gases, and provide less affluent and indigenous people with a chance to use their skills to teach tourists.

THINGS TO REMEMBER

> GEOGRAPHIC INSIGHT 1

• **Environment** Deforestation in this region contributes significantly to global climate change through the removal of trees, which as living plants naturally absorb carbon dioxide, a greenhouse gas. Once the trees are cut or burned, they release large amounts of carbon dioxide into the atmosphere. Additionally, despite the region's overall abundant water resources, some areas are experiencing a water crisis related to climate change, inadequate water infrastructure, and intensified use of water.

FIGURE 3.9 Ecotourism in the Amazon.

(A) A tourist poses at the bottom of a giant ceiba tree in the Ecuadorian Amazon.

(B) An Amazon river dolphin being fed by an ecotour guide.

(C) A tourist explores a walkway suspended in the canopy of tall rainforest trees in the Ecuadorian Amazon.

FIGURE 3.10 PHOTO ESSAY: Vulnerability to Climate Change in Middle and South America

Adapting to the multiple stresses that climate change is bringing to this region is proving to be quite a challenge. Water-related troubles, such as drought, hurricanes, flooding, and glacial melting, are combining with growing populations and persistent poverty to create a complex landscape of vulnerability to climate change.

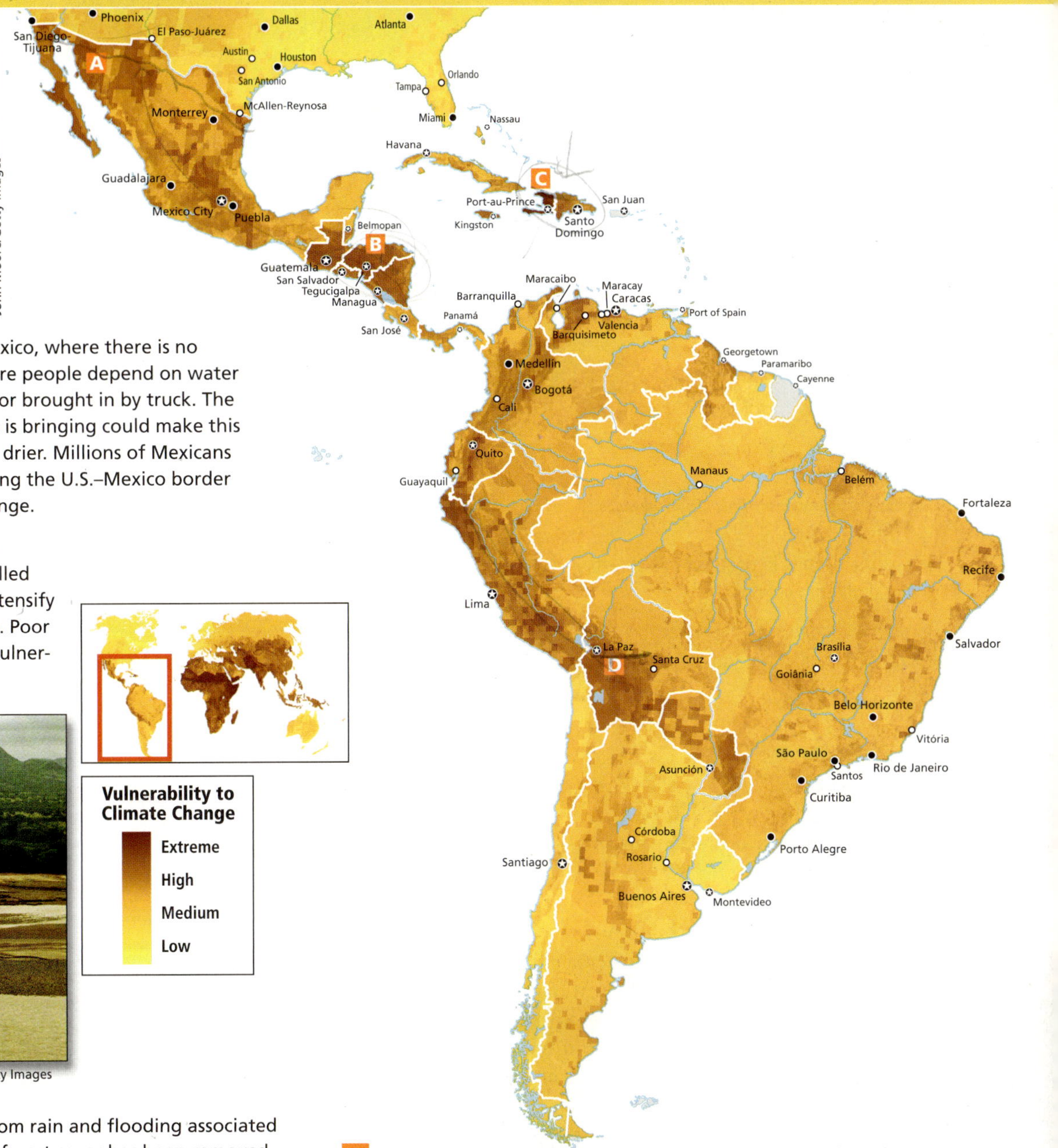

A A poor neighborhood in Nogales, Mexico, where there is no centralized water infrastructure and where people depend on water gathered off of the roofs of their homes or brought in by truck. The higher temperatures that climate change is bringing could make this area, where water is already scarce, even drier. Millions of Mexicans who have moved to work in factories along the U.S.–Mexico border are thus highly vulnerable to climate change.

B A major bridge in Honduras that was washed out by Hurricane Mitch, which killed 18,000 people. Hurricanes are likely to intensify as temperatures rise with climate change. Poor countries like Honduras are particularly vulnerable to the damage these storms bring.

Vulnerability to Climate Change
- Extreme
- High
- Medium
- Low

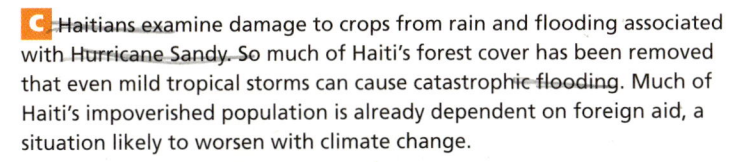

C Haitians examine damage to crops from rain and flooding associated with Hurricane Sandy. So much of Haiti's forest cover has been removed that even mild tropical storms can cause catastrophic flooding. Much of Haiti's impoverished population is already dependent on foreign aid, a situation likely to worsen with climate change.

D La Paz and many smaller cities and towns in Bolivia's drought-prone highlands receive much of their drinking water from glaciers in the Andes. Higher temperatures are causing these glaciers to melt rapidly. Many have already disappeared and the rest could be gone in 15 years.

A Teotihuacan, Mexico, once home to an estimated 150,000 to 250,000 people.
[Bjorn Holland/The Image Bank/Getty Images]

B Machu Picchu, an estate built for the Incan emperor of the fifteenth century.
[Lluís Vinagre/Flickr/Getty Images]

23,000 B.C.E.–12,000 B.C.E. Bering land bridge

200 B.C.E.–800 C.E. Teotihuacan flourishes

1325 C.E. Aztec capital of Tenochtitlán founded

FIGURE 3.11 VISUAL HISTORY OF MIDDLE AND SOUTH AMERICA

- One of the biggest impacts on rain forests is the clearing of land to raise cattle and grow crops.
- Hardwood logging and the extraction of underlying minerals, including oil, gas, and precious stones, are also contributing to deforestation.
- Governments in the region are beginning to embrace both economic development and environmental protection. One example of this new approach is ecotourism.
- Lax environmental policies in much of Middle and South America have allowed industries and cities to pollute both air and water with few restraints.

HUMAN GEOGRAPHY

HUMAN PATTERNS OVER TIME

The conquest of Middle and South America by Europeans set in motion a series of changes that helped create the ways of life found in this region today. The conquest wiped out most of the indigenous civilizations, and the conquerors set up new societies in their place. Many cultural features from the time of European colonialism endure to this day, as do some vestiges of the precolonial era.

The Peopling of Middle and South America

Recent evidence suggests that between 14,000 and 25,000 years ago, groups of hunters and gatherers from northeastern Asia spread throughout North America after crossing the Bering land bridge on foot, or moving along shorelines in small boats, or both. Some of these groups ventured south across the Central American isthmus, reaching the tip of South America by about 13,000 years ago.

By 1492, there were 50 to 100 million indigenous people in Middle and South America. In some places, population densities were high enough to threaten sustainability. People altered the landscape in many ways. They modified drainage to irrigate crops, they terraced hillsides, and they built paved walkways across swamps and mountains. They constructed cities with sewer systems, freshwater aqueducts, and huge earthen and stone ceremonial structures that rivaled the pyramids of Egypt (Figure 3.11A).

The indigenous people also practiced the system of **shifting cultivation** that is still common in wet, hot regions in Central America and the Amazon Basin. In this system, small plots are cleared in forestlands, the brush is dried and burned to release nutrients in the soil, and the clearings are planted with multiple crop species. Each plot is used for only 2 or 3 years and then abandoned for several decades—long enough to allow the forest to regrow. If there is sufficient land, this system is highly productive per unit of land and labor, and is sustainable for long periods of time. However, if population pressure increases to the point that a plot must be used before it has fully regrown and its fertility has been restored, its yields decrease drastically.

The **Aztecs** of the high central valley of Mexico had some technologies and social systems that rivaled or surpassed those of Asian and

shifting cultivation a productive system of agriculture in which small plots are cleared in forestlands, the dried brush is burned to release nutrients, and the clearings are planted with multiple species; each plot is used for only 2 or 3 years and then abandoned for many years of regrowth

Aztecs indigenous people of high-central Mexico noted for their advanced civilization before the Spanish conquest

122

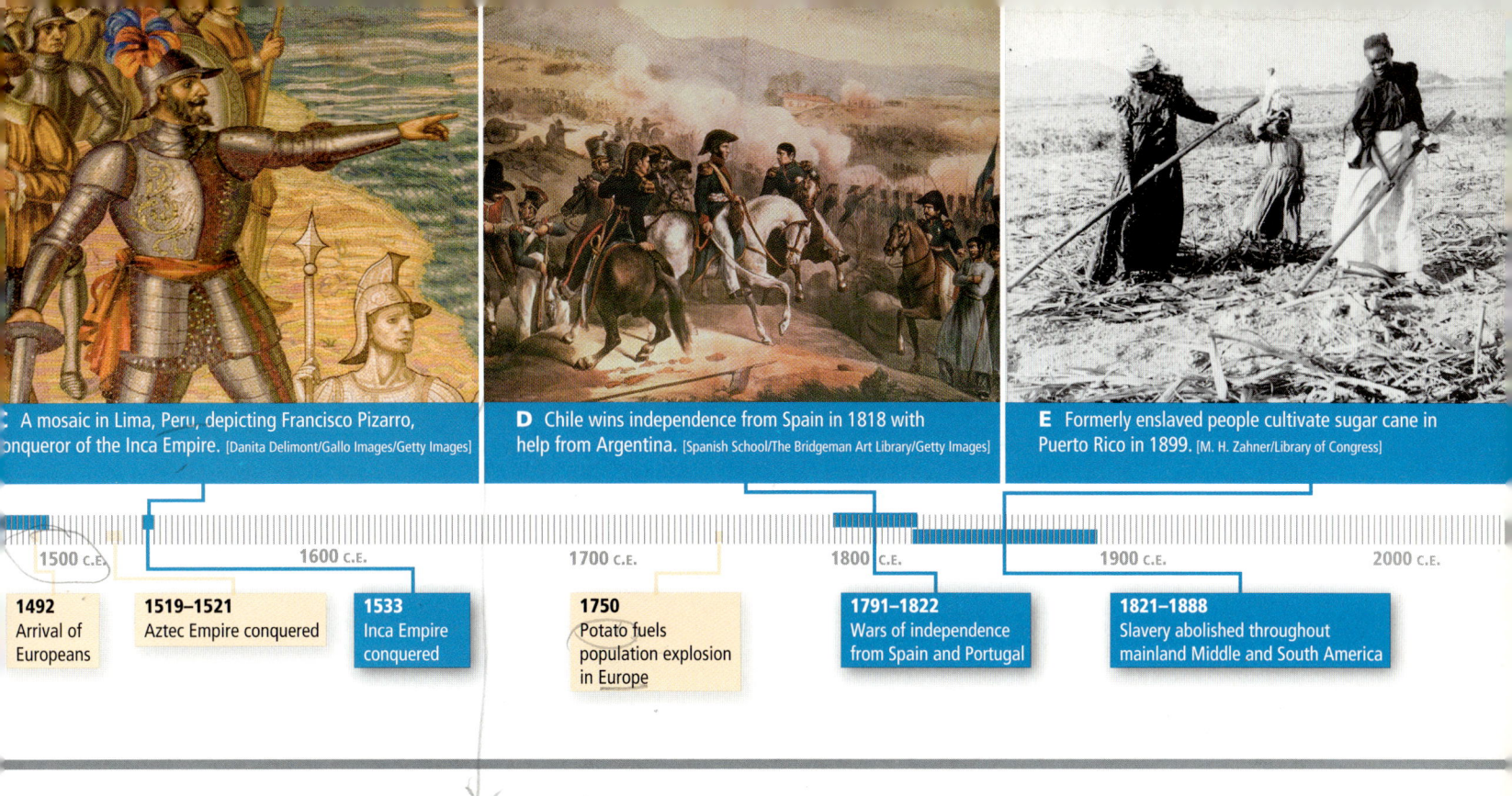

A A mosaic in Lima, Peru, depicting Francisco Pizarro, conqueror of the Inca Empire. [Danita Delimont/Gallo Images/Getty Images]

D Chile wins independence from Spain in 1818 with help from Argentina. [Spanish School/The Bridgeman Art Library/Getty Images]

E Formerly enslaved people cultivate sugar cane in Puerto Rico in 1899. [M. H. Zahner/Library of Congress]

1492	Arrival of Europeans
1519–1521	Aztec Empire conquered
1533	Inca Empire conquered
1750	Potato fuels population explosion in Europe
1791–1822	Wars of independence from Spain and Portugal
1821–1888	Slavery abolished throughout mainland Middle and South America

European civilizations of the time. Particularly well developed were urban water supplies, sewage systems, and elaborate marketing systems. Historians have concluded that by 1500 C.E., Aztecs probably lived more comfortably on the whole than their contemporaries in Europe.

In 1492, the largest state in the region was that of the **Incas**, stretching from what is now southern Colombia to northern Chile and Argentina. The main population clusters were in the Andes highlands, where the cooler temperatures at these high altitudes eliminated the diseases of the tropical lowlands, while proximity to the equator guaranteed mild winters and long growing seasons. For several hundred years, the Inca Empire was one of the most efficiently managed empires in the history of the world. Highly organized systems of labor were used to construct paved road systems, elaborate terraces, irrigation systems, and great stone cities in the Andean highlands (see Figure 3.11B). Incan agriculture was advanced, particularly in the development of crops, which included numerous varieties of potatoes and grains.

Incas indigenous people who ruled the largest pre-Columbian state in the Americas, with a domain stretching from what is now southern Colombia to northern Chile and Argentina

European Conquest

The European conquest of Middle and South America was one of the most significant events in human history (see Figure 3.11C). It rapidly altered landscapes and cultures and, through disease and slavery, ended the lives of millions of indigenous people.

Columbus established the first Spanish colony in 1492 on the Caribbean island of Hispaniola (which is now occupied by Haiti and the Dominican Republic). After learning of Columbus's exploits, other Europeans, mainly from Spain and Portugal on Europe's Iberian Peninsula, conquered the rest of Middle and South America.

The first part of the mainland to be invaded was Mexico, home to several advanced indigenous civilizations, most notably the Aztecs. The Spanish were unsuccessful in their first attempt to capture the Aztec capital of Tenochtitlán, but they succeeded a few months later after a smallpox epidemic decimated the native population. The Spanish demolished the grand Aztec capital in 1521 and built Mexico City on its ruins.

A tiny band of Spaniards, again aided by a smallpox epidemic, conquered the Incas in South America. Out of the ruins of the Inca Empire, the Spanish created the Viceroyalty of Peru, which originally encompassed all of South America except Portuguese Brazil. The newly constructed capital of Lima flourished, in large part as a transshipment point for enormous quantities of silver extracted from mines in the highlands of what is now Bolivia.

Diplomacy by the Roman Catholic Church prevented conflict between Spain and Portugal over the lands of the region. The Treaty of Tordesillas of 1494 divided Middle and South America at approximately 46° W longitude (**Figure 3.12**). Portugal took all lands to the east and eventually acquired much of what is today Brazil; Spain took all lands to the west.

The superior military technology of the Spanish and Portuguese sped the conquest of Middle and South America. A larger factor, however, was the vulnerability of the indigenous people to diseases carried by the Europeans. In the 150 years following 1492, the total population of Middle and South America was reduced by more than 90 percent to just 5.6 million. To obtain a new supply of labor to replace the dying indigenous people, the Spanish initiated the first shipments of enslaved Africans to the region in the early 1500s.

By the 1530s, a mere 40 years after Columbus's arrival, all major population centers of Middle and South America had been conquered and were rapidly being transformed by Iberian colonial

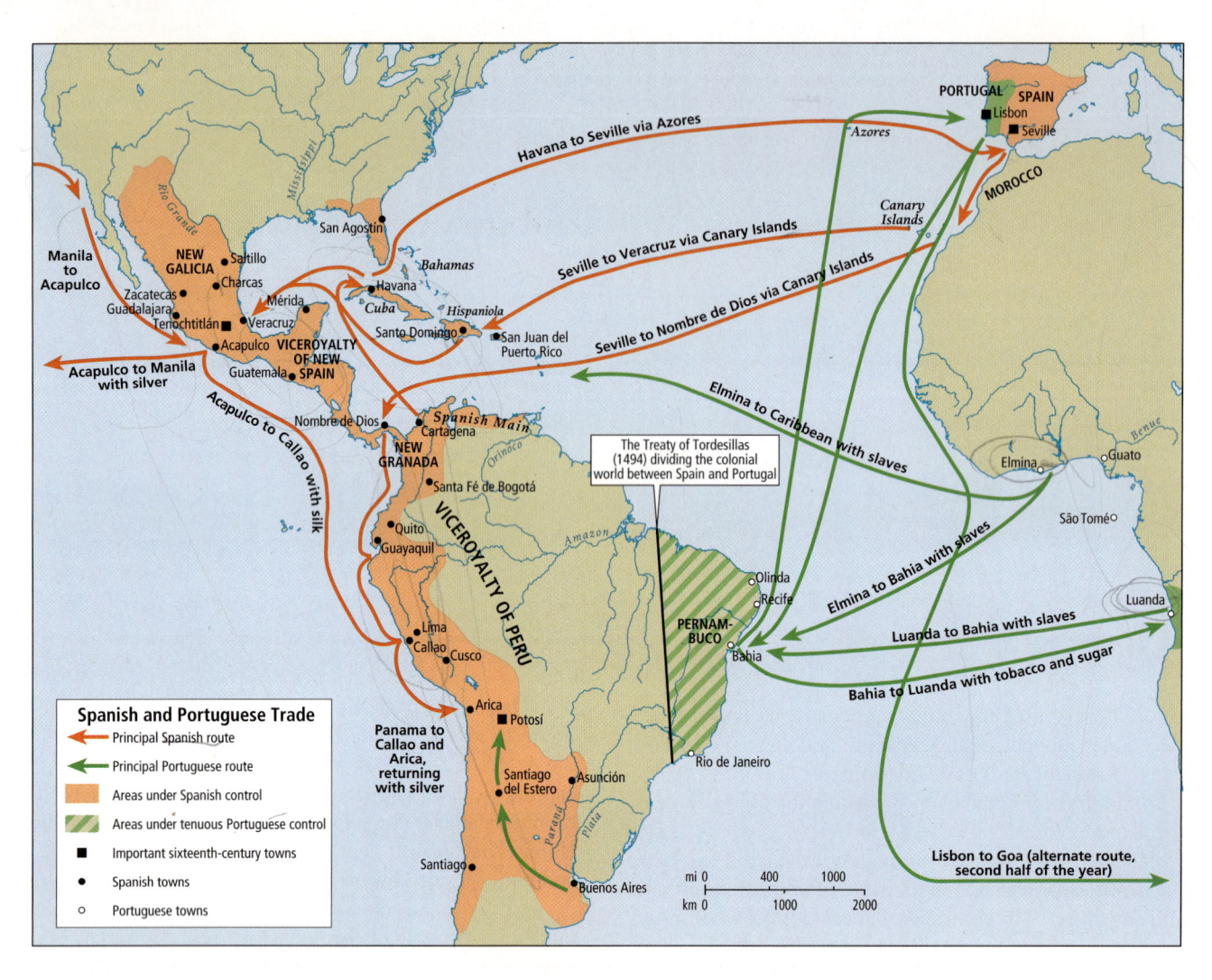

FIGURE 3.12 Spanish and Portuguese trade routes and territories in the Americas, circa 1600. The major trade routes from Spain to its colonies led to the two main centers of its empire, Mexico and Peru. The Spanish colonies could trade only with Spain, not directly with one another. By contrast, there were direct trade routes from Portuguese colonies in Brazil to Portuguese outposts in Africa. Many millions of Africans were enslaved and traded to Brazilian plantation and mine owners (as well as to Spanish, British, French, and Dutch colonies in Middle and South America). [Source consulted: *Hammond Times Concise Atlas of World History* (Maplewood, NJ: Hammond, 1994), pp. 66–67]

policies. The colonies soon became part of extensive regional and global trade networks, the latter with Europe, Africa, and Asia.

A Global Exchange of Crops and Animals

From the earliest days of the conquest, plants and animals were exchanged between Middle and South America, Europe, Africa, and Asia via the trade routes illustrated in Figure 3.12. Many plants essential to agriculture in Middle and South America today—rice, sugarcane, bananas, citrus, melons, onions, apples, wheat, barley, and oats, for example—were all originally imports from Europe, Africa, or Asia. When disease decimated the native populations of the region, the colonists turned much of the abandoned land into pasture for herd animals imported from Europe, including sheep, goats, oxen, cattle, donkeys, horses, and mules.

Just as consequential were plants first domesticated by indigenous people of Middle and South America. These plants have changed diets everywhere and have become essential components of agricultural economies around the globe. The potato, for example, had so improved the diet of the European poor by 1750 that it fueled a population explosion. Manioc (cassava) played a similar role in West Africa. Corn, peanuts, vanilla, and cacao (the source of chocolate) are globally important crops to this day, as are peppers, pineapples, and tomatoes (Table 3.1).

The Legacy of Underdevelopment

In the early nineteenth century, wars of independence left Spain with only a few colonies in the Caribbean (see Figure 3.11D). Figure 3.13 shows the European colonizing countries and the

TABLE 3.1	Globally important domesticated plants that originated in the Americas
Type	**Names and places of origin**
Seeds Quinoa	Amaranth—*Amaranthus cruentus*, S. Mexico, Guatemala Beans—*Phaseolus* (4 species), S. Mexico Maize (corn)—*Zea mays*, valleys of Mexico Peanut—*Arachis hypogaea*, central lowlands of S. America Quinoa—*Chenopodium quinoa*, Andes of Chile and Peru Sunflower—*Helianthus annuus*, southwest and southeast N. America
Tubers Potato	Manioc (cassava)—*Manihot esculenta*, lowland of Middle and S. America Potato (numerous varieties)—*Solanum tuberosum*, Lake Titicaca region of Andes Sweet potato—*Ipomoea batatas*, S. America Tannia—*Xanthosoma sagittifolium*, lowland tropical America
Vegetables Tomato	Chayote (christophene)—*Sechium edule*, S. Mexico, Guatemala Peppers (sweet and hot)—*Capsicum* (various species), many parts of Middle and S. America Squash (including pumpkin)—*Cucurbita* (4 species), tropical and subtropical America Tomatillo (husk tomato)—*Physalis ixocarpa*, Mexico, Guatemala Tomato (numerous varieties)—*Lycopersicon esculentum*, highland S. America
Fruit Pineapple	Avocado—*Persea americana*, S. Mexico, Guatemala Cacao (chocolate)—*Theobroma cacao*, S. Mexico, Guatemala Papaya—*Carica papaya*, S. Mexico, Guatemala Passion fruit—*Passiflora edulis*, central S. America Pineapple—*Ananas comosus*, central S. America Prickly pear cactus (huna)—*Opuntia* (several species), tropical and subtropical America Strawberry (commercial berry)—*Fragaria* (various species), genetic cross of Chilean berry and wild berry from N. America Vanilla—*Vanilla planifolia*, S. Mexico, Guatemala, perhaps Caribbean
Ceremonial and drug plants Coca	Coca (cocaine)—*Erythroxylon coca*, eastern Andes of Ecuador, Peru, and Bolivia Tobacco—*Nicotiana tabacum*, tropical America

dates of independence for the various Middle and South American countries. The supporters of the nineteenth-century revolutions were primarily **Creoles** (people of mostly European descent born in the Americas) and relatively wealthy **mestizos** (people of mixed European, African, and indigenous descent). The Creoles' access to the profits of the colonial system had been restricted by **mercantilism**, and the mestizos were excluded by racist colonial policies. Once these groups gained power, however, they became a new elite who controlled the state and monopolized economic opportunity. Because they did little to expand economic development or access

> **Creoles** people mostly of European descent born in the Americas
>
> **mestizos** people of mixed European, African, and indigenous descent
>
> **mercantilism** the policy by which European rulers sought to increase the power and wealth of their realms by managing all aspects of production, transport, and commerce in their colonies

to political power for the majority of their populations (see Figure 3.11E), the revolutions were incomplete.

Today, the economies of Middle and South America are much more complex and technologically sophisticated than they once were. Nevertheless, disparities persist: about 30 percent of the population is poor and lacks access to land, adequate food, shelter, water and sanitation, and basic education. Meanwhile, a small elite class has levels of affluence equivalent to those of the very wealthy in the United States. These conditions are in part the lingering result of colonial economic policies that favored the

126 CHAPTER 3 Middle and South America

FIGURE 3.13 The colonial heritage of Middle and South America. Most of Middle and South America was colonized by Spain and Portugal, but important and influential small colonies were held by Britain, France, and the Netherlands. Nearly all the colonies had gained independence by the late twentieth century. Those for which no date appears on the map are still linked in some way to the colonizing country. [Source consulted: *Hammond Times Concise Atlas of World History* (Maplewood, NJ: Hammond, 1994), p. 69]

export of raw materials and fostered privileges for wealthy elites and outside investors who often spent their profits elsewhere rather than reinvesting them within the region. These policies are further discussed in the "Globalization and Development" section below.

THINGS TO REMEMBER

- Recent evidence suggests that between 14,000 and 25,000 years ago, groups of hunters and gatherers from northeastern Asia spread throughout North America after crossing the Bering land bridge on foot, or moving along shorelines in small boats, or both.
- Shifting cultivation of small plots is a traditional and potentially sustainable agricultural strategy still used by some throughout the region.
- Spain and Portugal were the primary colonizing countries of Middle and South America, with smaller colonies held by Britain, France, Denmark, and the Netherlands. By the mid-nineteenth century, most of the larger colonies had gained independence.
- Exploitative development, incomplete revolutions, a dependence on raw materials exports, and the failure of elites to reinvest profits all led to underdevelopment in the region.

GLOBALIZATION AND DEVELOPMENT

> **GEOGRAPHIC INSIGHT 2**
>
> **Globalization and Development:** The integration of this region with the global economy has left it with the widest gap between rich and poor in the world. Long-term failures to address poverty, economic instability, and the flow of resources and money out of the region have resulted in many conflicts and inspired numerous efforts at reform. Nevertheless, in recent years several countries in this region have emerged as global economic leaders.

For decades, Middle and South America has been one of the poorer world regions—though not as poor, on average, as sub-Saharan Africa, South Asia, or Southeast Asia—and has had serious economic challenges. The unique colonial and postcolonial history of the region has resulted in wide economic inequality and income disparity, as well as persistent political disempowerment of poor people. In recent years, Middle and South America has emerged from a long series of efforts to change its economic trajectory. Several countries are now widely seen as the economic powerhouses of the future.

ECONOMIC INEQUALITY AND INCOME DISPARITY

With the exception of a few small countries, the **income disparity**—the gap between rich and poor—in this region is one of the biggest in the world. According to the most recent United Nations (UN) figures, the richest 10 percent of the population was between 19 and 94 times richer than the poorest 10 percent, depending on which country was being analyzed (Table 3.2). In recent years, disparities have been shrinking in Brazil, Chile, Mexico, and Venezuela, primarily because of new government economic and social policies aimed at reducing wide disparities. However, in Bolivia, Guatemala, Colombia, and Peru, disparities have recently increased. The poverty rate for the region as a whole is around 33 percent, but in the poorer countries (Haiti, Guatemala, Honduras, Nicaragua, Peru, Bolivia, and Paraguay), more than half the population now lives in poverty.

> **income disparity** the gap in income between rich and poor

Phases of Economic Development

The current economic and political situation in Middle and South America derives from the region's history, which can be divided into three major phases: the *early extractive phase*, the import substitution industrialization (ISI) phase, and the current structural adjustment and marketization phase, which includes sharply rising investment from abroad. All three phases have helped entrench wide income disparities despite a consensus that more egalitarian development is desirable.

Early Extractive Phase From the time European conquerors first arrived, economic development was guided by a policy of mercantilism in which Europeans extracted resources and controlled much of the economic activity in the American colonies in order to increase the power and wealth of their "mother" country. Even after the colonies became independent countries, many extractive enterprises were owned by foreign investors who had few incentives to build stable local economies.

A small flow of foreign investment and manufactured goods entered the region, while a vast flow of raw materials left for Europe and beyond. The money to fund the farms, plantations, mines, and transportation systems that enabled the extraction of resources for export came from abroad, first from Europeans and later from North Americans and other international sources.

TABLE 3.2 Income disparities in selected countries*

Ratio of wealth of richest 10% to poorest 10% of the population[†]

Country	HDI rank, 2001	1987–1995[‡] Richest 10% to poorest 10%	HDI rank, 2003	1998–2000[‡] Richest 10% to poorest 10%	HDI rank, 2009	2004–2007[‡,§] Richest 10% to poorest 10%
Middle and South America						
Bolivia	104	91:1	114	25:1	113	94:1
Brazil	69	49:1	65	66:1	75	41:1
Chile	39	34:1	43	43:1	44	26:1
Colombia	62	43:1	64	43:1	77	60:1
Guatemala	108	29:1	119	29:1	122	34:1
Mexico	51	26:1	55	35:1	53	21:1
Peru	73	23:1	82	22:1	78	26:1
Venezuela	61	24:1	69	44:1	58	19:1
Other selected countries						
China	87	13:1	104	13:1	92	13:1
France	13	9:1	17	9:1	8	9:1
Jordan	88	9:1	90	9:1	96	10:1
Philippines	70	16:1	85	17:1	105	14:1
South Africa	94	33:1	111	65:1	129	35:1
Thailand	66	12:1	64	13:1	87	13:1
Turkey	82	14:1	96	13:1	79	17:1
United States	6	17:1	7	17:1	13	16:1

*The UN used data from 1987–1995, 1988–2000, and 2004–2007 on either income or consumption to calculate an approximate representation of how much richer the wealthiest 10 percent of the population is than the poorest 10 percent. The lower the ratio, the more equitable the distribution of wealth in the country.

[†]Decimals rounded up or down.

[‡]Survey years fall within this range.

[§]Ratios are from the *United Nations Development Report 2009*.

Sources: *United Nations Human Development Report 2001*, Table 12; UNHDR 2003, Table 13; UNHDR 2009, Table M.

One example is the still highly lucrative Panama Canal. The French first attempted to build the canal in 1880. It was later completed and run by the United States and finally turned over to Panamanian control in 1999. The profits from these ventures were usually banked abroad, depriving the region of investment funds and tax revenues that could have made it more economically independent. Industries were slow to develop in the region, so even essential items, such as farm tools and household utensils, had to be purchased from Europe and North America at relatively high prices. Many people simply did without.

A number of economic institutions arose in Middle and South America to supply food and raw materials to Europe and North America. Large rural estates called **haciendas** were granted to colonists as a reward for conquering territory and people for Spain. For generations these essentially feudal estates were then passed down through the families of those colonists. Over time, the owners, who often lived in a distant city or in Europe, lost interest in the day-to-day operations of the haciendas. Productivity on these lands was generally low and hacienda laborers remained extremely poor. Nevertheless, haciendas produced a diverse array of products (cattle, cotton, rum, sugar) for local consumption and export.

Plantations were large factory farms. In addition to growing crops such as sugar, coffee, cotton, or (more recently) bananas, crops were often processed on site for shipment. Plantation owners made larger investments in equipment, and these farms were more efficient and profitable than haciendas. However, plantations had relatively little local economic impact. Instead of employing local populations, plantation owners imported enslaved people from Africa. The equipment that the plantations used was usually imported from Europe, which was also where plantation owners preferred to invest their profits. As a result, little money was available to support the development of local industries that could have grown up around the plantations.

First developed by the European colonizers of the Caribbean and northeastern Brazil in the 1600s, plantations became more common throughout Middle and South America by the late nineteenth century. Unlike haciendas, which were often established in the continental interior in a variety of climates, plantations were for the most part situated in tropical coastal areas with year-round growing seasons. Their coastal and island locations gave them easier access to global markets via ocean transport.

As markets for meat, hides, and wool grew in Europe and North America, the livestock ranch emerged, specializing in raising cattle and sheep. Today, commercial ranches serving such global markets as the fast-food industry are found in the drier grasslands and savannas of South America, Central America, and northern Mexico, and even in the wet tropics on freshly cleared rain forest lands.

Mining was another early extractive industry (Figure 3.14). Important mines (primarily gold and silver at first) were located on the island of Hispaniola and in north-central Mexico, the Andes, the Brazilian Highlands and in many other locations. Extremely inhumane labor practices were common in all of these mines. Today, oil and gas have been added to the mineral extraction industry, but rich mines throughout the region continue to produce gold, silver, copper, tin, precious gems, titanium, bauxite, and tungsten.

Profits from the region's mines, ranches, plantations, and haciendas continued to leave the region, even after the countries gained independence in the nineteenth century. One of the main reasons for this was that wealthy foreign investors retained control of many of the extractive enterprises.

FIGURE 3.14 A copper mine in northern Chile, where some of the largest copper mines in the world are located. Copper accounts for 13 percent of Chile's GDP, and Chile produces one-third of the world's copper, more than any other country. Chile's copper mines are dependent on imported machinery, such as large dump trucks that are made in the United States.

hacienda a large agricultural estate in Middle or South America, more common in the past; usually not specialized by crop and not focused on market production

plantation a large factory farm that grows and partially processes a single cash crop

import substitution industrialization (ISI) policies that encourage local production of machinery and other items that previously had been imported at great expense from abroad

Import Substitution Industrialization Phase After World War II, there were numerous political movements against the continuing domination of the economy and society by local elites and a few foreign investors. Some governments, notably those of Mexico and Argentina, tried to boost their economic independence by keeping money and resources within their borders through policies of **import substitution industrialization (ISI)**. Subsidies, among other measures, helped support the local production of machinery and other commonly imported items.

To encourage local people to buy manufactured goods from local suppliers, governments placed high tariffs on imported, usually higher-quality manufactured goods. The money and resources kept within each country were expected to provide the basis for further industrial development. These policies were intended to create well-paying jobs, raise living standards for the majority of people, and ultimately replace the extractive industries as the backbone of country economies.

The state-owned manufacturing sectors on which the success of ISI depended were never able to produce goods of a high-enough quality to compete with those produced in Asia, Europe, and North America. This was largely because the design as well as the technological and managerial skills needed to run globally competitive factories were lacking. Moreover, local consumers were not numerous or prosperous enough to support these industries. With demand for ISI products weak among domestic and foreign consumers alike, employment in these industries stagnated, tax revenues remained low, and social programs could not be adequately funded.

Not all state-owned corporations were losing propositions, however. ISI still survives in some countries, and periodically ISI schemes are considered for reimplementation in such places as Bolivia and Venezuela. Brazil—with its aircraft, armament, oil exports, and auto industries—and Mexico—with its oil and gas industries—both have had success with some ISI programs. Interest in state-supported manufacturing industries continues as a way to keep profits in-country; however, the general trend is now toward more market-oriented management and global competitiveness. Most countries still depend on the export of raw materials.

Beginning in the early 1970s, sharp increases in oil prices and decreases in global prices of raw materials ended a period of relative prosperity that had begun in the early 1950s. Reluctant to let go of the dream of rapid development, governments and private interests across the region continued to pursue ambitious plans to modernize and industrialize their national economies. They did this even as prices for raw material exports—their main source of income—fell. With false optimism, they paid for dam, road, and factory projects by borrowing millions of dollars from major international banks, most of which were in North America or Europe. When, in 1980, a global economic recession hit, not only were development plans halted, but a number of governments in the region also found that they were unable to repay their loans (Figure 3.15).

Structural Adjustment and the Marketization Phase Alarmed over the mounting debt of their clients, foreign banks that had made loans to governments in the region took action. The International Monetary Fund (IMF) developed and enforced policies that mandated profound changes in the organization of national economies. To ensure that sufficient money would be available to repay loans taken out from foreign banks to finance the now-discredited ISI phase, the IMF required **structural adjustment programs (SAPs)**—belt-tightening measures. These SAPs were based on concepts of **privatization** (the selling of formerly government-owned industries and firms to private investors), **marketization** (the development of a free market economy in support of free trade), and *globalization* (the opening of national economies to global investors; see Chapter 1, pages 31–34). At the time, these concepts were considered the soundest ways for countries to expand economically and thus to repay their debts to banks in North America, Europe, and Asia.

In the case of privatization, the investors to whom government firms were sold were multinational corporations located in North America, Europe, and Asia. Thus the SAP era resulted in industries across the region being returned to foreign ownership. The other main SAP policy, marketization, required that governments remove tariffs on imported goods of all types in order to obtain further loans. The removal of tariffs then caused many local industries to fail.

Crucially, SAPs also reversed the ISI-era trend of expanding government social programs and building infrastructure. To free up funds for debt repayment, governments were required to fire many civil servants and drastically reduce spending on public health, education, job training, day care, water systems, sanitation, and infrastructure building and maintenance. In other words, SAPs took government services away from the poor and middle-class workers. While severely cutting these badly needed government programs, SAPs encouraged the expansion of industries that were already earning profits by lowering taxes on their activities, thus further shrinking government revenues to pay for services. In the countries of Middle and South America, as in most developing countries, the most profitable industries remained those based on the extraction of raw materials for export. **61. PERUVIANS STRUGGLE TO GAIN HEALTH CARE ACCESS**

> **structural adjustment programs (SAPs)** policies that require economic reorganization toward less government involvement in industry, agriculture, and social services; sometimes imposed by the World Bank and the International Monetary Fund as conditions for receiving loans
>
> **privatization** the selling of formerly government-owned industries and firms to private companies or individuals
>
> **marketization** the development of a free market economy in support of free trade

VIGNETTE

In August of 2003, Orbalin Hernandez returned to his self-built shelter in the town of Mexicali on the border between Mexico and the United States. Recently fired for taking off his safety goggles while loading TV screens onto trucks at Thomson Electronics, he had just gone to the personnel office to ask for his job back. Because there were no previous problems with him, he was rehired at his old salary of $300 per month ($1.88 per hour). Thomson is a French-owned electronics firm that took advantage of NAFTA (see page 132) when it moved to Mexicali from Scranton, Pennsylvania, in 2001. There, its 1100 workers had been paid an average of $20 per hour. Twenty percent of the Scranton workers were unable to find work after Thomson left, and many still feel bitter toward the Mexicali workers.

In 2006, Orbalin and his fellow Mexicali workers were told that their wages were too high to allow their employers to compete with companies located in China, where in 2005, workers with the same skills as Orbalin earned just $0.35 an hour. Indeed, that year, 14 Mexicali plants closed and moved to Asia. Those firms remaining in Mexicali cut wages and reduced benefits.

By 2009, it appeared that the global recession was affecting Chinese–Mexican relations in new ways. Rising costs for fuel and storage and some questions about the quality of Chinese goods meant that it made less sense to move a Mexican factory to China. In fact, in 2012, China began to invest in Mexico in order to produce such things as motorcycles and trucks for the Mexican market. It is not yet clear whether this trend will be short-lived or the wave of the future.

Though Orbalin's salary at Thomson-Mexicali is barely enough for him to support his wife Mariestelle and their four children, he is grateful for the job. He and Mariestelle are originally from a farming community in the Mexican state of Tabasco, where the minimum wage in 2014 is $4.91 a day. [Source: NPR, ZNet, Bloomberg Businessweek, Maquiladora Portal. For detailed source information, see Text Sources and Credits.] ■

130 CHAPTER 3 Middle and South America

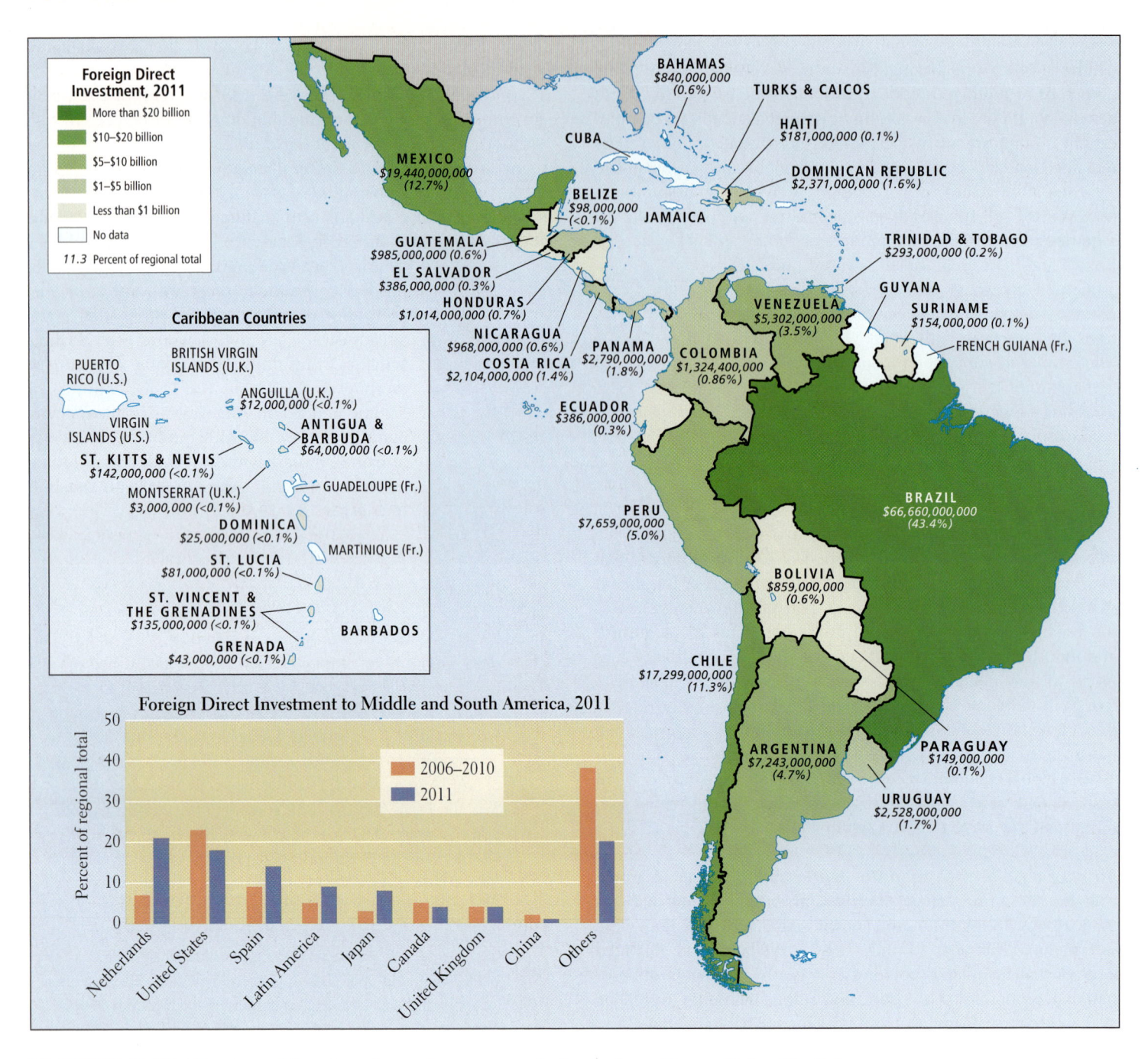

FIGURE 3.15 Origins of foreign direct investment in Latin America and the Caribbean, 2006–2011. In 2011, just a few years after the global financial crisis of 2008, FDI flows into the region reached their highest total ever, at U.S.$153.448 billion. The Netherlands was the leading investor in 2011, in large part because it is a conduit for investments from third-party countries. The United States was the second-leading investor, with 18 percent of the total; Spain was third; and Latin American countries were fourth in investing in other countries within the region. Brazil was the largest recipient of FDI in 2011, getting nearly half of the total, followed by Mexico and Chile. [Sources consulted: "Foreign Direct Investment in Latin America and the Caribbean, 2011," Figure 1.8, p. 38, Economic Commission for Latin America and the Caribbean, United Nations, April 16, 2012, at http://www.cepal.org/publicaciones/xml/2/46572/2012-182-LIEI-WEB.pdf]

Export Processing Zones A major component of SAPs was the expansion of manufacturing industries in **Export Processing Zones (EPZs)**, also known as *free trade zones*—specially created areas within a country where, in order to attract foreign-owned factories, taxes on imports and exports are not charged. The main benefit to the host country is the employment of local labor, which eases unemployment and brings money into the economy. Products are often assembled strictly for export to foreign markets.

> **Export Processing Zones (EPZs)** specially created legal spaces or industrial parks within a country where, to attract foreign-owned factories, duties and taxes are not charged

Economic Inequality and Income Disparity **131**

There are EPZs in nearly all countries on the Middle and South American mainland and on some Caribbean islands (in many other world regions also). However, the largest of the EPZs is the conglomeration of assembly factories, called **maquiladoras**, that are located along the Mexican side of the U.S.–Mexico border. Although these factories do provide employment, living conditions are difficult and exactly what role they will ultimately play in Mexican economic development is unclear, as the vignette on page 129 illustrates.

The Backlash Against SAPs The SAP era did not produce the sustained economic growth that was expected to relieve the debt crisis and create broader prosperity. SAPs failed to stimulate economic growth in large part because they encouraged more dependence on exports of raw materials just when prices for these items were falling in the global market.

Since the late 1990s, millions of voters in Middle and South America have registered their opposition to SAPs. Starting in 1999, presidents who explicitly opposed SAPs have been elected in eight countries in the region: Argentina, Bolivia, Brazil, Chile, Ecuador, Nicaragua, Uruguay, and Venezuela. Both Brazil and Argentina made considerable sacrifices to pay off their debts and liberate themselves from the restrictions of SAPs. Venezuela, under President Hugo Chávez, emerged as a leader of the SAP backlash. In 2005, the Chávez administration **nationalized** foreign oil companies operating in Venezuela. Since then it has also used its own considerable oil wealth to help other countries, such as Argentina, Bolivia, and Nicaragua, pay off their debts. Bolivia, led by Evo Morales (often an ally of Chávez), reversed the SAP policy of privatization by nationalizing that country's natural gas industry. Loans taken from the IMF dropped from $48 billion in 2003 to just $1 billion in 2008. During this same period, the IMF reversed its policies from promoting SAPs to promoting *Poverty Reduction Strategies Papers* (PRSPs) that attempt to combine marketization with strong investment in social programs. This new IMF position has been maintained across the globe now for several years, suggesting that voters in Middle and South America have had an important impact on a major global financial institution (see pages 33–34 in Chapter 1).

63. NEW BOLIVIAN ENERGY POLICY CAUSES CONCERN

67. U.S.–VENEZUELA ENERGY TIES ENDURE DESPITE DETERIORATING POLITICAL RELATIONS

The Current Era of Foreign Direct Investment

The most recent development in this region is the emergence of several countries—most notably Brazil, Mexico, and Chile—that are not only improving in economic health but are stepping into global economic leadership roles. These three have received notably increased flows of **foreign direct investment (FDI)**. FDI is defined as investment funds coming in to enterprises from outside the country (see Figure 3.15). FDI flows to Middle and South America in 2012 were on par with those coming into North America, Europe, or East Asia. In Brazil specifically, FDI increased by more than 80 percent in just 9 months of 2011, with the money flowing primarily into the telecommunications, food, metal works, and petroleum industries.

Brazil is part of a global foursome known as the BRIC countries: Brazil, Russia, India, and China. These are emerging economic powers, rich in resources and poised for transformation into highly developed, highly productive powerhouses.

One measure of the region's rising reputation as financially stable is the fact that in 2012, Christine Lagarde, director of the IMF, asked for help from Brazil, Mexico, and Peru in designing and funding a bailout package for overly indebted countries in the European Union—the first time countries in Middle and South America have played such a role.

Interest in investing in the region has risen significantly and shifted, as shown in the table in Figure 3.15. The Netherlands now plays the biggest role, primarily because it acts on behalf of a consortium of investor countries. Spain has had a major FDI role for a number of years. The absolute amounts invested by the United States are rising, but because of the strong interest of others, the U.S. proportion has decreased. China quietly invested in the region in relatively small ways for several decades, but recently it has increased its investments and appears to be most interested in buying or leasing (for an extended period) large swatches of agricultural land in Brazil, Argentina, and Chile.

The Role of Remittances Migration for the purpose of supporting family members back home with remittances is an extremely important economic activity throughout this region (see further discussion on page 139). Each year, remittances amount to roughly double the U.S. foreign aid to the region. Most such remittance migrations are within countries. While the amounts of cash sent home are small, these remittances are crucial for families that may not have much other income.

One report (by geographer Dennis Conway and anthropologist Jeffrey H. Cohen) suggests that couples who migrate to the United States from indigenous villages in Mexico typically work at menial jobs and live frugally in order to save a substantial nest egg. Then they may return home for several years to build a house (usually a family self-help project) and buy furnishings. When the money runs out, the couple, or just the husband or wife, may migrate again to save up another nest egg.

The Informal Economy

For centuries, low-profile businesspeople throughout Middle and South America have operated in the informal economy—they have supported their families through inventive entrepreneurship but their work is not officially recognized in the statistics and they do not pay business, sales, or income taxes. Most are small-scale operators involved with street vending or recycling used items such as clothing, glass, or waste materials.

---

maquiladoras foreign-owned, tax-exempt factories, often located in Mexican towns just across the border from U.S. towns, that hire workers at low wages to assemble manufactured goods which are then exported for sale

nationalize to seize private property and place it under government ownership, with some compensation

foreign direct investment (FDI) investment funds that come in to enterprises from outside the country

Regional Trade and Trade Agreements

The growth in international trade and foreign investment in the region, in part the result of SAPs, has been joined by growth in regional free trade agreements. Such agreements reduce tariffs and other trade barriers among a group of neighboring countries. The two largest free trade agreements within the region are NAFTA and UNASUR.

The **North American Free Trade Agreement (NAFTA)**, established a free trade bloc in 1994, consisting of the United States, Mexico, and Canada—and containing more than 450 million people. The main goal of NAFTA is to reduce barriers to trade, thereby creating expanded markets for the goods and services produced in the three countries. Since 1994, the value of the economies of these three countries has grown steadily; by 2014, it was worth over $19 trillion, making it the world's largest trade bloc. A subsequent effort by the United States to create a NAFTA-like trade bloc for all of the Americas (FTAA) has stalled in recent years, largely because of dissatisfaction with the effects of globalization.

UNASUR, a union of South American nations, was organized in May of 2008; it supersedes **Mercosur** and the Andean Community of Nations, two previous customs unions. The 12 member countries (Argentina, Bolivia, Brazil, Chile, Colombia, Ecuador, Guyana, Paraguay, Peru, Suriname, Uruguay, and Venezuela) have nearly 400 million people, with a combined economy worth nearly $8 trillion per year. UNASUR appears to want to emulate the European Union in that it has adopted resolutions on a multitude of economic, political, military, immigration, and trade issues, not the least of which is to reduce disparities of wealth and opportunity. 📹 **62. MANY VENEZUELANS UNCERTAIN ABOUT CHÁVEZ'S TWENTY-FIRST-CENTURY SOCIALISM**

The overall record of regional free trade agreements so far is mixed. While they have increased the amount of trade, the benefits of that trade usually are not spread evenly among regions or among all sectors of society. In Mexico, for instance, the benefits of NAFTA have gone mainly to the wealthy investors who are concentrated in the northern states that border the United States. There, the agreement facilitated the growth of maquiladoras by easing cross-border finances and transit. But significantly, as many as one-third of small-scale farmers throughout Mexico have lost their jobs because of the increased competition from subsidized corporate farms in the United States, which, with NAFTA, now have unrestricted access to Mexican markets. In particular, corn imports from the United States have driven down the price of corn in local markets, pushing small farmers out of business.

ON THE BRIGHT SIDE

Contributions from the Informal Economy

Throughout this region, nearly everyone depends on the informal economy in some way, as either a buyer or a seller. Critics argue that informal workers are just treading water, making too little to ever expand their businesses significantly. Moreover, the bribes they have to pay to avoid arrest or fines are much less beneficial to the economy as a whole than the taxes paid by legitimate businesses. Work in the informal economy is also risky because there is no protection of workers' health and safety, nor sick leave, retirement, or disability benefits (also often lacking in the formal economy). Nevertheless, the informal economy can be a lifesaver during times of economic recession. After a recession hit Peru in 2000, for example, 68 percent of urban workers were in the informal sector. They generated 42 percent of the country's total GDP, mostly through street-vending work.

In some instances, the informal economy can serve as an incubator for new businesses that may expand, eventually providing legitimate jobs for family and friends. Those who work in the informal economy as well as those who migrate and send remittances are often unfairly overlooked as the major contributors to the economies of the region that they are.

North American Free Trade Agreement (NAFTA) a free trade agreement made in 1994 that added Mexico to the 1989 economic arrangement between the United States and Canada

UNASUR a union of South American nations that was organized in May of 2008; it supersedes Mercosur and the Andean Community of Nations, two previous customs unions

Mercosur a free trade zone created in 1991 that links the economies of Argentina, Bolivia, Brazil, Paraguay, Uruguay, and Venezuela to create a common market

---

The record of UNASUR is too short to be evaluated, but the shift in the global economy since UNASUR's creation in 2008 may be creating conditions that will place this trade bloc in an advantageous position vis-à-vis markets far outside the region: China, India, Europe, and even North America.

Rapid Regional Expansion of Internet Use Overall, Middle and South America are more advanced in terms of information technology than are many other developing regions of the world—a fact that could open up many new economic opportunities for the region in the near future. Brazil ranks fifth in the world in terms of total number of Internet users, and has two world-class technology hubs in the environs of São Paulo.

Mexico, with 36.9 percent of its population connected, recently launched a program to give its citizens access to training and higher education via the Internet (Figure 3.16). Similar efforts throughout the region are part of a long-term shift, especially in urban areas, toward more technologically sophisticated and better-paid service sector employment. In 2000, only 3.2 percent of the region's population used the Internet, but that figure had risen to nearly 40 percent by 2011, an increase of more than 1000 percent.

Food Production and Development

Food production throughout Middle and South America is shifting away from small-scale, often subsistence-level production toward large-scale, green revolution agriculture that is aimed at earning cash. This shift has forced many small-scale farmers who cannot afford the investment in machinery or chemicals off their land and into cities or to wealthier countries where they often work illegally. In significant numbers, ordinary people have begun to effectively protest policies that have left them landless and poor.

For centuries, while people labored for very low wages on haciendas, the hacienda owners would at least allow them a bit of land to grow a small garden or some cash crops. SAPs encouraged a shift to green revolution agriculture, which was based on

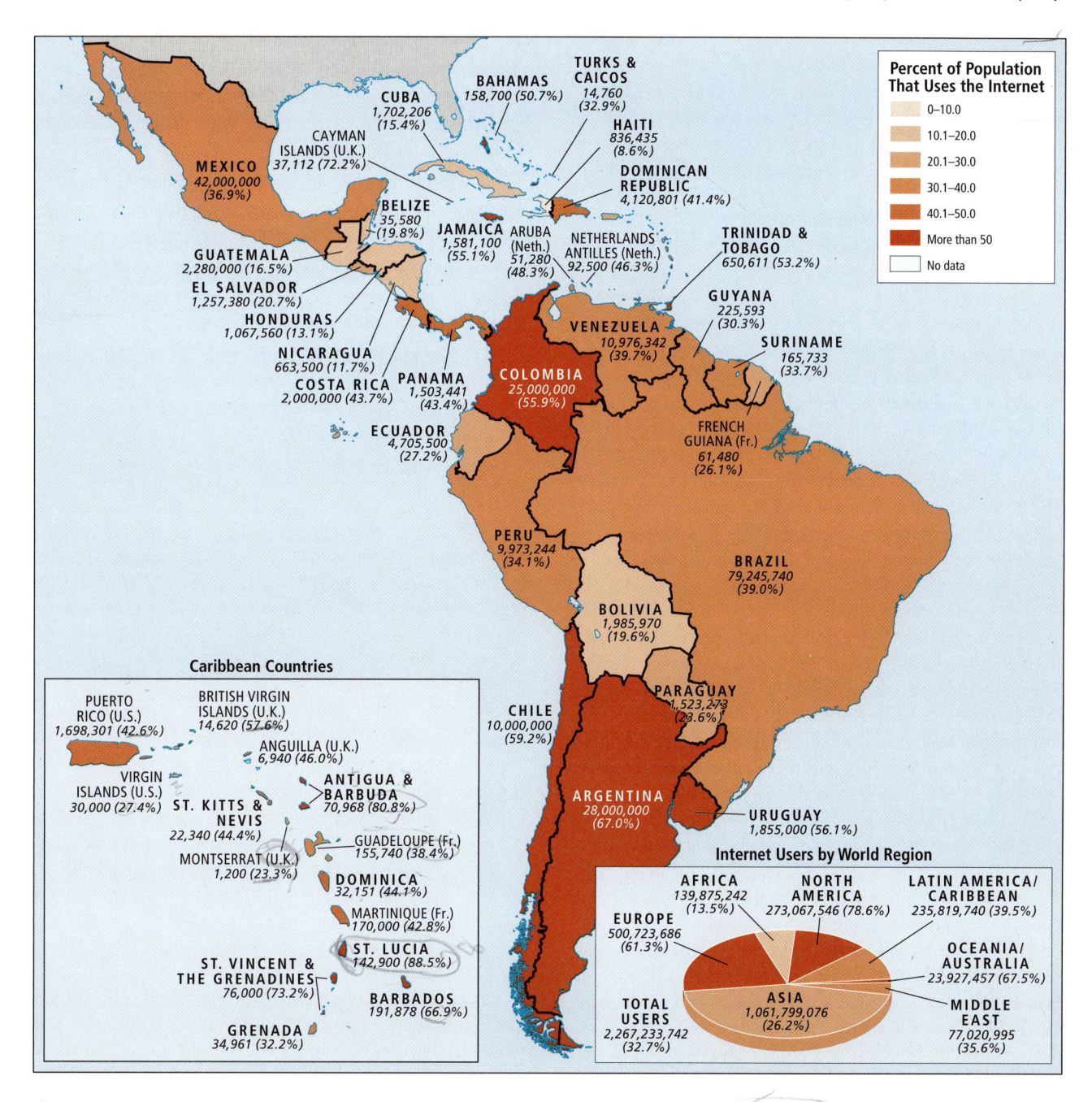

FIGURE 3.16 Internet use in Middle and South America, December 2011. The numbers on the map indicate the number of Internet users in each country; percents indicate the percentage of the country's population that uses the Internet. By December 2011, more than 235 million people (39.5 percent of the population) in Middle and South America were using the Internet; the number of users had risen slightly more than 1200 percent since 2000. Argentina has the highest percentage of users (67.0 percent) in South America; St. Lucia (88.5 percent) the highest in the Caribbean; and Costa Rica (43.7 percent) the highest in Middle America. [Sources consulted: Data from Internet World Stats: "Internet Users in the World: Distribution by World Regions—2012 Q2," at http://www.internetworldstats.com/stats.htm; "Latin American Internet Usage Statistics," at http://www.internetworldstats.com/stats10.htm; "Internet Usage and Population in the Caribbean," at http://www.internetworldstats.com/stats11.htm]

crops that could be exported for cash. This type of production, it was hoped, could help countries pay off debts faster. SAPs also made it easier for foreign multinational corporations such as Del Monte to use their financial resources to buy up many haciendas and other farms, converting them to plantations. This pattern of foreign corporate expansion into agriculture has continued into the modern FDI era. Many rural people have been forced off lands they once cultivated, and onto plantations, where they now work as migrant laborers for low wages, as the following story illustrates.

134 CHAPTER 3 Middle and South America

VIGNETTE

Aguilar Busto Rosalino used to work on a Costa Rican hacienda. He had a plot on which to grow his own food and in return worked 3 days a week for the hacienda. Since a banana plantation took over the hacienda, he rises well before dawn and works 5 days a week from 5:00 A.M. to 6:00 P.M., stopping only for a half-hour lunch break. Because he now lacks the time and land to farm, he must buy most of his food.

Aguilar places plastic bags containing pesticide around bunches of young bananas. He prefers this work to his last assignment of spraying a more powerful pesticide, which left him and 10,000 other plantation workers sterile. He works very hard because he is paid according to how many bananas he treats. Usually he earns between $5.00 and $14.50 a day.

It is common practice for these banana operations to fire their workers every 3 months so that they can avoid paying the employee benefits that Costa Rican law mandates. Although Aguilar makes barely enough to live on, he has no plans to press for higher wages because if he did he would be put on a "blacklist" of people that the plantations agree not to hire. [Source: Andrew Wheat. For detailed source information, see Text Sources and Credits.] ∎

Because of the moneymaking potential and political power of large-scale agriculture, conflicts are only rarely resolved in favor of small farmers and agricultural workers. Moreover, in rapidly urbanizing countries, green revolution agriculture is seen as the only way to supply cheap food for the millions of city dwellers. Ironically, many of these new urbanites were once farmers capable of feeding themselves and only moved to the cities because they were unable to compete with the new green revolution systems.

Many areas have had large farms, such as the haciendas mentioned on page 128, for centuries. These are now being converted to green revolution agriculture, similar to farms that can be found in the Midwestern United States. There is a wide belt of large-scale farming and ranching from the Argentine pampas into Brazil (Figure 3.17). And, as we saw in the earlier discussion of environmental issues (see page 124), rice, corn, soybeans, and meat animals, which are often produced by large green revolution farms on once-forested lands in and around the Amazon Basin, are all major exports for countries like Brazil and Argentina.

At a medium scale are zones of modern mixed farming (those that produce meat, vegetables, and specialty foods for sale in urban centers), located on large and small plots around most major urban centers. Figure 3.18 illustrates how the foodways of the area are based largely on indigenous plants (corn, potatoes, tomatoes) and methods of cooking (baking, grilling), and involve beef, pork, rice, seasonings, and cooking methods (frying) imported from Europe, Asia, and Africa. 🎥 **57. HAITI'S RISING COST OF FOOD WORRIES AID GROUPS**

✂ **THINGS TO REMEMBER**

> **GEOGRAPHIC INSIGHT 2**
> • **Globalization and Development** The integration of this region with the global economy has left it with the widest gap between rich and poor in the world.

Long-term failures to address poverty, economic instability, and the flow of resources and money out of the region have resulted in many conflicts and inspired numerous efforts at reform. Nevertheless, in recent years several countries in this region have emerged as global economic leaders.

• Three phases of economic development—extractive, import substitution industrialization (ISI), and structural adjustment programs (SAPs)—dominated the region into the twenty-first century.

• Since 1999, presidents who explicitly opposed SAPs have been elected in eight countries in the region: Argentina, Bolivia, Brazil, Chile, Ecuador, Nicaragua, Uruguay, and Venezuela.

• Brazil is part of a global foursome known as the BRIC countries: Brazil, Russia, India, and China. These are emerging economic powers, rich in resources and poised for transformation into highly developed, highly productive powerhouses.

• Since the 1990s, free trade policies have been adopted in some industries, and regional trading blocs have been formed and re-formed, with mixed results.

• Food production throughout this region is undergoing a shift away from small-scale, often subsistence-level production toward large-scale, green revolution agriculture that is aimed at earning cash. ∎

POWER AND POLITICS

> **GEOGRAPHIC INSIGHT 3**
>
> **Power and Politics:** After decades of rule by elites and the military, punctuated by disruptive foreign military interventions, political freedoms in Middle and South America are expanding. Almost all countries now have multiparty political systems and democratically elected governments. However, the international illegal drug trade continues to be a source of violence and corruption in the region.

In the last 30 years, there have been repeated peaceful and democratic transfers of power in countries once dominated by rulers who seized power by force. These **dictators** often claimed absolute authority, governing with little respect for the law or the rights of their citizens. Their authority was based on alliances between the military, wealthy rural landowners, wealthy urban entrepreneurs, foreign corporations, and even foreign governments such as the United States.

Although a slow expansion of political freedoms has transformed the politics of the region, problems remain (as seen in Figure 3.19 on page 137). Elections are sometimes poorly or unfairly run and their results are frequently contested. Elected governments are sometimes challenged by citizen protests or threatened with a **coup d'état**, in which the military takes control of the government by force. Such coups are usually a response to policies that are unpopular with large segments of the population, powerful elites, the military, or the United States. In the last decade, coups have been

dictator a ruler who claims absolute authority, governing with little respect for the law or the rights of citizens

coup d'état a military- or civilian-led forceful takeover of a government

FIGURE 3.17 Agricultural and mineral zones in Middle and South America. [Source consulted: Adapted from *Goode's World Atlas,* 21st ed. (Chicago: Rand McNally, 2005), p. 137, Minerals and Economic map]

threatened in Honduras, Venezuela, Colombia, Ecuador, and Bolivia; but more recently, peaceful democratic elections have become the norm in this region.

Hugo Chávez and Venezuelan Politics In Venezuela, the landslide election of Hugo Chávez as president in 1998 brought profound change to the country. Elites had long dominated Venezuelan politics, and profits from the country's rich oil deposits had not reached the poor. One-third of the population lived on U.S.$2 (PPP) a day, and per capita GNI (PPP) was lower in 1997 than it had been in 1977. Chávez campaigned as a champion of the poor, calling for the government to subsidize job creation, community health care, and food for the large underclass. He also redirected the profits of the largely foreign-managed oil industry to support government social programs. These policies, all of which were enacted soon after he was elected, eroded Chávez's original support from the small middle class and elites, who began to fear a turn toward broad government control. When the U.S. government expressed alarm, Chávez became a major critic of U.S. and other foreign influence in the region. Chávez helped lead the region-wide backlash against SAPs and condemned the United States' history of intervention in the internal affairs of countries across the region. **66. SOCIAL PROGRAMS AT ROOT OF CHÁVEZ'S POPULARITY**

Chávez was reelected in 2000, and then briefly deposed in a coup d'état in 2001 in which the United States participated covertly. He was quickly reinstated, however, by a groundswell of popular support among the large underclass who stood to benefit from his policies. His presidency was sustained in a referendum in 2004 and again in the landslide election of 2006. In 2009, voters approved a constitutional amendment removing presidential term limits, thereby allowing Chávez to run for president indefinitely. Some saw this as a triumph of the *populist movement* (see the definition on page 146), others as a loss for democracy and a drift back toward dictatorship (see the Figure 3.19 map). Hugo Chávez died in March of 2013 from cancer after having been treated numerous times by specialists in Cuba.
68. WILL CHÁVEZ INHERIT CASTRO'S REVOLUTIONARY MANTLE?
69. VENEZUELANS REJECT CONSTITUTIONAL CHANGES

The Drug Trade and Conflict

The international illegal drug trade is a major source of violence and corruption throughout the region. The primary drugs being traded are cocaine, heroin, and marijuana. Most drugs are produced in or pass through northwestern South America, Central America, and Mexico. Figure 3.20 on page 138 illustrates the geographic distribution of cocaine seizures in 2008. (Note: Such seizures may not reveal activity in areas where law enforcement is lax or entirely co-opted by drug traders.)

Production of cocaine and heroin is illegal in all of Middle and South America. However, public figures, from the local police on up to high officials, are paid to turn a blind eye to the industry. Most coca growers are small-scale farmers of indigenous or mestizo origin, working in remote locations where they can make a better income for their families from these plants than from other cash crops. Many rent land from the *drug cartels* that get the drugs to global markets. The cartels are monopolistic and violent, and those who oppose them—whether farmers, journalists, or law enforcement officials—often end up kidnapped, tortured, and brutally murdered.

In Colombia, the illegal drug trade has financed all sides of a continuing civil war that has threatened the country's democratic traditions and displaced more than 1.5 million people over the past several decades (see Figure 3.19B). Mexico also has faced increasing threats to its political stability from drug cartels that control a hugely profitable U.S.-oriented drug trade (see Figure 3.19A). **72. THOUSANDS KIDNAPPED IN COLOMBIA IN LAST DECADE**

FIGURE 3.18 LOCAL LIVES
Foodways in Middle and South America

A Corn tortillas are prepared in Mexico City. Corn has been a pillar of diets in Mexico and the countries of Central America for thousands of years. Now cultivated throughout the world, corn was first domesticated from a wild grass in southern Mexico. [Greg Elms/Lonely Planet Images/Getty Images]

B An Argentine *asado*, or barbeque. Here beef is roasted over an open bed of coals, using a special *parilla*, or grill. This method of cooking is a variant of methods first developed by the indigenous peoples of the grasslands of Argentina. Asado is the national dish of Argentina; it is prepared widely throughout Uruguay, Paraguay, and southern Brazil. [Virgina Sherwood/Bravo/NBCU Photo Bank via Getty Images]

C *Lomo saltado*, a Peruvian dish with Asian influences. First developed in restaurants started by Chinese immigrants, it is a mix of Chinese and Peruvian ingredients and culinary traditions. It consists of strips of beef or pork marinated in vinegar and stir-fried with onions, parsley, tomatoes, and other vegetables, and then served over rice and french fries. [Juanmonino/E+/Getty Images]

U.S. policy has emphasized stopping the production of illegal drugs in Middle and South America and interrupting trade flows rather than curtailing the demand for drugs in the United States. As a result, the U.S. *war on drugs* has led to a major U.S. presence in the region that is focused on supplying intelligence, herbicides used to destroy drug crops, military equipment, and training to military forces in the region. One consequence of this is that U.S. military aid to Middle and South America is now about equal to U.S. aid for education and other social programs in the region. Drug production levels in Middle and South America, though, are higher than ever, exceeding demand in the U.S. market, where street prices for many drugs have fallen in recent years.

Concerned and informed citizens of Middle and South America have been calling for radical changes in international drug policies, including the legalization of drug use in the main (U.S.) market, which they assert would cause criminals to lose interest in the drug trade. In 2013, Uruguay legalized and regulated marijuana use, cultivation, sale, and distribution in an effort to reduce drug-related violence and deprive criminals of a major source of income.

Foreign Involvement in the Region's Politics

Interventions in the region's politics by outside powers have frequently compromised political freedoms and human rights. Although the former Soviet Union, Britain, France, and other European countries have wielded much influence, by far the most active foreign power in this region has been the United States.

In 1823, the United States introduced the Monroe Doctrine to warn Europeans that no further colonization would be tolerated in the Americas. Subsequent U.S. administrations interpreted this policy more broadly to mean that the United States itself had the sole right to intervene in the affairs of the countries of Middle and South America, and it has done so many times. The official goal for such interventions was usually to make countries safe for democracy, but in most cases the driving motive was to protect U.S. political and economic interests.

At various times during the past 150 years, U.S.-backed unelected political leaders, many of them military dictators, have been installed in many countries in the region. After World War II, worried that Communism would infiltrate Middle and South America, the United States focused special military scrutiny on a number of countries. Since the 1960s, the United States has funded armed interventions in Cuba (1961), the Dominican Republic (1965), Nicaragua (1980s), Grenada (1983), and Panama (1989). Perhaps the most infamous intervention took place in Chile in 1973. With U.S. aid, the elected socialist-oriented government of Salvador Allende was overthrown. Allende was killed, and a military dictator, General Augusto Pinochet, was installed in his place. Over the next 17 years, the Pinochet regime imprisoned and killed thousands of Chileans who protested the loss of their political freedoms. In 1998, while recuperating from surgery in England, Pinochet was arrested under a Spanish judicial order and would have stood trial in Spain but was judged too mentally and physically impaired to be tried and was sent home to die.

The Cold War and Post–Cold War Eras Since 1959, the Caribbean island of Cuba has been governed by a radically socialist government led first by Fidel Castro and then, starting in 2006, by Fidel's brother Raúl. The revolution that brought Fidel Castro to power transformed a plantation and tourist economy once known for its extreme income disparities into one of the most egalitarian in the region. However, because Castro adopted socialism and allied Cuba with the Soviet Union, Cuba's relations with the United States became extremely hostile. The United States has funded many efforts to destabilize Cuba's government, and it actively discourages other countries from trading with Cuba. Since 1960, the United States has maintained an embargo

FIGURE 3.19 PHOTO ESSAY: Power and Politics in Middle and South America

An expansion of political freedoms is well under way in Middle and South America, although there are still many barriers to this process. Decades of elite and military rule, combined with rampant corruption and political repression, have at times inspired resistance in the form of popular revolutionary movements that attempt to take control of governments by force. The ensuing wars and brutal repression by governments and militants has cost many lives. The drug trade is also a major political force, with huge amounts of illicit cash being used to pay off elected officials, civil servants, police forces, and the military. Nevertheless, in the region as a whole, there has been a decline in political violence in recent decades.

A Federal agents search cars at a checkpoint in Juárez in Chihuahua, Mexico. Mexico's ongoing drug war has led to between 60,000 and 100,000 deaths over the past several years and corruption that has eroded local political institutions, especially near the U.S. border. Most of Mexico's drug production supplies the United States.

Democratization and Conflict
Democratization index
- Full democracy
- Flawed democracy
- Hybrid regime
- Authoritarian regime
- No data

Armed conflicts and genocides with high death tolls since 1990
- Ongoing conflict
- 1000–20,000 deaths
- 20,000–50,000 deaths
- 50,000–100,000 deaths
- 100,000–200,000 deaths

B Colombian police officers, crippled in combat with a rebel group, undertake a protest march destined for the capital in Bogotá. Popular revolutionary movements and a drug war drive this complex conflict, the worst in the region. Political institutions and the rule of law have been badly damaged, with many high-ranking officials linked to extreme human rights and civil liberties abuses.

C Cuban revolutionaries Che Guevara (left), Camilo Cienfuegos (center), and Juan Antonio Mella are immortalized in an aging mural in Havana. Since coming to power in 1959, Cuba's communist government has been criticized for jailing and sometimes executing leaders of the political opposition.

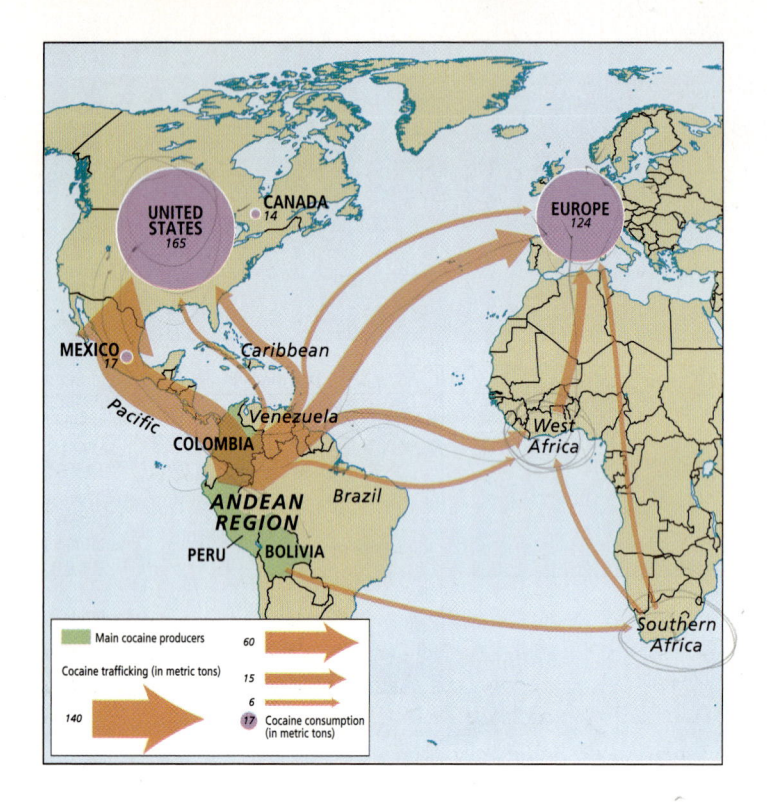

FIGURE 3.20 Linkages: Cocaine sources, trafficking routes, and seizures worldwide, 2008. Colombia, Peru, and Bolivia are the most important sources of cocaine cultivation and production in the world. The big cocaine markets are in the Americas; use has declined by 36 percent in the United States since 1998 and increased in South America. Cocaine use in Western Europe and West Africa is on the rise, due in part to new trade routes through West Africa to Europe. [Source consulted: *World Drug Report 2010*, United Nations Office on Drugs and Crime, at http://www.unodc.org/unodc/en/drug-trafficking/index.html]

against Cuba, and so far international efforts to lift the embargo have failed. **73. EU SPLIT OVER DEVELOPMENT COMMISSIONER'S PROPOSAL TO NORMALIZE RELATIONS WITH CUBA**

With help from the Soviet Union, Castro managed to dramatically improve the country's literacy and infant mortality rates and the life expectancy of its population. Unfortunately, he also imprisoned or executed thousands of Cubans who disagreed with his policies, many of whom fled to southern Florida. Following the demise of the Soviet Union, Castro opened the country to foreign investment. Many countries responded, and Cuba is now a major European tourist destination. However, political repression in Cuba persists, and relations with the United States remain distant.

Recent Revolutionary Movements

In response to continued poverty and inequality, and building on Middle and South America's long history of revolutionary politics, two major broad-based rural political movements have arisen in recent years. While neither poses any direct threat to national governments, they are building a strong network of support for political and economic change at the local level. The areas in which these movements take place are examples of what geographers call **contested space**, where various groups are in conflict over the right to use a specific territory as each sees fit.

The Zapatista Rebellion In the southern Mexican state of Chiapas, indigenous farmers have mobilized against the economic and political systems that have left them poor and powerless. The Mexican government redistributed some hacienda lands to poor farmers early in the twentieth century, but most fertile land in Chiapas is still held by a wealthy few who use green revolution agriculture to grow cash crops for export. The poor majority farm tiny plots on infertile hillsides. In 2000, about three-fourths of the rural population was malnourished, and one-third of the children did not attend school.

The Zapatista rebellion (named for the hero of the 1910 Mexican revolution, Emiliano Zapata) began on the day the North American Free Trade Agreement took effect in 1994. The Zapatistas view NAFTA as a threat because it diverts the support of the Mexican federal government from land reform to large-scale, export-oriented green revolution agriculture. Starting as an armed rebellion in Chiapas, the movement was largely suppressed by the Mexican national army. However, the Zapatistas' support in local communities allowed them to continue on in many parts of Chiapas. Their extensive use of the Internet to promote their cause earned the Zapatistas a worldwide following of supporters.

In 2003, after 9 years of armed resistance and largely unsuccessful negotiations with the Mexican government, the Zapatista movement redirected its energies toward local nonviolent political campaigns, mainly in Chiapas, to set up people's governing bodies parallel to local official governments. More than a decade into this process, the Zapatistas have made significant improvements in education, women's rights, and poverty reduction, thus deepening their support base in the areas they control. While the Zapatistas continue to focus more locally, the movement still has a national voice via campaigns against the political establishment during election years, and has a global reach through the Internet and numerous training programs for supporters of the movement from around the world.

contested space any area that two or more groups claim or want to use in different and often conflicting ways, such as the Amazon or Palestine

Brazil's Landless Movement Sixty-five percent of Brazil's arable and pasture land is owned by wealthy farmers who make up just 2 percent of the population. Since 1985, more than 2 million small-scale farmers have been forced to sell their land to larger farms that practice green revolution agriculture. Because the larger farms specialize in major export items, such as cattle and soybeans, they have been favored by governments wishing to increase exports. As a result, many poor farmers have been forced to migrate.

To help these farmers, organizations such as the Movement of Landless Rural Workers (MST) began taking over unused portions of some large farms. Since the mid-1980s, the MST has coordinated the occupation of more than 51 million acres of Brazilian land (an area about the size of Kansas). Some 250,000 families have gained land titles, while the elite owners have been paid off by the

Brazilian government and have moved elsewhere. There are now 1.5 million members of the MST spread across 23 of Brazil's 26 states, and movements with similar goals have been launched in Ecuador, Venezuela, Colombia, Peru, Paraguay, Mexico, and Bolivia.

ON THE BRIGHT SIDE

Children and the Landless Movement
In Brazil, the children of those in the MST movement find living in the landless protestor encampments a source of pride and positive identity. They learn both subsistence and leadership skills, gain knowledge of the natural environment, practice communitarian values, and are proud of the stance their parents have taken, as the video by Michalis Kontopodis, "Landless Children/Sem Terrinha," at http://tinyurl.com/9pxeclw, attests.

THINGS TO REMEMBER

> **GEOGRAPHIC INSIGHT 3**
• **Politics and Power** After decades of elite and military rule punctuated by disruptive foreign military interventions, political freedoms in Middle and South America are expanding. Almost all countries now have multiparty political systems and democratically elected governments. However, the international illegal drug trade continues to be a source of violence and corruption in the region.

• Elected governments are at times challenged by citizen protests or with a coup d'état because policies are unpopular with large segments of the population, powerful elites, or the military.

• The drug trade has emerged as a major obstacle to democracy, with huge amounts of illicit cash being used to pay off elected officials, civil servants, police forces, and the military.

• Interventions in the region's politics by outside powers have frequently compromised political freedoms and human rights.

• In response to continued poverty and inequality, and building on this region's long history of revolutionary politics, two major broad-based rural political movements have arisen in recent years.

URBANIZATION

> **GEOGRAPHIC INSIGHT 4**
>
> **Urbanization:** Since the early 1970s, Middle and South America has experienced rapid urban growth as rural people migrate to cities and towns. A lack of planning to accommodate the massive rush to the cities has created densely occupied urban landscapes that often lack adequate support services and infrastructure.

More than 70 percent of the people in the region live in settlements of at least 2000 people. Increasingly, one city, known as a **primate city**, is vastly larger than all the others, accounting for a large percentage of the country's total population (Figure 3.21). Examples in the region are Mexico City (with 23 million people, a bit more than 21 percent of Mexico's total population); Managua, Nicaragua (33 percent of that country's population); Lima, Peru (29 percent); Santiago, Chile (34 percent); and Buenos Aires, Argentina (36 percent).

The concentration of people into just one or two large cities in a country leads to uneven spatial development and to government policies and social values that favor the largest urban areas. Wealth and power are concentrated in one place, while distant rural areas and even other towns and cities have difficulty competing for talent, investment, industries, and government services. Many provincial cities languish as their most educated youth leave for the primate city.

Migration and Urbanization

It always takes some resourcefulness to move from one place to another. Those people who already have some years of education and strong ambition are the ones who migrate to cities. The loss of these resourceful young adults in which the community has invested years of nurturing and education is referred to as **brain drain**. Brain drain happens at several scales in Middle and South America: through rural-to-urban migration from villages to regional towns and from towns and small cities to primate cities. There is also an international brain drain when migrants move to North America and Europe; for example, one in four U.S. doctors are foreign born. Migrants often want better access to jobs and education, and to help their families with money they send back home.

Urban Landscapes A lack of planning to accommodate the massive rush to the cities has created dense urban landscapes with often inadequate infrastructure. The arrangement of urban landscapes is also very different from the common U.S. pattern. In the United States, a poor, older inner city is usually surrounded by affluent suburbs, with clear and planned spatial separation of residences by income as well as separation of residential, industrial, and commercial areas. In Middle and South America, by contrast, both affluent and working-class areas have become unwilling neighbors to unplanned slums filled with poor migrants. Too destitute even to rent housing, these "squatters" occupy parks and small patches of land wherever they can be found, as depicted in the diagram in Figure 3.22.

primate city a city, plus its suburbs, that is vastly larger than all others in a country and in which economic and political activity is centered

brain drain the migration of educated and ambitious young adults to cities or foreign countries, depriving the communities from which the young people come of talented youth in whom they have invested years of nurturing and education

favelas Brazilian urban slums and shantytowns built by the poor; called *colonias*, barrios, or *barriadas* in other countries

Unplanned Neighborhoods The best known of these unplanned communities are Brazil's **favelas** (Figure 3.23). In other countries they are known as slums, shantytowns, *colonias*, barrios, or *barriadas*. The settlements often spring up overnight on vacant land after a coordinated collective occupation by poor homeless families. They rarely occur with the landowner's permission, and are without city-supplied water, electricity, or sewer service. Housing is hastily constructed with whatever materials are available. Once the settlements are

FIGURE 3.21 PHOTO ESSAY: Urbanization in Middle and South America

Urbanization is at the heart of many transformations in this region. The last several decades have seen cities grow extremely fast, and 77 percent of the region's population now lives in cities. New opportunities have opened for migrants from rural areas, but new stresses have emerged as well. The low-skilled jobs they can get often don't pay well, and many people are forced to live in unplanned neighborhoods on the outskirts of huge cities.

Population living in urban areas
- 83%–100%
- 65%–82%
- 47%–64%
- 29%–46%
- 11%–28%
- No data

Population of metropolitan areas 2013
- 20 million
- 10 million
- 5 million
- 3 million

Note: Symbols on map are sized proportionally to metro area population

① Global rank (population 2013)

A Santiago is one of the region's primate cities, home to 34 percent of Chile's population. It dominates Chile's economy, generating 40 percent of the country's GDP.

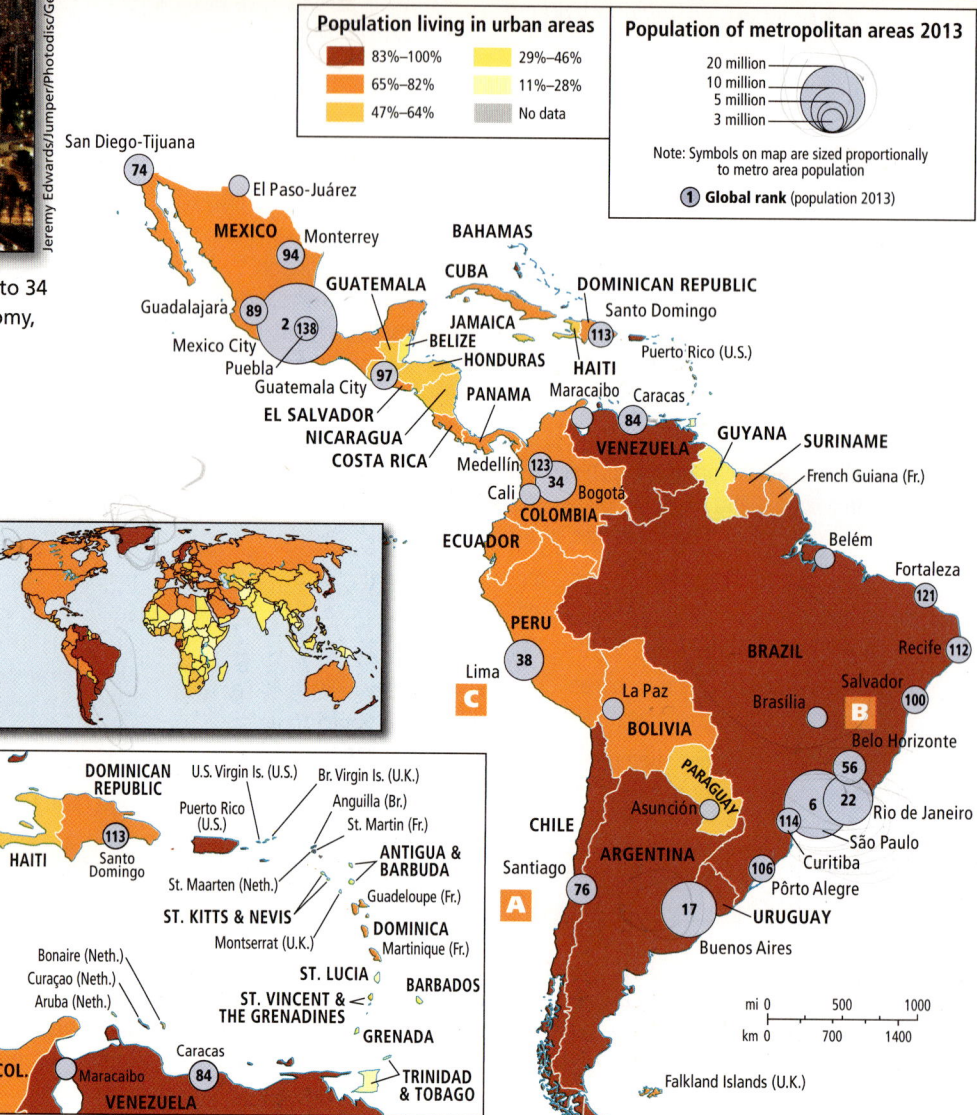

B A child works as a street vendor in Salvador, Brazil. Many new urban migrants work in similar informal sector jobs that generally pay less and are much less secure than formal sector jobs.

C Police evict 2500 people from an informal squatter settlement recently formed outside of Lima, Peru. Structures made out of woven grass mats are the first to go up, followed by wooden shacks and eventually cement or brick houses. Water and electricity may eventually be extended to such settlements, but only after years or decades of occupation.

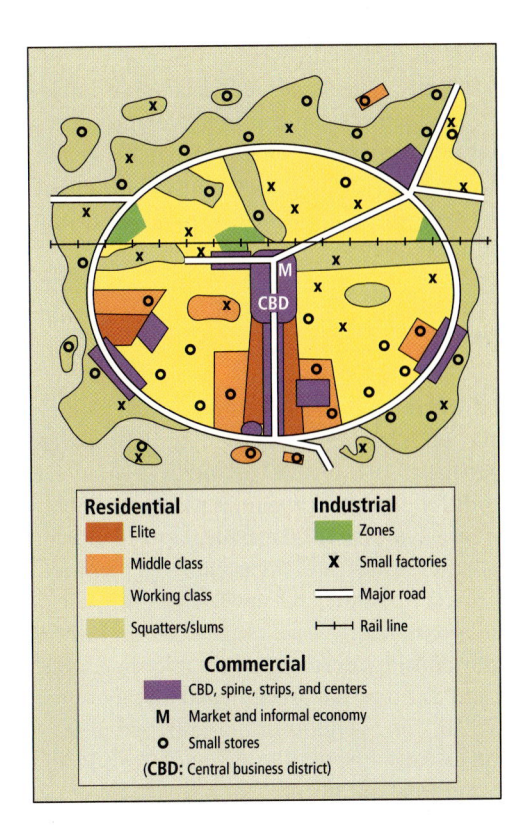

FIGURE 3.22 Crowley's model of urban land use in mainland Middle and South America. William Crowley, an urban geographer who specializes in Middle and South America, developed this model to depict how residential, industrial, and commercial uses are mixed together, with people of widely varying incomes living in close proximity to one another and to industries. Squatters and slum-dwellers ring the city in an irregular pattern.
[Source consulted: *Yearbook of the Association of Pacific Coast Geographers* 57(28): 1995; printed with permission]

Residential
- Elite
- Middle class
- Working class
- Squatters/slums

Industrial
- Zones
- X Small factories
- Major road
- Rail line

Commercial
- CBD, spine, strips, and centers
- M Market and informal economy
- o Small stores
- (CBD: Central business district)

FIGURE 3.23 A favela in Rio de Janeiro. A favela in Rio de Janeiro clings to a once-vacant hillside that was considered too steep for apartment buildings. Now home to thousands of people, it is still served mainly by footpaths and narrow alleyways. Water is only available sporadically, so many residents store water in tanks on their roof.

> **THINKING GEOGRAPHICALLY** Why are favelas often located on steep hills, and why do residents often store water on their roofs?

established, efforts to eject squatters usually fail, though police are often called in to attempt to evict them (see Figure 3.21C). The impoverished are such a huge portion of the urban population that even those in positions of power will not challenge them directly. Nearby wealthy neighborhoods simply barricade themselves with walls and security guards.

The squatters are frequently enterprising people who work hard to improve their communities. They often organize to press governments for social services. Some cities, such as Fortaleza in northeastern Brazil, even contribute building materials so that favela residents can build more permanent structures with basic indoor plumbing. Over time, as shacks and lean-tos are transformed through self-help initiatives into crude but livable suburbs, the economy of favelas can become quite vibrant, with much activity in the informal sector. Housing may be intermingled with shops, factories, warehouses, and other commercial enterprises. Favelas and their counterparts can become centers of pride and support for their residents, where community work, folk belief systems, crafts, and music (for example, Favela Funk) flourish. Many of the best steel bands of Port of Spain, Trinidad, have their homes in the city's shantytowns.

urban growth poles locations within cities that are attractive to investment, innovative immigrants, and trade, and thus attract economic development like a magnet

Planned Elite Landscapes In contrast to colorful but problematic favelas are districts in nearly every city that are modern, planned, and similar to elite urban districts across the world. Examples of such urban landscapes can be found in Rio de Janeiro, São Paulo, and other cities in Brazil's southeastern industrial heartland. These cities have districts that are elegant and futuristic showplaces, resplendent with the very latest technology and graced with buildings and high-end shops that rival those in New York, Singapore, and even Tokyo. Known as **urban growth poles**, these areas attract investment, trade, educated immigrants, and overall economic development—all of which make them stand out from surrounding landscapes.

Planners throughout the region now acknowledge, however, that in the rush to develop modern urban landscapes, municipal

VIGNETTE

Favelas are everywhere in Fortaleza, Brazil. The city grew from 30,000 to 300,000 people during the 1980s. By 2006, there were more than 3 million residents, most of whom had fled drought and rural poverty in the interior of the country. City parks of just a square block or two in middle-class residential areas were suddenly invaded by squatters. Within a year, 10,000 or more people were occupying crude, stacked concrete dwellings in a single park, completely changing the ambience of the surrounding upscale neighborhood. In the early days of the migration, lack of water and sanitation often forced the migrants to relieve themselves on the street.

One day, while strolling on the Fortaleza waterfront, Lydia Pulsipher chanced to meet a resident of a beachfront favela who invited her to join him on his porch. There, he explained how he and his wife had come to the city 5 years before, after being forced to leave the drought-plagued interior when a newly built irrigation reservoir flooded the rented land their families had cultivated for generations. With no way to make a living, they set out on foot for the city. In Fortaleza, they constructed the building they used for home and work from objects they collected along the beach. Eventually they were able to purchase roofing tiles, which gave the building an air of permanency. At the time of the visit, he maintained a small refreshment stand and his wife a beauty parlor that catered to women from the beach settlements. *[Source: Lydia Pulsipher's field notes. For detailed source information, see Text Sources and Credits.]* ∎

governments and developers neglected to underwrite the parallel development of a sufficient urban infrastructure (for an exception, see the discussion of the southern city of Curitiba below).

In short supply are sanitation and water systems, an up-to-date electrical grid, transportation facilities, schools, adequate housing, and medical facilities—all necessary to sustain modern business, industry, and a socially healthy urban population. In recent decades, there have been massive investments in these systems, especially in the larger and wealthier countries, such as Brazil, Chile, Argentina, and Mexico, that can afford the improvements. These investments are done in recognition that planned development of human capital through education, health services, and community development, whether privately or publicly funded, is a primary part of building strong urban economies.

Urban Transport Issues Transportation in large, rapidly expanding, partially unplanned urban areas with millions of poor migrants can be a special challenge. Favelas are often on the urban fringes, far from available low-skill jobs and ill-served by roads and public transport (see Figure 3.22). Workers must make lengthy, time-consuming, and expensive commutes to jobs that pay very little. The entire urban population gets caught up in endless traffic jams that cripple the economy and pollute the environment.

Yet the southern Brazilian city of Curitiba has carefully oriented its expansion around a master plan (dating from 1968) that includes an integrated transportation system funded by a public-private collaboration. Eleven hundred minibuses making 12,000 trips a day bring 1.3 million passengers from often remote neighborhoods to terminals, where they meet express buses to all parts of the city. The large pedestrian-only inner-city area is a boon for merchants because minibus commuters spend time shopping rather than looking for parking. Being able to get to work quickly and cheaply has helped the poor find and keep jobs. The reduced use of cars means that emissions and congestion are lowered and urban living is more pleasant.

Transportation in large, rapidly expanding, partially unplanned urban areas with millions of poor migrants can be a special challenge.

Gender and Urbanization Interestingly, rural women are just as likely as rural men to migrate to the city. This is especially true when employment is available in foreign-owned factories that produce goods for export. Companies prefer women for such jobs because they are a low-cost and usually passive labor force. Factors that push people out of rural areas can affect women disproportionately. For example, the shift toward green revolution agriculture has had a particularly hard impact on rural women because they are rarely considered for such jobs as farm-equipment operators or mechanics. In urban areas, unskilled migrant women usually find work as street vendors or domestic servants.

Male urban migrants tend to depend on short-term, low-skill day work in construction, maintenance, small-scale manufacturing, and petty commerce. Many work in the informal economy as street vendors, errand runners, car washers, and trash recyclers, and some turn to crime. In an urban context, both men and women feel sorely the loss of family ties and village life; the chances for recreating the family life they once knew are extremely low.

Once relocated to a city, migrant families may disintegrate because of extreme poverty and malnutrition, long commutes for both working parents, and poor quality of day care for children. At too young an age, children are left alone or sent into the streets to scavenge for food or to earn money for the family as street vendors (see Figure 3.21B).

THINGS TO **REMEMBER**

> **GEOGRAPHIC INSIGHT 4**
>
> • **Urbanization** Since the early 1970s, Middle and South America has experienced rapid urban growth as rural people migrate to cities and towns. A lack of planning to accommodate the massive rush to the cities has created densely occupied urban landscapes that often lack adequate support services and infrastructure.

• Increasingly, one city, known as a primate city, is vastly larger than all the others, accounting for a large percentage of the country's total population.

• Settlements known as slums, shantytowns, *colonias,* barrios, or *barriadas* often spring up overnight on vacant land after a coordinated collective occupation by poor homeless families.

• The southern Brazilian city of Curitiba has carefully oriented its expansion around a master plan that includes an integrated transportation system funded by a public-private collaboration.

• Rural women are just as likely as rural men to migrate to the city. ∎

POPULATION AND GENDER

> **GEOGRAPHIC INSIGHT 5**
>
> **Population and Gender:** During the early twentieth century, the combination of cultural and economic factors and improvements in health care created a population explosion. By the late twentieth century, improved living conditions, expanded access to education and medical care, urbanization, and changing gender roles were all working together to reduce population growth.

By 2013, about 606 million people were living in Middle and South America, close to ten times the highest estimated population of the region in 1492. Today, populations in Middle and South America continue to grow but are doing so at an ever-slower pace because of diminishing birth rates.

Slowing Population Growth

Birth rates stayed high for much of the twentieth century for a variety of reasons. In agricultural areas, children were seen as sources of wealth because they could do useful farm and household work at a young age and eventually would care for their aging elders. Worry over infant death rates persisted, so some parents had four or more children to be sure of raising at least a few to adulthood. Another factor was the Roman Catholic Church's opposition to systematic family planning.

By the 1970s, the region was undergoing a demographic transition (see Figure 1.28 on page 49). Between 1975 and 2013, the annual rate of natural increase (the birth rate minus the death rate) for the entire region fell from about 1.9 percent to 1.3 percent—a rate of growth only slightly more than the world average. Between 1990 and 2011, the rate of natural increase fell in every country in the region, as shown in Figure 3.24.

These changes happened for a variety of reasons. Better access to food, shelter, sanitation, and medical care reduced infant mortality, encouraging couples to have only two or three children instead of five or more. As populations urbanized and economies grew, women gained better access to education and employment outside the home, and so delayed childbearing and had smaller families. This trend is continuing as women move into high-level jobs such as management positions in corporations and government that traditionally went to men. Perhaps the most visible example of this is the women who have become presidents of Chile, Argentina, and Brazil in the last decade.

However, because 28 percent of the region's population is under the age of 15, even if couples have only one or two children, the population will continue to grow as this large group reaches the age of reproduction. Population projections for the year 2050 are for 780 million people, and supplying this extra 180 million people with food, water, homes, schools, and hospitals will be a challenge.

Population Distribution

The population density map of this region reveals a very unequal distribution of people. Comparing the population density map (Figure 3.25) with the landforms map (see Figure 3.1 on page 108), you will find areas of high population density in a variety of environments. Some of the places with the highest densities, such as those around Mexico City and in Colombia and Ecuador, are in highland areas. But high concentrations are also found in lowland zones along the Pacific coast of Central America, and especially along the Atlantic coast of South America. The cool uplands (*tierra templada*) were densely occupied even before the European conquest. Most coastal lowland concentrations, in the *tierra caliente*, are near seaports with vibrant globalizing economies and a cosmopolitan social life that attracts people.

THINGS TO REMEMBER

> **GEOGRAPHIC INSIGHT 5**
>
> • **Population and Gender** During the early twentieth century, the combination of cultural and economic factors and improvements in health care created a population explosion. By the late twentieth century, improved living conditions, expanded access to education and medical care, urbanization, and changing gender roles were all working together to reduce population growth.
>
> • Because 28 percent of the region's population is under age 15, even if couples choose to have only one or two children, the population will continue to grow as this large group reaches the age of reproduction.

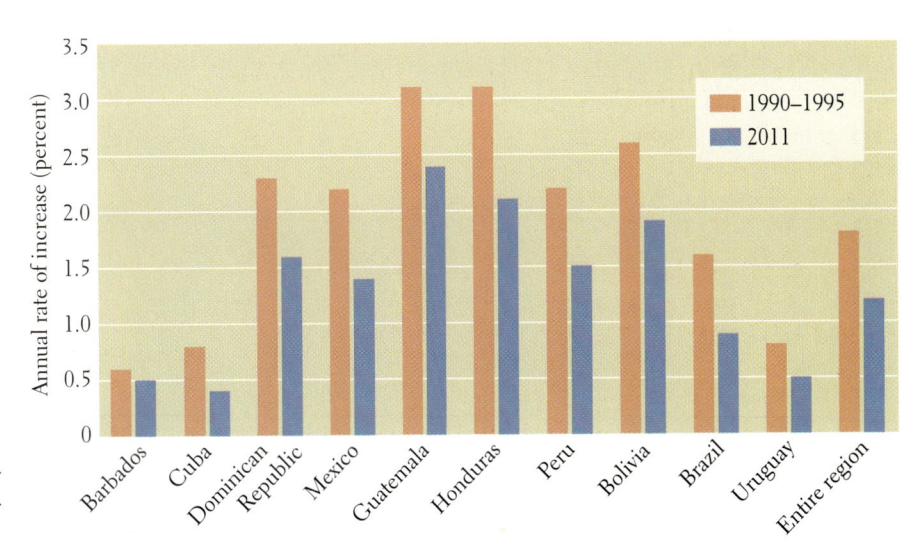

FIGURE 3.24 Trends in natural population increases, 1990–2011. The orange and blue columns show that rates of natural increase have declined steadily throughout the region. They are projected to continue to do so into the future. Nevertheless, in many countries, natural population increase remains high enough to outstrip efforts to improve standards of living. Note that the rates of natural increase are for a few selected countries and for the entire region. [Source consulted: *2011 World Population Data Sheet*, Population Reference Bureau, at http://www.prb.org/pdf11/2011population-data-sheet_eng.pdf]

144 CHAPTER 3 Middle and South America

FIGURE 3.25 Population density in Middle and South America.

SOCIOCULTURAL ISSUES

Under colonialism, a series of social structures evolved that guided daily life—standard ways of organizing the family, community, and economy. They included rules for gender roles, race relations, and religious observance. While these traditional social structures are now changing in response to a variety of forces discussed earlier in this chapter, much continuity with the past is also apparent. Here we look at the region's cultural diversity, family and gender roles, the impact of the Internet, and religion.

FIGURE 3.26 LOCAL LIVES
Festivals in Middle and South America

A A member of a *bateria*, or percussion band, is parading during Rio de Janeiro's *Carnaval*, the largest in the world. Carnaval, which is celebrated in much of the United States as Mardi Gras, takes place in the days and weeks before Lent, the 40-day period in Christian traditions of fasting and abstinence that precedes Easter. Elaborate celebrations take place in Barranquilla in Colombia, Port-au-Prince in Haiti, Port of Spain in Trinidad, and Montevideo in Uruguay. [Viviane Ponti/Lonely Planet Images/Getty Images]

B A procession commemorating Jesus walks on an *alfombra*, or carpet, made of colored sawdust, painstakingly drawn by hand on the streets of Antigua, Guatemala, during *Semana Santa*, or Holy Week, the last week of Lent. The processions will destroy the alfombras. This tradition is an adaptation of the Mayan practice of making elaborate ground designs with flowers and feathers for Mayan royalty to walk on as they made their way to important ceremonies. [Danita Delimont/Gallo Images/Getty Images]

C A richly decorated grave in Tijuana, Mexico, during *Día de los Muertos*—Day of the Dead—a holiday celebrated throughout Mexico on November 1. Families gather in cemeteries, where they build private altars and other decorations dedicated to deceased loved ones. The holiday is a "Christianized" version of an Aztec festival dedicated to the goddess Mictecacihuatl, Queen of the Underworld, whose job is to watch over the bones of the dead. [Claudio Cruz/LatinContent/Getty Images]

Cultural Diversity

The region of Middle and South America is culturally complex because many distinct indigenous groups were already present when the Europeans arrived, and then many cultures were introduced during and after the colonial period. From 1500 to the early 1800s, some 10 million people from many different parts of Africa were brought to plantations on the islands and in the coastal zones of Middle and South America. After the emancipation of African slaves in the British-controlled Caribbean islands in the 1830s, more than half a million Asians were brought there from India, Pakistan, and China as indentured agricultural workers. Their cultural impact remains most visible in Trinidad and Tobago, Jamaica, and the Guianas. In some parts of Mexico, Central America, the Amazon Basin, and the Andean Highlands, indigenous people have remained numerous. To the unpracticed eye, they may appear little affected by colonization, but this is not the case.

In the Caribbean and along the east coast of Central America as well as the Atlantic coast of Brazil, mestizos (people who have a mixture of African, European, and some indigenous ancestry) make up the majority of the population. Some of the most colorful festivals in the Americas, such as versions of Mardi Gras, known here as Carnival, are based on a melding of cultural practices from these three heritages (Figure 3.26A). In some areas, such as Argentina, Chile, and southern Brazil, people of Central European descent are also numerous. The Japanese, though a tiny minority everywhere in the region, increasingly influence agriculture and industry, especially in Brazil, the Caribbean, and Peru. Alberto Fujimori, a Peruvian of Japanese descent, campaigned on his ethnic "outsider" status to become president of Peru during a time of political upheaval.

acculturation adaptation of a minority culture to the host culture enough to function effectively and be self-supporting; cultural borrowing

assimilation the loss of old ways of life and the adoption of the lifestyle of another culture

In some ways, diversity is increasing as the media and trade introduce new influences from abroad. At the same time, the processes of **acculturation** and **assimilation** are also accelerating, thus erasing diversity to some extent as people adopt new ways. This is especially true in the biggest cities, where people of widely different backgrounds live in close proximity to one another.

Race and the Social Significance of Skin Color People from Middle and South America, especially those from Brazil, often proudly claim that race and color are of less consequence in this region than in North America. They are only partially right. In all of the Americas, skin color is now less associated with status than in the past. By acquiring an education, a good job, a substantial income, the right accent, and a high-status mate, a person of any skin color may become recognized as upper class.

Nevertheless, the ability to reduce the significance of skin color through one's actions is not quite the same as race having no significance at all. Overall, those who are poor, less educated, and of lower social standing tend to have darker skin than those who are educated and wealthy. And while there are poor people of European descent throughout the region, most light-skinned people are middle and upper class. Indeed, race and skin color have not disappeared as social factors in the region. In some countries—Cuba, for example, where overt racist comments are socially unacceptable—it is common for a speaker to use a gesture (tapping his or her forearm with two fingers) to indicate that the person referred to in the conversation is of African descent.

The Family and Gender Roles

The basic social institution in the region is the *extended family*, which may include cousins, aunts, uncles, grandparents, and more distant relatives. It is generally accepted that the individual should sacrifice many of his or her personal interests to those of the extended family and community and that individual well-being is best secured by doing so.

The arrangement of domestic spaces and patterns of socializing illustrate these strong family ties. Families of adult siblings, their mates and children, and their elderly parents frequently live together in domestic compounds of several houses surrounded by walls or hedges. Social groups in public spaces are most likely to be family members of several generations rather than unrelated groups of single young adults or married couples, as would be the case in Europe or the United States. A woman's best friends are likely to be her female relatives. A man's social or business circles will include male family members or long-standing family friends.

Gender roles in the region have been strongly influenced by the Roman Catholic Church. The Virgin Mary is held up as the model for women to follow through a set of values known as *marianismo*, which emphasizes chastity, motherhood, and service to the family. In this still-widespread tradition, the ideal woman is the day-to-day manager of the house and of the family's well-being. She trains her sons to enter the wider world and her daughters to serve within the home. Over the course of her life, a woman's power increases as her skills and sacrifices for the good of all are recognized and enshrined in family lore.

Her husband, the official head of the family, is expected to work and to give most of his income to his family. Still, men have much more autonomy and freedom to shape their lives than women because they are expected to move about the larger community and establish relationships, both economic and personal. A man's social network is considered just as essential to the family's prosperity and status in the community as is his work.

There are very different sexual standards for males and females. While all expect strict fidelity from a wife to a husband in mind and body, a man is much freer to associate with the opposite sex. Males measure themselves by the model of *machismo*, in which manliness is considered to consist of honor, respectability, fatherhood, household leadership, attractiveness to women, and the ability to be a charming storyteller. Traditionally, the ability to acquire money was secondary to other symbols of maleness. Increasingly, however, a new, market-oriented culture prizes visible affluence as a desirable male attribute.

While a general inclination to keep all kinds of sexual relationships "in the closet" persists, over the last decade attitudes have begun to change, due in part to television dramas (known as *telenovelas*) that have explored many formerly taboo sexual issues, especially homosexuality. Respect for the rights of gays and lesbians has grown significantly in recent years; Argentina and Mexico City legalized same-sex marriage in 2010, and Brazil and Uruguay did so in 2013.

Religion in Contemporary Life

Judaism, Islam, Hinduism, and indigenous beliefs are found across the region, but the majority of the people are at least nominal Christians, most of them Roman Catholic. While the Roman Catholic Church remains highly influential and relevant to the lives of believers, it has had to contend with popular efforts to reform it, as well as with increasing competition from other religious movements. From the beginning of the colonial era, the church was the major partner of the Spanish and Portuguese colonial governments. It received extensive lands and resources from colonial governments, and in return built massive cathedrals and churches throughout the region, sending thousands of missionary priests to convert indigenous people. The Roman Catholic Church encouraged these people to accept their low status in colonial society, to obey authority, and to postpone rewards for their hard work until heaven.

> **extended family** a family that consists of related individuals beyond the nuclear family of parents and children
>
> **marianismo** a set of values based on the life of the Virgin Mary, the mother of Jesus, that defines the proper social roles for women in Middle and South America
>
> **machismo** a set of values that defines manliness in Middle and South America
>
> **populist movements** popularly based efforts, often seeking relief for the poor
>
> **liberation theology** a movement within the Roman Catholic Church that uses the teachings of Jesus to encourage the poor to organize to change their own lives and to encourage the rich to promote social and economic equity

People throughout the region converted to the faith, some willingly and some only after the threats, intimidation, and torture of the Spanish Inquisition were brought to the Americas. Many indigenous people put their own spin on Catholicism, creating multiple folk versions of the Mass with folk music, more participation by women in worship services, and interpretations of Scripture that vary greatly from European versions. A range of African-based belief systems (Candomblé, Umbanda, Santería, Obeah, and Voodoo) combined with Catholic beliefs can be found in Brazil, northern South America, Middle America, and the Caribbean—wherever the descendants of Africans have settled. These African-based religions have attracted adherents of European or indigenous backgrounds as well, especially in urban areas.

The power of the Roman Catholic Church began to erode in the nineteenth century in places such as Mexico. **Populist movements**, aimed at addressing the needs of the poor masses, seized and redistributed some church lands. They also canceled the high fees the clergy had been charging for simple rites of passage such as baptisms, weddings, and funerals. Over the years, the Catholic Church became less obviously connected to the elite and more attentive to the needs of poor and non-European people. By the mid-twentieth century, the church was abandoning many of its earlier racist policies and ordaining clergy of African and indigenous descent. Women were also given more of a role in religious ceremonies.

In the 1970s, a Catholic movement known as **liberation theology** was begun by a small group of priests and activists. They were working to reform the church into an institution that could combat the extreme inequalities in wealth and power common in the region. The movement portrayed Jesus Christ as a social revolutionary who symbolically spoke out for the redistribution of wealth when he divided the loaves and fish among the multitude.

The perpetuation of gross economic inequality and political repression was viewed as sinful, and social reform as liberation from evil.

The legacy of liberation theology is now fading. At its height in the 1970s and early 1980s, liberation theology was the most articulate movement for region-wide social change. It had more than 3 million adherents in Brazil alone. But the Vatican objected to this popularized version of Catholicism, and today its influence is diminished. In countries such as Guatemala and El Salvador, governments targeted liberation theology participants, vilifying them as communist collaborators; hundreds were tortured and assassinated. Liberation theology has also had to compete with newly emerging evangelical Protestant movements. **79. POPE OPENS TOUR OF LATIN AMERICA IN BRAZIL**

evangelical Protestantism a Christian movement that focuses on personal salvation and empowerment of the individual through miraculous healing and transformation; some practitioners preach to the poor the "gospel of success"—that a life dedicated to Christ will result in prosperity for the believer

Evangelical Protestantism has spread from North America into Middle and South America, and is now the region's fastest-growing religious movement. About 10 percent of the population, or at least 50 million people, are adherents. The movement is growing rapidly in Brazil, Chile, the Caribbean, Mexico, and Middle America, especially among poor and middle classes in both rural and urban settings. It does not, however, share liberation theology's emphasis on combating the region's extreme inequalities in wealth and power. Some evangelical Protestants teach a "gospel of success," stressing that those true believers who give themselves to a new life of hard work and clean living will experience prosperity of the body (wealth) as well as of the soul.

The movement is *charismatic*, meaning that it focuses on personal salvation and empowerment of the individual through miraculous healing and psychological transformation. Evangelical Protestantism is not hierarchical in the same way as the Roman Catholic Church, and there is usually no central authority; rather, there are a host of small, independent congregations led by entrepreneurial individuals who may be either male or female.

Perhaps two of the most important contributions of evangelical Protestantism are the focus (as in Umbanda and other African-based religions) on helping people, especially urban migrants, cope with modern life; and the emphasis placed on involving women in leadership roles in church activities and services.

THINGS TO REMEMBER

- The region of Middle and South America is culturally complex because many distinct indigenous groups were already present when the Europeans arrived, and then many cultures were introduced during and after the colonial period.
- The basic social institution in the region is the extended family, which may include cousins, aunts, uncles, grandparents, and more distant relatives.
- Gender roles in the region have been strongly influenced by the Catholic Church through the ideals of *marianismo* and machismo.
- Evangelical Protestantism came to the region from North America and is now the fastest-growing religious movement in the region, though Catholicism is still the dominant religion.

GEOGRAPHIC INSIGHTS

Middle and South America: Review and Self-Test

1. Environment: Deforestation in this region contributes significantly to global climate change through the removal of trees, which as living plants naturally absorb carbon dioxide, a greenhouse gas. Once the trees are cut or burned, they release large amounts of carbon dioxide into the atmosphere. Additionally, despite the region's overall abundant water resources, some areas are experiencing a water crisis related to climate change, inadequate water infrastructure, and intensified use of water.

- What factors are driving the cycle of deforestation in this region? How is the market economy a factor?
- How are the processes of global warming and deforestation linked?
- What are some possibly sustainable alternatives to deforestation?
- Why are parts of this region experiencing a water crisis?

2. Globalization and Development: The integration of this region with the global economy has left it with the widest gap between rich and poor in the world. Long-term failures to address poverty, economic instability, and the flow of resources and money out of the region have resulted in many conflicts and inspired numerous efforts at reform. Nevertheless, in recent years several countries in this region have emerged as global economic leaders.

- What are some factors that are hampering the equitable distribution of wealth?
- Why did import substitution largely fail as a strategy to develop manufacturing and service-based industries?
- Why has there been so much dissatisfaction with SAPs?
- How have regional trading blocs faired in this region?
- What changes are occurring in this region with regard to food production?

3. Power and Politics: After decades of rule by elites and the military, punctuated by disruptive foreign military interventions, political freedoms in Middle and South America are expanding. Almost all countries now have multiparty political systems and democratically elected governments. However, the international illegal drug trade continues to be a source of violence and corruption in the region.

- What kinds of challenges do elected governments in this region face?
- How has the drug trade influenced the political process in this region?
- How have foreign powers impacted politics in this region?
- What factors have shaped the two rural political movements discussed in this chapter?

4. Urbanization: Since the early 1970s, Middle and South America has experienced rapid urban growth as rural people migrate to cities and towns. A lack of planning to accommodate the massive rush to the cities has created densely occupied urban landscapes that often lack adequate support services and infrastructure.

- How have primate cities impacted countries in this region?
- What kinds of problems do this region's slums, shantytowns, *colonias*, barrios, and *barriadas* have?
- Why is the Brazilian city of Curitiba unique in this region?
- Why might a rural woman decide to migrate to a city?

5. Population and Gender: During the early twentieth century, the combination of cultural and economic factors and improvements in health care created a population explosion. By the late twentieth century, improved living conditions, expanded access to education and medical care, urbanization, and changing gender roles were all working together to reduce population growth.

- What accounts for the population explosion of the last century?
- How has urbanization resulted in smaller family size and more options for women?
- How does the opportunity to gain an education affect a woman's fertility?
- Why are populations likely to grow in this region in the near future?

CRITICAL THINKING QUESTIONS

1. If the European colonists had come to Middle and South America in a different frame of mind—if, say, they were simply looking for a new place to settle and live quietly—how do you think the human and physical geography of the region would be different today?

2. Explain two main ways in which tectonic processes account for the formation of mountains in Middle and South America.

3. Reflecting on the whole chapter, pick some locations that impressed you with regard to the ways in which people are dealing with either environmental issues or with issues of income and wealth disparity or human well-being. Explain your selections.

4. Discuss the ways in which you see the historical circumstances of colonization affecting modern approaches to economic problems in Mexico, Bolivia, Brazil, Venezuela, or Cuba.

5. Describe the main patterns of migration in this region and discuss the effects of migration on both the sending and receiving societies.

6. Name three factors you see as important in increasing political freedoms in this region.

7. Explain how the Amazon Basin and its resources constitute an example of contested space.

8. Argue for or against the proposition that free trade blocs such as NAFTA and UNASUR (previously Mercosur) help the lowest-paid workers have some upward mobility.

9. How would you respond to someone who suggested that Middle and South America were helped toward development and modernization by the experience of European colonization?

› THINKING GEOGRAPHICALLY

Now that you have read about Middle and South America, you should be able to answer the following questions about the photos in this chapter. To answer these questions online, go to Geography LaunchPad.

Human Impacts on the Biosphere (page 119)

A What environmental impact is clearly visible in this photo?

D What land use likely preceded cattle grazing on the land shown in this photograph?

E Where are the soybeans grown on this land likely to end up?

Vulnerability to Climate Change (page 121)

A What clues can be seen in this photo that the neighborhood lacks a centralized water distribution system?

B From which direction are hurricanes most likely to hit Honduras?

C In addition to exposure to tropical storms and hurricanes, what else contributes to Haiti's vulnerability to climate change?

D Why is glacial melting of particular concern to cities in Bolivia?

Visual History of Middle and South America (page 122)

A How does this image of the ruins of Teotihuacan lend credence to the assertion that some indigenous groups may have been better off than their contemporaries in Europe?

B What about Machu Picchu in the Andean highlands indicates that it was more than a mere summer residence?

C Describe the mood of this depiction of the Spanish conquest of the Incas.

D In what way does this painting of Creole Argentinians and Chileans celebrating independence suggest that they were unlikely to found egalitarian societies?

E Even after the end of slavery, who constituted the labor force in sugar cultivation?

Power and Politics (page 137)

A What evidence can you find in the photo of a violent situation in Mexico?

B What are the indications that these men were once involved in violent combat?

C Why would the Cuban government lionize these men in a mural?

Urbanization (page 140)

A What about this picture suggests that there may be considerable wealth in Santiago?

B Why might work in the informal economy be considered insecure?

C What are some defining characteristics of "informal" squatter settlements?

CHAPTER KEY TERMS

- acculturation 145
- assimilation 145
- Aztecs 122
- biodiversity 114
- brain drain 139
- contested space 138
- coup d'état 134
- Creoles 125
- dictator 134
- ecotourism 118
- El Niño 116
- evangelical Protestantism 147
- Export Processing Zones (EPZs) 130
- extended family 146
- favelas 139
- foreign direct investment (FDI) 131
- hacienda 128
- import substitution industrialization (ISI) 128
- Incas 123
- income disparity 127
- indigenous 111
- isthmus 111
- liberation theology 146
- machismo 146
- maquiladoras 131
- *marianismo* 146
- marketization 129
- mercantilism 125
- Mercosur 132
- mestizos 125
- Middle America 111
- nationalize 131
- North American Free Trade Agreement (NAFTA) 132
- plantation 128
- populist movements 146
- primate city 139
- privatization 129
- shifting cultivation 122
- silt 114
- South America 111
- structural adjustment programs (SAPs) 129
- subduction zone 112
- temperature-altitude zones 114
- trade winds 114
- UNASUR 132
- urban growth poles 141

Map of Europe

chapter 4

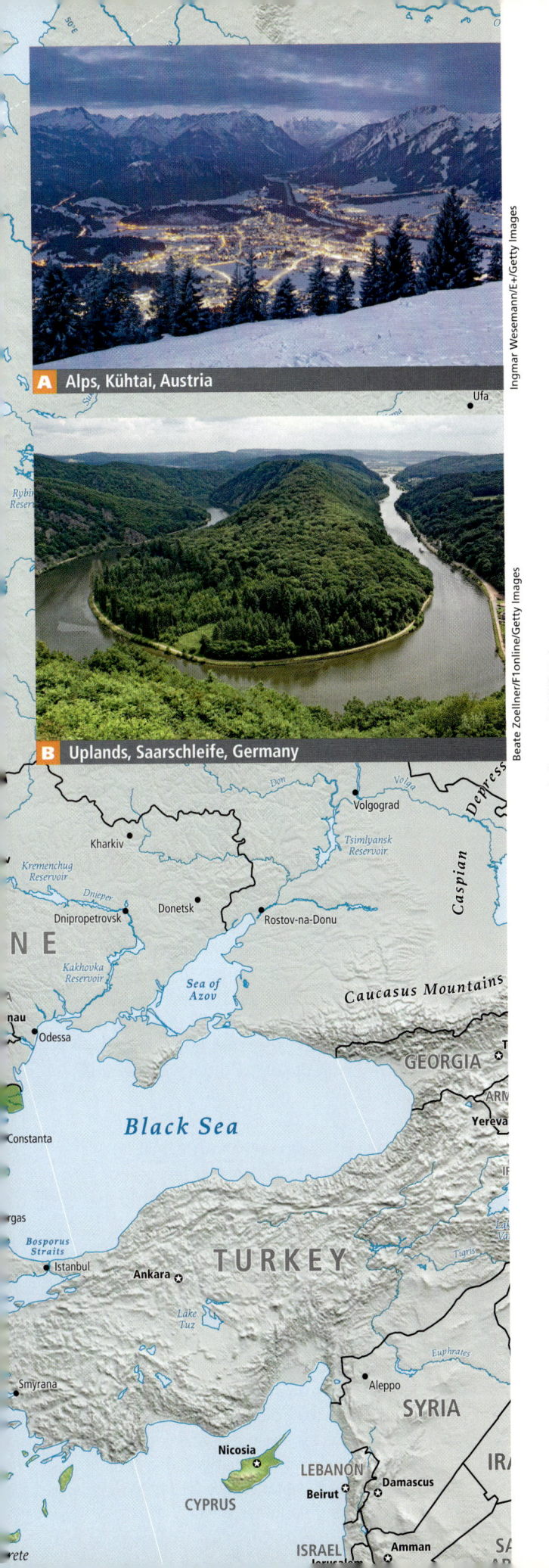

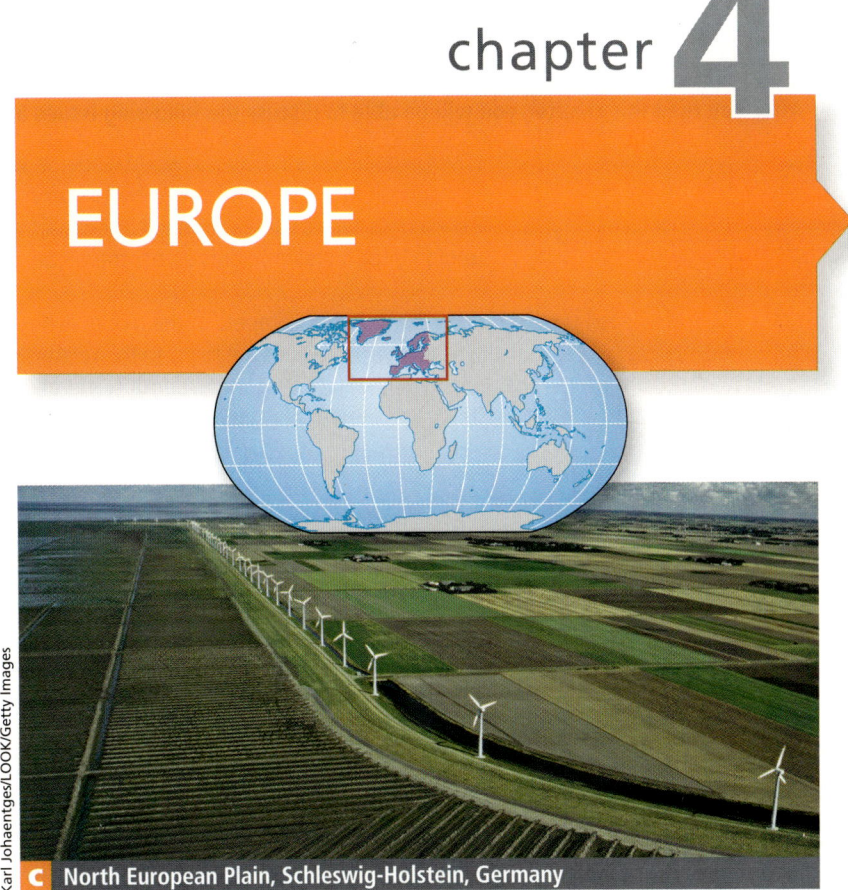

EUROPE

A Alps, Kühtai, Austria

B Uplands, Saarschleife, Germany

C North European Plain, Schleswig-Holstein, Germany

D Rhine River, Basel, Switzerland

E Danube River, Budapest, Hungary

FIGURE 4.1 Regional map of Europe.

151

CHAPTER 4 Europe

> GEOGRAPHIC INSIGHTS

After you read this chapter, you will be able to discuss the following issues as they relate to the five thematic concepts:

1. Environment: The European Union (EU) is a world leader in responding to climate change. Its goals for cutting greenhouse gas emissions are complemented by its many other strategies for saving energy and resources. However, much of Europe's air and many of its seas and rivers are very polluted, and consumption patterns in the region impact environments across the globe.

2. Globalization and Development: In order to better compete in the global economy, the EU has shifted labor-intensive industries from western Europe, where wages are high, to the relatively poorer, lower-wage member states of Central Europe. Growth in services has made up for some of the loss of industrial jobs in western Europe, though the unemployment rate is high in some countries. The EU continues to struggle to integrate the economies of its member states, especially those that use the EU's common currency, the euro.

3. Power and Politics: Following World War II, political freedoms have grown and strong welfare states have been established in this region. Since the fall of the Soviet Union in 1991, international politics within Europe have centered on the expansion of the EU into Central Europe and the development of EU political institutions. Having grown so much in the past several decades, the European Union, which is already a global economic power, could become a global counterforce to the United States in political and military affairs.

4. Urbanization: Europe's cities are both ancient and modern, with old town centers now surrounded by modern high-rise suburbs and supported by world-class urban infrastructures. Cities are the heart of Europe's economy, politics, and culture, with more than 70 percent of the EU population living in urban areas.

5. Population and Gender: Europe's population is aging as its fertility rates decline, which they are doing for a number of reasons, including the fact that career-oriented working women are choosing to have only one or two children. Fearing that aging populations could slow economic growth, European countries have tried to boost their fertility rates with a variety of policies, the most successful of which have addressed the needs of working mothers.

The European Region

Over the last 500 years, the region of Europe (shown in Figure 4.1) has had a central role in world history and geography as a colonizer of distant territories and as a major player in the development of capitalism. In the twentieth century, Europe brought the world to the edge of disaster with two massive wars. Since then it has revived, prospered, and designed a model for regional economic and political integration that has been copied in other regions. Europe is now at another important juncture: Can it perfect its union and extend it, or will the union dissolve in the face of growing subregional financial and social disparities? The five thematic concepts in this book are explored as they arise in the discussion of these regional issues. Scattered throughout the chapter, vignettes illustrate one or more of the themes as they are experienced in individual lives.

GLOBAL PATTERNS, LOCAL LIVES

Grigore Chivu (a pseudonym) wanders through the empty pig pens on his farm near Lugoj in western Romania. For generations, his family has made a meager but rewarding living by raising hogs and then processing and selling the meat. A few years ago, just before Romania joined the European Union (EU), he had more than 250 pigs. At Christmas, he and thousands of pig farmers across the country would slaughter a certain number of their pigs and preserve the meat using time-honored methods. Pig slaughtering was a time of high spirits, celebration, and community cooperation as the farmers contemplated the coming feast and their profits. Flavorful sausages, which they smoked and hung high in the rafters of their kitchens for further drying, were slowly parceled out to select customers, providing a steady income well into summer.

When Romania entered the European Union in 2007, all farmers were required to conform to EU standards for processing meat. The old methods were no longer allowed and for some the new standards were prohibitively expensive.

At first the farmers were uncertain just how to respond. Before they could organize a butchering cooperative that conformed to EU standards and for which development funds were available, an American company (based in Virginia) stepped into the breach. Smithfield, the world's largest pork producer, was expanding operations into Central Europe, where it planned to produce pork and pork products on an industrial scale and market them globally. Eager to enter the EU market, Smithfield enlisted the help of Romanian politicians and got permission (and even EU subsidies) to establish a conglomerate that included feed production, pig breeding, modernized sanitary barns for fattening thousands of hogs in small cages, and slaughterhouses.

FIGURE 4.2 Pig farms in Romania and the United States.

(A) Pigs on a farm in Romania that raises small animals.

(B) An industrial hog farm in Michigan, U.S., capable of raising thousands of pigs at a time.

> **THINKING GEOGRAPHICALLY** What about (A) suggests small-scale rearing of pigs?

In September of 2013, Smithfield was bought by a Chinese company, Shuanghui, which planned to redirect much of the company's global production toward supplying China's growing demand for pork. Shortly after Shuanghui purchased Smithfield, Romanian officials signed agreements with their Chinese counterparts to export more than 3 million pigs to China each year. With Romania already importing half of its own pork, the recently established Smithfield facilities are now seen as key to boosting production to keep up with the Chinese demand.

As the old, picturesque Romanian agricultural landscape is transformed into huge factory farms, the number of independent pig farmers is dropping; it has been reduced by more than 90 percent (Figure 4.2). Unable to compete with the lower prices Smithfield can charge, Grigore Chivu, like thousands of his fellow pig farmers, was considering migrating to western Europe, where stronger economies meant better-paying jobs. However, because of his age and traditional farming background, Chivu would likely find only low-wage work. Meanwhile, in Romania, he is too young for a pension. [Sources: Margareta Lelea; New York Times. For detailed source information, see Text Sources and Credits.]

Grigore Chivu is faced with disruptive change as his country adjusts to new circumstances in the European Union and the global economy, but he is also part of a global revolution in food production and marketing that is changing food patterns across the world. For example, Smithfield pork, produced and processed at low cost in Romania, is now marketed in China, a country whose reliance on imported food and factory farming is growing. Most small-scale Chinese pig farmers have already been driven out of business by the lower prices that large Chinese factory farms can charge for pork. Importing more pork from the rest of the world, including from Shuanghui's newly acquired Smithfield operations in Romania, will lower pork prices in China even further. The video *Pig Business* explores the many sides of the global shift from small farm production of hogs to large, corporate farm production. The section on Romania begins at 42 minutes (see http://tinyurl.com/96vh9ja).

The **European Union (EU)** is a supranational organization that unites most of the countries of West, South, North, and Central Europe (Figure 4.3). In principle, people, goods, and money can move freely throughout the European Union. Many people, like Grigore Chivu, decide to migrate only reluctantly, and do so only because poorer regions do not have the political power to push for useful reforms.

There are now 28 countries in the European Union; Croatia joined in 2013 and a few more hope to join in the next several years. The newest members (with the exceptions of Malta and Cyprus) are formerly Communist countries in Central Europe and generally have lower standards of living. Hundreds of thousands of workers from Central Europe have moved to other EU states to find better jobs. More recently, with rising job losses throughout the EU, many have returned home.

Since 2008, the European Union has been under growing financial stress as a number of the less industrialized and less prosperous countries (Greece, for example) have found it so hard to make loan payments that they face the possibility of national bankruptcy. Residents of the original and more prosperous

> **European Union (EU)** a supranational organization that unites most of the countries of West, South, North, and Central Europe

154 CHAPTER 4 Europe

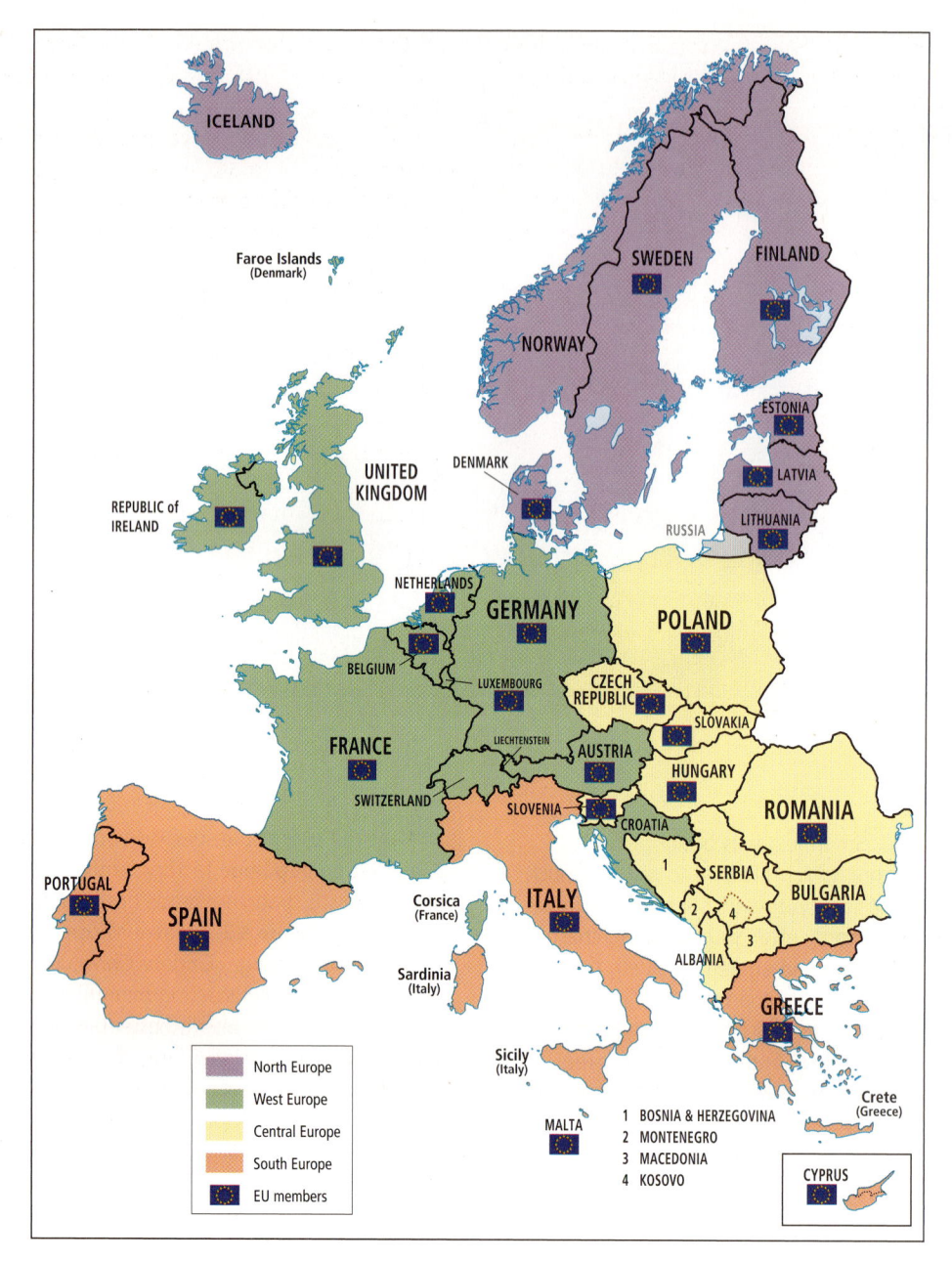

FIGURE 4.3 Political map of Europe.

EU members—such as France and especially Germany—resent having to support these failing economies and also fear that the migration of workers from the EU's south and east is bringing very different people into close contact with each other, resulting in political tensions and higher costs for social and educational services. Others, especially the migrants themselves, lament the **cultural homogenization**—the tendency toward uniformity of ideas, values, technologies, and institutions among associated culture groups—that is wiping out traditional ways. Meanwhile, economic planners and employers across Europe argue that allowing workers to migrate is crucial to Europe's economic growth and competitiveness on a global scale.

> **cultural homogenization** the tendency toward uniformity of ideas, values, technologies, and institutions among associated culture groups

THINGS TO REMEMBER

- Throughout the European Union, smaller family-run farms are giving way to larger farms run by corporations. This change is strongest in those Central Europe countries that have been recently admitted to the European Union and has resulted in many farm family members migrating to other EU countries in search of work.

What Makes Europe a Region?

Physically, Europe is a peninsula extending off the western end of the huge Eurasian continent. There are many peninsular appendages, large and small, on this giant peninsula of Europe. Norway and Sweden share one of the larger appendages. Other large peninsulas are the Iberian Peninsula (occupied by Portugal and Spain), as well as those of Italy and Greece. All extend into either the Atlantic Ocean or into the various seas that are adjacent to the Atlantic (see the Figure 4.1 map).

The perimeter of the region of Europe to the north, west, and south is primarily oceanic coastline. This unique access to the world ocean facilitated European exploration and colonization of distant territories. The eastern physical and cultural border of Europe has been difficult to define. Just where Europe ends and Asia begins has been a source of contention for millennia. In this book, at this time, the eastern limit of the European region is taken to be the border of the European Union. Other potential parts of Europe—Ukraine, Belarus, western Russia, Moldova, and the Caucasus—are covered in Chapter 5, and Turkey in Chapter 6.

Terms in This Chapter
This book divides Europe into four subregions (see Figure 4.3)—*North*, *West*, *South*, and *Central Europe*. *Central Europe* refers to all those countries formerly in the Soviet sphere that are now in the European Union, as well as the countries that were formerly in Yugoslavia, plus Albania.

For convenience, we occasionally use the term *western Europe* to refer to all the countries that were *not* part of the experiment with communism in the Soviet sphere and in Yugoslavia. That is, *western Europe* comprises the combined subregions of North Europe (except Estonia, Latvia, and Lithuania), West Europe (except the former East Germany), and South Europe. When we refer to the countries that were part of the Soviet sphere up to 1989, we use the pre-1989 label *eastern Europe*. EU-28 is the term for just the 28 member countries of the European Union. When we refer to the group of countries collectively known to some as the *Balkans* (Albania, Bosnia and Herzegovina, Bulgaria, Croatia, Macedonia, Montenegro, Romania, Serbia, and Slovenia), we use the term *southeastern Europe*.

PHYSICAL GEOGRAPHY AND ENVIRONMENTAL ISSUES

PHYSICAL PATTERNS

Europe's physical geography is shaped by its many peninsulas which reach into surrounding oceans and seas, and by its variable landforms, all of which affect the region's climates and vegetation.

Landforms

Although European landforms are fairly complex, the basic pattern is mountains, uplands, and lowlands, all stretching roughly west to east in wide bands. As you can see in Figure 4.1 (in the map and in photo A), Europe's largest and highest mountain chain, the *Alps*, runs west to east through the middle of the continent, from southern France through Switzerland and Austria. The Alps extend into the Czech Republic and Slovakia, and curve southeast as the Carpathian Mountains into Romania. This network of mountains is mainly the result of pressure from the collision of the northward-moving African Plate with the southeasterly moving Eurasian Plate (see Figure 1.8 on page 16). Europe lies on the westernmost extension of the Eurasian Plate.

South of the main Alps formation, lower mountains extend into the peninsulas of Iberia and Italy and along the Adriatic Sea through Greece to the southeast. The northernmost mountainous formation is shared by Scotland, Norway, and Sweden. These northern mountains are old (about the age of the Appalachians in North America) and have been worn down by glaciers and millions of years of erosion.

Extending northward from the central Alpine zone is a band of low-lying hills and plateaus curving from Dijon (France) through Frankfurt (Germany) to Krakow (Poland). These uplands (see Figure 4.1B) form a transitional zone between the high mountains and lowlands of the *North European Plain*, the most extensive landform in Europe (see Figure 4.1C). The plain begins along the Atlantic coast in western France and covers a wide band around the northern flank of the main European peninsula, reaching across the English Channel and the North Sea to take in southern England, southern Sweden, and most of Finland. The plain continues east through Poland, then broadens to the south and north to include all the land east to the Ural Mountains in Russia.

Crossed by many rivers and holding considerable mineral deposits, the coastal lowland of the North European Plain is an area of large industrial cities and densely occupied rural areas. Over the past thousand years, people have transformed the natural seaside marshes and vast river deltas into farmland, pastures, and urban areas by building dikes and draining the land with wind-powered pumps. This is especially true in the low-lying Netherlands, where concern over climate change and sea level rise is considerable.

The rivers of Europe link its interior to the surrounding seas. Several of these rivers are navigable well into the upland zone, and Europeans have built large industrial cities on their banks. The Rhine carries more traffic than any other European river, and the course it has cut through the Alps and uplands to the North Sea also serves as a route for railways and motorways (see Figure 4.1D). The area where the Rhine flows into the North Sea is considered the economic core of Europe. Rotterdam, Europe's largest port, is located here. The Danube River, larger and much longer than the Rhine, flows southeast from Germany, connecting the center of Europe with the Black Sea. As the European Union expands to the east, the economic and environmental roles of the Danube River basin, including the Black Sea, are getting more attention (see Figure 4.1E).

Vegetation and Climate

Nearly all of Europe's original forests are gone, some for more than a thousand years. Today, forests with very large and old trees exist only in scattered areas, especially on the more rugged mountain slopes (see Figure 4.1A, B) and in the northernmost parts of Norway, Sweden, and Finland. Forests are intensively managed and are regenerating on abandoned farmland. Although forests now cover about one-third of Europe, the dominant vegetation is crops and pasture grass.

Europe has three main climate types: temperate midlatitude, Mediterranean, and continental (Figure 4.4). The **temperate midlatitude climate**, characterized by year-round moisture, relatively mild winters, and long, mild to hot summers, dominates in western Europe, where the Atlantic Ocean moderates temperatures (see Figure 4.4A). To minimize the effects of heavy precipitation runoff, people in these areas have developed elaborate drainage systems for their houses and communities. Forests are both evergreen and deciduous.

A broad, warm-water ocean current called the **North Atlantic Drift** brings large amounts of warm water to the coasts of Europe. It is really just the easternmost end of the Gulf Stream, which carries warm waters from the Gulf of Mexico north along the eastern coast of North America and across the North Atlantic to Europe (see Figure 2.4 on page 63). While the air above the North Atlantic Drift is relatively warm and wet, the eastward-blowing winds that push this air over North and West Europe and the North European Plain have an equally large impact on climate in this region. These winds bring moderate temperatures and rain deep into the Eurasian continent, creating a climate that, although still fairly cool, is much warmer than elsewhere in the world at similar latitudes. There is some concern that global climate change could eventually weaken the North Atlantic Drift, leading to a significantly cooler Europe.

Farther to the south, the **Mediterranean climate** prevails—warm, dry summers and mild, rainy winters (see Figure 4.4B). In the summer, warm, dry air from North Africa shifts north over the Mediterranean Sea as far north as the Alps, bringing high temperatures and clear skies. Crops grown in this climate, such as olives, grapes, citrus, wheat, apples, and other fruits, must be drought-resistant or irrigated. In the fall, this warm, dry air shifts to the south and is replaced by cooler temperatures and rainstorms sweeping in off the Atlantic. Overall, the climate here is mild, and houses along the Mediterranean coast are often open and airy to afford comfort in the hot, sunny summers. During the short, mild winters, life moves indoors, where wood fires heat one or two small rooms.

In Central Europe, without the moderating influences of the Atlantic Ocean and the Mediterranean Sea, the climate is more extreme. In this region of **continental climate**, summers are fairly hot, and the winters become longer and colder the farther north or deeper into the interior of the continent one goes (see Figure 4.4C). Here, houses tend to be well insulated with small windows, low ceilings, and steep roofs that can shed snow. Crops must be adapted for much shorter growing seasons, and include corn and other grains plus fruit trees and a wide variety of vegetables adapted to the cold, especially root crops and cabbages.

> **temperate midlatitude climate** as in south-central North America, China, and Europe, a climate that is moist all year with relatively mild winters and long, mild to hot summers
>
> **North Atlantic Drift** the easternmost end of the Gulf Stream, a broad warm-water current that brings large amounts of warm water to the coasts of Europe
>
> **Mediterranean climate** a climate pattern of warm, dry summers and mild, rainy winters
>
> **continental climate** a midlatitude climate pattern in which summers are fairly hot and moist, and winters become longer and colder the deeper into the interior of the continent one goes

THINGS TO REMEMBER

- Europe is a region of peninsulas upon peninsulas.
- Europe has three main landforms: mountain chains, uplands, and the vast North European Plain.
- Europe has three principal climates: the temperate midlatitude, the Mediterranean climate, and the continental climate. ■

ENVIRONMENTAL ISSUES

> **GEOGRAPHIC INSIGHT 1**
>
> **Environment:** The European Union (EU) is a world leader in responding to climate change. Its goals for cutting greenhouse gas emissions are complemented by its many other strategies for saving energy and resources. However, much of Europe's air and many of its seas and rivers are very polluted, and consumption patterns in the region impact environments across the globe.

Europeans are increasingly aware of just how dramatically they have transformed their environments over the past several thousand years and are now taking action on many fronts. Nevertheless, the European Union still has a long way to go to meet its goals of clean air and water; sustainable development in agriculture, industry, and energy use; and maintenance of biodiversity.

European Leadership in Response to Global Climate Change

Europe leads the world in responsiveness to global climate change, with EU governments having agreed to cut greenhouse gas emissions by 20 percent by 2020. Europe has been more willing than any other region to address climate change, largely because its governments and corporations see economic advantages to doing so. Recent research suggests that investments in energy conservation, alternative energy, and other measures would cost EU economies 1 percent of their GDP. By contrast, doing nothing about climate change could *shrink* GDP by 20 percent.

🎬 **83. GERMAN INVESTMENTS IN CLEAN ENERGY PAY OFF**

Europe's increasing concern about global warming may also be influenced by public alarm at recent extreme weather. The summers of 2003 and 2012 broke high-temperature records across Europe. Crops failed, freshwater levels sank, forests burned, and deaths soared. In 2003, three thousand people died in France alone. On the other hand, in 2002, 2006, and 2012, rainfall and snowfall in Central Europe reached record levels. In the spring of 2006, the rivers of Central Europe—the Elbe, the Danube, and the Morava—flooded for weeks. Unusual vacillations in weather still continue.

FIGURE 4.4

PHOTO ESSAY: Climates of Europe

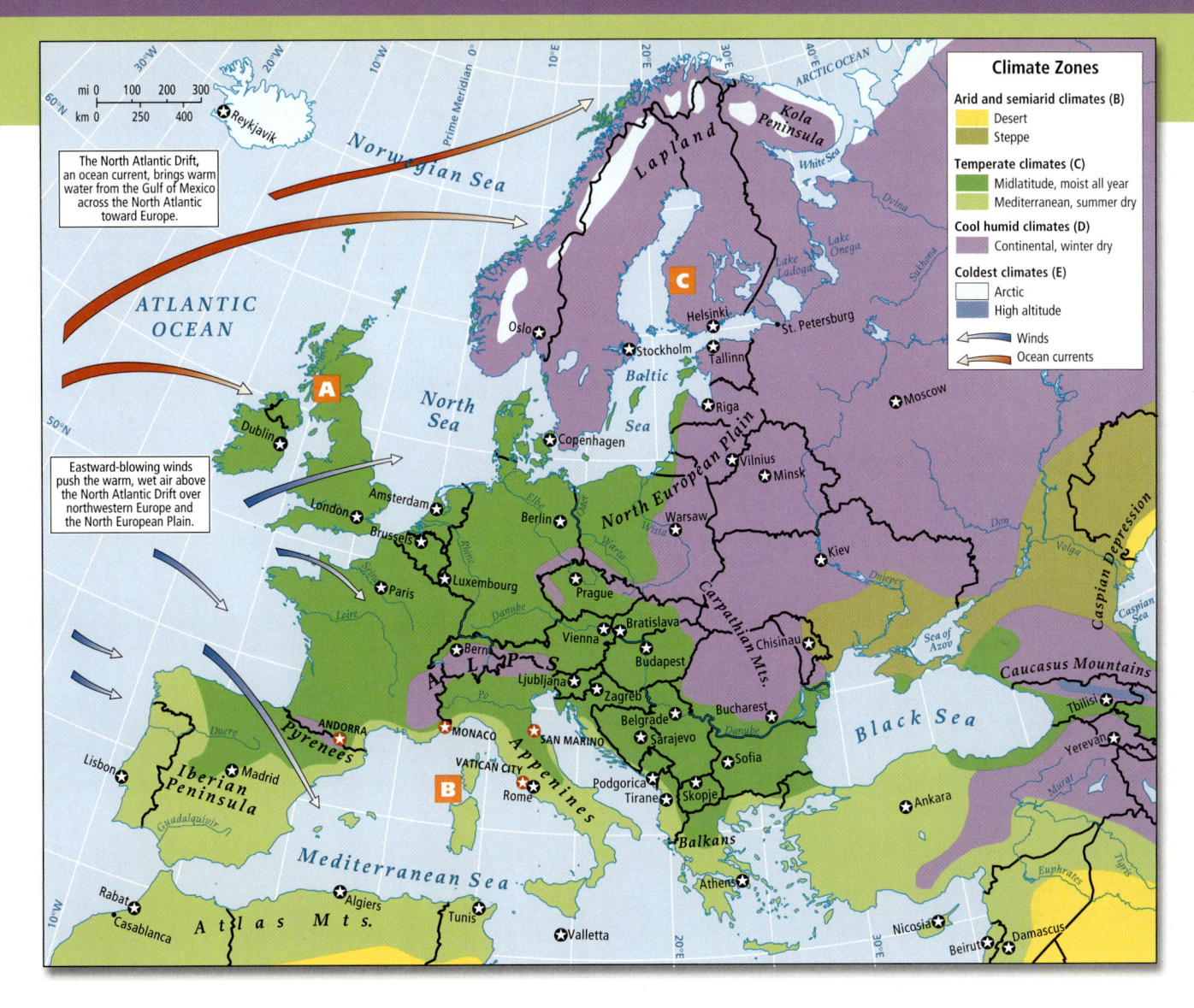

The North Atlantic Drift, an ocean current, brings warm water from the Gulf of Mexico across the North Atlantic toward Europe.

Eastward-blowing winds push the warm, wet air above the North Atlantic Drift over northwestern Europe and the North European Plain.

Climate Zones

Arid and semiarid climates (B)
- Desert
- Steppe

Temperate climates (C)
- Midlatitude, moist all year
- Mediterranean, summer dry

Cool humid climates (D)
- Continental, winter dry

Coldest climates (E)
- Arctic
- High altitude

→ Winds
→ Ocean currents

A Midlatitude, moist all year, Scotland

B Mediterranean, summer dry, Corsica

C Continental, winter dry, Finland

157

FIGURE 4.5   PHOTO ESSAY: Vulnerability to Climate Change in Europe

Despite Europe's wealth, technological sophistication, and well-developed emergency response systems, parts of the region are vulnerable to the effects of climate change. Many locations are exposed to drought, flooding, or sea level rise, and in some of these areas, people are too poor to afford appropriate precautions.

A As the water level in a reservoir in Catalonia, Spain, sinks, a once-submerged eleventh-century church reemerges. Rising temperatures will make Spain's climate drier, its evaporation rate increase, and its scarce water resources shrink further, threatening agriculture and drinking water, especially along the Mediterranean coast.

B Part of the massive Delta Works that protect the Netherlands from rising sea levels during storms. A response to massive flooding in 1953, the Delta Works significantly reduces the Netherlands' sensitivity to sea level rise, making the Netherlands only slightly more vulnerable to climate change than the rest of Europe, even though 60 percent of its population lives below sea level.

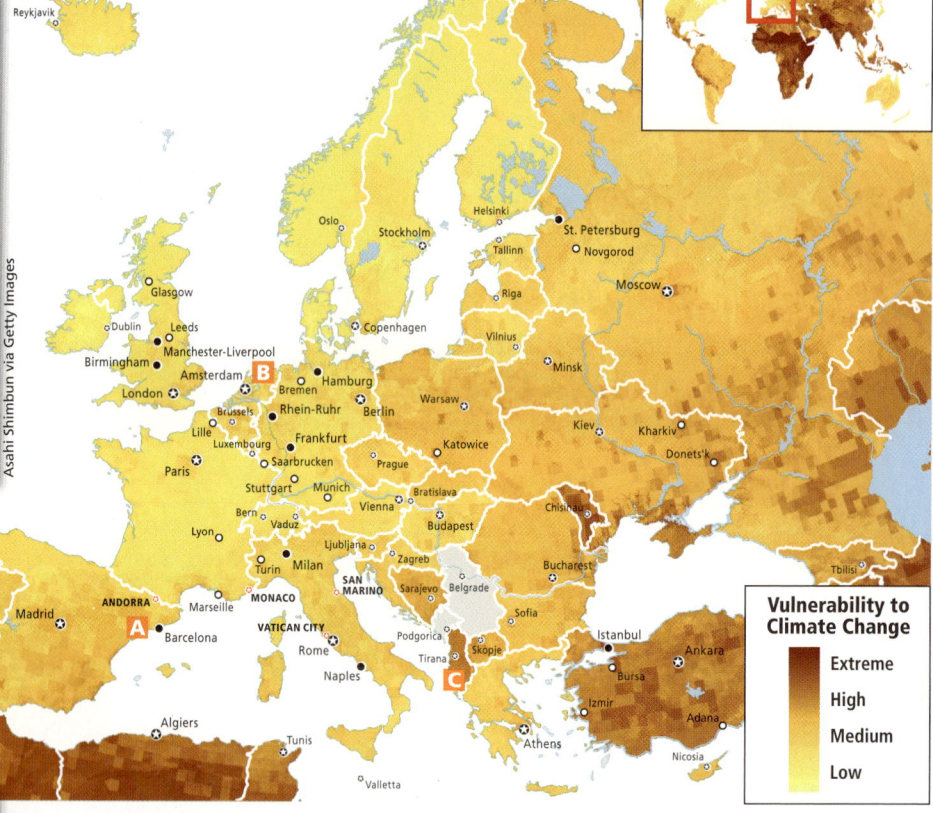

Vulnerability to Climate Change
- Extreme
- High
- Medium
- Low

C Albania's vulnerability to climate change is related to its low incomes and inadequate infrastructure of many kinds, including transportation. Albania's sensitivity to many potential disturbances is thus high and its resilience low. Many problems stem from misguided government policies. For example, until 1992 it was illegal for Albanians to own private automobiles. Even today, many poorer Albanians depend on horses and small, slow, horse-drawn carts as their main mode of transportation.

Environmental Issues **159**

Europe's Vulnerability to Climate Change Europe's wealth, technological sophistication, and well-developed emergency response systems make it more resilient to the consequences of climate change than are most regions of the world. Nevertheless, some areas are much more vulnerable than others because of their location, dwindling water resources (Figure 4.5A), and rising sea levels (see Figure 4.5B), or because of the effects of poverty. Figure 4.5 illustrates some of these vulnerabilities.

ON THE BRIGHT SIDE

Guerrilla Gardeners

Under cover of darkness, gardeners—who in the daytime are bureaucrats, stock traders, and computer jockeys—sneak into Central London to plant colorful flowers and foliage in traffic islands and roundabouts. These Green activists are part of a movement called Guerrilla Gardeners, which has thousands of activists in more than 30 countries, and a Web site, http://www.guerrillagardening.org/.

Bypassing the town councils (which tend to impose crippling rules), the Guerrilla Gardeners make quick assaults late at night, armed with trowels, spades, mulch, and watering cans. Authorities seem unable to stop the guerrillas from covering neglected urban land with blooming hyacinths, tulips, marigolds, shrubs, and even trees. Sneakier yet are the Seed Bombers, a group that packs flower seeds, soil, and water into compact parcels and tosses them into derelict patches of public land, where they shatter on impact, spewing forth seeds that produce plants capable of outcompeting the weeds. *[Source: NPR. For detailed source information, see Text Sources and Credits; see also Philip Booth's "Seed Bombers in Stroud," at http://www.youtube.com/watch?v=vY02FKd1Uco.]*

Europe's Green Behavior By global standards, Europeans use large amounts of resources and contribute about one-quarter of the world's greenhouse gas emissions. However, one European resident averages only one-half the energy consumption of the average North American resident. Europeans live in smaller dwellings that need less energy to heat or cool. They drive smaller, more fuel-efficient cars and often use public transportation. Because communities are more spatially compact, many people walk or bicycle wherever they need to go.

These energy-saving practices are related in part to the population densities and social customs of the region, and also to widespread explicit popular support for ecological principles. **Green** political parties influence national policies in all European countries, and Green policies are central to the agenda of the European Union. Results include strong regional advocacy for emission controls, well-entrenched community recycling programs, and grassroots work on local environmental concerns.

Green environmentally conscious

- 84. ICELAND: ENERGY TO SPARE
- 85. BRITAIN REQUIRES ENERGY INSPECTIONS FOR HOME SALES
- 86. MCD'S GARBAGE HEATS AND LIGHTS BRITISH CITY
- 87. LONDON LEADS BY EXAMPLE TO CURB POLLUTION, CLIMATE CHANGE

Changes in Transportation Although Europeans have for many years favored fast rail networks for both passengers and cargo (see Figure 4.20D on page 181) rather than private cars, trucks, and multilane highways, they have recently been drifting closer to the American model of private cars and trucks driven on sweeping freeways. This change is now happening in even the poorest parts of Europe, where cars have been scarce in the past (see Figure 4.5C). In response to rising fuel costs and CO_2 emissions, though, the European Union has developed long-term plans that reduce the emphasis on cars and trucks and involve designing *multimodal transport* to link high-speed rail to road, air, and water transportation (Figure 4.6). This is because highway transport is inefficient and polluting relative to rail or water transport. One gallon (3.7 liters) of gasoline can move one ton of cargo a distance of 59 miles (94 kilometers) by truck, 202 miles (323 kilometers) by rail, and 514 miles (822 kilometers) by waterway. An EU report shows that private cars used for passenger transportation produce three times more CO_2 emissions than does rail-based public transportation.

Europe's long, irregular coastline and the low cost of water transportation have been a boon for the development of links to global trade. Europe has numerous modern ocean ports that cater to container ships: Helsinki, Riga, Hamburg, Copenhagen, Antwerp, Rotterdam, Plymouth, Southampton, Le Havre, Marseille, Barcelona, and Koper. Soon, newly built Mediterranean ports in Morocco, Algeria, Tunisia, Malta, and Egypt, all of which cater at least in part to European markets, will join these ports. One-third of the world's container traffic now goes through the Mediterranean. The container ship industry is keenly attuned to the concerns about CO_2 and climate change, and is now using optimal (often slower) speeds that are carefully calculated to minimize fuel consumption and emissions while maximizing profits.

Europe's Impact on the Biosphere

There is a geographic pattern to the ways in which human activities have transformed Europe's landscapes over time. Western Europe shows the effects of dense population and heavy industrialization (Figure 4.7A), and Central Europe reveals the results of long decades of willful disregard for the environment. Central Europe continues to have a major impact on the biosphere through the air and water pollution it generates (see Figure 4.7B).

Europe's Energy Resources Europe's main energy sources have shifted over the years from wood to coal and, more recently, to petroleum and natural gas and in some countries to nuclear power. Europe is now leading the move to renewable energy sources in response to rising energy costs and commitments to cut greenhouse gas emissions.

The 28 members of the European Union (the EU-28) get a large portion of their fuel supplies from Russia—30 percent of their crude oil and 34 percent of their natural gas, as of 2011. Most of the gas now comes via pipelines through Belarus and Ukraine, but Turkey supplies Russian gas to Europe via the Black Sea and hosts a pipeline that carries gas from the Caspian Sea to Europe. Russia is negotiating for another trans-Turkey pipeline to carry oil and gas to the Mediterranean. While Europeans fear that this dependency will be used against them by Russia, which periodically withholds and releases flows to Europe through Ukraine and Belarus, Russia is actually also dependent on the EU fuel trade

160 CHAPTER 4 Europe

FIGURE 4.6 The Trans-European transport network. [Source consulted: European Commission Mobility and Transport, at http://ec.europa.eu/transport/themes/infrastructure/revision-t_en.htm]

because 80 percent of Russia's oil exports and 70 percent of its natural gas exports go to the European Union.

Another 30 percent of the gas and oil that the European Union consumes comes from various Middle Eastern producers. Large oil and gas deposits in the North Sea, most controlled by Norway (not an EU member), have alleviated Europe's dependence on "foreign" sources of energy, but the production of oil from the North Sea has already peaked and is expected to run out sometime in the next decade. **82. WINTER ON THE WAY**

The use of nuclear power to generate electricity has been more common in Europe than in North America. The EU-28 depends on nuclear power for 30 percent of its total needs. In France, 78 percent of the electricity is generated by nuclear power (compared with only 20 percent in the United States). However, support for nuclear power has declined, partly in response to the disastrous nuclear accidents in 1986 in Chernobyl, Ukraine, and in 2011 in Fukushima, Japan. The safe disposal of nuclear waste products is also a concern.

FIGURE 4.7 PHOTO ESSAY: Human Impacts on the Biosphere in Europe

Most of Europe has been transformed by human activity. Western Europe has some of the most heavily impacted landscapes and ecosystems, but the formerly communist countries of Central Europe also have severe environmental problems. Meanwhile, agricultural intensification is creating new problems in South Europe. The map shows some of the impacts on Europe's land, sea, and air.

Hans-Guenther Oed/STOCK4B/Getty Images

A The Garzweiler open-pit coal mine in Germany, one of the largest in the world, covers 25 square miles (66 square kilometers). Its operations forced the abandonment of 12 villages and towns, and now threaten local groundwater resources. The coal this mine produces is a heavily polluting variety called lignite. The high sulfur content of lignite is responsible for much of Europe's acid rain problem.

B A hotel in Lodz, Poland, in the shadow of a Soviet-era power plant. Central Europe has thousands of similar facilities, many of which are very polluting and in close proximity to human settlements. Most were built during or before the era of Soviet domination, when environmental safeguards were rare.

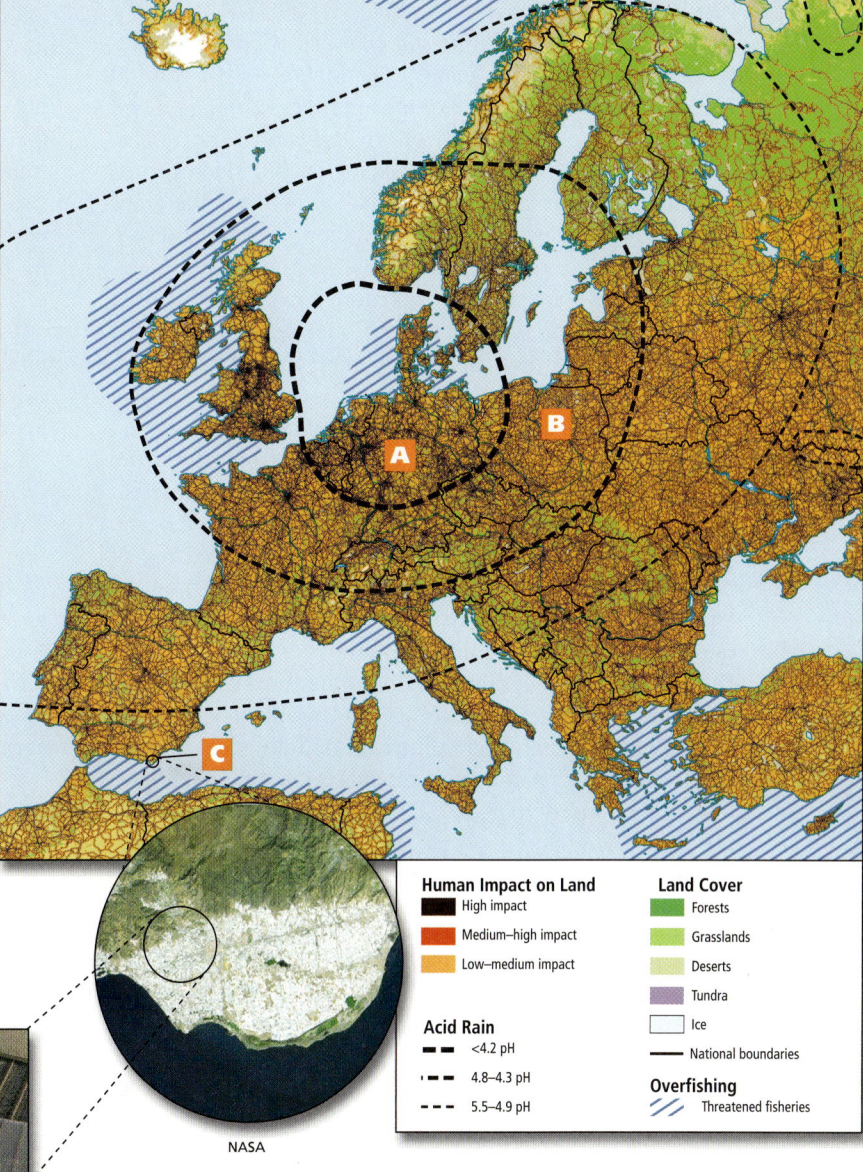

Ross McRoss/Flickr/Getty Images

Quim Llenas/Cover/Getty Images

NASA

Human Impact on Land
- High impact
- Medium–high impact
- Low–medium impact

Acid Rain
- <4.2 pH
- 4.8–4.3 pH
- 5.5–4.9 pH

Land Cover
- Forests
- Grasslands
- Deserts
- Tundra
- Ice
- National boundaries

Overfishing
- Threatened fisheries

C A sea of plastic greenhouses covers most of Campo de Dalía, a coastal plain in Spain's Almería province that has seen intense growth in year-round export-oriented vegetable production over the past several decades. Inside the greenhouses, pesticides and fertilizers are used so extensively that health problems are arising among the mainly Moroccan migrant workers and their families who work and live among the greenhouses. Meanwhile, water shortages are so severe that a desalination plant is being built nearby.

Air Pollution At present, there is significant air pollution in much of Europe, but it is particularly heavy over the North European Plain. This is a region of heavy industry, dense transportation routes, and large and affluent populations. The intense fossil fuel use associated with such lifestyles results not only in the usual air pollution but also in *acid rain*, which can fall far from where it was generated (see the Figure 4.7 map).

There is also much air pollution in the former communist states of Central and North Europe. Mines in Central Europe produce highly polluting soft coal that is burned in out-of-date factories and power plants (see Figure 4.7B). Central Europe produces high per capita emissions from burning oil and gas; it also receives air pollution blown eastward from western Europe. In Upper Silesia, Poland's leading coal-producing area, acid rain has destroyed forests, contaminated soils and the crops grown on them, and raised water pollution to deadly levels. Residents have higher rates of birth defects, high rates of cancer, and lower life expectancies than comparable populations in the rest of Europe. Industrial pollution was one of Poland's biggest obstacles to entry into the European Union; Poland barely met the EU's requirements.

Central Europe's severe environmental problems developed in part because the theories and policies promoted by the Soviet Union portrayed nature as existing only to serve human needs. During the Soviet era, little pollution data was collected and public protest against pollution was prohibited, but because of the recent shift toward democracy in places like Hungary and Bulgaria, more activism has been possible, and popular protest has resulted in reductions in air pollution.

The new market economies in the former Soviet bloc countries are improving energy efficiency and reducing emissions. Power plants, factories, and agriculture are polluting less, and the countries with the worst emissions records, such as Poland, have been making the most progress.

Freshwater and Seawater Pollution Sources of water pollution in Europe include insufficiently treated sewage, chemicals and silt in the runoff from agricultural plots and urban areas, consumer packaging litter, petroleum residues, and industrial effluent. Most inland waters contain a variety of such pollutants. Any pollutants that enter Europe's inland wetlands, rivers, streams, and canals eventually reach Europe's surrounding coastal waters. The Atlantic Ocean, the Arctic Ocean, and the northern reaches of the North Sea are better able to disperse chemical pollutants dumped into them because they are part of, or closely connected to, the circulating flow of the world ocean. In contrast, the Baltic, Mediterranean, and Black seas, along with the southern North Sea, are nearly landlocked bodies of water that do not have the capacity to flush themselves out quickly and are prone to accumulating chemical pollution (see the red areas in Figure 4.8).

ON THE BRIGHT SIDE

Energy Solutions

The European Union wants to increase its use of renewable energy in order to reduce fuel imports and thereby increase energy security, stimulate the economy with new energy-related jobs, and combat climate change. The EU's goals are to reduce CO_2 emissions by 20 percent, increase use of renewable energy by 20 percent, and power 60 percent of EU homes with renewably generated electricity, all by 2020. A wide array of alternative energy projects is now attracting significant investment. Wind power is generally the favored technology: Europe had 34 percent of the world's installed wind capacity in 2013, enough to supply about 7 percent of the EU's electricity. The use of solar power, while currently supplying only 2 percent of the EU's electricity, is surging, in part due to rapidly declining prices for Chinese-made solar panels. Europe has about 75 percent of the world's installed solar power capacity. In Germany, more than 60 percent of the country's electricity demand can be met by solar and wind power on a sunny summer day, which means that utilities must periodically shut down fossil fuel power plants to keep the power grid from overloading.

In the Mediterranean, the effect of the pollution that pours in from rivers, adjacent cities, industries, hotel resorts, and farms is exacerbated by the fact that the sea has just one tiny opening to the world ocean (see Figure 4.8). At the surface, seawater flows in from the Atlantic through the narrow Strait of Gibraltar and moves eastward. At the bottom of the sea, water exits through the same narrow opening, but only after it has been in the Mediterranean for 80 years. The natural ecology of the Mediterranean is attuned to this lengthy cycle, but the nearly 460 million people now living in the countries surrounding the sea have upset the balance. Their pollution stays in the Mediterranean for decades. As a result, fish catches have declined, beloved seaside resorts have become unsafe for swimmers, and agricultural workers have become sick.

There are 34 countries with coastlines on Europe's many seas, all with different economies, politics, and cultural traditions. Such diversity makes it difficult to cooperate to minimize pollution or even reduce the risk of severe pollution. Although most of the Mediterranean's pollution is generated by Europe, rapidly growing populations and economic development on the North African and eastern Mediterranean coasts of the sea also pose environmental threats because these countries still lack adequate urban sewage treatment and environmental regulations to control agricultural and industrial wastes.

The Wide Reach of Europe's Environmental Impact The term *virtual water*, introduced in Chapter 1 on page 26 as the volume of water required to produce, process, and deliver a good or service that a person consumes, has come into use as fresh water becomes more scarce. In Chapter 1, we discussed average per capita water needs per day (5 to 13 gallons [20 to 30 liters]) and noted that to arrive at a person's annual total water footprint, one must add the water required for basic needs to a person's virtual water footprint (see Table 1.1 on page 26). Although Europeans do not import as many of their consumer goods as do Americans, they still consume one-fifth of the world's imports; many of these goods have a high *virtual water component* (the water consumed in the production process). Nearly all EU countries import more virtual water than they export, and Europe as a region imports more virtual water than any other. The countries that produce Europe's imported virtual water often have very little water to begin with, and the costs of this water loss are not being adequately figured into the price of the products exported to Europe.

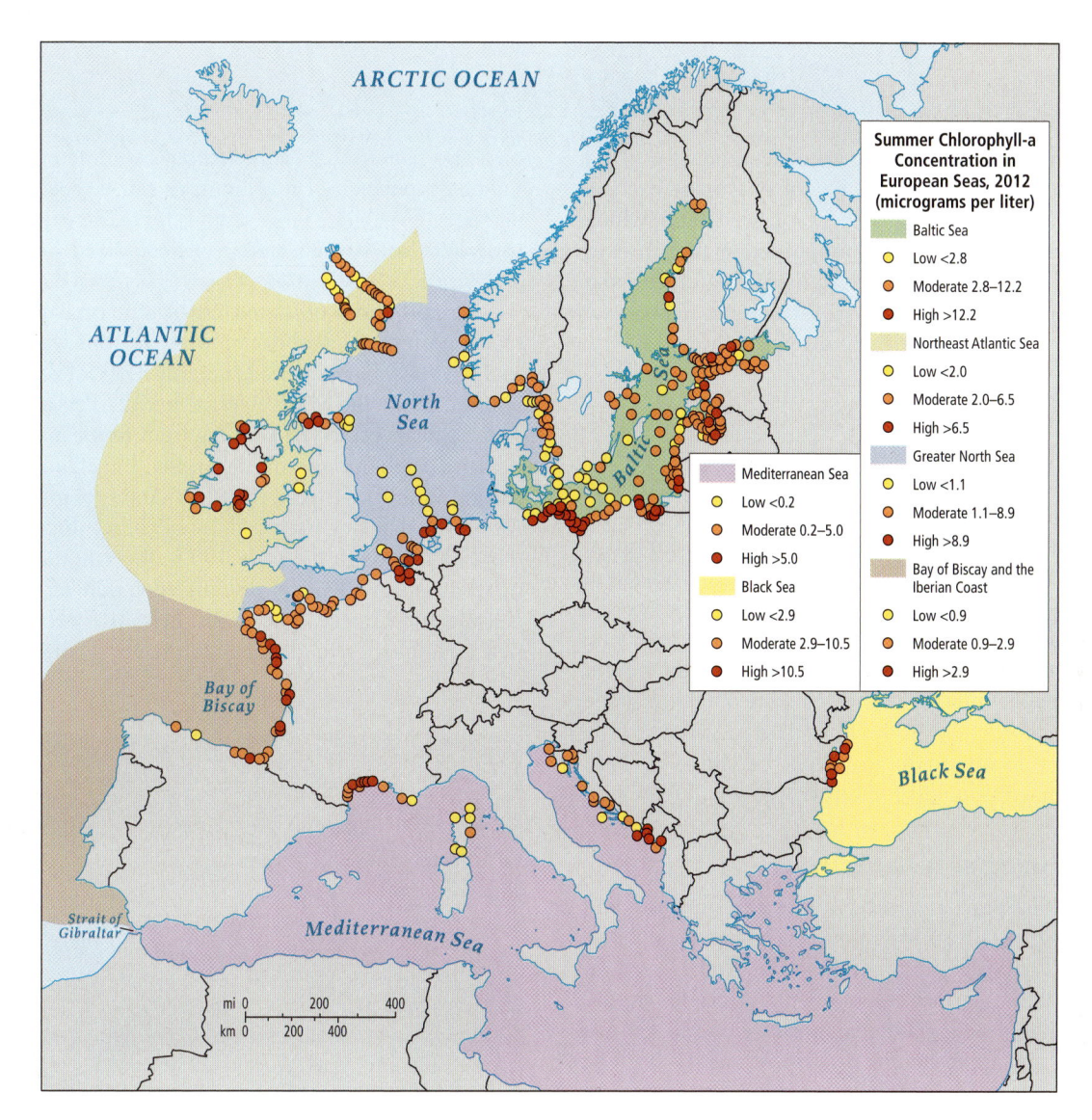

FIGURE 4.8 Pollution of the seas. Chlorophyll concentration is one measure of pollution levels in large bodies of water. Europe's exceptionally long and convoluted coast affords easy access to the world's oceans. However, pollution of the nearly landlocked Baltic, Black, and Mediterranean Seas (red and orange dots) is causing increasing concern. Lesser concentrations of chlorophyll are significant in the Atlantic and Arctic Oceans and in the Mediterranean and Black Seas. For this map, chlorophyll-a (Chla) was used to measure the amount of algae in bodies of water. Excessive algae growth is an indicator of pollution from fertilizers and sewage that fuel the growth of those organisms. [Source consulted: "Chlorophyll-a concentrations in European seas in 2010," European Environment Agency, at http://www.eea.europa.eu/data-and-maps/figures/map-of-summer-chlorophyll-a-concentrations-observed-in-1]

THINGS TO REMEMBER

> **GEOGRAPHIC INSIGHT 1**
- **Environment** The European Union (EU) is a world leader in responding to climate change. Its goals for cutting greenhouse gas emissions are complemented by its many other strategies for saving energy and resources. However, much of Europe's air and many of its seas and rivers are very polluted, and consumption patterns in the region impact environments across the globe.

- Recent research suggests that investments in energy conservation, alternative energy, and other measures would cost EU economies 1 percent of their GDP. By contrast, doing nothing about climate change could shrink GDP by 20 percent.

- European residents average only one-half the energy consumption of the average North American. Fuel costs and CO_2 emissions are increasingly being considered in the design of multimodal transport that links high-speed rail to road, air, and water transportation.

- Europe's main energy sources have shifted over the years from wood to coal and more recently to petroleum and natural gas, and in some countries to nuclear power.

- In the Mediterranean, the effect of the pollution that pours in from rivers, adjacent cities, industries, hotel resorts, and farms is exacerbated by the fact that there is just one tiny opening to the world ocean.

HUMAN GEOGRAPHY

For 500 years, Europe has profoundly influenced how the world trades, fights, thinks, and governs itself. Today Europe exerts a global influence primarily through the European Union, which is redefining both political and economic cooperation in this region and others. However, after years of increasing success and expansion, the European Union is now threatened by financial difficulties related to its currency, the euro.

HUMAN PATTERNS OVER TIME

The range of explanations for Europe's powerful and sustained impact on the world vary widely. One argument, for which there is very little evidence, is that Europeans are somehow a superior breed of humans. Another is that Europe's many bays, peninsulas, and navigable rivers have promoted commerce to a larger extent there than elsewhere. But, in fact, much of Europe's success is based on technologies and ideas it borrowed from elsewhere. For example, the concept of the peace treaty, so vital to current European and global stability, was first documented not in Europe but in ancient Egypt. In order to try to understand how Europe gained the leading role around the globe that it continues to have to this day, it is helpful to look at the broad history of this area.

Sources of European Culture

Starting about 10,000 years ago, the practice of agriculture and animal husbandry gradually spread into Europe from the uplands and plains associated with the Tigris and Euphrates rivers in Southwest Asia and from farther east in Central Asia and beyond. Mining, metalworking, and mathematics also came to Europe from these places and from various parts of Africa. All of these borrowed innovations increased the possibilities for trade and economic development in Europe.

The first European civilizations were ancient Greece (800 to 86 B.C.E.) and Rome (753 B.C.E. to 476 C.E.). Located in southern Europe, both Greece and Rome initially interacted more with the Mediterranean rim, Southwest Asia, and North Africa than with the rest of Europe, which then had only a small and relatively impoverished rural population. Later European traditions of science, art, and literature were heavily based on Greek ideas, which were themselves derived from yet earlier Egyptian and Southwest Asian (Arab and Persian) sources.

The Romans, after first borrowing from Greek culture, also left important legacies in Europe. Many Europeans today speak *Romance* languages, such as Spanish, Portuguese, Italian, French, and Romanian, all of which are largely derived from Latin, the language of the Roman Empire. European laws that determine how individuals own, buy, and sell land originated in Rome. Europeans then spread these legal customs throughout the world.

The practices that Romans used when colonizing new lands also shaped much of Europe. After a military conquest, the Romans secured control in rural areas by establishing large, plantation-like farms. Politics and trade were centered on new Roman towns built on a grid pattern that facilitated both commercial activity and the military repression of rebellions, as can be seen in Figure 4.11A on page 166). These same systems for taking and holding territory were later used when Europeans colonized the Americas, Asia, and Africa.

The influence of Islamic civilization on Europe is often overlooked. After the fall of Rome, while Europe was in a time known as the *early medieval period* (roughly 450 to 1300 C.E.; sometimes referred to as the *Dark Ages*), Turkish, Persian, Egyptian, and Arab scholars preserved learning from Rome and Greece in large libraries, such as those in Alexandria, Egypt, and Constantinople (now Istanbul), Turkey. Muslim Arabs originally from North Africa ruled Spain from 711 to 1492 (see Figure 4.11B). From the 1400s through the early 1900s, the Ottoman Empire (based in what is now Turkey) dominated much of southern Central Europe and Greece. The Arabs, Persians, and Turks all brought new technologies, food crops, architectural principles, and textiles

FIGURE 4.9 Predjama Castle in Slovenia. Predjama Castle, perched in the mouth of a cave in southwestern Slovenia, exemplifies the feudalism from which modern Europe eventually emerged. First mentioned in the historical record in the thirteenth century, Predjama became legendary as an impregnable fortress in the fifteenth century, when the Austrian Imperial Army laid siege to it for over a year, ignorant of a passage through the cave that kept the castle supplied. The Austrians finally succeeded in taking the castle and then the region when Predjama's owner, Erazem Leuger, was betrayed by one of his own men, who had informed the Austrians of the castle's greatest vulnerability. A cannonball killed Erazem in Predjama's unprotected outhouse.

FIGURE 4.10 LOCAL LIVES
People and Animals in Europe

A The bloodhound, a scent-tracking dog, may have been developed by the Romans in the third century C.E. and later brought to England from France before the fourteenth century. Originally bred to track wild game, bloodhounds are renowned for their ability to track individual people by their body scent. Sometimes called "sleuth hounds," they are still used today by law enforcement. [Rudi Von Briel/Peter Arnold/Getty Images]

B A rose-ringed parakeet of the type that has been kept as a pet in Europe also since the days of the Romans. Prized for their sociable nature and ability to mimic human speech, parakeets and parrots are popular pets throughout Europe, with some having escaped captivity to form wild populations in many European cities. [Manfred Pfefferle/Oxford Scientific/Getty Images]

C Originally bred by nomads along the Dalmatian (Croatian) coast as a guard dog, dalmatians became popular throughout Europe and North America during the nineteenth century. In the days of horse-drawn fire carriages, dalmatian "firehouse dogs" cleared the path for oncoming fire carriages and made the horses run faster by nipping at the their legs. [davespilbrow/Getty Images]

to Europe from Arabia, Persia, Anatolia, China, India, and Africa. Arabs also brought Europe its numbering system, mathematics, and significant advances in medicine and engineering, building on ideas they picked up in South Asia.

Beginning 500 years ago, Europe also began to draw on a host of cultural features from the various colonies that were established in the Americas, Asia, and Africa. For example, many food crops now popular in Europe came mostly from the Americas (potatoes, corn, peppers, tomatoes, beans, and squash; see Table 3.1 on page 125) and Southwest and Central Asia (wheat, leafy greens, garlic, onions, and apples).

The Inequalities of Feudalism

As the Roman Empire declined, a social system known as *feudalism* evolved during the *medieval period* (450–1500 C.E.). This system originated from the need to defend rural areas against local bandits and raiders from Scandinavia and the Eurasian interior. The objective of feudalism was to have a sufficient number of heavily armed, professional fighting men, or knights, to defend a much larger group of *serfs*, who were legally bound to live on and cultivate plots of land for the knights. Over time, some of these knights became a wealthy class of warrior-aristocrats, called the *nobility*, who controlled certain territories. Some nobles gained power over other knights, amassing vast kingdoms.

The often-lavish lifestyles and elaborate castles of the wealthier nobility were supported by the labors of the serfs (Figure 4.9). Most serfs lived in poverty outside castle walls and, much like slaves, were legally barred from leaving the lands they cultivated for their protectors.

The Role of Urbanization in the Transformation of Europe

While rural life followed established feudal patterns, new political and economic institutions were developing in Europe's towns and cities. Here, thick walls provided defense against raiders, and commerce and crafts supplied livelihoods, allowing the people more independence from feudal knights and kings. And Europeans, like people in other world regions, continued to develop distinctive cultures, including preferred working animals and pets (Figure 4.10).

Located along trade routes, people in Europe's urban areas were exposed to new ideas, technologies, and institutions from Southwest Asia, India, and China. Some institutions, such as banks, insurance companies, and corporations, provided the foundations for Europe's modern economy. Over time, citizens of Europe's urban areas established a pace of social and technological change that left the feudal rural areas far behind.

Urban Europe flourished in part because of laws that granted basic rights to urban residents. With adequate knowledge of these laws set forth in legal documents called *town charters*, people with few resources could protect their rights even if challenged by those who were more wealthy and powerful. Town charters provided a basis for European notions of *civil rights*, which have proved hugely influential throughout the world. With strong protections for their civil rights, some of Europe's townsfolk grew into a small middle class whose prosperity moderated the feudal system's extreme divisions of status and wealth (Figure 4.11C). A related outgrowth of urban Europe was a philosophy known as **humanism**, which emphasizes the dignity and worth of the individual, regardless of wealth or social status.

The liberating influences of European urban life transformed the practice of religion. Since late Roman

humanism a philosophy and value system that emphasizes the dignity and worth of the individual, regardless of wealth or social status

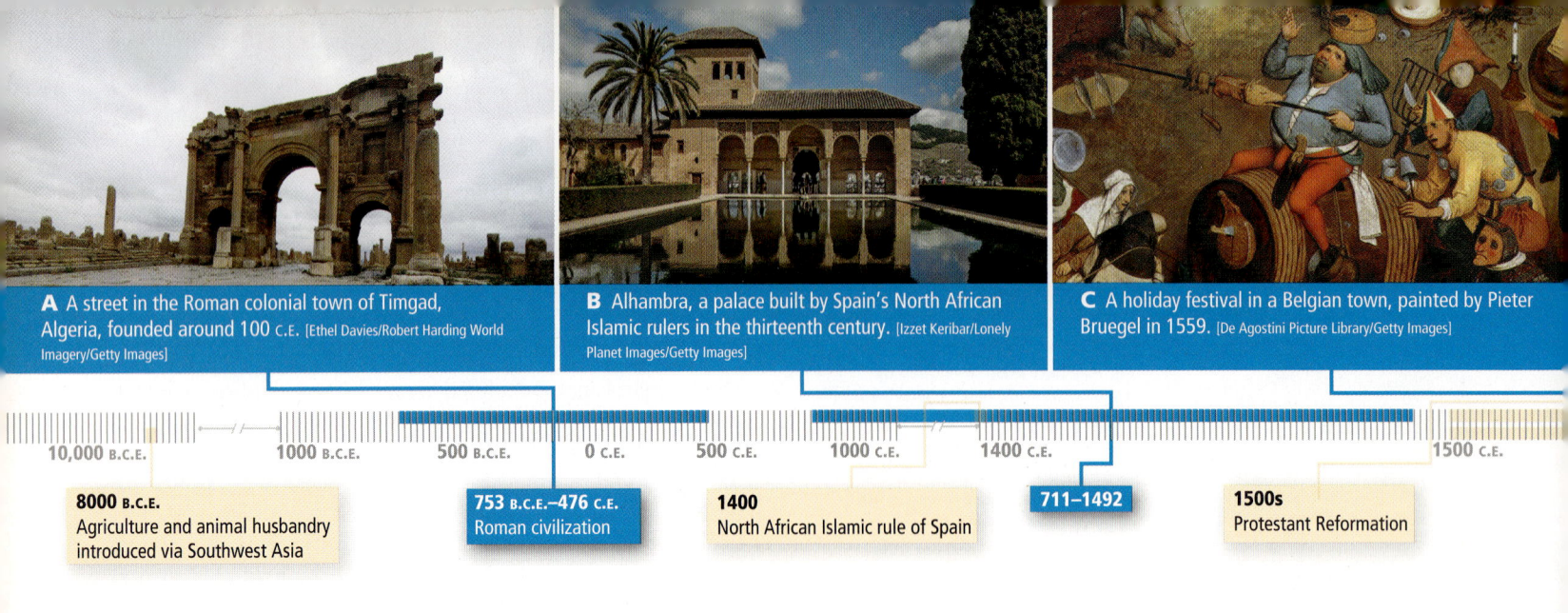

A A street in the Roman colonial town of Timgad, Algeria, founded around 100 C.E. [Ethel Davies/Robert Harding World Imagery/Getty Images]

B Alhambra, a palace built by Spain's North African Islamic rulers in the thirteenth century. [Izzet Keribar/Lonely Planet Images/Getty Images]

C A holiday festival in a Belgian town, painted by Pieter Bruegel in 1559. [De Agostini Picture Library/Getty Images]

8000 B.C.E. Agriculture and animal husbandry introduced via Southwest Asia

753 B.C.E.–476 C.E. Roman civilization

1400 North African Islamic rule of Spain

711–1492

1500s Protestant Reformation

FIGURE 4.11 VISUAL HISTORY OF EUROPE

times, the Catholic Church had dominated not just religion but also politics and daily life throughout much of Europe. In the 1500s, however, a movement known as the *Protestant Reformation* arose in the urban centers of the North European Plain. Reformers, among them Martin Luther, challenged Catholic practices—such as holding church services in Latin, which only a tiny educated minority understood—that stifled public participation in religious discussions. Protestants also promoted individual responsibility and more open public debate of social issues, altering the perception of the relationship between the individual and society. These ideas spread faster with the invention of the European version of the printing press, which enabled widespread literacy.

European Colonialism: The Founding and Acceleration of Globalization

A direct outgrowth of the greater openness and connectivity of urban Europe was the exploration and subsequent colonization of much of the world by Europeans. Spain, Portugal, the United Kingdom, and the Netherlands conquered vast overseas territories and created global trade relationships that transformed Europe economically and culturally and laid the foundation for the modern global economy. The increased commerce and cultural exchange began a period of accelerated *globalization* that persists today (see the discussion in Chapter 1 on pages 31–34).

In the fifteenth and sixteenth centuries, Portugal took advantage of advances in navigation, shipbuilding, and commerce to set up a trading empire in Asia and a colony in Brazil (Figure 4.11D). Spain soon followed, founding a vast and profitable empire in the Americas and the Philippines. By the seventeenth century, however, England, the Netherlands, and France had seized the initiative from Spain and Portugal. All European powers implemented **mercantilism**, a strategy for increasing a country's power and wealth by acquiring colonies and managing all aspects of their production, transport, and trade for the colonizer's benefit (Figure 4.12).

> **mercantilism** a strategy for increasing a country's power and wealth by acquiring colonies and managing all aspects of their production, transport, and trade for the colonizer's benefit

Mercantilism supported the Industrial Revolution in Europe (page 167) by supplying cheap resources from around the globe for Europe's new factories. The colonies also became markets for European manufactured goods. **88. ART EXHIBIT ENCOMPASSES THE WORLD OF PORTUGUESE EXPLORATIONS**

By the mid-1700s, wealthy merchants in London, Amsterdam, Paris, Berlin, and other western European cities were investing in new industries. Workers from rural areas poured into urban centers in England, the Netherlands, Belgium, France, and Germany to work in the manufacturing industries and mining. Wealth and raw materials flowed in from colonial ports in the Americas, Asia, and Africa. Some cities, such as Paris and London, were elaborately rebuilt in the 1800s to reflect their roles as centers of global empires.

By 1800, London and Paris, each of which had a million inhabitants, were Europe's largest cities, a status that eventually brought them to their present standing as *world cities* (cities of worldwide economic or cultural influence). London is a global center of finance, and Paris is a cultural center that has influence over global consumption patterns, from food to fashion to tourism.

By the nineteenth century, the English, French, and Dutch empires overshadowed those of Spain and Portugal. These empires extended European influence into Asia and Africa. By the twentieth century, European colonial systems had strongly influenced nearly every part of the world. Within Europe, the overseas empires of England, the Netherlands, and eventually France shifted wealth, investment, and general economic development toward western Europe and away from southern Europe and the Mediterranean.

Urban Revolutions in Industry and Politics

The wealth derived from Europe's colonialism helped fund two of the most dramatic transformations in a region already characterized by rebirth and innovation: the Industrial and democratic revolutions. Both first took place in urban Europe.

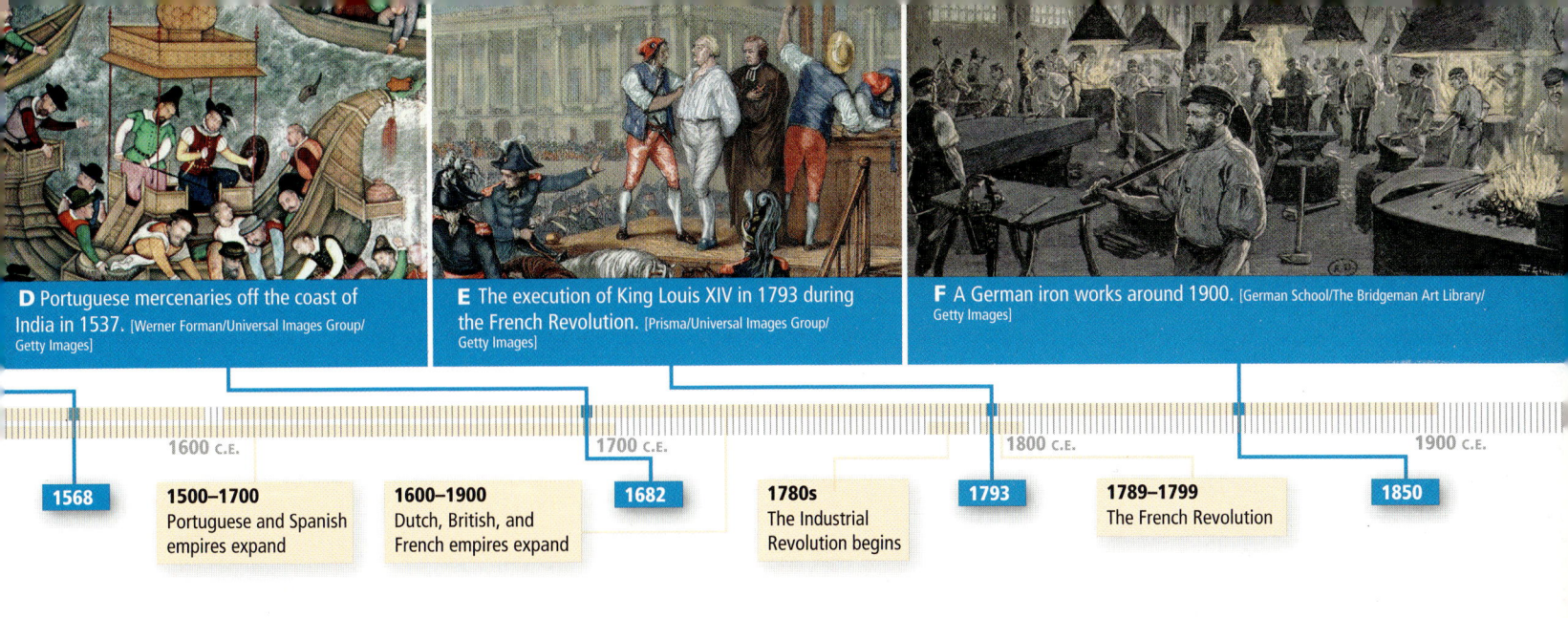

D Portuguese mercenaries off the coast of India in 1537. [Werner Forman/Universal Images Group/Getty Images]

E The execution of King Louis XIV in 1793 during the French Revolution. [Prisma/Universal Images Group/Getty Images]

F A German iron works around 1900. [German School/The Bridgeman Art Library/Getty Images]

1568	
1500–1700	Portuguese and Spanish empires expand
1600–1900	Dutch, British, and French empires expand
1682	
1780s	The Industrial Revolution begins
1793	
1789–1799	The French Revolution
1850	

The Industrial Revolution Europe's Industrial Revolution—particularly Britain's ascendancy as the leading industrial power of the nineteenth century—was intimately connected with colonial expansion and sugar production. In the seventeenth century, Britain

Transfers of Wealth to Europe, 1500–1840 (in billions of 1990 U.S. dollars)
- Profits from slave trade (British ships)
- Profits from slave trade (French ships)
- Silver exports to Spain and Portugal (1531–1810)
- Gold exports to Spain and Portugal (1503–1800)
- Sugar profits from use of slave labor in North America and Caribbean (18th century)
- Profits from Dutch spice trade (1650–1780)
- Profits from India–China opium trade (1800–1840)
- Land rents, taxes from small farmers (1760–1810)

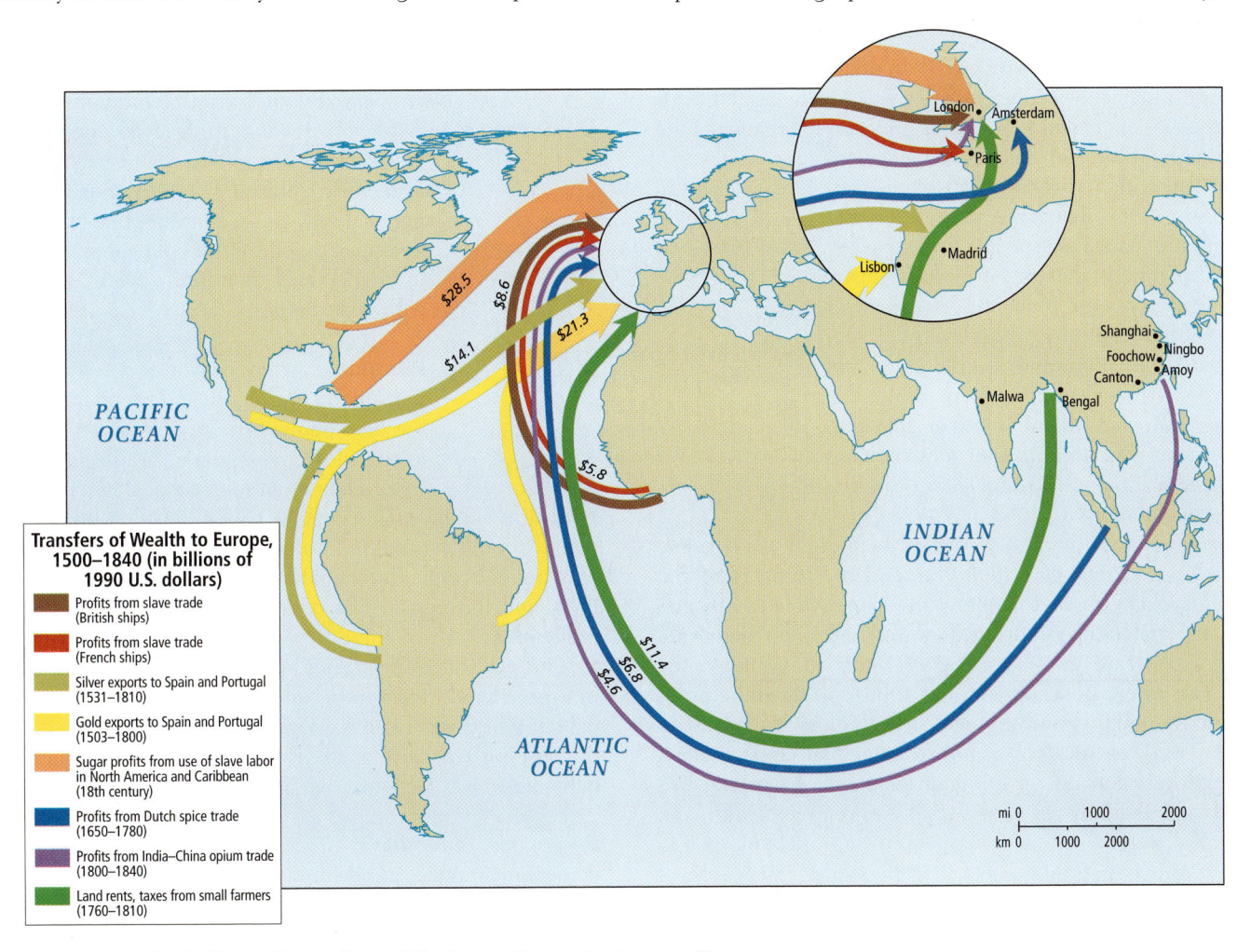

FIGURE 4.12 Transfers of wealth from the colonies to Europe. During the period of mercantilism, Europe received billions of dollars of income from its overseas colonies. [Source consulted: Alan Thomas, *Third World Atlas* (Washington, DC: Taylor & Francis, 1994), p. 29]

167

developed a small but growing trading empire in the Caribbean, North America, and South Asia, which provided it with access to a wide range of raw materials and to markets for British goods.

Sugar, produced by British colonies in the Caribbean, was an especially important trade crop (see Figure 4.12). Sugar production is a complex process that requires major investments in equipment, and for which a great deal of labor was once needed. Slaves were forcibly brought from Africa to help grow and process sugar. The skilled management and large-scale organization needed for sugar later provided a production model for the Industrial Revolution. The mass production of sugar also generated enormous wealth that helped fund industrialization.

By the late eighteenth century, Britain was introducing mechanization into its industries, first in textile weaving and then in the production of coal and steel. By the nineteenth century, Britain was the world's greatest economic power, with a huge and growing empire, expanding industrial capabilities, and the world's most powerful navy. Industrial technologies developed in Britain spread throughout continental Europe (see Figure 4.11F), North America, and elsewhere, thereby transforming millions of lives in the process.

Urbanization and Politics

Industrialization led to massive growth in urban areas in the eighteenth and nineteenth centuries. Extremely low living standards in Europe's cities created tremendous pressures for change in the political order, ultimately leading to democratization. Most early industrial jobs were dangerous and unhealthy, demanding long hours and offering little pay. Poor people were packed into tiny apartments that they shared with many others. Children were often weakened and ill because of poor nutrition, industrial pollution, inadequate sanitation and health care, and from being overworked as laborers. Water was often contaminated, and most sewage ended up in the streets. Opportunities for advancement through education were restricted to the tiny few who could afford it.

The ideas and information gained by the few poor people who did learn to read gave them the incentive to organize and protest for change. After lengthy and often violent struggles against the rich and politically powerful, Europe's huge and growing working class won greater political freedoms, such as the right to vote in elections. Continuing throughout the nineteenth and early twentieth centuries, these struggles also gained urban industrial workers the right to form unions that could bargain with employers and the government for higher wages and better working conditions. Eventually, political power, wealth, and opportunity were distributed more evenly throughout society. However, it is important to note that the road to these sorts of changes in Europe was rocky and often violent, just as it is now in many other parts of the world.

In 1789, the French Revolution led to the first major inclusion of common people in the political process in Europe. Angered by the extreme disparities of wealth in French society, and inspired by news of the popular revolution in North America, the poor rebelled against the monarchy and the elite-dominated power structure that still controlled Europe (see Figure 4.11E). As a result, the general populace, especially in urban areas, became involved in governing through democratically elected representatives. The political freedoms created by the French Revolution ultimately proved short-lived, as governments dominated by the elite soon regained control in France. Nevertheless, the French Revolution provided crucial inspiration to later urban democratic political movements in France and many other parts of the world.

The Impact of Communism During the struggles of the nineteenth century, popular discontent took the form of new revolutionary political movements that proposed radical changes to the established economic and political order. The political philosopher and social revolutionary Karl Marx framed the mounting social unrest in Europe's cities as a struggle between socioeconomic classes. His treatise *The Communist Manifesto* (1848) helped social reformers across Europe articulate ideas about how wealth could be more equitably distributed. East of Europe in Russia, Marx's ideas inspired the creation of a revolutionary communist state in 1917, the Union of Soviet Socialist Republics, often referred to as the USSR or the Soviet Union. Eventually, the Soviet Union extended its ideology and state power throughout Central Europe.

In West, North, and South Europe, communism gained some popularity, but this tended to lead to the formation of Communist political parties that operated peacefully within the context of democratizing political systems. In countries like France, Italy, and Spain, Communist political parties are still influential and often form governing coalitions with other parties.

Popular Democracy and Nationalism During the nineteenth and twentieth centuries, the development of democracy was also linked to the idea of **nationalism**, or allegiance to the state. The notion spread that all the people who lived in a certain area formed a nation and that loyalty to that nation should supersede loyalties to family, clan, or individual monarchs. Eventually, the whole map of Europe was reconfigured, and the mosaic of kingdoms gave way to a collection of *nation-states*. All of these new nations were, at varying paces, transformed into democracies by the political movements arising in Europe's industrial cities. Nationalism was a major component of both World War I and II, so in the period after the wars—called the *post-war era*—and as the European Union was constructed, nationalism was recognized as a problem and deemphasized.

Democracy and the Welfare State Channeled through the democratic process, public pressure for improved living standards moved most European governments toward becoming **welfare states** by the mid-twentieth century. Such governments accept responsibility for the well-being of their people, guaranteeing basic necessities such as education, employment, affordable food, and health care for all citizens. In time, government regulations on wages, hours, safety, and vacations established more

> **nationalism** devotion to the interests or culture of a particular country, nation, or cultural group; the idea that a group of people living in a specific territory and sharing cultural traits should be united in a single country to which they are loyal and obedient
>
> **welfare state** a government that accepts responsibility for the well-being of its people, guaranteeing basic necessities such as education, affordable food, employment, and health care for all citizens

harmonious relations between workers and employers. The gap between rich and poor declined, and overall civic peace increased. And although welfare states were funded primarily through taxes, industrial productivity and overall economic activity did not decline but rather increased, as did general prosperity.

Modern Europe's welfare states have yielded generally adequate to high levels of well-being for all citizens. However, just how much support the welfare state should provide is still a subject of intense debate—one that is currently being addressed in different ways across Europe.

Two World Wars and Their Aftermath

Despite Europe's many advances in industry and politics, by the beginning of the twentieth century, the region still lacked a system of collective security that could prevent war among its rival nations. Between 1914 and 1945, two extremely destructive world wars left Europe in ruins, no longer the dominant region of the world. At least 20 million people died in World War I (1914–1918) and 70 million in World War II (1939–1945). During World War II, Germany's Nazi government killed 15 million civilians during its failed attempt to conquer the Soviet Union. Eleven million civilians died at the hands of the Nazis during the Holocaust, a massive execution of 6 million Jews and 5 million gentiles (non-Jews), including ethnic Poles and other Slavs, Roma (Gypsies), disabled and mentally ill people, gays, lesbians, transgendered people, and political dissidents. **89. THE WORLD REMEMBERS VICTIMS OF HOLOCAUST**

The war divided Europe and much of the world into two battling camps. While a long list of countries participated, the main combatants were the *Axis* powers of Germany, Italy, and Japan; and the *Allies*—Russia, the United Kingdom, the United States, and France (except during the German occupation of 1940–1944). After World War II ended in 1945 with an Allied victory, finding a lasting peace became a primary goal that eventually led to the formation of the European Union. But first, a number of enduring changes took place. Germany was divided into two parts. West Germany became an independent democracy allied with the rest of western Europe—especially Britain and France—and the United States. Russia controlled East Germany and the rest of Central Europe (Latvia, Lithuania, Estonia, Poland, Czechoslovakia, Hungary, Romania, Bulgaria, Ukraine, Moldova, and Belarus). After the war, under Russia's dominant leadership, these war-devastated countries solidified into an ideologically communist alliance known as the *Soviet bloc*. The line between East and West Germany was part of what was called the Iron Curtain, a long, fortified border zone that separated western Europe from Central Europe (then called Eastern Europe). Yugoslavia did not become part of the Soviet bloc but also kept itself insulated from the West, partly to protect its socialized economy.

> **Holocaust** during World War II, a massive execution by the Nazis of 6 million Jews and 5 million gentiles (non-Jews), including ethnic Poles and other Slavs, Roma (Gypsies), disabled and mentally ill people, gays, lesbians, transgendered people, and political dissidents
>
> **Roma** the now-preferred term in Europe for Gypsies
>
> **Iron Curtain** a long, fortified border zone that separated western Europe from (then) eastern Europe during the Cold War
>
> **Cold War** a period of conflict, tension, and competition between the United States and the Soviet Union that lasted from 1945 to 1991
>
> **capitalism** an economic system characterized by privately owned businesses and industrial firms that adjust prices and output to match the demands of the market
>
> **communism** an ideology and economic system, based largely on the writings of the German revolutionary Karl Marx, in which, on behalf of the people, the state owns all farms, industry, land, and buildings (a version of socialism)
>
> **central planning** a communist economic model in which a central bureaucracy dictates prices and output, with the stated aim of allocating goods equitably across society according to need

The Cold War The division of Europe created a period of conflict, tension, and competition between the United States and the Soviet Union known as the Cold War, which lasted from 1945 to 1991. During this time, once-dominant Europe, and indeed the entire world, became a stage on which the United States and the Soviet Union competed for dominance. The central issue was the competition between capitalism—characterized by privately owned businesses and industrial firms that adjust prices and output to match the demands of the market—and communism (actually a version of socialism), in which the state owns all farms, industry, land, and buildings.

After 1945, in most of what we now call Central Europe, the Soviet Union forcibly implemented a communist-inspired economic model known as central planning, in which a central bureaucracy dictated prices and output, with the stated aim of allocating goods and services equitably across society according to need. This system was successful in alleviating long-standing poverty and illiteracy among the working classes and for women, but it was rife with waste, corruption, and bureaucratic bungling. It ultimately collapsed in the 1990s due to inefficiency, insolvency, high levels of environmental pollution, and public demands for more political freedoms.

The rest of post-war Europe, especially western Europe, developed capitalist economies with financial support from the United States under the *Marshall Plan* and from the governments of the involved countries. Basic facilities, such as roads, housing, and schools, were rebuilt. Economic reconstruction proceeded rapidly in the decades after World War II and included special attention to the social welfare needs of the general public. **91. MARSHALL PLAN'S 61ST ANNIVERSARY**

Decolonization Europe's decline during the two world wars also led to the loss of its colonial empires. Many European colonies had participated in the wars, with some (India, the Dutch East Indies, Burma, and Algeria) suffering extensive casualties, and almost all emerging economically devastated. After World War II, Europe was less able to assert control over its colonies as their demands for independence grew. By the 1960s, most former European colonies had gained independence, often after bloody wars fought against European powers and their local allies.

The Birth of the European Union

At the end of World War II, European leaders concluded that the best way to prevent the kinds of hostilities that had led to two world wars would be to forge closer economic ties among the European nation-states. The first step toward developing an economic union in Europe began in 1951 with the creation of a common market for coal and steel (two resources essential for war industries) that, in the rebuilding of Europe, enabled the monitoring of Germany's industrial rehabilitation. The next major step took place in 1958, when Belgium, Luxembourg, the Netherlands, France, Italy, and West Germany formed the *European Economic Community (EEC)*. The members of the EEC agreed to eliminate certain tariffs against one another and to promote mutual trade and cooperation. Denmark, Ireland, and the United Kingdom joined in 1973; Greece, Spain, and Portugal in the 1980s; and Austria, Finland, and Sweden in 1995, bringing the total number of member countries to 15. In 1992, the EEC became the European Union, with the intent to work toward a higher level of economic, social, and political integration that would make possible the free flow of goods and people across national borders.

THINGS TO REMEMBER

- During the medieval period, political and economic transformations in Europe's towns and cities challenged the feudal system that dominated rural areas.

- The vast overseas colonies of various European states created trade relationships that continue in the modern global economy. Europe was transformed and continues to benefit.

- Most of the countries in the world have been ruled by a European colonial power at some point in their history.

- Industrialization and the growth of cities set the stage for increasing the political power of ordinary people.

- The expansion of political freedoms in Europe has proceeded at different rates across the region and was interrupted by two world wars.

- The European Union was created with the intent to work toward a higher level of economic, social and political integration that would make possible the free flow of goods and people across national borders.

GLOBALIZATION AND DEVELOPMENT

GEOGRAPHIC INSIGHT 2

Globalization and Development: In order to better compete in the global economy, the EU has shifted labor-intensive industries from western Europe, where wages are high, to the relatively poorer, lower-wage member states of Central Europe. Growth in services has made up for some of the loss of industrial jobs in western Europe, though the unemployment rate is high in some countries. The EU continues to struggle to integrate the economies of its member states, especially those that use the EU's common currency, the euro.

The European Union uses a number of economic development strategies designed to ensure its ability to compete with the United States, Japan, and the developing economies of Asia, Africa, and South America. To increase Europe's export potential, the EU has focused on lowering the cost of producing goods in Europe. One strategy is to shift labor-intensive industries from the wealthiest EU countries in western Europe, where wages are high, to the relatively poorer, lower-wage member states of Central Europe (Figure 4.13). This strategy has generally worked well, helping poorer European countries prosper while keeping the costs of doing business low enough to restrain European companies from moving away from Europe to places where costs are lower still. However, while the economies of Central Europe have grown, in Europe's wealthiest countries, the resultant reduction of industrial capacity (*deindustrialization*) has led to higher unemployment rates. And despite these efforts, some EU firms have moved abroad to cut costs.

Europe's Growing Service Economies

As industrial jobs have declined across the region, most Europeans (about 70 percent) have found jobs in the service economy. Services, such as the provision of health care, education, finance, tourism, and information technology, are now the engine of Europe's integrated economy, drawing hundreds of thousands of new employees to the main European cities. For example, financial corporations that are located in London and serve the entire world play a huge role in the British economy. Many multinational companies are headquartered in London in part to take advantage of the city's financial sector.

Opportunities for economic growth provided by Europe's highly educated population have transformed the service sector, with almost 40 percent of all employment in the EU now in *knowledge-intensive services*. These kinds of jobs require considerable formal education, such as a college or graduate degree, and usually extensive training on the job. Nevertheless, in recent years, there has been concern within the EU about the declining profitability of European high-tech companies, such as the cell phone manufacturer Nokia, which was bought by the U.S. company Microsoft in 2014. Many of these companies have been unable to compete either with Asian companies that can build products more cheaply or with more innovative U.S. companies. While the EU is making little effort to compete with Asian companies on labor costs, it is now investing more in research and development to better compete with U.S. companies.

While the EU lagged behind North America in the development and use of personal computers and the Internet, it still leads the world in cell phone use. The information economy is especially advanced in North and West Europe, though some in South European countries have pioneered high-tech solutions for travel- and tourism-related industries.

A major component of Europe's service economy is tourism. Europe is the most popular tourist destination in the world, and one job in eight in the European Union is related to tourism. Tourism generates 13.5 percent of the EU's GDP and 15 percent of its taxes, although this varies with global economic conditions. Europeans are themselves enthusiastic travelers, frequently visiting one another's countries to attend local festivals (Figure 4.14) as well as traveling to many distant locations throughout the world. This travel is made possible by the long paid annual vacations—usually 4 to 6 weeks—that Europeans are granted by employers. Vacation days can be used a few at a time so that people can take numerous short trips. The most popular holiday destinations among EU members in 2011 were France, Spain, Italy, the United Kingdom, Germany, and Austria.

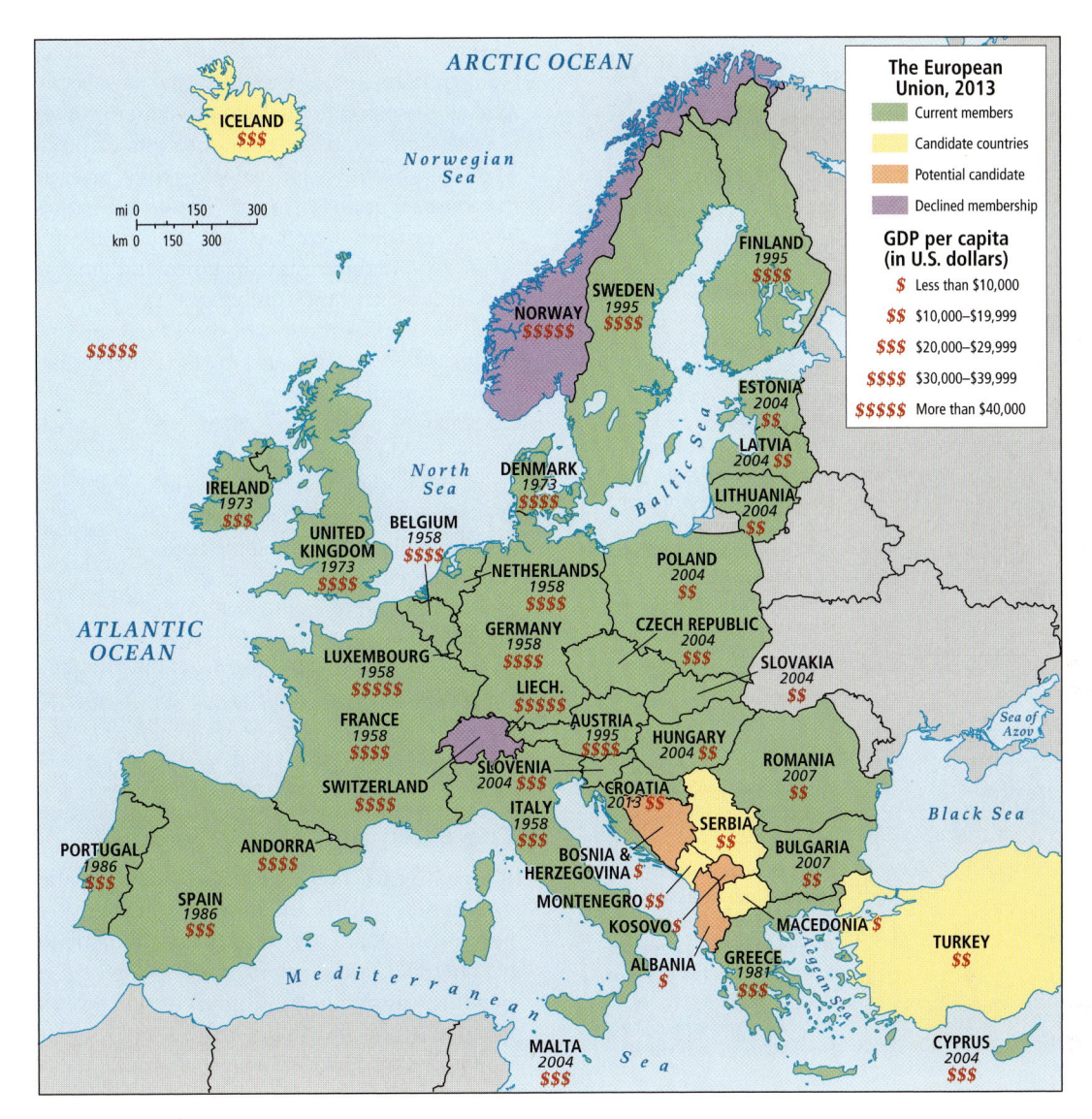

FIGURE 4.13 The current (2013) members of the EU and their dates of joining. The European Union, formed as the EEC in 1958 with the initial goal of economic integration, became the EU in 1992, and has led the worldwide movement toward greater regional cooperation. It is older and more deeply integrated than its closest competitor, the North American Free Trade Agreement (NAFTA). [Sources consulted: *United Nations Human Development Report 2009*, Table H, United Nations Development Programme, at http://hdr.undp.org/en/reports/global/hdr2009/; *Human Development Report 2011*, Sustainability and Equity: A Better Future for All, United Nations Development Programme, at http://hdr.undp.org/en/reports/global/hdr2011/]

Economic Comparisons of the EU and the United States The EU economy now encompasses more than 500 million people (out of a total of 540 million in the whole of Europe)—roughly 200 million more than live in the United States. Collectively, the EU countries are wealthy, with a joint economy slightly larger than that of the United States, making the EU the largest economy in the world and one that exerts a powerful influence on the global trading system (**Figure 4.15**). Even so, trade between countries within the EU amounts to about twice the monetary value of trade between the EU as a whole and the outside world.

In contrast to the United States, which usually imports far more than it exports—resulting in a trade deficit—the European Union often has maintained a trade balance in which the values of imports and exports are roughly equal. According to the World Bank, in terms of income, the average 2013 GNI per capita (PPP) for the European Union ($33,609) was significantly less than that of the United States ($50,120), although the disparity of wealth in the European Union was also lower by about one-third than in the United States, resulting in lower levels of poverty in the EU.

Careful regulation of EU economies contributes to generally slower economic growth than in the United States. These regulations also make the European Union more resilient to global financial crises, such as the crisis beginning in 2008, largely because of stronger control (than in the United States) of the banking industry and financial markets.

FIGURE 4.14 LOCAL LIVES
Festivals in Europe

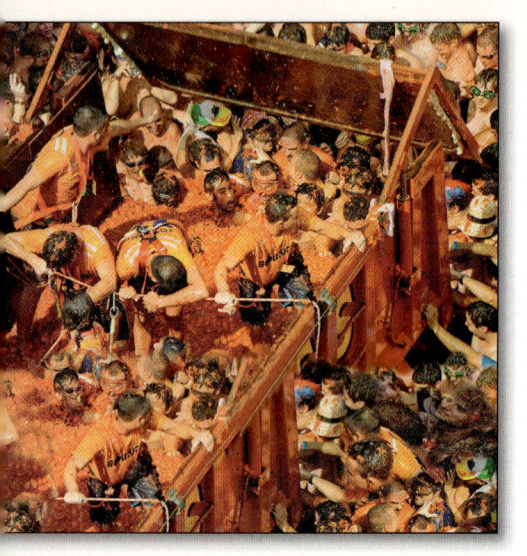

A Every year in Buñol, Spain, participants in "La Tomatina" throw tomatoes at each other. Supposedly, this festival began in 1945 when a group of young men, not allowed in a parade, grabbed tomatoes from a nearby vegetable stand and hurled them. The young men returned the next year and repeated the brawl, though they brought their own tomatoes. Today, truckloads of tomatoes are brought in, as shown here. [Denis Doyle/Getty Images]

B Riders in *Il Palio*, a horse race held each year in Siena, Italy. Jockeys from each of the city's 17 major neighborhoods circle the main plaza; they ride bareback and frequently get thrown from their horses. A colorful pageant precedes the race, which has been run since 1656, though similar races were run earlier. [Mike Hewitt/Allsport/Getty Images]

C Glastonbury Festival of Contemporary Performing Arts (known as "Glasto") is held each year in Somerset, England. First organized in 1970 by a farmer who was inspired by a Led Zeppelin performance, Glastonbury has grown to become one of Europe's largest music festivals. [Peter Still/Redferns/Getty Images]

Economic Integration and a Common Currency Economic integration has addressed a number of problems in Europe. Individual European countries have relatively small populations, which means that they have smaller internal markets for their products. As a result, their companies earn lower profits than those in large countries. The European Union solved this problem by joining European national economies into a common market. Companies in any EU country now have access to a much larger market and can potentially make larger profits through **economies of scale** (reductions in the unit costs of production that occur when goods or services are efficiently mass-produced, resulting in increased profits per unit). Before the European Union, when businesses sold their products to neighboring countries, their earnings were diminished by tariffs and other regulations, as well as by fees for currency exchanges.

The official currency of the European Union is the **euro** (€). Eighteen EU countries now use the euro: Austria, Belgium, Cyprus, Estonia, Finland, France, Germany, Greece, Ireland, Italy, Latvia, Luxembourg, Malta, the Netherlands, Portugal, Slovakia, Slovenia, and Spain. Countries that use the euro have a stronger voice in the creation of EU economic policies, and the use of a common currency greatly facilitates trade, travel, and migration within the European Union. All non-euro member states, except Sweden and the United Kingdom, have currencies whose value is determined by that of the euro. Depending on global financial conditions, either the euro or the U.S. dollar is the preferred currency of international trade and finance.

> **economies of scale** reductions in the unit cost of production that occur when goods or services are efficiently mass produced, resulting in increased profits per unit
>
> **euro** the official (but not required) currency of the European Union as of January 1999

The Euro and Debt Crises

In recent years, there have been major challenges to the euro, most significantly in the form of a debt crisis that has tested the mechanisms that governments use to maintain economic stability. The debt crisis first emerged in Greece, where government spending was significantly outpacing tax revenues. This was compounded by the growth of Greece's trade deficit during the global economic slowdown starting (in Europe) in 2008, which affected tourism and reduced demand for Greece's exports. Similar conditions in other countries led to similar debt crises in other euro zone countries, especially Ireland, Spain, and Portugal.

Normally, governments respond to high levels of debt by borrowing from other countries and then reducing the value of their currency, relative to that of the lenders, which makes these debts easier to pay. However, this was not an option for any member of the euro zone because no country could by itself manipulate the value of the euro. The EU has the European Central Bank (ECB), which controls the value of the euro, but it is strongly influenced by the larger euro zone economies, such as those of Germany and France. Banks in these countries had loaned large amounts of money to Greece and the other indebted governments of the euro zone, and they would lose money if the value of the euro declined. However, because all the economies of the euro zone were threatened by the crises, Germany and France had to act.

In addition to making large, inexpensive loans available to Greece and the other indebted governments through the ECB over the short term, the European Union, with crucial support by Germany and France, developed longer-term mechanisms to limit government debt in all countries that use the euro. This involved strengthening the political and financial ties between EU countries so that economic policies could be better developed and enforced by EU institutions. While many euro zone countries—even those

Globalization and Development **173**

(A) Shares in the World Market for Exports, 2008
(% share of world exports)
- EU-27* 16.7%
- China, excluding Hong Kong 12.4%
- United States 11.2%
- Japan 6.8%
- Canada 4.0%
- Rest of the world 48.9%

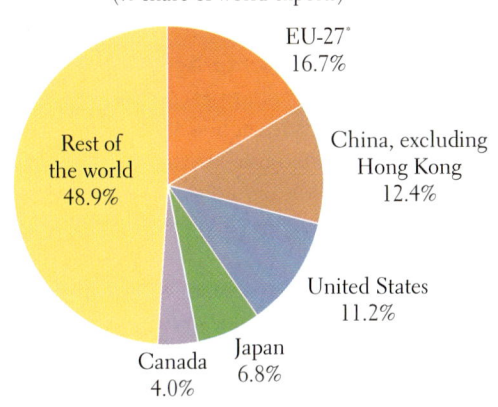

*External trade flows with extra EU-27.

(B) Shares in the World Market for Imports, 2008
(% share of world imports)
- EU-27* 19.1%
- United States 17.9%
- China, excluding Hong Kong 9.4%
- Japan 6.3%
- Canada 3.9%
- Rest of the world 43.8%

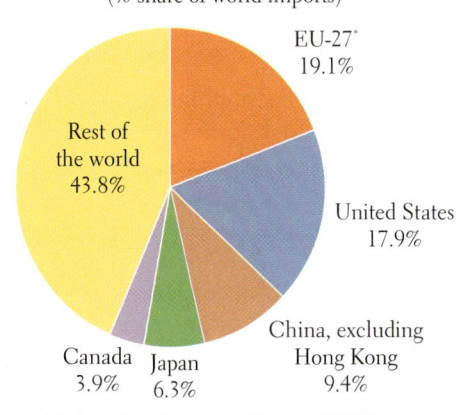

*External trade flows with extra EU-27.

(C) Main Trading Partners for Exports, 2009
(% share of extra EU-27 exports)
- United States 18.7%
- Switzerland 8.1%
- China, excluding Hong Kong 7.5%
- Russian Federation 6.0%
- Turkey 4.0%
- Norway 3.4%
- Japan 3.3%
- Rest of the world 49.0%

(D) Main Trading Partners for Imports, 2009
(% share of extra EU-27 imports)
- China, excluding Hong Kong 17.9%
- United States 13.3%
- Russian Federation 9.6%
- Switzerland 6.2%
- Norway 5.7%
- Japan 4.7%
- Turkey 3.0%
- Rest of the world 39.6%

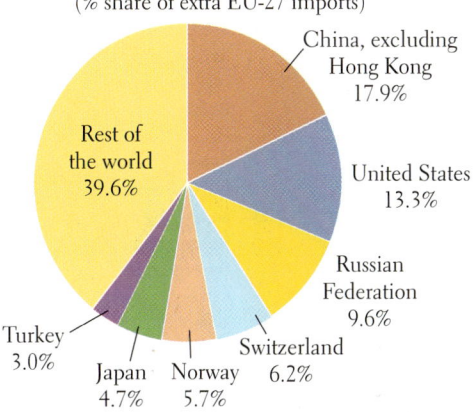

FIGURE 4.15 The EU's main trading partners and shares in world trade, 2008 and 2009. [Sources consulted: *Europe in Figures: Eurostat Yearbook 2009* (Eurostat Statistical Books, Luxembourg: Office for Official Publications of the European Commission, 2009), pp. 388 and 392, Figures 10.5, 10.6, 10.9, and 10.10, at http://epp.eurostat.ec.europa.eu/cache/ITY_OFFPUB/KS-CD-09-001/EN/KS-CD-09-001-EN.PDF; "International Trade in Goods," Figures 2, 3, 6, and 7, Eurostat, at http://epp.eurostat.ec.europa.eu/statistics_explained/index.php/International_trade_in_goods#Analysis_of_main_trading_partners]

in financial difficulty, such as Greece, Cyprus, and Slovenia—strenuously object to giving EU institutions more control over their financial affairs, their need for loans from the ECB has overridden these objections. Although short- and long-term solutions have been implemented, the euro zone crisis is still ongoing.

Food Production and the European Union

Concerns about food security in Europe have led to heavily subsidized and regulated agricultural systems. Subsidies to agricultural enterprises account for more than 40 percent of the budget of the European Union. Although large-scale food production by agribusiness is now dominant, there has also been a revival of organic and small-scale sustainable techniques.

Food security is especially important to Europeans, probably because stories of post–World War II food shortages are still so vivid. Most food is now produced on large, efficient mechanized farms. These farms require less labor and are more productive per acre than were farms before the 1970s. One result is that only about 2.3 percent of Europeans are now engaged in full-time farming. A second result of the efficiencies of agricultural mechanization is that the percentage of land in crops has declined since the mid-1990s, while forestlands have increased.

The Common Agricultural Program (CAP) The drastic decline of labor and land in farming has had an emotional effect on Europeans, who see it as endangering their cultural heritage and their goal of food self-sufficiency. To address these worries, the European Union established its wide-ranging **Common Agricultural Program (CAP)**, meant to guarantee secure and safe food supplies at affordable prices, provide a secure living for farmers, and preserve the quaint rural landscapes nearly everyone associates with Europe (Figure 4.16).

The CAP has long been criticized for undermining its own goals with its use of tariffs on imported agricultural goods and for giving **subsidies** (in this case, payments to

> **Common Agricultural Program (CAP)** an EU program, meant to guarantee secure and safe food supplies at affordable prices, that places tariffs on imported agricultural goods and gives subsidies to EU farmers
>
> **subsidies** monetary assistance granted by a government to an individual or group in support of an activity, such as farming, that is viewed as being in the public interest

farmers) to underwrite the costs of food production. Tariffs effectively raise food costs for millions of EU consumers. The subsidies tend to favor large, often corporate-owned farms because payments are based on the amount of land under cultivation. It was this aspect of the CAP that the Smithfield company (with the help of its European partners) took advantage of in setting up its giant pig farms in Romania (recall the opening vignette).

Protective agricultural policies like tariffs and subsidies—also found in North America, Japan, and elsewhere—can hurt farmers in the developing world. Tariffs lock farmers in poorer countries out of major markets. And subsidies encourage overproduction in rich countries (in order to collect more payments). The result is occasional gluts of farm products that are then sold cheaply on the world market. This practice, called *dumping*, lowers global prices to the point where farmers in developing countries are driven out of business.

The Growth of Corporate Agriculture and Food Marketing

As small family farms disappear in the European Union—just as they did several decades ago in the United States—smaller farms are being consolidated into larger, more profitable operations run by European and foreign corporations (see Figure 4.2A). These farms tend to employ very few laborers and use more machinery and chemical inputs.

The move toward corporate agriculture is strongest in Central Europe. When Communist governments gained power in the mid-twentieth century, they consolidated many small, privately owned farms into large collectives. After the breakup of the Soviet Union, these farms were rented to large corporations, which in turn further mechanized the farms and laid off all but a few laborers. Rural poverty rose. Small towns shrank as farmworkers and young people left for the cities. With EU expansion, the CAP has provided even more incentives for large-scale mechanized agriculture in Central Europe.

FIGURE 4.16 LOCAL LIVES
Foodways in Europe

A Tortellini, stuffed ring- or navel-shaped pasta from the region around Modena and Bologna, Italy. One legend claims that the gods Venus and Jupiter arrived at a tavern in Bologna one night, weary from an ongoing battle between Bologna and Modena. When they retired to their room, the innkeeper peeked through the keyhole and saw only Venus's navel. Inspired, he rushed off to the kitchen to create the first tortellini. [Jonathan Gelber/Getty Images]

B Swedish *gravlax*, raw salmon cured in salt and dill, is cut into thin slices. Gravlax was invented by fishermen who preserved salmon by salting it and then burying it in the sand on a beach above the high tide line. This process gave the dish its name: *grav* means grave and *lax* means salmon. [Dorling Kindersley/Getty Images]

C In French *confit* of duck legs, the meat is salted, seasoned, and slowly cooked in its own fat. The meat is then stored in the fat, which can preserve it for several weeks. Confit is considered a regional specialty of southwestern France. [Image Source/Getty Images]

A CASE STUDY
Green Food Production in Slovenia

During the Communist era, Slovenia was unlike most of the rest of Central Europe in that the farms were not collectivized. As a result, the average farm size is just 8.75 acres (3.5 hectares). Although Slovenia has plenty of rich farmland, as standards of living have risen, it has become a net importer of food. Nonetheless, Slovenia's new emphasis on private entrepreneurship, combined with a growing demand throughout Europe for organic foods, has encouraged some Slovene farmers to carve out an organic niche for themselves, first in local markets and eventually as exporters. The case of Vera Kuzmic is illustrative (Figure 4.17). ■

FIGURE 4.17 Vera Kuzmic in her market stall in Ljubljana. [Lydia Pulsipher]

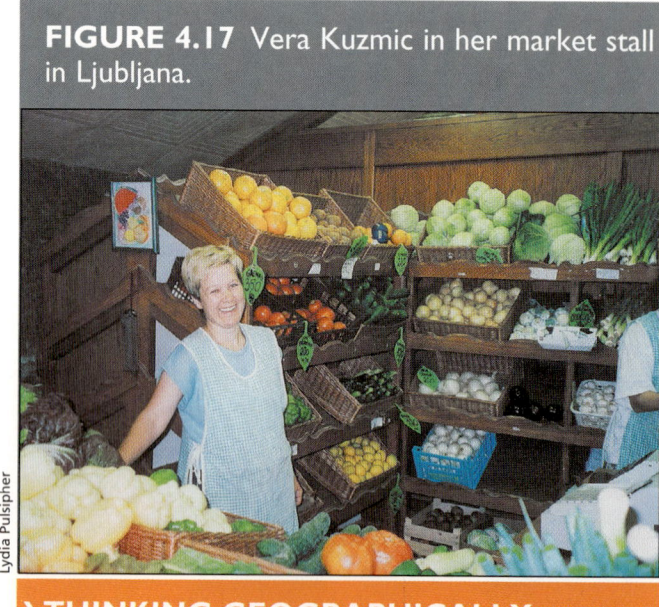

> **THINKING GEOGRAPHICALLY** How can you tell that Vera is involved in small-scale agriculture?

VIGNETTE

Vera Kuzmic (a pseudonym) lives 2 hours by car south of Ljubljana, Slovenia's capital. For generations, her family has farmed 12.5 acres (5 hectares) of fruit trees near the Croatian border. In the economic restructuring that took place after Slovenia became independent in 1991, Vera and her husband lost their government jobs. The Kuzmic family decided to try earning its living in vegetable-market gardening because vegetable farming could be more responsive to market changes than fruit tree cultivation. By 2000, the adult children and Mr. Kuzmic were working on the land, and Vera was in charge of marketing their produce and that of neighbors whom she had also convinced to grow vegetables.

Vera secured market space in a suburban shopping center in Ljubljana, where she and one employee maintained a small vegetable and fruit stall (see Figure 4.17). Her produce had to compete with less expensive, Italian-grown produce sold in the same shopping center—all of it produced on large corporate farms in northeastern Italy and trucked in daily. But Vera gained market share by bringing her customers special orders and by guaranteeing that only animal manure and no pesticides or herbicides were used on the fields. For a while, her special customer services and her organically grown produce kept her in business. But when Slovenia joined the European Union in 2004, she had to do more to compete with produce growers and marketers from across Europe who now had access to Slovene customers.

Anticipating the challenges to come, the Kuzmics' daughter Lili completed a marketing degree at the University of Ljubljana. The family incorporated their business and Lili is now its Ljubljana-based director, while Vera manages the farm. Lili's market research showed the wisdom of diversification. The Kuzmics continue to focus on Ljubljana's expanding professional population, who are willing to pay extra for fine organic vegetables and fruits. But now, in a banquet facility on the farm built with CAP funds, Vera also prepares special dinners for bus-excursion groups from across Europe interested in witnessing traditional farm life and in tasting Slovene ethnic dishes made from homegrown organic crops. [Source: Lydia Pulsipher. For detailed source information, see Text Sources and Credits.] ∎

THINGS TO REMEMBER

> **GEOGRAPHIC INSIGHT 2**

- **Globalization and Development** In order to better compete in the global economy, the EU has shifted labor-intensive industries from western Europe, where wages are high, to the relatively poorer, lower-wage member states of Central Europe. Growth in services has made up for some of the loss of industrial jobs in western Europe, though the unemployment rate is high in some countries. The EU continues to struggle to integrate the economies of its member states, especially those that use the EU's common currency, the euro.

- Collectively, the EU countries are wealthy, with a joint economy slightly larger than that of the United States, making the European Union the largest economy in the world.

- In recent years, there have been major challenges to the euro, most significantly in the form of a debt crisis that has tested the mechanisms that governments use to maintain economic stability.

- Concerns about food security in Europe have led to heavily subsidized and regulated agricultural systems. Subsidies to agricultural enterprises account for more than 40 percent of the budget of the European Union. ∎

POWER AND POLITICS

> **GEOGRAPHIC INSIGHT 3**
>
> **Power and Politics:** Following World War II, political freedoms have grown and strong welfare states have been established in this region. Since the fall of the Soviet Union in 1991, international politics within Europe have centered on the expansion of the EU into Central Europe and the development of EU political institutions. Having grown so much in the past several decades, the European Union, which is already a global economic power, could become a global counterforce to the United States in political and military affairs.

After World War II, most of West, South, and North Europe strengthened their commitment to the expansion of political freedoms. This transition was most pronounced in Germany, where a new post-war constitution gave strong protections to political freedoms in an effort to make the authoritarianism of Nazi Germany impossible in the future. A similar post-war constitution was drawn up in Italy.

In several other parts of Europe, the expansion of political freedoms came later, or was compromised by violence. Spain remained a dictatorship until 1975, and Portugal formally democratized only after a revolution related to decolonization in Africa in 1974 (Figure 4.18D). Until the late 1990s, Northern Ireland had sectarian violence between Catholics and Protestants so severe that free and fair elections were not possible (see Figure 4.18C). Central Europe did not significantly expand political freedoms until the Soviet Union dissolved in 1991 as its former satellites declared independence, one by one. Yugoslavia, a large Communist republic in southern Central Europe, always firmly outside the Soviet bloc, also dissolved beginning in the early 1990s. There the growth of political freedoms was hampered by a powerful wave of ethnic xenophobia and violence (see Figure 4.18A, B).

The Politics of EU Expansion

After the fall of the Soviet Union, the European Union expanded into Central Europe, transforming the region and setting the stage for future expansions that could reach well beyond the borders of Europe as it is currently defined.

As early as the 1980s, Soviet control over Central Europe began to falter in the face of a workers' rebellion in Poland, known as *Solidarity*. By 1990, East Germany had reunited with West Germany. The dissolution of the Soviet Union in 1991 brought the collapse of many economic and political relationships in

176 CHAPTER 4 Europe

Central Europe. Thousands of workers lost their jobs. In some of the poorest countries, including Romania, Bulgaria, and Serbia, social turmoil and organized crime threatened stability. Membership in the European Union became especially attractive to Central European political leaders and citizens who thought it would spur economic development for them that would be a result of investment by the wealthier EU member countries in West and North Europe. **90. POLES CELEBRATE THE 25TH ANNIVERSARY OF SOLIDARITY**

Standards for EU membership, however, are rather demanding and specific. A country must have both political stability and a democratically elected government. Each country has to adjust its constitution to EU standards that guarantee the rule of law, human rights, and respect for minorities. Each must also have a functioning market economy that is open to investment by foreign-owned companies and that has well-controlled banks. Finally, farms and industries must comply with strict regulations governing the finest details of their products and the health of environments. Meeting these requirements has been a challenge for members such as Romania, Bulgaria, and Hungary.

Two very wealthy countries chose not to join the European Union: Switzerland and Norway. They, plus Iceland, have long treasured their neutral role in world politics. Moreover, they were concerned about losing control over their domestic affairs. Iceland has now applied to join the European Union, primarily for financial security. During the recession that in Europe began in 2008, Iceland lost enormous wealth, in large part because of banking speculation in foreign markets. Iceland also became very indebted to the International Monetary Fund.

Several countries on the perimeter of Europe are candidate EU countries. In addition to Iceland, they include Turkey, Macedonia, Montenegro, and Serbia. The prospect of Turkey joining the EU is especially complex. Turkey has strained relations with the island country of Cyprus (which was admitted to the European Union in 2004), and Turkey has a history of human rights violations against minorities (especially against its large Kurdish population). There are also issues regarding the separation of religion and state. Turkey would be the first majority Muslim country to join the European Union. Turkey itself has some reservations. Its economy over the last decade has grown faster than the EU average, leading some to question the need to join the EU at all. Moreover, Turkey's political advantage may lie not in Europe but as a leader in the Middle East. Since the revolutions collectively referred to as the *Arab Spring*, and during the crisis in Syria (see Chapter 6, page 254), Turkey has been a voice of calm and reason, despite its somewhat heavy-handed response to a wave of Arab Spring–type protests that took place in Turkey in 2013.

FIGURE 4.18 PHOTO ESSAY: Power and Politics in Europe

Patterns of political power and conflict in Europe and in former European colonies generally indicate that countries with fewer political freedoms and lower levels of democratization also suffer the most from violent conflict (see Figure 1.22 in Chapter 1 on pages 40–41 for more). Since World War II, the most violent conflicts in Europe have taken place in southern Central Europe, where repressive governments frequently denied their populations political freedoms. Outside Europe, many long and brutal wars were fought to gain independence from European colonial powers. In part, these wars stemmed from economic exploitation and the extraction of wealth and resources from the "colonies" by Europeans, but the denial of political freedoms to the local non-European population also was a major impetus for independence. The map shows how low levels of democratization persist in the countries that fought wars of independence, decades after the formal end of colonial rule. Other factors also worked against democratization, but European imperialism helped set many of these countries on an authoritarian trajectory that has only recently started to change. [Source consulted: Democratization Index adapted from "The Democracy Index 2011: Democracy Under Stress," *Economist* Intelligence Unit, at http://www.eiu.com/public/topical_report.aspx?campaignid=DemocracyIndex2011]

D A group of Angolan rebels, captured during Angola's war of independence from Portugal, which fought three major wars in Africa during the 1960s and 1970s to hold on to its colonies there. The rising human and financial costs of these wars eventually brought about a revolution in Portugal itself in 1974, which at the time was an undemocratic authoritarian dictatorship. The revolution resulted in independence for the colonies and the eventual establishment of democracy in Portugal. None of Portugal's colonies have well-protected political freedoms, in part because of the authoritarian governing structures put in place by the Portuguese and because of the lengthy civil wars that followed decolonization.

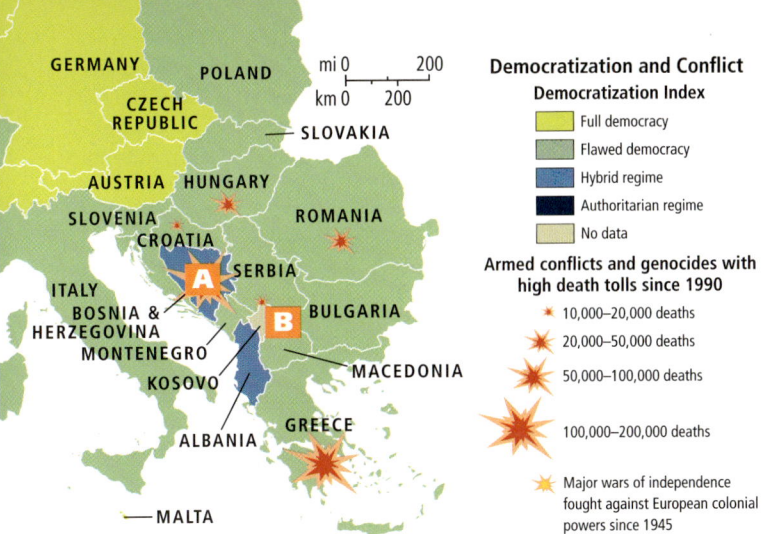

Democratization and Conflict

Democratization Index
- Full democracy
- Flawed democracy
- Hybrid regime
- Authoritarian regime
- No data

Armed conflicts and genocides with high death tolls since 1990
- 10,000–20,000 deaths
- 20,000–50,000 deaths
- 50,000–100,000 deaths
- 100,000–200,000 deaths
- Major wars of independence fought against European colonial powers since 1945

A The aftermath of Bosnia's civil war in a suburb of Sarajevo in 1996. Attempts by Serbia to limit democratic decision making within the former Yugoslavia sparked several wars during the 1990s. In Bosnia, a civil war that pitted different ethnicities against each other resulted in 175,000 deaths.

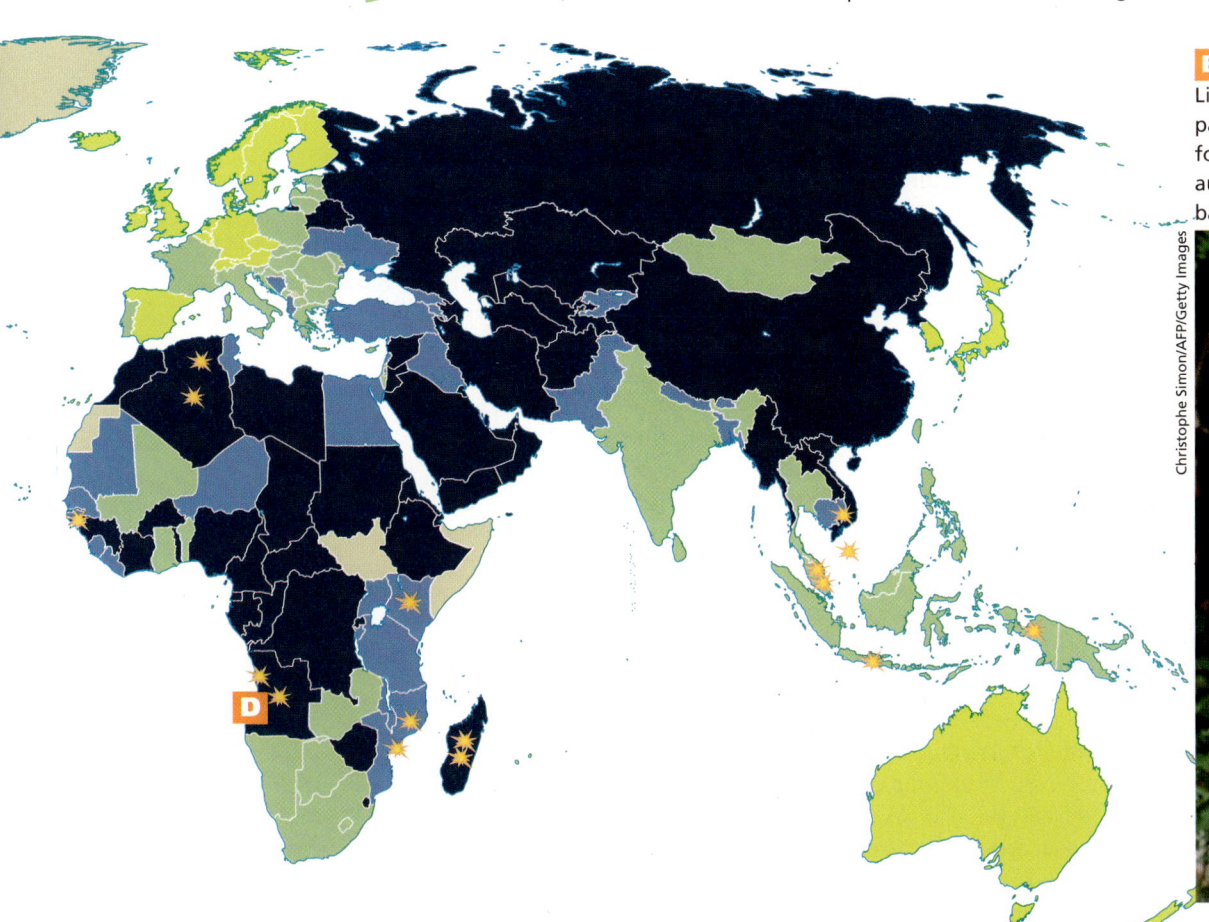

B A young member of the Kosovo Liberation Army in 1999. Like other parts of the former Yugoslavia, Kosovo fought for independence after Serbia's authoritarian policies denied people basic political freedoms.

C Two murals in Derry, Northern Ireland, mark an entrance to a "nationalist" neighborhood. The mural on the right depicts a boy wearing a gas mask and holding a homemade petrol bomb. Both murals were created during "the Troubles," a time of violent conflict from 1969 to 1998. One aspect of the still-sensitive conflict is competing claims about Northern Ireland's democratic legitimacy. The "nationalists" (who are usually Catholic) claim that Northern Ireland was undemocratically separated from the rest of Ireland in 1920 by the government of the United Kingdom during Ireland's war of independence from the U.K. The "loyalists" (usually Protestant) see Northern Ireland as an expression of the will of the voters in the state, most of whom are Protestant and want to remain part of the U.K.

178 CHAPTER 4 Europe

A few other countries—Ukraine, Moldova, and perhaps even the Caucasian republics (Armenia, Azerbaijan, and Georgia)—may at some time be invited to join. However, there is strong opposition to this within Europe, and Europe's huge and potentially powerful neighbor, Russia, opposes the expansion of the European Union into what it considers its sphere of influence.

🎥 93. EU TELLS TURKEY TO DEEPEN REFORMS

ON THE BRIGHT SIDE

The Nobel Peace Prize

Europe's long history of institution building has facilitated the healing of wounds. After both world wars, social and economic development have helped the countries of the region work cooperatively with each other to create and maintain the European Union. This accomplishment was recognized in October of 2012 with the awarding of the Nobel Peace Prize, a prize that normally goes to an individual, to the entire European Union.

EU Governing Institutions

Somewhat similar to the United States, the European Union has one executive branch and two legislative bodies. The *European Commission* acts like an executive branch of government, proposing new laws and implementing decisions. Each of the 28 member states gets one commissioner, who is appointed for a 5-year term, subject to the approval of the *European Parliament*. Commissioners are expected to uphold common interests and not those of their own countries. The entire commission must resign if censured by Parliament. The European Commission also includes about 25,000 civil servants who work in Brussels to administer the European Union on a day-to-day basis.

EU citizens directly elect the European Parliament. Each country elects a proportion of seats based on its population, much like the U.S. House of Representatives. The Parliament elects the president of the European Commission, who serves for 2½ years as a head of state and head of foreign policy. Laws must be passed in Parliament by 55 percent of the member states, which must contain 65 percent of the EU total population. In other words, a simple majority does not rule. The *Council of the European Union* is similar to the U.S. Senate in that it is the more powerful of the two legislative bodies. However, its members are not elected but consist of one minister of government from each EU country.

NATO and the Rise of the European Union as a Global Peacemaker

A new role for the European Union as a global peacemaker and peacekeeper is developing through the **North Atlantic Treaty Organization (NATO)**, which is based in Europe. During the Cold War, European and North American countries cooperated militarily through NATO to counter the influence of the Soviet Union. NATO originally included the United States, Canada, the countries of western Europe, and Turkey; it now includes almost all the EU countries as well.

Since the breakup of the Soviet Union, NATO has focused mainly on providing the international security and cooperation needed to expand the European Union. When the United States invaded Iraq in 2003, most EU members opposed the war. As worldwide opposition to the United States built, the global status of the European Union rose. Thereafter, with the United States preoccupied in Iraq and Afghanistan, NATO and the EU assumed more of a role as a global peacemaker. For example, NATO forces fought off attempts by pirates to seize merchant ships during the 2009 Somali pirate crisis off the northeast coast of Africa, and undertook a number of operations in the Mediterranean related to the Arab Spring in Algeria, Tunisia, Libya, and Syria.

🎥 94. NATO'S FUTURE ROLE DEBATED
🎥 95. NATO TO PROJECT DIFFERENT PHILOSOPHY
🎥 96. CONCERN OVER COMMON VALUES AT THE U.S.–EU SUMMIT
🎥 97. NATO LEADERS, PUTIN MEET IN BUCHAREST

Social Welfare Systems and Their Outcomes

In nearly all European countries, tax-supported systems of **social welfare** or **social protection** (the EU term) provide all citizens with basic health care; free or low-cost higher education; affordable housing; old age, survivor, and disability benefits; and generous unemployment and pension benefits. These systems developed during the nineteenth and twentieth centuries, and were broadly expanded in the post–World War II era that coincided with more protection of political freedoms. In recent decades, the cost of European welfare systems has become a political issue in some countries as economies have had slower growth rates and more unemployment.

Europeans generally pay much higher taxes than North Americans (the rate for EU countries is about 40 percent of GDP; for the United States, 27 percent; and for Canada, 30 percent); in return, they expect more in terms of services. In some cases, European governments are able to deliver services more cheaply than the so-called free market does elsewhere. For example, the European Union spends on average about $3000 per person for health care, while the United States spends more than $7000. Even at this much lower cost, the European Union has more doctors, more acute-care hospital beds per citizen, and better outcomes than the United States in terms of life expectancy and infant mortality.

European welfare systems can be classified into four basic categories (**Figure 4.19**). *Social democratic welfare systems*, common in Scandinavia, are the most generous systems. They attempt to create equality across gender and class lines by providing extensive health care, education, housing, and child and elder care benefits to all citizens from cradle to grave. Child care is widely available, in part to help women enter the labor market. But early childhood training, a key feature of this system, is also meant to ensure that in adulthood every citizen will be able to contribute to the best of his or

North Atlantic Treaty Organization (NATO) a military alliance between European and North American countries that was developed during the Cold War to counter the influence of the Soviet Union; since the breakup of the Soviet Union, NATO has expanded membership to include much of eastern Europe and Turkey, and is now focused mainly on providing the international security and cooperation needed to expand the European Union

social welfare (in the European Union, **social protection**) in Europe, tax-supported systems that provide citizens with benefits such as health care, affordable higher education and housing, pensions, and child care

her capability, and that citizens will not develop criminal behavior or abuse drugs. While finding comparable data is very difficult, surveys of crime victims in Scandinavia and the European Union show that the crime rate in Scandinavia is generally lower than in other parts of Europe.

The goal of *conservative* and *modest welfare systems* is to provide a minimum standard of living for all citizens. These systems are common in the countries of West Europe. The state assists those in need but does not try to assist upward mobility. For example, college education is free or heavily subsidized for all, but strict entrance requirements in some disciplines can be hard for the poor to meet. State-supported health-care and retirement pensions are available to all, but although there are movements to change these systems, they still reinforce the traditional "housewife contract" by assuming that women will stay home and take care of children, the sick, and the elderly. The "modest" system in the United Kingdom is considered slightly less generous than the "conservative" systems found elsewhere in West Europe. The two are combined here but shown separately in Figure 4.19.

In countries with *rudimentary welfare systems*, citizens are not considered to inherently have the right to government-sponsored support. These systems are found primarily in South Europe and in Ireland. Here, local governments provide some services or income for those in need, but the availability of such services varies widely, even within a country. The state assumes that when people are in need, their relatives and friends will provide the necessary support. The state also assumes that women work only part time, and thus are available to provide child care and other social services for free. Such ideas reinforce the custom of paying women lower wages than men.

Post-Communist welfare systems prevail in the countries of Central Europe. During the Communist era, these systems were comprehensive, resembling the cradle-to-grave social democratic system in Scandinavia, except that women were pressured to work outside the home. Benefits often extended to nearly free apartments, health care, state-supported pensions, subsidized food and fuel, and early retirement. However, in the post-Communist era, state funding has collapsed, forcing many to do without basic necessities. In many post-Communist countries, welfare systems are now being revised, but usually with an eye to reducing benefits and to extending work lives past age 65.

The Future of the Welfare State in Europe Europeans do not agree on the goals of their welfare systems, or on just how generous they should be. Some argue that Europe can no longer afford high taxes if it is to remain competitive in the global market. Others say that Europe's economic success and high standards of living are the direct result of the social contract to take care of basic human needs for all. The debate has been resolved

FIGURE 4.19 European social welfare/protection systems. The basic categories of social welfare systems shown here and described in the text should be taken as only an informed approximation of the existing patterns. In 2007 (the latest data available from the EU), the expenditures for social protection in the EU-27 were about 26.7 percent of GDP. The map shows the percent of GDP spent on social protection by the EU-28 countries in 2007. The 2007 data do not include Croatia. [Source consulted: *Europe in Figures: Eurostat Yearbook 2011*, Ch. 6 (especially Table 6.4), Eurostat, 2011, at http://epp.eurostat.ec.europa.eu/cache/ITY_OFFPUB/CH_06_2011/EN/CH_06_2011-EN.PDF]

differently in different parts of Europe, and the resulting regional differences have become a source of concern in the European Union. With open borders, unequal benefits can encourage those in need to flock to a country with a generous welfare system and overburden the taxpayers there. While many European countries are reforming their welfare systems, most remain among the most expansive and well-funded in the world.

THINGS TO REMEMBER

> **GEOGRAPHIC INSIGHT 3**
> • **Power and Politics** Following World War II, political freedoms have grown and strong welfare states have been established in this region. Since the fall of the Soviet Union in 1991, international politics within Europe have centered on the expansion of the EU into Central Europe and the development of EU political institutions. Having grown so much in the past several decades, the European Union, which is already a global economic power, could become a global counterforce to the United States in political and military affairs.

• After the fall of the Soviet Union, the European Union expanded into Central Europe in a series of enlargements that have transformed this region and set the stage for future expansions that may reach well beyond the borders of Europe as it is currently defined.

• Somewhat similar to the United States, the European Union has one executive branch and two legislative bodies.

• A new role for the European Union as a global peacemaker and peacekeeper is developing through the North Atlantic Treaty Organization (NATO), which is based in Europe.

• In nearly all European countries, tax-supported systems of social welfare or social protection (the EU term) provide all citizens with basic health care; free or low-cost higher education; affordable housing; old age, survivor, and disability benefits; and generous unemployment and pension benefits. ■

URBANIZATION

> **GEOGRAPHIC INSIGHT 4**
>
> **Urbanization:** Europe's cities are both ancient and modern, with old town centers now surrounded by modern high-rise suburbs and supported by world-class urban infrastructures. Cities are the heart of Europe's economy, politics, and culture, with more than 70 percent of the EU population living in urban areas.

Though historically based on trade and manufacturing, most European cities are now primarily service oriented. With many ancient educational and research institutions, urban Europe now has numerous hubs of innovation in the knowledge-intensive service sector that has become so crucial to the modern European economy. With thousands of years of history and culture visible in their architecture and well-funded museums, concert halls, theaters, and other cultural institutions, many cities in Europe also draw a disproportionate share of the world's tourists. In most years, Europe has 7 or 8 of the 20 most visited cities on the planet, a very large share given the region's size and population.

Many European cities began as trading centers more than a thousand years ago and still bear the architectural marks of medieval life in their historic centers (Figure 4.20A). These old cities are located either on navigable rivers in the interior or along the coasts because water transportation figured prominently (as it still does) in Europe's trading patterns.

Since World War II, nearly all the cities in Europe have expanded around their perimeters in concentric circles of apartment blocks (see Figure 4.20B). Usually, well-developed rail and bus lines link the blocks to one another, to the old central city, and to the surrounding countryside. Land is scarce and expensive in Europe, so only a small percentage of Europeans live in single-family homes, although the number is growing. Even single-family homes tend to be attached or densely arranged on small lots. Except in public parks, which are common, one rarely sees the sweeping lawns familiar to many North Americans. Because publicly funded transportation is widely available, many people live without cars, in apartments near city centers (see Figure 4.20C, D). However, many others commute daily by car, bus, or train from suburbs or ancestral villages to work in nearby cities.

Although deteriorating housing and slums do exist, especially in the poorer cities and towns of Central Europe, substantial public and private investments in housing, public transportation, sanitation, water, and utilities mean that most people maintain a generally high standard of urban living.

THINGS TO REMEMBER

> **GEOGRAPHIC INSIGHT 4**
> • **Urbanization** Europe's cities are both ancient and modern, with old town centers now surrounded by modern high-rise suburbs and supported by world-class urban infrastructures. Cities are the heart of Europe's economy, politics, and culture, with more than 70 percent of the EU population living in urban areas.

• Though once based on industry and manufacturing, most European cities are now primarily service oriented.

• Many older cities are located either on navigable rivers in the interior or along the coasts because water transportation figured prominently (as it still does) in Europe's trading patterns.

• Since World War II, nearly all the cities in Europe have expanded around their perimeters in concentric circles of apartment blocks.

• Land is scarce and expensive in Europe, so only a small percentage of Europeans live in single-family homes, although the number is growing. ■

POPULATION AND GENDER

> **GEOGRAPHIC INSIGHT 5**
>
> **Population and Gender:** Europe's population is aging as its fertility rates decline, which they are doing for a number of reasons, including the fact that career-oriented working women are choosing to have only one or two children. Fearing that aging populations could slow economic growth, European countries have tried to boost their fertility rates with a variety of policies, the most successful of which have addressed the needs of working mothers.

FIGURE 4.20

PHOTO ESSAY: Urbanization in Europe

Europe's cities are famous throughout the world for their architecture, economic dynamics, and cultural variety. Many are quite ancient but have expanded in recent decades with large apartment blocks connected to the old city centers by public transportation. Despite their global draw, many European cities will shrink in the next several decades, due to declining national populations (as in Italy and parts of Central Europe), as well as to deliberate efforts to shift growth to other urban centers, as in the case of London.

A The old urban core of Prague, Czech Republic, founded in the ninth century c.e. on the banks of the Vltava River.

B High-rise apartment blocks dot the periphery of Berlin's old urban core.

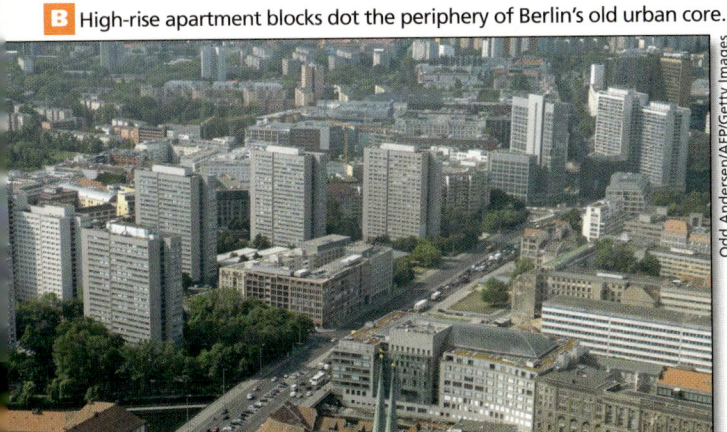

C La Rambla, a tree-lined pedestrian mall in Barcelona, Spain, is a popular strolling place among locals and tourists.

D A tram in Paris, France. The city's metropolitan transit system moves roughly 6 million people per day, and is the second-largest system in Europe, after London's.

Population Living in Urban Areas
- 83%–100%
- 65%–82%
- 47%–64%
- 29%–46%
- 11%–28%
- No data

Population of Metropolitan Areas, 2013
- 20 million
- 10 million
- 5 million
- 3 million

Note: Symbols on map are sized proportionally to metro area population

① Global rank (population 2013)

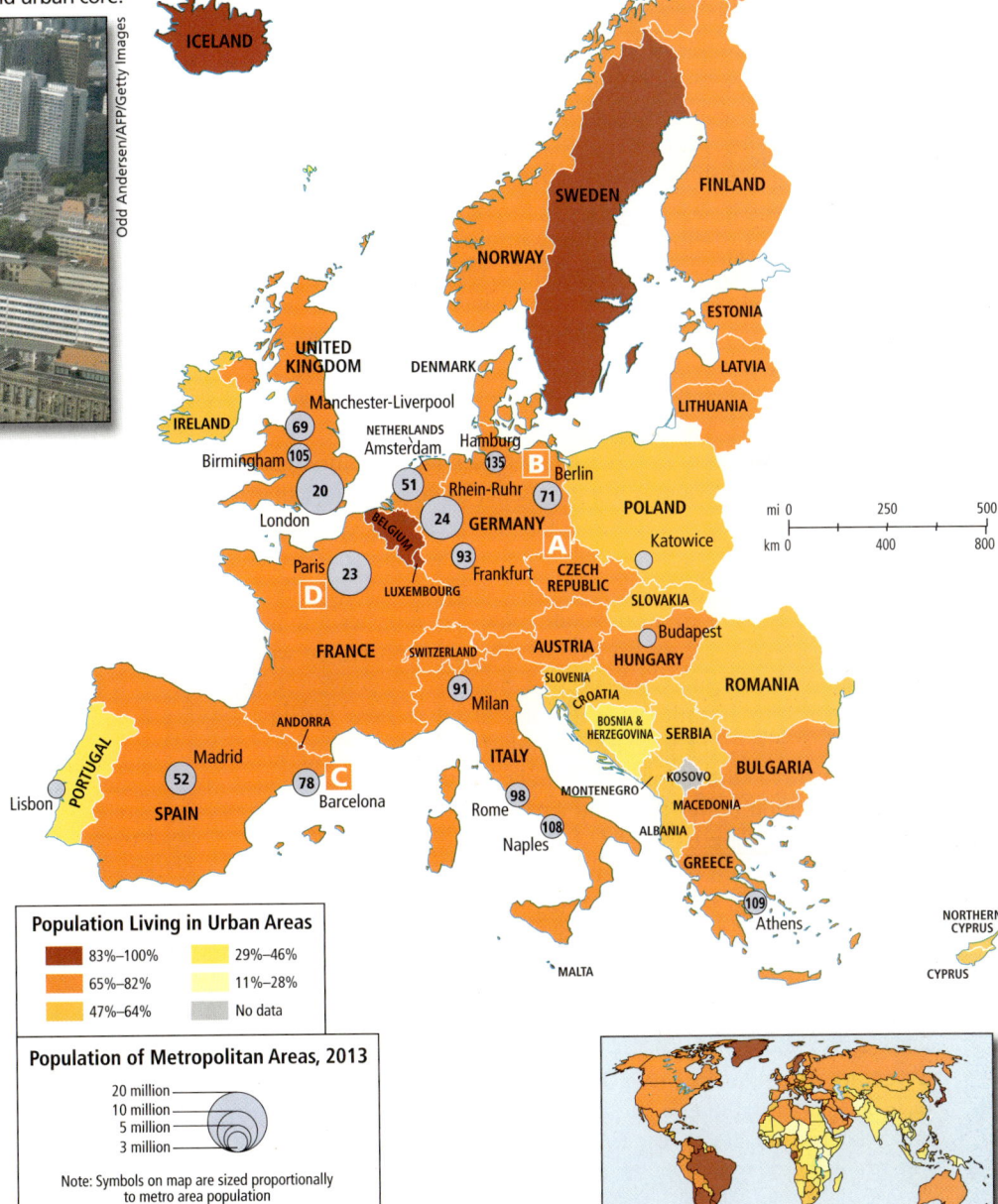

A stable population with a low birth rate has several economic consequences. Because fewer consumers are being born, economies may contract over time. Demand for new workers, especially highly skilled ones, may go unmet. Furthermore, the number of younger people available to provide expensive and time-consuming health care for the elderly, either personally or through tax payments, may decline. Currently, for example, there are just two German workers for every retiree.

Population aging is well advanced in Europe. Between 1960 and 2014, the proportion of the European population that is 14 years and under declined from 27 percent to 16 percent (well below the global average of 26 percent), while those over 65 increased from 9 percent to 16 percent (well above the global average of 8 percent). Longer life expectancies have influenced this trend; the average life expectancy of an EU citizen is 77 for men and 83 for women. However, a larger contributor to the aging of the population is the declining fertility rate.

Overall, Europe is now close to a negative rate of natural increase (<0.0), the lowest in the world. Already low fertility rates are shrinking across Europe, and they are lowest in Germany, the wealthiest country. The one-child family is increasingly common throughout Europe.

The declining birth rate is illustrated in the population pyramids of European countries (Figure 4.21), which look more like lumpy towers than pyramids. Examples are the population pyramid of Sweden and of the whole European Union (see Figure 4.21B, C). The pyramids' narrowing base indicates that for the last 35 years, far fewer babies have been born than in the 1950s and 1960s, when there was a post-war baby boom across Europe. By 2000, twenty-five percent of Europeans were having no children at all.

The reasons for these trends are complex. Urbanization and strong economic development play a role, as does the fact that more and more women want professional careers. In Germany, for example, birth rates are lowest among the most educated women, a group that is growing in size every year. Twenty-five percent of all German women are now choosing to remain unmarried well into their thirties, which means that they will likely have only one or two children, if any.

European governments have tried to boost their fertility rates with a variety of policies, the most successful of which have addressed the problems faced by working mothers. Current research on fertility policy suggests that career-minded working mothers are better aided by help with child care than by the generous full-pay maternity leave of several months to a year common in this region. This may relate to the professional consequences of taking long periods off from work, such

FIGURE 4.21 Population pyramids for Germany, Sweden, and the EU in 2012. These population pyramids have quite different shapes, but all exhibit a narrow base, indicating low birth rates. The EU pyramid **(C)** is at a different scale due to the much larger population. [Sources consulted: "Population Pyramid of Germany," "Population Pyramid of Sweden," and "Population Pyramid of the European Union," International Data Base, U.S. Census Bureau, 2012, at http://www.census.gov/population/international/data/idb/informationGateway.php]

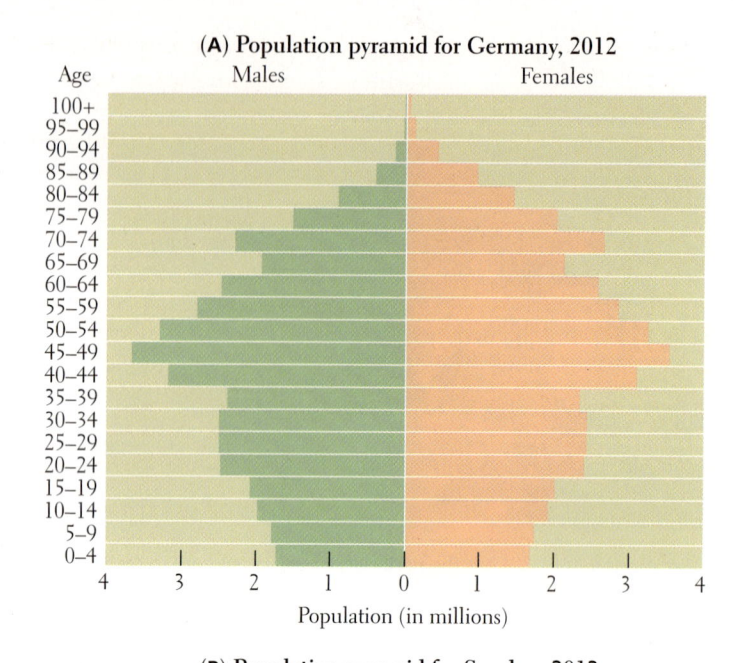

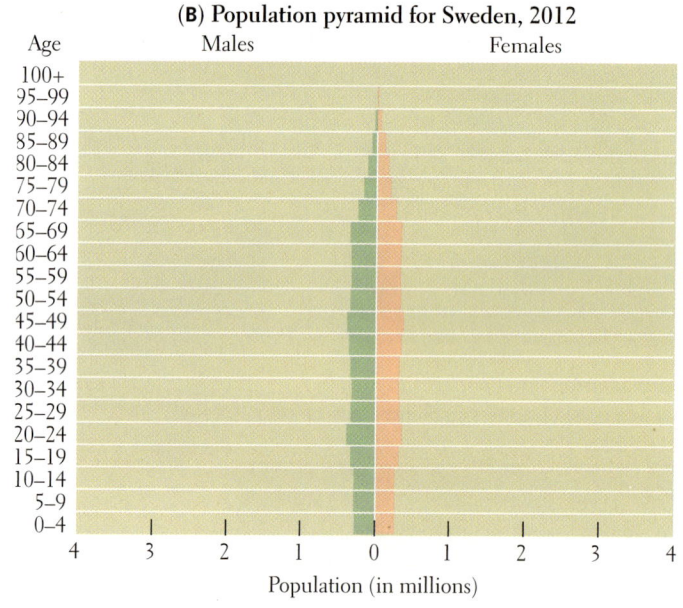

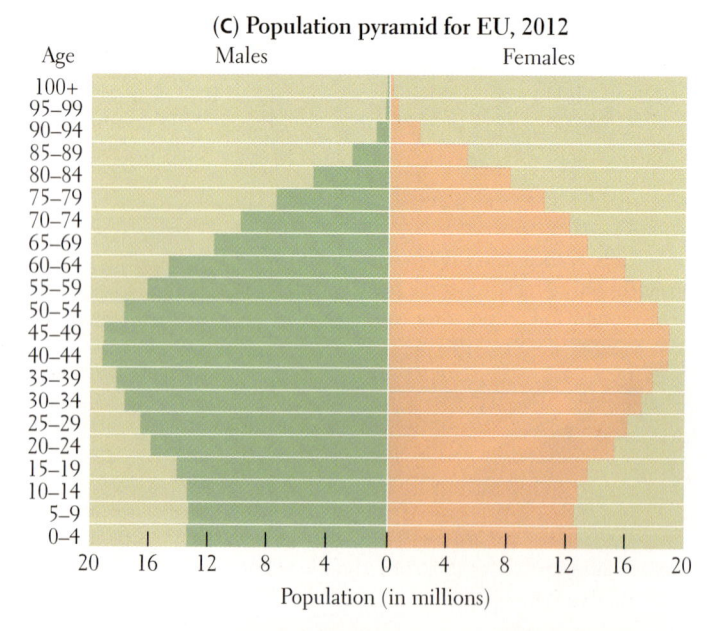

as missing opportunities for promotions. Fertility rates in both France and Sweden increased years ago after each country began providing free child care and preschools. In Germany, where there has historically been little government assistance for day care or preschools, fertility rates remain much lower. There is now a movement within the European Union to provide subsidized child care and more preschools and to lengthen school days.

Immigration provides only a partial solution to the dwindling number of young people. Immigrants from outside the region are the major source of population growth today in the EU, and they are slowing the trend toward population aging. However, to completely counter the low fertility rates common throughout this region, tens of millions of new immigrants would be required. This prospect is politically unpopular in Europe, where many voters are alarmed by the potential cultural changes brought by so many immigrants, often from very distant places with different value systems. Moreover, after a generation, most immigrants choose to have smaller families, making immigration a poor long-term solution to the aging of the population.

Population Distribution There are currently about 540 million Europeans. Of these, 500 million live within the European Union. The highest population densities stretch in a discontinuous band from the United Kingdom south and east through the Netherlands and central Germany into northern Switzerland (**Figure 4.22**). Northern Italy is another zone of high density, as

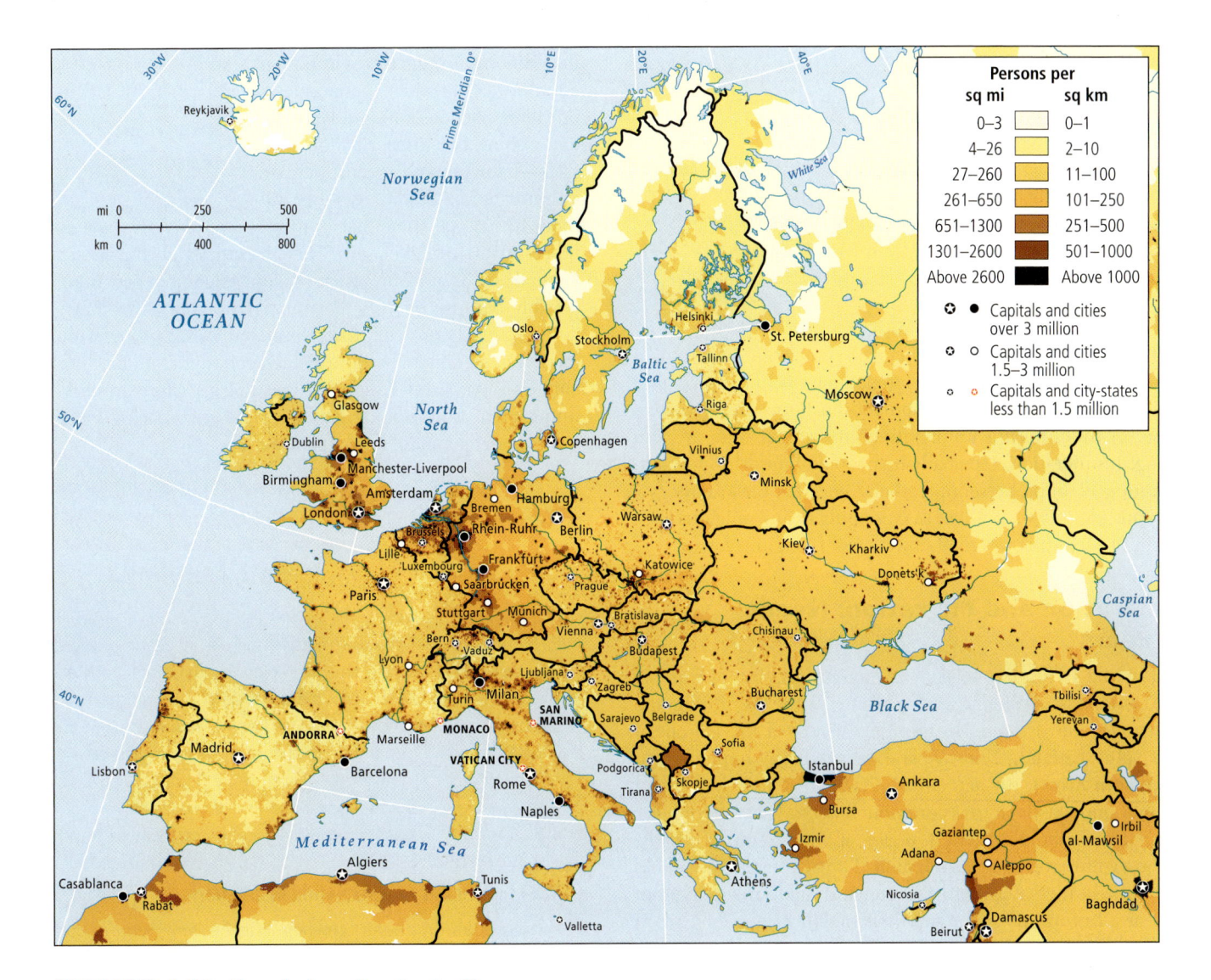

FIGURE 4.22 Population density in Europe. Europe's population is not growing, a fact that raises concerns about future economic conditions as the population ages. Many governments now encourage large families, with generous maternity and paternity leave (up to 10 months with full pay), free day care, and other incentives. However, few countries have seen much change in their population growth rate. A major reason for this is that more women now either work or want to work than ever before, so they are delaying childbearing and families remain small. As this trend shows no signs of decreasing, some officials are looking toward increased immigration as a possible solution. On the other hand, Europe has been attracting large numbers of immigrants for decades, and Europeans are increasingly wary of hosting large populations of foreigners who might not share their values.

are pockets in many countries along the Mediterranean coast. Overall, Europe is one of the more densely occupied regions on Earth, as shown on the world population density map in Figure 1.26 on pages 46–47.

> ### THINGS TO REMEMBER

> **GEOGRAPHIC INSIGHT 5**
> • **Population and Gender** Europe's population is aging as its fertility rates decline, which they are doing for a number of reasons, including the fact that career-oriented working women are choosing to have only one or two children. Fearing that aging populations could slow economic growth, European countries have tried to boost their fertility rates with a variety of policies, the most successful of which have addressed the needs of working mothers.

> • Because Europe's population is not growing and the number of consumers is staying the same or decreasing, economies may contract over time.

> • Immigration provides only a partial solution to the dwindling number of young people. There is political opposition to immigration on the scale that would be needed, and immigrants tend to have smaller families after a generation, making this only a short-term solution.

SOCIOCULTURAL ISSUES

The European Union was conceived primarily to promote economic cooperation and free trade, but its programs have social implications as well. As population patterns change across Europe, attitudes toward immigration and gender roles are also evolving. Immigration, especially by people of the Muslim faith, is a source of apprehension, perhaps because these days few Europeans identify themselves as belonging to any religious faith and they are left uncomfortable by shows of religious fervor.

Immigration and Migration: Needs and Fears

Until the mid-1950s, the net flow of migrants was out of Europe, to the Americas, Australia, and elsewhere. By the 1990s, the net flow was into Europe. In the 1990s, most of the European Union (plus Iceland, Norway, and Switzerland) implemented the **Schengen Accord**, an agreement that allows free movement of people and goods across common borders. The accord has facilitated trade, employment, tourism, and most controversially, migration within the European Union. The Schengen Accord has also indirectly increased both the demand for immigrants from outside the European Union and their mobility once they are in the European Union (Figure 4.23). **92. AFTER 50 YEARS, EUROPE STILL COMING TOGETHER**

Attitudes Toward Internal and International Migrants and Citizenship Like citizens of the United States, Europeans have ambivalent attitudes toward migrants. The internal flow of migration is mostly from Central Europe into North, West, and South Europe. These Central European migrants are mostly treated fairly, although prejudices against the supposed backwardness of Central Europe are still evident. Immigrants from outside Europe, so-called *international immigrants*, meet with varying levels of acceptance.

International immigrants often come both legally and illegally from Europe's former colonies and protectorates across the globe. Many Turks and North Africans come legally as **guest workers** who are expected to stay for only a few years, fulfilling Europe's need for temporary workers in certain sectors. Other immigrants are refugees from the world's trouble spots, such as Afghanistan, Iraq, Haiti, and Sudan. Many also come in illegally from all of these areas.

> **Schengen Accord** an agreement signed in the 1990s by the European Union and many of its neighbors that allows for free movement across common borders
>
> **guest workers** legal workers from outside a country who help fulfill the need for temporary workers but who are expected to return home when they are no longer needed
>
> **assimilation** the loss of old ways of life and the adoption of the lifestyle of another country

While some Europeans see international immigrants as important contributors to their economies—providing needed skills and labor and diminishing the impact of low birth rates—many oppose recent increases in immigration. Studies of public attitudes in Europe show that immigration is least tolerated in areas where incomes and education are low. Central and South Europe are the least tolerant of new immigrants, possibly because of fears that immigrants may drive down wages that are already relatively low. North and West Europe, with higher incomes and generally more stable economies, are the most tolerant. The European Union is increasing its efforts to curb illegal immigration from outside Europe while encouraging EU citizens to be more tolerant of legal migrants.

> ### A CASE STUDY
>
> #### The Rules for Assimilation: Muslims in Europe
>
> In Europe, culture plays as much of a role in defining differences between people as race and skin color. An immigrant from Asia or Africa may be accepted into the community if he or she has gone through a comprehensive change of lifestyle. **Assimilation** in Europe usually means giving up the home culture and adopting the ways of the new country. If minority groups—such as the Roma (see page 169)—who have been in Europe for more than a thousand years—maintain their traditional ways, it is nearly impossible for them to blend into mainstream society.
>
> **101. THE ART OF INTEGRATION IN GERMANY**
>
> Europe's small but growing Muslim immigrant population (Figure 4.24) is currently the focus of assimilation issues in the European Union. Muslims come from a wide range of places and cultural traditions, including North Africa, Turkey, and South Asia. Some of these immigrants maintain traditional dress, gender roles, and religious values, while others have assimilated into European culture.

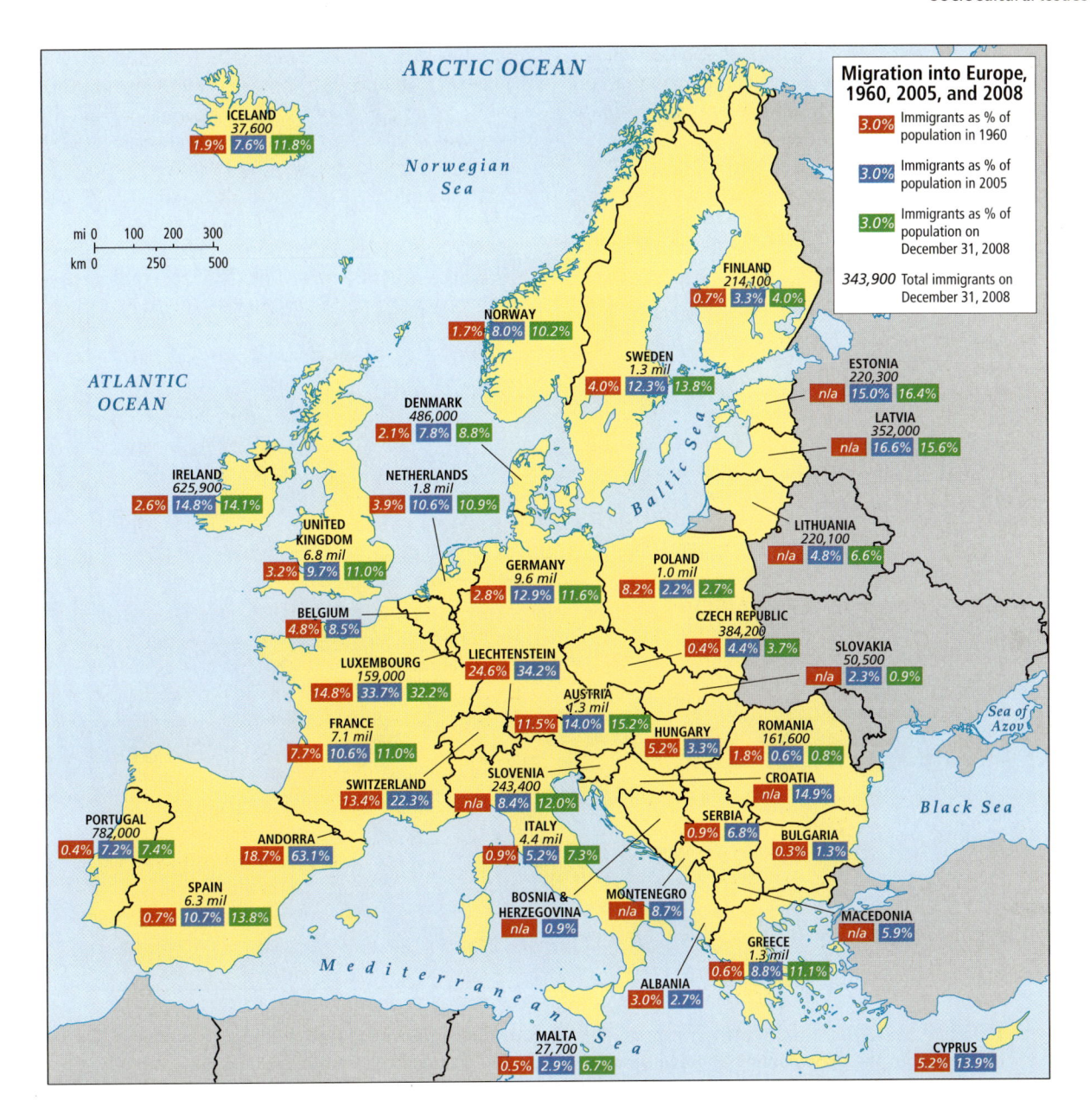

FIGURE 4.23 Migration into Europe, 1960, 2005, and 2008. Migration into Europe increased from 1960 through the 1990s and continued to increase into the twenty-first century for most countries. Migration continues to be a crucial issue in EU debates. The percentages of total population are shown in the three colored boxes for each country: the red box for 1960, the blue box for 2005, and the green box for 2008. The number of immigrants living in each country at the end of 2008 is above the two boxes. [Sources consulted: *Human Development Report 2009*, United Nations Development Programme, at http://hdrstats.undp.org/en/indicators/6.html; data in table is from *Migrants in Europe: A Statistical Portrait of the First and Second Generation*, Eurostat Statistical Books, 2011, Table 1, at http://epp.eurostat.ec.europa.eu/cache/ITY_OFFPUB/KS-31-10-539/EN/KS-31-10-539-EN.PDF]

The deepening alienation that has boiled over in recent years among some Muslim immigrants and their children—resulting in protests, riots, and sometimes even terrorism—relates primarily to the systematic exclusion of these less assimilated Muslims from meaningful employment, from some social services, and from higher education. In the wake of protests by Muslim residents, investigations by the French media revealed that the protestors' complaints were indeed legitimate. However, many Europeans, unaware of the extent to which their own societies discriminate against and stigmatize Muslims, have come to view Muslim protesters as simple malcontents. Meanwhile, young Muslims born in Europe who have never known life in any other place and have been schooled in the lofty ideals of the European Union harbor the greatest resentment against the constricted opportunities they face.

186 CHAPTER 4 Europe

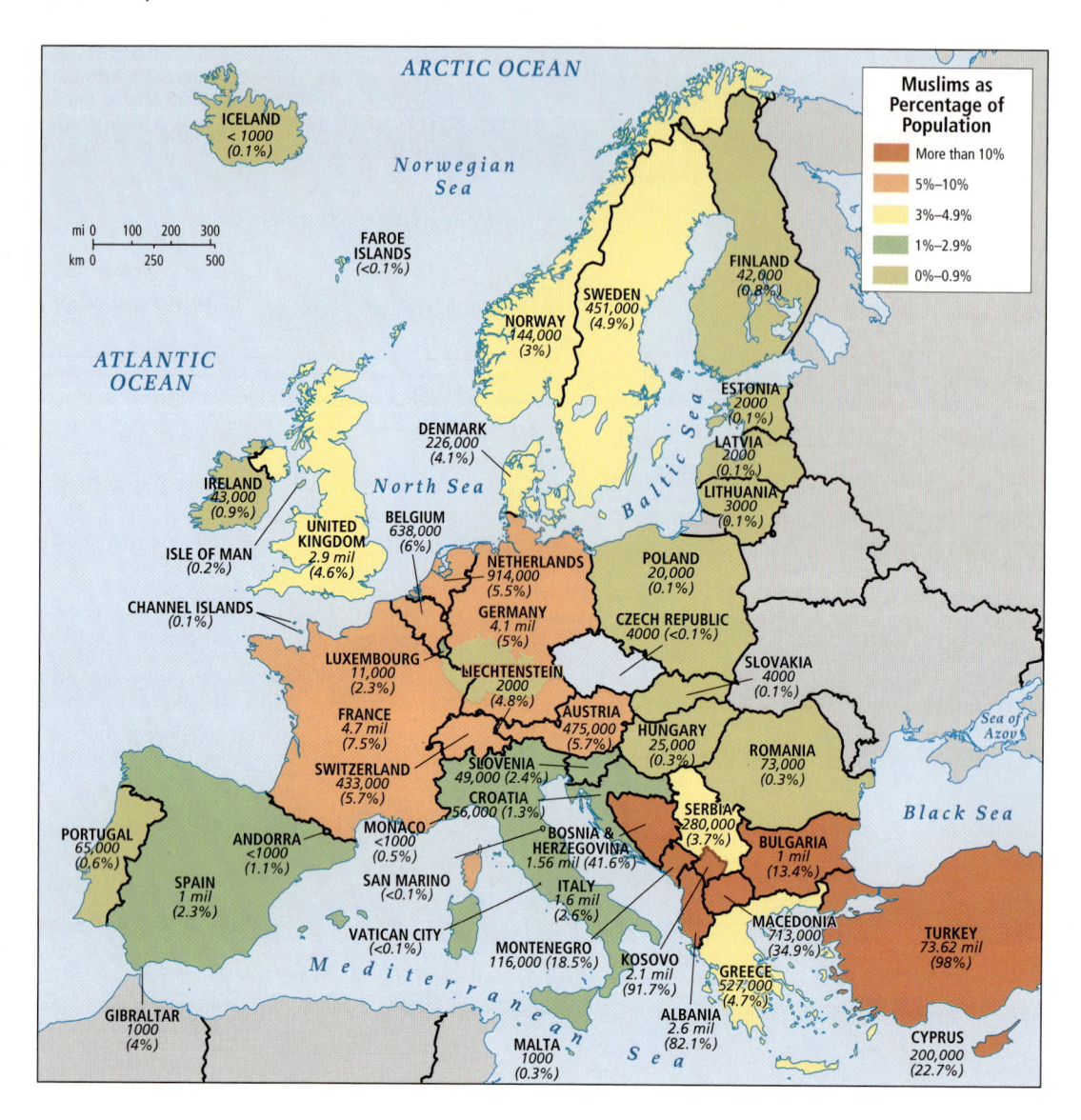

FIGURE 4.24 Muslims in Europe. Muslims are a small minority in most European countries, but their population has grown from a total of 29.6 million in 1990 to 44.1 million in 2010—amounting to about 6 percent of the total population—and is expected to rise to more than 58 million by 2030. Many intend to make the region their permanent home. Just how, or if, the assimilation of Muslims into largely secular Europe will proceed is a major topic of public debate. [Source consulted: "The Future of the Global Muslim Population," Religion and Public Life Project, Pew Research Center, January 27, 2011, at http://www.pewforum.org/2011/01/27/future-of-the-global-muslim-population-regional-sub-saharan-africa/]

In some cases, conflicts have also arisen over European perceptions of cultural aspects of Islam, such as the *hijab* (one of several traditional coverings for women). For example, in France in 2004, wearing of the hijab by observant Muslim schoolgirls became the center of a national debate about civil liberties, religious freedom, and national identity. French authorities wanted to ban the hijab but were wary of charges of discrimination. Eventually all symbols of religious affiliation were declared illegal in French schools (including crosses and yarmulkes, the Jewish head covering for men).

102. RELIGIOUS TOLERANCE FACING TEST IN BRITAIN
103. LONDON'S MEGA MOSQUE STIRS CONTROVERSY

Changing Gender Roles

Gender roles in Europe have changed significantly from the days when most women married young and worked in the home or on the family farm. Growing numbers of European women are working outside the home, and the percentage of women in professional and technical fields is rising rapidly (Figure 4.25). Nevertheless, European public opinion among both women and men largely holds that women are less able than men to perform the types of work typically done by men and that men are less skilled at domestic, caregiving, and nurturing duties. In most places, men have greater social status, hold more managerial positions, earn on average about 15 percent more pay for

Sociocultural Issues **187**

doing the same work, and have more autonomy in daily life than women—more freedom of movement, for example. These advantages for men retain a stronger hold in Central and South Europe today than they do in West and North Europe.

Certainly, younger men now assume more domestic duties than did their fathers, but women who work outside the home usually still face what is called a **double day** in that they are expected to do most of the domestic work in the evening in addition to their job outside the home during the day. UN research shows that in most of Europe, women's workdays, including time spent in housework and child care, are 3 to 5 hours longer than men's. (Iceland and Sweden reported that women and men there share housework equally.) Women burdened by the double day generally operate with somewhat less efficiency in a paying job than do men. They also tend to choose employment that is closer to home and that offers more flexibility in the hours and skills required. These more flexible jobs (often erroneously classified as part time) almost always offer lower pay and less opportunity for advancement, though not necessarily fewer working hours.

> **double day** the longer workday of women with jobs outside the home who also work as caretakers, housekeepers, and/or cooks for their families

Many EU policies encourage gender equality. Managerial posts in the EU bureaucracy are increasingly held by women, and well over half the university graduates in Europe are now women. Despite this, the political influence and economic well-being of European women lag behind those of European men. In most European national parliaments, women make up less than one-third of elected representatives. Only in North Europe, where several women have served as heads of government, do women come anywhere close to filling 50 percent of the seats in the legislature (Figure 4.26).

Numerous women have held high offices in the United Kingdom, but elsewhere in West Europe, this trend is only beginning. In 2009, Germany reelected Angela Merkel as its first woman chancellor (prime minister); in France, in 2007, Nicolas Sarkozy defeated his female opponent Ségolène Royal but then appointed women to a number of French cabinet-level positions. In 2012, fully one-half of newly elected French President François Hollande's cabinet was female. In 2013, Slovenia chose Alenka Bratušek as prime minister and for the first time elected a parliament that is nearly one-third women. Still, across Europe, women generally serve only in the lower ranks of government bureaucracies, where they implement policy but have limited power to formulate policy.

Although change is clearly underway in the European Union, economic empowerment for women has been slow on many fronts. For example, in 2006, unemployment was higher among women than among men in all but a few countries (the United Kingdom, Germany, the Baltic Republics, Ireland, Norway, and Romania), where the differences were slight—and 32 percent of women's jobs were part time, as opposed to only 7 percent of men's jobs. Throughout the European Union, women are paid less than men for equal work, despite the fact that young women tend to be more highly educated than young men.

Norway is a recognized global leader in redefining gender in society. It does this by directing much of its most innovative work

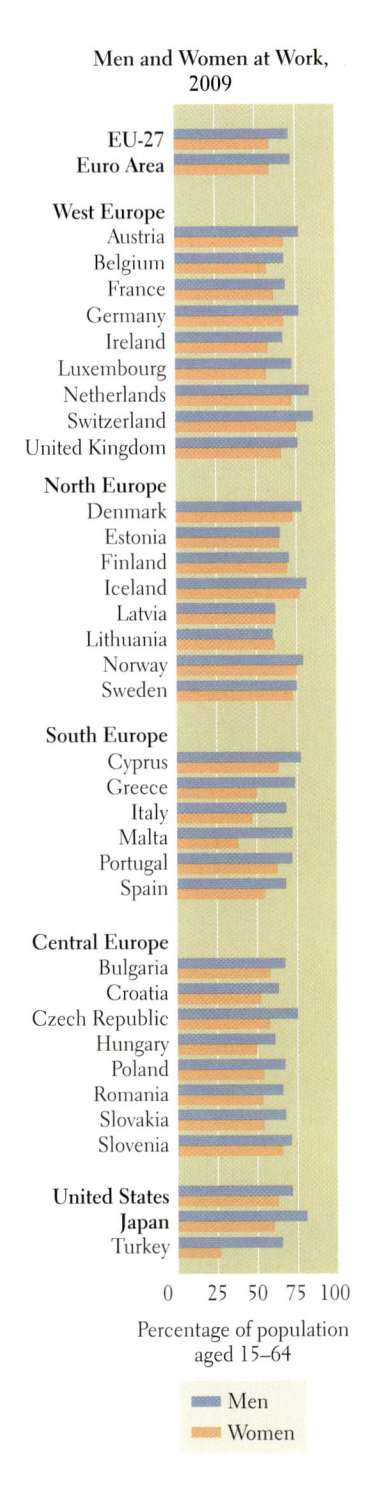

FIGURE 4.25 Women at work in Europe, 2009.
A majority of women in Europe work outside the home, and their numbers increased from 52 percent in 1998 to 58.6 percent in 2009. The exceptions are in Greece, Hungary, Italy, Malta, and the Republic of Macedonia, where less than 50 percent of women work outside the home. Included for comparison is Turkey, where slightly less than 25 percent of women work outside the home. These statistics, however, may not include some of the many women in Europe who work in the informal economy. In countries where the rate of women in the paid labor market is increasing, there is usually a concurrent decline in birth rates. [Source consulted: *Europe in Figures: Eurostat Yearbook 2011*, Figure 5.3, Eurostat, 2011, at http://epp.eurostat.ec.europa.eu/cache/ITY_OFFPUB/KS-CD-11-001/EN/KSCD-11-001-EN.PD]

Women in Parliaments, 2012

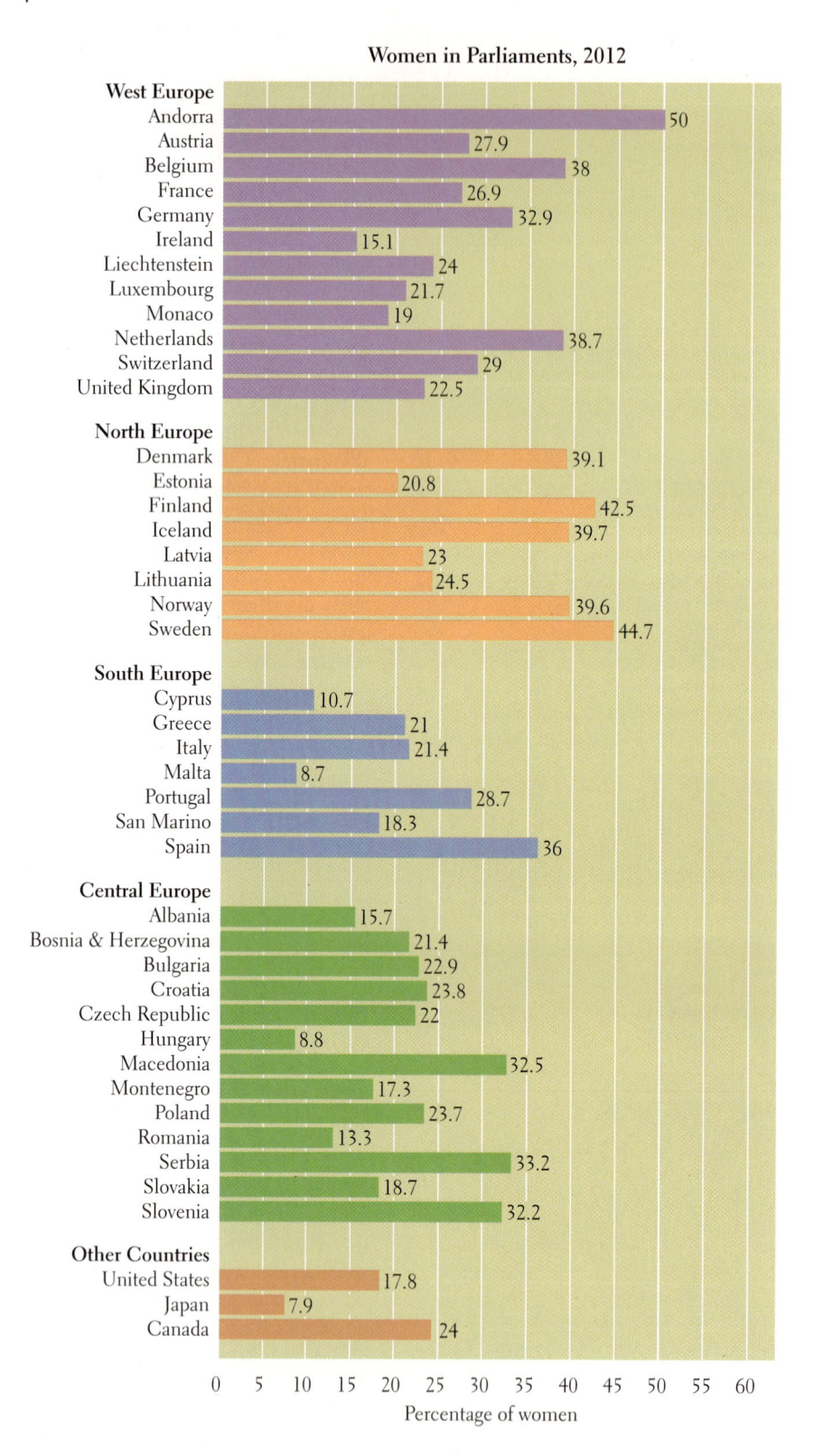

West Europe
- Andorra: 50
- Austria: 27.9
- Belgium: 38
- France: 26.9
- Germany: 32.9
- Ireland: 15.1
- Liechtenstein: 24
- Luxembourg: 21.7
- Monaco: 19
- Netherlands: 38.7
- Switzerland: 29
- United Kingdom: 22.5

North Europe
- Denmark: 39.1
- Estonia: 20.8
- Finland: 42.5
- Iceland: 39.7
- Latvia: 23
- Lithuania: 24.5
- Norway: 39.6
- Sweden: 44.7

South Europe
- Cyprus: 10.7
- Greece: 21
- Italy: 21.4
- Malta: 8.7
- Portugal: 28.7
- San Marino: 18.3
- Spain: 36

Central Europe
- Albania: 15.7
- Bosnia & Herzegovina: 21.4
- Bulgaria: 22.9
- Croatia: 23.8
- Czech Republic: 22
- Hungary: 8.8
- Macedonia: 32.5
- Montenegro: 17.3
- Poland: 23.7
- Romania: 13.3
- Serbia: 33.2
- Slovakia: 18.7
- Slovenia: 32.2

Other Countries
- United States: 17.8
- Japan: 7.9
- Canada: 24

Percentage of women

FIGURE 4.26 Women in national parliaments of Europe, 2013. Women comprise about half the adult population of the EU, but they have nowhere near their share of representation in European legislatures; hence, their influence on legislation is seriously restricted. Parliamentary elections in 2013 resulted in a rise in female representation in most countries. Notice that some regions are closer to equity than others, and also notice how the United States compares. [Source consulted: Women in National Parliaments, Inter-Parliamentary Union, February 1, 2014, at http://www.ipu.org/wmn-e/classif.htm]

on gender equality toward advancing men's rights in traditional women's arenas. For example, men and women are allowed to share the year of paid parental leave with a newborn or adopted child. This policy recognizes a father's responsibility in child rearing and provides a chance for father and child to bond early in life.

ON THE BRIGHT SIDE

European Women in Leadership Positions

In 1986, Gro Harlem Brundtland, a physician, became Norway's first female prime minister; she appointed women to 44 percent of all cabinet posts, thus making Norway the first country in modern times to have such a high proportion of women in important government policy-making positions. By 2013, women made up 39.6 percent of Norway's parliament (Sweden had 44.7 percent; Finland 42.5 percent; and Iceland 39.7 percent). Norway also requires a minimum of 40 percent representation by each sex on all public boards and committees. Even though the rule has not yet been implemented in all cases, female representation in state and municipal agencies averages over 35 percent. In addition, employment ads are required to be gender neutral, and all advertising must be nondiscriminatory.

THINGS TO REMEMBER

- Until the mid-1950s, the net flow of migrants was out of Europe, to the Americas, Australia, and elsewhere. By the 1990s, the net flow was into Europe.

- While some Europeans see international immigrants as important contributors to their economies—providing needed skills and labor and diminishing the impact of low birth rates—many oppose recent increases in immigration.

- Increasing numbers of European women are working outside the home, and the percentage of women in professional and technical fields is growing rapidly.

- Although change is clearly underway in the European Union, economic empowerment for women has been slow on many fronts.

GEOGRAPHIC INSIGHTS

Europe: Review and Self-Test

1. Environment: The European Union (EU) is a world leader in responding to climate change. Its goals for cutting greenhouse gas emissions are complemented by its many other strategies for saving energy and resources. However, much of Europe's air and many of its seas and rivers are very polluted, and consumption patterns in the region impact environments across the globe.

- How does Europe's physical geography complicate problems of river and sea pollution? What could the European Union do to reduce its contribution to the pollution of Europe's seas?

- Why might it make economic sense for Europe to address climate change now?

- On a more local level, what evidence is there that European ways of life contribute less to global warming than do North American ways?

- Which technological advances are being implemented in Europe to address climate change?

2. Globalization and Development: In order to better compete in the global economy, the EU has shifted labor-intensive industries from western Europe, where wages are high, to the relatively poorer, lower-wage member states of Central Europe. Growth in services has made up for some of the loss of industrial jobs in western Europe, though the unemployment rate is high in some countries. The EU continues to struggle to integrate the economies of its member states, especially those that use the EU's common currency, the euro.

- What challenges to the euro have emerged in recent years?

- What impact do Europe's heavily subsidized and regulated agricultural systems have on the rest of the world? What worries is the European Union addressing with its large investment in regulations of, and subsidies to, agricultural enterprises?

- What are some of the ways that the colonial empires created by European countries affect life in Europe today? How do colonial relationships still affect modern trading patterns?

3. Power and Politics: Following World War II, political freedoms have grown and strong welfare states have been established in this region. Since the fall of the Soviet Union in 1991, international politics within Europe have centered on the expansion of the EU into Central Europe and the development of EU political institutions. Having grown so much in the past several decades, the European Union, which is already a global economic power, could become a global counterforce to the United States in political and military affairs.

- How has the expansion of the EU into Central Europe transformed this region?

- In what ways is the European Union similar to the United States politically?

- What new role for the European Union is developing through the North Atlantic Treaty Organization (NATO)?

- What are the major types of tax-supported systems of social welfare (or social protection) found in Europe?

4. Urbanization: Europe's cities are both ancient and modern, with old town centers now surrounded by modern high-rise suburbs and supported by world-class urban infrastructures. Cities are the heart of Europe's economy, politics, and culture, with more than 70 percent of the EU population living in urban areas.

- How did the Industrial Revolution lead to urbanization in Europe?

- What is the connection between urbanization and democratization in Europe?

- Which economic sectors are most European cities now oriented toward?

- Why are so many cities in Europe located either on navigable rivers in the interior or along the coasts?

- Why do only a small percentage of Europeans live in single-family homes?

5. Population and Gender: Europe's population is aging as fertility rates decline for a number of reasons, including the fact that career-oriented working women are choosing to have only one or two children. Fearing that aging populations could slow economic growth, European countries have tried to boost their fertility rates with a variety of policies, the most successful of which have addressed the needs of working mothers.

- What are some of the potential consequences to Europe of aging populations and negative population growth rates?
- How do improvements in education and work opportunities for European women affect population growth?
- Which policy interventions have been most successful in countering low birth rates in Europe?
- How do immigration policies influence aging in Europe? What about these policies worries Europeans?

CRITICAL THINKING QUESTIONS

1. How do the issues of immigration in Europe compare with those in the United States? If you were a poor, undocumented immigrant searching for a way to support your family, would you choose the United States or Europe as a possible destination? Why?

2. What are the ways in which Europe is still linked to its former colonies?

3. How is the organization of urban space (housing, transportation, shopping) better suited to energy efficiency in Europe than in the United States?

4. What are some of the potential societal consequences of low birth rates in Europe?

5. How might the EU's status as the world's largest economy, and the rising importance of NATO, lead to changes in global power relationships?

6. What does the evolution of democracy in Europe suggest about current efforts to promote democracy in Iraq or Egypt?

7. Why might a pig farmer in West Africa criticize the EU's generous support for European agribusiness farmers?

8. Which European state-supported systems have offered women the best opportunities for employment outside the home?

9. Cite the evidence that Europe is seeking ways of life that contribute less to global warming.

10. Where is Europe's contribution to water-body pollution most severe? What could the European Union do to reduce its contribution to the pollution of Europe's oceans and seas?

› THINKING GEOGRAPHICALLY

Now that you have read about Europe, you should be able to answer the following questions about the photos in this chapter. To answer these questions online, go to Geography LaunchPad.

Vulnerability to Climate Change (page 158)

A What clues in this photo, other than the caption, indicate that the reservoir is at a very low level?

B Judging from the way these structures are arranged, how would you guess they work?

C How might owning a car decrease a family's sensitivity to flooding?

Human Impacts on the Biosphere (page 161)

A How might this large, open pit threaten groundwater resources?

B What clue(s) in the picture shows that large numbers of people may be exposed to pollution from the power plant?

C How might the large areas covered by plastic contribute to a water shortage?

Visual History (page 166)

A What about the photo of the ruins of Timgad reflects a particular Roman strategy for organizing space?

B What about the architecture of Alhambra suggests technological sophistication and an astute understanding of climate?

C What about this picture suggests relative prosperity in sixteenth-century Belgium?

D What about this picture suggests how the Portuguese felt about the value of trade with India?

E How is the execution of Louis XIV related to the evolution of democracy in Europe?

F What does this photo indicate about the impact of the Industrial Revolution on employment in Europe?

Power and Politics (page 176)

A What clues in this photo suggest war and not some other calamity?

C What are the political implications of the sign announcing entry into Free Derry? What might be the reasons for showing a child wearing a gas mask?

D Why could it be said that independence movements in Portugal's African colonies brought democracy to Portugal?

Urbanization (page 181)

A What are the likely reasons why Prague is located along a river?

B What can you see in this picture that illustrates what we know about urban life in Europe?

C Describe the people you see walking in this picture. How might walking as a mode of transportation affect the environment and urban spaces of Barcelona?

D How does the amount of greenhouse gas produced by rail-based public transportation compare to that of private automobile–based transportation?

CHAPTER KEY TERMS

assimilation 184
capitalism 169
central planning 169
Cold War 169
Common Agricultural Program (CAP) 173
communism 169
continental climate 156
cultural homogenization 154
double day 187
economies of scale 172

euro 172
European Union (EU) 153
Green 159
guest workers 184
Holocaust 169
humanism 165
Iron Curtain 169
Mediterranean climate 156
mercantilism 166
nationalism 168

North Atlantic Drift 156
North Atlantic Treaty Organization (NATO) 178
Roma 169
Schengen Accord 184
social welfare (in the European Union, social protection) 178
subsidies 173
temperate midlatitude climate 156
welfare state 168

A North European Plain, Belarus

B Caucasus Mountains, Georgia

C Volga River, Russia

D Urals, Russia

chapter 5

RUSSIA AND THE POST-SOVIET STATES

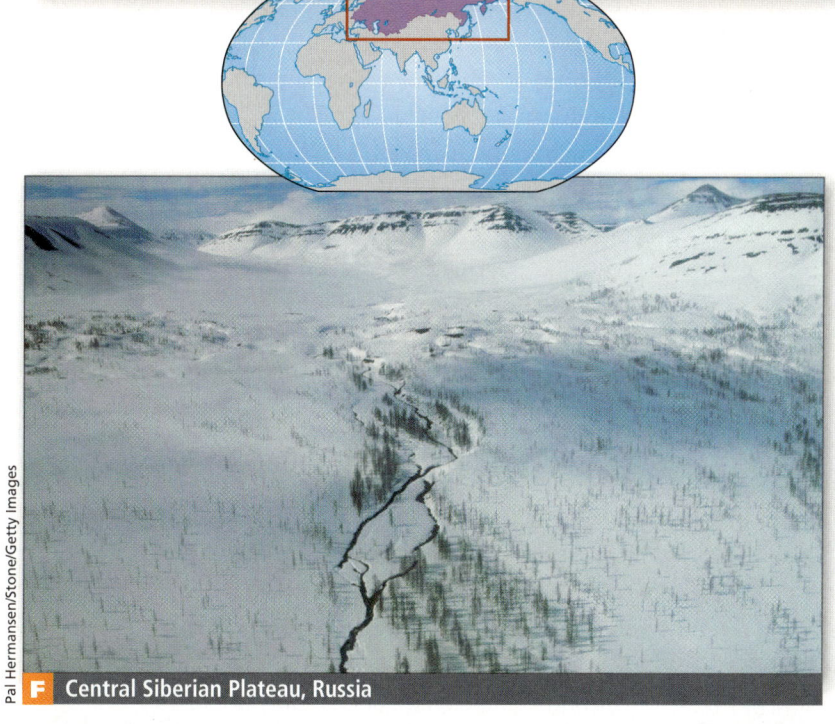

F Central Siberian Plateau, Russia

G Pacific Mountain Zone, Russia

E West Siberian Plain, Russia

FIGURE 5.1 Regional map of Russia and the post-Soviet states.

193

GEOGRAPHIC INSIGHTS

After you read this chapter, you will be able to discuss the following issues as they relate to the five thematic concepts:

1. Environment: Economic development has taken precedence over environmental concerns in Russia and the post-Soviet states for decades, resulting in numerous environmental problems such as severe pollution of the air and water by toxic industrial waste. This region's contribution to climate change is increasing as more people use automobiles for transportation. In parts of this region, food production systems and water resources are highly vulnerable to the impacts of climate change.

2. Globalization and Development: After the fall of the Soviet Union, economic reforms and globalization changed patterns of development in this region. Wealth disparity increased and jobs were lost as many Communist-era industries were closed or sold to the rich and well connected. The region is now largely dependent on its role as a leading global exporter of energy resources.

3. Power and Politics: This region has a long history of authoritarianism. While there is some pressure to expand political freedoms, there are still few opportunities for the public to influence the political process. Elected representative bodies often act as rubber stamps for very strong presidents and exercise only limited influence on policy making.

4. Urbanization: A few large cities in Russia and Central Asia are growing fast, fueled by the expansion of energy exports. Elsewhere, many cities are suffering from a lack of investment and maintenance as their economies struggle in the post-Soviet era.

5. Population and Gender: Populations are shrinking in many parts of the region due largely to lower birth rates and declines in life expectancy. These changes relate to consistently high levels of participation by women in the workforce and to the economic decline in the 1990s that followed the collapse of the Soviet Union. Men and women have been affected differently in the post-Soviet era, with men much more likely to suffer from alcoholism and related health problems.

The Russia and Post-Soviet States Region

Russia and the post-Soviet states (Figure 5.1) make up a large portion of Eurasia, the world's largest continent. Despite its vastness, the region is not highly populated and has a climate that, in many places, is unforgiving.

The five thematic concepts are explored as they arise in the discussion of regional issues. Vignettes, such as the one below, illustrate one or more of the themes as they are experienced in individual lives.

GLOBAL PATTERNS, LOCAL LIVES

It is February 2012 and a cold, snowy day in Moscow, the capital of the Russian Federation. Tens of thousands of protesters hold hands to form an enormous circle around the city, along one of the traffic-clogged ring roads that Moscow is known for. National elections are coming up soon, and the demonstrators are trying to defeat the favorite for the presidential post, Vladimir Putin. Marina Segupova, a 28-year-old interior decorator, participates in the demonstration, which she hopes will spur further protest. Using a metaphor apt for the moment, she describes the protest movement: "We are a snowball and we are rolling." At the same time, she acknowledges, change will not come easily. Another middle-class professional, the 56-year-old accountant Galina Venediktova, came to voice her opinion about corruption and poor public services. During that same month, other forms of protest were less conventional. The feminist punk band Pussy Riot staged an unauthorized "guerilla performance" at a Moscow cathedral, in which they decried the policies of Vladimir Putin and became a worldwide media sensation.

Putin had already served 8 years as president, from 2000 to 2008. Prevented by the constitution from running again for a third consecutive term, he had an ally become president for 4 years, after which Putin ran for a third and completely legal nonconsecutive term. Despite the efforts of Segupova, Venediktova, Pussy Riot, and other demonstrators, Putin is again serving as president of Russia after having decisively won the March 2012 election. The outcome of the election was hardly in doubt, due to the lack of other serious contenders for the presidency and Putin's dominance in state media reporting.

On the face of it, these events suggest that the political status quo has been maintained and that power is still concentrated within a small group of elites who have little interest in the expansion of political freedoms. However, the upwelling of demonstrations indicates that the political landscape is slowly changing in Russia. A few months earlier, a contested election in Russia's parliament, the Federal Assembly, also resulted in widespread protests by people who felt that the election had been rigged to ensure the victory of Putin's party. After Putin's election, protests continued in the streets and squares of central Moscow (Figure 5.2). In 2012, three members of Pussy Riot were put on trial and sentenced to prison for "hooliganism," which has reinforced the suspicion among many that political dissent in Russia will be repressed by the authorities.

As Russia slowly becomes wealthier, its people are demanding more political freedoms. This is especially true of the growing middle class, the urbanites in large cities (particularly those in Moscow and St. Petersburg), and the country's youth. But so far, politicians like Vladimir Putin have the support of the majority of Russians, who perceive them as strong leaders. People in hardscrabble industrial cities, small towns, and rural areas, along with the elderly, view Putin as a force for stability, which is favored by many, considering the political turbulence Russia has had during the last two decades. Putin being reelected also has larger global geopolitical ramifications because the relationship between Russia and the West sometimes involves suspicion on each side and sometimes, when it suits both regions, is one of cooperation. [Sources: Christian Science Monitor, New York Times, and Sky News. For detailed source information, see Text Sources and Credits.]

FIGURE 5.2 Political demonstration in Moscow. Riot police in Moscow detain journalist and civil rights activist Alexandr Podrabinek during a rally protesting the inauguration of Vladimir Putin as president of Russia in 2012. The region of Russia and post-Soviet states is transitioning from a political culture in which authority was rarely questioned to one in which protests are increasingly common. Nevertheless, many leaders continue to be highly authoritarian.

> **THINKING GEOGRAPHICALLY** Where in Russia are people more likely to support the policies of Vladimir Putin?

Russia is the largest country in a region that has entirely changed its political and economic systems in a short period of time. Barely two decades ago, the **Union of Soviet Socialist Republics (USSR)**, more commonly known as the **Soviet Union**, was the largest political unit in the world, stretching from Central Europe to the Pacific Ocean. It covered one-sixth of Earth's land surface. In 1991, the Soviet Union broke apart, ending a 70-year era of nearly complete governmental control of the economy, society, and politics. Over the course of a few years, attempts were made to substitute the Communist Party's economic control with capitalist systems similar to those of Western countries, which are based on competition among private businesses. The transition has proven difficult.

Politically, the Soviet Union has been replaced by a loose alliance known as the *Commonwealth of Independent States*. It consists of Russia and 11 independent post-Soviet states—the European states of Ukraine, Belarus, and Moldova; the Caucasian states of Georgia, Armenia, and Azerbaijan; and the Central Asian states of Kazakhstan, Kyrgyzstan, Tajikistan, Turkmenistan, and Uzbekistan (Figure 5.3). Three former Soviet republics, the Baltic states of Lithuania, Latvia, and Estonia, are now part of the European Union (see Chapter 4). Russia, which was always the core of the Soviet Union, remains predominant in the Commonwealth of Independent States and is influential in the world because of its size (at roughly three-quarters the size of the former Soviet Union, it is still the largest country in the world), population, military, and huge oil and gas reserves.

Union of Soviet Socialist Republics (USSR) the multinational union formed from the Russian empire in 1922 and dissolved in 1991; commonly known as the *Soviet Union*

Soviet Union see *Union of Soviet Socialist Republics*

FIGURE 5.3 Political map of Russia and the post-Soviet states.

196 CHAPTER 5 Russia and the Post-Soviet States

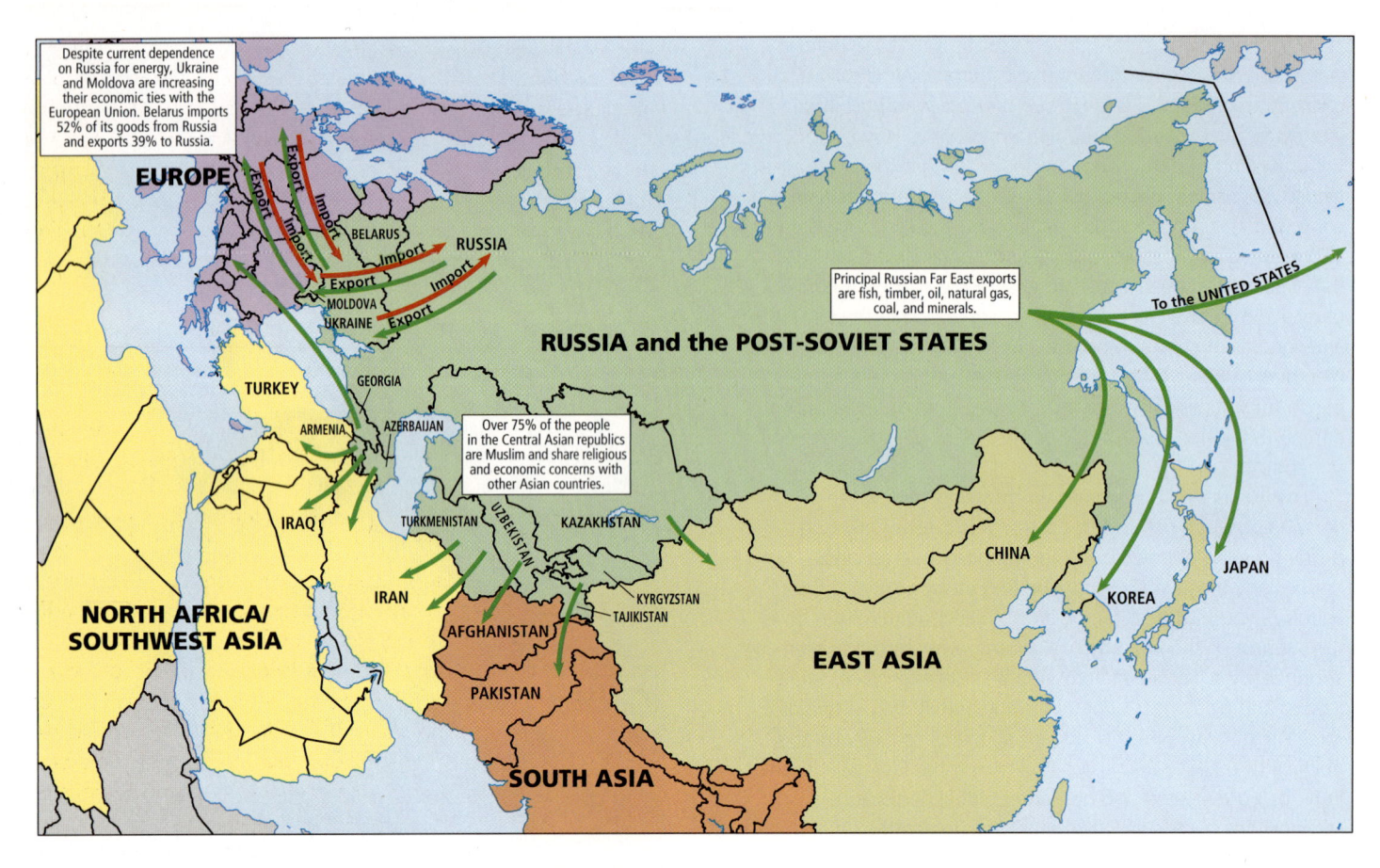

FIGURE 5.4 Russia and the post-Soviet states: Contacts with other world regions. The region of Russia and the post-Soviet states is in flux as all of the component countries rethink their geopolitical positions relative to one another and to adjacent regions. The arrows indicate various types of contact, ranging from economic to religious and cultural.

Geopolitically, this region is still sorting out its relationships. The Cold War between the Soviet Union and the United States and its allies is over. Most former Soviet allies in Central Europe have already joined the European Union, and some other countries in the region may eventually do the same. Trade with Europe, especially in oil and gas, is booming, if fraught with conflict. In the future, the Central Asian states currently allied with Russia may align themselves with neighbors in Southwest Asia or South Asia. The far eastern parts of Russia—and Russia as a whole—are already finding common trading ground with East Asia and Oceania (Figure 5.4).

After 70 years of authoritarian rule, elections are becoming the norm throughout this region. Whether these are actually free and fair elections is debatable, as opposition candidates are marginalized by lack of access to both print and broadcast media. Meanwhile, an explosion of crime and corruption has many wondering if the new political systems of this region are strong enough to endure.

THINGS TO REMEMBER

- Politically, the Soviet Union has been replaced by a loose alliance known as the Commonwealth of Independent States that consists of Russia and 11 independent post-Soviet states.

- A process of geopolitical realignment is taking place, in which some post-Soviet states seek closer relations with Europe, while others are developing ties with nearby Asian countries. However, most maintain strong ties with Russia, which dominates this region.

- Elections have become common in the region, although whether they are free and fair is uncertain. Political systems in the region are also threatened by crime and corruption.

What Makes Russia and the Post-Soviet States a Region?

Russia and the post-Soviet states are today associated as a region primarily because of their history from the nineteenth century onward. Imperial Russia underwent a convulsive revolution in 1917 that resulted in a new political and economic experiment with socialism that lasted until 1991. The leaders of this revolution continued with the imperial practice of territorial expansion that ultimately enveloped countries from the Baltic Sea to the Pacific and from the Arctic Sea to the mountains of Caucasia and southern Central Asia (see the Figure 5.1 map). Control of the economic and political systems of what eventually became known as the Soviet Union was exercised primarily from the Russian capital Moscow through authoritarianism reinforced with militarism. After the collapse of the Soviet Union in 1991, Moscow's dominance diminished somewhat as western parts of the region oriented more toward Europe and the Central Asian republics became more autonomous (see Figure 5.4). Nevertheless, Russia remains the dominant power in this region.

Terms in This Chapter

There is no entirely satisfactory new name for the former Soviet Union. In this chapter we use *Russia and the post-Soviet states*. One reason we do so is that economic, political, and social developments from the Soviet days still shape these countries today. Russia continues to be closely associated with these states economically, but they are independent countries, and their governments are legally separate from Russia's. Russia itself is formally known as the **Russian Federation** because it includes more than 21 (mostly) ethnic *internal republics* (also called *semiautonomous regions*)—such places as Chechnya, Ossetia, Tatarstan, and Buryatia—which all together constitute about one-tenth of the Federation's territory and one-sixth of its population. *Federation* should not be taken to mean that the internal republics share power equally with the central Russian government in Moscow.

PHYSICAL GEOGRAPHY AND ENVIRONMENTAL ISSUES

PHYSICAL PATTERNS

The physical features of Russia and the post-Soviet states vary greatly over the huge territory they encompass. The region bears some resemblance to North America in size, topography, climate, and vegetation.

Landforms

Because the region is relatively complex physically, a brief summary of its landforms is useful (see the Figure 5.1 map). Moving west to east, there is first the eastern extension of the North European Plain, including the plain of the Volga River (see Figure 5.1A, C), followed by the Ural Mountains (see Figure 5.1D), which are often considered the dividing line between European Russia and **Siberia**. East of the Urals are the West Siberian Plain (see Figure 5.1E), followed by an upland zone called the Central Siberian Plateau (see Figure 5.1F), and finally, in the Far East, a series of mountain ranges bordering the Pacific (see Figure 5.1G). To the south there are mountains and uplands (the Caucasus, shown in Figure 5.1B, and Central Asia, depicted in Figure 5.8A on page 203) as well as, in western Central Asia, semiarid grasslands, or **steppes** (see Figure 5.5B).

The eastern extension of the North European Plain rolls low and flat from the Carpathian Mountains in Ukraine and Romania, 1200 miles (about 2000 kilometers) east to the Ural Mountains (see Figure 5.1A, C, D). The part of Russia west of the Urals is often called *European Russia* because the Ural Mountains are traditionally considered part of the indistinct border between Europe and Asia. European Russia is the most densely settled part of the entire region (see Figure 5.22 on page 223) and its agricultural and industrial core. Its most important river is the Volga, which flows into the Caspian Sea. The Volga River and its tributaries form a major transportation route that connects many parts of the North European Plain, including Moscow, to St. Petersburg and the Baltic and White seas in the north and to the Black and Caspian seas in the south (see Figure 5.1C).

The Ural Mountains extend in a fairly straight line south from the Arctic Ocean into Kazakhstan (see the Figure 5.1 map). A low-lying range similar in elevation to the Appalachians, the Urals are not much of a barrier to humans and are only partially a barrier to nature (some European tree species do not extend east of the Urals). There are several easy passes across the mountains, and winds carry moisture all the way from the Atlantic into Siberia. Much of the Urals' once-dense forest has been felled to build and fuel new industrial cities.

The West Siberian Plain, east of the Urals, is the largest plains in the world (see Figure 5.1E). A vast, mostly marshy lowland about the size of the eastern United States drains toward the north into the Arctic Ocean. Long, bitter winters mean that in the northern half of this area, a layer of permanently frozen soil (**permafrost**) lies just a few feet beneath the surface. Permafrost is formed when the ground warms up during the short summer but the top layer of material insulates against this warming effect, leaving the subsurface always frozen. In the far north, the permafrost comes to within a few inches of the surface. Because water does not percolate down through this frozen layer, surface water accumulates to create wetlands that form above the permafrost. In the summer months these wetlands provide habitats for many migratory birds.

In the far north, a treeless area called the **tundra** lies where only mosses and lichens can grow because of the extreme cold, the shallow soils, and the permafrost. The West Siberian Plain is one of the world's largest oil and natural gas producers, although the harsh climate and permafrost make extraction difficult.

> **Russian Federation** Russia and its political subunits, which include 21 internal republics
>
> **Siberia** a region of Russia that is located east of the Ural Mountains
>
> **steppes** semiarid, grass-covered plains
>
> **permafrost** permanently frozen soil that lies just a few feet beneath the surface
>
> **tundra** a treeless area, between the ice cap and the tree line of arctic regions, where the subsoil is permanently frozen

The Central Siberian Plateau and the Pacific Mountain Zone, farther to the east, together equal the size of the United States (see Figure 5.1F, G). Permafrost prevails except along the Pacific coast. There, the ocean moderates temperatures; the many active volcanoes created as the Pacific Plate sinks under the Eurasian Plate supply additional heat. Lightly populated places like the Kamchatka Peninsula, Sakhalin Island, and Sikhote-Alin on Russia's southeastern Pacific coast are havens for wildlife.

To the south of the West Siberian Plain, steppes and deserts stretch from the Caspian Sea to the Chinese border. To the east of these grasslands are the Caucasus Mountains (see Figure 5.1B), and to the southeast, facing China and Mongolia, are a series of other mountain chains, including Pamir and Tien Shan. The rugged terrain has not deterred people from crossing these mountains. For tens of thousands of years, from the Caucasus to the Pamir and Tien Shan, people have exchanged plants (apples, onions, citrus, rhubarb, wheat), animals (horses, sheep, goats, cattle), technologies (cultivation, animal breeding, portable shelter construction, rug and tapestry weaving), and religious belief systems (principally Islam and Buddhism, but also Christianity, Hinduism, and Judaism).

Climate and Vegetation

The climates and associated vegetation in this large region are varied but less so than in other regions because so much territory here is taken up by expanses of midlatitude grasslands and by northern forest and tundra that are cold much of the year. No inhabited place on Earth has as harsh a climate as the northern part of the Eurasian landmass occupied by Russia and, particularly, by Siberia (shown in Figure 5.5C and the figure map). Winters are long and cold, with only brief hours of daylight. Summers are short and range from cool to hot, with long days. Precipitation is moderate, coming primarily from the west. In the northernmost areas, the natural vegetation is tundra grasslands. The major economic activities here are the extraction of oil, gas, and some minerals, as well as reindeer herding by the local indigenous population. Just south of the tundra is a vast, cold-adapted coniferous forest known as **taiga** that ranges from northern European Russia to the Pacific (and from a global perspective, this coniferous belt also includes much of Alaska, Canada, and Scandinavia). The largest portion of taiga lies east of the Urals, and here forestry—often unrestrained by ecological concerns—is a dominant economic activity. The short growing season and the large areas of permafrost generally limit crop agriculture, except in the southern West Siberian Plain, where people grow grain.

taiga subarctic coniferous forests

Because massive mountain ranges to the south block access to warm, wet air from the Indian Ocean, most rainfall in the entire region comes from storms that blow in from the Atlantic Ocean far to the west (see the Figure 5.5 map). But by the time these air masses arrive, most of their moisture has been squeezed out over Europe. A fair amount of rain does reach Ukraine, Belarus, European Russia (see Figure 5.5A), and the Caucasian republics; these regions are especially important areas for food production (vegetables, fruits, and grain). The natural vegetation in these western zones is open woodlands and grassland, though in ancient times, forests were common.

East of the Caucasus Mountains, the lands of Central Asia have semiarid to arid climates (see Figure 5.5B), influenced by their location in the middle of a very large continent. The summers are scorching and short and the winters are intense. In the desert zones, daytime-to-nighttime temperatures can vary by 50°F (28°C) or more. Northern Kazakhstan produces grain and grazing animals. The more southern areas (southern Kazakhstan, Uzbekistan, and Turkmenistan) have grasslands (steppes), which are also used for herding. In modern times, these grasslands have been used for irrigated commercial agriculture that is dependent upon glacially fed rivers, but most land is not useful for farming. The climates in the more mountainous area where Kazakhstan, Uzbekistan, Tajikistan, and Kyrgyzstan meet are varied and support a number of small-scale agricultural activities, some of them commercial.

The construction of land transportation systems has been held back by the climate of the region. Especially to the north, it is difficult to build during the long, harsh winters, which eventually give way to a spring period called the *rasputitsa*, or the "quagmire season," when melting permafrost (or, to a lesser extent, autumn rain) turns many roads and construction sites into impassable mud pits. Huge distances between populated places as well as the complex topography, especially in Siberia, add to the problem. As a result, few roads or railroads have ever been built outside of western Russia.

THINGS TO REMEMBER

- The part of Russia west of the Urals is often called European Russia because the Ural Mountains are traditionally considered part of the indistinct border between Europe and Asia.

- In the far north, a treeless area called the tundra lies where only mosses and lichens can grow because of the extreme cold, the shallow soils, and the permafrost.

- Just south of the tundra is a vast, cold-adapted coniferous forest known as taiga that ranges from northern European Russia to the Pacific (and from a global perspective, this coniferous belt also includes much of Alaska, Canada, and Scandinavia).

- Because massive mountain ranges to the south block access to warm, wet air from the Indian Ocean, most rainfall in the entire region comes from storms that blow in from the Atlantic Ocean far to the west (see the Figure 5.5 map).

- The construction of land transportation systems has been held back by the climate of the region. Especially to the north, it is difficult to build during the long, harsh winters, which eventually give way to a spring period called the *rasputitsa*, or the "quagmire season," when melting permafrost (or, to a lesser extent, autumn rain) turns many roads and construction sites into impassable mud pits.

ENVIRONMENTAL ISSUES

GEOGRAPHIC INSIGHT 1

Environment: Economic development has taken precedence over environmental concerns in Russia and the post-Soviet states for decades, resulting in numerous environmental problems such as severe pollution of the air and water by toxic industrial waste. This region's contribution to climate change is increasing as more people use automobiles for transportation. In parts of this region, food production systems and water resources are highly vulnerable to the impacts of climate change.

Soviet ideology held that nature was the servant of industrial and agricultural progress, and that humans should dominate nature on a grand scale. While this sentiment was common throughout much of the world at the time of the formation of the USSR, it does seem to have been taken further here than elsewhere. Joseph Stalin, leader of the USSR from 1922 to 1953 and a major architect of Soviet policy, is famous for having said, "We cannot expect charity from Nature. We must tear it from her." During the Soviet years, huge dams, factories, and other industrial facilities were built without regard for their effect on the environment or on public health. Russia and the post-Soviet states now have some of the worst environmental problems in the world. By 2000, more than 35 million people in the region (15 percent of the population) were living in areas where the soil was poisoned and the air was dangerous to breathe (see Figure 5.6A–C).

Since the collapse of the Soviet Union, the region's governments, beset with myriad problems, have been both reluctant to address environmental issues and incapable of doing so. As one Russian environmentalist put it, "When people become

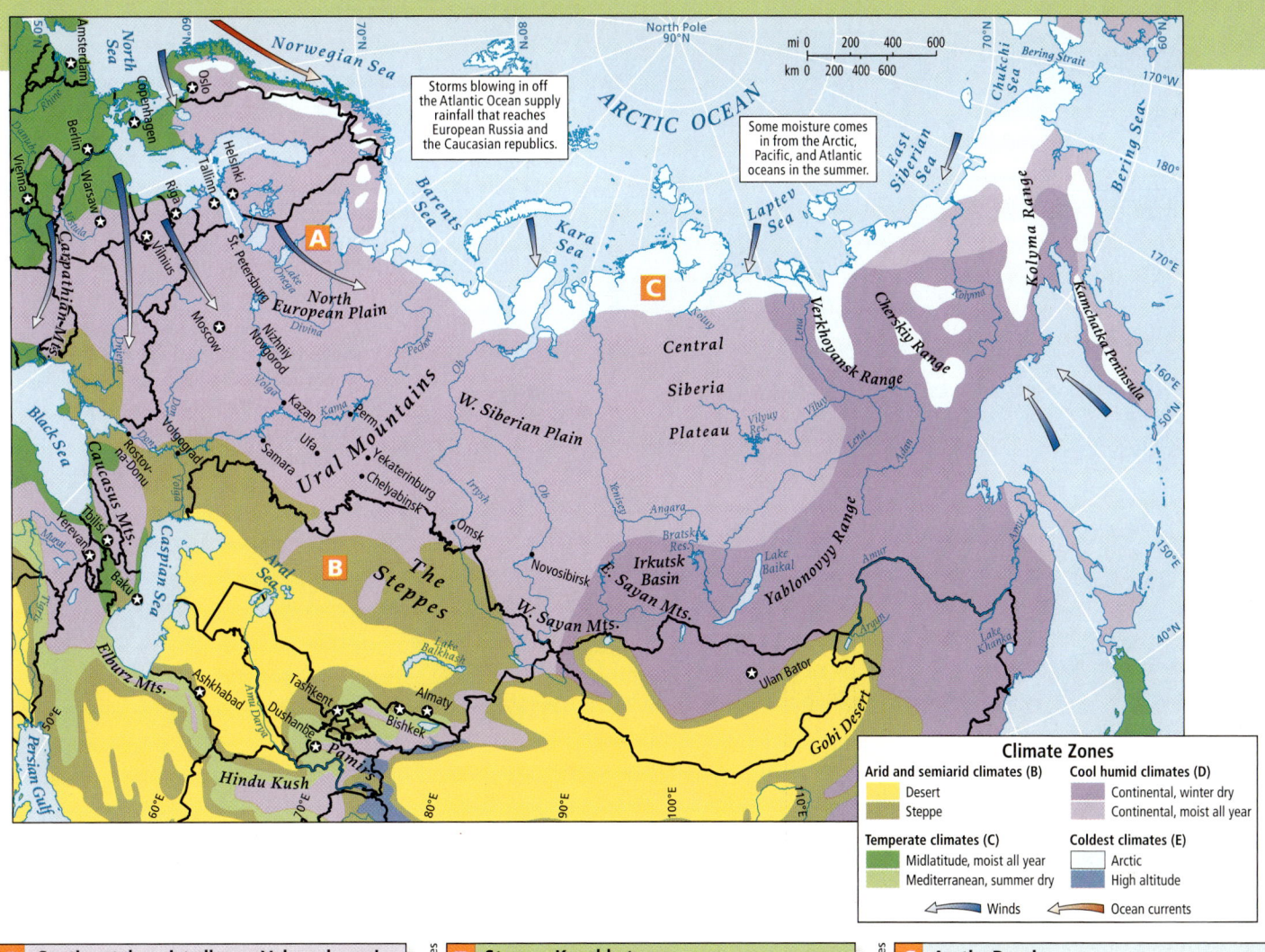

FIGURE 5.5   PHOTO ESSAY: Climates of Russia and the Post-Soviet States

Storms blowing in off the Atlantic Ocean supply rainfall that reaches European Russia and the Caucasian republics.

Some moisture comes in from the Arctic, Pacific, and Atlantic oceans in the summer.

Climate Zones

Arid and semiarid climates (B)
- Desert
- Steppe

Temperate climates (C)
- Midlatitude, moist all year
- Mediterranean, summer dry

Cool humid climates (D)
- Continental, winter dry
- Continental, moist all year

Coldest climates (E)
- Arctic
- High altitude

⇐ Winds   ⇐ Ocean currents

A Continental, moist all year, Vologada region

B Steppe, Kazakhstan

C Arctic, Russia

FIGURE 5.6 PHOTO ESSAY: Human Impacts on the Biosphere in Russia and the Post-Soviet States

This region is home to some of the worst pollution in the world. More than 70 years of industrial development with few environmental safeguards have wreaked havoc on ecosystems. Some attempts to become cleaner and less polluting are being made, but decades may pass before significant improvements are realized.

A A cloud of pollution from a nearby nickel smelter hangs over the city of Norilsk. The smelter emits about 1 percent of all global emissions of sulfur dioxide, which helps form acid rain. Enough heavy metals have accumulated in the soils around Norilsk that they can now be mined commercially. The company that runs the smelter has promised improvements but estimates that there will not be any significant changes in emissions until between 2015 and 2020.

B This false-color satellite image shows Norilsk. Purple indicates areas where vegetation has been killed by pollution. The city is located just below the main lake seen in the upper part of the image.

Human Impact, 2002

Land cover:
- Forests
- Grasslands
- Deserts
- Tundra
- Ice

Human impact on land:
- High impact
- Medium–high impact
- Low–medium impact

C The abandoned city of Pripyat in Ukraine, once home to 50,000 people, was evacuated after the explosion of the Chernobyl nuclear power plant, visible on the horizon.

D Oil production in the Russian Arctic. With such a short growing season, many arctic environments are extremely fragile. Some of the largest oil spills ever have happened here and have caused severe damage to ecosystems and to the indigenous people who depend on them. [Alexander Zemlianichenko Jr./Bloomberg via Getty Images]

more involved with their stomachs, they forget about ecology." Pollution controls are complicated by a lack of funds and by an official unwillingness to correct past environmental abuses. Figure 5.6 shows just a few examples of human impacts on the region's environment, most of which are related to ongoing industrial pollution (as in Norilsk, in Figure 5.6A, B), nuclear contamination (as in Pripyat, Ukraine, in Figure 5.6C), or the extraction of fossil fuels. The sale of oil and gas on the global market is now a major source of income for Russia and several of the post-Soviet states, yet relatively little attention is being paid even today to the associated environmental impacts (see Figure 5.6D).

Urban and Industrial Pollution

Urban and industrial pollution was often ignored during Soviet times as cities expanded quickly—with workers flooding in from the countryside—to accommodate the new industries that often generated lethal levels of pollutants. Citizens' concerns about pollution were suppressed and environmental movements such as those in North America and western Europe never developed.

It is often difficult to link urban pollution directly to health problems because the sources of contamination are multiple and difficult to trace. Such **nonpoint sources of pollution** include untreated automobile exhaust, raw sewage, and agricultural chemicals that drain from fields into water supplies. In all urban areas of the region, air pollution resulting from the burning of fossil fuels is skyrocketing as more people purchase cars and as the industrial and transport sectors of the economy continue to grow.

> **nonpoint sources of pollution** diffuse sources of environmental contamination, such as untreated automobile exhaust, raw sewage, and agricultural chemicals that drain from fields into water supplies

Some cities were built around industries that produce harmful by-products. The former chemical weapons–manufacturing center of Dzerzhinsk has been listed by the nonprofit Blacksmith Institute as one of the ten most polluted cities in the world. It is competing with the city of Norilsk, where much of the vegetation has been killed off around the city's metal-smelting complex—the largest facility of its kind in the world (see Figure 5.6B). Norilsk is widely considered to be the most polluted city in Russia.

Nuclear Pollution

Russia and the post-Soviet states are also home to extensive nuclear pollution, the effects of which have spread globally. The world's worst nuclear disaster occurred in Ukraine in 1986, when the Chernobyl nuclear power plant exploded. The explosion severely contaminated a vast area in northern Ukraine, southern Belarus, and Russia. It spread a cloud of radiation over much of Central Europe, Scandinavia, and eventually the entire planet. In the area surrounding Chernobyl, more than 300,000 people were evacuated from their homes, and Pripyat, which used to have a population of 50,000 people, has been an abandoned ghost town since the disaster (see Figure 5.6C). The ultimate health effects are impossible to assess exactly. The United Nations estimates that 6000 people developed cancer due to the accident, although many believe the number to be much higher. **278. SAFETY AT THE CENTER OF NUCLEAR POWER OPERATIONS WORLDWIDE**

When the Soviet Union collapsed, many of the post-Soviet states inherited nuclear facilities and even nuclear weapons. One such country is Kazakhstan (see Figure 5.3). Remote areas of eastern Kazakhstan used to serve as a testing ground for Soviet nuclear devices. The residents were sparsely distributed and no one took the trouble to protect them from nuclear radiation. A museum in Kazakhstan now displays the preserved remains of hundreds of deformed human fetuses and newborns.

On the positive side, Kazakhstan's government has disarmed the nuclear warheads that it inherited from the Soviet Union and has worked with U.S. authorities to safeguard its remaining nuclear weapons material. Kazakhstan plans to use its nuclear technology to become a global player in the nuclear energy sector. It is already the single biggest producer of uranium in the world, generating 35 percent of the global production total.

In addition to mining and exporting the raw material, Kazakhstan's government wants to produce nuclear fuel both for global markets and domestic nuclear power plants and to develop commercial repositories for radioactive waste from other countries. However, the global appetite for nuclear expansion remains uncertain after the 2011 Fukushima accident in Japan, and no reliably safe system has yet been found for storing nuclear waste until it is no longer radioactive. Environmental and human rights NGOs have raised strong opposition, pointing to the region's poor environmental track record. Kazakhstan's population may also be suspicious of nuclear power because they have lived for so long with land poisoned by radioactive waste.

The Globalization of Resource Extraction and Environmental Degradation

Russia and the post-Soviet states have considerable natural and mineral resources (see Figure 5.14 on page 211). Russia itself has the world's largest natural gas reserves, major oil deposits, and forests that stretch across the northern reaches of the continent. Russia also has major deposits of coal and industrial minerals such as iron ore and nickel. The Central Asian states share substantial deposits of oil and gas, which are centered on the Caspian Sea and extend east toward China. The value of all these resources is determined in the global marketplace, and their extraction affects the global environment.

Despite obvious environmental problems, cities like Norilsk (see Figure 5.6A), which sits on huge mineral deposits, continue to attract investment crucial to the new Russian economy. The area is rich in nickel—used in steel and other industrial products—and other minerals, such as copper. Norilsk Nickel, a company that was privatized at the end of communism and is now the largest producer of nickel in the world, dominates the city. With an extensive base of mineral resources, and unconstrained by meaningful environmental regulations, Russian-controlled Norilsk Nickel has used its enormous profits to become a global player in the mining industry. Despite its poor environmental record, it has purchased mining operations in Australia, Botswana, Finland, the United States, and South Africa and now employs more than 80,000 people worldwide.

Rivers, Irrigation, and the Loss of the Aral Sea To the south of Russia in the Central Asian states, the landlocked Aral Sea, once the fourth-largest lake in the world, was fed by the Syr Darya and Amu Darya rivers for millions of years. These rivers, which bring water from melting glaciers and snow in the mountains of Kyrgyzstan, Tajikistan, and Afghanistan, have been used to irrigate agriculture for thousands of years.

In the 1960s, the Soviet leadership ordered that the two rivers be diverted to irrigate millions of acres of cotton in naturally arid Kazakhstan and Uzbekistan. So much water was consumed by these projects that within a few years, the Aral Sea had shrunk measurably (Figure 5.7A). By the early 1980s, no water at all from the two rivers was reaching the Aral Sea, and by 2001, the sea had lost 75 percent of its volume and had shrunk into three smaller lakes. The region's huge fishing industry, which accounted for one-sixth of the fish catch in the entire USSR, died out because of the salinity of the water. Once-active port cities were marooned many kilometers from the water. The decline and disappearance of the Aral Sea has been described as the largest human-made ecological disaster in history.

The shrinkage of the Aral Sea has caused many human health problems. Winds that sweep across the newly exposed seabed pick up salt and chemical residues, creating poisonous dust storms. At the southern end of the Aral Sea in Uzbekistan, 69 percent of the people report chronic illnesses caused by air pollution, a lack of clean water, and underdeveloped sanitation.

Efforts to increase water flows to the Aral Sea have been hampered by politics because the now-independent countries of Uzbekistan and Kazakhstan share the sea. Uzbekistan, in particular, wants to continue its massive irrigation programs, which have made it the world's seventh-largest cotton grower. Nevertheless, some restorative actions have been taken. Kazakhstan, which

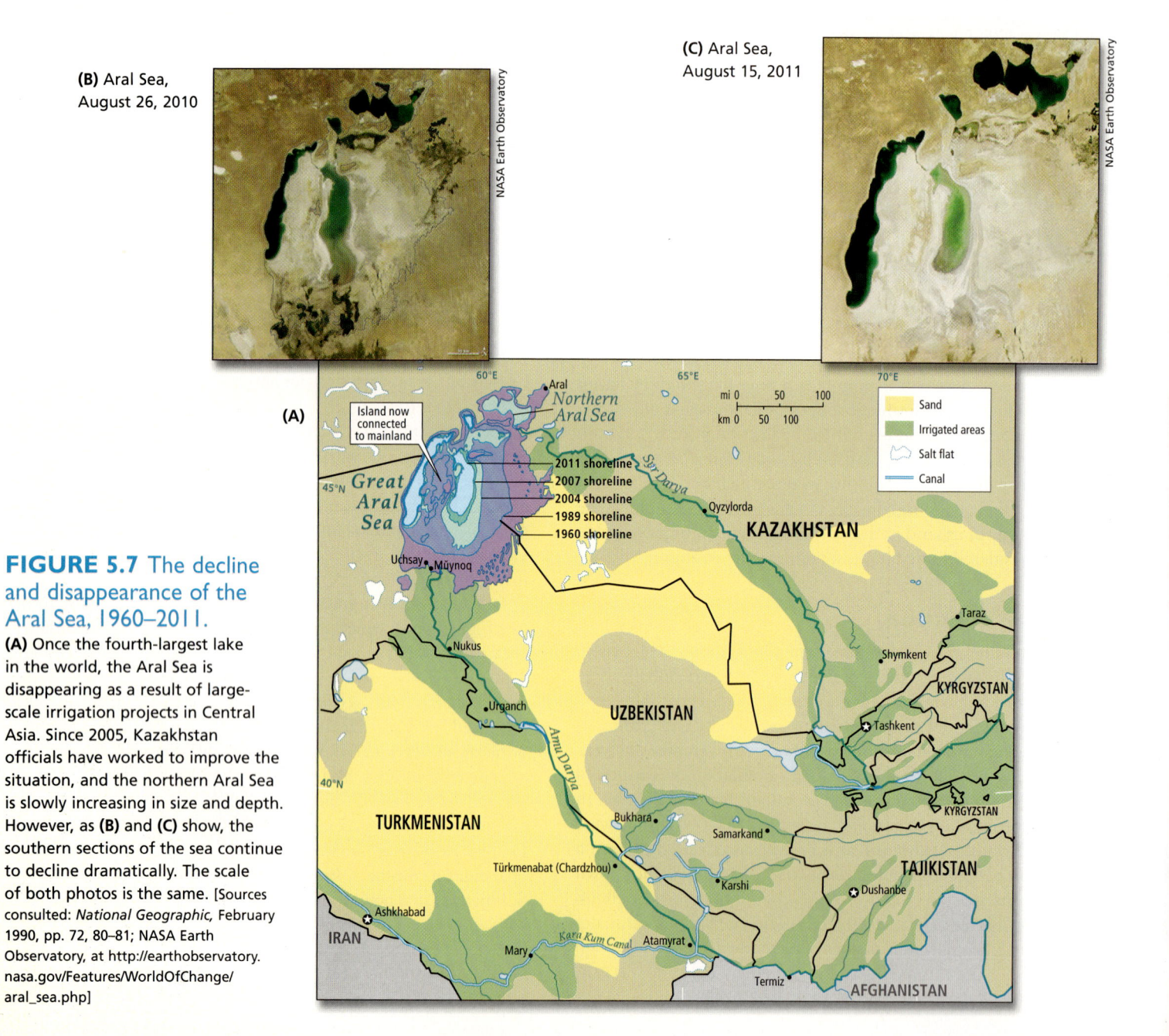

FIGURE 5.7 The decline and disappearance of the Aral Sea, 1960–2011.

(A) Once the fourth-largest lake in the world, the Aral Sea is disappearing as a result of large-scale irrigation projects in Central Asia. Since 2005, Kazakhstan officials have worked to improve the situation, and the northern Aral Sea is slowly increasing in size and depth. However, as **(B)** and **(C)** show, the southern sections of the sea continue to decline dramatically. The scale of both photos is the same. [Sources consulted: *National Geographic,* February 1990, pp. 72, 80–81; NASA Earth Observatory, at http://earthobservatory.nasa.gov/Features/WorldOfChange/aral_sea.php]

(B) Aral Sea, August 26, 2010

(C) Aral Sea, August 15, 2011

FIGURE 5.8 PHOTO ESSAY: Vulnerability to Climate Change in Russia and the Post-Soviet States

Much of the region has a medium level of vulnerability to climate change, but Central Asia has high and extreme levels of vulnerability. In Central Asia, vulnerability will increase as temperatures rise and glaciers in the Pamirs and other mountains melt. The dependence of so many Central Asians on agriculture makes them very sensitive to drought and water scarcity. Widespread poverty, combined with poorly developed disaster response systems, reduces the overall resilience to these and other disturbances.

A A river fed by glacial melting in Tajikistan. Most of Central Asia depends on rivers for drinking and irrigation water. If these rivers receive less water from glaciers, there will not be enough water to supply people and their crops.

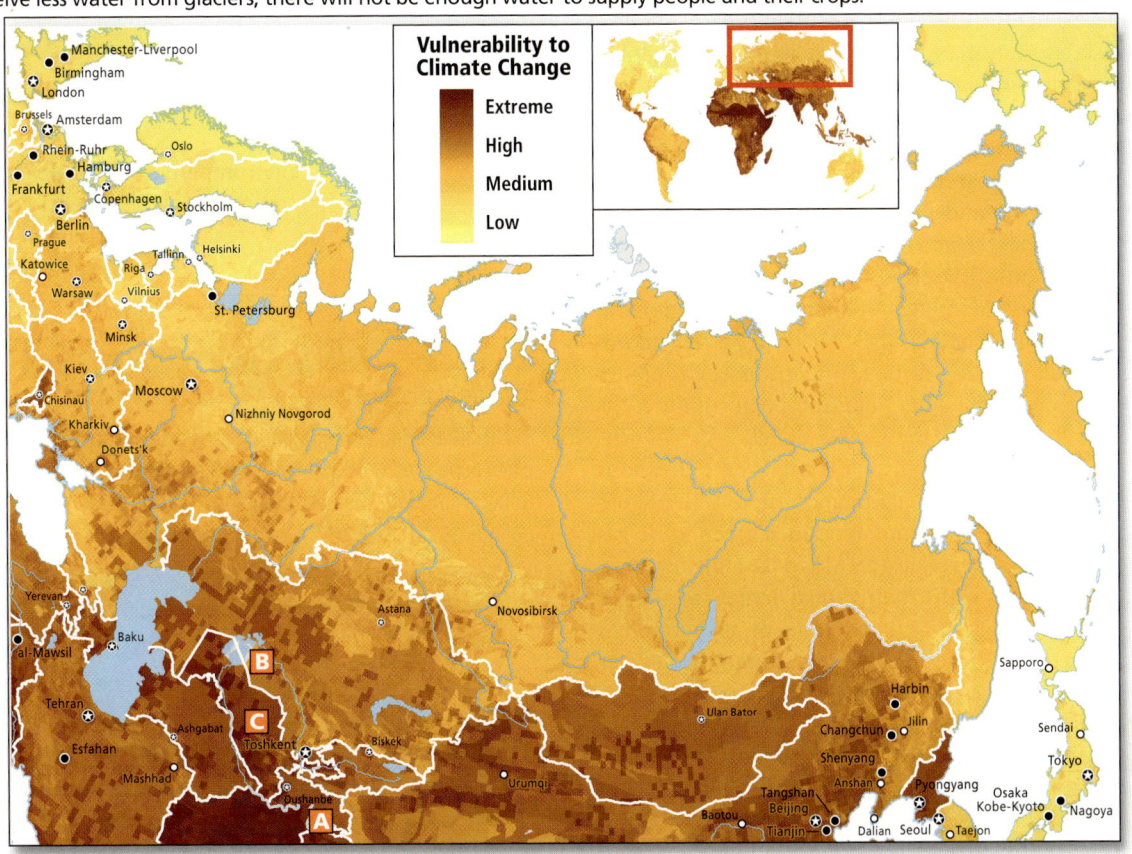

B A fishing boat stranded in what used to be part of the Aral Sea, which has shrunk dramatically because the river waters that feed the sea have been diverted to irrigate cotton and food crops. Even less water will reach the Aral Sea as glaciers continue melting.

C An irrigated cotton field in Uzbekistan. Forty-four percent of Uzbeks work in agriculture, and cotton is the country's leading export. If irrigation water were to become less available, the livelihoods of millions of Uzbeks, 33 percent of whom live below the poverty line, would be threatened.

relies on oil more than cotton for income, built a dam in 2005 to keep fresh water in the northern Aral Sea. By the following year, the water had risen 10 feet, and the fish catch was improving. Kazakh fishers note that there is now more open public debate about what to do next regarding the Aral Sea. In the southern Aral Sea, however, water levels remain very low (see Figure 5.7B, C). In Uzbekistan, for instance, where 44 percent of the workforce is employed in agriculture, the limitation of irrigation has forced economic diversification, and cotton production now makes up just 11 percent of the country's exports, far less than the 45 percent it represented in 1990. **107. DYING SEA MAKES COMEBACK**

Climate Change

Climate change is an issue in this region for many reasons. Although high levels of CO_2 and other greenhouse gases are produced here, Russia's forests play a major role in absorbing global CO_2. And while portions of the region are only somewhat vulnerable to the negative effects of a changing climate (shown on the map in Figure 5.8), Central Asia, which is prone to drought and water scarcity, has high to extreme levels of vulnerability.

Even though its CO_2 and other greenhouse gas emissions have declined since the Soviet era, Russia still has the fifth-highest rate of greenhouse gas emissions in the world. This is due in part to Russia's use of energy-inefficient technologies and practices such as burning off or "flaring" of natural gas that comes to the surface when oil wells are drilled. While Russia has higher greenhouse gas emissions per capita than European countries, its levels are not as high as those in the United States.

Russia has shown some willingness to limit its own emissions by, for example, signing international treaties, such as the Kyoto Protocol and the Copenhagen Accord, to reduce greenhouse gas emissions. Such treaties, however, use 1990 as a baseline for emissions. Russia's compliance with international targets to reduce greenhouse emissions is largely an outcome of the country's post-1990 economic decline after the end of communism. Lower levels of industrial output have meant lower emissions.

Central Asia is particularly vulnerable to the impacts of climate change because of its dependence on water from rivers that are fed by glaciers in the Pamirs and other mountains in Kyrgyzstan and Tajikistan (see Figure 5.8). In recent decades, the glaciers have shrunk because of more extreme temperature ranges and decreases in rain and snowfall. In this area, most precipitation falls in the mountains in the winter and spring. The glaciers act as water-storage systems, supplying melted ice flow to the rivers during the summer, when irrigation is most needed for growing crops. If current trends continue, the glaciers will melt earlier in the year, resulting in lower river flows in the summer. Central Asia's agricultural systems would either have to adapt to a spring growing season (winters are too cold), or farmers would have to store water for use in the summer. Either proposition would demand complex and expensive changes on a massive scale.

ON THE BRIGHT SIDE

Better Times Ahead?

While environmental pollution remains serious throughout this region, recent surveys suggest that pro-environmental attitudes are increasing. Awareness of environmental issues is rising, as is the willingness to act. For example, a survey by Alinga Consulting Group, based in Moscow, showed that the proportion of Russians willing to work to solve environmental problems went from 26 percent in 2005 to 30 percent in 2009. The proportion unwilling to take part in such activities went down from 56 percent to 47 percent over the same period. These and other surveys also indicate that Russians are most willing to take action on environmental issues at the local level.

THINGS TO REMEMBER

› **GEOGRAPHIC INSIGHT 1**
- **Environment** Economic development has taken precedence over environmental concerns in Russia and the post-Soviet states for decades, resulting in numerous environmental problems such as severe pollution of the air and water by toxic industrial waste. This region's contribution to climate change is increasing as more people use automobiles for transportation. In parts of this region, food production systems and water resources are highly vulnerable to the impacts of climate change.

- Urban and industrial pollution was often ignored during Soviet times as cities expanded quickly—with workers flooding in from the countryside—to accommodate the new industries that often generated lethal levels of pollutants.

- The world's worst nuclear disaster occurred in Ukraine in 1986, when the Chernobyl nuclear power plant exploded, severely contaminating a vast area in northern Ukraine, southern Belarus, and Russia.

- Massive water diversions from rivers that once fed the Aral Sea began in the 1960s when the Soviet leadership ordered millions of acres to be irrigated for cotton production in naturally arid Kazakhstan and Uzbekistan.

- Central Asia is especially vulnerable to climate change, given its dependence on irrigated agriculture that uses water from rivers fed by glacial melt.

HUMAN GEOGRAPHY

The goals of the Soviet experiment begun in 1917 were unique in human history: to quickly and totally reform an entire human society. The Soviet Union's collapse brought an even more rapid shift from centrally planned economies to market economies. The transition remains incomplete and for many, life is proceeding amid economic, political, and social uncertainty.

HUMAN PATTERNS OVER TIME

The core of the entire region has long been European Russia, the large and densely populated homeland of the ethnic Russians. Expanding gradually from this center, the Russians conquered a large area inhabited by a variety of other ethnic groups. These conquered territories remained under Russian control as part of the Soviet Union (1917–1991), which attempted to create an integrated social and economic unit out of the disparate territories. The breakup of the Soviet Union has diminished Russian domination throughout this region, though Russia's influence is still considerable.

The Rise of the Russian Empire

For thousands of years, the militarily and politically dominant people in the region were **nomadic pastoralists** who lived on the meat and milk provided by their herds of sheep, horses, and other grazing animals, and used animal fiber to make yurts, rugs, and clothing. As possibly the first people to domesticate horses, their movements followed the changing seasons across the wide grasslands that stretch from the Black Sea to the Central Siberian Plateau. The nomads would often take advantage of their superior horsemanship and hunting skills to plunder settled communities. To defend themselves, permanently settled peoples gathered in fortified towns.

Towns arose in two main areas: the dry lands of greater Central Asia and the forests of Caucasia, Ukraine, and Russia. As early as 5000 years ago, Central Asia had settled communities that were supported by irrigated croplands (see Figure 5.10A). These communities were enriched by their locations, which were central for trade along what became known as the Silk Road, that vast, ancient, interwoven ribbon of major and minor trading routes between China and the Mediterranean that also had lesser connections to other places (Figure 5.9). Many commodities were traded along the route, but the eastern tradition of silk fabric became highly valued in the West, which is why the term Silk Road was applied by later historians.

About 1500 years ago, the **Slavs**, a group of farmers, including those known as the Rus (possibly of Scandinavian origin), emerged in what is now Poland, Ukraine, and Belarus. They moved east, founding numerous settlements, including the towns of Kiev in about 480 C.E. and Moscow in 1100. By 600, Slavic trading towns were located along all the rivers west of the Ural Mountains. The Slavs prospered from a lucrative trade route over land and along the Volga River that connected Scandinavia (North Europe) and Central and Southwest Asia (via Constantinople, modern-day Istanbul). Powerful kingdoms developed in Ukraine and European Russia. In the mid-800s, Greek missionaries introduced both Christianity (now known as "Orthodox Christianity") and the Cyrillic alphabet that we now associate with the Russian language (Figure 5.10B). The Cyrillic alphabet is still used in most of the region's countries.

In the twelfth century, the Mongol armies of Genghis Khan conquered the forested lands of Ukraine and Russia. The **Mongols** were a loose confederation of nomadic pastoral people centered in East and Central Asia. Moscow's rulers became tax gatherers for the Mongols, dominating neighboring kingdoms

> **nomadic pastoralists** people whose way of life and economy are centered on tending grazing animals who are moved seasonally to gain access to the best pasture
>
> **Slavs** a group of people who originated between the Dnieper and Vistula Rivers in modern-day Poland, Ukraine, and Belarus
>
> **Mongols** a loose confederation of nomadic pastoral people centered in East and Central Asia, who by the thirteenth century had established by conquest an empire that stretched from Europe to the Pacific

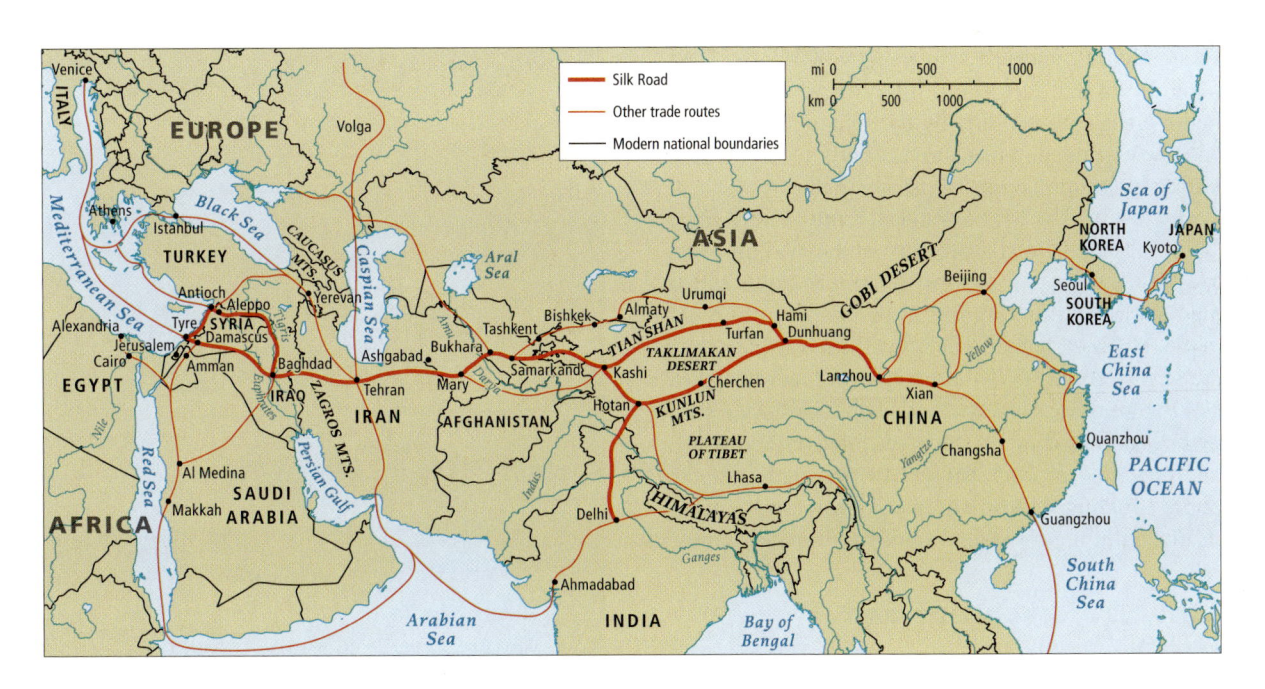

FIGURE 5.9 The ancient Silk Road and related trade routes. Merchants who worked the Silk Road rarely traversed the entire distance. Instead, they moved back and forth along part of the road, trading with merchants to the east and west. [Source consulted: "Maps of the Silk Road," the Silk Road Project and the Stanford Program on International and Cross-Cultural Education, at http://www.silkroadproject.org/tabid/177/default.aspx]

A The fort at Bukhara, Uzbekistan, founded in 500 B.C.E. along the Silk Road. [placchic/Flickr/Getty Images]

B An Orthodox Christian monastery in Russia, founded around 1500 C.E. [Nick Laing/AWL Images/Getty Images]

C The summer palace of Czar Peter the Great in St. Petersburg, built from 1714 to 1755. [Martin Child/Photodisc/Getty Images]

3000 B.C.E. Settled communities in Central Asia, supported by irrigated agriculture and trade on the Silk Road

500 B.C.E. Fort at Bukhara founded along the Silk Road

500 C.E. Slavs emerge

1500 C.E.

1598–1689 Russian annexation of western Siberia

FIGURE 5.10 VISUAL HISTORY OF RUSSIA AND THE POST-SOVIET STATES

and eventually growing powerful enough to challenge local Mongol rule. In 1552, the Slavic ruler Ivan IV ("Ivan the Terrible") conquered the Mongols, marking the beginning of the Russian empire. St. Basil's Cathedral, a major landmark in Moscow, commemorates the victory (Figure 5.11).

By 1600, Russians centered in Moscow had conquered many former Mongol territories, integrating them into the growing empire and extending it to the east, as shown in Figure 5.12. The first major non-Russian area to be annexed was western Siberia (1598–1689). Russian expansion into Siberia (and even into North America, along its Pacific coast) resembled the spread of European colonial powers throughout Asia and the Americas. Russian colonists took land and resources from the Siberian populations and treated the people of those cultures as inferiors. Moreover, migrations of laborers from Russia to Siberia meant that indigenous Siberians were vastly outnumbered by the eighteenth century. By the mid-nineteenth century, Russia had also expanded its reach to the south in order to gain control of Central Asia's cotton crop, its major export.

The Russian empire was ruled by a powerful monarch, the **czar**, who lived in splendor (along with a tiny aristocracy) while the vast majority of the people lived short, brutal lives in poverty (see Figure 5.10C, D). Many Russians were *serfs* who were legally bound to live on and farm land owned by an aristocrat. If the land was sold, the serfs were transferred with it. Serfdom was legally ended in the mid-nineteenth century. However, the inequities of Russian society persisted into the twentieth century, fueling opposition to the czar. By the early twentieth century, a number of violent uprisings were underway.

czar the title of the ruler of the Russian empire

The Communist Revolution and Its Aftermath

Periodic rebellions against the czars and elite classes took place over the centuries. By the mid-nineteenth century, these rebellions were being led not only by serfs, but also by members of the tiny educated

FIGURE 5.11 St. Basil's Cathedral and the Red Square, Moscow. Now the most recognized building in Russia, St. Basil's Cathedral was built between 1555 and 1561 to commemorate the defeat of the Mongols by Ivan the Terrible, the first Russian czar. A central chapel, capped by a pyramidal tower, stands amid eight smaller chapels, each with a colorful "onion dome." Each chapel commemorates a saint on whose feast day Czar Ivan had won a battle.

D Poor men and women, known as Burlaks, hauling freight along the Volga River in 1873, as depicted by the Russian painter Ilya Repin. [Ilya Efimovich Repin/The Bridgeman Art Library/Getty Images]

E A World War II–era poster of a Soviet soldier. [Galerie Bilderwelt/Getty Images]

F A journey to the International Space Station begins at Baikonur Cosmodrome in 2012. [Bill Ingalls/NASA/Getty Images]

1714–1755

1796–1914 Russian annexation of Central Asia and Caucasia

1873

1917 Bolshevik Revolution

1922–1953 Stalin era

1941–1945

1945–1991 Cold War

2012

---

and urban middle class. In 1917, at the height of Russian suffering during World War I, Czar Nicholas II was overthrown in a revolution. What followed was a civil war between rival factions with different ideas about how the revolution should proceed. Eventually the revolution brought a complete restructuring of the Russian economy and society, according to what at first promised to be a more egalitarian model.

Of the disparate groups that coalesced to launch the Russian revolution, the **Bolsheviks** succeeded in gaining control during the post-revolution civil war. The Bolsheviks were inspired by the principles of **communism** as explained

Bolsheviks a faction of Communists who came to power during the Russian Revolution

communism an ideology, based largely on the writings of the German revolutionary Karl Marx, that calls on workers to unite to overthrow capitalists and establish an egalitarian society where workers share what they produce

capitalists usually a wealthy minority that owns the majority of factories, farms, businesses, and other means of production

by the German revolutionary philosopher Karl Marx. Marx criticized the societies of Europe as inherently flawed because they were dominated by **capitalists**—the wealthy minority who owned the factories, farms, businesses, and other means of production. Marx pointed out that the wealth of capitalists was actually created in large part by workers, who nonetheless remained poor because the capitalists undervalued their work. Under communism, workers were called upon to unite to overthrow governments that supported the capitalist system, take over the means of production (farms, factories, services), and establish a completely egalitarian society

FIGURE 5.12 Russian imperial expansion, 1300–1945. A series of powerful rulers expanded Russia's holdings across Eurasia to the west and east. Expansion was particularly vigorous after 1700, when Russia acquired Siberia. Note that the centuries-long imperial expansion was reversed in 1991, when the Soviet Union was disbanded and the new post-Soviet states were created. [Source consulted: Robin Milner-Gulland with Nikolai Dejevsky, *Cultural Atlas of Russia and the Former Soviet Union*, rev. ed. (New York: Checkmark Books, 1998), pp. 56, 74, 128–129, 177]

207

without government or currency. The philosophy held that people would work out of a commitment to the common good, sharing whatever they produced so that each had their basic needs met.

One Bolshevik leader, Vladimir Lenin, argued that the people of the former Russian empire needed a transition period in which to realize the ideals of communism. Accordingly, Lenin's Bolsheviks formed the **Communist Party**, which set up a powerful government based in Moscow. Lenin believed that the government should run the economy, taking land and resources from the wealthy and using them to benefit the poor majority.

The Stalin Era After Lenin suffered a series of strokes in 1922, Joseph Stalin stepped in to reorganize the society and economy in a highly authoritarian style. His 31-year rule of the Soviet Union brought a mixture of brutality and revolutionary change that set the course for the rest of the Soviet Union's history. He sought to increase general prosperity through rapid industrialization made possible by a **centrally planned**, or **socialist, economy**. The state owned all real estate and means of production, and government officials in Moscow managed all economic activity. This system became known as the *command economy*. In this economy, central control was used to determine where all factories, apartment blocks, and transport infrastructure should be located, and to manage all production, distribution, and pricing of products. The idea was that under a socialist system, the economy would grow more quickly, which would hasten the transition to the idealized communist state in which everyone shared equally. This notion was reflected in the new name chosen for the country: the Union of Soviet Socialist Republics ("Soviet" means, roughly, "council").

The centerpiece of Stalin's vision was massive government investment in gigantic development projects, such as factories, dams, and chemical plants, some of which are still the largest of their kind in the world. To increase agricultural production, he forced farmers to join large, government-run collectives.

Stalin's strategy of government control met many of his expectations. It brought rapid economic development and higher standards of living. Schools were provided for previously uneducated poor and rural children, and contributed to major technological and social advances. During the Great Depression of the 1930s, the Soviet Union's industrial productivity grew steadily, allowing for the development of a large military, even while the economies of other countries stagnated.

However, Stalin's economic development model also had deep flaws. With production geared largely toward heavy industry (the manufacture of machines, transportation equipment, military vehicles, and armaments), less attention was paid to the demand for consumer goods and services that could have further improved living standards. Ultimately, the Soviet Union proved to be less innovative than non-communist countries in western Europe and less able to develop advanced technologies to sustain economic growth over time.

The most destructive aspect of Stalin's rule, however, was his ruthless use of the secret police and mass executions to silence anyone who opposed him. Those who resisted were forced to work in prison labor camps, the so-called *gulags*, many of which were located in Siberia. Prisoners there mined for minerals and also built and worked in newly constructed industrial cities. Millions were executed or died from overwork under harsh conditions. Estimates vary greatly, but between 20 and 60 million people were killed as a result of Stalin's policies.

World War II and the Cold War

During World War II, the Soviet Union did more than any other country to defeat the armies of Nazi Germany (see Figure 5.10E). A failed attempt to conquer the Soviet Union exhausted Hitler's war machine. In the process of defeating Germany on the Eastern Front, 23 million Soviets were killed, more than all the other European casualties combined. After the war, Stalin was determined to erect a buffer of allied communist states in Central Europe that would be the battleground of any future war with Europe. Because they were unwilling to risk another war, the leaders of the United States and the United Kingdom ceded control of much of Central Europe to the Soviet Union while they busied themselves with reconstructing western Europe. However, when Stalin's intention to utterly dominate Central Europe became clear, the United States and its allies organized to stop the Soviets from extending their power even further into Europe and elsewhere. The result was a politically and economically divided Europe separated by an "Iron Curtain" and the **Cold War**, a nearly 50-year-long global geopolitical rivalry that pitted the Soviet Union and its allies against the United States and its allies around the world (**Figure 5.13**).

A wide variety of economic and political problems led to the eventual collapse of the Soviet Union. Fundamental flaws with Soviet central planning led to innumerable inefficiencies and chronic mismanagement compared to most market economies. This was recognized during Mikhail Gorbachev's tenure as president of the USSR (1985 to 1991), and a private market economy began to be permitted. Under an overall policy of **perestroika**, or "restructuring," government decision making was decentralized and private entrepreneurs were allowed to legally sell goods and services. Gorbachev also introduced policies collectively known as **glasnost**, or "openness," which encouraged more transparency, openness, and publicity in the workings of all levels of the Soviet government. However, many problems remained, such as the diversion of scarce resources to the military, including the Soviet space program, and away from much-needed economic and social development (see Figure 5.10F).

Communist Party the political organization that ruled the USSR from 1917 to 1991; other communist countries, such as China, Mongolia, North Korea, and Cuba, also have ruling Communist parties

centrally planned, or **socialist, economy** an economic system in which the state owns all land and means of production, while government officials direct all economic activity, including the locating of factories, residences, and transportation infrastructure

Cold War the contest that pitted the United States and western Europe, who were espousing free market capitalism and democracy, against the USSR and its allies, who were promoting a centrally planned economy and a socialist state

perestroika literally, "restructuring"; the restructuring of the Soviet economic system that was done in the late 1980s in an attempt to revitalize the economy

glasnost literally, "openness"; the policies instituted in the late 1980s under Mikhail Gorbachev that encouraged more transparency and openness in the workings of all levels of the Soviet government

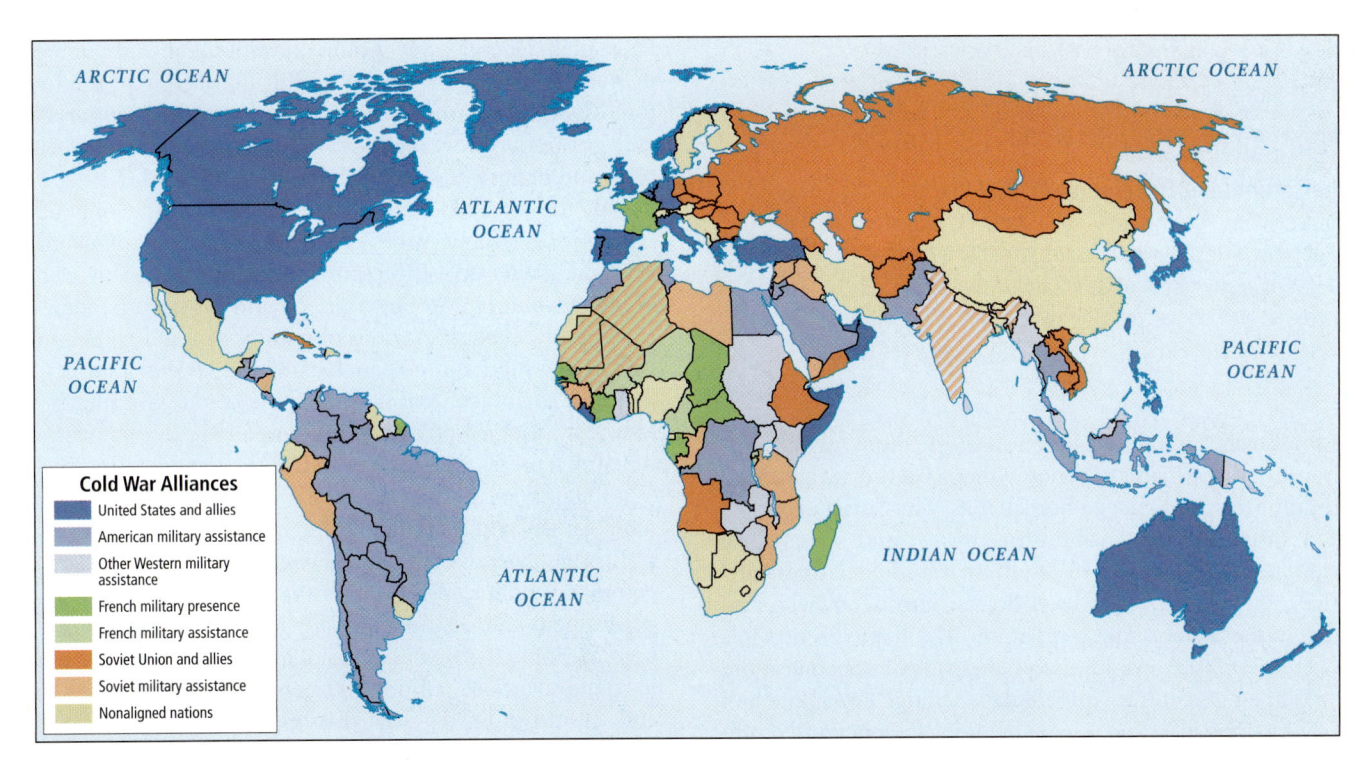

FIGURE 5.13 The Cold War in 1980. In the post–World War II contest between the Soviet Union and the United States, both sides enticed allies through economic and military aid. The group of countries militarily allied with the Soviet Union was known as the Warsaw Pact. Some countries remained nonaligned. [Source consulted: Clevelander, at http://en.wikipedia.org/wiki/Image:New_Cold_War_Map_1980.png]

Soviet efforts to promote communism in less developed countries were also expensive, as was extensive political, economic, and military coercion in Central Europe and Central Asia. As problems multiplied, so did resistance to Soviet control, especially in Central Europe in the late 1980s. Because Gorbachev did not resort to violence to control the countries of Central Europe, they were able to abandon communism and instead establish democracy. Eventually, this dissension spread to the republics that made up the Soviet Union itself, and by 1991, the USSR had formally dissolved. **104. FORMER SOVIET UNION LAUNCHED SPACE AGE 50 YEARS AGO**

A war in Afghanistan also helped to stretch the USSR past the breaking point (see Figure 5.16A on page 215). In the 1970s in Afghanistan, which lies just to the south of the Central Asian states, there was contention among a small group who favored a Western-style democratic government, those who favored communism, and Muslims who favored a theocratic state based on Islamic law. The Soviets feared that the strongly anti-communist Muslim movement in Afghanistan would influence nearby Central Asian Soviet republics to unite under Islam and rebel against the USSR. In response, the USSR backed an unpopular communist government that allowed Soviet military bases in Afghanistan. This brought strong resistance from the Muslim fundamentalist mujahedeen movement, and in 1979, Russia launched a war to support the Afghan government. A bloody, decade-long conflict ensued, concluding with the Soviets being defeated by Afghan guerillas who were financed, armed, and trained by the United States. Within 2 years of the Soviet withdrawal from Afghanistan, the USSR disintegrated. The Soviet–Afghan conflict and the aftermath are discussed further in Chapter 8.

THINGS TO REMEMBER

- As early as 5000 years ago, Central Asia had settled communities that were supported by irrigated croplands and enriched by trade along what became known as the Silk Road.

- The Russian empire (1598–1917) was ruled by a powerful monarch, the czar, who lived in splendor (along with a tiny aristocracy) while the vast majority of the people lived in poverty. Many Russians were serfs who were legally bound to live and work on farm land owned by an aristocrat.

- The Bolsheviks, a group inspired by the principles of communism, were the dominant leaders of the Russian Revolution of 1917. Their goal, which was never achieved, was an egalitarian society where people would work out of a commitment to the common good, sharing whatever they produced.

- Taking control in 1922, Joseph Stalin brought a mixture of brutality and revolutionary change that set the course for the rest of the Soviet Union's history. He established a centrally planned, or socialist, economy in which the state owned all property and means of production, while government officials in Moscow directed all economic activity. This system became known as a command economy.

- After World War II, a nearly 50-year-long global geopolitical rivalry known as the Cold War pitted the Soviet Union and its allies against the United States and its allies.

GLOBALIZATION AND DEVELOPMENT

> **GEOGRAPHIC INSIGHT 2**
>
> **Globalization and Development:** After the fall of the Soviet Union, economic reforms and globalization changed patterns of development in this region. Wealth disparity increased and jobs were lost as many Communist-era industries were closed or sold to the rich and well connected. The region is now largely dependent on its role as a leading global exporter of energy resources.

When the Soviet Union disbanded in 1991, there was a disorderly transition to a market economy that gave the advantage to a small number of well-placed bureaucrats. The Soviet economy consisted almost entirely of industries owned and operated by the government. In the early 1990s, these were sold at drastically low prices, often to the high-level Soviet bureaucrats who had run them in the old command economy. The hope was that this process of **privatization** would enable industries to operate more efficiently in a competitive "free market" setting. Two and a half decades later, there have been some gains in efficiency and profitability, mostly only in a few key export-oriented industries. The most widespread effect of privatization was to make a small number of bureaucrats fabulously rich very quickly. Many became what are known as **oligarchs**—individuals who are so wealthy and politically connected that they wield enormous, often clandestine power. These oligarchs continue to exercise power in government and private enterprise.

The transition to a market economy took place so rapidly that it was known as "shock therapy." A wide range of policies attempted to revitalize the sluggish economy by moving as quickly away from the old command economy as possible. Today, approximately 65 percent of Russia's economy is in private hands, a dramatic change from the virtually 100 percent state-owned economy of 1991.

Unfortunately, shock therapy came at steep cost for many Russians. A major part of the reforms was the abandonment of government price controls that once kept goods affordable to all. Instead, newly privatized businesses determined prices in the environment of great uncertainty created by shock therapy. This led to skyrocketing prices for the many goods that were in high demand but also in short supply. While a few people became rich, many lost their savings because of inflation and could barely pay for basic necessities such as food. Toward the end of the 1990s, as opportunities opened and competition developed, the supply of goods increased and prices fell. But even today, the effects of shock therapy are still being felt. For example, Russian economists estimate that even though average incomes have grown by 45 percent since 1991, almost all of this gain has gone to the wealthiest 20 percent of the population; the bottom 20 percent of the population is earning about half what they did in 1991.

Oil and Gas Development: Fueling Globalization

Natural gas and crude oil have emerged as the region's most lucrative exports, introducing a new wave of globalization and fossil fuel–dependent economic development. Although control of Russia's oil and gas resources still lies largely in the hands of the Russian government, the struggle to control Central Asia's oil and gas resources is ongoing, involving multinational corporations and many foreign governments (Figure 5.14).

By the year 2000, after an initial decade of economic instability, rising revenues from oil and gas began to finance Russia's economic recovery. In response, Russia's central government tightened control over the entire oil and gas sector. Today, Russia is the world's largest exporter of natural gas, and taxes on energy companies fund more than half of the federal budget. While this does guarantee that at least some of the revenues from the new fossil fuel–based industries benefit Russia's population, the remaining profits are concentrated in the hands of a few oligarchs.

Russia's relationship with Europe is intertwined with the lucrative oil and gas trade. The majority state-owned **Gazprom** is Russia's largest company as well as the world's largest gas company, and because Gazprom is state controlled, the Russian government can use the company to promote the country's geopolitical interests. Russia occasionally tries to use its gas exports to manipulate the politics and economies of Ukraine, Belarus, and countries in Central Europe. These countries are all much more reliant on gas from Gazprom than is the rest of Europe. EU member states receive about 25 percent of their natural gas from Gazprom, with Central Europe much more dependent on Gazprom than other parts of the EU. Russia is also very reliant on this trade; approximately 30 percent of Gazprom's profits come from sales to the European Union.

> **privatization** the sale of industries that were formerly owned and operated by the government to private companies or individuals
>
> **oligarchs** in Russia, those who acquired great wealth during the privatization of Russia's resources and who use that wealth to exercise power
>
> **Gazprom** in Russia, the state-owned energy company; it is the largest gas entity in the world

In 2009, Russia curtailed Ukraine's access to Gazprom's gas in a dispute over prices and as a response to Ukraine moving closer to the EU rather being a Russian ally. Because pipelines that supply Europe with gas run through Ukrainian territory, gas shortages also developed in parts of Europe. The EU became concerned about its reliance on Russian energy and is now rapidly turning to other sources of fossil fuels as well as renewable energy sources. The political dimensions of these issues are discussed further on page 214.

Central Asia has yet to find stability because a tug-of-war has evolved between Russia and foreign multinational energy corporations over the right to develop, transport, and sell Central Asian oil and gas resources. The new Central Asian states do not have the capital to develop the resources themselves, yet individuals and special interests in the Central Asian countries are reaping enormous financial rewards from selling the rights to exploit oil and gas to outsiders.

Because Central Asia is landlocked, the only way to get oil and gas from there to world markets is via pipelines (see Figure 5.14), the routes of which have become a major source of contention. Russia, the United States, the European Union, Turkey, China, and India have all developed or proposed various routes. A 1900-mile (3000-kilometer) pipeline from the Caspian Sea to China serves the growing demand for energy in China.

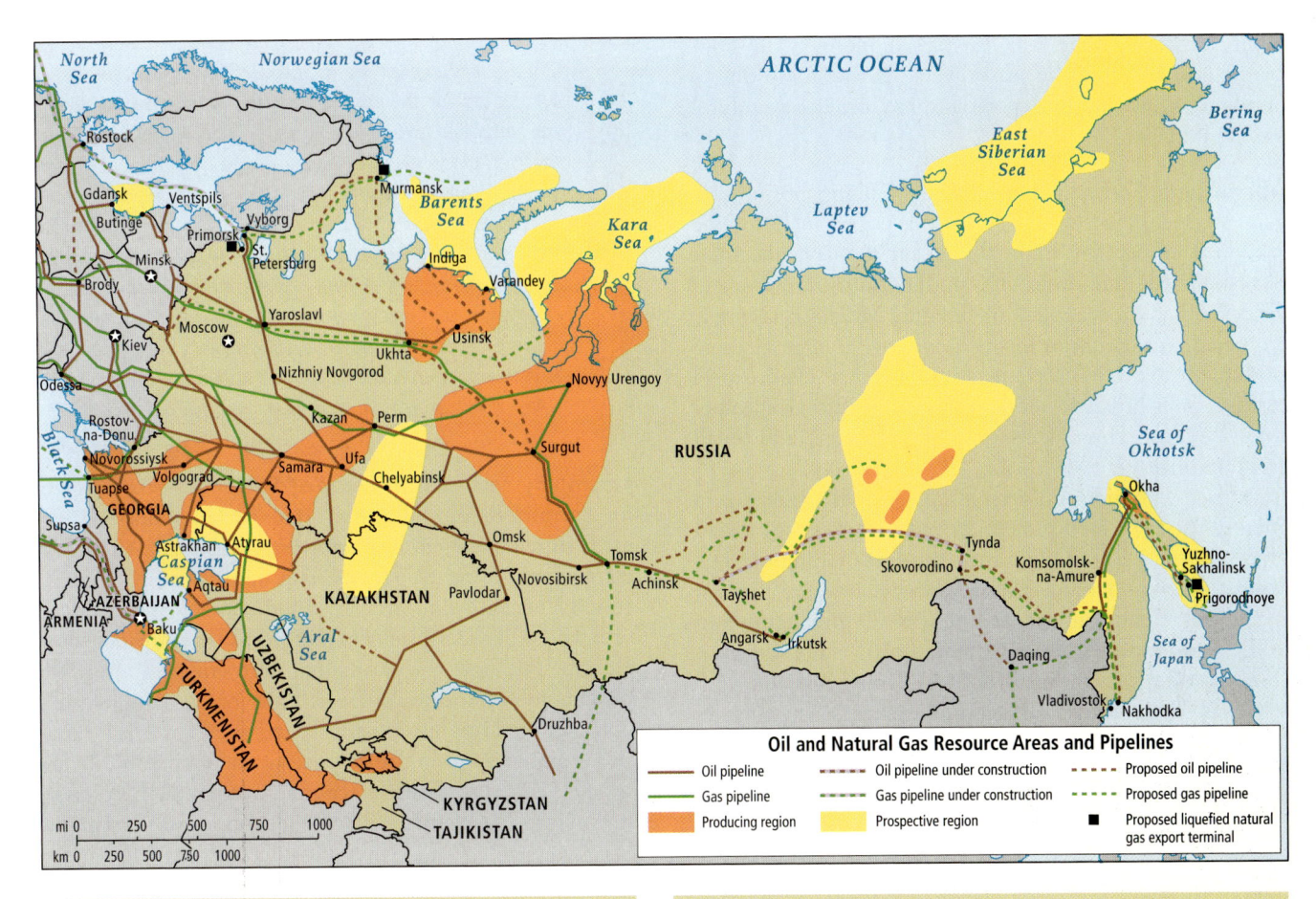

Oil and Natural Gas Resource Areas and Pipelines

Oil in Russia (2011–2012)	
Oil reserves	60 billion barrels
Oil reserves as percentage of world	4.07 percent
Saudi Arabian reserves	267.02 billion barrels
U.S. reserves	22.3 billion barrels (proved 2009)
Oil production	10,228 million barrels per day
Oil production as percentage of world	11.75 percent
U.S. oil production	5512 million barrels per day (2010)
Oil exports	7.083 million barrels per day
Oil exporter, rank	8
Oil exports to U.S.	580,000 barrels per day (2009)

Natural Gas in Russia (2011–2012)	
Gas reserves	1680 trillion cubic feet
Gas reserves as percentage of world	25.05 percent
Iranian reserves	1046 trillion cubic feet
U.S. reserves	272.5 trillion cubic feet
Gas production	23,647 billion cubic feet
Gas production as percentage of world	21.14 percent
U.S. gas production	1103.2 billion cubic feet (2009)
Gas exports	5753 billion cubic feet
Gas exporter, rank	90
Gas exports, to Europe	4948 billion cubic feet

FIGURE 5.14 Oil and natural gas: Russia and the post-Soviet states' resources and pipelines. For additional information and maps, see http://www.eia.gov/countries/. [Source consulted: U.S. Department of Energy, Energy Information Administration, "Russia Country Analysis Brief," May 2008, at http://www.eia.doe.gov/emeu/cabs/Russia/images/Russian%20Energy%20at%20a%20Glance%202007.pdf]

The United States maintains a military presence to protect its interests in countries such as Georgia and Uzbekistan, both of which, along with Turkey, have pipelines to Europe. Russia has built pipelines that traverse its own territory and reach into Europe. Each of these parties has made numerous efforts to encourage Central Asian countries to use their pipelines instead of those of their competitors. **108. ENERGY REVENUES AND CORRUPTION INCREASE IN RUSSIA**

Russia's Relations with the Global Economy Both the European Union and the United States want to ensure that Russia's oil and gas wealth does not finance a return to the hostile relations of the Cold War. For this reason, Russia was invited to join the World Trade Organization (WTO) and the **Group of Eight (G8)**, an organization of eight affluent countries with large economies.

Looking to form an economic counterweight to the United States and the European Union, Russia has recently been meeting with Brazil, India, and China to form a trade consortium known as BRIC. All four countries are large, populous, and have had spectacular economic growth over the last decade, making the alliance a symbol of the growing power of the

Group of Eight (G8) an organization of eight countries with large economies: France, the United States, Britain, Germany, Japan, Italy, Canada, and Russia

"developing" world. Of the four, Russia's economy is the slowest growing and arguably the least diverse in the group due to its dependence on fossil fuel exports, but it is also the wealthiest on a per capita basis.

Institutionalized Corruption The cost of doing business in Russia is affected by a high level of corruption. In 2013, the organization Transparency International rated Russia the most corrupt of the BRIC countries and among the most corrupt in the world. With the exceptions of Belarus, Moldova, Georgia, and Azerbaijan, all countries in this region were rated as more corrupt than Russia. Civil servants and the police often expect businesses to offer bribes to receive licenses and to avoid other regulatory hurdles. This affects large corporations and small entrepreneurs alike.

Even those who make it through the labyrinth of setting up a business must then face protection racketeers, as organized crime has become a fact of life during the post-Communist era. Many oligarchs have become closely connected to the *Russian Mafia*, a highly organized criminal network dominated by former *KGB* (Russia's intelligence agency during the Cold War, now known as the Federal Security Service, or FSB) and military personnel who control the thugs on the streets. The Russian Mafia has extended its influence into nearly every corner of the post-Soviet economy, especially in illegal activities and the arms trade. These crime syndicates have also developed international networks through which they control illicit activities abroad, especially in Europe.

> **underemployment** the condition in which people are working too few hours to make a decent living or are highly trained but working at menial jobs

The Growing Informal Economy A large share of the economic activity in the region takes place in the informal sector. To some extent, the new informal economy is an extension of the old one that flourished under communism. The black market of that time was based on currency exchange and the sale of hard-to-find luxuries. In the 1970s, for example, savvy Western tourists could enjoy a vacation on the Black Sea paid for by a pair or two of smuggled Levi's blue jeans and some Swiss chocolate bars. Today, many people who have lost stable jobs due to privatization now depend on the informal economy for their livelihood.

Workers in the informal economy tend to be young, unskilled adults; retirees on tiny pensions; and those with only a low level of education. The majority of these workers operate out of their homes by selling, among other things, cooked foods, vodka made in their bathtubs, and clothing and electronics that have been smuggled into the country. The World Bank has estimated that the informal sector makes up about 40 percent of the Russian economy. In the Caucasian countries, the number is as high as 50 to 60 percent. Such large percentages indicate that people may actually be better off financially than official GDP per capita figures suggest.

Despite the fact that the informal economy helps people survive, it is not popular with governments. Unregistered enterprises do not pay business and sales taxes and usually are so underfinanced that they tend not to grow into job-creating formal sector businesses. In many cases, informal businesspeople must pay protection money to local gangsters (the so-called *Mafia tax*) to keep from being reported to the authorities.

> **VIGNETTE**
>
> Natasha is an engineer in Moscow. She has managed to keep her job and the benefits it carries, but in order to better provide for her family, she sells secondhand clothes in a street bazaar on the weekends. Asked about her customers, Natasha says, "Many are former officials and high-level bureaucrats who just can't afford the basics for their families any longer." Some are older retired people whose pensions are so low that they resort to begging on Moscow's elegant shopping streets close to the parked Rolls-Royces of Moscow's rich. These often highly educated and only recently poor people buy used sweaters from Natasha and eat in nearby soup kitchens. [Source: This composite story is based on work by Alessandra Stanley, David Remnick, David Lempert, and Gregory Feifer.]

Unemployment and the Loss of Social Services Since becoming privatized, most formerly state-owned industries have cut many jobs in an attempt to compete against more-efficient foreign companies. Losing a job is especially devastating in a former Soviet country because one also loses the subsidized housing, health care, and other social services that were often provided along with the job. The new companies that have emerged rarely offer full benefits to employees. There is little job security because most small private firms appear quickly and often fail. Discrimination is also a problem, given the absence of equal opportunity laws. Job ads often contain wording such as "only attractive, skilled people under 30 need apply."

Unemployment figures for the region vary widely. Since 2013, official unemployment rates have been under 6 percent in Russia, and under 9 percent in Ukraine. In Belarus, unemployment is officially only about 1 percent, but actual unemployment may be higher because many remaining state-owned firms cannot pay employees still listed as workers. Many people do not bother to register as unemployed because of the absence of unemployment benefits. The rate of **underemployment**, which measures the number of people who are working too few hours to make a decent living or who are highly trained but are working at low-paying jobs, is even higher in all of these countries.

Food Production in the Post-Soviet Era

Across most of the region, agriculture is precarious at best, either hampered by a short growing season and boggy soils or requiring expensive inputs of labor, water, and fertilizer. Because of Russia's harsh climates and rugged landforms, only 10 percent of its vast expanse is suitable for agriculture. The Caucasus mountain zones are some of the only areas in the region where rainfall adequate for agriculture coincides with a relatively warm climate and long growing seasons. Together with Ukraine and European Russia, this area is the agricultural backbone of the region (**Figure 5.15**). The best soils are in an area stretching from Moscow south toward the Black and Caspian seas, and extending west to include much of Ukraine and Moldova. In Central Asia, irrigated agriculture is extensive, especially where long growing seasons support cotton, fruit, and vegetables.

Changing Agricultural Production During the 1990s, agriculture went into a general decline across the entire region. In most of the former Soviet Union, yields dropped by 30 to 40 percent

FIGURE 5.15 Agriculture in Russia and the post-Soviet states. Agriculture in this part of the world has always been a difficult proposition, partly because of the cold climate and short growing seasons and partly because soil fertility or lack of rainfall are problems in all but a few places (Ukraine, Moldova, and Caucasia). [Source consulted: Robin Milner-Gulland with Nikolai Dejevsky, *Cultural Atlas of Russia and the Former Soviet Union*, rev. ed. (New York: Checkmark Books, 1998), pp. 186–187, 198–199, 204–205, 216–217]

compared to previous production levels. This was due mostly to the collapse of the subsidies and internal trade arrangements of the Soviet Union. Many large and highly mechanized collective farms were suddenly without access to equipment, fuel, or fertilizers. Since that time, agricultural output has rebounded but even today it remains below Soviet levels. Russia's commercial agricultural sector is still characterized by many large and inefficient farms, most of which are now run by corporations. Russians are now heavily dependent on "homegrown" food—small household plots, gardens, and family farms that produce 59 percent of the total agricultural output on only 20 percent of the arable land.

In Central Asia, agriculture has been reorganized with more emphasis on smaller, food-producing family farms. Grain and livestock farming (for meat, eggs, and wool) are increasing. Production levels per acre and per worker also are increasing. China, through the Asian Development Bank, has provided assistance in reducing the use of agricultural chemicals, which were used extensively during Soviet times.

Farmers in Georgia, favored with warm temperatures and abundant moisture from the Black and Caspian seas, can grow citrus fruits and even bananas; they do so primarily on family, not collective, farms. Before 1991, most of the Soviet Union's citrus and tea came from Georgia, as did most of its grapes and wine. Georgia still exports some food to Russia, but because of political tensions between the two countries, Georgia is increasing the amount of food and wine it sells to Europe and other markets outside this region.

THINGS TO REMEMBER

> **GEOGRAPHIC INSIGHT 2**
>
> • **Globalization and Development** After the fall of the Soviet Union, economic reforms and globalization changed patterns of development in this region. Wealth disparity increased and jobs were lost as many Communist-era industries were closed or sold to the rich and well connected. The region is now largely dependent on its role as a leading global exporter of energy resources.

• Russian economists estimate that even though average incomes have grown by 45 percent since 1991, almost all of this gain has gone to the wealthiest 20 percent of the population; the bottom 20 percent of the population is earning about half what they did in 1991.

• Natural gas and crude oil have emerged as the region's most lucrative exports, introducing a new wave of globalization and fossil fuel–dependent economic development.

- The World Bank has estimated that the informal sector makes up about 40 percent of the Russian economy and 50 to 60 percent of the economies of the Caucasian countries.

- Russians are heavily dependent on "homegrown" food—small household plots, gardens, and family farms that produce 59 percent of the total agricultural output on only 20 percent of the arable land.

POWER AND POLITICS

> **GEOGRAPHIC INSIGHT 3**
>
> **Power and Politics:** This region has a long history of authoritarianism. While there is some pressure to expand political freedoms, there are still few opportunities for the public to influence the political process. Elected representative bodies often act as rubber stamps for very strong presidents and exercise only limited influence on policy making.

Since 1991, the politics in this region have remained very authoritarian and largely dominated by Russia, which sees itself as the successor to the Soviet Union. For almost two decades, Russia's political scene has been dominated by one person, Vladimir Putin, a former high official in the KGB. Putin was elected president in 2000 and has been the country's de facto leader as either president or prime minister ever since. During this time, he has exercised tight control over political and economic policy and consolidated political power in Moscow. He has been popular for bringing more peace and prosperity to Russia and for restoring Russia's image of itself as a world power. Putin has also been criticized for preventing meaningful reforms at the local, state, and national levels, and for extending government control over the press and the media (see page 217). While criticism of the government is now formally allowed, most fear that such criticism would be met with retribution from the *securocrats* or *siloviki* (former FSB functionaries loyal to Putin) who control many governing institutions.

Putin is the most dominant political person in the region, and is likely to remain in power for the near future. But long after he goes, authoritarianism will likely hold sway in this region. For one thing, Russia's current constitution, written in 1993, gives sweeping powers to the president, which will mean that a strong, and likely authoritarian, leader will succeed Putin. Many scholars of this region have argued that Russia's geography, featuring vast territory and few mountain or other barriers to prevent invasion by outsiders, favors a single powerful authoritarian ruler with a strong military.

117. PUTIN CONFIRMATION
123. NEW RUSSIAN LEADER

Elsewhere in the region, authoritarianism and the limited political freedoms it allows are the norm. Belarus, Ukraine, Moldova, the Caucasian republics (including Chechnya, discussed on pages 216–217), and Central Asia are still most often run by authoritarian leaders who are unaccustomed to criticism or to sharing power (see the map in Figure 5.16). Elections often involve systematically intimidating voters and arresting the political opposition.

The Color Revolutions and Political Freedoms In a few of the post-Soviet states, there have been a number of transitions to somewhat greater political freedoms. A series of so-called *color revolutions* took place, led by coalitions of educated young adults who rallied under various symbolic colors meant to unite them despite their differences. Funded largely by foreign NGOs and governments, including the United States, that wanted to hasten democratization, it appeared for a time that these movements would result in significant reforms. However, while elections have been held and some freedom of expression is now tolerated, many forms of authoritarian control and practices that constrain political freedoms remain.

The color revolutions include the Rose Revolution in Georgia (2003–2004), the Orange Revolution in Ukraine (January 2005; see Figure 5.16B), and the Tulip Revolution in Kyrgyzstan (April 2005). In Russia itself, similar protest movements have been funded by agencies of foreign governments, notably USAID (the United States Agency for International Development). In 2012, as part of a general crackdown on protests during Putin's bid for the presidency, USAID was banned in Russia for funding various political groups opposed to Putin.

In Ukraine, the Orange Revolution took place during the presidential elections of 2004. A candidate who favored closer ties with Russia, Viktor Yanukovych, won the election, but the results were so obviously rigged by the government that widespread protests went on for months. The protests, combined with international pressure, resulted in a new vote in which Victor Yuschenko, who favored closer ties with the European Union and Ukrainian membership in NATO, won.

In subsequent elections, however, the pro-Russian Yanukovych, returned to power. Ukraine today is divided in terms of its geographic orientation. Areas to the west, which border Central Europe, tend to favor closer ties with Europe, while areas to the east, where many of Ukraine's Russian-speaking minority resides, lean toward Russia.

Crisis in Ukraine In early 2014, one of the most challenging crises this region has seen since the fall of the Soviet Union emerged in Ukraine, pitting Russia against the EU and the United States in a seeming throwback to the days of the Cold War. After Yanukovych backed away from previous commitments to increase economic ties between Ukraine and the EU, a wave of protests developed in Ukraine's capital of Kiev. Like the color revolutions, these protests were supported by U.S. and European governments and pro-democracy NGOs. After the Yanukovych regime attempted to forcibly repress the protests, which resulted in more than 80 deaths, the regime itself was toppled by an alliance of demonstrators and political groups opposed to the Yanukovych regime and in favor of closer integration with the EU. Shortly thereafter, Crimea, an autonomous republic within Ukraine that is dominated by ethnic Russians, voted in a referendum to secede from Ukraine. The referendum was declared illegal by Ukraine, which only allows referendums at the national scale. Russian military forces, already present in large numbers in Crimea as part of various treaties with Ukraine, began massing on the borders of Ukraine's eastern provinces, several of which have large ethnic Russian populations that support closer ties with Russia. Throughout eastern Ukraine, administrative and police buildings were taken over by local pro-Russian forces. In response, Ukraine's military retook some of these buildings and began preparations to repel a potential Russian invasion.

FIGURE 5.16 PHOTO ESSAY: Power and Politics in Russia and the Post-Soviet States

Since World War II, most conflicts in this region have been in Caucasia and Central Asia. Most occurred just before or shortly after the breakup of the Soviet Union. Since then, popular movements pushing for greater political freedoms have emerged in several countries. However, violent conflict continues to be a major source of political instability in this region.

A Children in Afghanistan play on a tank dating from the Soviet Union's 10-year war in that country. The Soviets intervened in 1979 to prop up an authoritarian and unpopular Afghan government that had initiated radical reforms and murdered much of the political opposition. The war resulted in at least 1.5 million deaths and a humiliating defeat for the USSR. Fourteen thousand Soviet soldiers died, billions of dollars were spent, and practically nothing was gained. Public outrage over the Afghan war played a major role in the breakup of the Soviet Union. Since the fall of the USSR, the expansion of political freedoms has proceeded at a much more rapid pace throughout the region.

Democratization and Conflict

Armed conflicts and genocides with high death tolls since 1945
- Ongoing conflict
- 1000–5000 deaths
- 5000–50,000 deaths
- 50,000–300,000 deaths
- 300,000–1,000,000 deaths

Democratization index
- Full democracy
- Flawed democracy
- Hybrid regime
- Authoritarian regime
- No data

B Ukrainian protesters pushing for a transition from an authoritarian to a more democratic political system during the Orange Revolution of 2004–2005. The movement followed a model pioneered in Serbia and Georgia, where student groups, funded by NGOs and governments in Europe and the United States, spearheaded grassroots political campaigns that featured massive nonviolent demonstrations. A similar color revolution took place in Kyrgyzstan.

C Chechen police in a show of force in Grozny, Chechnya, where rebels have fought two unsuccessful wars for independence from Russia and have conducted ongoing terror campaigns against Russia since the fall of the Soviet Union. Political instability in Caucasia is one of the greatest factors working against the expansion of political freedoms in this region.

216 CHAPTER 5 Russia and the Post-Soviet States

Regardless of the eventual outcome of the ongoing confrontation between Russia and Ukraine, underlying tensions will remain. Ukraine's government will be challenged to balance the demands of citizens in the western parts of the country, many of whom want closer ties with the EU, with those of ethnic Russians in the east who want closer ties and possibly even political union with Russia. Russia will have to weigh the economic benefits of its gas trade with Europe against its long-term opposition to the expansion of the EU into Ukraine. Additionally, the EU and the United States will have to decide whether supporting color revolutions or other political movements in this region is worth risking much more tense relations with Russia.

Russification the czarist and Soviet policy of encouraging ethnic Russians to settle in non-Russian areas as a way to assert political control

Cultural Diversity and Russian Domination Russia's long history of expansion into neighboring lands has left it and neighboring countries with exceptionally complex internal political geographies. As the Russian czars and then the Soviets pushed the borders of Russia eastward toward the Pacific Ocean over the past 500 years, they conquered a number of non-Russian areas. Russians are now the main inhabitants of many of those areas, some of which are now within countries neighboring Russia, such as Ukraine. Other territories have been designated semiautonomous internal republics (Figure 5.17) within Russia and have significant ethnic minority populations that are descendant from indigenous people or trace their origins to long-ago migrations from Germany, Turkey, or Persia.

Both the czars and the USSR had a policy of **Russification**, whereby large numbers of ethnic Russian migrants were settled in non-Russian ethnic areas and given the best jobs and most powerful positions in regional governments. The goal was to force potentially rebellious regional minorities to conform to the state's goals. However, even during the Soviet period, but especially after, minorities organized to resist Russification and to enhance local ethnic identities.

The Conflict in Chechnya Shortly after the breakup of the Soviet Union, several internal republics demanded more autonomy, and two of them, Tatarstan and Chechnya, declared outright independence.

FIGURE 5.17 Ethnic character of Russia and percentage of Russians in the post-Soviet states. Russia, with all of its internal republics and administrative regions, is formally called the Russian Federation. The ethnic character of many internal republics was changed by the policy of central planning, so Russians now form significant minorities in the republics. The pie charts for each country indicate the percentages of Russians in the post-Soviet states. In 2002, the ethnic makeup of Russia was 79.8 percent Russian, 3.8 percent Tatar, 2 percent Ukrainian, 1.2 percent Bashkir, 1.1 percent Chuvash, and 12.1 percent other. [Sources consulted: James H. Bater, *Russia and the Post-Soviet Scene* (London: Arnold, 1996), pp. 280–281; Graham Smith, *The Post-Soviet States* (London: Arnold, 2000), p. 75; "Ethnic Groups," *The World Factbook 2012*, Central Intelligence Agency, at https://www.cia.gov/library/publications/theworld-factbook/fields/2075.html#rs]

Tatarstan has since been placated with greater economic and political autonomy. However, Chechnya's stronger resistance to Moscow's authority has led to the worst bloodshed of the post-Soviet era and raised many doubts about Russia's commitment to human rights.

Chechnya, located on the fertile northern flanks of the Caucasus Mountains (see the inset map in Figure 5.17), is home to 800,000 people. Partially in response to Russian oppression, the Chechens converted from Orthodox Christianity to the Sunni branch of Islam in the 1700s. Since then, Islam has served as an important symbol of Chechen identity and as an emblem of resistance to the Orthodox Christian Russians, who annexed Chechnya in the nineteenth century.

In 1942, during World War II, as the Germans invaded Russia, a group of Chechens rebels simultaneously waged a guerilla war against the Soviets. Near the end of the war, Stalin exacted his revenge by deporting the majority of the Chechen population (as many as 500,000 people) to Kazakhstan and Siberia. Here they were held in concentration camps, and many died of starvation. The Chechens were finally allowed to return to their villages in 1957, but a heavy propaganda campaign portrayed them as traitors to Russia, a designation that greatly affected their daily lives.

In 1991, as the Soviet Union was dissolving, Chechnya declared itself an independent state. Russia saw this as a dangerous precedent that could spark similar demands by other cultural enclaves throughout its territory. Russia also wished to retain the agricultural and oil resources of the Caucasus; it had planned to build pipelines across Chechnya to move oil and gas to Europe from Central Asia. Russia responded to acts of terrorism by Chechen guerrillas with bombing raids and other military operations that killed tens of thousands of people and created 250,000 refugees.

Most Chechen guerillas have since given up, most Russian combat troops have been pulled out of Chechnya, and Russia has begun making substantial investments in rebuilding the capital city of Grozny. While this shift is a welcome change, some Chechen rebels have continued their struggle by carrying out brutal terrorist attacks, many of which now take place in Moscow and other areas outside Chechnya. Chechnya remains highly militarized (see Figure 5.16C), and the ongoing conflict there continues to raise doubts about Russia's ability to peacefully address internal political dissent. **116. CHECHNYA HANGS ON TO UNEASY PEACE**

Conflict in Georgia Just south of Chechnya, conflicts between Russia and Georgia over the ethnic republics of South Ossetia and Abkhazia have worsened in the post-Soviet era. Both republics became part of Georgia at the behest of Joseph Stalin (a native Georgian). He then moved Georgians and Russians into Ossetia and Abkhazia so that the native people became a minority in their own place. More recently, Russia has strategically supported the ethnic Ossetian and Abkhazian populations' agitation for secession from Georgia, even granting Russian citizenship to between 60 and 70 percent of the non-Georgian population. Russia may have done this in retaliation for Georgia's increasingly close relations with the United States and the European Union, as evidenced by its candidacy for NATO membership. A pipeline that links the oil fields of Azerbaijan with the Black Sea via Georgia is another source of contention (see Figure 5.14). The Black Sea route is important because it is an expedient way to ship oil to world markets, as the Black Sea connects to the Mediterranean Sea and, ultimately, to the Atlantic Ocean. Russia would like to enhance its control over the oil resources of the Caspian Sea region by routing all pipelines through Russian-controlled territory.

In 2008, Georgia's military, attempting to gain control over rebelling parts of South Ossetia, engaged with the Russian army on Russian territory, but that conflict ended without clear resolution. It is not clear at this time whether Abkhazia and South Ossetia will break away and become independent countries, become provinces within the Russian Federation, or be satisfied by offers of more autonomy within Georgia's federal structure.

The Media and Political Reform During the Soviet era, all communication media were under government control. There was no free press, and public criticism of the government was a punishable offense. Between 1991 and the early 2000s, the communications industry was a center of privatization, and several media tycoons emerged to challenge the authorities. Privately owned newspapers and television stations regularly criticized the policies of various leaders of Russia and the other states. It appeared that a free press was developing.

In Vladimir Putin's rise to power, however, the most outspoken newspapers and TV stations in Russia were shut down. Since then, critical analysis of the government has become rare throughout Russia. Journalists openly critical of government policies have been treated to various forms of punishment: censorship, exile, or violence. Since 2000, about 200 journalists have been killed under suspicious circumstances.

ON THE BRIGHT SIDE

Controlling Nuclear Material

In the post–9/11 world, concern spread that military corruption in this region could put nuclear weapons from the former Soviet arsenal into the hands of terrorists. After huge military-funding cuts, weapons, uniforms, and even military rations were routinely sold on the black market. In 2007, Russian smugglers were caught trying to sell nuclear materials on the black market. Recent developments are more encouraging. In an effort to fight corruption and the temptation to sell nuclear materials in the black market, the Russian government has given impoverished military personnel long-delayed pay raises or termination compensation. In addition, all countries in the region are now cooperating with the International Atomic Energy Agency in controlling nuclear material (see the discussion on page 201).

THINGS TO REMEMBER

> **GEOGRAPHIC INSIGHT 3**
> • **Power and Politics** This region has a long history of authoritarianism. While there is some pressure to expand political freedoms, there are still few opportunities for the public to influence the political process. Elected representative bodies often act as rubber stamps for very strong presidents and exercise only limited influence on policy making.

- A series of so-called *color revolutions* took place throughout this region, led by coalitions of educated young adults, and funded largely by foreign NGOs and governments, including that of the United States, that wanted to hasten democratization.

- In early 2014, one of the most challenging crises this region has seen since the fall of the Soviet Union emerged in Ukraine, pitting Russia against the EU and the United States, in a seeming throwback to the days of the Cold War.

- Russia's long history of expansion into neighboring lands has left it and surrounding countries with exceptionally complex internal political geographies. ■

URBANIZATION

> **GEOGRAPHIC INSIGHT 4**
>
> **Urbanization:** A few large cities in Russia and Central Asia are growing fast, fueled by the expansion of energy exports. Elsewhere, many cities are suffering from a lack of investment and maintenance as their economies struggle in the post-Soviet era.

Even though Russia and the post-Soviet states have the largest land area of any world region, and vast areas of lightly populated or uninhabited land, this is still largely a region of cities. This is especially true in the more European countries of Ukraine, Belarus, and Russia, where 70 percent or more of the population lives in cities. The Caucasian republics of Georgia, Armenia, and Azerbaijan are somewhat less urbanized (around 60 percent), and Central Asia is only 47 percent urban.

Russia's capital city of Moscow, with 11.5 million people, is a primate city, as are all other capital cities of the nations in this region. The dominance of primate cities reflects the importance of politics and government in determining where economic development would happen during the Soviet and post-Soviet eras.

Urban life for most remains shaped by the Communist-era legacies of central planning. Giant apartment blocks, designed by bureaucrats and built for industrial workers, dominate most cities, especially on the fringes of the older, pre-Soviet urban cores. In cities that are growing, such as Moscow and St. Petersburg, the housing shortages of the Soviet era have continued since 1991. Cramped, drab apartments, badly in need of repair, with shared kitchens and bathrooms, are common. At the community scale, inadequate sewage, garbage, and industrial waste management pose serious long-term health and environmental threats.

In Moscow, housing shortages are irregular but particularly severe because the city has grown so rapidly since 1991, when it began receiving much of the new investment in the region. Due to high demand, the cost of housing has increased so dramatically that Moscow has one of the highest costs of living in the world. Shortages are also exacerbated by the ability of wealthy people to buy up multiple apartments and refurbish them into one luxurious dwelling.

Outside of Moscow, St. Petersburg, and a few other cities that have been growing because they are receiving revenues from energy exports (Figure 5.18C), many urban areas are suffering from general deterioration and the absence of funds for maintenance due to slow or absent economic growth. In some cases, this is causing overall housing shortages (see Figure 5.18B).

Urban Legacies of Soviet Regional Economic Development

One of the more difficult legacies of the Soviet era involves the huge industrial cities in the lightly inhabited expanses of Siberia and the Pacific coast. Founded originally as trading posts and fortress towns during the pre-Soviet Russian empire, these towns became the focus of Soviet planners, who were eager to solidify their control of the vast interior east of Moscow. They became home to major extractive industries and defense-related infrastructure, but further growth was always hampered by the vast distances and challenging physical geography that separate them from the main population centers in the west of the region. Even so, nearly 90 percent of Siberia's people are concentrated in the few large urban areas. Life in these cities is vastly different from the more traditional rural lifestyles that characterized this region only a few generations ago (Figure 5.19).

Because the region's rivers run mainly north–south, there is a need for land transportation systems, such as railroads and highways, that run east to west to connect these cities with the west. The harsh climate of the region makes this infrastructure exceedingly expensive to build and maintain, and much has fallen into disrepair in the post-Soviet era. Only one poorly paved road and one main rail line—the Trans-Siberian Railroad (Figure 5.20 on page 221)—runs the full east–west length, connecting Moscow with Vladivostok, the main port city on the Russian Pacific coast.

Urban Central Asia

While Central Asia has historically had a relatively low rate of urbanization, recent expansion in the energy industries has fueled new urban growth. Cities here have also grown because of the region's relatively high overall population growth. Tashkent, the capital of Uzbekistan, has 2.3 million people, making it the fourth-largest city in the region, after Moscow, St. Petersburg, and Kiev. Almaty, the former capital of Kazakhstan, is the next-largest city in Central Asia, at 1.3 million, followed by Astana, the new capital of Kazakhstan, which has 750,000 people.

VIGNETTE Yernar Zharkeshov is a 24-year-old university-educated man who is moving back home to Kazakhstan after having spent time in Britain and Singapore. He is seeking new opportunities in his homeland, which are increasing because of its extensive oil exports. The natural choice for him is Astana, a city that was designated the capital of Kazakhstan in 1997. Yernar quickly found a job as a government economist; he is just one of thousands of young people who have migrated to the growing capital.

The idea of Astana came from Nursultan Nazarbayev, who has been president of Kazakhstan since its independence from the Soviet Union in 1991. Without much public debate, he moved the capital from Almaty in the south to the remote and sparsely populated steppe in the north that is known for its harsh climate. Here he realized his vision for a grand and impressive new capital, built more or less from scratch. Astana is meant to symbolize a new, forward-looking Kazakhstan, and Nazarbayev has spared no expense. The city is full of stunning buildings, either designed by world-renowned architects or by Nazarbayev himself (see Figure 5.18C).

FIGURE 5.18 PHOTO ESSAY: Urbanization in Russia and the Post-Soviet States

An uneven pattern of urbanization is developing in this region. In several countries, the largest cities are growing rapidly, as they are centers of new development and globalization. Elsewhere, however, many cities are struggling economically and even shrinking in terms of population because they are less able to attract new investment.

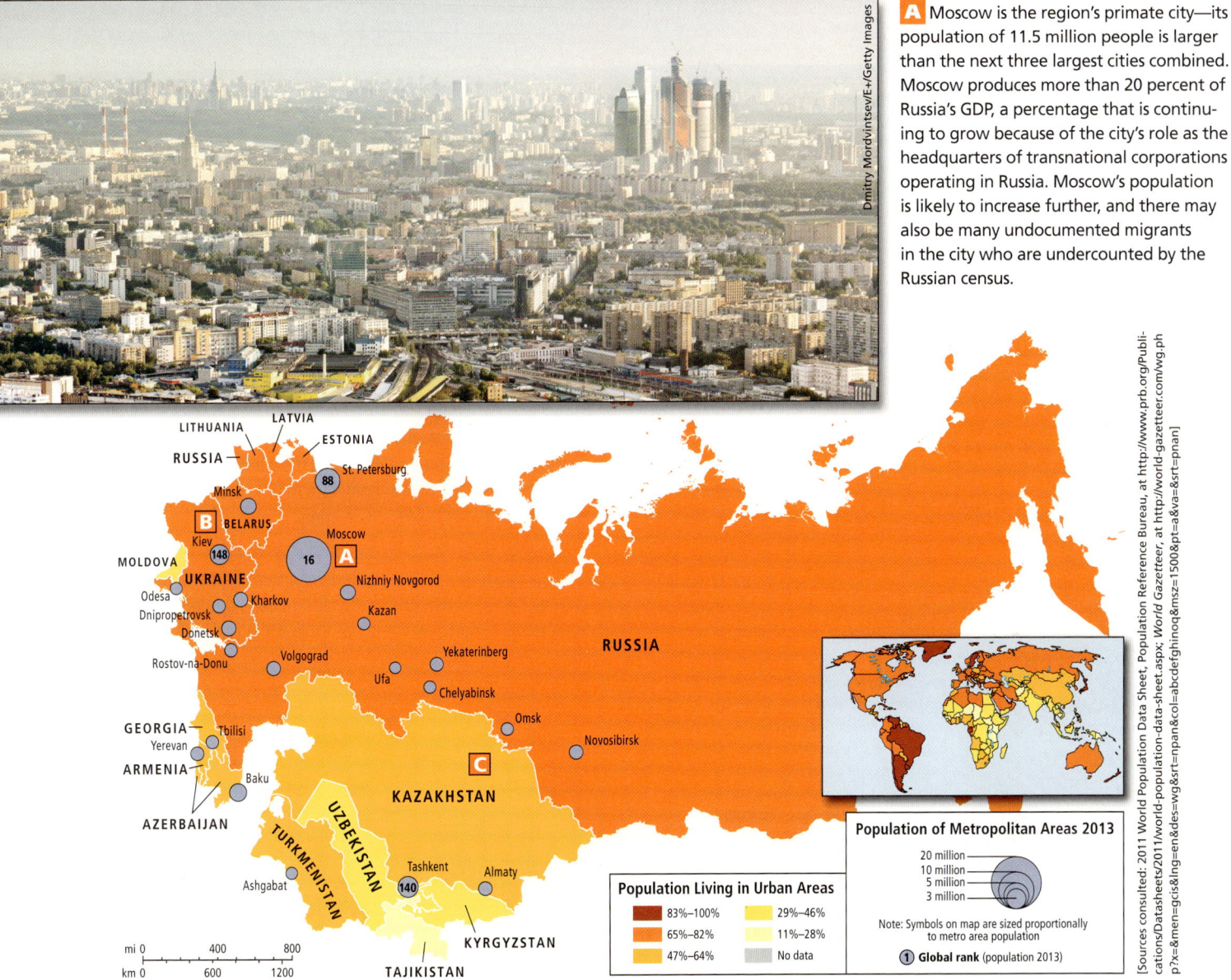

A Moscow is the region's primate city—its population of 11.5 million people is larger than the next three largest cities combined. Moscow produces more than 20 percent of Russia's GDP, a percentage that is continuing to grow because of the city's role as the headquarters of transnational corporations operating in Russia. Moscow's population is likely to increase further, and there may also be many undocumented migrants in the city who are undercounted by the Russian census.

Population Living in Urban Areas
- 83%–100%
- 65%–82%
- 47%–64%
- 29%–46%
- 11%–28%
- No data

Population of Metropolitan Areas 2013
- 20 million
- 10 million
- 5 million
- 3 million

Note: Symbols on map are sized proportionally to metro area population

① Global rank (population 2013)

B Abandoned buildings in Kiev, Ukraine, where the population is projected to decline by 0.02 percent by 2020. This is much less than the projected decline of 5 percent by 2020 for Ukraine as a whole.

C Astana, Kazakhstan's capital, has undergone massive new development that has largely been supported by the increased exploitation of nearby oil resources by local and international companies. The population of Astana doubled in the 10 years after Astana was designated the new capital.

FIGURE 5.19 LOCAL LIVES
People and Animals in Russia and the Post-Soviet States

A A trained hunting eagle is released by its master in Kazakhstan. Training an eagle requires great devotion over a long period of time. Raised by hand by their master, eagles are kept blindfolded from birth to make them depend on and trust their human companions. A Kazakh proverb says that "as the man trains his eagle, so does the eagle train his man." A golden eagle is on the flag of Kazakhstan. [Vyacheslav Oseledko/AFP/Getty Images]

B Siberian huskies take a break from training for a long-distance sled race. One of the oldest dog breeds, huskies were bred several thousand years ago by the Chukchi people in far eastern Siberia, near Alaska (see Figure 5.1 on page 193) to pull heavy loads on sleds over long distances. With their legendary strength and stamina, huskies helped the Chukchi and other peoples explore and populate new territories. They were also central to U.S. penetration into Alaska in the early twentieth century. [Dmitry Kostyukov/AFP/Getty Images]

C Reindeer are tended by a group of the indigenous Nenets people in western Siberia. Reindeer have been raised throughout northern Russia for thousands of years for their meat, hides, antlers, milk, and to pull sleds. Having weathered the collapse of Soviet state-run herding and meat-production economies, reindeer herders in some areas now find themselves in conflict with Russia's oil and gas producers, whose facilities often block important migration routes. [WIN Initiative/The Image Bank/Getty Images]

Like many other migrants to Astana, Yernar Zharkeshov is an ethnic Kazakh. President Nazarbayev's unspoken motive for moving the capital is to consolidate the country's northern territories, which are currently dominated by ethnic Russians. The presence of these Russians is the result of the practice of Russification during the days of the Soviet Union. Astana has been designed to promote the opposite—the "Kazakhification" of the country's north. [Source: National Geographic. For detailed source information, see Text Sources and Credits.]

THINGS TO REMEMBER

> **GEOGRAPHIC INSIGHT 4**

- **Urbanization** A few large cities in Russia and Central Asia are growing fast, fueled by the expansion of energy exports. Elsewhere, many cities are suffering from a lack of investment and maintenance as their economies struggle in the post-Soviet era.

- Russia's capital city of Moscow, with 11.5 million people, is a primate city, as are all other capital cities of the nations in the region.

- Urban life for most remains shaped by the Communist-era legacies of central planning. Giant apartment blocks, designed by bureaucrats and built for industrial workers, dominate most cities, especially on the fringes of the older, pre-Soviet urban cores.

- One of the more difficult legacies of the Soviet era involves the huge industrial cities in the lightly inhabited expanses of Siberia and the Pacific coast, where urban growth has long been hampered by spatial isolation from the main population centers of the region.

- While Central Asia has historically had a relatively low rate of urbanization, recent expansion in the energy industries has fueled new urban growth.

POPULATION AND GENDER

> **GEOGRAPHIC INSIGHT 5**

Population and Gender: Populations are shrinking in many parts of the region due largely to lower birth rates and declines in life expectancy. These changes relate to consistently high levels of participation by women in the workforce and to the economic decline in the 1990s that followed the collapse of the Soviet Union. Men and women have been affected differently in the post-Soviet era, with men much more likely to suffer from alcoholism and related health problems.

With high death rates and low birth rates, this region is undergoing a unique, late-stage variant of the demographic transition (see Chapter 1, page 48). In all but the Caucasian and Central Asian states, populations are shrinking faster than in any world region other than Europe. During the Soviet era, the increase in women's opportunities to become educated and work outside the home curtailed population growth. Free health care and adequate retirement pensions also helped lower incentives for large families, as people didn't have to depend on their children

FIGURE 5.20 Principal industrial areas and land transport routes of Russia and the post-Soviet states. The industrial, mining, and transportation infrastructure is concentrated in European Russia and adjacent areas. The main trunk of the Trans-Siberian Railroad and its spurs link industrial and mining centers all the way to the Pacific, but the frequency of these centers decreases with distance from the borders of European Russia. [Sources consulted: Robin Milner-Gulland with Nikolai Dejevsky, *Cultural Atlas of Russia and the Former Soviet Union*, rev. ed. (New York: Checkmark Books, 1998), pp. 186–187, 198–199, 204–205, 216–217; http://www.travelcenter.com.au/russia/images/trans-sib-map-v3.jpg]

for support in old age. Severe housing shortages were an additional disincentive to having children, and many families chose to have only one or two children. All of these factors were leading to slower population during the last years of the Soviet Union, but the population didn't decline until the economic collapse and the "shock therapy" of the 1990s, which had major social repercussions.

By the mid-2000s, the rate of shrinkage had slowed down considerably across the region, and by 2013 Russia's population was again growing. This growth may not last, as the population decline of the 1990s means that fewer people will be in their reproductive years in the coming decade. However, the dark predictions of demographic collapse that are still common regarding this region probably need to be revised.

Russia has made considerable efforts to fight population decline. Couples are now offered a bonus for having a second child that is equivalent to more than 2 years' worth of wages for the average Russian. However, similar policies in Europe have been less successful than efforts to address the concerns of career-minded working mothers, who often value access to inexpensive, high-quality day care and longer school days more than they do money. Another questionable policy involves attracting back Russians and their dependents who live abroad. In 2007 and 2008, Russia spent $300 million to send emissaries to places like Brazil, Egypt, Germany, and all the post-Soviet states to lure "returnees." Only 10,300 were recruited.

Gender and Life Expectancy

Much of the reason for the population decline, and possibly the recent slowing of this decline, relates to life expectancy. Life expectancy in the region differs dramatically according to gender, with women living 10 to 12 years longer than men. By comparison, around the world, women live on average only 5 years longer than men. Life expectancy for women changed only marginally after the collapse of the Soviet Union, but life expectancy for men dropped considerably. In Russia, for example, between 1987 and 1994, male life expectancy dropped from 64.9 to 57.6 years. Since then, male life expectancy has recovered somewhat, but at 64 years, it is still the shortest in any developed country. Female life expectancy has held steady at 75 to 76 years throughout this period.

Major causes of low life expectancies in the region are the loss of health care, which was usually tied to employment, and the physical and mental distress caused by lost jobs and social disruption. These factors, though, might not explain much of the gender difference in life expectancy since the fall of the Soviet Union, given that women worked outside the home almost as much as men during the Soviet era and seem to have been more likely to have lost jobs during the 1990s. Another possible cause for low male life expectancy is alcohol abuse and alcohol-related accidents and suicides, which are much more common among men. Russia has by far the world's highest rate of alcohol-related health disorders (1277 per 100,000 people in Russia, compared to 600 per 100,000 people in the United States), and more than half of all deaths among the working-age population are alcohol-related. On the positive side, recent increases in male life expectancy may be a result of declining rates of alcohol poisoning since the mid-2000s. This change is likely an outcome of government and private campaigns to reduce alcohol consumption and improve treatment of alcoholism.

Population pyramids for several countries (Figure 5.21) show the overall population trends in the region and reflect geographic

FIGURE 5.21 Population pyramids for Russia, Belarus, Kyrgyzstan, and Kazakhstan. Note that the pyramid for Russia is at a different scale (millions) than the other three pyramids (thousands) because of Russia's larger total population. This difference does not significantly affect the pyramid's shape. [Sources consulted: "Population Pyramids of Russia," "Population Pyramids of Belarus," "Population Pyramids of Kyrgyzstan," and "Population Pyramids of Kazakhstan," *International Data Base*, U.S. Census Bureau, 2012, at http://www.census.gov/population/international/data/idb/informationGateway.php]

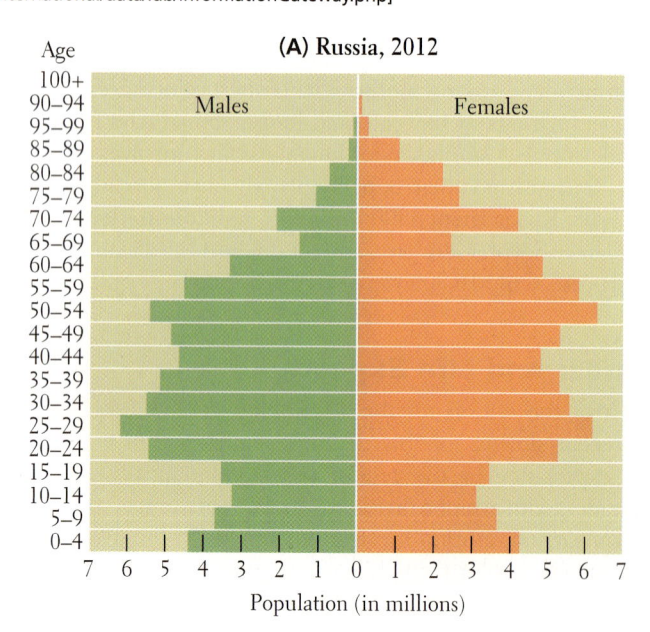

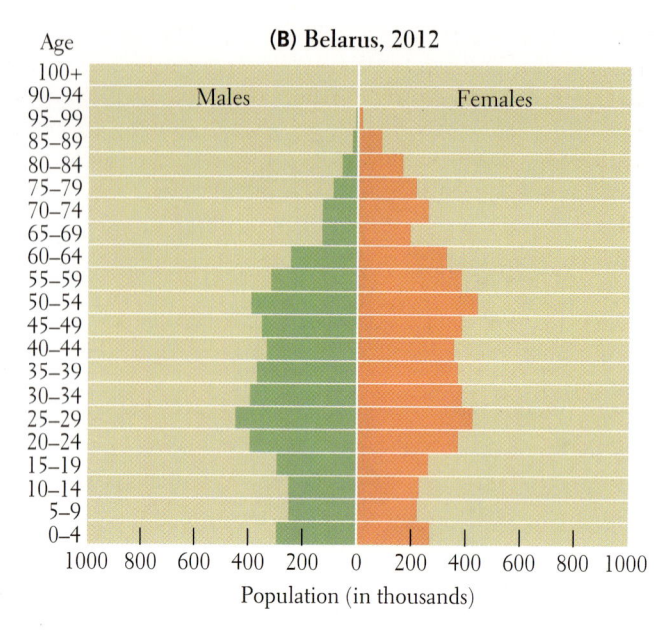

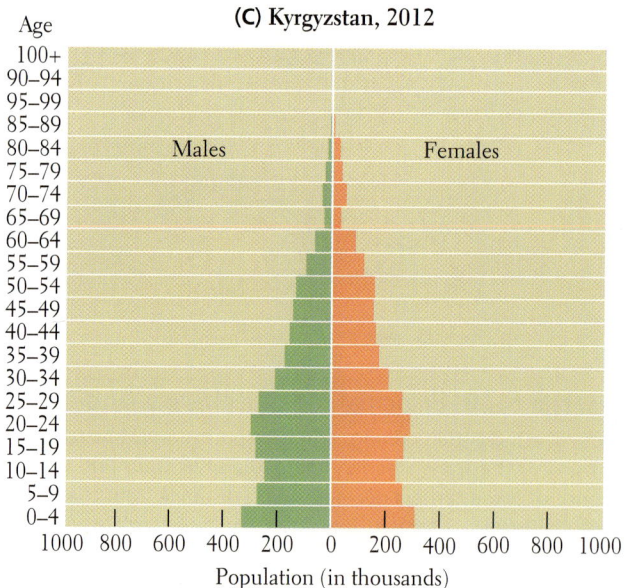

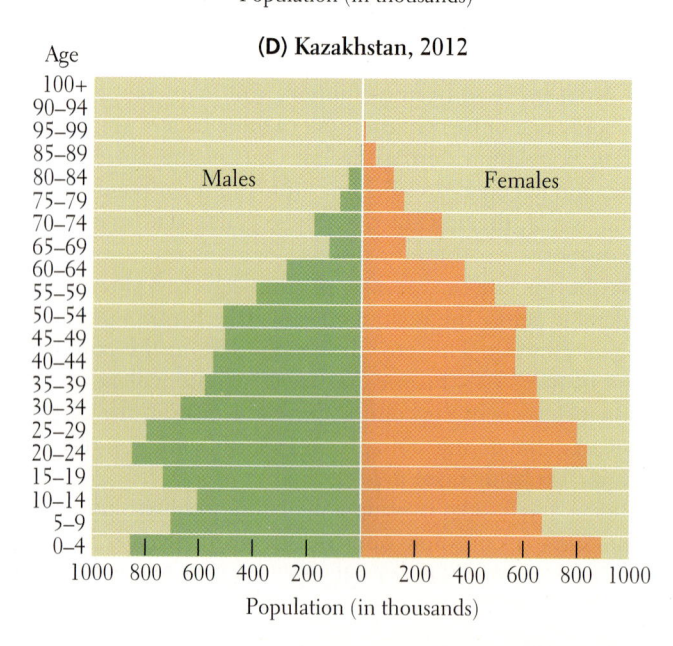

differences in life expectancy and fertility. The pyramids for Belarus and Russia resemble those of European countries (for example, Germany, as shown in Figure 4.21A on page 182), but there are also differences. First, they are significantly narrower at the bottom, indicating that birth rates in the last several decades have declined sharply. Because so few children are born, in 20 years there will be few prospective parents, which means that low birth rates and population decline will probably continue into the foreseeable future. Also, the narrower point at the top for males shows their much shorter average life span compared to women.

In all countries, there has been a recent rebound in birth rates in the youngest age group, ranging from a slight increase in Russia to a significant rebound in Kazakhstan. Kyrgyzstan's pyramid indicates a younger population structure, which is common in less affluent economies. In the Central Asian countries, population will likely expand during the next decades.

Population Distribution

A broad area of moderately dense population forms an irregular triangle that stretches from Ukraine on the Black Sea north to St. Petersburg on the Baltic Sea and east to Novosibirsk, the largest city in Siberia. In this triangle, settlement is highly urbanized, but the cities are widely dispersed (Figure 5.22).

A secondary spur of dense settlement extends south from Russia into **Caucasia**, the mountainous region between the Black Sea and the Caspian Sea, where there are several primate cities of well over 1 million people each. In Central Asia, another patch of relatively dense settlement is centered on the cities of Tashkent. Along major Central Asian rivers during the Soviet period, the development of irrigated cotton farming and mineral extraction resulted in patches of high rural density, fueled partly by ethnic Russian immigration.

> **Caucasia** the mountainous region between the Black Sea and the Caspian Sea

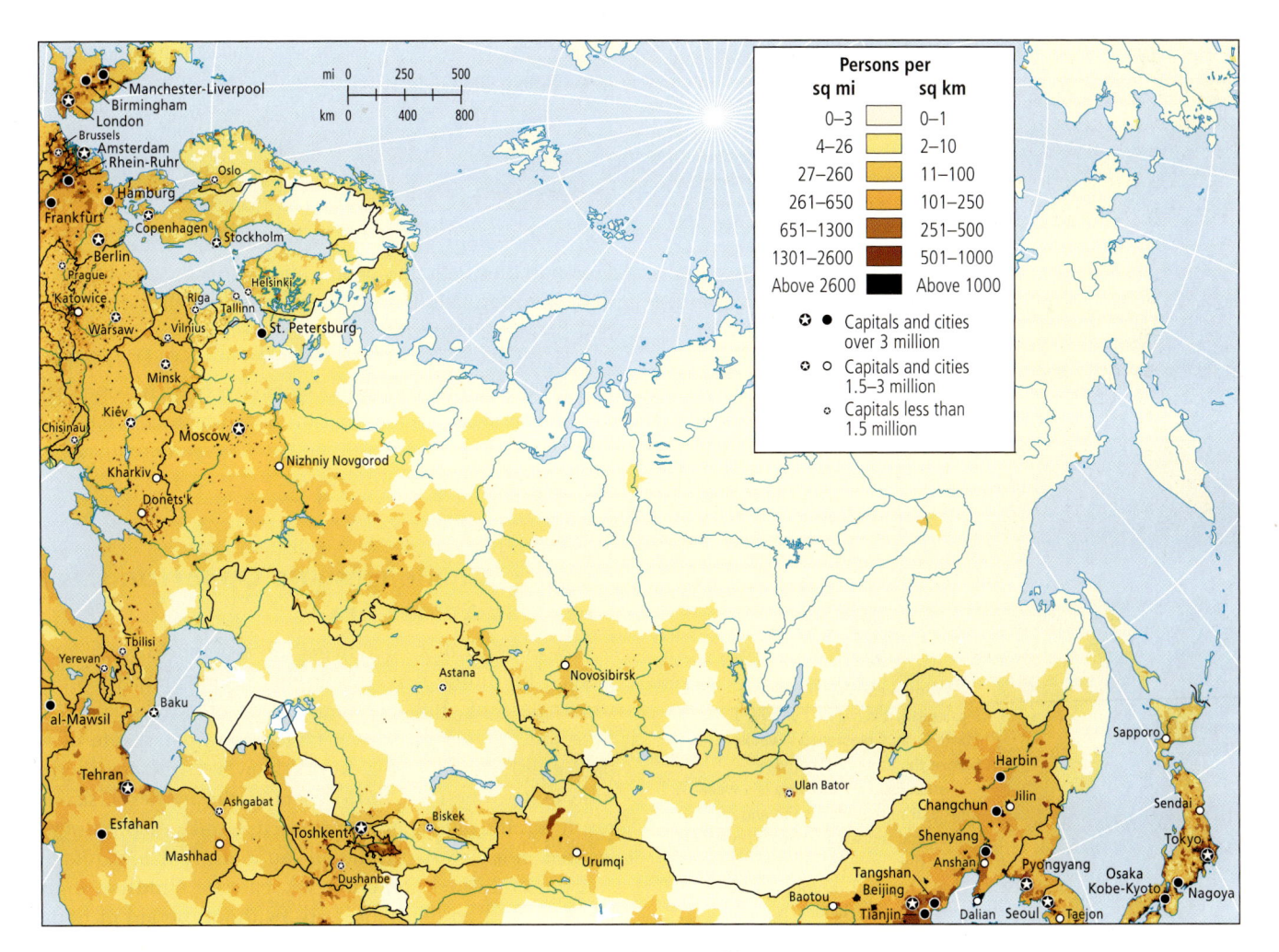

FIGURE 5.22 Population density in Russia and the post-Soviet states. Population trends in this region are quite uneven. While Central Asia and some countries in Caucasia are growing, populations in the rest of the region are shrinking. Some of this unevenness may be related to cultural differences or to varying levels of dependence on social welfare institutions that collapsed with the end of the Soviet Union.

FIGURE 5.23 LOCAL LIVES
Festivals in Russia and the Post-Soviet States

A A man performs stunts on horseback during the festival of Nauryz in Kyrgyzstan. Nauryz is an ancient Zoroastrian festival, originating in Iran, that marks the New Year and the first day of spring on March 21. Forbidden during the Soviet era, Nauryz celebrations are now a source of pride throughout Central Asia. [Vyacheslav Oseledko/AFP/Getty Images]

B Revelers at KaZantip, a 6-week-long electronic dance music festival held on Ukraine's Crimean Peninsula during the summer. More than 300 DJs entertain as many as 150,000 revelers each year at 14 festival venues, with electronic music playing 21 hours a day. A celebrated tradition is the notion that KaZantip is an independent republic, complete with its own constitution prohibiting a wide variety of "chauvinistic" behavior and promoting love and fun. [Oleg Nikishin/Getty Images]

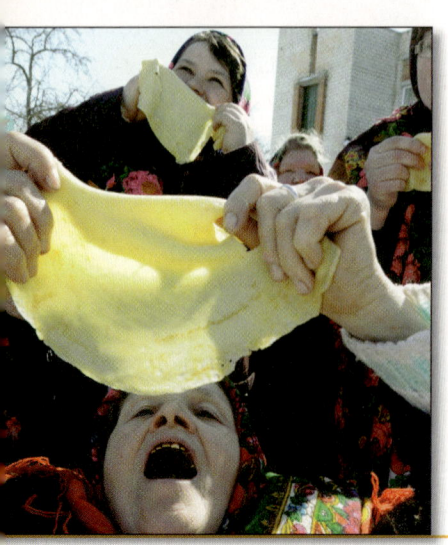

C Women eat pancakes during Maslenitsa, this region's answer to Mardi Gras. Celebrated 7 weeks before the Eastern Orthodox Christian Easter, Maslenitsa is a week-long festival that marks the end of winter and the last chance to revel before the onset of Lent (see page 101 in Chapter 2). During Maslenitsa, eating meat is forbidden, but dairy products and pancakes are consumed in large quantities. [Viktor Drachev/AFP/Getty Images]

THINGS TO REMEMBER

> **GEOGRAPHIC INSIGHT 5**
>
> • **Population and Gender** Populations are shrinking in many parts of the region due largely to lower birth rates and declines in life expectancy. These changes relate to consistently high levels of participation by women in the workforce and to the economic decline in the 1990s that followed the collapse of the Soviet Union. Men and women have been affected differently in the post-Soviet era, with men much more likely to suffer from alcoholism and related health problems.

• By the mid-2000s, the rate of population decline had slowed down considerably across the region, and by 2013 Russia's population was again growing.

• Life expectancy in the region differs dramatically according to gender, with women living 10 to 12 years longer than men. By comparison, around the world, women live on average only 5 years longer than men.

• Recent increases in male life expectancy may be a result of declining rates of alcohol poisoning since the mid-2000s. This change is likely an outcome of government and private campaigns to reduce alcohol consumption and improve treatment of alcoholism.

SOCIOCULTURAL ISSUES

When the winds of change began to blow through the Soviet Union in the 1980s, few people anticipated the rapidity and depth of the transformations or the social instability that resulted. On the one hand, new freedoms have encouraged self-expression, individual initiative, and cultural and religious revivals (Figure 5.23); on the other hand, the post-Soviet era has brought very hard times to many as access to jobs and the social safety net have become more difficult. Men and women have been differently affected by the massive changes in this region.

Gender: Challenges and Opportunities in the Post-Soviet Era

Soviet policies that encouraged all women to work for wages outside the home have transformed this region. By the 1970s, ninety percent of able-bodied women of working age had full-time jobs, giving the Soviet Union the highest rate of female paid employment in the world. However, the traditional attitude that women are the keepers of the home persisted. The result was the *double day* for women. Unlike men, most women worked in a factory or office or on a farm for 8 or more hours and then returned home to cook, care for children, shop daily for food, and do the housework (without the aid of household appliances). Because of shortages (the result of central-planning miscalculations), they also had to stand in long lines to procure necessities for their families.

These special pressures on working women affected diets because there was so little time or space for cooking and there

Sociocultural Issues **225**

FIGURE 5.24 LOCAL LIVES
Foodways in Russia and the Post-Soviet States

A A bowl of borscht, or beet soup, a Ukrainian dish with many variants, is popular throughout this region. Beets are extremely nutritious root vegetables that can tolerate cold weather well, enabling farmers to make the most out of the short growing season found throughout much of this region. Borscht is served warm in the winter and chilled in the summertime. [Alexandra Grablewski/Lifesize/Getty Images]

B Pirozhki are baked yeast buns stuffed with meat, vegetables, or fruit, and glazed with egg to give them a golden color. Shown here are cabbage, egg, and dill pirozhki. They are a popular fast food throughout this region, and even in neighboring Mongolia. They are often made at home and, depending on the stuffing, can be an appetizer, a main dish, or a dessert. [Lara Hata/Photodisc/Getty Images]

C A girl in a refugee camp prepares *non*—a kind of bread cooked throughout Central, South, and Southwest Asia (where it is also known as *naan*). Non is considered sacred by many Central Asians and is traditionally served with great reverence. In Uzbekistan, for example, it is never cut with a knife, but rather is broken by hand and put near each place setting at a table, always with the flat side down (serving it flat side up is considered insulting). One tradition holds that when an Uzbek person leaves a house, he or she should bite off a piece of bread, which will be kept for that person to eat upon return. [Viktor Drachev/AFP/Getty Images]

were so few helpful appliances. The everyday cuisine in most of the region was limited during the Communist era; only recently have market forces made a wider range of food available. Traditional recipes are now being revived. Figure 5.24 shows several distinct dishes of the region that are prepared at home or in restaurants. The popularity of hearty soups and stews and the common use of root vegetables and grains (bread is an essential part of every meal) reflect the region's climate, soil, and agricultural systems.

When the first market reforms in the 1980s reduced the number of jobs available to all citizens, President Gorbachev encouraged women to go home and leave the increasingly scarce jobs to men. Many women lost their jobs involuntarily. By the late 1990s, women made up 70 percent of the registered unemployed, despite the fact that because of illness, death, or divorce, many if not most were the sole support of their families. Consequently, many had to find new jobs.

On average, the female labor force in Russia is now more educated than the male labor force. The same pattern is emerging in Belarus, Ukraine, Moldova, and parts of Muslim Central Asia. In Russia, the best-educated women commonly hold professional jobs, but they are unlikely to hold senior supervisory positions and are paid less than their male counterparts. In 2007, the wages of women professional workers averaged 36 percent less than those of men. Still, this region ranks higher in gender income equity than many others.

The Trade in Women During the economic boom stimulated by marketization and oil and gas wealth, the "marketing" of women became one of the less savory entrepreneurial activities. One part of this market is the Internet-based mail-order bride services aimed at men in Western countries. A woman in her late teens or early twenties, usually seeking to escape economic hardship, pays about $20 to be included in an agency's catalog of pictures and descriptions. (One Internet agency advertises 30,000 such women.) She is then assessed via e-mail or Facebook by the prospective groom, who then travels—usually to Russia or Ukraine—to meet women he has selected from the catalog and to potentially have one of them accompany him to the West to marry.

> **trafficking** the recruiting, transporting, and harboring of people through coercion for the purpose of exploiting them

Sex work has also increased in recent years. Precise numbers are hard to come by, but a 2010 UN report estimates that 140,000 women have been smuggled into Europe, where they become part of a sex trade that generates $3 billion annually. The post-Soviet states comprise the primary source region for Europe's sex workers. Most of the women are supplied by members of the Russian Mafia, who have been known to kidnap schoolgirls or deceive women who are looking for jobs as maids or waitresses in Europe, and then force them to work as prostitutes. This is a form of **trafficking**, which is defined by the United Nations as the recruiting, transporting, and harboring of people through coercion for the purpose of exploiting them.

The Political Status of Women One way for women to address institutionalized discrimination is to increase their political power. Although women were granted equal rights in the Soviet constitution, they never held much power. In 1990, women accounted for 30 percent of Communist Party membership but

Women in Legislatures, 2003, 2009, and 2013

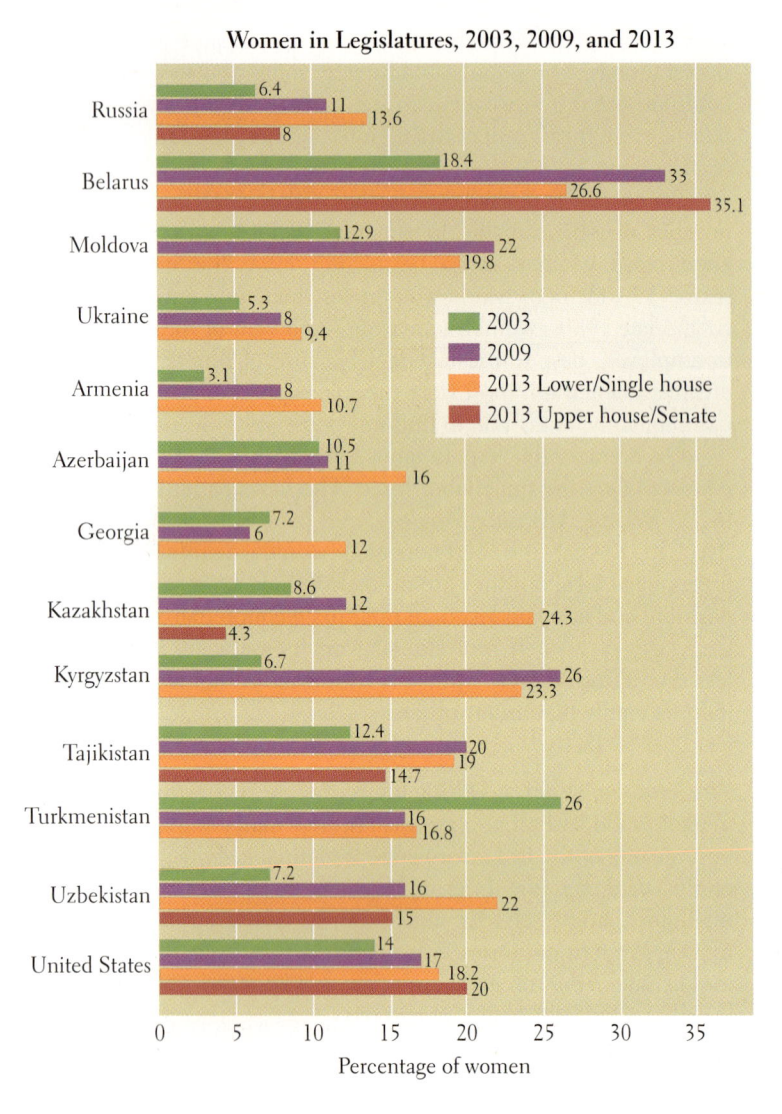

Country	2003	2009	2013 Lower/Single house	2013 Upper house/Senate
Russia	6.4	11	13.6	8
Belarus	18.4	33	26.6	35.1
Moldova	12.9	22	19.8	
Ukraine	5.3	8	9.4	
Armenia	3.1	8	10.7	
Azerbaijan	10.5	11	16	
Georgia	7.2	6	12	
Kazakhstan	8.6	12	24.3	4.3
Kyrgyzstan	6.7	26	23.3	
Tajikistan	12.4	20	19	14.7
Turkmenistan	26	16	16.8	
Uzbekistan	7.2	16	22	15
United States	14	17	18.2	20

FIGURE 5.25 Women legislators, 2003, 2009, and 2013. This graph shows the percentage of legislators in Russia and the post-Soviet states (with the United States for comparison) who are women. Notice that over the last decade in some countries, the number of women in parliament has expanded and then contracted. Also note that the countries with the largest percentage of female legislators are not necessarily those with the most open democratic participation. [Source consulted: *Human Development Report 2011 Statistical Annex*, Table 4, United Nations Development Programme, at http://www.beta.undp.org/content/dam/undp/library/corporate/HDR/2011%20Global%20HDR/English/HDR_2011_EN_Tables.pdf]

just 6 percent of the governing Central Committee. The very few in party leadership often held these positions at the behest of male relatives.

Since the fall of the Soviet Union, the political empowerment of women has grown somewhat, but long-standing cultural biases against women in positions of power remain. In many countries, the number of women in legislatures has risen (Figure 5.25), but often this is because male leaders want to appear more progressive in the eyes of international donors. Frequently they promote women who are the least likely to work for change. Support among women themselves for women's political movements is not widespread, as many fear being seen as anti-male or against traditional feminine roles.

Religious Revival in the Post-Soviet Era

The official Soviet ideology was atheism, and religious practice and beliefs were seen as obstacles to revolutionary change. Orthodox Christianity, which was the official religion of the Russian Empire, was tolerated, but few people went to church, in part because the open practice of religion could be harmful to one's career. Now, religion is a major component of the general cultural revival across the former Soviet Union.

The overall level of religiosity is moderate in Russia by global standards but a number of people, especially those from indigenous ethnic minority groups, are turning back to ancient religious traditions. For example, the Buryats from east of Lake Baikal in Siberia, who are related to the Mongols, are relearning the prayers and healing ceremonies of the Buryat version of Tibetan Buddhism, which they adopted in the eighteenth century. The shamans who lead them have organized into a guild to give official legitimacy to their spiritual work. They now pay taxes on their clergy income.

In Russia, Ukraine, Moldova, Belarus, Georgia, and Armenia, most people have some ancestral connection to Orthodox Christianity. Those with Jewish heritage form an ancient minority, mostly in the western parts of Russia and Caucasia, where they trace their heritage back to 600 B.C.E. Religious observance by both groups increased markedly in the 1990s, and many sanctuaries that had been destroyed or used for nonreligious purposes by the Soviets were rebuilt and restored.

A countertrend to the robust revival of Orthodox Christianity is the spread of evangelical Christian sects from the United States (Southern Baptists, Adventists, and Pentecostals). Evangelical Christianity first came to Russia in the eighteenth century, but after 1991, American missionary activity increased markedly. The notion often promoted by this movement—that with faith comes economic success—may be particularly comforting both to those struggling with hardship and to those adjusting to new prosperity.

In Central Asia, Islam was repressed by the Soviets, who feared Islamic fundamentalist movements would cause rebellion against the dominance of Russia. Today, Islam is openly practiced and is increasingly important politically across Central Asia, Azerbaijan, and some of the Russian Federation's internal ethnic republics, such as Chechnya and Tatarstan. Especially in the Central Asian states, however, the return to religious practices is often a subject of contention. Some local leaders still view traditional Muslim religious practices as obstacles to social and economic reform.

Militant Islamic movements have often been violently repressed by Central Asian governments, in part out of fear of influences from nearby countries where Islam plays a greater role in politics. In some cases, these influences are clear—as in Tajikistan's civil war (1992–1997), which involved Afghan Taliban fighters, and the conflict in Chechnya in which Saudi militants fought. At times, government repression has backfired, as when Uzbekistan and Kyrgyzstan joined forces to eliminate an extremist Islamic movement in 2000. In their fervor to eliminate radicals, these governments persecuted ordinary devout Muslims, especially men, which actually helped recruit *more* Islamic militants instead. Human Rights Watch, an organization that monitors human rights abuses worldwide, reports that Uzbekistan's government jails and tortures believers who worship independently outside the strict supervision of the state. In part, this is done out of concern that the Arab Spring (see page 254) will spread and lead to demands for more human rights and religious freedom in Uzbekistan too.

THINGS TO REMEMBER

- By the 1970s, ninety percent of able-bodied women of working age had full-time jobs, giving the Soviet Union the highest rate of female paid employment in the world.

- By the late 1990s, women made up 70 percent of the registered unemployed, despite the fact that because of illness, death, or divorce, many if not most were the sole support of their families.

- The official Soviet ideology was atheism, and religious practice and beliefs were seen as obstacles to revolutionary change. Now, religion is a major component of the general cultural revival across the former Soviet Union.

- In Central Asia, Islam was repressed by the Soviets, who feared Islamic fundamentalist movements would cause rebellion against the dominance of Russia. Today, Islam is openly practiced and increasingly important politically across Central Asia, Azerbaijan, and some of the Russian Federation's internal ethnic republics, such as Chechnya and Tatarstan.

GEOGRAPHIC INSIGHTS

Russia and the Post-Soviet States: Review and Self-Test

1. Environment: Economic development has taken precedence over environmental concerns in Russia and the post-Soviet states for decades, resulting in numerous environmental problems such as severe pollution of the air and water by toxic industrial waste. This region's contribution to climate change is increasing as more people use automobiles for transportation. In parts of this region, food production systems and water resources are highly vulnerable to the impacts of climate change.

- How did Soviet authorities regard the problem of urban and industrial pollution?

- Describe two cities discussed in this region that are particularly contaminated by pollution, and describe the sources of pollution in each case.

- What factors led to the shrinking of the Aral Sea?

- Why is Central Asia especially vulnerable to climate change?

2. Globalization and Development: After the fall of the Soviet Union, economic reforms and globalization changed patterns of development in this region. Wealth disparity increased and jobs were lost as many Communist-era industries were closed or sold to the rich and well connected. The region is now largely dependent on its role as a leading global exporter of energy resources.

- Who has gained the most from recent economic growth in Russia?

- Discuss the evidence that Russia's natural resources are becoming the foundation of a new role for Russia in the global economy and in regional politics.

- How does Central Asia's geographic situation affect the export of its energy resources?

- Why is the informal sector of the economy important in this region?

- What kinds of spaces produce most of the agricultural output of the region?

3. Power and Politics: This region has a long history of authoritarianism. While there is some pressure to expand political freedoms, there are still few opportunities for the public to influence the political process. Elected representative bodies often act as rubber stamps for very strong presidents and exercise only limited influence on policy making.

- Who led the color revolutions, who funded them, and why?

- Why might the conflict in Ukraine in 2014 be seen as a revival of Cold War tensions? In what ways is it different from Cold War conflicts?

- How has Russia's history of expansion into neighboring lands influenced the political geography of the region?

- What impact did Vladimir Putin's rise to power have on the media?

4. Urbanization: A few large cities in Russia and Central Asia are growing fast, fueled by the expansion of energy exports. Elsewhere, many cities are suffering from a lack of investment and maintenance as their economies struggle in the post-Soviet era.

- How are the reasons for housing shortages today different from those of the Soviet era?

- What type of city are most capitals of this region?
- What relics of Communist-era central planning can be found in cities in the region?
- What factors hold back economic growth in the huge industrial cities located in Siberia and along the Pacific coast?
- What has fueled recent urbanization in Central Asia?

5. Population and Gender: Populations are shrinking in many parts of the region due largely to lower birth rates and declines in life expectancy. These changes relate to consistently high levels of participation by women in the workforce and to the economic decline in the 1990s that followed the collapse of the Soviet Union. Men and women have been affected differently in the post-Soviet era, with men much more likely to suffer from alcoholism and related health problems.

- What Soviet-era developments helped curtail population growth? What is being done in the post-Soviet era to encourage population growth?
- How does life expectancy vary according to gender in this region?
- How does alcohol use relate to male life expectancy in Russia?
- How and why has access to health care and other social services changed for so many in this region since the fall of the USSR?

CRITICAL THINKING QUESTIONS

1. What were the social circumstances that gave rise to the Russian Revolution? Explain why the revolutionaries adopted a form of communism grounded in rapid industrialization and central planning.

2. In general terms, discuss the ways in which the change from a centrally planned economy to a more market-based economy has affected career options and standards of living for the elderly, for young professionals, for unskilled laborers, for members of the military, for women, and for former government bureaucrats.

3. Discuss the changing birth rate in Russia and other countries in the region. What are the major causes of change? What do you think the impact of this decline can have on neighboring countries and regions? Where are similar declines happening elsewhere?

4. Given Russia's crucial role in World War II and the defeat of Germany, discuss what led to the Cold War between the Soviet Union and the West.

5. Given the changing options for women in the region, what do you anticipate will be the trend in population growth in the future? Will birth rates remain low or will they increase? Why?

6. Ethnic and national identities are affecting how the internal republics of Russia envision their futures. How might the geography of Russia change if these feelings were to intensify? In what ways might the global economy be affected?

› THINKING GEOGRAPHICALLY

Now that you have read about Russia and the post-Soviet states, you should be able to answer the following questions about the photos in this chapter. To answer these questions online, go to Geography LaunchPad.

Human Impacts on the Biosphere (page 200)

A What is the life expectancy for men in Norilsk?

C Why was Pripyat evacuated?

D In addition to oil pollution, from what other form of pollution has the Arctic portion of this region suffered?

Vulnerability to Climate Change (page 203)

A Why would rivers like the one in this photo receive less water over the long term due to climate change?

B What major resource did the Aral Sea once provide in abundance?

C Why would a decline in Uzbekistan's cotton industry hurt the country's economy?

Visual History (page 206)

A In addition to trade along the Silk Road, what supported Central Asian cities?

B From where was Orthodox Christianity introduced into this region?

D How does this painting convey the poverty of the Burlaks?

E How many Soviets were killed during World War II?

F How did the Cold War strain the USSR?

Power and Politics (page 215)

A How did the USSR's war in Afghanistan ultimately contribute to democratization in the Soviet Union?

B Who funded the color revolutions?

C How does Russia's unwillingness to let Chechnya secede relate to fossil fuel resources?

Urbanization (page 219)

A Why is Moscow growing so rapidly?

B What larger-scale population trend does Ukraine's decline relate to?

C What is responsible for the growth of Astana?

CHAPTER KEY TERMS

Bolsheviks 207
capitalists 207
Caucasia 223
centrally planned, or socialist, economy 208
Cold War 208
communism 207
Communist Party 208
czar 206
Gazprom 210
glasnost 208
Group of Eight (G8) 211
Mongols 205
nomadic pastoralists 205
nonpoint sources of pollution 201
oligarchs 210
perestroika 208
permafrost 197
privatization 210
Russian Federation 197
Russification 216
Siberia 197
Slavs 205
Soviet Union 195
steppes 197
taiga 198
trafficking 225
tundra 197
underemployment 212
Union of Soviet Socialist Republics (USSR) 195

Land Elevations

meters	feet
4877	16,000
3353	11,000
2134	7,000
914	3,000
305	1,000
152	500
0	0

mi 0 100 200 300 400 500
km 0 200 400 600 800
1:24,500,000
Lambert Azimuthal Equal Area Projection

A Sahara, Morocco

ATLANTIC OCEAN

Madeira Islands (Portugal)

PORTUGAL — Lisbon
Iberian Peninsula
SPAIN — Toledo, Barcelona
Guadalquivir
Balearic Islands
Corsica
Rome — VATICAN CITY
ITALY — Naples
Tyrrhenian Sea
Sardinia
MONTENEGRO
Tiranë — ALBANIA
MACEDONIA — Skopje
BULGARIA — Sofia
Aegean Sea
GREECE — Athens
Ionian Sea
Sicily
MALTA — Valletta
Crete

Black Sea
Istanbul — Bosporus Straits
TURKEY — Ankara
Anatolian Plateau
Lake Tuz
Izmir
Adana
Aleppo
Al Ladhiqiyah
SY
Nicosia — CYPRUS
LEBANON — Beirut
Damascus
ISRAEL — Tel Aviv, Jerusalem
oPT — Gaza
JORD — Amman

Mediterranean Sea

Tanger, Rabat, Casablanca, Fes — MOROCCO
Oran, Algiers, Constantine — Beskra **B**
Atlas Mountains
Tunis — TUNISIA
El Oued, Gafsa, Sfax
Touggourt
Marrakech
Ain Sefra
Ouargla
Tripoli
Banghazi
Tobruk
Sidi Barrani
Alexandria
As Sallum
Port Said
Cairo
Al Jizah — Suez
Al Aqabah **C**
Tabuk

Canary Islands (Spain)
Las Palmas
El Aaiún
Western Sahara (Morocco)
Erfoud **G**
A

ALGERIA
LIBYA
Libyan Desert
EGYPT
Al Uqsur (Luxor) **F**
Aswan, Aswan High Dam
Lake Nasser

Sahara

Ahaggar Mountains
Tamanrasset
Zillah
Waw An Namus
Al Jawf
Bir Misahah
Nubian Desert
Dunqulah

Nouakchott
MAURITANIA
MALI
Tombouctou
NIGER
CHAD
SUDAN
Khartoum
Al Fashir
Blue Nile

NEGAL
THE GAMBIA
Bamako — Niger
GUINEA
Conakry
Freetown — SIERRA LEONE
Monrovia — LIBERIA
CÔTE D'IVOIRE
Yamoussoukro
Abidjan
Ivory Coast
Ouagadougou — BURKI
White Volta
Black Volta
Lac de Kossou
GHANA
Accra
Lomé — Porto Novo
Gold Coast
Bight of Benin
Gulf of Guinea

B Atlas Mountains, Algeria

CAMEROON
Malabo — EQUATORIAL GUINEA
Douala
Yaoundé
Bangui
CENTRAL AFRICAN REPUBLIC
Libreville — GABON
Prime Meridian
SOUTH SUDAN
Shambe
Juba
White Nile
DEMOCRATIC REPUBLIC OF THE CONGO
UGANDA
Kampala
TANZ AN

C Gulf of Aqaba, Red Sea

D Rub'al Khali dunes, Saudi Arabia

Vilnius, Minsk — BELARUS — RUSSIA
Kiev — UKRAINE — Kharkiv
Dnipropetrovsk
MOLDOVA — Chisinau
Donetsk
Rostov-na-Donu
Sea of Azov
ROMANIA — Bucharest
Danube
Odesa
Ordu

chapter 6

NORTH AFRICA AND SOUTHWEST ASIA

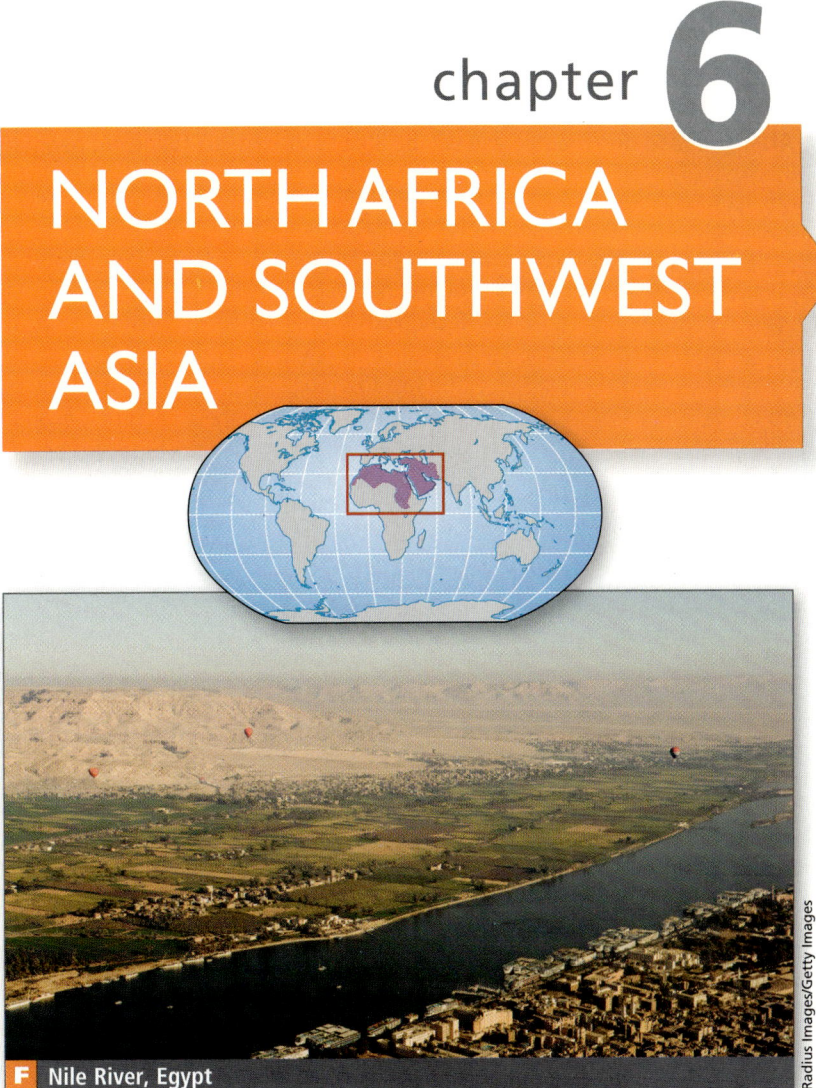

E Mt. Ararat, Turkey

F Nile River, Egypt

G Wadi el Mellah, Morocco

FIGURE 6.1 Regional map of North Africa and Southwest Asia.

231

GEOGRAPHIC INSIGHTS: NORTH AFRICA AND SOUTHWEST ASIA

After you read this chapter, you will be able to discuss the following geographic insights as they relate to the five thematic concepts:

1. **Environment:** The population of this predominantly dry region is growing faster than is access to cultivable land and water for agriculture and other human uses. Many countries here are dependent on imported food, and because global food prices are unstable, the region often has food insecurity. Climate change could reduce food output both locally and globally and make water, already a scarce commodity, a source of conflict. New technologies offer solutions to water scarcity but are expensive and unsustainable.

2. **Globalization and Development:** The vast fossil fuel resources of a few countries in this region have transformed economic development and driven globalization. In these countries, economies have become powerfully linked to global flows of money, resources, and people. Politics have also become globalized, with Europe and the United States strongly influencing many governments.

3. **Power and Politics:** Despite the presence of elected bodies of government, authoritarian power structures prevail throughout North Africa and Southwest Asia. Beginning with the Arab Spring of 2010, waves of protests swept the region, resulting in the overthrow of several authoritarian governments. Within the region, official responses to the Arab Spring have included both repression and reform.

4. **Urbanization:** Two patterns of urbanization have emerged in the region, both tied to the global economy. In the oil-rich countries, the development of spectacular new luxury-oriented urban areas is based on global flows of money, credit, goods, and skilled people. In the oil-poor countries, cities are old and have little capacity to handle the masses of poor, rural migrants attracted by the promise of jobs. In these older, poorer cities, jobs are often scarce and many people live in overcrowded slums with few services.

5. **Population and Gender:** This region has the second-highest population growth rate in the world, after that of sub-Saharan Africa. Part of the reason for this is that women are generally poorly educated and tend not to work outside the home. Childbearing thus remains crucial to a woman's status, a situation that encourages large families. The role and social standing of women in the region is one of the issues being addressed by current political movements.

The North Africa and Southwest Asia Region

The region of North Africa and Southwest Asia (shown in Figures 6.1 and 6.2) contains 21 countries plus the occupied Palestinian Territories. Physically, the region is part of two continents: Africa and Eurasia. The North African countries of this region stretch from Western Sahara and Morocco on the Atlantic Ocean, through Algeria, Tunisia, Libya, and Egypt along the Mediterranean Sea, plus Sudan, which lies south of Egypt. Eurasia begins on the eastern side of the Red Sea with the Arabian Peninsula, which consists of Saudi Arabia, Yemen, Oman, the United Arab Emirates, Qatar, Bahrain, and Kuwait (see Figure 6.1C); the countries of the eastern Mediterranean *littoral* (shoreline), including Jordan, Israel, the occupied Palestinian Territories, Lebanon, and Syria; and Turkey, Iraq, and Iran. This region is often referred to as the *Middle East*, a term we do not use in this book because it arose during the colonial era and describes the region from a limited Euro-American geographic perspective. To someone in China or Japan, the region lies to the far west, and to someone in Russia, it lies to the south.

The five thematic concepts in this book are explored as they arise in the discussion of regional issues; the interactions between two or more themes are often featured. Vignettes, like the one on the next page about the changing role of women in this region, illustrate one or more of the themes as they are experienced in individual lives.

The Arab Spring, a movement that will likely continue evolving for decades, has been brought on by deep and structural problems. Large numbers of the poor, the uneducated, the educated unemployed, minorities, and women have been pushed further and further into poverty and powerlessness by political and economic systems that have privileged the wealthy and politically connected. Movements like the Arab Spring arise when enough people decide they can no longer tolerate these systems.

The issue of gender equity in the region of North Africa and Southwest Asia is controversial because many people are convinced that only Westerners (meaning people of European origin) raise this issue, and that they do so in order to belittle Muslim culture for having a set of values that places females in a special, and subservient, category. But Mona Eltahawy's article and the response it has received across the region suggests that gender equity is becoming a central theme in the Arab Spring movement and that women are leading a profound transformation of this region.

What Makes North Africa and Southwest Asia a Region?

To most outsiders, North Africa and Southwest Asia is a region characterized by five qualities: It is the center of the religion of Islam; it holds a great deal of Earth's petroleum resources; water is generally scarce here; Arab culture dominates most countries; and women are discriminated against more intensely than in most of the rest of the world. While these five features are all present, they alone provide a far too simplistic picture of the region.

The vast majority of people practice **Islam**, a monotheistic religion that emerged between 601 and 632 C.E. Islam is a faith that is interpreted in many different ways. Most Muslims are moderate in their thinking and accept the validity of other beliefs, especially that of Christianity, because Mary and Jesus both play important roles in Islam. Only a vocal minority of Muslims is drawn to ultra-fundamentalist versions of Islam known as *Islamism*. The term **Islamism** refers to a Muslim religious activist movement that seeks to curb secular influences that are spreading as a result of globalization. Some Islamists characterize Western influence as corrupt and destructively self-serving, while others are more open to Western ideas. Stronger in some parts of the region than in others, Islamism has tended to gain popularity during economic recessions and to wane with economic booms.

Fossil fuel reserves—made up of oil and natural gas formed over millions of years from the fossilized remains of dead plants and animals—are highly uneven in their distribution (see Figure 6.18 on page 251) and are found mainly around the Persian Gulf. These fossil fuels are extracted and exported throughout the world at tremendous profit, but the profits are not equitably distributed to the people in the countries from which they are generated. For example, in the oil-rich Gulf states (see the discussion of wealth disparities), profits go primarily to the members of a few large, privileged families, and only modest amounts are spread to the rest of the citizens.

This region is generally arid but the degree of aridity and its impact on development varies widely. Newly revealed underground water resources may change development options.

Arab culture and language is widespread, but many people in the region are not Arab. The population of Turkey and Iran, the second and third most populous countries in the region, are of non-Arab ethnicities, as are many minority populations, such as the Kurds, Berbers, Christians, and Jews.

Finally, the role and status of women, a point of contention for many years, is in transition. The experiences of women vary widely from country to country. In Turkey, women have many more options for education and work than in Saudi Arabia or Yemen. Gender issues also vary between more rural and urban areas. Urban women tend to be more educated, outspoken, and active in commerce, public life, and government, while rural women tend to lead more secluded domestic lives with few educational opportunities. The movement toward greater gender equity is generally strongest in urban areas.

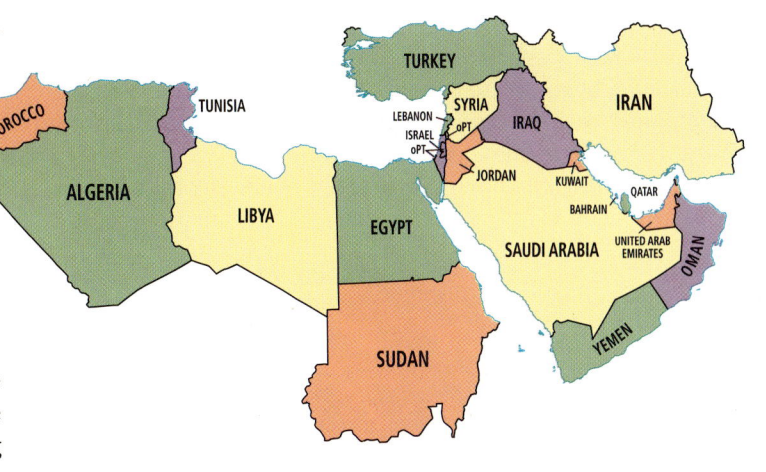

FIGURE 6.2 Political map of North Africa and Southwest Asia.

Terms in This Chapter

In this book, as mentioned earlier, we choose to not use the common term *Middle East*. The term *Arab world* is used only where it applies, since many people in the region are not of Arab ethnicity.

We use the term **occupied Palestinian Territories (oPT)** to refer to Gaza and the West Bank, those areas where Israel still exerts control despite treaty agreements; the word "occupied" is lowercase to show the supposed temporary quality of the occupation. The U.S. Department of State uses the term *Palestinian Territories*. The United Nations, upon recently agreeing to give the territory observer status, now refers to it as the *State of Palestine*.

> **Islam** a monotheistic religion that emerged in the seventh century C.E. when, according to tradition, the archangel Gabriel revealed the tenets of the religion to the Prophet Muhammad
>
> **Islamism** a grassroots religious revival in Islam that seeks political power to curb what are seen as dangerous secular influences; also seeks to replace secular governments and civil laws with governments and laws guided by Islamic principles
>
> **fossil fuel** a source of energy formed from the remains of dead plants and animals
>
> **occupied Palestinian Territories (oPT)** Palestinian lands occupied by Israel since 1967

GLOBAL PATTERNS, LOCAL LIVES

Mona Eltahawy, a Muslim Egyptian-American journalist, was angry when she wrote an essay for *Foreign Policy* magazine entitled "Why Do They Hate Us?" in the spring of 2012. It is easy to see why. The *Arab Spring*—a political movement that washed across North Africa and into Egypt, beginning in 2010—held the promise of real change in her native land. For a time, the tens of thousands of largely peaceful male and female demonstrators in Cairo's Tahrir Square shared a common purpose: the resignation of the authoritarian president, Hosni Mubarak (Figure 6.3). He had remained in power for nearly 30 years, aided by a powerful and oppressive military, and supported by internal corruption and large foreign aid packages from the United States.

For the many thousands of women—students and young professionals—who participated in the Cairo demonstrations, it was exhilarating to be welcomed by male compatriots, many of whom were members of the Muslim Brotherhood, a large and amorphous organization of men opposed to the corrupt Mubarak regime and its affiliations with the West. The Muslim Brotherhood, however, is also known for its stance against liberalizing women's rights. For a time, the common purpose of ousting Mubarak united everyone.

Eltahawy, who grew up in Saudi Arabia, the United Kingdom, and Egypt, and who has focused her career on social justice and women's rights in Egypt, joined the throng of demonstrators. As the protests went on, not only did retaliation by the military and police toward male and female demonstrators become increasingly brutal, but antagonism toward women by the Muslim Brotherhood demonstrators and even from young, more liberal male demonstrators grew, leading to random violence by men against women. Mubarak's resignation in February 2011 left undetermined the way in which a new government would be established; so as the military asserted control, the demonstrations continued. One day in the fall of 2011, Eltahawy, who continued to report on the increasingly frustrated demonstrators in Tahrir Square, was arrested along with some others. While in custody, both her arms were broken in a beating, and she was sexually assaulted, as were many other women demonstrators.

In her essay, which she wrote some months after her release, Eltahawy raged against repressive political authorities and against Muslim men as a whole, who she expected to be in solidarity with women demonstrators. Her hope that movement toward gender equity would be one of the end products of the Arab Spring was dashed by what she saw and experienced, so she lashed out with a diatribe against the many large and small ways in which she saw women being belittled—not by the Qur'an (see page 237), but by religious social practice in Egypt and elsewhere (Figure 6.4). In so expressing her rage, Eltahawy seems to have invigorated a cross-regional discussion on gender equity that is attracting serious male and female participants. They are expressing a surprising level of agreement that the Arab Spring must include the human rights of women. [Source: Foreign Policy, Al Jazeera. For more detailed information, see Text Sources and Credits.]

FIGURE 6.3 Women and the Arab Spring. An Egyptian woman holds up a widely circulated photo that shows riot police stripping and beating a woman during an Arab Spring protest in Cairo in 2011. Women have been very active in protests in Egypt and elsewhere in the region in recent years, and some groups, including parts of Egypt's government, have felt threatened by this. Some female protestors have become a target for sexual assaults designed to make others perceive them as dangerous, "dirty," or immoral.

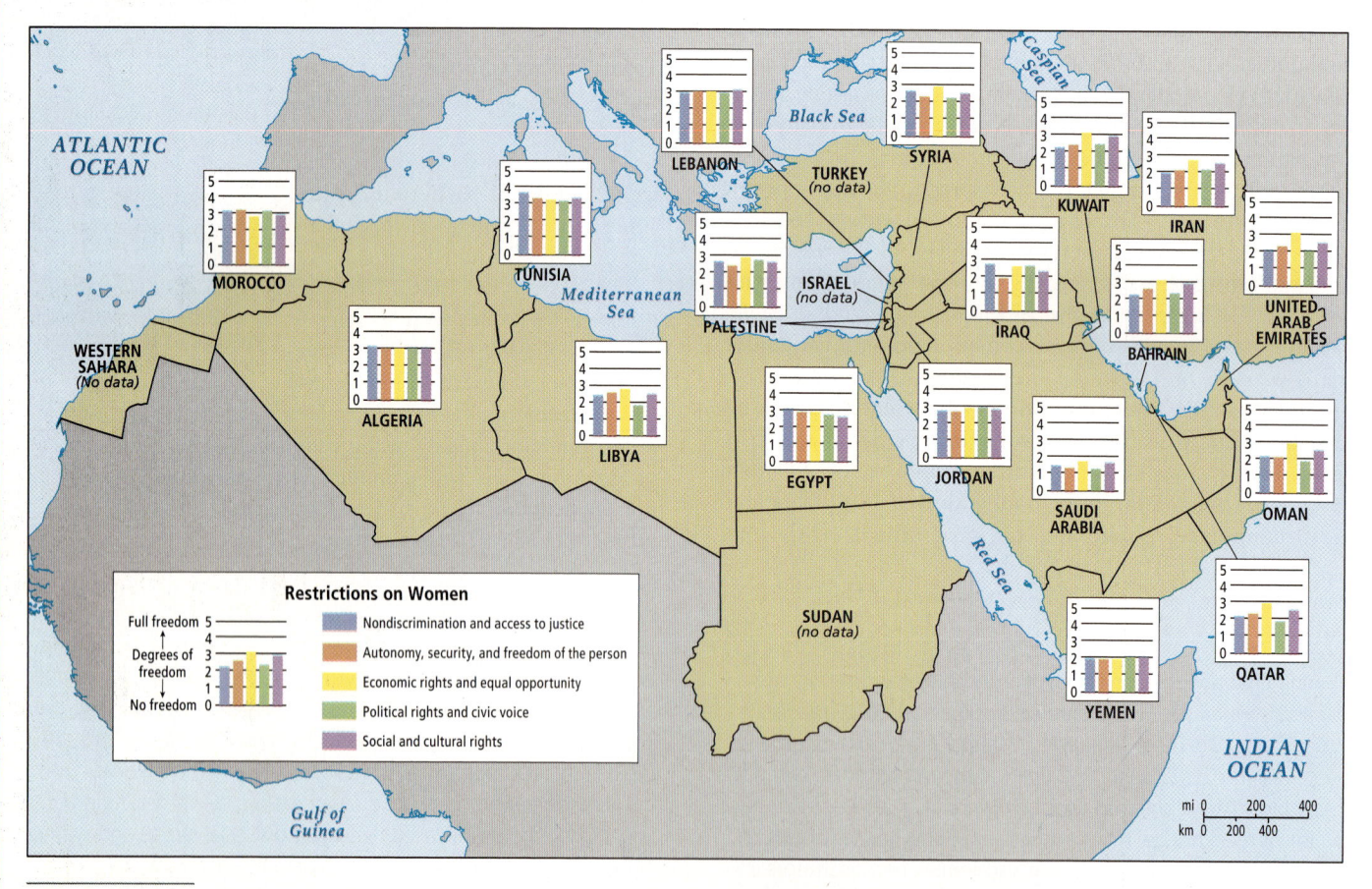

FIGURE 6.4 Variations in women's freedom to exercise their human rights in five categories. Each country was evaluated on a scale of 1 to 5, with 1 being the lowest degree of freedom and 5 being the highest. A variety of factors was assessed for each nation: actions taken and omitted; the legal system; the political environment; actual implementation of laws and policies; and the role that non-state individuals and institutions played in strengthening or weakening women's rights and freedoms. [Sources consulted: Sanja Kelly and Julia Breslin, eds., Women's Rights in the Middle East and North Africa (New York: Freedom House); Freedom House reports on Israel and Turkey, at http://www.freedomhouse.org]

> **THINGS TO REMEMBER**
>
> - The Arab Spring, a broad-based but not centrally organized movement against corruption and toward more political freedom, has spread across the region since 2010.
> - Women are increasingly participating in political movements but continue to face difficulties in doing so.
> - Issues of gender equity are becoming central to the public dialogue.

PHYSICAL GEOGRAPHY AND ENVIRONMENTAL ISSUES

PHYSICAL PATTERNS

Landforms and climates are particularly closely related in this region. The climate is dry and hot in the vast stretches of the relatively low, flat land; it is somewhat moister where mountains capture orographic rainfall. The lack of vegetation perpetuates aridity. Without plants to absorb and hold moisture, the rare but occasionally copious rainfall simply runs off, evaporates, or sinks rapidly into underground aquifers, of which there are many across the region.

Climate

No other region in the world is as dry as North Africa and Southwest Asia (Figure 6.5A). A belt of dry air that circles the planet between roughly 20° N and 30° N creates desert climates in the Sahara of North Africa, the eastern Mediterranean, the Arabian Peninsula, Kuwait, Iraq, and Iran.

The Sahara's size and location under this high-pressure belt of dry air make it a particularly hot desert region. In some places, temperatures can reach 130°F (54°C) in the shade at midday (see Figure 6.1A). With little water or moisture-holding vegetation to retain heat, nighttime temperatures can drop quickly to below freezing. Nevertheless, in even the driest zones, humans survive as traders and nomadic herders at scattered oases, where they maintain groves of drought-resistant plants such as date palms. Desert inhabitants often wear light-colored, loose, flowing robes that reflect the sunlight, retain body moisture during the day, and provide warmth at night.

In the uplands and at the desert margins, enough rain falls to nurture grass, some trees, and limited agriculture. Such is the case in northwestern Morocco; along the Mediterranean coast in Algeria, Tunisia, and Libya; in the highlands of Sudan, Yemen, and Turkey; and in the northern parts of Iraq and Iran (see Figure 6.5B–D). The rest of the region, generally too dry for cultivation, has for generations been the prime herding lands for nomads, such as the Kurds of Southwest Asia, the Berbers and Tuareg (a branch of Berbers) in North Africa, and the Bedouin of the steppes and deserts on the Arabian Peninsula. Recently, most nomads have been required to settle down; some of their lands are now irrigated for commercial agriculture, but the general aridity of the region means that sources of irrigation water (rivers and aquifers) are scarce.

Landforms and Vegetation

The rolling landscapes of rocky and gravelly deserts and steppes cover most of North Africa and Southwest Asia (see the Figure 6.1 map and Figure 6.5). In a few places, mountains capture sufficient moisture to allow plants, animals, and humans to flourish. In northwestern Africa, the Atlas Mountains, which stretch from Morocco on the Atlantic coast to Tunisia on the Mediterranean coast, lift damp winds from the Atlantic Ocean, creating orographic rainfall of more than 50 inches (127 centimeters) per year on windward slopes (see Figure 6.1B). In some Atlas Mountain locations, there is enough snowfall to support a ski industry.

A rift that formed between two tectonic plates—the African Plate and the Arabian Plate—separates Africa from Southwest Asia (see Figure 1.8 on page 16). The rift, which began to form about 12 million years ago, is now filled by the Red Sea. The Arabian Peninsula lies to the east of this rift. Mountains bordering the rift in the southwestern corner rise to 12,000 feet (3658 meters). They capture enough rainfall to sustain agriculture, which has been practiced there for thousands of years. Irrigation has been used for millennia to increase yields.

Behind these mountains, to the northeast of the Arabian Peninsula, lies the great desert region of the Rub'al Khali. Like the Sahara, it has virtually no vegetation. The sand dunes of the Rub'al Khali, which are constantly moved by strong winds, are among the world's largest, some as high as 2000 feet (610 meters; see Figure 6.1D).

The landforms of Southwest Asia are more complex than those of North Africa. The Arabian Plate is colliding with the Eurasian Plate and pushing up the mountains and plateaus of Turkey and Iran (see Figure 1.8). Turkey's mountains lift damp air that passes over Europe and the Mediterranean from the Atlantic, resulting in considerable rainfall (see Figure 6.1E). Only a little rain makes it over the mountains to the interior of Iran, which is very dry. The tectonic movements that create mountains also create earthquakes, a common hazard in Southwest Asia.

There are only three major river systems in the entire region, and all have attracted human settlement for thousands of years. The Nile flows north from the moist central East African highlands (see Figure 6.1F). It crosses arid Sudan and desert Egypt and forms a large delta on the Mediterranean. The Euphrates and Tigris rivers both begin with the rain that falls in the mountains of Turkey. The rivers then flow southeast to the Persian Gulf. A much smaller river, the Jordan, starts as snowmelt in the uplands of southern Lebanon and flows through the Sea of Galilee to the Dead Sea. Most other streams are for most of the year dry riverbeds, or *wadis*, carrying water only after the generally light rains that fall between November and April (Figure 6.1G shows a wadi in Morocco).

North Africa and Southwest Asia were home to some of the very earliest agricultural societies. Today, rain-fed agriculture is practiced primarily in the highlands and along parts of the Mediterranean

FIGURE 6.5   PHOTO ESSAY: Climates of North Africa and Southwest Asia

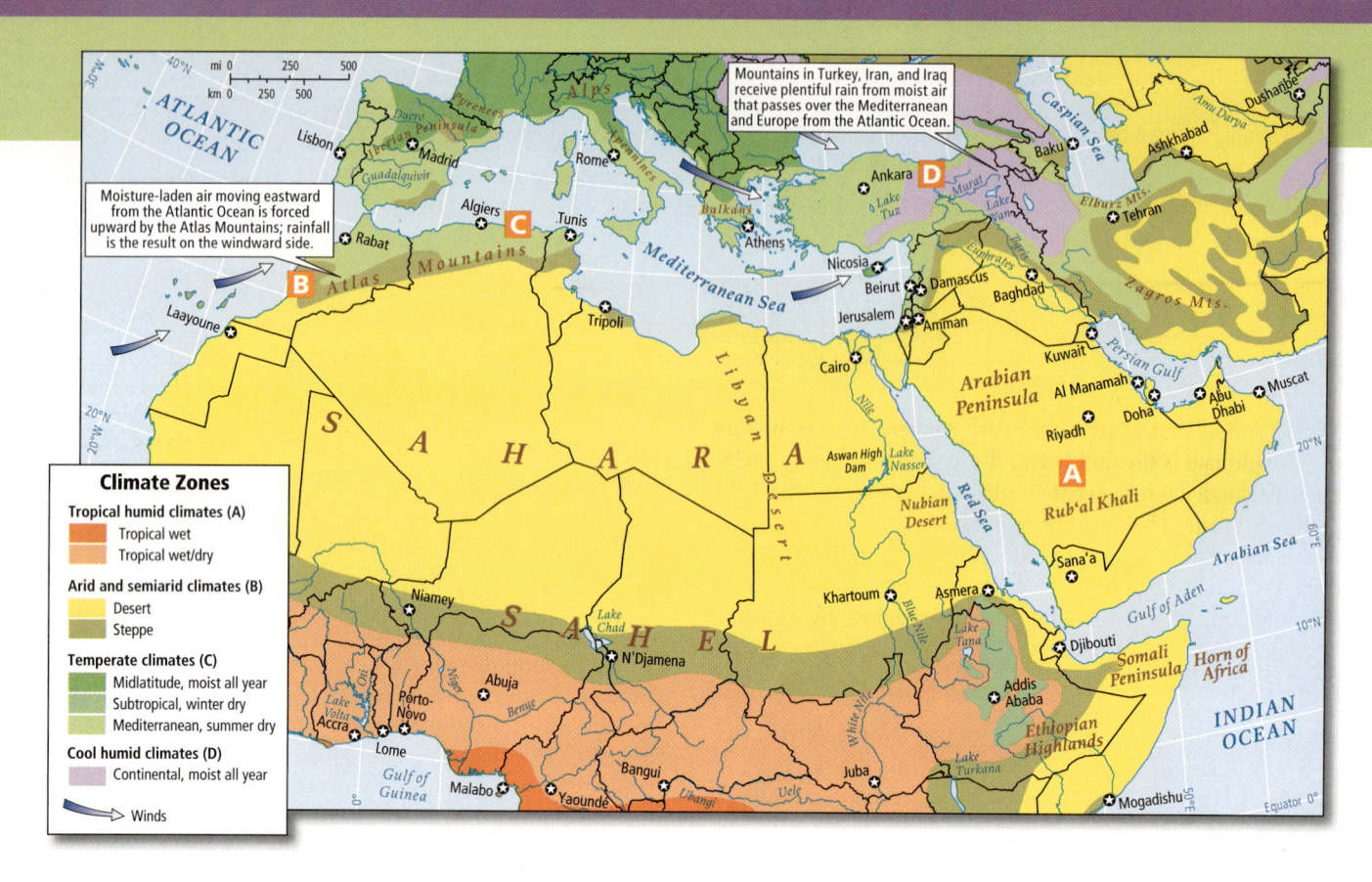

Moisture-laden air moving eastward from the Atlantic Ocean is forced upward by the Atlas Mountains; rainfall is the result on the windward side.

Mountains in Turkey, Iran, and Iraq receive plentiful rain from moist air that passes over the Mediterranean and Europe from the Atlantic Ocean.

Climate Zones

Tropical humid climates (A)
- Tropical wet
- Tropical wet/dry

Arid and semiarid climates (B)
- Desert
- Steppe

Temperate climates (C)
- Midlatitude, moist all year
- Subtropical, winter dry
- Mediterranean, summer dry

Cool humid climates (D)
- Continental, moist all year

→ Winds

A Desert, Saudi Arabia

B Steppe, Morocco

C Mediterranean, summer dry, Algeria

D Continental, moist all year, Turkey

236

coast, where there is enough precipitation to grow citrus fruits, grapes, olives, and many vegetables, though supplemental irrigation is often needed. While irrigated agriculture is an age-old type of farming, it has become increasingly significant in modern times, both in the valleys of the major rivers, where seasonal flooding fills irrigation channels, and where aquifers are tapped by wells to provide water for cotton, wheat, barley, vegetables, and fruit trees.

ENVIRONMENTAL ISSUES

Environmental concerns are only beginning to be an overt focus in this region. This is in part because for thousands of years people here have confronted the challenges of a naturally arid environment and have been quite attuned to its scarcities, which exceed those in most other places on Earth (Figure 6.6).

An Ancient Heritage of Water Conservation

> **GEOGRAPHIC INSIGHT 1**
>
> **Environment:** The population of this predominantly dry region is growing faster than is access to cultivable land and water for agriculture and other human uses. Many countries here are dependent on imported food, and because global food prices are unstable, the region often has food insecurity. Climate change could reduce food output both locally and globally and make water, already a scarce commodity, a source of conflict. New technologies offer solutions to water scarcity but are expensive and unsustainable.

The Qur'an (or Koran), the holy book of Islam, guides believers to avoid spoiling or degrading human and natural environments and to share resources, especially water, with all forms of life. In actual practice, the residents of this region conserve water better than most people in the world. Daily bathing is a religious requirement and is often done in public baths, where water use is minimized. For millennia, mountain snowmelt has been captured and moved to dry fields and villages via constructed underground water conduits, called *qanats*, that minimize evaporation. Likewise, traditional architectural designs that maximize shade and airflow are used to create buildings that stay cool.

Qur'an (or Koran) the holy book of Islam, believed by Muslims to contain the words Allah revealed to Muhammad through the archangel Gabriel

Despite their history of water-conserving technologies and practices, however, this region's 477 million residents now have such limited water resources that even clever combinations of ancient and modern measures are no longer sufficient to ensure an adequate supply of water. Growing populations, water pollution, and unwise modern usages of water virtually guarantee that water shortages will be more extreme in the future (Figure 6.7A–C), especially with the likelihood of climate change–related drier conditions. As shown in the map in Figure 6.6, Turkey has the most plentiful water resources, followed by Iran, Iraq, and Syria. However, all other countries in the region suffer from *freshwater scarcity*, meaning that they have less water than the minimum the United Nations considers necessary to support basic human development—1000 cubic meters per person per year.

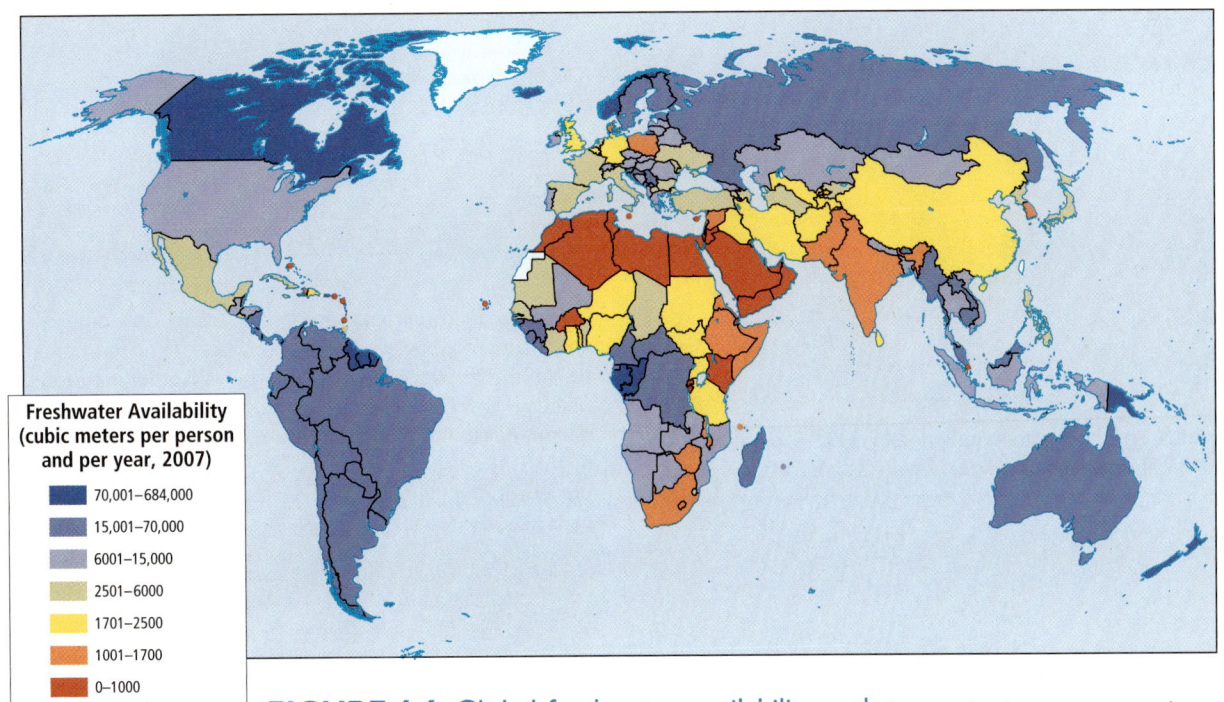

FIGURE 6.6 Global freshwater availability and stress. Freshwater stress and scarcity occur when so much water is withdrawn from rivers, lakes, and aquifers that not enough water remains to meet human and ecosystem requirements for sustainability. [Source consulted: United Nations Environmental Programme, *Vital Water Graphics: An Overview of the State of the World's Fresh and Marine Waters*, 2nd ed., 2008, at http://www.unep.org/dewa/vitalwater/article69/html]

FIGURE 6.7 PHOTO ESSAY: Human Impacts on the Biosphere in North Africa and Southwest Asia

War and water scarcity have resulted in major impacts on the biosphere in this region. The growing population threatens to intensify these stresses on ecosystems.

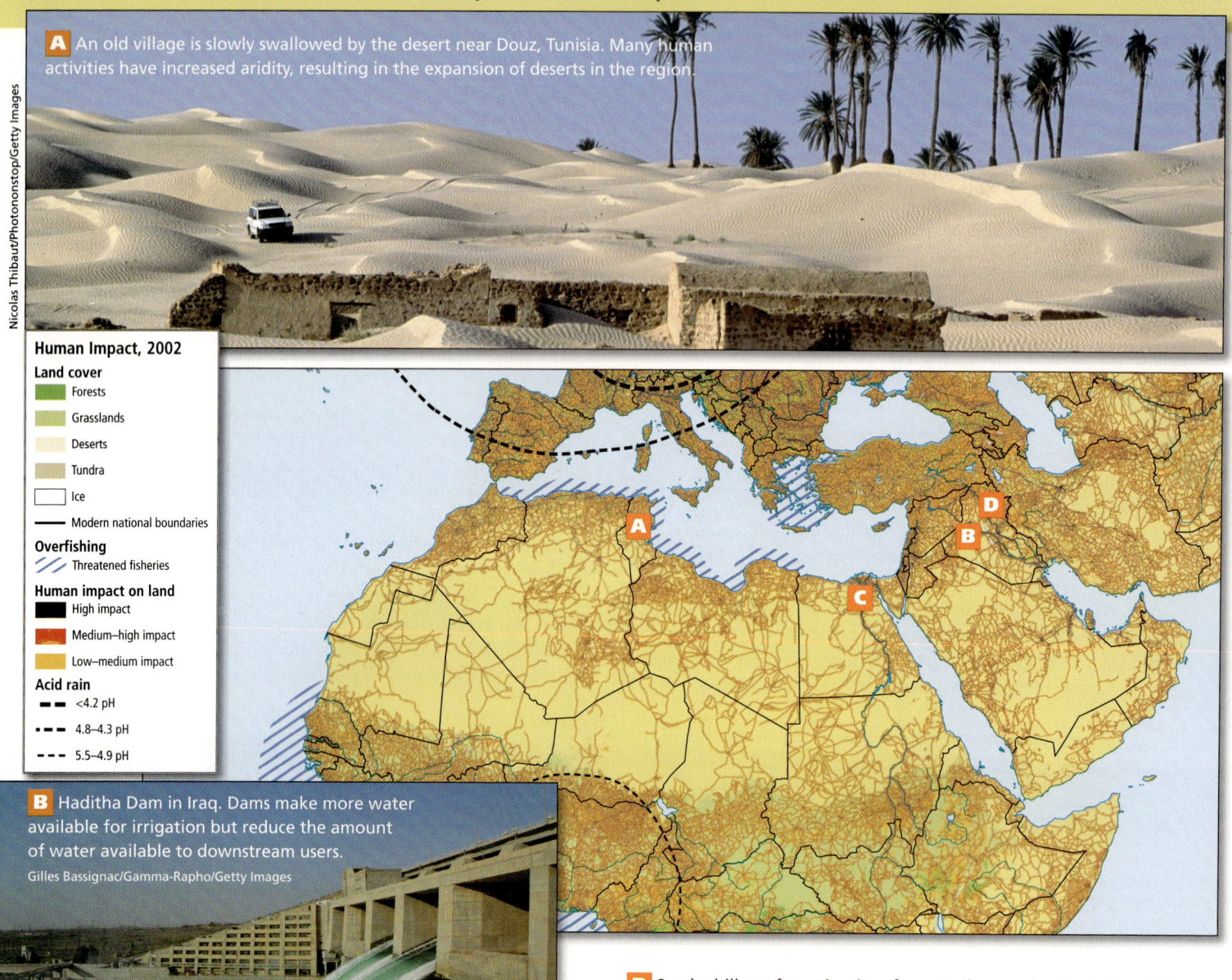

A An old village is slowly swallowed by the desert near Douz, Tunisia. Many human activities have increased aridity, resulting in the expansion of deserts in the region.

Human Impact, 2002

Land cover
- Forests
- Grasslands
- Deserts
- Tundra
- Ice
- Modern national boundaries

Overfishing
- Threatened fisheries

Human impact on land
- High impact
- Medium–high impact
- Low–medium impact

Acid rain
- <4.2 pH
- 4.8–4.3 pH
- 5.5–4.9 pH

B Haditha Dam in Iraq. Dams make more water available for irrigation but reduce the amount of water available to downstream users.

C Outside Alexandria, Egypt, an automated irrigation system moves through a potato plantation located on what would otherwise be desert land.

D Smoke billows from the site of an attack on a pipeline in Iraq. War and insurgency have been major sources of environmental degradation in this region. The largest oil spill in history occurred during the first Gulf War (1991) when the Iraqi government spilled 300 million gallons of oil into the Persian Gulf in order to thwart a land invasion by the United States. Lebanon suffered a smaller, though similarly devastating, spill in 2006 as a result of Israeli bombing raids.

Could There Be New Sources of Water?

A recent reassessment of groundwater storage in North Africa by a consortium of British scientists revealed that deep aquifers in Algeria, Libya, Egypt, and Sudan may hold up to 100 times the water available through renewable freshwater resources (Figure 6.8). This *fossil water*, deposited thousands of years ago, is unevenly distributed and so deep that special pumps are required to bring it to the surface. There probably isn't enough to meet the expected increases in demand for agricultural irrigation and drinking water for rapidly growing urban populations, but this stored groundwater may help alleviate scarcities caused by climate variability.

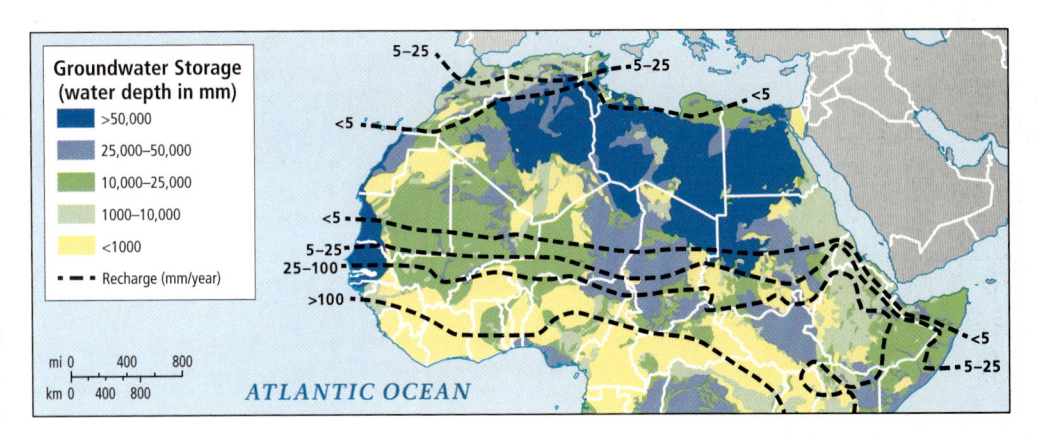

FIGURE 6.8 Water for the future? A map of newly discovered fossil groundwater deep below the deserts in North Africa. This ancient water deposit, already tapped in Libya, is expensive to extract and is unlikely to meet future needs because it is not being replenished. [Source consulted: A. M. MacDonald, H. C. Bonsor, B. E. O. Dochartaigh, and R. G. Taylor. (2012) Quantitative maps of groundwater resources in Africa. *Environmental Research Letters* 7. 024009 doi:10.1088/1748-9326/7/2/024009]

Water and Food Production

The predominant use of water in North Africa and Southwest Asia is for irrigated agriculture, even though agriculture does not contribute significantly to national economies. In Tunisia, for example, agriculture accounts for just 10 percent of GDP but 86 percent of all the water used. Just 13 percent of water is used in homes, and only 1 percent by industry. Most national governments actually subsidize irrigated agriculture because they are worried about depending completely on imported food. Irrigated agriculture also supports jobs and family economies that are crucial to rural communities. Some of this region's food specialties appear in Figure 6.9.

Until the twentieth century, agriculture was confined to a few coastal and upland zones where rain could support cultivation, and to river valleys (such as the Nile, Tigris, and Euphrates valleys) where farms could be irrigated with simple gravity-flow technology. However, to accommodate population growth and development, Libya, Egypt, Saudi Arabia, Tunisia, Syria, Turkey, Israel, and Iraq all now have ambitious mechanized irrigation schemes that have expanded agriculture deep into formerly uncultivable desert environments (see Figure 6.7B, C).

> **salinization** a process that occurs when large quantities of water are used to irrigate areas where evaporation rates are high, leaving behind dissolved salts and other minerals
>
> **desertification** a set of ecological changes that converts arid lands into deserts

Over time, irrigation projects damage soil fertility through **salinization**. When irrigation is used in hot, dry environments, the water evaporates, leaving behind a salty residue of minerals. When too much residue accumulates, the plants are unable to grow or even survive. This human-induced loss of fertility is one of the largest environmental water issues that the world faces today in arid and semiarid regions.

Israel has developed relatively efficient techniques of *drip irrigation* that dramatically reduce the amount of water used, thereby limiting salinization and freeing up water for other uses. Until very recently, however, poorer states have been unable to afford this somewhat complex technology, which requires an extensive network of hoses and pipes to deliver water to each plant.

Vulnerability to Climate Change

North Africa and Southwest Asia are especially vulnerable to the multiple effects of climate change. Higher temperatures are already increasing evaporation rates where water scarcity has been a problem for many years (Figure 6.10B, C). These climatic changes, along with people's intensified use of water, are transforming more nondesert lands into deserts. Also, as the climate warms, a sea level rise of a few feet could severely impact the Mediterranean coast, especially the Nile Delta, one of the poorest and most densely populated lowland areas in the world (see Figure 6.10A), and the Persian Gulf coasts, where vast new high-tech cities that lie at sea level would likely be flooded (see Figure 6.24A on page 264). Conversely, under some climate-change scenarios, periods of unusually intense rainfall could increase flooding.

Independent of any environmental effects of climate change, global efforts to reduce dependencies on fossil fuel consumption could devastate oil- and gas-based economies and transform the region's geopolitics, leaving countries vulnerable to losing some of their global power and to having lower overall income levels. Despite the enormous wealth from fossil fuel sales that has flowed into some of these countries over the past 40 years, few countries have undertaken the significant economic diversification needed to prepare for the reduced global consumption of the fossil fuels they sell.

Climate Change and Desertification Climate change could accelerate **desertification**, the conversion of nondesert lands into deserts (see Figure 6.7A). As temperatures increase, soil moisture decreases because of evaporation. As a result, plant cover is reduced, which causes less protection for arid soil. Bare patches of soil can become badly eroded by wind, and eventually sand dunes can blow onto formerly vegetated land.

In the grasslands (steppes) that often border deserts, a wide array of land use changes contribute to the general drying. For example, as groundwater levels fall because water is being

FIGURE 6.9 LOCAL LIVES
Foodways in North Africa and Southwest Asia

A A stew is prepared in a Moroccan *tagine*, a heavy clay pot with two parts: a round bottom and a cone-shaped lid with a hole in the top. The bottom holds the food while the hole in the lid allows excess steam to escape, preventing the food from overcooking. The cone shape of the lid encourages condensation of moisture, thus reducing water loss and keeping flavors in the stew. Tagines are placed directly on hot coals or a grill and left for 3 or more hours. [Gary Yeowell/The Image Bank/Getty Images]

B Falafel is fried in Gaza, Palestine. Either fava beans or chickpeas are mixed with parsley, scallions, garlic, and spices and then formed into a ball or patty that is deep-fried. Falafel is often made into a flatbread sandwich and served with tomatoes, cucumbers, lettuce, and sauces based on tahini, a paste made from ground sesame seeds. The origin of this popular vegetarian street food, which has spread throughout this region and much of the world, is unclear. Egyptian Coptic Christians say they invented it as a substitute for meat during Lent. Others claim it originated in ancient Egypt, or even in South Asia. [Abid Katib/Getty Images]

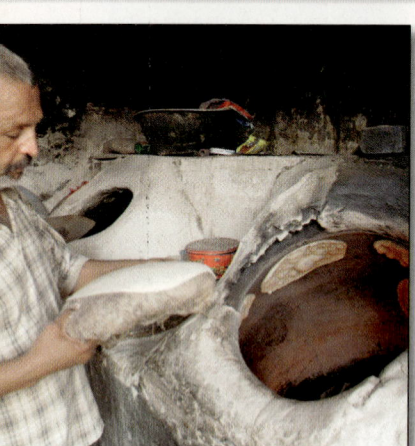

C *Khubz*, or flatbread, is made in a *tannur*, a wood- or charcoal-fired oven. Possibly originating in ancient Sumer, khubz consists of a dough of flour, water, and salt that is rolled flat and then pressed onto the interior side walls of the tannur, where the radiant heat from the fire rapidly bakes it. Khubz is so central to diets in this region that in recent decades most governments have subsidized it, keeping prices artificially low to avoid political uprisings among the urban poor. [Sabah Arar/AFP/Getty Images]

pumped out for cities and irrigated agriculture, some plant roots no longer reach sources of moisture, causing the plants to die. In other cases, international development agencies have inadvertently contributed to desertification by encouraging nomadic herders to take up settled cattle ranching of the type practiced in western North America. Agencies encourage settlement because the mobility of nomads is viewed as complicating management for modern state bureaucracies where records must be kept, taxes collected, and children sent to school with regularity. However, ranching on fragile grasslands can lead to overgrazing, which can kill the grasses and increase evaporation. If irrigation is used to support ranching, such projects can also deplete groundwater resources, resulting in long-term desertification. Incidentally, encouraging nomads to settle may actually increase their vulnerability to climate change, because it is their very mobility and skills at cycling seasonally through multiple environments that have helped nomads adapt to climate variability over many generations.

Imported Food and Virtual Water Because agriculture is so difficult due to the aridity of this region, the diets of nearly all people here include imported food. The total per capita water use of this region must thus include the water used to produce this imported food. *Virtual water* is the volume of water used to produce all that a person consumes in a year (see Chapter 1, page 26). For example, 1 kilogram (2.2 pounds) of beef requires 15,500 liters (4094 gallons) of water to produce, while 1 kilogram of goat meat requires just 4000 liters (1056 gallons). Goat meat, which is popular in this region and produced locally, has a much less significant water component than beef, but beef consumption is on the rise. One kilogram of corn requires 900 liters (238 gallons) of water, and 1 kilogram of wheat, 1350 liters (357 gallons). Beef, corn, and wheat are common imports; in fact, North Africa and Southwest Asia import more wheat than any other world region.

Strategies for Increasing Access to Water Some strategies have been developed for increasing supplies of fresh water, but each presents a set of difficulties. All of them are expensive and some have enormous potential to cause more wasting of water.

Seawater desalination The fossil fuel–rich countries of the Persian Gulf have invested heavily in **seawater desalination** technologies that remove the salt from seawater, making it suitable for drinking or irrigation. Subsidized desalination plants supply 70 percent of Saudi Arabia's drinking water and some of its wheat field irrigation. The process of desalination uses huge amounts of energy, which is produced by burning fossil fuels and results in emissions that contribute to global warming. If all the costs of producing food with desalinated water were counted—the fuel burned, the cost of irrigation and other equipment, and the fact that irrigated soil inevitably loses productivity due to soil salinization—the wheat so produced would be far too expensive for anyone to buy. Government subsidies for desalination have kept the wheat artificially cheap, though there are now plans to remove these subsidies so that the water can be used by urban populations.

> **seawater desalination** the removal of salt from seawater—usually accomplished through the use of expensive and energy-intensive technologies—to make the water suitable for drinking or irrigating

Groundwater pumping Many countries pump groundwater from underground aquifers to the surface for irrigation or drinking water. Under Muammar el-Qaddafi, Libya had invested some of its earnings from fossil fuels into one of the world's largest groundwater pumping projects, known as the *Great Man-Made River*. This project draws on part of the ancient fossil water deposit mentioned earlier to supply almost 2 billion gallons of

FIGURE 6.10   PHOTO ESSAY: Vulnerability to Climate Change in North Africa and Southwest Asia

Water scarcity, sea level rise, desertification, food scarcity, and political instability are some of the factors that make parts of this region highly vulnerable to climate change.

A Alexandria, Egypt, located on the Mediterranean coast of the Nile Delta region, is a low-lying city that is exposed to rising sea levels. Efforts to improve the sea wall around the city, visible at the right of the photo, may be outpaced by rising sea levels. The entire Nile Delta is struggling to adapt to salt water that is moving up rivers and permeating soils, making agriculture extremely difficult and polluting freshwater resources that cities depend on. Half of Egypt's population lives in the Nile delta, and 80 percent of the country's imports and exports run through Alexandria.

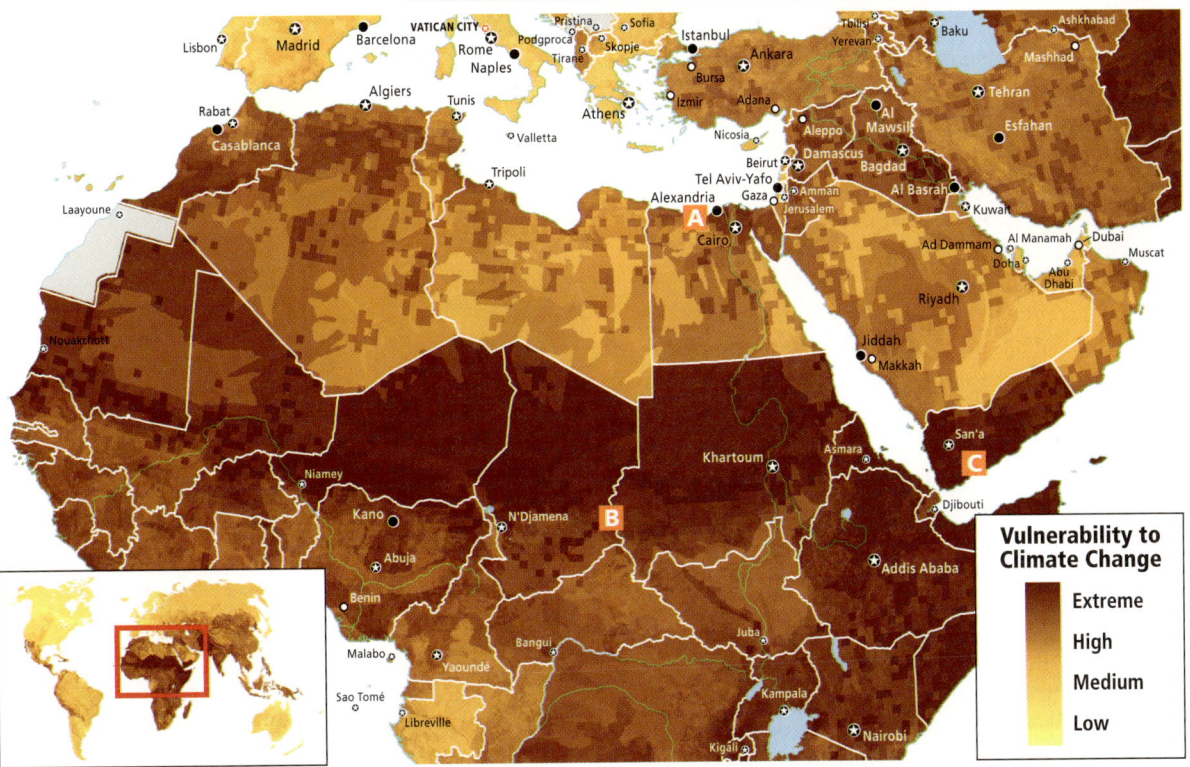

Vulnerability to Climate Change
- Extreme
- High
- Medium
- Low

B Refugees from Darfur, Sudan, line up for food and water in neighboring Chad. Sudan is significantly exposed to drought, and widespread poverty has left the population very sensitive to any disruption in food or water supplies. Meanwhile, political instability is uprooting people from their homes and complicating emergency response and longer-term planning efforts that might lend greater resilience.

C Residents of Wadi Mur, Yemen, ride out a sandstorm. The same factors that make Sudan so vulnerable to climate change also affect Yemen. However, Yemen also receives large numbers of refugees from Somalia and other parts of Africa, which further stretches its resources and frustrates planning efforts.

water per day to Libya's coastal cities and to 600 square miles of agricultural fields. With this irrigated agriculture, Libya had hoped to grow enough food to end its current food imports and even supply food to the European Union. However, the rate of natural replenishment of the aquifer is far slower than the rate of extraction, making this use of groundwater unsustainable. Hydrologists report fissures in the land surface above the aquifer that they associate with land sinking, or *subsidence*, as the water is withdrawn.

Dams and reservoirs Dams and reservoirs built on the regions' major river systems to increase water supplies have created new problems. In Egypt, for example, the natural cycles of the Nile River have been altered by the construction of the Aswan dams. Downstream of these two dams, water flows have been radically reduced, making it necessary to use supplemental irrigation, which leads to soil salinization (see Figure 6.7C), and the lack of flooding means that fertility-enhancing silt is no longer deposited on the land. As a result, expensive fertilizers are now necessary. Moreover, with less water and silt coming downstream, parts of the Nile Delta are sinking into the sea or are experiencing saltwater intrusion because the seawater is flowing into aquifers and wells in coastal zones, rendering once-fresh water unusable. Upstream, the artificial reservoir created by the dams has flooded villages, fields, wildlife habitats, and historic sites, and created still-water pools that harbor parasites. Dams can also cause hardship across borders. The Aswan High Dam sits on Egypt's border with Sudan, an area that has been a source of contention for many years.

Dams as water-acquisition strategies strain cross-border relations The need to share the water of rivers that flow across or close to national boundaries often impedes national development efforts. Turkey's Southeastern Anatolia Project, which involves the construction of several large dams on the upper reaches of the Euphrates River for hydropower and irrigation purposes, has reduced the flow of water to the downstream countries of Syria and Iraq (Figure 6.11). In negotiations over who should get Euphrates water, for example, Turkey argues that it should be allowed to keep more water behind its dams because the river starts in Turkey and most of its water originates there as mountain rainfall. Meanwhile, Iraq's claim to the water is linked to the fact that the Euphrates travels the longest distance in Iraq. 285. THE JORDAN RIVER IS DYING

THINGS TO REMEMBER

- Landforms and climates are particularly closely related in this driest region in the world. Rolling deserts and steppes cover most of the land. In a few places, mountains capture moisture, allowing plants, animals, and humans to flourish.

> **GEOGRAPHIC INSIGHT 1** • **Environment** The population of this predominantly dry region is growing faster than is access to cultivable land and water for agriculture and other human uses. Many countries here are dependent on imported food, and because global food prices are unstable, the region often has food insecurity. Climate change could reduce food output both locally and globally and make water, already a scarce commodity, a source of conflict. New technologies offer solutions to water scarcity but are expensive and unsustainable. ■

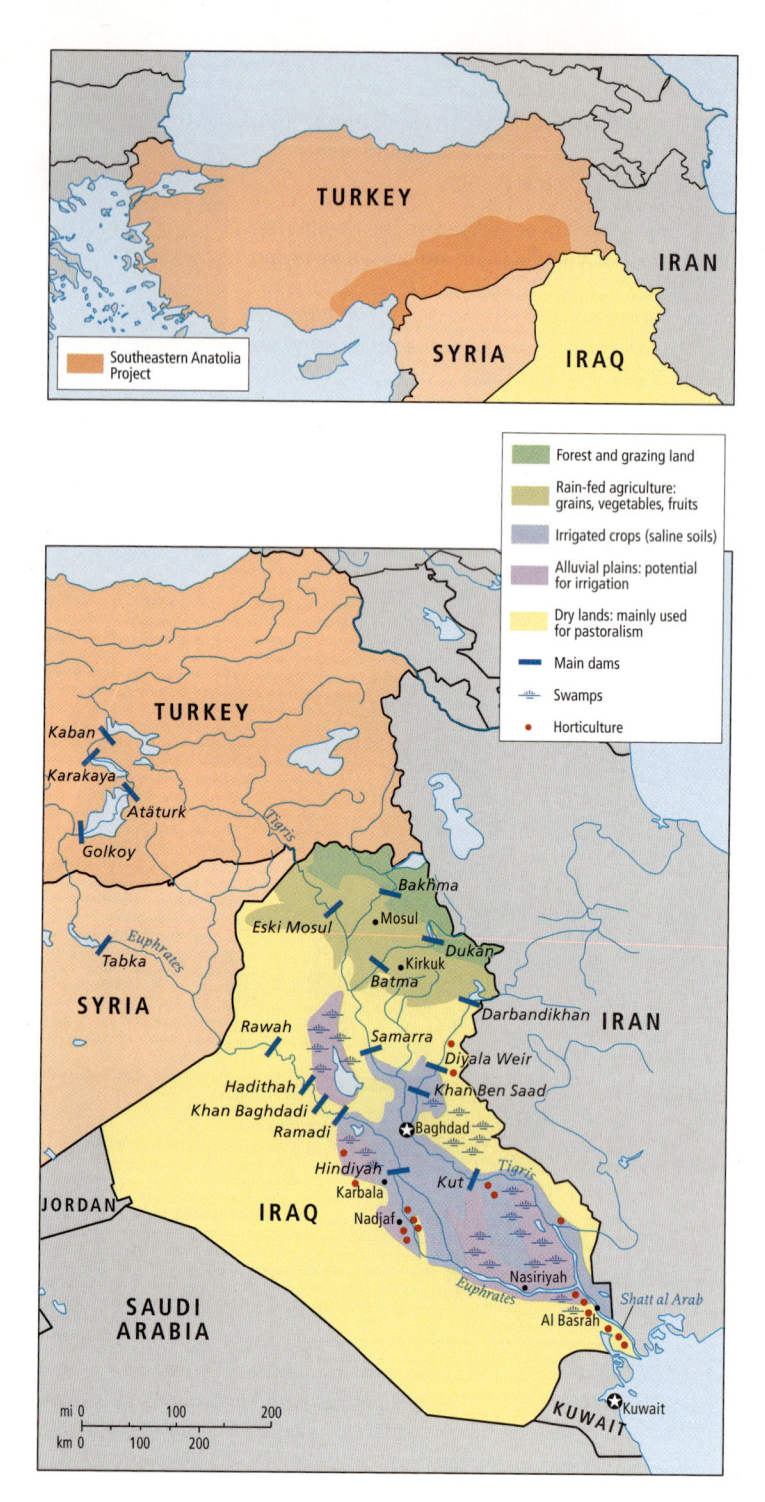

FIGURE 6.11 Dams on the Tigris and Euphrates drainage basins. Turkey's projects to manage the Tigris and Euphrates river basins through dam construction have international implications. Water that is retained in Turkey will not reach its neighbors. The main map shows Turkey's dams on the headwaters of the two rivers, as well as dams built in Syria and Iraq, all of which will have environmental effects, especially on the lower reaches of both rivers in Iraq. The smaller map shows the full extent of Turkey's Southeastern Anatolia Project. [Source consulted: United Nations Environmental Programme, *Vital Water Graphics: Problems Related to Freshwater Resources,* "Turning the Tides" map, at http://www.unep.org/dewa/assessments/ecosystems/water/vitalwater/22.htm]

HUMAN GEOGRAPHY

Often heralded as a "cradle of civilization" and the birthplace of three major world religions, the North Africa and Southwest Asia regions have more recently struggled with the impacts of outsiders. In the current era, social and political change has lagged behind economic development, which itself varies widely around the region. Why has the wealth generated by oil not resulted in a spreading of opportunities, such as broad public education, an opening up of public discourse, and an expansion of educational and political opportunities for women? As indicated in the opening vignette, change is now underway across this region and is gaining momentum, but there are countervailing forces that could inhibit political reform, social change, and economic development.

Fertile Crescent an arc of lush, fertile land formed by the uplands of the Tigris and Euphrates river systems and the Zagros Mountains, where nomadic peoples began the earliest known agricultural communities

HUMAN PATTERNS OVER TIME

Important developments in agriculture, societal organization, and urbanization took place long ago in this part of the world. Also, three of the world's great religions were born here: Judaism, Christianity, and Islam.

Agriculture and the Development of Civilization

Between 8000 and 10,000 years ago, formerly nomadic peoples founded some of the earliest known agricultural communities in the world. These communities were located in an arc formed by the uplands of the Tigris and Euphrates river systems (in modern Turkey and Iraq) and the Zagros Mountains of modern Iran (see Figure 6.16 on pages 248–249). This zone is often called the **Fertile Crescent** because of its once-plentiful fresh water; its fertile soil that is seasonally replenished by flooding; its open forests and grasslands; its abundant wild grains; and its fish, goats, sheep, wild cattle, camels, horses, and other large animals (Figures 6.12 and 6.13).

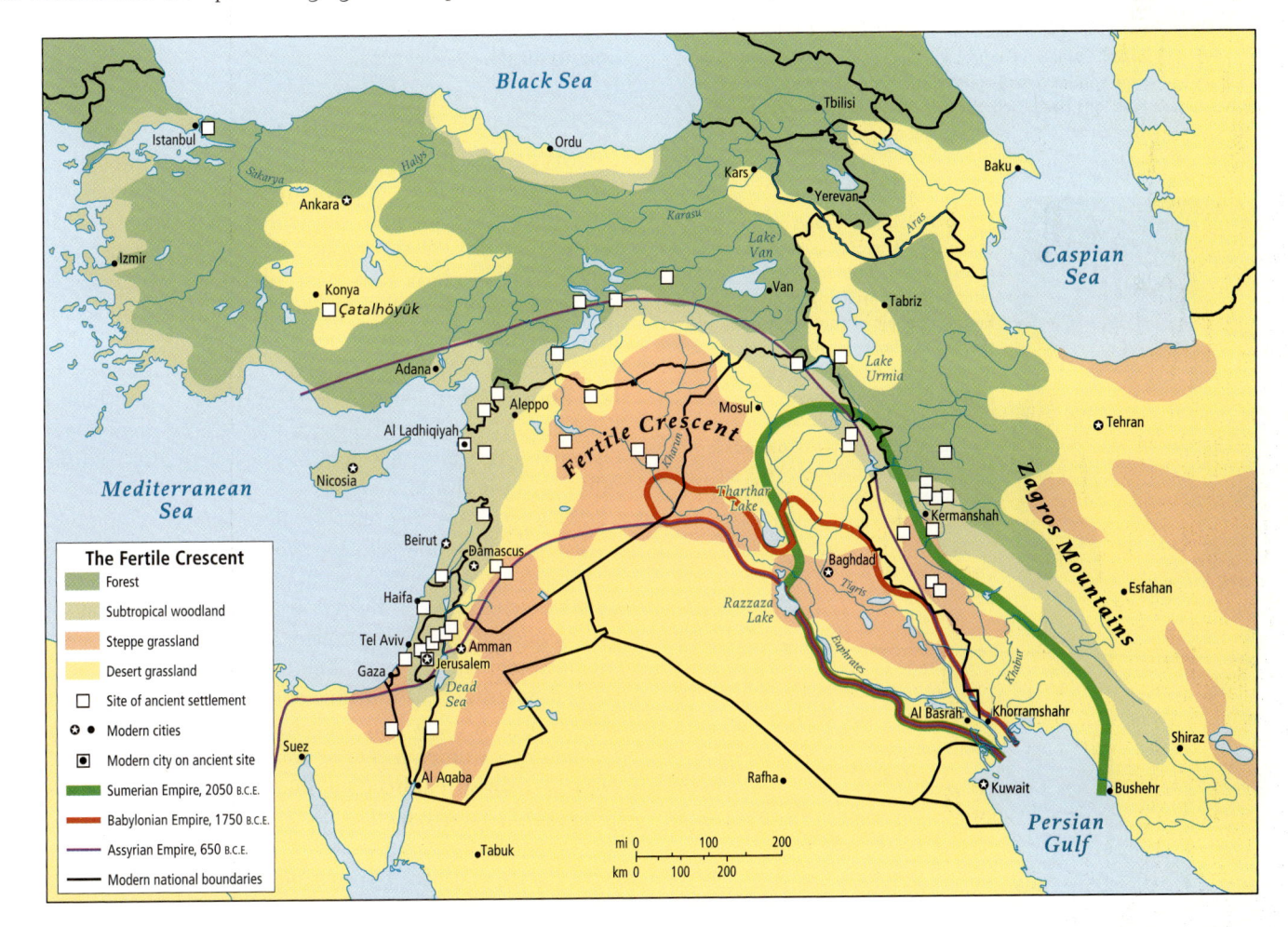

FIGURE 6.12 The Fertile Crescent, one of the earliest known agricultural sites. About 10,000 years ago, people in the Fertile Crescent began domesticating cereal grains, legumes, and animals, especially sheep and goats. The uses of domesticated animals spread into Europe and Africa as agricultural peoples traded their surpluses for other goods or moved into other regions. Three major empires developed successively in the eastern part of the Fertile Crescent: the Sumerian, the Babylonian, and the Assyrian. [Source consulted: Bruce Smith, *The Emergence of Agriculture* (New York: Scientific American Library, 1995), p. 50]

The skills of these early people in domesticating plants and animals allowed them to build ever more elaborate settlements. The settlements eventually grew into societies based on widespread irrigated agriculture along the foot of the mountains and in river basins, especially along the Tigris and Euphrates. Nomadic herders living in adjacent grasslands traded animal products for the grain and other goods produced in the settled areas.

Over several thousand years, agriculture spread to the Nile Valley, west across North Africa, north and west into Europe, and east to the mountains of Persia (modern Iran). Ultimately, other cultivation systems across the world were influenced by developments in the Fertile Crescent. Eventually, small settlements associated with agriculture took on urban qualities: dense populations, specialized occupations, concentrations of wealth, and centralized government and bureaucracies. For example, the agricultural villages of Sumer (in modern southern Iraq), which existed 5000 years ago, gradually turned into city-states that extended their influence over the surrounding territory. The Sumerians developed wheeled vehicles, oar-driven ships, and irrigation technology.

From time to time, nomadic tribes who had adopted the horse as a means of conquest banded together and, with devastating cavalry raids, swept over agricultural settlements. They then set themselves up as a ruling class, but soon these former nomads adopted the settled ways and cultures of the peoples they conquered and thus themselves became vulnerable to attack.

Agriculture and Gender Roles

Increasing research evidence suggests that the dawning of agriculture may have marked the transition to markedly distinct roles for men and women. Archaeologist Ian Hodder reports that at the 9000-year-old site of Çatalhöyük (see Figure 6.12), near modern Konya in south-central Turkey, where the economy was primarily based on hunting and gathering, there is little evidence of gender differences. Families were small and men and women performed similar chores in daily life. Both had comparable status and power, and both played key roles in social and religious life. Archaeological evidence elsewhere indicates that gender roles were also egalitarian in other preagricultural societies.

Scholars think that after the development of agriculture, as the accumulation of wealth and property became more important in human society, concerns about family lines of descent and inheritance emerged. This led in turn to the idea that women's bodies needed to be controlled so that a woman could not become pregnant by a man other than her mate and thus confuse lines of inheritance. From this core concern about secure lines of descent grew many practices aimed at reinforcing the idea that the mating of women had to be controlled and, by extension, that women's daily spatial freedom, interaction with men, and sexuality needed to be curtailed.

The Coming of Monotheism: Judaism, Christianity, and Islam

The very early religions of this region were founded on a belief in many gods who controlled natural phenomena; such was the case through the Greek era and into the Roman period.

FIGURE 6.13 LOCAL LIVES
People and Animals in North Africa and Southwest Asia

A A camel race in Jordan. Camels were probably domesticated between 4500 and 5000 years ago in the southern Arabian Peninsula. Extensively adapted to arid climates, camels have long been used to carry people and heavy loads across deserts. They are also prized for their lean meat and highly nutritious milk. [Salah Malkawi/Getty Images]

B An Arabian horse and his trainer at the stud farm of a Saudi prince in Saudi Arabia. Bred for war and raids by the nomadic desert-dwelling Bedouin people of Southwest Asia, the Arabian horse is known for its endurance, intelligence, speed, stealth, and ability to survive on relatively little food and water. Arabian horses are one of the oldest horse breeds, with evidence of their domestication going back 4500 years. Now they are a part of most major horse breeds throughout the world. [Reza/Getty Images]

C A mummified cat from ancient Egypt, where cats were revered and sometimes worshiped as symbols of the cat goddess Bastet. Part of a wider tradition of animal worship in Egypt, cats were often treated with the same respect accorded humans, including mummification before burial. Though cat worship was officially banned in 390 B.C., cats have been kept as pets and for pest control. [Clive Streeter/Dorling Streeter/Getty Images]

Then, several thousand years ago, **monotheism**—the belief system based on the idea that there is only one god—began to emerge. The three major monotheistic world religions—Judaism, Christianity, and Islam—all have connections to the eastern Mediterranean; there the city of Jerusalem is sacred to all three. (Muslims also revere Makkah (Mecca) and Al Madinah (Medina) in Saudi Arabia.) All three religions have a connection to a sacred text: the Torah (the first five books in what Christians refer to as the Old Testament) for Jews; the Old and New Testaments of the Bible for Christians; and for Muslims both the Bible and the Qu'ran are sacred.

Judaism was founded approximately 4000 years ago. According to tradition, the patriarch Abraham led his followers from Mesopotamia (modern Iraq) to the shores of the eastern Mediterranean (modern Israel and the occupied Palestinian Territories), where he founded Judaism. Judaism is characterized by the belief in one God, a strong ethical code summarized in the Ten Commandments, and an enduring ethnic identity reinforced by dietary and religious laws.

After rebelling against the Roman Empire, which culminated in their expulsion in 73 C.E. from the eastern Mediterranean, some Jews were enslaved by the Romans and most migrated to other lands in a movement known as the **diaspora** (the dispersion of an originally localized people). Many Jews migrated across North Africa and Europe, and others went to various parts of Asia. After 1500, Jews were among the earliest European settlers in all parts of the Americas.

Christianity is based on the teachings of Jesus of Nazareth, a Jew who, claiming to be the son of God, gathered followers in the area of Palestine about 2000 years ago. Jesus, who became known as Christ (meaning *anointed one* or *Messiah*), taught that there is one God, who primarily loves and supports humans but who will judge those who do evil. This philosophy grew popular, and both Jewish religious authorities and Roman imperial authorities of the time saw Jesus as a dangerous challenge to their power.

After Jesus' execution in Jerusalem in about 32 C.E., his teachings were written down (the Gospels) by those who followed him, and his ideas as interpreted by these writers spread and became known as Christianity. Centuries of persecution ensued, but by 400 C.E., Christianity had become the official religion of the Roman Empire. However, following the spread of Islam after 622 C.E., only remnants of Christianity remained in Southwest Asia and North Africa.

Islam, now the overwhelmingly dominant religion in the region, emerged in the seventh century C.E., after the Prophet Muhammad transmitted the Qur'an to his followers by writing down what was conveyed to him by Allah ("God" in Arabic). Born in about 570 C.E., Muhammad was a merchant and caravan manager in the small trading town of Makkah (Mecca) on the Arabian Peninsula near the Red Sea (see Figure 6.16B). Followers of Islam, called **Muslims**, believe that Muhammad was the final and most important in a long series of revered prophets, which includes Abraham, Moses, and Jesus.

Unlike the many versions of Christianity, Islam has virtually no central administration and only an informal religious hierarchy (this is somewhat less true of the Shi'ite version of Islam; see the discussion on page 246). The world's one billion Muslims may communicate directly with God (Allah). A clerical intermediary is not necessary, though there are numerous clerical leaders who help their followers interpret the Qur'an. An important effect of the lack of a central authority is that the interpretation of Islam varies widely within and among countries, from group to group, and from individual to individual.

Today, 93 percent of the people in the region are followers of Islam; for them, the Five Pillars of Islamic Practice embody the central teachings of Islam. Not all Muslims are fully observant, but the Pillars have had an enormous impact on daily life, including on festivals and religious holidays, since the time of Muhammad (Figure 6.14).

> **monotheism** the belief system based on the idea that there is only one god
>
> **Judaism** a monotheistic religion characterized by the belief in one god (Yahweh), a strong ethical code summarized in the Ten Commandments, and an enduring ethnic identity
>
> **diaspora** the dispersion of Jews around the globe after they were expelled from the eastern Mediterranean by the Roman Empire beginning in 73 C.E.; the term can now refer to other dispersed culture groups
>
> **Christianity** a monotheistic religion based on the belief in the teachings of Jesus of Nazareth, a Jew, who described God's relationship to humans as primarily one of love and support, as exemplified by the Ten Commandments
>
> **Muslims** followers of Islam
>
> **hajj** the pilgrimage to the city of Makkah (Mecca) that all Muslims are encouraged to undertake at least once in a lifetime

The Pillars of Muslim Practice

1. A testimony of belief in Allah as the only God and in Muhammad as his messenger (prophet).

2. Daily prayer at five designated times (daybreak, noon, midafternoon, sunset, and evening). Although prayer is an individual activity, Muslims are encouraged to pray in groups and in mosques. The call to prayer, broadcast five times a day in all parts of the region, is a constant reminder to all people to reflect on their beliefs.

3. Obligatory fasting (no food, drink, or smoking) during the daylight hours of the month of Ramadan, followed by a light celebratory family meal after sundown. Ramadan falls in the ninth month of the Islamic (lunar) calendar.

4. Obligatory almsgiving (*zakat*) in the form of a "tax" of at least 2.5 percent. The alms are given to Muslims in need. *Zakat* is based on the recognition of the injustice of economic inequity. Although it is usually an individual act, the practice of government-enforced *zakat* is returning in certain Islamic republics.

5. Pilgrimage (**hajj**) at least once in a lifetime to the Islamic holy places, especially the Masjid Al-Haram and the Kaaba in Makkah (Mecca), during the twelfth month of the Islamic calendar.

Saudi Arabia occupies a prestigious position in Islam, as it is the site of two of Islam's three holy shrines: Makkah, the birthplace of the Prophet Muhammad and of Islam, and Al Madinah (Medina), the site of the Prophet's mosque and his burial place. (The third holy shrine is in Jerusalem.) The fifth pillar of Islam has placed Makkah and Al Madinah at the heart of Muslim religious geography for more than 1400 years. Today, a large private sector service industry, owned and managed by members of the huge

FIGURE 6.14 LOCAL LIVES
Festivals of North Africa and Southwest Asia

A Colombian singer Shakira performs at Mawazine, a world music festival held each year in Rabat, the capital of Morocco. Part of many conscious efforts to portray Morocco and Rabat as "open to the world," the festival has featured great traditional and international artists such as Cheb Khaled and Stevie Wonder. It has also come under criticism from Islamist politicians who see it as encouraging immoral behavior. [Abdelhak Senna/AFP/Getty Images]

B A banquet is set in Gaza, Palestine, for Iftar, the sunset meal that Muslims eat to break the fast each night in the holy month of Ramadan. During Ramadan, observant Muslims refrain from eating, drinking, quarreling, or having sex between sunrise and sunset. They offer extra prayers each day as a demonstration of their submission to God. Wealthy Muslims often sponsor public banquets, such as the one shown above. [Mohammed Abed/AFP/Getty Images]

C An Iranian woman celebrates Nauryz, an ancient New Year holiday of Zoroastrian origin celebrated throughout Iran, parts of Turkey, and much of Central Asia (see Figure 5.23 on page 224). People jump over bonfires while singing a verse of purification that is meant to remove sickness and problems, replacing them with warmth and energy. [Behrouz Mehri/AFP/Getty Images]

the Qur'an. But there are many interpretations of the Qur'an, several renderings of shari'a, and a wide variety of versions of observant Muslim life. Some Muslims believe that no other legal code is necessary in an Islamic society, because shari'a provides guidance in all matters of life, including worship, finance, politics, marriage, sex, diet, hygiene, war, and crime. Other Muslims think that secular law is more useful in modern societies that are increasingly multicultural because secular law makes allowances for different religious sensibilities. The debate about whether shari'a or secular law is best has raged for hundreds of years, and recently flared up again as part of the Arab Spring.

Insofar as the interpretation of shari'a is concerned, the Muslim community is split into two major groups that formed after the death of Muhammad, when divisions arose over who should succeed the Prophet and have the right to interpret the Qur'an for all Muslims: **Sunni** Muslims, who today account for 85 percent of the world community of Islam, and **Shi'ite** (or **Shi'a**) Muslims, who live primarily in Iran but also in southern Iraq, Syria, and southern Lebanon. While Sunnis have a relatively decentralized religious system for interpreting shari'a, Shi'ites recognize an authoritative priestly class whom they call *mullahs*.

This division continues today, with disagreements about the interpretation of shari'a exacerbated by countless local disputes over land, resources, and philosophies. In Iraq, for example, long-standing conflict between Sunnis and Shi'ites intensified after the U.S. invasion in 2003, as rivalries arose over which group should control political power and fossil fuel resources.

shari'a literally, "the correct path"; Islamic religious law that guides daily life according to the interpretations of the Qur'an

Sunni the larger of two major groups of Muslims who have different interpretations of shari'a

Shi'ite (or Shi'a) the smaller of two major groups of Muslims who have different interpretations of shari'a; Shi'ites are found primarily in Iran and southern Iraq

The Spread of Islam

Among the first converts to Islam were the Bedouin—nomads of the Arabian Peninsula. By the time of Muhammad's death in 632 C.E., they were already spreading the faith and creating a vast Islamic sphere of influence. Over the next century, Muslim armies built an Arab–Islamic empire over most of Southwest Asia, North Africa, and the Iberian Peninsula of Europe (Figure 6.15).

While most of Europe was stagnating during the medieval period (450–1500 C.E.), the Arab–Islamic empire nurtured learning and economic development. Muslim scholars traveled throughout Asia and Africa, advancing the fields of architecture, history, mathematics, geography, and medicine. Centers of learning flourished from Baghdad (Iraq) to Toledo (Spain). During the early Arab–Islamic era, the development of banks, trusts, checks, receipts, and bookkeeping fostered vibrant economies and wide-ranging trade. The traders founded settlements and introduced new forms of living spaces. The architectural legacy of Arabs and Muslims lives on in Spain, India, Central Asia, the Americas, and in countless buildings across the world.

Saud family that rules Saudi Arabia, organizes and oversees the 5- to 7-day hajj for more than 2.5 million devout foreign visitors.

Islamic Religious Law and Variable Interpretations Beyond the Five Pillars, Islamic religious law, called **shari'a**, or "the correct path," guides daily life according to the principles of

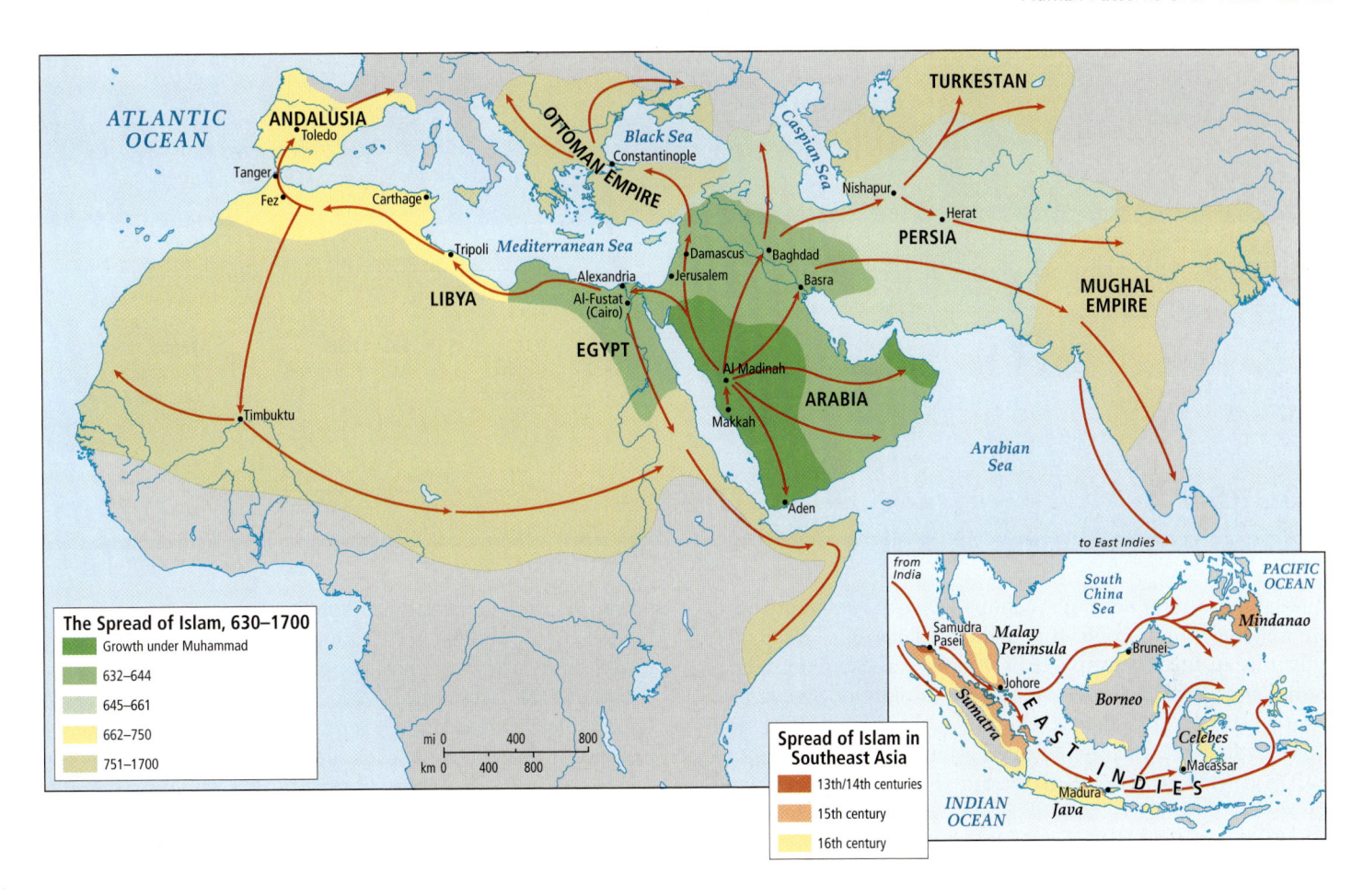

FIGURE 6.15 The spread of Islam, 630–1700. In the first 120 years after the death of the Prophet Muhammad in 632, Islam spread, primarily by conquest. Over the next several centuries, Islam was carried to distant lands by both traders and armies. [Source consulted: Richard Overy, ed., *The Times History of the World* (London: Times Books, 1999), pp. 98–99]

By the end of the tenth century, the Arab–Islamic empire had begun to break apart. From the eleventh to the fifteenth centuries, Mongols from eastern Central Asia (eventually converting to Islam by 1330) conquered parts of the Arab-controlled territory, forming the Muslim Mughal Empire, centered in what is now north India. Meanwhile, beginning in the 1200s, nomadic Turkic herders from Central Asia began to converge in western Anatolia (Turkey) where they eventually forged the **Ottoman Empire**, which became the most influential Islamic empire the world has ever known.

By the 1300s, the Ottomans had become Muslim and by the 1400s they had defeated the Christian Byzantine Empire centered in Constantinople, which was the successor to the Roman Empire. The Ottomans took over Constantinople, renamed it Istanbul, and soon controlled most of the eastern Mediterranean, Egypt, and Mesopotamia. By the late 1400s, they also controlled much of southeastern and central Europe and by the 1600s had taken over parts of coastal North Africa from the Arabs. Previously, in the 1490s, the Arab Muslims had lost their control of the Iberian Peninsula to Christian kingdoms. Islam currently still dominates in a huge area that stretches from Morocco to western China and includes northern South Asia, as well as Malaysia, Brunei, and Indonesia in Southeast Asia (see the inset of Figure 6.15).

Once a location was completely conquered, the Ottoman Empire, like the Arab–Islamic empire before it, encouraged religious tolerance toward the conquered peoples so long as they adhered to a religion with a sacred text. Jews, Christians, Buddhists, and Zoroastrians were allowed to practice their religions, although there were attractive economic and social advantages to converting to Islam. Multicultural urban life in Ottoman cities facilitated vast trading networks spanning the known world, and Istanbul became a cosmopolitan capital with elaborate buildings and lavish public parks that outshined anything in Europe until the nineteenth century (Figure 6.16C).

> **Ottoman Empire** the most influential Islamic empire the world has ever known; begun in the 1200s when nomadic Turkic herders from Central Asia converged in western Anatolia (Turkey)

Western Domination, State Formation, and Antidemocratic Practices

The Ottoman Empire ultimately withered in the face of a Europe made powerful by colonialism and the Industrial Revolution.

A Mud-brick houses in Harran, Turkey, have been inhabited for at least 5000 years. [Wu Swee Ong/Flickr/Getty Images]

B Masjid Al-Haram, built in Makkah in 630. [Rabi Karim Photography/Flickr/Getty Images]

C Istanbul's "Blue Mosque," completed in 1616. [Ayhan Altun/Flickr/Getty Images]

8000 B.C.E.	Nomads founded agricultural communities in the Fertile Crescent
3000 B.C.E.	
400 C.E.	Roman Empire, officially Christian
630	
632	Death of Muhammad
632–1500s	Arab–Islamic empires
1300s	Rise of the Ottoman Empire
1616	

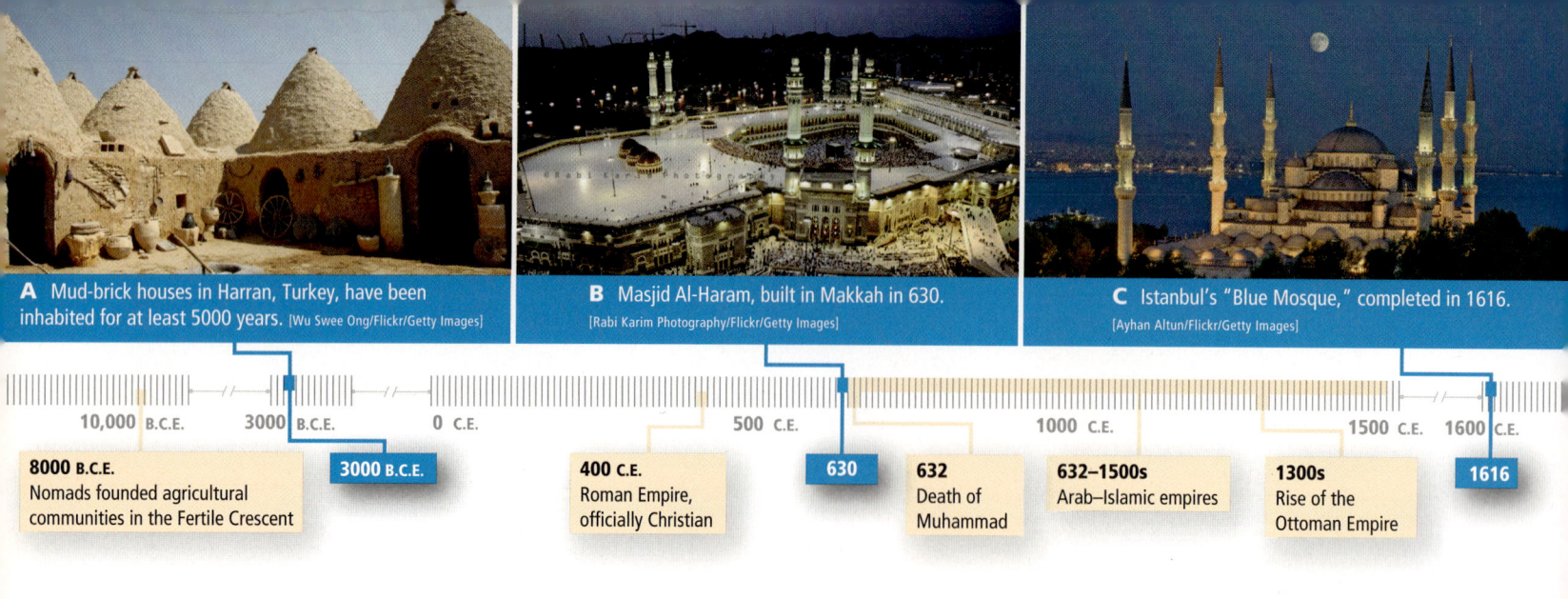

FIGURE 6.16 VISUAL HISTORY OF NORTH AFRICA AND SOUTHWEST ASIA

Throughout the nineteenth century, North Africa provided raw materials for Europe in a trading relationship dominated by European merchants. By 1830, France was exercising direct control over parts of the North African territory of Algeria (see Figure 6.16D). France took control of Tunisia in 1881 and Morocco in 1912 (Spain controlled a bit of the Mediterranean coast and what is now called Western Sahara); Britain gained control of Egypt in 1882 and Sudan in 1898; and Italy took control of Libya in 1912 (Figure 6.17A on page 250).

World War I (1914–1918) brought the fall of the Ottoman Empire, which had allied itself with Germany. At the end of the war, the victorious powers dismantled the Ottoman Empire and all of the former Ottoman territories; only Turkey was recognized as an independent country. The rest of the formerly Ottoman-controlled territory was allotted to France and Britain as protectorates (see Figure 6.17B). On the Arabian Peninsula, Bedouin tribes were consolidated under Sheikh Ibn Saud in 1932, and Saudi Arabia began to emerge as an independent country.

World War II (1939–1945) further affected the political development of North Africa and Southwest Asia. Most significantly, in the aftermath of the Holocaust in Europe, the Jewish state of Israel was created in the eastern Mediterranean on land inhabited by Arab farmers and nomadic herders as well as by some Jews (see the discussion on pages 259–262; see also Figure 6.16E).

By the 1950s, European and U.S. energy companies played a key role in influencing who ruled Iran and Saudi Arabia—countries where vast oil deposits were to become especially lucrative. In Egypt, a major cotton producer, European textile companies played a similar role. Officials in the governments of these countries received financial benefits from foreign companies and showed their loyalty to these companies with low taxes on oil and cotton exports and easy access to land. While a tiny ruling elite grew fabulously wealthy, even the modest revenues generated by oil and other taxes were not invested in creating opportunities for the vast majority of poor people. Over time, ever more political power accrued, especially to foreign energy companies. The United States and Western Europe supported those autocratic local leaders who were most sympathetic to the interests of their companies and their Cold War strategies vis-à-vis the Soviet Union (see Figure 6.16F). As a result of these concerns, both the Europeans and the Americans supported undemocratic governments and stood in the way of reforms that would have resulted in a more educated populace able to participate in the full range of democratic institutions.

THINGS TO REMEMBER

- About 10,000 years ago in the Fertile Crescent, formerly nomadic peoples founded some of the world's earliest known agricultural communities. The domestication of plants and animals allowed them to build ever more elaborate settlements that eventually grew into societies that were based on widespread irrigated agriculture.

- The three major monotheistic world religions—Judaism, Christianity, and Islam—all have their origins in this region in the eastern Mediterranean. Islam is by far the largest in numbers of adherents in the region, and it is the principal faith in all of the region's countries except Israel.

- For Muslims (93 percent of the population in this region), the Five Pillars of Islamic Practice embody the central teachings of Islam. Some Muslims are fully observant, some are not; but the Pillars have had an impact on daily life for 1400 years.

- Beginning in the nineteenth century and continuing through the end of World War II, European colonial powers ruled or controlled most countries in the region.

- Following World War II, the state of Israel was created, and in countries with oil deposits and other useful assets, the United States and Western Europe supported those autocratic local leaders who were most likely to maintain a friendly attitude toward U.S. and European business and geopolitical strategic interests. ■

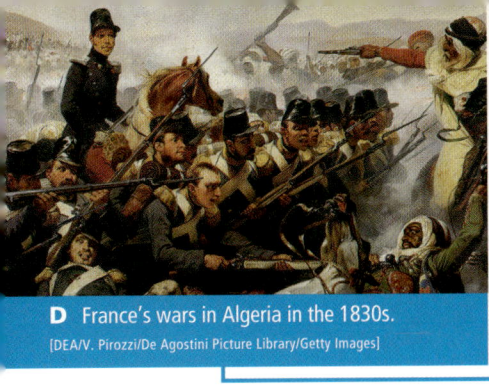

D France's wars in Algeria in the 1830s. [DEA/V. Pirozzi/De Agostini Picture Library/Getty Images]

E Jewish, Arab, and UN representatives negotiate a cease-fire during the 1948 war. [AFP/Getty Images]

F The king of Saudi Arabia meets with the U.S. secretary of defense in 2011. [Chip Somodevilla/Getty Images]

1700 C.E. — 1800 C.E. — 1900 C.E. — 2000 C.E.

- **1830s**
- **1830–1918** European colonial expansion
- **1918** Fall of the Ottoman Empire
- **1948** War, resulting in the creation of the state of Israel
- **1950s–present** Strong European and U.S. influence on politics
- **2011**

GLOBALIZATION AND DEVELOPMENT

There are major economic barriers to peace and prosperity within North Africa and Southwest Asia. Wealth from fossil fuel exports remains in the hands of an elite few, while most people work at low-wage urban jobs or are relatively poor farmers or herders. The economic base is unstable because the main resources are fossil fuels and agricultural commodities, both of which are subject to wide price fluctuations on world markets. Meanwhile, in many poorer countries, large national debts and the need to stay competitive in the global market are forcing governments to streamline production and cut jobs and social services.

Fossil Fuel Exports

> **GEOGRAPHIC INSIGHT 2**
>
> **Globalization and Development:** The vast fossil fuel resources of a few countries in this region have transformed economic development and driven globalization. In these countries, economies have become powerfully linked to global flows of money, resources, and people. Politics have also become globalized, with Europe and the United States strongly influencing many governments.

This region, which has two-thirds of the earth's known reserves of oil and natural gas, is the major fossil fuel supplier to the world. Most oil and gas reserves are located around the Persian Gulf (Figure 6.18 on page 251) in the countries of Saudi Arabia, Kuwait, Iran, Iraq, Oman, Qatar, and the United Arab Emirates (UAE). The North African countries of Algeria, Libya, and Sudan also have oil and gas reserves. It is important to note, however, that several countries in this region—Morocco, Tunisia, Syria, Lebanon, Jordan, Israel, and Turkey—have minimal oil and gas resources or are net importers of fossil fuels. Meanwhile, Egypt and Sudan produce enough to supply most of their own needs, but export little. Regardless of their own petroleum resources, all countries in the region are affected by the geopolitics of oil and gas.

European and North American companies were the first to exploit the region's fossil fuel reserves early in the twentieth century. These companies paid governments a small royalty for the right to extract oil (natural gas was not widely exploited in this region until the 1960s, though gas extraction has grown rapidly since then). Oil was processed at on-site refineries owned by the foreign companies and sold at very low prices, primarily to Europe and the United States and eventually to other countries, such as Japan.

Global oil and gas prices have risen and fallen dramatically since 1973, when governments in the Gulf states raised the price of oil and gas for several reasons. For one, the Gulf states were imposing a kind of penalty in response to U.S. support for Israel in the 1973 Yom Kippur War between Israel and neighboring Arab countries (discussed further below). The oil-producing states were also reacting to a long-term trend of rising prices on products exported from European countries and the United States to the Gulf states. The oil price increase in 1973 was made possible by the founding in 1960 of a **cartel**—a group of producers strong enough to control production and set prices for products—known as **OPEC (Organization of the Petroleum Exporting Countries)**. For a few months after the Yom Kippur War, there was a total halt of oil shipments from the Gulf states to the United States and to a few other countries that supported Israel. When the Gulf states resumed their shipments, they raised the price of oil. OPEC remains the main organization of oil-producing states; membership, which changes from time to time, now includes all of the states indicated in Figure 6.19 on page 253. OPEC members cooperate to periodically restrict or increase oil

cartel a group of producers strong enough to control production and set prices for products

OPEC (Organization of the Petroleum Exporting Countries) a cartel of oil-producing countries—currently, Algeria, Angola, Iran, Iraq, Kuwait, Libya, Nigeria, Qatar, Saudi Arabia, the United Arab Emirates, Ecuador, and Venezuela—that was established to regulate the production and price of oil and natural gas

FIGURE 6.17 Colonial regimes in North Africa and Southwest Asia.

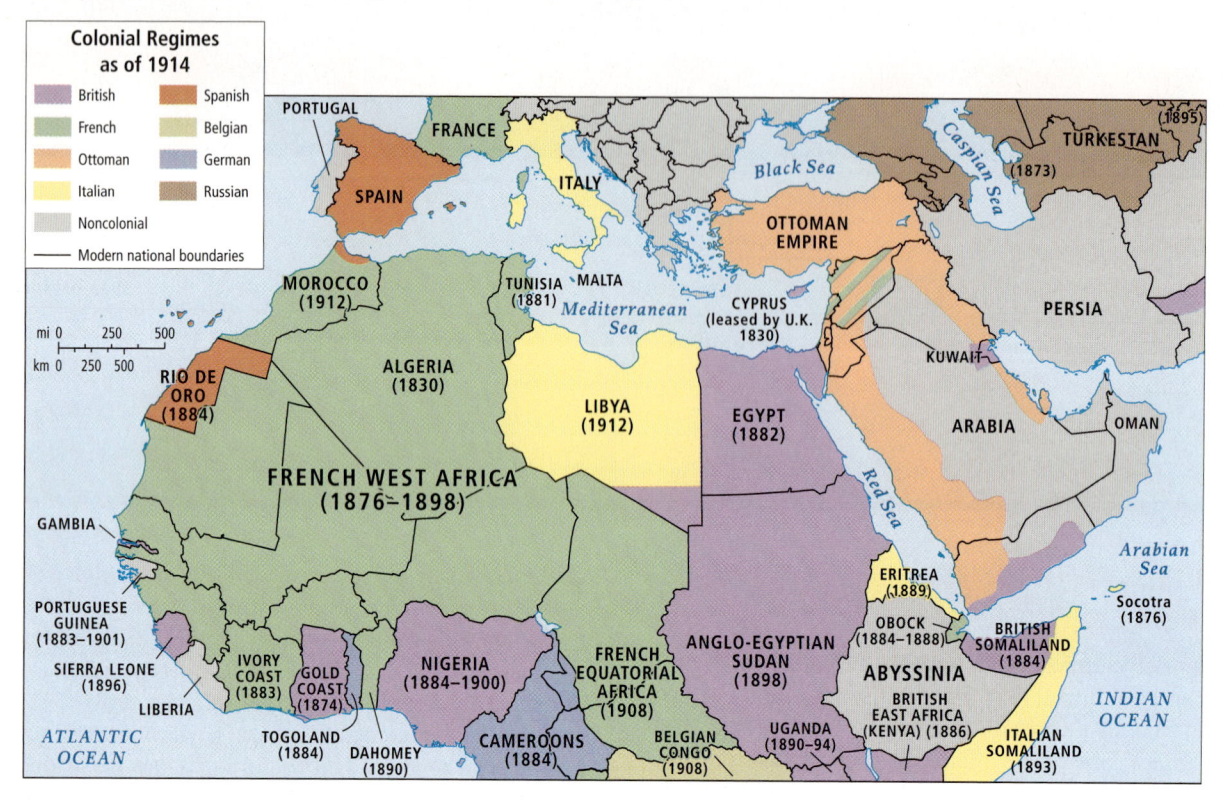

(A) European powers began influencing the affairs of the region in the nineteenth century and expanded their control by 1914 at the beginning of World War I. The dates on the map indicate when the Europeans took control of each country. [Source consulted: *Hammond Times Concise Atlas of World History* (Maplewood, NJ: Hammond, 1994), pp. 100–101]

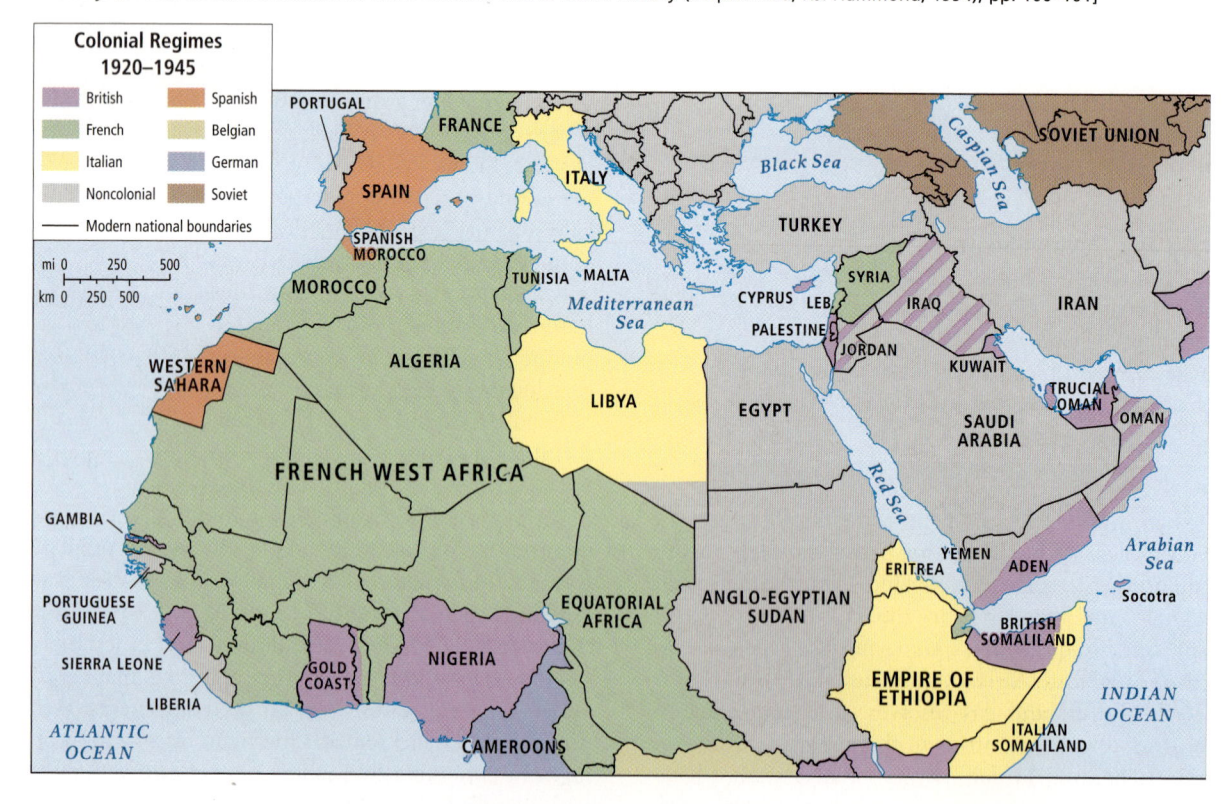

(B) Between 1920 and 1945, what was left of the Ottoman Empire in the eastern Mediterranean became protectorates administered by the British and French. The striped areas reflect the British colonial practice of allowing local rulers to govern while the British retained control of many of the colonies' policies and actions. [Sources consulted: *Rand McNally Historical Atlas of the World* (Chicago: Rand McNally, 1965), pp. 36–37; *Cultural Atlas of Africa* (New York: Checkmark Books, 1988), p. 59]

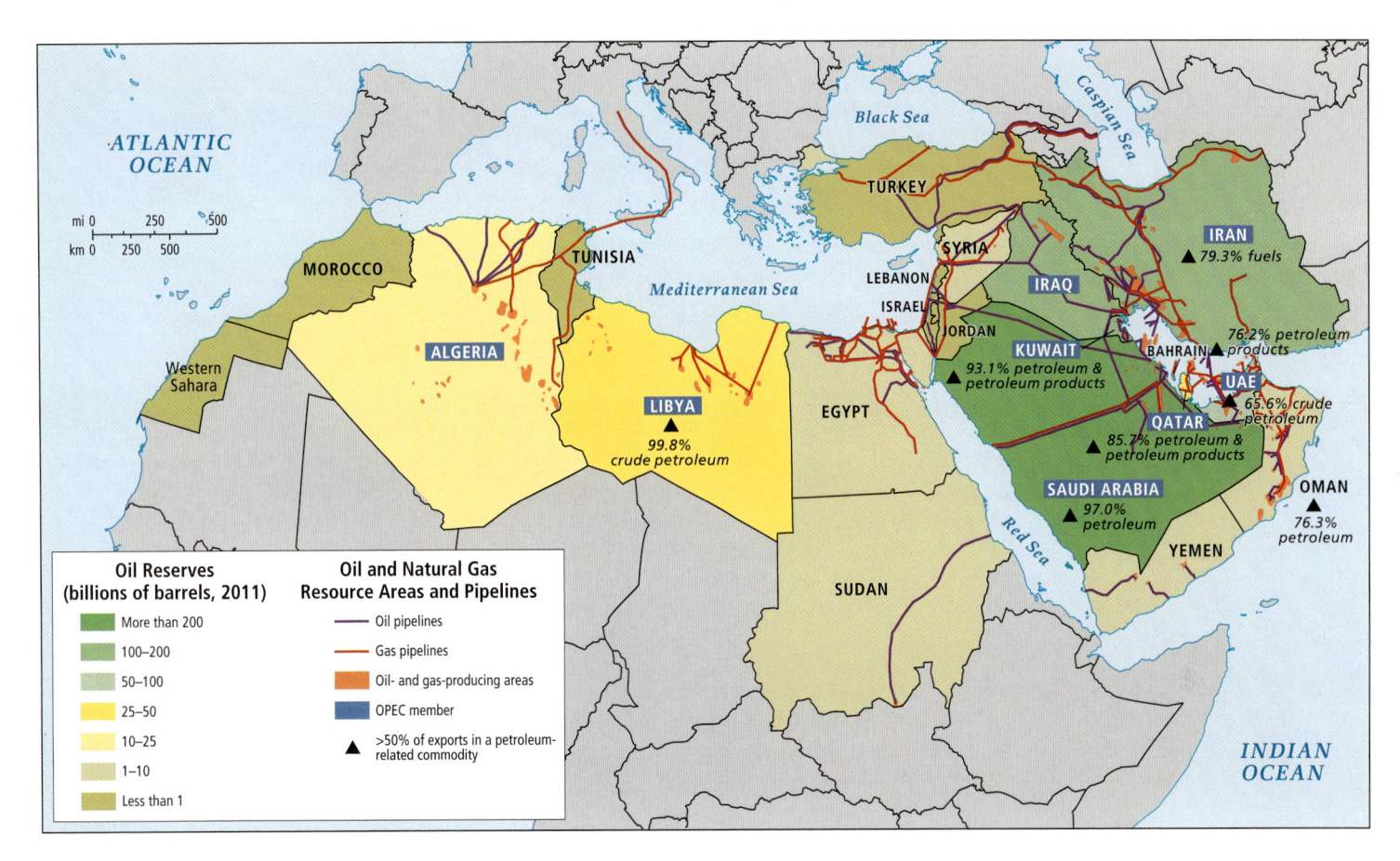

FIGURE 6.18 Economic issues: Oil and gas resources in North Africa and Southwest Asia. (A) The map shows the oil reserves, oil and gas resource areas, and pipelines in the region. [Sources consulted: U.S. Department of Energy, Country Analysis Briefs, at http://www.eia.doe.gov/emeu/cabs/Region_me/html; U.S. Department of Energy, "Selected Oil and Gas Pipeline Infrastructure in the Middle East," at http://www.eia.doe.gov/cabs/Saudi_Arabia/images/Oil%20and%20Gas%20Infrastructure%20Persian%20Gulf%20(large)%20(2).gif]

production, thereby significantly influencing the price of oil on world markets. A move to create an OPEC-like cartel for natural gas is currently underway.

Many OPEC countries, which were exceedingly poor just 40 years ago, have become much wealthier since 1973, but they have also become more vulnerable to economic downturns. For example, oil income in Saudi Arabia shot up from U.S.$2.7 billion in 1971 to U.S.$110 billion in 1981, to U.S.$592.4 billion in 2012. But this wealth has not been widely shared within OPEC countries or within the region. The Gulf states were slow to invest their fossil fuel earnings in basic human resources at home, and they only gradually began to explore other economic strategies in case oil and gas ran out. Like their poorer neighbors, OPEC countries remain very dependent on the industrialized world for their technology, manufactured goods, skilled labor, and expertise.

Recently, the Gulf states have begun to invest heavily in roads, airports, new cities, irrigated agriculture, and petrochemical industries. Saudi Arabia is investing in six new cities in different parts of the country to spur investment and advance economic and social development. A good example of the scale of projects in the Gulf states is Palm Jumeirah (Palm Island) in Dubai, UAE (see Figure 6.24A on page 264), which is intended to be the centerpiece of Dubai's planned tourism economy—insurance against the day when the regional oil economy fails for whatever reason. The recession that started in 2008 badly damaged the economies of the Gulf states. Most of the massive building projects that were underway, including Palm Island and similar projects in the UAE, screeched to a halt. The recession showed that Dubai's tourism economy is highly vulnerable to economic downturns. It also affected remittances sent back to home countries because, with drastically fewer jobs available in the recession, those countries that have many temporary workers in the Gulf states have had a steep decline in remittances sent home.

Resources have been poured primarily into visible physical development; much less oil wealth has been invested in education, social services, public housing, and health care. Libya, before the Arab Spring, and Kuwait are exceptions to this pattern, having developed ambitious plans for addressing all social service needs, with especially heavy investment in higher education. Iran is another possible exception, though because the Iranian government shares very little of its data with the rest of the world, it is hard to be certain how much it is investing in social services.

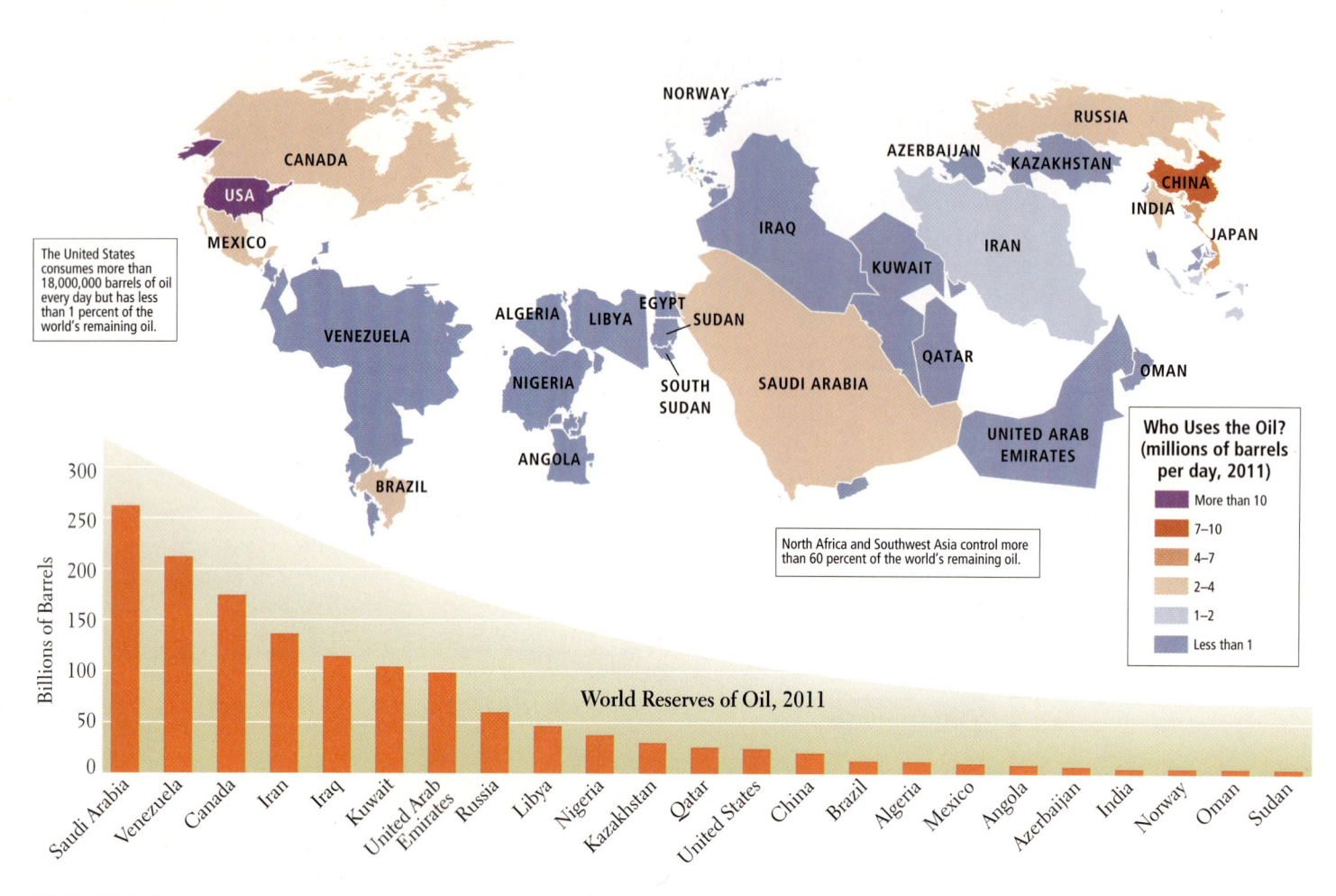

FIGURE 6.18 **(B)** The graph shows the oil reserves by country; in the map, the size of countries corresponds to the amount of oil reserves held, and the color corresponds to the amount used. [Sources consulted: "Who Has the Oil? A Map of the World Oil Reserves," at http://gcaptain.com/who-has-the-oil-a-map-of-world-oil-reserves/; "Countries: Proved Reserves," U.S. Department of Energy, Energy Information Administration, at http://www.eia.gov/countries/index.cfm?view=reserves.]

Meanwhile, the non-OPEC countries (those that do not produce fossil fuels) do not share directly in the wealth of the Persian Gulf states, and for the most part, the oil-rich countries have not helped their poorer neighbors develop.

Many factors beyond OPEC's control strongly influence world oil prices. Recent rapid industrialization in China and India has driven up their use of fuel and other petroleum products, creating long-term pressures that are raising the price of oil. Employing a theory known as "peak oil," some experts argue that in an era of diminished oil reserves—which these experts say has already begun—prices will be driven higher. On the other hand, efforts to combat climate change with renewable energy sources have been increasingly successful and could dramatically reduce the demand for oil. Current global oil flows are summarized in Figure 6.19.

Economic Diversification and Growth

Greater **economic diversification**—expansion into a wider range of economic development strategies—could have a significant impact on the region, bringing economic growth and broader prosperity and thereby limiting the damage caused by a drop in the prices of oil, gas, or other commodities on the world market.

By far the most diverse economy in the region is that of Israel, which has a large knowledge-based service economy and a particularly solid manufacturing base. Israel's goods and services and the products of its modern agricultural sector are exported worldwide. Turkey is the next most diversified, in part because—like Israel—it has never had oil income to fall back on. Egypt, Morocco, Libya, and Tunisia are also starting to move into many new economic activities. Some fossil fuel–rich countries tried to diversify into other industries. This was true in Syria until the dictator Bashar al-Assad refused to respond to civil protests from an economically stressed citizenry, which precipitated a revolution (see page 254). A somewhat more responsive government in the UAE developed an economy in which only 25 percent of GDP is based directly on fossil fuels, and where trade, tourism, manufacturing, and financial services are now dominant. Even so, most Gulf states are still very dependent on fossil fuel exports and the spin-off industries they generate.

economic diversification the expansion of an economy to include a wider array of activities

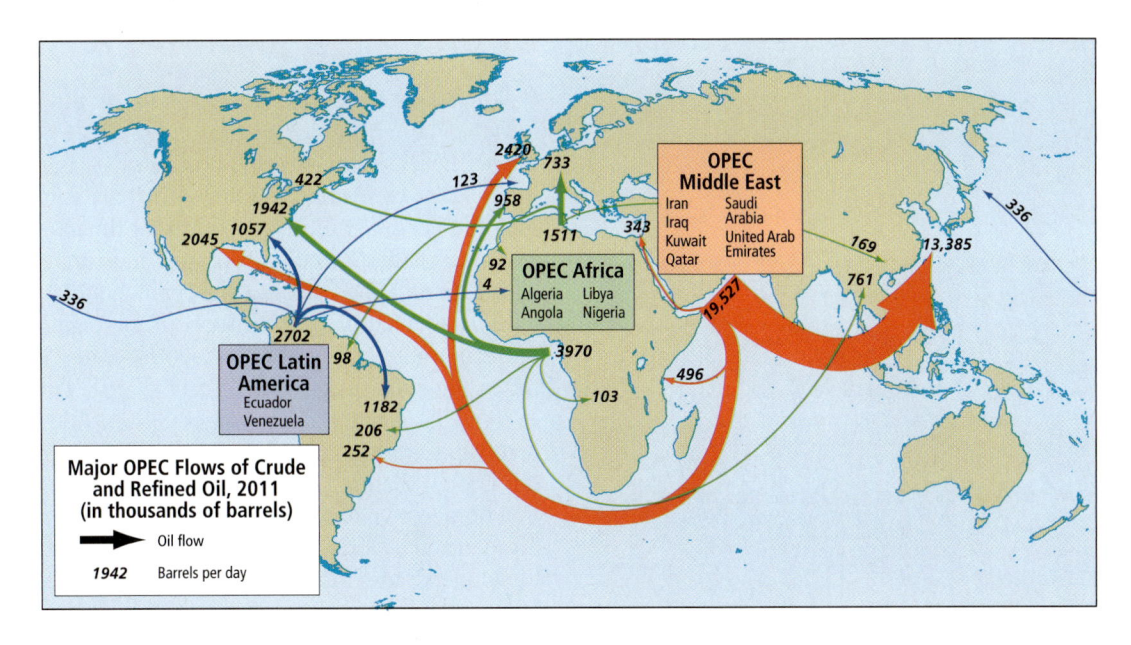

FIGURE 6.19 Major OPEC oil flows in 2011. The map shows the average number of barrels (in thousands) of crude oil and petroleum products distributed per day by OPEC to regions of the world. [Sources consulted: Organization of the Petroleum Exporting Countries, *Annual Statistical Bulletin, 2008, 2009*; *Annual Statistical Bulletin, 2012* (Vienna: OPEC, 2012), pp. 47–48, 53–54, at http://www.opec.org/opec_web/static_files_project/media/downloads/publications/ASB2012.pdf]

Diversification has also been limited by economic development policies that favored *import substitution* (see pages 128–129 in Chapter 3). Beginning in the 1950s, many governments, such as those of Turkey, Egypt, Iraq, Israel, Syria, Jordan, Tunisia, and Libya, established state-owned enterprises to produce goods for local consumption. Among the major products were machinery and metal items, textiles, cement, processed food, paper, and printing. These enterprises were protected from foreign competition by tariffs and other trade barriers. With only small local markets to cater to, profitability was low and the goods were relatively expensive. Without competitors, the products tended to be shoddy and unsuitable for sale in the global marketplace. Meanwhile, the extension of government control into so many parts of the economy nurtured corruption and bribery and squelched entrepreneurialism. Many of the state-owned import substitution enterprises have subsequently either gone bankrupt or been sold to private investors.

Economic diversification and export growth were also limited by a lack of financing, private and public, from within the region. For example, rather than invest locally in mundane but needed consumer products, members of the Saudi royal family generally invest their wealth in more profitable private firms in Europe, North America, and Southeast Asia. Even when private and public investment has stayed in the region, it has gone into lavish projects, such as those in Dubai that give jobs to skilled Asians and may not necessarily prove to be economically profitable. Only recently have governments recognized that they need to invest locally in order to plan wisely for a time when oil and gas run out.

Finally, both international and domestic military conflicts and the ensuing political tensions have stymied economic diversification because they have resulted in some of the highest levels (proportionate to GDP) of military spending in the world. Military spending diverts funds from other types of development, such as health care and education (Figure 6.20). The top four spenders—Saudi Arabia, Oman, Israel, and Jordan—lead the world in military expenditures as a percentage of GDP, and the percentage in all countries in the region, except Tunisia and Libya, is above the global average of 2 percent.

THINGS TO REMEMBER

> **GEOGRAPHIC INSIGHT 2**
> • **Globalization and Development** The vast fossil fuel resources of a few countries in this region have transformed economic development and driven globalization. In these countries, economies have become powerfully linked to global flows of money, resources, and people. Politics have also become globalized, with Europe and the United States strongly influencing many governments.

• Profits in the oil- and gas-producing countries were low until OPEC initiated price increases in 1973. Since then, wealth has accrued mainly to a privileged few.

• Until very recently, countries in the region, with the exception of Turkey and Israel, had not diversified their economies. In the Gulf states, huge profits have not been invested at home, but rather abroad, thus limiting diversification. When import substitution policies have been tried, they have failed to lift countries out of poverty.

Military, Health, and Education Expenditures as Percent of GDP

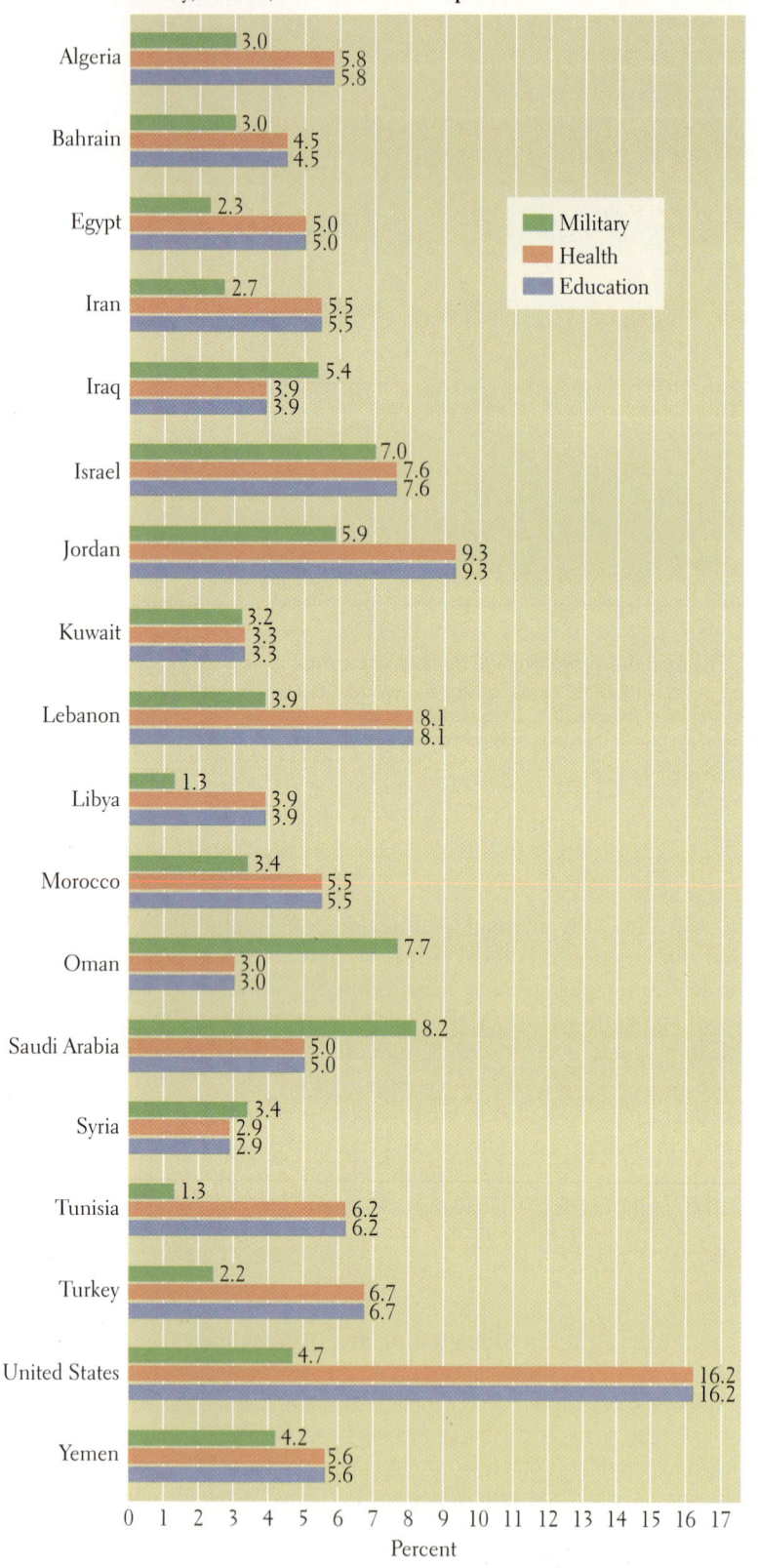

Country	Military	Health	Education
Algeria	3.0	5.8	5.8
Bahrain	3.0	4.5	4.5
Egypt	2.3	5.0	5.0
Iran	2.7	5.5	5.5
Iraq	5.4	3.9	3.9
Israel	7.0	7.6	7.6
Jordan	5.9	9.3	9.3
Kuwait	3.2	3.3	3.3
Lebanon	3.9	8.1	8.1
Libya	1.3	3.9	3.9
Morocco	3.4	5.5	5.5
Oman	7.7	3.0	3.0
Saudi Arabia	8.2	5.0	5.0
Syria	3.4	2.9	2.9
Tunisia	1.3	6.2	6.2
Turkey	2.2	6.7	6.7
United States	4.7	16.2	16.2
Yemen	4.2	5.6	5.6

FIGURE 6.20 Military, health, and education expenditures as a percent of GDP, 2008–2009. These graphs show those countries in the region that have data for each variable. Note that the data is from 2008–2009 and expenditures may have changed due to the Arab Spring. The United States is included for comparative purposes. [Sources consulted: *Human Development Report 2010*, Table 15, and *Human Development Report 2011*, Table 10 (New York: United Nations Development Programme, 2010)]

POWER AND POLITICS

Political and economic cooperation in the region has been thwarted by a complex tangle of hostilities between neighboring countries. Many of these hostilities are the legacy of outside interference by Europe and the United States in regional politics, including colonial intrusions in earlier times, and more recently, by the activities of global oil, gas, industrial, and agricultural corporations. The Israeli–Palestinian conflict, which has profoundly affected politics throughout the region, is rooted in the persecution of the Jews in nineteenth-century Europe, culminating in the Holocaust during World War II. The Iran–Iraq war of 1980–1988 and the Gulf War of 1990–1991 were instigated in part by pressures on petroleum resources, again from Europe and the United States. The siege of violence in Iraq, though begun by a homegrown dictator, came to a head when the United States invaded and occupied Iraq, beginning in March of 2003. Though the United States formally withdrew on December 15, 2011, violence continues to plague Iraq.

The Arab Spring

> **GEOGRAPHIC INSIGHT 3**
>
> **Power and Politics:** Despite the presence of elected bodies of government, authoritarian power structures prevail throughout North Africa and Southwest Asia. Beginning with the Arab Spring of 2010, waves of protests swept the region, resulting in the overthrow of several authoritarian governments. Within the region, official responses to the Arab Spring have included both repression and reform.

The political protests that swept across North Africa in 2010 and into Syria by 2011 eventually resulted in the toppling of long-standing dictatorships in Tunisia, Egypt, and then Libya. The protest movement also presented an increasingly serious threat to President Assad in Syria. The early demonstrations were precipitated by high unemployment and rising food prices caused by the global recession that began in 2008; poor living conditions; government corruption; and the absence of freedom of speech and other political freedoms. An undercurrent of dissent among women was also palpable in every country. Movements for women's rights specifically and human rights generally were rejuvenated (see the opening vignette).

The political disquiet reflected that, for much of their history, most governments in this region gave citizens very little ability to influence how decisions were made. Constitutions were not constructed to facilitate widespread participation in public discourse or to protect the rights of women and minorities. Laws were simply interpreted to suit the factions that held power. Elections were either nonexistent or were rigged to reelect those already in office or their chosen successors. Meanwhile, freedom of speech and of the press was strictly limited, especially if it involved criticism of the government. As a result, most people across the region saw their governments as unresponsive and corrupt and viewed themselves as powerless to influence government in any way other than by massive protests.

The role that Islam should play in society has been a contentious issue, and it relates directly to determining which system of laws should be adopted as well as which protections women should be afforded.

> **ON THE BRIGHT SIDE**
>
> **The Arab Spring as a Conduit for New Ideas**
> The potential for the Arab Spring to bring political and social change quickly to this region was at first overstated. However, the demonstrations and the debates that surround them suggest that things will not return to the former status quo of elitist control. The public shift toward support for democracy has the potential to reduce violence because the possibility of having a voice tends to curb anger. Furthermore, officials in many countries have now seen the power of public protests and are motivated to engage in dialogue with those who want to see changes.

Which First: Elections or Constitutions?

The Arab Spring movement began in Tunisia and spread to Egypt, then to Libya, and eventually to every country in the region—including Iran—with the exceptions of Israel, Qatar, Turkey, and the UAE. The size and focus of the protests varied, but they all raised hopes for real political reforms. However, there were no viable reform models extant within the region to follow. Should constitutions be revised first to guarantee freedoms and equal participation for all, or should elections come first, with the winning majority framing the new constitutions? Tunisia and Egypt chose to have elections first, but turnout was so low that not even one-third of the eligible voters participated. In both cases, Islamist-oriented governments were elected. These governments then proceeded to adopt constitutions that did not adequately protect the rights of more secular political groups, minorities, or women. In the case of Egypt, political turmoil and violence resulted as the Islamists governed in ways that outraged their numerous political opponents, women, and minorities. In 2013, the Islamists were removed from power by Egypt's military.

More incremental approaches to reform were chosen by the kings of Jordan and Morocco, resulting in less chaos and violence. In both cases, the respective kings negotiated with reformists well before full-blown Arab Spring mass movements developed, bargaining to keep some of their royal powers while agreeing to give up others. They both supported the idea that lasting political reform should begin with constitutional reforms hammered out not by elected bodies but by constitutional councils made up of a broad range of people. These councils were charged with making the political systems more democratic and inclusive, and in both cases women were significantly represented, as were religious minorities and those with secular points of view. When elections were held, candidates from a wide range of political perspectives won positions. In Morocco, for example, an Islamist party won enough seats in parliament (107 out of 395 seats) to form a government with allies; but with only a little over a quarter of the total seats, the Islamists had to accept cooperation and compromise as a necessity, an important precedent for the future.

In Egypt, by contrast, where elections were the first priority, the highly organized and Islamist Muslim Brotherhood mobilized their supporters and swept to power with only a minority of the electorate voting. Having won, the Muslim Brotherhood then controlled the writing of an authoritarian constitution that gave the government sweeping powers, such as the ability to curtail freedom of speech, especially that of journalists. The custom that all women required the guardianship of a male was revived and the rights of religious minorities were undermined. Shortly before he was removed from office, President Morsi, a Muslim Brotherhood stalwart, eventually declared that he had powers that superseded those of the Egyptian courts, thus appearing to give himself absolute power.

Islamism in a Globalizing World Ironically, the Arab Spring, with its emphasis on political freedoms, actually opened up space for Islamist groups, especially Salafists, who took momentary advantage of the new openness to assert a tightly closed agenda. **Salafism** is an extreme, purist Qur'an-based version of Islam that has little room for adaptation to modern times. Globalization is a particular problem for these conservative Muslims, who lament what they see as the global spread of open sexuality, consumerism, and hedonism, transmitted in part by TV, movies, and popular music. Salafists were once explicitly peaceful, focusing mostly on the much-needed social services they provided to the poor. But since the 1990s, the idea of *jihad* (a fight) against the West has gained popularity among young Salafists. Many conservative Muslims—Salafists or not—object to what they see as the Western emphasis on the liberalization of women's roles, especially the idea that women should be educated and active outside the home (see Figure 6.27 on page 268).

> **Salafism** an extreme, purist Qur'an-based version of Islam that has little room for adaptation to modern times
>
> **theocratic states** countries that require all government leaders to subscribe to a state religion and all citizens to follow rules decreed by that religion
>
> **secular states** countries that have no state religion and in which religion has no direct influence on affairs of state or civil law

- 139. WHAT MOTIVATES A TERRORIST?
- 140. NEW POLL OF ISLAMIC WORLD SAYS MOST MUSLIMS REJECT TERRORISM
- 144. JIHADIST IDEOLOGY AND THE WAR ON TERROR

Historically, Islamism is rooted in the interaction between religious and governmental authority (Figure 6.21). Governments in Saudi Arabia, Yemen, the UAE, Oman, and Iran are **theocratic states** in which Islam is the official religion and political leaders are considered to be divinely guided by both Allah and the teachings of the Qur'an. Elsewhere, such as in Turkey as well as in Tunisia, Libya, and even Syria (at least before the 2011 rebellion began), governments are officially **secular states**, where religious parties are not allowed and the law is neutral on matters of religion. In practice, even in this region's secular states, Islamic ideas influence government policies. Combined with an authoritarian political culture, this has meant that political freedoms, such as free speech and the right to hold public meetings, were severely limited *outside of Islam*. In countries such as Egypt, for decades the only public spaces in which people were allowed to gather were mosques, and the only public discussions free of censorship by the government were religious discourses. In this context many political movements became rooted in Islam that might not have done so otherwise.

The militancy often associated with Islamism is characteristic of many popular political movements in this region, where challenges to the authority of governments are frequently met

FIGURE 6.21 PHOTO ESSAY: Power and Politics in North Africa and Southwest Asia

Political freedoms are relatively few in this region, which is one of the least democratized in the world. Many countries suffer from a violent political culture, as conflicts that might have been resolved peacefully through a free media and fair elections have instead been prolonged, embittering and enraging otherwise reasonable people. Meanwhile, wars and terrorism have given authoritarian regimes an excuse to forcibly repress legitimate political opposition groups.

with violent repression (see the Figure 6.21 map). In both secular and theocratic governments, political freedoms are sometimes so weak that minorities have been denied the right to engage in their own cultural practices and to speak their own languages. Journalists and private citizens have been harassed, jailed, or even killed for criticizing governments or exposing corruption. While the Arab Spring protests were in part a response to these conditions, the extent to which the new governments and political reforms will protect political freedoms remains to be seen.

153. TURKEY VOTES FOR STABILITY

Will There Be an Iranian Spring? When the political protests known as the Arab Spring erupted across North Africa, many wondered if Iran (which is not Arab) would join in the protests. A seeming prelude occurred in June 2009, when many Iranians suspected the presidential elections had been rigged. Hundreds of thousands of people in Tehran took to the streets in protest. The demonstrations first centered on election fraud, but when the Iranian police killed several unarmed demonstrators, the brutality of the autocratic, theocratic state became a major focus. The Iranian street protests persisted into 2010, when the activities of the Arab Spring in North Africa began. However, there were no major protests during the 2013 Iranian elections, in large part because the leaders of the 2009 protests were in jail or in exile.

The Role of the Press, Media, and Internet in Political Change

In some countries—Turkey, Israel, and Morocco, for example—the press is reasonably free and opposition newspapers are aggressive in their criticism of the government. But in much of the region, journalism can be a risky career, often leading to imprisonment. Egypt has a checkered history where the press is concerned. For years, the Mubarek government harassed Hisham Kassem, editor of the independent English-language weekly, *Cairo Times*. When he became too critical of the government, he lost his license to publish in Egypt. For a time he took great pains to write and print his paper outside the country and smuggle it into Egypt, always risking arrest. Business leaders in Cairo began to provide backing for Kassem, enabling him to publish *Al-Masry Al-Youm*, a daily newspaper that specialized in domestic issues of corruption, election fraud, and the need for an independent judiciary. Kassem and many independent journalists like him participated actively in the 2011 protests in Egypt that brought an end to the 30-year regime of Hosni Mubarak. In this he shared a common cause with Egypt's formerly repressed Islamist opposition group, the Muslim Brotherhood, which won control of the national government in elections in 2012. However, Kassem and many in the media were highly critical of efforts by the Muslim Brotherhood–led government to censor print and broadcast journalists' criticisms of the government.

In Saudi Arabia, a dozen newspapers are available every morning, but all are owned or controlled by the royal Saud family, and all journalists are constrained by the fact that they may

A U.S. Navy warships patrol the Persian Gulf. U.S. and European militaries have had a large presence in this region for a number of years, especially because of the importance of the region's oil exports to the global economy. The U.S.-led invasion of Iraq in 2003 was justified in part as an effort to ensure greater political freedoms for the people of Iraq. However, many U.S. officials have openly admitted that safeguarding foreign access to the region's oil reserves was a larger motivation.

E Thousands celebrate in Tripoli, Libya, on the second anniversary of the end of Libya's civil war of 2011. Like the war in Syria, this conflict, in which 25,000–30,000 people were killed, was ignited by the government's brutal repression of Arab Spring protests.

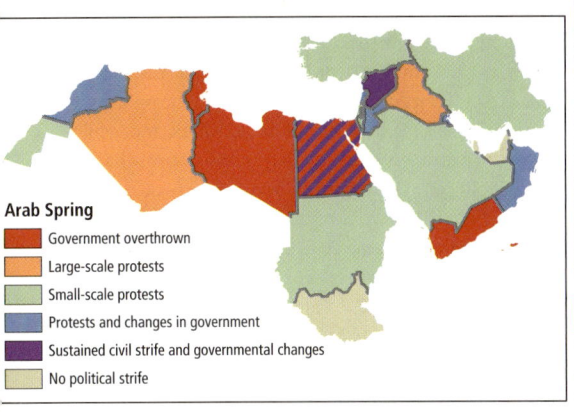

Arab Spring
- Government overthrown
- Large-scale protests
- Small-scale protests
- Protests and changes in government
- Sustained civil strife and governmental changes
- No political strife

B An Egyptian man displays a Star of David painted on his shoe, an act of symbolic desecration, during an anti-Israel rally. Ongoing tensions between Israel and its Arab neighbors have worked against political freedom throughout the region. Many Arab governments use the conflict to justify repressing legitimate political parties. Meanwhile, Israel recently banned anyone who has visited "enemy countries" (Lebanon, Syria, Iraq, Iran, Saudi Arabia, Sudan, Libya, and Yemen) from running for national office for 7 years. [Khaled Desouki/AFP/Getty Images]

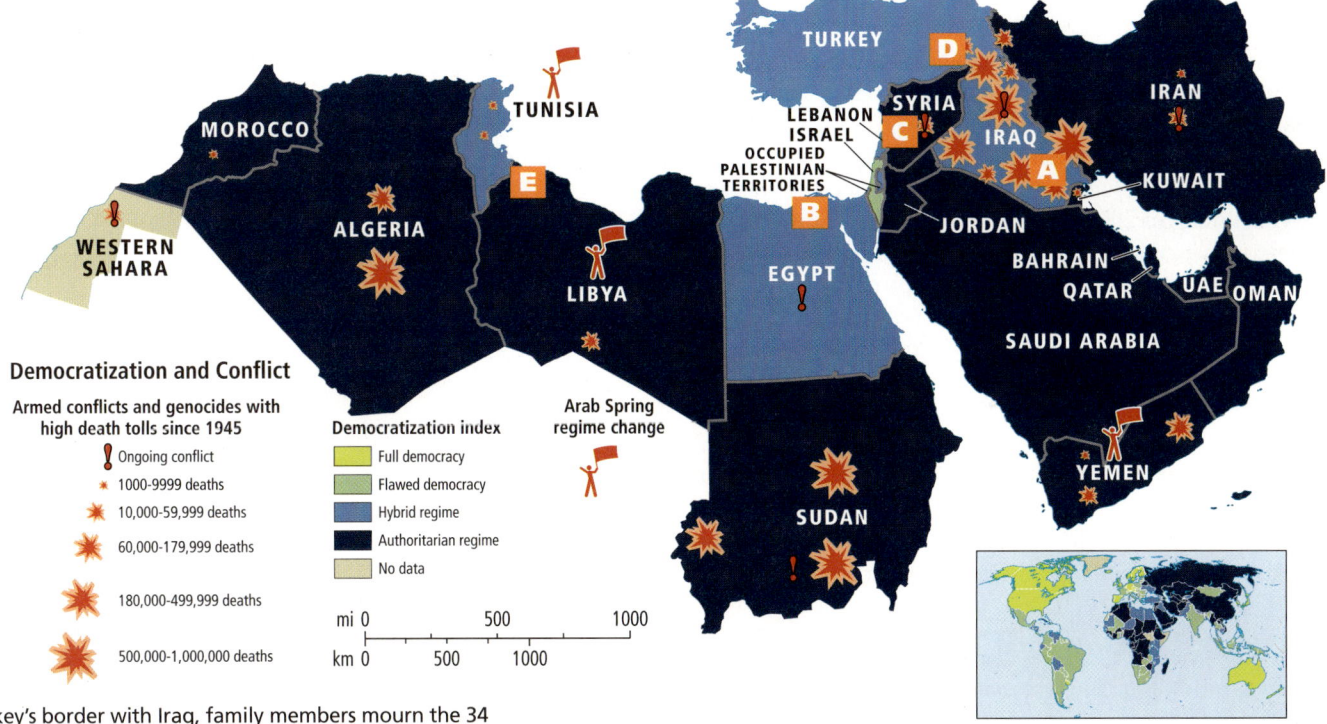

Democratization and Conflict

Armed conflicts and genocides with high death tolls since 1945
- Ongoing conflict
- 1000–9999 deaths
- 10,000–59,999 deaths
- 60,000–179,999 deaths
- 180,000–499,999 deaths
- 500,000–1,000,000 deaths

Democratization index
- Full democracy
- Flawed democracy
- Hybrid regime
- Authoritarian regime
- No data

Arab Spring regime change

D Near Turkey's border with Iraq, family members mourn the 34 Kurdish civilians killed by the Turkish air force, who mistook them for rebels. Kurds form a majority of the population in southeastern Turkey and in neighboring parts of Iraq and Iran. Kurds have long been denied political freedoms and in some cases even the right to speak their language. Though Kurdish militias still exist, democratization has recently reduced public support for violence, especially in northern Iraq.

C A rebel sniper in the city of Aleppo, Syria. A civil war was sparked by the Syrian government's harsh crackdown on Arab Spring protests, beginning in 2011. More than 100,000 people have been killed and millions have been displaced by the conflict.

not print anything critical of Islam or of the Saud family, which numbers in the tens of thousands. When accidents happen or some malfeasance by a public official is revealed, the story is blandly reported with little effort to explore the causes of events or their effects, or to hold responsible officials accountable.

A beacon of journalistic change is the broadcasting network Al Jazeera, founded by the emir of Qatar, Sheikh Hamad bin Khalifa al-Thani, but now privately owned and independent of the Qatar government. Al Jazeera is credited not only with changing the climate for public discourse across the region but with changing public opinion outside of the region through Al Jazeera English, the source of some of the information used in the opening vignette. The various versions of Al Jazeera now openly cover controversy, including reforms proposed by the most radical Iranian and Arab Spring protesters, with no apparent censorship. **147. AL JAZEERA LAUNCHES GLOBAL BROADCAST OPERATION**

The role of the Internet and cell phone technology as a force for political change emerged first during the Iranian protests beginning in 2009, when the fact that so many Iranians had video-capable cell phones meant that via the Internet, the world saw instantly the brutality of the Iranian police and army. Use of the Internet (e-mail, Facebook, crowdsourcing, and Twitter) became commonplace during the Arab Spring, as ordinary Egyptians with nothing but an inexpensive cell phone were able to give real-time reports about activities in the streets of Cairo, Tripoli, and Benghazi. Twitter became the chief medium because it is free, mobile, and quick. The data from many sources could then be compiled via software such as *Ushahidi*, which was invented and made popular by several Kenyan engineers (discussed in the opening vignette of Chapter 7).

Democratization and Women

Most countries in this region now allow women to vote, and two countries—Israel and Turkey—have elected female heads of state (prime ministers) in the past. Nevertheless, women who want to actively participate in politics still face many barriers. In the Gulf states, where women's political status is the lowest, circumstances vary from country to country. Oman gave all women the right to vote in 2003. In Kuwait in 2009, four highly educated women were elected to parliament; they quickly energized the pace of law making, often publicly criticizing their male colleagues, who were frequently absent for crucial votes. In Saudi Arabia, by contrast, women will be given the right to vote only in local elections in 2015, and only one Saudi woman serves as a public official. Across the region, women average less than 12 percent of national legislatures, the lowest of any world region and half the world average. By 2011, only in Bahrain, Iraq, Israel, Libya, Morocco, and Sudan did women make up more than 15 percent of parliaments. Women lost ground in Egypt when the post–Arab Spring elections in 2012 brought fewer women to office than there were before. Moreover, there is a tendency—even among women—to not support female candidates.

ON THE BRIGHT SIDE

Reforms for Women Could Address Labor Problems

An important impetus for change in the region is the increasing number of women who are becoming educated and employed outside the home. In a number of countries, women outnumber men in universities; most notably, women make up 70 percent of university students (but only 5 percent of the workforce) in Saudi Arabia. While Saudi women activists have characterized their country as practicing a sort of apartheid with its own women, circumscribing or forbidding all manner of public activities, and while laws there still require a male guardian to accompany an adult woman when she leaves the house, change is in the wind. Recently, a Saudi prince, known for his interests in reform, suggested that one way to reduce dependence on foreign workers is to allow Saudi women to become drivers of taxis and trucks, as is now allowed in Jordan. [Source: Ulf Laessing, "Saudi Prince Asks Why Women Are Not in the Driver's Seat," http://www.trust.org/item/20110309172500-r8i3d.]

THINGS TO REMEMBER

> **GEOGRAPHIC INSIGHT 3**

- **Power and Politics** Despite the presence of elected bodies of government, authoritarian power structures prevail throughout the region. Beginning with the Arab Spring of 2010, waves of protests swept this region, resulting in the overthrow of several authoritarian governments. Within the region, official responses to the Arab Spring have included both repression and reform.

- While the immediate causes of the Arab Spring demonstrations may be economic, activists who have for years promoted increased political freedoms as a path to peace in this region have been reinvigorated by the protests and continue their work to end authoritarian rule. Women participants have thus far been both energized and disappointed by the results.

- Demand for more political freedoms is growing and public spaces for debate, other than mosques, are emerging. The press is reasonably free in some countries and severely curtailed in others. Al Jazeera is credited with changing the climate for public discourse across the region and with changing public opinion outside the region.

- Most countries now allow women to vote, and the two most developed countries have elected female heads of state. Even though women who want to actively participate in politics still face barriers, patterns are changing as more women are becoming educated and are working outside the home. ∎

THREE WORRISOME GEOPOLITICAL SITUATIONS IN THE REGION

The region of North Africa and Southwest Asia is known as a center of especially troublesome political issues that command the attention of major global powers. These issues can all be related in one way or another to the politics of the international oil trade (see Figure 6.21A), and at least one may be related to global climate change.

Situation 1: Fifty Years of Trouble Between Iraq and the United States

The origins of the U.S. war with Iraq, beginning in 2003, lie in 1963, when the United States backed a coup that installed a pro-U.S. government that evolved into the regime of Saddam Hussein. For decades, the United States had an amicable relationship with the Iraqi government, driven in large part by Iraq's considerable oil reserves, the fourth largest in the world after those of Saudi Arabia, Canada, and Iran. The United States publicly supported Iraq in the 1980–1988 war between Iraq and Iran, but secretly supplied Iran with weapons during the Reagan administration (1981–1989) when it appeared that Iraq might become more troublesome if it won the war. Relations with Iraq took a dramatic turn for the worse in 1990 when Saddam Hussein, Iraq's dictator, invaded Kuwait. The United States, under President George H. Bush, forced Iraq's military out of Kuwait in the Gulf War of 1990–1991, and afterward placed Iraq under crippling economic sanctions, but failed to tame Saddam Hussein (see Figure 6.7D).

After the terrorist attacks of September 11, 2001, the U.S. administration of President George W. Bush (2001–2009) first launched a war on Afghanistan, but by 2003, shifted its focus to Iraq and its autocratic president, Saddam Hussein (see Chapter 2, pages 86–87). The Bush administration claimed that Iraq had or was creating an arsenal of weapons of mass destruction, and used this pretext to declare war on March 20, 2003, with the goals of confiscating Iraq's weapons, removing Saddam from power, and turning Iraq into a democracy. Since then, several senior U.S. officials involved with the planning and execution of the Iraq war have stated that a major motivation was to increase U.S. control of Iraq's oil resources. This explanation for the war is now widely accepted throughout the region.

After the initial invasion met little resistance, President Bush declared the war won on May 1, 2003. However, terrorist bombs and insurgent attacks soon erupted, with violence peaking between 2006 and 2007, after which it gradually decreased but did not cease. By March 2013, a total of 4422 U.S. troops had been killed and 31,926 had been wounded. Furthermore, more than 150,000 veterans of the conflict had been diagnosed with some form of serious mental disorder, including posttraumatic stress disorder (PTSD). The death toll for Iraqis, including civilians, was much higher, estimated as at least 111,407 and possibly as high as 1.3 million, when counting Iraqi deaths caused by the harsh conditions created by the war.

The failure to understand the simmering tensions between Iraq's major religious and ethnic groups crippled the U.S. intervention. Sunnis in the central northwest had dominated the country under Saddam, although they constituted only 32 percent of the population. Shi'ites in the south, with 60 percent of the population, have dominated politics since the fall of Saddam. This has given more influence to neighboring Iran, whose mostly Shi'ite population has an affinity with Iraqi Shi'as. Meanwhile, Kurds in the northeast, once brutally suppressed under Saddam Hussein, are allied with Kurdish populations in Turkey, Iran, and Syria, and resist cooperating with the Iraqi national government in Baghdad. **134. KURDISH NATIONALISTS IN IRAQ, TURKEY SEEK LAND OF THEIR OWN**

Recent studies of Iraqi public opinion indicate that a majority of Iraqis want a strong central government that can protect them from violent insurgents and that can maintain control of the country's large fossil fuel reserves. Polls also show that the vast majority of Iraqis want all fighting to stop and all U.S. and allied military forces to leave. All U.S. "combat" forces left Iraq in 2012, though thousands of personnel will remain for years to come as trainers and technical support for the Iraqi military.

Situation 2: The State of Israel and the "Question of Palestine"

The Israeli–Palestinian conflict has lasted more than 60 years and has included several major wars and innumerable skirmishes. It is a persistent obstacle to political and economic cooperation within the region and it complicates the relations between many countries in this region and the rest of the world (see Figure 6.21B).

Israel's excellent technical and educational infrastructure, its diverse and prospering economy, and the large aid contributions (public and private) it receives from the United States and elsewhere have made it one of the region's wealthiest, most technologically advanced, and militarily powerful countries.

The Palestinian people, by contrast, are severely impoverished and undereducated after years of conflict, inadequate government, and meager living (as noted in Table 6.1), often in refugee camps. Through a series of events over the past 60 years, Palestinians have lost most of the lands on which they used to live. They now live in two main areas—the West Bank (home to 2 million Palestinians) and the Gaza Strip (1.1 million), with

TABLE 6.1	Circumstances and human well-being among Palestinians and Israelis, 2012					
Statistics show some stark differences in well-being between Palestinians and Israelis, including differences in overall satisfaction with life.						
	Population (in millions) (mid-2012)	GNI PPP in U.S.$ (2012)	Life expectancy at birth (2012)	Infant mortality rate (2012)	HDI ranks (2012)*	Overall life satisfaction (2011)[†]
Palestinians	4.3	3359	73	21	110	4.7
Israelis	7.9	27,660	82	3.4	16	7.4

*See http://hdrstats.undp.org/en/countries/profiles/PSE.html
[†]Ranked on a scale of 1 to 10, with 0 representing least satisfied and 10 representing most satisfied
Source: Population Reference Bureau 2012 Data Sheet
Human Rights Watch report: http://www.hrw.org/middle-eastn-africa/israel-palestine

another 2 million living as refugees in Jordan, where they outnumber ethnic Jordanians. The West Bank and Gaza are both highly dependent on Israel's economy. Israel often takes military action in these two zones in retaliation for Palestinian suicide bombings and rocket fire launched primarily from Gaza. The West Bank Palestinian territories continue to be encircled by security walls built by Israel, partly to defend against violence, but also to curtail Palestinian access.

154. 60 YEARS AFTER ISRAEL'S FOUNDING, PALESTINIANS ARE STILL REFUGEES

131. ISSUES FROM 1967 ARAB–ISRAELI WAR REMAIN UNRESOLVED

The Creation of the State of Israel In the late nineteenth century, as a response to centuries of discrimination and persecution in Europe and Russia, a small group of European Jews, known as **Zionists**, began to purchase land in a part of the Ottoman Empire known at the time as Palestine. Most sellers were wealthy non-Palestinian Arabs and Ottoman Turks living outside of Palestine. These absentee landowners had leased their lands for many years to Palestinian tenant farmers and herders or had granted them the right to use the land freely. Such rights were no longer recognized by the Zionists who wanted the land to become farmed and occupied by Jews. Historians still debate whether or not the Zionists or the former landlords adequately compensated the displaced indigenous inhabitants.

> **Zionists** those who have worked, and continue to work, to create a Jewish homeland (Zion) in Palestine
>
> **intifada** a prolonged Palestinian uprising against Israel

On their newly purchased lands, the Zionists established communal settlements called *kibbutzim*, into which a small flow of Jews came from Europe and Russia. While Jewish and Palestinian populations intermingled in the early years, tensions emerged as Zionist land purchases displaced more and more Palestinians.

In 1917, the British government adopted the Balfour Declaration, which favored the establishment of a national home for Jews in the Palestine territory, and which explicitly stated that "nothing shall be done which may prejudice the civil and religious rights of existing non-Jewish communities in Palestine, or the rights and political status enjoyed by Jews in any other country." It was at this time that the word *Palestine*, with roots far back in history, began to be used officially. In 1923 after World War I, the United Kingdom received what is called a "mandate" from the League of Nations to administer the territory of Palestine (Figure 6.22A).

By 1946, following the genocide of 6 million Jews in Nazi Germany's death camps during World War II, strong sentiment had grown throughout the world in favor of a Jewish homeland in the space in the eastern Mediterranean that Jews had shared in ancient times with Palestinians and other Arab groups. Hundreds of thousands of Jews began migrating to Palestine and, against British policy, many took up arms to support their goal of a Jewish state.

Decades of Conflict and the Two-State Solution In November 1947, after an intense debate in the UN General Assembly, the UN adopted the "Plan of Partition with an Economic Union" that ended the earlier mandate and called for the creation of both Arab and Jewish states (the *two-state solution* still under discussion today) with special international status for the city of Jerusalem (see Figure 6.22B).

The Palestinians and neighboring Arab countries fiercely objected to the establishment of a Jewish state and feared that they would continue to lose land and other resources. Then, as now, the conflict between Jews and Palestinian Arabs was less about religion than control of land, settlements, and access to water. The sequence of the changes in allotments of land to Israel and to the Palestinians over the last 90 years can be followed in Figure 6.22.

On the same day that the British reluctantly ended the mandate and withdrew their forces, May 14, 1948, the Jewish Agency for Palestine unilaterally declared the state of Israel on the land designated to them by the Plan of Partition. Warfare between the Jews and Palestinians began immediately. Neighboring Arab countries—Lebanon, Syria, Iraq, Egypt, and Jordan—supporting the Palestinian Arabs, invaded Israel the next day. Fighting continued for several months, during which Israel prevailed militarily. An armistice was reached in 1949, and as a result, the Palestinians' land shrank still further, with the remnants incorporated into Jordan and Egypt (see Figure 6.22C).

In the repeated conflicts over the next decades—such as the Six-Day War in 1967 and the Yom Kippur War in 1973—Israel again defeated its larger Arab neighbors, expanding into territories formerly controlled by Egypt (Sinai), Syria (Golan Heights), and Jordan (West Bank of the Jordan River; see also Figure 6.22D). Since 1948, hundreds of thousands of Palestinians have fled the war zones, with many forcibly removed to refugee camps in nearby countries. Some Palestinians stayed inside Israel and became Israeli citizens, but they have not been treated by the state as equal to Jewish Israelis.

128. HEZBOLLAH—SERVING MUSLIMS WITH GOD AND GUNS

133. LEBANESE OIL SPILL—COLLATERAL DAMAGE OF THE BOMBINGS

In 1987, the Palestinians mounted the first of two prolonged uprisings, known as the **intifada**, characterized by escalating violence. The first ran until 1993, when the Oslo Peace Accords provided that Israel withdraw from parts of the Gaza Strip and the West Bank, and that the Palestinian Authority be the entity that would enable Palestinians to govern themselves in their own state. The second intifada began in 2000 and continues into the present, primarily fueled by the expansion of Israeli settlements—in breach of the Oslo Accords—into Palestinian territories in Gaza, the West Bank, and the Golan Heights (see Figure 6.22F). Both sides have suffered substantial trauma and casualties. According to B'Tselem, an Israeli human rights group, far more Palestinians have died in this violence (since 2000, there have been 6630 Palestinian deaths versus 1097 Israeli deaths).

Over the years since 1947, the vision of a two-state solution has not died. In 2012 the Palestinians, led by their president, Mahmoud Abbas, officially petitioned the UN General Assembly for status as a nonmember observer state. The UN Assembly voted 138 to 9 to approve this petition (41 nations abstained), and thus the Palestinians gained official recognition that they had never before achieved.

FIGURE 6.22 Israel and Palestine, 1923–1949, and Israel and the Palestinian Territory after 1949.

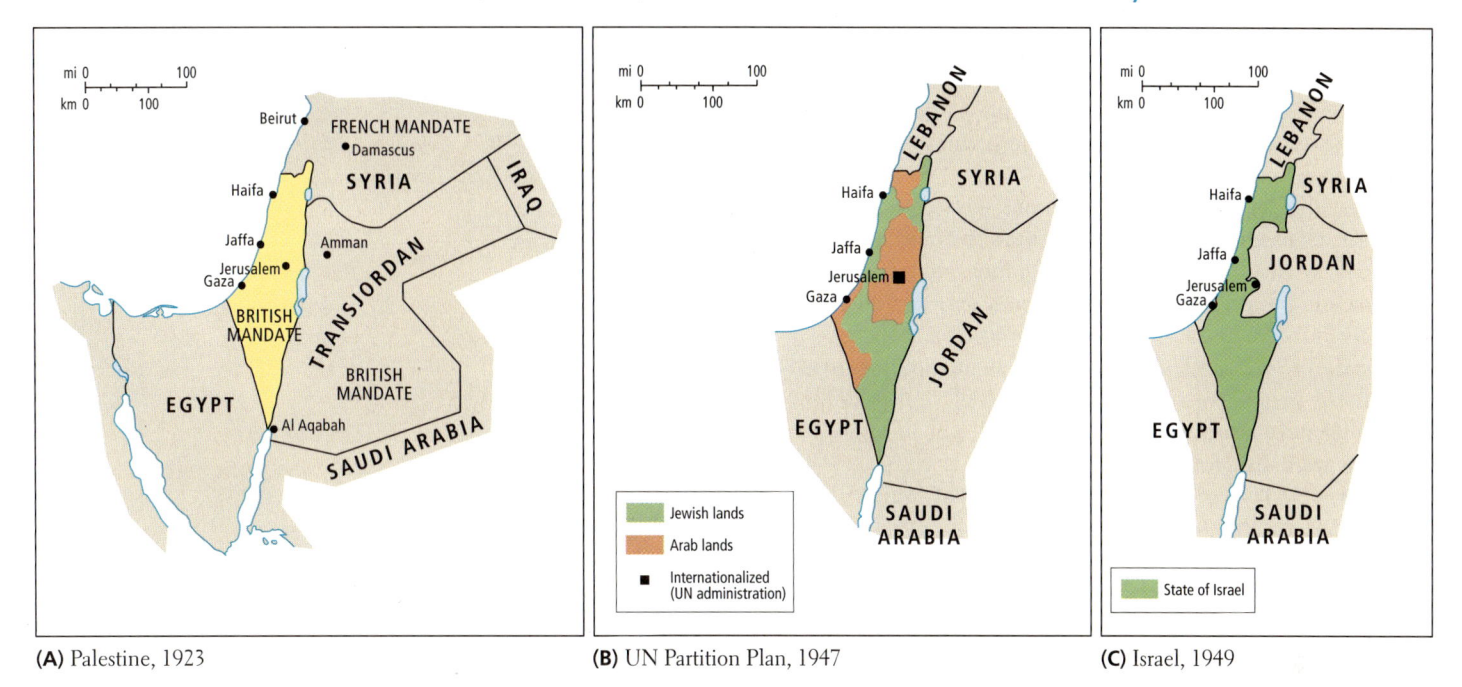

(A) Palestine, 1923

(B) UN Partition Plan, 1947

(C) Israel, 1949

Israel and Palestine, 1923–1949. **(A)** Palestine, 1923. Following World War I, Britain controlled what was called Palestine and is now Israel and Jordan (Transjordan was the precursor to Jordan). **(B)** The UN Partition Plan, 1947. After World War II, the United Nations developed a plan for separate Jewish and Palestinian (Arab) states. **(C)** Israel, 1949. The Jewish settlers did not agree to the partition plan; instead, they fought and won a war, creating the state of Israel. [Sources consulted: Colbert C. Held, *Middle East Patterns—Places, Peoples, and Politics* (Boulder, CO: Westview Press, 1994), p. 184; *The Israeli Settlements in the Occupied Territories,* 2002, Foundation for Middle East Peace, at http://www.firstpr.com.au/nations]

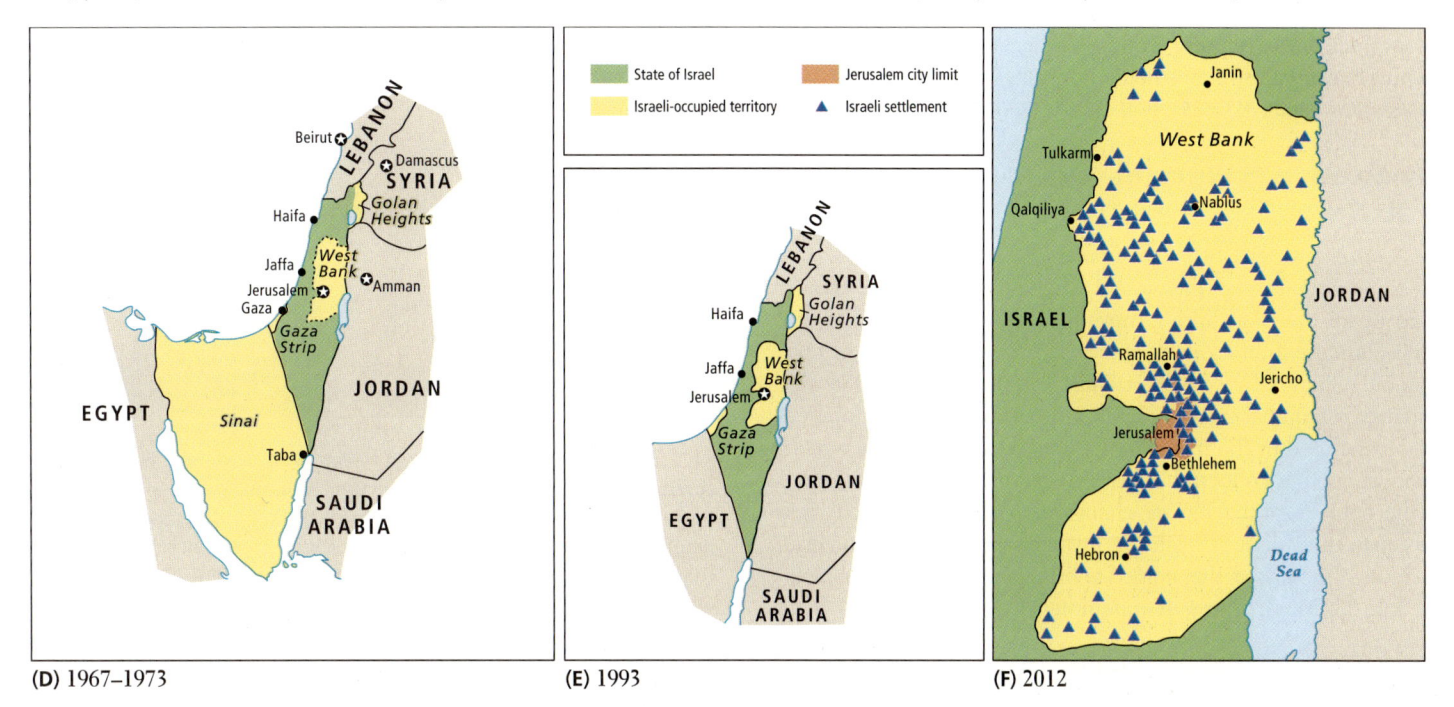

(D) 1967–1973

(E) 1993

(F) 2012

Israel and the Palestinian Territory after 1949. When the state of Israel was created in 1949, its Arab neighbors were opposed to a Jewish state. **(D)** In 1967, Israel soundly defeated combined Arab forces and took control over Sinai, the Gaza Strip, the Golan Heights, and the West Bank. **(E)** In subsequent peace accords, Sinai was returned to Egypt, but Israel maintained control over the Golan Heights and the West Bank, claiming that they were essential to Israeli security. **(F)** Although the Palestinians were granted some autonomy in the Gaza Strip and the West Bank, during the 1990s the Israelis, contrary to verbal agreements in the Oslo Accords, continued to build Jewish settlements in the West Bank and Golan Heights. By 2012, there were hundreds of extra-legal Israeli settlements in these areas. This map shows only West Bank settlements. [Sources consulted: "Map: Golan Heights," at http://www.fmep.org/reports/archive/vol.-22/no.-6/map-golan-heights; Geoffrey Aronson, "The Occupation Returns to Center Stage," *Foundation for Middle East Peace Settlement Report* 22 (6), at http://www.fmep.org/reports/archive/vol.-22/no.-6/the-occupation-returns-to-center-stage; http://thinkprogress.org/security/2012/05/14/483860/eu-settlements-threaten-two-states/?mobile=nc; http://thinkprogress.org/wp-content/uploads/2012/05/APN-Settlement-Map1.png]

Territorial Disputes When Israel occupied Palestinian lands in 1967, the UN Security Council passed a resolution requiring Israel to return those lands, known as the occupied Palestinian Territories (oPT), in exchange for peaceful relations between Israel and neighboring Arab states. This *land-for-peace* formula, which set the stage for an independent Palestinian state, has been only partially fulfilled.

Despite the land-for-peace agreement, between 1967 and 2013, Israel secured ever more control over the land and water resources of the occupied territories. Israel took what appeared to be a major step toward peace in 2005 when it removed all Jewish settlements from the Gaza Strip, but this progress was negated by the blockade of Gaza's economy and the significantly stepped-up settlement in the West Bank. Between 2005 and mid-2013, some 95,000 new Israeli settlers were added and thousands of Palestinians were displaced.

The **West Bank barrier**, a high containment wall (Figure 6.23A) built beginning in 2003, encircles Jewish settlements on the West Bank and separates approximately 30,000 Palestinian farmers from their fields. It also blocks roads that once were busy with small businesses, effectively annexes 6 to 8 percent of the West Bank to Israel, and severely limits Palestinian access to much of the city of Jerusalem, most of which is now on the Israeli side of the barrier. The barrier, declared illegal by the World Court and the United Nations, and opposed by the United States, is nonetheless very popular among Israelis because it has reduced the number of Palestinian suicide bombings. **132. WEST BANK BARRIER, NEW DIVIDE IN PALESTINIAN–ISRAELI CONFLICT**

West Bank barrier a 25-foot-high concrete wall in some places and a fence in others that now surrounds much of the West Bank and encompasses many of the Jewish settlements there

Situation 3: Failure of the Arab Spring in Syria

The worries raised by the rebellion in Syria (beginning in 2011 and continuing through the present) are basically the same as those raised by all the Arab Spring movements. Can autocratic governments continue to block democratic reforms? Will Islamist factions more often than not succeed in taking control? Will women be able to change patriarchal customs? Will gains in economic development be overwhelmed by the material destruction caused by civil war? In Syria, yet another concern has been raised that carries implications for the entire region: Climate change and the water scarcity that accompanies it (discussed later in this section) are now thought to be among the stressors that helped to bring on the Syrian rebellion.

The political causes of rebellion in Syria are many. Syria used to be one of the more developed countries in the region. Several factions of Muslims, as well as Christians and even a few Jews, lived together peaceably. Markets were full of life, women could have careers and go out alone past midnight, old men played board games while children played along community streets, tourists flocked to ancient historic sites, and Syrian TV dramas disseminated the Syrian hospitable way of life across the region. But after the Assad family swept to power in a bloodless coup in 1970, Syria slowly morphed into an authoritarian state. By the time Bashar al-Assad took over from his father Hafez in 2000, Syria had gained a reputation for backing assassinations in

FIGURE 6.23 Conflict and cooperation between Israelis and Palestinians. Both conflict and cooperation between Israelis and Palestinians have increased in recent years.

(A) Palestinian women on their way to Jerusalem to celebrate Ramadan pass through a heavily guarded checkpoint at the West Bank barrier in Bethlehem, Palestine. The barrier has hindered the flow of people and goods between Israel and the West Bank, severely damaging the latter's economy.

(B) Palestinian women working in a factory in Jerusalem. Israeli companies have access to the expertise, equipment, and investment capital that make the growth of manufacturing industries possible, while Palestinians supply cheap labor. While many economic partnerships are developing between Israelis and Palestinians, they suffer when violence forces border closures.

Lebanon and bombings in Iraq, and for supporting *Hezbollah*—a Lebanese-based radical Shi'ite group that has close links to Iran, opposes Israel, and uses terrorist tactics.

Although some Syrians enjoyed the way the Assads thumbed their noses at the United States and Europe (the West), more educated and secular Syrians could see that their access to decent lives and participation in civic life was being ever more curtailed by the Assad regime. The trappings of a free society, such as shopping malls, Coca-Cola, Internet cafés, and Facebook were not able to pacify people when debate, elections, and creativity were stifled by an authoritarian government and a society rife with corruption even in the civil courts.

The Assad family belongs to the minority Alawite sect of Shi'ites, while the majority of Syrians are Sunni Muslims. To keep political equilibrium, the Assads favored a secular state similar to Turkey's. However Sunnis were kept out of governing circles. Sunni antagonism against Alawites grew because Alawites led particularly privileged lives, with many advantages coming to them from the Assad family.

Arab Spring protests were met with brutal repression in Syria. Several pro-democracy protestors in Damascus who demanded the release of political prisoners were shot dead during March 2011. The protestors persisted, and from the spring of 2011 well into 2013, the Syrian government, backed by the army and aided by weapons and food assistance from Russia and Iran, escalated its attacks on the dissidents. In time, it became apparent that the rebel factions came from opposing points of view—some favoring pro-democratic secular changes, others wanting more Islamist or Salafist conservative reforms. The dissenters were unable to coalesce around common goals for Syria. Although for the most part the Syrian army remained loyal to Assad, some soldiers from Sunni parts of the country defected to the rebels; yet despite all the destruction and loss of life, at no time did the Assad administration agree to peace-making concessions.

By early 2013, more than 100,000 people had died from violence and related causes, with children making up at least one-third of the dead. Aid agencies estimated that by March of 2013, a million Syrians lacked adequate food, 2.5 million were internally displaced, and 1 million had fled across borders to Turkey, Iraq, Lebanon, and Jordan. Western governments have been reluctant to intervene.

THINGS TO REMEMBER

- The United States has had a long and complicated relationship with Iraq that dates back to the 1960s, when the United States tried to influence Iraq's internal affairs.
- While the modern conflict between Israel and the Palestinians is decades old, the situation remains dynamic and complex and continues to be an obstacle to widespread political and economic cooperation in the region.
- Economic cooperation is already a fact of life for Israelis and Palestinians, and more such interaction is a crucial component of any solution to the current conflict. The entire region would benefit from the peace dividend that comes from broader economic cooperation.
- Political protests against the autocratic Assad regime in Syria have become complicated by the involvement of competing points of view among Islamist, Salafist, secularist, and pro-democracy dissident factions.

ON THE BRIGHT SIDE

Economic Interdependence as a Peace Dividend

Though rarely covered in the media, economic cooperation is a fact of life for Israelis and Palestinians and is a crucial component for any two-state solution. Economic ties between Palestine and Israel have been essential to both for many years. Israel is the largest trading partner of the West Bank and the Gaza Strip, and provides Palestinians with a currency (the Israeli shekel), electricity, and most imports. Israel needs the labor of tens of thousands of Palestinian workers in fields, factories, and homes (see Figure 6.23B).

Currently, there are several cooperative industrial parks that have been established jointly by the Palestinian Authority and the Israeli government in the West Bank and the Gaza Strip. Dubbed "peace parks," the goal of these parks is to use economic development to overcome conflict. Some people are critical of them because Israel tightly controls Palestinian access to the world market. It may take some time for the peace parks to earn the trust of Palestinians, given Israel's history of discouraging industrial development in the occupied territories. Nevertheless, neighboring Israeli and Palestinian cities, such as Gilboa in Israel and Jenin in the West Bank, are pursuing the peace park idea. In 2012, Israelis and Palestinians quietly reached new agreements to expand Palestinian access to jobs in, and trade with, Israel. The agreements address tax procedures, increase the number of entrance permits, and help ease the movement of people through roadblocks.

URBANIZATION

> **GEOGRAPHIC INSIGHT 4**
>
> **Urbanization:** Two patterns of urbanization have emerged in the region, both tied to the global economy. In the oil-rich countries, the development of spectacular new luxury-oriented urban areas is based on global flows of money, credit, goods, and skilled people. In the oil-poor countries, cities are old and have little capacity to handle the masses of poor, rural migrants attracted by the promise of jobs. In these older, poorer cities, jobs are often scarce and many people live in overcrowded slums with few services.

Urbanization is transforming this region. Two globalized patterns of urbanization with distinct geographic signatures are now apparent: one pattern in the newly oil-rich countries, and another in those countries where development has primarily resulted in rural-to-urban migration, with its associated crowded slum housing.

Most people lived in small rural settlements until the 1970s, when people began to migrate in significant numbers from villages to urban areas in response to economic forces driven by oil wealth and globalization, and by agricultural modernization that displaced small farmers. In locations other than the Gulf states, the shift toward green revolution export-oriented agriculture reduced the amount of labor required to produce crops headed for the world market. Even though the green revolution created needed export income for the countries, people who became part of the excess labor force had to leave the countryside for the cities. In North Africa, drought also instigated rural-to-urban migration (Figure 6.24B). By 2013, more than 70 percent of the region's people lived in urban areas (see the Figure 6.24 map); more than 434 cities had populations of at least 100,000; and 37 cities had more than 1 million people. The largest metropolitan area in the region, and one of the largest in the world, is Cairo, with 17 million residents in 2012.

The petroleum-rich Gulf states are now highly urbanized. Between 70 and 100 percent of the (still small) population

FIGURE 6.24 PHOTO ESSAY: Urbanization in North Africa and Southwest Asia

Globalization has brought two distinct patterns of urbanization to this region. In fossil fuel–rich countries, populations are more urbanized and city governments have undertaken lavish building booms, drawing in laborers and highly skilled workers from across the globe. In countries without fossil fuel wealth, cities are receiving massive flows of poor, rural migrants. This is partially a result of economic reforms (emphasizing export agriculture and industrialization) aimed at making rural and urban economies more globally competitive.

A Palm Jumeirah is an artificial palm-shaped island in Dubai, UAE. Built by more than 40,000 workers, mostly low-wage migrants from South Asia, Palm Jumeirah will eventually house 65,000 people, mostly in villas and condos that cost millions of dollars each. Palm Jumeirah and several similar developments are central to Dubai's efforts to build a globalized tourism-based economy that will prosper long after the region's fossil fuel resources are exhausted.

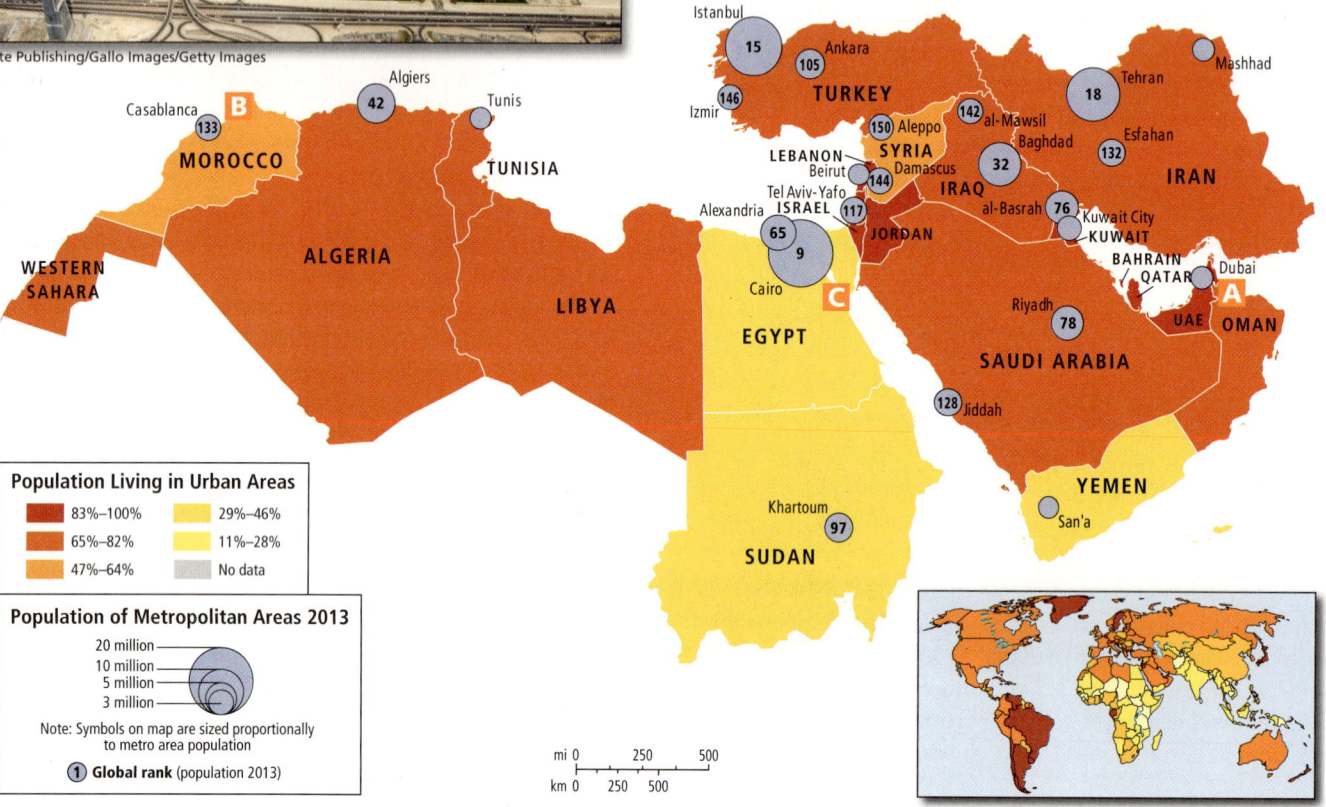

Population Living in Urban Areas
- 83%–100%
- 65%–82%
- 47%–64%
- 29%–46%
- 11%–28%
- No data

Population of Metropolitan Areas 2013
- 20 million
- 10 million
- 5 million
- 3 million

Note: Symbols on map are sized proportionally to metro area population

① Global rank (population 2013)

B A Moroccan man searches for valuables in a garbage dump outside Casablanca. The city's population is growing quickly due largely to migration from Morocco's drought-stricken interior, but the lack of jobs in the city is resulting in escalating unemployment and crime.

C A market in Cairo. After decades of migration from rural areas, Cairo is the region's largest city; it is severely overcrowded.

now lives in urban areas, which are extravagant in design. These modern Gulf cities draw investment in high-tech ventures and high-end tourism. The new ventures and the construction of office and living space require a wide range of skilled workers from all over the world. For example, Dubai (one of the United Arab Emirates) has built elaborate (and as yet largely unoccupied) new high-rise condominiums for the rich. On Palm Jumeirah, a fanciful palm tree–shaped island and peninsula off the Gulf coastline, huge homes await buyers (see Figure 6.24A).

Outside the Gulf states, there has been far less planning and financing for urban growth. For example, in 1950, Cairo had about 2.4 million residents, while today it is home to over 17 million people. These millions of new residents live in huge makeshift slums. Meanwhile, much of Cairo's middle class occupies the medieval interiors of the old city, where streets are narrow pedestrian pathways and plumbing and other services are chronically dysfunctional (see Figure 6.24C).

Internal and International Migration The prospect of better education opportunities, jobs, and living conditions pull rural internal migrants into Cairo and other cities outside the Gulf states. However, because stable, well-paying jobs are scarce, many migrants end up working in the informal economy (see pages 131–132 in Chapter 3), sometimes as street vendors, as casual laborers, or as menial service providers. Unfortunately, education also does little to ensure employment; for decades this region has been noted for its large numbers of unemployed and underemployed university graduates.

In the Gulf states, on the other hand, there has been a deficit of trained native young people willing and able to work in white-collar jobs, yet surplus trained workers from neighboring countries have not been welcomed. Instead, immigrants come from all over the world to be temporary guest workers. Some work as laborers on construction sites and as low-wage workers throughout the service economy, but many work in shops and professions. In some countries, such as the UAE, guest workers make up 85 percent of the labor force. Most employers prefer Muslim guest workers, and over the last two decades, several hundred thousand Muslim workers have arrived from Palestinian refugee camps in Lebanon and Syria, as well as from Egypt, Pakistan, and India. Some female domestic and clerical workers come from Muslim countries in Southeast Asia, including the Philippines. Overwhelmingly, these immigrant workers are temporary residents with no job security and no right to become citizens. They remit most of their income to their families at home and often live in stark conditions alongside the opulence of those enriched by oil and gas.

Refugees comprise a major category of migrants, as this region has the largest number of refugees in the world. Usually they are escaping human conflict, but environmental disasters such as earthquakes or long-term drought also displace many people. When Israel was created in 1949, many Palestinians were placed in refugee camps in Lebanon, Syria, Jordan, the West Bank, and the Gaza Strip. Palestinians still constitute the world's oldest and largest refugee population, numbering at least 5 million. Elsewhere, Iran is sheltering more than a million Afghans and Iraqis because of continuing violence and instability in their home countries. Across the region, even more people are refugees within their own countries: 1.7 million Iraqis are internal refugees, and in Sudan, about 2.3 million internal refugees are living in camps. The conflict against the Assad regime in Syria has recently produced several million internal and external refugees.

Refugee camps often become semipermanent communities of stateless people in which whole generations are born, mature, and die. Although residents of these camps can show enormous ingenuity in creating a community and an informal economy, the cost in social disorder is high. Tension and crowding create health problems. Because birth control is generally unavailable, refugee women have an average of 5.78 children each. Disillusionment is widespread. Years of hopelessness, extreme hardship, lack of education or well-paid employment take their toll on youth and adults alike. Because of these factors, it is easy for Islamists and jihadists to find new recruits, which they do by providing basic services to camp residents. Moreover, even though international organizations also provide for refugees, the refugees constitute a huge drain on the resources of their host countries. In Jordan, for example, native Jordanians are a minority in their own country because the more than 3 million Palestinian refugees and their children account for well over half the total population of the country, and their presence has changed life for all Jordanians. They have been joined by over 750,000 Iraqis, and as of 2013, more than 250,000 Syrians have crossed into Jordan.

THINGS TO REMEMBER

› **GEOGRAPHIC INSIGHT 4** • **Urbanization** Two patterns of urbanization have emerged in the region, both tied to the global economy. In the oil-rich countries, the development of spectacular new luxury-oriented urban areas is based on global flows of money, credit, goods, and skilled people. In the oil-poor countries, cities are old and have little capacity to handle the masses of poor, rural migrants attracted by the promise of jobs. In these older, poorer cities, jobs are often scarce and many people live in overcrowded slums with few services. ∎

POPULATION AND GENDER

› **GEOGRAPHIC INSIGHT 5**

Population and Gender: This region has the second-highest population growth rate in the world, after sub-Saharan Africa. Part of the reason for this is that women are generally poorly educated and tend not to work outside the home. Childbearing thus remains crucial to a woman's status, a situation that encourages large families. The role and social standing of women in the region is one of the issues being addressed by current political movements.

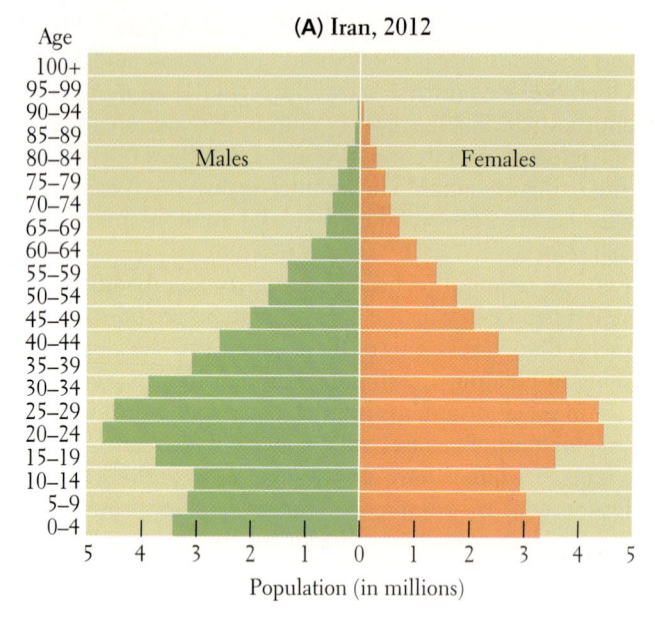

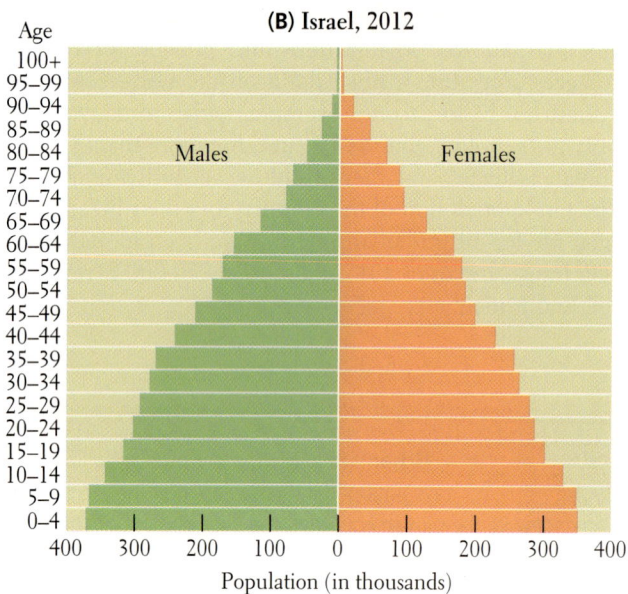

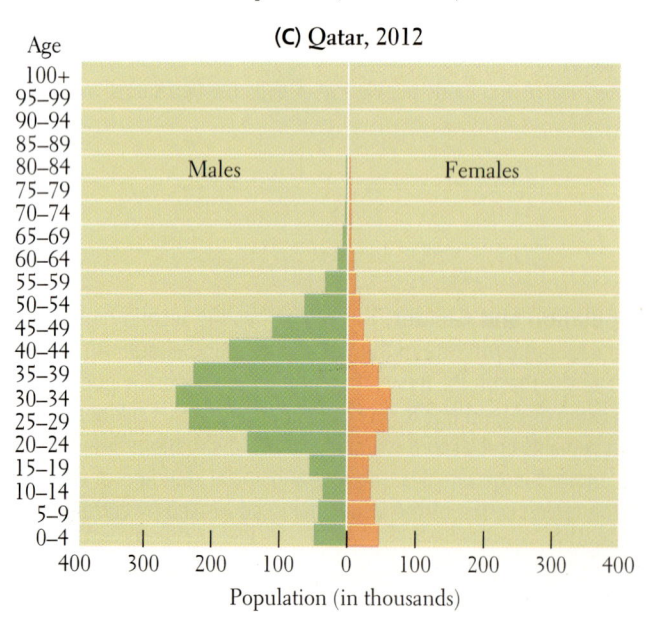

Although fertility rates have dropped significantly since the 1960s, to 3.1 children per adult woman in 2012, they are still higher than the world average of 2.6. Only sub-Saharan Africa, at 5.1 children per woman, is growing faster. At current growth rates, the population of the region will be about 540 million by 2025. This growth will severely strain supplies of fresh water and food, and worsen shortages of housing, jobs, medical care, and education.

As noted in many world regions, population growth rates are higher in societies where women are not accorded basic human rights, are less educated, and work primarily inside the home. In places where women have opportunities to work or study outside the home, they usually choose to have fewer children. Figure 6.29 shows that as of 2011, considerably less than 50 percent of women across the region (except in Israel, Kuwait, and Qatar) worked outside the home at jobs other than farming. Moreover, on average only 74.8 percent of adult females can read—a fact that limits their employability—whereas 88 percent of adult males can read.

For uneducated women who work only at home or in family agricultural plots, children remain the most important source of personal fulfillment, family involvement, and power. This may partially explain why in 2011 a good deal less than half of women in this region were using modern methods of contraception, and only 54 percent were using any method of contraception at all. Both of these numbers are well below the world average of 55 and 62 percent, respectively. Other factors resulting in low use of contraception are male dominance over reproductive decisions and the unavailability or high cost of effective birth control products.

The deeply entrenched cultural preference for sons in this region is both a cause and a result of women's lower social and economic standing. It also contributes to population growth, as families sometimes continue having children until they have a desired number of sons. Moreover, some young females may not survive because of malnutrition and associated illnesses or because female fetuses are sometimes aborted. The result is that males slightly outnumber females in several age cohorts of the population pyramids in Figure 6.25 (see also the discussion in Chapter 1 on page 47). In Qatar and the UAE, gender imbalance is extreme (see Figure 6.25C); in these countries,

FIGURE 6.25 Population pyramids for Iran, Israel, and Qatar. The population pyramid for Iran **(A)** is at a different scale (millions) from those for Israel **(B)** and Qatar **(C)** (thousands). The imbalance of Qatar's pyramid in the 25–54 age groups is caused by the presence of numerous male guest workers. Note, too, that pyramids A and C show missing females in the younger age groups. (This is most easily observed by drawing lines from the ends of the male and female age bars to the scale at the bottom of the pyramid and comparing the numbers.) [Source consulted: "Population Pyramids for Iran," *International Data Base*, U.S. Census Bureau, 2012, at http://www.census.gov/population/international/data/idb/informationGateway.php]

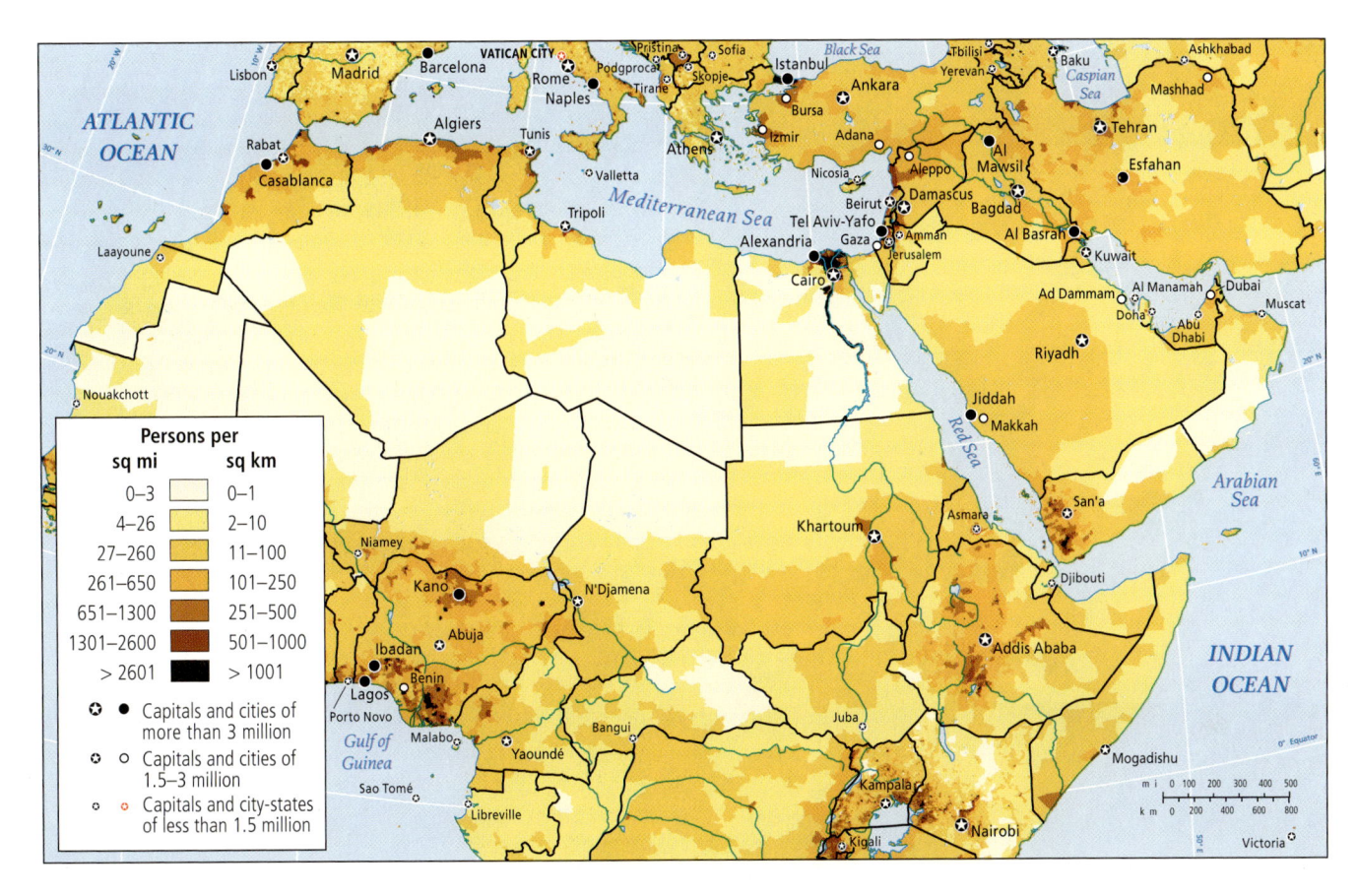

FIGURE 6.26 Population density in North Africa and Southwest Asia.

the unusually large numbers of males over the age of 15 is the result of the presence of numerous male guest workers. Opportunities for women are opening and this change is likely to reduce women's interest in having large families. Slower population growth will mean that fewer resources will have to be invested in housing, schools, and services (see "On the Bright Side" on page 258).

Changing Population Distribution

Although the region as a whole is nearly twice as large as the United States, most of the population is concentrated in the few areas that are useful for agriculture. Vast tracts of desert are virtually uninhabited, while the region's 477 million people are packed into coastal zones, river valleys, and mountainous areas that capture orographic rainfall (compare Figure 6.26 with Figure 6.5). Population densities in these areas can be quite high. For example, some of Egypt's urban neighborhoods have more than 260,000 people per square mile (100,000 people per square kilometer), a density 4 times higher than that of New York City, the densest city in the United States.

SOCIOCULTURAL ISSUES

This section explores the broad social changes occurring in this region with regard to families, gender, and space. Like most institutions in this region, the family remains predominantly patriarchal. Related to this are divisions between the public spaces of society, which are usually occupied by men, and the private spaces of the home, where women spend most of their time. More legal limits are imposed on women here than in any other region, with various countries restricting women's political freedoms, how they dress, and their ability to occupy public spaces. Change on these fronts is happening, as women push for more equitable treatment, but the pace is slow.

Families and Gender

The family is the most important institution in this region. Although the role of the family is changing, a person's family is still such a large component of personal identity that the idea of individuality is almost a foreign concept. Each person is first and foremost part of a family, and the defining parameter of one's role in the family is gender identity (see Figures 6.3 and 6.4, both of which accompany the opening vignette).

The head of the family is nearly always a man; even when a woman is widowed or divorced, she is under the tacit supervision of a male, perhaps her father, her adult son, or another male relative. An educated unmarried woman with a career outside the home is likely to live in the home of her parents or a brother and to defer to them out of respect in decision making. Traditionally, men are considered more stable and

FIGURE 6.27 Domestic spaces. Many of the older cities and buildings in this region reflect the division between public space and private or domestic space.

(A) As seen from above, nearly every house in this old part of Baghdad has a courtyard with a garden in it.

(B) Projecting windows covered with lattice or louvers, known as a *mashrabiya*, in Jeddah, Saudi Arabia. Mashrabiyas allow women to look out on the street below and catch breezes without being seen.

(C) An interior view of a courtyard in Cairo. Courtyards are places where women can do chores and manage children while remaining secluded from the outside world.

capable of making decisions and thus, it is thought that they should be in charge. Women usually agree or at least comply with minimal argument. These **patriarchal** ideas are beginning to change as the result of modernization. During the Arab Spring, a few brave women have challenged convention and taken a political stance (see the opening vignette; see also the discussion below).

Gender Roles and Gendered Spaces

Carefully specified gender roles are common in many cultures, and there is often a spatial component to these roles. In the region of North Africa and Southwest Asia, in both rural and urban settings, the ideal is for men and boys to go forth into *public spaces*—the town square, shops, the market. Women are expected to inhabit primarily *private spaces*. But there are many possible exceptions to those ideals.

To facilitate this ideal of public/private spaces for the sexes, traditional family compounds included a courtyard that was usually a private, female space within the home (Figure 6.27A); the only men who could enter it were relatives. For the urban upper classes, female space was an upstairs set of rooms with latticework or shutters at the windows, which increased the interior ventilation and from which it was possible to look out at street life without being seen (see Figure 6.27B, C). Today, the majority of people in the region live in urban apartments, yet even in these there is a demarcation of public and private space. One or two formally furnished reception areas are reserved for nonfamily visitors, and rooms deeper into the dwelling are for family-only activities. When guests are present, women in the family are usually absent or present only briefly. Customs vary not only from country to country but also from rural to urban settings, by social class, and by personal preference. Today, many women as well as men go out into public spaces, but the conditions under which women enter these spaces remain an issue and a woman who challenges convention does so at her own peril and that of her family's reputation.

The requirement that women stay out of public view (also known as **female seclusion**) is most strictly enforced in the more conservative Muslim countries of the **Gulf states** (Saudi Arabia, Kuwait, Bahrain, Oman, Qatar, and the United Arab Emirates). Women in these countries are generally expected to remain in private spaces except when on important business; then they are to be accompanied by a male relative. In the more secular Islamic countries—Morocco, Tunisia, Libya, Egypt, Turkey, Lebanon, and Iraq—women regularly engage in activities that place them in public spaces. In these countries, a group of women may go out together for social events such

patriarchal relating to a social organization in which the father is supreme in the clan or family

female seclusion the requirement that women stay out of public view

Gulf states Saudi Arabia, Kuwait, Bahrain, Oman, Qatar, and the United Arab Emirates

FIGURE 6.28 Variations on the veil as portable seclusion. There is an almost infinite variety of interpretations of the veil.

(A) An Iraqi woman wears a head scarf.

(B) Schoolgirls in Iran wear a uniform that covers most of their hair and a suit that covers most of their body.

(C) Tunisian women are covered except for their eyes and hands.

> **THINKING GEOGRAPHICALLY** Why do these women wear a veil, no matter what style of veil they choose?

as a high-spirited evening dinner at a restaurant. Some women wear conservative religious clothing; others dress in Western styles. Increasingly, female doctors, lawyers, teachers, and businesspeople are found in even the most conservative societies. The map in Figure 6.4, accompanying the opening vignette, compares the various levels of restrictions on women across the region.

Affluent urban women may observe seclusion either rarely (especially if they are highly educated), or, if they are relatively affluent but less educated, even more strictly than do rural women. Although rural women are often more traditional in their outlook, they have many tasks that they must perform outside the home: agricultural work, carrying water, gathering firewood, and buying or selling food in markets. Meanwhile, upper-class women who can afford servants to perform daily tasks in public spaces can more easily stay secluded and afford the amenities (TV, DVDs, cell phones) that relieve the boredom and isolation of seclusion. It should also be noted that seclusion customs are in flux, being relaxed in some countries, such as Morocco, Turkey, and Tunisia, but becoming more strictly enforced in other countries, such as post–Arab Spring Egypt and parts of North Africa where **jihadists** (especially militant Islamists) are newly active.

Many women in this region use clothing as a way to create private space (Figure 6.28). This is done with the many varieties of the **veil**, which may be a garment that totally covers the person's body and face or just a scarf that covers the person's hair. In some cultures, even prepubescent girls wear the veil; in others, they go unveiled until their transition into adulthood is observed. The veil allows a devout Muslim woman to preserve a measure of seclusion when she enters a public space, thus increasing the territory she may occupy with her honor preserved. A modern young woman may choose to wear a headscarf with jeans and a t-shirt in order to signal to the public that she is both an up-to-date woman and an observant Muslim.

There is considerable debate about the origin and validity of female seclusion and whether veiling and seclusion are specifically Muslim customs. Scholars say that these ideas, as well as the custom of having more than one wife, actually predate Islam by thousands of years and do not derive from the teachings of the Prophet Muhammad. In fact, Muhammad may have been reacting against such customs when he advocated equal treatment of males and females. Muhammad's first wife, Khadija, did not practice seclusion, and worked as an independent businesswoman whose counsel Muhammad often sought.

jihadists especially militant Islamists

veil the custom of covering the body with a loose dress and/or of covering the head—and in some places the face—with a scarf

146. EDUCATION, ECONOMIC EMPOWERMENT ARE KEYS TO A BETTER LIFE FOR MUSLIM WOMEN

The Rights of Women

This region has notably more restrictive customary and legal limits on women than any other. In the Gulf states, women cannot travel independently or even drive a car or shop without male supervision. Yet even here changes have recently been made. In the decade of the 2000s, some women in the Gulf states became more active in public life, education, and business. In Qatar, Sheikha Moza, the second of the three wives of the emir (Muslim ruler) of Qatar, Sheikh Hamad bin Khalifa Al-Thani, has implemented remarkable changes for women. They can now drive, attend university, be elected to political office, and work side by side with men; the sheikha has founded a battered women's shelter and serves as a UNESCO envoy. In the UAE, a few female activists, along with other female lawyers and journalists, have begun to question persistent patriarchal attitudes. Saudi Arabia remains the most restrictive country, but even there it is now possible for a woman to register a business without first proving that she has hired a male manager. A few young Saudi women even staged mini-demonstrations by posting on YouTube videos of themselves driving. Although they were eventually punished by the state for this behavior, they inspired further challenges to convention.

The Arab Spring in North Africa demonstrated just how elusive political equality for women can be. In Tunisia, Libya, and Egypt, some women were active as public protesters against undemocratic regimes, but military and civilian supporters of the various regimes singled them out for particularly harsh retribution. Among their fellow male demonstrators, it was usually the younger men who supported the women's rights and physically defended them when supporters of the various regimes attacked.

A source of contention within this region and abroad is the practice of polygyny (see Chapter 7, page 310): when a man takes more than one wife at a time. Although the Qur'an allows a man up to four wives, it generally does not encourage this and imposes financial limits on the practice by requiring that each wife be given separate and equal living quarters and support. While legal in most of the region (not Tunisia), polygyny is relatively rare, with less than 4 percent of males in North Africa practicing it. In Southwest Asia, polygyny—not covered in any reliable statistical surveys, but only in secondary reports—may be somewhat more common. About 5 percent in Jordan and 8 to 12 percent of marriages in Kuwait are polygamous. The social justifications given for polygyny further illustrate traditional ideas about women's rights in Islam. In common legal practice in many countries, an unmarried woman under 40 must be regarded as a minor requiring protection. Therefore, if a man takes her as a second or third wife, she gains support and safety. Similarly, a widow is saved from disgrace and poverty if her dead husband's brother takes her as an additional wife.

Female genital mutilation (FGM, defined and discussed in Chapter 7 on page 310) is not widely practiced in this region except in Egypt, where although it is illegal, as of 2013 according to the World Health Organization, more than 90 percent of women have undergone the practice, mostly before age 15.

Political and social equality with men can only improve with more economic independence for women (Figure 6.29). With the sole exception of Israel, women in this region generally do not work outside the home; when they do, they are paid, on average, only about 40 percent of what men earn for comparable work. Only India, Pakistan, and other South Asian countries have a similarly large wage gap based on gender.

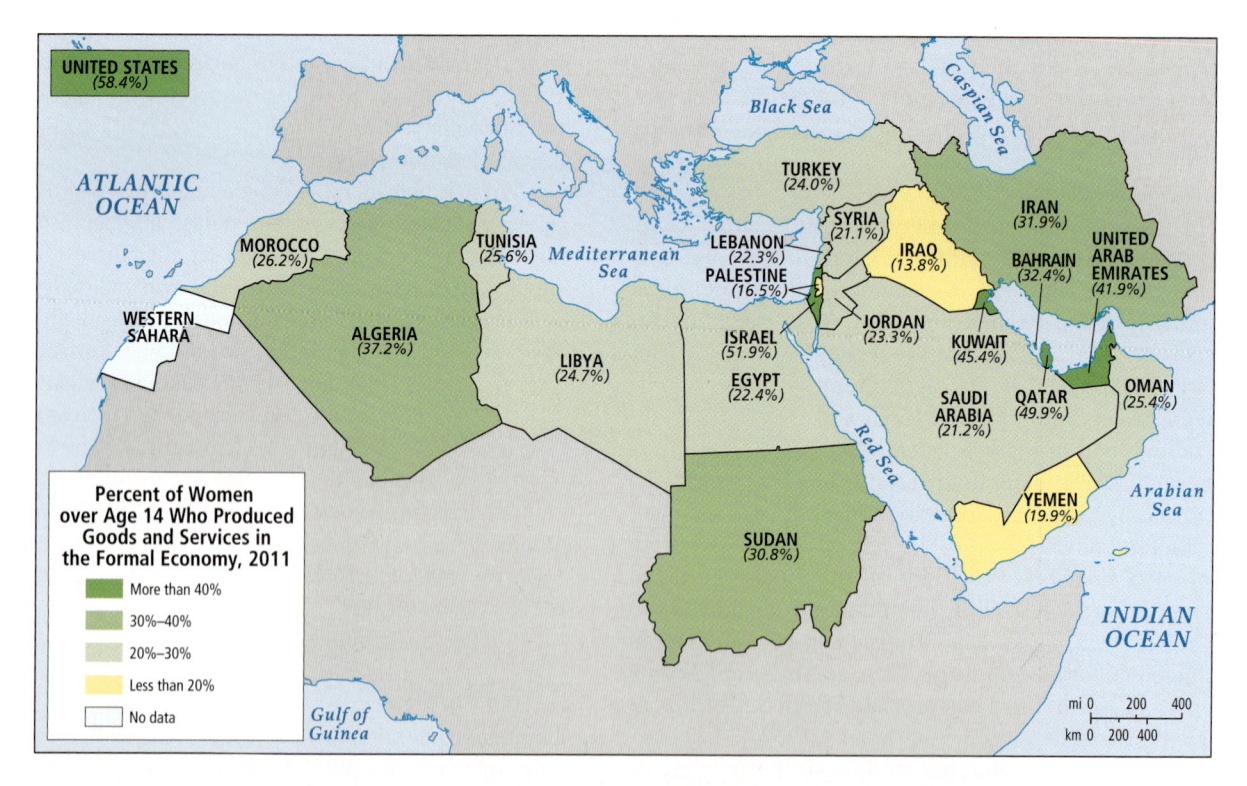

FIGURE 6.29 Percentage of the region's women who are in the labor force (2011). Between 2005 and 2011, the percentage of women who were earning wages in the labor force increased in 10 of the 22 countries and territories in the region, with the largest increase, more than 13%, in Qatar. The largest decreases (more than 10%) of women working were in Lebanon, Syria, and Yemen and may be a result of major unrest in those countries. [Source consulted: *Human Development Report 2011*, Sustainability and Equity: A Better Future for All, United Nations Development Programme, at http://hdr.undp.org/en/reports/global/hdr2011/]

The Lives of Children

Three observations can be made about the lives of children in the Islamic cultures of North Africa and Southwest Asia. First, in most families children contribute to the welfare of the family starting at a very young age. In cities, they run errands, clean the family compound, and care for younger siblings. In rural areas, they do all these chores and also tend grazing animals, fetch water, and tend gardens. Second, their daily lives take place overwhelmingly within the family circle. Both girls and boys spend their time within the family compound or in urban areas in adjacent family apartments. Their companions are adult female relatives and siblings and cousins of both sexes. Even teenage boys in most parts of the region identify more with family than with age peers.

In rural areas, prepubescent girls can move around in public space as they go about their chores in the village. After puberty they may be restricted to the family compound and required to wear the veil. The U.S. geographer Cindi Katz found that until puberty, rural Sudanese Muslim girls have considerably more spatial freedom than do girls of similar ages in the United States, who are rarely allowed to range alone through their own neighborhoods.

The third observation is that school, television, and the Internet increasingly influence the lives of children and introduce them to a wider world. Most children go to school; many boys go for a decade or more, and increasingly girls go for more than a few years; and in some countries (Algeria, Israel, Jordan, Lebanon, Libya, the occupied Palestinian Territories, Oman, Qatar, Tunisia, and the UAE), a larger percentage of girls than boys go to school. Like educated women everywhere, these girls will make life choices different from those of their mothers.

Even in rural areas, it is fairly common for the poorest families to have access to a television, which is often on all day, in part because it provides a window on the world for secluded women. Television can serve either to reinforce traditional cultural values or as a vehicle for secular perspectives, depending on which channels are watched.

THINGS TO REMEMBER

> **GEOGRAPHIC INSIGHT 5**
> • **Population and Gender** This region has the second-highest population growth rate in the world, after that of sub-Saharan Africa. Part of the reason for this is that women are generally poorly educated and tend not to work outside the home. Childbearing thus remains crucial to a woman's status, a situation that encourages large families. The role and social standing of women in the region is one of the issues being addressed by current political movements.

- The family is the most important societal institution in the region.

- Most Muslim families in this region are patriarchal in structure. The role and status of women, the spaces they occupy, and the clothes they wear are often carefully defined, circumscribed, and in some cases rigidly imposed.

GEOGRAPHIC INSIGHTS

North Africa and Southwest Asia: Review and Self-Test

1. Environment: The population of this predominantly dry region is growing faster than is access to cultivable land and water for agriculture and other human uses. Many countries here are dependent on imported food, and because global food prices are unstable, the region often has food insecurity. Climate change could reduce food output both locally and globally and make water, already a scarce commodity, a source of conflict. New technologies offer solutions to water scarcity but are expensive and unsustainable.

- How will people get enough water to grow food in the future, especially if climate change makes this dry region even drier?

- What factors could make it difficult to obtain enough water for agriculture? What technologies could provide potential solutions? How crucial is imported food for this region?

2. Globalization and Development: The vast fossil fuel resources of a few countries in this region have transformed economic development and driven globalization. In these countries, economies have become powerfully linked to global flows of money, resources, and people. Politics have also become globalized, with Europe and the United States strongly influencing many governments.

- How have the huge fossil fuel reserves of some countries transformed economic development and driven globalization in this region?

- What flows of money, resources, and people characterize globalization in the Gulf states? How important are outside political influences?

- How might changes in job opportunities for women in this region relieve the dependency on imported labor? Where is this likely to be most important?

3. Power and Politics: Despite the presence of elected bodies of government, authoritarian power structures prevail throughout North Africa and Southwest Asia. Beginning with the Arab Spring of 2010, waves of protest swept the region, resulting in the overthrow of several authoritarian governments. Within the region, official responses to the Arab Spring have included both repression and reform.

- What would you say are the chief signs that this region may be democratizing?

- How have recent events changed the political landscape of this region?

- What has repression of the political opposition often resulted in? What has been the effect of integrating political opposition movements into the democratic process?

4. Urbanization: Two patterns of urbanization have emerged in the region, both tied to the global economy. In the oil-rich countries, the development of spectacular new luxury-oriented urban areas is based on global flows of money, credit, goods, and skilled people. In the oil-poor countries, cities are old and have little capacity to handle the masses of poor, rural migrants attracted by the promise of jobs. In these older, poorer cities, jobs are often scarce and many people live in overcrowded slums with few services.

- How has globalization shaped different patterns of urbanization throughout the region?

- Why is urbanization currently so different in the Gulf states than in the rest of the region? How have those economic reforms that are aimed at improving global competitiveness, especially in export-oriented agriculture, influenced urbanization outside the Gulf states?

5. Population and Gender: This region has the second-highest population growth rate in the world, after that of sub-Saharan Africa. Part of the reason for this is that women are generally poorly educated and tend not to work outside the home. Childbearing thus remains crucial to a woman's status, a situation that encourages large families. The role and social standing of women in the region is one of the issues being addressed by current political movements.

- How does the low status of women contribute to this region's high population growth?

- Relative to other regions, how high is this region's population growth rate? How do women's education levels influence their decisions about how many children to have?

CRITICAL THINKING QUESTIONS

1. Which social forces in North Africa and Southwest Asia modify the power of religion?

2. To what extent is the present-day map of North Africa and Southwest Asia related to the dismantling of the Ottoman Empire after World War I?

3. Considering the various factors that encourage relatively high fertility in this region, design themes for a public education program that would effectively encourage lower birth rates. Which population groups would you target? How would you incorporate cultural sensitivity into your project?

4. Consider the new forces that are affecting urban landscapes: immigration and globalization. Identify some of the expected effects on ordinary people of these abrupt changes to traditional living spaces.

5. Compare and contrast the public debate over the proper role of religion in public life in your country and in one country in this region (for example, Turkey, Morocco, Egypt, or Saudi Arabia). Contrast the roles of religious fundamentalists in the debates in your country and in the country chosen.

6. Gender is a complex subject in this region. Choose a rural location and an urban location and make a list of the forces in each that would affect the future of a 20-year-old woman. Describe those hypothetical futures objectively; that is, without using any judgmental terminology.

7. Why is it important to know that in some countries of this region agriculture may produce only a small amount of the GDP yet employ 40 percent or more of the people? What are some of the things such a relationship would indicate about the state of development in that country? What public policies would be appropriate in these circumstances—for example, should agriculture be deemphasized? How might this deemphasis affect food security?

8. Describe the circumstances that led to support in Europe and the United States for the formation of the state of Israel. Why did the West overlook the Palestinian people in this political undertaking?

9. Discuss the possibilities that scarcity of water is or will become a cause of violence in the region. What is the evidence against this happening?

> THINKING GEOGRAPHICALLY

Now that you have read about North Africa and Southwest Asia, you should be able to answer the following questions about photos in this chapter. To answer these questions online, go to Geography LaunchPad.

Human Impacts on the Biosphere (page 238)

B What about a dam creates soil moisture problems for downstream users of river water once the dam is operational?

C How might a dam like the Aswan on the upper part of the Nile River in Egypt contribute to the salinity and lowered fertility in lowland regions that were once naturally flooded but are now irrigated by pumped water?

Vulnerability to Climate Change (page 241)

A What about Alexandria's location makes it particularly vulnerable to sea level rise?

B What are some of the factors in the region that complicate emergency responses to climate change, thus limiting resilience?

C How does Yemen's situation as a haven for refugees from nearby countries make it more vulnerable to climate change?

Visual History (page 248)

A What allowed for urban settlements like Harran to develop?

B What ceremonial purpose is filled by the Masjid Al-Haram and the Kaaba in Makkah?

C Which Turkish city had lavish buildings and parks that outshined anything in Europe until the nineteenth century?

D When did France first assert control over territory in what is now Algeria?

E The state of Israel was formally created in the aftermath of what event in Europe?

F Why did European and U.S. companies become involved in deciding who ruled Iran and Saudi Arabia?

Power and Politics (page 256)

A What circumstances made many suspicious of the U.S. claim that the 2003 war with Iraq was undertaken to bring democracy to Iraq?

B Why is the 1973 war between Israel and Egypt cited as one of the causes of political repression within countries of the region?

C What were the issues that motivated the Syrians to take to the streets in prolonged demonstrations in 2011?

D What has changed for Kurds in Iraq that has lessened their use of violent protests?

E What accounts for the higher death toll in Libya's Arab Spring episode than that in Egypt's?

Urbanization (page 264)

A What circumstances dealt a blow to high-tech and tourism development in the Gulf states, such as Dubai?

B By 2013, what percent of the population in this region lived in urban areas?

C What has driven the migration of rural people to the region's old established cities, such as Cairo?

CHAPTER KEY TERMS

cartel 249
Christianity 245
desertification 239
diaspora 245
economic diversification 252
female seclusion 268
Fertile Crescent 243
fossil fuel 233
Gulf states 268
hajj 245
intifada 260
Islam 233

Islamism 233
jihadists 269
Judaism 245
monotheism 245
Muslims 245
occupied Palestinian Territories (oPT) 233
OPEC (Organization of the Petroleum Exporting Countries) 249
Ottoman Empire 247
patriarchal 268
Qur'an (or Koran) 237
Salafism 255

salinization 239
seawater desalination 240
secular states 255
shari'a 246
Shi'ite (or Shi'a) 246
Sunni 246
theocratic states 255
veil 269
West Bank barrier 262
Zionists 260

Map of Africa (Western, Central, and Southern)

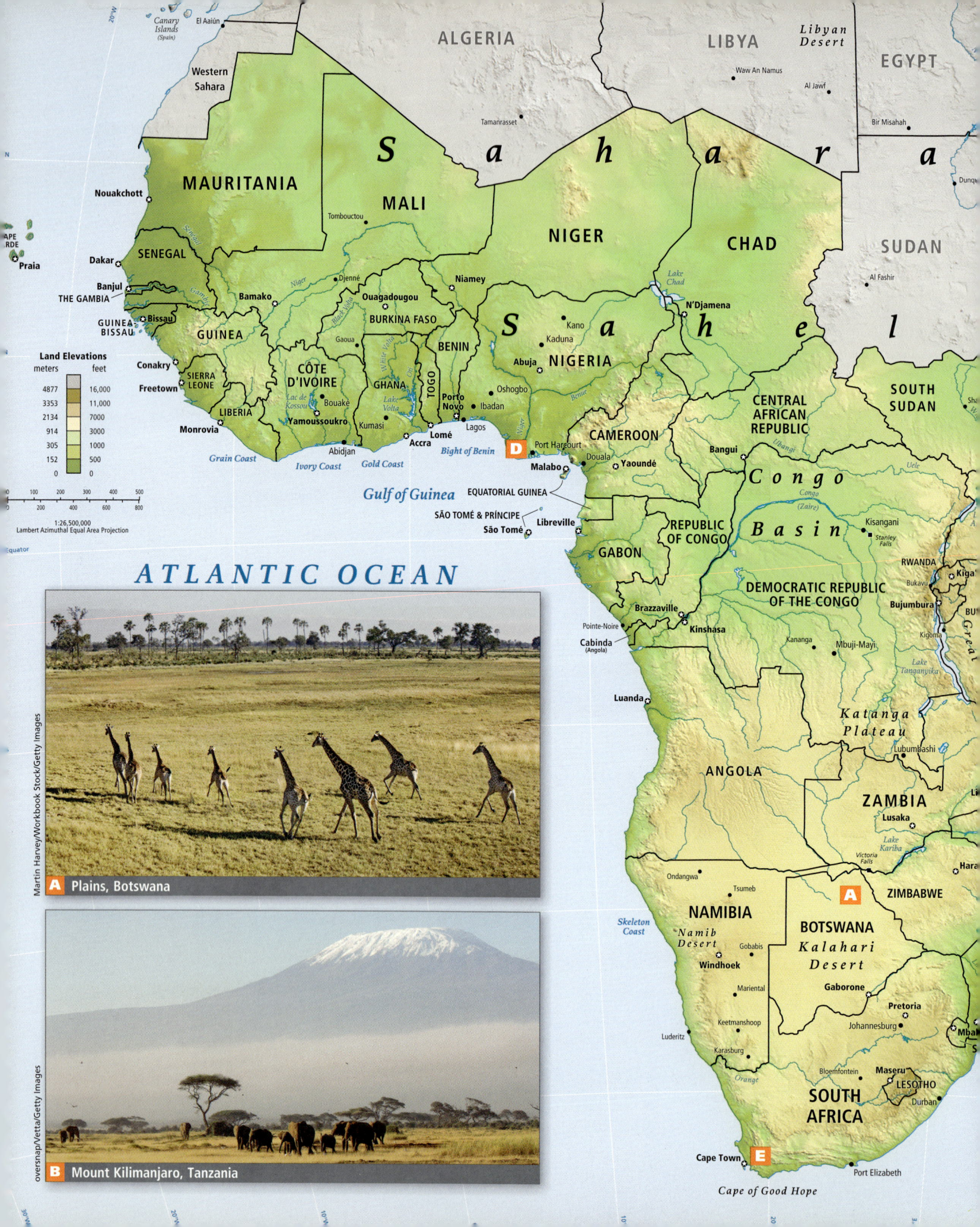

Countries and Regions Labeled:

- Western Sahara
- Mauritania
- Mali
- Niger
- Chad
- Algeria
- Libya
- Egypt
- Sudan
- South Sudan
- Senegal
- The Gambia
- Guinea-Bissau
- Guinea
- Sierra Leone
- Liberia
- Côte d'Ivoire
- Burkina Faso
- Ghana
- Togo
- Benin
- Nigeria
- Cameroon
- Central African Republic
- Equatorial Guinea
- São Tomé & Príncipe
- Gabon
- Republic of Congo
- Democratic Republic of the Congo
- Rwanda
- Angola
- Zambia
- Zimbabwe
- Namibia
- Botswana
- South Africa
- Lesotho

Physical Features:
- Sahara
- Sahel
- Libyan Desert
- Canary Islands (Spain)
- Cape Verde
- Grain Coast, Ivory Coast, Gold Coast
- Bight of Benin
- Gulf of Guinea
- Congo Basin
- Katanga Plateau
- Lake Tanganyika
- Lake Chad
- Lake Kariba
- Victoria Falls
- Skeleton Coast
- Namib Desert
- Kalahari Desert
- Cape of Good Hope
- Atlantic Ocean

Cities (selected):
El Aaiún, Nouakchott, Dakar, Banjul, Bissau, Conakry, Freetown, Monrovia, Yamoussoukro, Abidjan, Bamako, Tombouctou, Ouagadougou, Gaoua, Bouaké, Kumasi, Accra, Lomé, Porto Novo, Lagos, Ibadan, Oshogbo, Abuja, Kaduna, Kano, Niamey, Djenné, N'Djamena, Al Fashir, Port Harcourt, Douala, Malabo, Yaoundé, Bangui, São Tomé, Libreville, Brazzaville, Kinshasa, Pointe-Noire, Cabinda (Angola), Luanda, Kananga, Mbuji-Mayi, Kisangani, Bukavu, Bujumbura, Kigoma, Lubumbashi, Lusaka, Harare, Ondangwa, Tsumeb, Windhoek, Gobabis, Mariental, Keetmanshoop, Lüderitz, Karasburg, Gaborone, Pretoria, Johannesburg, Maseru, Bloemfontein, Durban, Cape Town, Port Elizabeth, Tamanrasset, Waw An Namus, Al Jawf, Bir Misahah

Land Elevations

meters	feet
4877	16,000
3353	11,000
2134	7,000
914	3,000
305	1,000
152	500
0	0

Scale 1:26,500,000
Lambert Azimuthal Equal Area Projection

A Plains, Botswana
Martin Harvey/Workbook Stock/Getty Images

B Mount Kilimanjaro, Tanzania
oversnap/Vetta/Getty Images

chapter 7
SUB-SAHARAN AFRICA

C Great Rift Valley, Kenya

D Niger River Delta, Nigeria

E Escarpment at Cape Town, South Africa

FIGURE 7.1 Regional map of sub-Saharan Africa.

275

GEOGRAPHIC INSIGHTS: SUB-SAHARAN AFRICA

After you read this chapter, you will be able to discuss the following geographic insights as they relate to the five thematic concepts:

1. **Environment:** Sub-Saharan Africa's poverty and political instability leave its people less able to adapt to climate change than the inhabitants of other world regions, though many Africans are developing strategies to cope with increasing variability in rainfall and temperature. As a whole, this region has contributed little to the buildup of greenhouse gases in the atmosphere, but deforestation by rural Africans and by multinational logging companies is intensifying global climate change.

2. **Globalization and Development:** Most sub-Saharan African economies are dependent on the export of raw materials. This pattern, a legacy of the era of European colonialism, results in economic instability because prices for raw materials can vary widely from year to year. While a few countries are diversifying and industrializing, most still sell raw materials and all rely on expensive imports of food and manufactured goods.

3. **Power and Politics:** Governments in sub-Saharan Africa tend to be authoritarian, though there is a general shift in the region toward the expansion of political freedoms. Free and fair elections have brought about dramatic changes in some countries, while elections tainted by widespread suspicions of fraud have often been followed by surges of violence in others.

4. **Urbanization:** Sub-Saharan Africa is undergoing a massive wave of rural-to-urban migration; it has the fastest rate of urbanization in the world. Much of this growth is unplanned and 70 percent of the urban population now lives in impoverished slums characterized by inadequate sanitation; poor access to clean water or food; and crowded, fire-prone housing.

5. **Population and Gender:** Of all the world regions, populations are growing fastest in sub-Saharan Africa. However, growth is slowing as the demographic transition takes hold, especially in a few of the more prosperous countries where better health care and lower levels of child mortality, along with more economic and educational opportunities, are encouraging smaller families. In particular, women are better able to pursue careers and to choose to have fewer children.

The Sub-Saharan Region

Sub-Saharan Africa (Figure 7.1) contains 48 countries and occupies a space bigger than North America and Europe combined. However, it is not particularly densely populated. This region is the ancient home of modern humans; it has megafauna that most of us will see only in zoos and has a broad variety of landscapes. But it is not a region that is well known to most and is often erroneously thought of as desperately poverty-stricken and barely capable of helping itself.

The five thematic concepts in this book are explored as they arise in the discussion of regional issues; the interactions between two or more themes are often featured. Vignettes, like the one that follows about Juliana Rotich, illustrate one or more of the themes as they are experienced in individual lives.

Although sub-Saharan Africa (see Figure 7.1) is often portrayed as a place mired in seemingly unsolvable problems related to poverty, disease, corruption, and conflict, it actually has many success stories. Highly competent women and men, such as Juliana Rotich, use a powerful language of opportunity, optimism, and innovation to characterize the Africa they know today. In 2012, six of the ten fastest-growing economies in the world were in sub-Saharan Africa. World Bank studies (2010) indicate that a number of countries, such as Congo (Brazzaville), with some of the world's richest deposits of oil, gold, platinum, copper, and other strategic minerals, are taking control of mineral resources and using the profits to reduce poverty.

During the era of European colonialism (the 1850s to the 1950s), this region's massive wealth of human talent and natural resources flowed out of Africa to Europe and the Americas. Even after African countries became politically independent in the 1950s, 1960s, and 1970s, wealth continued to flow out of Africa. Now, wealth is being increasingly created in Africa for Africans. While much of region is impoverished and parts are in armed conflict, there are many hopeful signs that discord is diminishing and well-being increasing.

What Makes Sub-Saharan Africa a Region?

Sub-Saharan Africa is a region distinct from North Africa for physical, cultural, and historical reasons. The Sahara Desert and the **Sahel**—that grassy transition zone between the desert and wetter climes to the south—present major physical obstacles to human habitation (though trading caravans have traversed the Sahara and Sahel for thousands of years). Because of this, sub-Saharan Africa developed largely separate from North Africa. For millennia, sub-Saharan Africa was known to the outside world only by the accounts of a few travelers, such as those of the Arab explorer Ibn Battuta,

> **Sahel** a band of arid grassland, where steppe and savanna grasses grow, that runs east-west along the southern edge of the Sahara

GLOBAL PATTERNS, LOCAL LIVES

Few people in the East African country of Kenya have computers, but nine of ten Kenyans have mobile phones. They use these simple and inexpensive devices (not smart phones) with texting (SMS) capabilities and perhaps a camera for a multitude of tasks that are done on computers in Europe or North America. One of the most consequential tasks is mobile banking—sending and receiving money via mobile phone. More than 20 percent of the Kenyan gross domestic product (GDP) flows through the mobile banking system. This means that owners of even the smallest businesses can order materials and pay bills and employees without having to pay for expensive equipment and accounting systems or making an expensive time-consuming trip to a bank in a distant city. Small-scale farmers with crops to sell can use SMS to find out where the best market is at the moment and can maximize their profits by bypassing middlemen who charge high fees.

Juliana Rotich (Figure 7.2) grew up in a small town in Kenya, just as Africa was first experiencing the mobile phone and Internet revolutions in communication. She was admitted to a U.S. university, where she became an IT major. She then collaborated with friends in the online community to found and run Ushahidi, a web-based, open source reporting system with interactive mapping capabilities that is now widely used throughout the world. Ushahidi, which means "testimony" in Swahili, was originally designed for quickly sharing information about the violence that broke out in Kenya during the 2008 elections. A user can connect to Ushahidi through a phone or a Web site to instantly learn specific details about a situation. Information gathered from hundreds or even thousands of users (in a process called *crowdsourcing*) can be verified, mapped, and disseminated in real time.

Many people benefit when the barriers to the use of technology are lowered and suddenly virtually everyone has new and useful tools—in this case, a particular software tool and

FIGURE 7.2 Juliana Rotich. Juliana Rotich (standing) is a young Kenyan telecommunications specialist with a computer science degree from the University of Missouri. She founded and directs the company that created *Ushahidi*, a crowd sourcing platform that enables mobile phone users to map crisis information in order to be able to quickly respond to the crises. Ushahidi is now used around the world. Rotich, who also writes a respected blog, *Afromusing*, is part of a growing number of educated Africans who focus on applying technology to African development needs.

mobile phones—in their hands. The businesspeople and farmers of Kenya have clearly been beneficiaries; but rescuers have also been able to use the Ushahidi software—with its mapping capabilities—to locate people buried by earthquakes in Haiti, Chile, and Japan. Ushahidi enabled crucial communication between dissidents during and after the Arab Spring revolutions in Libya and Egypt and during the Syrian civil war (Figure 7.3). Within sub-Saharan Africa, Ushahidi is now being used to aid in disease control, health care, political reform, wildlife inventorying, and the control of corruption. *[Source: "99 Faces." For detailed source information, see Text Sources and Credits.]*

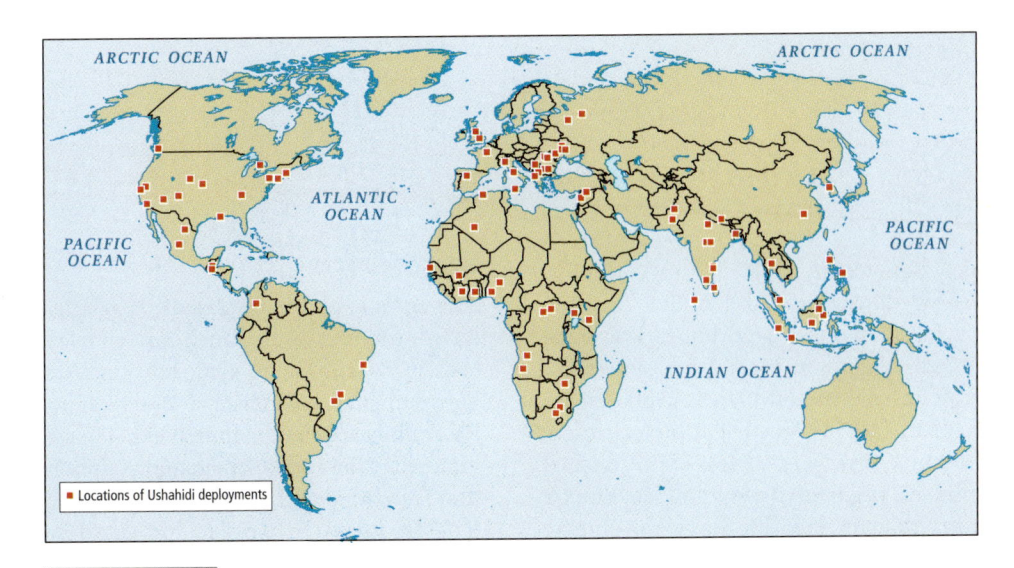

FIGURE 7.3 Ushahidi platform users worldwide. The map shows a selection of the 132 countries in which Ushahidi has been deployed.

who traveled widely throughout the region in the 1300s. In the mid-1400s, however, the Portuguese sent their exploratory fleets down the west coast of Africa, beginning a 600-year-long period of European colonialism that transformed sub-Saharan Africa.

The Europeans noted what for them were exotic qualities: dark-skinned people with unique and varied ways of life and cultures, who possessed valuable trade items, such as precious minerals, exotic plants and animals, and fine textiles. For centuries, however, Europeans discounted the sophistication and complexity of sub-Saharan Africa's cultural and historical heritage of horticulture, weaving, mining, and metalwork. Everything south of the Sahara was lumped into "Black Africa," and for some this term is still used to distinguish the region from Africa as a whole.

Terms in This Chapter

In this chapter, we refer only occasionally to the whole continent, and then we refer to it simply as Africa. Sub-Saharan Africa is defined by the countries shown in Figure 7.4. For cultural and historical reasons, the new country of South Sudan is included in this region. For similar reasons, the country of Sudan is not included because its location on the Nile River and the strong Arab influence in the capital of Khartoum have brought Sudan into closer association with North Africa and Southwest Asia. Sudan is instead discussed in Chapter 6.

The names of African countries can often be confusing. For example, there are two neighboring countries called Congo—the Democratic Republic of the Congo and the Republic of Congo. Because these designations are both lengthy and easily confused, we abbreviate them in this text. The Democratic Republic of the Congo (formerly Zaire) carries the name of its capital in parentheses: Congo (Kinshasa). The Republic of Congo is called Congo (Brazzaville). Check the regional map in Figure 7.1 to see the locations of these countries and capitals.

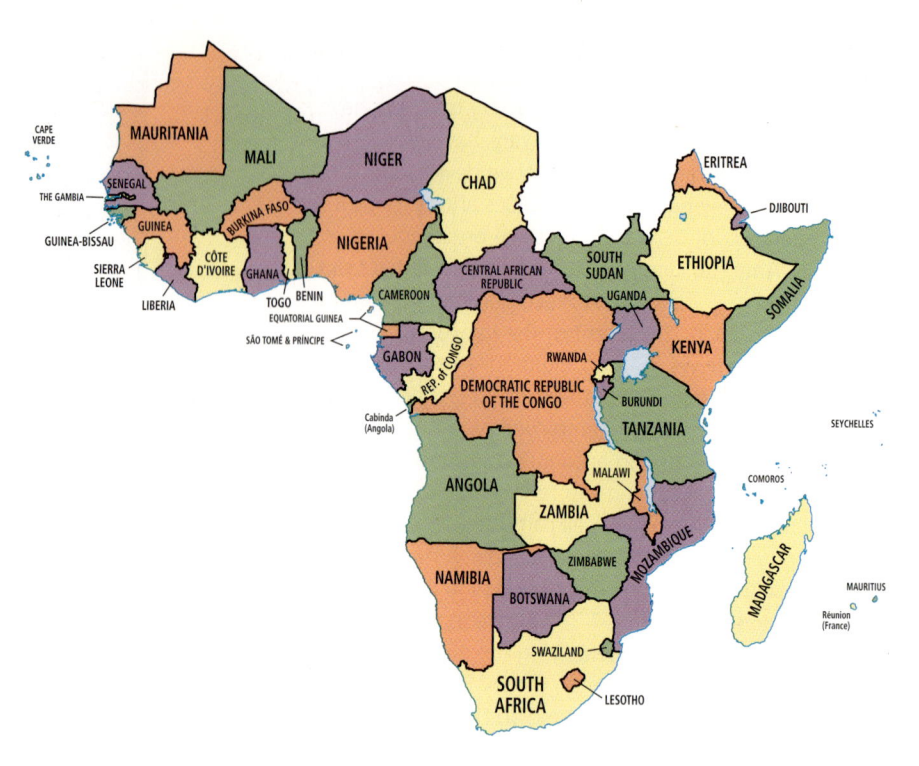

FIGURE 7.4 Political map of sub-Saharan Africa.

THINGS TO REMEMBER

- The many stereotypes of sub-Saharan Africa—that it is a region of conflict, disease, corruption, and poverty—are being countered by young, educated Africans who are constructing new public agencies and private businesses, often grounded in modern technology, in their home countries.

- Some sub-Saharan countries now have economies that are among the fastest-growing in the world.

PHYSICAL GEOGRAPHY AND ENVIRONMENTAL ISSUES

PHYSICAL PATTERNS

The African continent is big—the second largest after Asia and about three times the size of the United States. But Africa's great size is not matched by its surface complexity. Africa has no major mountain ranges, but it does have several high volcanic peaks, including Mount Kilimanjaro (19,324 feet [5890 meters] high; see Figure 7.1B) and Mount Kenya (17,057 feet [5199 meters] high). At their peaks, both have permanent snow and ice, though these features have shrunk dramatically due to global climate change and associated local environmental changes. More than one-fourth of the African continent is covered by the Sahara Desert, which reaches to the edge of the Mediterranean Sea.

Landforms

The surface of the continent of Africa can be envisioned as a raised platform, or plateau, bordered by fairly narrow and uniform coastal lowlands. The platform slopes downward to the north; it has an upland region with several high peaks in the southeast, and lower uplands in the northwest. The steep escarpments (long, high cliffs) between plateau and coast have obstructed transportation and hindered connections to the outside world. Africa's lengthy, uniform coastlines provide few natural harbors; Cape Town in South Africa being an exception (see Figure 7.1E).

Geologists usually place Africa at the center of the ancient supercontinent Pangaea (see Figure 1.8 on page 16). As landmasses broke off from Africa and moved away—North America to the northwest, South America to the west, and India to the northeast—Africa readjusted its position only slightly. Because its shift was so small, it did not pile up long, linear mountain ranges, as did the other continents when their plates collided with one another (see page 13).

Africa continues to break apart along its eastern flank. There, the Arabian Plate has already split away and drifted to the northeast, leaving the Red Sea, which separates Africa and Asia. Another split, known as the Great Rift Valley, extends south from the Red Sea more than 2000 miles (3200 kilometers) (see Figure 7.1C). In the future, Africa is expected to split again along these rifts.

Climate and Vegetation

Most of sub-Saharan Africa has a tropical climate (see the Figure 7.5 map). Average temperatures generally stay above 64°F (18°C) year-round everywhere except at the more temperate southern tip of the continent and in the cooler upland zones (hills, mountains, high plateaus). Seasonal climates in Africa differ more by the amount of rainfall than by temperature.

Most rainfall comes to Africa by way of the **intertropical convergence zone (ITCZ)**, a band of atmospheric currents that circle the globe roughly around the equator (see the inset map in Figure 7.5). At the ITCZ, warm winds converge from both the north and the south and push against each other. This causes the air to rise, cool, and release moisture in the form of rain. The rainfall produced by the ITCZ is most abundant in Africa near the equator, where dense tropical rainforests flourish in places such as the Congo Basin (see Figure 7.5A, D). The ITCZ shifts north and south seasonally, generally following the area of Earth's surface that has the highest average temperature at any given time. Thus, during the height of summer in the Southern Hemisphere in January, the ITCZ might bring rain far enough south to water the dry grasslands, or steppes, of Botswana. During the height of summer in the Northern Hemisphere in August, the ITCZ brings rain as far north as the southern fringes of the Sahara, to the area called the Sahel, a band of arid grassland 200 to 400 miles (320 to 640 kilometers) wide that runs east-west along the southern edge of the Sahara (see the Figure 7.5 map). At roughly 30° N latitude (the Sahara) and 30° S latitude (the Namib and Kalahari deserts), the air, which has dried out during its passage, descends, forming a subtropical high-pressure zone that shuts out lighter, warmer, moister air. As a result of this system (which is in no way precise), deserts tend to be found in Africa (and on other continents) in these zones about 30 degrees north and south of the equator.

The tropical wet climates that support equatorial rain forests are bordered on the north, east, and south by seasonally wet/dry subtropical woodlands (see Figure 7.5B). These give way to moist tropical savannas or steppes, where tall grasses and trees intermingle in a semiarid environment. These tropical wet, wet/dry, and steppe climates have provided suitable land for different types of agriculture for thousands of years, but the amount of moisture available in each of these climates can vary greatly, in cycles that are decades long. Farther to the north and south lie the true desert zones of the Sahara, the Namib, and the Kalahari (see Figure 7.5C). This banded pattern of African ecosystems is modified in many areas by elevation and wind patterns.

Without mountain ranges to block them, wind patterns can have a strong effect on climate in Africa. Winds blowing north along the east coast keep ITCZ-related rainfall away from the **Horn of Africa**, the triangular peninsula that juts out from northeastern Africa below the Red Sea. As a consequence, the Horn of Africa is one of the driest parts of the continent. Along the west coast of the Namib Desert, moist air from the Atlantic is blocked from moving over the desert by cold air above the northward-flowing water of the Benguela Current. Like the cool Peru Current off South America, the Benguela is an oceanic current that is chilled by its passage past Antarctica. Rich in nutrients, it supports a major fishery along the west coast of Africa (see the Figure 7.5 map).

ENVIRONMENTAL ISSUES

> ### GEOGRAPHIC INSIGHT 1
>
> **Environment:** Sub-Saharan Africa's poverty and political instability leave its people less able to adapt to climate change than the inhabitants of other world regions, though many Africans are developing strategies to cope with increasing variability in rainfall and temperature. As a whole, this region has contributed little to the build-up of greenhouse gases in the atmosphere, but deforestation by rural Africans and by multinational logging companies is intensifying global climate change. **158. DISAPPEARING GLACIERS ON MT. KILIMANJARO RAISE ENVIRONMENTAL CONCERNS**

Deforestation and Climate Change

Sub-Saharan Africa contributes to CO_2 emissions and potential climate change primarily through deforestation. Trees absorb CO_2 as they photosynthesize, thus removing carbon from the air and storing it as biomass, a process known as **carbon sequestration**. When trees are burned or when they decompose after they die, they release the stored CO_2 into the atmosphere. Because of this, deforestation is a major contributor to carbon buildup in the atmosphere and ultimately to climate change.

African countries lead the world in the *rate* of deforestation, the percentage of total forest area lost. Of the eight countries that had the world's highest rates of deforestation between 1990 and 2005, six are in sub-Saharan Africa (Burundi, Togo, Nigeria, Benin, Uganda, and Ghana). This deforestation constitutes a major human environmental impact across the region (Figure 7.6). The countries that have the most emissions from deforestation are Brazil and Indonesia, but Nigeria and Congo (Kinshasa) have the third and fourth most, respectively. **168. PLAN TO CLEAR-CUT UGANDAN FOREST RESERVE FOR GROWING SUGARCANE SPARKS CONTROVERSY**

---

intertropical convergence zone (ITCZ) a band of atmospheric currents that circle the globe roughly at the equator; warm winds from both north and south converge at the ITCZ, pushing air upward and causing copious rainfall

Horn of Africa the triangular peninsula that juts out from northeastern Africa below the Red Sea and wraps around the Arabian Peninsula

carbon sequestration the removal and storage of carbon taken from the atmosphere

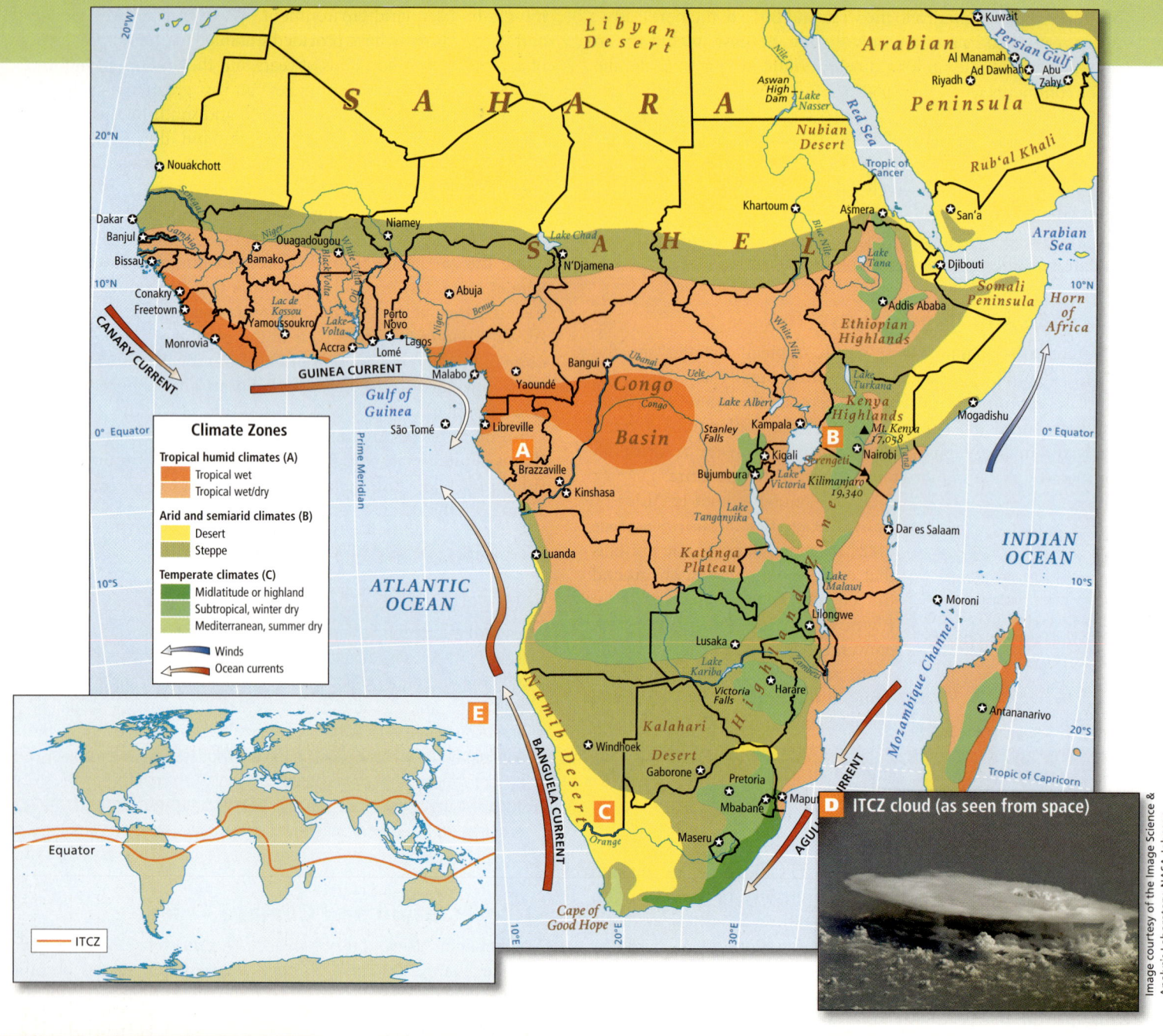

FIGURE 7.5

PHOTO ESSAY: Climates of Sub-Saharan Africa

Climate Zones

Tropical humid climates (A)
- Tropical wet
- Tropical wet/dry

Arid and semiarid climates (B)
- Desert
- Steppe

Temperate climates (C)
- Midlatitude or highland
- Subtropical, winter dry
- Mediterranean, summer dry

→ Winds
→ Ocean currents

E — ITCZ (world map inset)

D ITCZ cloud (as seen from space)
Image courtesy of the Image Science & Analysis Laboratory, NASA Johnson Space Center

A Tropical wet, Gabon 
Jacques Jangoux/Photo Researchers/Getty Images

B Subtropical, winter dry, Kenya 
Nigel Hicks/Dorling Kindersley/Getty Images

C Desert, Namibia 
David Yarrow Photography/The Image Bank/Getty Images

FIGURE 7.6   PHOTO ESSAY: Human Impacts on the Biosphere in Sub-Saharan Africa

Deforestation, desertification, and increasing water use have had major impacts on sub-Saharan Africa's ecosystems.

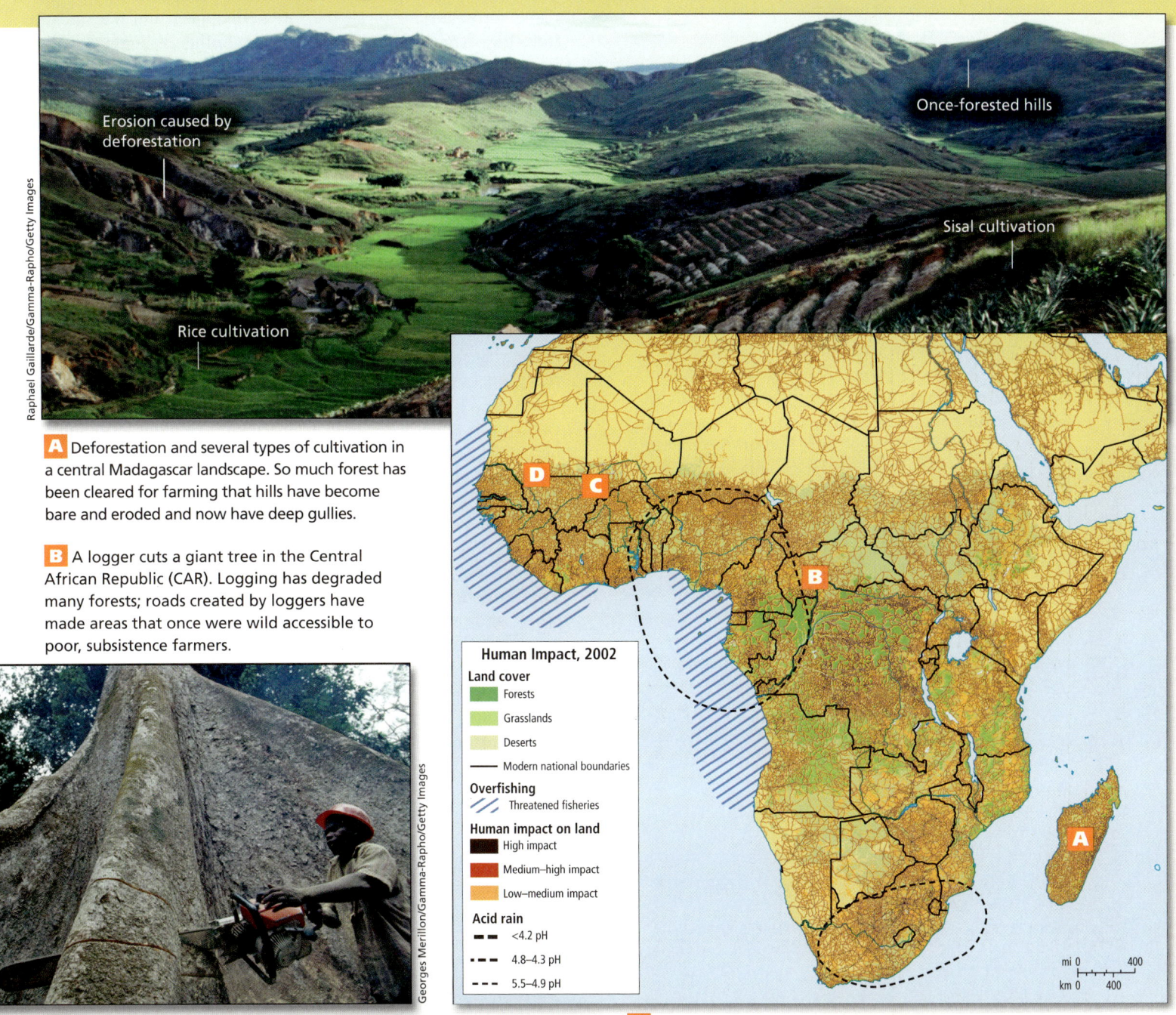

Erosion caused by deforestation

Once-forested hills

Sisal cultivation

Rice cultivation

A Deforestation and several types of cultivation in a central Madagascar landscape. So much forest has been cleared for farming that hills have become bare and eroded and now have deep gullies.

B A logger cuts a giant tree in the Central African Republic (CAR). Logging has degraded many forests; roads created by loggers have made areas that once were wild accessible to poor, subsistence farmers.

Human Impact, 2002

Land cover
- Forests
- Grasslands
- Deserts
- Modern national boundaries

Overfishing
- Threatened fisheries

Human impact on land
- High impact
- Medium–high impact
- Low–medium impact

Acid rain
- <4.2 pH
- 4.8–4.3 pH
- 5.5–4.9 pH

C A fisher casts his net in the Niger River wetlands, where carefully synchronized resource-use patterns have been developed over millennia.

D A dam at Manantan in western Mali. A similar dam planned for upstream of the Niger River wetlands would alter the river's flow in order to supply farmers with irrigation water. The project would increase food security for farmers elsewhere in Mali but threaten the livelihoods of fishers, farmers, and pastoralists in the Niger River wetlands. [Matt Fletcher/Lonely Planet Images/Getty Images]

VIGNETTE

Liberian environmental activist Silas Siakor is an affable and unassuming fellow. But his casual style belies his remarkable sleuthing ability and fierce dedication to his homeland. At great personal risk, Siakor uncovered evidence that 17 international logging companies were bribing Liberia's then-president, Charles Taylor, with cash and guns. Taylor allowed the companies to illegally log Liberia's forests, which are home to many endangered species, including forest elephants and chimpanzees. In return, the companies paid cash and provided weapons that Taylor used to equip his personal armies. Made up largely of kidnapped and enslaved children, Taylor's armies fought those of other Liberian warlords in a 14-year civil war that took the lives of 150,000 civilians. The war also spilled over into neighboring Sierra Leone, where another 75,000 people died. Meanwhile, the logging companies—based in Europe, China, and Southeast Asia—reaped huge fortunes from the tropical forests.

Silas Siakor pulled together publicly available information that had been previously ignored by the international community and information provided by ordinary citizens in ports, villages, and lumber companies. He prepared a clear, well-documented report proving the massive logging fraud. In response to Siakor's report, the UN Security Council voted to impose sanctions to stop the timber trade and prosecute some of the people involved. Charles Taylor fled to Nigeria, but in 2006 he was turned over to face a war crimes tribunal in The Hague, Netherlands. In April of 2012, Taylor received a 50-year sentence.

In Liberia, democratic elections followed in 2006, after which Ellen Johnson-Sirleaf took office as Africa's first elected female president. In a bold move—given the poverty and political instability of her country—she cancelled all contracts with timber companies pending a revision of Liberian forestry law. By 2012, President Johnson-Sirleaf was herself being audited by the global Extractive Industries Transparency Initiative (EITI). Their report showed that $8 billion worth of illegal resource contracts were made under Johnson-Sirleaf's administration. Liberia, a country that is now thought of as rich in resources, is the first African country to submit to the EITI anticorruption audits. [Sources: Goldman Environmental Prize, National Public Radio, and Silas Siakor. For detailed source information, see Text Sources and Credits.]

175. 'EZRA,' TRAGIC TALE OF CHILD SOLDIERS IN AFRICA

---

major oil producer, most people use fuelwood because they cannot afford petroleum products. Not only are charcoal prices rising across the region as the forests disappear, but the smoke from the charcoal is causing asthma and other respiratory problems, as well as creating greenhouse gases.

In Maputo, the capital of Mozambique, a group of small-scale farmers are engaged in a for-profit project that combines new technology with traditional crops to create a sustainable fuel and food preparation system. The goal is to shift the 1.2 million inhabitants of Maputo from relying on charcoal—made from the disappearing old-growth forests in the north of the country—to using ethanol made from the roots of the cassava plant grown by the farmers on their own land. Newly developed small, metal cookstoves run on the cassava-based ethanol.

The multifaceted stove/ethanol/cassava pilot project in Maputo is funded by four American firms that hope to replicate it elsewhere around the world. Their goal is to save the forests, relieve fuel shortages, halt rising fuelwood prices, and alleviate air pollution. Other benefits include improving food security, since cassava is a major food crop in Africa, and creating local small businesses that can convert cassava into ethanol and build the innovative cookstoves.

Elsewhere, African governments that are trying to slow deforestation are encouraging **agroforestry**—the farming of economically useful trees in conjunction with the usual subsistence and cash crops. Using the farmed trees for fuel and construction helps reduce dependence on old-growth forests and can provide income through the sale of farmed wood. For example, by practicing agroforestry on the fringes of the Sahel, a family in Mali can produce fuelwood, fencing and building materials, medicinal products, and food—all on the same piece of land. This would double what could be earned from a single crop or from animal herding. However, critics caution that agroforestry sometimes introduces *invasive species*, nonnative plants that threaten many African ecosystems by outcompeting indigenous species.

> **agroforestry** growing economically useful crops of trees on farms, in conjunction with the usual plants and crops, to reduce dependence on trees from nonfarmed forests and to provide income to the farmer
>
> **subsistence agriculture** farming that provides food for only the farmer's family and is usually done on small farms

Agricultural Systems, Food, Water, and Vulnerability to Climate Change

Agricultural systems in Africa are undergoing a rapid transition from subsistence farming to commercial farming. Unfortunately, this change is happening at a time when the advantages of traditional agriculture are just beginning to be appreciated. The social and environmental effects of turning rapidly to commercial production are not fully understood at this point, especially given the as-yet little-known ecological impacts of climate change.

Traditional Agriculture Most sub-Saharan Africans practice **subsistence agriculture**—farming that provides food for only the farmer's family—usually on small farms of about 2 to 10 acres

---

The story of Silas Siakor and Ellen Johnson-Sirleaf illustrates how, like many other countries in this region, Liberia is still in the process of building strong political institutions that can extract and use the country's resources sustainably and for the benefit of its citizens.

Most of Africa's deforestation is driven by the growing demand for farmland and fuelwood, although logging by international timber companies is also increasing. Africans use wood (or charcoal made from wood) to supply nearly all their domestic energy (see Figure 7.6A, B). Wood remains the cheapest fuel available, in part because of African traditions that consider forests to be a free resource held in common. Even in Nigeria, a

(1 to 4 hectares). Most subsistence farmers also practice **mixed agriculture**, which involves raising a diverse array of crops and a few animals as livestock. Many also fish, hunt, herd, and gather some of their food from forest or grassland areas. The wide variety of food produced means that the cuisine of Africa is also varied (Figure 7.7).

ON THE BRIGHT SIDE

The Value of Agroforestry

Agroforestry has long been promoted by the Green Belt Movement, founded by Dr. Wangari Maathai of Kenya, which helps rural women plant trees for use as fuelwood and to reduce soil erosion. The movement grew from Maathai's belief that a healthy environment is essential for democracy to flourish. Thirty years and 30 million trees after she began, Maathai was awarded the 2004 Nobel Peace Prize for her contributions to sustainable development, democracy, and peace.

FIGURE 7.7 LOCAL LIVES
Foodways in Sub-Saharan Africa

A Ethiopian and Eritrean cuisine features spicy vegetable and meat stews and salads served with *injera*, a type of savory pancake made from fermented grains. Pieces of the injera are torn off and used to scoop up the rest of the meal. Injera is traditionally cooked on a clay plate over an open fire. Because cooking injera on an open fire requires a lot of wood, people also use more efficient stoves, including electric models that are popular in cities, to cook injera. [Photostock Israel/Oxford Scientific/Getty Images]

B Recently harvested cassava root is transported in Nigeria. Originally domesticated in Brazil, cassava was brought to Africa in the seventeenth century as part of the slave trade. Because it grows abundantly in a wide variety of soils and environments and is easy to cultivate, it has been a popular crop, despite its relatively low nutritional value. Sub-Saharan Africa produces more cassava than any world region, with Nigeria as the world's largest producer. [Lynn Johnson/National Geographic/Getty Images]

C A *potjiekos*, or "small pot stew," is prepared in South Africa. Cast iron pots for cooking were brought to South Africa by migrants from the Netherlands, who developed the potjiekos during their expansion north from Capetown into the interior. These "trekkers" made stews of wild game, vegetables, and spices each evening, keeping the leftovers in the pot for the next day's move. The trekkers also added fresh ingredients as they travelled. Making a potjiekos is a social activity, with fireside conversation throughout the 3 to 6 hours that it takes to cook the stew. [Graeme Williams/Gallo Images/Getty Images]

To maintain soil quality in a tropical wet or wet/dry climate, subsistence farmers have for generations used **shifting cultivation**, in which small patches of forest are cleared, with the detritus either burned or left to decay. The clearings are cultivated with a wide variety of plants for 2 or 3 years and then are abandoned and left to regrow. After a few decades, the soil naturally replenishes its organic matter and nutrients and is ready to be cultivated again. However, when fallow periods are reduced, as is happening in many rural areas that have increasingly high population densities, the soil can become degraded.

These traditional, subsistence, mixed, and shifting cultivation food-acquisition techniques have advantages and disadvantages in the present era of climate change. All of these closely related forms of agriculture provide a diverse array of strategies for coping with the changes in temperature and rainfall that climate change may bring. As in many other parts of sub-Saharan Africa, traditional farmers in Nigeria (usually women) grow complex tropical gardens, often with 50 or more species of plants. Some of the plants can handle drought, while others can withstand intense rain or heat.

However, the subsistence nature of most African farming can also leave families without much cash. If harvests are too low to provide surplus that can be sold, and hunting and gathering fail to provide supplementary food for the family, there may not be enough money to buy food. While such situations can lead to famine, it is important to note that the most serious famines in Africa have occurred not because of low harvests but rather because of political instability that disrupts economies and food growing and distribution systems. **162. FELLOWSHIP PROGRAM AIMS TO OPEN DOORS FOR AFRICAN WOMEN IN AGRICULTURAL RESEARCH**

mixed agriculture farming that involves raising a diverse array of crops and animals on a single farm, often to take advantage of several environmental riches

shifting cultivation a productive system of agriculture in which small plots are cleared in forestlands, the dried brush is burned to release nutrients, and the clearings are planted with multiple species; each plot is used for only 2 or 3 years and then abandoned for many years of regrowth

commercial agriculture farming in which crops are grown deliberately for cash rather than solely as food for the farm family

Commercial Agriculture Much of Africa is now shifting over to **commercial agriculture**, in which crops are grown deliberately for cash rather than solely as food for the farm family. This type of production also has advantages and disadvantages with respect to climate change. If harvests are good and prices for crops are adequate, farmers can earn enough cash to get them through a year or two of poor harvests. Having some cash income can also allow poor families to invest in other means of earning a living, such as opening a grinding mill to help process other farmers' harvests or building a stone oven to bake neighbors' bread for a fee.

However, some of the most common commercial crops, such as peanuts, cacao beans (used for making chocolate), rice, tea, and coffee, are often less adapted to environments outside their native range (Figure 7.8D). They are more likely to produce smaller yields if temperatures increase or water becomes scarce. Moreover, to maximize profits, these crops are often grown in large fields containing only a single plant species. This can leave crops vulnerable to pests; if crops fail, farmers have no other garden foods to rely on. The potential for commercial agriculture to help farmers adapt to the uncertainties of global climate change is also limited by the instability of prices for commercial crops. Prices can rise or fall dramatically from year to year because of overproduction or crop failures both in Africa and abroad.

Commercial agriculture—whether small scale and locally managed, or large scale and owned and operated by international corporations (agribusiness)—often requires permanently cleared fields. Because soil fertility declines rapidly once the trees are removed, however, commercial crops in the tropics are more likely to fail even under ideal climatic conditions. Chemical fertilizers can compensate for this loss, but are often too expensive for ordinary farmers, and may lose their effectiveness when used repeatedly. Moreover, rains almost always wash much of the fertilizer into nearby waterways, thus polluting them. This may ultimately hurt farmers and fishers by spoiling drinking water and reducing the quantity of available fish, which are important sources of protein for rural people.

These aspects of commercial agricultural systems make them a risky choice even under the most ideal climatic conditions, let alone in the hotter and more drought-prone environments that climate change may bring. Foreign agricultural "experts" who often push commercial systems can be woefully ignorant of local conditions and the value of local cultivation techniques. In Nigeria, for example, women grow most of the food for family consumption and are guardians of the knowledge that makes Nigeria's traditional farming systems work. However, women are rarely included or even consulted by the foreign experts who promote commercial agricultural development projects. At best, women are employed as field laborers. Consequently, diverse subsistence agricultural systems based on numerous plant species and deep knowledge about the agriculture have been replaced by less stable commercial systems that are based on a single plant species and that do not take into account the skills and knowledge base of women.

Recently, agricultural scientists have begun to recognize past mistakes and have been trying to incorporate the traditional knowledge of African farmers, female and male, into commercial agriculture systems. For example, scientists at Nigeria's International Institute for Tropical Agriculture are developing cultivation systems that, like traditional systems, use many species of plants that help each other cope with varying climatic conditions. Most of these systems are designed for both subsistence and commercial agriculture, and so can give families both a stable food supply and cash to help them ride out crop failures and pay school fees for their children (see Figure 7.8C).

Water Resources, Irrigation, and Water Management Alternatives

Although sub-Saharan Africa has a large wet tropical zone, much of the region is seasonally dry (see the Figure 7.5 map). Freshwater shortages are expected to increase, as is the number of refugees (see Figure 7.8A). In water-deficit conditions, many people look to irrigation to provide more stability for agricultural systems (see Figure 7.6D).

New Groundwater Discoveries Is there sufficient groundwater in Africa? Until recently, the answer was a resounding no. However, a recent comprehensive review and quantitative mapping of all available data on groundwater for the African continent (Figure 7.9) revealed that the quantity of **groundwater** (water naturally stored in aquifers as many as 5000 years ago, during wetter climate conditions) was nearly 100 times that of previous estimates. In many cases, the water was close enough to the surface to be accessed via inexpensive bore well pumps. This does not mean that Africa suddenly has an inexhaustible supply of water. Nor does it mean that it will be easy to get this water to the people who need it. The largest reservoirs of groundwater are in lightly populated areas (such as North Africa and Botswana; see Figure 7.1A). The most heavily populated country, Nigeria, has only a relatively small supply of groundwater. Nonetheless, this news on groundwater resources is welcome because climate change is likely to reduce precipitation in irregular and unpredictable patterns across the continent, and because per capita water use is likely to expand exponentially as this region develops economically (see Chapter 1, page 26).

> **groundwater** water naturally stored in aquifers as many as 5000 years ago during wetter climate conditions

New Large- and Small-Scale Irrigation Projects Government officials continue to favor large-scale irrigation projects to address the need for increased food supplies in the face of current and future water scarcity. Africa has a number of major rivers—the Nile, the Congo, the Zambezi, and the Limpopo—and climate change will affect all of them. Here we focus on the Niger, one of Africa's most important rivers, which flows through several very different ecological zones.

The Niger rises in the tropical wet Guinea Highlands and carries summer floodwaters northeast into the normally arid lowlands of Mali and Niger (see the Figure 7.5 map and Figure 7.6C, D). There the waters spread out into lakes and streams that nourish wetlands. For a few months of the year (June through September), the wetlands ensure the livelihoods of millions of fishers, farmers, and pastoralists. These people share the territory in carefully synchronized patterns of land use that have survived for millennia. Wetlands along the Niger produce 8 times more plant matter per acre than the average wheat field does. They provide seasonal pasture for millions of domesticated animals, and serve as an important habitat for wildlife.

The governments of Mali and Niger now want to dam the Niger River and channel its water into irrigated agribusiness projects that they hope will help feed the more than 26 million people in both countries. However, the dams would forever change the seasonal rise and fall of the river. The irrigation systems may also pose a threat to human health. Systems that rely on surface storage not only lose a great deal of water through evaporation and leakage, but the standing pools of water they create often breed mosquitoes that spread tropical diseases such as malaria,

FIGURE 7.8 PHOTO ESSAY: Vulnerability to Climate Change in Sub-Saharan Africa

Much of this region is highly exposed to drought, flooding, and other climatic disturbances. Poverty and having little access to cash increase sensitivity to climate change, while political instability makes it harder for governments to effectively use disaster management to help boost resilience.

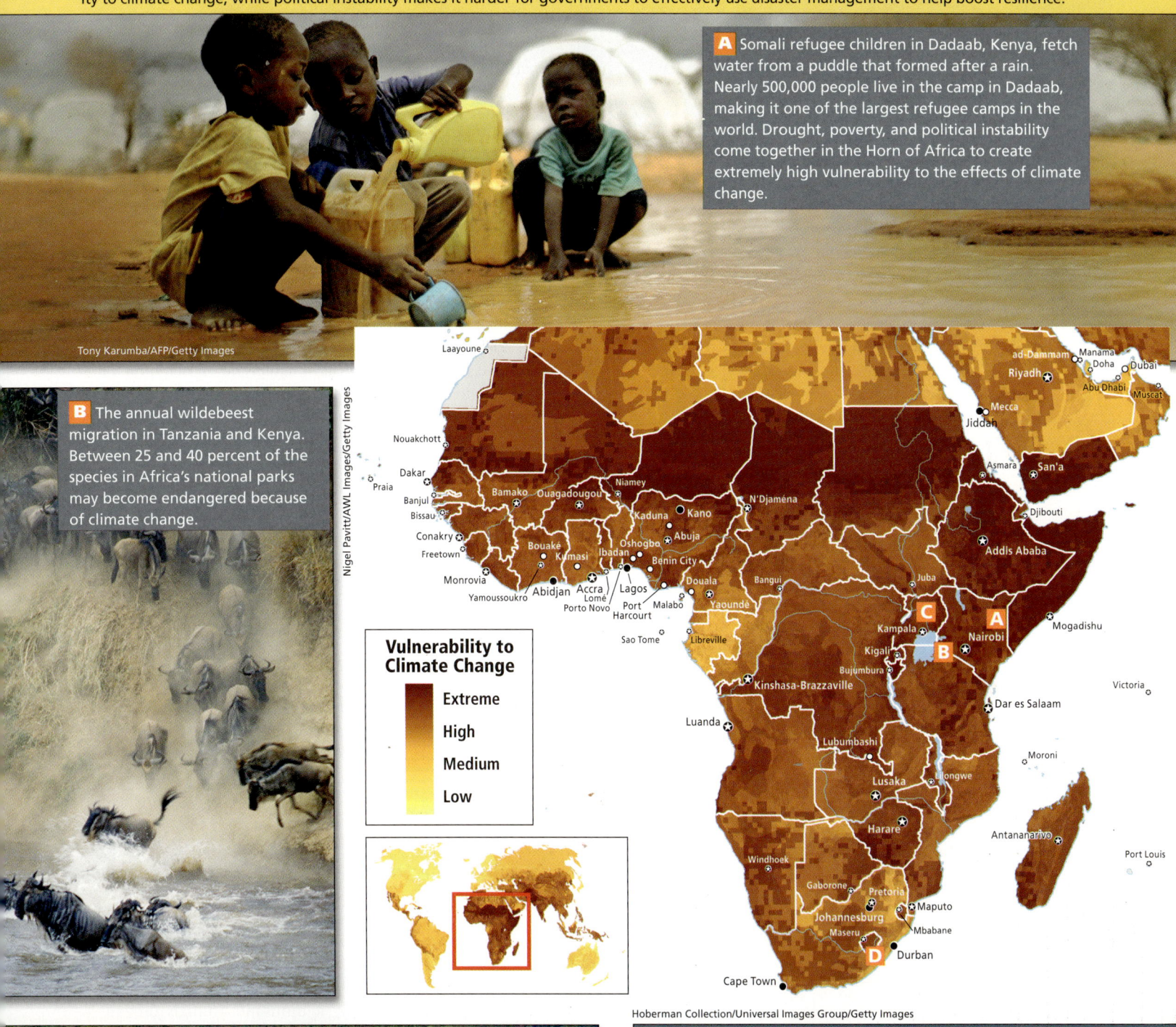

A Somali refugee children in Dadaab, Kenya, fetch water from a puddle that formed after a rain. Nearly 500,000 people live in the camp in Dadaab, making it one of the largest refugee camps in the world. Drought, poverty, and political instability come together in the Horn of Africa to create extremely high vulnerability to the effects of climate change.

B The annual wildebeest migration in Tanzania and Kenya. Between 25 and 40 percent of the species in Africa's national parks may become endangered because of climate change.

Vulnerability to Climate Change
- Extreme
- High
- Medium
- Low

C A landscape of subsistence and mixed agriculture in Uganda. These diversified systems make farmers somewhat more resilient to climatic disturbances, but they provide little cash to help farmers buy other food they might need.

D Tea harvested for export in South Africa. Commercial crops like tea can increase the farmers' resilience to climate change by providing cash to buy any needed food. But tea, like many commercial crops, requires expensive and polluting fertilizers, and prices for tea are unstable.

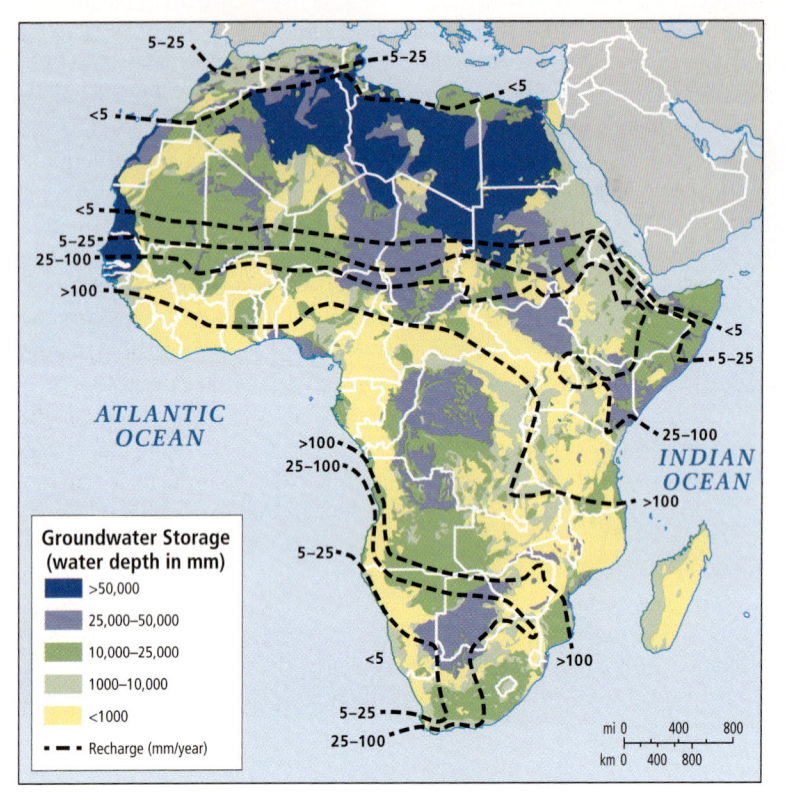

FIGURE 7.9 Newly exposed water resources. It was long thought that Africa had only shallow and sparse deposits of groundwater. But recent research has revealed that for the African continent, the quantity of water stored in aquifers is nearly 100 times that of previous estimates. Some is near the surface and relatively accessible, but because much is fossil water deposited as many as 5000 years ago, it cannot be *recharged* (replenished) quickly under present drier climate conditions. Also, many densely populated areas have only small deposits. Compare this map of the depth of groundwater deposits and number of years needed to recharge with the Figure 7.22 population map. [Sources consulted: A. M. MacDonald, H. C. Bonsor, B. E. O. Dochartaigh, and R. G. Taylor, 2012, "Quantitative Maps of Groundwater Resources in Africa," *Environmental Research Letters* 7. doi:10.1088/1748-9326/7/2/024009; boundaries of surficial geology of Africa courtesy of the U.S. Geological Survey; country boundaries sourced from ArcWorld © 1995–2011 ESRI. All rights reserved.]

Herding and Desertification Herding, or **pastoralism**, is practiced by millions of Africans, primarily in savannas, on desert margins, and in the mixture of grass and shrubs called *open bush*. Herders live off the milk, meat, and hides of their animals. They typically circulate seasonally through wide areas, taking their animals to available pasturelands and trading with settled farmers for grain, vegetables, and other necessities.

Many traditional herding areas in Africa are now undergoing *desertification*, the process by which arid conditions spread to areas that were previously moist (see Chapter 6, pages 239–240). This drying out is often the result of the loss of native vegetation; traditional herding may be partially to blame, but economic development schemes that encourage cattle raising are also at fault. Cattle need more water and forage than do traditional herding animals and therefore place greater stress on native grasslands than goats or camels do. Agricultural intensification in the Sahel also contributes to desertification, as scarce water resources are diverted to irrigation, leaving dry, vegetation-less soils exposed to wind erosion.

> **pastoralism** a way of life based on herding; practiced primarily on savannas, on desert margins, and in the mixture of grass and shrubs called *open bush*

Over the last century, desertification has shifted the Sahel to the south. For example, the *World Geographic Atlas* in 1953 showed Lake Chad situated in a forest well south of the southern edge of the Sahel. By 1998, the Sahara itself was encroaching on Lake Chad (see the Figure 7.5 map) and the lake had shrunk to a tenth of the area it occupied in 1953.

Wildlife and Climate Change

Africa's world-renowned wildlife faces multiple threats from both human and natural forces, all of which could become more severe with increases in global climate change. The Intergovernmental Panel on Climate Change, a body of scientists tasked by the United Nations with assessing scientific information relevant to understanding climate change, estimates that 25 to 40 percent of the species in Africa's national parks may become endangered as a result of climate change.

Wildlife managers are developing new management techniques to help animals survive. For example, one of the greatest wildlife spectacles on the planet took a tragic turn in 2007. The annual 1800-mile-long (2900-kilometer-long) natural migration of more than a million wildebeests, zebras, and gazelles in Kenya's Maasai Mara game reserve requires animals to traverse the Mara River. In the best of times, this is a difficult migration that usually results in roughly a thousand animals drowning. In the past, the park's managers have taken a hands-off approach to the migration, considering the losses normal. However, in 2007, extremely heavy rains, possibly related to global climate change, swelled the Mara River to record levels. When the animals tried to cross, 15,000 drowned (see Figure 7.8B). Park managers are now taking a more active role in helping the migrating animals cope with unusual climatic conditions that may worsen with climate change. This may involve stopping animals from attempting a river crossing or directing them to a safer crossing. **169. PROTECTING NATURE IN GUINEA COLLIDES WITH HUMAN NEEDS**

and harbor the snails that host schistosomiasis, a debilitating parasite that enters the skin of humans who spend time standing in still water to fish or do other chores.

Many smaller-scale alternatives are available. In some parts of Senegal, for example, farmers are using hand- or foot-powered pumps to bring water from rivers or ponds directly to the individual plants that need it. This is in some ways a more modern version of traditional African irrigation practices whereby water is delivered directly to the roots of the plants by human, often female, water brigades. Smaller-scale projects provide the same protection against drought that larger systems offer, but they are much cheaper and simpler to operate for small farmers. They also avoid the social dislocation and ecological disruption of larger projects. Already used successfully for 15 years, these low-tech pumps will help farmers adjust to the drier conditions that may come with climate change.

Farmers' dependence on hunting wild game (*bushmeat*) for part of their food and income is already a major threat to wildlife in much of Africa. For example, farmers who need food or extra income are killing endangered species such as gorillas, chimpanzees, and especially elephants in record numbers (Figure 7.10C). If crop harvests are diminished by global climate change, many farmers will become even more dependent on bushmeat and on income from selling contraband, such as ivory. The threat to wild populations of various species has led to calls to expand and establish new protected areas for wildlife.

Africa's national parks constitute one-third of the world's preserved national parkland. The parks are struggling to deal with *poaching* (illegal hunting) within the park boundaries by members of surrounding communities. Poaching is often fueled by demand outside Africa for exotic animal parts (tusks, hooves, penises) as medicines and aphrodisiacs, especially in Asia, where efforts to educate consumers about the damage done by their purchases are only beginning. In 2011, the western black rhino was declared extinct, due largely to poaching driven by demand for its horns by practitioners of Chinese medicine.

Some parks are now using profits from ecotourism to fund development in nearby communities that previously depended in part on poaching. For example, in 1985, wildlife poaching threatened the animal population in Zambia's Kasanka National Park. Park managers decided to generate employment for the villagers through tourism-related activities. They built tourist lodges and wildlife-viewing infrastructure and started cottage industries to make products to sell to the tourists. Funding also goes to local clinics and schools, and students are included in research projects within the park. Local farmers have expanded into alternative livelihoods, such as beekeeping and agroforestry. Today, poaching in Kasanka is rare, its wildlife populations are booming, tourism is growing, and local communities have an ongoing stake in the park's success. Sub-Saharan Africa's rich array of animals has long played an economic role in daily life. Three species are discussed in Figure 7.10.

FIGURE 7.10 LOCAL LIVES
People and Animals in Sub-Saharan Africa

A The Rhodesian ridgeback, or the African lion hound, helps lion hunters by distracting the lions. Very intelligent and capable hunters themselves (they can kill animals as large as baboons), ridgebacks are now used mostly as guard dogs. The nomadic KhoiKhoi people developed the ridgebacks by interbreeding indigenous dogs of southern Africa with a variety of European dogs. [Hans Surfer/Flickr/Getty Images]

B African grey parrots, highly intelligent and able to mimic human speech, are popular pets worldwide. One bird famously greeted Dr. Jane Goodall, known for her research with chimpanzees, by saying, "Got a chimp?" (The bird had seen a photo of Goodall with a chimpanzee.) As many as 21 percent of the wild African grey parrots are caught each year, and the International Union for the Conservation of Nature (IUCN) lists the parrots as a vulnerable species. [Enrique R Aguirre Aves/Getty Images]

C Elephants are a major attraction at Africa's many national parks and nature reserves, but park space cannot sustain large numbers of elephants whose natural impulse is to pull down trees to eat leaves. In response, some parks have reduced their elephant populations. The parks, though, may be the only hope for the survival of elephants that are hunted for their ivory tusks. Hunters and poachers have for years tended to take large-tusked elephants, leading to natural selection for elephants with smaller tusks or with no tusks at all—a once-rare trait that is now found in about 30 percent of the population. [Massimo Pizzotti/Photographer's Choice/Getty Images]

THINGS TO REMEMBER

- The surface of the continent of Africa can be envisioned as a raised platform, or plateau, bordered by fairly narrow and uniform coastal lowlands.

- Most of sub-Saharan Africa has a tropical wet or wet/dry climate, except at the southern tip and in the cooler uplands. Seasonal climates in Africa differ more by the amount of rainfall than by temperature.

- Most rainfall comes to Africa by way of the intertropical convergence zone (ITCZ), a band of atmospheric currents that circle the globe roughly around the equator. The effects of desertification are most dramatic in the region called the Sahel.

> **GEOGRAPHIC INSIGHT 1** • **Environment** Sub-Saharan Africa's poverty and political instability leave its people less able to adapt to climate change than the inhabitants of other world regions, though many Africans are developing strategies to cope with increasing variability in rainfall and temperature. As a whole, this region has contributed little to the build-up of greenhouse gases in the atmosphere, but deforestation by rural Africans and by multinational logging companies is intensifying global climate change.

- The primary way in which sub-Saharan Africa contributes to CO_2 emissions and potential climate change is through deforestation. Much of Africa's deforestation is driven by the growing demand for farmland and fuelwood, but corruption on the part of local authorities and international timber companies also plays a role.

- Long-standing physical challenges, such as deforestation, desertification, and increasing water scarcity, have had a major impact on sub-Saharan Africa's ecosystems. Commercial food production for Africa's cities has many ecological drawbacks. These challenges are likely to increase with climate change.

- Newly discovered groundwater resources hold promise for alleviating some of Africa's water scarcities.

A A museum diorama of early humans (*Homo erectus*) in southern Africa. [Peter V. Bianchi/National Geographic/Getty Images]

B The ruins of Great Zimbabwe. [Andrew McConnell/Robert Harding World Imagery/Getty Images]

C Sankore Mosque, Tombouctou, Mali, built in 1327. [Amar Grover/AWL Images/Getty Images]

2 million years ago First human species evolve in eastern Africa

5000 B.C.E. Early agriculture in the Sahel

1400 B.C.E. Iron- and steelmaking in northeastern Africa

700–1500 C.E.

1250–1600 C.E. Mali Empire

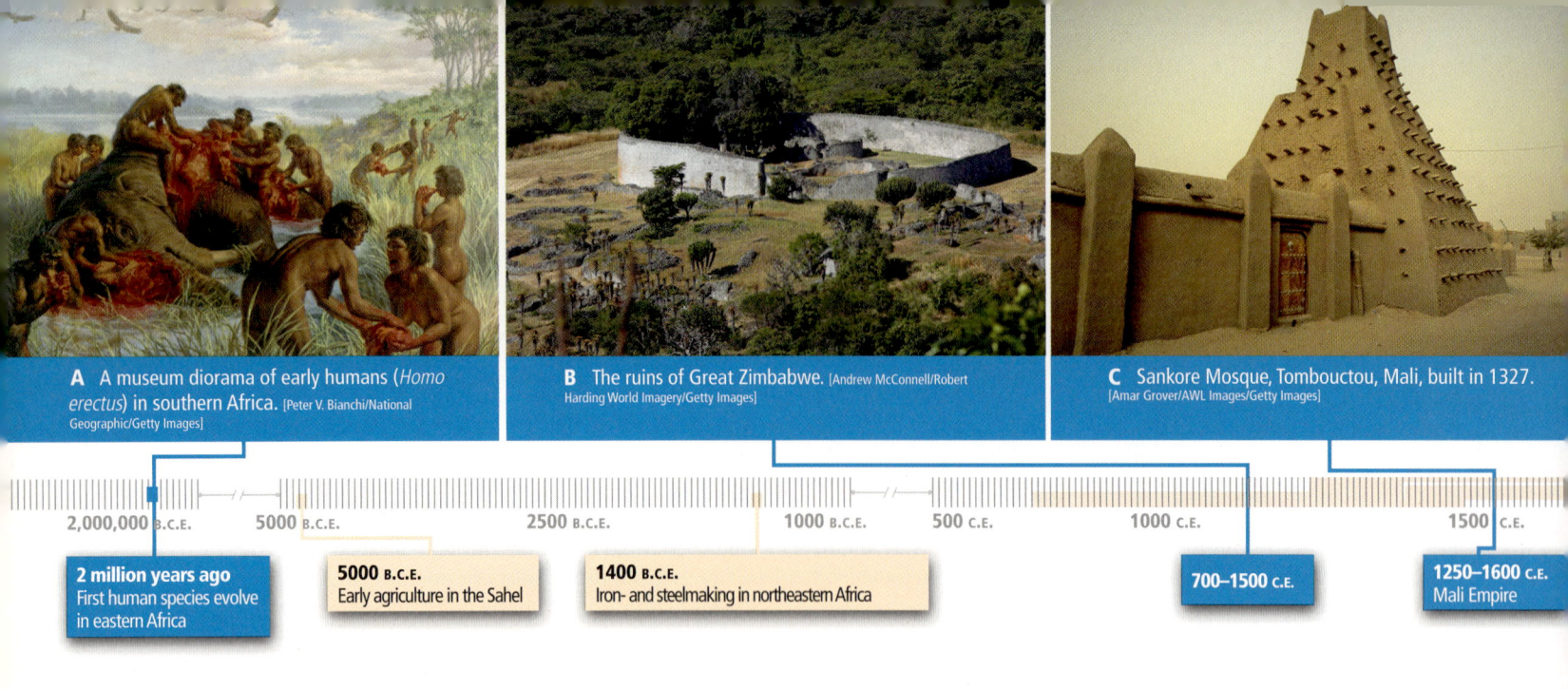

FIGURE 7.11 VISUAL HISTORY OF SUB-SAHARAN AFRICA

HUMAN GEOGRAPHY

Africa's rich and ancient history is often overshadowed by the dramatic changes that came with European colonialism, beginning about 600 years ago. These changes were so powerful that even today many countries are still struggling with the lingering effects of the colonial era.

HUMAN PATTERNS OVER TIME

Africa's rich past has often been misunderstood and dismissed by people from outside the region. European slave traders and colonizers called Africa the *Dark Continent* and assumed it was a place where little of significance had occurred in human history. The substantial and elegantly planned cities of Benin in western Africa, Djenné in the Niger River basin, and Loango in the Congo Basin, which European explorers encountered in the 1500s, never became part of Europe's image of Africa. Even today, most people outside the continent are unaware of Africa's internal history or its contributions to world civilization, let alone its role in the emergence of humankind as a species.

The Peopling of Africa and Beyond

Africa is the original home of humans (Figure 7.11A). It was probably in eastern Africa (in what are today the highlands of Ethiopia, Kenya, and Uganda) that the first human species (*Homo erectus*) evolved more than 2 million years ago. *Homo erectus* differed anatomically from humans today. These early, tool-making humans ventured out of Africa, reaching north of the Caspian Sea and beyond as early as 1.8 million years ago. Anatomically, modern humans (*Homo sapiens*) evolved about 250,000 years ago from earlier hominoids (probably *Homo erectus*) in eastern Africa. Like the migrations of earlier human species, those of modern humans radiated out of Africa, first toward the eastern Mediterranean (*Homo sapiens* reached the eastern Mediterranean about 90,000 years ago), then spreading across mainland and island Asia, and only later turning west into Europe. After coexisting with earlier dispersions of *Homo erectus* and other *Homo* species (all originating in Africa) in Eurasia and Europe, *Homo sapiens* eventually outcompeted them all.

Early Agriculture, Industry, and Trade in Africa

In Africa, people began to cultivate plants as far back as 7000 years ago in the Sahel and the highlands of present-day Sudan and Ethiopia. Agriculture was brought south to equatorial Africa 2500 years ago and to southern Africa about 1500 years ago. Trade routes spanned the African continent, extending north to Egypt and Rome and east to India and China. Gold, elephant tusks, and timber from tropical Africa were exchanged for salt, textiles, beads, and a wide variety of other goods.

About 3400 years ago, people in the vicinity of Lake Victoria learned how to smelt iron and make steel. By 700 C.E., when Europe was still recovering from the collapse of the Roman Empire, a remarkable civilization with advanced agriculture, iron production, and gold-mining technology had developed in the highlands of southeastern Africa in what is now Zimbabwe. This empire, now known as the Great Zimbabwe Empire, traded with merchants from Arabia, India, Southeast Asia, and China, exchanging the products of its mines and foundries for silk, fine porcelain, and exotic jewelry. The Great Zimbabwe Empire collapsed around 1500 for reasons not yet understood (see Figure 7.11B).

Complex and varied social and economic systems existed in many parts of Africa well before the modern era. Several influential centers, made up of dozens of linked communities, developed

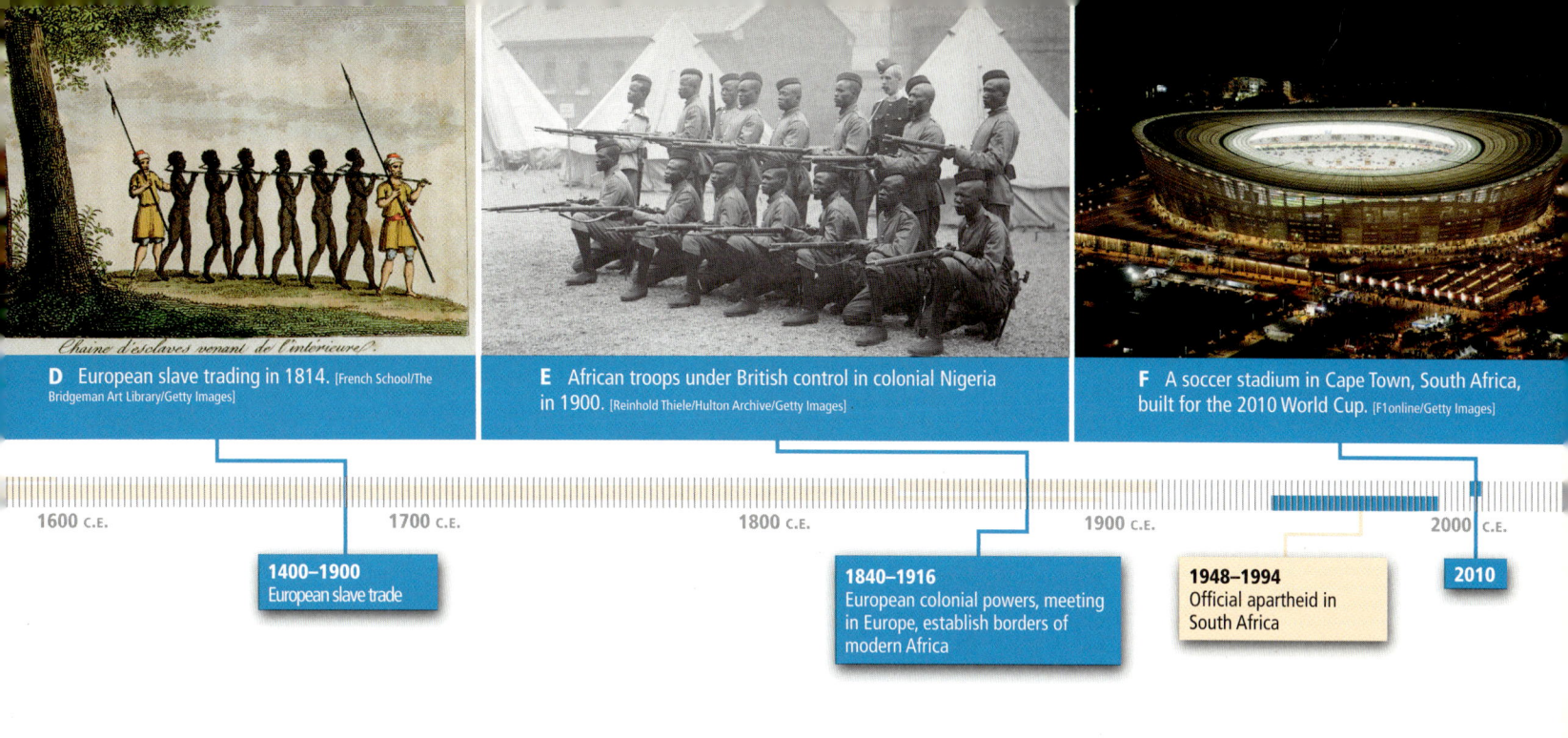

D European slave trading in 1814. [French School/The Bridgeman Art Library/Getty Images]

E African troops under British control in colonial Nigeria in 1900. [Reinhold Thiele/Hulton Archive/Getty Images]

F A soccer stadium in Cape Town, South Africa, built for the 2010 World Cup. [F1online/Getty Images]

1600 C.E. — 1700 C.E. — 1800 C.E. — 1900 C.E. — 2000 C.E.

1400–1900 European slave trade

1840–1916 European colonial powers, meeting in Europe, establish borders of modern Africa

1948–1994 Official apartheid in South Africa

2010

---

in the forest and the savanna of the western Sahel. There, powerful kingdoms and empires rose and fell, such as Djenné; Ghana (700–1000 C.E.), centered in what is now Mauritania and southwestern Mali; and the Mali Empire (1250–1600 C.E.), centered a bit to the east on the Niger River in what became the famous Muslim trading and religious center of Tombouctou (Timbuktu), originally founded about 1100 C.E. by Tuareg nomads as a seasonal camp (see Figure 7.11C). Some rulers periodically sent large trade caravans carrying gold and salt through Tombouctou on to Makkah (Mecca), where their opulence was a source of wonder.

Africans also traded slaves. Long-standing customs of enslaving people captured during war fueled this trade. The treatment of slaves within Africa was sometimes brutal and sometimes reasonably humane. Long before the arrival of Islam, a slave trade developed with Arab and Asian lands to the east. After the spread of Islam began, around 700 C.E., the slave trade continued, and Muslim traders exported close to 9 million African slaves to parts of Southwest, South, and Southeast Asia (Figure 7.12). When slaves were traded to non-Africans, indigenous checks on brutality were abandoned. For example, to ensure sterility and to help promote passivity, Muslim traders often preferred to buy castrated male slaves.

The course of African history shifted dramatically in the mid-1400s, when Portuguese sailing ships began to appear off Africa's west coast. The names given to stretches of this coast by the Portuguese and other early European maritime powers reflected their interest in Africa's resources: the Gold Coast, the Ivory Coast, the Pepper Coast, and the Slave Coast.

By the 1530s, the Portuguese had organized a slave trade with the Americas. The trading of slaves by the Portuguese, and then by the British, Dutch, and French, was more widespread and brutal than any trade of African slaves that preceded it. African slaves became part of the elaborate production systems supplying the raw materials and money that fueled Europe's Industrial Revolution.

To acquire slaves, the Europeans established forts on Africa's west coast and paid nearby African kingdoms with weapons, trade goods, and money to make slave raids into the interior. Some slaves were taken from enemy kingdoms in battle. Many more were kidnapped from their homes and villages in the forests and savannas. Most slaves traded to Europeans were male because they brought the highest prices and the raiding kingdoms preferred to keep captured women for their reproductive capabilities. Between 1600 and 1865, about 12 million captives were packed aboard cramped and filthy ships and sent to the Americas. One-quarter or more of them died at sea. Of those who arrived in the Americas, about 90 percent went to plantations in South America and the Caribbean. Between 6 and 10 percent were sent to North America (see Figure 7.12).

The European slave trade severely drained parts of Africa of human resources and set in motion a host of damaging social responses within Africa that even today are not completely understood (see Figure 7.11D). The trade enriched the African kingdoms that could successfully conquer their neighbors and impoverished or enslaved more peaceful or less powerful kingdoms. It also encouraged the slave-trading kingdoms to become dependent on European trade goods and technologies, especially guns.

Slavery persists in modern Africa and is a growing problem that some argue exceeds the transatlantic slave trade of the past. Today, slavery is most common in the Sahel region, where several countries have made the practice officially illegal only in the past few years (Figure 7.13). People may become enslaved during war. They may be sold by their parents or relatives to pay off debts or forced into slavery when migrating to find a job in a city. Some are even enslaved by gangsters in Europe, where they become street vendors, domestic servants, or prostitutes. In Côte d'Ivoire, Burkina Faso, and Nigeria, forced child labor is used increasingly in commercial agriculture such as cacao (chocolate) production. In Liberia, Sierra Leone, Cameroon, and Congo, young boys have been enslaved as miners and soldiers. It is hard to know exactly how many Africans are currently enslaved, but estimates range from several million to more than 10 million.

290 CHAPTER 7 Sub-Saharan Africa

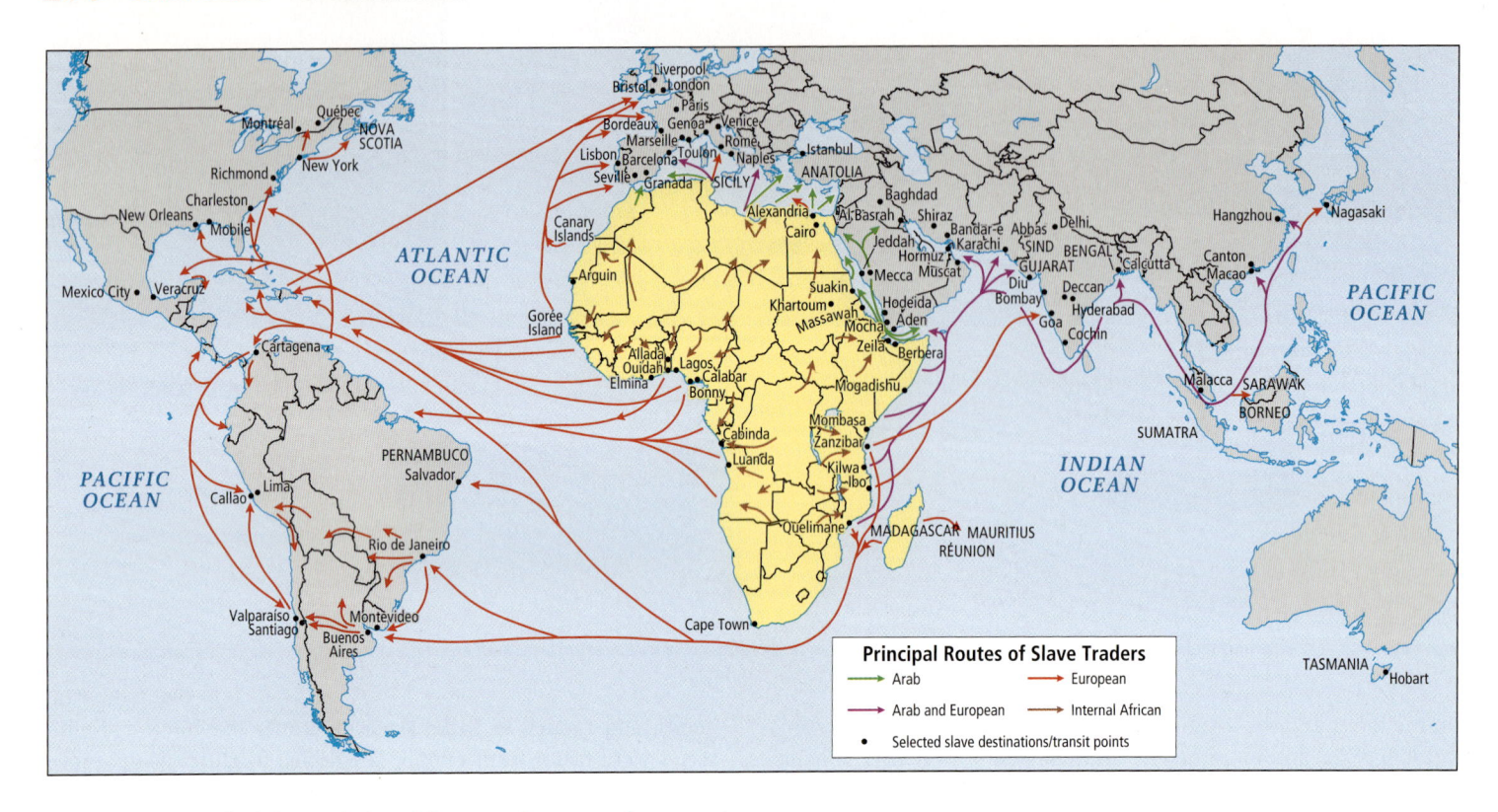

FIGURE 7.12 Map of the African slave trade. The origin of the African slave trade is indigenous and predates Islam. Outside of Africa, many slave markets were established around the Mediterranean and the Indian Ocean and eventually, after 1500 C.E., millions of slaves were brought to the Americas. [Sources consulted: Joseph H. Harris, in Monica B. Visona et al., *A History of Art in Africa* (New York: Harry N. Abrams, 2001), pp. 502–503]

FIGURE 7.13 Modern slavery in Niger. A man born as a slave in Niger is presented with a document that declares his freedom in 2008. Though slavery was "abolished" in Niger in 1960, and made a criminal offense in 2003, it remains deeply embedded in society. At least 43,000 people are enslaved in Niger, most of them born into the status. Slaves have no rights, little access to education, and will spend most of their lives herding cattle, tending fields, or working in their "master's" house.

› **THINKING GEOGRAPHICALLY** What does this photo suggest about the social relationships of the man being freed from slavery in Niger?

The Scramble to Colonize Africa

The European slave trade wound down by about the mid-nineteenth century, as Europeans found it more profitable to use African labor within Africa to extract raw materials for Europe's growing industries.

European colonial powers competed avidly for territory and resources, and by World War I, only two areas in Africa were still independent (Figure 7.14): Liberia on the west coast (populated by former slaves from the United States) and Ethiopia (then called Abyssinia) in East Africa. Ethiopia managed to defeat early Italian attempts to colonize it. Otto von Bismarck, the German chancellor who convened the 1884 Berlin Conference at which the competing European powers formalized the partitioning of Africa, revealed the common arrogant European attitude when he declared, "My map of Africa lies in Europe." With some notable exceptions, the boundaries of most African countries today derive from the colonial boundaries set up between 1840 and 1916 by European administrators and diplomats to suit their purposes (see Figure 7.11E). These territorial divisions lie at the root of many of Africa's current problems.

In some cases, the boundaries were purposely drawn to divide tribal groups and thus weaken them. In other cases, groups who used a wide range of environments over the course of a year were forced into smaller, less diverse lands or forced to settle down entirely, thus losing their traditional livelihoods. As access to resources shrank, hostilities between competing groups developed. Colonial officials often encouraged these hostilities, purposely replacing strong leaders with leaders who could be manipulated. Food production, which was the mainstay of most African economies

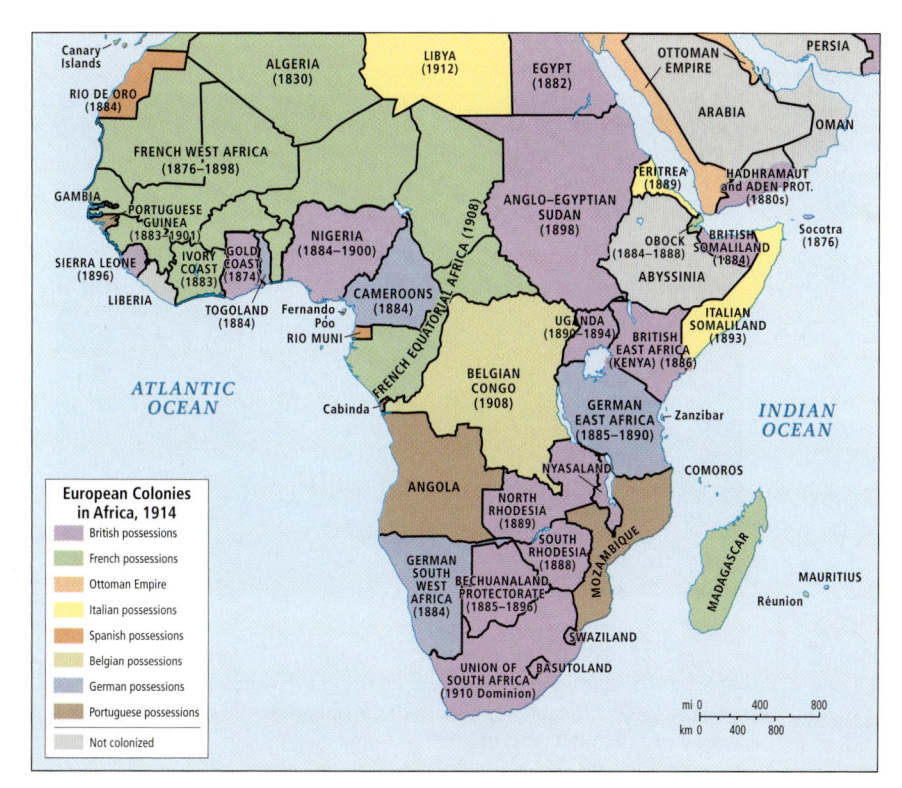

FIGURE 7.14 The European colonies in Africa in 1914. The dates on the map indicate the beginning of officially recognized control by the European colonizing powers. Countries without dates were informally occupied by colonial powers for a few centuries. [Source consulted: Alan Thomas, *Third World Atlas* (Washington, DC: Taylor Francis, 1994), p. 43]

until the colonial period, was discouraged in favor of activities that would support European industries: cash-crop production (cotton, rubber, palm oil) and mineral and wood extraction. Eventually, many formerly prosperous Africans became hungry and poor.

One of the main objectives of European colonial administrations in Africa, in addition to extracting as many raw materials as possible, was to create markets in Africa for European manufactured goods. The case of South Africa provides insights into the European expropriation of African lands and the subjugation of African peoples. In this case, these aims ultimately led to the infamous system of racial segregation known as **apartheid** (see the Case Study on the next page).

> **apartheid** a system of laws mandating racial segregation. South Africa was an apartheid state from 1948 until 1994

Power and Politics in the Aftermath of Independence

The era of formal European colonialism in Africa was relatively short. In most places, it lasted for about 80 years, from roughly the 1880s to the 1960s. In 1957, Ghana became the first sub-Saharan African colonial state to become independent. The last sub-Saharan African country to gain independence was South Sudan in 2011, not from a European power, but from its neighbor Sudan, after more than 40 years of civil war.

Africa entered the twenty-first century with a complex mixture of enduring legacies from the past and looming challenges for the future. Although it has been liberated from colonial domination, most old colonial borders remain intact (compare the colonial borders in Figure 7.14 with the modern country borders in Figure 7.4 on page 278). Often these borders exacerbate conflicts between hostile groups by joining them into one resource-poor political entity. Other borders divide potentially powerful ethnic groups, thereby diminishing their influence; or cut off nomadic people from the resources they use on a seasonal basis.

For many years and with few exceptions, governments continued to mimic the colonial bureaucratic structures and policies that distanced them from their citizens. Corruption and abuse of power by bureaucrats, politicians, and wealthy elites stifled individual initiative, civil society, and entrepreneurialism, creating instead frustration and suspicion. Too often, coups d'état were used to change governments. Democracy, where it existed, was often weakly connected only to voting and not to true participation in policy making at the community, regional, and national levels.

Even today African policy makers, like their colonial predecessors, are often insensitive to the widespread poverty and hardship around them. This was demonstrated during the summer of 2010 when South Africa hosted the 2010 World Cup, the world's largest sporting event and a first for this region, which up to that point had never hosted a global sporting event of similar scale. The preparations involved demolishing the homes of tens of thousands of poor, urban residents who were relocated to make way for the elegant new soccer stadiums (see Figure 7.11F). More recently, in an August 2012 crackdown reminiscent of those under apartheid, police shot dead 34 miners during a strike at a platinum mine in Marikana in Northwest Province. (More strikes followed.)

Other relics of the colonial era can be seen in the dependence of African economies on relatively expensive imported food and manufactured goods as well as on the production of agricultural and mineral raw materials for which profit margins are low and prices on the global market highly unstable. Many sub-Saharan African countries often end up competing against each other and remain economically entwined in trade relationships that rarely work to their advantage. The wealthiest country in the region remains South Africa, where the economy has highly profitable manufacturing and service sectors as well as extractive sectors.

THINGS TO REMEMBER

- Anatomically modern humans (*Homo sapiens*) evolved from earlier hominoids in eastern Africa about 250,000 years ago. By about 90,000 years ago, *Homo sapiens* had reached the eastern Mediterranean. They joined and eventually replaced other hominid species from Africa that had migrated throughout Eurasia starting 1.8 million years ago.

- Agriculture, industry, and international trade have an early history in Africa; agriculture began as early as 7000 years ago in northeast sub-Sahara; iron smelting 3400 years ago; and trade extended across the Indian Ocean to Asia long before Europeans came to east Africa.

CASE STUDY

The Colonization of South Africa

The Cape of Good Hope is a rocky peninsula that shelters a harbor on the southwestern coast of South Africa. Portuguese navigators seeking a sea route to Asia first rounded the Cape in 1488. The Portuguese remained in nominal control of the Cape until the 1650s, when the Dutch took possession with the intention of establishing settlements. Dutch farmers, called *Boers*, expanded into the interior, bringing with them herding and farming techniques that used large tracts of land and depended on the labor of enslaved Africans. The British were also interested in the wealth of South Africa, and in 1795, they seized control of areas around the Cape of Good Hope. When slavery was outlawed throughout the British Empire in 1834, large numbers of slave-owning Boers migrated to the northeast in order to keep their slaves. There, in what became known as the Orange Free State and the Transvaal, the Boers often came into violent conflict with African inhabitants.

In the 1860s, extremely rich deposits of diamonds and gold were unearthed in these areas, securing the Boers' economic future. African men were forced to work in the diamond and gold mines for minimal wages under extreme hardship and unsafe conditions. They lived in unsanitary compounds that travelers of the time compared to large cages.

Britain, eager to claim the wealth of the mines, invaded the Orange Free State and the Transvaal in 1899, waging the Boer War. This brutal war briefly gave the British control of the mines until resistance by Boer nationalists forced the British to grant independence to South Africa in 1910. This independence, however, applied to only a small minority of whites: the Boers (who have since been known as Afrikaners) and some British people who chose to remain. Not until 1994 would full political rights (voting, freedom of expression, freedom of assembly, freedom to live where one chose) be extended to black South Africans—who made up more than 80 percent of the population—and to Asians and mixed race or "coloured" people (2 percent and 8 percent of the population, respectively).

In 1948, the long-standing segregation of South African society was reinforced by *apartheid*, a system of laws that required everyone except whites to carry identification papers at all times, to live in racially segregated areas, and to use segregated transportation and other infrastructure (Figure 7.15). Eighty percent of the land was reserved for the use of white South Africans, who at that time made up just 10 percent of the population. Black, Asian, and "coloured" people were assigned to ethnically based "homelands." The South African government considered the homelands to be independent enclaves within the borders of, but not legally part of, South Africa. Nevertheless, the South African government exerted strong influence in the homelands. Democracy theoretically existed throughout South Africa, but nonwhites were allowed to vote only in the homelands.

FIGURE 7.15 Apartheid in South Africa. A South African woman protests apartheid by using a "whites only" car on a train in 1952. Africa's long history of European domination officially ended only in 1994 with the first national elections in which black South Africans were allowed to vote.

The African National Congress (ANC) was the first and most important organization that fought to end racial discrimination in South Africa. Formed in 1912 to work nonviolently for civil rights for all South Africans, the ANC grew into a movement with millions of supporters, most of them black, but some of them white. Its members endured decades of brutal repression by the white minority. One of the most famous members to be imprisoned was Nelson Mandela, a prominent ANC leader, who was jailed for 27 years and finally released in 1990.

Violence increased throughout South Africa until the late 1980s, when it threatened to engulf the country in civil war. The difficulties of maintaining order, combined with international political and economic pressure, forced the white-dominated South African government to initiate reforms that would end apartheid. A key reform was the dismantling of the homelands. Finally, in 1994, the first national elections took place in which black South Africans could participate. Nelson Mandela, the long-jailed ANC leader, was elected the country's president and proved to be an extraordinary leader. His efforts at reconciliation between the country's different racial groups earned him a Nobel Peace Prize in 1993. Today in South Africa, finding new leaders with the unselfish, clear-eyed vision Mandela had has proven difficult. The thorny process of dismantling systems of racial discrimination and corruption continues, with varying amounts of success. ■

- Slavery was practiced in Africa within several powerful kingdoms, and then pre-Islam Arab traders extended the custom around the Mediterranean and to Asia. Muslims continued the practice. The European slave trade to the Americas severely drained the African interior of human resources and set in motion a host of damaging social responses within Africa. Slavery persists in parts of modern Africa.

- The main objectives of European colonial administrations in Africa were to extract as many raw materials as possible; to create markets in Africa for European manufactured goods and food exports; and to divide and conquer by splitting powerful ethnic groups or by placing groups hostile to one another under the same jurisdiction, governed by complicit indigenous leaders.

- During the era of European colonialism, sub-Saharan Africa's massive wealth of human talent and natural resources flowed out of Africa. Even after the countries of the region became politically independent, wealth continued to flow out of Africa because of fraud and unfavorable terms of trade.
- South Africa trod a rocky road through colonial oppression and apartheid to its present relatively precarious position as the wealthiest country in sub-Saharan Africa.
- Governments have continued to mimic colonial policies, including corruption, in ways that have kept sub-Saharan African countries economically dependent into the twenty-first century. This has stifled individual initiative, civil society, and entrepreneurialism. Political freedoms are often poorly protected. ■

GLOBALIZATION AND DEVELOPMENT

Sub-Saharan Africa emerged from the exploitation and dependence of the colonial era just in time to get sucked into the vortex of globalization, where small or weak countries are often left with little control over their fates.

> **GEOGRAPHIC INSIGHT 2**
>
> **Globalization and Development:** Most sub-Saharan African economies are dependent on the export of raw materials. This pattern, a legacy of the era of European colonialism, results in economic instability because prices for raw materials can vary widely from year to year. While a few countries are diversifying and industrializing, most still sell raw materials and all rely on expensive imports of food and manufactured goods.

For many centuries, sub-Saharan Africa has contributed human labor and raw materials to the global economy, but the profits from turning these human resources and raw materials into higher-value manufactured and processed products have gone to wealthier countries. Raw materials that are traded (usually to other countries) for processing or manufacturing into more valuable goods are called **commodities**. Sub-Saharan commodities include cotton, cacao beans, coffee, timber, palm oil, unrefined oil, gas, precious stones, and metals. The profits of commodity production are usually too low to lift poor countries out of poverty. Because many other poor countries are producing the same commodities, competition between them on the world market often drives down profits.

Commodity prices are also subject to wide fluctuations that create economic instability. This is because commodities are of more or less uniform quality and are traded as such on a global basis. For example, on the major commodities exchanges, a pound of copper may sell for U.S.$3 to $5, regardless of whether it is mined in South Africa, Chile, or Papua New Guinea. Because of the global scale of commodity trading, a change anywhere in the world that influences the supply or demand for a particular commodity can cause immediate instability in the pricing of that commodity. For example, an earthquake in Chile could damage a major copper mine, restricting the supply of copper and sending up the price on world markets. In response, governments that make money from sales of raw copper may overestimate their future revenues and commit to expensive infrastructure projects. But the demand for copper can be decreased for many reasons; for example, innovations in the construction industry can reduce the use of copper in wiring and plumbing, resulting in a rapid drop in the price of copper on world markets. Governments that depend on revenues from copper may suddenly need to halt infrastructure projects and cut essential services such as electricity, road maintenance, or health care.

Since 2000, there has been a noticeable shift in many African countries away from reliance on just one or two commodities. Nearly all have diversified to some extent, a trend that is discussed in the section "The Current Era of Diverse Globalization" on page 295.

Successive Eras of Globalization

Successive eras of globalization have transformed Africa over the past several centuries. They include the era of colonial exploitation; the transition from colonial status to political independence; and the current era, in which some African countries are moving away from a dependence—known as **commodity dependence**—on raw materials exports, and toward new types of economic development and regional integration.

Most early European colonial administrations in Africa (Britain, France, Germany, Belgium) evolved directly out of private resource-extracting corporations, such as the German East Africa Company. The welfare of Africans and their future development was a low priority for most colonial administrators. Education and health care for Africans were generally neglected by colonial governments, which were guided by *mercantilism* (see page 125 in Chapter 3)—the idea that colonies existed to benefit their colonizers. Laborers and farmers were paid poorly and the colonial governments strongly discouraged any economic activities that might compete with those of Europe. African economies were thus hindered from making a transition to the more profitable manufacturing-based industries that were transforming Europe and North America.

For years, commodity dependence, widespread poverty, and the lack of internal markets for local products and services characterized all sub-Saharan African economies, with the partial exception of South Africa, the only sub-Saharan African country with a manufacturing base. Early on, profits from its commodity exports (mainly minerals) were reinvested in the manufacturing of mining and railway equipment. In the late twentieth century, South Africa developed a large service sector with particular strengths in finance and communications that support the mining and manufacturing industries. Today, with only 6 percent of sub-Saharan Africa's population, South Africa produces 30 percent of the region's economic output.

Unfortunately, the relics of colonialism and apartheid have left most sub-Saharan countries, including South Africa, with a **dual economy**—one part rich and industrialized, the other below

---

commodities raw materials that are traded, usually to other countries, for processing or manufacturing into more valuable goods

commodity dependence economic dependence on exports of raw materials

dual economy an economy in which the population is divided by economic disparities into two groups, one prosperous and the other near or below the poverty level

294 CHAPTER 7 Sub-Saharan Africa

the poverty level and primarily reliant on low-wage labor and small-scale informal enterprises. By the 1940s, even though the labor of black South Africans was essential to the country's prosperity, 84 percent of black South Africans lived at a bare subsistence level. By 2009, fifty percent of black South Africans still lived below the poverty line (compared with just 7 percent of white South Africans), and only 22 percent of black South Africans had finished high school (compared with 70 percent of white South Africans).

The Era of Structural Adjustment By the 1980s, most African countries remained poor and dependent on their volatile and relatively low-value commodity exports. Attempts at investing in manufacturing industries failed (discussed below); and governments struggled to make payments on the large loans taken out for these projects. Infrastructure (roads, water, and utilities) was either not built or not maintained. A breaking point came in the early 1980s, when an economic crisis swept through the region and much of the rest of the developing world, leaving most countries unable to repay their debts at all.

In response, the IMF and the World Bank designed *structural adjustment programs* (SAPs; see page 129 in Chapter 3) to enforce repayment of the loans. SAPs did have some useful results. They tightened bookkeeping procedures and thereby curtailed corruption and waste in bureaucracies. They closed some corrupt state-owned industrial and service monopolies, opened some sectors of the economy to medium- and small-scale business entrepreneurs, and made tax collection more efficient. But overall, SAPs had many unintended consequences and failed at their primary objective—reducing debt (Figure 7.16).

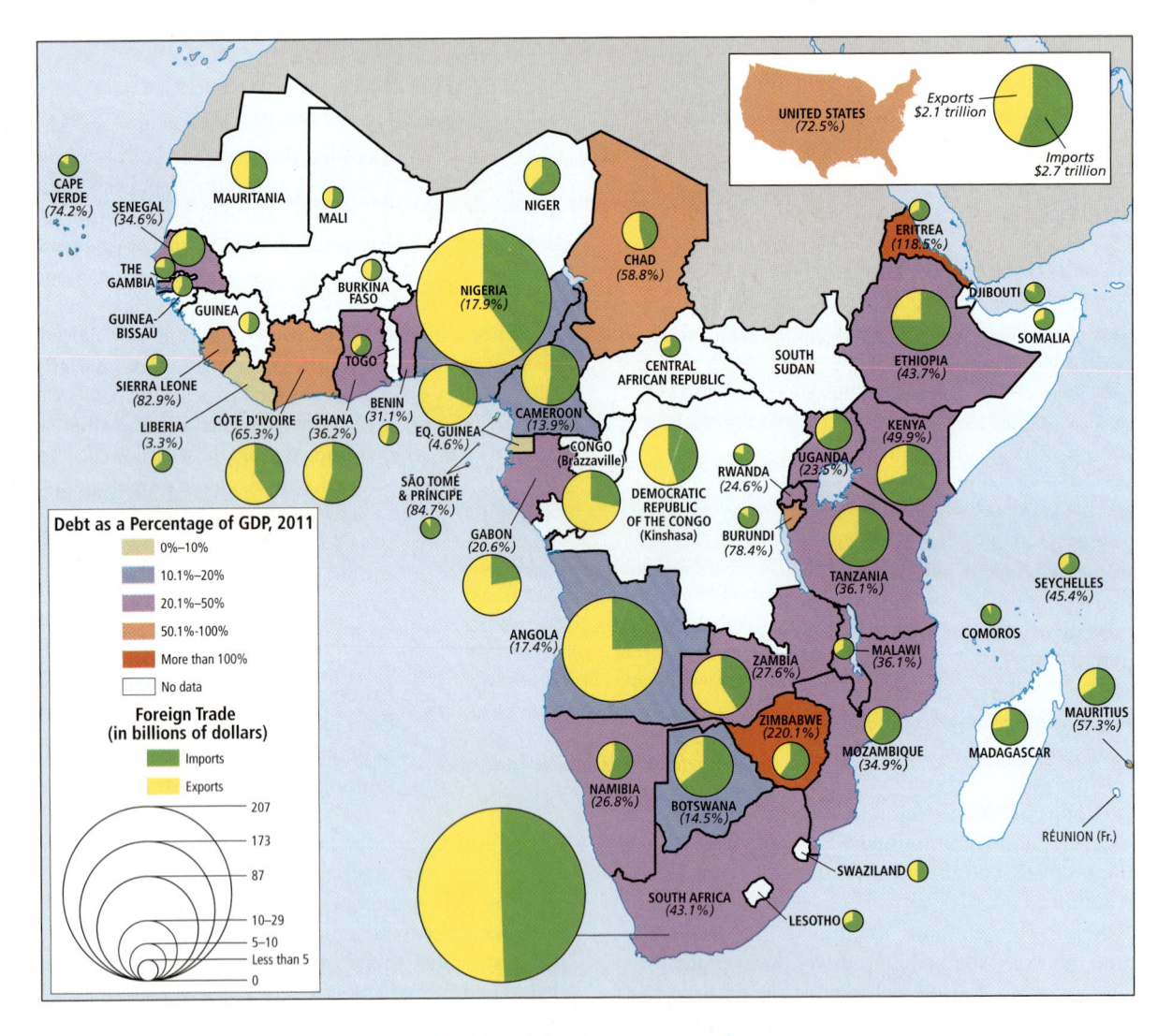

FIGURE 7.16 Economic issues: Public debt, imports, and exports. Public debt is increasing across Africa, partly as a result of borrowing to fund development projects; but the U.S. public debt (see inset) far exceeds that of most sub-Saharan countries. In all but ten of the countries in sub-Saharan Africa, imports exceed exports. These ten are Angola, Chad, Democratic Republic of the Congo (indicated in the text as Congo [Kinshasa]), Republic of Congo (indicated in the text as Congo [Brazzaville]), Equatorial Guinea, Gabon, Nigeria, South Africa, Zambia, and Côte d'Ivoire. Minerals (copper and cobalt) are the main export earners for Zambia; cocoa, coffee, and timber for Côte d'Ivoire; and the export earnings of the other eight are from oil and petroleum products. [Sources consulted: *World Fact Book 2012: Africa: South Africa*, Central Intelligence Agency, at https://www.cia.gov/library/publications/the-world-factbook/geos/sf.html]

To facilitate loan repayment, SAPs required governments to sell off inefficient government-owned enterprises, often at bargain-basement prices. In addition, government jobs in social services, education, health, and agricultural programs were slashed so that tax revenues could be devoted to loan repayment. If countries refused to implement SAP requirements, the international banks cut off any future lending for economic development.

As unemployment rose, so did political instability. The deteriorating infrastructure reduced the quality of the remaining social services, transportation, and financial services, which scared away potential investors. SAPs also reduced food security because agricultural resources were shifted toward the production of cash crops for export. Between 1961 and 2009, per capita food production in sub-Saharan Africa actually decreased by 14 percent, making it the only region on Earth where people were not eating as well as they had in the past (see Figure 1.18 on page 30).

Informal Economies to the Rescue Ultimately, Africa's informal economies provided relief from the hardships created by SAPs. Informal economies in Africa are ancient and varied, providing employment and useful services and products. People grow and sell garden produce, prepare food, vend a wide array of products on the street, sell time cards (or credits) for mobile phones, make craft items and utensils, or do child and elder care. Others in the informal economy, however, earn a living at sex work, distilling liquor, or smuggling scarce or illegal items, including drugs, weapons, endangered animals, bushmeat, and ivory. Because most of these activities take place "under the radar," informal jobs may involve criminal acts, wildly unsafe activities, and hazardous substances.

In most African cities, the role of the informal economy has grown from one-third of all employment to more than two-thirds. Such jobs are a godsend to the poor, but they create problems for governments because the informal economy typically goes untaxed, so less money is available to pay for government services or to repay debts. Moreover, as the informal sector grows, profits have declined as more people compete to sell goods and services to those with little disposable income. And although women typically dominate informal economies, when large numbers of men lose their jobs in factories or the civil service, they may crowd into the streets and bazaars as vendors, displacing the women and young people.

In 2000, in response to the now widely recognized failures of SAPs, and the overemphasis on the power of markets to guide development, the IMF and the World Bank replaced SAPs with *Poverty Reduction Strategy Programs*, or *PRSPs*, and with *Sustainable Structural Transformation (SST)* strategies. These policies are similar to SAPs in that they push market-based solutions intended to reduce the role of government in the economy, but they differ in several ways. They focus on reducing poverty and diversifying economies into manufacturing, rather than on just "development" and debt repayment per se, and are generally promoting more democratic reforms. They also include the possibility that a country may have all or most of its debt "forgiven" (paid off by the IMF, the World Bank, or the African Development Bank) if the country follows the PRSP rules. Forty sub-Saharan countries had qualified for and been approved to receive debt relief as of July 2010.

The Current Era of Diverse Globalization The current wave of globalization promises a different role for Africa, and is resulting in new sources of investment that bring jobs and strengthen infrastructure. As discussed at the beginning of the chapter, one significant trend is that young African professionals at home and abroad are supporting development in a multitude of ways. Another trend is foreign direct investment (FDI) at the corporate level. While Europe and the United States are still the largest sources of investment in sub-Saharan Africa, Asia's influence on African economies is increasing significantly, but there are potential problems involved in Asia's role.

In recent years, China and, to a lesser extent, India have begun to view sub-Saharan Africa as a new frontier for their large and growing economies. These two countries now exert a powerful influence on the region through their demand for Africa's export commodities, direct investment in agribusiness, mining and industry, and the sale of their manufactured goods. It is not clear whether Asian investments will ultimately prove beneficial or hurtful to sub-Saharan African economies.

Together, China and India consume about 15 percent of Africa's exports, but their share is growing twice as fast as is that of any of Africa's other trading partners. Improved infrastructure (roads, ports, utilities, and technology) could ultimately facilitate intra-African trade, linking countries that have never before been able to trade. China's and India's investment in African agriculture, undertaken to meet rising demands for food in China and India, could result in more efficient production overall through *technology transfer*, and thus create higher earnings for Africa and more food supplies for African internal urban markets. Overall, African food security could increase.

It is important, however, to keep an eye on the extent to which Asian investments in Africa actually benefit Africans rather than simply replicate the injustices of colonialism. For example, Chinese investments in mining and infrastructure development could have boosted local African economies by providing construction jobs for African laborers and professional design and management experience for mid-level educated Africans, but this benefit never materialized because African governments agreed to China's demand that all work be done by Chinese companies using Chinese workers (**Figure 7.17**). Most controversial has been the willingness of Chinese companies to deal with brutal and corrupt local leaders, such as Liberia's now-convicted and imprisoned Charles Taylor (diamonds and timber extraction) and Zimbabwe's Robert Mugabe (mining), who bartered away their countries' resources at bargain prices and used the profits to enrich themselves and to wage war against their own citizens.

The expansion of communication technology within this region would not have been possible without numerous innovations supplied by Africans. For example, in order to make service available to the very poor, prepayment credits are sold in very small units. In fact, many people now make a living selling these small credits. Other Africans have created ways to charge mobile phones in remote villages where there is no electricity, as the Malawian William Kamkwamba figured out how to do (for more on William Kamkwamba's story, see the vignette on page 298).

Foreign private investment is also transforming the communications environment in ways that encourage regional economic development in Africa. The India-based cell phone company Airtel operates a borderless network that now covers 18 countries: Burkina Faso, Chad, Congo (Kinshasa), Congo (Brazzaville), Gabon, Ghana, Kenya, Madagascar, Malawi, Niger, Nigeria, Rwanda, Seychelles, Sierra Leone, Tanzania, South Sudan, Uganda, and Zambia. Airtel's customers can make calls across the network at local rates without incurring surcharges, and recently, the company added Internet access, SMS, international roaming, and portal applications to the local rate. The borderless aspect of Airtel's network, and others like it, is helping increase trade and interaction between African countries, which have long been goals of African regional economic development efforts.

160. SOMALILAND EXPATRIATES RETURN HOME TO HELP NATIVE LAND DEVELOP

161. AFRICAN UNION APPEALS TO DIASPORA TO AID HOMELANDS

Regional and Local Economic Development

As they create alternatives to past development strategies, many African governments have been focusing on regional economic integration similar to that of the European Union. Hoping to utilize existing talent and consumer demand for African-made products, local agencies and public and private donors are pursuing grassroots development designed to foster very basic innovation at the local level that can then be marketed across the region.

The Potential of Regional Integration According to the World Trade Organization (WTO), less than 20 percent of the total trade of sub-Saharan Africa in 2011 was conducted between African countries. This is true partly because so many countries produce the same raw materials for export. And everywhere except South Africa, industrial capacity is so low that the raw materials cannot be absorbed within the continent, so African countries compete with each other and with all other global producers to sell to the main customers, which are currently in Europe, North America, and industrialized Asia. This failure to trade with each other can be traced to divisive colonial policies, lack of transportation and communication grids, and arcane bureaucratic regulations.

African Regional Integration Over the last several decades, regional trading blocs have been formed to encourage neighboring countries to trade with each other and cooperate in the production of manufactured (value-added) exports. The trade blocs are somewhat fluid; they form and then reorganize, as is depicted in Figure 7.19, a list and simplified map of the evolving African Regional Economic Communities (RECS). By combining the markets of several countries (as do NAFTA and the EU), regional trade blocs can create a market size sufficient to foster industrialization and entrepreneurialism. Africa's many different regional trade blocs share several goals: reducing tariffs between members, forming common currencies, reestablishing peace in war-torn areas, cooperating to upgrade transportation and communication infrastructure, and building regional industrial capacity. Building a full-scale, continent-wide economic union along the lines of the European Union is a long-term goal. According to studies by the WTO, if sub-Saharan countries could increase trade with each

---

ON THE BRIGHT SIDE

Africans Investing in Africa

The role of African professionals in globalization is growing. As was noted in the opening vignette, perhaps the most dramatic sign of change for African economies is that educated Africans who are skilled in IT and other high-tech fields are staying home and starting businesses instead of migrating to richer regions. Their incomes, along with *remittances* (money sent to family members from Africans working and living abroad, primarily in Europe and North America), allow them and those they help support to buy mobile phones, start small businesses, fund education for children, build houses, and help the needy. Remittances are a more stable source of investment than *foreign direct investment*, in that they tend to come regularly from committed donors who will continue their support for years. They are also much more likely to reach poorer communities. But payments will stop if remitters lose their jobs in America or Europe.

Young, educated adults are profoundly changing the trajectory of Africa's development. They have been largely responsible for the rapid spread of mobile phones and have developed a wide array of new and important uses of cell phone technology. Figure 7.18 is a map of mobile phone subscriptions by country as of 2012. Between 1998 and 2010, mobile phone subscribers in sub-Saharan Africa increased by more than 100 times, from about 4 million to about 450 million. Google estimates that by 2016 there will be a billion mobile phones in sub-Saharan Africa, or one for every person.

FIGURE 7.17 China in Angola. The man on the right is one of the estimated 20,000 Chinese workers in Angola. China has given loans and aid to Angola in excess of U.S.$4 billion since 2004, and in return, China has been guaranteed a large portion of Angola's future oil production. In addition, 70 percent of Angola's development projects, such as building and upgrading railways, have been given to Chinese companies, most of which import workers from China.

> **THINKING GEOGRAPHICALLY** Why might China be particularly interested in investments in Africa's transportation infrastructure?

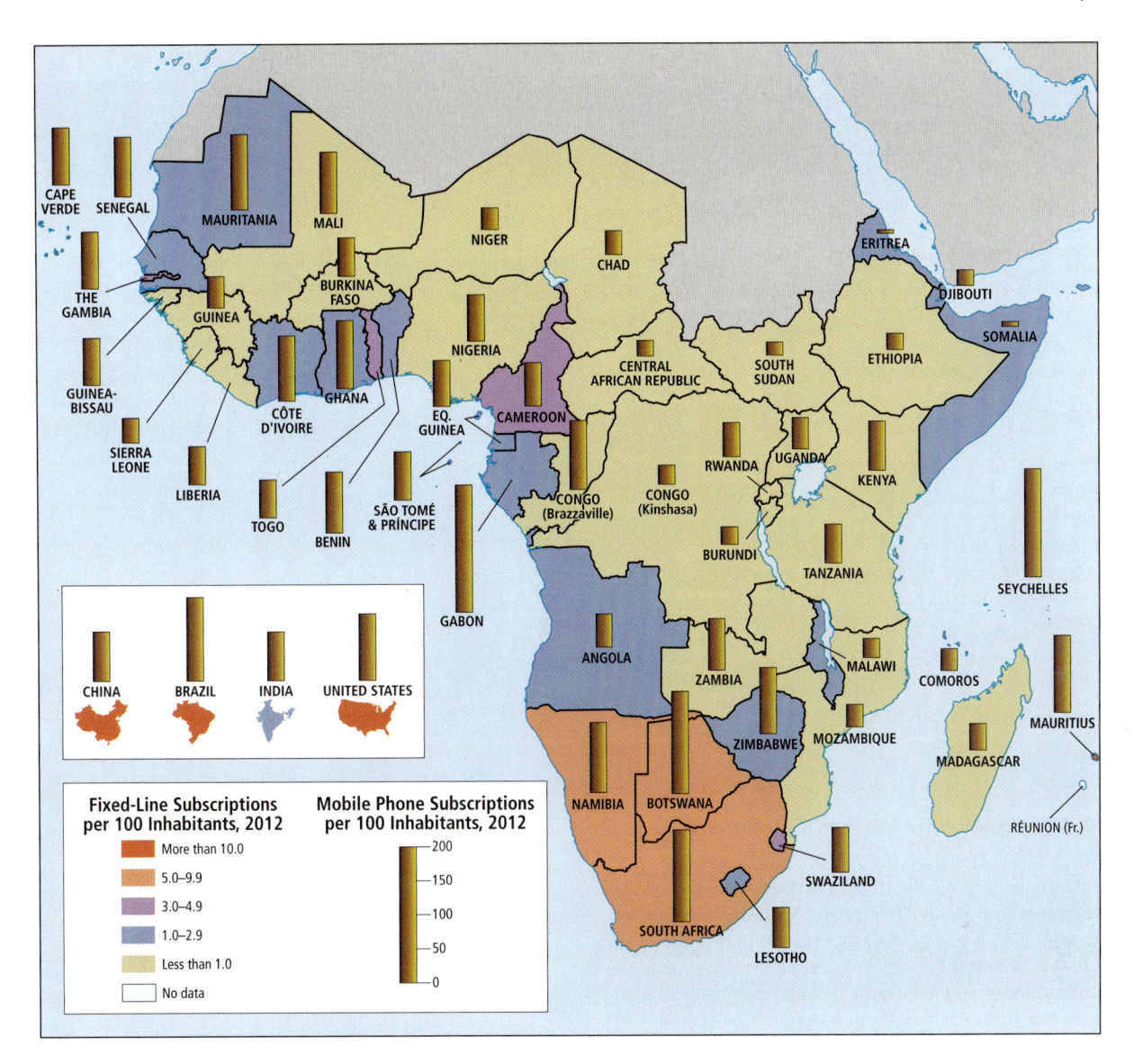

FIGURE 7.18 Mobile phone users in sub-Saharan Africa. In the past decade, mobile phone use in sub-Saharan Africa has boomed. It is estimated that by 2016, there will be more than a billion mobile phones in the region, including in the hands of even very poor citizens, because of innovations that make it possible to buy service access in tiny amounts. One of the most popular uses of phones in Africa is to move money from one account to another.

other by just 1 percent, the region would show a total added income of $200 billion per year—a substantial internal contribution toward the alleviation of poverty.

The goals of regional and local economic development are being boosted by the creation of *value chains*. These link parts of the production chain of a final product in order to maximize the regional or local economic impact. For example, farmers in Endau, Kenya, were able to increase their profits five-fold by switching from growing corn to producing sorghum after a local nongovernmental organization (NGO) negotiated a deal with a brewery in Nairobi to use the sorghum in their beer. Some farmers have used their boosted income to increase their profits further by buying milling equipment and grinding the sorghum of other local farmers into sorghum flour, which is more valuable than unprocessed sorghum.

Countries and regions with strong value chains are more able to keep the profits their industries create from going elsewhere. Such value chains are one of the fastest-growing aspects of economic development worldwide.

Local Development An increasingly common strategy for improving living standards in this region is **grassroots economic development**. Projects using this strategy are designed to provide sustainable livelihoods in rural and urban areas, and

> **grassroots economic development**
> economic development projects designed to provide sustainable livelihoods in rural and urban areas; these often use simple technology that requires minimal or no investment in imported materials

298 CHAPTER 7 Sub-Saharan Africa

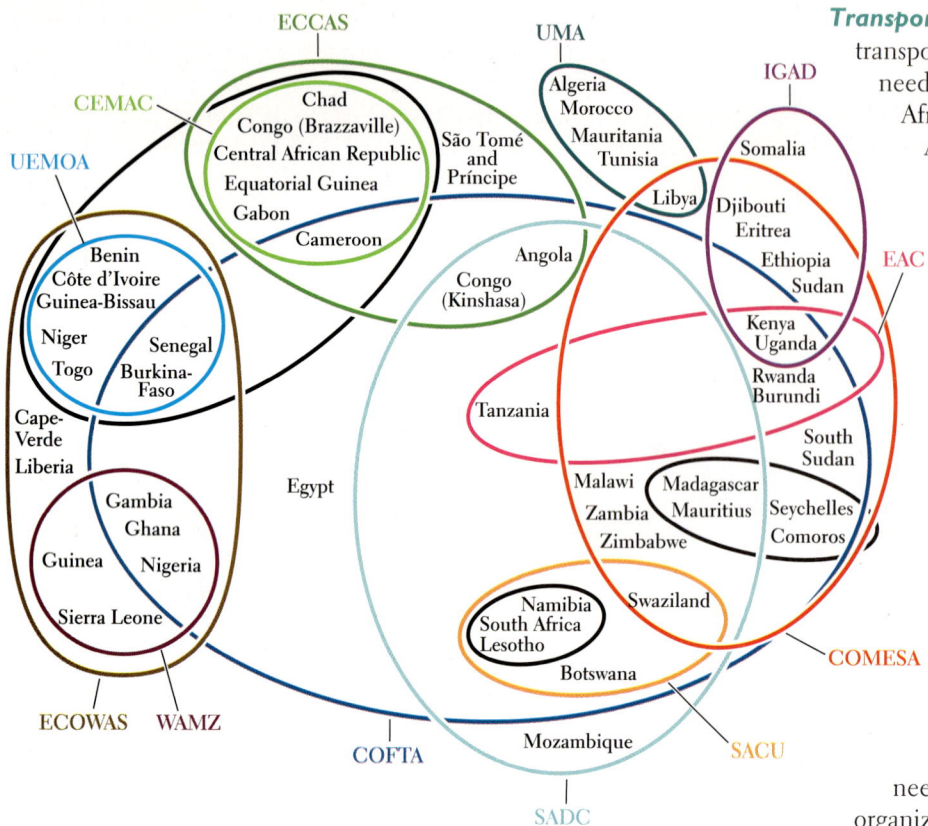

FIGURE 7.19 Principal trade organizations in sub-Saharan Africa. There is a lot of overlap between the countries and the regional trade organizations, but often one country is dominant in each group. South Africa and Nigeria, for example, are dominant in three regional trade organizations; but in fact, South Africa uses its many well-established relationships to trade with nearly all countries in the region. Because political stability is essential to trade, some of the strongest trade organizations, such as the Economic Community of West African States (ECOWAS), have become involved in the amelioration of conflicts. [Source consulted: UN Conference on Trade and Development, *Economic Development in Africa Report 2009: Strengthening Regional Economic Integration for Africa's Development* (New York: United Nations, 2009), p. 12, Figure 1, at http://www.unctad.org/en/docs/aldcafrica2009_en.pdf]

often use simple technology that requires minimal or no investment in imported materials. One such approach is **self-reliant development**, which consists of small-scale self-help projects that use local skills to create products or services for local consumption. One of the most important aspects of this type of development is that control of the projects remains in local hands, so that participants retain a sense of ownership and commitment in difficult economic times. One district in Kenya has more than 500 such self-reliant groups. Most members are women who build water tanks, plant trees, and terrace land so that it can be farmed for food. They also build schools and form credit societies.

self-reliant development small-scale development in rural areas that is focused on developing local skills, creating local jobs, producing products or services for local consumption, and maintaining local control so that participants retain a sense of ownership over the process

Transportation Needs The issue of improving rural transportation illustrates how a focus on local African needs can generate unique solutions. When non-Africans learn that transportation facilities in Africa are in need of development, they usually imagine building and repairing roads for cars and trucks. But a recent study that analyzed village transportation on a local level found that 87 percent of the goods moved are carried via narrow footpaths on the heads of women. Women "head up" (their term) firewood from the forests, crops from the fields, and water from wells. An average adult woman spends about 1.5 hours each day moving the equivalent of 44 pounds (20 kilograms) more than 1.25 miles (2 kilometers).

Unfortunately, the often-treacherous footpaths trodden by Africa's load-bearing women have been virtually ignored by African governments and international development agencies, which tend to focus solely on roads for motorized vehicles (which are also badly needed). Grassroots-oriented nongovernmental organizations are now making far less expensive but equally necessary improvements to Africa's footpaths. Some women have been provided with bicycles, donkeys, and even motorcycles that can travel on the footpaths. This saves time and energy for women, who can direct more of their efforts to becoming educated and generating income.

Energy Needs Africa exports oil but its own energy needs, which are currently unmet even at the most basic level of home electricity, can be addressed by local solutions, as the vignette illustrates.

VIGNETTE

In Malawi, 14-year-old William Kamkwamba was forced to drop out of school when a famine struck his country in 2001 and his family could no longer afford the $80 school fee. Depressed at the prospect of having no future, he went to a local library when he was able. There he found a book in English called *Using Energy* that described an electricity-generating windmill. With an old bicycle frame, tractor fan blades, PVC pipes, and scraps of wood, he built a windmill that generated enough power (stored in a car battery) to light his home, run a radio, and charge neighborhood cell phones. More elaborate energy projects followed.

Now known as "the boy who harnessed the wind," William appeared on Jon Stewart's *Daily Show* in the United States in 2009 to explain how he plans to start his own windmill company and other ventures that will bring power to remote places across Africa. He graduated from the first pan-African prep school in South Africa and now attends Dartmouth College in the United States. William's web page is http://www.williamkamkwamba.com. ■

THINGS TO REMEMBER

> **GEOGRAPHIC INSIGHT 2**

- **Globalization and Development** Most sub-Saharan African economies are dependent on the export of raw materials. This pattern, a legacy of the era of European colonialism, results in economic instability because prices for raw materials can vary widely from year to year. While a few countries are diversifying and industrializing, most still sell raw materials and all rely on expensive imports of food and manufactured goods.

- Prospective investors in sub-Saharan Africa have been discouraged by problems that structural adjustment programs (SAPs) either ignored or worsened.

- Recently, Africa has become a new frontier for Asia's large and growing economies, especially those of China and India, which both buy and sell in sub-Saharan Africa.

- Africa's informal economies have provided some relief from the hardships created by SAPs. These economies are ancient, varied, and agile; they provide employment and useful services and products.

- A major and enduring source of investment funds in Africa comes from members of the African diaspora, who send regular remittances to friends and families.

- Many African governments are focusing on regional economic integration along the lines of the European Union. Grassroots economic development is also being pursued.

POWER AND POLITICS

> **GEOGRAPHIC INSIGHT 3**
>
> **Power and Politics:** Governments in sub-Saharan Africa tend to be authoritarian, though there is a general shift in the region toward the expansion of political freedoms. Free and fair elections have brought about dramatic changes in some countries, while elections tainted by widespread suspicions of fraud have often been followed by surges of violence in other countries.

After years of being ruled by corrupt elites and the military, Africa is now shifting toward systems with greater political freedoms. Yet change is often blocked by conflict, and even when reforms are enacted and free elections established, violence often accompanies these elections.

Ethnic Rivalry Africa has suffered from frequent civil wars that are in many ways the legacy of colonial era policies of **divide and rule**. Divisions and conflicts between ethnic or religious groups were deliberately aggravated by European colonial powers. To make it difficult for Africans to unite against colonial rule, the borders and administrative units of the African colonies were designed so that different and sometimes hostile groups were put together under the same jurisdiction (Figure 7.20 shows the situation in Nigeria).

divide and rule the deliberate intensification of divisions and conflicts by potential rulers; in the case of sub-Saharan Africa, by European colonial powers

After independence, when Africans took over, governance was complicated because African officials inevitably belonged to one local ethnic group or another and so were not seen as impartial in their attempts to resolve conflicts. Moreover, during the colonial era, older indigenous traditions for ensuring ethical behavior, punishing greed by leaders, and resolving ethnic

CASE STUDY
Conflict in Nigeria

Nigeria was and remains in part a creation of British divide-and-rule imperialism. Many disparate groups—speaking 395 indigenous languages—have been joined into one unusually diverse country.

The British reinforced a north-south dichotomy that mirrored the physical north (dry)–south (wet) patterns. Among the Hausa and Fulani ethnic groups in the north, the British ruled via local Muslim leaders who did not encourage public education. In the south, the Yoruba and Igbo ethnic groups, who were primarily a combination of animist and Christian, were ruled more directly by the British, who encouraged attendance at Christian missionary schools that were open to the public. At independence, the south had more than 10 times as many primary and secondary school students as the north. It was more prosperous, and southerners also held most government civil service positions. Yet the northern Hausa dominated the top political posts, in part because of their long association with British colonial administrators. Over the years, bitter and often violent disputes erupted between the southern Yoruba–Igbo and northern Hausa–Fulani regarding the distribution of increasingly scarce clean water, development funds, jobs, and oil revenues, and the severe environmental damage done by oil extraction.

The politics of oil have complicated the troubles in Nigeria. Nigeria's oil is located on lands occupied by the Ogoni people, which lie in the south along the edges of the Niger River delta (see Figure 7.1D). The Ogoni are a group of about 300,000 who are distinct from other ethnicities in Nigeria. Virtually none of the profits from oil production and export and very little of the oil itself goes to Ogoniland. While receiving few benefits from oil extraction, Ogoniland has suffered gravely from the resulting pollution (see the photo in Figure 1.4 on page 8). Oil pipelines crisscross Ogoniland, and spills and blowouts are frequent; between 1985 and 2009 there were hundreds of spills, many larger than that of the Exxon Valdez disaster in Alaska. Natural gas, a by-product of oil drilling, is burned off, even though it could be used to generate electricity—something many Ogoni lack. Royal Dutch Shell, a multinational oil company that is very active in the Niger River delta, acknowledges that while historically it has netted $200 million in profit yearly from Nigeria, it has paid a total of just $2 million to the Ogoni community in 40 years.

Geographic strategies have been used to reduce tensions in Nigeria. One approach has been to create more political states (Nigeria now has 30) and thereby reallocate power to smaller local units with fewer ethnic and religious divisions. However, when large, wealthy states were subdivided, reputedly to spread oil profits more evenly, the actual effect of the subdivision was to mute the voice and power of the public.

178. NEW NIGERIAN PRESIDENT INHERITS TURBULENT NIGER DELTA

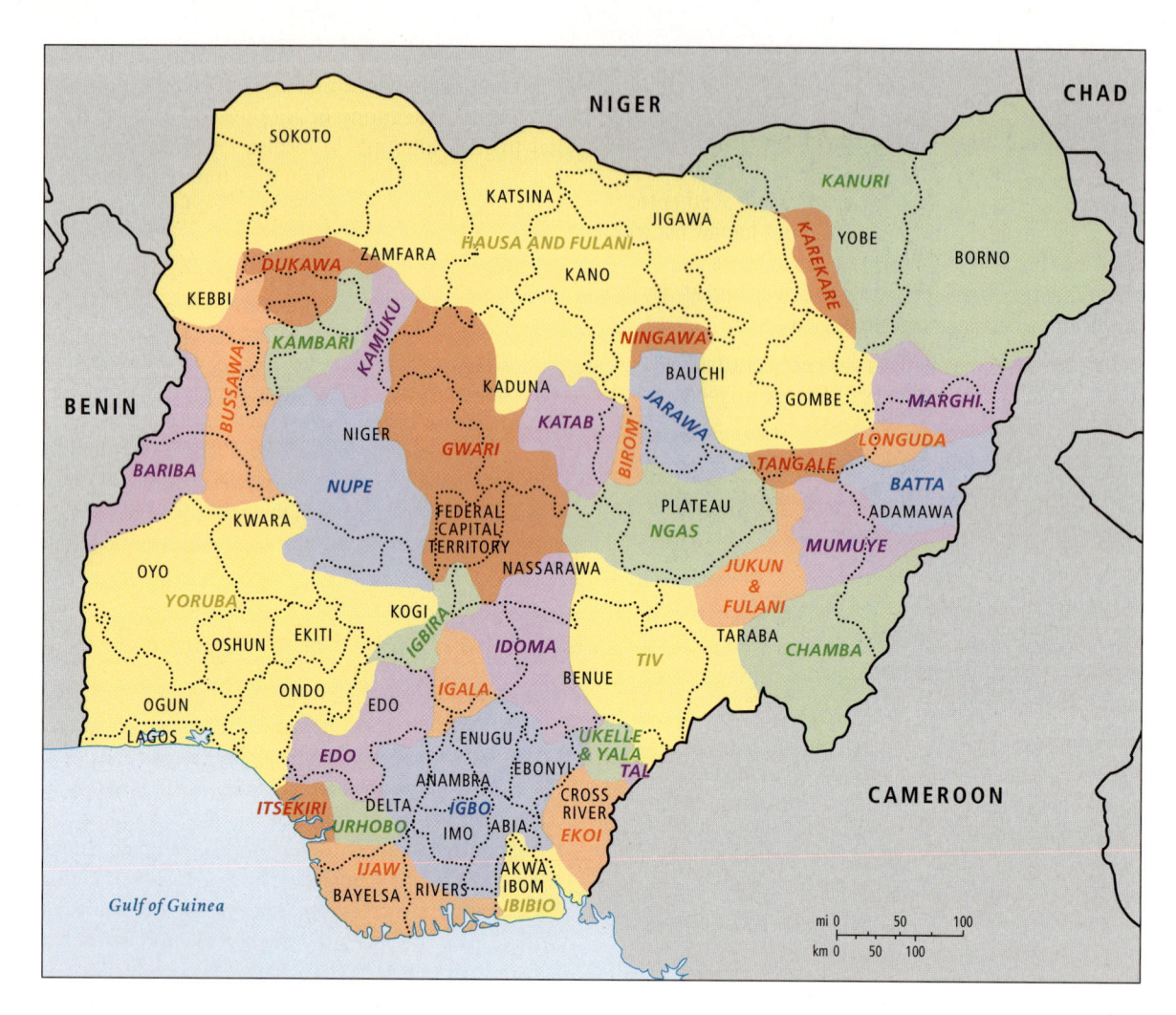

FIGURE 7.20 Nigeria's ethnic geography. There are thousands of ethnic groups distributed across sub-Saharan Africa. Nigeria alone has more than 400; but when Europeans set up political units, they either ignored ethnic boundaries or purposely divided large ethnic groups. [Source consulted: *Ethnic Map of Nigeria*, Online Nigeria: Community Portal of Nigeria, at www.onlinenigeria/com/mapethnic.asp#]

conflict had been weakened. The political institutions that replaced them were often not adept at conflict resolution, resulting in hostilities between ethnic groups that have at times devolved into civil wars (Figure 7.21C).

The Role of Geopolitics Cold War geopolitics between the United States and the former Soviet Union deepened and prolonged African conflicts that grew out of divide-and-rule policies. After independence, some sub-Saharan African governments turned to socialist models of economic development, often receiving economic and military aid from the Soviet Union. Other governments became allies of the United States, receiving equally generous aid (see Figure 5.13 on page 209). Both the United States and the USSR tried to undermine each other's African allies by arming and financing rebel groups.

In the 1970s and 1980s, southern sub-Saharan Africa became a major area of East–West tension. The United States aided South Africa's apartheid government in military interventions against Soviet-allied governments in Namibia, Angola, and Mozambique. Another area of Cold War tension was the Horn of Africa, where Ethiopia and Somalia fought intermittently throughout the 1960s, 1970s, and 1980s. At different times, the Soviets and Americans funded one side or the other. The failure by both sides in the Cold War to anchor the aid they dispensed to any requirements that it be used for sustainable development became a major stimulus for militarization and corruption. Once this source of cash was removed at the end of the Cold War, corrupt officials turned to selling off natural resources and commodities.

genocide the deliberate destruction of an ethnic, racial, religious, or political group

Conflict and the Problem of Refugees Conflicts create refugees and refugees are commonplace across sub-Saharan Africa. Refugees pour back and forth across borders throughout this region and are displaced within their own countries. Often, they are trying to escape **genocide** (the deliberate destruction of an ethnic, racial, religious, or political group). With only 11 percent of the world's

FIGURE 7.21 PHOTO ESSAY: Power and Politics in Sub-Saharan Africa

This region has long been plagued by political violence. A long-term shift toward democratization may help reduce bloodshed as governments become more responsive to their citizens. As in other regions, the most democratized countries have had fewer violent conflicts since 1945. Nevertheless, in any country, the transition to democracy can be tumultuous, and elections are often marked by violence.

A A man in Nairobi, Kenya, runs from police at a demonstration where he has been protesting demands by legislators for an increase in their salaries. That this demonstration could take place at all is a step forward, as is the presence of the media. However, the fact that participants in this nonviolent protest were beaten, tear-gassed, and arrested shows how limited the political freedoms are in Kenya.

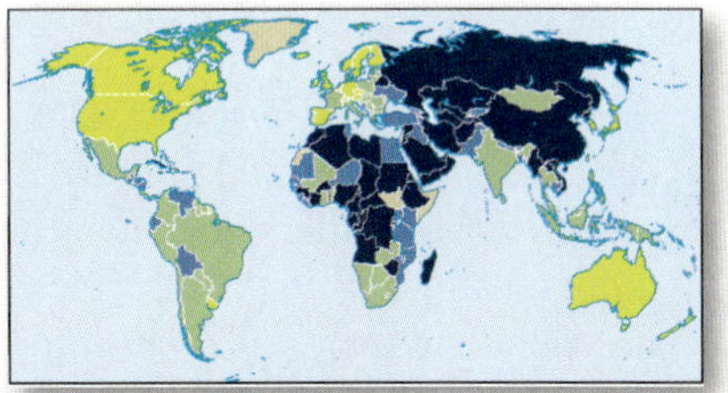

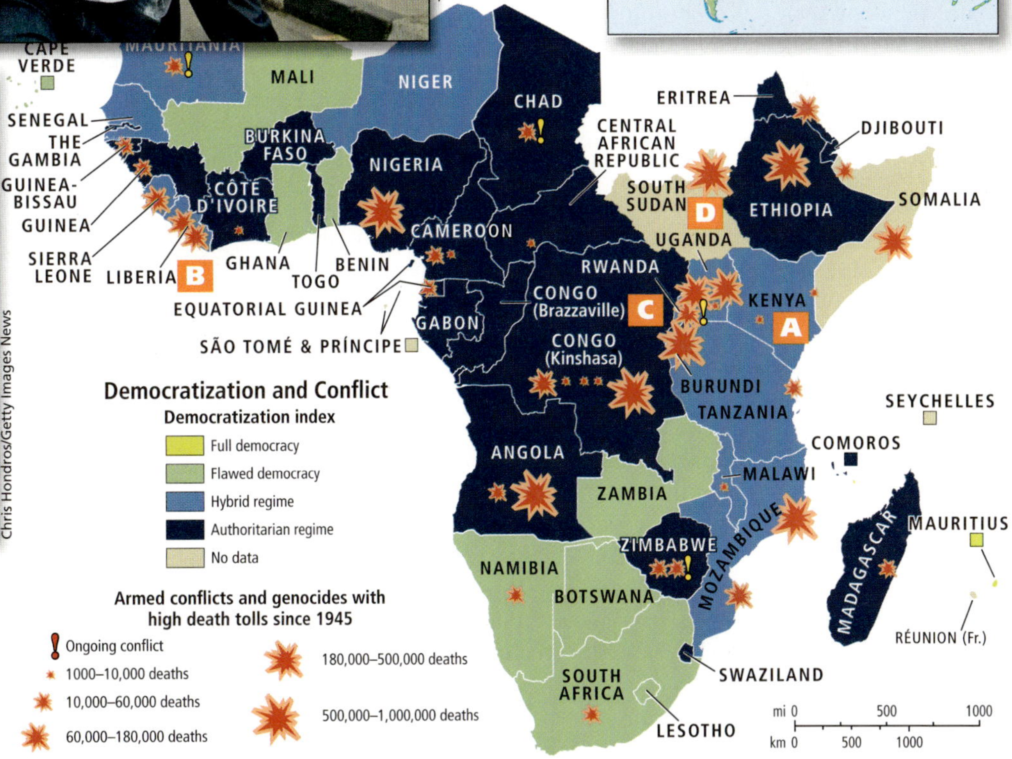

Democratization and Conflict

Democratization index
- Full democracy
- Flawed democracy
- Hybrid regime
- Authoritarian regime
- No data

Armed conflicts and genocides with high death tolls since 1945
- Ongoing conflict
- 1000–10,000 deaths
- 10,000–60,000 deaths
- 60,000–180,000 deaths
- 180,000–500,000 deaths
- 500,000–1,000,000 deaths

B Ellen Johnson-Sirleaf of Liberia is the first woman to be elected head of state in this region. Her election in 2005 was considered free and fair by international observers and helped Liberia begin to resolve tensions that had produced two devastating civil wars.

C Members of a rebel militia in eastern Congo that is battling the government for control of the area's gold, tin, and coltan (a mineral used in cell phones) resources.

D Refugees from Sudan relocate to South Sudan after a referendum in 2011 established the latter's independence from the former. South Sudan is plagued by political corruption and is involved in continuing hostilities with Sudan; both of these problems have constrained the export of its main resource, oil.

population, this region contains about 19 percent of the world's refugees, and if people displaced within their home countries are also counted, the region has about 28 percent of the world's refugee population (see Figure 7.21D). Women and children constitute 75 percent of Africa's refugees because many adult men who would be refugees are either combatants, jailed, or dead.

As difficult as life is for these refugees, they also place a severe burden on the areas that host them. Even with help from international agencies, the host areas find their own development plans deferred by the arrival of so many distressed people who must be fed, sheltered, and given health care. Large portions of economic aid to Africa have been diverted to deal with the emergency needs of refugees.

Successes and Failures of Democratization In sub-Saharan Africa, efforts to expand political freedoms have produced mixed results. The number of elections held in the region has increased dramatically. In 1970, only 11 states had held elections since independence. By 2006, twenty-five out of the then 44 sub-Saharan African states had held open, multiparty, secret ballot elections, with universal suffrage. By 2014, every country had held some form of election, though not all could be considered "free and fair" (see the Figure 7.21 map).

While the implementation of democratic elections has increased (see Figure 7.21B), the growth of other political freedoms—such as freedom of speech, the freedom to assemble in public, and the ability to participate in policy formation at the local and national levels—has been more irregular and uneven. Public frustration with suspicious election outcomes, authoritarian policies, and violent repression of peaceful protests (see Figure 7.21A) has often led to rebellion (see Figure 7.21C).

At times, flawed elections have brought about massive violence. In 2008, after several peaceful election cycles, Kenya had an election so corrupt that the country erupted in deadly protest riots. More than 1000 people died and 600,000 were displaced by mobs of enraged voters. By 2010, a new Kenyan constitution gave some hope that political freedoms would be better protected. In 2008, local elections in Nigeria sparked similar violence, and in the Congo (Kinshasa) in 2006, the first elections held there in 46 years resulted in violence that left more than 1 million Congolese people refugees within their own country. In the 2012 parliamentary elections, the will of the voting public became more evident when the ruling party lost more than 40 percent of its legislative seats to opposition parties.

Zimbabwe may represent the worst case of the failed attempts at democratization. In the 1970s, Robert Mugabe became a hero to many for his successful guerilla campaign in what was then called Rhodesia (now Zimbabwe). The defeated white minority government had been allied with apartheid South Africa but was not formally recognized by any other country because of its extreme racist policies. Mugabe was elected president in 1980, following relatively free multiparty elections. Over the years, however, his authoritarian policies have impoverished and alienated more and more Zimbabweans while increasing his personal wealth. In the 1990s, he implemented a highly controversial land-redistribution program that resulted in his supporters gaining control of the country's best farmland, much of which had been in the hands of white Zimbabweans. This move decimated agricultural production and contributed to a chronic food shortage and a massive economic crisis. The resulting political violence created 3 to 4 million refugees who fled mostly to neighboring South Africa and Botswana. Mugabe held on to his office through the rigged elections of 2002 and 2008, but was then forced to share power in 2009 with opposition leader Morgan Tsvangirai, who became prime minister. After some of Tsvangirai's policies were implemented, Zimbabwe began to have some modest economic growth. In 2012, though, Mugabe called for early elections and appeared to be financing his campaign with money from state-owned diamond mines. The elections, which occurred in 2013 and returned Mugabe to power, were widely condemned as fraudulent. **159. ZIMBABWE'S ROBERT MUGABE—A PROFILE**

In other places, however (most recently in South Sudan and Rwanda and some years ago in South Africa), elections have helped end civil wars and resettle refugees as the possibility of becoming respected elected leaders induced former combatants to lay down their arms. In Sierra Leone and Liberia, public outrage against corrupt ruling elites resulted in elections that brought fortuitous changes of leadership. Over the long term, fair and regular elections can help enable people to have more input at the policy level, which forces governments to be more responsive to the needs of their citizens.

Gender, Power, and Politics

The education of African women and their economic and political empowerment are now recognized as essential to development. As girls' education levels rise, population growth rates fall; as the number of female political leaders increases, so does their influence on policies designed to reduce poverty.

One major characteristic of the expansion of political freedoms has been the increase in the number of women across Africa who are assuming positions of political influence. This chapter opened with the story of Juliana Rotich, a businesswoman in Kenya whose road to success was built on policy changes pushed by a number of female Kenyan political figures over the last decade. In Nigeria, Ngozi Okonjo-Iweala, the first female finance minister, now fights for the empowerment of women entrepreneurs as a director at the World Bank. In Liberia—the country devastated by the logging fraud, child-soldier recruitment, and general brutality of the dictator Charles Taylor (see the vignette on Silas Siakor on page 282)—President Ellen Johnson-Sirleaf (also a former World Bank economist) began to work on democratic and environmental reforms immediately after she took office (see Figure 7.21B). In Rwanda, racked by genocide and mass rapes in the 1990s, women now make up 51 percent of the national legislature, the highest percentage in the world. And Rwandan women are also leaders at the local level, where they make up 40 percent of the mayors. In Mozambique, 39.2 percent of the parliament is female; in Burundi, 36 percent; and in South Africa, 42.7 percent. Altogether, there are 17 African countries where the percentage of women in national legislatures is well above the world average of 17.7 percent (the U.S. figure is 16.8 percent). **170. WOMEN HAVE STRONG VOICE IN RWANDAN PARLIAMENT**

There has been a sea change in attitudes toward women in politics. The policies that establish quotas for the number of female members of parliament are usually a reflection of a larger, post-conflict commitment to the empowerment of women and to involving women in all aspects of political and civil society. Sometimes the quotas are written into a country's constitution, but it is important to note that many female leaders in Africa, including Ellen Johnson-Sirleaf and at least half of Rwanda's female parliamentarians, were elected without the aid of quotas.

THINGS TO REMEMBER

> **GEOGRAPHIC INSIGHT 3**
>
> • **Power and Politics** Governments in sub-Saharan Africa tend to be authoritarian, though there is a general shift in the region toward the expansion of political freedoms. Free and fair elections have brought about dramatic changes in some countries, while elections tainted by widespread suspicions of fraud have often been followed by surges of violence in other countries.

• Africa has suffered from frequent civil wars that are in many ways the legacy of colonial era policies of divide and rule.

• With only 11 percent of the world's population, this region contains about 19 percent of the world's refugees, and if people displaced within their home countries are also counted, the region has about 28 percent of the world's refugee population.

• Over the long term, fair and regular elections can help enable people to have more input at the policy level, which forces governments to be more responsive to the needs of their citizens.

• One major characteristic of the expansion of political freedoms has been the increase in the number of women across Africa who are coming into positions of political power. ■

URBANIZATION

> **GEOGRAPHIC INSIGHT 4**
>
> **Urbanization:** Sub-Saharan Africa is undergoing a massive wave of rural-to-urban migration; it has the fastest rate of urbanization in the world. Much of this growth is unplanned and 70 percent of the urban population now lives in impoverished slums characterized by inadequate sanitation; poor access to clean water or food; and crowded, fire-prone housing.

In the 1960s, only 15 percent of sub-Saharan Africans lived in cities; now about 37 percent do (see the Figure 7.22 map). Estimates are that by 2030, half of all Africans will be urban dwellers. In 1960, just one sub-Saharan African city—Johannesburg, South Africa—had more than 1 million people; in 2009, fifty-two did. The largest sub-Saharan African city is Lagos, Nigeria, where various estimates put the population at between 11 and 13 million; by 2020, Lagos is projected to have 20 million people. Much of this growth is taking place in primate cities (see Chapter 3, page 139). For example, Kampala, Uganda, with 1.8 million people, is almost 10 times the size of Uganda's next-largest city, Gulu. Migration from rural areas to cities is the biggest factor behind sub-Saharan cities' growth rate of 5 percent per year, the highest in the world.

Most of this urbanization is happening in an uncontrolled fashion that makes urban life extremely difficult. Because governments and private investors have paid little attention to the need for affordable housing, most migrants have to construct their own dwellings using found materials. The vast, unplanned, one-story slums surrounding older urban centers house 72 percent of Africa's urban population. Transportation in these huge and shapeless settlements is a jumble of government buses and private vehicles, and boats in waterfront locales (see Figure 7.22A). Parents often have to travel long hours through extremely congested traffic to reach distant jobs, getting much of their sleep while sitting on a crowded bus and leaving their children unsupervised and susceptible to the influence of gangs (see Figure 7.22B).

This rapid and uncontrolled urbanization has major implications for the region, economically, politically, demographically, and in relation to gender. Economically, further development of manufacturing industries and the service sector will be needed to provide jobs to the growing urban population. Politically, governments will have to become more responsive to the needs of urban slum dwellers who are already a large and volatile political force that will become more powerful as their numbers grow.

In terms of population, urbanization is slowing growth. This is because urban life strengthens all the factors that influence the demographic transition—increased education, better health care, and more opportunities for women to work outside the home. Hence, while rural sub-Saharan African women give birth to about six children, urban African women give birth to about four. While this is still almost double the world average, the demographic transition has only just begun; urban sub-Saharan Africa's birth rates will continue to decline.

Part of the reason that urban fertility rates are still as high as they are is that compared to cities in other regions, sub-Saharan Africa's cities are still plagued by health risks that compel families to have large numbers of children so that a few may survive into adulthood (see Figure 7.22A, C). Although improved over rural conditions, poverty and disease are still relatively widespread in cities and clean water and sanitation are generally lacking. Urban families also tend to have many children because of low levels of education and economic opportunities for women and limited access to birth control (when compared to global levels).

Rural-and-Urban Migration The migration of sub-Saharan people looking for work in North Africa, Europe, Turkey, or the United States is now a familiar pattern, but the vast majority of African migrants are moving between rural and urban areas within Africa. People generally move to the cities for jobs and educational opportunities. However, the volatility of many African commodity-based economies, combined with the low living standards found in many slums, means that there is a great deal of what geographer Deborah Potts describes as *circular migration*, or back-and-forth migration between rural and urban areas. When urban economies stagnate and jobs become scarce, the higher cost of living in cities combines with the hazards of slum life to convince many migrants to move back to the countryside at least temporarily. This constant movement makes it hard to estimate urban or rural populations, and it is

FIGURE 7.22   PHOTO ESSAY: Urbanization in Sub-Saharan Africa

Urban populations are exploding because of migration from rural areas and the relatively high birth rates in cities. By 2030, half of the people in the region will live in cities, most in slums that are plagued by violence and inadequate access to water, sanitation, and education. Largely ignored by most governments, people in slums survive by helping themselves. [Sources consulted: *2011 World Population Data Sheet*, Population Reference Bureau, at http://www.prb.org/pdf11/2011population-data-sheet_eng.pdf; *World Gazetteer*, at http://world-gazetteer.com/wg.php?x=&men=gcis&lng=en&des=wg&srt=npan&col=abcdefghinoq&msz=1500&pt=a&va=&srt=pnan]

A A part of Makoko slum in Lagos, Nigeria, the largest city in this region and one of the fastest growing in the world. Makoko is home to more than 100,000 people. Waterborne diseases such as malaria are common, as is flooding, which kills and displaces residents on a regular basis. Because of these hazards and the fact that the community sits on potentially valuable waterfront property, the municipal government of Lagos periodically evicts some of Makoko's residents and demolishes their homes.

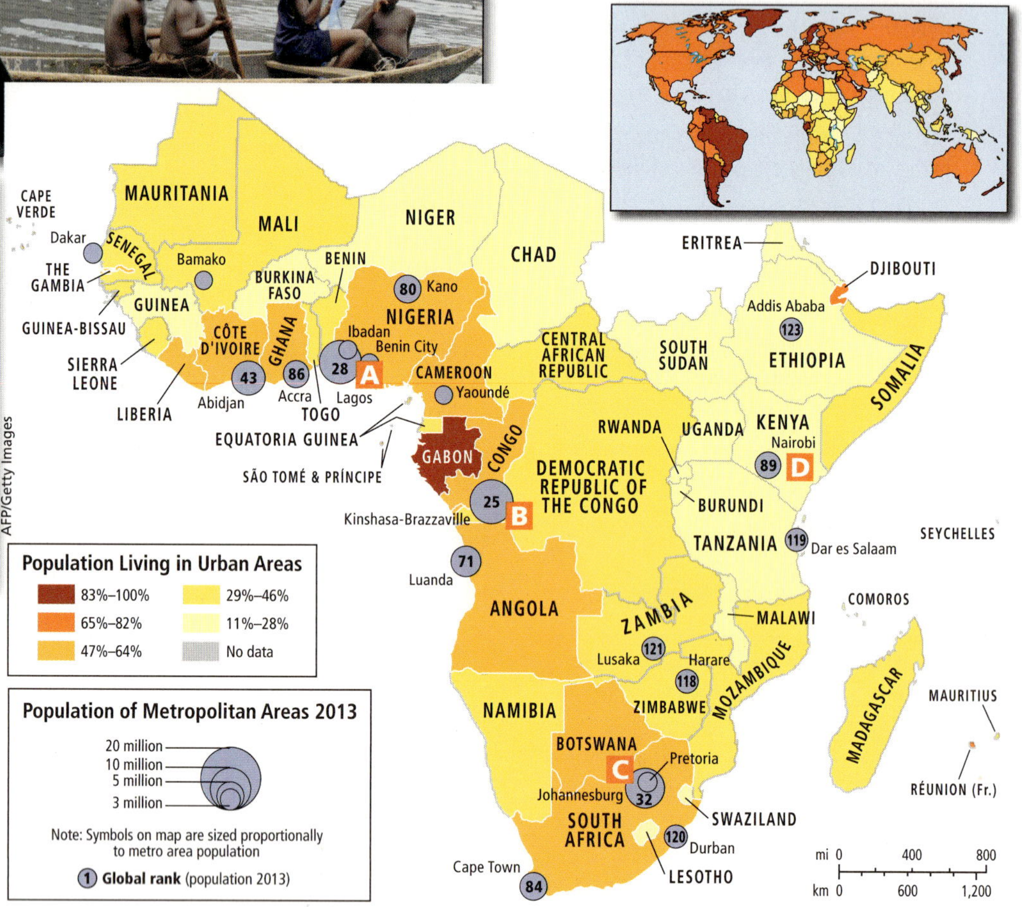

B Opposing gangs battle for control of territory in Kinshasa, Congo. Gangs control most slum neighborhoods in this region. Violence is widespread and includes organized attacks and robberies of passersby, the police, and other gangs.

C A girl drinks from a public water tap in Soweto, South Africa. Slum houses rarely have indoor plumbing, and people obtain water from public spigots. Latrines provide the only sanitation.

D A "sack garden," one of many innovative self-help projects found in Kibera slum in Nairobi, Kenya. High food costs and high unemployment rates make gardening attractive to many urban Africans.

Population Living in Urban Areas
- 83%–100%
- 65%–82%
- 47%–64%
- 29%–46%
- 11%–28%
- No data

Population of Metropolitan Areas 2013
- 20 million
- 10 million
- 5 million
- 3 million

Note: Symbols on map are sized proportionally to metro area population

① Global rank (population 2013)

possible that urban populations and their growth rates are being overestimated. However, even accounting for circular migration, sub-Saharan Africa still has the highest urbanization rate in the world.

Much of the migration between African cities and rural areas crosses international borders, which makes estimating urban and rural populations even harder. In 2005, the United Nations estimated that there were 17 million such migrants, most going to West Africa and Southern Africa, because jobs are more numerous in both areas. Migrants often go to rural areas as agricultural laborers, or to the cities to work in various jobs in the informal sector. The five sub-Saharan African countries with the most immigrants in 2005 were: Côte d'Ivoire (2.4 million), Ghana (1.7 million), South Africa (1.1 million), Nigeria (1.0 million), and Tanzania (0.8 million). Together, these countries have 40 percent of the migrants in Africa. Even though the migrants typically make few demands, the impact on receiving areas, especially on crowded cities that are unable to provide adequately for existing residents, is substantial. Like migrants who leave the continent, internal migrants live frugally and send much of their earnings home to families.

Food Security in Urban Areas Food supplies for urban residents can be brought in from the countryside, imported from distant lands, or, as is increasingly common, be grown in cities by urban residents. In some cities, residents have taken the initiative to produce their own food in the tiny spaces between houses (see Figure 7.22D) and in the wastelands that surround urban shantytowns.

The Scarcity of Clean Water in Sub-Saharan Cities Public health is a major urban concern, as many water distribution systems are susceptible to being contaminated by harmful bacteria from untreated sewage and garbage (see Figure 7.22C). Only the largest sub-Saharan African cities have sewage treatment plants, and few of these extend to the slums that surround them. The result is frequent outbreaks of waterborne diseases such as cholera, dysentery, and typhoid. It is estimated that unsafe water and sanitation facilities result in an annual loss of $28.4 billion from illness and premature death in Africa alone.

Although the safety of city water supplies is improving, there are frequent periodic water shortages in as many as 11 out of 14 sub-Saharan cities. To control usage during a crisis, city officials limit service to a few hours a day, causing people to store water in vessels that may become contaminated. Recent reassessments of the availability of groundwater resources may eventually alleviate these urban water shortages, but recall that the natural distribution of groundwater does not match up well with the locations of the big cities (see page 284; see also Figure 7.9 on page 286).

ON THE BRIGHT SIDE

Urban Food Gardens

A 2012 UN Food and Agricultural Organization (FAO) report, "The Greening of African Cities," emphasizes that the growing of produce in and around cities, on even the tiniest scraps of land, has an advantage over both rural market gardening and imported food in supplying urban people with safe, nutritious food. Fruits and vegetables can be highly perishable, so if urban consumers can produce their own food, they are likely to eat better, to create less waste, and to have lower food transport costs. Also, food waste can be immediately recycled as compost, reducing or eliminating the need for fertilizers; the greenbelts created by urban gardens can reduce urban air and water pollution.

If urban gardening is to become a viable food security solution, some adjustments will be necessary. At present, urban farmers often use *brownfields*—land where the soil is contaminated by past industrial activities. Moreover, urban farmers frequently use wastewater that may not be safe to irrigate their crops. But these problems can be corrected. Raised cultivation beds can be sealed off from the contaminated soil of brownfields and for irrigation, rainwater can be harvested off nearby roofs, or, alternatively, wastewater can be treated before it is used.

Because few urban gardeners own the land they cultivate, their gardens can be confiscated for development without warning or compensation. One adaptation to this uncertainty is semi-mobile "sack gardens" (see Figure 7.22D) that can be relocated if necessary. Finally, to avoid the overuse of fertilizers and pesticides, some education of urban farmers on more organic techniques may be necessary.

THINGS TO REMEMBER

› GEOGRAPHIC INSIGHT 4 • **Urbanization** Sub-Saharan Africa is undergoing a massive wave of rural-to-urban migration; it has the fastest rate of urbanization in the world. Much of this growth is unplanned and 70 percent of the urban population now lives in impoverished slums characterized by inadequate sanitation; poor access to clean water or food; and crowded, fire-prone housing.

• In the 1960s, only 15 percent of sub-Saharan Africans lived in cities; now about 37 percent do. Estimates are that by 2030, half of all Africans will be urban dwellers.

• When urban economies stagnate and jobs become scarce, the higher cost of living in cities combine with the hazards of slum life to convince many migrants to move back to the countryside at least temporarily, resulting in *circular migration*, or back-and-forth migration between rural and urban areas.

• In some cities, residents have taken the initiative to produce their own food in the tiny spaces between houses and in the wastelands that surround urban shantytowns.

• Public health is a major urban concern, as many water distribution systems are susceptible to being contaminated by harmful bacteria from untreated sewage and garbage. ∎

POPULATION AND GENDER

› GEOGRAPHIC INSIGHT 5

Population and Gender: Of all the world regions, populations are growing fastest in sub-Saharan Africa. However, growth is slowing as the demographic transition takes hold, especially in a few of the more prosperous countries where better health care and lower levels of child mortality, along with more economic and educational opportunities, are encouraging smaller families. In particular, women are better able to pursue careers and to choose to have fewer children.

In fewer than 50 years, sub-Saharan Africa's population has more than quadrupled, growing from about 200 million in 1960 to 828 million in 2009 to slightly under 1 billion in 2013. By 2050, the population of this region could reach just over 2 billion. Contrary to the perception outsiders often hold, however, Africa as a whole is much less densely populated than most of Europe and Asia—36 people per square kilometer, compared to the global average of 51 people per square kilometer. If fertility rates remain high though, places that are relatively uncrowded now may change dramatically over the next few decades (Figure 7.24; see also Figure 1.26 on pages 46–47). In some rural areas, the population density may far exceed the carrying capacity of the land, which will lead to the impoverishment of subsistence farmers. Increasingly, however, Africa's population is becoming concentrated in a few urban areas, such as the cities along the coast of the Gulf of Guinea in West Africa, discussed later.

At the moment, sub-Saharan Africa's annual rate of natural increase (2.6 percent) is the highest in the world; even so, overall rates are slowing in nearly every country, especially in urban areas. Nonetheless, people continue to have more children than would be necessary to simply maintain population numbers. Birth rates are as high as they are because many Africans view children as both an economic advantage and a spiritual link between the past and the future. Childlessness is considered a tragedy, as children ensure a family's genetic and spiritual survival.

Large families are also viewed as having economic value, as children and young adults still perform important work on family farms and in family-scale industries. Moreover, in this region of generally poor health care and resultant high incidence of infant mortality, parents may have extra children in the hope of raising a few to maturity. In all but a few countries, the *demographic transition*—the sharp decline in births and deaths that usually accompanies economic development (see Figure 1.28 on page 49)—has barely begun, and more people now survive long enough to reproduce than did in the past. If Africa is on the brink of a development surge, however, fertility rates could decline quite sharply as women choose roles other than motherhood.

At least five countries in the region have gone through the demographic transition. In South Africa, Botswana, Seychelles, Réunion, and Mauritius (the last three being small island countries off Africa's east coast), circumstances have changed enough to make smaller families desirable and attainable. In all five countries, per capita incomes are five to ten times the sub-Saharan average of U.S.$2000. Advances in health care have cut the infant mortality rate to less than half the regional average of 88 infant deaths per 1000 live births. Because of this, parents can have only two or three children and expect them to live to adulthood. The circumstances of women in these five countries has also improved, as reflected in female literacy rates of about 80 to 90 percent, compared to the regional average of 54 percent. There are more opportunities for women to work outside the home at decent-paying jobs than there are in these countries than in the rest of the region. Thus, many women are choosing to use contraception because they have life options beyond motherhood. Indeed, the percentage of married women using contraception in these five countries is double or even triple the rate for sub-Saharan Africa as a whole, which is only 22 percent (the world average is 61 percent).

The population pyramids in Figure 7.25 show the contrast between countries that are growing rapidly, such as Nigeria, Africa's most populous country (with 170.1 million people and a natural increase rate of 2.6 percent), and countries that have already gone through the demographic transition, such as South Africa (with 51.1 million people and a natural increase rate of 0.9 percent). Nigeria's pyramid is very wide at the bottom because over half the population is under the age of 20. In just 15 years, this entire group will be of reproductive age. Only 15 percent of Nigerian women use any sort of birth control.

In contrast to Nigeria, South Africa's pyramid has contracted at the bottom because its birth rate has dropped from 35 per 1000 to 21 per 1000 over the last 20 years. This decrease is primarily an effect of economic and educational improvements as well as social changes that have come about since the end of apartheid

ON THE BRIGHT SIDE

The TivaWater Jug

The problem of unsafe water for households may have a solution that is simple, affordable, and may create local jobs. A for-profit group, working with grassroots organizations in Uganda but subsidized by American investors, has designed a simple plastic jug, called the TivaWater jug, with an interior sand and clay filtering device that purifies water (Figure 7.23). The jug is made to be affordable to urban poor people in Ugandan shantytowns. While initially made in the United States, the jug is now manufactured in Uganda, is sold and distributed by small businesses there, and will eventually be exported by Ugandans to other African countries.

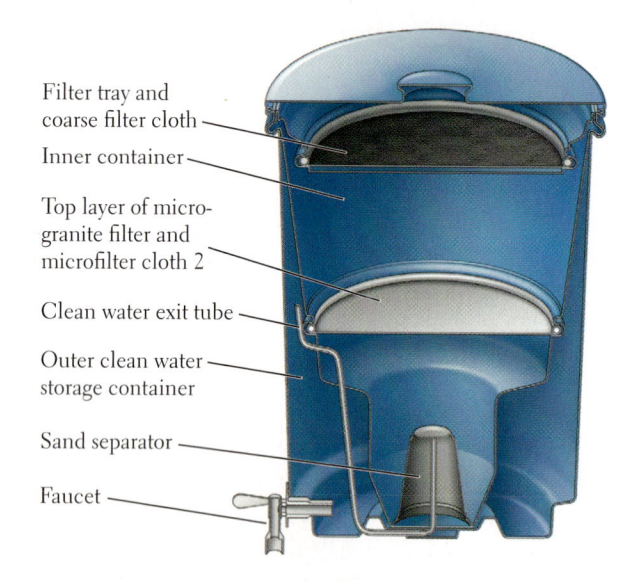

FIGURE 7.23 The TivaWater jug. This specially designed water-filtering jug, made of unbreakable food-grade plastic, removes dirt, bacteria, and viruses, using a new version of biosand technology. It is affordable through subsidies to the very poor, is easy to maintain, and has a faucet that dispenses the water.

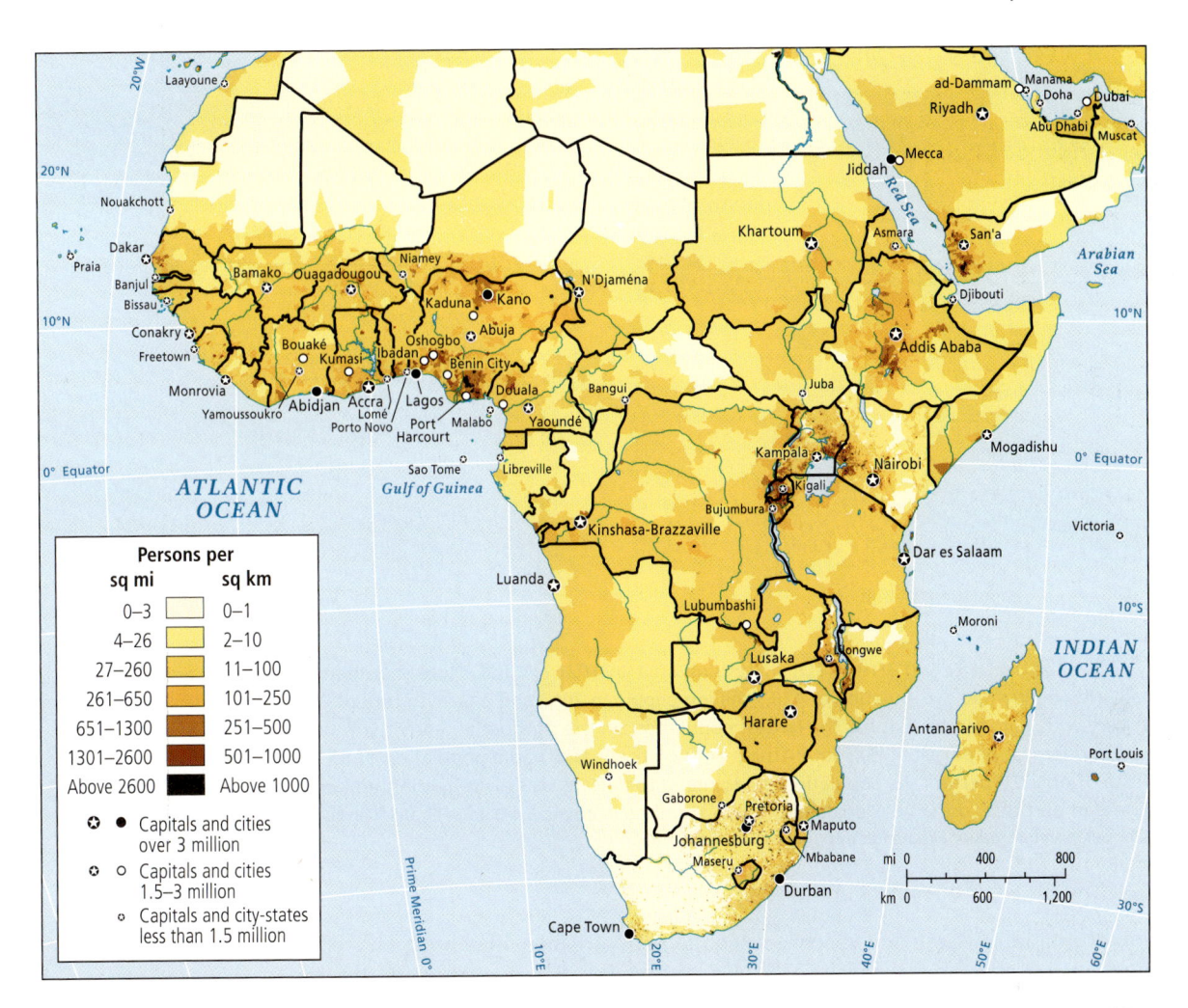

FIGURE 7.24 Population density in sub-Saharan Africa. Of all the world regions, populations are growing fastest in sub-Saharan Africa. The demographic transition is taking hold in a few of the more prosperous countries. In these countries, better health care and more economic and educational opportunities for women are enabling women to pursue careers and to choose to have fewer children. [Sources consulted: Deborah Balk, Gregory Yetman, et al., Center for International Earth Science Information Network, Columbia University, at http://www.ciesin.columbia.edu]

(1994). It is likely to persist as the advantages of smaller families, become clear, especially to women. The birth rate decrease is all the more remarkable because contraception is used by only 60 percent of South African women. Part of the birth rate decrease in South Africa is probably a consequence of the high rate of HIV among young adults there (see pages 308–309).

Population and Public Health Infectious diseases, including HIV-AIDS, are by far the largest killers in sub-Saharan Africa and are responsible for about 50 percent of all deaths. Some diseases are linked to particular ecological zones. For example, people living between the 15th parallels north and south of the equator are most likely to be exposed to sleeping sickness (trypanosomiasis), which is spread among people and cattle by the bites of tsetse flies. The disease attacks the central nervous system and results in death if untreated. Several hundred thousand Africans suffer from sleeping sickness, and most of them are not treated because they cannot afford the expensive drug therapy.

Africa's most common chronic tropical diseases, schistosomiasis and malaria, are linked to standing fresh water. Their incidence has thus increased with the construction of dams and irrigation projects. Schistosomiasis is a debilitating, though rarely fatal, disease that affects about 170 million sub-Saharan Africans. It develops when a parasite carried by a particular freshwater snail enters the skin of a person standing in water. Malaria, spread by the anopheles mosquito (which lays its eggs in standing water), is more deadly. The disease kills at least 1 million sub-Saharan Africans annually, most of them children under the age of 5. Malaria also infects millions of adult Africans who are left feverish, lethargic, and unable to work efficiently because of the disease.

Until recently, relatively little funding was devoted to controlling the most common chronic tropical diseases. Now, however, major international donors are funding research in Africa and elsewhere. More than 60 research groups in Africa are working on a vaccine that will prevent malaria in most people.

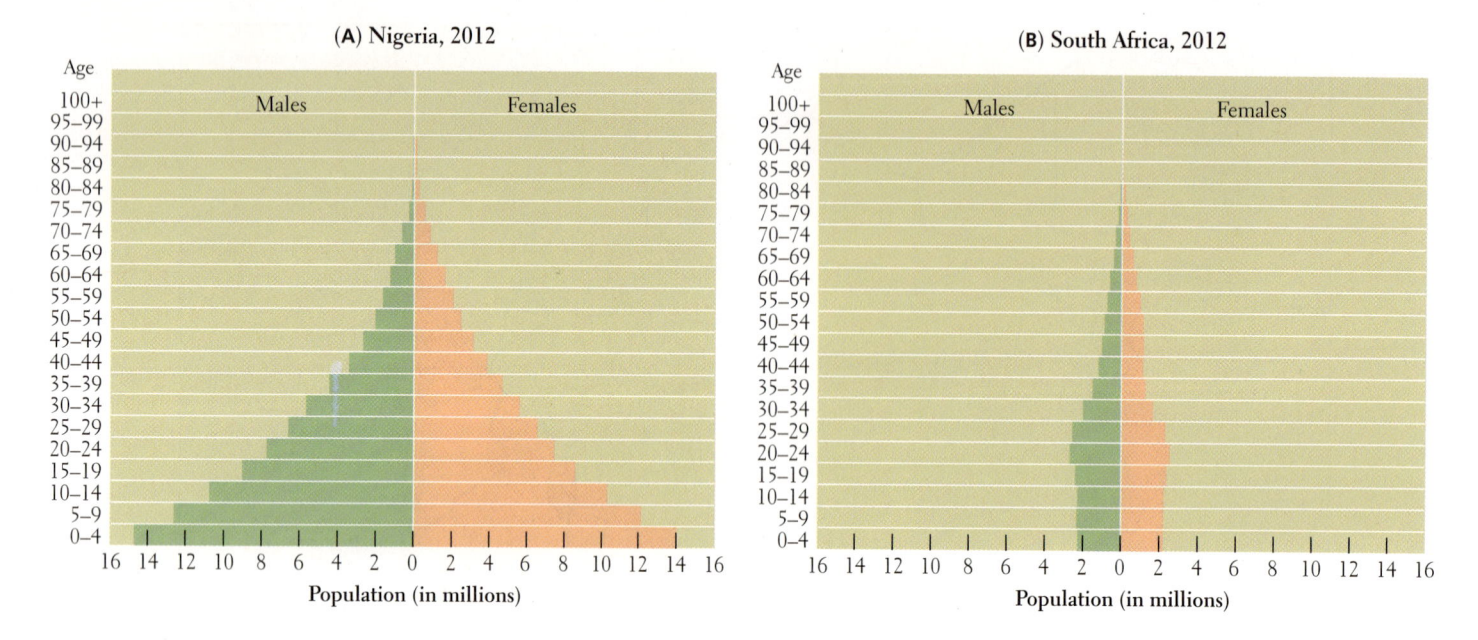

FIGURE 7.25 Population pyramids for Nigeria and South Africa. Note that the two pyramids are at the same scale. **(A)** Nigeria had a population of 170.1 million people in 2012 and a growth rate of 2.6 percent, as indicated by its very wide pyramid base. It has the largest population on the continent, nearly twice that of Ethiopia, which has the second-largest population. **(B)** South Africa, on the other hand, had a population of 51.1 million in 2012 and a growth rate of only 0.9 percent; its population is beginning to decline, as shown by the shrinking of the pyramid base. The decline is in part due to the AIDS epidemic. [Source consulted: International Data Base, U.S. Census Bureau, at http://www.census.gov/population/international/data/idb/informationGateway.php]

The distribution of simple, low-cost, and effective mosquito nets is also reducing the transmission rates of malaria.

HIV-AIDS in Sub-Saharan Africa A leading cause of death in Africa, and the leading cause of death for women of reproductive age, is acquired immunodeficiency syndrome (AIDS), caused by the human immunodeficiency virus (HIV) (Figure 7.26). As of 2010, more than 16 million sub-Saharan Africans had died of the disease, and as many as 80 million more AIDS-related deaths are expected by 2025. In 2010, sub-Saharan Africa had two-thirds of the estimated worldwide total of 33 million people living with HIV. In Botswana, one of the richest countries in Southern Africa, nearly 29.2 percent of the adult female population is infected. The Southern Africa region alone accounted for 31 percent of global AIDS deaths in 2010. While the epidemic is subsiding a bit, it has significantly constrained economic development.

Worldwide, women account for half of all people living with HIV, but in Africa, HIV-AIDS affects women more than men: 59 percent of the region's HIV-infected adults are women. The reasons for this pattern are related to the social status of women in sub-Saharan Africa and elsewhere.

The rapid urbanization of Africa has hastened the spread of HIV, which is much more prevalent in urban areas. It is not uncommon for poor, new urban migrants removed from their village support systems to become involved in the sex industry in order to survive economically. In some cities, virtually all sex workers are infected. Meanwhile, many urban men, especially those with families back in rural villages, visit prostitutes on a regular basis. These men often bring HIV back to their rural homes. As transportation between cities and the countryside has improved, bus and truck drivers have also become major carriers of HIV to rural villages.

A number of myths and social taboos surround HIV-AIDS, exacerbating the problem of controlling the spread of the disease and making education a key component in combating the epidemic. For example, some men think that only sex with a mature woman can result in infection, so very young girls are increasingly sought as sex partners (sometimes referred to as "the virgin 'cure'"). Elsewhere, infection is considered such a disgrace that even those who are severely ill refuse to get tested, yet they remain sexually active.

The Social Costs of HIV-AIDS Across the continent, the consequences of the HIV-AIDS epidemic are enormous. Children contract HIV primarily in utero or from nursing; just over 400 children under the age of 15 die each day from AIDS-related causes. Millions of parents, teachers, skilled farmers, craftspeople, and trained professionals have been lost. More than 15 million children have been orphaned and many have no family left to care for them or to pass on vital knowledge and life skills. The disease has severely strained the health-care systems of most countries. Demand for treatment is exploding, drugs are prohibitively expensive, and many health-care workers themselves are infected. Because so many young people are dying of AIDS, decades of progress in improving the life expectancy of Africans have been erased. For example, in 1990, adult life expectancy in South Africa was 63 years, but in 2012 it was 54 years.

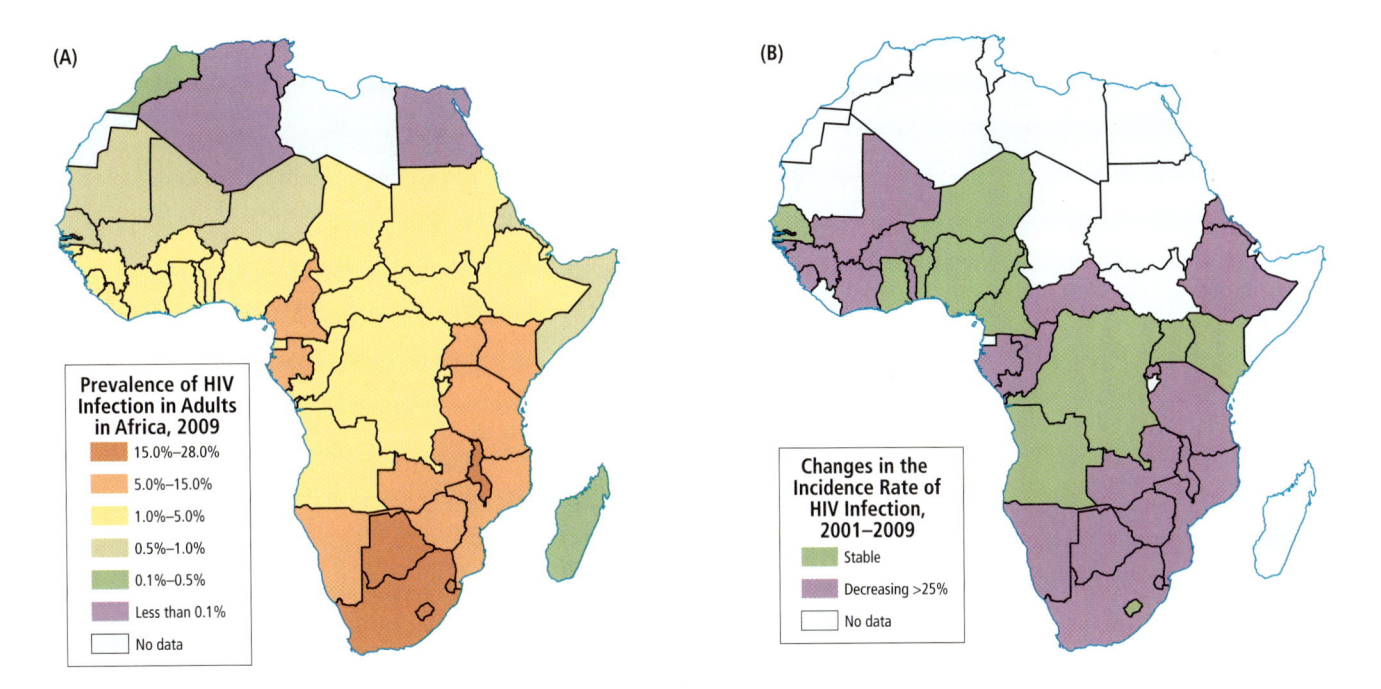

FIGURE 7.26 HIV-AIDS in Africa. New HIV infections fell in sub-Saharan Africa from 2.2 million in 2001 to 1.8 million in 2009. In this period, rates declined more than 25 percent in 22 of the countries. The rate has remained stable elsewhere in the region. The number of deaths from AIDS has declined in sub-Saharan Africa, in large part because of the increased availability of antiretroviral medications. [Sources consulted: "Global Report 2010 Fact Sheet," UNAIDS, at http://www.unaids.org/documents/20101123_FS_SSA_em_en.pdf; *Global Report: UNAIDS Report on the Global AIDS Epidemic/2010*, UNAIDS, at http://www.unaids.org/globalreport/documents/20101123_GlobalReport_full_en.pdf]

HIV-AIDS Education and Medication The most effective means of prevention are the massive education programs that lower rates of infection among those who can read and understand explanations about how HIV is spread. For example, Senegal started HIV-AIDS education in the 1980s, and levels of infection there have so far remained low. Major education campaigns in Uganda lowered the incidence of new HIV infections from 15 percent to just 4.1 percent between 1990 and 2004. By contrast, infection rates have soared in areas such as South Africa, where a few top politicians have put forth untenable theories about the causes of HIV infection or have denied that HIV-AIDS is a problem.

HIV-AIDS medications (known as antiretroviral therapy) are prohibitively expensive for most families in sub-Saharan Africa. The United Nations and a number of private philanthropies cover some of the costs. The medications are not a cure and must be taken daily for the rest of one's life, but they do make it possible to live with the disease. Pharmaceutical firms in India, China, and elsewhere have come out with generic antiretroviral versions that have reduced the cost to about $350 per year, about one-sixth of the average annual income in the region. Approximately five out of every six recipients worldwide are in sub-Saharan Africa, and most of those are in Southern Africa. As yet, only half of those who need the medicines are receiving them; the goal is for all of those with HIV-AIDS to be covered by 2015. The cost to the world of keeping parents alive with antiretroviral therapy is far less than that of dealing with the social effects of millions of orphans.

THINGS TO REMEMBER

> **GEOGRAPHIC INSIGHT 5**
• **Population and Gender** Of all the world regions, populations are growing fastest in sub-Saharan Africa. However, growth is slowing as the demographic transition takes hold, especially in a few of the more prosperous countries where better health care and lower levels of child mortality, along with more economic and educational opportunities, are encouraging smaller families. In particular, women are better able to pursue careers and to choose to have fewer children.

• In fewer than 50 years, sub-Saharan Africa's population has more than quadrupled, growing from about 200 million in 1960 to 828 million in 2009, to slightly under 1 billion in 2013. By 2050, the population of this region could reach just over 2 billion.

• The overall low population density figures in sub-Saharan Africa—36 people per square kilometer, compared to the global average of 51 people per square kilometer—can be misleading. Densities are extremely high in some places and very low in less habitable areas.

• Sub-Saharan Africa has higher rates of infectious diseases than any other region in the world, with the world's worst epidemics of malaria, schistosomiasis, sleeping sickness, and HIV-AIDS. ■

SOCIOCULTURAL ISSUES

To the casual observer, it may appear that a majority of sub-Saharan Africans live traditional lives in rural villages. It is true that with just 37 percent of its population in urban areas, Africa is the world's second-most-rural region. A closer look, however, reveals that change is occurring rapidly even in rural areas.

Gender Issues

Long-standing African traditions dictate a fairly strict division of labor and responsibilities between men and women. In general, women are responsible for domestic activities, including raising the children, attending to the sick and elderly, and maintaining the house. Women collect water and firewood and prepare nearly all the food. Men are usually responsible for preparing land for cultivation. In the fields intended to produce food for family use, women sow, weed, and tend the crops, as well as process them for storage. In the fields where cash crops are grown, men perform most of the work and retain control of the money earned. When husbands in search of cash income migrate to work in mines or in urban jobs, women take over nearly all of the agricultural work, often using simple hand tools, not machines or even draft animals. When there are small agricultural surpluses or handcrafted items to trade, it is women who transport and sell them in the market. Throughout Africa, married couples often keep separate accounts and manage their earnings as individuals, but when a wife sells her husband's produce at the market, she usually gives the proceeds to him.

Ideas in Africa about how males and females should relate to each other come from a variety of sources. Even in the ancient past, social controls tempered gender relationships. Most marriages were social alliances between families; therefore, husbands and wives spent most of their time doing their tasks with family members of their own sex rather than with each other. Then as today, women primarily influenced other women and men influenced men. It is important to note that having multiple wives—the practice of **polygyny** (Figure 7.27)—is more common in sub-Saharan Africa (where it has ancient pre-Muslim, pre-Christian roots) than in Muslim North Africa (as discussed in Chapter 6 on page 270). In a 2012 study of the historical causes, distribution, and persistence of polygyny in Africa, economist James Fenske (2013) produced the map in Figure 7.27. He noted that while polygyny is on a marked decline, the reasons for this decline are unclear. Current improvements in women's education don't seem to reduce its incidence, but economic development and declining infant mortality do.

Women face significant sexual and physical abuse in the home and during civil unrest, when rape and beatings are more common. Many of the female politicians who have come to power in the last decade are now addressing these issues. Most have emphasized the need for more women to be in the position to influence and enforce policies to reduce violence against women. In Liberia, for instance, President Ellen Johnson-Sirleaf has defined women's rights as a national security issue. Liberia has recruited women to serve in both the police and the armed forces.

A practice known as **female genital mutilation (FGM)** (formerly called female circumcision) has been documented in 27 countries in the central portion of the African continent (Figure 7.28). It also is known in Yemen, India, Indonesia, Iraq, Israel, Malaysia, the United Arab Emirates, and, in some cases, in the United States, Canada, Australia, and New Zealand. The practice predates Islam and Christianity, and is performed today among all social classes and in Christian, Muslim, and animist (see the definition on page 312) religious traditions. In the procedure, which is usually performed without anesthesia, a young girl's entire clitoris and parts of her labia are removed. In the most extreme cases (called *infibulation*), her vulva is stitched nearly shut. This mutilation far exceeds that of male circumcision, eliminating any possibility of sexual stimulation for the female and making urination and menstruation difficult. Intercourse is painful and childbirth is particularly devastating because the flesh scarred by the mutilation is inelastic. A 2006 medical study conducted with the help of 28,000 women in six African countries showed that women who had undergone FGM were 50 percent more likely to die during childbirth, and their babies were at similarly high risk. The practice also leaves women exceptionally susceptible to infection, especially HIV infection. **171. FEMALE GENITAL MUTILATION STILL COMMON IN SOMALILAND**

The practice is probably intended to ensure that a female is a virgin at marriage and that she has a low interest in intercourse other than for procreation thereafter. While in decline today, FGM is still widespread among some groups. Among the Kikuyu of

> **polygyny** the practice of having multiple wives
>
> **female genital mutilation (FGM)** removing the labia and the clitoris and sometimes stitching the vulva nearly shut

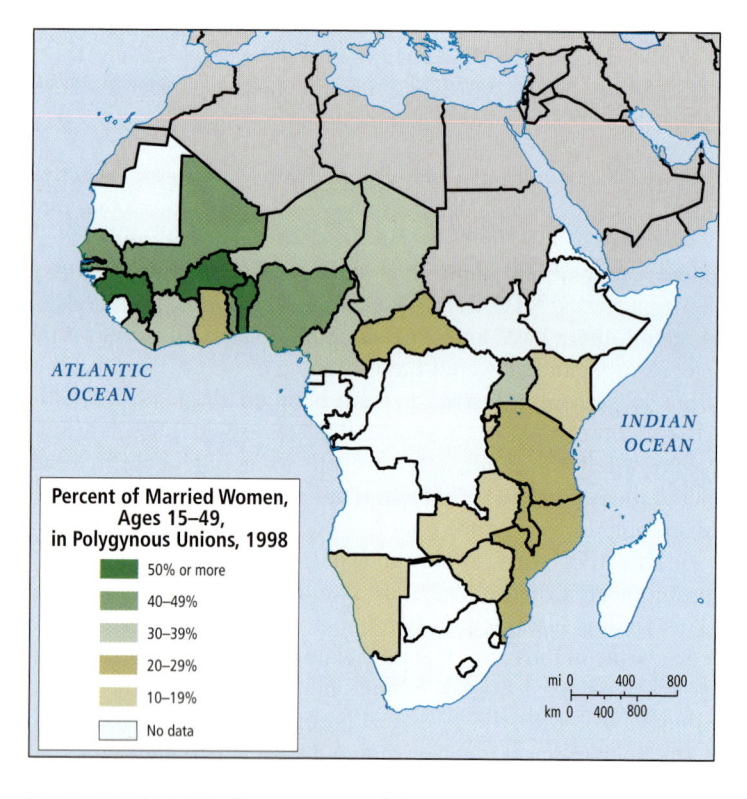

FIGURE 7.27 Polygyny in Africa. The map depicts the spatial distribution of polygyny in sub-Saharan Africa. Seventy-two percent of the women indicated they are the only wife but that their husbands have relations with other women; 19 percent are in two-wife marriages; 7 percent in three-wife marriages; and the remaining 2 percent are in marriages with four or more wives. [Source consulted: James Fenske, "African Polygamy: Past and Present," *Editorial Express*, February 15, 2012, at https://editorialexpress.com/cgi-bin/conference/download.cgi?db_name=CSAE2012&paper_id=115]

Percent of Girls and Women Living in African Countries Who Have Undergone FGM

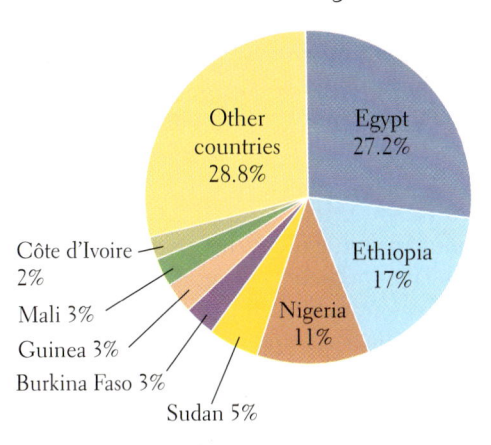

- Other countries 28.8%
- Egypt 27.2%
- Ethiopia 17%
- Nigeria 11%
- Sudan 5%
- Burkina Faso 3%
- Guinea 3%
- Mali 3%
- Côte d'Ivoire 2%

FIGURE 7.28 Percent of girls and women living in African countries who have undergone FGM. The World Health Organization estimates that there are 91.5 million girls and women living in Africa today who have undergone some form of FGM. Of those, 12.4 million are girls between 10 and 14 years of age. Close to half of them are in two countries: Egypt and Ethiopia. Incidences of FGM have been documented in 27 countries in Africa and in 8 other countries. There have also been some cases documented in Europe, North America, Australia, and New Zealand, which are probably due to migration. [Source consulted: "An Update on WHO's Work on Female Genital Mutilation (FGM), Progress Report," World Health Organization, 2011, at http://www.who.int/reproductivehealth/publications/fgm/rhr_11_18/en/index.html]

Kenya, nearly all females would have had FGM 40 years ago, but today only about 40 percent do. Kikuyu women's rights leaders have had some success in curbing FGM by replacing it with right-of-passage ceremonies that joyously mark a girl's transition to puberty.

Many African and world leaders have concluded that the practice is an extreme human rights abuse, and it is now against the law in 16 countries. The most successful eradication campaigns are those that emphasize the threat FGM poses to a woman's health, thus making it socially acceptable to not undergo this ritual. The World Health Organization in 2008 took a strong stand against FGM, saying:

Female genital mutilation has been recognized as discrimination based on sex because it is rooted in gender inequalities and power imbalances between men and women and inhibits women's full and equal enjoyment of their human rights. It is a form of violence against girls and women, with physical and psychological consequences.

WORLD HEALTH ORGANIZATION, *Eliminating Female Genital Mutilation—An Interagency Statement* (GENEVA: WORLD HEALTH ORGANIZATION PRESS, 2008), P. 10.

Religion

Africa's rich and complex religious traditions derive from three main sources: indigenous African belief systems (many of them animist), Islam, and Christianity.

Indigenous Belief Systems Traditional African religions, found in every part of the continent, may be the most ancient on Earth, since this is where human beings first evolved. Figure 7.29 highlights the countries in which traditional beliefs remain particularly strong. Traditional beliefs and rituals often are used to bring departed ancestors into contact with living people, who in turn are the connecting links in a timeless spiritual community that stretches into the future. The future is reached only if family members procreate and perpetuate the family heritage through storytelling.

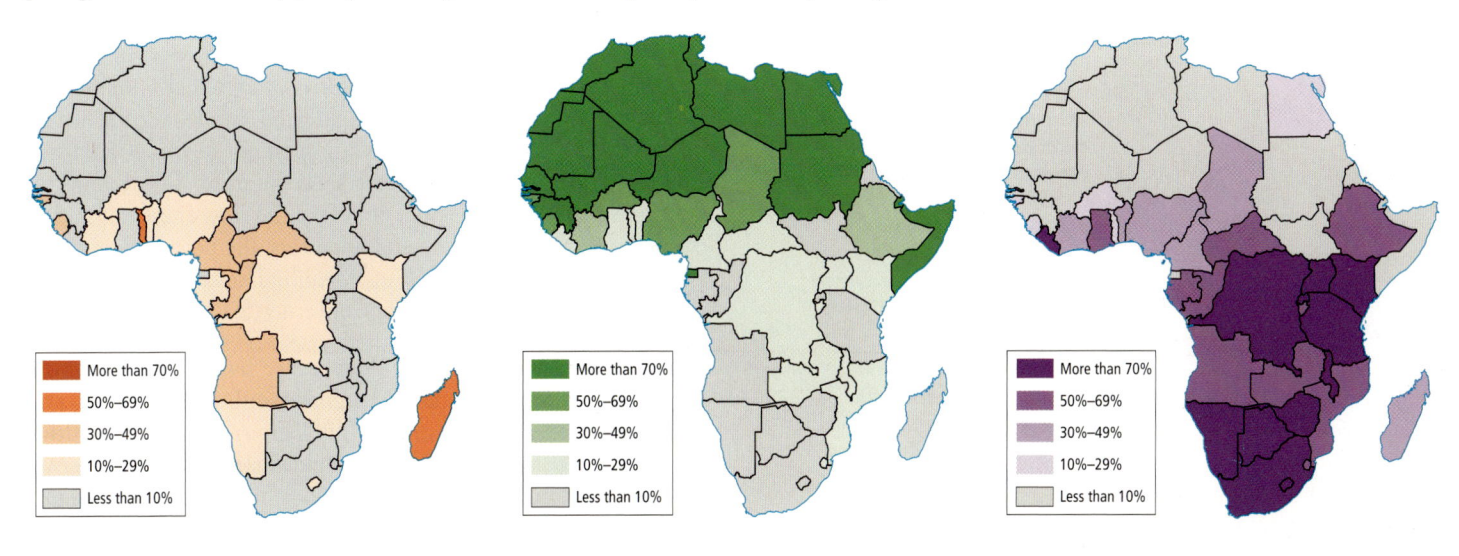

(A) Distribution of traditional religions.

(B) Distribution of Islam.

(C) Distribution of Christianity (Judaism is diminishing).

FIGURE 7.29 Religions in Africa. Notice that the various religions in Africa overlap in distribution; in many countries, one is dominant but others are present. [Sources consulted: Matthew White, "Religion in Africa," in *Historical Atlas of the Twentieth Century* (October 1998), at http://users.erols.com/mwhite28/aforelg.htm; revised with new data from the CIA, *The World Factbook*, 2009, at https://www.cia.gov/cia/publications/factbook/index.html]

Most traditional African beliefs can be considered **animist** in that spirits, including those of the deceased, are thought to exist everywhere—for example, in trees, streams, hills, and art. In return for respect (expressed through ritual), these spirits offer protection from sickness, accidents, and the ill will of others. African religions tend to be fluid and adaptable to changing circumstances. For example, in West Africa, Osun, the god of water traditionally credited with healing powers, is now also invoked for those suffering economic woes.

Religious beliefs in Africa, as elsewhere, continually evolve as new influences are encountered. If Africans convert to Islam or Christianity, they commonly retain aspects of their indigenous religious heritage. The three maps in Figure 7.29 show a spatial overlap of belief systems, but they do not convey the philosophical blending of two or more faiths, which is widespread. In the Americas, the African diaspora has influenced the creation of new belief systems developed from the fusion of Roman Catholicism and African beliefs. Voodoo in Haiti, Santería in Cuba, and Candomblé in Brazil (see Chapter 3, page 146) are examples of this fusion.

Islam and Christianity Islam began to extend south of the Sahara in the first centuries after Muhammad's death in 632 C.E., and today, about one-third of sub-Saharan Africans are Muslim (see Figure 7.29B). Islam is now the predominant religion throughout the Sahel and parts of East Africa, where Muslim traders from North Africa and Southwest Asia brought the religion. Powerful Islamic empires have arisen here since the ninth century. The latest of these empires challenged European domination of the region in the late nineteenth century.

Today, about half of Africans are Christian. Christianity first came to the region via Egypt and Ethiopia shortly after the time of Christ, well before it spread throughout Europe (see Figure 7.29C). Christianity did not come to the rest of Africa until nineteenth-century missionaries from Europe and North America became active along the west coast. Many Christian missionaries provided the education and health services that colonial administrators had neglected. In the twentieth century, old-line established churches began to gain adherents in Africa. The Anglican Church (Church of England) grew so rapidly that by 2000, Anglicans in Kenya, Uganda, and Nigeria outnumbered those in the United Kingdom. The Anglican Church in Africa attracts the educated urban middle class.

> **animism** a belief system in which spirits, including those of the deceased, are thought to exist everywhere and to offer protection to those who pay their respects

Festivals across the continent celebrate the variety of African religious traditions. As is the case with festivals everywhere, such events offer an opportunity to represent ideal versions of culture, a chance to revel in joy and renewed religious dedication, and a chance to make important social contacts, reinforce identity, and perform community service (Figure 7.30).

Evangelical Christianity and the Gospel of Success In contrast to the Anglican Church, modern evangelical versions of Christianity appeal to the less-educated, more recent urban migrants—the most rapidly growing segment of the population. These independent, evangelical sects interpret Christianity as being aligned with traditional beliefs in the importance of sacrificial gifts to spiritual leaders and ancestors, the power of miracles, and the idea that devotion brings wealth. The result is one of the world's fastest-growing Christian movements.

Ethnicity and Language Ethnicity, as we have seen throughout this book, refers to the shared language, cultural traditions, and political and economic institutions of a group. The map of Africa in Figure 7.31 shows a rich and complex mosaic of languages. Yet despite its complexity, this map does not adequately depict Africa's cultural diversity (compare it with Figure 7.20 of Nigeria on page 300).

Most ethnic groups have a core territory in which they have traditionally lived, but very rarely do groups occupy discrete and exclusive spaces. Often several groups share a space, practicing different but complementary ways of life and using different resources. For example, one ethnic group might be subsistence cultivators, another might herd animals on adjacent grasslands, and a third might be craft specialists working as weavers or metalsmiths.

FIGURE 7.30 LOCAL LIVES
Festivals in Sub-Saharan Africa

A During Timkat, a festival that marks the baptism of Jesus Christ, Ethiopian Orthodox Christian priests in Addis Ababa carry replicas of the Ten Commandments, wrapped in cloth, on their heads. [Matjaz Krivic/Getty Images]

B South African Muslims spot the new moon that marks the end of Ramadan, a month of fasting during which observant Muslims do not eat, drink, quarrel, or have sex between sunrise and sunset. Throughout Africa, many Muslims mark the end of Ramadan with prayers for peace and unity as well as with drumming, singing, dancing and general celebration. [Gianluigi Guercia/AFP/Getty Images]

C The 8-day festival of the Umhangla (the "Reed Dance" ceremony) protects the virginity of young unmarried women, who gather in the king of Swaziland's village and repair the surrounding fence with tall reeds. Later, in costumes, the women dance for the royal family, foreign dignitaries, tourists, and other spectators. Traditionally, the king also chooses a new wife. The women who wear red feathers in their hair are the daughters of the king. [Denny Allen/Gallo Images/Getty Images]

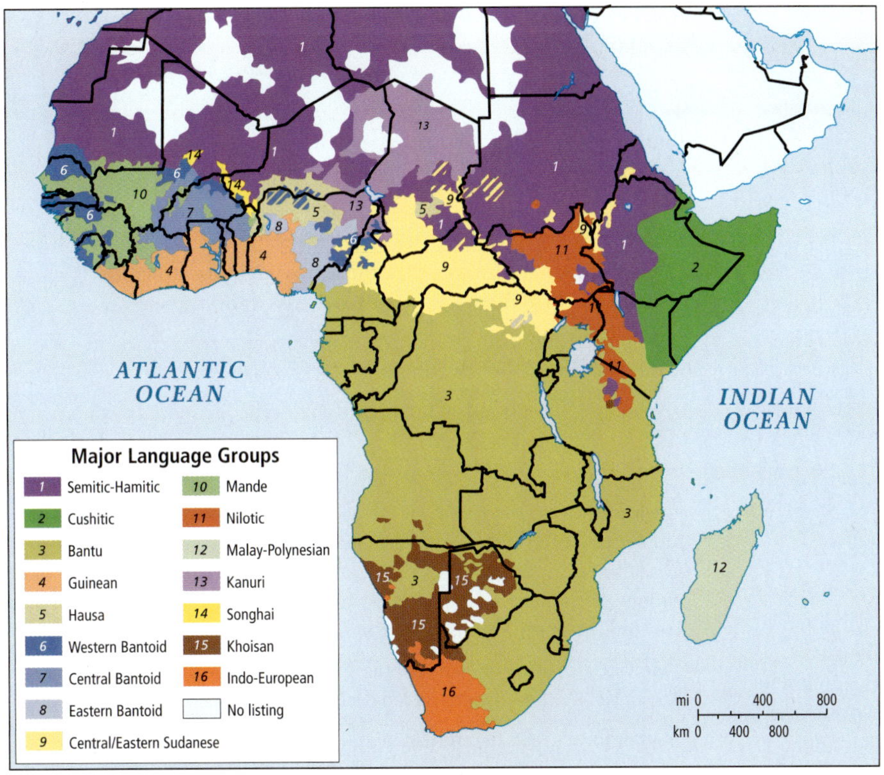

Sociocultural Issues **313**

as a method of control, assigning a higher status to the Tutsis. Hutu and Tutsi now think of themselves as having very different ethnicities and abilities. In the 1990s and again in 2004, the Hutu-run government encouraged attacks on the minority Tutsi people. Hutus were also killed, but those killed tended to be people who opposed the genocide. The best estimates indicate that more than 1 million people, most of them Tutsis, died during the two genocide catastrophes.

Some African countries have only a few ethnic groups; others have hundreds. Cameroon, sometimes referred to as a microcosm of Africa because of its ethnic complexity, has 250 different ethnic groups. Different groups often have extremely different values and practices, making the development of cohesive national policies difficult. Nonetheless, the vast majority of African ethnic groups have peaceful and supportive relationships with one another.

To a large extent, language correlates with ethnicity. More than a thousand languages, falling into more than a hundred language groups, are spoken in Africa; Bantu is the largest such group (see Figure 7.31). Most Africans speak their native tongue and a **lingua franca** (language of trade). Some languages are spoken by only a few dozen people, while other languages, such as Hausa, are spoken by millions of people from Côte d'Ivoire to Cameroon. Lingua francas—such as Hausa, Arabic, and Swahili— are taking over because they better suit people's needs or have become politically dominant; some African languages are dying out. Former colonial languages such as English, French, and Portuguese (all classed as Indo-European languages in Figure 7.31) are also widely used in commerce, politics, education, and on the Internet.

lingua franca a language of trade

FIGURE 7.31 Major language groups of Africa.
[Sources consulted: Edward F. Bergman and William H. Renwick, *Introduction to Geography—People, Places, and Environment* (Englewood Cliffs, NJ: Prentice-Hall, 1999), p. 256; Jost Gippert, *TITUS Didactica*, at http://titus.uni-frankfurt.de/didact/karten/afr/afrikam.htm]

People may also be very similar culturally and occupy overlapping spaces but identify themselves as being from different ethnic groups. Hutu farmers and Tutsi cattle raisers in Rwanda share similar occupations, languages, and ways of life. However, European colonial policies purposely exaggerated ethnic differences

VIGNETTE

Preachers at the Miracle Center in Kinshasa describe the Gospel of Success forcefully: "The Bible says that God will materially aid those who give to Him. . . . We are not only a church, we are an enterprise. In our traditional culture you have to make a sacrifice to powerful forces if you want to get results. It is the same here."

Generous gifts to churches are promoted as a way to bring divine intervention to alleviate miseries, whether physical or spiritual. Practitioners donate food, television sets, clothing, and money. One woman gave 3 months' salary in the hope that God would find her a new husband.

Like all religious belief systems, the Gospel of Success is best understood within its cultural context. Many of the believers new to the city feel isolated and are looking for a supportive community to replace the one they left behind. In return for material contributions to the church and volunteer services, members receive social acceptance and community assistance in times of need. ■

THINGS TO REMEMBER

- Gender relationships in sub-Saharan Africa are complex and variable, but in this region as in others, women are generally subordinate and are often physically abused. Nevertheless, change is underway.

- Female genital mutilation has terrible consequences for women's health. It has been recognized as sexual discrimination because it is rooted in gender inequalities and power imbalances between men and women.

- Traditional African religions are among the most ancient on Earth and are found in every part of the continent. Today, however, about one-third of sub-Saharan Africans are Muslim and about half are Christian. Many in the evangelical movement, a subset of Christianity, promote the Gospel of Success as a way of life.

- Sub-Saharan Africa is a mosaic of more than 1000 languages that are roughly similar to the mosaic of ethnic groups. Hausa, Arabic, and Swahili are now the most commonly used languages of trade. ■

GEOGRAPHIC INSIGHTS

Sub-Saharan Africa: Review and Self-Test

1. Environment: Sub-Saharan Africa's poverty and political instability leave its people less able to adapt to climate change than the inhabitants of other world regions, though many Africans are developing strategies to cope with increasing variability in rainfall and temperature. As a whole, this region has contributed little to the build-up of greenhouse gases in the atmosphere, but deforestation by rural Africans and by multinational logging companies is intensifying global climate change.

- Why are many African food production systems particularly vulnerable to climate change?
- Why are African farmers and herders particularly sensitive to changes in temperature and water availability?
- How do poverty and having limited access to cash influence the vulnerability of many Africans to climate change?

2. Globalization and Development: Most sub-Saharan African economies are dependent on the export of raw materials. This pattern, a legacy of the era of European colonialism, results in economic instability because prices for raw materials can vary widely from year to year. While a few countries are diversifying and industrializing, most still sell raw materials and all rely on expensive imports of food and manufactured goods.

- How did European colonization affect economic development and food production in Africa?
- Why does being a supplier of raw materials put a country at a disadvantage in the global economy?
- What are the current positive and negative trends in food production in sub-Saharan Africa?
- Are there any exceptions in the region to the general pattern of economic dependence on exports of cheap raw materials?
- Why is the development of manufacturing industries in sub-Saharan Africa considered such an important step in raising standards of living?

3. Power and Politics: Governments in sub-Saharan Africa tend to be authoritarian, though there is a general shift in the region toward the expansion of political freedoms. Free and fair elections have brought about dramatic changes in some countries, while elections tainted by widespread suspicions of fraud have often been followed by surges of violence in others.

- What forces are working for and against the expansion of political freedoms in sub-Saharan Africa?

- To what extent can the tendency toward authoritarianism in sub-Saharan Africa be traced to institutions put in place during the colonial era?
- What role have elections played in both diffusing and inciting violence?
- What are some examples of authoritarian political structures that have given way to more democratic systems of government?
- Why have national crises led, in some cases, to an increase in women's political empowerment?
- What is the relationship between rights and education for women and the alleviation of poverty?

4. Urbanization: Sub-Saharan Africa is undergoing a massive wave of rural-to-urban migration; it has the fastest rate of urbanization in the world. Much of this growth is unplanned and 70 percent of the urban population now lives in impoverished slums characterized by inadequate sanitation; poor access to clean water or food; and crowded, fire-prone housing.

- How is urbanization influencing population growth in sub-Saharan Africa? Relate these trends to the demographic transition.
- What forces lead to circular migration?
- Why is urban gardening an important innovation in sub-Saharan Africa's modernization?
- Why are water distribution systems a public health concern in this region's slums?

5. Population and Gender: Of all the world regions, populations are growing fastest in sub-Saharan Africa. However, growth has been slowing as the demographic transition has begun taking hold, especially in a few of the more prosperous countries where better health care and lower levels of child mortality, along with more economic and educational opportunities, are encouraging smaller families. In particular, women are better able to pursue careers and to choose to have fewer children.

- On what grounds would the claim be made that sub-Saharan Africa is not densely populated?
- What do population pyramids tell us about a country's population?
- What special spiritual role do children play in family and community life in sub-Saharan Africa?
- What role do infectious diseases play in sub-Saharan African development, life expectancies, and use of resources?
- Why do women bear the brunt of the HIV-AIDS epidemic in sub-Saharan Africa?

CRITICAL THINKING QUESTIONS

1. If you were to investigate the origins of lumber and lumber products sold where you live, where would you start? What would make you think that you are or are not participating in the African timber trade? How is this trade related to the concept of *neocolonialism*, the revival of colonial-like resource extraction in the modern period?

2. Describe the ways in which the environment is linked to the spread of the chronic communicable diseases of Africa. Why might some scientists argue that malaria is a worse threat to Africa's development than is HIV-AIDS?

3. Describe the trajectory of population growth in Africa. What are the factors that contribute to this pattern? How might the developed countries help the situation?

4. Reflect on the many passages throughout this chapter regarding gender roles in Africa, then state the points that you think are the most influential in the economic development process.

5. Describe the important ways in which women are exerting an influence on African political processes.

6. Technology is changing life in Africa. What do you think are the most crucial ways in which technology will modify the ability of ordinary people to better their lives and to participate in civil society?

7. What are the various circumstances that make young, heterosexual women so susceptible to HIV-AIDS in sub-Saharan Africa? How does the social and economic status of the average woman (compared to that of the average man) influence the spread of HIV among women?

8. Looking at the various African locations where food is in short supply, explain the main factors that account for the shortages. How would you respond to those who say that drought and climate are the principal causes of famine?

9. If you were asked to make a speech to the Rotary Club in your town about hopeful signs out of Africa, what would you include in your talk? Which pictures from this book (or elsewhere) would you show?

› THINKING GEOGRAPHICALLY

Now that you have read about sub-Saharan Africa, you should be able to answer the following questions about the photos in this chapter. To answer these questions online, go to Geography LaunchPad.

Human Impacts on the Biosphere (page 281)

A Other than the caption, what clues are there to suggest that this landscape has been cleared of forest?

B What about this photo suggests that the trees being logged are from old-growth forests?

C What about the fishing method depicted in the photo suggests that catches are fairly small in size?

Vulnerability to Climate Change (page 285)

A How is poverty evident in this photo?

B What caused the death of many wildebeests in 2007?

C How is agricultural diversity illustrated in this photo?

D What kind of climate change in Africa most often causes a reduced yield for a crop like tea?

Visual History (page 288)

A What are the clues in the picture that food may have been a source of conflict between early humans and between humans and other animals?

B Aside from the central fortlike structure, describe any further evidence of human habitation.

C What does the existence of a mosque in Mali in 1327 indicate about the geography of trading patterns in this part of Africa?

D What is the evidence of European involvement in the slave trade in this image?

E What about this photo suggests hierarchies of status and power in British colonial Africa?

F What about this stadium conveys the wealth and Westernization of South Africa?

Power and Politics (page 301)

A In what ways does this picture show both the strengths and limits of political freedoms in Kenya?

B Why might it be significant that Ellen Johnson-Sirleaf was elected as Liberia's president without the help of quotas that reserve certain elected positions for women?

C What does this photo suggest about the strength of this rebel militia and its ability to acquire uniforms and arms?

D What does this photo convey about the general well-being of the people depicted?

Urbanization (page 304)

A What do the types of houses, the water, and the boats suggest about the location of this slum?

B What are some factors that contribute to the prevalence of gangs in African slums?

C How might the lack of clean water and sanitation create an incentive for people to have large families?

D How does this project in Nairobi embody the idea of "self-reliant development"?

CHAPTER KEY TERMS

agroforestry 282
animism 312
apartheid 291
carbon sequestration 279
commercial agriculture 283
commodities 293
commodity dependence 293
divide and rule 299

dual economy 293
female genital mutilation (FGM) 310
genocide 300
grassroots economic development 297
groundwater 284
Horn of Africa 279
intertropical convergence zone (ITCZ) 279
lingua franca 313

mixed agriculture 283
pastoralism 286
polygyny 310
Sahel 276
self-reliant development 298
shifting cultivation 283
subsistence agriculture 282

NORTHERN AFGHAN PROVINCES
1. JOWZJAN
2. BALKH
3. SAMANGAN
4. KONDUZ
5. BAGHLAN
6. PARVAN
7. VARDAK
8. TAKHAR
9. KAPISA
10. KABUL
11. LOWGAR
12. BADAKHSHAN
13. LAGHMAN
14. KONARHA
15. NANGARHAR

A Coastal lowlands, South India

B Himalaya Mountains, Nepal, as seen from space

chapter 8

SOUTH ASIA

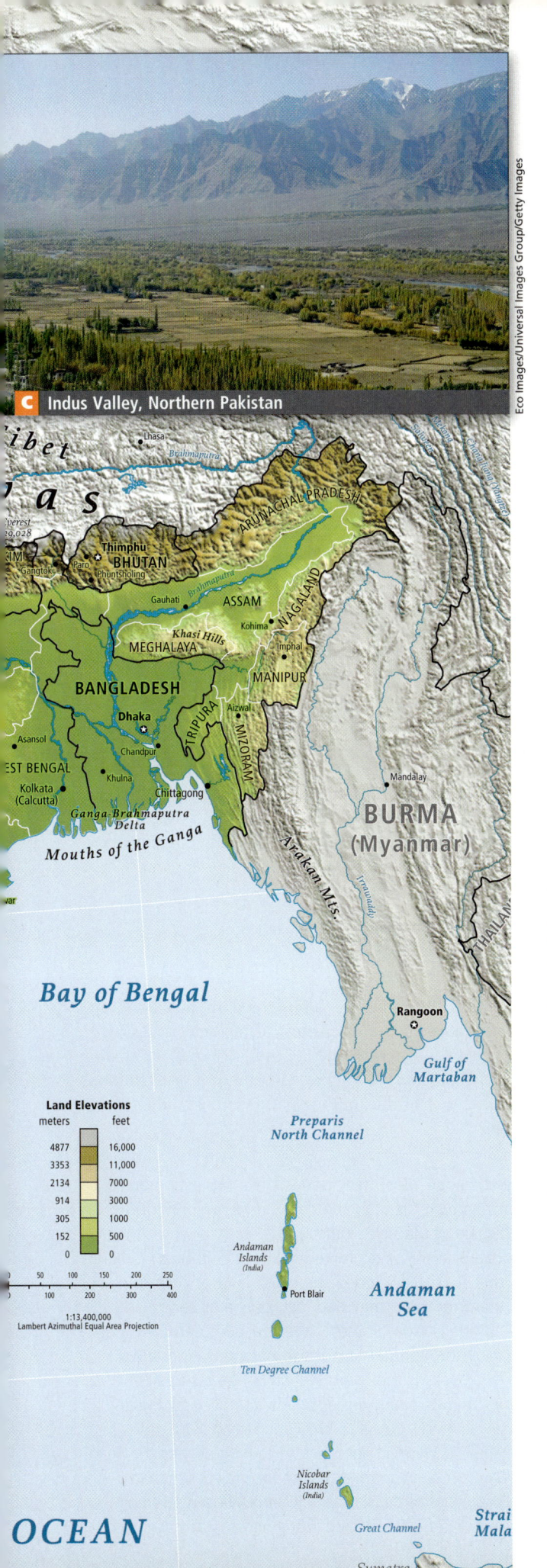

C Indus Valley, Northern Pakistan

D Western Ghats, Kerala, India

E Deccan Plateau, Karnataka, India

F Indo-Gangetic Plain, North India

FIGURE 8.1 Regional map of South Asia.

317

GEOGRAPHIC INSIGHTS: SOUTH ASIA

After you read this chapter, you will be able to discuss the following geographic insights as they relate to the five thematic concepts:

1. Environment: Climate change puts more lives at risk in South Asia than in any other region in the world, primarily due to water-related issues. Over the short term, droughts, floods, and the increased severity of storms imperil many urban and agricultural areas. Over the longer term, sea level rise may profoundly affect coastal areas and glacial melting poses a threat to rivers and aquifers.

2. Globalization and Development: Globalization benefits some South Asians more than others. Educated and skilled South Asian workers with jobs in export-connected and technology-based industries and services are paid more and sometimes have better working conditions. Less-skilled workers in both urban and rural areas are left with demanding but very low-paying jobs.

3. Power and Politics: India, South Asia's oldest, largest, and strongest democracy, has shown that the expansion of political freedoms can ameliorate conflict. Across the region, when people have been able to participate in policy-making decisions and implementation—especially at the local level—seemingly intractable conflict has been diffused and combatants have been willing to take part in peaceful political processes.

4. Urbanization: South Asia has two general patterns of urbanization: one for the rich and the middle classes and one for the poor. The areas that the rich and the middle classes occupy include sleek, modern skyscrapers bearing the logos of powerful global companies, universities, upscale shopping districts, and well-appointed apartment buildings. The areas that the urban poor occupy are chaotic, crowded, and violent, with overstressed infrastructures and menial jobs. These two patterns often coexist in very close proximity.

5. Population and Gender: In this most densely populated of world regions, population growth is slowing as the demographic transition takes hold. Birth rates are falling due to rising incomes, urbanization, better access to health care, and the fact that women are finding more opportunities to study and work outside the home, and thus are delaying childbearing and having fewer children. However, a severe gender imbalance is developing in this region due to age-old beliefs that males are more useful to families than are females. As a result, adult males significantly outnumber adult females.

The South Asian Region

South Asia (Figure 8.1), like so many of today's world regions, began its modern history as the result of European colonization. During that time, economic, political, and social policies rarely put the needs of South Asian people first. Toward the end of the colonial era, major upheavals precipitated by departing colonists left the region with a legacy of distrust and difficult political borders that still lingers.

The five thematic concepts highlighted in this book are explored as they arise in the discussion of regional issues, with interactions between two or more themes featured, as in the geographic insights above. Vignettes, like the one that follows, illustrate one or more of the themes as they are experienced in individual lives.

GLOBAL PATTERNS, LOCAL LIVES

Narendra Modi, the chief minister of the state of Gujarat in western India (see Figure 8.1), is passionate about securing water for his state. In 2006, draped in garlands by well-wishers, he embarked on a flamboyant hunger strike to protest a decision by India's national government in New Delhi to limit the height of the Sardar Sarovar Dam on the Narmada River. Four Indian states lie in the drainage basin of the Narmada. Modi wanted to trap more water behind a higher dam so that as many as 50 million of his constituents in Gujarat would have more access to water, primarily for irrigation. Gujarat produces much of the food needed by India's west coast cities.

At the same time as Modi's hunger strike, far away in New Delhi, Medha Patkar, the leader of the "Save the Narmada River" movement, was in day 18 of her hunger strike protesting against the same dam. As water rose in the dam's reservoir, 320,000 farmers and fishers in Madhya Pradesh, the state in which the reservoir is located, were being forced to move. Although Indian law requires that these "evacuees" be given land or cash to compensate them for what they had lost, only a fraction had received any compensation. Some of the evacuees demonstrated their objection by forfeiting their right to compensation and refusing to move even as the rising waters of the reservoir consumed their homes. Many were forcibly removed by Indian police and have since relocated to crowded urban slums.

Environmental and human rights problems have been at the heart of the Sardar Sarovar Dam controversy since the project began in 1961. The natural cycles of the Narmada River have been severely disturbed. Once a placid, slow-moving river—and one of India's most sacred—the disruption of flows and the flooding upstream of the dam have caused massive die-offs of aquatic life, resulting in high unemployment levels among farmers and fishers. Many people have also lost their homes to the rising reservoir waters (Figure 8.2A).

In addition to cost overruns that are more than three times the original cost estimates, the ultimate benefits of the dam have been called into question. A major justification for the dam was that in addition to irrigation water, it would provide drinking water for 18 million people in the greater river basin as well as 1450 megawatts of electricity. But the demand for irrigation waters to serve drought-prone areas in the adjacent states of Gujarat and Rajasthan is now so strong that there will not be sufficient water to generate even 10 percent of the 1450 planned megawatts on a sustained basis. Furthermore, 80 percent of the areas in Gujarat most vulnerable to drought are not yet connected to the project by the necessary canals. Therefore, 90 percent of the water planned for irrigation still flows into the sea.

Concerned that the economic benefits would be small and easily negated by the environmental costs, the World Bank withdrew its funding of the dam some years ago. Ecologists say that far less costly water-management strategies such as rainwater harvesting, intensified groundwater recharge (artificially assisted replenishment of the aquifer), and watershed management would be better options for the farmers of Gujarat and Rajasthan.

Minister Modi quickly ended his hunger strike when the Indian Supreme Court ruled that the Sardar Sarovar Dam could be raised higher (it now stands at 121.92 meters, or 400 feet). The following day, Medha Patkar ended her fast as well, because, in the same decision, the Supreme Court ruled that all people displaced by the dam must be adequately relocated. Furthermore, the court decision confirmed that human impact studies are required for all dam projects. Up to then, no such study had been done for the Sardar Sarovar Dam. Nonetheless, by September of 2012, the dam project was 8 years behind schedule and had stalled again in court. Both sides used the Internet to promote their positions. Meanwhile, Narendra Modi achieved national prominence and began an unsuccessful campaign to be elected prime minister of India.

[Sources: India eNews; Frontline; Friends of River Narmada (http://tinyurl.com/bhhanz3); Environmental Justice, Issues, Theories, and Policy; and Wall Street Journal. For detailed source information, see Text Sources and Credits.] ■

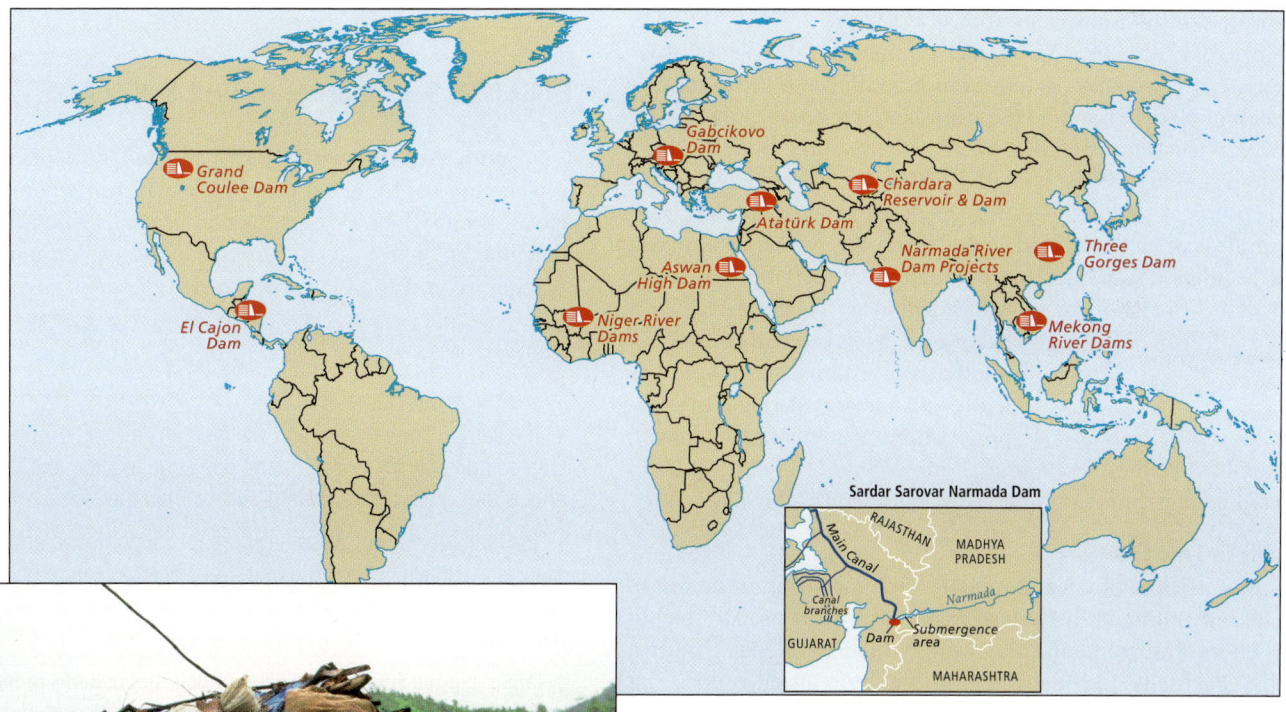

FIGURE 8.2 The Sardar Sarovar Dam on the Narmada and other major dams around the world. The Sardar Sarovar Dam on the Narmada River is only one of hundreds of thousands of dam projects throughout the world that together have displaced between 40 and 80 million people. The map shows some of the major dams around the world. **(A)** Before they were displaced by the reservoir of the Sardar Sarovar Dam, rural people stand outside homes that are now submerged.

320 CHAPTER 8 South Asia

The recent history of water management in the Narmada River valley highlights some key issues now facing South Asia and other regions that have developing economies. Across the world, large, poor populations depend on increasingly overtaxed environments. Improving the standard of living for the poor nearly always requires increases in water and energy use. Often, misplaced efforts to meet these urgent needs make neither economic nor environmental sense, but are driven to completion by political and social pressures. In the case of the Sardar Sarovar Dam, the wealthier, more numerous, and more politically influential farmers of Gujarat have tipped the scale—over the protests of the evacuees along the Narmada River—in favor of a project that may be creating more problems than it is solving.

The role of water in South Asian life, including the ways water use intersects with other central issues, is a recurring topic in this chapter.

What Makes South Asia a Region?

The countries that make up the South Asia region in this book are Afghanistan and Pakistan in the northwest; the Himalayan states of Nepal and Bhutan; Bangladesh in the northeast; India (including the Indian territories of Lakshadweep, Andaman, and Nicobar Islands); and the island countries of Sri Lanka and the Maldives (Figure 8.3; see also Figure 8.1). Physically, these countries occupy territory known as the *Asian subcontinent*—the portion of a tectonic plate that joined the Eurasian continent nearly 60 million years ago (discussed below). Historically and culturally, these modern countries have a shared patchwork of religious, social, and political features linked to ancient conquerors from Central Asia, explorers and traders from the Arabian Peninsula and Southeast Asia, and more recent colonizers from Europe.

This region can be compared to a patchwork, as its extensive history of internal and external influences has left it with a fragmented and overlapping pattern of religions, languages, ethnicities, economic theories, forms of government, attitudes toward gender and class, and ideas about land and resource use.

> **subcontinent** a term often used to refer to the entire Indian peninsula, including Nepal, Bhutan, India, Pakistan, and Bangladesh

Terms in This Chapter

Because its clear physical boundaries set it apart from the rest of the Asian continent, the term **subcontinent** is often used to refer to the entire Indian peninsula, which includes Nepal, Bhutan, India, Pakistan, and Bangladesh (the term usually does not include Afghanistan).

FIGURE 8.3 Political map of South Asia.

South Asians have recently adopted new place names to replace the names given them during British colonial rule. The city of Bombay, for example, is now officially *Mumbai*, Madras is *Chennai*, Calcutta is *Kolkata*, Benares is *Varanasi*, and the Ganges River is the *Ganga River*.

THINGS TO REMEMBER

• The Sardar Sarovar Dam and similar dams throughout the world are created primarily for irrigation and power generation, but they have side effects that often create more problems than they solve.

• The management of water—where it is apportioned and whether there is too much or too little of it—has been a feature of South Asian life for a long time. Climate change now makes water management more problematic.

• Improving the standard of living of poor populations in many parts of the world nearly always requires increases in water and energy use.

PHYSICAL GEOGRAPHY AND ENVIRONMENTAL ISSUES

PHYSICAL PATTERNS

Many of the landforms and even climates of South Asia are the result of huge tectonic forces. These forces have positioned the Indian subcontinent along the southern edge of the Eurasian continent, where it is surrounded by the warm Indian Ocean and shielded from cold airflows from the north by the massive mountains of the Himalayas (see Figure 8.1).

Landforms

The Indian subcontinent and the territory surrounding it are a dramatic illustration of what can happen when two tectonic plates collide. The Indian-Australian Plate, which carries India, broke free from the eastern edge of the African continent millions of years ago and drifted to the northeast (see Figure 1.8 on page 16). When this plate collided with the Eurasian Plate about 60 million years ago, it left a giant peninsula jutting into the Indian Ocean. As the relentless pushing from the south continued, both the leading (northern) edge of what became South Asia and the southern edge of Eurasia crumpled and buckled. The result is the world's highest mountains—the Himalayas, which rise more than 29,000 feet (8800 meters)—as well as other mountain ranges that curve away from the central impact zone—in the west, the Hindu Kush and Karakoram Range and in the east, the lesser-known ranges that border China and Burma (see Figure 8.1B and the figure map). The continuous compression also lifted the Plateau of Tibet, which rose up behind the Himalayas to an elevation of more than 15,000 feet (4500 meters) in some places. The compression and mountain-building process continues into the present.

The three great rivers of South Asia—the Indus, the Ganga, and the Brahmaputra—all begin within 100 miles (160 kilometers) of one another in the Himalayan highlands near the Tibet, Nepal, and India borders; they are largely fed by glacial meltwater (see the Figure 8.1 map and Figure 8.5). The main river basins of the Indus and Ganga lie to the southwest and south of the Himalayas in what is called the Indo-Gangetic Plain (see Figure 8.1C, F). The Indus flows southwest through Pakistan and ultimately joins the Arabian Sea. The Ganga flows southeast to the Bay of Bengal. The Brahmaputra flows east through Tibet and then turns south through Arunachal Pradesh and Assam into Bangladesh. The mouths of the Ganga and Brahmaputra are blended together in a giant, ever-changing delta.

Lying south of the Indo-Gangetic Plain is the Deccan Plateau, an area of modest uplands (1000–2000 feet [300–600 meters] in elevation) interspersed with river valleys (see Figure 8.1E). This upland region is bounded on the east and west by two moderately high mountain ranges, the Eastern and Western Ghats (see Figure 8.1D). The Ghat mountain ranges descend to long but narrow coastlines interrupted by extensive river deltas and floodplains (see Figure 8.1A). The river valleys and coastal zones are densely occupied; the uplands are only slightly less so.

Because of its continual high degree of tectonic activity and deep crustal fractures, South Asia is prone to devastating earthquakes, such as the magnitude 7.7 quake that shook the state of Gujarat in western India in 2001; the 7.6 quake that hit the India–Pakistan border region in 2005; and the 6.4 quake in Quetta Province, Pakistan, in 2008 that left 120,000 people homeless. Coastal areas are also vulnerable to tidal waves, or *tsunamis*, that are caused by undersea earthquakes. In 2004, a massive tsunami originating off Sumatra in Southeast Asia wrecked much of coastal Sri Lanka and southern India, killing tens of thousands of people there and many more in Southeast Asia.

Climate and Vegetation

The climate of South Asia is characterized by a seasonal reversal of winds known as monsoons (Figure 8.4). These monsoon

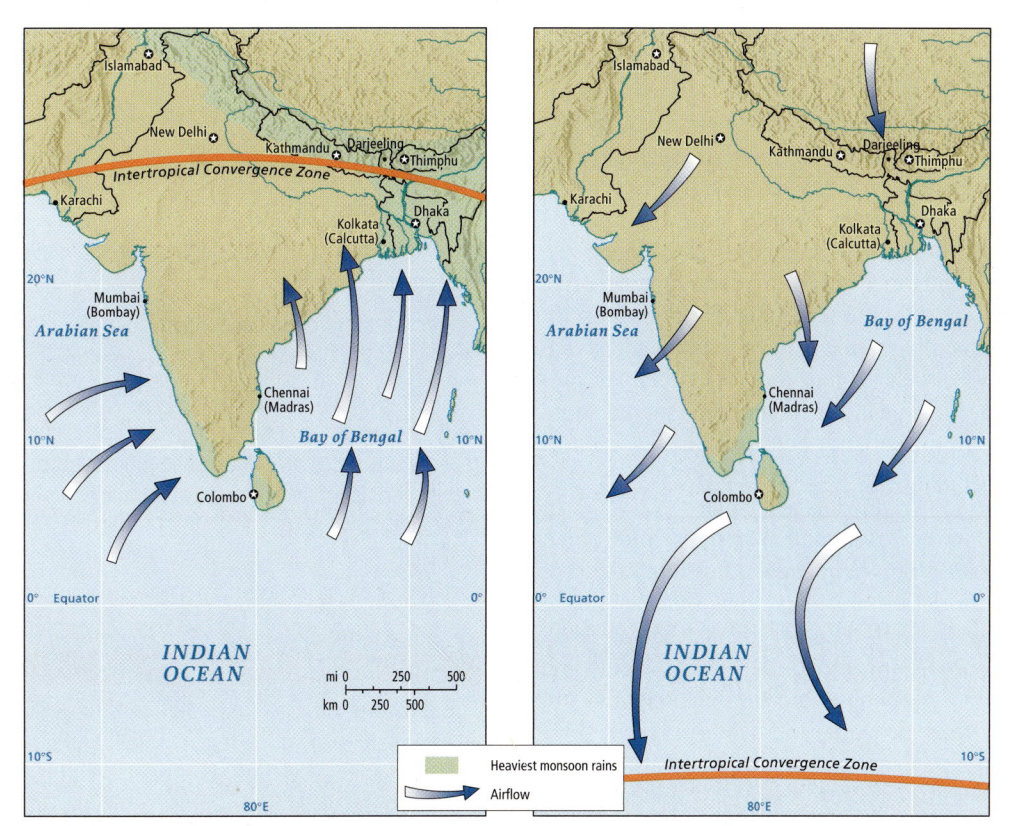

FIGURE 8.4 Summer and winter monsoons in South Asia.

(A) In the summer, the ITCZ moves across northern India, picking up huge amounts of moisture from the ocean, which are then deposited over India and Bangladesh.

(B) In the winter, cool, dry air blows from the Eurasian continent south across India toward the ITCZ, which in winter lies far to the south.

winds are affected by the *intertropical convergence zone (ITCZ)*, which, as the name suggests, is a zone created when air masses moving south from the Northern Hemisphere and north from the Southern Hemisphere converge near the equator. As the warm air rises and cools, it drops copious amounts of precipitation. As described in Chapter 7 (see page 279), the ITCZ shifts north and south seasonally. The intense rains of South Asia's summer monsoons are likely caused by the ITCZ being sucked onto the land by a vacuum created when huge air masses over the Eurasian continent heat up and rise into the upper troposphere.

The **summer monsoon** (see Figure 8.4A) begins in early June, when the warm, moist ITCZ air first reaches the mountainous Western Ghats. The rising air mass cools as it moves over the mountains, releasing rain that nurtures patches of dense tropical rain forests and tropical crops on the upper slopes of the Western and Eastern Ghats and in the central uplands. Once on the other side of India, the monsoon gathers additional moisture and power in its northward sweep up the Bay of Bengal, sometimes turning into tropical cyclones.

As the monsoon system reaches the hot plains of Bangladesh and the Indian state of West Bengal in late June, columns of warm, rising air create massive, thunderous cumulonimbus clouds that drench the parched countryside. Monsoon rains sweep east to west, parallel to the Himalaya Mountains, moving across northern India to Pakistan and finally petering out over the Kabul Valley in eastern Afghanistan by July. Rainfall is especially intense in the east, north of the Bay of Bengal, where the town of Darjeeling holds the world record for annual rainfall—about 35 feet, even though no rain falls for half the year. These patterns of rainfall are reflected in the varying climate zones (Figure 8.5) and agricultural zones of South Asia (see Figure 8.17 on page 341). Although sufficient rain falls in central India to support forests, most land has long been cleared of forest (see the discussion on page 329) and planted in crops. Much of the remaining forests are small and so separated from each other that they no longer provide suitable habitat for India's wildlife. Forest vegetation is more common in the Himalayan highlands and foothills, but even in the mountains there is widespread deforestation, often leaving only patchy areas of formerly richly forested land.

Periodically, the monsoon seasonal pattern is interrupted and serious drought ensues. This happened in July and August of 2009, when the worst drought in 40 years struck much of South Asia (see the discussion below). Then in September of 2009, heavy rains came to central south India, causing crop-damaging floods that killed hundreds of people. Increasingly, scientists are concluding that the extreme droughts and floods of recent years are not an anomaly but instead are part of global climate change.

The **winter monsoon** (see Figure 8.4B) is underway by November each year, when the cooling Eurasian landmass sends the cooler, drier, heavier air over South Asia. This cool, dry air from the north sinks down to the lower elevations and pushes the warm, wet air back south to the Indian Ocean. Very little rain falls in most of the region during this winter monsoon. However, as the ITCZ retreats southward across the Bay of Bengal, it picks up moisture that is then released as early winter rains over parts of southeastern India and Sri Lanka.

The monsoon rains deposit large amounts of moisture over the Himalayas, much of it in the form of snow and ice that add to the existing mass of glaciers (see page 23 in Chapter 1). Meltwater from these glaciers feeds the headwaters of the Indus, the Ganga, and the Brahmaputra. These rivers carry enormous loads of sediment, especially during the rainy season. When the rivers reach the lowlands, their velocity slows and much of the sediment settles out as silt which is then repeatedly picked up and redeposited by successive floods. As illustrated in the diagram of the Brahmaputra River in Figure 8.6, the movement of silt constantly rearranges the floodplain landscape, complicating human settlement and agricultural efforts. However, the seasonal deposit of silt nourishes much of the agricultural production in the densely occupied plains of Bangladesh. The same is true on the Ganga and Indus plains.

> **summer monsoon** rains that begin every June when the warm, moist ITCZ air first reaches the mountainous western Ghats
>
> **winter monsoon** a weather pattern that begins by November, when the cooling Eurasian landmass sends the cooler, drier, heavier air over South Asia

THINGS TO REMEMBER

- Because of the high degree of tectonic activity and deep crustal fractures, all the countries of South Asia are prone to devastating earthquakes.
- The three major rivers of this region originate from glacial meltwater high in the northwest Himalaya Mountains.
- The monsoons have a major influence on South Asia's climate. In June, the warm, moist air of the summer monsoon reaches India, where coastal mountains and rising hot air force the moist monsoon air up, cooling it and causing rain. Monsoon rains run parallel to the Himalayas in a variable band that reaches across northern India and Pakistan, finally petering out over the Kabul Valley in eastern Afghanistan by July.
- Periodic interruptions of the monsoon rainfall patterns can result in devastating droughts.
- Precipitation from monsoons is especially intense in the eastern foothills of the Himalayas.
- The three main rivers—the Indus, the Ganga, and the Brahmaputra—carry enormous loads of sediment and deposit it as silt, a process that is repeated by successive floods. The movement of silt constantly rearranges the floodplain landscape, complicating human settlement and agricultural efforts.

ENVIRONMENTAL ISSUES

GEOGRAPHIC INSIGHT 1

Environment: Climate change puts more lives at risk in South Asia than in any other region in the world, primarily due to water-related issues. Over the short term, droughts, floods, and the increased severity of storms imperil many urban and agricultural areas. Over the longer term, sea level rise may profoundly affect coastal areas and glacial melting poses a threat to rivers and aquifers.

FIGURE 8.5 **PHOTO ESSAY: Climates of South Asia**

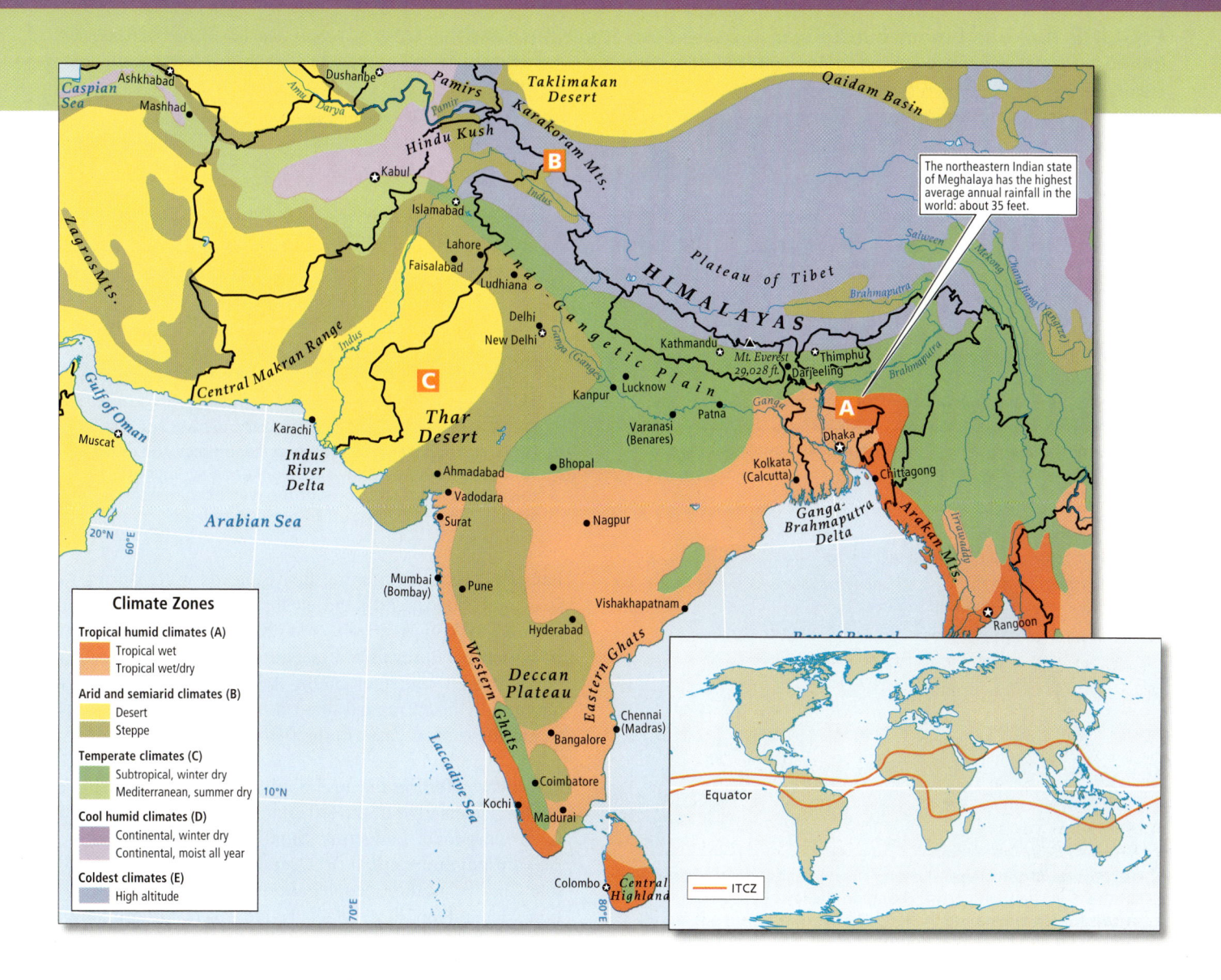

The northeastern Indian state of Meghalaya has the highest average annual rainfall in the world: about 35 feet.

Climate Zones

Tropical humid climates (A)
- Tropical wet
- Tropical wet/dry

Arid and semiarid climates (B)
- Desert
- Steppe

Temperate climates (C)
- Subtropical, winter dry
- Mediterranean, summer dry

Cool humid climates (D)
- Continental, winter dry
- Continental, moist all year

Coldest climates (E)
- High altitude

ITCZ

A Tropical wet, Meghalaya, India
Tushar S. Chowdhury/Flickr/Getty Images 

B High altitude, Kashmir, India 
Tauseef Mustafa/AFP/Getty Images

C Desert, Jaisalmer, India 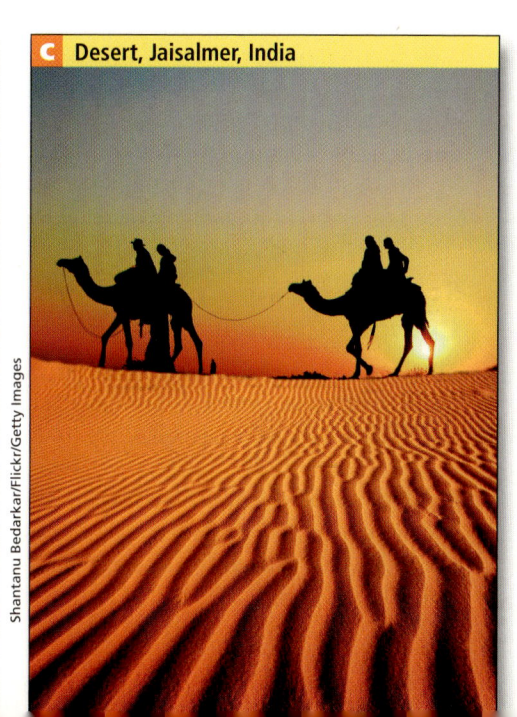
Shantanu Bedarkar/Flickr/Getty Images

FIGURE 8.6 The Brahmaputra River in Bangladesh at various seasonal stages. People who live along the river have learned to adapt their farms to a changing landscape, and along much of the river farmers are able to produce rice and vegetables nearly year-round. [Source consulted: *National Geographic*, June 1993, p. 125]

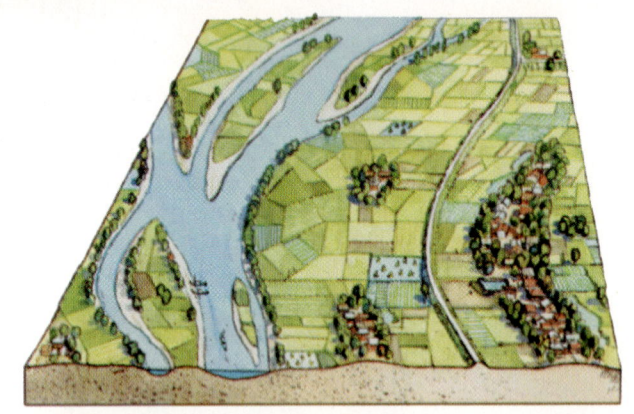

(A) Pre-monsoon stage: The river normally flows in multiple channels across the flat plain.

(B) Peak flood stage: During peak flood times, the great volume of water overflows the banks and spreads across fields, towns, and roads. The force of the water carves new channels, leaving some places cut off from the mainland.

(C) Post-monsoon stage: The river returns to its banks, but some of the new channels persist, changing the lay of the land and access for those who use the land. As the river recedes, it leaves behind silt and algae that nourish the soil. New ponds and lakes form and fill with fish.

Humans have lived in South Asia for at least 50,000 years, but as recently as 1700 C.E. (just before British colonization), population density and human environmental impacts were relatively light compared to the population and impacts of the period that followed; since 1700, they have grown exponentially.

South Asia's Vulnerability to Climate Change

All across South Asia, people face a wide range of challenges made more serious by climate change. Here we focus on eight of these challenges, all of which relate to water in some way.

1. Water Scarcity and Flooding Significant sections of South Asia have lived with water scarcity problems for millennia, mostly due to low average rainfall, and also because of extreme seasonal variability in rainfall. The Indus Valley of Pakistan, for example, has many ancient structures that captured water during wet seasons and floods and held it for use during dry times (see Figure 8.10A on page 332). Now, the shifting rainfall patterns and rising temperatures linked to climate change are creating drier conditions in much of South Asia, and occasionally causing abnormally heavy rainfall that can result in widespread flooding, as was the case in Pakistan in 2010.

2. Glacial Melting Because South Asia's three largest rivers are fed by glaciers high in the Himalayas, the issue of glacial melting is of particular concern. (Smaller glacial-melt rivers serve Afghanistan and Central Asia; see Figure 5.8A on page 203). As many as 703 million people, almost half of the region's population, depend on these glacially fed rivers for drinking water, domestic and industrial uses, and irrigation. (Irrigation requires more water than industrial and domestic uses combined.) While the immediate effect of glacial melting may be flooding, the real threat is that as the glaciers shrink, they will provide less water each year to recharge the ancient aquifers beneath the heavily populated Indo-Gangetic plains south of the Himalayan Range. The amount of water pumped for multiple uses on these plains already exceeds the rate of recharge; the long-term effect of the pumping, therefore, will be severe water shortages.

Both water conservation and increased water storage will be needed for supplies to last through the dry winter monsoon. High Himalayan communities in Ladakh, on the India–China border, are attempting to retain autumn glacial melt in stone catchments where the melt refreezes over the winter and is available for spring irrigation.

3. Sea Level Rise Tens of millions of impoverished farmers and fishers live near sea level in South Asia, most of them in Bangladesh, which has more people vulnerable to sea level rise (see page 23 in Chapter 1) than any other country in the world (**Figure 8.7A**). With 162 million people already squeezed into

FIGURE 8.7   **PHOTO ESSAY: Vulnerability to Climate Change in South Asia**

Climate change is putting more lives at risk in South Asia than in any other region. However, many responses to climate change are being developed that may increase resilience to climate-related hazards.

A A girl in coastal Bangladesh washes fish outside her home, which was flooded and partially ruined after a cyclone destroyed the river embankment that once protected her village from high tides. Bangladesh is extremely exposed to sea level rise, increased storm intensity, and flooding related to glacial melting. Because it has a large population of poor people, Bangladesh is very sensitive to these climate hazards.

B In Afghanistan, a man makes improvements to irrigation infrastructure that helps reduce the sensitivity of nearby areas to water shortage. Widespread poverty and political instability contribute to Afghanistan's extremely high vulnerability to climate hazards.

Vulnerability to Climate Change
- Extreme
- High
- Medium
- Low

C A high tide during monsoon rains floods part of Mumbai's rail system. Low-lying Mumbai (formerly Bombay) is at significant risk for sea level rise; while the city's wealth and well-developed emergency response systems increase its resilience to climate hazards, much of Mumbai remains quite sensitive.

D Workers add boulders to reinforce an embankment that guards against erosion south of Kolkata (formerly Calcutta) on an island that is very exposed to sea level rise and storm surge. Such embankments can reduce sensitivity to water-level changes, but they are expensive to build and require maintenance.

a country the size of Iowa, as many as 17 million Bangladeshis might have to find new homes if sea levels rise by 3 to 5 feet. The biggest economic impacts of sea level rise would come from the submergence of parts of South Asia's largest cities, such as Dhaka in Bangladesh, Mumbai (see Figure 8.7C and pages 23–24 in Chapter 1) and Kolkata in India, and Karachi in Pakistan.

Also threatened by continuing sea level rise are the Maldives Islands in the Indian Ocean, 80 percent of which lie 1 meter or less above the sea. Beach erosion is so severe that homes built only a few years ago in this richest of South Asia's countries are falling into the sea.

All of these places are developing responses to climate change, though the resources available to do so are often very limited in this region (see Figure 8.7D).

4. Rising per Capita Use of Water As we learned in Chapter 1, humans require an average of 5 to 13 gallons (20 to 50 liters) of clean water per day for basic domestic needs: drinking, cooking, and bathing/cleaning. In South Asia's poorest urban and rural areas, per capita domestic water consumption is about 5 to 6 gallons per day. Consumption increases as incomes rise, in part because people add conveniences to homes and businesses that consume large amounts of water (a flush of a toilet, for instance, can use several gallons). The middle and upper classes consume closer to 13 gallons per day, and some much more than that. As more and more South Asians join the middle class, they will consume more water per capita. Add population growth to that and it becomes easy to see that the demand for water is bound to increase across this region. A window into exactly how much water would be needed is provided in Delhi (the ancient city and New Delhi are now commonly referred to simply as Delhi). Seventeen five-star hotels, serving a few thousand guests, most of whom are in India on business from various parts of the developed world, use about 210,000 gallons (800,000 liters) of water daily. This would be enough to serve the needs of 42,000 people living in Delhi's slums.

5. The Role of Virtual Water We must also consider that domestic water consumption is only a fraction of a person's actual water consumption. Much of South Asia's water is tied up as virtual water used to produce the manufactured goods and agricultural products that South Asians themselves consume as well as those that are made for export. The concept of *virtual water*, introduced in Chapter 1 (page 26) is especially useful in assessing the sustainability of water resources in the drier regions of South Asia. Afghanistan, Pakistan, and northwest India are all naturally dry regions that have been made more arid by thousands of years of human use.

The water that goes into the production of all agricultural and industrial products of these regions (whether used locally or exported) is drawn from supplies that are so scarce ordinary citizens often must survive on less water than is considered healthy. In the past, when production was mostly for local consumption, less virtual water was used and more remained in the region. Now, export crops—cotton, grain, rice, fruits, and nuts—are all produced with irrigation; thus all use a large amount of virtual water (Table 8.1). Because these crops are exported and the virtual water is consumed elsewhere in the global marketplace, these water-scarce regions are in effect subsidizing the water use of wealthy global consumers. The same is true for manufactured goods with high virtual water inputs, such as textiles, garments, hides, leather goods, and sporting goods. Furthermore, in these dry regions, water for crops and industry is frequently drawn down faster than it is naturally replenished, so little of this water use is sustainable. Most importantly, the costs of depleted water used in production are not sufficiently accounted for in the pricing of goods exported from these regions.

6. Inefficient Allocation and Use of Water There is wide spatial and temporal variation in the availability of water across the region, and this is only likely to increase with climate change. South Asia has more than 20 percent of the world's population but only 4 percent of its freshwater, often making access to water difficult. Disputes are increasingly common, yet few citizens understand the gravity of the problem because political rivalries and bureaucratic inertia have delayed the development and implementation of national water policies. As a result, there are cases, as described below, when scarce water is used to purposely create a flood and in so doing threatens the well-being of millions.

When two or more states or countries share a water basin, conflicts can become geopolitical in scale. During the dry season, India occasionally diverts as much as 60 percent of the Ganga River flow to Kolkata to flush out channels where silt is accumulating and hampering river traffic (see the Ganga-Brahmaputra Delta on map in Figure 8.1 and the map in Figure 8.5). These diversions temporarily deprive Bangladesh of normal freshwater flow. Reductions in the freshwater levels in the Ganga-Brahmaputra Delta allow saltwater from the Bay of Bengal to penetrate inland, ruining agricultural fields. The diversion has also caused major alterations in Bangladesh's coastline, damaging its small-scale fishing industry. Thus, the livelihoods of 40 million rural Bangladeshis have been put at risk to serve the needs of Kolkata's 16 million people, triggering protests in

TABLE 8.1 Virtual water component table

Produce	Virtual water content (in liters)
1 apple	125
1 kg bread	1608
1 kg milled rice	2500
1 kg cotton fabric	10,000
1 kg cow leather	17,000
1 U.S. dollar worth of industrial goods	80

Source: Water Footprint Product Gallery, Water Footprint Network Web site, at http://www.waterfootprint.org/?page=files/productgallery

Bangladesh. In the late 1990s, India signed a treaty with Bangladesh promising a fairer distribution of water, but as of 2010, Bangladesh was still receiving a considerably reduced flow. India's draft water policy of 2012 merely mentions the need to resolve the issue.

📹 **182. COCA-COLA BLAMED FOR INDIA'S WATER PROBLEMS**

ON THE BRIGHT SIDE

Technological Solutions

To address the issues of climate warming and drought, public and private entities are using alternative energy sources to reduce CO_2 emissions in the atmosphere. For example, India, which contributes the seventh-largest amount of greenhouse gases in the world, is also home to the world's largest producer of plug-in electric cars. The Mahindra Reva E20 sells in India for about U.S.$15,000 with a government subsidy. India's small but surging middle class now has the disposable income to afford these economy cars. Even factoring in emissions from the plants that generate the electricity used to charge the cars, electric cars produce substantially lower levels of CO_2 emissions than do gasoline- or diesel-powered cars.

Solar and wind energy are the focus of public investment for several South Asian countries because they have the potential to decrease greenhouse gas emissions. In north and west South Asia, where there is less cloud cover and energy is in greatest demand by industries and high-tech firms, the development of solar power is being emphasized. Wind energy is most efficiently produced in the wind-prone state of Tamil Nadu in southern India. Nationally, the use of wind power increased 22 percent per year from 1992 to 2010; by 2011, it constituted 70 percent of India's renewable energy generation.

7. Water Pollution First, it should be noted that most people do not drink water out of the tap in South Asia. Widespread contamination within most urban and rural water supplies means that water must be boiled or otherwise purified before it is consumed. Sewage, and garbage that is dumped into rivers and streams, is a major source of contamination (**Figure 8.8A**). Most sewage enters these bodies of water in raw form because city sewage systems exceeded their capacity long ago. Only 5 percent of cities or towns have any sewage treatment at all.

As is often the case in South Asia, religion plays a role in the resolution of serious problems such as water pollution. In 2008, Veer Bhadra Mishra, who was a Brahmin priest and professor of hydraulic engineering at Banaras Hindu University in Varanasi, received approval from India's central government to build a series of processing ponds that use India's heat and monsoon rains to clean Varanasi's sewage at half the cost of other methods. The technology has since been mired in bureaucratic delay, but that didn't stop Mishra from preaching a contemporary religious message to the thousands who visited his temple on the banks of the Ganga River, which is sacred to Hindus: No longer is it valid to believe that the Ganga is a goddess who purifies all she touches while assuming that it is impossible to cause her damage. Rather, Mishra said, because the Ganga is their symbolic mother, it would be a travesty for Hindus to smear her with sewage and other wastes.

Basic water safety is especially crucial in historic religious pilgrimage towns, such as Varanasi, where each year millions of Hindus come to die, be cremated, and have their ashes scattered over the Ganga River. The number of such final pilgrimages has increased with population growth and affluence, causing wood for cremation fires to become scarce. As a result, incompletely cremated bodies are being dumped into the river, where they pollute water used for drinking, cooking, and ceremonial bathing (see Figure 8.8B). In an attempt to deal with this problem, the government recently installed an electric crematorium on the riverbank. It is attracting considerable business, as a cremation in this facility costs 30 times less than a traditional funeral pyre.

8. Crop Failures and Food Insecurity Insufficient water threatens two essential agricultural activities in South Asia: the production of export crops and the production of food for domestic consumption. The primary use of water in South Asia is for irrigation, and although old and new ways of conserving water in fields are being tried (see Figure 8.7B), the demand for irrigation water is not likely to decrease. For most countries in the region, agriculture supplies jobs for close to 50 percent of the workforce and the export crops they produce contribute 17 percent or more of the GNI. Water conservation measures will contribute to crop success and food security by preserving the possibility of irrigating; meanwhile, the work being done to reduce CO_2 levels will help alleviate climate warming and drying trends.

Since ancient times, both India and Pakistan have pioneered methods of increasing the rate at which water deposited during the summer monsoon percolates through the soil and into underground aquifers rather than evaporating. This practice has made more water available for irrigation during the dry season. Drip irrigation technology, now known to be the most efficient way to irrigate, would also help conserve water. Because agriculture uses the most water of any human activity (more than 70 percent in some places), drip irrigation could free up enormous amounts of water for other uses. However, the relatively high cost of modern efficient drip equipment has hampered widespread implementation of drip systems.

Responses to Water Issues Related to Global Climate Change

Unless effective action is taken against climate change, Himalayan glaciers could eventually disappear, causing South Asia's largest rivers to run nearly dry, at least during the winter when cool, dry air flows off the Eurasian continent and rainfall is sparse (see Figure 8.4 and the Figure 8.5 map). South Asia is pioneering some innovative responses to the multiple threats posed by global climate change. India, by far the largest country in the region, has some experience in developing and implementing emergency plans. In 2012, India adopted a national water-use plan designed to address the many water crises arising from climate change. However, the government has already had to back down on the most controversial aspects of the plan, such as the proposal that all government water

FIGURE 8.8 PHOTO ESSAY: Human Impacts on the Biosphere in South Asia

South Asia's huge population and growing industries have had major impacts on water and air quality, as well as on the extent and health of remaining ecosystems, such as forests.

A Students walk on a bridge across a very polluted and garbage-choked stream in a slum area in Rawalpindi, Pakistan. Nationally, only 8 percent of urban sewage is treated before it is discharged. Meanwhile, 40 percent of premature deaths in Pakistan are attributed to water-related diseases. [Farooq Naeem/AFP/Getty Images]

B A Hindu woman washes her baby in the polluted Ganga River in Varanasi, India. Many Hindus consider this a sacred act of ritual purification.

Human Impact, 2002

Land cover
- Forests
- Grasslands
- Deserts
- Tundra
- Ice
- Modern national boundaries

Overfishing
- Threatened fisheries

Human impact on land
- High impact
- Medium–high impact
- Low–medium impact

Acid rain
- Early threat of acid rain

C A public bus billows smoke in New Delhi, India, which now rivals Beijing, China, for having the worst air pollution in the world. Sixty-seven percent of air pollution in New Delhi comes from vehicles. [Prashanth Vishwanathan/Bloomberg/Getty Images]

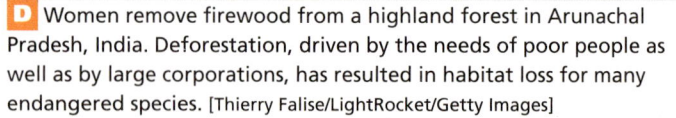

D Women remove firewood from a highland forest in Arunachal Pradesh, India. Deforestation, driven by the needs of poor people as well as by large corporations, has resulted in habitat loss for many endangered species. [Thierry Falise/LightRocket/Getty Images]

Environmental Issues 329

FIGURE 8.9 LOCAL LIVES
People and Animals in South Asia

A A Bengal tiger (*Panthera tigris tigris*) crosses a stream in Nepal. Fewer than 2500 individuals remain in the wild, many of them in national parks and reserves in India, Bangladesh, Nepal, and Bhutan. Despite ongoing conservation efforts since the 1970s, Bengal tiger populations are in a steady decline due to demand for tiger body parts by some practitioners of traditional Chinese medicine. An endangered species, tigers are the national animal of both India and Bangladesh. [Tom Brakefield/Stockbyte/Getty Images]

B A cow lounges in a fabric store in Varanasi, India. Cattle are allowed to roam free in some cities in India, due to the reverence Hindus widely hold for cows. The killing of cattle for any purpose is banned in 6 of India's 28 states and allowed without restriction in only 4. Scholars have found much evidence to suggest that cows have not always been sacred, and were eaten by Hindus in the distant past. [Buena Vista Images/The Image Bank/Getty Images]

C Asian elephants at work in Nagarhole National Park in Karnataka, India, carrying fallen branches and brush. Asian elephants (*Elephas maximus*) are a separate species distinct from African elephants (*Loxodonta africana*) and have, among several other unique characteristics, the largest volume of cerebral cortex of any land animal. This portion of the brain is used for memory, attention, perceptual awareness, thought, language, and consciousness. Asian elephants can be trained to follow instructions, a quality that has made them useful to humans for work and warfare for thousands of years. They are particularly prized for their ability to cross difficult terrain while carrying heavy objects such as logs. [Dibyangshu Sarkar/AFP/Getty Images]

services be privatized and that all water users, including even the poorest of the poor, be charged a fee for the water they use.

Deforestation

Deforestation has been occurring in South Asia since the first agricultural civilizations developed between 5000 and 10,000 years ago. Ecological historians have shown that as the forests vanished, the northwestern regions of the subcontinent (from India to Afghanistan) became increasingly drier. The pace of deforestation has increased dramatically over the past 200 years. By the mid-nineteenth century, perhaps a million trees a year were felled for use in building railroads alone. Such radical deforestation jeopardized the well-being both of people and of animals in South Asia (Figure 8.9).

South Asia's forests are shrinking because of commercial logging and expanding village populations that use wood for building and for fuel. Many of South Asia's remaining forests are in mountainous or hilly areas, where forest clearing dramatically increases erosion during the rainy season. In addition to the loss of CO_2-absorbing forests, one result of deforestation is massive landslides that can destroy villages and close roads. With fewer trees and less soil to retain the water, rivers and streams become clogged with runoff, mud, and debris. The effects can reach so far downstream that increased flooding in the plains of Bangladesh is now linked to deforestation in the Himalayas.

Resistance to Deforestation Unlike China and many other nations facing similar problems, the countries of South Asia have a healthy and vibrant culture of environmental activism that has alerted the public to the consequences of deforestation. In 1973, for example, in the Himalayan district of Uttarakhand (then known as Uttaranchal), India, a sporting-goods manufacturer planned to cut down a grove of ash trees so that the factory, in the distant city of Allahabad, could use the wood to make tennis racquets. The trees were sacred to nearby villagers, however, and when their protests were ignored, a group of local women took dramatic action. When the loggers came, they found the women hugging the trees and refusing to let go until the manufacturer decided to find another grove.

The women's action grew into the *Chipko movement* (literally, "hugging"), which is also known as the *social forestry movement*. The movement has spread to other forest areas and has been responsible for slowing deforestation and increasing ecological awareness. Proponents of the movement argue that the management of forest resources should be turned over to local communities. They say that people living at the edges of forests possess complex local knowledge of those ecosystems that has been gained over generations—knowledge about which plants are useful for food, medicines, and fuel, and as building materials. Those who live in forested areas are more likely to manage the forests carefully because they want their descendants to benefit from forests for generations to come.

Industrial Air Pollution

In many parts of South Asia, the air as well as the water is endangered by industrial activity. Emissions from vehicles and

CASE STUDY

Nature Preserves in the Nilgiri Hills

The Mudumalai Wildlife Sanctuary and neighboring national parks in the Nilgiri Hills (part of the Western Ghats) harbor some of the last remaining forests in southern India. Here, in an area of about 600 square miles, live a few of India's last wild tigers and a dozen or more other rare species, such as sloth bears and barking deer (see Figure 8.9A). Even much smaller forest reserves play an important role in conservation. At 287 acres, Longwood Shola is a tiny remnant of the ancient tropical forests that once covered the Nilgiris.

Phillip Mulley, a naturalist, Christian minister, and leader of the Badaga ethnic group, points out that the indigenous peoples of the Nilgiris must now compete for space with a growing tourist industry (1.7 million visitors in 2005). In addition, huge tea plantations were cut out of forestlands by the Tamil Nadu state government to provide employment for Tamil refugees from the conflict in Sri Lanka (see Figure 8.20B). So while the forestry department and citizen naturalists are trying to preserve forestlands, the social welfare department, faced with a huge refugee population, is cutting them down. [Source: Lydia and Alex Pulsipher and the government of Tamil Nadu. For detailed source information, see Text Sources and Credits.]

180. TRIBAL PEOPLE IN INDIA WANT TO PROTECT INDIGENOUS WAYS OF LIFE

181. INDIA'S NATIONAL SYMBOL BECOMING MORE DIFFICULT TO SPOT

coal-burning power plants are so bad that breathing Delhi's air is equivalent to smoking 20 cigarettes a day (see Figure 8.8C on page 328). The acid rain caused by industries up and down the Ganga River basin is destroying good farmland and renowned heritage monuments such as the Taj Mahal.

M. C. Mehta, a Delhi-based lawyer, became an environmental activist partly in response to the condition of the Taj Mahal, whose white marble was becoming pitted by acid rain. For more than 20 years, he has successfully promoted environmental legislation that has removed hundreds of the most polluting factories from India's river valleys. His efforts are also a response to a disastrous event that took place in central India in 1984: An explosion at a pesticide plant in Bhopal produced a gas cloud that killed at least 15,000 people and severely damaged the lungs of 50,000 more. The explosion was largely the result of negligence on the part of the U.S.-based Union Carbide Corporation (which owned the plant) and the local Indian employees who ran the plant. To help address the tragedy, the Indian government launched an ambitious campaign to clean up poorly regulated factories, a project that is far from complete.

THINGS TO REMEMBER

> **GEOGRAPHIC INSIGHT 1**
> • **Environment** Climate change puts more lives at risk in South Asia than in any other region in the world, primarily due to water-related issues. Over the short term, droughts, floods, and the increased severity of storms imperil many urban and agricultural areas. Over the longer term, sea level rise may profoundly affect coastal areas and glacial melting poses a threat to rivers and aquifers.

• Unless effective action against climate change is taken, Himalayan glaciers could eventually disappear, causing South Asia's largest rivers to run nearly dry, at least during the winter. As many as 703 million people, almost half of the region's population, depend on these glacially fed rivers for drinking water, domestic and industrial uses, and irrigation.

• Pakistan, India, and Bangladesh have major air and water pollution problems related to industrial waste and lack of sewage treatment facilities.

• South Asia's forests are shrinking because of commercial logging and expanding village populations that use wood for building and for fuel.

• The *Chipko* ("tree hugging") *social forestry movement* in Uttarakhand, India, has spread to other forest areas, slowing deforestation and increasing ecological awareness.

HUMAN GEOGRAPHY

Home to some of the earth's oldest and most influential civilizations, South Asia faces the world's great transitions with an enviable ability to innovate while maintaining ancient traditions. Tens of thousands of years of continuous human occupation, and the integration of numerous influences from outside the region, have given South Asia astounding cultural diversity to draw on in this time of change. In the current era of globalization, South Asia is integrating powerful global economic forces while struggling to achieve an economic transformation that benefits more than a privileged few. The politics of this region are also in transition; people are demanding more political freedom, challenging the authoritarian systems of the past. In this most densely populated of world regions, the demographic transition is underway as population growth slows while changing gender roles give women more freedom. Throughout these great transitions, the people of South Asia often manage to create a synthesis of tradition and innovation that smooths over the most jarring changes.

HUMAN PATTERNS OVER TIME

A variety of groups have migrated into South Asia, many of them as invaders who conquered peoples already there. Despite much blending over the millennia, the continued coexistence and interaction of many of these groups make South Asia both a richly diverse and an extremely contentious place.

The Indus Valley Civilization

There are indications of early humans in South Asia as far back as 200,000 years ago, but the first real evidence of modern humans in the region is about 38,000 years old. The first large agricultural communities, known as the **Indus Valley civilization** (or **Harappa culture**), appeared about 4500 years ago along the Indus River in what is modern-day Pakistan and northwest India. The architecture and urban design of this civilization were quite advanced for the time. Homes featured piped water and sewage disposal. Towns were well planned, with wide, tree-lined boulevards laid out in a grid. Evidence of a trade network that extended to Mesopotamia and eastern Africa has also been found.

Vestiges of the Indus Valley civilization's agricultural system survive to this day in parts of the valley, including infrastructure for storing monsoon rainfall to be used for irrigation in dry times (shown in **Figure 8.10A**). Possible cultural and linguistic remnants of the Indus Valley civilization survive today among the Dravidian peoples of southern India, who originally migrated from the Indus region beginning about 2000 years ago.

Scholars have long debated the reasons for the decline of the Indus Valley civilization. Some believe that complex geologic (seismic) and ecological changes (drier climate) brought about a gradual demise. Others argue that foreign invaders brought a swift collapse, instigating out-migration.

A Series of Invasions

The first recorded invaders to join the indigenous people of South Asia came from Central and Southwest Asia into the rich Indus Valley and Punjab about 3500 years ago. Many scholars believe that these people, referred to as Indo-European (a linguistic term; see Figure 8.11), in conjunction with those of the Harappa and other indigenous cultures, instituted some of the early elements of classical Hinduism, the predominant religion of India today. One of those elements was the still-influential caste system (discussed on pages 332–334), which divides society into hereditary hierarchical categories.

Wave after wave of other invaders arrived, including the Persians, the armies of the Greek general Alexander the Great, and numerous Turkic and Mongolian peoples. Defensive structures against these invaders can still be found across northwest South Asia. Jews came to the Malabar Coast of Southwest India more than 2500 years ago, Christians shortly after the time of Jesus. Arab traders came by land and sea to India long before the emergence of Islam; and then starting about 1000 years ago, Arab traders and religious mystics introduced Islam to what are now Afghanistan, Pakistan, and northwest India. By sea, the Arabs brought Islam to the coasts of southwestern India and Sri Lanka. In 1526, the **Mughals**, a group of Turkic Persian people from Central Asia, invaded from the north, intensifying the growth of Islam. The Mughals reached the height of their power and influence in the seventeenth century, controlling the north-central plains of South Asia. The last great Mughal ruler (Aurangzeb) died in 1707, but the cultural legacy of the Mughals remained, even as the power and range of the dynasty declined. One aspect of this legacy is the 520 million Muslims now living in South Asia. The Mughals also left a unique heritage of architecture, art, and literature that includes the Taj Mahal, miniature painting, and a rich tradition of lyric poetry (see Figure 8.10B). The Mughals contributed to the evolution of the Hindi language, which became the language of trade of the northern subcontinent and which is still used by more than 400 million people.

Language and Ethnicity

One result of the numerous invasions of South Asia over the millennia is that today there are many distinct ethnic groups, each with its own language or dialect. In India alone, 18 languages are officially recognized, but there are actually hundreds of distinct languages. This complexity results partly from strong cultural traditions that allow groups to maintain distinct identities in the midst of foreign cultures and withstand invasion or even factors that could force an entire group to relocate. As shown in **Figure 8.11**, the Dravidian language-culture group, represented by numbers 14 through 19, is an ancient group, thought to have originated in Pakistan, that predates the Indo-European invasions by a thousand years or more. Today, because of ancient migrations from Pakistan, Dravidian languages are found mostly in southern India, but a remnant of the extensive Dravidian past can still be found in the Indus Valley in south-central Pakistan.

> **Indus Valley civilization** the first substantial settled agricultural communities, which appeared about 4500 years ago along the Indus River in modern-day Pakistan and northwest India
>
> **Harappa culture** see Indus Valley civilization
>
> **Mughals** a dynasty of Central Asian origin that ruled India from the sixteenth century to the nineteenth century
>
> **Hinduism** a major world religion practiced by approximately 900 million people, 800 million of whom live in India; a complex belief system, with roots both in ancient literary texts (known as the *Great Tradition*) and in highly localized folk traditions (known as the *Little Tradition*)

By the seventeenth century, Hindi—an amalgam of Persian-based and Sanskrit-based Indo-European languages—was the dominant language throughout northern India and what is now Pakistan, where it is known as Urdu. Today, variants of Hindi serve as national languages for both India and Pakistan, though it is the first, or native, language of only a minority. English is a common second language throughout the region. As the language of the British colonial bureaucracy, English remains a language used at work by professional people of all categories. Between 10 and 15 percent of South Asians speak, read, and write English. Many others use a version of spoken English.

Religious Traditions

The main religious traditions of South Asia are Hinduism, Buddhism, Sikhism, Jainism, Islam, and Christianity (**Figure 8.12**). (For a discussion of Islam, Christianity, and Judaism, see pages 244–246 in Chapter 6.)

Hinduism is a major world religion practiced by approximately 900 million people, 800 million of whom live in India. It is a complex belief system, with roots in both ancient literary texts (known as the *Great Tradition*) and in highly localized folk traditions (known as the *Little Tradition*).

The Great Tradition is based on 4000-year-old scriptures called the *Vedas*. Its major tenet is that all gods are merely illusory manifestations of the ultimate divinity, which is formless and infinite. Some devout Hindus worship no gods at all, and

A A water storage tank at Mohenjo-daro, in Pakistan.
[Luca Tettoni/Robert Harding World Imagery/Getty Images]

B The Taj Mahal, a Mughal mausoleum in Agra, India.
[Mlenny Photography/E+/Getty Images]

| 2000,000 B.C.E. | 5000 B.C.E. | 0 C.E. | 1500 C.E. | 1600 C.E. | 1700 C.E. |

200,000 B.C.E. Early humans enter South Asia

2500 B.C.E. Mohenjo-daro

1526–1857 C.E. Mughal Empire

1643–1653 Taj Mahal

FIGURE 8.10 VISUAL HISTORY OF SOUTH ASIA

Languages

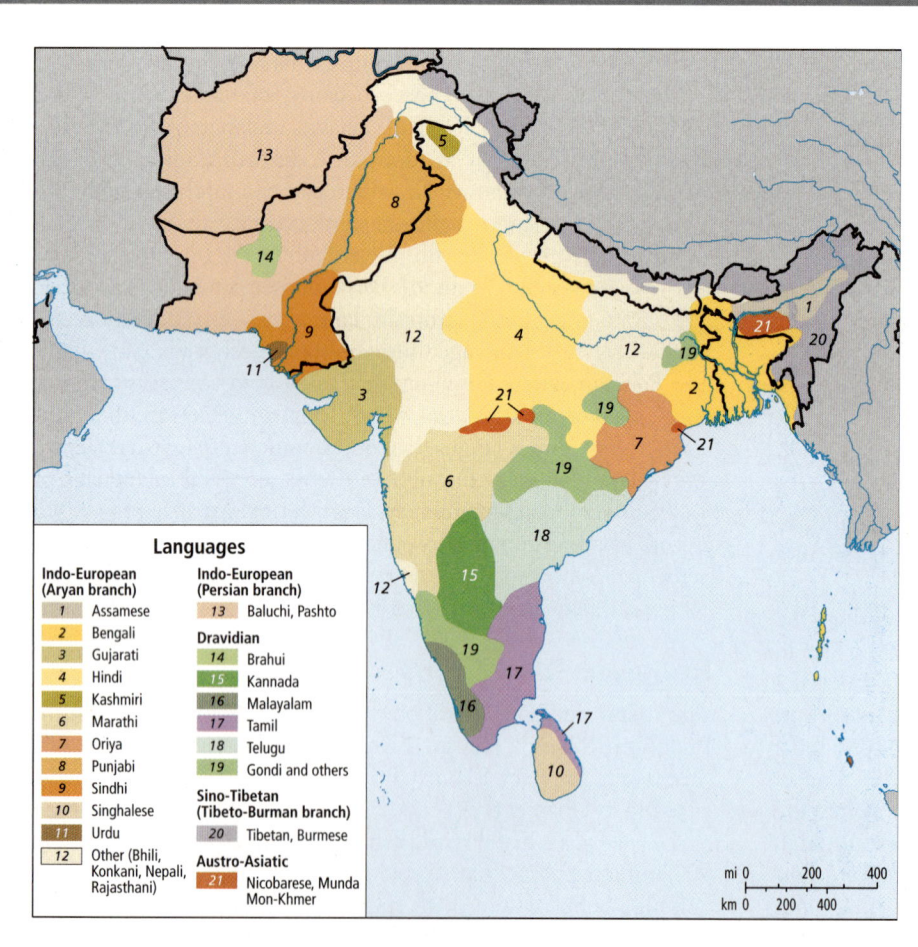

Indo-European (Aryan branch)
1. Assamese
2. Bengali
3. Gujarati
4. Hindi
5. Kashmiri
6. Marathi
7. Oriya
8. Punjabi
9. Sindhi
10. Singhalese
11. Urdu
12. Other (Bhili, Konkani, Nepali, Rajasthani)

Indo-European (Persian branch)
13. Baluchi, Pashto

Dravidian
14. Brahui
15. Kannada
16. Malayalam
17. Tamil
18. Telugu
19. Gondi and others

Sino-Tibetan (Tibeto-Burman branch)
20. Tibetan, Burmese

Austro-Asiatic
21. Nicobarese, Munda Mon-Khmer

FIGURE 8.11 Major language groups of South Asia.
The modern pattern of language distribution in South Asia is testimony to the fact that this region has long been a cultural crossroads. [Source consulted: Alisdair Rogers, ed., *Peoples and Cultures* (New York: Oxford University Press, 1992), p. 204]

instead engage in meditation, yoga, and other spiritual practices designed to bring people to a state described as "infinite consciousness." The average person, however, is thought to need the help of personified divinities in the form of gods and goddesses (the Little Tradition). While all Hindus recognize some deities (such as Vishnu, Shiva, Ganesh, and Krishna), many deities are found only in one region, one village, or even one family.

Some beliefs are held in common by nearly all Hindus. One is the belief in reincarnation, the idea that any living thing that desires the illusory pleasures (and pains) of life will be reborn after it dies. A reverence for cows, which are seen as only slightly less spiritually advanced than humans, also binds all Hindus together (see Figure 8.9B). This attitude, along with the Hindu prohibition on eating beef, may stem from the fact that cattle have been tremendously valuable in rural economies as the primary source of transport, field labor, dairy products, fertilizer, and fuel (animal dung is often burned).

Caste: An Explanation Hinduism includes the **caste system**, a complex and ancient way of dividing society into hereditary hierarchical categories (see Figure 8.30 on page 357). One is born into a given subcaste, or community (called a *jati*), that traditionally defined much of one's life experience—where one would live, where and what one

caste system a complex, ancient Hindu system for dividing society into hereditary hierarchical classes

jati in Hindu India, the subcaste into which a person is born, which traditionally defines the individual's experience for a lifetime

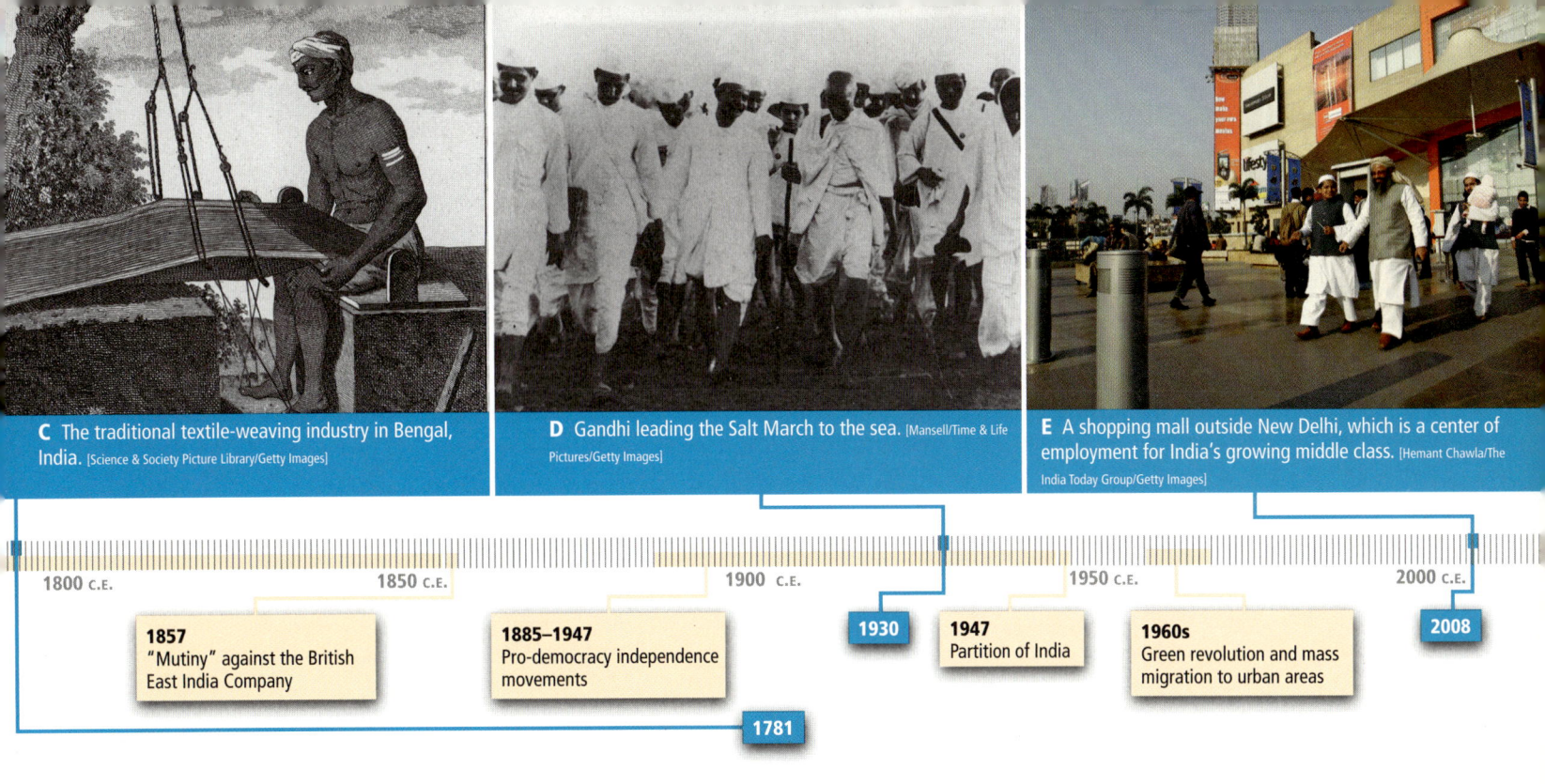

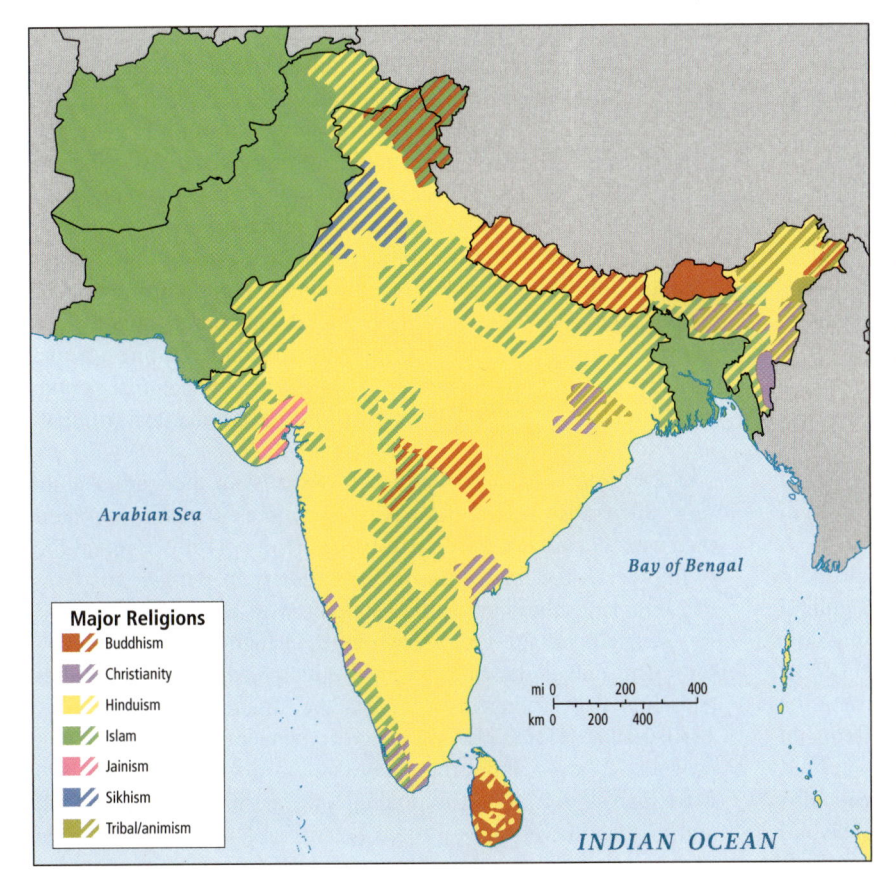

C The traditional textile-weaving industry in Bengal, India. [Science & Society Picture Library/Getty Images]

D Gandhi leading the Salt March to the sea. [Mansell/Time & Life Pictures/Getty Images]

E A shopping mall outside New Delhi, which is a center of employment for India's growing middle class. [Hemant Chawla/The India Today Group/Getty Images]

1800 C.E. – 2000 C.E.	
1857	"Mutiny" against the British East India Company
1885–1947	Pro-democracy independence movements
1930	
1947	Partition of India
1960s	Green revolution and mass migration to urban areas
2008	
1781	

FIGURE 8.12 Major religions in South Asia. Notice that while Hinduism dominates in India, there are variable and overlapping (striped) patterns of other religious traditions in India, Nepal, and Sri Lanka. On the other hand, Bhutan, Bangladesh, Pakistan and Afghanistan have very little religious diversity. [Source consulted: Gordon Johnson, *Cultural Atlas of India* (New York: Facts on File, 1996), p. 56]

Major Religions:
- Buddhism
- Christianity
- Hinduism
- Islam
- Jainism
- Sikhism
- Tribal/animism

could eat and drink, with whom one would associate, one's marriage partner, and often one's livelihood. The classical caste system has four main divisions or tiers, called **varna**, within which are thousands of *jatis* and sub-*jatis*, which vary from place to place.

Brahmins, members of the priestly caste, are the most advantaged in caste hierarchy. Thus they must conform to those behaviors that are considered most ritually pure (for example, strict vegetarianism and abstention from alcohol). As is the case with many castes, Brahmins are found in many occupations outside their place as priests in the *varna* system (as barbers and hairdressers, for example). Below Brahmins, in descending rank, are *Kshatriyas*, who are warriors and rulers; *Vaishyas*, who are landowning (small-plot) farmers and merchants; and *Sudras*, who are low-status laborers and artisans. A fifth group, the *Dalits*—"the oppressed," or untouchables—is actually considered to be so lowly as to have no caste. Dalits perform those tasks that caste Hindus consider the most despicable and ritually polluting: killing animals, tanning hides, cleaning, and disposing of refuse. A sixth group, also outside the caste system, is the *Adivasis*, who are thought to be descendants of the region's ancient aboriginal inhabitants.

Although *jatis* are associated with specific sub-categories of occupations, in modern economies, this aspect of caste is more symbolic than real. Members of a particular *jati* do, however, follow the same social and cultural customs, dress in a similar manner, speak the same dialect, and tend to live in particular neighborhoods or villages.

varna the four hierarchically ordered divisions of society in Hindu India underlying the caste system: Brahmins (priests), Kshatriyas (warriors/kings), Vaishyas (merchants/landowners), and Sudras (laborers/artisans)

333

This spatial separation arises from the higher-caste communities' fears of ritual pollution, such as through physical contact or the sharing of water or food with lower castes. When one stays in the familiar space of one's own *jati*, one is enclosed in a comfortable circle of families and friends that becomes a mutual aid society in times of trouble. The social and spatial cohesion of *jatis* helps explain the persistence and respect paid to a system that, to outsiders, seems to put a burden of shame and poverty on the lower ranks.

It is important to note that caste and class are not the same thing. Class refers to economic status, and there are class differences within caste groups because of differences in wealth. Historically, upper-caste groups (Brahmins and Kshatriyas) owned or controlled most of the land, and lower-caste groups (Sudras) were the laborers, so caste and class tended to coincide. There were many exceptions, however. Today, as a result of legally mandated expanding educational and economic opportunities, caste and class status are less connected. Some Vaishyas and Sudras have become large landowners and extraordinarily wealthy businesspeople, while some Brahmin families struggle to achieve a middle-class standard of living. By and large, however, Dalits remain very poor.

Geographic Patterns of Religious Beliefs The geographic pattern of religion in South Asia is complex and overlapping. As Figure 8.12 shows, there is a core Hindu area in central India, with other faiths more common on the fringes of the region.

The approximately 520 million Muslims in South Asia form the majority in Afghanistan, Pakistan, Bangladesh, and the Maldives; and the 140 million Muslims in India are a large and important minority, comprising 12 percent of the population. They live mostly in the northwestern and central Ganga River plain.

Buddhism began about 2600 years ago as an effort to reform and reinterpret Hinduism. Its origins are in northern India, where it flourished early in its history before spreading eastward to East and Southeast Asia. About 10 million people—only 1 percent of South Asia's population—are Buddhists. They are a majority in Bhutan and Sri Lanka.

Jainism, like Buddhism, originated as a reformist movement within Hinduism more than 2000 years ago. Jains (about 6 million people, or 0.6 percent of the region's population) are found mainly in cities and in western India. They are known for their educational achievements, promotion of nonviolence, and strict vegetarianism.

Sikhism was founded in the fifteenth century as a challenge to both Hindu and Islamic systems. Sikhs believe in one god, hold high ethical standards, and practice meditation. Philosophically, Sikhism accepts the Hindu idea of reincarnation but rejects the idea of caste. (In everyday life, however, caste plays a role in Sikh identity.) The 21 million Sikhs in the region live mainly in Punjab, in northwestern India. Their influence in India is greater than their numbers because throughout India many Sikhs hold positions in the government, in the military, and in police forces.

The first Christians in the region are thought to have arrived in the far southern Indian state of Kerala with St. Thomas, the apostle of Jesus, in the first century C.E. Today, Christians and Jews are influential but tiny minorities along the west coast of India. A few Christians live on the Deccan Plateau and in northeastern India near Burma.

Animism, the most ancient religious tradition, is practiced throughout South Asia, especially in central and northeastern India, where there are indigenous people whose occupation of the area is so ancient that they are considered aboriginal inhabitants. (For a discussion of animism, see page 312 in Chapter 7.)

Globalization and the Legacies of British Colonial Rule

As Mughal rule declined, a number of regional states and kingdoms rose and fought with one another (Figure 8.13). The absence of one strong power created an opening for yet another invasion. By the late 1700s, several European trading companies were competing to gain a foothold in the region. Of these, Britain's East India Company was the most successful. By 1857, the East India Company, acting as an extension of the British government, repressed a rebellion against European intrusion and became the dominant power in the region.

The British controlled most of South Asia from the 1830s through 1947 (Figure 8.14). By making the region part of the British Empire, the British accelerated the process of globalization in South Asia, transforming the region politically, socially, and economically. Even areas not directly ruled by the British felt the influence of their empire. Afghanistan repelled British attempts at military conquest, but the British continued to intervene there, trying to make Afghanistan a "buffer state" between British India and Russia's expanding empire. Nepal remained only nominally independent during the colonial period, and Bhutan became a protectorate of the British Indian government.

The Deindustrialization of South Asia As in their other colonies, the British used South Asia's resources primarily for their own benefit, often with disastrous results for South Asians. One example was the fate of the textile industry in Bengal (modern-day Bangladesh and the Indian state of West Bengal).

Bengali weavers, long known for their high-quality muslin cotton cloth, initially benefited from the increased access that traders gave them to overseas markets in Asia, the Americas, and Europe. By 1750, South Asia had an advanced manufacturing economy that produced 12 to 14 times more cotton cloth than Britain alone and more than all of Europe combined. However, during the eighteenth and nineteenth centuries, Britain's own highly mechanized textile industry—based on cotton grown in India, various other colonies, and the American South—developed cheaper cloth that then replaced Bengali muslin. This shift happened first in the British colonies in the Americas, then in Europe, and eventually throughout South Asia. Potential competition for the British textile industry was hindered by the British East India Company, which began severely punishing Bengalis who continued to run their own looms. As a

---

Buddhism a religion of Asia that originated in India in the sixth century B.C.E. as a reinterpretation of Hinduism; it emphasizes modest living and peaceful self-reflection leading to enlightenment

Jainism a religion of Asia that originated as a reformist movement within Hinduism more than 2000 years ago; Jains are found mainly in western India and in large urban centers throughout the region and are known for their educational achievements, nonviolence, and strict vegetarianism

Sikhism a religion of South Asia that combines beliefs of Islam and Hinduism

But rural South Asia already had an abundance of agricultural labor, so many migrated to emerging urban centers. In the 1830s, a drought worsened an already difficult situation and more than 10 million people starved to death. Throughout the nineteenth century, similar events forced millions of South Asian workers to join a stream of indentured laborers migrating to other British colonies in the Americas, Africa, Asia, and the Pacific, where their descendants can still be found.

183. CARIBBEAN BEAT PULSES WITH INDIAN ACCENTS

Democratic and Authoritarian Legacies

Contemporary South Asian governments retain institutions put in place by the British to administer their vast empire. These governments inherited many of the shortcomings of their colonial forebears, such as authoritarian bureaucratic procedures that tend to resist change. While the governments have proved functional over time, there have been numerous major civil disturbances in virtually every country. Democratic governments were not instituted on a large scale until the final days of the Empire, but since independence in 1947 (see below), people have been able to use the political freedoms afforded by democracy to voice their concerns and to make many peaceful transitions of elected governments. Still, the struggle to retain and build democratic institutions continues.

Independence and Partition
The tremendous changes brought by the British inspired many resistance movements among South Asians. Some of these were militant movements intent on pushing the British out by force, such as the unsuccessful rebellion of South Asian soldiers employed by the British East India Company in 1857. Other movements, such as the Indian National Congress (founded in 1885), used political means to agitate for more democracy, which they saw as the route to South Asian political independence. Although both militant and political actions were brutally repressed, the democracy movements gained worldwide attention and, after decades of struggle, were successful.

Nonviolence as a Political Strategy
In the early twentieth century, Mohandas Gandhi, a young lawyer from Gujarat—who would later become known by the honorific bestowed on him, "Mahatma," meaning "venerable"—emerged as a central political leader in South Asia's independence movement. He used tactics of **civil disobedience** to nonviolently defy laws imposed by the British that discriminated against South Asians. Gathering a large group of peaceful protesters, he would notify the government that the group was about to break a discriminatory law. If the authorities ignored the act, the demonstrators would have made their point and the law would be weakened. If the government instead used force against the peaceful demonstrators, it

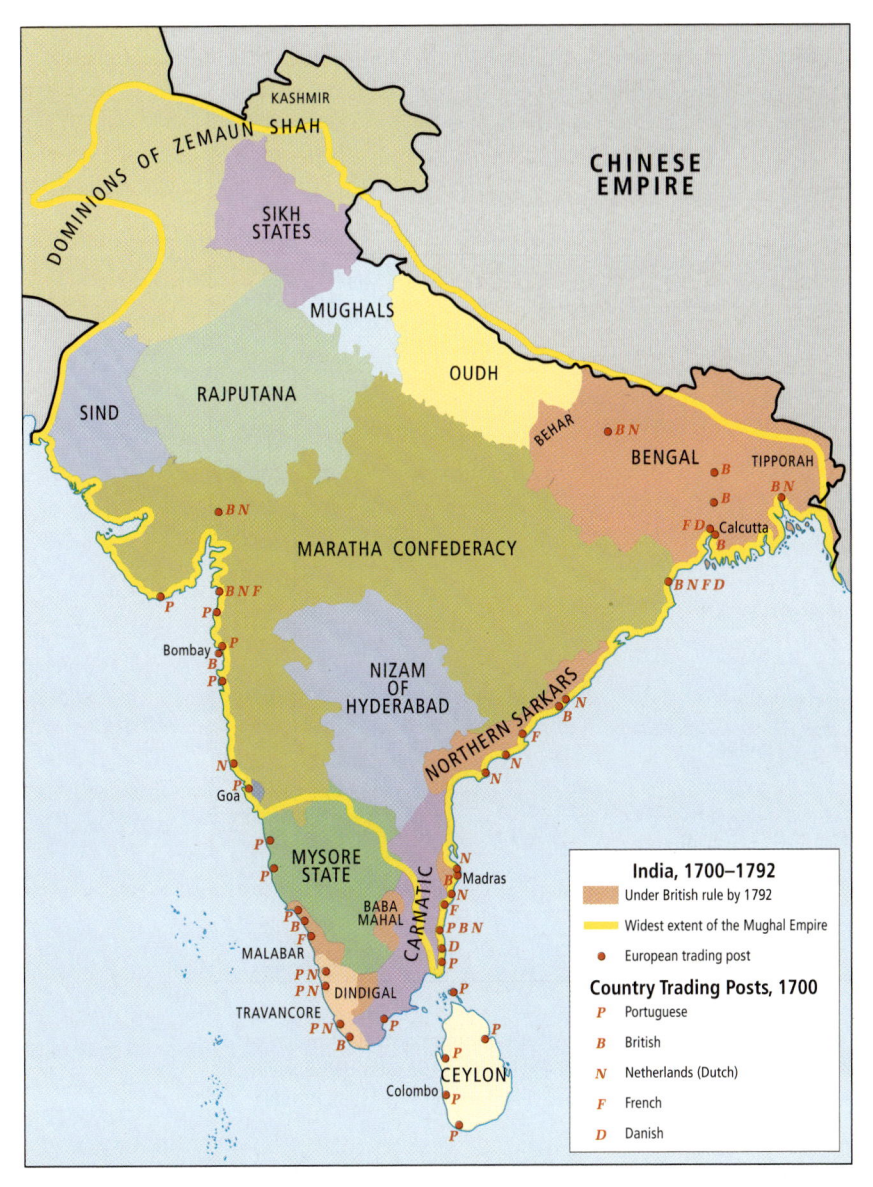

FIGURE 8.13 Precolonial South Asia. By 1700, several European nations had established trading posts along the coast of India and Ceylon (now Sri Lanka). After the death of the Mughal ruler Aurangzeb in 1707, the Mughals' ability to assert strong central rule throughout South Asia declined. A number of emergent regional states competed with one another for territory and power. Among the strongest was the Maratha Confederacy, composed of a number of small states dominated by the Maratha peoples, who were known for their martial skills. Weakness of administrative control at the center and constant rivalries between these South Asian states paved the way for British conquest by the end of the eighteenth century.
[Sources consulted: William R. Shepherd, *The Historical Atlas* (New York: Henry Holt, 1923–1926), p. 137; Gordon Johnson, *Cultural Atlas of India* (New York: Facts on File, 1996), p. 111]

result, the venerable South Asian textile industry survived in only a few places (see Figure 8.10C on page 333). As Lord Bentinck, a British colonial official put it, "While the mills of Yorkshire prospered, the bones of Bengali weavers bleached on the plains of India."

Many people who were pushed out of their traditional livelihood in textile manufacturing were compelled to find work as landless laborers.

civil disobedience protesting of laws or policies by peaceful direct action

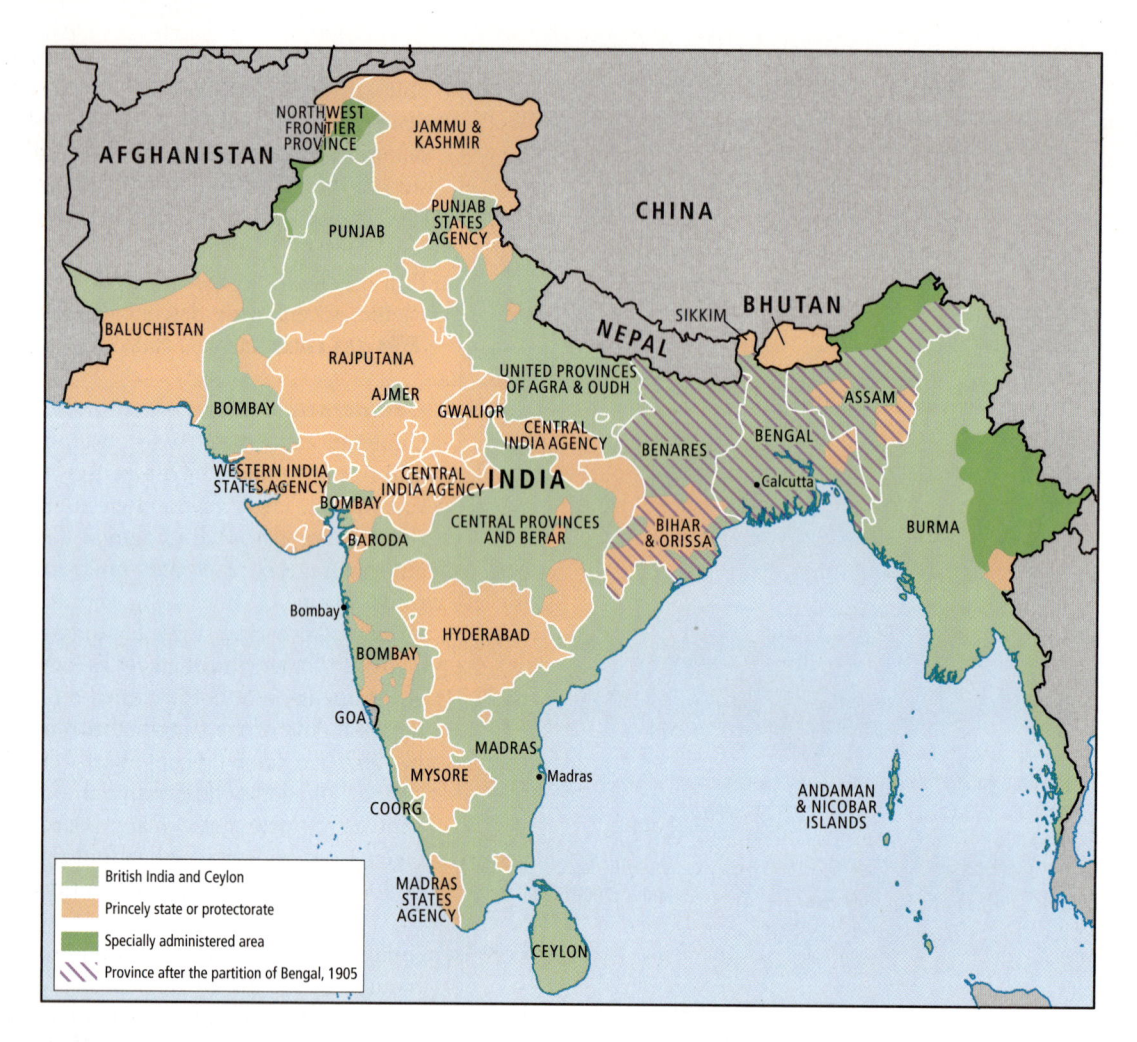

FIGURE 8.14 The British Indian Empire, 1860–1920. After winning control of much of South Asia, Britain controlled lands from Baluchistan to Burma, including Ceylon and the islands between India and Burma. [Source consulted: Gordon Johnson, *Cultural Atlas of India* (New York: Facts on File, 1996), p. 158]

would lose the respect of the masses. Throughout the 1930s and 1940s, this technique was used to slowly but surely undermine British authority across South Asia. The most famous example was the Salt March of 1930, when Gandhi led tens of thousands of people on a march to the sea, where they made salt by evaporating seawater, thus breaking the law that made it illegal for South Asians to produce salt (see Figure 8.10D on page 333). The purpose of the law had been to facilitate British rule by controlling a vital human nutritional necessity. Breaking the law on such a massive scale created an international media frenzy that catapulted Gandhi to global notoriety; it moved millions of South Asians to support independence from Britain. The hunger strikes described in the vignette that opens this chapter about the Sardar Sarovar Dam on the Narmada are another strategy that reflect Gandhi's legacy of nonviolent political protest.

The Partition of India in 1947 When British India was granted independence by Britain in 1947, it was divided into two independent countries: India, which was predominantly Hindu; and West and East Pakistan, which was predominantly Muslim (Figure 8.15). (Afghanistan, Bhutan, and Nepal were never officially British colonies; Ceylon [now Sri Lanka] became independent in 1948.) This division—called **Partition**—which Gandhi greatly lamented, was perhaps the most enduring and damaging outcome of colonial rule.

Muslim political leaders, concerned about the fate of a minority Muslim population in a united India with a Hindu majority, first suggested the idea of two nations. Though Partition was highly controversial, it became part of the independence agreement between the British and the Indian National Congress (India's principal nationalist party). Northwestern and northeastern India, two very different places historically and culturally, but where the populations were predominantly Muslim, became a single country consisting of two parts known as West and East Pakistan, separated by northern India (see Figure 8.15). Although both India and Pakistan maintained secular constitutions with no official religious affiliation, the general understanding was that Pakistan would have a Muslim majority and India a Hindu majority. Fearing that

> **Partition** the breakup following Indian independence that resulted in the establishment of Hindu India and Muslim Pakistan

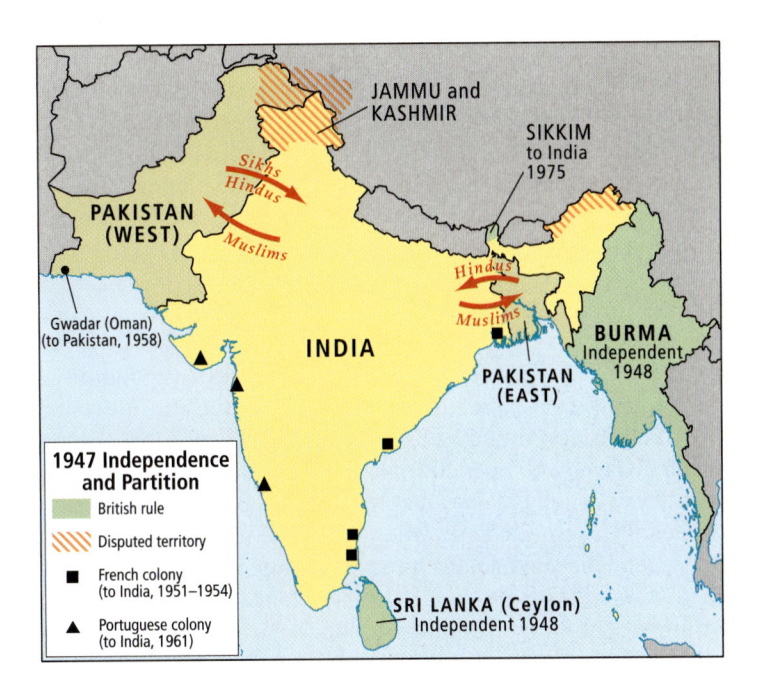

FIGURE 8.15 Independence and Partition. India became independent of Britain in 1947, and by 1948, the old territory of British India was partitioned into the independent states of India and East and West Pakistan. The Jammu and Kashmir region was contested space, and remains so today. Sikkim went to India, and both Burma and Sri Lanka became independent. Following additional civil strife, East Pakistan became the independent country of Bangladesh in 1971. [Source consulted: *National Geographic*, May 1997, p. 18]

they would be persecuted if they did not move, more than 7 million Hindus and Sikhs migrated to India from their ancestral homes in what had become West or East Pakistan. A similar number of Muslims left their homes in India for one of the parts of Pakistan. In the process, civil society broke down: Families and communities were divided, looting and rape were widespread, and between 1 and 3.4 million people were killed in numerous local outbreaks of violence. In 1971, after a bloody civil war, Pakistan was divided into Bangladesh (formerly East Pakistan) and Pakistan (formerly West Pakistan). **184. 60TH ANNIVERSARY OF INDIA–PAKISTAN PARTITION ON AUGUST 15TH**

Partition was the tragic culmination of the divide-and-rule approach the British used throughout their empire (see pages 299–300 in Chapter 7). This approach heightened tensions between South Asian Muslims and Hindus, thus creating a role for the British as seemingly indispensable and benevolent mediators. Instead of relieving tensions, the partition of India and Pakistan laid the groundwork for the wars, skirmishes, strained relations, and ongoing arms race between India and Pakistan that persist to this day.

The Post-Independence Period In the more than 60 years since the departure of the British, South Asians have experienced both progress and setbacks. Democracy has expanded steadily, albeit somewhat slowly. India is now the world's most populous democracy. It is gradually dismantling age-old traditions that hold back women, poor, low-caste Hindus, and other disadvantaged groups, and a vibrant, if still small, middle class is emerging (see Figure 8.10E on page 333). Pakistan has held a number of elections, and while some of its governments have been effective, more often they have been corrupt, militaristic, authoritarian regimes. The long feud between Pakistan and India about the border between the two countries (discussed on page 344), meanwhile, has sapped resources and ruined innumerable lives. Bangladesh, despite the damage it sustained in its war of independence from Pakistan in 1971, has actually had a more stable, responsive, and less militaristic (though not trouble-free) government than Pakistan.

Under British rule, agricultural modernization lagged, and during World War II, the Bengal Famine took an estimated 4 million lives. After independence, progress in agricultural production was slow until the late 1960s, when the green revolution brought marked improvements (see pages 340–341). The move to modernized farming on large tracts of land with far fewer agricultural workers brought prosperity to some South Asians, relieved food scarcities, and made food exports possible; but it also forced millions to migrate to the cities. This migration intensified the already vast economic disparities in urban South Asia.

Industrialization became a main goal after independence, partly in response to the dismantling of industry during the colonial period. The emphasis on technical training has produced several generations of highly skilled engineers whose talents are in demand around the world. In most countries in the region, urban-based industrial and service economies now constitute a far larger share of GNI than agriculture (which nonetheless continues to employ 50 percent or more of the workforce). The information technology (IT) sector is growing especially rapidly in India.

THINGS TO REMEMBER

- The Indus Valley/Harappa civilization, which began 4500 years ago, was remarkable for its innovative developments in water management and agriculture.

- South Asia has been invaded many times, primarily from the north by groups stretching from Greece to Mongolia.

- There are many distinct ethnic groups in South Asia, each with its own language or dialect. Today, variants of Hindi are the principal languages of India and Pakistan, while Bengali is the official language of Bangladesh. English is a common second language throughout South Asia.

- Caste and class are not the same thing: "Class" refers to economic status, and "caste" to hereditary hierarchical social categories. There are class differences within caste groups because of differences in wealth. Even though India's constitution bans caste discrimination, caste is still hugely influential in Indian society.

- There is a geographic pattern to religion in South Asia, but it is not absolute and people of different religions often live in close proximity. Relations between Muslims and Hindus can be quite tense, occasionally resulting in violent confrontation. Other religious traditions also play prominent local roles in the life of South Asia.

- Following some 300 years of Mughal rule, the British controlled most of South Asia from the 1830s through 1947, profoundly influencing the region politically, socially, and economically.

- British colonial rule in South Asia channeled resources to Europe, and in so doing depressed flourishing industries and inhibited development.

- In the early twentieth century, Mohandas "Mahatma" Gandhi emerged as a central political leader of India's independence movement, using nonviolent civil disobedience that eventually led to independence.

- Instead of relieving tensions, Partition—the 1947 division of British India into two countries, India and Pakistan—laid the groundwork for the repeated wars, skirmishes, strained relations, and ongoing arms race between India and Pakistan. ■

GLOBALIZATION AND DEVELOPMENT

> **GEOGRAPHIC INSIGHT 2**
>
> **Globalization and Development:** Globalization benefits some South Asians more than others. Educated and skilled South Asian workers with jobs in export-connected and technology-based industries and services are paid more and sometimes have better working conditions. Less-skilled workers in both urban and rural areas are left with demanding but very low-paying jobs.

South Asia is a region of startling economic incongruities. India is a good example: It is home to hundreds of millions of desperately poor people, but it is also a global leader in the computer software industry and in engineering innovation. It is celebrated for being among the world's largest emerging economies, with a growing middle class and even a robust space program. And yet its poor often see little improvement in their own lives from this economic development.

Economic Trends

Rapid economic development, national self-sufficiency, and poverty reduction have been the main goals for all parts of the region since the end of the colonial era in the 1940s. Over the past several decades, a wave of globalized economic development has reduced the emphasis on self-sufficiency, with strong growth in the IT and other high-tech industries, as well as in the automotive industry in some parts of the region. Less successful, but still promising, is the development of modernized agribusiness, which involves the three major economic sectors: agriculture, industry, and services. Especially in India, but also in Pakistan and Bangladesh, agribusiness has drastically increased the amount of food available while channeling the profits from growing food to an ever-smaller number of farmers.

Agriculture still employs more than 50 percent of the workers in South Asia, but the contribution of agriculture to most national economies is much lower, averaging less than 20 percent of the GNI for the region. The industrial sector employs far fewer people but produces between one-quarter and one-third of GNI in all countries except Nepal, Bhutan, and Afghanistan. The service sector has expanded more rapidly than either agriculture or industry and now produces more than half of the GNI in every country except Afghanistan and Bhutan.

Industry over Agriculture: A Vision of Self-Sufficiency After independence from Britain in 1947, the new leaders in India, Pakistan, Bangladesh, and Sri Lanka tended to favor industrial development over agriculture. Influenced by socialist ideas and industrial success in the Soviet Union, they believed that government involvement in industrialization was necessary to ensure high levels of job creation that would reduce poverty. They also wanted their respective countries to become self-sufficient, avoiding dependence on manufactured goods imported from the industrialized world (see pages 128–129 in Chapter 3 for a discussion of import substitution industrialization). To reach their goal of self-sufficiency, governments took over the industries they believed to be the linchpins of a strong economy: steel, coal, transportation, communications, and a wide range of manufacturing and processing industries.

South Asian industrial policies in the decades after independence generally failed to meet their goals. The emphasis on industrial self-sufficiency was not suited to countries that had such large agricultural populations. In India, for example, governments invested huge amounts of money in a relatively small industrial sector—even today, industry employs only 19 percent of the population, while agriculture employs 53 percent. Since such a small portion of the population directly benefited from this investment, industrialization failed to significantly increase South Asia's overall prosperity.

Another problem was that the measures intended to boost employment often contributed to inefficiency. One policy encouraged industries to employ as many people as possible, even if they were not needed. So until recently, for example, it took more than 30 Indian workers to produce the same amount of steel as 1 Japanese worker. Consequently, Indian steel was not competitive in the world market for years. In addition, as in the former Soviet Union, decisions about which products should be produced were made by ill-informed government bureaucrats and were not driven by consumer demand. Until the 1980s, items that would improve daily life for the poor majority, such as cheap cooking pots or simple tools, were produced only in small quantities and were of inferior quality. At the same time, there was a relative abundance of large kitchen appliances and large cars that only a tiny minority could afford.

Economic Reforms

During the 1990s, much of South Asia began to undergo economic reforms aimed at increasing competitiveness and creating secure jobs in the private sector. In many other world regions, similar reforms took place as part of structural adjustment programs (SAPs; see page 129 in Chapter 3) which were mandated by the International Monetary Fund (IMF) and the World Bank.

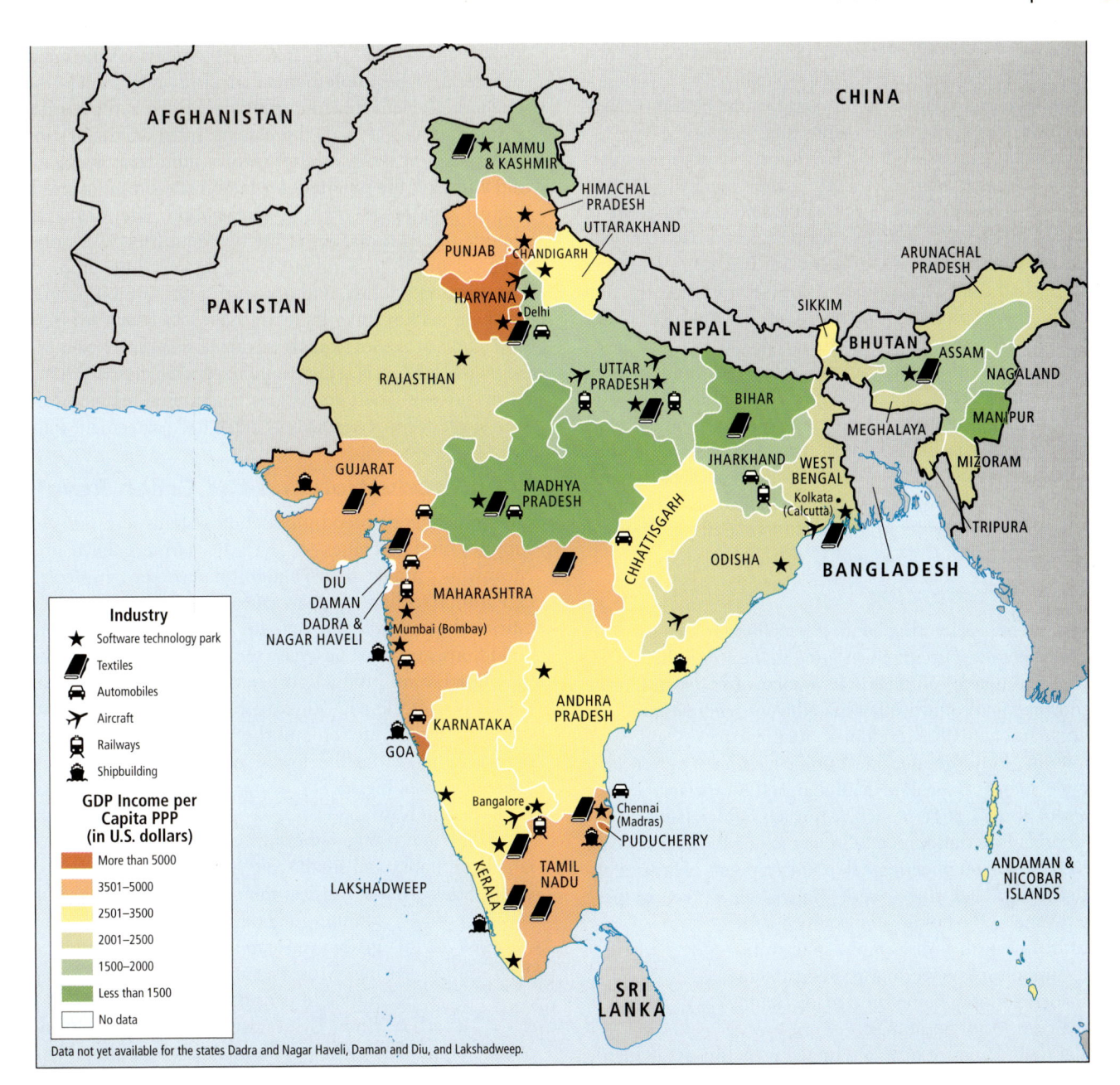

FIGURE 8.16 GDP income per capita (PPP) and industrial and IT centers in India. The presence of industry is often associated with higher incomes, and because of this, planners sometimes try to bring industry to low-income places. In some areas, poverty may be so great that even fairly intensive industrial development is able to raise average incomes above base levels only slowly. (The per capita income PPP information since 2009 for India's states is not available.) [Sources consulted: Directorate of Economics & Statistics of each of the respective state governments, 2006; GDP data from the Reserve Bank of India, Table 8, "Per Capita Net State Domestic Product at Factor Cost—State-Wise (at Current Prices)," at http://rbidocs.rbi.org.in/rdocs/Publications/PDFs/008T_BST130913.pdf; data from "Comparing Indian States and Territories with Countries: An Indian Summary," *Economist*, at http://www.economist.com/content/indiansummary]

In India, by contrast, as a response to an earlier financial crisis in the 1980s, the government itself initiated economic reforms. These consisted of, among other things, the privatization of government-run industries and banks and the relaxation of tariffs on imports. India's economic reforms have been arguably more successful than SAPs and similar reforms imposed by the IMF and the World Bank in other countries.

India's economic reforms have freed many private companies from a maze of regulations, enabling both foreign and Indian companies to invest heavily in manufacturing and other industries (Figure 8.16). Some Indian companies, such as Reva, which builds economical electric cars for India's middle class, are quite innovative. Foreign auto companies are also flocking to India, drawn by its large and cheap workforce, its excellent educational infrastructure (for the middle and upper classes), and especially by its large demand for manufactured goods of all sorts. Nearly every major global automobile company is currently establishing significant manufacturing facilities somewhere in India—producing everything from economical first cars for Indian families to luxury brands such as Mercedes-Benz for wealthier Indians. So many global manufacturers have flocked to India in recent years that the country is challenging China as an exporter of manufactured goods.

One possible outcome of all this growth is that as manufacturing jobs increase, many of India's urban poor will see their incomes rise. They will then increase their consumption of Indian-made products, thus fueling further growth. By 2009 (the latest date for which comparable figures are available), GDP per capita (PPP) had risen in all Indian states (see Figure 8.16), and the average per capita income (PPP) in India in 2011 was U.S.$3468, nearly ten times higher than it was in 1992.

So far, most of the benefits from India's self-implemented economic reforms seem to be going to the highly skilled and educated urban upper and middle classes. India's middle class now stands at 50 million—still a tiny minority of India's 1.2 billion—but is likely to grow dramatically to perhaps as many as 600 million people by 2030. However, the new economic policies are producing wider urban/rural income disparities.

Offshore Outsourcing

India's current manufacturing boom is benefiting from a previous boom in offshore outsourcing that started in the 1990s. In **offshore outsourcing**, a company contracts to have some of its business functions performed in a country other than the one where its products or services are actually developed, manufactured, and sold. Companies in North America and Europe have been outsourcing jobs to cities such as Bangalore, Mumbai, and Ahmadabad to take advantage of India's large, college-educated, and relatively low-cost workforce. Jobs outsourced to India include those in IT, data entry, Web design, engineering, telephone support, pharmaceutical research, and "back office" work. Many of the workers in these jobs are women. By holding outsourced jobs with North American firms, South Asian workers have gained experience that positions them perfectly for jobs now being created by South Asian–owned firms in South Asia.

> **offshore outsourcing** the contracting of certain business functions or production functions to providers where labor and other costs are lower

Major global finance firms are increasingly hiring the highly skilled workers on Dalal Street—Mumbai's "Wall Street"—to provide finance and accounting services. Stiff global competition makes cost cutting imperative for finance firms, and India's relatively low salaries provide a solution. While a junior analyst from an Ivy League school is paid $150,000 a year in the United States, a graduate of a top Indian business school is paid only $35,000 a year in India. That Indian employee's salary, however, buys a much higher standard of living than the U.S. employee would have. In fact, well-educated Indian migrants in the United States are now returning home to take these seemingly low salaries because they can still live well and join this exciting development phase in their home country. This trend of return-migration could eventually happen across the South Asian region.

Free Trade Within South Asia

To facilitate trade between countries and to lure extra-regional investors into this huge market, South Asia has attempted to create region-wide free trade agreements. Agreements were scheduled to be implemented in 2012, but progress has been slow and uneven. Because of its size, India is the biggest player in South Asia's free trade discussions. India is a major supplier for Sri Lanka and Bangladesh but buys little from them in return. And although the potential for trade between India and Pakistan is great and could include cars, cotton, chemicals, and food imported from India, as yet the two countries have only a small trade relationship, with India buying very little from Pakistan. The ongoing political and border disputes inhibit progress.

Even within India there are blocks to freer trade across state borders. The maze of regulations mentioned earlier extends down to the interstate level, and there is little reconciliation of the states' varying tax policies. Trade across state borders is so hampered that it has been suggested that India needs a free trade agreement with itself.

Food Production and the Green Revolution

Due to changes in food production systems in South Asia, food supplies and incomes of wealthy farmers have increased. Meanwhile, impoverished agricultural workers, displaced by new agricultural technology, have been forced to seek employment in cities, where they often end up living in crowded, unsanitary conditions, and must purchase food instead of growing their own.

Agricultural production per unit of land has increased dramatically over the past 50 years, especially in parts of India; nevertheless, agriculture remains the least efficient economic sector, meaning that it has the lowest return on investments of land, labor, and cash. Figure 8.17 shows the distribution of agricultural zones in South Asia.

Until the 1960s, agriculture across South Asia was based largely on traditional small-scale systems (the average farm holding in South Asia is still less than 2 acres) that managed to feed families in good years, but often left them hungry or even starving in years of drought or flooding. Moreover, these systems did not produce sufficient surpluses for the region's growing cities. Even now, cities must import food from outside South Asia. Much of South Asia's agricultural land is still cultivated by hand, and for decades, agricultural development was neglected in favor of industrial development, especially in India (discussed below). Nevertheless, by the 1970s, important gains in agricultural production had begun.

Beginning in the late 1960s, the green revolution (see page 29 in Chapter 1) boosted grain harvests dramatically through the use of new agricultural tools and techniques. Such innovations included seeds bred for high yield and resistance to disease and wind damage; fertilizers; mechanized equipment; irrigation; pesticides; herbicides; and double-cropping (producing two crops consecutively per year). To a lesser extent, the increase in the amount of land under cultivation also contributed to a larger yield. Where the new techniques were used, yield per unit of farmland improved by more than 30 percent between 1947 and 1979 and both India and Pakistan became food exporters.

Despite the achievements of the green revolution, the benefits have been uneven. The prosperity of some Indian states, such as Punjab and Haryana, which have extensive irrigation networks, has increased tremendously, but the successes are precarious. Many impoverished farmers who were unable to afford the special seeds, fertilizers, pesticides, and new equipment could not compete and had to give up farming. Most migrated to the cities, where their skills matched only the lowest-paying jobs.

South Asia needs alternatives to standard green revolution strategies because it will have difficulty maintaining current levels of food production over the long term. The green revolution's dependency on chemical fertilizers, pesticides, and high levels of irrigation all contribute to aquifer depletion, waterway pollution, increased erosion, and the loss of soil fertility through the buildup of salt in soils. *Soil salinization* (see page 240 in Chapter 6) is already reducing yields in many areas, such as the Pakistani Punjab, which is Pakistan's most productive—but highly irrigated—agricultural zone.

192. TECHNOLOGY KEY TO PRODUCING MORE FOOD

The methods of agroecology are a potential remedy for some of the failings of green revolution agriculture. **Agroecology** often involves the revival and use of traditional methods, such as

agroecology the practice of traditional, nonchemical methods of crop fertilization and the use of natural predators to control pests

ON THE BRIGHT SIDE

Microcredit: A Bangladeshi Innovation

So far, the Grameen Bank has been an enormous success in Bangladesh, where it has loaned over U.S.$8 billion to more than 8 million borrowers. Similar microcredit projects have been established in India and Pakistan and throughout Africa, Middle and South America, North America, and Europe. In 2006, Dr. Yunus was awarded the Nobel Peace Prize for his work in microcredit. In 2009, he received the Presidential Medal of Freedom from President Obama.

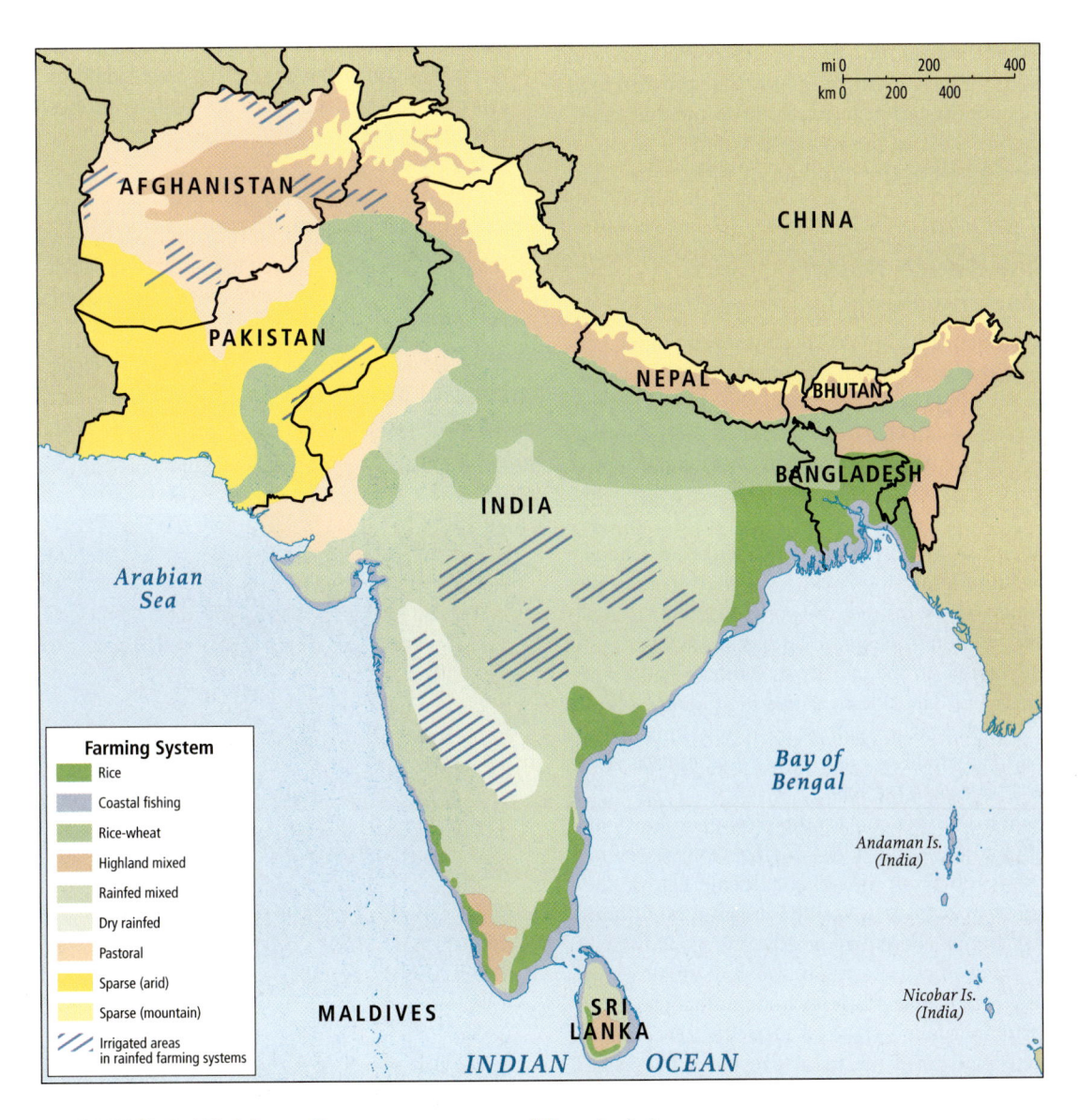

FIGURE 8.17 Major farming systems of South Asia. [Sources consulted: John Dixon and Aidan Gulliver with David Gibbon, *Farming Systems and Poverty: Improving Farmers' Livelihoods in a Changing World* (Rome and Washington, DC: FAO and World Bank, 2001), at http://www.fao.org/farmingsystems/FarmingMaps/SAS/01/FS/index.html]

fertilizing crops with animal manure, intercropping (planting several species together) with legumes to add nitrogen and organic matter, water conservation, and using natural predators to control pests. Although the knowledge required is extensive, unlike green revolution techniques, the methods of agroecology are advantageous to poor farmers because the necessary resources are readily available in most rural areas and the knowledge can be handed down orally from generation to generation.

One reason that the green revolution has managed to increase food supplies but not eliminate hunger and malnutrition is that food tends to go to those who have money to spend. As noted previously, agricultural modernization usually pushes unskilled farm workers off the land and into underemployment. Between 1970 and 2001, the amount of food produced per capita in South Asia increased 18 percent and the proportion of undernourished people dropped from 33 percent of the population to 22 percent. Nonetheless, 22 percent of nearly 1.6 billion is 352 million people—roughly half of the world's total undernourished population. Because of corruption and social discrimination, government programs that provide food to the poor have generally failed to reach those most in need. For example, despite massive resources devoted to improve child nutrition, about half of the children in India show signs of malnutrition.

Microcredit: A South Asian Innovation for the Poor

Over the past four decades, a highly effective strategy for lifting people out of extreme poverty and the hunger and malnutrition that so often accompanies it has been pioneered in South Asia. **Microcredit** makes very small loans (generally under U.S.$100) available to very low-income, would-be business owners. Throughout South Asia, as in most of the world, banks are generally not interested in administering the small loans that poor people, especially poor women, need. Instead, the poor must rely on small-scale moneylenders, who often charge interest rates as high as 30 percent or more *per month*. In the late 1970s, Muhammad Yunus, an economics professor in Bangladesh, started the Grameen Bank, or "Village Bank," which makes small loans, mostly to people in rural villages who wish to start businesses. **195. NOBEL PEACE PRIZE GOES TO BANGLADESH'S "BANKER TO THE POOR"**

The microcredit loans often pay for the start-up costs of small enterprises, which include such wide-ranging ventures as cell phone–based services, chicken raising, small-scale egg production, the construction of pit toilets, and the distribution of home water purification systems. Potential borrowers (more than 90 percent of whom are women) are organized into small groups that are collectively responsible for repaying any loans to group members. If one member fails to repay a loan, then everyone in the group is denied loans until the loan is repaid. This system, reinforced with weekly meetings, creates incentives for members to repay their loans (Figure 8.18). The repayment rate on the loans is extremely high, averaging around 98 percent—much higher than the rate most banks achieve.

> **microcredit** a program based on peer support that makes very small loans available to very low-income entrepreneurs

THINGS TO REMEMBER

> **GEOGRAPHIC INSIGHT 2**

- **Globalization and Development** Globalization benefits some South Asians more than others. Educated and skilled South Asian workers with jobs in export-connected and technology-based industries and services are paid more and sometimes have better working conditions. Less-skilled workers in both urban and rural areas are left with demanding but very low-paying jobs.

- The service sector is growing rapidly in all countries and dominates most economies in terms of the contributions made to GNI (or GDP), but it still employs a relatively small proportion of workers.

- During the 1990s, much of South Asia began to undergo economic reforms aimed at increasing competitiveness and creating secure jobs in the private sector.

- India's economic reforms have freed many private companies from a maze of regulations, enabling both foreign and Indian companies to invest heavily in manufacturing and other industries. However, the new economic policies are producing wider income disparities between rural and urban areas.

- Due to changes in food production systems in South Asia, food supplies and incomes of wealthy farmers have increased. Meanwhile, impoverished agricultural workers, displaced by new agricultural technology, have been forced to seek employment in cities, where they often end up living in crowded, unsanitary conditions and must purchase food instead of growing their own.

- The Grameen Bank and its strategy of microcredit have been very successful in Bangladesh; this method of small-loan financing has spread around the world, including to the United States.

FIGURE 8.18 Microcredit for small-scale entrepreneurs. Borrowers from the Grameen Bank operate a business manufacturing and repairing fishing nets in rural Bangladesh. Following a model established in 1974 by Bangladeshi economist Mohammad Yunus, such groups gather weekly to discuss business and pay their loan installments. For many, this is a treasured social outing.

POWER AND POLITICS

> **GEOGRAPHIC INSIGHT 3**
>
> **Power and Politics:** India, South Asia's oldest, largest, and strongest democracy, has shown that the expansion of political freedoms can ameliorate conflict. Across the region, when people have been able to participate in policy-making decisions and implementation—especially at the local level—seemingly intractable conflict has been diffused and combatants have been willing to take part in peaceful political processes.

Throughout South Asia's history, supporters of opposing political ideologies—for instance, religious freedom versus a religious state—commonly tried to resolve their differences through violence. The most successful paths out of these conflicts have been those in which opposing groups managed to compete peacefully in democratic elections. Nevertheless, every South Asian country and every foreign country that has intervened in the region's politics have missed opportunities to resolve conflicts through democratic means, resorting instead to the use of force (see Figure 8.20A, B on page 345).

A particularly virulent source of conflict within all South Asian countries is corruption, often linked to purposeful bureaucratic inefficiency, especially the soliciting of bribes to perform a service. A number of movements to address conflict and corruption are becoming popular among the increasingly politically aware middle class. Because more people are now employed in the private sector instead of in government-owned industries, more people have the courage to challenge government officials.

The Hindu–Muslim Relationship Indian independence leaders like Mohandas Gandhi and Jawaharlal Nehru (first prime minister of India, 1947–1964) emphasized the common cause of throwing off British rule that once united Muslim and Hindu Indians. Since independence, members of the Muslim upper class have been prominent in Indian national government and the military. Muslim generals have served India willingly, even in its wars with Pakistan after Partition. Hindus and Muslims often interact amicably, and they occasionally marry each other. Both groups have influenced the region's cuisine (see Figure 8.28).

But there is a dark side to the Hindu–Muslim relationship. At the community level in South Asia, relations between the region's two largest religious groups are often quite tense. In some Indian villages, Hindus may regard Muslims as members of low castes. Religious rules about food are often the source of discord because dietary habits are a primary means of distinguishing caste. After Partition in 1947, some wealthy Hindu landowners remained in what was then East Pakistan (now Bangladesh). In some Bangladeshi villages today, Hindus are often somewhat wealthier even though Muslims may be a majority. Although the two groups may coexist amicably for many years, they view each other as different, and conflict resulting from religious or economic disputes—euphemistically called **communal conflict**—can erupt over seemingly trivial events, as described in the following vignette.

communal conflict a euphemism for religiously based violence in South Asia

VIGNETTE

The sociologist Beth Roy, who studies communal conflict in South Asia, recounts an incident that she refers to as "Some trouble with cows" in the village of Panipur (a pseudonym) in Bangladesh (Figure 8.19). The incident started when a Muslim farmer either carelessly or provocatively allowed one of his cows to graze in the lentil field of a Hindu. The Hindu complained, and when the Muslim reacted complacently, the Hindu seized the offending cow. By nightfall, Hindus had allied themselves with the lentil farmer and Muslims with the owner of the cow. More Muslims and Hindus converged from the surrounding area, and soon there were thousands of potential combatants lined up facing each other. Fights broke out. The police were called. In the end, a few people died when the police fired into the crowd of rioters. [Source: Beth Roy. For detailed source information, see Text Sources and Credits.] ■

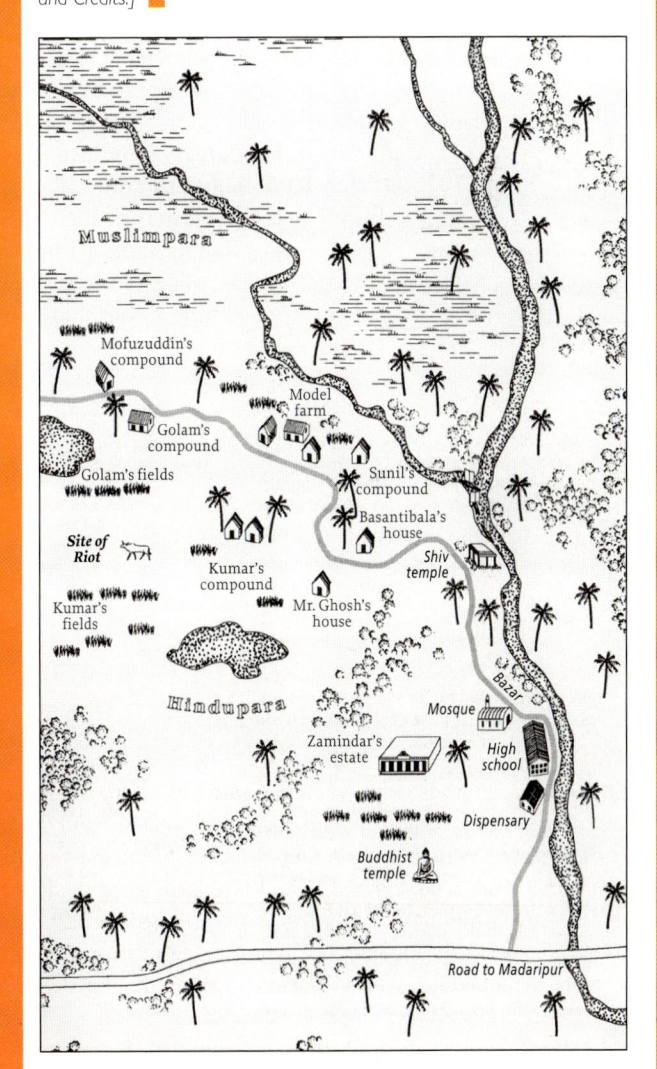

FIGURE 8.19 Some trouble with cows. This map of the village of Panipur (a pseudonym) illustrates how intimately the separate Muslim and Hindu communities were connected. The map shows Muslim and Hindu areas, the area where the riot took place, and numerous other features of village life. [Beth Roy/University of California Press Books]

At the state and national level, Hindu–Muslim conflict has taken on strong political overtones, aspects of which are discussed below.

Religious Nationalism

The association of a particular religion with a particular territory or political unit—be it a neighborhood, a city, or an entire country—to the exclusion of other religions, is commonly called **religious nationalism**. The ultimate goal of such movements is often political control over a given territory.

Although both India and Pakistan were formally created as secular states, religious nationalism has been a reality in both countries, shaping relations between people and their governments. Rejecting the idea of multiculturalism, and referring back to the days of Partition, India is increasingly thought of as a Hindu state, while Pakistan calls itself an Islamic Republic, and Bangladesh a People's Republic. In each country, many people in the dominant religious group strongly associate their religion with their national identity.

In India, urban men from middle- and upper-caste groups are the predominant supporters of Hindu nationalism (sometimes called *Hindutva*). Hindutva proponents not only promote Hinduism, they fear the erosion of their castes' political influence and particularly resent the extension of the quota system for government jobs and admission to universities to lower-caste groups (see pages 332–334). Conflict results because politically mobilized lower castes and members of other religious groups are no longer willing to follow the dictates of the dominant castes.

religious nationalism the association of a particular religion with a particular territory or political unit to the exclusion of other religions

regional conflict a conflict created by the resistance of a regional ethnic or religious minority to the authority of a national or state government; currently these are the most intense armed conflicts in South Asia

Political parties based on religious nationalism have gained popularity throughout South Asia. Although their members think of these parties, such as the Bharatiya Janata Parishad (BJP) in India, as forces that will purge their country of corruption and violence, they are usually no less corrupt or violent than secular parties.

Movements Against Government Inefficiency and Corruption

In recent years, reform movements have been developed by the many South Asians who have been frustrated by government inefficiency, corruption, religious nationalism, caste politics, and the failure of governments to deliver on their promises of broad-based prosperity. Bureaucrats who demand bribes have lately been the focus of such activism. Confronted by a bureaucrat who asked for nearly $200 to issue a legitimate income tax refund, one Indian couple in Bangalore launched the Web site I Paid a Bribe, aimed at collecting information about crooked officials. The idea caught on quickly and similar sites now exist in more than 17 countries. Anna Hazare of Maharashtra, a former military man who advocates that those convicted of corruption lose a hand as punishment, is a more militant anticorruption crusader in India. Although most Indians quickly backed off such extreme measures, Hazare, despite his extremism, has attracted a following in the new middle class, especially among educated women, who are increasingly active in politics.

Two aspects of the high-tech revolution in South Asia are likely to make corruption more difficult for bureaucrats and elected officials. The first is the spread of cell phones and their use in banking and money transfers, which will now be more traceable. The second is the *Aadhaar Project*, an ambitious effort to provide all 1.2 billion Indians with a unique photo/digitized ID card (UID). By October of 2012, a total of 208 million cards had been issued, with complete coverage of India expected before 2020. The UID project is designed to facilitate the distribution of government benefits of all types, much of which currently fails to reach the poorest people because there are few formal records that can be used to identify them. Currently, much of the money intended to reach these people ends up in the hands of corrupt officials.

The Growing Influence of Women and Young Voters

In the state elections of 2012, women and young voters were particularly active, coming to the polls in large numbers, with specific issues and candidates in mind. In several of the largest states, voter turnout was 50 percent higher than it had been in the past. The anticorruption movements mentioned above, widely covered in the press and by Web sites, apparently motivated voters. Large numbers of new voters, at least a third of whom are 18 to 19 years old, have recently registered to vote. These educated and urbanized voters, male and female, who have far more access to information than voters have had in the past, could usher in a new political era.

Regional Conflicts

The most intense armed conflicts in South Asia today are **regional conflicts** in which nations dispute territorial boundaries or a minority actively resists the authority of a national or state government. Most of these hostilities arise from the authoritarian tendencies of governments that at times work against the growth of political freedoms that might otherwise defuse political tensions.

Conflict in Kashmir Since 1947 and the post-independence dividing of India and Pakistan, between 60,000 and 100,000 people have been killed in violence in Kashmir. At the root of the violence is a struggle for territory between India and Pakistan, neither of which is willing to let the people of Kashmir resolve the dispute democratically (see the Figure 8.20 map).

For many years, Kashmir has been a Muslim-dominated area, and in 1947, some Kashmiris believed that it should be turned over to Pakistan for this reason. Although the Hindu maharaja (king) of Kashmir wanted Kashmir to remain independent at the time, the most popular Kashmiri political leader and significant portions of the populace favored joining India. They preferred India's stated ideals of having a secular government to Pakistan's less robust safeguards for religious freedom. When Pakistan-sponsored raiders invaded western Kashmir in 1947, the maharaja quickly agreed to join India. A brief war between Pakistan and India resulted in a cease-fire line that became a tenuous boundary.

Pakistan attempted to invade Kashmir again in 1965 but was defeated. India and Pakistan are technically still waiting for a UN decision about the final location of the border (Figure 8.20C). The Ladakh region of Kashmir (see the Figure 8.1 map) is the object of a more limited border dispute between India and China.

FIGURE 8.20

PHOTO ESSAY: Power and Politics in South Asia

Most armed conflicts in South Asia have been sparked by governmental authoritarianism that has eroded political freedoms. Supporters of political opposition groups have faced disenfranchisement, imprisonment, and sometimes execution. However, in some cases, growing respect for political freedoms has paved the way for peaceful reconciliation between former combatants.

A Local residents watch a burning supply convoy in Pakistan that was attacked by militants as it attempted to carry fuel to U.S.-led NATO forces in Afghanistan. Pakistan no longer openly supports U.S. and NATO efforts in Afghanistan and these forces have reduced their roles and now officially only provide support for Afghan forces. The conflict in Afghanistan was largely the U.S.'s response to the September 11th, 2001, attacks on the United States, though one of the stated goals of the war was to support political freedoms in Afghanistan.

B Several years after the official end of Sri Lanka's civil war, refugees continue to arrive in neighboring countries, such as Indonesia (shown here). The war was sparked in part by policies that discriminated against Tamils, politically and economically. Tamil insurgents went on to disrupt numerous elections, though some of their supporters have now become peaceful elected politicians. In July 2013, Australia banned Sri Lankan refugees and relocated them to Papua New Guinea, which agreed to take them.

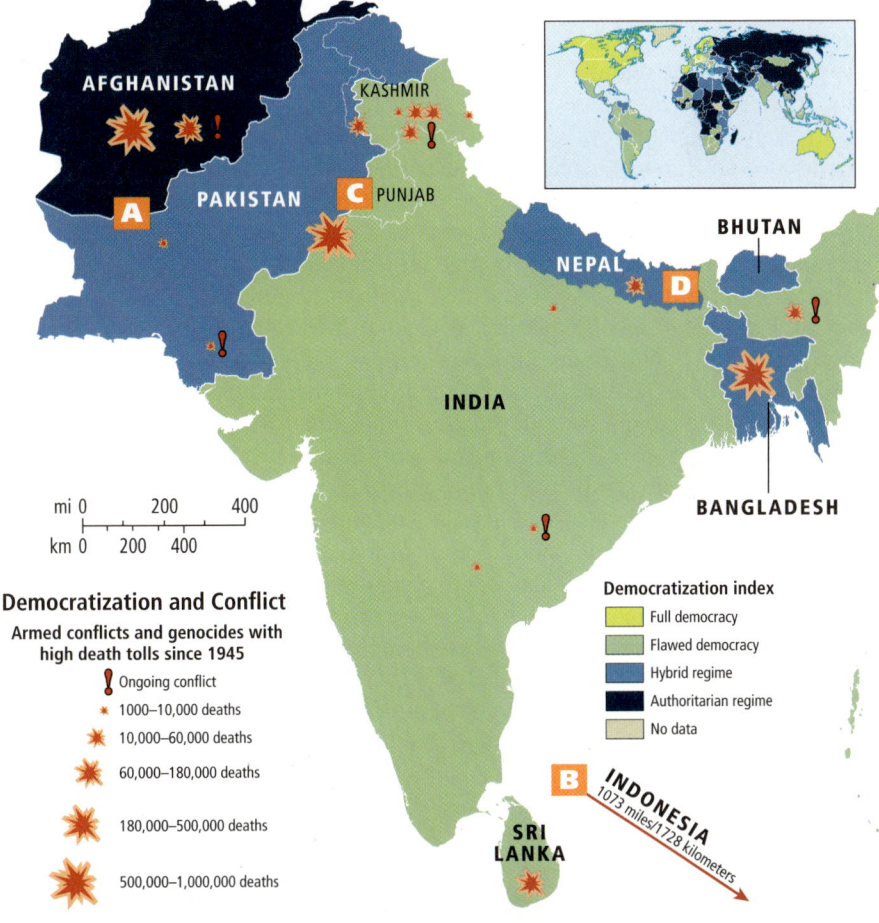

Democratization and Conflict
Armed conflicts and genocides with high death tolls since 1945

- ! Ongoing conflict
- 1000–10,000 deaths
- 10,000–60,000 deaths
- 60,000–180,000 deaths
- 180,000–500,000 deaths
- 500,000–1,000,000 deaths

Democratization index
- Full democracy
- Flawed democracy
- Hybrid regime
- Authoritarian regime
- No data

C The India–Pakistan border can be seen from space, thanks to the floodlights illuminating the fence that stretches for much of its length. The fence was built to discourage arms smuggling related to the region's many conflicts, especially the one in Kashmir.

D The former monarch of Nepal, King Gyanendra, whose removal from power paved the way for the growth of political freedoms and the end of Nepal's civil war. Since then, former combatants, such as the Maoists, have become peaceful political parties. [R. Fox Photography/Lonely Planet Images/Getty Images]

After years of military occupation, most Kashmiris now support independence from both India and Pakistan. However, neither country is willing to hold a vote on the matter. Anti-Indian Kashmiri guerrilla groups equipped with weapons and training from Pakistan have carried out many bombings and assassinations. Blunt counterattacks launched by the Indian government have killed large numbers of civilians and alienated many Kashmiris.

Another complication in the Kashmir dispute is the fact that both India and Pakistan—which came close to war against each other in 1999 and again in 2002—have nuclear weapons. Because of the nationalistic fervor of the protagonists, many see the conflict in Kashmir as more likely to result in the use of nuclear weapons than any other conflict in the world. Analysts agree that any use at all of nuclear weapons would have severe repercussions for everyone on Earth.

War and Reconstruction in Afghanistan In the 1970s, political debate in Afghanistan became polarized. On one side were several factions of urban elites, who favored modernization and varying types of democratic reforms. Opposing them were rural, conservative religious leaders, whose positions as landholders and ethnic leaders were threatened by the proposed reforms. Divisions intensified as successive governments, all of which came to power through military coups, became more and more authoritarian. Political opponents were imprisoned, tortured, and killed by the thousands, resulting in a growing insurgency outside the major cities.

In 1979, fearing that a civil war in Afghanistan would destabilize neighboring Soviet republics in Central Asia, the Soviet Union invaded Afghanistan. Rural conservative leaders (often erroneously called "warlords") and their followers formed an anti-Soviet resistance group, the *mujahedeen*. As Afghan resistance to Soviet domination increased, the mujahedeen became ever more strongly influenced by militant Islamist thought and by Persian Gulf Arab activists who provided funding and arms. At the time, the United States, still searching for Cold War allies against the Soviet Union, joined with Pakistan in supporting the mujahedeen. Moderate, educated Afghans who favored democratic reforms fled the country during this turbulent time, hoping to go back eventually when peace returned.

In 1989, after heavy losses—14,000 Soviet soldiers killed and billions of dollars wasted—the Soviets gave up and left Afghanistan. Anarchy prevailed for a time as mujahedeen factions fought one another, adding to the 1.5 million civilians and combatants killed in the war with the Soviets.

In the early 1990s, a radical religious, political, and military movement called the *Taliban* emerged from among the mujahedeen. The Taliban wanted to control corruption and crime and minimize Western ways—especially those related to the role, status, and dress of women—that had been introduced in earlier decades by the urban elites and reinforced or made more extreme by the Russian occupation. The Taliban wanted to strictly enforce *shari'a*, the Islamic social and penal code (see page 246 in Chapter 6). Efforts by the Taliban to purge Afghan society of non-Muslim influences included greatly restricting women (see pages 355–357), promoting only fundamentalist Islamic education, and publicly banning the production of opium, to which many Afghan men had become addicted, all the while privately promoting its sale to raise funds for their side. By 2001, the Taliban controlled 95 percent of the country, including the capital, Kabul.

198. TALIBAN INSURGENCY FUELED BY POPPY CULTIVATION

Following the events of September 11, 2001, the United States and its allies focused on removing the Taliban, who were giving shelter to Osama bin Laden and his international Al Qaeda network. By late 2001, the Taliban were overpowered by an alliance of Afghans supported heavily by the United States, the United Kingdom, and eventually by NATO (see the Figure 8.20 map).

In 2003, the United States launched the war in Iraq that diverted national attention, troops, and financial resources away from Afghanistan. Almost immediately the Taliban were back again, effectively thwarting the ability of Afghanistan's new government to ensure security and to meet the needs of people outside Kabul. Based in rural areas in both Pakistan and Afghanistan, the Taliban are now aided by widespread distrust of the government in Kabul, which is seen as corrupt (see Figure 8.20A). It appears that most people in Afghanistan favor a democratic government based on Muslim principles, but functionally flawed elections beginning in 2004 have resulted in an ever-lower voter turnout, diminishing from 7.4 million in 2004 to 3.2 million in 2010. The next presidential elections were scheduled for 2014.

In May of 2011, bin Laden was killed in a raid by U.S. forces in the town of Abbottabad, Pakistan. Another raid killed the next-highest Al Qaeda commander, and sporadic drone attacks killed both combatants and civilians. The United States took these actions without the knowledge or consent of Pakistani authorities, which suggests a weakening of the U.S.–Pakistani alliance in the so-called War on Terror. With the threat from Al Qaeda seemingly diminished, calls within the United States for a faster withdrawal from Afghanistan increased; withdrawal is now underway and scheduled to be completed in 2014 (see Chapter 2, pages 84–85), though the withdrawal is unlikely to be total.

Sri Lanka's Civil War The Singhalese have dominated Sri Lanka since their migration from Northern India several thousand years ago. Today they make up about 74 percent of Sri Lanka's population of 20.5 million people. Most Singhalese are Buddhist. Tamils, a Hindu ethnic group from South India, make up about 18 percent of the total population of Sri Lanka. About half of these Tamils have been in Sri Lanka since the thirteenth century, when a Tamil Hindu kingdom was established in the northeastern part of the island. The other half were brought over by the British in the nineteenth century to work on tea, coffee, and rubber plantations. Some Tamils have done well, especially in urban areas, where they dominate the commercial sectors of the economy. However, many others have remained poor laborers isolated on rural plantations.

Upon its independence in 1948, Sri Lanka had a thriving economy led by a vibrant agricultural sector and a government that made significant investments in health care and education. It was poised to become one of Asia's most developed economies. But, driven by nationalism, Singhalese was made the only official language and Tamil plantation workers were denied the right to vote. Efforts were also made to deport hundreds of thousands of Tamils to India. In the 1960s, the government shifted investment away from agricultural development and toward urban manufacturing and textile industries, which were dominated by Singhalese.

By 1983, the Tamil minority, lacking political power and influence, chose to use guerilla warfare against the Singhalese, mounting an army known as the Tamil Tigers.

For more than 30 years, the entire island was subjected to repeated terrorist bombings and kidnappings. Peace agreements were attempted several times, but in the end it was an overwhelming military victory by the government, combined with an effective crackdown on international funding for the Tamils, that forced the Tamil surrender in May of 2009. However, dissatisfaction with the peace process has resulted in an ongoing flow of Tamil and other refugees from Sri Lanka (see Figure 8.20B).

Despite many years of violence that severely curtailed the tourist industry, economic growth in other sectors has been surprisingly robust in Sri Lanka. Driven by strong growth in food processing, textiles, and garment making, Sri Lanka is today one of the wealthiest nations in South Asia on a per capita GDP basis, and it provides for the human well-being of its citizens.

Nepal's Rebels After a civil war that ended generations of monarchical rule, Nepal has endured years of political turmoil. An elected legislature and multiparty democracy were introduced into Nepal in 1990, but until 2008, a royal family governed with little respect for the political freedoms of the Nepalese people.

In 1996, inspired by the ideals of the late Chinese leader Mao Zedong (but with no apparent support from China), Maoist revolutionaries took advantage of public discontent and waged a "people's war" against the Nepalese monarchy. Following a decade of civil war, during which 13,000 Nepalese died, the Maoists had both military control of much of the countryside and strong political support from most Nepalese. Persistent poverty and lack of the most basic development under the dictatorial rule of the latest monarch, King Gyanendra (see Figure 8.20D), led to massive protests that forced Gyanendra to step down in 2006. Soon thereafter, the Maoists declared a cease-fire with the government.

In 2008, the Maoists won sweeping electoral victories that gave them a majority in parliament and made their former rebel leader prime minister. Then, in May of 2009, when his many conditions for reforming Nepalese society remained unmet, the Maoist prime minister resigned in a tactical parliamentary move and took his party into opposition against a new prime minister and his weak 22-party coalition. Although the Maoist opposition agreed to participate in writing a new constitution, divisive issues relating to power sharing and dividing the country into ethnic states have obstructed the creation of a broadly acceptable constitution. Observers concur, however, that if the various factions can see that peaceful democratic processes will give them a voice, Nepal will probably not return to civil war. **200. FUTURE OF NEPAL'S KING GYANENDRA IN QUESTION**

Caste and Politics Despite decades of effort to fight the influence of caste, politics in India are still dramatically influenced by caste. In the twentieth century, Mohandas Gandhi began an organized effort to eliminate discrimination against "untouchables." As a result, India's constitution bans caste discrimination. However, in recognition that caste is still hugely influential in society, India began an affirmative action program upon independence from Britain. The program reserves a portion of government jobs, places in higher education, and parliamentary seats for Dalits and Adivasis. Together, the two groups now constitute approximately 23 percent of the Indian population and are guaranteed 22.5 percent of government jobs. Extended in 1990, this program now includes other socially and educationally "backward castes" (the term used in India), such as disadvantaged *jatis* of the Sudras caste, allotting them an additional 27 percent of government jobs. However, reserving nearly half of government jobs in this way has resulted in considerable controversy. In 2006, medical students successfully protested against quotas in elite higher-education institutions for lower-caste applicants. The Indian Supreme Court ruled in favor of the students.

At the local level, most political parties design their vote-getting strategies to appeal to subcaste loyalties. They often secure the votes of entire *jati* communities with political favors such as new roads, schools, or development projects. These arrangements fly in the face of the official ideologies of the major political parties and of Indian government policies, which actively work to eliminate discrimination on the basis of caste. Nevertheless, the role of caste in politics seems to be increasing at this time; several new political parties that explicitly support the interests of low castes have emerged.

Among educated people in urban areas, the campaign to eradicate discrimination on the basis of caste may appear to have succeeded, but the reality is more complex. Throughout the country, there are now some Dalits serving in powerful government positions. Members of high and low castes ride city buses side by side, eat together in restaurants, and attend the same schools and universities. For some urban Indians—especially educated professionals who meet in the workplace—caste is deemphasized as the crucial factor in finding a marriage partner. However, less than 5 percent of marriages cross even *jati* lines, let alone the broader gulf of *varna*. Nearly everyone notices the tiny social clues that reveal an individual's caste, and in rural areas, where the majority of Indians still reside, the divisions of caste remain prevalent.

THINGS TO REMEMBER

> **GEOGRAPHIC INSIGHT 3**
> • **Power and Politics** India, South Asia's oldest, largest, and strongest democracy, has shown that the expansion of political freedoms can ameliorate conflict. Across the region, when people have been able to participate in policy-making decisions and implementation—especially at the local level—seemingly intractable conflict has been diffused and combatants have been willing to take part in peaceful political processes.

• Religious nationalist movements are increasingly attractive to people frustrated by government inefficiency, corruption, and caste politics, and by the failure of governments to deliver on their promises of broad-based prosperity.

• New populations of voters—women and educated youth—are changing the outcomes of elections across the region.

• The most intense armed conflicts in South Asia today are regional conflicts in which nations dispute territorial boundaries or a minority actively resists the authority of a national or state government.

• Despite decades of effort to fight its influence, caste still dramatically affects politics in India.

URBANIZATION

> **GEOGRAPHIC INSIGHT 4**
>
> **Urbanization:** South Asia has two general patterns of urbanization: one for the rich and the middle classes and one for the poor. The areas that the rich and the middle classes occupy include sleek, modern skyscrapers bearing the logos of powerful global companies, universities, upscale shopping districts, and well-appointed apartment buildings. The areas that the urban poor occupy are chaotic, crowded, and violent, with overstressed infrastructures and menial jobs. These two patterns often coexist in very close proximity.

Although today only 30 percent of the region's population lives in urban areas, South Asia has several of the world's largest metropolitan areas. By 2012, Mumbai had 22.9 million people; Kolkata, 16 million; Delhi, 19.5 million; Dhaka, 15.4 million; and Karachi, 11.1 million. All of these cities have grown quickly and all now have massive slums (Figure 8.21D). This growth is likely to continue, and South Asia's current urban population of about 460 million people could expand to as many as 712 million by 2025.

Many middle-class South Asians move to cities for education, training, or business opportunities (Figure 8.21C). Because people come to cities where schooling is more available, large cities have higher literacy rates than rural areas. Mumbai and Delhi both have literacy rates above 80 percent, approximately 25 percent above the average for the country; Dhaka, in Bangladesh, and Karachi, in Pakistan, both have 63 percent literacy, more than 15 percent above the rate for each country as a whole.

Mumbai

Bombay is the name by which most Westerners know South Asia's wealthiest city. It is now called Mumbai, after the Hindu goddess Mumbadevi. Mumbai has the largest deepwater harbor on India's west coast. With more than 21 million people, its metropolitan area hosts India's largest stock exchange, has multiple IT firms, and is the home of the nation's central bank. It pays about a third of the taxes collected in the entire country and brings in nearly 40 percent of India's trade revenue. Its annual per capita GDP is three times that of India's next wealthiest city, the capital Delhi.

Mumbai's wealth also extends into the realm of culture through the city's flourishing creative arts industries, including its internationally known film industry. The city is known as "Bollywood" because it produces popular Hindi movies portraying love, betrayal, and family conflicts. (The term *Bollywood* is a combination of *Bombay* and *Hollywood*.) The stories, played out on lavish sets and accompanied by popular music and dance, serve to temporarily distract their huge audiences from the physical difficulties of daily life.

Mumbai's wealth is most evident when one looks up at the elegant high-rise condominiums built for the city's rapidly growing middle class. But at street level, the urban landscape is dominated by large numbers of people living on the sidewalks, in narrow spaces between buildings, and in large, rambling shantytowns (see Figure 8.21A). The largest of the shantytowns is Dharavi, which houses up to a million people in less than 1 square mile. It is said to contain 15,000 one-room factories turning out thousands of products that are sold in the global marketplace, many of them from India's recycled plastic and metal. Dharavi rivals Orangi Township in Karachi, Pakistan, for the title of Asia's most populous slum, but it is also generating future members of the middle class, at least among the factory owners.

There is yet a third aspect to urban life in South Asia. Were one to carefully observe places as widely separated and culturally different as Mumbai, Chennai, Kathmandu, and Peshawar, one would discover that beyond the main avenues and shantytowns, these cities also contain thousands of tightly compacted, reconstituted villages where daily life is intimate and familiar, not anonymous as in Western cities. Koli is one such place (see the vignette on page 350).

As it is doing in sub-Saharan Africa, the cell phone is bringing urban-style communication into the South Asian countryside. Mumbai-based reporter Anand Giridharadas found that over half the population of South Asia now has access to a cell phone. Even though this access is unevenly distributed, for many the cell phone has become a means of creating the type of social privacy that the bedroom stands for in Western life. With a cell phone, a person can conduct relationships, keep secrets, and access information, all while avoiding interference from authority figures. One measure of how rapidly cell phone technology has taken over is that many more South Asians now have a cell phone than have the use of a flush toilet. Indeed, many young South Asians now view IT access as essential.

Risks to Children in Urban Settings In rural areas, a child can perform useful tasks by age 5, thereby contributing to family well-being. But what about children in urban areas? A report by the International Labor Organization (2008) found that 23 million children between the ages of 5 and 14 are working in South Asia, many of them in urban areas. They work primarily in the informal sector as domestic servants; in export-oriented factories like those small enterprises in urban slums; and as dump scavengers. Some are forced into the sex trade. Clearly, many of these occupations are entirely unsafe and inappropriate for children, who should be protected from such exploitation.

But if we look more closely at one occupation that children have, that of carpet weaving, some interesting issues arise. Carpet weaving is an ancient artistic and economic enterprise in South Asia; traditionally, it has been a family-run enterprise, with women and children as weavers and men as merchants. For thousands of years, young children have learned weaving skills from their parents and have become proud members of the family's home-based production unit. But in today's rapidly modernizing society, the role of children as family workers must be balanced with the need for children to attend school. One of the chief benefits of urban life is education, where children learn the skills that will enable them to survive and prosper in a modern economy, such as the ability to read, write, and do basic math.

FIGURE 8.21

PHOTO ESSAY: Urbanization in South Asia

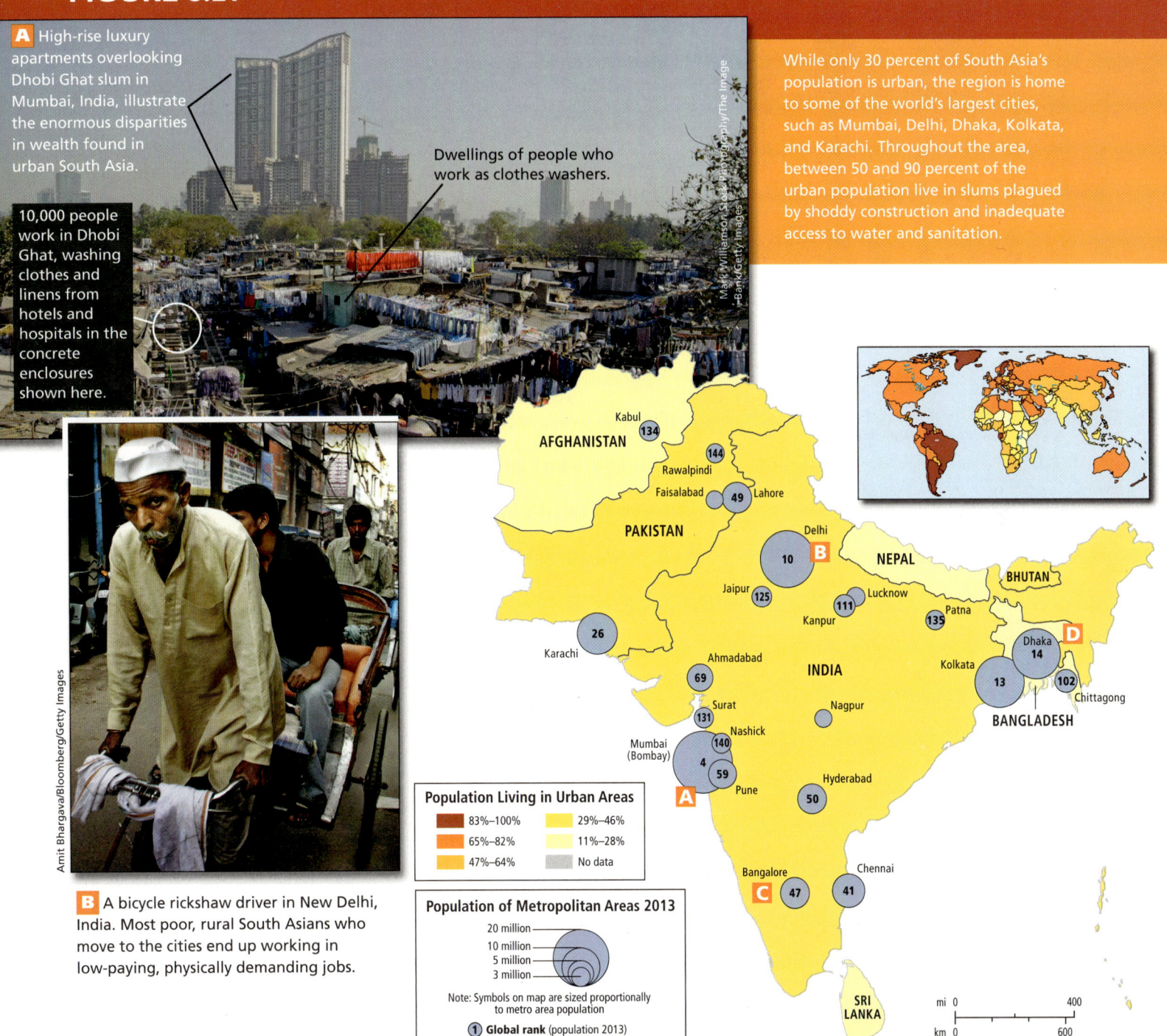

A High-rise luxury apartments overlooking Dhobi Ghat slum in Mumbai, India, illustrate the enormous disparities in wealth found in urban South Asia.

10,000 people work in Dhobi Ghat, washing clothes and linens from hotels and hospitals in the concrete enclosures shown here.

Dwellings of people who work as clothes washers.

While only 30 percent of South Asia's population is urban, the region is home to some of the world's largest cities, such as Mumbai, Delhi, Dhaka, Kolkata, and Karachi. Throughout the area, between 50 and 90 percent of the urban population live in slums plagued by shoddy construction and inadequate access to water and sanitation.

B A bicycle rickshaw driver in New Delhi, India. Most poor, rural South Asians who move to the cities end up working in low-paying, physically demanding jobs.

Population Living in Urban Areas
- 83%–100%
- 65%–82%
- 47%–64%
- 29%–46%
- 11%–28%
- No data

Population of Metropolitan Areas 2013
- 20 million
- 10 million
- 5 million
- 3 million

Note: Symbols on map are sized proportionally to metro area population.

① Global rank (population 2013)

Cities on map:
- Kabul 134 (Afghanistan)
- Rawalpindi 144
- Faisalabad
- Lahore 49 (Pakistan)
- Delhi 10 (B)
- Jaipur 125
- Kanpur
- Lucknow 111
- Patna 135
- Dhaka 14 (D)
- Chittagong 102
- Karachi 26
- Ahmadabad 69
- Kolkata 13
- Surat 131
- Nashick 140
- Nagpur
- Mumbai (Bombay) 4 (A)
- Pune 59
- Hyderabad 50
- Bangalore 47 (C)
- Chennai 41
- Sri Lanka

C A computer class in Mysore, India. Many middle-class South Asians come to cities and stay there in order to take advantage of educational opportunities and the availability of jobs.

D A slum in Dhaka, Bangladesh, that grew up in a railroad right-of-way. Many slums develop in areas that have clear risks to human habitation and few safeguards against potentially deadly hazards.

As global trade has increased the demand for fine, hand-woven carpets, most of the profit has gone not to the weaving families but to South Asian middlemen and foreign traders from Europe and America. The possibility of making large profits has led some unscrupulous carpet merchants to set up factories where kidnapped children are forced to produce carpets. Obviously, kidnapping and enslavement must be stopped, but the question remains: Is there room for cottage industry employment of children within the family circle?

ON THE BRIGHT SIDE

Can Child Labor Be "Fair Labor"?
The United Nations, South Asian governments, and NGOs like RugMark are now addressing this issue of child labor in the hand-woven carpet industry. They have instituted an active program to curb child labor abuses while remaining open to the positive experience for a child of learning a skill and being part of a family production unit. India has established a national system to certify that exported carpets are made in shops where the children go to school, have an adequate midday meal, and receive basic health care. Such carpets bear the label "Kaleen" or "RugMark."

VIGNETTE

Koli is an ancient fishing village that predates the city of Mumbai and is now squeezed between Mumbai's elegant coastal high-rises and the Bay of Mumbai (Figure 8.22). Ringed by fishing boats, Koli is a labyrinth of low-slung, tightly packed homes. Some villagers still fish every day. The screeching of taxis and buses is soon lost in quiet calm as one ducks into a narrow covered passageway that winds through the village and branches in multiple directions. At first Koli appears impoverished, but inside, well-appointed homes, some with marble floors, TVs, and computers, open onto the dimly lit but pleasant alleys. The visitor soon learns that this is no slum or warren of destitute shanties, but rather a community of educated bureaucrats, tradespeople, and artisans who constitute South Asia's rising urban middle class. [Source: From the field notes of Lydia and Alex Pulsipher and their colleague, Tom Osmand]. ∎

FIGURE 8.22 The village of Koli. Now surrounded by the vibrant city of Mumbai, in 1534 when the Portuguese colonists arrived, Koli was a small fishing settlement on a beautiful bay. The village remains and there are still some fishers in Koli, but this exterior view of Koli is misleading. In the interior, out of sight to strangers, dwellings have been rebuilt and refurbished. Most residents are educated and work as bureaucrats or in the service sector of the city. [Alex Pulsipher]

Risks to Women and Men in Urban Settings When rural adults move to cities, the need to find a way to make a living, the crowded conditions, the lack of housing, and the welter of new experiences can leave them vulnerable to exploitation.

Women's experience with purdah (seclusion) can leave them especially at risk. It is inexperienced and undereducated rural women and girls who are most often recruited or forced into **sex work**, the selling of sexual acts for a fee. The vast, urban, slum-based brothels in which sex workers work are legendary in the cities of South Asia and have been documented in, among other places, *Half the Sky*, a book and video by Nicholas Kristof and Sheryl WuDunn.

sex work the provision of sexual acts for a fee

Brothels are notoriously exploitative of women wherever they are found; but the age of the mobile phone is changing the geography of sex work, at once allowing sex workers to move out of brothels, control those who will become their clients, and manage their incomes, which used to be seized by pimps and madams. However, being spatially autonomous also leaves them less likely to have access to condoms or to learning about how to avoid HIV exposure, for which sex workers are at high risk.

Rural men new to cities are also at risk: On-the-job accidents, exposure to HIV, extreme pressure to support families on tiny wages by working long hours at physically demanding jobs (see Figure 8.21B), and the loss of camaraderie with fellow villagers are just a few of the hardships felt by men. **188. GLOBAL POPULATION BOOM PUTS 'MEGA' PRESSURE ON CITIES IN DEVELOPING WORLD**

THINGS TO REMEMBER

> **GEOGRAPHIC INSIGHT 4**

- **Urbanization** South Asia has two general patterns of urbanization: one for the rich and the middle classes and one for the poor. The areas that the rich and the middle classes occupy include sleek, modern skyscrapers bearing the logos of powerful global companies, universities, upscale shopping districts, and well-appointed apartment buildings. The areas that the urban poor occupy are chaotic, crowded, and violent, with overstressed infrastructures and menial jobs. These two patterns often coexist in very close proximity.

- Mumbai is South Asia's wealthiest city and is a showcase of the economic disparity that characterizes South Asian cities.

- Some South Asian urban neighborhoods share many of the intimate qualities of village life.

- When rural people move to cities, the loss of village constraints on behavior, the need to find a way to make a living, the crowded conditions, the lack of housing, and the welter of new experiences can leave them vulnerable to exploitation. ∎

POPULATION PATTERNS

> **GEOGRAPHIC INSIGHT 5**
>
> **Population and Gender:** In this most densely populated of world regions, population growth is slowing as the demographic transition takes hold. Birth rates are falling due to rising incomes, urbanization, better access to health care, and the fact that women are finding more opportunities to study and work outside the home, and thus are delaying childbearing and having fewer children. However, a severe gender imbalance is developing in this region due to age-old beliefs that males are more useful to families than are females. As a result, adult males significantly outnumber adult females.

South Asia is the most densely populated region in the world (Figure 8.23; see also Figure 1.26 on page 46). The region already has more people (1.68 billion) than China (1.35 billion), which has almost twice the land area of South Asia. With 1.46 billion people, India alone will have overtaken China's 1.40 billion by 2025. By 2050, China's population (if the one-child policy persists) will be shrinking at the rate of about 20 million people per decade, while India's may still be growing. However, a notable trend is that current rates of growth have slowed to about 1.6 percent and fertility is declining significantly. Only in Afghanistan and Pakistan are fertility rates above 3 children per woman. In Bangladesh and all other countries of the region, women average just 2.5 children (Figure 8.24). Families are choosing to have fewer children for a variety of reasons, such as improved educational and employment opportunities for women, better health care, and urbanization.

Population densities in this region are highest in the cities that lie just south of the Hindu Kush and Himalayas, and are at their peak in the Ganga-Brahmaputra Delta (see Figure 8.23), much of which is a densely occupied rural agricultural and fishing area.

Geographic Patterns in the Status of Women On average, women's literacy rates, social status, earning power, and welfare are generally lowest in the belt that stretches from the northwest in Afghanistan across Pakistan, western India, Nepal, Bhutan, and the Ganga Plain into Bangladesh. Women fare somewhat better in eastern and central India and considerably better in southern India

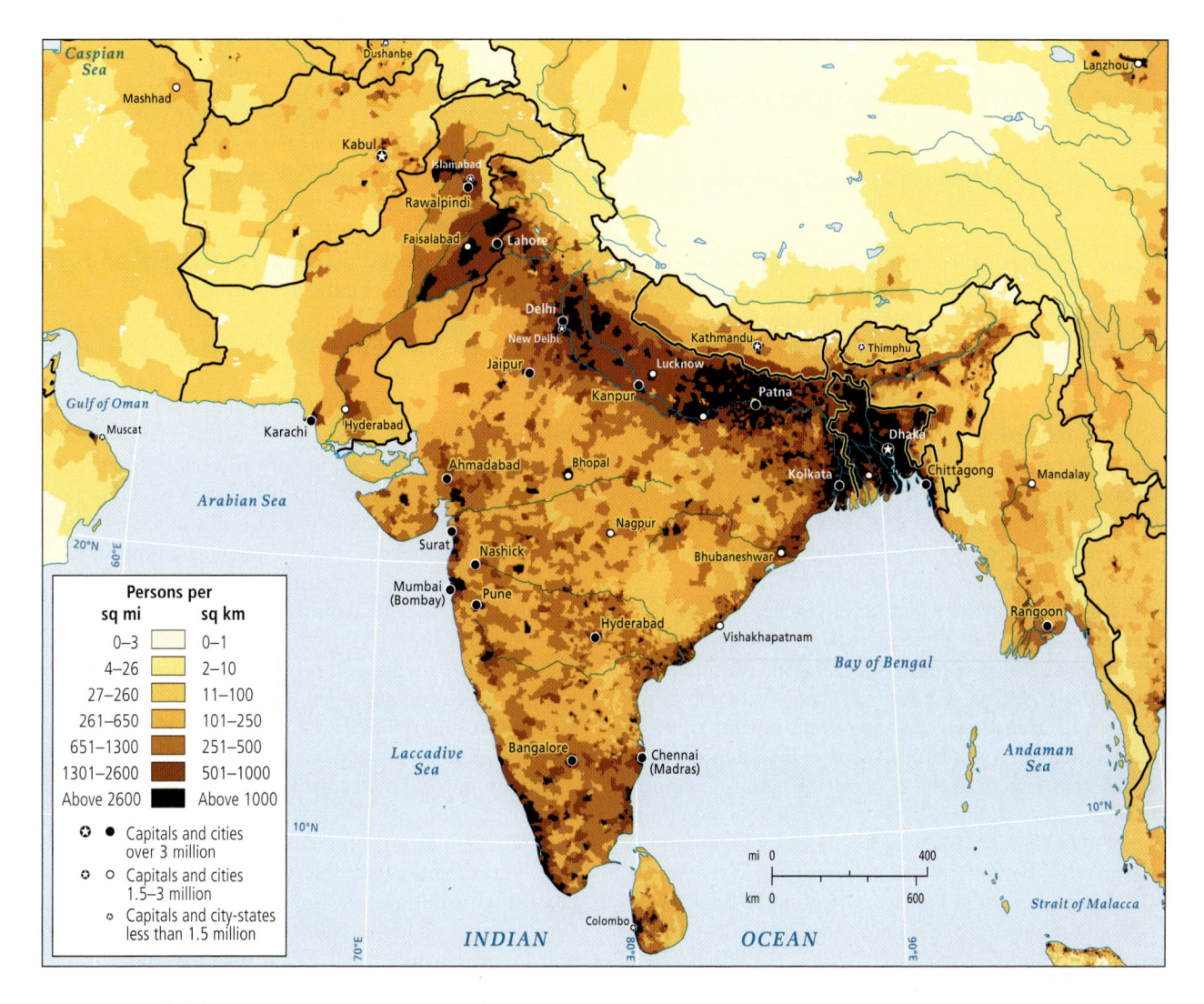

FIGURE 8.23 Population density of South Asia.

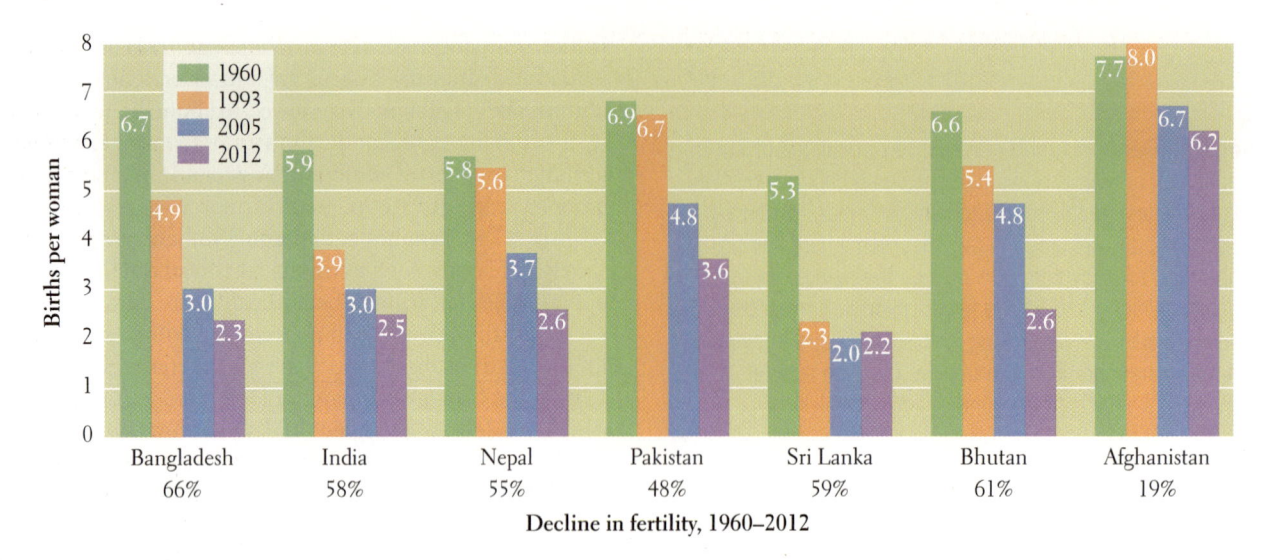

FIGURE 8.24 Total fertility rates in South Asia, 1960–2012. All South Asian countries except Afghanistan have had a substantial decline in fertility since 1960. Afghanistan's rate has declined less than 20 percent over the past 50 years, probably because of almost continuous conflicts, the overwhelmingly rural population (77 percent), and the low literacy rates for both men (43.1 percent) and women (12.6 percent). [Sources consulted: *2012 World Population Data Sheet*, Population Reference Bureau, at http://www.prb.org/pdf12/2012-population-datasheet_eng.pdf; "Bhutan—Fertility Rate: Fertility Rate, Total (births per woman)," Index Mundi, at http://www.indexmundi.com/facts/bhutan/fertility-rate; "Afghanistan—Fertility Rate: Fertility Rate, Total (births per woman)," Index Mundi, at http://www.indexmundi.com/facts/afghanistan/fertility-rate; "Afghanistan," *The World Fact Book 2012*, Central Intelligence Agency, at https://www.cia.gov/library/publications/the-world-factbook/geos/af.html]

and in Sri Lanka. In these latter regions, where literacy rates are higher (Figure 8.25), different marriage, inheritance, and religious practices give women better access to education and resources.

Slowing Population Growth: Health Care, Urbanization, and Gender Efforts to slow population growth through better health care have been underway for more than 50 years. Today, urbanization and the rising status of women are helping to reduce the birth rate.

With improved health care—such as rehydration to control the effects of diarrhea in infants—far fewer babies are dying in infancy. In 1992, infant mortality rates were 91 per 1000 live births in India, 109 in Pakistan, and 120 in Bangladesh. By 2012, those rates were reduced to 47 in India, 68 in Pakistan, and 43 in Bangladesh. With better assurance that their babies will survive to adulthood and be able to care for their elderly parents, couples now choose more often to have just two children.

At the same time that improved access to health care has helped propel the reduction in family size, the shift in population from rural to urban areas has reduced the economic incentive for having a large family (see the Figure 8.21 map). Whereas children in rural areas (starting at an early age) contribute their labor to farming, thus increasing the family income, the labor of children in urban areas is less likely to boost the family income. Children in urban areas are more likely to go to school, and thus represent a cost to the family for school fees, books, and uniforms.

Improvements in the status of women have also slowed population growth. As in other parts of the world, women in South Asia who have the opportunity to go to school and/or to work at a paying job tend to delay childbearing and have fewer children.

A comparison of the population pyramids as well as some other statistics for Sri Lanka and Pakistan helps illustrate the influence of health care and the status of women on birth rates (Figure 8.26; see also Figures 8.24 and 8.25). As a country that is far along but not completely through the demographic transition (see Figure 1.28 on page 49), Sri Lanka has a much higher GNI per capita (U.S.$4943) than Pakistan (U.S.$2550). Health care is generally better in Sri Lanka, as indicated by its much lower infant mortality rate (in 2012, Sri Lanka had 12 deaths per 1000 live births versus Pakistan's 68 per 1000). Sri Lanka has also worked harder to educate and economically empower its women. Three key indicators of women's education and empowerment that are associated with reduced fertility rates are much higher in Sri Lanka than in Pakistan: female literacy (90 percent versus 30 percent), the percentage of young women who attend high school (89 percent versus 19 percent), and the percentage of women who work outside the home (36 percent versus 16 percent). Surprisingly, Sri Lanka has achieved all this with an urbanization rate of just 15 percent, far lower than that of Pakistan's 35 percent.

Despite Pakistan's rather dismal statistics compared to those of Sri Lanka, its population pyramid does indicate that birth rates have slowed significantly (notice that the bottom of the pyramid in Figure 8.26B narrows noticeably). If this trend continues, Pakistan, as well as all South Asian countries except Afghanistan, will enjoy what is called the *demographic dividend*: a bulge of young people entering the workforce and choosing to have fewer children, who will thus have the time and energy to study, work, and contribute to the economy and civil society. Down the road, however, South Asian countries that have had a demographic dividend will have to deal with an aging population for which there will be far fewer younger caregivers and taxpayers to support the elderly. **185. CHILD LABOR PERSISTS IN INDIA DESPITE NEW LAWS**

Population Patterns **353**

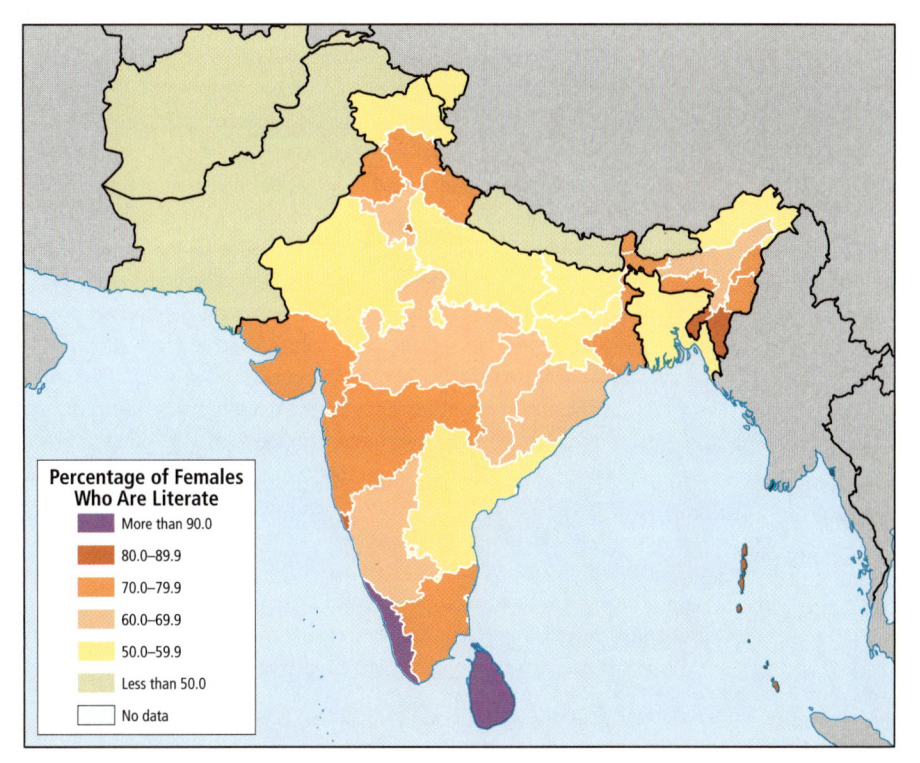

FIGURE 8.25 Female literacy in South Asia. Female literacy lags behind male literacy in all countries in South Asia, and is below 50 percent in four countries. Female literacy is crucial to improving the lives not only of women, but also of children: Women who can read often seize opportunities to earn an income and nearly always use this income to help their children. In India, overall literacy has risen in the past decade, and in 2011, female literacy there reached 65 percent, up from 54.5 percent in 2005. [Sources consulted: "District Wise Female Literacy Rate of India," Maps of India, at http://www.mapsofindia.com/census2001/femaleliteracydistrictwise.htm; *United Nations Human Development Report 2009* (New York: United Nations Development Programme), Table J, "Gender-Related Development Index and Its Components," at http://hdr.undp.org/en/reports/global/hdr2009/; "Map of Literacy Rate in India," Maps of India, at http://www.mapsofindia.com/census2011/literacy-rate.html; "Literacy," *The World Fact Book*, Central Intelligence Agency, at https://www.cia.gov/library/publications/the-world-factbook/ields/2103.html#af]

Gender Imbalance South Asian populations have a significant gender imbalance because of cultural customs that make sons more likely than daughters to contribute to a family's wealth. A popular toast to a new bride is "May you be the mother of a hundred sons." Many middle-class couples who wish to have sons hire high-tech laboratories that specialize in identifying the sex of a fetus. The intention is to abort female fetuses. This practice is now illegal, but it persists. Poorer South Asians may neglect the health of female children, with some even committing female infanticide (the dowry implications are discussed on page 355). 📹 **186. GIRLS PAY THE PRICE FOR INDIA'S PREFERENCE FOR BOYS**

Gender imbalance of the magnitude India is facing could create serious problems, such as surges in crime and drug abuse related to the presence of so many young men with no prospect of having a family. In all cultures, the possibility of having a family is a stabilizing influence for young men. The efforts of the state of Kerala to overcome gender imbalance by paying more attention to women's development are worth watching. Kerala's government has made women's development a priority, funding education for women well beyond basic levels. The results are reflected in its female literacy rate, which is the highest in India at 87.8 percent (the average for India as a whole is 54.5 percent; see Figure 8.25 for the female literacy map), and in the number of women who work outside the home. In Kerala, far from being viewed as an economic liability to the family, daughters are seen as an asset. Women can travel the public streets alone and in groups and commonly work in public places, many in high positions in government, education, health care, IT industries, and other professions.

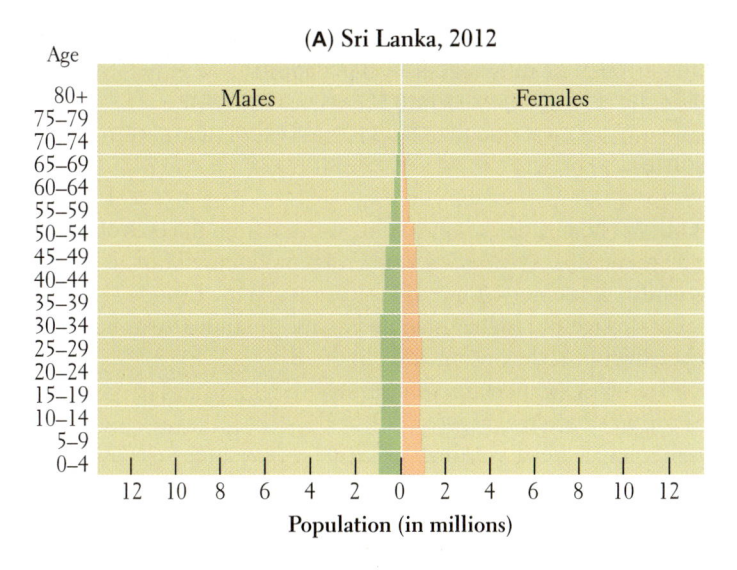

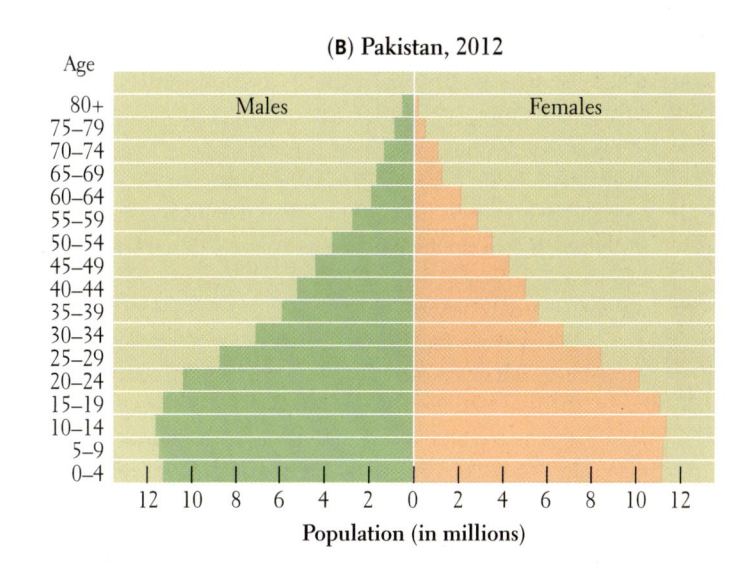

FIGURE 8.26 Population pyramids for Sri Lanka and Pakistan. [Source consulted: *International Data Base*, U.S. Census Bureau, at http://www.census.gov/population/international/data/idb/informationGateway.php]

THINGS TO REMEMBER

› GEOGRAPHIC INSIGHT 5

- **Population and Gender** In this most densely populated of world regions, population growth is slowing as the demographic transition takes hold. Birth rates are falling due to rising incomes, urbanization, better access to health care, and the fact that women are finding more opportunities to study and work outside the home, and thus are delaying childbearing and having fewer children. However, a severe gender imbalance is developing in this region due to age-old beliefs that males are more useful to families than are females. As a result, adult males significantly outnumber adult females.

- There is a geographic pattern to the status of women, especially with regard to their literacy and earning power. Generally speaking, women are less restricted in the southern and eastern sections of this region.

- While population growth is slowing across the region, momentum provided by a young population means that growth will continue for the foreseeable future. ∎

SOCIOCULTURAL ISSUES

Within the life of one South Asian village or urban neighborhood, there can be considerable cultural variety. Differences based on caste, economic class, ethnic background, gender, religion, and even language are usually accommodated peacefully by long-standing customs, such as religious and ethnic festivals and foodways (Figure 8.27 and Figure 8.28), that guide cross-cultural interaction. However, South Asia is also undergoing a number of social changes, political shifts, and cultural transitions that challenge many traditional practices, especially those related to gender.

The Texture of Village Life

The vast majority—about 70 percent—of South Asians live in the hundreds of thousands of villages. Even many of those now living in South Asia's giant cities were born in a village or occasionally visit an ancestral rural community, so for most people village life is a common experience.

VIGNETTE

The writer Richard Critchfield, who studied village life in more than a dozen countries, wrote that the village of Joypur (Bangladesh) in the Ganga-Brahmaputra Delta is set in "an unexpectedly beautiful land, with a soft languor and gentle rhythm of its own." In the heat of the day, the village is sleepy: Naked children play in the dust, women meet to talk softly in the seclusion of courtyards, and chickens peck for seeds.

In the early evening, mist rises above the rice paddies and hangs there "like steam over a vat." It is then that the village comes to life, at least for the men. The men and boys return from the fields, and after a meal in their home courtyards, the men come "to settle in groups before one of the open pavilions in the village center and talk—rich, warm Bengali talk, argumentative and humorous, fervent and excited in gossip, protest, and indignation" as they discuss their crops, an upcoming marriage, or national politics. [Source: Richard Critchfield, Villages.] ∎

VIGNETTE

The anthropologist Faith D'Aluisio and her colleague Peter Menzel offer another peek into village life as night falls in Ahraura, a village in the state of Uttar Pradesh in north-central India. In the enclosed women's quarters of a walled compound, Mishri is finishing her day by the dying cooking fire as her 1-year-old son tunnels his way into her sari to nurse himself to sleep. Mishri, who is 27, lives in a tiny world bounded by the walls of the courtyard she shares with her husband, five children, and several of her husband's kin. Like many villages in northern India, her village observes the practice of *purdah*, in which women keep themselves apart from men (see the discussion below).

That Mishri can observe purdah is a mark of status because it shows she need not help her husband in the fields. Within the compound, she works from sunup to sundown, chatting only momentarily with two women who cover their faces and scurry from their own courtyards to hers for the short visit. Mishri is devoted to her husband, who was chosen for her by her family when she was 10; out of respect, she never says his name aloud. [Source: Adapted from Faith D'Aluisio and Peter Menzel, Women in the Material World, 1996.] ∎

Social Patterns in the Status of Women

The status of women in South Asia varies significantly along rural/urban and religious divides. Generally speaking, rural women have far less freedom than do urban women. In rural India, middle- and upper-caste Hindu women are often more restricted in their movements than are lower-caste women because they have a status to maintain. Meanwhile, lower-caste women who go into public spaces may have to contend with sexual harassment and exploitation from upper-caste men. The socioeconomic status of Muslim women in South Asia is notably lower than that of their Hindu and Christian counterparts. In India, some of this is related to the generally lower incomes and standard of living for Muslims, which also usually means lower educational levels. Muslim women also work outside the home less than non-Muslim women in India. Low rates of education and workforce participation for women also prevail in Muslim-dominated countries such as Pakistan and Afghanistan, though rates are significantly higher in Bangladesh.

🎬 **187. SUFI ROCK SINGER FALU BLENDS OLD WITH NEW**

🎬 **191. REMEMBERING PAKISTAN'S FORMER PRIME MINISTER BENAZIR BHUTTO**

Purdah The practice of concealing women from the eyes of nonfamily men, especially during women's reproductive years, is known as **purdah**. It is observed in various ways across the region. The practice is strongest in Afghanistan and across the Indo-Gangetic Plain, where within both Muslim and Hindu communities, women are often secluded within structures (Figure 8.29) and wear veils or head coverings. Purdah is less strict in central and southern India, but even there, separation between unrelated men and women is maintained in public spaces. In general, low-caste Hindus do not observe this custom, but that is changing. In recent decades, as some low-status households have increased their incomes, they have adopted purdah as a sign of their rising wealth.

> **purdah** the practice of concealing women from the eyes of nonfamily men

FIGURE 8.27   LOCAL LIVES
Festivals in South Asia

A A village festival in Pakistan features *kabaddi*, a popular South Asian sport in which teams take turns sending a "raider" across a field center line. That person must tag, or in some cases wrestle to the ground, members of the other team and then return to his or her own side without taking a breath. Kabaddi has been played at the Indian National Games since 1939 and at the Asian Games since 1991. [Amir Mukthar/FlickrVision/Getty Images]

B Celebrants in Kolkata, India, during Holi—a festival celebrating the end of winter and beginning of spring. Holi evolved from temple worship practices involving the application of color to statues. In a riotous and celebratory atmosphere, people of different ages, genders, castes, and economic backgrounds temporarily disregard their differences and hurl the colors of the coming spring at each other. [Sanjay Kanojia/AFP/Getty Images]

C Pilgrims during the 2010 Kumbh Mela bathe in the Ganga River at Haridwar. During this event, which is held every 3 years, Hindus purify themselves by bathing in the sacred waters of the river. In 2013, the 45-day-long event attracted over 100 million participants. [Pedro Ugarte/AFP/Getty Images]

Purdah practices have influenced the architecture of South Asia. Homes are often in walled compounds that seclude kitchens and laundries as women's spaces. In grander homes, windows to the street are usually covered with lattice screens, known as *jalee*, that allow in air and light but shield women from the view of outsiders.

Marriage, Motherhood, and Widowhood Throughout South Asia, most marriages are arranged by the parents of the prospective bride and groom. Particularly in wealthier, better-educated families, the wishes of the bride and groom are considered, but in some cases they are not (Figure 8.30). This is especially true of child marriage, when a young girl (often as young as 12 and in some places even younger) is married off to a much older man.

Usually a bride goes to live in her husband's family compound, where she becomes a source of domestic labor for her mother-in-law. Most brides work at domestic tasks for many years until they have produced enough children to have their own crew of small helpers, at which point they gain some prestige and a measure of autonomy.

Motherhood in South Asia determines much about a woman's status within her community. A woman's power and mobility increase when she has grown children and becomes a mother-in-law herself. On the other hand, in some communities, the death of a husband, regardless of cause, is a disgrace to a woman and can completely deprive her of all support and even of her home, children, and reputation. Widows may be ritually scorned and blamed for their husband's death. Widows of higher caste rarely remarry, and in some areas, they become bound to their in-laws as household labor or may be asked to leave the family home. Most simply become marginalized "aunties" in extended families and help with household duties of all sorts.

Dowry and Violence Against Females A **dowry** is a sum of money paid by the bride's family to the groom's family at the time of marriage. Dowries originated as an exchange of wealth between Muslim landowners or high-caste families that practiced purdah. With her ability to work reduced by purdah, a woman was considered a liability for the family that took her in. Changing dowry customs appear to be a cause of the growing incidence of various kinds of domestic violence against females in Pakistan, India, and Bangladesh.

> **dowry** a price paid by the family of the bride to the groom (the opposite of *bride price*); formerly a custom practiced only by the rich

Until the last several decades, only wealthy families gave the groom a dowry—in this case, a substantial sum that symbolized the family's wealth, meant to give a daughter a measure of security in her new family.

Ironically, the increases in affluence and education have reinforced the custom of dowry and made it much more common for all families. As more men became educated, their families felt that their diplomas increased their worth as husbands and gave them the power to demand larger and larger dowries. Soon, the practice spread through lower-caste families wanting to upgrade their status. Now the dowries they must pay to get their daughters married can cripple poor families. A village proverb captures this dilemma: "When you raise a daughter, you are watering another man's plant."

Gender, Politics, and Power

As countries in South Asia have moved toward greater respect for political freedoms, the status of women in the region has risen. India, Pakistan, Bangladesh, and Sri Lanka have all had female heads of state (prime ministers) in the past. However, it is important to note that all of these women were either wives or daughters of previous heads of state. Women have been notably less successful

FIGURE 8.28 LOCAL LIVES
Foodways in South Asia

A Boys in Hampi, Karnataka, eat a south Indian *thali*, or feast, featuring rice, lentils, and various other vegetarian dishes served on a banana leaf. Most South Asian cuisine is eaten with the hands; in South India and Sri Lanka, banana leaves often serve as plates. [Huw Jones/Lonely Planet Images/Getty Images]

B *Chapatis* are cooked on a griddle at a wedding in Rajasthan in northwest India. Chapati is a kind of bread made of stone-ground wheat flour and cooked on a griddle or curved iron *tava* located above a fire or oven. The bread is unleavened—it is made of dough that doesn't rise because it doesn't contain yeast. Most popular in northern South Asia, where wheat is more widely grown than rice, chapatis are torn into pieces and used to scoop up vegetable or meat dishes. [Dario Mitidieri/Photonica World/Getty Images]

C Street food in a crowded marketplace in Dhaka, Bangladesh, during Ramadan. On offer are whole chickens, lamb, and various vegetable dishes, including potatoes and lentils. Usually the food is taken home and shared with family members. [AFP/Getty Images]

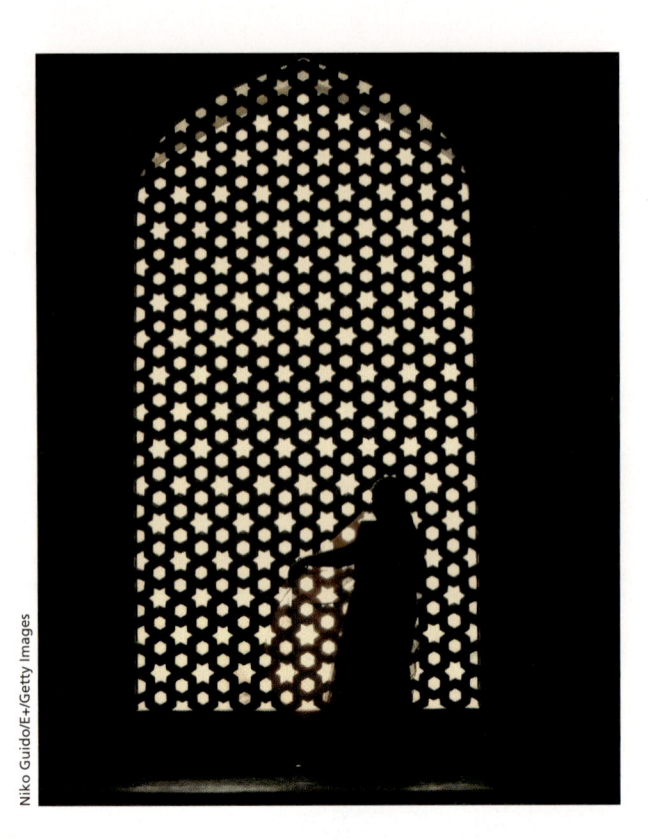

FIGURE 8.29 Material culture of purdah. A lattice screen in Fatehpur Sikri, India. Known as *jalee*, lattice screens are often found in parts of South Asia where women are secluded. Like the louvers and latticed windows of Southwest Asia (see Figure 6.27B on page 268), *jalee* allow ventilation and let in light but shield women from the view of strangers.

TABLE 8.2	Percentages of South Asian women in Parliament, 2013
Country	Percentage in Parliament
Sri Lanka	5.8
Bhutan	8.5
Pakistan	19.5
Nepal	33.2
Bangladesh	19.7
India	11.0
Afghanistan	27.7

Source: Women in National Parliaments, Inter-Parliamentary Union Web site at: http://www.ipu.org/wmn-e/classif.htm

in local elections, and at the parliamentary level Indian and Sri Lankan women remain very poorly represented (Table 8.2).

The very low percentage of women in India's parliament inspired a confederation of Muslim and Hindu women's groups to lobby for legislation that would temporarily (for a 15-year trial period) reserve one-third of the seats in the lower house of Parliament and in state assemblies for women. Such one-third quotas are already in place in Pakistan (19.5 percent), Nepal (33.2 percent), and Bangladesh (19.7), but are so far close to being met only in Nepal.

If recent voter turnout and political activism trends continue, as discussed on page 344, there is likely to be a major improvement in female representation in the region's national parliaments and in local offices.

Women and the Taliban in Afghanistan Women in Afghanistan have frequently suffered brutal repression since a conservative Islamist

Sociocultural Issues **357**

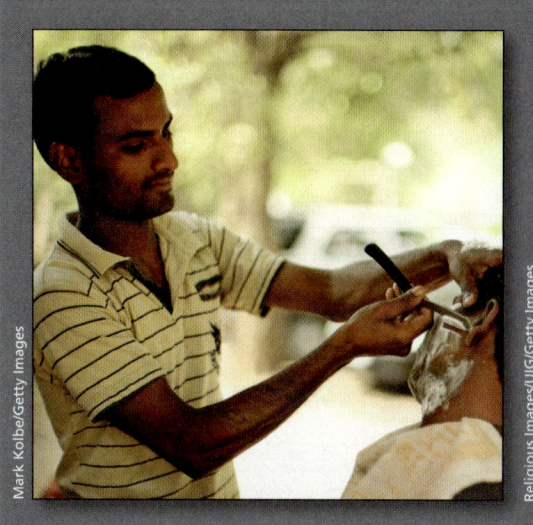

FIGURE 8.30 Caste and marriage. Caste remains a particularly powerful force with respect to marriage.

(A) A barber, often a member of a "barber caste" or *jati*, practices his trade in Chandigarh, India. In many areas, members of the barber caste act as go-betweens when families are arranging marriages for their children.

(B) A Brahmin priest officiates at a Hindu wedding ceremony in Varanasi, India. The specifics of the ceremony vary considerably, depending on the caste of the bride and groom.

(C) A *hijra* performs at a wedding ceremony in Rawalpindi, Pakistan. *Hijras* are men who leave their families, often at a young age, to join a "*hijra* family." Here they undergo a gradual transition toward femininity, sometimes culminating in castration. They also must learn the *hijra* trades, which include singing and dancing for weddings and birth ceremonies, sometimes fortune telling, and often sex work. Because of their unique identity, *hijras* are allowed to move between different castes with greater ease than most.

movement, the **Taliban**, gained control of the government there in the mid-1990s. Prior to that time, rights for women in Afghanistan were slowly but steadily improving, and upper-class women had many freedoms; they could dress in Western styles and they had the right to attend gender-integrated university courses. The Taliban support strict and radical interpretations of Islamic law, forcing females, including urban professional women, to live in seclusion. In regions where the Taliban retain control, girls and women are not allowed to work outside the home or attend school. In virtually all parts of the country, despite the decline of Taliban control, women must wear a heavy, completely concealing garment, called a *burqa* (or burka), whenever they are out of the house. (Men also must follow a dress code, though a less restrictive one.) Although the Taliban were driven from official power in November of 2001, they maintain control of the rural southern provinces and mountainous zones near Pakistan, where cultural and religious conservatism continues to adversely affect Afghan women. Even efforts to provide women and girls with a basic education, such as that described in the following vignette about Radio Sahar, run into hostility, and oftentimes violence.

Taliban an archconservative Islamist movement that gained control of the government of Afghanistan in the mid-1990s

189. REPORT: DOMESTIC VIOLENCE WIDESPREAD IN AFGHANISTAN

190. FRONTRUNNER DOCUMENTARY TELLS STORY OF AFGHAN POLITICIAN WHO INSPIRES WOMEN

VIGNETTE

From behind her microphone at Radio Sahar ("Dawn"), Nurbegum Sa'idi speaks to a female audience on a wide range of topics. Located in the city of Herat, Radio Sahar is one in a network of independent women's community radio stations that has sprung up in Afghanistan since early 2003. Radio Sahar provides 13 hours of daily programming consisting of educational items that address cultural, social, and humanitarian matters as well as music and entertainment. For example, one recent broadcast aimed at informing women of their legal rights followed the life of a young woman who was physically abused by her husband and his entire family. The woman took the brave step of asking for a divorce. As a result, she was forced into hiding, where she was counseled on the steps she might take next. A reported 600,000 Afghan women and youth listen to Radio Sahar while they do their chores.

Girls on the Air, a film by Valentina Monti, reveals the diversity of the ideas and hopes for the future of the young journalists who founded Radio Sahar. A clip can be seen at http://www.youtube.com/watch?v=c6KxtDHbuuY. [*Source: Internews Afghanistan. For detailed source information, see Text Sources and Credits.*]

Men and the Taliban in Afghanistan

The extreme repression of male individuality under the Taliban bears recognition for the hardships it causes. Hypermasculine societies, such as that of the Taliban, tend to repress creativity and emotional sensitivity in males, and that repression can lead to closed minds and even violence toward others, particularly women. Nevertheless, despite the many expressions of hypermasculinity in Afghanistan, Pakistan, and elsewhere in this region, there is also a long history of alternative expressions of gender (see Figure 8.30C).

> **ON THE BRIGHT SIDE**
>
> **Cell Phones and Literacy**
>
> In order to reach women held in deep seclusion, the Afghan government is now making available basic reading and writing lessons on special mobile phones distributed free to Afghan women. The reading/writing software was developed by an Afghan IT firm with USAID assistance. As a woman achieves literacy, she can add other subjects with free apps.

THINGS TO REMEMBER

- Purdah is practiced in both Muslim and Hindu households, but the status of Muslim women is significantly lower than that of their Hindu, Sikh, Jain, Buddhist, and Christian counterparts.
- Changing dowry customs appear to be a cause of the growing incidence of various kinds of domestic violence against females in Pakistan, India, and Bangladesh.
- Motherhood in South Asia determines much about a woman's status within her community.
- South Asia has had a number of women in very high positions of power, and young women today on average have more educational and employment opportunities than women of a generation ago. However, the overall status of women in the region is notably lower than the status of men.

GEOGRAPHIC INSIGHTS

South Asia: Review and Self-Test

1. Environment: Climate change puts more lives at risk in South Asia than in any other region in the world, primarily due to water-related issues. Over the short term, droughts, floods, and the increased severity of storms imperil many urban and agricultural areas. Over the longer term, sea level rise may profoundly affect coastal areas and glacial melting poses a threat to rivers and aquifers.

- Why is the issue of melting glaciers in the Himalayas so important in this region? Compare the short- and long-term effects of this change.
- What additional threats do coastal areas of South Asia face?

2. Globalization and Development: Globalization benefits some South Asians more than others. Educated and skilled South Asian workers with jobs in export-connected and technology-based industries and services are paid more and sometimes have better working conditions. Less-skilled workers in both urban and rural areas are left with demanding but very low-paying jobs.

- What kinds of workers in South Asia have gained from the recent boom in foreign investment? What about South Asian workers is especially attractive to foreign investors in high-tech industries?
- How are jobs in South Asia linked to the global economy?
- What kinds of workers in South Asia are benefiting the least from these new foreign investments?

3. Power and Politics: India, South Asia's oldest, largest, and strongest democracy, has shown that the expansion of political freedoms can ameliorate conflict. Across the region, when people have been able to participate in policy-making decisions and implementation—especially at the local level—seemingly intractable conflict has been diffused and combatants have been willing to take part in peaceful political processes.

- Which conflicts in South Asia have been made worse by an unwillingness on the part of governments and warring parties to allow free and fair elections?
- What democratic strategy has been used to diffuse some conflicts in the region, at least for the short term?
- What are some signs that political participation is catching on in South Asia?

4. Urbanization: South Asia has two general patterns of urbanization: one for the rich and the middle classes and one for the poor. The areas that the rich and the middle classes occupy include sleek, modern skyscrapers bearing the logos of powerful global companies, universities, upscale shopping districts, and well-appointed apartment buildings. The areas that the urban poor occupy are chaotic, crowded, and violent, with overstressed infrastructures and menial jobs. These two patterns often coexist in very close proximity.

- What has been the impact on food production of the introduction of new seeds, fertilizers, pesticides, and equipment in South Asia? How has this shift resulted in the growth of cities?
- What kinds of jobs and housing do people in South Asia usually find in urban areas?

5. Population and Gender: In this most densely populated of world regions, population growth is slowing as the demographic transition takes hold. Birth rates are falling due to rising incomes, urbanization, better access to health care, and the fact that women are finding more opportunities to study and work outside the home, and thus are delaying childbearing and having fewer children. However, a severe gender imbalance is developing in this region due to age-old beliefs that males are more useful to families than are females. As a result, adult males significantly outnumber adult females.

- What has created the strong preference for sons among many South Asian families? Where in South Asia is the preference for sons the weakest?
- What is the link between fertility and education for women of South Asia?

CRITICAL THINKING QUESTIONS

1. Dams and the reservoirs they create are sources of irrigation water and generators of electricity. Name some of the factors that have rendered so many places in South Asia, as well as elsewhere around the world, in need of markedly more water and electricity.

2. Explain why some would say that India has been part of globalization for thousands of years. Tie this history to what is happening in the present.

3. What are some of the lingering features of the British colonial era in South Asia? To what extent is globalization reinforcing or erasing these features?

4. How are changes in agriculture resulting in urbanization?

5. Describe microcredit and its impact on extreme poverty in South Asia.

6. Describe how South Asia is vulnerable to climate change. Identify some responses to the climate crisis that are emerging from this region.

7. Describe the main challenges to democracy in South Asia.

8. Identify the factors that have led to a recent boom in manufacturing industries in South Asia.

9. What factors have encouraged high population growth in South Asia in the past? What factors may encourage slower population growth in the future?

10. Describe the main factors that create South Asia's gender imbalance.

> THINKING GEOGRAPHICALLY

Now that you have read about South Asia, you should be able to answer the following questions about photos in this chapter. To answer these questions online, go to Geography LaunchPad.

Vulnerability to Climate Change (page 325)

A What suggests that this girl comes from a very poor family?

B Why is the use of simple irrigation technology a sensible strategy in a country like Afghanistan?

C List all the indications in this photo that Mumbai has drainage problems.

D How might such embankments help reduce sensitivity to storm surge and sea level rise?

Human Impacts on the Biosphere (page 328)

A What are the indications that water in this stream is polluted?

B What is the greater source of pollution for the Ganga River, incompletely cremated dead bodies or industrial and urban pollution?

C What does this photo suggest about sources of air pollution in New Delhi?

D What effects can deforestation have on humans and the environment?

Visual History (page 332)

A This structure was designed to store water from what source?

B Suggest some principal features of the structure and grounds of the Taj Mahal.

C What is significant about the fact that as of 1750, before the onset of British colonialism, South Asia produced 12 to 14 times more cotton than Britain and the rest of Europe combined?

D What was the immediate result of the Salt March?

E What does this photo suggest about India's modern economy?

Power and Politics (page 345)

A What aspects of this photo suggest that the convoy is part of a government-funded army?

C What direction was the camera facing when this photo was taken?

D How did the Maoists describe the war they waged against King Gyanendra?

Urbanization (page 349)

A and **D** What circumstances have brought South Asians into the crowded urban districts in these two photos of Mumbai and Dhaka?

C What is notable about the members of this classroom?

CHAPTER KEY TERMS

agroecology 341
Buddhism 334
caste system 332
civil disobedience 335
communal conflict 343
dowry 355
Harappa culture 331
Hinduism 331
Indus Valley civilization 331

Jainism 334
jati 332
microcredit 342
Mughals 331
offshore outsourcing 340
Partition 336
purdah 354
regional conflict 344
religious nationalism 344

sex work 350
Sikhism 334
subcontinent 320
summer monsoon 322
Taliban 357
varna 333
winter monsoon 322

A: Plateau of Tibet, China

chapter 9
EAST ASIA

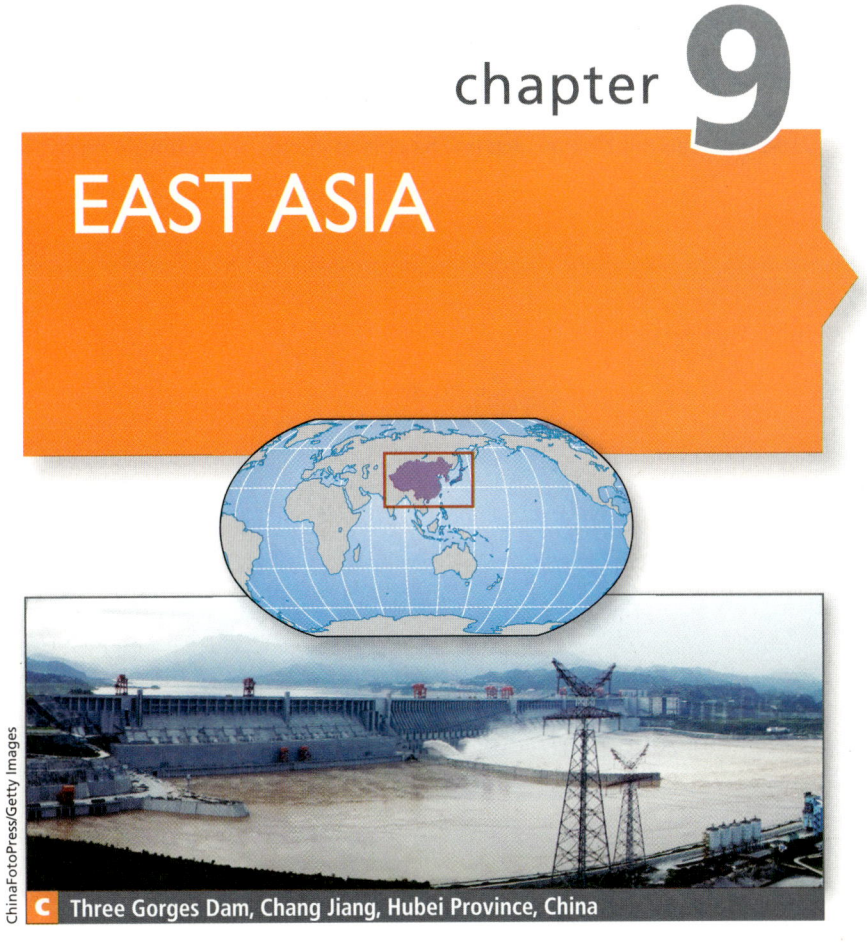

C Three Gorges Dam, Chang Jiang, Hubei Province, China

D Hong Kong Island and the South China Sea

E Mount Fuji, Japan

B Mongolia

FIGURE 9.1 Regional map of East Asia.

361

CHAPTER 9 East Asia

GEOGRAPHIC INSIGHTS: EAST ASIA

After you read this chapter, you will be able to discuss the following geographic insights as they relate to the five thematic concepts:

1. **Environment:** East Asia's most serious environmental problems result from its high population density combined with its rapid urbanization and environmentally unsustainable economic development. Climate change may intensify the droughts and floods that have long plagued this region.

2. **Globalization and Development:** East Asia pioneered a spectacularly successful economic development strategy that has transformed economies across the globe. Governments in Japan and then Taiwan, South Korea, and eventually China intervened strategically in the economy to encourage the production of manufactured goods destined for sale abroad, primarily to the large economies of North America and Europe.

3. **Power and Politics:** As East Asia has developed economically, the pressure for more political freedoms has grown. Japan, South Korea, and Taiwan are now among the more politically free places in the world, and demands for political change in China are increasing, especially in urban areas.

4. **Urbanization:** Across East Asia, cities have grown rapidly over the last century, fueled by export-oriented manufacturing industries. China has recently undergone the most massive and rapid urbanization in the history of the world. Its urban population, now more than 720 million people, has tripled since China initiated economic reforms in the 1980s.

5. **Population and Gender:** Although East Asia remains the most populous world region, families here are having far fewer children than in the past, resulting in populations that are aging. Meanwhile, the legacy of China's now largely abandoned "one child" policy, combined with an enduring cultural preference in China for male children, has created a shortage of females.

The East Asia Region

East Asia (Figure 9.1), with 1.6 billion people, is the most populous world region. The region is dominated by China, which has a population of 1.35 billion. China's dominance is not due only to its population, but also to its physical size and its economic role in the global economy. Just as the striking economic rise of Japan, South Korea, and Taiwan has been the model for developing countries over the past 50 years, the opening of China to the global economy will continue to transform East Asia and the world for 50 years or more. The changes underway in this region today are immense. Incomes have risen and cities have boomed, but so have air and water pollution and greenhouse gas emissions. At the same time, there are signs that new, cleaner technologies may mitigate these pollution problems.

The five thematic concepts in this book are explored as they arise in the discussion of regional issues, with interactions between two or more themes featured. Vignettes, like the one that follows about the situation of workers in the East Asian economy, illustrate one or more of the themes as they are experienced in individual lives.

GLOBAL PATTERNS, LOCAL LIVES

In 2000, at age 18, Li Xia (Li is her family name) left her farming village in China's Sichuan Province (Figure 9.2A) for Dongguan, then a city of 6.4 million, in the Guangdong Province of southern China. She was accompanied by two friends. A few months earlier, the government had taken their families' farmland for an urban real estate project, paying compensation of only U.S.$2000 per family. The three young women accepted an offer to work in a Dongguan toy factory so they could send money back to their families, who now have to pay cash for food and housing (see Figure 9.2B).

When the young women arrived in the city, they joined 4.3 million other recent internal migrants, who formed 63 percent of Dongguan's population. Like Li Xia, they had migrated illegally without government-approved residency rights, and thus were dependent on their employers for housing.

As did many others, Li Xia and her friends soon found that the labor recruiters had lied about their wages. They would be paid U.S.$30 per month, not U.S.$45 as they had been promised. In addition, they would work 12-hour shifts in 100°F heat, receive no overtime pay, and have only 1 day off a month. But there was no point in protesting, as the recruiters purposely brought in thousands of extra workers to replace any complainers.

Xia felt better when she saw that the toy factory was a clean, modern building known locally as the "Palace of Girls" (nearly all 3500 employees were female). Within a day, she had completed her training, signed a 3-year contract, and mastered her task of putting eyes on stuffed animals destined for toddlers in the United States and Europe. Her enthusiasm faded, however, when she learned that she and her friends would be spending much of their money on the expensive but low-quality food provided by the company and would be sharing one small room and a tiny bath with eight other women and a rat or two.

FIGURE 9.2 Workers, development, and urbanization in the new Chinese economy. In 2009, more than 145 million Chinese workers—over one-third of them women—were migrant workers who had moved from rural areas to urban areas to find work, usually without the government's permission and hence without the full rights of citizenship they would be entitled to had they stayed in their rural homes. By 2025, another 243 million rural migrants will join them, giving China an overall urban population of more than 1 billion people. Geographer Kam Wing Chan observed that the number of workers moving long distances has steadily increased since the 1990s, with the majority moving to Guangdong Province, which some now call the "world factory," and into the lower Chang Jiang region around Shanghai. [Source consulted: Kam Wing Chan, "Internal Migration in China: Trends, Geography and Policies," *Population Distribution, Urbanization, Internal Migration and Development: An International Perspective* (New York: United Nations Department of Economic and Social Affairs, Population Division, 2011), pp. 81–109.]

(A) A village in rural Sichuan province.

(B) Workers in Dongguan, Guangdong Province, check stuffed toys for defects.

After her contract was up, Xia returned home to her village in Sichuan. There, using ideas and assertiveness she had gained from her time in Dongguan, she opened a snack stand. In just 1 month, she made ten times her investment of U.S.$12. But this was still less than she could earn in the city, so after a year Xia returned to Dongguan to try again for a well-paid factory job.

Since her first arrival, the city had grown by 20 percent and now had 1400 foreign companies trying to hire thousands of workers. Through connections, Xia and her sister found jobs requiring midlevel skills in a Taiwanese-owned fiber optics factory at twice the wages Xia had earned at the toy factory. One year after her first trip to Dongguan, Xia was making a bit more than U.S.$100 a month, enough to live relatively comfortably with only three roommates. She had prospects for a raise, and she was sending money home. [Sources: Washington Post, National Public Radio, and Wall Street Journal. For detailed source information, see Text Sources and Credits.]

The experiences of Li Xia illustrate first how the needs of rural areas are being subverted to the needs of China's burgeoning cities. Developers are increasingly targeting rural land, and farmers are rarely given a fair price for their land. The reason for this is that the individual Chinese farmer does not own farmland; he or she only leases it from the local government, which may be inclined to sell the land to high-bidding developers.

Urbanization, globalization, and changes in gender roles are transforming East Asia as millions of rural young adults are flocking to East Asia's coastal cities to work in factories. There they produce goods for sale on global markets, learn new skills, and gain new confidence (see the Figure 9.2 map).

Until 2014, most rural-to-urban migration in China was illegal because the **hukou** (household registration) **system**, an ancient practice reinforced in the Maoist era (see page 378) effectively tied rural people to the place of their birth. Today the system is being reformed (see page 390) and migrants are being allowed to move to approved

hukou system the system in China that ties people to their place of birth; each person's permanent residence is registered and any person who wants to migrate must obtain permission from authorities to do so

cities, as long as they obtain permission from the authorities. This is a major change for the more than 160 million people like Li Xia who have been ignoring the *hukou* system for decades, becoming part of what is called the **floating population**, a term used in China to describe those who have no rights to subsidized housing, schools, or health care in the place to which they have migrated. These migrants generally work in menial, low-wage jobs and make agonizing sacrifices to send money home to children and spouses living in rural areas. Between 2000 and 2010, the number of people migrating from rural to urban areas in China in search of work more than doubled (see Figure 9.20C on page 391). This amounts to approximately half of the working people in China's cities.

> **floating population** the Chinese term for people who live in a place other than their household registration location; many are jobless or underemployed people who have left economically depressed rural areas for the cities

For decades, the Chinese government only weakly enforced the *hukou* system because urban industries were in need of a growing workforce. The lax enforcement enabled the mobility of the labor force while keeping that labor vulnerable and exploitable. The current reforms to the *hukou* system are designed both to encourage more migration to cities, which have had labor shortages in recent years, and to control the places that migrants can relocate to, thus keeping the biggest cities from being overwhelmed by new residents. A major benefit for migrants is that they now have better access to housing, schools, and health care.

THINGS TO REMEMBER

- The migration of people from rural China to cities includes 160 million workers who have moved without legal papers and who now constitute half of China's urban labor force.
- Most rural workers who have moved to China's cities without official permission are in low-paying jobs with little or no access to social services or education.
- The current reforms to the *hukou* system are designed to better control the places that migrants can relocate to, thus keeping the biggest cities from being overwhelmed by new residents.

What Makes East Asia a Region?

Part of the region of East Asia is China, home to nearly one-fourth of humanity. The vast East Asian territory stretches from the Taklimakan Desert in far western China to Japan's rainy Pacific coastline, and from the frigid mountains of Mongolia in the north to the tropical landscapes of Hainan, China's southernmost province, in the south (see Figure 9.1; see also Figure 9.4B on page 366). East Asia (**Figure 9.3**) is comprised of the countries of China, Mongolia, North Korea, South Korea, Japan, and Taiwan (the last has operated as an independent country since World War II but is claimed by China as a province). These countries are grouped together because of their cultural and historical roots, many of which are in China. Because of China's great size, historical influence, enormous population, and huge economy, it is given particular emphasis in this chapter. Japan, whose large and prosperous economy makes it a major player on the world stage, is also emphasized.

FIGURE 9.3 Political map of East Asia.

Terms in This Chapter

East Asian place-names can be very confusing to those unfamiliar with East Asian languages. We give place-names in English transliterations of the appropriate Asian language, taking care to avoid redundancies. For example, *he* and *jiang* are both Chinese words for river. Thus the Yellow River is the Huang He, and the Long River, also called the Yangtze in English, is the Chang Jiang in China; it is redundant to add the term *river* to either name. The word *shan* appears in many place-names and usually means mountain.

Pinyin (a spelling system based on Chinese sounds) versions of Chinese place-names are now commonplace. For example, the city once called Peking in English is now Beijing, and Canton is Guangzhou.

The region historically known as Manchuria is here referred to as China's Far Northeast to emphasize its geographical location. Although China refers to Tibet as Xizang, people around the world who support the idea of Tibetan self-government avoid using that name. This text uses Tibet for the region (with Xizang in parentheses), and Tibetans for the people who live there.

PHYSICAL GEOGRAPHY AND ENVIRONMENTAL ISSUES

PHYSICAL PATTERNS

A quick look at the regional map of East Asia (see Figure 9.1) reveals that the topography here is perhaps the most rugged of any world region. East Asia's varied climates result from the meeting of huge warm and cool air masses and the dynamic interaction between land and oceans. The region's large human population has affected the variety of ecosystems that have evolved there over the millennia and that still contain many important and unique habitats.

Landforms

The complex topography of East Asia is partially the result of the slow-motion collision of the Indian subcontinent with the southern edge of Eurasia over the past 60 million years. This tremendous force created the Himalayas and lifted up the Plateau of Tibet (see Figure 9.1A; also depicted in gray and gold in the Figure 9.1 map), which can be considered the highest of four descending steps that define the landforms of mainland East Asia, moving roughly west to east.

The second step down from the Himalayas is a broad arc of basins, plateaus, and low mountain ranges (depicted in yellowish tan in the Figure 9.1 map). These landforms include the broad, rolling highland grasslands and deep, dry basins and deserts of western China (such as the Taklimakan Desert) and Mongolia (see Figure 9.1B), as well as the Sichuan Basin and the rugged Yunnan–Guizhou Plateau to the south, which is dominated by a system of deeply folded mountains and valleys that bend south through the Southeast Asian peninsula.

The third step, directly east of this upland zone, consists mainly of broad coastal plains and the deltas of China's great rivers (shown in shades of green in the Figure 9.1 map). Starting from the south, this step is defined by three large lowland river basins: the Zhu Jiang (Pearl River) basin, the massive Chang Jiang basin (see Figure 9.1C), and the lowland basin of the Huang He on the North China Plain. Each of these rivers has a large delta. Despite the deltas being subject to periodic flooding, they have historically been used for agriculture. However, coastal cities have now spread into many of these deltas, filling in wetlands and farms with zones of dense population and industrialization (see Figure 9.20 on page 391). Low mountains and hills (shown in light brown in the Figure 9.1 map) separate these river basins. China's Far Northeast and the Korean Peninsula are also part of this third step.

The fourth step consists of the continental shelf, covered by the waters of the Yellow Sea, the East China Sea, and the South China Sea. Numerous islands—including Hong Kong, Hainan, and Taiwan—are anchored on this continental shelf; all are part of the Asian landmass (see Figure 9.1D).

The islands of Japan have a different geological origin: they are volcanic, not part of the continental shelf. They rise out of the waters of the northwestern Pacific in the tectonically active zone where the Pacific, Philippine, and Eurasian plates grind against one another. Lying along a portion of the Pacific Ring of Fire (see Figure 1.9 on page 17), the entire Japanese island chain is particularly vulnerable to disastrous volcanic eruptions, earthquakes, and **tsunamis** (seismic sea waves). Volcanic Mount Fuji, the highest peak in the country and a recognizable national symbol (see Figure 9.1E), last erupted in 1707. However, the mountain is still classed as active, and deep internal rumblings have been detected since 2001. In March of 2011, the largest earthquake in recorded Japanese history (registering 9.2 on the Richter scale) hit off the coast of Honshu, Japan, near the city of Sendai. The quake and subsequent tsunami killed tens of thousands of people and damaged several nuclear reactors located on the coast, resulting in the second-worst nuclear accident ever in the world (discussed further on page 369).

> **tsunami** a large sea wave caused by an earthquake

The East Asian landmass has few flat portions, and most flat land is either very dry or very cold. Consequently, the large numbers of people who occupy the region have had to be particularly inventive in creating spaces for agriculture. They have cleared and terraced entire mountain ranges, until recently using only simple hand tools (see Figure 9.5A on page 368). They have irrigated drylands with water from melted snow, drained wetlands using elaborate levees and dams, and applied their complex knowledge of horticulture and animal husbandry to help plants and animals flourish in difficult conditions.

Climate

East Asia has two principal contrasting climate zones, shown in the Figure 9.4 map: the dry interior west and the wet (monsoon) east. Recall from Chapter 8 that the term *monsoon* refers to the seasonal reversal of surface winds that flow from the Eurasian continent to the surrounding oceans during winter and from the oceans inland during summer.

The Dry Interior Because land heats up and cools off more rapidly than water does, the interiors of large landmasses in the midlatitudes tend to be intensely cold in winter and extremely hot in summer. Western East Asia, roughly corresponding to the first two topographic steps described earlier, is an extreme example of such a midlatitude continental climate because it is very dry. In fact, this area is farther away from an ocean than any other place on Earth's surface (it is also known as a *pole of inaccessibility*). With little vegetation or cloud cover to retain the warmth of the sun after nightfall, summer daytime and nighttime temperatures may vary by as much as 100°F (55°C).

Grasslands and deserts of several types cover most of the land in this dry region (see Figure 9.4C). Only scattered forests grow on the few relatively well-watered mountain slopes and in protected valleys supplied with water by snowmelt. In all of East Asia, humans and their impacts are least conspicuous in the large, uninhabited portions of the deserts of Tibet (Xizang), the Tarim Basin in Xinjiang, and the Mongolian Plateau.

The Monsoon East The monsoon climates of the east are influenced by the extremely cold conditions of the huge Eurasian landmass in the winter and the warm temperatures of the surrounding seas and oceans in the summer. During the dry winter monsoon, descending frigid air sweeps south and east through East Asia, producing long, bitter winters on the Mongolian Plateau, on the North China Plain, and in China's Far Northeast (see Figure 9.4D). While occasional freezes may reach as far as southern China, winters there are shorter and less severe. The cold air of the dry winter monsoon is partially deflected by the east–west mountain ranges of the Qin Ling, and the warm waters of the South China Sea moderate temperatures on land.

As the continent warms during the summer monsoon, the air above it rises, pulling in wet, tropical air from the adjacent seas. The warm, wet air from the ocean deposits moisture on the land in the form of seasonal rains. As the summer monsoon moves northwest, it must cross numerous mountain ranges and

FIGURE 9.4
PHOTO ESSAY: Climates of East Asia

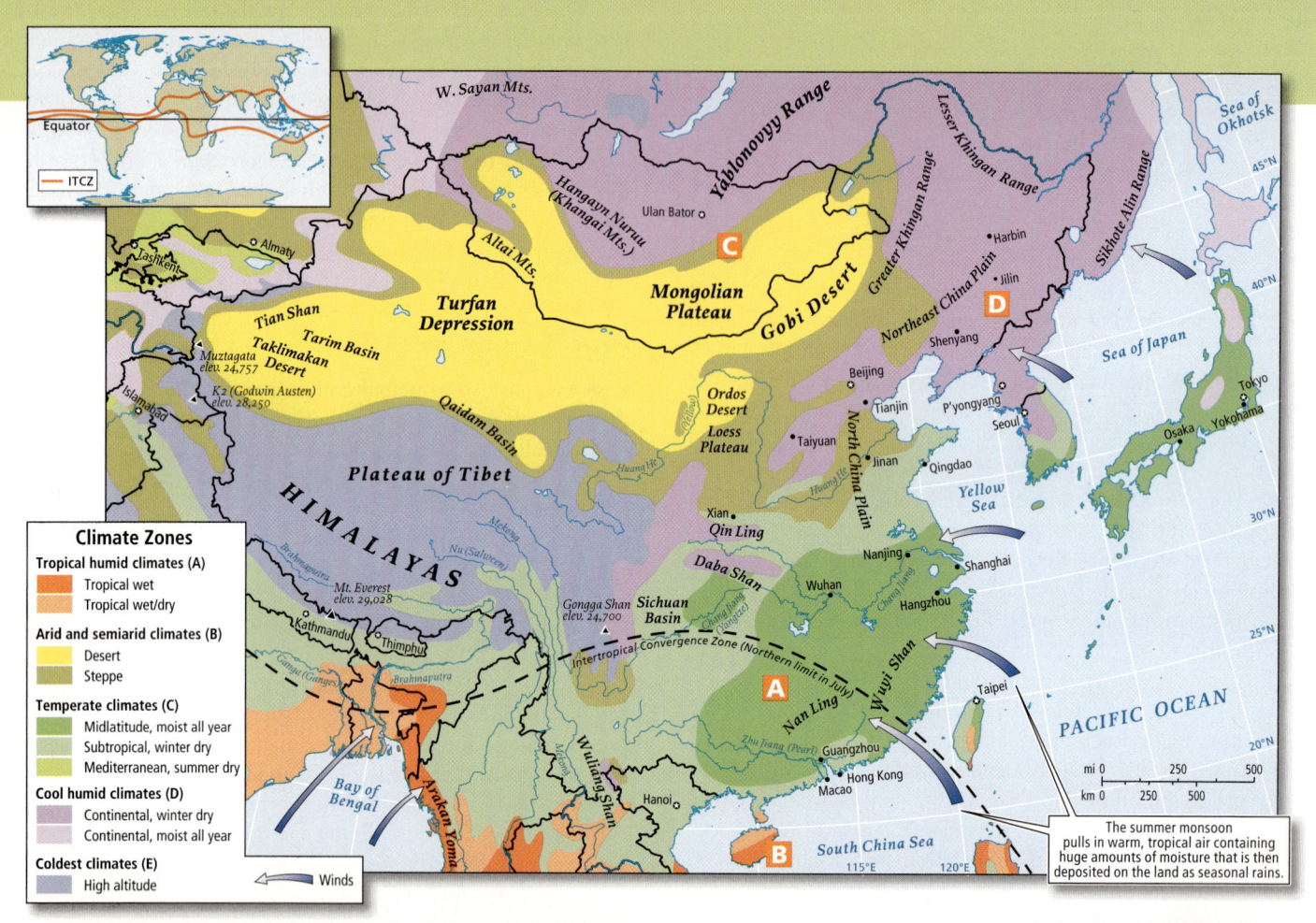

Climate Zones

Tropical humid climates (A)
- Tropical wet
- Tropical wet/dry

Arid and semiarid climates (B)
- Desert
- Steppe

Temperate climates (C)
- Midlatitude, moist all year
- Subtropical, winter dry
- Mediterranean, summer dry

Cool humid climates (D)
- Continental, winter dry
- Continental, moist all year

Coldest climates (E)
- High altitude

→ Winds

The summer monsoon pulls in warm, tropical air containing huge amounts of moisture that is then deposited on the land as seasonal rains.

A Midlatitude, moist all year, Guilin, Guangxi Zhuang Autonomous Region, China

B Tropical wet, Hainan, China

C Steppe, Mongolia

D Continental, winter dry, Jilin, China

displace cooler air. Consequently, its effect is weakened toward the northwest. Thus, the Zhu Jiang basin in the far southeast is drenched with rain and has warm weather for most of the year (see Figure 9.4A), whereas the Chang Jiang basin, which lies in central China to the north of the Nan Ling range, has about 5 months of summer monsoon weather. The North China Plain, north of the Qin Ling and Dabie Shan ranges, receives even less monsoon rain—about 3 months of monsoon each year. Very little monsoon rain reaches the dry interior.

Korea and Japan have wet climates year-round, similar to those found along the Atlantic Coast of the United States, because of their proximity to the sea. They still have hot summers and cold winters because of their northerly location and exposure to the continental effects of the huge Eurasian landmass. Japan and Taiwan actually receive monsoon rains twice: once in spring, when the main monsoon moves toward the land, and again in autumn, as the winter monsoon forces warm air off the continent. This retreating warm air picks up moisture over the coastal seas, which is then deposited on the islands. Much of Japan's autumn precipitation falls as snow, in particular in northern latitudes and at higher altitudes.

Natural Hazards The entire coastal zone of East Asia is intermittently subject to **typhoons** (tropical cyclones or hurricanes). Japan's location along the northwestern edge of the Pacific Ring of Fire (see Figure 1.9 on page 17) results in volcanic eruptions, earthquakes, and tsunamis. These natural hazards are a constant threat in Japan; the heavily populated zone from Tokyo southwest through the Inland Sea (between Shikoku and southern Honshu) is particularly endangered. Earthquakes are also a serious natural hazard in Taiwan and in China's mountainous interior.

> **typhoon** a tropical cyclone or hurricane in the western Pacific Ocean

THINGS TO REMEMBER

- There are four main topographical zones, or "steps," that form the East Asian continent.
- Japan was created by volcanic activity along the Pacific Ring of Fire.
- East Asia has two principal contrasting climates: the dry continental interior (west) and the monsoon east.
- East Asia faces a wide range of natural hazards, including earthquakes, tsunamis, volcanic eruptions, and tropical storms (typhoons). ■

ENVIRONMENTAL ISSUES

> **GEOGRAPHIC INSIGHT 1**
>
> **Environment:** East Asia's most serious environmental problems result from its high population density combined with its rapid urbanization and environmentally unsustainable economic development. Climate change may intensify the droughts and floods that have long plagued this region.

Climate change, especially global warming, is a growing concern in East Asia. China is the world's largest overall producer of greenhouse gases (but still produces less than a third of U.S. emissions on a per capita basis; see page 21 in Chapter 1). China has recently been working to limit its emissions for a variety of reasons, but even though it has significantly increased its efficiency through the use of new technologies, its emissions still make up almost a quarter of the world's total. This amount may double by 2030 as China's urban households use more cars, air conditioning, appliances, and computers. Japan, Korea, and Taiwan are also major greenhouse gas emitters.

China's Vulnerability to Glacial Melting Glaciers on the Plateau of Tibet capture monsoon moisture and store it as frozen ice, which is then slowly released as meltwater. Two of China's largest rivers, the Huang He and the Chang Jiang, are partially fed by these glaciers, which are now melting so rapidly that scientists predict they will eventually disappear. One effect could be significantly lower flows in these two great rivers during the winter, when little rain falls. Both have already begun to run low during winter, and trade has been affected because riverboats and barges have become stranded on sandbars. Another effect could be a reduction in the amount of irrigation water available for dry-season farming.

Water Shortages Nearly every year, abnormally low rainfall or abnormally high temperatures create a drought somewhere in China. These droughts often cause more suffering and damage than any other natural hazard.

Droughts can be worsened by human activity on a local or regional scale. When people begin to live or farm in dry environments, as many millions have done in China and Mongolia during the twentieth century, *desertification* (see page 239 in Chapter 6) can result. People clear vegetation to grow crops, which require more water than the natural vegetation, so the crops must be irrigated with water pumped to the surface from underground aquifers (Figure 9.5A). Many dry areas in China are subject to strong winds that can blow away topsoil once the vegetation is removed. Furthermore, irrigated crops are often less able than natural vegetation to hold the soil. This has resulted in huge dust storms much like those that plagued the central United States in the 1930s Dust Bowl (see page 73 in Chapter 2). High dunes of dirt and sand have appeared almost overnight in some parts of China that border the desert, threatening crops, roads, and homes. Dust storms can also move over long distances from western China and affect large cities, such as Beijing, near the coast. Particulate matter from these dust storms even circles the entire globe in the upper atmosphere.

VIGNETTE

In Ningxia Huizu Autonomous Region on the Loess Plateau in northwest China, Wang Youde squints out at what are now sand-colored low hills barren of vegetation, thinking about how these vast tracts of former farmland have been transformed into deserts by a combination of human error and climate change.

The Loess Plateau was already prone to dust storms during times of drought (*loess* means "wind-deposited soil"). Agricultural expansion into loess areas led to the removal of thick, natural, deep-rooted grasses which once helped hold down the soil. One can still see the agricultural terraces on the arid slopes where now

FIGURE 9.5 — PHOTO ESSAY: Vulnerability to Climate Change in East Asia

China is particularly vulnerable to drought, desertification, flooding, and other hazards that climate change may intensify.

A The terraced hillsides of China's eastern Gansu Province are in a dry upland zone, where agriculture depends on rainfall or irrigation water taken from rivers or underground aquifers. Rainfall is already unpredictable and could become more so with climate change. Melting glaciers on the Plateau of Tibet threaten to reduce river flows, and overuse of groundwater for irrigation is depleting aquifers. Large impoverished rural populations are increasingly being forced to relocate to other parts of China.

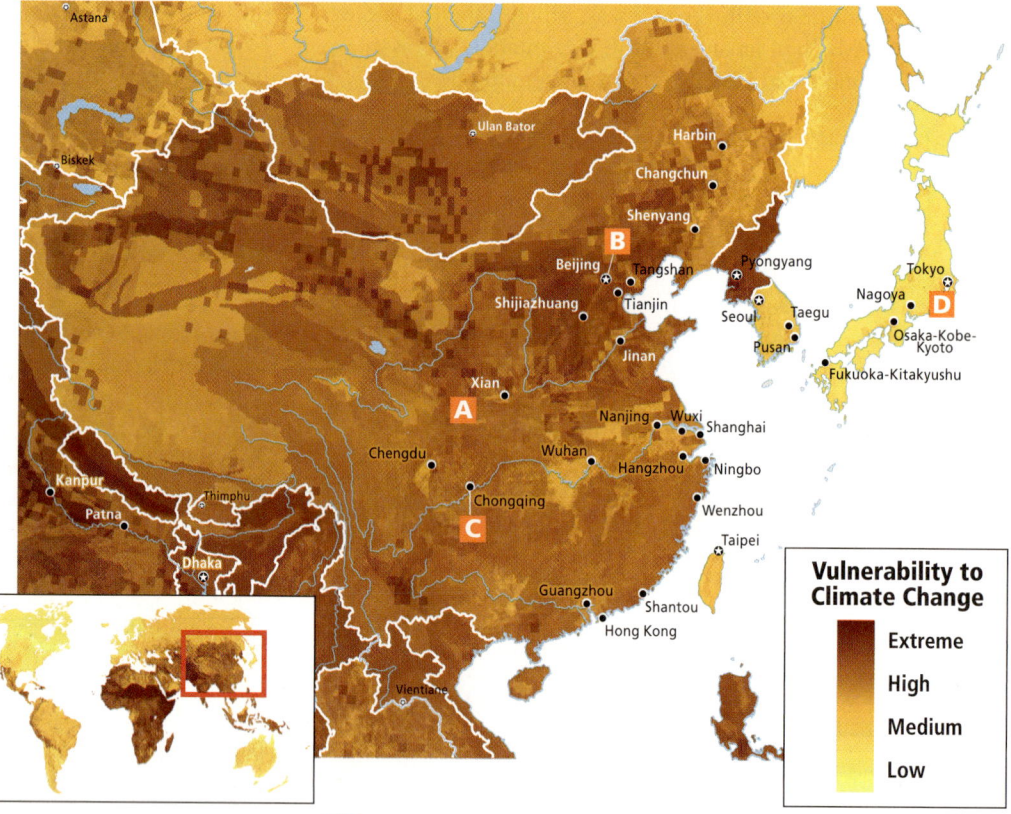

B A canal outside of Beijing, China, that is subject to severe water shortages and the effects of massive increases in population. China's national government is creating huge canals and other infrastructure to divert water from distant rivers in southern and western China to Beijing and surrounding areas.

C Flooding hits the Chang Jiang at Chongqing, China, carrying off boats and barges. Climate change could result in increased flooding. This is especially likely in eastern China, where there is much rainfall during the summer monsoon.

D A solar power station in Kawasaki, Japan. While it has made only small reductions in its greenhouse gas emissions, Japan has the third-largest installed solar power–generating capacity in the world (behind those of Germany and Italy).

Vulnerability to Climate Change
- Extreme
- High
- Medium
- Low

not even grass grows. Wang Youde's family and 30,000 others fled the area when Youde was 10 years old because one day a sand dune covered their village. From his early childhood, he remembers flowers, birdsong, and occasional snowfalls; all have vanished. Now Youde is back, heading up a project to revegetate thousands of hectares with drought-resistant plants that will hold the soil. By hand, squares of braided straw or stones are laid down to keep water from running off the land and to protect planted seedlings. The seedling survival rate is only 20 to 30 percent. Yet Youde says, "Every time we see an oasis that we have created we are very satisfied . . . [b]ecause we have poured sweat and blood into our work." His adult children are helping him, hoping to remain with the family and escape the hardships of migrating to find urban factory work. [Source: Asia Pacific News. For detailed source information, see Text Sources and Credits.]

Water shortages are particularly intense in the North China Plain, which produces half of China's wheat and a third of its corn. Here the water table is falling more than 10 feet per year because of the increased use of groundwater for irrigation and urban needs. Meanwhile, withdrawals of water from the Huang He often make the river's lower sections run completely dry during the winter and spring. A gigantic development called the South–North Water Transfer Project is underway which will attempt to rectify the problem (see Figure 9.5B). The goal is to divert water from the Chang Jiang basin in central China to the Huang He basin in the north using dams, canals, and pipelines. Multiple locations in the area from the Plateau of Tibet in the west to the coastal provinces in the east will receive water from the project. The project is not expected to be completed until 2050.

China is also trying to avoid water shortages, focusing on water conservation to stretch existing supplies. Already, 30 percent of China's urban water is recycled, and many cities are trying to raise this percentage. China is also making a major effort to remove pollutants from wastewater discharged by industry and farming, which together account for 85 percent of water use.

Japan, the Koreas, and Taiwan have monsoon rainfall patterns, which give them a generally wetter climate and make them less vulnerable to drought than China and Mongolia.

Flooding in Central China The same shifting patterns of rainfall that may worsen droughts can also worsen flooding. Under usual conditions, the huge amounts of rain deposited on eastern China during the summer monsoon periodically can cause catastrophic floods along the major rivers. If global warming leads to even slight changes in rainfall patterns, flooding could be much more severe (see Figure 9.5C). Engineers have constructed elaborate systems of dikes, dams, reservoirs, and artificial lakes to help control flooding. However, these systems failed in 1998, when heavy rains in the Chang Jiang basin caused some of the worst flooding in decades, resulting in the death of approximately 4000 people. The pattern is now repeating frequently on the Chiang Jiang and other rivers in central and southern China (significant floods happened in 2010, 2011, and 2012). Every year there are large numbers of casualties from drowning and mudslides.

Natural Hazards and Energy Vulnerability Until the Sendai earthquake and tsunami of 2011, Japan, along with many other countries, planned to increase its use of nuclear power as a way to reduce greenhouse gas emissions. However, the tsunami washed over the Fukushima nuclear plant and destroyed its cooling systems, which caused a meltdown of several of the plant's nuclear reactors. The severity of the nuclear disaster that followed in the wake of the tsunami halted or curtailed many plans for expanding nuclear power in Japan and across the world. The accident forced the evacuation of more than 200,000 people and temporarily contaminated the water supply of Tokyo. For miles around the reactor, the ground, food, and livestock were contaminated and thousands of gallons of water, containing 10,000 times the normal level of iodine-131 (a radioactive substance), were released into the Pacific Ocean. Enough radiation was released into the atmosphere that scientists detected measurable radiation increases in the United States and Europe. On the positive side, the accident is expected to boost reliance both on renewable resources and on fossil fuel in the near future.

Food Security and Sustainability

East Asia's **food security**—the capacity of the people in a geographic area to consistently provide themselves with adequate food—is increasingly linked to the global economy. The wealthiest countries in the region can buy food on the global market, but they are less able to produce sufficient food domestically for two main reasons. Much agricultural land has been lost to urban and industrial expansion, and rising levels of affluence have led to a demand for more meat, which is either imported itself or requires the importation of animal feed.

food security when people consistently have access to sufficient amounts of food to maintain a healthy life

About 75 percent of the food consumed in Japan, South Korea, and Taiwan is imported. In China, on the other hand, self-sufficiency in grain production is important to national identity because devastating famines were a recurring problem before and during the Communist Revolution. China is now nearly self-sufficient with regard to basic necessities, but it relies on imports of commodities to supply it with grain and soybeans for animal feed and luxury food items.

Recent dramatic increases in global food prices illustrate the perils of dependence on food imports, especially for the poor. In 2007–2008, the world market price for grain and other basic food commodities shot up, and it did so again in 2010–2011. The reasons for the increases in food prices include high oil prices (because producing food is often energy intensive), the use of corn to produce ethanol fuel (ethanol is used as a replacement for gasoline in the Americas and elsewhere) rather than as a food source, and the rise in demand from China and other economically expanding countries for agricultural commodities. **217. FOOD PRICES SKYROCKETING IN CHINA**

Food Production Just over half of East Asia's vast territory can support agriculture (Figure 9.6). In much of this area, food production has been pushed well beyond what can be sustained over the long term. As a result, East Asia's fertile zones are shrinking. In China, roughly one-fifth of the agricultural land has been lost since the Communist Revolution of 1949, largely because of urban and industrial expansion and agricultural mismanagement that created soil erosion and desertification (see the vignette on page 367).

As urban populations become more affluent, they consume more meat and other animal products that require more land and resources than the plant-based national diet of the past. Soybeans are one example of this change in diet. Soybeans are no longer used mainly for human food but for animal feed, for farmed fish food, and especially for high-grade cooking oil. More than 70 percent of the soybeans used in China are now imported, primarily from the United States, Brazil, and Argentina.

Even with the increase in meat consumption, vegetable dishes with an ancient history remain popular throughout East Asia (Figure 9.7A), and meat and dairy dishes are still prepared in traditional ways in places where raising farm animals has been the norm (see Figure 9.7C).

ON THE BRIGHT SIDE

Responses to the Climate Crisis

East Asia is increasingly responding to the warming aspect of global climate change. Japan has led these efforts for decades. In 1997, the Japanese city of Kyoto hosted the meeting in which countries first committed to reduce their greenhouse gas emissions. Japanese automakers such as Toyota were among the first to develop and sell hybrid gas-electric vehicles, and Japan is now third only to Germany and Italy in its installed solar power–generating capacity (see Figure 9.5D). China is now the world leader in manufacturing photovoltaic cells used to generate electricity from solar energy and has one of the world's largest solar projects.

Rice Cultivation The region's most important grain is rice, and over the millennia its cultivation has transformed landscapes throughout central and southern China, Japan, Korea, and Taiwan. In these areas, rainfall is sufficient to sustain **wet rice cultivation**, which can be highly productive.

Wet rice cultivation requires elaborate systems of water management, as the roots of the plants must be submerged in water early in the growing season. Centuries of painstaking human effort have channeled rivers into intricate irrigation systems, and whole mountainsides have been transformed into descending terraces that evenly distribute the

wet rice cultivation a prolific type of rice production that requires the plant roots to be submerged in water for part of the growing season

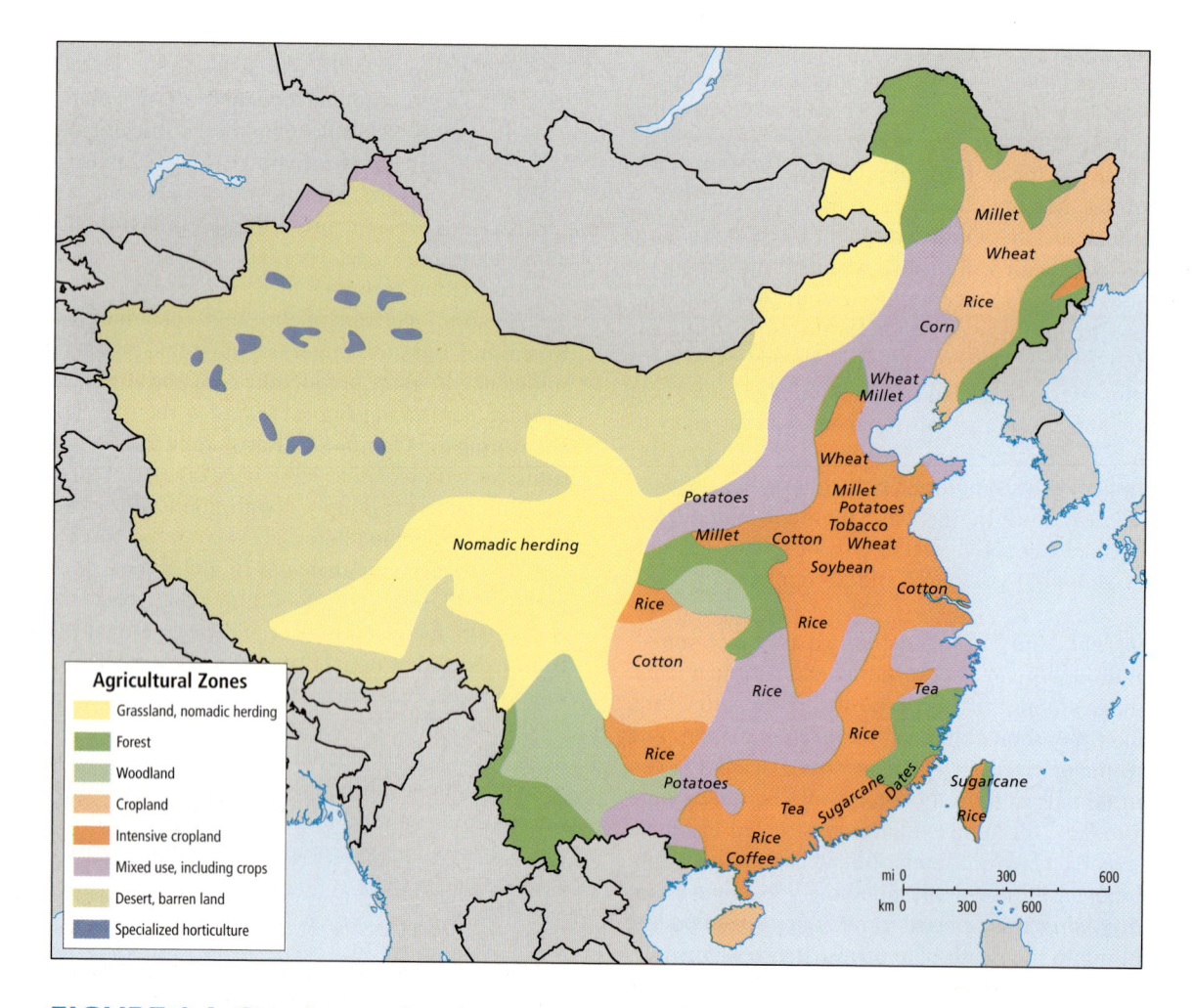

FIGURE 9.6 China's agricultural zones. The economic reforms instituted in recent years in China include more regional specialization in agricultural products. [Sources consulted: "World Agriculture," *National Geographic Atlas of the World,* 8th ed. (Washington, DC: National Geographic Society, 2005), p. 19; "China: Economic, Minerals" map, *Goode's World Atlas,* 21st ed. (New York: Rand McNally, 2005), pp. 39, 207]

FIGURE 9.7 LOCAL LIVES
Foodways in East Asia

A Kimchi, a Korean dish made of fermented and often highly seasoned vegetables, was developed at least 3000 years ago as a food storage and preservation technique. Specific kinds of kimchi are made at different times of the year because fermentation happens at variable rates, depending on the particular vegetables available for fermenting at that point in the year and on the temperature at which each vegetable ferments. [Martin Moos/Lonely Planet Images/Getty Images]

B A wealthy restaurant owner in Japan poses with a 448-pound (220-kilogram) tuna that he bought for U.S.$1.8 million at the Tsukiji fish market in Tokyo, Japan. All of this fish will be served raw as sushi, which is a combination of cooked, vinegared rice and other ingredients, usually raw seafood. With roughly 2 percent of the world's population, Japan consumes 15 percent of the global fish catch. [Yoshikazu Tsuno/AFP/Getty Images]

C A woman in rural Mongolia prepares *byaslag*, a cheese made in Mongolia of yak, cow, sheep, or goat milk. Unlike most cheeses originating in Europe, byaslag is not aged but rather is desiccated in the dry air of Mongolia. It is one of many unique products made with animal milk in Mongolia. [Bruno Morandi/The Image Bank/Getty Images]

Fisheries and Globalization Many East Asians depend heavily on ocean-caught fish for protein (see Figure 9.7B). People in Japan, South Korea, and Hong Kong eat more than twice as much fish and shellfish as do Americans. The Japanese in particular have had a huge impact on the seas of not only this region, but also of the entire world. With less than 2 percent of the world's population, Japan consumes 15 percent of the global wild fish catch. There are some 4000 coastal fishing villages in Japan, sending out tens of thousands of small crafts to work nearby waters each day.

There are also many large Japanese fishing vessels, complete with onboard canneries and freezers, that harvest oceans around the world. They can do so because waters at a certain distance away from land, typically 200 nautical miles (230 miles or 370 kilometers), are considered international waters and outside the jurisdiction of any individual country. Environmentalists have criticized Japan for overfishing. For example, Japanese fishing off western Africa has reduced the catches of local fishers so much that many have been forced to migrate to Europe for work. Today there is little room for Japan to expand its fish imports because the global wild fish catch has been static or in decline since 1990 because of overfishing.

China is today both the largest producer and consumer of fish in the world. Unlike Japan, the world's second-largest consumer of fish, most of China's production is based on aquaculture rather than on wild-caught fish.

People and Animals Animals are used in East Asia not only for food, but also for medicinal, aesthetic, and cultural purposes. For example, traditional Chinese medicine involves, among other things, the capturing of wild animals. Such animals are found in markets around China because they are believed to have healing properties (Figure 9.8A). Unfortunately, due to this practice, some species are now endangered.

Particularly in Japan, aquaculture provides living ornamentation to gardens and other carefully tended landscapes. Carp have been domesticated and bred to have a variety of colors and patterns (see Figure 9.8C) and are a central part of many Japanese gardens. As an adaptive species that does well under a wide variety of aquatic conditions, ornamental carp are now found around the world.

Animals are often central to a culture, as is the case in Mongolia, where many species of animals are fundamental to the nomadic culture that historically dominated there. A way of life once common in areas of dry grasslands around the world (see Figure 9.8B), nomadic living is becoming less common than it was in the past as places like Mongolia become more urbanized.

THINGS TO REMEMBER

› **GEOGRAPHIC INSIGHT 1** • **Environment** East Asia's most serious environmental problems result from its high population density combined with its rapid urbanization and environmentally unsustainable economic development. Climate change may intensify the droughts and floods that have long plagued this region.

• China's main river systems are being affected by the melting of its glaciers, which ultimately leads to water shortages.

• The region's most important grain is rice, and over the millennia, rice cultivation has transformed landscapes throughout central and southern China, and Japan, the Koreas, and Taiwan.

water. Writing about wet rice cultivation in Sichuan Province, geographer Chiao-Min Hsieh describes how "[e]verywhere one can hear water gurgling like music as it brings life and growth to the farms." However, these same cultivation techniques, combined with extensive forestry and mining, have led to the loss of most natural habitats in all but the most mountainous, dry, or remote areas. With so little suitable agricultural land left, further expansion of the area under wet rice cultivation is unlikely.

FIGURE 9.8 LOCAL LIVES
People and Animals in East Asia

A A stall in a night market in Beijing sells dried seahorses, lizards, and starfish, all of which are used in traditional Chinese medicine (TCM). Millions of animals are captured each year for use in TCM, and some are being pushed dangerously close to extinction. [Blue Jean Images/Getty Images]

B Sheep shearing in Mongolia, where livestock raising occupies 75 percent of the land. It was the backbone of the economy for thousands of years, until the 1970s, when mining and services took the lead. Sheep, goats, yaks (a long-haired Central Asian bovine), horses, and cattle are the most commonly raised animals. Many of these animals are still raised by nomadic and seminomadic herders. However, most of the people who were once herders are now moving to the city because of economic downturns and an increasingly variable climate. [Jacques Marais/Gallo Images/Getty Images]

C *Nishikigoi* (also known as koi), a variety of carp, in a garden in Himeji, Japan. Carp were domesticated in both ancient China and Rome as a food source. Nishikigoi, which means "brocaded carp," are an ornamental variety developed in Japan in the 1820s. Currently, the trade in ornamental nishikigoi is larger than the trade of carp raised for food. [Alexander Safonov/Flickr/Getty Images]

- About three-quarters of the food consumed in Japan, South Korea, and Taiwan is imported. So far, China is self-sufficient with regard to basic necessities such as rice, but it is highly dependent on imports for other foods.
- People in East Asia consume much fish and seafood. Some of this food is harvested in waters far away from East Asia. ■

Three Gorges Dam: The Power of Water

The Three Gorges Dam (Figure 9.9A) is at this time the largest dam in the world, at 600 feet (183 meters) high and 1.4 miles (2.3 kilometers) wide. It was designed to improve navigation on the Chang Jiang and control flooding, but it is most lauded for its role in generating hydroelectricity for China, a country that uses considerable amounts of energy and is working to reduce its greenhouse gas emissions. But the dam also comes with a social cost. The Chinese environmental activist Dai Qing notes that China has 22,000 large dams, all of which have displaced people—perhaps as many as 60 million—without any attention to their rights as stakeholders in the projects.

Many experts involved with the Three Gorges project see serious design flaws. Of greatest concern is the dam's position above a seismic fault. The dam was built at the east end of the Three Gorges because the deep canyons provided a prodigious reservoir for water to power turbines, providing electricity for all of central China, from the sea to the Plateau of Tibet. Unfortunately, the enormous weight of the water and its percolation through geological fissures in the 370-mile-long (600-kilometer-long) reservoir behind the dam could trigger earthquakes. Already at issue are huge landslides along the gorges, lubricated by the rising reservoir water. Meanwhile, cracks in the dam raise doubts about its structural integrity. Similar defects led to the failure of China's much smaller Banqiao Dam during a 1975 typhoon, which caused 150,000 deaths from the flooding and an ensuing famine. Even if the Three Gorges dam holds, its potential to generate power will probably be reduced by the buildup of eroded silt behind the dam.

Any failure of the dam would be a financial as well as human disaster. Official construction costs are $25 billion, but the real costs may end up being three times this figure, due in part to unforeseen negative environmental and social impacts and to theft from the project by corrupt officials.

Also of concern is the incalculable cost associated with relocating the 1.2 million people who once lived where the dam now forms a reservoir. Thirteen major cities have been submerged, along with 140 large towns, hundreds of small villages, 1600 factories, and 62,000 acres (25,000 hectares) of farmland. In some cases, communities were rebuilt immediately uphill from the submerged location and people only had to relocate a short distance; others had to migrate to distant parts of the country. The reservoir has also destroyed important archaeological sites, as well as some of China's most spectacular natural scenery. There are significant environmental costs as well. The giant sturgeon, for example, a fish that can weigh as much as three-quarters of a ton and is as rare as China's giant panda, may become extinct. Sturgeon used to swim more than 1000 miles (1600 kilometers) up the Chang Jiang past the location of the dam to spawn. Now the sturgeon's migration, and consequently their reproductive process, has been irretrievably altered. **209. THREE GORGES DAM LEAVES SOME CHINESE SWAMPED**

The plan to build the Three Gorges Dam came just as UN development specialists were deciding that the benefits dams could bring do not sufficiently outweigh the many problems they cause. Decades ago, international funding sources such as the World Bank withdrew their support for the Three Gorges Dam

FIGURE 9.9   PHOTO ESSAY: Human Impacts on the Biosphere in East Asia

Economic development in East Asia has often proceeded without adequate environmental protection or safeguards for human health. Here we focus on the Three Gorges Dam, air pollution, and energy issues.

A China's Three Gorges Dam is the largest dam in the world. It is capable of supplying about 3 percent of China's electricity needs, resulting in significantly less air pollution than the several large coal-fired power plants that it replaces. However, the dam has also caused significant environmental damage related to its 370-mile-long (600-kilometer-long) reservoir, which has altered the habitats of endangered species; submerged the dwellings of 1.2 million people; and become a cesspool of untreated sewage, industrial waste, and garbage that once flowed downstream.

B Pollution from vehicles chokes the motorcade of a political campaign in Taipei, Taiwan.

Human Impact, 2002

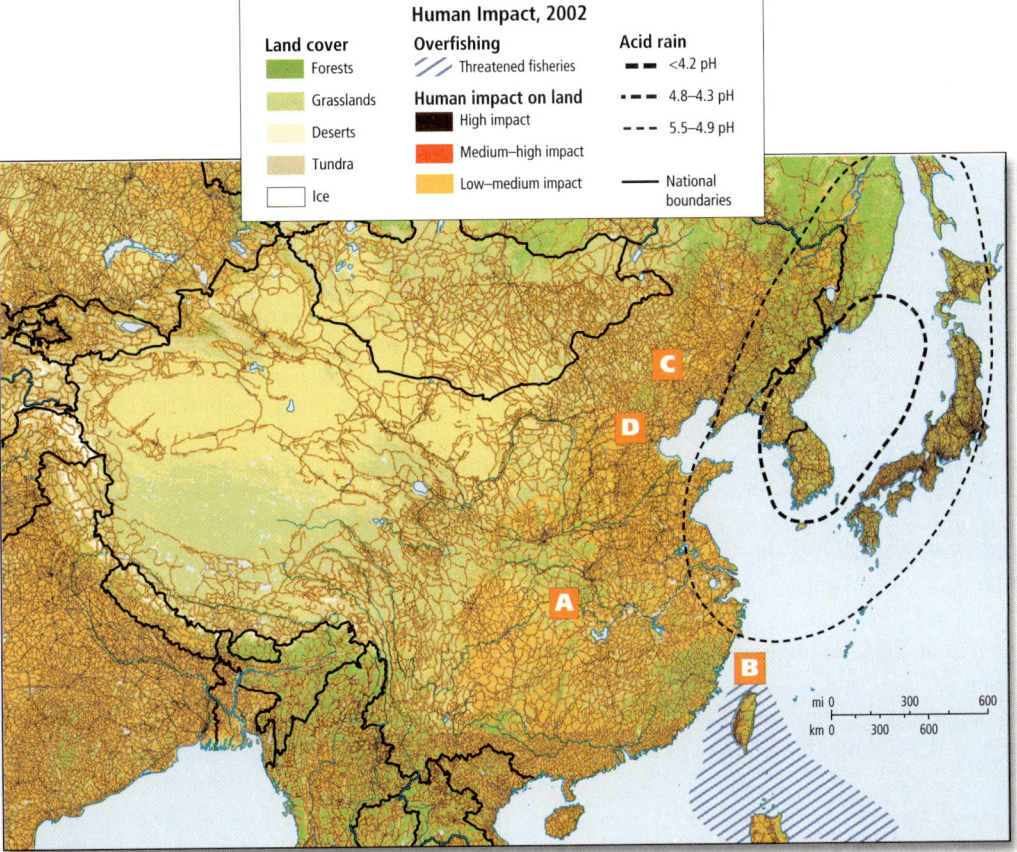

Land cover:
- Forests
- Grasslands
- Deserts
- Tundra
- Ice

Overfishing:
- Threatened fisheries

Human impact on land:
- High impact
- Medium–high impact
- Low–medium impact

Acid rain:
- <4.2 pH
- 4.8–4.3 pH
- 5.5–4.9 pH
- National boundaries

C An open-pit coal mine in China's province of Inner Mongolia. China is already the world's biggest producer and consumer of coal, and its demand for coal is likely to increase in the future. China is the largest contributor of greenhouse gases in the world.

D A toxic haze of air pollution engulfs Beijing and the North China Plain, often blowing eastward to reach Korea, Japan, and even the United States and Europe. China's coal-fired power plants, industries, and vehicles are emerging as major global environmental concerns.

because of concerns over the social and environmental costs and other shortcomings of the project. However, Chinese industrialists who need the energy, construction companies that have prospered from building the dam and its many ancillary projects, and government officials eager to impress the world and leave their mark on China continue not only to support the *Da Ba* (the Big Dam), but also to look for other locations around the world where China can gain influence and profit by building dams.

Air Pollution: Choking on Success

Air pollution is often severe throughout East Asia, but the air quality in cities is particularly poor. After South Asia, the pollution in East Asian cities ranks among the worst in the world. China's pollution problems are particularly important because its demand for energy is growing so fast. Globally, coal burning is a major source of air pollution, and China is the world's largest consumer and producer of coal, accounting for 46 percent of all the coal burned in the world each year (see Figure 9.9C, D). Between 1980 and 2011, China's coal consumption grew sixfold. By 2006, China was bringing a new coal-fired power plant online every week. The pace of expansion has slowed somewhat since then in favor of other energy sources. Nevertheless, at any given point in time, there are still hundreds of new coal-fired plants in the planning stages in China.

The combustion of coal releases high levels of pollutants—suspended particulates, smog-causing ozone, and sulfur dioxide—all of which can cause respiratory ailments. In Chinese cities, these emissions can be many times higher than the World Health Organization (WHO) defines as safe. Figure 9.10 shows that almost no Chinese cities have air quality that is classified as "good" with regard to one type of pollutant—particulates. The worst pollution is often in cities, where homes are in close proximity to industries that depend heavily on coal for fuel. The use of coal to heat homes can also be a source of pollution, as the following vignette shows.

FIGURE 9.10 China's most polluted and least polluted cities in terms of air quality. Based on a 2011 report from the WHO, the map shows the 20 cities in China that have the most air pollution and the 10 cities with the least air pollution, as measured by the number of PM_{10} particles per cubic meter. PM_{10} particles are microscopic; by definition, they are less than 10 micrometers in diameter. These tiny particles can get into people's lungs and potentially cause serious health problems. The cities of northern and central China suffer from higher levels of air pollution than those in the south. [Sources consulted: "China's Most Polluted Cities—WHO Index," *China Briefing*, September 28, 2011, at http://www.china-briefing.com/news/2011/09/28/chinas-most-polluted-cities-who-index.html; "Particle Pollution and Your Health," *AirNow*, at http://airnow.gov/index.cfm?action=particle_health.page11]

Sulfur dioxide from coal burning also contributes to acid rain, which is displaced to the northeast by prevailing winds that reach the Koreas, Japan, Taiwan, and beyond. Figure 9.9D and the map above it show the zones of heavy pollution. Japan and Taiwan are particularly afflicted (see Figure 9.9B). Particulates from China's coal burning are transported globally by high-altitude, west-to-east flowing jet streams, thus affecting air quality in North America and Europe. Locally, toxic mercury from coal plants accumulates in aquatic life and, ultimately, in humans.

Air pollution from vehicles is also severe even though the use of cars for personal transportation in China is in its early stages. For years, China's vehicles have had very high rates of lead and carbon dioxide emissions. The government has begun to work on the problem. As of 2008, new Chinese cars had to meet EU standards for mileage and emissions—which are much stricter than those for U.S. cars. Even so, to provide decent air quality during the 2008 Summer Olympics, China removed half of Beijing's cars from the road. A number of polluting industries were temporarily shut down as well. The actions had the intended effect; Beijing's air was noticeably better during the Olympic summer.

VIGNETTE Every winter, an elderly couple got through the biting Beijing winters by feeding 1200 one-kilogram coal bricks into a small iron stove. The ashes and coal dust blackened their belongings, and they worried about carbon monoxide poisoning. Then, in 2009 they were given a new electric space heater by the city government. In a move to clean up Beijing's air and reduce the city's contribution to China's greenhouse gas emissions, the city replaced nearly 100,000 old coal stoves with electric heaters and cut electric nighttime rates to just 3 cents per kilowatt-hour. The couple was astonished with the change the new cheap heater made in their lives. But of course, 100,000 more homes were now dependent on electrical power, most of it generated by coal-fired power plants. [Source: *New York Times/International Herald Tribune Global Edition*. For detailed source information, see Text Sources and Credits.] ■

203. WORLDWATCH INSTITUTE: 16 OF 20 OF WORLD'S MOST POLLUTED CITIES IN CHINA

Air Pollution Elsewhere in East Asia Public health risks related to air pollution are also serious in the largest cities of Japan (Tokyo and Osaka), Taiwan (Taipei), Mongolia (Ulan Bator), and South Korea (Seoul), and in adjacent industrial zones. Even with antipollution legislation and increased enforcement, high population densities and rising expectations about better living standards make it difficult to improve environmental quality. Taiwan is a case in point.

Taiwan has some of the dirtiest air in the world. Some of the main causes are the island's extreme population density of 250 people per square mile (646 per square kilometer), its high rate of industrialization, and its close proximity to industrialized south China. There are now 4 motor vehicles (cars or motorcycles) in Taiwan for every five residents—more than 18.5 million exhaust-producing vehicles on this small island. In addition, there are nearly eight registered factories per square mile (three per square kilometer), all emitting waste gases. The government of Taiwan acknowledges that the air is six times dirtier than that of the United States or Europe.

Both North and South Korea depend on hydro- or nuclear-generated electricity to run factories and heat buildings, unlike elsewhere in East Asia. North Korea has relatively few industries and uses very few cars, so its air pollution levels are thought to be low for the region (although few reliable data are available for North Korea). South Korea uses fossil fuels in its numerous industries and has many gasoline-powered cars. These are the sources of most of its internally generated air pollution. Mongolia generally has the region's cleanest air, but in Ulan Bator, the pollution from coal heaters has inspired some imaginative projects. For example, whole sections of Ulan Bator are now heated by centrally located boilers, which supply hot water to apartment buildings and individual dwellings.

THINGS TO REMEMBER

- In an effort to address droughts, floods, and its burgeoning need for electricity, China has built many dams, the largest of which is the Three Gorges Dam, which now poses many potential problems itself.
- Rapid urban economic development in East Asia, combined with weak environmental protection, has resulted in some of the most polluted cities on the planet.

HUMAN GEOGRAPHY

East Asia is home to some of the most ancient civilizations on Earth. Settled agricultural societies have flourished in China for more than 7000 years, which makes China among the oldest continuous civilizations in the world. After centuries of leading the world technologically and economically, East Asia was gradually eclipsed by European powers beginning in the seventeenth century. Outright domination by Europeans occurred only in certain places and not until the nineteenth century. The twentieth century was violent and tumultuous, and after World War II, the countries of East Asia adopted entirely new economic systems. Three of the countries—China, Mongolia, and North Korea—became communist, and stayed so until the 1980s, when all but North Korea began to allow more capitalist practices. The other three countries—Japan, South Korea, and Taiwan—took a more capitalist route after World War II and today have among the highest standards of living in the world. China is now a global economic superpower that rivals the United States and the European Union.

HUMAN PATTERNS OVER TIME

Chinese civilization evolved from several hearths, including the North China Plain, the Sichuan Basin, and the lands of interior Asia that were inhabited by Mongolian nomadic pastoralists. On East Asia's eastern fringe, the Korean Peninsula and the islands of Japan and Taiwan were profoundly influenced by the culture of China, but they were isolated enough that each developed a distinctive culture and maintained political independence most of the time. In the early twentieth century, Japan industrialized rapidly by integrating the European influences that China disdained. As Figure 9.11 shows, both ancient and contemporary traditions are celebrated in East Asia today.

Bureaucracy and Imperial China

Although humans have lived in East Asia for hundreds of thousands of years, the region's earliest complex civilizations appeared in China about 4000 years ago. Written records exist only from the civilization that was located in north-central China. There, a small, militarized, feudal aristocracy (see page 165 in Chapter 4 for a discussion of feudalism) controlled vast estates on which the majority of the population lived and worked as semi-enslaved farmers and laborers. The landowners usually owed allegiance to one of the petty kingdoms that dotted northern China. These kingdoms were relatively self-sufficient and well-defended with private armies.

An important move away from feudalism came with the Qin empire (beginning in 221 B.C.E.), which instituted a trained and salaried bureaucracy in combination with a strong military to extend the monarch's authority into the countryside. One part of the legacy of the Qin empire is shown in Figure 9.12B on page 378.

The Qin system proved more efficient than the old feudal allegiance system it replaced. The estates of the aristocracy were divided into small units and sold to the previously semi-enslaved farmers. The empire's agricultural output increased because the people worked harder to farm the land they now owned. In addition, the salaried bureaucrats were more responsible than the aristocrats they replaced, especially about building and maintaining levees, reservoirs, and other tax-supported public works that reduced the threat of flood, drought, and other natural disasters. Although the Qin empire was short-lived, subsequent empires maintained Qin bureaucratic ruling methods, which have proved essential in governing a united China.

FIGURE 9.11 LOCAL LIVES
Festivals in East Asia

A A man dressed as a samurai—a member of Japan's preindustrial-era warrior class—for a festival in Fukushima, Japan. Thousands of parades, processions, and street festivals are held each year in Japan, often celebrating events in local history or legend. [Toru Yamanaka/AFP/Getty Images]

B Participants in the Boryeong Mud Festival, an annual event created in 1996 to celebrate the medicinal properties of the mud found near the town of Boryeong. The 2-week-long festival draws about 2.2 million people each year and features a mud pool, mud slides, a mud prison, and cosmetics and treatments that are based on the medicinal qualities of the mud. [Jung Yeon-Je/AFP/Getty Images]

C A man participates in *Naadam*, or "Games," which form the basis of festivals held throughout Mongolia in early July to celebrate the country's nomadic history. There are three games practiced: archery, horseracing, and wrestling. Both men and women compete in archery, though separately. [Bruno Morandi/Robert Harding World Imagery/Getty Images]

Confucianism Molds East Asia's Cultural Attitudes

The philosophy of **Confucianism** is closely related to China's bureaucratic ruling tradition. Confucius, who lived from 551 to 479 B.C.E., was an idealist who was interested in reforming government and eliminating violence from society. He thought human relationships should involve a set of defined roles and mutual obligations. Confucian values include courtesy, knowledge, integrity, and respect for and loyalty to parents and government officials. These values diffused across the region and are still widely shared throughout East Asia (see Figure 9.12A).

> **Confucianism** a Chinese philosophy that teaches that the best organizational model for the state and society is a hierarchy based on the patriarchal family

The Confucian Bias Toward Males The model for Confucian philosophy was the patriarchal extended family. The oldest male held the seat of authority and was responsible for the well-being of everyone in the family. All other family members were aligned under the patriarch according to age and gender. Beyond the family, the Confucian patriarchal order held that the emperor was the grand patriarch of all China, charged with ensuring the welfare of society. Imperial bureaucrats were to do his bidding and commoners were to obey the bureaucrats.

Over the centuries, Confucian philosophy penetrated all aspects of East Asian society. Concerning the ideal woman, for example, a student of Confucius wrote: "A woman's duties are to cook the five grains, heat the wine, look after her parents-in-law, make clothes, and that is all! When she is young, she must submit to her parents. After her marriage, she must submit to her husband. When she is widowed, she must submit to her son." These concepts about limited roles for women affected society at large, where the idea developed that sons were the more valuable offspring, with public roles, while daughters were primarily servants within the home.

The Bias Against Merchants For thousands of years, Confucian ideals were used to maintain the power and position of emperors and their bureaucratic administrators at the expense of merchants. In parable and folklore, merchants were characterized as a necessary evil, greedy and disruptive to the social order. At the same time, the services of merchants were sorely needed. Such conflicting ideas meant the status of merchants waxed and waned. At times, high taxes left merchants unable to invest in new industries or trade networks. At other times, however, the anti-merchant aspect of Confucianism was less influential, and trade and entrepreneurship flourished. Under communism, merchants again acquired a negative image. Later, when a market economy was encouraged, the social status of merchants rose once again.

Cycles of Expansion, Decline, and Recovery Although the Confucian bureaucracy at times facilitated the expansion of imperial China (Figure 9.13 on page 378), its resistance to change also led to periods of decline. Heavy taxes were periodically levied on farmers, bringing about farmer revolts that weakened imperial control. Threats from outside, particularly invasions by nomadic people from what are today Mongolia and western China, inspired the creation of massive defenses such as the Great Wall (see Figure 9.12C) built along China's northern border. Nevertheless, the Confucian bureaucracy always recovered from invasions. After a few generations, the nomads were indistinguishable from the Chinese. At the same time, Chinese culture and civilization absorbed a tremendous mixture of different influences and, as a result, has itself changed.

One nomadic invasion did result in important links between China and the rest of the world. In the 1200s, the Mongolian military leader Genghis Khan and his descendants were able to conquer all of China. They then pushed west across Asia as far as Hungary and Poland (see page 205 in Chapter 5). It was during the time of this Mongol empire (also known as the Yuan Empire) that traders such as the Venetian Marco Polo made the first direct contacts between China and Europe. These connections proved much more significant for Europe, which was dazzled by China's wealth and technologies, than for China, which saw Europe as backward and barbaric.

Indeed, from 1100 to 1600, China remained the world's most developed region, despite enduring several cycles of imperial expansion, decline, and recovery. It had the largest economy, the highest living standards, and the most magnificent cities. Improved strains of rice allowed dense farming populations to expand throughout southern China and supported large urban industrial populations. Nor was innovation lacking: Chinese inventions included paper making, printing, paper currency, gunpowder, and improved shipbuilding techniques.

Why Did China Not Colonize an Overseas Empire?

During the well-organized Ming dynasty, 1368–1644, Zheng He, a Chinese Muslim admiral in the emperor's navy, directed an expedition that could have led to China conquering a vast overseas empire. From 1405 to 1433, Zheng He sailed 250 ships—the biggest and most technologically advanced fleet that the world had ever seen. Zheng He took his fleet throughout established Chinese trade routes to Southeast Asia, across the Indian Ocean, and all the way to the east coast of Africa.

The lavish voyages of Zheng He were funded for almost 30 years, but they never resulted in an overseas empire like those established by European countries a century or two later. These newly explored regions simply lacked much that China needed or wanted. Moreover, back home the empire was continually threatened by the armies of nomads from Mongolia, so any surplus resources were needed for upgrading the Great Wall. Eventually the emperor decided that Zheng He's explorations were not worth the effort. In the years following Zheng He's voyages, China reduced its contacts with the rest of the world as the emperor focused on repelling the Mongols. As a result, the pace of technological change slowed, leaving China ill prepared to respond to growing challenges from Europe after 1600. China did, however, become a regional colonizing power by extending control to include territories in Central Asia and Southeast Asia (see Figure 9.13).

European and Japanese Imperialism

By the mid-1500s, during Europe's Age of Exploration, Spanish and Portuguese traders interested in acquiring China's silks, spices, and ceramics found their way to East Asian ports. They brought a number of new food crops from the Americas to exchange such as corn, peppers, peanuts, and potatoes. These new sources of nourishment contributed to a spurt of economic expansion and population growth during the Qing, or Manchurian, dynasty (1644–1912), and by the mid-1800s, China's population was more than 400 million; at that same time, Europe had 270 million people.

By the nineteenth century, European merchants gained access to Chinese markets and European influence increased markedly. In exchange for Chinese silks and ceramics, British merchants supplied opium from India, which was one of the few things that Chinese merchants would trade for. The emperor attempted to crack down on this drug trade because of its debilitating effects on Chinese society. The result was the Opium Wars (1839–1860), in which Britain badly defeated China. Hong Kong became a British possession, and British trade, including its opium trade, expanded throughout China well into the twentieth century (see Figure 9.12D).

The final blow to China's long preeminence in East Asia came in 1895, when a rapidly modernizing Japan won a spectacular naval victory over China in the Sino-Japanese War (and to solidify its emerging regional dominance, Japan also defeated Russia). After this first defeat by the Japanese, the Qing dynasty made only halfhearted attempts at modernization, and in 1912, it was overthrown by an internal revolt and collapsed. From the time of the decline of the Qing empire (1895) until China's Communist Party took control in 1949, much of the country was governed by provincial rulers in rural areas and by a mixture of Chinese, Japanese, and European administrative agencies in the major cities.

China's Turbulent Twentieth Century

In response to the absence of a central state authority, two rival reformist groups arose in China in the early twentieth century. The Nationalist Party, known as the Kuomintang (KMT), was an urban-based movement that appealed to workers as well as the middle and upper classes. The Chinese Communist Party (CCP), on the other hand, found its base among the rural poor. At first the KMT gained the upper hand, uniting the country in 1924. However, Japan's invasion of China in 1931 changed the dynamic.

By 1937, Japan had control of most major Chinese cities. The KMT did not resist the Japanese effectively and were confined to the few deep interior cities not under Japanese control. The CCP, however, waged a constant guerilla war against the Japanese throughout rural China, where they gained widespread support. Japan's brutal occupation caused 10 million Chinese deaths, including those of 250,000 to 300,000 civilians in the city of Nanjing, then China's capital. Similar acts of wartime brutality, as well as the keeping of local "comfort women" (forced prostitution) by the Japanese army in the occupied territories during World War II, still complicate the relationship between Japan and its East Asian neighbors today. When Japan finally withdrew in 1945, defeated at the end of World War II by the United States, Russia, and other Allied forces, the vastly more popular CCP pushed the KMT out of the country and into exile in Taiwan. In 1949, the CCP, led by Mao Zedong, proclaimed the country the "People's Republic of China," with Mao as president.

A The oldest Confucian temple, built in Confucius's hometown of Qufu, China, in 478 B.C.E. [Imagemore Co., Ltd./Getty Images]

B Part of a terra cotta army of thousands buried with the first Qin emperor in 210 B.C.E. [Robert G. Brown/Design Pics/Perspectives/Getty Images]

C A section of the Great Wall built in 1570. [Melanie Stetson Freeman/The Christian Science Monitor/Getty Images]

5000 B.C.E. Agricultural societies develop in China

551–478 B.C.E. Confucian philosophy begins

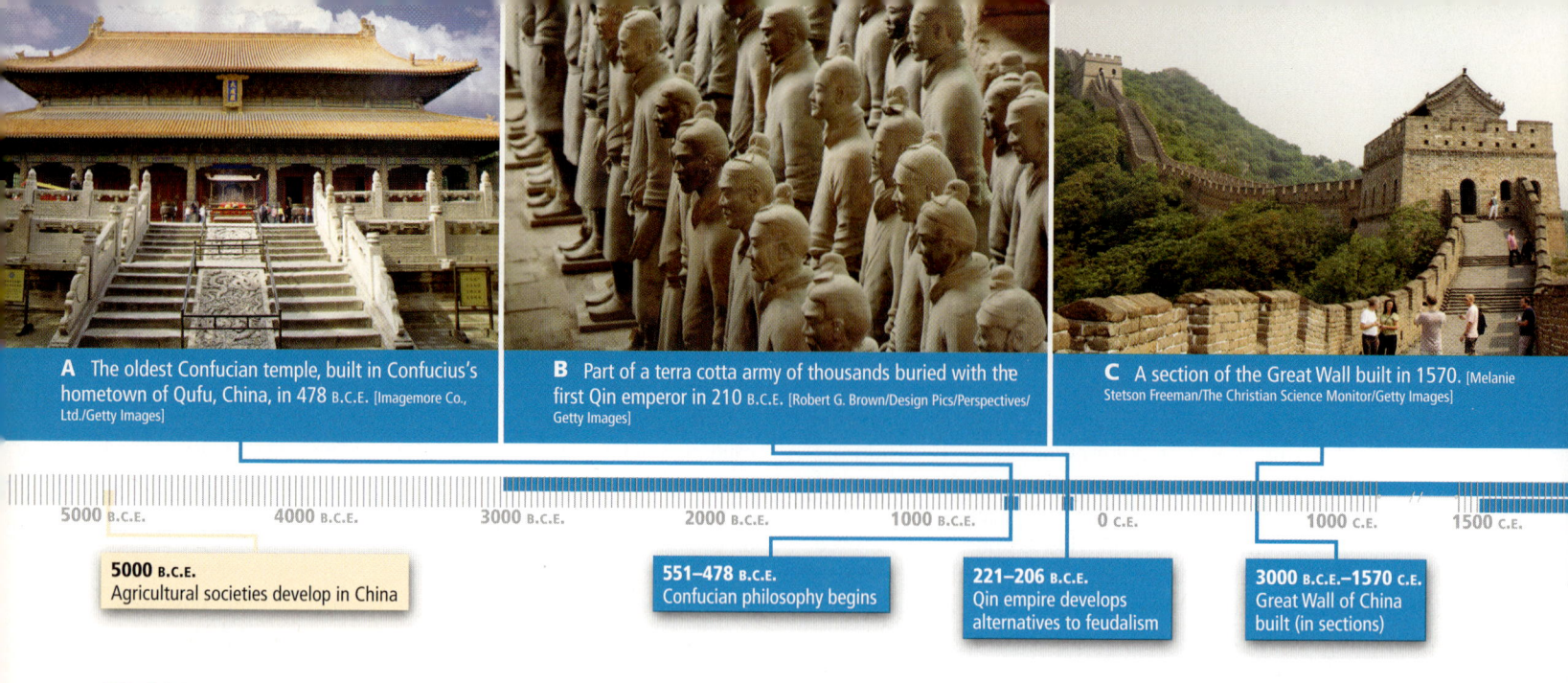

221–206 B.C.E. Qin empire develops alternatives to feudalism

3000 B.C.E.–1570 C.E. Great Wall of China built (in sections)

FIGURE 9.12 VISUAL HISTORY OF EAST ASIA

Mao's Communist Revolution Mao Zedong's revolutionary government became extremely powerful, dominating all the outlying areas of China—the Far Northeast, Inner Mongolia, and western China (Xinjiang)—and launched a brutal occupation of Tibet (Xizang). The People's Republic of China was in many ways similar to past Chinese empires. The Chinese Communist Party replaced the Confucian bureaucracy and Mao Zedong became a sort of emperor with unquestioned authority. China received support from the Soviet Union, but the two communist states remained wary of each other for decades.

Among the early beneficiaries of the revolution were the masses of Chinese farmers and landless laborers. On the eve of the revolution, huge numbers lived in abject poverty. Famines were frequent, infant mortality was high, and life expectancy was low. The vast majority of women and girls held low social status and spent their lives in unrelenting servitude.

The revolution drastically changed this. All aspects of economic and social life became subject to central planning by the Communist Party. Land and wealth were reallocated, often resulting in an improved standard of living for those who needed it most. Major efforts were made to improve agricultural production and to reduce the severity of floods and droughts. Everyone, regardless of age, class, or gender, was mobilized to construct almost entirely by hand huge public

FIGURE 9.13 The extent of Chinese empires, 221 B.C.E.–1850 C.E. The Chinese state has expanded and contracted throughout its history. [Source consulted: *Hammond Times Concise Atlas of World History* (Maplewood, NJ: Hammond, 1994)]

Greatest Extent of Chinese Empires:
- Qin 221 B.C.E.
- Han 2 C.E.
- Tang 907
- Tang zone of cultural dominance
- Ming 1644
- Qing 1850
- Qing tributary states
- Modern national boundaries
- Modern provincial borders

378

D British warships attack near Guangzhou during the First Opium War in 1840. [Hulton Archive/Getty Images]

E An official Chinese-government postcard made during the Cultural Revolution in 1969, showing peasants reading a book of the thoughts of Mao Zedong. [AFP/Getty Images]

F A "bullet train" passes Mt. Fuji in Japan in 2013. [DAJ/Amana Images/Getty Images]

1500s–1949	European and Japanese imperialism throughout East Asia
1945–Present	Japan rebuilds after WWII
1949–Present	Communist China
1980s–Present	Economic reforms in China
2006	

works projects—roads, dams, canals, whole mountains terraced into fields for rice and other crops. "Barefoot doctors" with rudimentary medical training dispensed basic medical care, midwife services, and nutritional advice to people in even the most remote locations. Schools were built in the smallest of villages. Opportunities for women became available, and some of the worst abuses against them, such as the crippling binding of feet to make women's feet small and childlike, were stopped. Most Chinese people who are old enough to have witnessed these changes say that the revolution did a great deal to improve overall living standards for the majority.

Mao's Missteps Progress, however, came at enormous human and environmental costs. During the **Great Leap Forward** (a government-sponsored program of massive economic reform initiated in the 1950s), 30 million people died, many from famine brought on by poorly planned development; others because they were persecuted for opposing the reforms. Meanwhile, deforestation, soil degradation, and agricultural mismanagement became widespread. In the aftermath of the Great Leap Forward, some Communist Party leaders tried to correct the inefficiencies of the centrally planned economy only to be demoted or jailed as Mao Zedong remained in power.

In 1966, partially in response to the failures of the Great Leap Forward, a political movement known as the **Cultural Revolution** enforced support for Mao and punished dissenters. Everyone was required to study the "Little Red Book" of Mao's sayings (see Figure 9.12E). Educated people and intellectuals were a main target of the Cultural Revolution because they were thought to instigate dangerously critical evaluations of Mao and the central planning of the Communist Party. Tens of millions of Chinese scientists, scholars, and students were sent out of the cities to labor in mines and industries or to jail, where as many as 1 million died. Children were encouraged to turn in their parents. Petty traders were punished for being capitalists, as were those who adhered to any type of organized religion. The Cultural Revolution so disrupted Chinese society that by Mao's death in 1976, the Communists had been seriously discredited.

Changes After Mao Two years after Mao's death, a new leadership formed around Deng Xiaoping. Limited market reforms were instituted, but the Communist Party retained tight political control. In 2009, after more than 30 years of reform and remarkable levels of economic growth, China's economy became the third largest in the world, behind those of the European Union and the United States. It passed Japan (although Japan's per capita wealth is still five times that of China's) when China managed to weather the world recession beginning in 2008 better than expected. However, the disparity of wealth in China has been increasing for some years, human rights are still often abused, and political activity remains tightly controlled even as discontent occasionally boils over into open protests against the government.

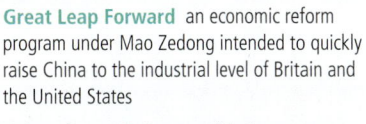

> **Great Leap Forward** an economic reform program under Mao Zedong intended to quickly raise China to the industrial level of Britain and the United States
>
> **Cultural Revolution** a political movement launched in 1966 to force the entire population of China to support the continuing revolution

👉 THINGS TO REMEMBER

- Settled agricultural societies have flourished in China for more than 7000 years, which makes China among the oldest continuous civilizations in the world.

- In China, an important move away from feudalism came with the Qin empire (beginning in 221 B.C.E.), which instituted a trained and salaried bureaucracy in combination with a strong military to extend the monarch's authority into the countryside.

- For six centuries, from 1100 to 1600, China was the world's most developed region, despite cycles of imperial expansion, decline, and recovery.

- Following the Opium Wars (1839–1860)—in which Britain badly defeated China—Hong Kong became a British possession, and British trade, including its opium trade, expanded throughout China, well into the twentieth century.

• In the wake of a brutal occupation by Japan during World War II, Mao Zedong's revolutionary communist government became extremely powerful, dominating all the outlying areas of China. ■

Japan Becomes a World Leader

Although China's influence was preeminent in East Asia for thousands of years, Japan, with only one-tenth the population and 5 percent of the land area of China, dominated East Asia economically and politically for much of the twentieth century. Japan's rise as a modern global power resulted largely from its response to challenges from Europe and North America.

Beginning in the mid-sixteenth century, active trade with Portuguese colonists brought new ideas and military technology that strengthened Japan's wealthier feudal lords (*shoguns*), allowing them to unify the country under a military bureaucracy. However, the shoguns monopolized contact with the European outsiders, refusing to allow Japanese people to leave the islands, on penalty of death.

A second period of radical change began when a small fleet of U.S. naval vessels arrived in Tokyo Bay in 1853. The foreigners, carrying military technology far more advanced than Japan's, forced the Japanese government to open the economy to international trade and political relations. In response, a group of reformers (the Meiji) seized control of the Japanese government, setting the country on a crash course of modernization and industrial development that became known as the *Meiji Restoration*. During this time, Japanese students were sent abroad and experts were recruited from around the world, especially from Western nations, to teach everything from foreign languages to modern military technology. Investments emphasized infrastructure development, especially in transportation and communication. The improvements that resulted enabled Japan's economy to grow rapidly, surpassing China's size in the early twentieth century.

Between 1895 and 1945, Japan fueled its economy with resources from a vast colonial empire. Equipped with imported European and North American military technology, its armies occupied first Korea, then Taiwan, then coastal and eastern China, and eventually Indonesia and much of Southeast Asia, as shown in Figure 9.14. Many people in these areas still harbor resentment about the brutality they suffered at Japanese hands. Japan's imperial ambitions ended with its defeat in World War II

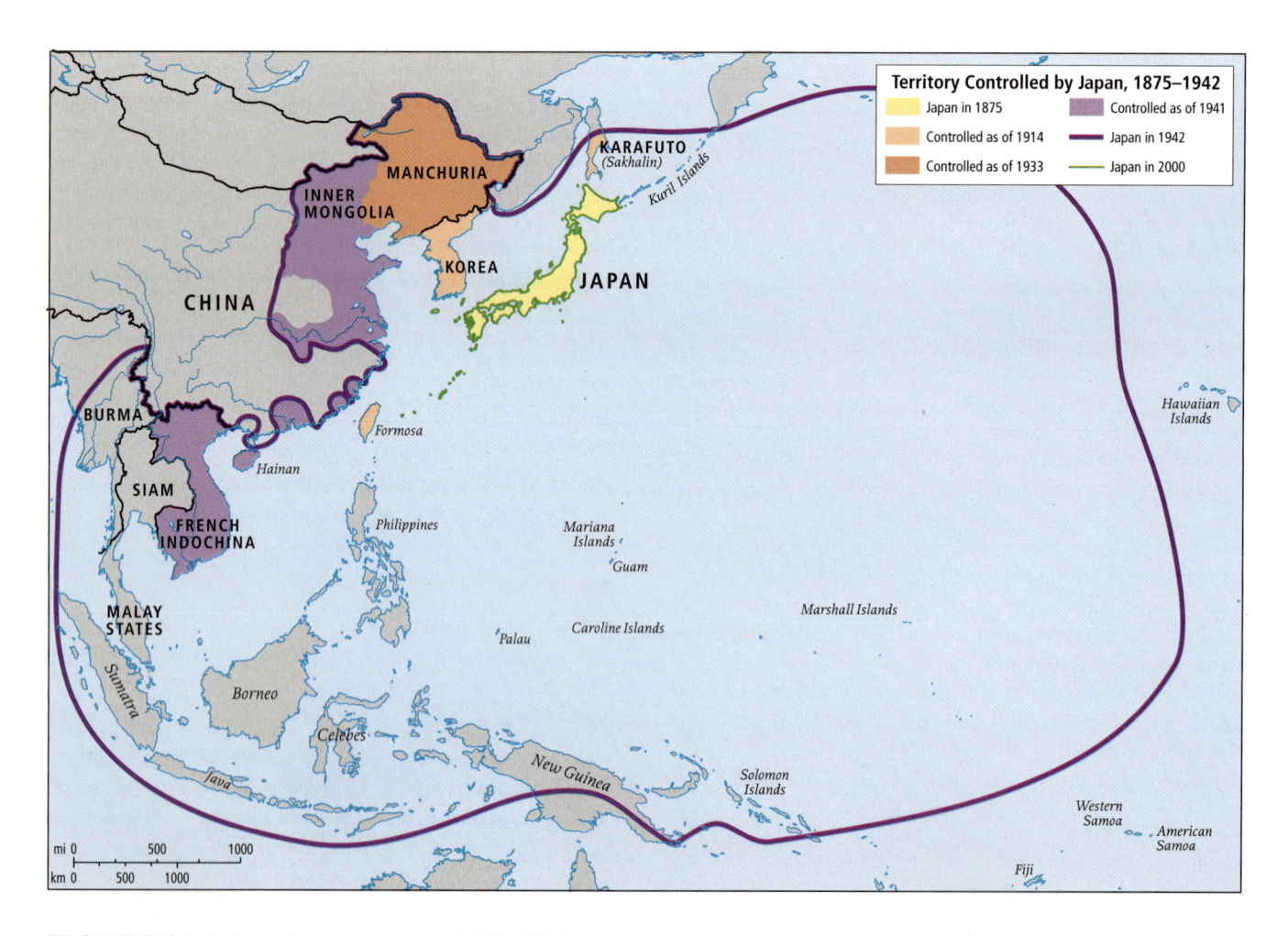

FIGURE 9.14 Japan's expansions, 1875–1942. Japan colonized Korea, Taiwan (then known as Formosa), Manchuria, China, parts of Southeast Asia, and several Pacific islands to further its program of economic modernization and to fend off European imperialism in the early twentieth century. [Source consulted: *Hammond Times Concise Atlas of World History* (Maplewood, NJ: Hammond, 1994)]

and its subsequent occupation by U.S. military forces until 1952.

204. JAPANESE STILL RESOLVING FEELINGS ABOUT THE WAR

Immediately following World War II, the U.S. government imposed many social and economic reforms on Japan. Japan was required to create a democratic constitution and reduce the emperor's role to a symbolic one. Its military was reduced dramatically, forcing it to rely on U.S. forces to protect it from attack. With U.S. support, Japan rebuilt rapidly after World War II, and it eventually became a giant in industry and global business, exporting automobiles, electronic goods, and many other products (see Figure 9.12F). Japan's economy is still among the world's largest, wealthiest, and most technologically advanced (see Figure 9.20D on page 391).

Chinese and Japanese Influences on Korea, Taiwan, and Mongolia

The history of the remaining East Asia region has largely been shaped by what transpired in China and Japan.

The Korean War and Its Aftermath Korea was a unified but poverty-stricken country until 1945. At the end of World War II, the United States and the Soviet Union, as victorious allies, agreed to divide the former Japanese colony. The Soviet Union occupied and established a communist regime to the north, while the United States took control of the southern half of the Korean peninsula, where it instituted reforms similar to those in Japan. After the United States withdrew its troops in the late 1940s, North Korea attacked South Korea. The United States returned to defend the south, leading a 3-year war against North Korea and its allies, the Soviet Union and Communist China.

After great loss of life on both sides and the devastation of the peninsula's infrastructure, the Korean War ended in 1953 in a truce. A demilitarized zone (DMZ) was established near the 38th parallel; it serves as the de facto border between North and South Korea. This border, which is 2.5 miles (4 kilometers) wide, is called "demilitarized" because it is a buffer zone not claimed by either side, although in the surrounding area, both sides are heavily armed.

North Korea closed itself off from the rest of the world, and to this day it remains isolated and impoverished, occasionally gaining international attention by hinting at its nuclear potential (see Figure 9.18D on page 387). South Korea, on the other hand, now has a prosperous and technologically advanced market economy. Relations between the two countries remain tense, with occasional skirmishes breaking out along the border.

201. INTER-KOREAN COOPERATION GROWS
210. HIGH-TECH KOREA EYES 'UBIQUITOUS' FUTURE

Taiwan's Uncertain Status For thousands of years, Taiwan was a poor, agricultural island on the periphery of China. Then, between 1895 and 1945, it became part of Japan's regional empire. In 1949, when the Chinese nationalists (the Kuomintang) were pushed out of mainland China by the Chinese Communist Party, they set up an anti-communist government in Taiwan, naming it the Republic of China (ROC). For the next 50 years, with U.S. aid and encouragement, the ROC became modern and industrialized. Its economy quickly dwarfed that of China and remained dominant until the 1990s, serving as a prosperous icon of capitalism right next door to massive Communist China.

Today, Taiwan remains an economic powerhouse. Taiwanese investors have been especially active in Shanghai and the cities of China's southeast coast. Yet mainland China has never relinquished its claim to Taiwan. As China's economic and military power has increased, the United Nations, the World Bank, and the United States, along with most other countries, agencies, and institutions, have judiciously tiptoed around the issue of whether Taiwan should continue to be treated as an independent country or as a rebellious province of China. Most Taiwanese feel that their country should hold on to its sovereignty (see Figure 9.18C on page 387).

Mongolia Seeks Its Own Way For millennia, Mongolia's nomadic horsemen posed periodic threats to China, so much so that the Great Wall was built, reinforced, and extended to combat them (see Figure 9.12C). China has long been obsessed with both deflecting and controlling its northern neighbor; and China did control Mongolia from 1691 until the 1920s. Revolutionary communism spread to Mongolia soon thereafter, and Mongolia continued as an independent communist country under Soviet, not Chinese, guidance until the breakup of the Soviet Union in 1989. Communism brought education and basic services. Literacy for both men and women rose above 95 percent. Deeply suspicious of both Russia and China, Mongolia has been on a difficult road to a market economy since 1989. In need of cash to participate in the modern world, many families have elected to abandon nomadic herding and permanently locate their portable *ger* homes near Ulan Bator and search for paid employment. While the economy has developed, the rapid and drastic change in lifestyle has also resulted in broken homes, increasing personal debt, and poverty, in a society that formerly took pride in an egalitarian if not prosperous standard of living.

THINGS TO REMEMBER

- For much of the twentieth century, Japan dominated East Asia economically and politically. Japan's rise as a modern global power resulted largely from its response to challenges from Europe and North America.
- North Korea remains a communist country and is impoverished and cut off from the wider world.
- Taiwan and South Korea emerged after World War II as rapidly industrializing countries.
- Until recently, Mongolia retained communist connections to the USSR.

GLOBALIZATION AND DEVELOPMENT

> **GEOGRAPHIC INSIGHT 2**
>
> **Globalization and Development:** East Asia pioneered a spectacularly successful economic development strategy that has transformed economies across the globe. Governments in Japan and then Taiwan, South Korea, and eventually China intervened strategically in the economy to encourage the production of manufactured goods destined for sale abroad, primarily to the large economies of North America and Europe.

After World War II, the countries of East Asia established two basic types of economic systems. The Communist regimes of China, Mongolia, and North Korea relied on central planning by the government to set production goals and to distribute goods among their citizens. In contrast, the governments of first Japan and then Taiwan and South Korea established **state-aided market economies** with the assistance and support of the United States and Europe. In this type of economic system, market forces, such as supply and demand and competition for customers, determine many economic decisions. However, the government intervenes strategically, especially in the financial sector, to make sure that certain economic sectors develop in a healthy fashion. Investment in the country by foreigners is also limited so that the government can retain more control over the direction of the economy and so that economic development benefits domestic interests. In the cases of Japan, South Korea, and Taiwan, government intervention was designed to enable **export-led growth**. This economic development strategy relies heavily on the production of manufactured goods destined for sale abroad—in this case primarily to North America and Europe—while limiting imports for local consumers.

More recently, the differences among East Asian countries have diminished as China and Mongolia have set aside strict central planning and adopted many aspects of state-aided market economies and export-led growth. China in particular relies heavily on exports of its manufactured goods to North America and Europe.

The Japanese Miracle

Throughout the nineteenth century, the economies of Japan, Korea, and Taiwan were minuscule compared with China's. During the twentieth century, though, all three grew tremendously, in part because of ideas that originated in Japan.

Japan Rises from the Ashes Japan's recovery after its crippling defeat at the end of World War II is one of the most remarkable tales in modern history. Except for Kyoto, which was spared because of its historical and architectural significance, all of Japan's major cities were destroyed by the United States. Most notably, the United States bombed Hiroshima and Nagasaki with nuclear weapons, and leveled Tokyo with incendiary bombs.

📹 **205. SURVIVORS RECALL THE NUCLEAR BOMBING OF HIROSHIMA**

Key to Japan's rapid recovery was its state-aided market economy, in which the government guided private investors in creating new manufacturing industries. The overall strategy was one of export-led economic growth, with the Japanese government negotiating trade agreements with the United States and Europe. These and other deals ensured that large and wealthy foreign markets would be willing to import Japanese manufactured goods. South Korea, Taiwan, countries in Southeast Asia, and eventually China and many other parts of the developing world have imitated Japan's model of a state-aided and export-oriented economy. Also central to Japan's recovery, but less imitated abroad, were arrangements between the government, major corporations, and labor unions that guaranteed lifetime employment by a single company for most workers in return for relatively modest pay.

The government-engineered trade and labor arrangements produced explosive economic growth of 10 percent or more annually between 1950 and the 1970s. The leading sectors were export-oriented automobile and electronics manufacturing. Japanese brand names such as Sony, Panasonic, Nikon, and Toyota became household words in North America and Europe. Products made by these companies sold at much higher volumes than would have been possible in the then relatively small Japanese and nearby Asian economies. Although growth slowed considerably during the 1990s and again beginning in 2007, Japan's postwar "economic miracle" continues to have an immense worldwide impact as a model, and Japan remains a significant actor in the world economy. Japan purchases resources from all parts of the world for its industries and domestic use. These purchases and investments in various local economies create jobs for millions of people around the globe.

Productivity Innovations in Japan Over the years, Japan has made two major innovations in manufacturing that have boosted its productivity and been diffused to other industrial economies, changing the spatial arrangements of industries. The **just-in-time system** clusters together companies that are part of the same production process so that they can deliver parts to each other when they are needed (Figure 9.15). For example, factories that make automobile parts are clustered around the final assembly plant, delivering parts literally minutes before they will be used. This saves money by making production more efficient and reducing the need for warehouses.

A related innovation is the *kaizen* system of continuous improvement in manufacturing and business management. In this system, production lines are constantly surveyed for errors, which helps ensure that fewer defective parts are produced. Production lines are also constantly adjusted and improved to save time and energy. Both the just-in-time and the *kaizen* systems have been imitated by companies around the world and have been taken overseas by Japanese companies that invest abroad. For example, Toyota uses *kaizen* and just-in-time in all of its U.S. plants.

Mainland Economies: Communists in Command

After World War II, economic development on East Asia's mainland proceeded on a dramatically different course than it did in Japan. Communist economic systems transformed poverty-ridden China, Mongolia, and North Korea. Private property was abolished, and the state took full control of the economy, loosely following the example of the Soviet centrally planned economy

---

state-aided market economy an economic system based on market principles such as private enterprise, profit incentives, and supply and demand, but with strong government guidance; in contrast to the free market (limited government) economic system of the United States and, to a lesser degree, Europe

export-led growth an economic development strategy that relies heavily on the production of manufactured goods destined for sale abroad

just-in-time system the system pioneered in Japanese manufacturing that clusters companies that are part of the same production system close together so that they can deliver parts to each other precisely when they are needed

***kaizen* system** a system of continuous manufacturing improvement, pioneered in Japan, in which production lines are constantly adjusted, improved, and surveyed for errors to save time and money and ensure that fewer defective parts are produced

FIGURE 9.15 Japan's just-in-time system. In this example of the just-in-time system, related industries are clustered together around a Toyota plant in Toyota City, Japan. The facilities supply each other with various automobile parts and other inputs needed for the production process.

(see Chapter 5, page 207). These sweeping changes transformed life for the poor majority but ultimately proved less resilient and successful than was hoped.

By design, most people in the communist economies could not consume more than the bare necessities. On the other hand, the policy—called the *iron rice bowl* in China—of guaranteeing nearly everyone a job for life, sufficient food, basic health care, and housing was better than what they had before. One drawback was that overall productivity remained low.

The Commune System When the Communist Party first came to power in China in 1949, its top priority was to make monumental improvements in both agricultural and industrial production. The Communist governments in North Korea and Mongolia held similar goals, though these countries had much smaller populations and resource bases to work with.

In the years following World War II, an aggressive agricultural reform program joined small landholders together into cooperatives so that they could pool their labor and resources to increase production. In time, the cooperatives became full-scale communes, with an average of 1600 households each. The communes, at least in theory, took care of all aspects of life. They provided health care and education, and built rural industries to supply such items as simple clothing, fertilizers, small machinery, and eventually even tractors. The rural communes also had to fulfill the ambitious expectations that the leaders in Beijing had for better flood control, expanded irrigation systems, and especially, more food production.

The Chinese commune system met with several difficulties. Rural food shortages developed because farmers had too little time to farm. They were required to spend much of their time building roads, levees, terraces, and drainage ditches, or working in the new rural industries. Local Communist Party administrators often compounded the problem by overstating harvests in their communes to impress their superiors in Beijing. The leaders in Beijing responded by requiring larger food shipments to the cities, which caused even greater food shortages in the countryside.

Although the Chinese agricultural communes were inefficient and created devastating food scarcities during the Great Leap Forward, they did eventually result in a stable food supply that kept Chinese people well fed.

In North Korea, the Communists have pushed military strength at the expense of broader economic development. Agriculture has been so neglected that production is precarious, with food aid from its immediate neighbors or the United States a yearly necessity and famine a constant threat. In recent years, North Korea has used rocket launchings, nuclear tests, and even the threat of war to intimidate its neighbors, and it is selling its military technology abroad to pay for its food imports. Intermittently, the U.S. government suspends food aid to encourage more peaceful policies from the North Korean leadership, but with limited results.

In Mongolia, Communist policy followed the Soviet model of collectivization, though it was tailored to Mongolia's economy, which at the time was largely one of herding and agriculture. There were minimal changes to the nomadic pastoralist way of life, but the role of mining and industry was emphasized, and the herding and forestry contributions to GDP declined to less than 20 percent.

Focus on Heavy Industry The Communist leadership in China, North Korea, and Mongolia believed that investment in heavy industry would dramatically raise living standards. Massive investments were made in the mining of coal and other minerals and in the production of iron and steel. Especially in China, heavy machinery was produced to build roads, railways, dams, and other infrastructure improvements that leaders hoped would increase overall economic productivity. However, much as in India (see pages 334–337 in Chapter 8), the vast majority of the population remained impoverished agricultural laborers who received little benefit from industries that created jobs mainly in urban areas. Not enough attention was paid to producing consumer goods (such as cheap pots, pans, hand tools, and other household items) that would have driven modest internal economic growth and improved living standards for the rural poor. Even in the urban areas, growth remained sluggish because, as also was the case in the Soviet command economy, small miscalculations by bureaucrats resulted in massive shortages and production bottlenecks that constrained economic growth.

Struggles with Regional Disparity For centuries, China's interior west has been poorer and more rural than its coastal east. The interior west has been locked into agricultural and herding economies, while the economies of the east have benefited from trade and industry in even the most restricted times. The first effort to address regional disparities, right after the revolution, was an economic policy that focused on a combination of investment from the central government and the development of

regional self-sufficiency. Each region was encouraged to develop as an independent entity with both agricultural and industrial sectors that would create jobs and produce food and basic necessities. This policy did little to lessen regional disparities and in fact constrained economic growth by inhibiting trade between regions.

Economic Reforms in China

In the 1980s, China's leaders enacted economic reforms that changed the country's economy in five ways. First, building on Communist-era methods of financial control but also integrating strategies pursued in Japan, China used the state-owned banking system to direct investment into export-oriented manufacturing industries. Second, at the local level, economic decision making was decentralized and given the name **responsibility system**. Under the responsibility system, farmers now could make household-level decisions, subject to the approval of the commune, about how to use land and which crops to grow. Third, the reforms allowed existing and newly established businesses to sell their produce and goods in competitive markets. Fourth, **regional specialization**, rather than regional self-sufficiency, was encouraged in order to take advantage of regional variations in climate, natural resources, and location, and thereby encourage national economic integration. In practice, this has meant that coastal urban areas focus on export-oriented manufacturing and advanced service production, while resource-based industries and production for domestic consumption are fostered in the interior provinces. Finally, the government allowed foreign direct investment in Chinese export-oriented enterprises and the sale of foreign products in China.

This five-part shift dramatically improved the efficiency with which food and goods were produced and distributed. On the other hand, regional disparities have widened and on the provincial level, rural–urban disparities are sometimes extreme (see Figure 9.16). Indeed, the floating population discussed at the beginning of this chapter is a direct consequence of this spatially uneven development.

China's economic reforms have transformed not only China's economy but also the economies of wider East Asia and indeed the whole world. Today China is the world's largest producer of manufactured goods, supplying consumers across the globe. Of the other Communist-led countries, Mongolia has participated in this revolution only moderately, mainly as a raw material exporter, and North Korea has not participated at all.

East Asia's Role as Mega-Financier As U.S. debt has mounted in recent decades (up to U.S.$12.1 trillion by 2013), so has concern about the fact that China and Japan now own much of this debt. China owns about 10 percent of U.S. debt, and Japan 9 percent (Figure 9.17); many people are worried that either country could destabilize the U.S. economy by selling or refusing to buy more U.S. debt. However, such an event is highly unlikely. Both countries' ownership of U.S. debt is a side effect of their huge trade relationship with the United States, which leaves them with large amounts of dollars. These amounts are large enough that, worldwide, the only relatively stable investment available that China and Japan can buy in sufficient quantities with all their dollars is U.S. debt. Moreover, dumping their vast holdings of U.S. debt and therefore destabilizing the U.S. economy would mean that both China and Japan would not be able to sell as many of their goods to consumers in the United States, which would destabilize their own economies. Hence China and Japan have strong incentives to continue buying U.S. government debt.

East Asians are able to use their reserves to finance development elsewhere in the world, thereby securing privileged trade deals. In recent years, China's lending to developing countries has outpaced even that of the World Bank. Often these loans are made in return for guarantees that the recipient country will supply China with needed industrial materials, especially fossil fuels.

> **regional self-sufficiency** an economic policy in Communist China that encouraged each region to develop independently in the hope of evening out the wide disparities in the national distribution of production and income
>
> **responsibility system** in the 1980s, a decentralization of economic decision making in China that returned agricultural decision making to the farm household level, subject to the approval of the commune
>
> **regional specialization** specialization (rather than self-sufficiency) in order to take advantage of regional variations in climate, natural resources, and location

---

👉 THINGS TO **REMEMBER**

› GEOGRAPHIC INSIGHT 2

- **Globalization and Development** East Asia pioneered a spectacularly successful economic development strategy that has transformed economies across the globe. Governments in Japan and then Taiwan, South Korea, and eventually China intervened strategically in the economy to encourage the production of manufactured goods destined for sale abroad, primarily to the large economies of North America and Europe.

- Japanese management innovations known as the just-in-time and *kaizen* systems were so successful that they diffused and created clusters of economic activity throughout the world.

- After World War II, Communist economic systems transformed poverty-ridden China, Mongolia, and North Korea. Private property was abolished and the state took full control of the economy, loosely following the example of the Soviet centrally planned economy.

- By design, most people in the Communist economies could not consume more than the bare necessities, but the policy—called the *iron rice bowl* in China—of guaranteeing nearly everyone a job for life, sufficient food, basic health care, and housing was better than what they had before.

- In the 1980s, China's leaders enacted reforms that changed the country's economy, enabling China to become the world's largest producer of manufactured goods, supplying consumers across the globe.

- Because China and Japan now own about 20 percent of U.S. public debt, many people are worried that either country could destabilize the U.S. economy by selling or refusing to buy more U.S. debt. However, such an event is highly unlikely. ■

POWER AND POLITICS

› GEOGRAPHIC INSIGHT 3

Power and Politics: As East Asia has developed economically, the pressure for more political freedoms has grown. Japan, South Korea, and Taiwan are now among the more politically free places in the world, and demands for political change in China are increasing, especially in urban areas.

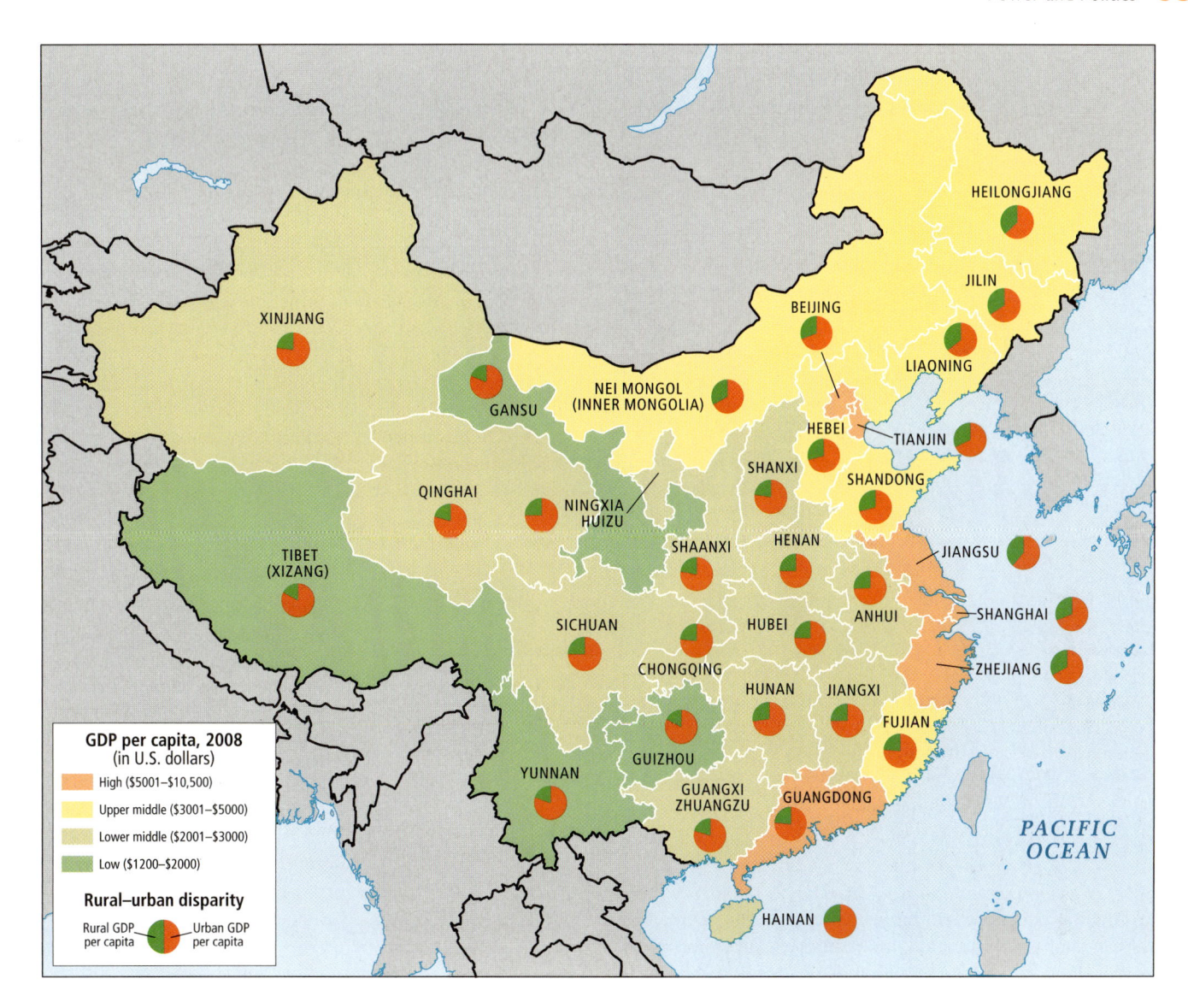

FIGURE 9.16 China's regional GDP and rural–urban income per capita disparities, 2008. Notice the disparity in GDP per capita across China, as indicated by the colors of the provinces, as well as the rural–urban income disparity in each province, as represented by the pie diagrams. [Sources consulted: Invest in China, "Per Capita Cash Income of Rural Households by Region (Third Quarter, 2009)," at http://www.stats.gov.cn/English/Statisticaldata/QuarterlyData/200911/t20091102_56720.html; "Income of Urban Households by Region (Third Quarter, 2009)," at http://www.fdi.gov.cn/pub/FDI_EN/Economy/Investment%20Environment/Macroeconomic%20Indices/ Population%20&%20GDP/t20091120_114779.htm]

Demands for more political freedom are growing throughout East Asia. Japan's current political structure was established after World War II, South Korea's in the late 1980s, and Taiwan's in the mid-1990s. All have steadily expanded political freedoms since their inception. Mongolia has had a dramatic expansion of political freedom since abandoning socialism in 1992. China, however, remains under the tight control of an authoritarian regime, and North Korea is even more tightly held. The Figure 9.18 map shows each country's level of democracy. The red starbursts indicate places where civil unrest has broken out since 1945.

Pressures for Political Change in China

With China now a globalized economy, many wonder how much longer the Communist Party can remain in control without allowing a significant expansion of political freedom throughout the entire country. The Communist Party officially says that China is a democracy, and indeed some elections have long been held at the village level and within the Communist Party. However, representatives of the National People's Congress, the country's highest legislative body, are appointed by the Communist Party elite, who maintain tight control throughout all levels of government. Most experts on China agree that while radical change is unlikely in the near future, a slow but steady shift toward more political freedom is underway. Change might be inevitable as the population becomes more prosperous, educated, and widely traveled, thereby becoming exposed to places that have more political freedoms. However, the Communist Party remains determined to repress any major reform movements that arise from outside the party's senior leadership.

Major Foreign Holders of U.S. Treasury Securities (August 2012)
(in billions of U.S. dollars)

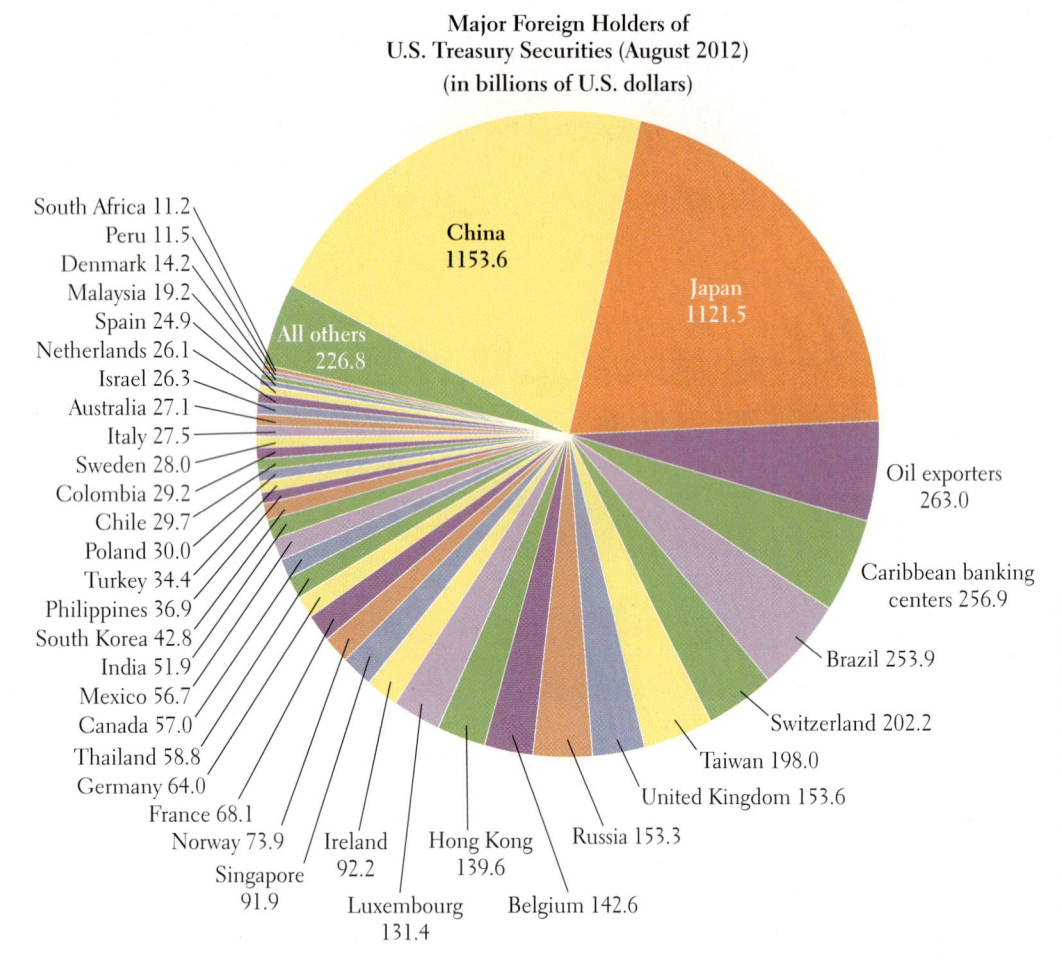

- China 1153.6
- Japan 1121.5
- All others 226.8
- Oil exporters 263.0
- Caribbean banking centers 256.9
- Brazil 253.9
- Switzerland 202.2
- Taiwan 198.0
- United Kingdom 153.6
- Russia 153.3
- Belgium 142.6
- Hong Kong 139.6
- Luxembourg 131.4
- Ireland 92.2
- Singapore 91.9
- Norway 73.9
- France 68.1
- Germany 64.0
- Thailand 58.8
- Canada 57.0
- Mexico 56.7
- India 51.9
- South Korea 42.8
- Philippines 36.9
- Turkey 34.4
- Poland 30.0
- Chile 29.7
- Colombia 29.2
- Sweden 28.0
- Italy 27.5
- Australia 27.1
- Israel 26.3
- Netherlands 26.1
- Spain 24.9
- Malaysia 19.2
- Denmark 14.2
- Peru 11.5
- South Africa 11.2

FIGURE 9.17 Foreign holders of U.S. Treasury securities, August 2012. The total dollar amount of U.S. securities held by foreign countries is more than $5 trillion; note, though, that the United States holds more than $6 trillion in securities from foreign countries. [Source consulted: "Major Foreign Holders of Treasury Securities (in Billions of Dollars)," U.S. Treasury, at http://www.treasury.gov/resource-center/data-chart-center/tic/Documents/mfh.txt]

Less than a decade after market reforms began, demands for political change culminated in a series of pro-democracy protests that drew hundreds of thousands of people to Beijing's Tiananmen Square in 1989 (see Figure 9.18A). These protests were brutally repressed, with thousands (the precise number is uncertain) of students and labor leaders massacred by the military. The Tiananmen Square event made clear that China's integration into the global economy would take place on political terms dictated by the state.

International Pressures Informed consumers and environmentalists in developed countries have criticized China for its "no holds barred" pursuit of economic growth. Much of China's growth has been built on environmentally destructive activities and harsh conditions for workers, both of which effectively reduce production costs so that the prices of Chinese goods are low on world markets.

The 2008 Olympics in Beijing highlighted a number of issues that have also created an impetus for increasing democracy. Before the games, the international media drew attention to the human rights abuses of workers, protesters, prisoners, ethnic minorities such as Tibetans and Uygurs (see pages 397–400), and spiritual groups. The most prominent of such groups is Falun Gong, a Buddhist-inspired movement that has been perceived as a threat by the Chinese government because of its independence and capacity to mount a worldwide media campaign for its cause. The extravagance of the 2008 Olympics also became a subject of criticism by the global media, with many journalists pointing out that no democracy would ever be able to devote so much tax revenue (an estimated $40 billion) to such an event, especially not in a country with as many pressing human and environmental problems as China has. Today, many of Beijing's Olympic venues are unused; at the same time, residents have benefited from the improvements to transportation and infrastructure that were part of the Olympic rebuilding of Beijing.

211. OLYMPIC RELAY CUT SHORT BY PARIS PROTESTS
213. BEIJING OLYMPICS: POLITICAL BATTLEGROUND?
221. REPORTS OF SALE OF EXECUTED FALUN GONG PRISONERS' ORGANS IN CHINA CALLED "SHOCKING"

Information Technology and Political Freedom The spread of information via the Internet has increased the push for political freedom from within China. Since the revolution in 1949, China's central government has controlled the news media in the country. By the late 1990s, though, the expanding use of electronic communication devices was loosening central control over information. By 2001, twenty-three million people were connected to the Internet in China; 538 million (or 40 percent of the population) were connected by 2012 (Figure 9.19). However, Internet access is not evenly distributed. A huge digital divide has

FIGURE 9.18 PHOTO ESSAY: Power and Politics in East Asia

Political freedoms are better protected in Japan, Taiwan, and South Korea than in China and North Korea, where people have fewer opportunities to directly influence their governments. Pressure for more substantive public participation in politics is growing throughout the region.

A The annual vigil in Hong Kong commemorating those who died in the 1989 Tiananmen Square massacre in Beijing has become a focus for China's pro-democracy movement. As many as 150,000 people have shown up for the vigil in recent years. Hong Kong is the only part of China where the right to engage in political protests is protected by law.

Democratization and Conflict
Democratization Index
- Full democracy
- Flawed democracy
- Hybrid regime
- Authoritarian regime
- No data

Armed Conflicts and Genocides with High Death Tolls Since 1945
- Ongoing conflict
- 1000–10,000 deaths
- 10,000–100,000 deaths
- 100,000–1,000,000 deaths
- 1,000,000–3,000,000 deaths
- 30,000,000 deaths

B Japanese politicians visit the Yasukuni Shrine in Tokyo, which honors Japan's war dead, including convicted war criminals from World War II. Condemned by China, Taiwan, and Korea, these visits spark protests throughout the region.

C A Taiwanese activist puts her country's flag on a model of one of the Senkaku/Diaoyu Islands, one of several disputed islands in the East China Sea that is claimed by Japan, China, and Taiwan. Many Taiwanese see this and other disputes as ways to reaffirm the country's continued independence from China, which considers Taiwan a rebellious province.

D North Korea's longest-range ballistic missile is paraded in P'yongyang. In 2006, North Korea tested this missile and others, along with several nuclear weapons. China, Japan, South Korea, Russia, and the United States all expressed varying levels of concern.

388 CHAPTER 9 East Asia

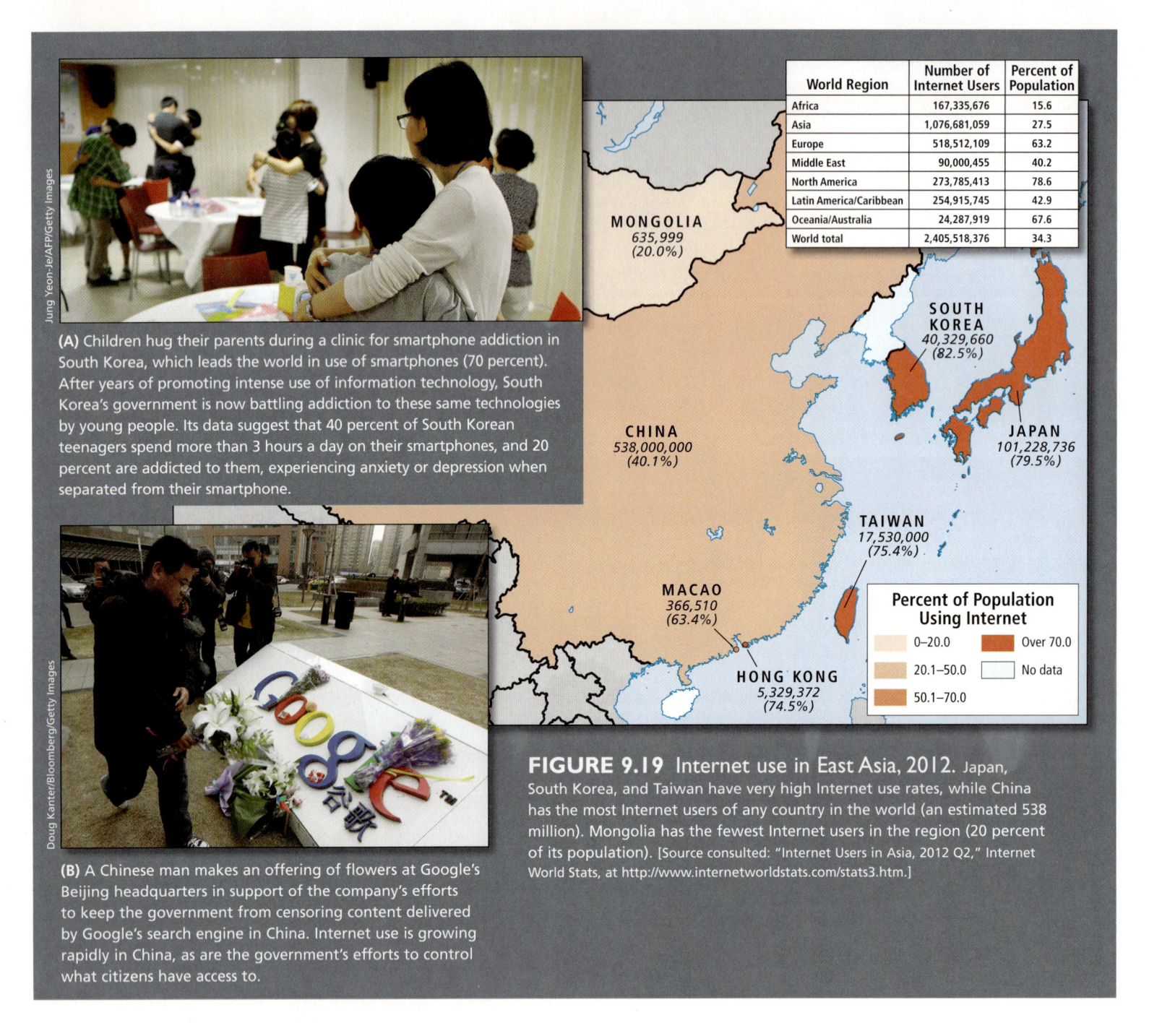

World Region	Number of Internet Users	Percent of Population
Africa	167,335,676	15.6
Asia	1,076,681,059	27.5
Europe	518,512,109	63.2
Middle East	90,000,455	40.2
North America	273,785,413	78.6
Latin America/Caribbean	254,915,745	42.9
Oceania/Australia	24,287,919	67.6
World total	2,405,518,376	34.3

MONGOLIA 635,999 (20.0%)
SOUTH KOREA 40,329,660 (82.5%)
JAPAN 101,228,736 (79.5%)
CHINA 538,000,000 (40.1%)
TAIWAN 17,530,000 (75.4%)
MACAO 366,510 (63.4%)
HONG KONG 5,329,372 (74.5%)

Percent of Population Using Internet
- 0–20.0
- 20.1–50.0
- 50.1–70.0
- Over 70.0
- No data

(A) Children hug their parents during a clinic for smartphone addiction in South Korea, which leads the world in use of smartphones (70 percent). After years of promoting intense use of information technology, South Korea's government is now battling addiction to these same technologies by young people. Its data suggest that 40 percent of South Korean teenagers spend more than 3 hours a day on their smartphones, and 20 percent are addicted to them, experiencing anxiety or depression when separated from their smartphone.

(B) A Chinese man makes an offering of flowers at Google's Beijing headquarters in support of the company's efforts to keep the government from censoring content delivered by Google's search engine in China. Internet use is growing rapidly in China, as are the government's efforts to control what citizens have access to.

FIGURE 9.19 Internet use in East Asia, 2012. Japan, South Korea, and Taiwan have very high Internet use rates, while China has the most Internet users of any country in the world (an estimated 538 million). Mongolia has the fewest Internet users in the region (20 percent of its population). [Source consulted: "Internet Users in Asia, 2012 Q2," Internet World Stats, at http://www.internetworldstats.com/stats3.htm.]

emerged between the cities, where more than 60 percent have Internet access, and the rural interior, where only about 23 percent do. As economic development spreads over time, it is likely that the divide will diminish.

Not too long ago, telephones were very rare and people had to obtain permission to use them, but now millions of Chinese people have access to an international network of information. It is much more difficult for the government to give inaccurate explanations for problems caused by inefficiency and corruption. Journalists can now check the accuracy of government explanations by phoning witnesses or the principal actors directly and then posting their findings on the Internet. Analysts, both inside and outside China, see the availability of the Internet to ordinary Chinese citizens as a watershed event that supports the expansion of political freedom.

Social networking technologies have emerged as a powerful tool for collective action in China. In recent years, the use of cell phone–based Twitter-like services known as *weibo* has grown exponentially. This type of microblogging has been instrumental in forcing major media coverage and public discussion of incidents that expose government corruption and mismanagement. For example, when local officials tried to cover up and censor media coverage of a fire in an apartment building in Shanghai in 2010 and a train wreck in Zhejiang province in 2011, people used weibo to make these become major national discussions.

Nevertheless, the Internet in China is not the open forum that it is in most Western countries. Since 2013, Chinese Internet users have been required to register their real names when using an online account. This is an attempt by the government to monitor free speech and anonymous criticisms against party officials. And if people writing blogs in China use the words "democracy," "freedom," or "human rights," they may receive the following reminder: "The title must not contain prohibited language, such as profanity. Please type a different title." Such censorship, informally dubbed the "Great Firewall," has been aided by U.S. technology firms such as Yahoo and Microsoft, which allow the Chinese government to use software that blocks access to certain Web sites for users in China. Google, in dispute with the Chinese government over state-sponsored censorship, invasion of human rights activists' email accounts, and cyber-attacks, moved its operations to Hong Kong, where more freedom is allowed (see Figure 9.19B). Meanwhile, the Web browser from the Chinese company Baidu has a much larger market share in China than Google does.

Protests and Political Freedom Protests by workers for better pay and living conditions, and by farmers and urban dwellers displaced by new real estate developments are becoming increasingly common. China's government reported that there were 58,000 public protests in the country in 2003. This number rose to 74,000 in 2004 and 87,000 in 2005. Since then, the government has stopped issuing complete statistics on protests. The nonprofit group Human Rights Watch estimates that 100,000 to 200,000 such protests are being held each year.

Many people have joined *interest groups* that pressure the government to take action on particular problems. One of the most publicized of such groups was formed by parents whose children died after schools collapsed during a massive earthquake (which registered 7.8 on the Richter scale) in Sichuan in May 2008. They demanded changes in policies that allowed schools to be poorly constructed and pushed for compensation for their lost children.

In December 2008, people seeking more political freedoms signed a manifesto called the *08 Charter*, which calls for a decentralized federal system of government (that is, more power to the provinces), democratic elections, and the end of the Communist Party's political monopoly. Liu Xiaobo, coauthor of the 08 Charter, was arrested, and in December 2009 he was sentenced to 11 years in prison, even as prominent international human rights activists lobbied on his behalf. In 2010, Liu Xiaobo was awarded the Nobel Peace Prize, though neither he nor his family was allowed to attend the ceremony in Norway.

In an effort to maintain control over China's increasingly articulate protestors, the government has allowed elections to be held for "urban residents' committees." The idea seems to be to provide a peaceful outlet for voicing frustrations and creating limited change at the local level. Many of these elections are hardly democratic, with candidates selected by the Communist Party. However, in cities where unemployment is high and where protests have been particularly intense, elections tend to be more free, open, and truly democratic. It could be that interest group protests are an important first step toward actual participatory democracy, even when some, such as those who signed the 08 Charter, are harshly stifled.

Japan's Recent Political Shifts

Significantly, the government that played such a central role during the post–World War II rise of the Japanese economy was controlled from 1955 to 2009 by one political party, the Liberal Democratic Party (LDP). This led to criticism that Japan lacked a meaningful democracy. In 2009, the Japanese people elected a government led by the opposition party, the Democratic Party of Japan (DPJ), only to reverse course again and bring the LDP back to power in 2012. A main difference between the two parties is that the more nationalistic LDP takes a hawkish stance against China and indeed all neighboring countries (see Figure 9.18B). Having closer relations with China may be advantageous for Japan, though, as China has the potential to be a major market for Japanese exports in the future.

Political Tensions Between East Asian Countries

In recent years, political tensions have been rising between the countries of East Asia even as their economies have become more linked. While these tensions are rooted in contemporary economic conditions and political calculations by governments, they are also fueled by the painful legacies of World War II and its aftermath.

Recently, two major disputes between East Asian countries have arisen over uninhabited islands that lie over oil and natural gas reserves. In the case of the Spratly Islands of the South China Sea (see page 430 in Chapter 10), both China and Taiwan are pursuing claims to the islands, as are several Southeast Asian countries.

In the East China Sea, the Senkaku Islands (in China, these are known as the Diaoyu Islands) are claimed by China, Japan, Taiwan, and Korea. The conflict over the Senkaku/Diaoyu Islands has escalated lately, mainly because of competing claims on them by China and Japan. Because China and Japan are the world's first- and second-largest importers of fossil fuels, respectively, they each have an interest in controlling nearby reserves. Within China, the conflict often inflames widespread resentment about Japan's brutal occupation of China before and during World War II. Japanese nationalists, on the other hand, eager to counter China's rising power and armed with several treaties that explicitly give it control over the islands, have staged protests and planted Japanese flags on the islands. Many now see this as East Asia's most risky potential military conflict of the post–World War II period.

THINGS TO REMEMBER

> **GEOGRAPHIC INSIGHT 3**
> • **Power and Politics** As East Asia has developed economically, the pressure for more political freedoms has grown. Japan, South Korea, and Taiwan are now among the more politically free places in the world, and demands for political change in China are increasing, especially in urban areas.

390 CHAPTER 9 East Asia

- By 2012, there were 538 million people in China using the Internet, representing 40 percent of the country's population. China has strict controls on Internet access and use, but social networking is being used to promote the expansion of political freedoms.

- Chinese citizens who share a grievance against the government are increasingly cooperating with one another in their protests. In December 2008, a manifesto for change, the 08 Charter, was signed and submitted to the government by 350 protestors.

- In Japan, one political party has controlled the government for all but 3 years since 1955, leading to criticism that Japan lacks a vigorous democracy.

- Recently, two major disputes between East Asian countries have arisen over islands that may lie over oil and natural gas reserves. ■

URBANIZATION

> **GEOGRAPHIC INSIGHT 4**
>
> **Urbanization:** Across East Asia, cities have grown rapidly over the last century, fueled by export-oriented manufacturing industries. China has recently undergone the most massive and rapid urbanization in the history of the world. Its urban population, now more than 720 million people, has tripled since China initiated economic reforms in the 1980s.

Throughout most of the twentieth century, the most dynamic East Asian cities were in Japan, Korea, and Taiwan. Tokyo, Japan, is the world's largest urban area, with 34 million inhabitants. The vast Kobe–Osaka–Kyoto area in Japan has 18 million (shown in the Figure 9.20 map). Korea is home to Seoul, East Asia's second-largest urban area, at 25 million people, and the fourth-largest urban economy in the world (after Tokyo, New York, and Los Angeles). The metropolitan area of Taiwan's capital of Taipei is home to almost 7 million and has the second-highest per capita income in East Asia ($48,000) after Tokyo ($65,000).

However, for the last 30 years, the strongest urban growth in East Asia, and indeed in the entire world, has been in China. In 2012, nineteen of the 20 fastest-growing cities in the world, as measured by increase in GDP per capita, were in China. Despite the rapid growth, urbanites still represent only 53 percent of China's total population, so more growth there is likely, whereas in every other country in the region—even North Korea—the majority of the population has been urban for at least two decades. Most East Asian cities are extremely crowded, and future growth poses a challenge for planners and residents (see Figure 9.20A, B).

Spatial Disparities China's focus on export-oriented manufacturing based in urban areas has led to rural areas lagging behind cities in access to jobs and income, education, and medical care. The map in Figure 9.16 on page 385 depicts the problem. GDP per capita is significantly lower in China's interior provinces than it is in coastal provinces, and within each province there is a disparity between rural and urban places, shown in pie diagrams on the map of rural versus urban incomes. Notice that rural–urban GDP disparities, while still significant, are less extreme in northeastern and coastal provinces than in interior and western provinces. Similar rural-to-urban disparity patterns are found in all other countries in East Asia, including North Korea.

📹 **206. OLD BEIJING MAKING WAY FOR MODERN DEVELOPMENT**
📹 **219. SOME CHINESE FEAR PRIVATE PROPERTY LAW WILL COST THEM THEIR HOMES**

International Trade and Special Economic Zones When China initiated economic reforms in the 1980s, many people were wary of the disruption that could result from abruptly opening the economy to international trade. To ease the transition, China first selected 5 coastal cities as sites where foreign technology and management could be imported to China and free trade established. These **special economic zones (SEZs)** (Figure 9.21) now operate like EPZs found in other world regions (see page 130 in Chapter 3). In the late 1990s, the program was expanded to 32 other cities, many of them in the interior. These new locations were designated *economic and technology development zones* (ETDZs). Like SEZs, the ETDZs provide footholds for international investors and multinational companies eager to establish operations in the country.

This program was successful. Today the SEZs and ETDZs are China's greatest **growth poles**, meaning that their development, like a magnet, is drawing yet more investment and migration. The first coastal SEZs were spectacularly successful. In just 25 years, many coastal cities, including Dongguan (the city Li Xia migrated to in the vignette that opens this chapter), grew from medium-sized towns or even villages into some of the largest urban areas in the world. Foreign direct investment remains concentrated on the eastern coast, which accounts for 94 percent of China's exports. As shown in Figure 9.16, GDP per capita (PPP) rates remain noticeably higher on the eastern coast relative to China's interior.

> **special economic zones (SEZs)** free trade zones within China, which are commonly called export processing zones (EPZs) elsewhere
>
> **growth poles** zones of development whose success draws more investment and migration to a region

Transportation Improvements

Recently, ETDZs in the interior have had higher rates of growth than those on the coast. One reason for this is that China spends 9 percent of its GDP on transportation improvements (by comparison, the United States spends 2 to 3 percent), which has made central and western China more accessible. New highways and railroads may eventually enable the interior to catch up with the economic development of the coast. The world's longest high-speed rail now takes a traveler from Beijing to southern China in 10 hours rather than 22, and it is routed through central China, not along the coast.

Urban Labor Shortages, Hukou System Reform, and Economic Transitions In 2013, China reached a major milestone that every rapidly developing country eventually reaches: its booming service sector now creates more jobs than its industrial sector. Many of these jobs offer much better pay and working conditions than industrial jobs do, which is creating a serious shortage of workers in the industrial sector.

FIGURE 9.20   PHOTO ESSAY: Urbanization in East Asia

Some of the largest, wealthiest, and fastest-growing urban areas in the world are in East Asia. China's cities are undergoing particularly rapid growth.

A A man searches for usable lumber in the remains of an old neighborhood in downtown Shanghai that is being demolished to make way for high-rise apartments for the city's burgeoning population of wealthy residents. The inhabitants of this neighborhood were forcibly evicted by the city government and resettled in cheaper apartments on the urban fringe. In the name of modernization, few attempts at renovating historic neighborhoods, called *hutongs*, are being made in urban China.

Population of Metropolitan Areas 2013
- 20 million
- 10 million
- 5 million
- 3 million

Note: Symbols on map are sized proportionally to metro area population.

① Global rank (population 2013)

Population Living in Urban Areas
- 83%–100%
- 65%–82%
- 47%–64%
- 29%–46%
- 11%–28%
- No data

B Large expanses of high-rise apartment buildings characterize much of the current urban development throughout East Asia. Shown here are the outskirts of Seoul, South Korea.

C Migrants from rural areas arrive in Beijing. Most of China's urban growth is based on migration from rural areas.

D People camp out to buy the latest iPhone in Ginza, Tokyo, the most lavish shopping district in the world's largest and wealthiest city.

[Sources consulted: 2011 World Population Data Sheet, Population Reference Bureau, at http://www.prb.org/pdf11/2011population-datasheet_eng.pdf; World Gazetteer, at http://world-gazetteer.com/wg.php?x=&men=gcis&lng=en&des=wg&srt=npan&col=abcdefghinoq&msz=1500&pt=a&va=&srt=pnan]

392 CHAPTER 9 East Asia

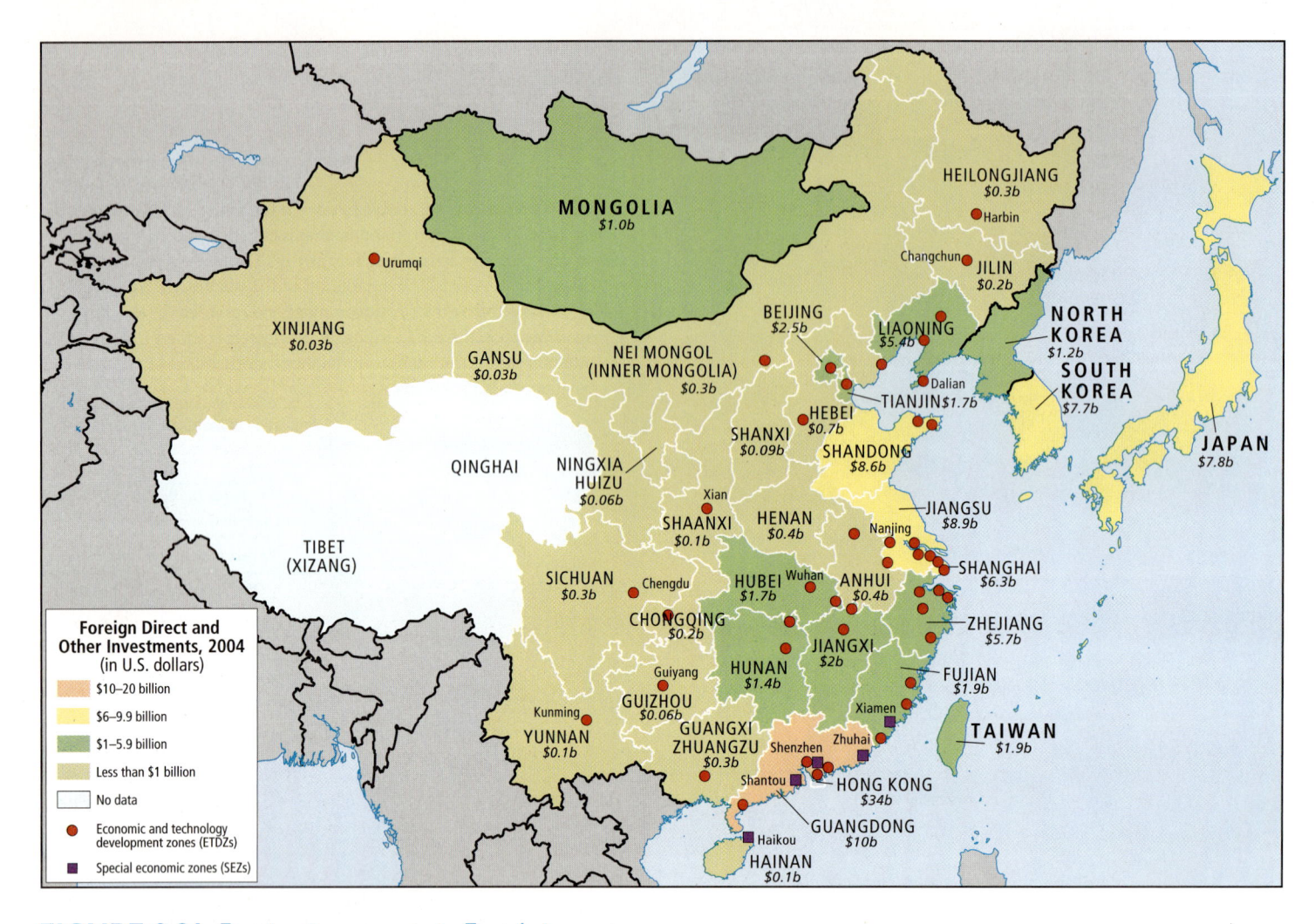

FIGURE 9.21 Foreign investment in East Asia. The map shows China's original special economic zones (SEZs) and its more recently designated economic and technology development zones (ETDZs). The colors on the map reflect levels of foreign investment in each of the countries of the region and in each of China's provinces. [Source consulted: Invest in China, at http://www.fdi.gov.cn/common/info.jsp?id5ABC00000000000022787. Specific Web site no longer available without registering at http://www.fdi.gov.cn/pub/FDI_EN/Statistics/default.htm.]

Until recently, China's spectacular urban growth was based on hundreds of millions of new urban migrants who were willing to put up with adverse conditions to earn a little cash. Fewer migrants are now willing to work in such jobs, creating periodic labor shortages. Some factory owners have been forced to offer higher pay, better working conditions, and shorter workdays or more time off. The government is also trying to offset labor shortages by making changes to the hukou system that will gradually give formal urban residency status to the vast floating population of people who have migrated illegally.

The extra costs these changes impose—especially the higher wages and the improved access to subsidized housing, education, and health care for millions of urban migrants—mean that China is no longer the cheapest place to manufacture products. Some factories have already moved to Vietnam, Bangladesh, and countries in Africa with even cheaper labor.

Some industrial employers hope that enough new migrants will be drawn to the cities by the hukou reforms that the labor shortages will disappear. The experiences of other countries, though, suggests that China's urban service sector will continue to attract more and more migrants. The rapid growth of the service sector is being fueled by the purchasing power and consumption of China's increasingly wealthy urban populations.

These changes in China's labor market added a new twist to the story of Li Xia, whom we met in the vignette that opens this chapter.

VIGNETTE

Li Xia and her sister returned home to rural Sichuan a second time in 2006. With the money she had saved from her second job in Dongguan, Xia tried to open a bar in the front room of her parents' house. She hoped to introduce the popular custom of karaoke singing she had enjoyed in the city, but people in her village could not stand the noise, and family tensions rose.

In early 2007, news that training was now available in Dongguan for skilled electronics assemblers convinced Xia to try the city again. Past experience with the bureaucracy and her knowledge of Dongguan helped Xia to sign up for the electronics training. The cost of tuition came out of her wages, which were only U.S.$350 a month rather than the U.S.$400 she had thought she would be paid. But this still left her with enough to live on, even after she sent money home.

In November of 2008, Xia heard rumors that a global recession was causing orders for electronics to be cut and that she might soon be laid off. But her factory limped along with a reduced staff. Then, miraculously, in June of 2009, orders picked up. Consumers in America had continued to buy electronics even as they downsized their homes and cars.

By 2011, Dongguan's export industries were once again struggling to meet orders, and this time their main obstacle was a shortage of labor. The cost of living in Dongguan had risen quickly, and not enough employers in the industrial sector were willing to raise wages accordingly. Some migrant workers were returning to their homes in the interior provinces, where the lower cost of living was more in line with the lower wages. Others were staying in the cities to take advantage of new legal residency status granted to some Dongguan workers as part of the hukou reforms. Among those who stayed, many were finding better-paying jobs in the now-blossoming service sector. Even with the higher salary that she was able to command because of her electronics training course, by 2012, Xia was eyeing jobs at hotels, an insurance company, and a nearby karaoke bar. *[Sources: USA Today, National Public Radio, Wall Street Journal, New York Times, China Daily, PlasticsNews.com, Dongguan Today. For detailed source information, see Text Sources and Credits.]*

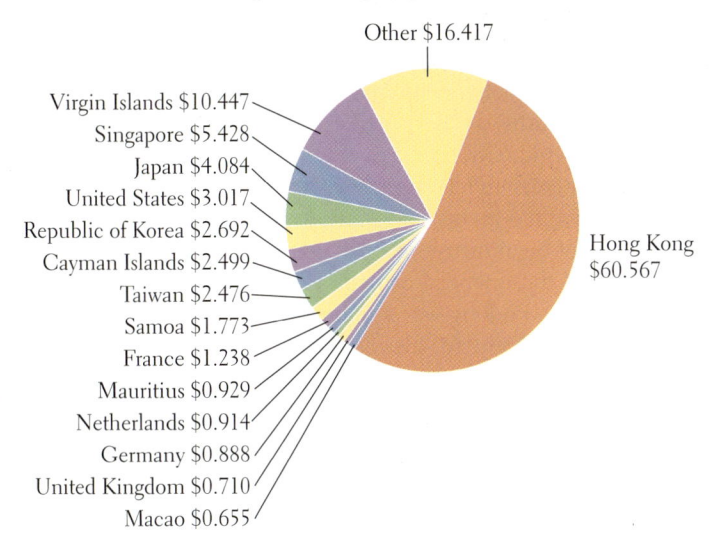

FIGURE 9.22 Foreign investment capital in China, 2010. Foreign investment in China reached nearly $115 billion in 2010, and more than $124 billion in 2011. Note that investment from Hong Kong, even though it is part of China, is considered "foreign" here. Many investments that actually originate in various countries across the globe are funneled through Hong Kong banks, as was the case when Hong Kong was a British territory. The same holds true for small banking centers such as in Taiwan, the Virgin Islands, and the Cayman Islands. [Source consulted: Invest in China, February 7, 2012, at http://www.fdi.gov.cn/1800000121_10000041_8.html]

Hong Kong's Special Role

Hong Kong, which was a British possession from the end of the first Opium War in 1842 until 1997, has long had a special role as a link to the global economy for China. Before 1997, some 60 percent of foreign investment in China was funneled through Hong Kong, and since then Hong Kong has remained the financial hub for China's booming southeastern coast (**Figure 9.22**).

Hong Kong is one of the most densely populated cities in the world; its 7.2 million people are packed into only 23 square miles (60 square kilometers), an area roughly the same size as the borough of Manhattan in New York City but with four times as many people. Hong Kong is also one of the richest cities in the world: its per capita GNI (adjusted for PPP) in 2013 was similar to that of the United States, at U.S.$53,050. Hong Kong also has the world's third-largest container port (after those of Shanghai and Singapore).

In July 1997, Britain's 99-year lease of Hong Kong ran out and Hong Kong became a special administrative region (SAR) of China. Hong Kong still operates under a political and legal system established by the British that allows for more political freedom. Even so, over a million residents, many of them quite wealthy, left before China took over in 1997, worried that their fortunes and freedoms would diminish. China allows more political freedom in Hong Kong, in part because of the island's important role as a link to global trade. Given the obvious success of many cities along China's southeastern coast, from Macao to Shanghai, Hong Kong will probably continue to be a financial hub as this very rapidly growing region develops. **220. AS HONG KONG ENTERS SECOND DECADE UNDER CHINA, CITY PONDERS PLACE IN ECONOMIC GIANT**

Shanghai's Latest Transformation Shanghai has historically been a trendsetting city. Its opening to Western trade in the early nineteenth century spawned a period of phenomenal economic growth and cultural development that led to it being called the "Paris of the East." As a result of China's recent reentry into the global economy, Shanghai is undergoing another boom, which has enriched some people and dislocated others (see Figure 9.20A). On average, residents of Shanghai are now as affluent as people in Central European countries like Poland and Hungary, and if current growth trends continue, their wealth will soon be on par with that of residents of Portugal or Slovenia. Shanghai today has the world's busiest cargo port, and the region around the city is now responsible for as much as a quarter of China's GDP.

In less than a decade, the city's urban landscape has been remade by the construction of more than a thousand business and residential skyscrapers; subway lines and stations; highway overpasses; bridges; and tunnels. Shanghai's shopping district on

Nanjing Road is as imposing as any such district around the world. For hundreds of miles into the countryside, suburban development linked to Shanghai's economic boom is gobbling up farmland, and displaced farmers have rioted.

Pudong, the city's new financial center, sits across the Huangpu River from the Bund—Shanghai's famous, elegant row of big brownstone buildings that served as the financial capital of China until half a century ago. Previously, Pudong was a maze of dirt paths and sprawling neighborhoods of simple, tile-roofed houses, but its former residents were pushed out to make way for soaring high-rises. New construction is restricted in the historic section of the Bund, which is another reason why the centrality of Pudong has been attractive as a new business center. The Oriental Pearl Tower has become a visual symbol for the economic vitality of Pudong, the city of Shanghai, and all of China.

ON THE BRIGHT SIDE

Shanghai's Pajama Culture

Shanghai's long-standing role as a window to the outside world has meant that it often is host to behavior that is less common elsewhere in the country. For example, for years the people of Shanghai have relaxed in public in their pajamas—light, loose cotton tops and bottoms stamped with images of puppies or butterflies. The city government has tried to squelch the custom, but the citizens have proved recalcitrant. The police seem to understand that images of them arresting old and young alike for wearing pajamas would be ridiculous, especially in the international media, so for now the issue is unresolved.

THINGS TO REMEMBER

> **GEOGRAPHIC INSIGHT 4**
> • **Urbanization** Across East Asia, cities have grown rapidly over the last century, fueled by export-oriented manufacturing industries. China has recently undergone the most massive and rapid urbanization in the history of the world. Its urban population, now more than 720 million people, has tripled since China initiated economic reforms in the 1980s.

• China's focus on export-oriented manufacturing based in urban areas has led to rural areas lagging behind cities in access to jobs and income, education, and medical care.

• China's SEZs and ETDZs have been spectacularly successful, becoming major growth poles that draw investment and migration.

• In 2013, China reached a major milestone that every rapidly developing country eventually reaches: its booming service sector now creates more jobs than its industrial sector.

• A British possession from the end of the first Opium War in 1842 until 1997, Hong Kong has long had a special role as a link to the global economy for China.

• Shanghai today has the world's busiest cargo port, and the region around the city is now responsible for as much as a quarter of China's GDP. ■

POPULATION AND GENDER

> **GEOGRAPHIC INSIGHT 5**
>
> **Population and Gender:** Although East Asia remains the most populous world region, families here are having far fewer children than in the past, resulting in populations that are aging. Meanwhile, the legacy of China's now largely abandoned "one child" policy, combined with an enduring cultural preference in China for male children, has created a shortage of females.

Of all world regions, only Europe has a lower rate of natural increase than that of East Asia (less than 0.0 percent increase per year, compared to East Asia's 0.4 percent). In China, this is partially due to the legacy of government policies that harshly penalized families for having more than one child. But in China as well as elsewhere, urbanization and changing gender roles are also resulting in smaller families, regardless of official policy. Only in the two poorest countries—Mongolia (with 2.9 million people) and North Korea (with 24.6 million)—are women still averaging two or more children each, but even there family size is shrinking.

Responding to an Aging Population

Low birth rates mean that fewer young people are being added to the population, and improved living conditions mean that people are living longer across East Asia. The overall effect is that the average age of the populations is rising. Put another way, populations are aging. For Mongolia, South Korea, North Korea, and Taiwan, it will be several decades before the financial and social costs of supporting numerous elderly people will have to be addressed. China faces especially serious future problems with elder care because the one-child policy and urbanization have so drastically reduced family kin-groups. Japan, on the other hand, has already been dealing with the problem of having a large elderly population that requires support and a reduced number of young people to do the job.

Japan's Options Japan's population is growing slowly and aging rapidly, raising concerns about economic productivity and humane ways to care for dependent people. The demographic transition (see page 48 in Chapter 1) is well underway in this highly developed country, where 86 percent of the population lives in cities. Japan has a negative rate of natural increase (−0.2), the lowest in East Asia and on par with that of Europe. If this trend continues, Japan's population is projected to plummet from the current 128 million to 95.5 million by 2050.

At the same time, the Japanese have the world's longest life expectancy, at 83 years (Figure 9.23). As a result, Japan also has the world's oldest population, 24 percent of which is over the age of 65. By 2055, this age group will account for approximately 40 percent of the population. By 2050, Japan's labor pool could be reduced by more than a third, but it would still need to produce enough to take care of more than twice as many retirees as it does now. Clearly, these demographic changes will have a momentous effect on Japan's economy, and the search for solutions is underway. One possibility is increasing immigration to bring in

FIGURE 9.23 Japan's aging population. An elderly woman in an adult day-care center on Gogo Island, Japan. Japan has the world's longest life expectancy (83 years) and also the world's oldest population: 24 percent of its people are over the age of 65.

younger workers who will fill jobs and contribute to the tax rolls, as the United States and Europe have done.

In Japan, recruiting immigrant workers from other countries is a very unpopular solution to the aging crisis. Many Japanese people object to the presence of foreigners, and the few small minority populations with cultural connections to China or Korea have for years faced discrimination. The children of foreigners born in Japan are not granted citizenship, and some who have been in the country for generations are still thought of as foreign. Today, immigrants must carry an "alien registration card" at all times.

Nonetheless, foreign workers are dribbling into Japan in a multitude of legal and illegal ways. Many are so-called guest workers from South Asia brought in to fill the most dangerous and lowest-paying jobs, with the understanding that they will eventually leave. Others are the descendants of Japanese people who once migrated to South America (Brazil and Peru). Regardless, Japan's foreign population remains tiny, making up only 1.7 percent (2.2 million people) of the total population. A recent UN report estimates that Japan would have to admit more than 640,000 immigrants per year just to maintain its present workforce and avoid a 6.7 percent annual drop in its GDP.

In a novel approach to Japan's demographic changes, the government has invested enormous sums of money in robotics over the past decade. Robots are already widespread in Japanese industries such as auto manufacturing, and their industrial use is growing. Now they are also being developed to care for the elderly, to guide patients through hospitals, to look after children, and even to make sushi. By 2025, the government plans to replace up to 15 percent of Japan's workforce with robots.

China's Options The proportion of China's population over 65 is now only 9 percent, but this will change rapidly as conditions improve and life expectancies increase by 5 to 10 years to become more like those of China's affluent East Asian neighbors. However, a crisis in elder care is already upon China for two other reasons: the high rate of rural-to-urban migration and the shrinkage of family support systems because of the one-child policy (discussed below).

When hundreds of millions of young Chinese people were lured into cities to work, most thought rural areas would benefit from remittances, and this has happened. However, few anticipated that the one-child family would mean that for every migrant, two aging parents would be left to fend for themselves, often in rural, underdeveloped areas. China's parliament passed a law in 2013 that requires family members to visit and support their elderly relatives. While such a law may be unenforceable, it shows how worried the government is about the social consequences of the prospect of an aging society, including the increasing spatial mismatch between the elderly who need care and their younger relatives who often have moved elsewhere in search of employment.

The Legacies of China's One-Child Policy

In response to fears about overpopulation and environmental stress, China had a one-child-per-family policy from 1979 to 2013. Since 2013, when China also instituted broader economic reforms, couples have been allowed to have two children, so long as one of the people in the couple is an only child, which is the case in most couples. The policy was enforced with rewards for complying and with large fines for not complying. As a result, China's rate of natural increase (0.5 percent) is the same as that of the United States, and less than half the world average (1.2 percent). The policy was relaxed because of growing concern about the forecasted shrinking of China's population, which should begin sometime between 2025 and 2050 (Figure 9.24B), creating many of the same economic and social problems that Japan is now facing. The one-child policy will be abandoned completely by 2020.

The one-child policy has transformed Chinese families and Chinese society as a whole. For example, an only child has no siblings, so within two generations, the kinship categories of brother, sister, cousin, aunt, and uncle have disappeared from most families, meaning that any individual has very few, if any, related age peers with whom to share family responsibilities. Most children are doted on by several adults and children are not taught to share by their siblings. Conscious efforts must be made to instill self-sufficiency in only children.

The one-child policy has sometimes been enforced brutally. At various times and in places, the government has waged a campaign of forced sterilizations and forced abortions for mothers who already have one child.

Gender Imbalance and the Cultural Preference for Sons: Missing Females and Lonely Males The one-child policy, combined with an ancient cultural preference for male children, resulted in a severe gender imbalance. For many couples, the prospect of their only child being a daughter, without the possibility of having a son in the future, was devastating. The preference for sons relates to deeply patriarchal Confucian values that have prevailed throughout this region for millennia (see page 376).

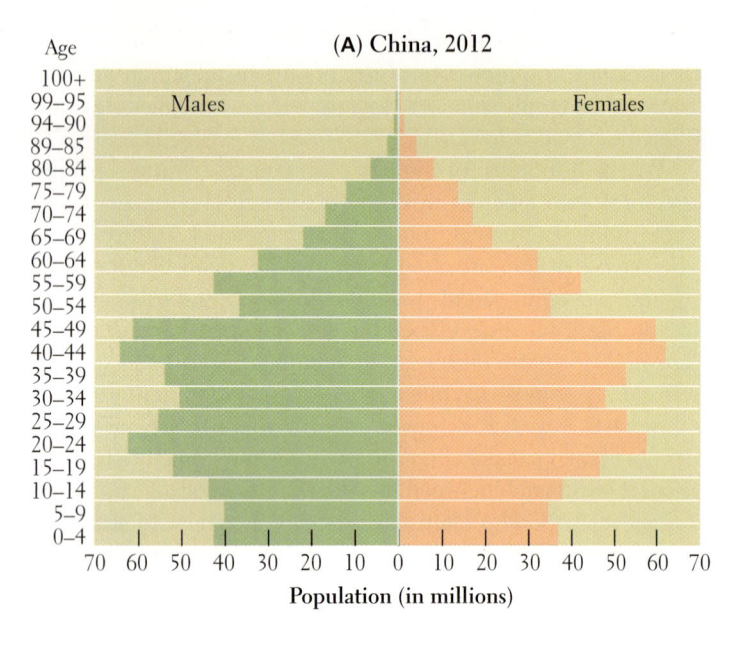

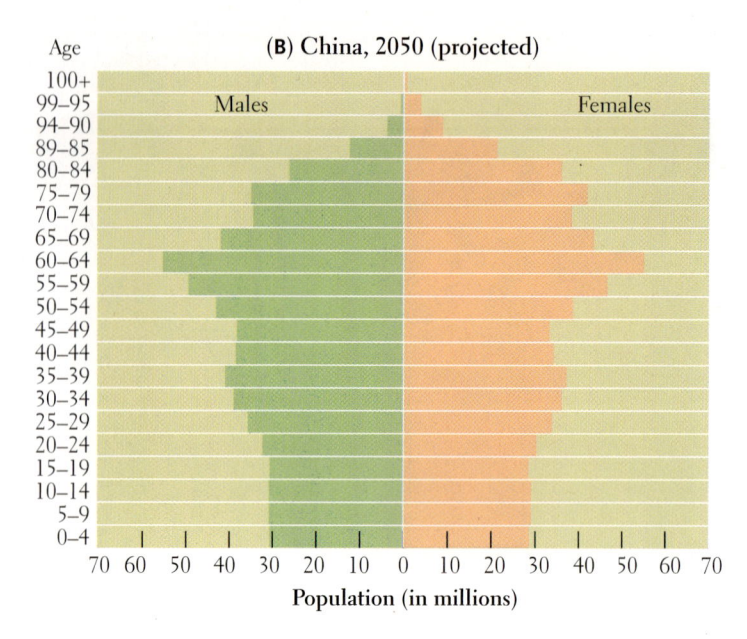

FIGURE 9.24 Population pyramids for China, 2012 and 2050 (projected). [Source consulted: *International Data Base*, U.S. Census Bureau, 2012, at http://www.census.gov/population/international/data/idb/informationGateway.php]

Indeed, gender imbalance has emerged in the Koreas and Taiwan even without the one-child policy, which suggests that the influence of Confucian values is strong.

China's more severe gender imbalance is illustrated by its population pyramid (Figure 9.24A). The average global sex ratio at birth is 105 boys to 100 girls, which evens out to 101 boys to 100 girls by the age of 5. In China, however, there are 113 boys for every 100 girls born. In fact, for nearly every category until age 70, males outnumber females. The census data show that there are already 50 million more men than women.

What happened to the missing girls? There are several possible answers. Given the preference for male children, the births of these girls may simply have gone unreported by families hoping to conceal their daughters as they tried to conceive a male child. There are many anecdotes of girls being raised secretly or even disguised as boys. Also, adoption records indicate that girls are given up for international adoption much more often than boys are. Or the girls may have died in early infancy, either through neglect or infanticide. Finally, some parents have access to medical tests that can identify the sex of a fetus. There is evidence that in China, as elsewhere around the world, many parents choose to abort female fetuses.

There is some evidence that attitudes may be changing. For example, in Japan, South Korea, and Mongolia, the percentage of women receiving secondary education equals or exceeds that of men. In Japan, there is a slight gender imbalance *in favor of females*. In China, the makers of social policy have tried for decades to eliminate the old Confucian bias toward men by empowering women economically and socially. In some ways, they are succeeding, as Chinese women now participate in the workforce to a large degree. Nevertheless, the preference for sons persists.

A Shortage of Brides A major side effect of the preference for sons is that there is now a growing shortage of women of marriageable age throughout East Asia. In 2012, China alone had an estimated deficit of 11 million women aged 20 to 35. Females are also effectively "missing" from the marriage rolls because many educated young women are too busy with career success to meet eligible young men.

Research suggests that at least 10 percent of young Chinese men will fail to find a mate; and poor, rural, uneducated men will have the most difficulty. The shortage of women will lower the birth rate yet further, which will contribute to the expected shrinkage of the population over the next century. Without spouses, children, or even siblings, there will be no one to care for single men when they age. Furthermore, China's growing millions of single young men are emerging as a potential threat to civil order, as they may be more prone to drug abuse, violent crime, HIV infection, and sex crimes. Cases of kidnapping and forced prostitution of young girls and women are already increasing.

Population Distribution

In East Asia, people are not evenly distributed across the land (Figure 9.25). China, with 1.35 billion people, has more than one-fifth of the world's population. However, 90 percent of these people are clustered on the approximately one-sixth of the total land area that is suitable for agriculture, and roughly half of these live in urban areas. The population is concentrated especially densely in the eastern third of China in the North China Plain, the coastal zone from Tianjin to Hong Kong that includes the delta of the Zhu Jiang (Pearl River) in the southeast, the Sichuan Basin, and the middle and lower Chang Jiang (Yangtze) basin.

The west and south of the Korean Peninsula are also densely settled, as are northern and western Taiwan. In Japan, settlement is concentrated in a band that stretches from the cities of

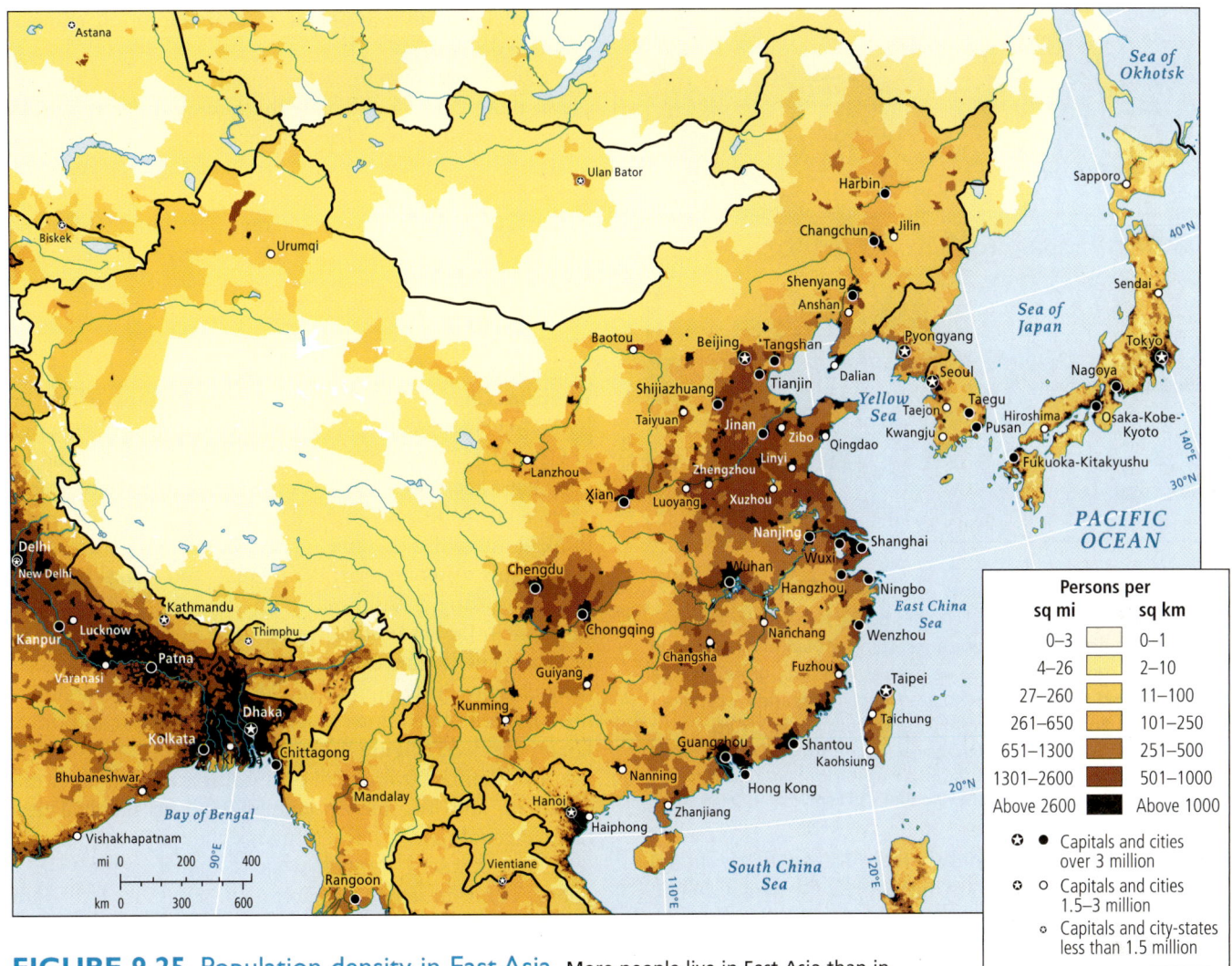

FIGURE 9.25 Population density in East Asia. More people live in East Asia than in any other world region, but population growth in East Asia is now slowing as the size of families is decreasing. Most of the region now faces the challenge of caring for large elderly populations. China is grappling also with an unexpected consequence of its one-child-per-family policy: a shortage of women.

Tokyo and Yokohama on the island of Honshu, south through the coastal zones of the Inland Sea to the islands of Shikoku and Kyushu. This urbanized region is one of the most extensive and heavily populated metropolitan zones in the world, accommodating 86 percent of Japan's total population. The rest of Japan is mountainous and more lightly settled. Mongolia is only lightly settled, with one modest urban area.

THINGS TO REMEMBER

> **GEOGRAPHIC INSIGHT 5**

- **Population and Gender** Although East Asia remains the most populous world region, families here are having far fewer children than in the past, resulting in populations that are aging. Meanwhile, the legacy of China's now largely abandoned "one child" policy, combined with an enduring cultural preference in China for male children, has created a shortage of females.

- The Japanese have the world's longest life expectancy, at 83 years, and the world's oldest population—24 percent of Japanese people are over the age of 65. By 2055, this age group is projected to account for 40 percent of the population.

- China faces a crisis in elder care for two reasons: the high rate of rural-to-urban migration and the shrinkage of family support systems because of the legacy of the one-child policy.

- Research suggests that at least 10 percent of young Chinese men will fail to find a mate; and poor, rural, uneducated men will have the most difficulty. ∎

SOCIOCULTURAL ISSUES

Most countries in East Asia have one dominant ethnic group, but all countries have considerable cultural diversity. In China, for example, 93 percent of Chinese citizens call themselves "people of the Han." The name harks back about 2000 years to the Han

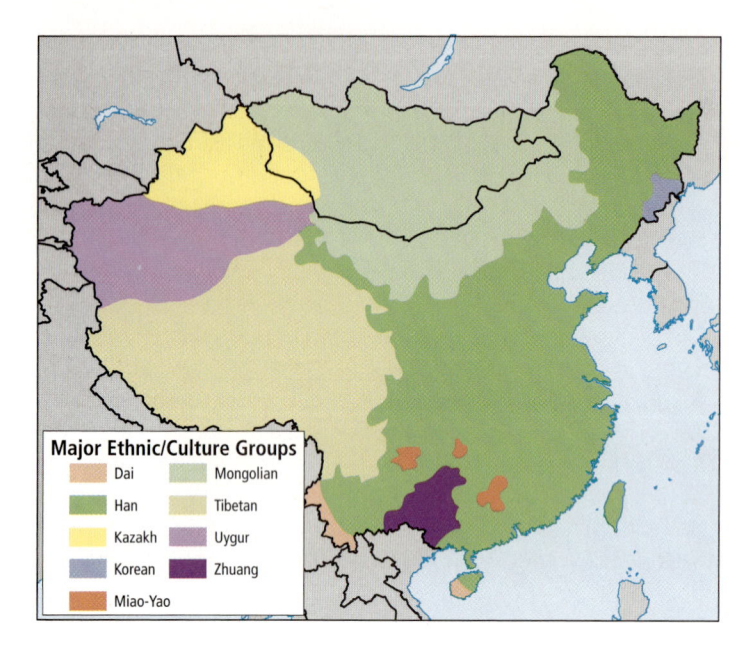

FIGURE 9.26 Major ethnic groups of China. This map shows the areas traditionally occupied by the Han and by ethnic minorities. It does not show the recent resettling of Han in Xinjiang and Tibet, nor does it show the Hui, people from many ethnic groups whose ancestors converted to Islam and who are found in disparate locations along the old Silk Road and in coastal southeastern China. [Source consulted: Chiao-min Hsieh and Jean Kan Hsieh, *China: A Provincial Atlas* (New York: Macmillan, 1995), p. 12. The Web site http://www.index-china.com/minority/minorityenglish.htm includes a comprehensive survey of minorities in China.]

empire but it gained currency only in the early twentieth century, when nationalist leaders were trying to create a mass Chinese identity. The term *Han* simply connotes people who share a general way of life and pride in Chinese culture. The main language spoken by the Han is Mandarin, although it is only one of many Chinese dialects.

China's non-Han minorities number about 117 million people in more than 55 different ethnic or culture groups scattered across the country. Most live outside the Han heartland of eastern China (Figure 9.26). Some of these areas have been designated *autonomous regions*, where minorities theoretically manage their own affairs. In practice, however, the Communist Party in Beijing controls the fate of these regions, especially those considered to have potentially rebellious populations and those that have resources of economic value. We profile a few of China's ethnic groups here.

Western China's Muslims Muslims of various ethnic origins have long been prominent minorities in China. All are originally of Central Asian origin and most are Turkic people who historically have been nomadic herders. Others specialized in trading. They tend to be concentrated in China's northwest and often think of themselves as quite separate from mainstream China.

Uygurs (pronounced WEE-gurs) and Kazakhs, who are Turkic-speaking Muslims, live in the autonomous region of Xinjiang in the far northwest (see the Figure 9.1 map). Though relatively few Uygurs or Kazakhs remain nomadic herders today, contact between them and similar peoples in Central Asia has been revived since China's market reforms began and the Soviet Union dissolved (Figure 9.27).

The Beijing government claims this area's oil, other mineral resources, and irrigable agricultural land for national development. Accordingly, it has sent troops and many millions of Han settlers to Xinjiang through what is known as the "Go West" policy. The Han settlers fill most managerial jobs in the bureaucracy, mineral extraction, the military, and power generation. An important secondary role of the Han is to dilute the power of Uygurs and Kazakhs within their own lands. In Xinjiang, there are now almost as many Han as Uygurs (8 versus 9 million), plus small numbers of other minorities.

The Beijing government has rushed to develop Xinjiang and its capital Urumqi by making special development zones (ETDZs) in the area, but the Uygurs have been left out of most policy-making roles and indeed have been excluded from participating in the economic boom. One young Uygur man in Urumqi writes of his discontent: "I am a strong man, and well-educated. But [Han] Chinese firms won't give me a job. Yet go down to the railroad station and you can see all the [Han] Chinese who've just arrived. They'll get jobs. It's a policy to swamp us."

Until recently, the Uygur people of Xinjiang expressed their resistance to Han dominance merely by reinvigorating their Islamic culture. Islamic prayers were increasingly heard publicly, more Muslim women were wearing Islamic dress, Uygur was spoken rather than Chinese, and Islamic architectural traditions were being revived. Then more active resistance groups, formed by Uygur separatists, began carrying out attacks on Chinese targets. In 2009, violence erupted in the streets of Urumqi between Uygurs and the Han, and about 200 people were killed. The Beijing government responded by harshly punishing the rioters and broadcasting the accusation that all Uygur separatists, even those committed to

FIGURE 9.27 The Silk Road market in Kashi. A Uygur man tests a horse before selling it at the Sunday market in Kashi, a city in China's Xinjiang Uygur Autonomous Region. Kashi has been an important trading center along the Silk Road for at least 2000 years.

nonviolence, are Islamic fundamentalists bent on terrorism. In 2013, China's government began arresting nonviolent Uygur Internet-based activists.

Distinct from the Uygurs and Kazakhs are the Hui, who altogether number about 10 million. The original Hui people were descended from ancient Turkic Muslim traders who traveled the Silk Road from Europe across Central Asia to Kashgar (now Kashi) and on to Xian in the east (see the map of the Silk Road in Figure 5.9 on page 205; see also Figure 9.27). Subgroups of Hui live in the Ningxia Huizu Autonomous Region and throughout northern, western, and southwestern China. There they continue as traders, farmers, and artisans. The tradition of commercial activity and adoption of the Chinese language among the Hui has facilitated their success in China. Many are active in the new free market economies of southeastern China as businesspeople, technicians, and financial managers, using their money not just to buy luxury goods but also to revive religious instruction and to fund their mosques, which are now more obvious in the landscape.

The Tibetans The Tibetans, in contrast to the prosperous and assimilated Hui, are an impoverished ethnic minority group of nearly 5 million people scattered thinly over a huge, high, mountainous region in western China. The history of Tibet's political status vis-à-vis China is long and complex, characterized by both cordial relations and conflict. During China's imperial era (prior to the twentieth century), Tibet maintained its own government but faced the constant threat of invasion and of Chinese meddling in its affairs. In the early 1900s, Tibet declared itself separate and free from China and conducted its affairs as an independent country. It was able to maintain this status until 1949–1950, when the Chinese Communist army "liberated" Tibet, promising a "one-country, two-systems" structure, which suggested a great deal of regional self-governance. A Tibetan uprising in 1959 led China to abolish the Tibetan government and violently reorder Tibetan society. Since the 1950s, the Chinese government has referred to Tibet as the Xizang Autonomous Region. The Chinese government suppressed Tibetan Buddhism—a religion that Tibetans rally behind as a symbol of independence—by destroying thousands of temples and

FIGURE 9.28 The Beijing-Tibet Railway. The western section of this railroad opened in 2006. It is the world's highest railway, with nearly 600 miles (1000 kilometers) of line that are at an altitude above 13,000 feet (4000 meters). The trip from Beijing to Lhasa now takes less than 48 hours, and it costs much less to transport goods and people via the railway than it does using the existing highway. However, many Tibetans worry that the railway will Tibet become a conduit for more Han migration to Tibet. [Source consulted: "Travel and Tours," http://www.chinatibettrain.com/index.html.]

(A) Han Chinese passengers aboard the train combat altitude sickness by sleeping and using supplemental oxygen supplied to every passenger.

(B) A view of the train as it passes through the mountains of the Tibetan Plateau.

monasteries and massacring many thousands of monks and nuns. In 1959, the spiritual and political leader of Tibet, the Dalai Lama, was forced into exile in India along with thousands of his followers.

216. DALAI LAMA CALLS FOR AUTONOMY BUT NOT INDEPENDENCE

By the 1990s, the Beijing government's strategy was to overwhelm the Tibetans with secular social and economic modernization and with Han Chinese settlers rather than outright military force (though China maintains a military presence in Tibet). To attract trade and quell foreign criticism of its treatment of Tibetans, China now spends hundreds of millions of dollars on housing and on roads, railroads, and a tourism infrastructure that capitalizes on European and American interest in Tibetan culture. China presents its actions in Tibet as part of its overall strategy to integrate the entire country economically and socially. Schools are being built and jobs opened up to young Tibetans. A new railway link connecting Tibet more conveniently to the rest of China was completed in 2006 (Figure 9.28). The Han in Tibet see the railway as a public service that will promote Tibetan development, but Tibetan activists see it as a conveyor belt for more Han dominance over the Tibetan economy and culture. 214. TIBETAN BUDDHIST NUNS' PROTEST SONGS BRING PUNISHMENT FROM CHINA

Within Tibetan culture, women have held a somewhat higher position than in other cultures of East Asia. Buddhism did introduce many patriarchal attitudes to Tibet, but these did not curtail other more equitable traditions. Among nomadic herders, women could have more than one husband, just as men were free to have more than one wife. At marriage, a husband often joins the wife's family. By comparison, Han Chinese culture has typically regarded the women of Tibet and other western minorities as barbaric precisely because their roles were not circumscribed. They rode horses, they worked alongside the men in herding and agriculture, and they were generally more assertive in daily life than their Chinese counterparts.

Indigenous Diversity in Southern China In Yunnan Province in southern China, more than 20 groups of ancient native peoples live in remote areas of the deeply folded mountains that stretch into Southeast Asia. These groups speak many different languages, and many have cultural and language connections to the indigenous people of Tibet, Burma, Thailand, or Cambodia. Gender relations are different here than among the Han. A crucial difference may be that among several groups, most notably the Dai, a husband moves in with his wife's family at marriage and provides her family with labor or income. A husband inherits from his wife's family rather than from his birth family.

Taiwan's Many Minorities In Taiwan and the adjacent islands, the Han account for 95 percent of the population, but Taiwan is also home to 60 indigenous minorities. Some have cultural characteristics—languages, crafts, and agricultural and hunting customs—that indicate a strong connection to ancient cultures in Southeast Asia and the Pacific. The mountain dwellers among these groups have resisted assimilation more than the plains peoples. Both groups may live on reservations set aside for indigenous minorities if they choose, but most are now being absorbed into mainstream urbanized Han-influenced Taiwanese life. The Han are themselves not homogenous in Taiwan. They are divided into subgroups based on their ancestral home in China, and the people (and their descendants) who fled to Taiwan around the time of the establishment of the People's Republic of China in 1949 are often called "mainlanders."

The Ainu in Japan There are several indigenous minorities in Japan and most have suffered considerable discrimination. A small and distinctive minority group is the **Ainu**, characterized by their light skin, heavy beards, and thick, wavy hair. Now numbering only about 30,000 to 50,000, the Ainu are a racially and culturally distinct group thought to have migrated many thousands of years ago from the northern Asian steppes. They once

FIGURE 9.29 The Ainu of Japan. An Ainu man creates an item to be used in ancestor worship in Tokyo. In 2008, Japan's government formally recognized the Ainu as an indigenous people of Japan, more than 500 years after the Ainu began to be marginalized by mainstream Japanese culture.

Ainu an indigenous cultural minority group in Japan characterized by their light skin, heavy beards, and thick, wavy hair, who are thought to have migrated thousands of years ago from the northern Asian steppes occupied Hokkaido and northern Honshu and lived by hunting, fishing, and some cultivation, but they are now being displaced by forestry and other development activities. Few full-blooded Ainu remain because, despite prejudice, they have been steadily assimilated into the mainstream Japanese population (Figure 9.29). Somewhat belatedly, the Japanese parliament officially recognized the Ainu as an indigenous minority in 2008.

East Asia's Most Influential Cultural Export: The Overseas Chinese

China has had an impact on the rest of the world not only through its global trade, but also through the migration of its people to nearly all corners of the world. The first recorded emigration by the Chinese took place more than 2200 years ago. Following that, China's contacts then spread eastward to Korea and Japan, westward into Central and Southwest Asia via the Silk Road, and by the fifteenth century, to Southeast Asia, coastal India, the Arabian Peninsula, and even Africa.

Trade was probably the first impetus for Chinese emigration. The early merchants, artisans, sailors, and laborers came mainly from China's southeastern coastal provinces. Taking their families with them, some settled permanently on the peninsulas and islands of what are now Indonesia, Thailand, Malaysia, and the Philippines. Today they form a prosperous urban commercial class known as the *Overseas Chinese*. The quintessential Overseas Chinese state in Southeast Asia is Singapore, where 77 percent of the population is ethnically Chinese (see page 440 in Chapter 10).

In the nineteenth century, economic hardship in China and a growing international demand for labor spawned the migration of as many as 10 million Chinese people to countries all over the world. By the middle of the twentieth century, many others fleeing the repression of China's Communist Revolution joined those from earlier migration. As a result, "Chinatowns" are found in most major world cities. The term *Overseas Chinese* has been extended to apply to Chinese emigrants and their descendants in all such locations.

THINGS TO REMEMBER

- There are several distinct minority groups in East Asia: the Uygurs, Kazakhs, and Tibetans in western China; the Hui, scattered throughout central China; and smaller indigenous groups in southern China, Taiwan, and Japan.

- There is resentment and sometimes open resistance to Han Chinese domination in the various minority homelands.

- Throughout East Asia, minorities have experienced discrimination.

- Millions of Overseas Chinese live in cities and towns around the world, especially in Southeast Asia. Most have settled permanently in their foreign locations and many (perhaps most) maintain relationships with China.

GEOGRAPHIC INSIGHTS

East Asia: Review and Self-Test

1. Environment: East Asia's most serious environmental problems result from its high population density combined with its rapid urbanization and environmentally unsustainable economic development. Climate change may intensify the droughts and floods that have long plagued this region.

- Why has East Asia suffered for so many years from enormously destructive droughts and devastating floods? What has been done to protect the population from such hazards?

- How might the melting of China's highest glaciers affect the country's rivers? Which areas might be more affected than others?

- In which ways is China more vulnerable to global warming than Japan, the Koreas, or Taiwan?

2. Globalization and Development: East Asia pioneered a spectacularly successful economic development strategy that has transformed economies across the globe. Governments in Japan and then Taiwan, South Korea, and eventually China intervened strategically in the economy to encourage the production of manufactured goods destined for sale abroad, primarily to the large economies of North America and Europe.

- East Asian countries are able to buy food on the global market but not produce the food they need at home. What problems might this create both for East Asia and the rest of the world?

- What is East Asia's, and especially Japan's, role in the global harvesting of wild-caught fish in the oceans?

3. Power and Politics: As East Asia has developed economically, the pressure for more political freedoms has grown. Japan, South Korea, and Taiwan are now among the more politically free places in the world, and demands for political change in China are increasing, especially in urban areas.

- How is the Internet emerging as a tool to promote the expansion of political freedoms in China?

- How has China's government responded to protests and other efforts to promote political change by Chinese citizens?

- Why do some criticize Japan as lacking a vigorous democracy?

- Why do so many countries claim control of uninhabited islands in or near the waters of East Asia?

4. Urbanization: Across East Asia, cities have grown rapidly over the last century, fueled by export-oriented manufacturing industries. China has recently undergone the most massive and rapid urbanization in the history of the world. Its urban population, now more than 720 million people, has tripled since China initiated economic reforms in the 1980s.

- How has China's focus on export-oriented manufacturing affected rural areas?

- What role did SEZs and ETDZs play in China's economic reforms?

- What major milestone did China's reach in 2013?

- What special role has Hong Kong played in China's transformation?
- What sets Shanghai apart among China's major cities?

5. Population and Gender: Although East Asia remains the most populous world region, families here are having far fewer children than in the past, resulting in populations that are aging. Meanwhile, the legacy of China's now largely abandoned "one child" policy, combined with an enduring cultural preference in China for male children, has created a shortage of females.

- Why do most families want a male child? What problems have been created by the resulting shortage of women?
- What problems of an aging society are arising in Japan that are related to its rapidly aging population? What might be the solutions to such problems?

CRITICAL THINKING QUESTIONS

1. Why is the interior west of continental East Asia (western China and Mongolia) so dry and subject to extremes in temperature?

2. Historically, China has been afflicted with recurring famines. While famine is no longer a concern, China's access to food has changed with economic development. How?

3. How might a less authoritarian political system have resulted in changes in the overall conception and design of China's Three Gorges Dam?

4. Why might a Confucian praise or criticize China's current political system and its approach to economic and political change?

5. In what ways might globalization have taken a different course if the voyages of Admiral Zheng He had inspired China to conquer a vast colonial empire, like that of Great Britain?

6. Contrast Japan's pre–World War II policies in East Asia with its current role in the region. Describe the principal similarities or differences.

7. In what ways has the one-child-per-family policy helped maintain stability in China? How might it affect China in the future in terms of gender balance and economic growth?

8. How has East Asia's spatial pattern of urbanization been shaped by integration into the global economy?

9. What information technologies are now available to the Chinese and what role can such technology have in the democratization of the country?

› THINKING GEOGRAPHICALLY

Now that you have read about East Asia, you should be able to answer the following questions about the photos in this chapter. To answer these questions online, go to Geography LaunchPad.

Vulnerability to Climate Change (page 368)

A What is one process that can result from people beginning to live or farm in dry environments?

B What is causing the water table to fall each year on the North China Plain?

C Flooding is especially likely in what part of China?

Human Impacts on the Biosphere (page 373)

A Which design flaw is most concerning to experts involved with the Three Gorges Dam project?

B Where is the worst air pollution in Taiwan?

C Of all the coal burned in the world each year, how much is burned in China?

D What allows particulates from China's coal burning to be transported globally?

Visual History (page 378)

A What is the model for Confucian philosophy?

B How did the Qin empire influence subsequent Chinese empires?

C The Great Wall was built in response to what?

D Which major Chinese city and port was a British possession from the time of the Opium Wars until recently?

E After the Communist Revolution in China, how did the Chinese government control all aspects of economic and social life?

F How did Japan's political and economic development proceed after World War II?

Power and Politics (page 387)

A What happened in Tiananmen Square in 1989?

C When was democracy in Taiwan established?

D What is North Korea purchasing with money earned from sales of its military technology abroad?

Urbanization (page 391)

A What happened to many of the former residents of Pudong, the new financial center of Shanghai?

C Migrants to urban areas who ignore the hukou system are considered part of what population?

D The urbanized region that stretches from the cities of Tokyo and Yokohama on Honshu south through the coastal zones of the Inland Sea to the islands of Shikoku and Kyushu is home to what percent of Japan's population?

CHAPTER KEY TERMS

Ainu 400
Confucianism 376
Cultural Revolution 379
export-led growth 382
floating population 364
food security 369
Great Leap Forward 379

growth poles 390
hukou system 363
just-in-time system 382
kaizen system 382
regional self-sufficiency 384
regional specialization 384
responsibility system 384

special economic zones (SEZs) 390
state-aided market economy 382
tsunami 365
typhoon 367
wet rice cultivation 370

A Gorges, Burma (Myanmar)

chapter 10
SOUTHEAST ASIA

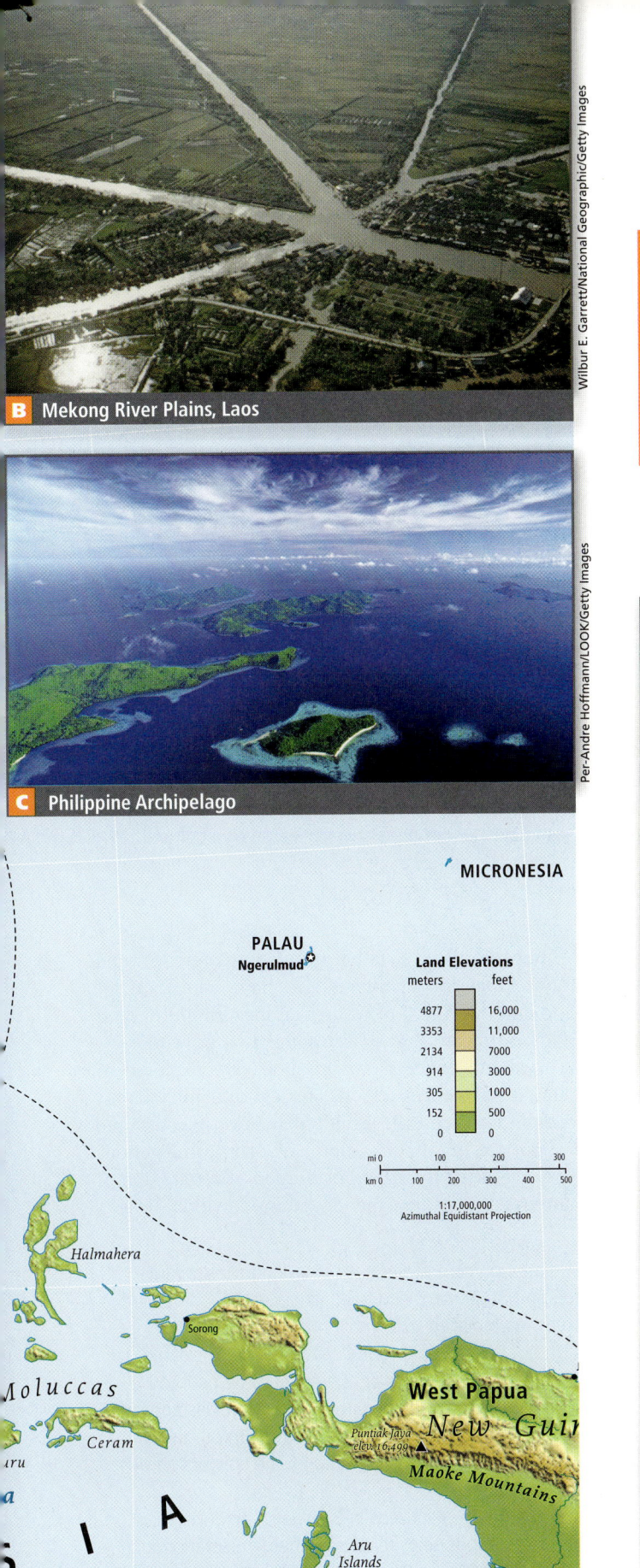

B Mekong River Plains, Laos

C Philippine Archipelago

D Volcanoes, Java, Indonesia

E Post-earthquake/tsunami, Sumatra, Indonesia

FIGURE 10.1 Regional map of Southeast Asia.

405

GEOGRAPHIC INSIGHTS: SOUTHEAST ASIA

After you read this chapter, you will be able to discuss the following geographic insights as they relate to the five thematic concepts:

1. Environment: Many of Southeast Asia's most critical environmental issues relate in some way to climate change. Deforestation is rapid in this region and is a major global source of greenhouse gas emissions, which intensify climate change. This region is also highly vulnerable to the impacts of climate change. Increased flooding and droughts threaten food production on land and rising ocean temperatures strain aquatic ecosystems.

2. Globalization and Development: Globalization has brought both spectacular successes and occasional declines to the economies of Southeast Asia. Key to the economic development of this region are strategies that were pioneered earlier in East Asia: the formation of state-aided market economies combined with export-led economic development.

3. Power and Politics: There has been a general expansion of political freedoms throughout Southeast Asia in recent decades, but authoritarianism, corruption, and violence have at times reversed these gains.

4. Urbanization: While Southeast Asia as a whole is only 43 percent urban, its cities are growing rapidly as agricultural employment declines and urban industries expand. The largest Southeast Asian cities, which are receiving most of the new rural-to-urban migrants, rarely have sufficient housing, water, sanitation, or jobs for all their people.

5. Population and Gender: Population dynamics vary considerably in this region because of differences in economic development, government policies, prescribed gender roles, and religious and cultural practices. With regard to gender, economic change has brought better job opportunities and increased status for women, who then often choose to have fewer children. Some countries also have gender imbalances because of a cultural preference for male children.

The Southeast Asia Region

The region of Southeast Asia (shown in Figure 10.1) has attracted global attention for several decades because so many of its countries emerged poverty-stricken from World War II only to embark on a rapid journey to relative prosperity. There have been ups and downs: difficulties in finding smooth paths to stable democratic processes and institutions; urbanization at too rapid a pace; and a tendency to rely on development strategies that carry grave environmental and social side effects. Some of this region's transitions are now emulated by other developing regions: rapid industrialization, expansion of the middle class, education for the masses, the empowerment of women, improvement of food security, public health care, and slower population growth.

The five thematic concepts in this book are explored as they arise in the discussion of regional issues, with interactions between two or more themes featured. Vignettes, like the one that follows about the rights of a group of people indigenous to the state of Sarawak on the island of Borneo, illustrate one or more of the themes as they are experienced in individual lives.

GLOBAL PATTERNS, LOCAL LIVES

In December 2005, a group of indigenous people in the Malaysian state of Sarawak, on the island of Borneo (see the Figure 10.1 map), attended a public meeting wearing orangutan masks. They carried signs informing onlookers that, although the government protects natural areas for orangutan, it ignores the basic right of indigenous people to live on their own ancestral lands and practice their forest skills (Figure 10.2A).

Over the years, Sarawak forest dwellers have tried many tactics to save their lands from the deforestation that logging companies do in preparation for expanding the oil palm plantations (see Figure 10.2B). Palm oil is a major food and cosmetic oil export; it is sold primarily to Europe, North America, Japan, and China. In the mid-twentieth century, the Sarawak state government began licensing logging companies to cut down tropical forests occupied by indigenous peoples (see Figure 10.3) and export the timber.

By the 1980s, ninety percent of Sarawak's lowland forests had been degraded and 30 percent had been clear-cut. Much of the cleared area was replaced by oil palm plantations. As a consequence, indigenous people found that even on uncleared land, their hunts declined. Streams and rivers became polluted with eroded sediment and by the fertilizers and pesticides applied to palm trees. Virtually none of the profits from palm oil, though, were returned to the communities affected by the conversion of forests to plantations.

In the 1990s, a group of citizens in Berkeley, California, who were concerned about news reports of the deforestation in Sarawak, organized the Borneo Project. They offered to become a "sister city" to one indigenous group in Sarawak, the Uma Bawang, giving help wherever it was needed. With the help of the Borneo Project, the Uma Bawang began a community-based mapping project in 1995. Using rudimentary compass-and-tape techniques, they began mapping both the extent and the biological

FIGURE 10.2 Indigenous people in Sarawak, Borneo.

(A) An indigenous man in Sarawak hunts in the forests of Malaysian Borneo.

(B) Palm oil plantations in Sarawak.

content of their forest home. Since then, indigenous people from across Sarawak have learned how to use global positioning systems (GPS), geographic information systems (GIS), and satellite imagery to make more sophisticated maps. The power of these maps was demonstrated in 2001 when they were used to help win a precedent-setting court case that protected indigenous lands from an encroaching oil palm plantation.

This favorable court ruling was challenged in 2005 when the government appealed, but in 2009, the Malaysian federal court ruled in favor of the Uma Bawang. The decision stated that native customary land rights had been protected since 1939, when British colonial officials had directed the district lands and survey departments to map the boundaries of native lands. Now indigenous people have the right to sue the government for past illegal leases to logging companies, and as of 2011, some 203 such cases were in litigation. The outcome of this litigation is as yet unclear, and in the Indonesian part of Borneo, deforestation for oil palm plantations is proceeding rapidly, as shown in the 2011 video "Indigenous Community Witnesses End of Forest for Palm Oil," at http://tinyurl.com/d5pmr49. [Source: The Borneo Wire. For detailed source information, see Text Sources and Credits.] ∎

This account of the tactics that indigenous groups are using to secure their rights to control and manage their ancestral lands mirrors the accounts of similar struggles in Middle and South America (pages 110–111) and highlights a number of themes in this book: the environment, globalization and development, and power and politics. In particular, it shows how issues that may seem to have only local significance can gain global attention. This connection of the local to the global presents both challenges and opportunities for the world's indigenous people (Figure 10.3). On one hand, the global market provides the demand for timber and palm oil; on the other, it is support from groups like the Borneo Project that help communities such as the Uma Bawang have their land claims validated. In fact, the use of mapping to help indigenous groups secure their legal rights to ancestral lands has now become a global phenomenon. Hundreds of indigenous groups throughout the world are collaborating with mapping specialists, many of them geographers, often resulting in their land rights being formally recognized by governments for the first time. In 2007, the United Nations' Department of Economic and Social Affairs passed the Declaration on the Rights of Indigenous People and issued a report, *State of the World's Indigenous Peoples*, that documents several indigenous community mapping efforts. The report is available online at http://tinyurl.com/ybpfgpp.

What Makes Southeast Asia a Region?

Southeast Asia is physically a manifestation of tectonic forces that are described in the "Physical Patterns" section (see page 409). Aside from physical commonalities, the peninsula and island countries of Southeast Asia today share a deep cultural past, related to but separate from the cultures of southwestern China. With the exception of Thailand, they all also share more recent experiences with European colonialism. During and after World War II, most countries went through severe political turmoil followed by the rapid modernization and industrialization that continues into the present.

408   CHAPTER 10   Southeast Asia

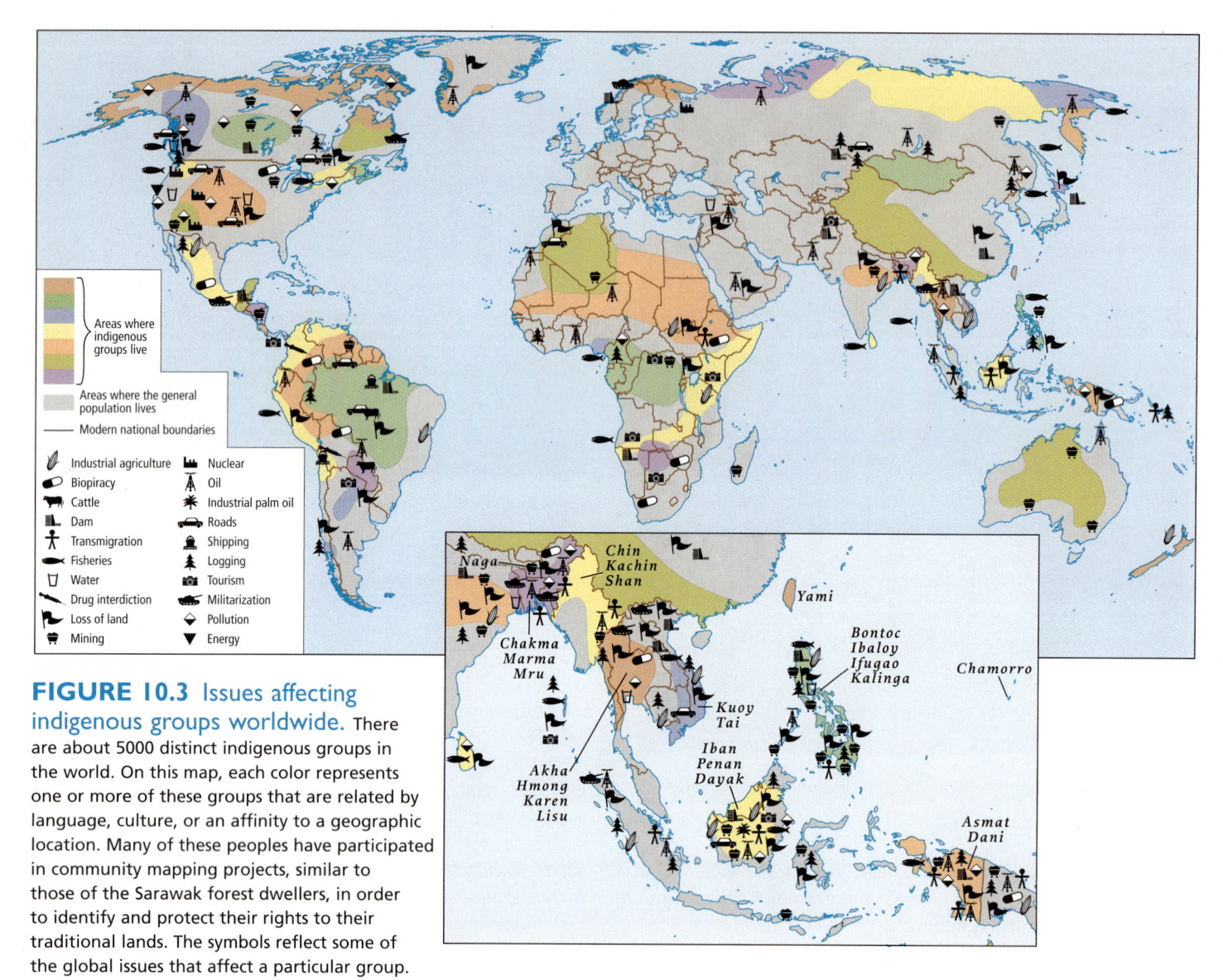

FIGURE 10.3 Issues affecting indigenous groups worldwide. There are about 5000 distinct indigenous groups in the world. On this map, each color represents one or more of these groups that are related by language, culture, or an affinity to a geographic location. Many of these peoples have participated in community mapping projects, similar to those of the Sarawak forest dwellers, in order to identify and protect their rights to their traditional lands. The symbols reflect some of the global issues that affect a particular group.

[Sources consulted: "Struggling Cultures," *National Geographic Atlas of the World*, 8th ed. (Washington, DC: National Geographic Society, 2005), p. 15; "Globalization: Effects on Indigenous Peoples" map, in Jerry Mander and Victoria Tauli-Corpuz, eds., *Paradigm Wars: Indigenous Peoples' Resistance to Economic Globalization* (San Francisco, CA: Sierra Club Books, 2006)]

Terms in This Chapter

The mainland Southeast Asian countries are Burma, Thailand, Laos, Cambodia, Vietnam, Malaysia, and Singapore; the island countries are Indonesia, Brunei, Timor-Leste (East Timor), and the Philippines. Malaysia occupies part of the peninsula and parts of the island of Borneo (Figure 10.4).

Many governments in Southeast Asia choose to dispense with place-names that originated in their colonial past. However, when the governments that make these changes lose respect in the international community by violating the human rights of their citizens, their chosen name may not be acknowledged. Such is the case with Burma (see Figure 10.4), where a military government seized control in a coup d'état in 1990, shortly after an election, and changed the country's name to Myanmar. As discussed later in the chapter, the government of Burma is once again in transition; in press reports, both names are used. In this text, we use the country's traditional name of Burma, not Myanmar.

Another potential point of confusion is Borneo, a large island that is shared by three countries. The part of the island known as Kalimantan is part of Indonesia; while Sarawak and Sabah on the north coast are part of Malaysia; and Brunei is a very small, independent, oil-rich country, also on the north coast.

👉 THINGS TO **REMEMBER**

• Indigenous land claims are no longer just local issues. Indigenous groups have shifted their efforts to secure their rights to ancestral lands to the global scale, highlighting a trend in which many issues that were once local or national are now becoming global concerns. ∎

PHYSICAL GEOGRAPHY AND ENVIRONMENTAL ISSUES

PHYSICAL PATTERNS

The physical patterns of Southeast Asia have a continuity that is not immediately obvious on a map. A map of the region shows a unified mainland region that is part of the Eurasian continent and a vast and complex series of islands arranged in chains and groups. These landforms are actually related in tectonic origin. Climate is another source of continuity; most of the region is tropical or subtropical.

Landforms

Southeast Asia is a region of peninsulas and islands (see Figure 10.1). Although the region stretches over an area larger than the continental United States, most of that space is ocean; the area of all the region's landmass amounts to less than half that of the contiguous United States. Burma, Thailand, Laos, Cambodia, and Vietnam occupy the large Indochina peninsula that extends south of China. This peninsula itself sprouts the long, thin Malay Peninsula that is shared by outlying parts of Burma and Thailand, a main part of Malaysia, and the city-state of Singapore, which is built on a series of islands at the southern tip. The **archipelago** (a series of large and small islands) that fans out to the south and east of the mainland is grouped into the countries of Indonesia, Malaysia, Brunei, Timor-Leste (East Timor), and the Philippines (see Figure 10.1C). Indonesia alone has some 17,000 islands, and the Philippines, 7000.

> **archipelago** a group, often a chain, of islands

The irregular shapes and landforms of the Southeast Asian mainland and archipelago are the result of the same tectonic forces that were unleashed when India split off from the African Plate and gradually collided with Eurasia (see Figure 1.8 on page 16). As a result of this collision, which is still underway, the mountainous folds of the Plateau of Tibet reach heights of almost 20,000 feet (6100 meters). These folded landforms, which bend out of the high plateau, turn south into Southeast Asia. There, they descend rapidly and then fan out to become the Indochina peninsula. Deep gorges widen out into valleys that stretch toward the sea, each containing hills of 2000 to 3000 feet (600 to 900 meters) and a river or two flowing from the mountains of the Yunnan–Guizhou Plateau of China to the Andaman Sea, the Gulf of Thailand, and the South China Sea (see Figure 10.1A and the figure map). The major rivers of the Indochina peninsula are the Irrawaddy and the Salween in Burma; the Chao Phraya in Thailand; the Mekong, which flows through Laos, Cambodia, and Vietnam; and the Black and Red rivers of northern Vietnam. Several of these rivers have major delta formations—especially the Irrawaddy, Chao Phraya, and the Mekong—that are intensively cultivated and settled (see the discussion on pages 417–418; see also Figure 10.1B).

The curve formed by Sumatra, Java, the Lesser Sunda Islands (from Bali to Timor), and New Guinea conforms approximately to the shape of the Eurasian Plate's leading southern edge (see Figure 1.8 on page 16). Where the Indian-Australian Plate plunges beneath the Eurasian Plate along this curve, hundreds of earthquakes occur and volcanoes abound, especially on the islands of Sumatra and Java (see Figure 10.1D, E). There are also volcanoes and earthquakes in the Philippines where the Philippine Plate pushes against the eastern edge of the Eurasian Plate and is in turn pushed by the Pacific Plate. The volcanoes of Southeast Asia are considered part of the Pacific Ring of Fire (see Figure 1.9 on page 17; see also Figure 10.1C, D).

Volcanic eruptions and the mudflows and landslides that follow eruptions complicate and endanger the lives of many Southeast Asians. Earthquakes are especially problematic because of the tsunamis they can set off (see Figure 10.1E). The tsunami of December 2004, triggered by a giant earthquake (9.1 in magnitude) just north of Sumatra, swept east and west across the Indian Ocean, taking the lives of 230,000 people (170,000 in Aceh Province, Sumatra) and injuring many more. It was one of the deadliest natural disasters in recorded history. There has been a series of strong earthquakes since, with several along coastal Sumatra in August and September of 2009, April and May 2010, and April of 2012; but they did not generate notable tsunamis. Over the long run, the volcanic material creates new land and provides minerals that enrich the soil for farming.

The now-submerged shelf of the Eurasian continent, known as Sundaland (Figure 10.5) extends under the Southeast Asian peninsulas and islands. It was above sea level during the recurring ice ages of the Pleistocene epoch, when much of the world's water was frozen in glaciers. The exposed shelf allowed ancient people and Asian land animals (such as elephants, tigers, rhinoceroses, and orangutans; Figure 10.6) to travel south to what became the islands of Southeast Asia when sea levels rose.

FIGURE 10.4 Political map of Southeast Asia.

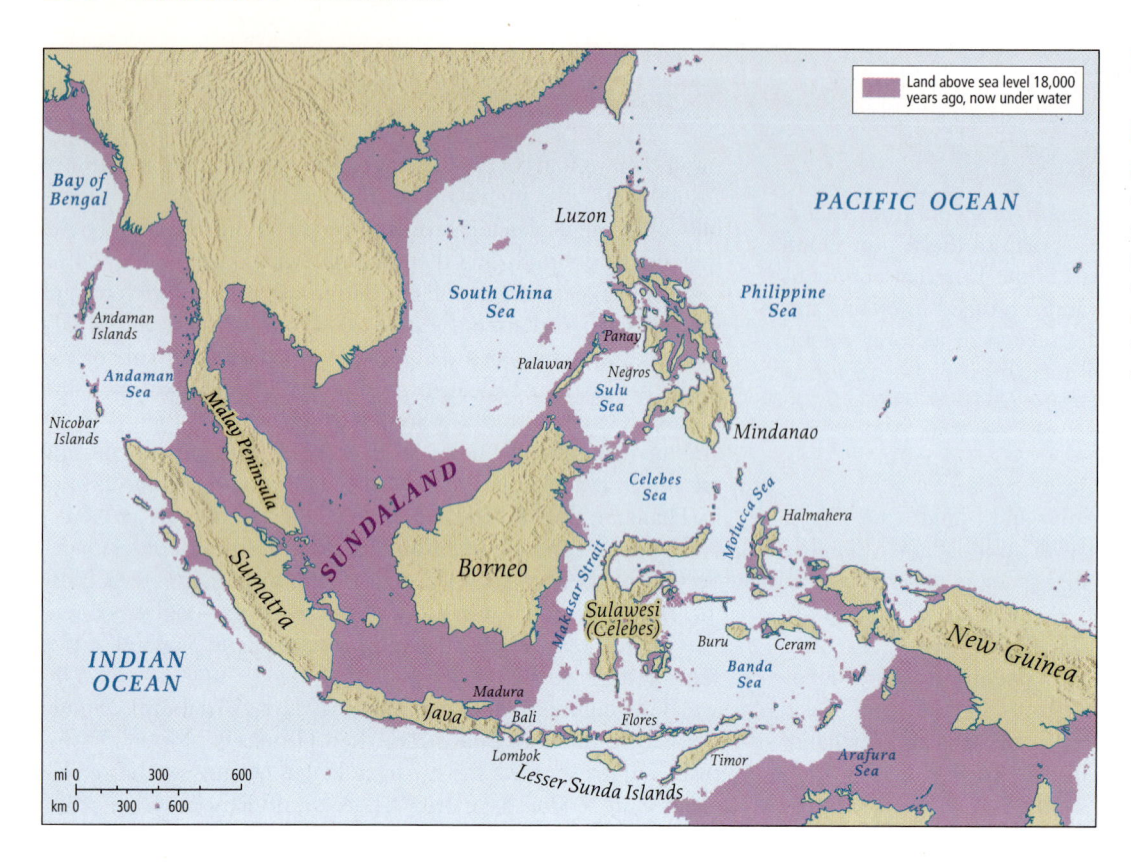

FIGURE 10.5 Sundaland 18,000 years ago, at the height of the last ice age. The now-submerged shelf of the Eurasian continent that extends under Southeast Asia's peninsulas and islands was exposed during the last ice age and remained above sea level until about 16,000 years ago, when that ice age was ending.

Climate and Vegetation

The largely tropical climate of Southeast Asia is distinguished by continuous warm temperatures in the lowlands—consistently above 65°F (18°C)—and heavy rain (Figure 10.7). The rainfall is the result of two major processes: the monsoons (seasonally shifting winds) and the movement of the intertropical convergence zone (ITCZ), an area centered roughly at the equator, where surface winds converge from the northern and southern hemispheres and rise upward, resulting in rainfall. The wet summer season extends from May to October, when the warming of the Eurasian landmass sucks in moist air from the surrounding seas and pulls the ITCZ northward (see the Figure 10.7 map). Between November and April, there is a long dry season on the mainland, when the seasonal cooling of Eurasia causes dry, cool air from the interior of the continent to flow out toward the sea, pushing the ITCZ southward (see Figure 11.5 on page 452). On the many islands, however, the winter can also be wet because the air that flows from the continent warms and picks up moisture as it passes south and east over the seas. The air releases its moisture as rain after ascending high enough over elevated landforms to cool. With rains coming from both the monsoon and the ITCZ, the island part of Southeast Asia is one of the wettest areas of the world.

Irregularly every 2 to 7 years, the normal patterns of rainfall are interrupted, especially in the islands, by the El Niño phenomenon (see Figure 11.6 on page 453). In an El Niño event, the usual patterns of air and water circulation in the Pacific are reversed. Ocean temperatures are cooler than usual in the western Pacific near Southeast Asia. Instead of warm, wet air rising and condensing as rainfall, cool, dry air sits on the ocean surface. The result is severe drought, often with catastrophic effects for farmers and for tinder-dry forests that regularly catch fire. The cool El Niño air can trap smoke and other pollutants at Earth's surface, creating unusually toxic smog and such low visibility that planes sometimes crash.

The soils in Southeast Asia are typical of the tropics. Although not particularly fertile, they will support dense and prolific vegetation when left undisturbed for long periods. The warm temperatures and damp conditions promote the rapid decay of **detritus** (dead organic material) and the quick release of useful minerals. These minerals are taken up directly by the roots of the living forest rather than enriching the soil. Because rainfall is usually abundant during the summer wet season (when drought is only episodic), this region has some of the world's most impressive forests in both tropical and subtropical zones (see Figure 10.7). On the mainland, human interference, coupled with the long winter dry season, has reduced the forest cover. The relationships between humans and the environment that affect forests elsewhere in the region are discussed in the opening vignette and in the next section.

detritus dead organic material (such as plants and insects) that collects on the ground

THINGS TO REMEMBER

- The irregular shapes and landforms of the Southeast Asian mainland and archipelago are the result of the same tectonic forces that were unleashed when India split off from the African Plate and crashed into Eurasia.

- Continuous warm temperatures in the lowlands and heavy rain distinguish the tropical climate of Southeast Asia, except during those irregular years of the El Niño phenomenon.

FIGURE 10.6 LOCAL LIVES
People and Animals in Southeast Asia

A A cat at a temple in Chiang Mai, Thailand. Considered the guardians of statues in temples throughout Thailand, cats are written about in ancient poetry and celebrated by Thailand's royal family. Stray cats are given refuge at certain temples. [John S. Lander/LightRocket/Getty Images]

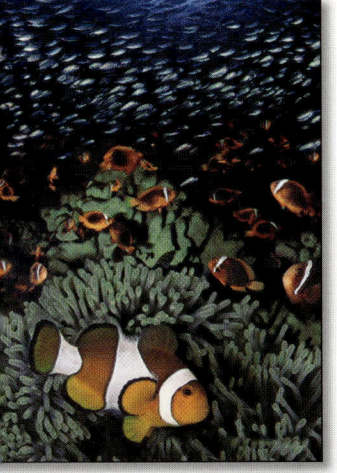

B Clownfish swim amid sea anemones that capture prey with paralyzing toxins. Immune to these toxins, the clownfish eat tiny animals that could harm the anemones; in return, they are provided with a safe home. Clownfish, whose natural habitat is this region's saltwater, have also been successfully bred for aquarium tanks. [Per-Andre Hoffmann/LOOK/Getty Images]

C A pig is decorated and paraded around a village in Vietnam during Tet, the Vietnamese New Year. Pork is especially prized in Vietnam and is an essential part of holiday meals. This pig will be served as an offering to village gods. [Hoang Dinh Nam/AFP/Getty Images]

ENVIRONMENTAL ISSUES

> ### GEOGRAPHIC INSIGHT 1
>
> **Environment:** Many of Southeast Asia's most critical environmental issues relate in some way to climate change. Deforestation is rapid in this region and is a major global source of greenhouse gas emissions, which intensify climate change. This region is also highly vulnerable to the impacts of climate change. Increased flooding and droughts threaten food production on land and rising ocean temperatures strain aquatic ecosystems.

Deforestation

Southeast Asia has the second-highest rate of deforestation of any world region, after sub-Saharan Africa. With only 5 percent of the world's forests, Southeast Asia has accounted for nearly 25 percent of deforestation around the world in the last 10 years. The forces driving deforestation in Southeast Asia are global. Figure 10.8A, B and the figure map show the supply chain that takes wood from various tropical regions to Europe and Asia, where it is turned into furniture and other products. Figures 10.8C, D show who is importing and exporting sawn wood; Figure 10.8E lists the different activities for which the land is deforested legally. It is extremely difficult to know the extent of illegal deforestation. The World Wildlife Fund estimated that more than 40 percent of the timber production in 2006 in Indonesia was illegal.

The regional environmental impacts of deforestation and the agriculture that often follows it are addressed in Figure 10.9 on page 414. One effect is that indigenous people of the forest lose their living space and resources when the land is given over to export agriculture. Another is the loss of habitat for orangutans, the Sumatran tiger, the Sumatran rhinoceros, and tens of thousands of other less well-known species that are displaced when rain forests are converted to agricultural uses. Finally, the emissions of greenhouse gases that accompany deforestation processes contribute to climate change. **222. INDONESIA'S POLLUTED ENVIRONMENT THREATENS HAWKSBILL TURTLE**

Climate Change and Deforestation

Deforestation is a major contributor to global climate change because enormous amounts of carbon dioxide (CO_2) are released when forests are removed and the underbrush is burned. Fewer trees also mean that less CO_2 is absorbed from the atmosphere. Every day, 13 to 19 square miles (34 to 50 square kilometers) of Southeast Asia's rain forests are destroyed, much of it done illegally. A great deal of this deforestation takes place in Malaysia and Indonesia, where the logging of tropical hardwoods for sale on the global market is a major activity (see Figure 10.8).

Significant amounts of CO_2 are also released when the forests are converted to oil palm plantations, especially when the forest is burned to clear the land for the plantations. For decades, these activities have made Indonesia among the world's biggest contributors to global climate change. Since 1950, the tropical forest cover on the Indonesian part of the island of Borneo alone has decreased by about 60 percent. Oil palm plantations now constitute the predominant use of land there.

FIGURE 10.7

PHOTO ESSAY: Climates of Southeast Asia

The presence of so much wet, tropical air, combined with the movements of the intertropical convergence zone (ITCZ) across the region twice a year, makes Southeast Asia one of the wettest regions in the world.

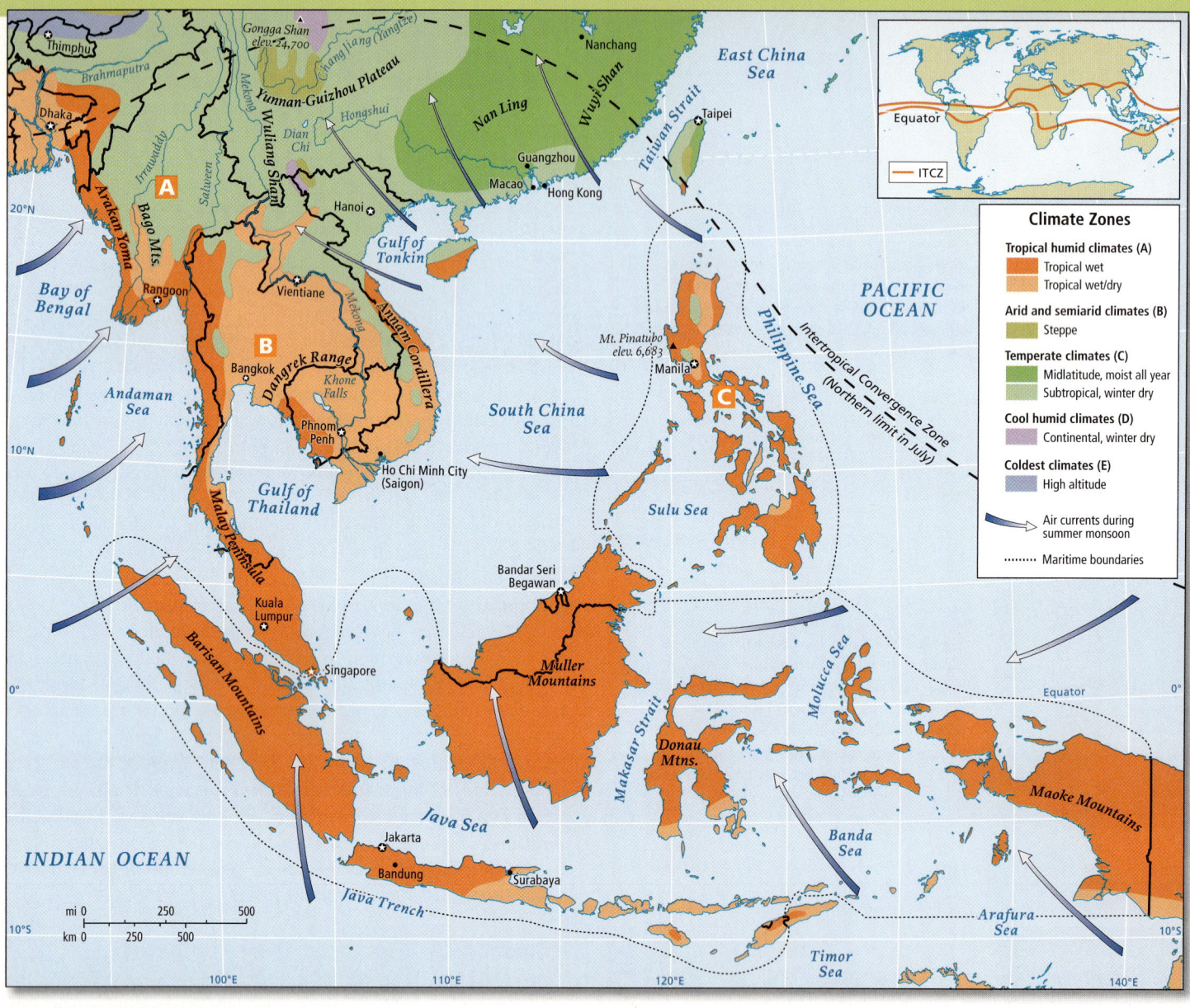

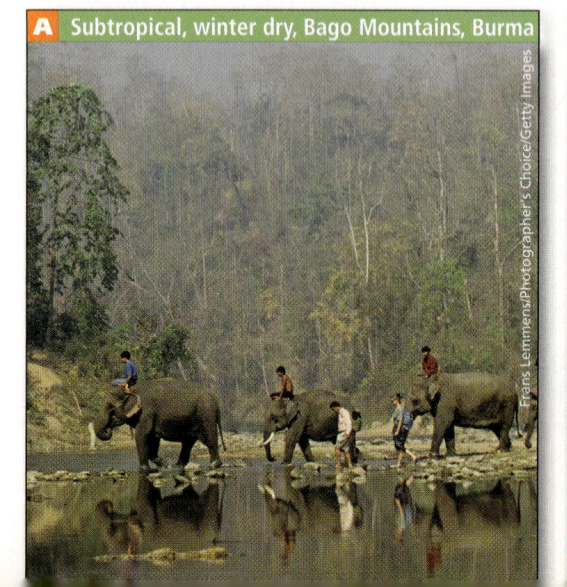

Climate Zones

Tropical humid climates (A)
- Tropical wet
- Tropical wet/dry

Arid and semiarid climates (B)
- Steppe

Temperate climates (C)
- Midlatitude, moist all year
- Subtropical, winter dry

Cool humid climates (D)
- Continental, winter dry

Coldest climates (E)
- High altitude

→ Air currents during summer monsoon
⋯⋯ Maritime boundaries

A Subtropical, winter dry, Bago Mountains, Burma

B Tropical, wet/dry, Khao Yai, Thailand

C Tropical, wet, Alaminos, Philippines

FIGURE 10.8 Trade in tropical timber. Most of the consumer demand for tropical timber is in North America, Europe, and Japan. Increasingly large quantities of tropical timber sold to China are made into furniture, plywood, and flooring, which are then sold to consumers in the developed world. The trade shown on the map is mostly legal, but much of the wood that fuels China's wood-processing industries is harvested illegally in Southeast Asia. [Sources consulted: *Annual Review and Assessment of the World Timber Situation 2010* (Yokohama, Japan: International Tropical Timber Organization, 2010), pp. 13–15, at http://www.itto.int/annual_review/; Sam Lawson and Larry MacFaul, "Illegal Logging and Related Trade: Indicators of the Global Response" (London: Chatham House, 2010), at http://awsassets.panda.org/downloads/chatham_house_illegallogging_2010.pdf]

(A) A logging road in Sabah, Borneo.

(B) Tropical hardwoods are made into toy train tracks in Dongguan, China.

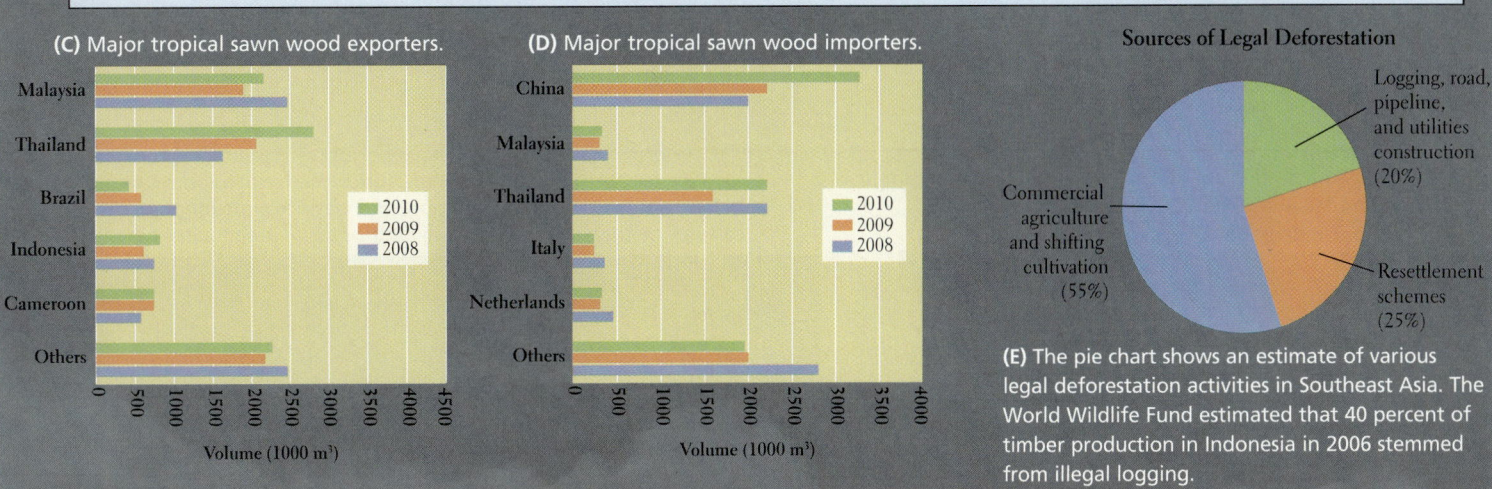

Major Trade Flows of Tropical Sawn Wood, 2009 (millions of cubic meters): 1.0, 0.6, 0.4, 0.3, 0.2, 0.1

* Malaysia-Taiwan (0.1)
** Thailand-Malaysia (0.1)
*** Malaysia-Singapore (0.1)
**** Indonesia-Malaysia (0.1)
★ Laos-Thailand (0.8)
★★ Malaysia-Thailand (0.4)
★★★ Vietnam-Hong Kong (0.1)
☆ Indonesia-Vietnam (0.1)
☆☆ Malaysia-Philippines (0.1)
☆☆☆ Malaysia-China (0.2)
☆☆☆☆ Philippines-China (0.3)

(C) Major tropical sawn wood exporters.

Malaysia, Thailand, Brazil, Indonesia, Cameroon, Others (2010, 2009, 2008) — Volume (1000 m³)

(D) Major tropical sawn wood importers.

China, Malaysia, Thailand, Italy, Netherlands, Others (2010, 2009, 2008) — Volume (1000 m³)

Sources of Legal Deforestation

- Commercial agriculture and shifting cultivation (55%)
- Resettlement schemes (25%)
- Logging, road, pipeline, and utilities construction (20%)

(E) The pie chart shows an estimate of various legal deforestation activities in Southeast Asia. The World Wildlife Fund estimated that 40 percent of timber production in Indonesia in 2006 stemmed from illegal logging.

> **THINKING GEOGRAPHICALLY** What happens to the atmosphere when there are fewer trees due to legal or illegal logging?

FIGURE 10.9 PHOTO ESSAY: Human Impacts on the Biosphere in Southeast Asia

Deforestation and the conversion of land to agricultural uses in Southeast Asia are having a profound impact both on regional ecosystems and on the entire biosphere. [Sources consulted: *United Nations Environment Programme, 2002, 2003, 2004, 2005, 2006* (New York: United Nations Development Programme), at http://maps.grida.no/go/collection/globio-geo-3]

A A worker sprays a palm oil plantation in Puntianai, on the island of Sumatra in Indonesia, with pesticides. Deforestation in much of Malaysia and Indonesia is now driven by the expansion of oil palm plantations.

Human Impact, 2002

Land cover:
- Forests
- Grasslands
- Deserts
- Tundra
- Ice

Acid rain:
- 5.5–4.9 pH

Human impact on land:
- High impact
- Medium–high impact
- Low–medium impact

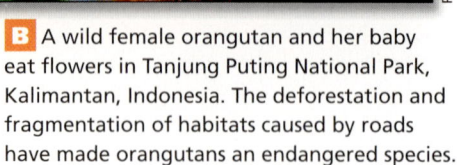

Overfishing:
- Threatened fisheries
- National boundaries
- Maritime boundaries

B A wild female orangutan and her baby eat flowers in Tanjung Puting National Park, Kalimantan, Indonesia. The deforestation and fragmentation of habitats caused by roads have made orangutans an endangered species.

C A satellite image of fires used to clear forests for agriculture in Indonesia and Malaysia. The CO_2 produced contributes to climate change and degrades local air quality.

D Rice terraces in the Philippines. Agricultural expansion is a major force behind deforestation in the region, and wet rice agriculture, because it releases methane, is a significant contributor of greenhouse gases.

The Costs of Logging, Legal and Illegal Deforestation costs Southeast Asian countries billions of dollars per year in lost forest services, such as renewable food, construction material, and fuel resources; rainwater capture and purification; flood amelioration; biodiversity preservation; and CO_2 absorption. Government regulations that limit logging have been constrained by rampant corruption, but recent efforts to enforce limits may be yielding results. With a vision of sustainable logging as a long-term source of income for his country, Indonesian president Susilo Bambang Yudhoyono has lead the fight to reduce deforestation (see "On the Bright Side: Indonesia Tries to Slow Deforestation and Reduce Greenhouse Gas Emissions").

ON THE BRIGHT SIDE

Indonesia Tries to Slow Deforestation and Reduce Greenhouse Gas Emissions

In 2011, as part of Indonesia's ongoing efforts to reduce greenhouse gas emissions, President Yudhoyono of Indonesia announced a 2-year moratorium on new logging and oil palm plantations on old-growth forests and peat swamps. The moratorium covered 65 million hectares (1.6 billion acres) of land, an area that is roughly the size of Japan. The agreement was facilitated by a pledge from Norway's government to provide $1 billion to help Indonesia reduce corruption and mismanagement in its forestry sector.

The moratorium was renewed for another 2 years in 2013 to enable the government, corporations, and foresters to adjust to a long list of changes aimed at reducing pressure on existing old-growth forests and peat swamps. While the moratorium will eventually be lifted, it has prompted a wide range of changes aimed at reducing deforestation, which is now the centerpiece of Indonesia's plans to lessen its greenhouse gas emissions.

Oil Palm Plantations Palm oil, which is used around the world in food products, as a cooking oil, in soaps and cosmetics, and as a machine oil, is at the center of debates over global climate change in Southeast Asia. The oil palm originated in West Africa and was brought to Southeast Asia early in the colonial era. Today, Indonesia and Malaysia are the world's largest palm oil producers, and the expansion of their oil palm plantations has resulted in extensive deforestation. Much of the deforestation is in lowland peat swamps which, in a natural state, are capable of storing large amounts of carbon in their soils. To create space for oil palm plantations, the swamps are drained and most of their vegetation is burned (see Figure 10.9A). Often the fires spread underground to the peat beneath the forests, smoldering for years after the surface fires have been extinguished. These subsurface fires release enormous amounts of carbon into the atmosphere. In recent years, smoke from burning forests and peat has periodically covered much of the region, dramatically reducing air quality and even making it necessary for people in rural areas to wear masks.

Palm oil has been promoted as a potential solution to global climate change because it can be converted into fuel for automobiles. The claim is that because the palm trees absorb CO_2 from the atmosphere just as the original forest did, palm oil could be considered a "carbon neutral" fuel. This is a fallacy because palm trees do not absorb carbon at a rate equivalent to natural rain forests—and if forests are burned to clear land for plantations, it can take decades to offset the initial carbon emissions produced by deforestation. If a peat swamp is cleared for an oil palm plantation, more carbon is released through the burning of subsurface peat than centuries of CO_2 absorption by growing oil palms can counteract.

Climate Change and Food Production

Food production contributes to global climate change in two ways. First, it is a major contributor to deforestation, and second, some types of cultivation actually produce significant greenhouse gases.

Shifting Cultivation Also known as *slash-and-burn* or *swidden* cultivation, shifting cultivation has been practiced sustainably for thousands of years in the hills and uplands of mainland Southeast Asia and in many parts of the islands (Figure 10.10 map). To maintain soil fertility in these warm, wet environments where nutrients are quickly lost to decay, traditional farmers move their fields every 3 years or so, letting old plots lie fallow for 15 years or more. The regrowth of forest on once-cleared fields not only regenerates the soil and inhibits runoff, it also absorbs significant amounts of CO_2 from the atmosphere. However, if fallow periods are shortened or are disrupted by logging, soil fertility can collapse, making future cultivation impossible. In some cases, fertility can be temporarily restored through the use of chemical or organic fertilizers, but these can be too expensive for most farmers and chemical fertilizers may be ineffective after a year or be too unhealthy for food cultivation. Tropical soils left bare of forest for too long eventually turn into hard, infertile, sunbaked clay called *laterite*.

Where population densities are relatively low, subsistence farmers can practice *sustainable shifting cultivation* indefinitely, but to allow for long fallow periods and still support human populations, this system requires larger areas than other types of agriculture. If population density increases, farmers are usually forced to shorten fallow periods, thus inhibiting forest regrowth and soil regeneration. Even though individual plots are small, the plots eventually become close to one other because shifting cultivation requires clearing forest at each move. Shifting cultivation now accounts for a significant portion of the region's deforestation. Moreover, because burning usually clears the forests, shifting cultivation can result in wildfires. This is especially true during an El Niño period, when rainfall is low. These wildfires have increased in recent years, particularly in Indonesia, further contributing to deforestation and carbon emissions there.

Wet Rice Cultivation Another major contributor to global climate change is Southeast Asia's most productive form of agriculture. *Wet rice cultivation* (sometimes called *paddy rice*) entails planting rice seedlings by hand in flooded and often terraced fields that are first cultivated with hand-guided plows pulled by water buffalo or tractors (see Figure 10.9D). Wet rice cultivation has transformed landscapes throughout Southeast Asia. It is practiced throughout this generally well-watered region,

(A) A rubber plantation in Cambodia.

(B) Slash-and-burn agriculture in Thailand.

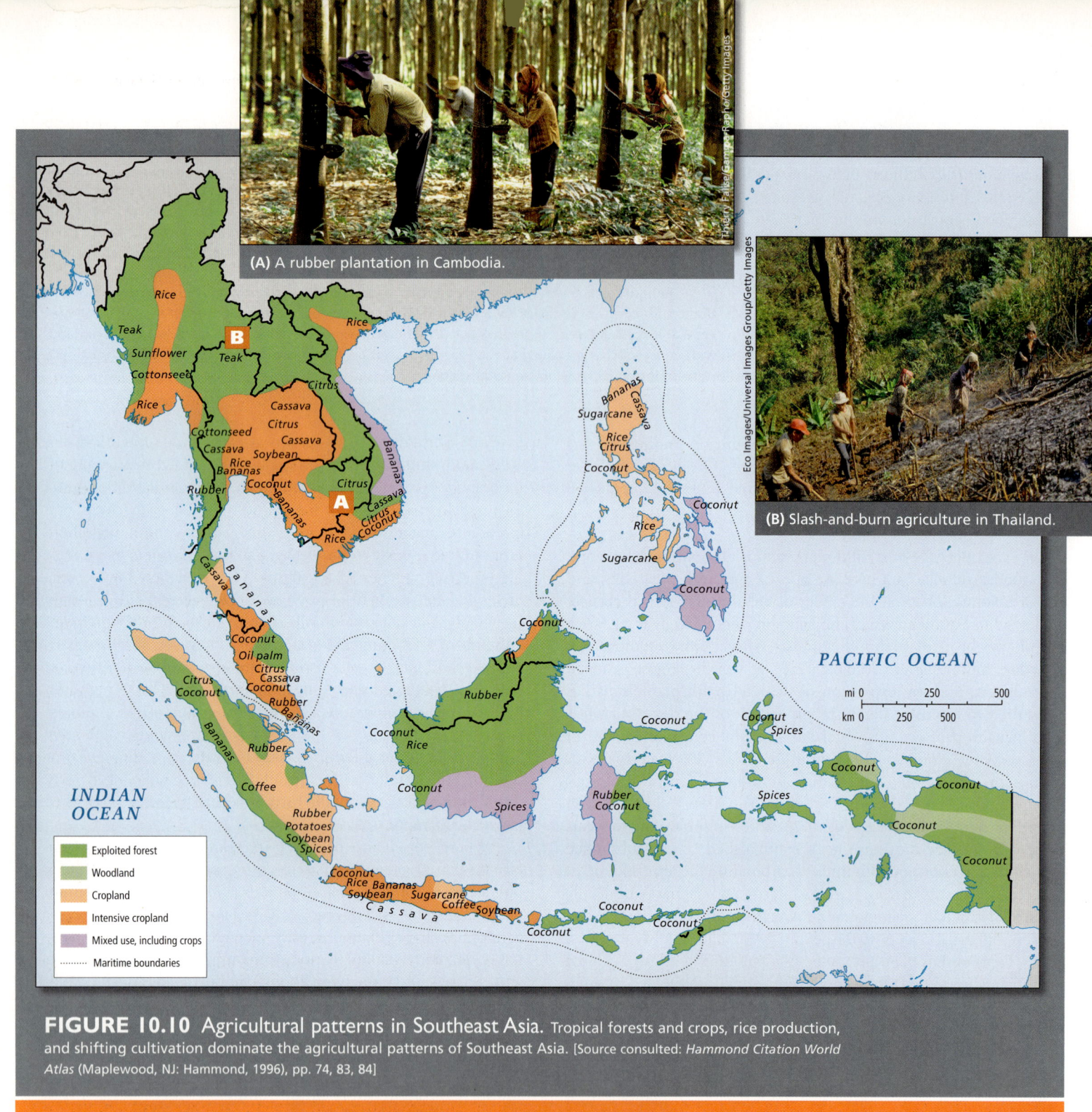

FIGURE 10.10 Agricultural patterns in Southeast Asia. Tropical forests and crops, rice production, and shifting cultivation dominate the agricultural patterns of Southeast Asia. [Source consulted: *Hammond Citation World Atlas* (Maplewood, NJ: Hammond, 1996), pp. 74, 83, 84]

Legend:
- Exploited forest
- Woodland
- Cropland
- Intensive cropland
- Mixed use, including crops
- Maritime boundaries

> **THINKING GEOGRAPHICALLY** On plantations of all types, native plants are replaced by marketable crop plants. What might be a way in which using slash-and-burn agriculture helps farmers avoids this particular impact?

especially on rich volcanic soils and in places where rivers and streams bring a yearly supply of silt.

The flooding of rice fields also results in the production of methane, a powerful greenhouse gas responsible for about 20 percent of global climate change. It is estimated that up to one-third of the world's methane is released from flooded rice fields, where organic matter in soil undergoes fermentation as oxygen supplies are cut off. Wet rice has been cultivated for thousands of years, but the increase in human population in the last 25 years has driven a 17 percent expansion of the area devoted to wet rice. The map in Figure 10.10 shows patterns of field and forest crops across the region.

416

Commercial Agriculture During the recent decades of relative prosperity, small farms once operated by families have been combined into large commercial farms owned by local or multinational corporations. These farms produce cash crops for export, such as rubber, palm oil, bananas, pineapples, tea, and rice (see Figure 10.10A and the figure map). Commercial farming on large tracts of deforested land reduces the need for labor by using mechanization, irrigation, and chemicals for fertilizer and pest control. The objectives of commercial farming are to generate big yields and quick profits, not to develop long-term sustainable agriculture. Many commercial farmers have achieved dramatic boosts in harvests (especially of rice) by using high-yield crop varieties, the result of green revolution research that has been applied in many parts of the world (see pages 340–341 in Chapter 8).

As we have noted in relation to other world regions, large-scale commercial farming has significant negative environmental effects, including the loss of wildlife habitat and hence the loss of biodiversity (see Figure 10.9B); increases in soil erosion, flooding, and chemical pollution; and depletion of groundwater resources. In addition, poor and subsistence farmers unable to afford the new technologies find they cannot compete with large commercial farms and are forced to migrate to cities to look for work.

Climate Change and Water

Many of Southeast Asia's potential vulnerabilities to global climate change are related to water resources. Four areas of vulnerability may affect the region's economy and food supply: glacial melting, increased evaporation, coral reef bleaching, and storm surge flooding (Figure 10.11).

Melting Glaciers and Increased Evaporation Like much of Asia, mainland Southeast Asia's largest rivers (the Irrawaddy, Salween, Mekong, and Red rivers) are fed during the dry season (November through March) by glaciers high in the Himalayas. These glaciers are now melting at a rate that could result in their eventual disappearance.

As glacial melting accelerates, the immediate risk is catastrophic flooding. While some areas have long been adapted to dramatic seasonal floods, which have taken place in this region for most of human history, many areas are unprepared for floods (see Figure 10.11D).

A longer-term concern is reduced dry-season flows in the rivers where glacial melt water has been the primary source of stream flow. As much as 15 percent of this region's rice harvest depends on the dry-season flows of the major rivers. The loss of these harvests would strain many farmers' incomes. In addition, there would likely be food shortages in cities. Coming on top of a global rise in food prices in recent years, this would place further strain on the incomes of poor people throughout the region.

The higher temperatures associated with current trends in global climate change mean evaporation rates will rise, resulting in drier conditions in fields, lower lake levels, and lower fish catches because of changing habitats for aquatic animals. Evaporation that results in reduced river and groundwater levels can also cause saltwater intrusions into estuaries and freshwater aquifers.

Coral Reef Bleaching Global climate change is expected to increase sea temperatures in Southeast Asia, threatening the coral reefs that sustain much of the region's fishing and tourism industries. A coral reef is an intricate structure that is composed of the calcium-rich skeletons of millions of tiny living creatures called coral polyps (see Figure 10.6B). The polyps are subject to **coral bleaching** (see Figure 10.11C), or color loss, which results when photosynthetic algae that live in the corals are expelled by a variety of human-instigated or naturally occurring changes. Both rising water temperature related to climate change and oceanic acidification resulting from the absorption of excess CO_2 in the atmosphere can cause bleaching. Ocean acidification is thought to have increased by about 30 percent over the last 340 years, a rate of change that may exceed the ability of ocean creatures to adapt. Bleaching also occurs in response to pollution from urban sources such as sewage discharges, stormwater runoff, or industrial water pollution.

> **coral bleaching** color loss that results when photosynthetic algae that live in corals are expelled

Overfishing can contribute to bleaching as well. Under normal conditions, coral that is assaulted with just one of the bleaching situations recovers within weeks or months. However, severe or repeated bleaching can cause corals to die. Unprecedented global coral bleaching events affecting roughly half of the world's coral reefs took place in 1998, 2002, 2004, and 2006, and regional events occur somewhere every year. **224. NEW SPECIES OF UNDERSEA LIFE FOUND NEAR INDONESIA**

Many of the fish caught in Southeast Asia's seas depend on healthy coral reefs for their survival. The thousands of rural communities throughout coastal Southeast Asia that rely on fish for food are thus also threatened by coral bleaching. So far, however, the greatest observable impacts on people have been in the tourism industry. In the Philippines, the coral bleaching of 1998 brought a dramatic decline in tourists who come to dive the country's usually spectacular reefs, resulting in a loss of about U.S.$30 million to the economy.

Storm Surges and Flooding Although the relationship is not yet entirely understood, violent tropical storms seem to be increasing in both number and intensity as the climate warms. When Typhoon Haiyan hit in 2013, it became the strongest typhoon to make its way onto land in recorded history. Haiyan struck the Philippines, killing at least 5600 people and leaving millions homeless. Climatologists studying data on tropical storms predict that the entire Southeast Asian region will have a somewhat higher number of typhoons over the next few years and that the duration of peak winds along coastal zones will increase. This is significant because many poor, urban migrants have crowded into precarious dwellings in low-lying coastal cities such as Manila, Bangkok, Rangoon,

FIGURE 10.11   PHOTO ESSAY: Vulnerability to Climate Change in Southeast Asia

Much of Southeast Asia's vulnerability to climate change relates to water. Here we explore vulnerabilities related to tropical storms, flooding, and coral reefs.

A A father and his children on Mindanao, Philippines, sift through the wreckage of Typhoon Bopha in search of coconuts. Typhoons (known in the Atlantic as hurricanes) are projected to increase in frequency as a result of the warmer temperatures created by climate change.

Vulnerability to Climate Change
- Extreme
- High
- Medium
- Low

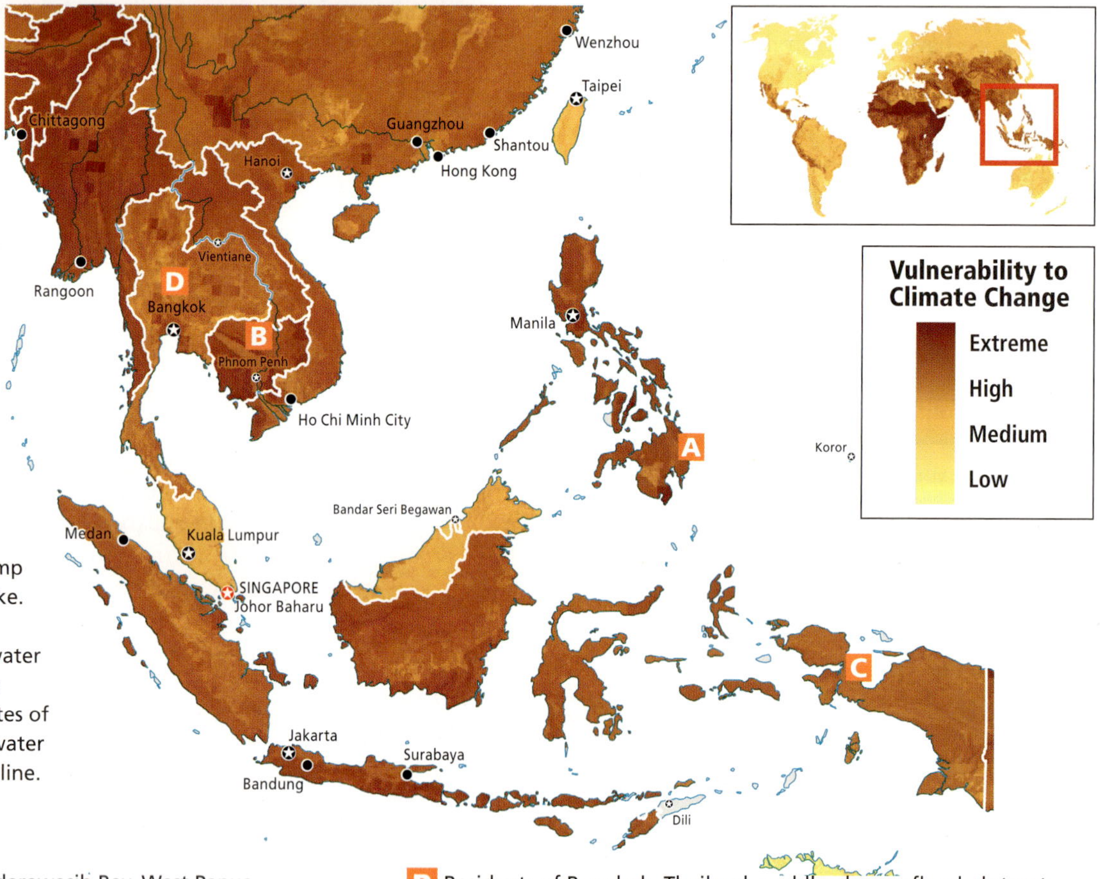

B A husband and wife harvest shrimp and fish on Cambodia's Tonlé Sap Lake. Sixty percent of Cambodia's protein intake comes from this inland freshwater lake. Higher temperatures related to climate change may be increasing rates of evaporation from the lake, causing water levels to drop and fish catches to decline.

C A partially bleached coral in Cenderawasih Bay, West Papua, Indonesia. Climate change is raising ocean temperatures, resulting in more coral bleaching events that severely degrade reefs, which are home to many aquatic life forms.

D Residents of Bangkok, Thailand, paddle along a flooded street during catastrophic flooding, which is becoming more common in the city. Climate change could increase flooding along rivers as patterns of rainfall and glacial melting change.

Responses to Climate Change

One goal of all governments in this region is to reduce the amount of fossil fuel consumption, and some are meeting this goal despite the frequently high start-up costs of doing so. Both the Philippines and Indonesia have significant potential for generating electricity from *geothermal energy* (heat stored in Earth's crust). This energy is particularly accessible near active volcanoes, which both countries have in abundance. The Philippines already generates 17 percent of its electricity from geothermal energy; it is second only to the United States in the amount of geothermal power it generates. By some estimates, geothermal energy could eventually provide a majority of Indonesia's energy needs. Solar energy is another attractive source, given that the entire region lies near the equator, the part of the planet that receives the most solar energy. For most countries, however, wind is the most cost-effective option, especially in Laos and Vietnam, where many population centers are in high-wind areas.

THINGS TO REMEMBER

> **GEOGRAPHIC INSIGHT 1**
> - **Environment** Many of Southeast Asia's most critical environmental issues relate in some way to climate change. Deforestation is rapid in this region and is a major global source of greenhouse gas emissions, which intensify climate change. This region is also highly vulnerable to the impacts of climate change. Increased flooding and droughts threaten food production on land and rising ocean temperatures strain aquatic ecosystems.

- The expansion of palm oil cultivation has resulted in extensive deforestation that has released enormous amounts of carbon into the atmosphere.

- Much of this region's rice harvest depends on the dry-season flows of the major rivers of mainland Southeast Asia, which are threatened by glacial melting.

- Climate change poses special challenges for people in low-lying cities and in typhoon zones.

HUMAN GEOGRAPHY

First settled in prehistory by migrants from the Eurasian continent, Southeast Asia was later influenced by Chinese, Indian, and Arab traders. Later still, it was colonized by Europe (from the 1500s to the early 1900s), and the Philippines were occupied by the United States from 1898 to 1946. During World War II, much of the region was occupied by Japan. By the late twentieth century, occupation and domination by outsiders had ended and the region was profiting from selling manufactured goods to its former colonizers.

Like Middle and South America, Africa, and South Asia, Southeast Asia is now expanding its links to the global economy. It has had more success than other areas in achieving widespread, if modest, prosperity. It has done so largely by following the example of some of East Asia's most successful countries, such as Japan, Korea, Taiwan, and more recently, China.

> **Australo-Melanesians** a group of hunters and gatherers who moved approximately 40,000 to 60,000 years ago from what are now the northern Indian and Burman parts of southern Eurasia to the exposed landmass of Sundaland and present-day Australia
>
> **Austronesians** a group of skilled farmers and seafarers from southern China who migrated south to various parts of Southeast Asia between 5000 and 10,000 years ago

HUMAN PATTERNS OVER TIME

The modern indigenous populations of Southeast Asia arose from two migrations widely separated in time. In the first migration, about 40,000 to 60,000 years ago, **Australo-Melanesians**, a group of hunters and gatherers from what is now northern India and Burma, moved to the exposed landmass of Sundaland and into present-day Australia. Their descendants still live in Indonesia's easternmost islands and in small, usually remote pockets on other islands and the Malay Peninsula, and in Australia, New Guinea, and other parts of Oceania (see the Figure 11.1 map on page 446).

In the second migration (from 6000 to 10,000 years ago, after the last ice age), people from southern China began moving into Southeast Asia. Their migration gained momentum about 5000 years ago, when a culture of skilled farmers and seafarers from southern China, the **Austronesians**, migrated first to what is now Taiwan, then to the Philippines, and then into the islands of Southeast Asia and the Malay Peninsula (see Figure 10.13A on page 422). Some of these adventurous sea travelers eventually moved westward to southern India and to Madagascar (off the east coast of Africa; see Chapter 7), and eastward to the far reaches of the Pacific islands (see Figure 11.13 on page 462).

Diverse Cultural Influences

Over the last several thousand years, Southeast Asia has been and continues to be shaped by a steady circulation of cultural influences, both internal and external. Merchants, religious teachers, and sometimes even invading armies from China, India, and Southwest Asia came to the region by overland trade routes and the surrounding seas. These newcomers brought religions, trade goods (such as cotton textiles), and food plants (such as mangoes and tamarinds) deep into the Indonesian and Philippine archipelagos and throughout the mainland. The monsoon winds, which blow from the west in the spring and summer, facilitated access for merchant ships from South Asia and the Persian Gulf. The ships sailed home on winds blowing from the east in the autumn and winter. These easterly winds carried ships with

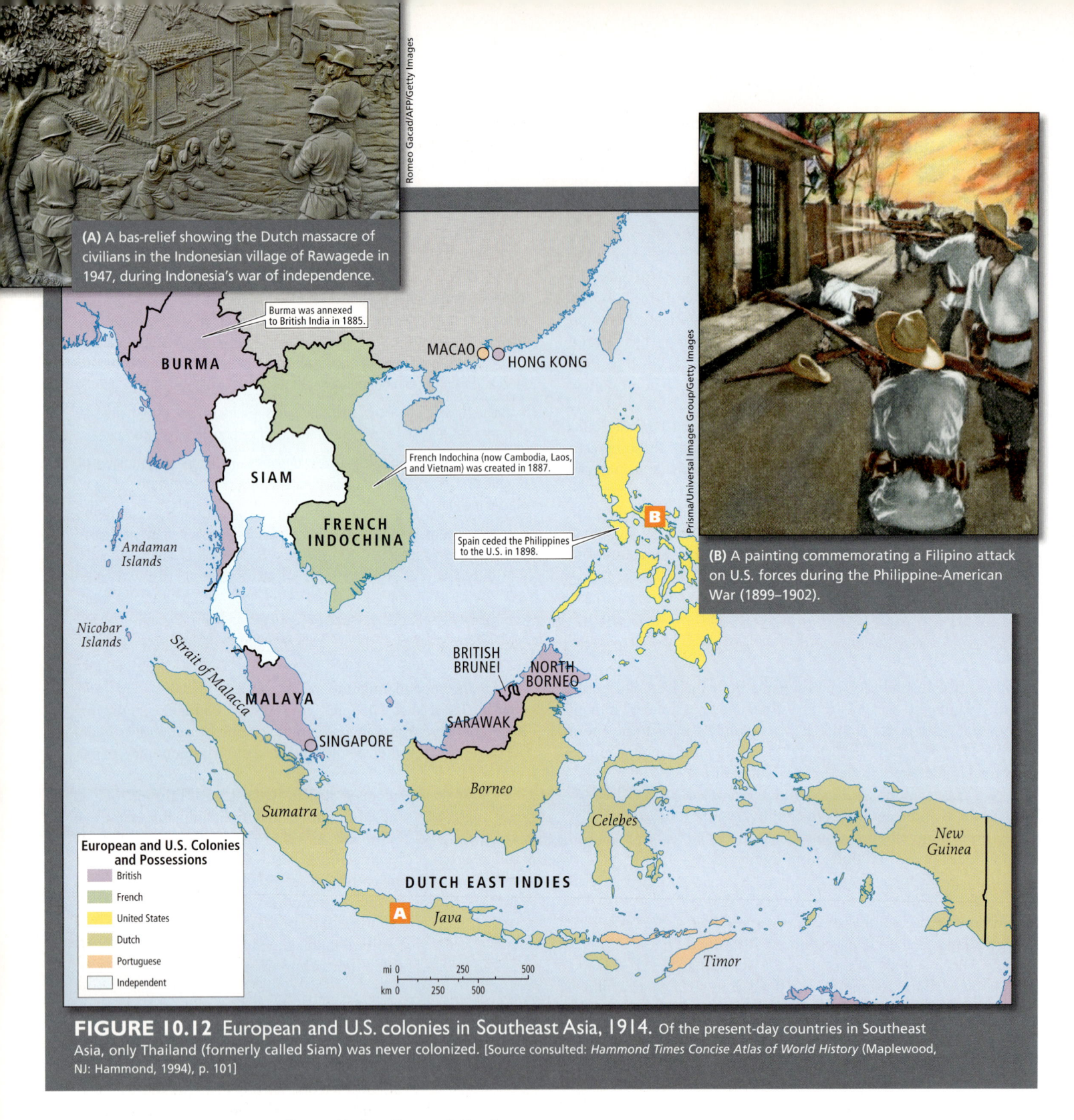

(A) A bas-relief showing the Dutch massacre of civilians in the Indonesian village of Rawagede in 1947, during Indonesia's war of independence.

(B) A painting commemorating a Filipino attack on U.S. forces during the Philippine-American War (1899–1902).

FIGURE 10.12 European and U.S. colonies in Southeast Asia, 1914. Of the present-day countries in Southeast Asia, only Thailand (formerly called Siam) was never colonized. [Source consulted: *Hammond Times Concise Atlas of World History* (Maplewood, NJ: Hammond, 1994), p. 101]

spices, bananas, sugarcane, silks, and other East and Southeast Asian items, as well as people, to the wider world.

Religious Legacies Spatial patterns of religion in Southeast Asia reveal an island–mainland division that reflects the history of influences from India, China, Southwest Asia, and Europe. Both Hinduism and Buddhism arrived thousands of years ago via monks and traders who traveled by sea and along overland trade routes that connected India and China through Burma (see Figure 10.13B). Many early Southeast Asian kingdoms and empires switched back and forth between Hinduism and Buddhism as their official religion. Spectacular ruins of these Hindu–Buddhist empires are scattered across the region; the most famous is the city of Angkor in what is now Cambodia. At its zenith in the 1100s, Angkor was among the largest cities in the world, and its ruins are now a World Heritage Site (see Figure 10.16A on page 427). Today, Buddhism dominates mainland Southeast Asia, while Hinduism remains dominant only on the Indonesian islands of Bali and Lombok.

420

In Vietnam, people practice a mix of Buddhist, Confucian, and Taoist customs that reflect the thousand years (ending in 938 C.E.) when Vietnam was part of various Chinese empires. China's traders and laborers also brought cultural influences to scattered coastal zones throughout Southeast Asia.

Islam is now dominant in the islands of Southeast Asia. Islam came mainly through South Asia after India was conquered by Mughals in the fifteenth century. Muslim spiritual leaders and traders converted many formerly Hindu–Buddhist kingdoms in Indonesia, Malaysia, and parts of the southern Philippines, where Islam is still dominant. Roman Catholicism is the predominant religion in Timor-Leste and in most of the Philippines, colonized respectively by Portugal and Spain.

European Colonization

Between the sixteenth and the twentieth centuries, several European countries established colonies or quasi-colonies in Southeast Asia (Figure 10.12). Drawn by the region's fabled spice trade, the Portuguese established the first permanent European settlement in Southeast Asia at the port of Malacca, Malaysia, in 1511. Although better ships and weapons gave the Portuguese an advantage, their anti-Islamic and pro-Catholic policies provoked strong resistance in Southeast Asia. Only on the small island of Timor-Leste did the Portuguese establish Catholicism as the dominant religion.

Arriving first in 1521, the Spanish had established trade links across the Pacific between the Philippines and their colonies in the Americas by 1540 (Figure 10.13C). Like the Portuguese, they practiced a style of colonial domination grounded in Catholicism, but they met less resistance because of their greater tolerance of non-Christians. The Spanish ruled the Philippines for more than 350 years; as a result, except for the southernmost islands, the Philippines is the most deeply Westernized and certainly the most Catholic part of Southeast Asia.

The Dutch were the most economically successful of the European colonial powers in Southeast Asia. From the sixteenth to the nineteenth centuries, they extended their control of trade over most of what is today called Indonesia, known at the time as the Dutch East Indies. The Dutch became interested in growing cash crops for export. Between 1830 and 1870, they forced indigenous farmers to leave their own fields and work part time without pay on Dutch coffee, sugar, and indigo plantations. The resulting disruption of local food production systems caused severe famines and provoked resistance that often took the form of Islamic religious movements. Resistance movements hastened the spread of Islam throughout Indonesia, where the Dutch had made little effort to spread their Protestant version of Christianity.

Beginning in the late eighteenth century, the British established colonies at key ports on the Malay Peninsula. They held these ports both for their trade value and to protect the Strait of Malacca, the shortest passage for sea trade between Britain's empire in India and China. In the nineteenth century, Britain extended its rule over the rest of modern Malaysia in order to benefit from Malaysia's tin mines and plantations. Britain also added Burma to its empire, which provided access to forest resources and overland trade routes into southwest China.

The French first entered Southeast Asia as Catholic missionaries in the early seventeenth century. They worked mostly in the eastern mainland area in the modern states of Vietnam, Cambodia, and Laos. In the late nineteenth century, spurred by rivalry with Britain and other European powers for access to the markets of nearby China, the French formally colonized the area, which became known as French Indochina.

In all of Southeast Asia, the only country not to be colonized by Europe was Thailand (then known as Siam). Like Japan, it protected its sovereignty through both diplomacy and a vigorous drive toward European-style modernization.

Struggles for Independence

Agitation against colonial rule began in the late nineteenth century when Filipinos fought first against Spain in 1896. They then resisted control by the United States, which began in 1898 after the Spanish-American War (see the map in Figure 10.12). However, the Philippines and the rest of Southeast Asia did not win independence until after World War II (see Figure 10.12A). By then, Europe's ability to administer its colonies had been weakened by the devastation of the war, during which Japan had conquered most of the parts of Southeast Asia that had been controlled by Europe and the United States (see Figure 10.13D). Japan held these lands until it was defeated by the United States at the end of the war. By the mid-1950s, the colonial powers had granted self-government to most of the region, and all of Southeast Asia was independent by 1975.

The Vietnam War The bitterest battle for independence took place in French Indochina (the territories of Vietnam, Laos, and Cambodia, acquired in the nineteenth century). Although all three became nominally independent in 1949, France retained political and economic power over them. Various nationalist leaders, most notably Vietnam's Ho Chi Minh, headed resistance movements against continued French domination. After failed diplomatic efforts in Europe and the United States, the resistance leaders accepted military assistance from Communist China and the Soviet Union, despite ancient antipathies toward China for its previous millennia of domination. In this way, the Cold War was brought to mainland Southeast Asia.

In 1954, Ho Chi Minh defeated the French at Dien Bien Phu in northern Vietnam. The United States stepped in because it had become worried about the spread of international communism should the anticolonial resisters, now supported by Communists, succeed. The **domino theory**—the idea that if one country "falls" to communism, other nearby countries will follow—was a major influence in this decision, because both North Korea and China had recently become communist. The Vietnamese resistance, which controlled the northern half of the country, attempted to wrest control of the southern half of Vietnam from the U.S.-supported and quite corrupt South Vietnamese government. The pace of the war

domino theory a foreign policy theory that used the idea of the domino effect to suggest that if one country "fell" to communism, others in the neighboring region would also fall

A Ifugao tribal people, one of several Austronesian ethnic groups in the Philippines. [John S. Lander/LightRocket/Getty Images]

B The ruins of Bagan, capital of many Buddhist kingdoms since at least 874 C.E. [Joe & Clair Carnegie/Libyan Soup/Getty Images]

C A monument to the landing of Ferdinand Magellan in the Philippines, 1521. [John S. Lander/LightRocket/Getty Images]

60,000–40,000 B.C.E. Austro-Melanesian migrations

8000 B.C.E. Austronesian migrations

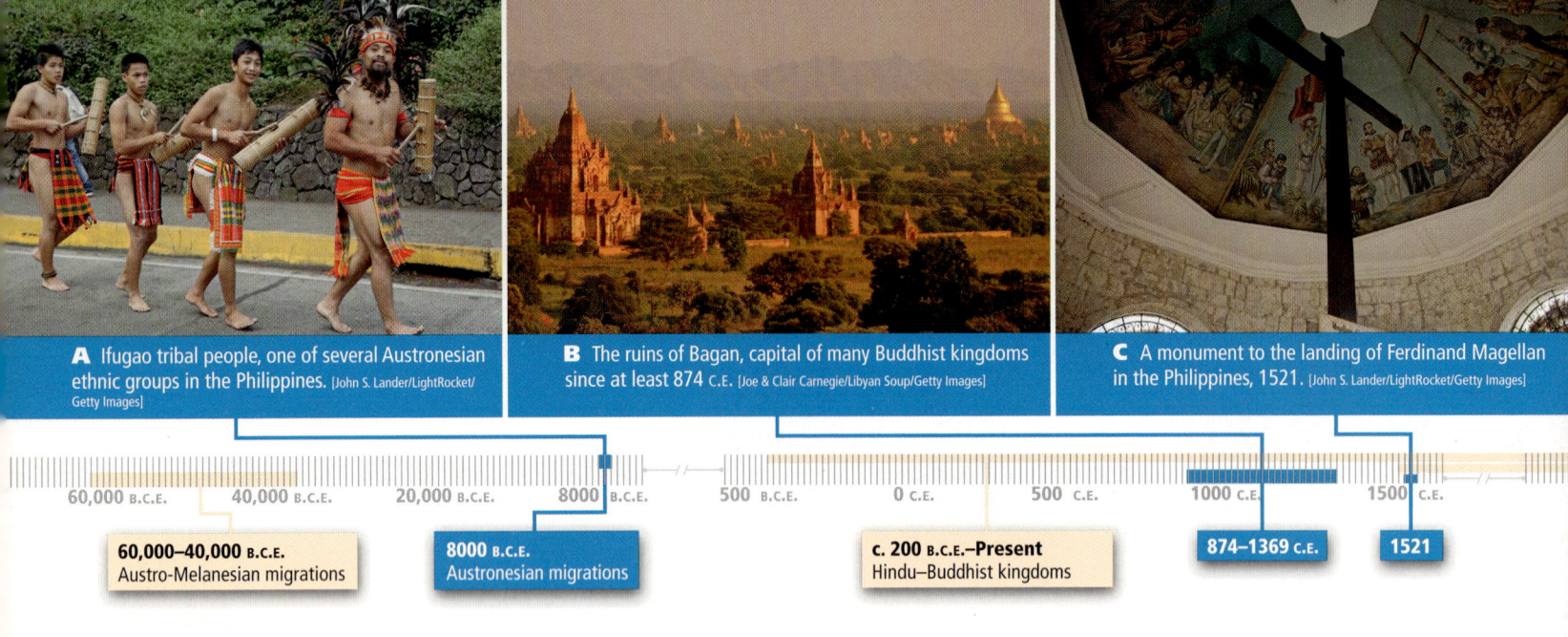

c. 200 B.C.E.–Present Hindu–Buddhist kingdoms

874–1369 C.E.

1521

FIGURE 10.13 VISUAL HISTORY OF SOUTHEAST ASIA

---

accelerated in the mid-1960s. After many years of brutal conflict, public opinion in the United States forced a U.S. withdrawal from the conflict in 1973. The civil war continued in Vietnam, finally ending in 1975, when the north defeated the south and established a new national government.

More than 4.5 million people died during the Vietnam War, including more than 58,000 U.S. soldiers. Another 4.5 million on both sides were wounded, and bombs, napalm, and chemical defoliants ruined much of the Vietnamese environment. Land mines continue to be a hazard to this day, and the effects of the highly toxic defoliant known as Agent Orange are still causing debilitating birth defects among many rural Vietnamese and Laotian people (see Figure 10.13E). The withdrawal from Vietnam in 1973 ranks as one of the most profound military defeats in U.S. history. After the war, the United States crippled Vietnam's recovery by imposing severe economic sanctions that lasted until 1993. Since then, however, the United States and Vietnam have developed a significant trade relationship.

The "Killing Fields" in Cambodia In Cambodia, where the Vietnam War had spilled over the border, a particularly violent revolutionary faction called the Khmer Rouge seized control of the government in 1975. Inspired by the vision of a rural communist society, they attempted to destroy virtually all traces of European influence. They targeted Western-educated urbanites in particular, forcing them into labor camps where more than 2 million Cambodians—one-quarter of the population—starved or were executed in what became known as the "killing fields."

In 1979, Vietnam deposed the Khmer Rouge and, until 1989, ruled Cambodia through a puppet government. A 2-year civil war then ensued. Despite major UN efforts throughout the 1990s to establish a multiparty democracy in Cambodia, the country remained plagued by political tensions between rival factions and by government corruption.

In March 2009, Kang Kek Iew, the first of the Khmer Rouge leaders to be tried for war crimes and genocide, was forced to listen to and watch lengthy accounts of the torture of men, women, and children that he supervised. He was convicted in 2010 and sentenced to 35 years in prison. Most operatives in the killing fields will never be prosecuted. **223. CAMBODIAN HIP-HOP ARTIST TELLS STORY THROUGH RAP**

THINGS TO REMEMBER

- The modern indigenous populations of Southeast Asia arose from two migrations, one about 40,000 to 60,000 years ago (the Australo-Melanesians), and one 6000 to 10,000 years ago (the Austronesians).
- Today, Buddhism dominates mainland Southeast Asia, while Hinduism remains dominant only on the Indonesian islands of Bali and Lombok.
- Islam is now dominant in the islands of Southeast Asia.
- Between the sixteenth and the twentieth centuries, several European countries and the United States and Japan established colonies or quasi-colonies that covered almost all of Southeast Asia.
- Colonial rule in Southeast Asia began in 1511 with the Portuguese and came to a violent end in Vietnam and Cambodia, where war took the lives of more than 4.5 million, including more than 58,000 U.S. soldiers.

GLOBALIZATION AND DEVELOPMENT

GEOGRAPHIC INSIGHT 2

Globalization and Development: Globalization has brought both spectacular successes and occasional declines to the economies of Southeast Asia. Key to the economic development of this region are strategies that were pioneered earlier in East Asia: the formation of state-aided market economies combined with export-led economic development.

D Japanese forces land in the Philippines during World War II. [Keystone/Getty Images]

E The defoliant Agent Orange was sprayed over Vietnam and Laos between 1962 and 1971 to destroy crops and expose North Vietnamese troops. [Dick Swanson/Time Life Pictures/Getty Images]

F Bangkok, Thailand, is emerging as a global center for manufacturing and trade. [Igor Prahin/Flickr/Getty Images]

1511–1975 European colonial era

1830–1870 Dutch forced-labor system causes famines in Indonesia

1896–1975 Struggles for independence

1941

1955–1975 Vietnam War

1962–1971

1975–1979 Khmer Rouge rules Cambodia

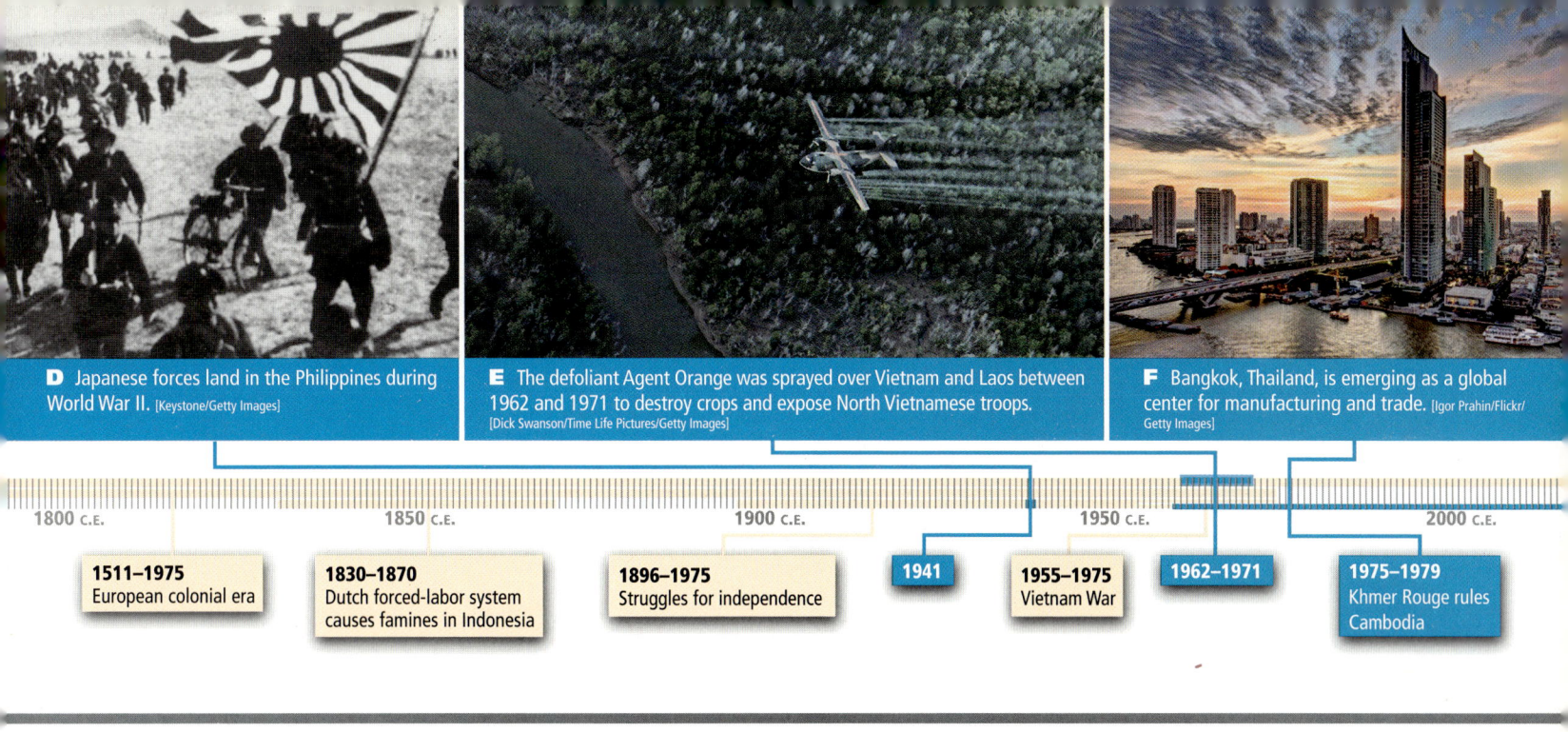

During and immediately after the era of European colonialism, trade among the countries within the region was inhibited by the fact that they all exported similar goods—primarily food and raw materials—and imposed tariffs against one another. They imported consumer products, industrial materials, machinery, and fossil fuels, mostly from the developed world.

Beginning in the 1960s, though, some national governments in Southeast Asia were able to create strong and sustained economic expansion by emulating two strategies for economic growth pioneered earlier by Japan, Taiwan, and South Korea (see pages 381–382 in Chapter 9). One was the formation of state-aided market economies. National governments in Indonesia, Malaysia, Thailand, and to some extent the Philippines intervened strategically in financial institutions to make sure that certain economic sectors developed; in addition, investment by foreigners was limited so that the governments could have more control over the direction of the economy. The other strategy was export-led economic development, which focused investment on industries that manufactured products for export, primarily to developed countries. These strategies amounted to a limited and selective embrace of globalization in that global markets for the region's products were sought but foreign sources of capital were not.

These approaches were a dramatic departure from those used in other developing areas. In Middle and South America and parts of Africa, postcolonial governments relied on import substitution industries that produced manufactured goods mainly for local use. By contrast, export-led growth allowed Southeast Asia's industries to earn much more money in the vastly larger markets of the developed world. Standards of living increased markedly, especially in Malaysia, Singapore, Indonesia, and Thailand. Other important results of Southeast Asia's success were a decrease in wealth disparities and improvements in vital statistics—lower infant and maternal mortality, longer life expectancy, and lower population growth. By the 1990s, Southeast Asian countries had some of the highest economic growth rates in the world. The growth was based on an economic strategy that emphasized the export of manufactured goods—initially clothing and then more sophisticated technical products.

Export Processing Zones In the 1970s, some governments in the region began adopting an additional strategy for encouraging economic development. This time, they sought foreign sources of capital, but the places in which those sources could invest were limited to specially designated free trade areas. Such export processing zones (EPZs; see the discussion in Chapter 3) are places in which foreign companies can set up their facilities using inexpensive, locally available labor to produce items only for export. (Maquiladoras in Mexico are an example of companies set up by foreign firms.) Taxes are eliminated or greatly reduced as long as the products are re-exported, not sold in-country. Since the 1970s, EPZs have expanded economic development in Malaysia, Indonesia, Vietnam, and the Philippines, and are now used widely in China and Middle and South America.

The Feminization of Labor Between 80 and 90 percent of the workers in the EPZs are women, not only in Southeast Asia but in other world regions as well (Figure 10.14). The **feminization of labor** has been a distinct characteristic of globalization over the past three decades (see pages 362–363 in Chapter 9). Employers prefer to hire young, single women because they are perceived as the least expensive, least troublesome employees. Statistics do show that, generally, women across the globe will work for lower wages than men, will not complain about poor and unsafe working conditions, will accept being restricted to certain jobs on the basis of sex, and are not as likely as men to agitate for promotions; though, of course, there are many exceptions. The reasons for this are complex and are discussed throughout this book.

feminization of labor the rising numbers of women in both the formal and the informal labor force

FIGURE 10.14 The feminization of labor. Women assemble circuit boards in a factory in Vietnam.

Steve Raymer/Asia Images/Getty Images

> **THINKING GEOGRAPHICALLY** Why do many employers prefer workers to be young, single women?

Working Conditions In general, the benefits of Southeast Asia's "economic miracle" have been unequally apportioned. In the region's new factories and other enterprises, it is not unusual for assembly-line employees to work 10 to 12 hours per day, 7 days per week for less than the legal minimum wage and without benefits. Labor unions that typically would address working conditions and wage grievances are frequently repressed by governments, and international consumer pressure to improve working conditions at U.S.-based companies has been only partially effective. For example, the U.S.-based shoe company Nike has tried to sidestep the entire issue of customer complaints for years by outsourcing its manufacturing to non-U.S. contractors in the region. U.S. fair labor NGOs pursued one Nike subcontractor in Jakarta, Indonesia, where for 19 years workers were forced to work an extra hour every day without pay. The NGOs filed a lawsuit against the subcontractor, which they won, resulting in a $1 million settlement that will give the 4500 workers $222 each. However, given that there are more than 160,000 workers for Nike subcontractors in Indonesia alone, this settlement is but a drop in the bucket.

A much more powerful force that is improving working conditions and driving up pay is the growth of the service sector throughout the region. Many service sector jobs require at least a high school education, and competition for hiring these educated workers means that wages and working conditions are already better than in manufacturing and are likely to improve faster.

In Singapore, the Philippines, and Thailand, the service sector already dominates the economy in employment and as a percentage of GDP. The service sector accounts for 40 percent of Cambodia's GDP and 55 percent of Timor-Leste's, though agriculture still employs the majority of workers in both countries. In other countries in the region, the service sector is approaching parity with the industrial or agricultural sectors. Only Brunei is still dominated by its industrial sector, which is entirely based on oil production.

Economic Crisis and Recovery: The Perils of Globalization

Periodically, economic crises have swept through this region, with varying effects on society. For example, a crisis in the late 1990s forced millions of people into poverty and changed the political order in some countries. A major cause of this crisis was the lifting of controls on Southeast Asia's once highly regulated financial sector. There were also some geographic aspects to it: the banks involved were located primarily in Singapore and the major cities of Thailand, Malaysia, and Indonesia. These were also the countries to feel the immediate effects of the crisis. While urban areas were hit hardest, eventually the effects filtered into the hinterland, and workers even in remote areas lost jobs.

Deregulating Investment As part of a general push by the International Monetary Fund (IMF) to open national economies to the free market, Southeast Asian governments relaxed controls on the financial sector in the 1990s. Soon, Southeast Asian banks were flooded with money from investors in the rich countries of the world who hoped to profit from the region's growing economies. Flush with cash and newfound freedoms, the banks often made reckless decisions. For example, bankers made risky loans to real estate developers, often for high-rise office building construction. As a result, many Southeast Asian cities soon had far too much office space, with millions of investment dollars tied up in projects that contributed little to economic development.

One of the forces that led bankers to make such bad decisions was a kind of corruption known as **crony capitalism**. In most Southeast Asian countries, as elsewhere, corruption is related to the close personal and family relationships between high-level politicians, bankers, and wealthy business owners. In Indonesia, for example, the most lucrative government contracts and business opportunities were reserved for the children of former president Suharto,

> **crony capitalism** a type of corruption in which politicians, bankers, and entrepreneurs, sometimes members of the same family, have close personal as well as business relationships

who ruled the country from 1967 to 1997. His children became some of the wealthiest people in Southeast Asia. This kind of corruption expanded considerably with the new foreign investment money, much of which was diverted to bribery or unnecessary projects that brought prestige to political leaders.

The cumulative effect of crony capitalism and the lifting of controls on banks was that many ventures failed to produce any profits at all. In response, foreign investors panicked, withdrawing their money on a massive scale. In 1996, before the crisis, there was a net inflow of U.S.$94 billion to Southeast Asia's leading economies. In 1997, inflows had ceased and there was a net outflow of U.S.$12 billion.

The IMF Bailout and Its Aftermath The IMF made a major effort to keep the region from sliding deeper into recession by instituting reforms designed to make banks more responsible in their lending practices. The IMF also required structural adjustment policies (SAPs; see pages 129–130 in Chapter 3), which required countries to cut government spending (especially on social services) and abandon policies intended to protect domestic industries.

After several years of economic chaos and much debate over whether the IMF bailout helped or hurt a majority of Southeast Asians, economies began to recover. In the largest of the region's economies, growth resumed by 1999; by 2006, while it was still an ominous reminder of the risks of globalization, the crisis had been more or less overcome.

The Impact of China's Growth During the Southeast Asian financial crisis of the late 1990s, Singapore, Malaysia, and Thailand lost out to China in attracting new industries and foreign investors. By the early 2000s, China attracted more than twice as much foreign direct investment (FDI) as Southeast Asia. However, China's growth also became an opportunity for Southeast Asia. Singapore, Malaysia, Indonesia, Thailand, and the Philippines "piggy-backed" on China's growth by winning large contracts to upgrade China's infrastructure in areas such as wastewater treatment, gas distribution, and shopping mall development.

> **Association of Southeast Asian Nations (ASEAN)** an organization of Southeast Asian governments that was established to further economic growth and political cooperation

Southeast Asia has also used its comparative advantages over China to attract investment. In poorer countries, such as Vietnam, wages have remained lower than in China. This has attracted investments in low-skill manufacturing that had been going to China. Meanwhile, some of the region's wealthier countries—Singapore, Malaysia, and Thailand—have been positioning themselves as locations that offer more highly skilled labor and better high-tech infrastructure than China does.

The Global Recession Beginning in 2007 The region was hit again by the effects of the global recession that began in 2007. Growth slowed markedly because high oil and food prices restricted disposable cash worldwide, and consumers in wealthy countries (especially in the United States and the EU countries, which are Southeast Asia's biggest trade partners) faced crippling debt that seriously curtailed their spending. By late 2009, East and Southeast Asia appeared to be recovering. In part, the recovery was based on the increased demand for goods and services within the domestic economies of China and Southeast Asia, where consumers now had some disposable cash. Southeast Asia's ability to respond to this demand was facilitated by policies that opened up intraregional trade and access to China.

But the positive developments of 2009 did not last. By 2012, financial troubles in Europe and slow recoveries in the United States and Japan all lowered demand for Southeast Asian products. Singapore, which has a highly sophisticated economy based on its banking services and the transfer of goods through its giant port, is particularly susceptible to slowdowns in rich countries. Manufacturing countries like Indonesia, Malaysia, and Thailand also fare best when there is demand for their products in international markets. Cambodia, Laos, and Burma still rely on raw material exports to places like China. However, in order to maintain its economic momentum, China has begun discouraging imports in favor of Chinese-produced goods, thus reducing its demand for raw materials and manufactured goods from Southeast Asia.

Regional Trade and ASEAN

During the 1980s and 1990s, Southeast Asian countries traded more with China and the rich countries of the world than they did with each other. This issue of insufficient trade between Southeast Asian countries was the reason behind the creation of the **Association of Southeast Asian Nations (ASEAN)**, an increasingly strong organization of Southeast Asian nations (see Figure 10.4 for locations).

Regional Integration ASEAN started in 1967 as an anti-Communist, anti-China association, but it now focuses on agreements that strengthen regional cooperation, including agreements with China. One example is the Southeast Asian Nuclear Weapons–Free Zone Treaty signed in December 1995. Another is the ASEAN Economic Community, or AEC, a trade bloc patterned after the North American Free Trade Agreement and the European Union. Now, ASEAN focuses on increasing trade between members of the association.

In January of 2010, ASEAN established the ASEAN Free Trade Area (AFTA) to facilitate regional trade. The total intraregional trade between ASEAN countries in 2010 was more than trade with any other outside country or region, with ASEAN countries representing 25.9 percent of total imports and 25 percent of total exports (Figure 10.15). By 2012, ASEAN was addressing a variety of other issues, such as the free flow of services, skilled labor, investment, and capital; consumer protection; intellectual property rights; and leveling the playing field for competition. ASEAN's overall goals are to develop a sustainable regional economic community and to decrease development gaps between and within ASEAN member states. Externally, ASEAN has ratified free trade agreements with Australia, New Zealand, China, India, Japan, and South Korea.

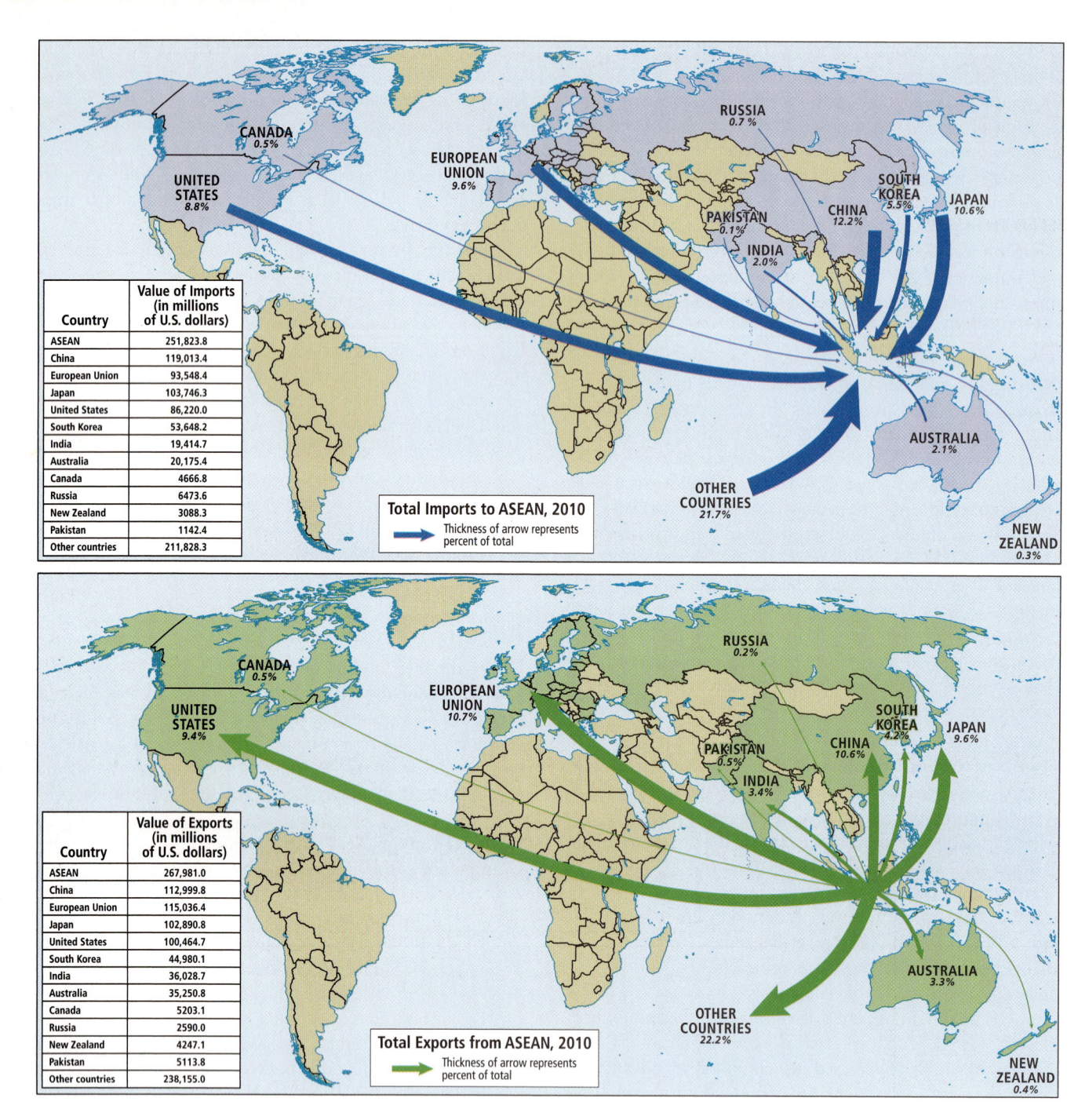

FIGURE 10.15 ASEAN imports and exports, 2010. The ASEAN countries are very active in world trade as importers and exporters. The largest share of trade, however, is among the ASEAN countries themselves, with imports at 25.9 percent and exports at 25 percent. [Source consulted: "ASEAN External Trade Statistics," Table 19, ASEAN, February 15, 2012, at http://www.aseansec.org/external-trade-statistics/]

Tourism International tourism is an important and rapidly growing economic activity in most Southeast Asian countries (Figure 10.16). Between 1991 and 2001, the number of international visitors to the region doubled to more than 40 million; in 2010, international visitors reached over 73.7 million. As in other trade matters, Southeast Asians are themselves increasingly touring neighboring countries; in 2010, for example, 34.8 million tourists to ASEAN countries were from within the region (Figure 10.17). This is a positive trend because familiarity between neighbors lays the groundwork for various forms of regional cooperation, such as infrastructure improvements.

In response to its popularity with global and regional tourists, ASEAN members have been working to improve the region's transportation infrastructure. One such project is the Asian Highway,

FIGURE 10.16 Development of the Asian Highway: Transportation infrastructure for tourism. Many of the Southeast Asian World Heritage Sites are located along the partially completed Asian Highway, which will eventually connect Europe to Indonesia. The UNESCAP report, "Development of the Asian Highway," describes the project and includes a map of the entire system. **(A)** Buddha heads sculpted from stone at the Angkor Wat temple complex in Cambodia, the largest religious monument in the world. [Sources consulted: "World Heritage List," UN World Heritage Convention, at http://whc.unesco.org/en/list/; "Tourism Attractions Along the Asian Highway," UN Economic and Social Commission for Asia and the Pacific, 2004, at http://www.unescap.org/our-work/transport/asian-highway/about; *Asia Times Online*, at http://www.atimes.com/atimes/Asian_Economy/images/highways.html]

a web of standardized roads that loops through the mainland and connects it with Malaysia, Singapore, and Indonesia (the latter via ferry; see the Figure 10.16 map). Eventually, the Asian Highway will facilitate ground travel through 32 Eurasian countries from Russia to Indonesia and from Turkey to Japan.

The surge of tourism in Southeast Asia has also raised concerns about becoming too dependent on an industry that leaves economies vulnerable to events that precipitously stop the flow of visitors (natural disasters, political upheavals), or that leave local people vulnerable to the sometimes destructive demands of tourists (see the discussion of sex tourism on pages 442–443). Examples of disasters are the tsunami of December 2004 that killed several thousand international tourists in Thailand and Indonesia, and various human-made disasters such as the terrorist bombings in Bali (2002, 2005) and in southern Thailand (2006, 2012). In addition, tourism can divert talent from contributing to national development, as the following vignette illustrates.

VIGNETTE

Tan Phuc waits patiently for the mechanic to mount the new rear tire on his Chinese-made 125cc motorcycle. He is a motorcycle tour guide in Vietnam who likes to take his clients to places along Highway 14 in the Central Highlands, close to the historic Ho Chi Minh Trail. Tan's current client, a Dutch tourist, has spent 3 days filling his digital camera with images of vast coffee plantations covered in snowy blossoms, of silkworms frying in oil while their cocoons are unwound and spun into silk thread, and of indigenous minority children who met the gaze of his camera with casual curiosity. Tan finds his clients in bus stations and backpacker hostels and negotiates a daily rate for tours—usually between U.S.$50 and U.S.$75, not insignificant in a country that has an average per capita annual income of U.S.$2790.

Tan Phuc's life has always been affected by global forces. As a child in the Mekong Delta, he sold produce to U.S. soldiers at a nearby base. During the war, he lost his brother, a soldier for

South Vietnam; after the war, his father spent 10 years in a Communist reeducation camp. Even with the hardships, Tran obtained an education. But when the annual inflation rate of 400 percent shrank his salary as a high school math teacher, Tran could not adequately meet his family's needs. Like other skilled Vietnamese people, Tran took advantage of Vietnam's transition to a market economy beginning in the 1980s and established a business that caters to tourists. This means that he, along with many other educated Vietnamese people, no longer works in occupations crucial to Vietnam's future, like education. [Source: The field work of Karl Russell Kirby. For detailed source information, see Text Sources and Credits.]

230. BANKERS, ANALYSTS SEE RESURGENT ASIA 10 YEARS AFTER ECONOMIC CRISIS

FIGURE 10.17 Visitor arrivals to ASEAN countries, 2010. [Source consulted: "ASEAN Tourism Marketing Strategy (ATMS) 2012–2015," Figures 2-2 and 2-3 (Jakarta: ASEAN Secretariat, March 2012), p. 13, at http://www.aseansec.org/tourism-statistics/]

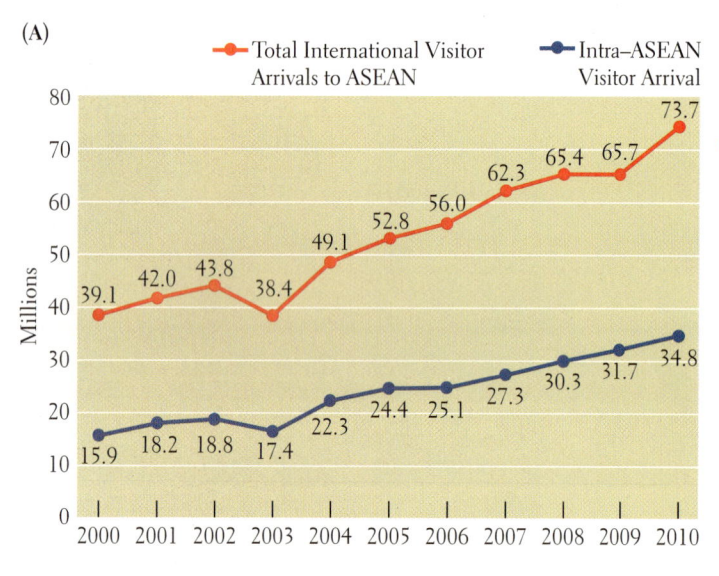

(A) Proportion of ASEAN visitors to total international visitor arrivals.

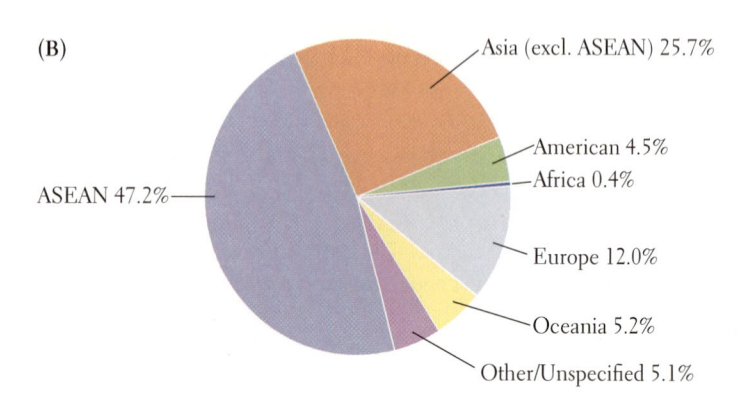

(B) Origin of international visitors to ASEAN countries.

THINGS TO REMEMBER

> **GEOGRAPHIC INSIGHT 2**

- **Globalization and Development** Globalization has brought spectacular successes as well as occasional declines to the economies of Southeast Asia. Key to the economic development of this region are strategies that were pioneered earlier in East Asia: the formation of state-aided market economies combined with export-led economic development.

- The region was hit by a devastating recession in the late 1990s that lasted as late as 2006 in some places, and by a less severe slowdown related to the global recession that began in 2007.

- One of the forces that contributed to the recessions of the last two decades was a kind of corruption known as crony capitalism.

- ASEAN started in 1967 as an anti-Communist, anti-China association, but it now focuses on agreements that strengthen regional cooperation, including agreements with China.

- Regional integration, including the tourism infrastructure of the region, is expected to foster growth and encourage cooperation with China.

POWER AND POLITICS

> **GEOGRAPHIC INSIGHT 3**

Power and Politics: There has been a general expansion of political freedoms throughout Southeast Asia in recent decades, but authoritarianism, corruption, and violence have at times reversed these gains.

While there is a general shift away from authoritarianism and toward more political freedoms, the Figure 10.18 map shows the wide variation in political freedoms in the region. Some people argue that authoritarianism has deep cultural roots in Southeast Asia and should be given greater respect. Others see a certain amount of authoritarianism as necessary to control corruption and political violence. Still others argue that respect for political freedoms is more likely to expose corruption and transform militant movements into peaceful political parties.

Southeast Asia's Authoritarian Tendencies

Some Southeast Asian leaders, such as Singapore's former prime minister, Lee Kuan Yew, have said that Asian values are not compatible with Western ideas of democracy and political freedom. Yew asserts that Asian values are grounded in the Confucian view that individuals should be submissive to authority; therefore Asian countries should avoid the highly contentious public debate of open electoral politics. Nevertheless, when confronted with governments that abuse their power, people throughout Southeast Asia have not submitted but rebelled (see Figure 10.18A, B). Even in Singapore, the Western-educated son of Lee Kuan Yew, Lee Hsien Loong, who is now prime minister,

FIGURE 10.18   PHOTO ESSAY: Power and Politics in Southeast Asia

There are significant barriers to public participation in politics in this region, including political violence and authoritarian political cultures. However, people have been pushing for more access to the political process, and some countries have recently enacted political reforms.

A Rohingya refugees at a camp in Burma. The Burmese government does not allow Rohingyas to become citizens. Burma has more than 100 distinct ethnic minority groups, many of whom are denied basic political freedoms. Burma's central government is dominated by the Burman ethnic group that forms more than two-thirds of the country's population. Several ethnically based violent insurgent groups are currently battling Burma's armed forces.

Democratization and Conflict
Armed conflicts and genocides with high death tolls since 1990
- Ongoing conflict
- 1000–5000 deaths
- 5000–50,000 deaths
- 50,000–300,000 deaths
- 300,000–1,00,000 deaths
- 1,000,000–3,000,000 deaths

Democratization index
- Full democracy
- Flawed democracy
- Hybrid regime
- Authoritarian regime
- No data

B The banned flag of the militant Free Papua Movement is carried at a demonstration in West Papua, Indonesia, shortly before police open fire. While political freedoms are growing in Indonesia, political violence and repression of political movements is still common.

C Members of the Moro Islamic Liberation Front (MILF) in Mindanao, Philippines, celebrate an accord between their organization and the government. In return for the MILF's cessation of militancy, the Muslim-dominated areas of Mindanao were granted more autonomy.

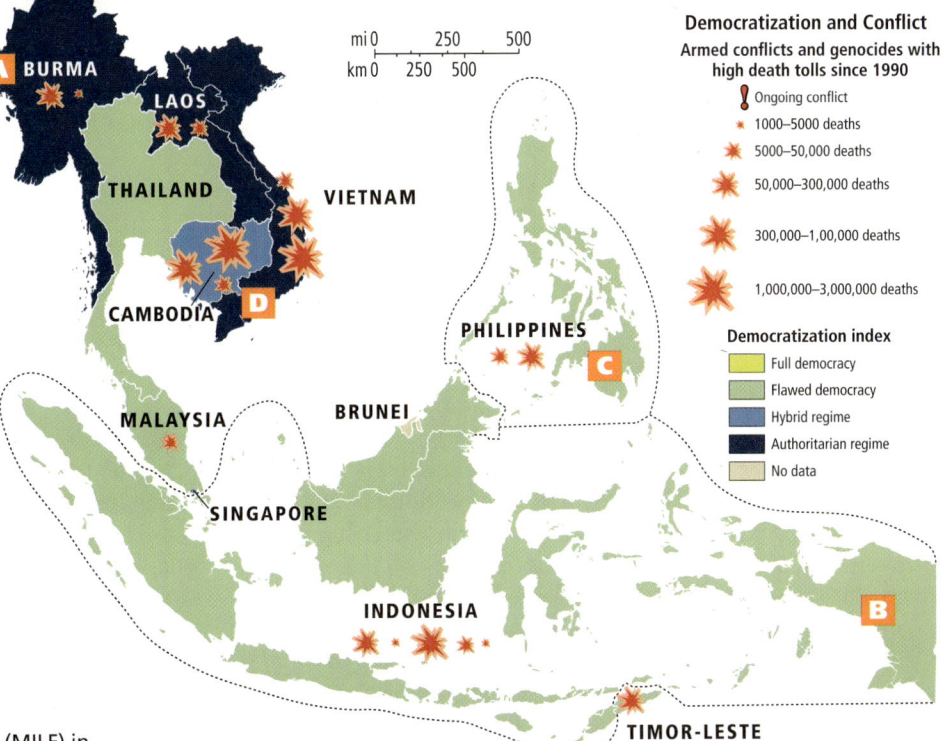

D Police and local militia members stop journalists from taking pictures outside a court in Ho Chi Minh City, Vietnam, where a pro-democracy activist was being tried. Vietnam's government remains communist and authoritarian and allows the population few political freedoms.

has expressed more interest in the growth of political freedoms than his father did.

While authoritarianism seems to be giving way to more political freedom, this process is often complicated by corruption and violent state repression of political movements. For example, after being plagued with a long line of dictators, the Philippines has now elected governments that have resolved some significant problems. However, militants in the southern islands continue an insurgency that periodically devolves into violence (see Figure 10.18C).

Thailand, long regarded as the most stable democracy in the region, is now dealing with deep divisions between those who favor authoritarianism (including many among the economic elite and the military) and those who are part of or support the large populist political movement. In 2006, the military took over Thailand's government after a corrupt but charismatic prime minister, Thaksin Shinawatra, was implicated in several scandals. Despite recent elections, the matter remains unresolved. Shinawatra is in exile in Europe and his many supporters periodically take over the streets of Bangkok. In 2011, Shinawatra's sister, Yingluck, was elected prime minister, with overwhelming support from her brother's supporters. In 2014, she was removed from power by Thailand's Constitutional Court, after which the military seized control of the government.

Other countries are also struggling to find a balance between meeting demands for more political freedoms and relying on authoritarianism to create stability. Cambodia's democracy is precarious, and violence there is common. The wealthier and usually stable countries of Malaysia and Singapore continue to use authoritarian versions of democracy that impose severe limitations on freedom of the press and freedom of speech.

When more democratically oriented governments fail to keep the peace, authoritarianism, often backed by a strong military, is seen as a justified counterforce to "too much" democracy. In virtually every country, the military has been called on to restore civil calm after periods of civil disorder. Military rank is highly regarded and many top elected officials have had military careers.

Authoritarian rule remains firmly entrenched in some countries such as Burma where numerous regional ethnic minorities and pro-democracy movements have been repressed with an iron fist for decades. Nevertheless, a gradual expansion of political freedoms is now taking place in Burma, as evidenced by the recent freeing of political prisoners, such as Aung San Suu Kyi, and a reduction in restrictions on the press.

Little political reform is occurring in Laos and Vietnam (see Figure 10.18D), where authoritarian socialist regimes have a firm grip on power, or in Brunei, which is governed by an authoritarian sultanate. For these and other reasons, authoritarianism is likely to remain a powerful force in the politics of this region for some time.

Militarism and China

While the countries of the region are unlikely to go to war against each other, they are keeping an eye on China, given its recent interest in the South China Sea. Newly discovered oil resources there, as well as fishing resources and vital shipping routes, have resulted in competing territorial claims by China, Taiwan, Vietnam, Malaysia, Brunei, Indonesia, and the Philippines. Most countries have recently invested heavily in military equipment, primarily from Europe and the United States. In the midst of this buildup, countries with no claims on the South China Sea are arming themselves. In recent years, tiny, wealthy, trade-centered Singapore has spent one quarter of its GDP on weapons. It has the largest military budget in the region and now produces armored troop carriers for use at home and for export. **322. THAILAND'S PROTESTERS HIGHLIGHT RIFTS, POLITICAL PARTICIPATION**

Can the Expansion of Political Freedoms Help Bring Peace to Indonesia?

All Southeast Asian countries have diverse multicultural populations that at times come into conflict with each other. Efforts to resolve differences take many forms, both peaceful and violent. Much attention is now focused on Indonesia as it works to find peaceful political solutions to ethnic conflicts.

An important but still tentative shift toward democracy took place in Indonesia in the wake of the economic crisis of the late 1990s. After three decades of semi-dictatorial rule by President Suharto, the economic crisis spurred massive demonstrations that forced Suharto to resign. Since then, democratic parliamentary and presidential elections have brought a new political era to the country.

Indonesia is the largest country in Southeast Asia and the most fragmented—physically, culturally, and politically. It is comprised of more than 17,000 islands (3000 of which are inhabited), stretching over 3000 miles (8000 kilometers) of ocean. It is also the most culturally diverse, with hundreds of ethnic groups and multiple religions. Although Indonesia has the largest Muslim population in the world, there are also many Christians, Buddhists, Hindus, and followers of various local religions and spiritual traditions. With all these potentially divisive forces, many wonder whether this multi-island country of 250 million might be headed for political disintegration.

Until the end of World War II, Indonesia was a loose assembly of distinct island cultures that Dutch colonists managed to hold together as the Dutch East Indies. When Indonesia became an independent country in 1945, its first president, Sukarno, hoped to forge a new nation out of these many parts, founded on a fairly strong communist ideology. To that end, he articulated a national philosophy known as *Pancasila*, which was aimed at holding the disparate nation together, primarily through nationalism and concepts of religious tolerance.

In 1965, during the height of the Cold War, Suharto, a staunchly anti-Communist general in the Indonesian army, ousted Sukarno in a coup and ruled the country for another 33 years. It is now clear that Suharto's regime was responsible for a purge of suspected Communists, during which as many as a million people were killed.

Despite his abrupt removal from office, Sukarno's unifying idea of Pancasila endures as a central theme of life in Indonesia (and less formally throughout the region). Pancasila embraces five precepts: *belief in God*, and the observance of *conformity*, *corporatism* (often defined as "organic social solidarity with the state"), *consensus*, and *harmony*. These last four precepts could be interpreted as discouraging dissent or even loyal opposition, and they seem to require a perpetual stance of boosterism. Pancasila has been criticized by some Indonesians as being too secular, while others see it as not going far enough to protect the freedom of people to believe in multiple deities, as Indonesia's many Hindus do, or freedom to not believe in any deity at all. Some praise Pancasila's other precepts for counteracting the extreme ethnic diversity and geographic dispersion of the country, while others say that the precepts have had a chilling effect on participatory democracy and on criticism of the government, the president, and the army.

The first orderly democratic change of government in Indonesia did not take place until national elections in 2004. Since then, there have been several peaceful elections, and the government is stable enough to allow citizens to publicly protest some policies. But corruption remains a threat that encourages feelings of nostalgia for the Suharto model of authoritarian government.

Separatist movements have sprouted in four distinct areas in recent years, demonstrating Indonesia's fragility. The only rebellion to succeed was in Timor-Leste, which became an independent country in 2002. However, its case is unique in that this area was under Portuguese control until 1975, when it was forcibly integrated into Indonesia. Two other separatist movements (in the Molucca Islands and in West Papua; see Figure 10.18B) developed largely in response to Indonesia's forced **resettlement schemes**. Also known as *transmigration schemes*, these programs have relocated approximately 20 million people from crowded islands such as Java to less densely settled islands since 1965. The policies were originally initiated under the Dutch in 1905 to relieve crowding and provide agricultural labor for plantations in thinly populated areas. After independence, Indonesia used resettlement schemes both to remove troublesome people and to bring outlying areas under closer control of the central government in Jakarta. Resettlement schemes continue today, though at a much smaller scale of roughly 60,000 people per year.

resettlement schemes government plans to move large numbers of people from one part of a country to another to relieve urban congestion, disperse political dissidents, or accomplish other social purposes; also called *transmigration*

The far-western province of Aceh, in the north of Sumatra, has long been troubled by political violence. However, recent expansions of political freedoms in Aceh may have charted a course to peace. Conflicts there originally developed because most of the wealth yielded by Aceh's resources, primarily revenues from oil extraction, was going to the central government in Jakarta. The Acehnese people protested what they saw as the expropriation of oil without compensation, and Jakarta sent the military to silence them. Many who spoke out against the military presence were accused of terrorism and arrested, jailed, forced into hiding, or killed. The conflict seemed unresolvable. Then in 2004, the earthquake and tsunami in Aceh, which killed more than 170,000 Acehnese, suddenly brought many outside disaster relief workers to Sumatra because the Indonesian government was unable to give sufficient aid to the victims. Global press coverage of the relief effort mentioned the recent political violence; this created a powerful incentive for separatists and the government to cooperate in order to receive outside aid. A resulting peace accord signed in 2005 brought many former combatants into the political process as democratically elected local leaders. Virtually all the separatists laid down their arms.

Terrorism, Politics, and Economic Issues Like authoritarianism, terrorism has been a counterforce to political freedoms in this region. Terrorist violence short-circuits the public debate that is at the heart of democratic processes and appears to reinforce the need for repressive authoritarian measures. However, it is important to recognize that terrorist movements often thrive in the context of both economic deprivation and political repression, two factors that are often interrelated. As is the case in the southern Philippines (see Figure 10.18C), terrorism seems to draw support from people who feel shut out of opportunities for economic advancement. The peaceful solution to terrorism seems to be to listen to both the political and economic desires of those who might be attracted to terrorism.

225. TERROR AND ISLAMIC STRUGGLE IN INDONESIA
232. VIOLENCE IN THAILAND'S MUSLIM SOUTH INTENSIFIES

THINGS TO REMEMBER

› **GEOGRAPHIC INSIGHT 3** • **Power and Politics** There has been a general expansion of political freedoms throughout Southeast Asia in recent decades, but authoritarianism, corruption, and violence have at times reversed these gains.

• Some Southeast Asian leaders, such as Singapore's former prime minister, Lee Kuan Yew, have said that Asian values are not compatible with Western ideas of democracy and political freedom.

• While the countries of the region are unlikely to go to war against each other, they are keeping an eye on China, given its recent interest in the South China Sea.

• All Southeast Asian countries have diverse multicultural populations that at times come into conflict with each other.

• Like authoritarianism, terrorism has been a counterforce to political freedoms in this region. ■

URBANIZATION

› **GEOGRAPHIC INSIGHT 4**

Urbanization: While Southeast Asia as a whole is only 43 percent urban, its cities are growing rapidly as agricultural employment declines and urban industries expand. The largest Southeast Asian cities, which are receiving most of the new rural-to-urban migrants, rarely have sufficient housing, water, sanitation, or jobs for all their people.

Farmers who were once able to produce enough food for their families are now forced to migrate to cities, where they must purchase their food, and where the only affordable housing is in slums that lack essential services.

Southeast Asia as a whole is only 43 percent urban, but the rural–urban balance is shifting steadily in response to declining agricultural employment and booming urban industries. The forces driving farmers into the cities are called the *push factors* in rural-to-urban migration. They include the rising cost of farming, which is related to the use of new technologies and competition with agribusiness. *Pull factors*, in contrast, are those that attract people to the city, such as abundant manufacturing jobs and education opportunities. In Southeast Asia, as in all other regions, these factors have come together to create steadily increasing urbanization. Malaysia is already 64 percent urban; the Philippines, 63 percent; Brunei, 72 percent; and Singapore, 100 percent.

Employment in agriculture has been declining throughout Southeast Asia since the introduction of new production methods that increase the use of labor-saving equipment and reduce the need for human labor. Chemical pesticide and fertilizer use has also grown. While such additives can increase harvests and the supply of food to cities, they also drive the cost of production higher than what most farmers can afford. Many family farmers have sold their land to more prosperous local farmers or to agribusiness corporations and have moved to the cities. These people, skilled at traditional farming but with little formal education, often end up in the most menial of urban jobs (Figure 10.19C) and live in circumstances that do not allow them to grow their own food (see Figure 10.19A).

Labor-intensive manufacturing industries (garment and shoe making, for instance) are expanding in the cities and towns of the poorer countries, such as Cambodia, Vietnam, and parts of Indonesia and Timor-Leste. In the urban and suburban areas of the wealthier countries—Singapore, Malaysia, Thailand, and parts of Indonesia and the northern Philippines—technologically sophisticated manufacturing industries are also growing. These include automobile assembly, chemical and petroleum refining, and computer and other electronic equipment assembly. Riding on this growth in manufacturing are innumerable construction projects that often provide employment to recent migrants (see Figure 10.19B, D).

Cities like Jakarta, Manila, and Bangkok, among the most rapidly growing metropolitan areas in the world, are *primate cities*—cities that, with their suburbs, are vastly larger than all others in a country. Bangkok is more than 20 times larger than Thailand's next-largest metropolitan area, Udon Thani; and Manila is more than 10 times larger than Davao, the second-largest city in the Philippines. Thanks to their strong industrial base and political power and the massive immigration they attract, primate cities can dominate whole countries; and in the case of Singapore, the city constitutes the entire country.

Rarely can such cities provide sufficient housing, water, sanitation, or decent jobs for all the new rural-to-urban migrants. Many millions of urban residents in this region live in squalor, often on floating raft-villages on rivers and estuaries. Of all the cities in Southeast Asia, only Singapore provides well for nearly all of its citizens (see Figure 10.19B). Even there, however, a significant undocumented non-citizen population lives in poverty on islands surrounding the city. The experience of rural-to-urban migrants who go to Bangkok or Jakarta is more typical; migrants there often live in slums on the banks of polluted, trash-ridden bodies of water (see Figure 10.19A).

Emigration Related to Globalization

The same push and pull factors behind urbanization are also driving millions to migrate out of Southeast Asia. These migrants are a major force of globalization, as they supply much of the world's growing demand for low- and middle-wage workers who are willing to travel or live temporarily in foreign countries. For example, 40 percent of the foreign workers in Taiwan are Indonesian. These migrants are also a globalizing force within their home countries because their remittances (monies sent home) boost family incomes and supply governments with badly needed **foreign exchange** (foreign currency) that countries use to purchase imports. Filipinos working abroad are their country's largest source of foreign exchange, sending home more than U.S.$6 billion annually and increasing household annual income by an average of 40 percent.

> **foreign exchange** foreign currency that countries need to purchase imports

The Merchant Marines Skilled male seamen from Southeast Asia make up a significant portion of the international merchant marines, where conditions are considerably better than they are for most migrant workers. Nonetheless, in the merchant marines, it is customary for workers to be paid according to their homeland's pay scales, which makes employing seamen from low-wage Southeast Asian countries attractive to ship owners. The seamen work aboard international freighters or on luxury cruise liners as deckhands, cooks, engine mechanics; a few become officers. Generally, seamen work for 6 months at a time—with only a few hours a day for breaks—saving nearly every penny. At the end of a tour of duty, they return home to their families for another 6 months, where they often contribute financially to the well-being of an extended group of kin and friends and send their children to receive an advanced education.

The Maid Trade Women constitute well over 50 percent of the more than 8 million emigrants from Southeast Asia. Many skilled nurses and technicians from the Philippines work in European, North American, and Southwest Asian cities. About 3 million participate in the global "maid trade" (shown in Figure 10.20A and the figure map). Most are educated women from the Philippines and Indonesia who work under 2- to 4-year contracts in wealthy homes throughout Asia. An estimated 1 to 3 million Indonesian maids now work outside of the country, mostly in the Persian Gulf.

The maid trade has become notorious for abusive working conditions and for having employers who often do not pay what they promise. In Saudi Arabia, the NGO Human Rights Watch is monitoring the cases of Muslim Indonesian women who were brutally abused—two were killed—by members of a privileged

FIGURE 10.19 PHOTO ESSAY: Urbanization in Southeast Asia

Southeast Asia is rapidly urbanizing as manufacturing and service sector industries pull in people from rural areas and as changes in agriculture push farmers to the cities. Many cities are struggling to cope with rapid growth. [Source consulted: *2011 World Population Data Sheet*, Population Reference Bureau, at http://www.prb.org/pdf11/2011population-data-sheet_eng.pdf]

A A slum area in Jakarta, Indonesia, where 62 percent of the population lives in slums. Jakarta will grow by almost 50 percent by 2020.

B Publicly built housing in Singapore, where 85 percent of the population lives in apartment buildings designed, built, and managed by the government.

C Migrant workers from rural areas pull carts to a construction site in Hanoi, Vietnam. Many migrants start out in low-wage, manual labor jobs like this one.

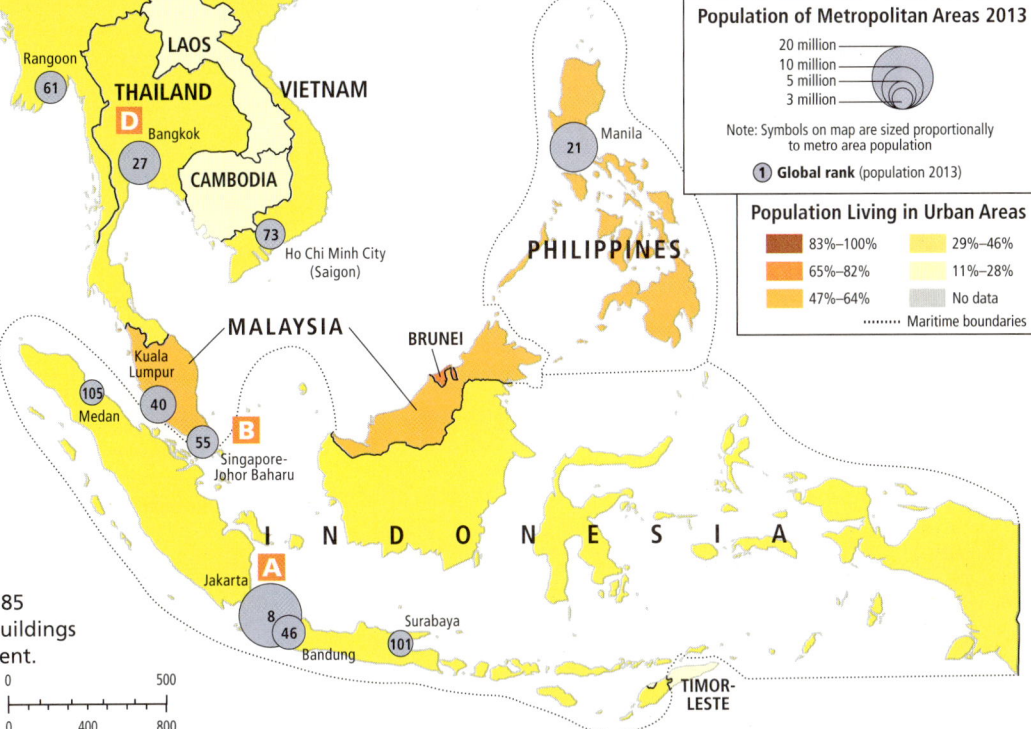

D A woman works at a construction site in Bangkok, Thailand. Gender roles are changing as more women move to the cities.

Population of Metropolitan Areas 2013
- 20 million
- 10 million
- 5 million
- 3 million

Note: Symbols on map are sized proportionally to metro area population

① Global rank (population 2013)

Population Living in Urban Areas
- 83%–100%
- 65%–82%
- 47%–64%
- 29%–46%
- 11%–28%
- No data
- .... Maritime boundaries

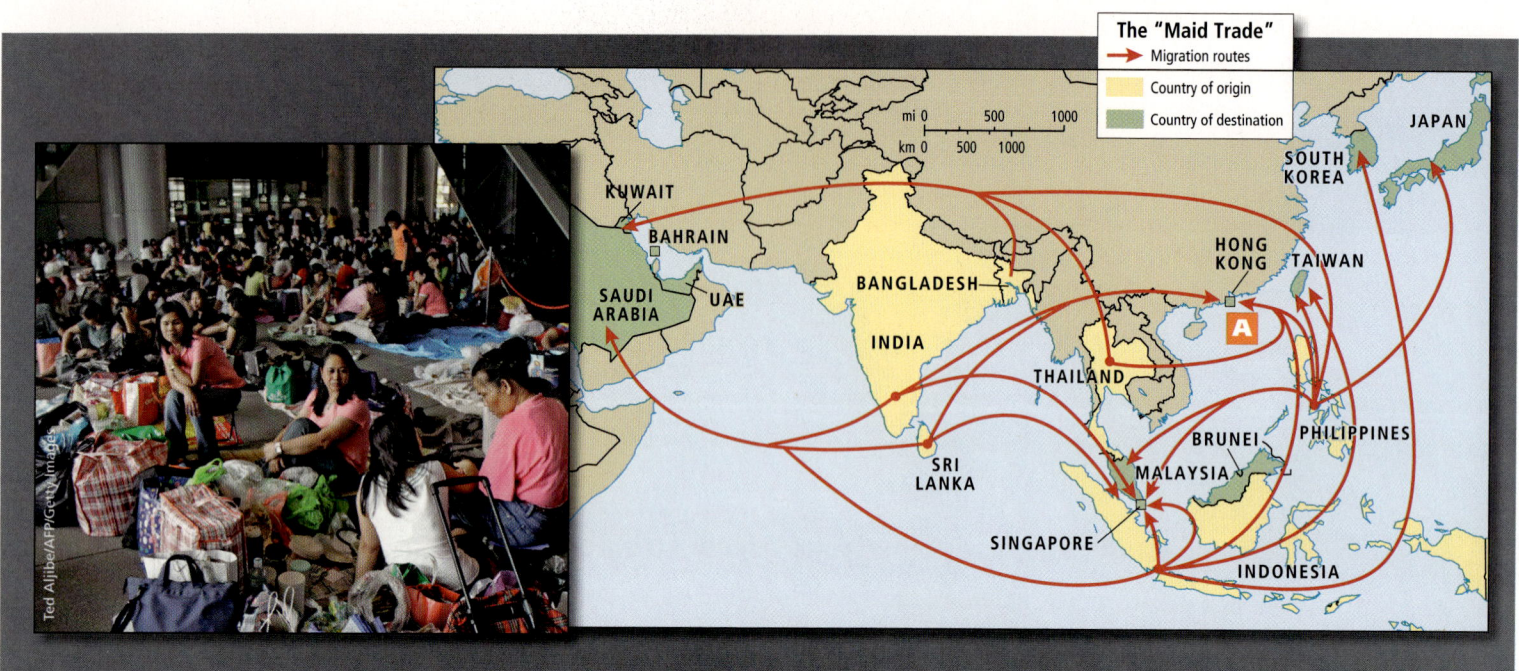

FIGURE 10.20 Globalization: The "maid trade." By 2009, about 3 million women, primarily from the Philippines, Indonesia, and Sri Lanka, were working as domestic servants across Asia. **(A)** Filipina housemaids and nannies in Hong Kong relax on blankets on their day off. The Philippine government mandates that they must have Sundays off, so many public areas in central Hong Kong are occupied on Sundays by small groups of women talking, trading goods, playing games, and packing up things to send home. **(B)** Map of the "maid trade," showing the origins and destinations of the maids. [Sources consulted: Joni Seager, *The Penguin Atlas of Women in the World* (New York: Penguin Books, 2003), p. 73, with updated information from the Migration Policy Institute, at http://www.migrationpolicy.org/regions/asia-and-pacific?qt-recent_program_activities=4#qt-recent_program_activities.]

> **THINKING GEOGRAPHICALLY** How does this photo portray the inventiveness of women in the maid trade?

Saudi family. The Philippines went so far as to ban the maid trade in 1988, but reestablished it in 1995 after better pay and working conditions were negotiated with the countries that receive the workers.

VIGNETTE Every Sunday is amah (nanny) day in Hong Kong. Gloria Cebu and her fellow Filipina maids and nannies stake out temporary geographic territory on the sidewalks and public spaces of the central business district (see Figure 10.20A). Informally arranging themselves according to the different dialects of Tagalog (the official language of the Philippines) they speak, they create room-like enclosures of cardboard boxes and straw mats, where they share food, play cards, give massages, and do each other's hair and nails. Gloria says it is the happiest time of her week, because for the other 6 days she works alone, caring for the children of two bankers.

Gloria, who is a trained law clerk, has a husband and two children back home in Manila. Because the economy of the Philippines has stagnated, she can earn more in Hong Kong as a nanny than in Manila in the legal profession. Every Sunday she sends most of her income (U.S.$125 a week) home to her family. [Sources: Kirsty Vincin and the Economist. For detailed source information, see Text Sources and Credits.]

THINGS TO REMEMBER

> **GEOGRAPHIC INSIGHT 4**
• **Urbanization** While Southeast Asia as a whole is only 43 percent urban, its cities are growing rapidly as agricultural employment declines and urban industries expand. The largest Southeast Asian cities, which are receiving most of the new rural-to-urban migrants, rarely have sufficient housing, water, sanitation, or jobs for all their people.

• Labor-intensive manufacturing industries (garment and shoe making, for instance) are expanding in the cities and towns of the poorer countries, such as Cambodia, Vietnam, and parts of Indonesia and Timor-Leste.

• In the urban and suburban areas of the wealthier countries—Singapore, Malaysia, Thailand, and parts of Indonesia and the northern Philippines—technologically sophisticated manufacturing industries are also growing.

• Cities like Jakarta, Manila, and Bangkok, among the most rapidly growing metropolitan areas in the world, are *primate cities*—cities that, with their suburbs, are vastly larger than all others in a country and are often the focus of rural-to-urban migration.

• Thousands of people emigrate from the region each year to seek temporary or long-term employment in the Middle East, Europe, North America, Asia, or on the world's oceans.

POPULATION AND GENDER

> **GEOGRAPHIC INSIGHT 5**
>
> **Population and Gender:** Population dynamics vary considerably in this region because of differences in economic development, government policies, prescribed gender roles, and religious and cultural practices. With regard to gender, economic change has brought better job opportunities and increased status for women, who then often choose to have fewer children. Some countries also have gender imbalances because of a cultural preference for male children.

Southeast Asia is home to well over 600 million people (almost double the U.S. population), who occupy a land area that is about one-half the size of the United States. At current rates, Southeast Asia's population, which is large and growing, is projected to reach more than 800 million by 2050, by which time much of this population will live in cities (Figure 10.21). However, population projections could be inaccurate, both because rates of natural increase are continuing to slow markedly and because many Southeast Asians are migrating to find employment outside the region.

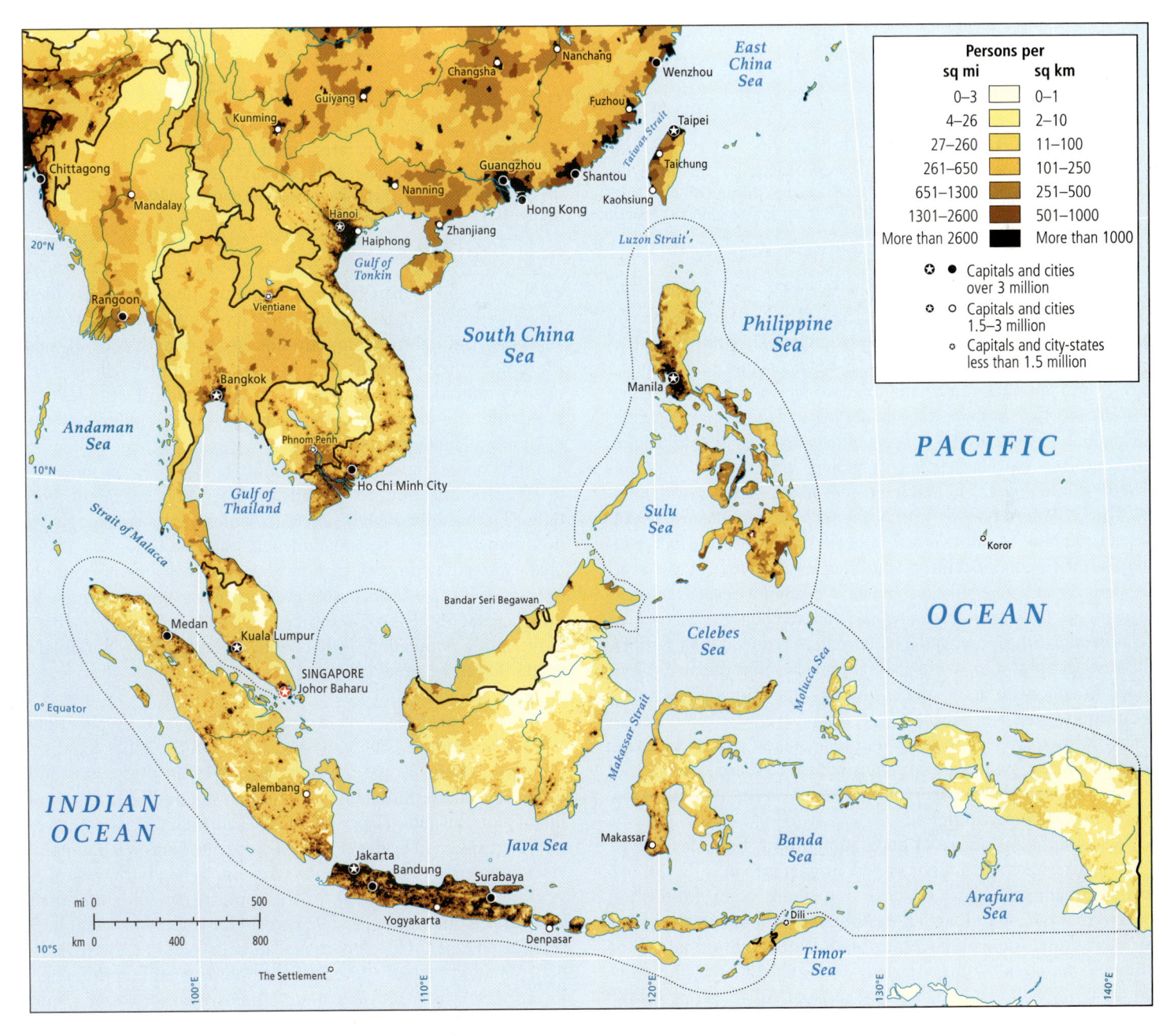

FIGURE 10.21 Population density in Southeast Asia. Population growth in Southeast Asia is slowing, largely related to economic development, urbanization, changing gender roles, and government policies. Fertility rates have declined sharply in all countries since the 1960s but are still high in the poorest areas.

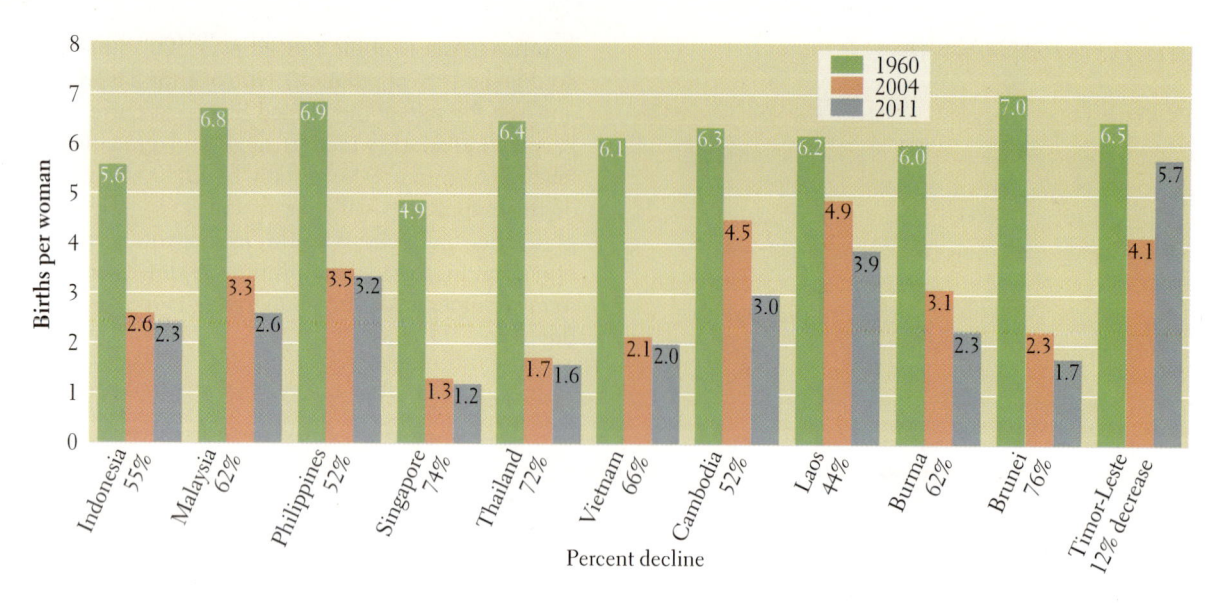

FIGURE 10.22 Total fertility rates, 1960, 2004, and 2011. Total fertility rates declined for all Southeast Asian countries from 1960 to 2011, but they did increase in Timor-Leste between 2004 and 2011. [Sources consulted: Sasha Loffredo, *A Demographic Portrait of South and Southeast Asia* (Washington, DC: Population Reference Bureau, 1994), p. 9; *2009 World Population Data Sheet* (Washington, DC: Population Reference Bureau, 2009)]

Population Dynamics Population dynamics vary considerably among the countries of Southeast Asia. The variation is due in part to differences in economic development, government policy, prescribed gender roles, and broader religious and cultural practices. Most countries are nearing the last stage of the demographic transition, where births and deaths are low and growth is minuscule or slightly negative. In the last several decades, overall fertility rates in Southeast Asia have dropped rapidly (Figure 10.22). Whereas women formerly had 5 to 7 children, they now have 1 to 3, with Laos (3.9) and Timor-Leste (5.7) being the major exceptions. However, because in most countries populations are still quite young (between one-quarter and one-third of the people are aged 15 years or younger), modest population growth is likely for several decades because so many are just coming into their reproductive years.

Brunei, Singapore, and Thailand have reduced their fertility rates so sharply—below replacement levels—that they will soon need to cope with aging and shrinking populations. Regionally, education levels are the highest in Singapore, where most educated women work outside the home at skilled jobs and professions. The Singapore government is now so concerned about the low fertility rate that it offers young couples various incentives for marrying and procreating.

An important source of population growth for Singapore is the steady stream of highly skilled immigrants that its vibrant economy attracts. Many of these immigrants come from elsewhere in the region, though the country also draws highly skilled workers from the United States and Europe. Singapore also attracts immigrants from elsewhere in Southeast Asia, primarily in the building trades and low-wage services. All countries in this region are losing population to emigration, with the exceptions of Singapore, Malaysia, and Brunei.

Thailand's low fertility rate of 1.6 children per adult woman was achieved in part via a government-sponsored condom campaign, and in part by rapid economic development, which made many couples feel that smaller families would be best. Also, as women gained more opportunities to work and study outside the home, they decided to have fewer children. High literacy rates for both men and women, along with Buddhist attitudes that accept the use of contraception, have also been credited for the decline in Thailand's fertility rate.

The poorest and most rural countries in the region show the usual correlation between poverty, high fertility, and infant mortality. Timor-Leste, which suffered a violent, impoverishing civil disturbance before it became independent from Indonesia, is still characterized by poverty, high fertility, and high infant mortality (64 per 1000 live births). However, its oil resources promise to bring prosperity if those resources are used to help the welfare of the entire population. If that happens, both fertility and infant mortality can be expected to fall.

In Cambodia and Laos, fertility rates average 3.0 and 3.9 children, respectively. Infant mortality rates are 58 per 1000 live births for both Cambodia and Laos. On the other hand, in Vietnam, where people are only slightly more prosperous and urbanized, the fertility rate (2.0 children per adult woman) and infant mortality rate (16 per 1000 births) are far lower. These lower rates are explained by the fact that as a socialist state, Vietnam provides basic education and health care—including birth control—to all, regardless of income. In Vietnam, literacy rates are more than 96 percent for men and 92 percent for women, whereas they are only 64 percent for women in Cambodia and in Laos. In addition, Vietnam's rapidly developing economy is pulling in foreign investment, which provides more employment for women, and careers are replacing child rearing as the central focus of many women's lives.

Out of concern that these changes weren't happening rapidly enough, and that a rapidly growing population would jeopardize

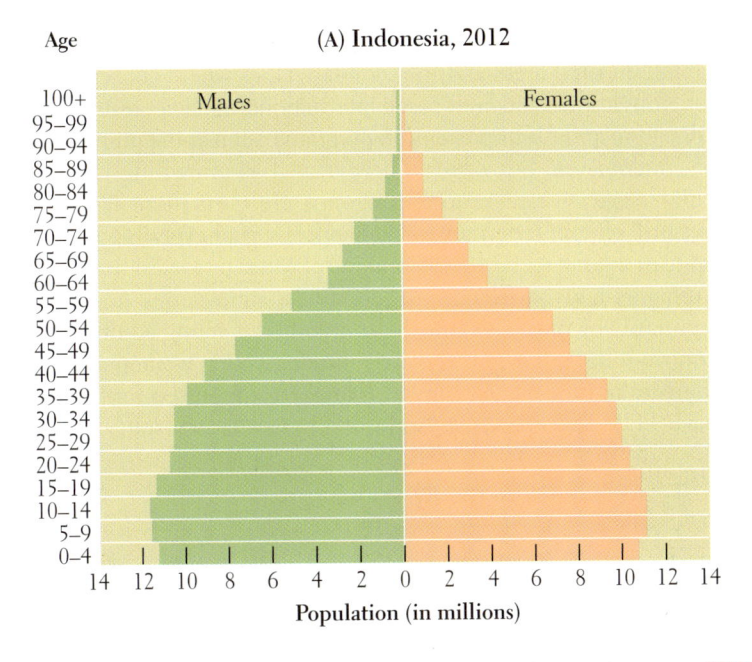

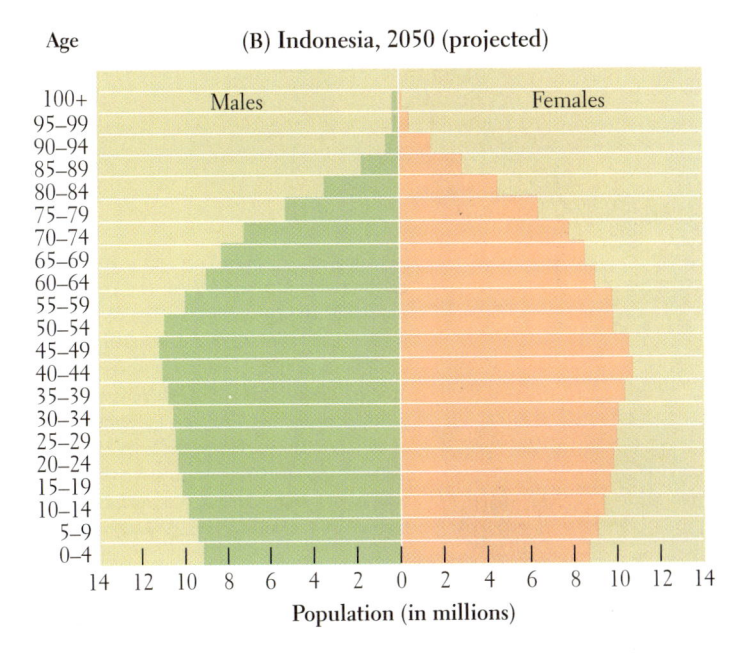

FIGURE 10.23 Population pyramids for Indonesia, 2012 and 2050 (projected). In 2012, Indonesia's population was 248.2 million. It is projected to be 313 million in 2050. [Source consulted: *International Data Base*, U.S. Census Bureau, 2012, at http://www.census.gov/population/international/data/idb/informationGateway.php]

Vietnam's upswing in economic development, the government of Vietnam recently reintroduced a two-child family policy that it has had on and off since the 1960s. This policy has already brought on gender imbalance, as some couples are choosing abortion if the fetus is female.

The Philippines, which has a higher per capita income than Laos, Cambodia, Timor-Leste, or Vietnam, is an anomaly in population patterns, primarily because it is predominantly Roman Catholic (83 percent, see the Figure 10.25 map on page 439), a religion that officially does not allow birth control. Fertility there is among the highest in the region (3.2 per adult female); infant mortality is in the medium range (22 per 1000 live births); and maternal childbirth deaths are high despite female literacy rates also being high (93 percent).

VIGNETTE In the Roman Catholic Philippines, Gina Judilla, who works outside the home, has had six children with her unemployed husband. They wanted only two, but because of the strong role of the Catholic Church and the political pressure it exerts, birth control was not available to them. Abortion is legal only to save the life of the mother, so with every succeeding pregnancy she tried folk methods of inducing an abortion. None worked. Now she can afford to send only two of her six children to school.

A move by family planners to provide national reproductive health services and sex education is underway. A recent survey showed that 48 percent of all pregnancies in the Philippines in 2010 were unintended, which often led to illegal "backstreet" abortions. In 2011, only one-third of Philippine women had access to modern birth control methods. [Source: Likhaan Center for Women's Health, Inc., and the New York Times. For detailed source information, see Text Sources and Credits.] ■

Population Pyramids The youth and gender features of Southeast Asian populations are best appreciated by looking at the population pyramids for Indonesia. Figure 10.23 shows the 2012 population and the projected population for 2050. The wide bottom of the 2012 pyramid indicates that most people are age 30 and under, but the projections to 2050 show that eventually, with declining birth rates, those 30 and under will be outnumbered by those 35 and older. Indonesia will accumulate ever-larger numbers in the upper age groups, and the pyramid will eventually be more box-shaped, as those for Europe are now. The gender disparities (more males than females) that have developed over the last 20 years can be seen by carefully examining the length of the bars on the male and female sides of the pyramid for those aged 15–19 and under. The difference is slight but significant because there is a rather consistent deficit of females, which indicates that they were selected out before birth, as is the case in Vietnam.

Southeast Asia's Encounter with HIV-AIDS As in sub-Saharan Africa (see Chapter 7), HIV-AIDS is a significant public health issue in Southeast Asia, although infection and death rates are now declining across the region. Cambodia, Thailand, and Burma currently have the highest infection rates.

Thailand is one of the few countries that has been able to drastically reduce the incidence of HIV-AIDS. It did so through a well-funded program that increased the use of condoms, decreased STDs dramatically, and reduced visits to sex workers by half. In Thailand in 2001, AIDS was the leading cause of death, overtaking stroke, heart disease, and cancer, but it is now just the third leading cause of death. Estimates are that about 500,000 Thais are infected, down from 1 million 12 years ago. Men between the ages of 20 and 40 have the highest rates of infection, but the rate of infection among women is rising.

FIGURE 10.24 LOCAL LIVES
Festivals in Southeast Asia

A Songkran, or New Year's Day, is celebrated in Thailand. Water is thrown on friends and passersby as a sign of respect and renewal. This developed from the tradition of gathering water used to wash statues of the Buddha and gently pouring that water on the shoulder of elderly relatives. However, since the festival happens during April, a very hot time of the year, a tradition developed of throwing water on anyone. [STR/AFP/Getty Images]

B A calligrapher at the Temple of Literature in Hanoi, Vietnam, elegantly writes down intentions for Tet (the New Year, celebrated in January or February). Exchanges of such calligraphy are hung in people's homes; the expressed intentions surround a particular idea or quality such as happiness, wealth, virtue, knowledge, talent, or long life. [Hoang Dinh Nam/AFP/Getty Images]

C An elaborately costumed dancer in the Ati-Atihan Festival in Kalibo, Aklan, in the Philippines. Originally celebrating the migration to the Philippines of a group of indigenous people from the island of Borneo, Ati-Atihan includes non-Christian traditions but is also celebrated by Christian Filipinos as a day honoring the infant Jesus. [Jeremy Jones Villasis/FlickrVision/Getty Images]

Elsewhere in the region, HIV rates are expected to increase rapidly in rural areas and in secondary cities where conservative religious leaders and faith-based international agencies restrict sex education and AIDS-prevention programs, such as the promotion of condom use. Sex education is viewed as promoting promiscuity. Aggressive prevention programs are even more essential because of popular customs that support sexual experimentation (at least among men), the reluctance of women to insist that their husbands and boyfriends use condoms, intravenous drug use (primarily by men), and the high mobility of young adults. Also contributing to the spread of HIV among young women are sex tourism and sex-related human trafficking (discussed further on page 442). **233. ACTIVISTS: AIDS LINKED TO WOMEN'S RIGHTS ABUSES**

THINGS TO REMEMBER

› **GEOGRAPHIC INSIGHT 5**
- **Population and Gender** Population dynamics vary considerably in this region because of differences in economic development, government policies, prescribed gender roles, and religious and cultural practices. With regard to gender, economic change has brought better job opportunities and increased status for women, who then often choose to have fewer children. Some countries also have gender imbalances because of a cultural preference for male children.

- While population growth rates have slowed across the region, they are slowing the most in the developed economies of Singapore, Brunei, and Thailand.

- An important source of population growth for Singapore is the steady stream of highly skilled immigrants that its vibrant economy attracts.

- The variations in population growth are linked to relative prosperity, gender roles, and availability of birth control.

- HIV-AIDS is a threat, to varying degrees, throughout the region. Preventative measures have proven quite successful in Thailand.

SOCIOCULTURAL ISSUES

Because of the region's long and complex history, the people of Southeast Asia have a great diversity of cultures and religious traditions. Now globalization and urbanization are adding yet more diversity as new cultural influences come in from abroad and as urban women gain more independence and pursue careers outside the home.

Cultural and Religious Pluralism

Southeast Asia is a place of **cultural pluralism** in that it is inhabited by groups of people from many different backgrounds. Over the past 40,000 years, migrants have come to the region from India, the Tibetan Plateau, the Himalayas, China, Southwest Asia, Japan, Korea, and the Pacific. Many of these groups have remained distinct, partly because they lived in isolated pockets separated by rugged topography or seas. However, the religious practices and traditions of many groups show diverse cultural influences (Figure 10.24).

cultural pluralism the cultural identity characteristic of a region where groups of people from many different backgrounds have lived together for a long time but have remained distinct

Sociocultural Issues **439**

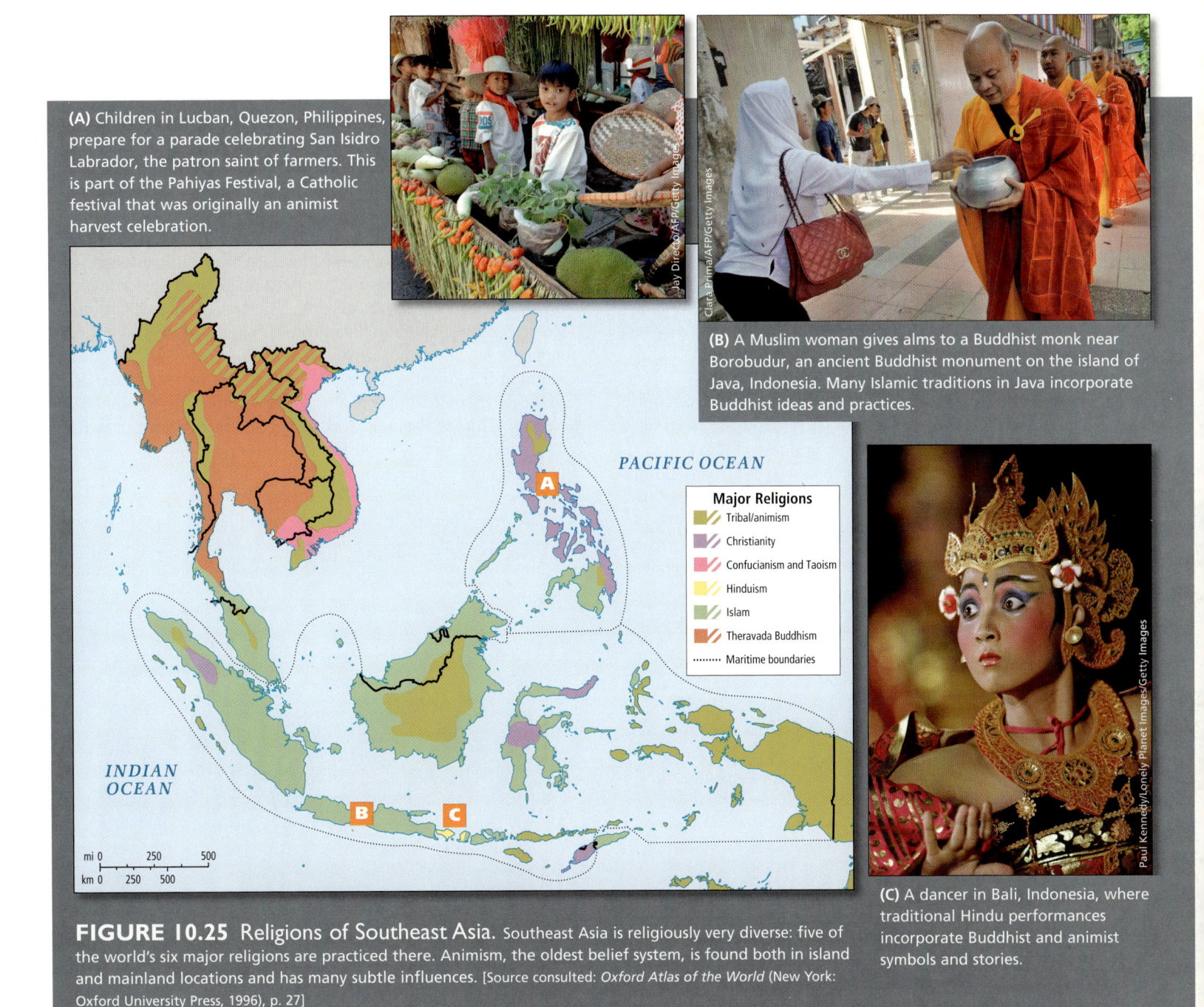

(A) Children in Lucban, Quezon, Philippines, prepare for a parade celebrating San Isidro Labrador, the patron saint of farmers. This is part of the Pahiyas Festival, a Catholic festival that was originally an animist harvest celebration.

(B) A Muslim woman gives alms to a Buddhist monk near Borobudur, an ancient Buddhist monument on the island of Java, Indonesia. Many Islamic traditions in Java incorporate Buddhist ideas and practices.

(C) A dancer in Bali, Indonesia, where traditional Hindu performances incorporate Buddhist and animist symbols and stories.

FIGURE 10.25 Religions of Southeast Asia. Southeast Asia is religiously very diverse: five of the world's six major religions are practiced there. Animism, the oldest belief system, is found both in island and mainland locations and has many subtle influences. [Source consulted: *Oxford Atlas of the World* (New York: Oxford University Press, 1996), p. 27]

As discussed in the "Religious Legacies" section (page 420), the major religious traditions of Southeast Asia include Hinduism, Buddhism, Confucianism, Taoism, Islam, Christianity, and animism (Figure 10.25). In animism, a belief system common among many indigenous peoples, natural features such as rocks, trees, rivers, crop plants, and the rains all carry spiritual meaning. These natural phenomena are the focus of festivals and rituals to give thanks for bounty and to mark the passing of the seasons, and these ideas have permeated all the imported religious traditions of the region.

The patterns of religious practice are complex; the patterns of distribution reveal an island–mainland division. All but animism originated outside the region and were brought primarily by traders, priests (Brahmin and Christian), and colonists. Buddhism is dominant on the mainland, especially in Burma, Thailand, and Cambodia. In Vietnam, people practice a mix of Buddhism, Confucianism, and Taoism that originated in China. Islam is dominant in Indonesia (the world's largest Muslim country), on the southern Malay Peninsula, and in Malaysia, and it is increasing in popularity in the southern Philippines. Roman Catholicism is the predominant religion in Timor-Leste and the Philippines, where it was introduced by Portuguese and Spanish colonists, respectively. Hinduism first arrived with

Indian traders thousands of years ago and was once much more widespread; now it is found only in small patches, chiefly on the islands of Bali and Lombok, east of Java. Recent Indian immigrants who came as laborers in the twentieth century (during the latter part of the European colonial period) have reintroduced Hinduism to Burma, Malaysia, and Singapore, but only as minority communities.

All of Southeast Asia's religions have changed as a result of exposure to one another. Many Muslims and Christians believe in spirits and practice rituals that have their roots in animism. Hindus and Christians in Indonesia, surrounded as they are by Muslims, have absorbed ideas from Islam, such as the seclusion of women. Muslims have absorbed ideas and customs from indigenous belief systems, especially ideas about kinship and marriage, as illustrated in the vignette below.

Globalization Brings Cultural Diversity and Homogeneity

Southeast Asia's cities are extraordinarily culturally diverse, a fact that is illustrated by their food customs (Figure 10.27). But in some ways, this diversity decreases as the many groups continue to be exposed to each other and to global culture. For example, Malaysian teens of many different ethnicities (Chinese, Tamil, Malay, Bangladeshi) spend much of their spare time following the same European soccer teams, playing the same video games, visiting the same shopping centers, eating the same fast food, and talking to each other in English. While rural areas retain more traditional influences—of the world's 6000 or so actively spoken languages, 1000 can be found in rural Southeast Asia—in cities, one main language usually dominates trade and politics.

One group that has brought cultural diversity to Southeast Asia is the Overseas (or ethnic) Chinese (see Chapter 9, page 401). Small groups of traders from southern and coastal China have been active in Southeast Asia for thousands of years; over the centuries, there has been a constant trickle of immigrants from China. The forebearers of most of today's Overseas Chinese, however, began to arrive in large numbers during the nineteenth century, when the European colonizers needed labor for their plantations and mines. Later, those who fled China's Communist Revolution after 1949 sought permanent homes in Southeast Asian trading centers. Today, more than 26 million Overseas Chinese live and work across Southeast Asia, primarily as shopkeepers and as small business owners. A few are wealthy financiers, and a significant number are still engaged in agricultural labor.

Chinese success at small-scale commercial and business activity throughout the region has reinforced the perception that the Chinese are diligent, clever, and extremely frugal, often

VIGNETTE

Although arranged marriages have historically been the norm across Southeast Asia, in most urban and rural areas, marriages are now love matches. Such is the case for Harum and Adinda, who live in Tegal, on the island of Java in Indonesia. They met in high school and some 10 years later, after saving a considerable sum of money, decided to formally ask both sets of parents if they could marry. Harum, an accountant, would normally be expected to pay the wedding costs, which could run to many thousands of dollars; however, Adinda was able to contribute from her salary as a teacher. Like nearly all Javanese people, both are Muslim. Because Islam does not have elaborate marriage ceremonies, colorful rituals from Christianity, Buddhism, and indigenous animism will enhance the elaborate and festive occasion.

In preparation, both bride and groom participate in unique Javanese rituals that remain from the days when marriages were arranged and the bride and groom did not know each other (Figure 10.26). A *pemaes*, a woman who prepares a bride for her wedding and whose role is to inject mystery and romance into the marriage relationship, bathes and perfumes the bride. She also puts on the bride's makeup and dresses her, all the while making offerings to the spirits of the bride's ancestors and counseling her about how to behave as a wife and how to avoid being dominated by her husband. The groom also takes part in ceremonies meant to prepare him for marriage. Both are counseled that their relationship is bound to change over the course of the decades as they mature and as their family grows older.

Despite the elaborate preparations for marriage, divorce in Indonesia (and also in Malaysia) is fairly common among Muslims, who often go through one or two marriages early in life before they settle into a stable relationship. Although the prevalence of divorce is lamented by society, it is not considered outrageous or disgraceful. Apparently, ancient indigenous customs predating Islam allowed for mating flexibility early in life, and this attitude is still tacitly accepted. [Source: Jennifer W. Nourse and Walter Williams. For detailed source information, see Text Sources and Credits.] ■

FIGURE 10.26 A Javanese wedding. In the *dahar klimah* phase of a traditional Javanese wedding, the bridegroom makes three small balls of the food and feeds them to the bride; she then does the same for him. The ritual reminds them that they should share joyfully whatever they have.

Sociocultural Issues **441**

working very long hours. Although few are actually wealthy, with their region-wide family and friendship connections and access to start-up money, some have been well positioned to take advantage of the new growth sectors in the globalizing economies of the region. Sometimes externally funded, new Chinese-owned enterprises have put out of business older, more traditional establishments

ON THE BRIGHT SIDE

LGBT Rights in Thailand and Vietnam

Lesbian, gay, bisexual, and transgender people are discriminated against throughout the world, and Southeast Asia is no exception. Both Thailand and Vietnam, though, are considering giving more legal recognition to same-sex couples in the arenas of property and child custody. Thailand is among the world's most open societies with regard to transgender people; transgender women are referred to as *kathoey*, and transgender men as *tom,* both of which are nonderogatory terms. Of the two, the more prominent are the kathoey, whose gender identity has become somewhat celebrated in recent years, though kathoeys often face discrimination in daily life. Both kathoeys and toms may soon benefit from legal protection as a "third gender." 318. **THAILAND'S "THIRD SEX" WANTS ACCEPTANCE, LEGAL SUPPORT**

FIGURE 10.27 LOCAL LIVES

Foodways in Southeast Asia

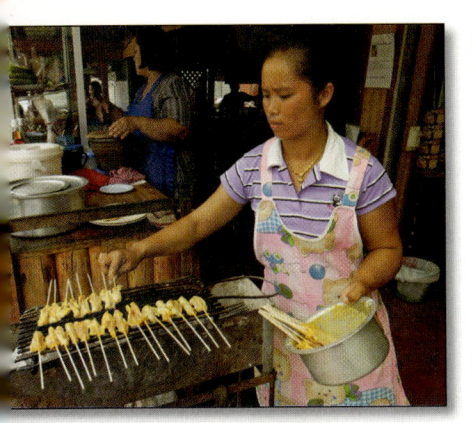

A Satay, or marinated and grilled meat or fish often served with a peanut sauce, is prepared at a market in Chiang Mai, Thailand. Satay first became popular in this region in the nineteenth century on the island of Java, Indonesia, after the arrival of Muslim merchants and other immigrants from South and Southwest Asia. Some food scholars argue that satay is related to kebab, a food found throughout South and Southwest Asia. [John Elk/Lonely Planet Images/Getty Images]

B Throughout Southeast Asia, cold desserts feature shaved ice, coconut milk, evaporated or condensed milk, sweeteners, beans, noodles, corn, jelly, and other toppings. They are called *ais kacang* in Malaysia, *halo-halo* in the Philippines, *cendol* in Indonesia, and *ching bo leung* in Vietnam. Ice machines on European ships in the early twentieth century turned these desserts from beverages into ice cream–like sundaes. [Seet Ying Lai Photography/Flickr/Getty Images]

C A *banh mi* sandwich, a street food in Vietnam and Laos, is a perfect example of the union of European (in this case, French) and Southeast Asian cuisine. European ingredients include French bread (a baguette), mayonnaise, and pâté. Vietnamese and Laotian ingredients include coriander, pickled vegetables, hot peppers, cucumber, sliced pork or headcheese, and fish sauce. This sandwich now has a U.S. audience and fans. [Rebecca Skinner/Lonely Planet Images/Getty Images]

---

that depend on local people of modest incomes (both ethnic Malay and ethnic Chinese).

In recent years, when low- and middle-income Southeast Asians were hurt by the recurring financial crises, some tended to blame their problems on the Overseas Chinese. Waves of violence resulted. Chinese people were assaulted, their temples desecrated, and their homes and businesses destroyed. Conflicts involving the Overseas Chinese have taken place in Vietnam, Malaysia, and in many parts of Indonesia (Sumatra, Java, Kalimantan, and Sulawesi) as well. Some Overseas Chinese have attempted to diffuse tensions through public education about Chinese culture. Others have shown their civic awareness by financing economic and social aid projects to help their poorer neighbors, usually of local ethnic origins (Malay, Thai, Indonesian).

Gender Patterns in Southeast Asia

Gender roles are being transformed across Southeast Asia by urbanization and the changes it brings to family organization and employment. Here we look at some surprising traditional patterns of gender roles in extended families. Moving to the city shifts people away from extended families and toward the nuclear family. The gains that women have made in political empowerment, educational achievement, and paid employment have not erased gender disparities.

Family Organization, Traditional and Modern Throughout the region, it has been common for a newly married couple to reside with, or close to, the wife's parents. Along with this custom is a range of behavioral rules that empower the woman in a marriage, despite some basic patriarchal attitudes. For example, a family is headed by the oldest living male, usually the wife's father. When he dies, he passes on his wealth and power to the husband of his oldest daughter, not to his own son. (A son goes to live with his wife's parents and inherits from them.) Hence, a husband may live for many years as a subordinate in his father-in-law's home. Instead of the wife being the outsider, subject to the demands of her mother-in-law—as is the case, for example, in South Asia—it is the husband who must show deference. The inevitable tension between the wife's father and the son-in-law is resolved by the custom of *ritual avoidance*—in daily life they simply arrange to not encounter each other much. The wife manages communication between the two men by passing messages and even money back and forth. Consequently, she has access to a wealth of information crucial to the family and has the opportunity to influence each of the two men.

Another traditional custom that empowers women is that women are often the family financial managers. Husbands turn over their pay to their wives, who then apportion the money to various household and family needs.

Urbanization and the shift to the nuclear family mean that young couples now frequently live apart from the extended family, an arrangement that takes the pressure to defer to his father-in-law off the husband. Because this nuclear family unit is often dependent entirely on itself for support, wives usually work for wages outside the home. Although married women lose the power they would have if they lived among their close kin, they are empowered by the opportunity to have a career and an income. The main drawback of this compact family structure, as many young families have discovered in Europe and the United States, is that there is no pool of relatives available to help working parents with child care and housework. Further, no one is left to help elderly parents maintain the rural family home.

Political and Economic Empowerment of Women Women have made some impressive gains in politics in Southeast Asia. Economically, they still earn less money than men and work less outside the home, but this will likely change if their level of education in relation to that of men continues to increase (see "On the Bright Side").

Southeast Asia has had several prominent female leaders over the years, most of whom have risen to power in times of crisis as the leaders of movements opposing corrupt or undemocratic regimes. In the Philippines, Corazon Aquino, a member of a large and powerful family, became president in 1986 after leading the opposition to Ferdinand Marcos, whose 21-year presidency was infamous for its corruption and authoritarianism. She is credited with helping reinvigorate political freedoms in the Philippines. Gloria Macapagal-Arroyo, from another powerful family, became president in 2001 after opposing a similarly corrupt president, and then she was accused of corruption herself. Still, her administration kept the economy sound throughout the global recession starting in 2007. In 2010, Corazon's son, Benigno Aquino III, became president.

In Indonesia, Megawati Sukarnoputri became president in 2001 after decades of leading the opposition to Suharto's notoriously corrupt 31-year reign. In Burma, for more than two decades, a woman, Aung San Suu Kyi, has led opposition to the military dictatorship. In Thailand, Yingluck Shinawatra was elected prime minister in 2011 as part of a popular political movement started by her brother.

All of these women leaders were wives, daughters, or siblings of powerful male political leaders, which raises some questions of nepotism. However, family favoritism cannot account for several countries where the percentage of female national legislators is well above the world average of 18 percent: Timor-Leste (29 percent), Laos (25 percent), Vietnam (24 percent), Singapore (22 percent), the Philippines (22 percent), and Cambodia (21 percent).

Despite their increasing successes in politics and their acknowledged role in managing family money, women still lag well behind men in terms of economic well-being. Throughout the region, men have a higher rate of employment outside the home than women, and are paid more for doing the same work. But changes may be on the way. In Brunei, Malaysia, the Philippines, and Thailand, significantly more women than men are completing training beyond secondary school. If training qualifications were the sole consideration for employment, women would appear to have an advantage over men. This advantage may be significant if service sector economies, which generally require more education, become dominant in more countries. The service economy already dominates in Singapore, the Philippines, Thailand, and Indonesia.

Globalization and Gender: The Sex Industry

Southeast Asia has become one of several global centers for the sex industry, supported in large part by international visitors willing to pay for sex. **Sex tourism** in Southeast Asia grew out of the sexual entertainment industry that served foreign military troops stationed in Asia during World War II, the Korean War, and the Vietnam War. Now, primarily civilian men arrive from around the globe to live out their fantasies during a few weeks of vacation. The industry is found throughout the region but is most prominent in Thailand. In 2013, twenty-six million tourists visited Thailand alone, up from 250,000 in 1965, and some observers estimate that as many as 70 percent were looking for sex. Even though the industry is officially illegal, some Thai government officials have even publicly praised sex tourism for its role in helping the country weather the economic crises, because it supported jobs. Some corrupt officials also favor sex tourism because it provides them with a source of untaxed income from bribes.

One result of the "success" of sex tourism is a high demand for sex workers, which has attracted organized crime. Estimates of the numbers of sex workers vary from 30,000 to more than a million in Thailand alone. Gangs often coerce girls and women into remaining in sex work once they have been forced into the trade. Demographers estimate that 20,000 to 30,000 Burmese girls taken against their will—some as young as 12—are working in Thai brothels. Their wages are too low to enable them to buy their own freedom. In the course of their work, they must service more than 10 clients per day, and they are routinely exposed to physical abuse and sexually transmitted diseases, especially HIV.

ON THE BRIGHT SIDE

Women's Improving Status in Southeast Asia

Building on strong traditional roles at the family level, women now occupy prominent political roles in many countries, and their rising educational levels suggest that even in the most conservative countries, changes in gender roles will be significant. One result of the rising political status of women is that there is more focus on the abuse of women and girls in the sex industry.

---

sex tourism the sexual entertainment industry that serves primarily men who travel for the purpose of living out their fantasies during a few weeks of vacation

VIGNETTE

Twenty-five-year-old Watsanah K. (not her real name) awakens at 11:00 every morning, attends afternoon classes in English and secretarial skills, and then goes to work at 4:00 P.M. in a bar in Patpong, Bangkok's red light district. There she will meet men from Europe, North America, Japan, Taiwan, Australia, Saudi Arabia, and elsewhere, who will pay to have sex with her. She leaves work at about 2:00 A.M., studies for a while, and then goes to sleep.

Watsanah was born in northern Thailand to an ethnic minority group made up of impoverished subsistence farmers. She married at 15 and had two children shortly thereafter. Several years later, her husband developed an opium addiction. She divorced him and left for Bangkok with her children. She found work at a factory that produced seatbelts for a nearby automobile plant. In 2008, Watsanah lost her job as the result of the global economic crisis. To feed her children, she became a sex worker.

Although the pay, between U.S.$400 and U.S.$800 a month, is much better than the U.S.$100 a month she earned in the factory, the work is dangerous and demeaning. Sex work, though widely practiced and generally accepted in Thailand, is illegal, and the women who do it are looked down on. As a result, Watsanah must live in constant fear of going to jail and losing her children. Moreover, she cannot always make her clients use condoms, which puts her at high risk of contracting AIDS and other sexually transmitted diseases. "I don't want my children to grow up and learn that their mother is a prostitute," says Watsanah. "That's why I am studying. Maybe by the time they are old enough to know, I will have a respectable job." [*Source: Field notes of Alex Pulsipher and Debbi Hempel; Kaiser Family Foundation; BBC News. For detailed source information, see Text Sources and Credits.*]

THINGS TO REMEMBER

- The major religious traditions of Southeast Asia include Hinduism, Buddhism, Confucianism, Taoism, Islam, Christianity, and animism. All originated outside the region, with the exception of the animist belief systems.

- Lesbian, gay, bisexual, and transgender people are discriminated against throughout Southeast Asia, but both Thailand and Vietnam are considering giving more legal recognition to same-sex couples.

- Women have made some impressive gains in politics in Southeast Asia, but they still earn less money than men and work less outside the home.

- Gender roles are being transformed across Southeast Asia by urbanization and the changes it brings to family organization and employment.

- Some countries have become centers for the global sex industry, which puts many women at risk of violence and disease.

GEOGRAPHIC INSIGHTS

Southeast Asia: Review and Self-Test

1. Environment: Many of Southeast Asia's most critical environmental issues relate in some way to climate change. Deforestation is rapid in this region and is a major global source of greenhouse gas emissions, which intensify climate change. This region is also highly vulnerable to the impacts of climate change. Increased flooding and droughts threaten food production on land and rising ocean temperatures strain aquatic ecosystems.

- How does deforestation lead to more greenhouse gas emissions?
- How is food production for global markets related to deforestation?
- How is deforestation related to increased risks of floods and droughts?

2. Globalization and Development: Globalization has brought both spectacular successes and occasional declines to the economies of Southeast Asia. Key to the economic development of this region are strategies that were pioneered earlier in East Asia: the formation of state-aided market economies combined with export-led economic development.

- How is globalization linked to the economic success of Southeast Asia?
- How is globalization linked to rapid urbanization and the growth of slums in this region?
- Why have global recessions caused such dramatic economic declines for the many workers in Southeast Asia?

3. Power and Politics: There has been a general expansion of political freedoms throughout Southeast Asia in recent decades, but authoritarianism, corruption, and violence have at times reversed these gains.

- How does the history of colonialism in this region influence the current dynamics of political power?
- What arguments have some Southeast Asian leaders made about Asian values not being compatible with Western ideas of democracy and political freedom?
- Why are some countries concerned about China's interest in the South China Sea?

- What are some of the explanations for the violence against ethnic or religious minorities that tends to occur during hard economic times?
- How is Indonesia's government working to alleviate ethnic tensions?

4. Urbanization: While Southeast Asia as a whole is only 43 percent urban, its cities are growing rapidly as agricultural employment declines and urban industries expand. The largest Southeast Asian cities, which are receiving most of the new rural-to-urban migrants, rarely have sufficient housing, water, sanitation, or jobs for all their people.

- How have development strategies, such as mechanized agriculture and agribusiness, changed the degree of food self-sufficiency in this region?
- How are development and urbanization linked to the decision of many to migrate abroad for employment?

5. Population and Gender: Population dynamics vary considerably in this region because of differences in economic development, government policies, prescribed gender roles, and religious and cultural practices. With regard to gender, economic change has brought better job opportunities and increased status for women, who then often choose to have fewer children. Some countries also have gender imbalances because of a cultural preference for male children.

- Where have population growth rates slowed the most in this region?
- What is a major source of population growth for Singapore other than natural increase?
- What are variations in population growth linked to in this region?
- Where have efforts at HIV-AIDS prevention been more successful?
- Describe the circumstances that have led to better job opportunities for women.
- How are the improving roles and status of women changing attitudes toward female sex work?

CRITICAL THINKING QUESTIONS

1. What do you think are the most serious threats to Southeast Asia posed by global climate change? What are the most promising responses to global climate change emerging from this region?

2. Which countries in the region have been most successful in controlling population growth? How do population issues differ among Southeast Asian countries? How do the population issues of the region as a whole compare with those of Europe or Africa?

3. Discuss the role of crony capitalism in the economic crisis of the 1990s in this region.

4. Did those responsible for resolving the economic crisis of the 1990s opt for reregulation or deregulation of financial institutions? Why did they make this choice?

5. In what ways did state aid to market economies and export-led growth amount to a strategic and limited embrace of globalization?

6. Which countries in this region have more political freedoms? Which have fewer? Why do you think that is the case?

7. Describe the spatial distribution of major religious traditions of this region. How have these religious traditions influenced each other over time?

8. What factors have produced major flows of refugees in Southeast Asia?

9. Relationships and gender roles in some traditional Southeast Asian families are not necessarily what an outsider might expect. Describe the characteristics of these relationships that interested you most. Why did they catch your attention? Discuss some of the ways in which gender roles vary from those typically found in South Asia or East Asia.

10. How has the role of the Overseas Chinese evolved over time? Why have some people in this region resented them?

› THINKING GEOGRAPHICALLY

Now that you have read about Southeast Asia, you should be able to answer the following questions about photos in this chapter. To answer these questions online, go to Geography LaunchPad.

Human Impacts on the Biosphere (page 414)

A How does this landscape of recently planted oil palms differ from that of the forest that was removed for this plantation?

B What are these orangutans eating? Do you think these animals would relish a meal of oil palm leaves? Why is this likely or unlikely?

C Can you identify the geographic locations barely visible through the smoke?

Vulnerability to Climate Change (page 418)

A Typhoons are particularly significant for people living on which landforms?

C The text describes several causes of coral bleaching. Explain the link between coral bleaching and human activity in the case of rising ocean temperatures.

D How might flood-resistant cities help people adapt to climate change?

Visual History (page 422)

A From where did those who populated this region in prehistory come?

B How did Hinduism and Buddhism arrive in Southeast Asia?

C Which European countries established colonies in Southeast Asia?

D During World War II, Japan conquered which parts of Southeast Asia?

E What has been the consequence of exposure to Agent Orange for many rural Vietnamese and Laotian people?

Power and Politics (page 429)
A What details in this picture are clues that these people are refugees?
B Why would a government ban a flag?
C What does this picture suggest about the MILF?

Urbanization (page 433)
A Why might these people be living on the water in Jakarta?
B If government-built apartment buildings in Singapore are intended for the middle class, what are the common facilities that house most of the undocumented, noncitizen population?
C What about this photo suggests that these men may be of rural origins?
D Why might the situation depicted in this photo be viewed as a triumph for women's rights in Bangkok?

CHAPTER KEY TERMS

archipelago 409
Association of Southeast Asian Nations (ASEAN) 425
Australo-Melanesians 419
Austronesians 419
coral bleaching 417
crony capitalism 424
cultural pluralism 438
detritus 410
domino theory 421
feminization of labor 423
foreign exchange 432
resettlement schemes 431
sex tourism 442

A Australia's Eastern Highlands, Blue Mountains

B Australian Desert

chapter 11

OCEANIA: AUSTRALIA, NEW ZEALAND, AND THE PACIFIC

C Tupai Island, French Polynesia

D New Zealand, Milford Sound, South Island

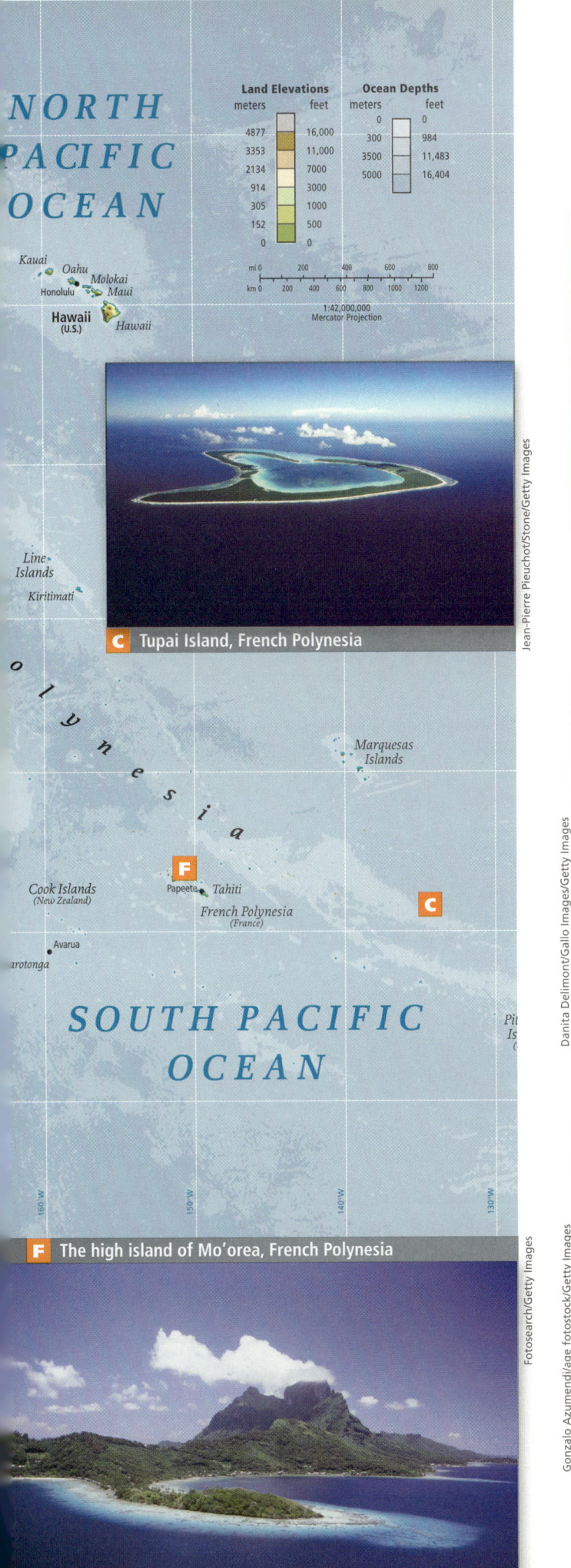

F The high island of Mo'orea, French Polynesia

E Great Barrier Reef

FIGURE 11.1 Regional map of Oceania.

447

GEOGRAPHIC INSIGHTS: OCEANIA

After you read this chapter, you will be able to discuss the following geographic insights as they relate to the five thematic concepts:

1. Environment: Oceania faces a host of environmental problems and public awareness of environmental issues is keen. Global climate change, primarily warming, has brought rising sea levels and increasingly variable rainfall. Other major threats to the region's unique ecology have come from the introduction of many nonnative species and the expansion of herding, agriculture, fishing, and human settlements.

2. Globalization and Development: Globalization, coupled with Oceania's stronger focus on neighboring Asia (rather than on its long-time connections with Europe), has transformed patterns of trade and economic development across Oceania. The transformation is being driven largely by Asia's growing affluence, its enormous demand for resources, and its similarly massive production of manufactured goods.

3. Power and Politics: In recent decades, stark divisions have emerged in Oceania over definitions of democracy—the system of government that dominates in New Zealand and Australia—versus the Pacific Way, a political and cultural philosophy based in the traditional cultures of the Pacific islands.

4. Urbanization: Oceania is only lightly populated but it is highly urbanized. The shift from agricultural and resource-based economies toward service economies is a major reason for the urbanization of the wealthiest parts of Oceania (Australia, New Zealand, Hawaii, Guam), where 80 to 100 percent of the population lives in cities. These trends are weakest in Papua New Guinea and many smaller Pacific islands.

5. Population and Gender: In this largest but least populated world region, there are two main patterns relevant to population and gender. Australia, New Zealand, and Hawaii have older and more slowly growing populations, and relatively more opportunities for women. The Pacific islands and Papua New Guinea have much more rural, younger, and rapidly growing populations, with fewer opportunities for women.

The Oceania Region

The region of Oceania, shown in Figure 11.1, is a vast and often idealized region that faces a number of difficult realities. The five thematic concepts in this book are explored as they arise in the discussion of these issues. Vignettes, like the one that follows about the Kiribati archipelago and the challenges its people face due to climate change, help illustrate the themes as they are experienced in individual lives.

GLOBAL PATTERNS, LOCAL LIVES

Aurora (a pseudonym) is a nursing student in Brisbane, Australia, who is undergoing a wrenching personal transition due to climate change. Aurora is from Kiribati, a Pacific nation of 33 tiny islands barely 6 and a half feet above sea level. Rising seas are swamping the Kiribati archipelago, which straddles the equator just west of the international date line (180° longitude). Aurora pines for her island's blue lagoons. While she studies for a new future and prepares to become her household's primary income earner, she does so without the daily close company of her raucous extended family.

Several years ago, Aurora and her fellow citizens (100,000 people live in Kiribati) began to notice alarming environmental changes. Their drinking water was getting brackish (salty), several shoreline villages had sunk below sea level, and the climate had become so dry that skilled gardeners could no longer raise their customary cabbages, tomatoes, cucumbers, and swamp taro (Figure 11.2A, B). Drought conditions were being worsened by rising salty tides that saturated formerly fertile soil. Scientists predicted that, with rising seas and more frequent droughts and storms, much of Kiribati will be uninhabitable in a few decades.

Aurora's nursing education, funded by a government program called AusAID, is one of several strategies developed by Kiribati president Anote Tong to encourage his people to gradually "migrate with dignity." President Tong sees climate change as a huge challenge for his country, and he hopes that other South Pacific islands, as well as Australia and New Zealand, will allow Kiribati people to resettle there permanently. Through the AusAID program, Aurora and other young people receive an education that will help them find employment outside Kiribati and support their large extended families after their family members have joined them.

But it is a tough transition. Another nursing student says he feels like he is losing his identity as his nation sinks into the sea. Some students' parents refuse to leave Kiribati. They see the changes in sea level and rainfall, but as Christians they believe that "God is not so silly to allow people to perish just like that."

In March 2012, the BBC (British Broadcasting Corporation) reported that President Tong had approached landowners on the island nation of Fiji, 1300 miles to the south (Figure 11.3), hoping to buy Fijian land. His goals are twofold: to produce food to export

FIGURE 11.2 Kiribati.

(A) Inhabitants of a now-relocated Kiribati village stand where their village used to be. Kiribati's coastline is eroding as sea level rises.

(B) A swamp taro pit that was rendered infertile by intruding salt water. Swamp taro is a major food source for inhabitants of low-lying islands. It is cultivated in a carefully managed pit that makes use of a layer of fresh (not salty) groundwater that exists along the coastline. As sea level rises, salt water intrudes into this layer of fresh groundwater, killing the plants or stunting their growth.

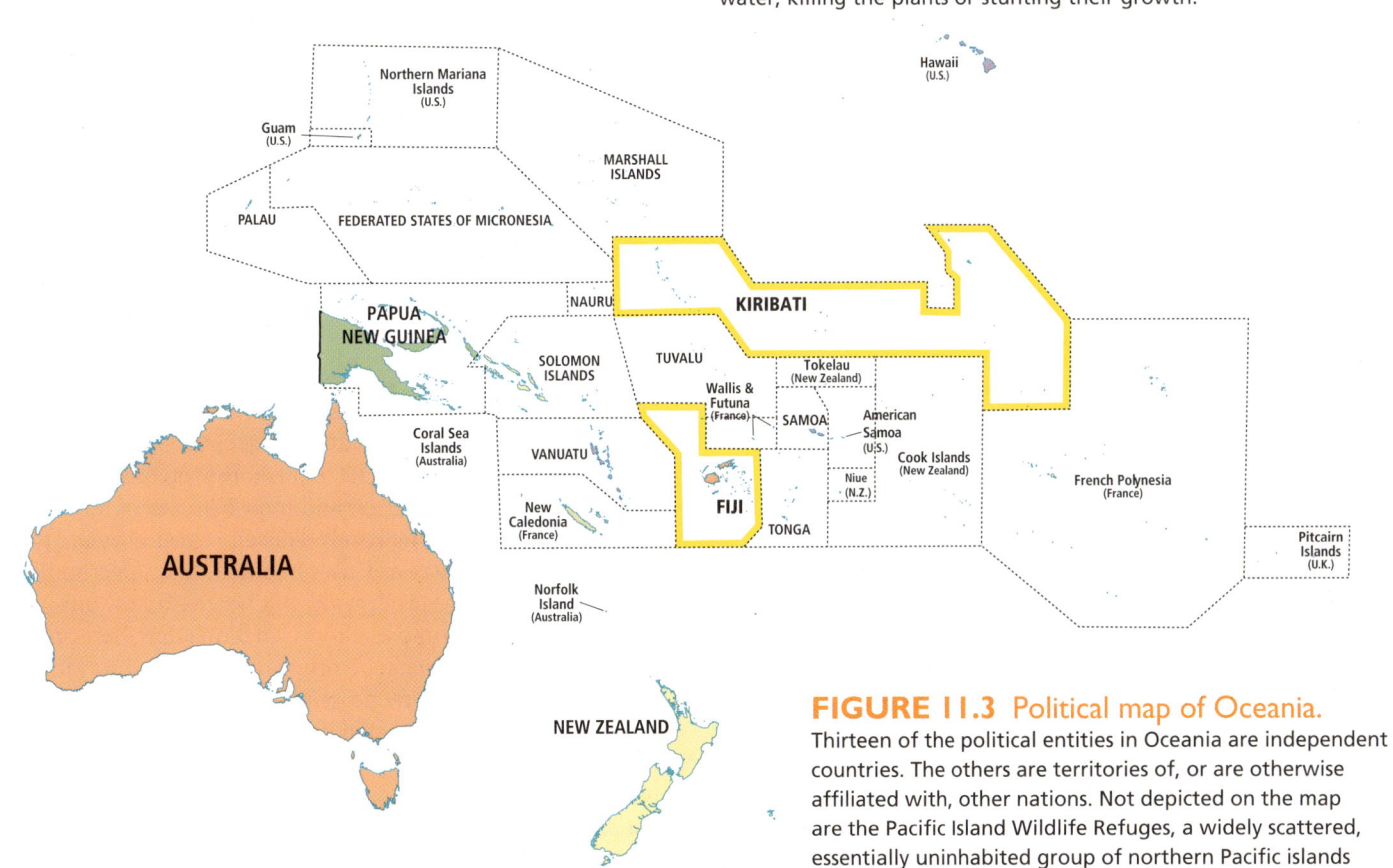

FIGURE 11.3 Political map of Oceania. Thirteen of the political entities in Oceania are independent countries. The others are territories of, or are otherwise affiliated with, other nations. Not depicted on the map are the Pacific Island Wildlife Refuges, a widely scattered, essentially uninhabited group of northern Pacific islands that constitute a U.S. territory.

to Kiribati and to import Fijian soil to replace the soil lost to sea level rise in Kiribati. This proposition will surely trigger controversy in Fiji, which already has its own social and political problems (see pages 468–470). [Sources: National Public Radio; BBC News Asia. For detailed source information, see Text Sources and Credits.] ■

Perhaps nowhere else on Earth are the effects of climate change as measurably real as they are in the low-lying islands of the Pacific, where changes in rainfall patterns and sea level are strikingly evident. These islands have contributed little to the greenhouse gases that have triggered rapid climate change. Because many people in the islands have fairly simple lifestyles, most Pacific islands have relatively low emissions of greenhouse gases on a per capita basis. Their emissions are also low as a whole because the islands' total population is small. But seemingly remote places in the Pacific are anything but remote from general global influences. Beyond climate change, people in this region are dealing with changes brought on by tourism; environmental degradation caused by modern mining techniques; freshwater scarcity; ocean pollution; and rapidly changing cultural and trading patterns.

PHYSICAL GEOGRAPHY AND ENVIRONMENTAL ISSUES

What Makes Oceania a Region?

Oceania is made up of Australia, New Zealand, Papua New Guinea, and the many small islands scattered across the Pacific Ocean (see Figure 11.3). It is a unique world region in that it is composed primarily of ocean and covers the largest area of Earth's surface of any region, yet it is home to only 38.7 million people, the vast majority of whom live in Australia (22.7 million), Papua New Guinea (6.9 million), and New Zealand (4.4 million). In this book, we also include the U.S. state of Hawaii (1.7 million) in Oceania. All together, the thousands of other Pacific islands are home to only 3 million people. The Pacific Ocean is the link that unites Oceania as a region, profoundly influencing life even on dry land, but across the region, the ocean also acts as a biological and cultural barrier.

Terms in This Chapter

Some maps in this chapter show different place-names for the same locations. This reflects the political evolution of the region, where some islands previously were grouped under one name that remains in everyday use but are now grouped into country units with another name. For example, the Caroline Islands, located north of New Guinea, still go by that name on maps and charts, but they have been divided into two countries: Palau (a small group of islands at the western end of the Caroline Islands) and the Federated States of Micronesia, which extends over 2000 miles west to east, from Yap to Kosrae. In addition, there are three names commonly used to refer to large cultural groupings of islands: Micronesia, Melanesia, and Polynesia. These groupings are not political units; rather, they are based on ancient ethnic and cultural links.

> **Gondwana** the great landmass that formed the southern part of the ancient supercontinent Pangaea
>
> **Great Barrier Reef** the longest coral reef in the world, located off the northeastern coast of Australia

PHYSICAL PATTERNS

The Pacific Ocean serves as both a link and a barrier in Oceania. Plants and animals have found their way from island to island by floating on the water, swimming, or flying, and humans have long used the ocean as a way to make contact with other peoples for visiting, trading, and raiding. Unfortunately, as we shall see, the ocean also serves as a conveyer of pollution, especially discarded plastics (see page 460). But the wide expanses of water also profoundly limit the natural diffusion of plant and animal species and keep Pacific Islanders spatially isolated from one another. The vast ocean has imposed solitude and fostered self-sufficiency and subsistence economies among its human occupants well into the modern era.

Continent Formation

The largest landmass in Oceania is the ancient continent of Australia at the southwestern edge of the region (see the Figure 11.1 map). The Australian continent is partially composed of some of the oldest rock on Earth and has been relatively stable for more than 200 million years, with very little volcanic activity and only an occasional mild earthquake. Australia was once a part of the great landmass called **Gondwana** that formed the southern part of the ancient supercontinent Pangaea (see Figure 1.8 on page 16). What became present-day Australia broke free from Gondwana and drifted until it eventually collided with the part of the Eurasian Plate on which Southeast Asia sits. That impact created the mountainous island of New Guinea to the north of Australia.

Australia is shaped roughly like a dinner plate with an irregularly broken rim with two bites taken out of it: one in the north (the Gulf of Carpentaria) and one in the south (the Great Australian Bight). The center of the plate is the great lowland Australian desert, with only two hilly zones and rocky outcroppings (see Figure 11.1B). The eastern rim of Australia is composed of uplands; the highest and most complex of these are the long, curving Eastern Highlands (labeled "Great Dividing Range" in the Figure 11.1 map; see also Figure 11.1A). Over millennia, the forces of erosion—both wind and water—have worn most of Australia's landforms into low, rounded formations. Some of these, like Uluru (Ayers Rock), are quite spectacular (Figure 11.4).

Off the northeastern coast of the continent lies the **Great Barrier Reef**, the largest coral reef in the world and a World Heritage Site since 1981 (see Figure 11.1E). It stretches along the coast of Queensland in an irregular arc for more than 1250 miles (2000 kilometers), covering 135,000 square miles (350,000 square kilometers). The Great Barrier Reef is so large that it influences Australia's climate by interrupting the westward-flowing ocean currents in the mid–South Pacific circulation pattern. Warm water is shunted to the south, where it warms the southeastern coast of Australia. Threats to the health of the Great Barrier Reef are discussed on page 455.

FIGURE 11.4 Uluru (Ayer's Rock). This land formation, near the center of Australia, is a smooth remnant of ancient mountains. The site is held sacred by central Australian Aborigines. It is also one of Australia's most popular tourist destinations.

Island Formation

The islands of the Pacific were (and are still being) created by a variety of processes related to the movement of tectonic plates. The islands found in the western reaches of Oceania—including New Guinea, New Caledonia, and the main islands of Fiji—are remnants of the Gondwana landmass; they are large, mountainous, and geologically complex. Other islands in the region are volcanic in origin and form part of the Ring of Fire (see Figure 1.9 on page 17). Many in this latter group are situated in boundary zones where tectonic plates are either colliding or pulling apart. For example, the Mariana Islands east of the Philippines are volcanoes that were formed when the Pacific Plate plunged beneath the Philippine Plate. The two islands of New Zealand were created when the eastern edge of the Indian-Australian Plate was thrust upward by its convergence with the Pacific Plate.

The Hawaiian Islands were produced through another form of volcanic activity associated with **hot spots**, places where particularly hot magma moving upward from Earth's core breaches the crust in tall plumes. Over the past 80 million years, the Pacific Plate has moved across these hot spots, creating a string of volcanic formations 3600 miles (5800 kilometers) long. The youngest volcanoes, only a few of which are active, are on or near the islands known as Hawaii.

Volcanic islands exist in three forms: volcanic high islands, low coral atolls, and *makatea*, which are coral platforms raised or uplifted by volcanism. High islands are usually volcanoes that rise above the sea into mountainous rocky formations which, because of their varying height and rugged landscapes, contain a rich variety of environments. New Zealand, the Hawaiian Islands, Mo'orea, and Easter Island are examples of high islands (see Figure 11.1D, F). An **atoll** is a low-lying island or chain of islets, formed of coral reefs that have built up on the circular or oval rim of a submerged volcano (see Figure 11.1C). These reefs are arranged around a central lagoon that was once the volcano's crater. Because of their low elevation, atoll islands tend to have only a small range of environments and very limited supplies of fresh water.

> **hot spots** individual sites of upwelling material (magma) that originate deep in Earth's mantle and surface in a tall plume; hot spots tend to remain fixed relative to migrating tectonic plates
>
> **atoll** a low-lying island or chain of islets, formed of coral reefs that have built up on the circular or oval rim of a submerged volcano
>
> **Maori** Polynesian people indigenous to New Zealand
>
> **Roaring Forties** powerful air and ocean currents at about 40° S latitude that speed around the far Southern Hemisphere virtually unimpeded by landmasses

Climate

Although the Pacific Ocean stretches almost from pole to pole, most of the land of Oceania is situated within the Pacific's tropical and subtropical latitudes. The tepid water temperatures of the central Pacific bring mild climates year-round to nearly all the inhabited parts of the region (Figure 11.5). The southernmost reaches of Australia and New Zealand have the widest seasonal variations in temperature.

Moisture and Rainfall With the exception of the vast arid interior of Australia, much of Oceania is warm and humid nearly all the time. New Zealand and the high islands of the Pacific receive copious rainfall; before human settlement, they supported dense forest vegetation. Now, after 1000 years of human impact, much of that forest is gone (see Figure 11.5B; see also Figure 11.9D on page 458).

Travelers approaching New Zealand, either by air or by sea, often notice a distinctive long white cloud that stretches above the north island. Seven hundred years ago, early **Maori** settlers (members of the Polynesian group) also noticed this phenomenon and they named that place *Aotearoa*, "land of the long white cloud," a name that is now applied to all of New Zealand.

The legendary **Roaring Forties** (named for the 40th parallel south) are powerful air and ocean currents that speed around the far Southern Hemisphere virtually unimpeded by landmasses. These *westerly winds* (winds that blow west to east), which are responsible for Aotearoa's distinctive bodies of moist air, deposit a drenching 130 inches (330 centimeters) of rain per year in the New Zealand highlands and more than 30 inches (76 centimeters)

FIGURE 11.5
PHOTO ESSAY: Climates of Oceania

The fringes of the Asian winter monsoon push the ITCZ south, bringing moist conditions to New Guinea, across the Australian north coast, and around to the far southeast during the southern summer.

El Niño events reverse the normal water and air temperature patterns, bringing drought and cooler temperatures to New Guinea.

Intertropical Convergence Zone – Southern limit in January

The Eastern Highlands block moist, westward-moving air so that rain does not reach the interior, but is abundant along the eastern slopes of the highlands.

Powerful westerly air currents—the Roaring Forties—speed around the far Southern Hemisphere, virtually unimpeded by landmasses.

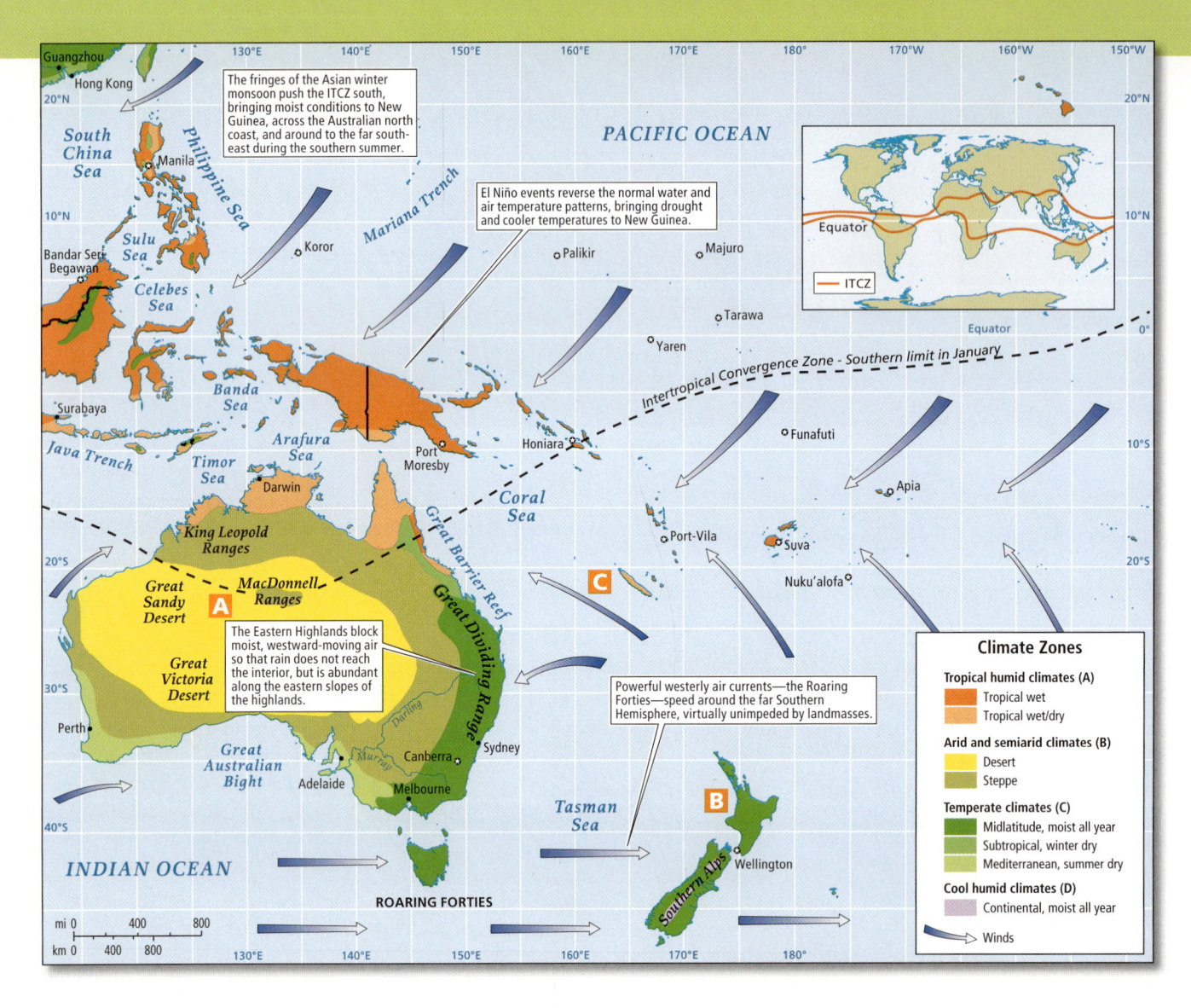

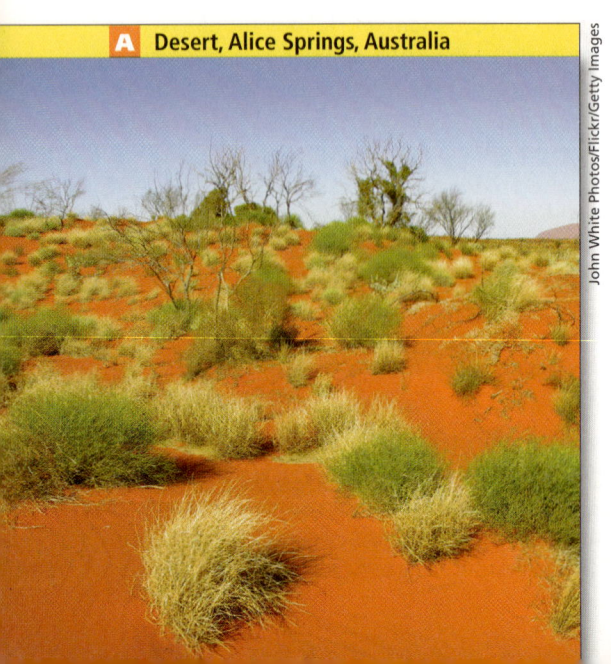

ROARING FORTIES

Climate Zones

Tropical humid climates (A)
- Tropical wet
- Tropical wet/dry

Arid and semiarid climates (B)
- Desert
- Steppe

Temperate climates (C)
- Midlatitude, moist all year
- Subtropical, winter dry
- Mediterranean, summer dry

Cool humid climates (D)
- Continental, moist all year

→ Winds

A Desert, Alice Springs, Australia

B Midlatitude, moist all year, New Zealand

C Tropical wet/dry, New Caledonia

per year on the coastal lowlands (see Figure 11.5B). At the southern tip of New Zealand's North Island, the wind averages more than 40 miles per hour (64 kilometers per hour) nearly 120 days per year. Farmers in the area stake their cabbages to the ground so the plants will not blow away.

By contrast, two-thirds of Australia is overwhelmingly dry (see Figure 11.5A). The dominant winds affecting Australia are the north and south *easterlies* (winds that blow east to west) that converge east of the continent. The Great Dividing Range blocks the movement of moist, westward-moving air so that rain does not reach the interior (an orographic pattern; see Figure 1.12 on page 20). As a result, a large portion of Australia receives less than 20 inches (50 centimeters) of rain per year, and humans have found rather limited uses for this interior territory. But the eastern (windward) slopes of the highlands receive more abundant moisture. This relatively moist eastern rim of Australia was favored as a habitat by both indigenous people and the Europeans who displaced them after 1800. During the southern summer, the fringes of the monsoon that passes over Southeast Asia and Eurasia bring moisture across Australia's northern coast. There, annual rainfall varies from 20 to 80 inches (50 to 200 centimeters).

Overall, Australia is so arid that it has only one major river system, which is in the temperate southeast where most Australians live. There, the Darling and Murray rivers drain one-seventh of the continent, flowing west and south into the Indian Ocean near Adelaide. One measure of Australia's overall dryness is that the entire average *annual* flow of the Murray-Darling river system is equal to just *one day's* average flow of the Amazon in Brazil.

In the island Pacific, mountainous high islands also exhibit *orographic rainfall* patterns, with a wet windward side and a dry leeward side (see Figure 11.5C). Rainfall amounts on the low-lying islands vary considerably across the region. Some of the islands lie directly in the path of trade winds, which usually deliver between 60 and 120 inches of rain per year. These islands support a remarkable variety of plants and animals. Other low-lying islands, particularly those on the equator, receive considerably less rainfall and are dominated by grasslands that support little animal life.

El Niño Recall from Chapter 3 the *El Niño* phenomenon, a pattern of shifts in the circulation of air and water in the Pacific that occurs irregularly every 2 to 7 years. Although these cyclical shifts, or oscillations, are not yet well understood, scientists have worked out a model of how the oscillations may occur (Figure 11.6).

The El Niño event of 1997–1998 illustrates the effects of this phenomenon. By December of 1997, the island of New Guinea (north of Australia; see the Figure 11.1 map) had received very little rainfall for almost a year. Crops failed, springs and streams dried up, and fires broke out in tinder-dry forests. The cloudless sky allowed heat to radiate up and away from elevations above 7200 feet (2200 meters), so temperatures at high elevations dipped below freezing at night for stretches of a week or more. Tropical plants died, and people unaccustomed to chilly weather became ill. Meanwhile, at the other end of the system, along the Pacific coasts of North, Central, and South America, the warmer-than-usual weather brought unusually strong storms, high ocean surges, and damaging wind and rainfall (see Figure 11.6C).

FIGURE 11.6 A model of the El Niño phenomenon.

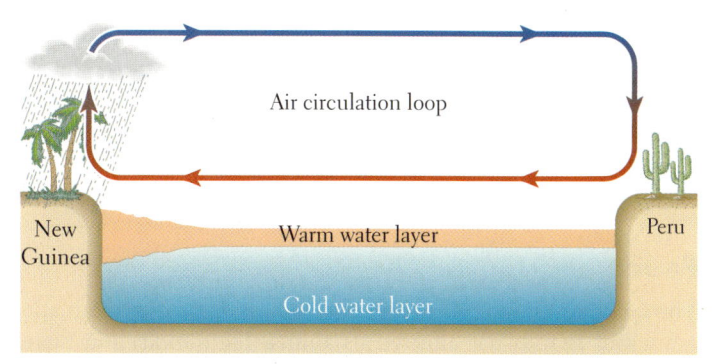

(A) Normal equatorial conditions. Water in the equatorial western Pacific (the New Guinea/Australia side) is warmer than water in the eastern Pacific (the Peru side). Due to prevailing wind patterns, the warm water piles up in the west. Warm air rises above the warm-water bulge in the western Pacific and forms rain clouds. The rising air cools and, once in the higher atmosphere, moves in an easterly direction. In the east, the dry, cool air descends, bringing little rainfall to Peru.

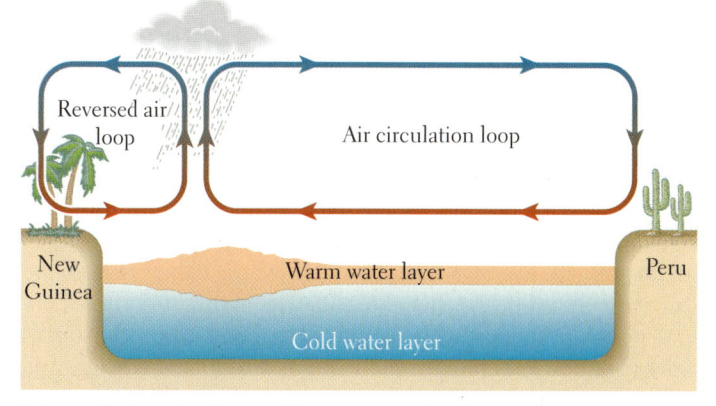

(B) Developing El Niño conditions. As an El Niño event develops, the ocean surface's warm-water bulge (orange) begins to move east. The air rising above it splits into two formations, one circling east to west in the upper atmosphere and one west to east.

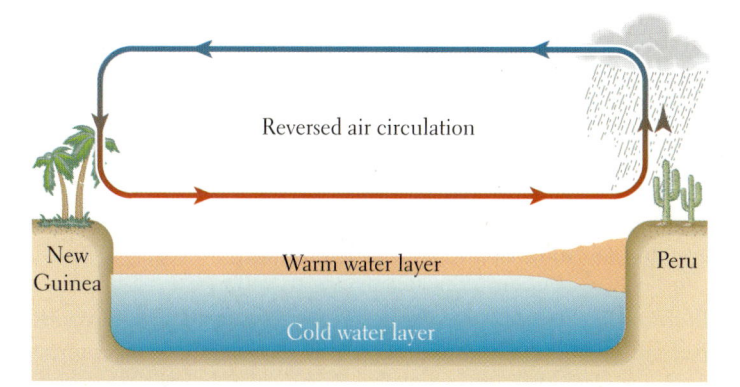

(C) Fully developed El Niño. Slowly, as the bulge of warm water at the surface of the ocean moves east, it forces the whole system into the fully developed El Niño, with air at the surface and in the upper atmosphere flowing in reverse of normal **(A)**. Instead of warm, wet air rising over the mountains of New Guinea and condensing as rainfall, cool, dry, cloudless air descends to sit at Earth's surface. Meanwhile, in the east, the normally dry, clear coast of Peru has clouds and rainfall.
[Sources consulted: Environmental Dynamics Research, Inc., 1998; Ivan Cheung, George Washington University, Geography 137, Lecture 16, October 29, 2001]

In the 1980s, an opposite pattern in which normal weather conditions become intensified was identified and named *La Niña*. It is now understood that La Niña patterns can bring unusually severe precipitation events (including tornadoes and blizzards) to places from the Indian Ocean to North America. La Niña is thought to have played a role in the major floods of 2010–2011 in Queensland, Australia, which were especially damaging because they followed a lengthy El Niño–connected drought.

Fauna and Flora

The fact that Oceania is comprised of an isolated continent and numerous islands has affected its animal life (*fauna*) and plant life (*flora*). Many of its species are **endemic**, meaning that they exist in a particular place and nowhere else on Earth. This is especially true in Australia (of the 750 species of birds known in Australia, more than 325 species are endemic), but many Pacific islands also have endemic species.

Animal and Plant Life in Australia

The uniqueness of Australia's animal and plant life is the result of the continent's long physical isolation, large size, relatively homogeneous landforms, and arid climate. Since Australia broke away from Gondwana more than 65 million years ago, its animal and plant species have evolved in isolation. One spectacular result of this isolation is the presence of more than 144 living species of endemic marsupial animals. **Marsupials** are mammals whose babies at birth are still at a very immature stage; the marsupial then nurtures them in a pouch equipped with nipples. The best-known marsupials are kangaroos; other marsupials include wombats, koalas, and bandicoots. The **monotremes**, egg-laying mammals that include the duck-billed platypus and the spiny anteater, are endemic to Australia and New Guinea.

Most of Australia's endemic plant species are adapted to dry conditions. Many of the plants have deep taproots to draw moisture from groundwater, and small, hard, pale green or shiny leaves to reflect heat and to hold moisture. Much of the continent is grassland and scrubland with bits of open woodland; there are only a few true forests, found in pockets along the Eastern Highlands and the southwestern tip and in Tasmania (Figure 11.7). Two plant genera account for nearly all the forest and woodland plants: *Eucalyptus* (450 species, often called gum trees) and *Acacia* (900 species, often called wattles).

Plant and Animal Life in New Zealand and the Pacific Islands

Naturalists and evolutionary biologists have had a great interest in the species that inhabited the Pacific islands before humans arrived. Charles Darwin formulated many of his ideas about evolution after visiting the Galápagos Islands of the eastern Pacific (see the Figure 3.1 map on page 108) and the islands of Oceania.

Islands gain plant and animal populations from the sea and air around them as organisms are carried from larger islands and continents by birds, storms, or ocean currents. Once these organisms "colonize" their new home, they may evolve over time into new species that are unique to one island. High, wet islands generally contain more varied species because their more complex environments provide niches for a wider range of wayfarers and thus more opportunities for evolutionary change.

Once they arrive, human inhabitants modify the flora and fauna of islands. In prehistoric times, Asian explorers in oceangoing canoes brought plants such as bananas and breadfruit and animals such as pigs, chickens, and dogs to Oceania. European settlers later brought grains, vegetables, fruits, invasive grasses, cattle, sheep, goats, rabbits, housecats, and rats. Today, human activities from tourism to military exercises to urbanization continue to change the flora and fauna of Oceania. **240. BREADFRUIT ADVOCATES SAY IT COULD SOLVE HUNGER IN TROPICAL REGIONS**

Generally, the diversity of land animals and plants is richest in the western Pacific, near the larger landmasses. It thins out to the east, where the islands are smaller and farther apart. The natural rain forest flora is rich and abundant in New Zealand and New Guinea, and also on the high islands of the Pacific. However, the natural fauna is much more limited on these islands. While New Guinea has fauna comparable to Australia, to which it was once connected via Sundaland (see Figure 10.5 on page 410), New Zealand and the Pacific islands have no indigenous land mammals, almost no indigenous reptiles, and only a few indigenous species of frogs, as they were never connected to Australia and New Guinea by a land bridge that land animals could cross. Two indigenous birds in New Zealand, the kiwi and the huge moa (a bird that grew

> **endemic** belonging or restricted to a particular place
>
> **marsupials** mammals whose babies at birth are still at a very immature stage; the marsupial then nurtures them in a pouch equipped with nipples
>
> **monotremes** egg-laying mammals, such as the duck-billed platypus and the spiny anteater

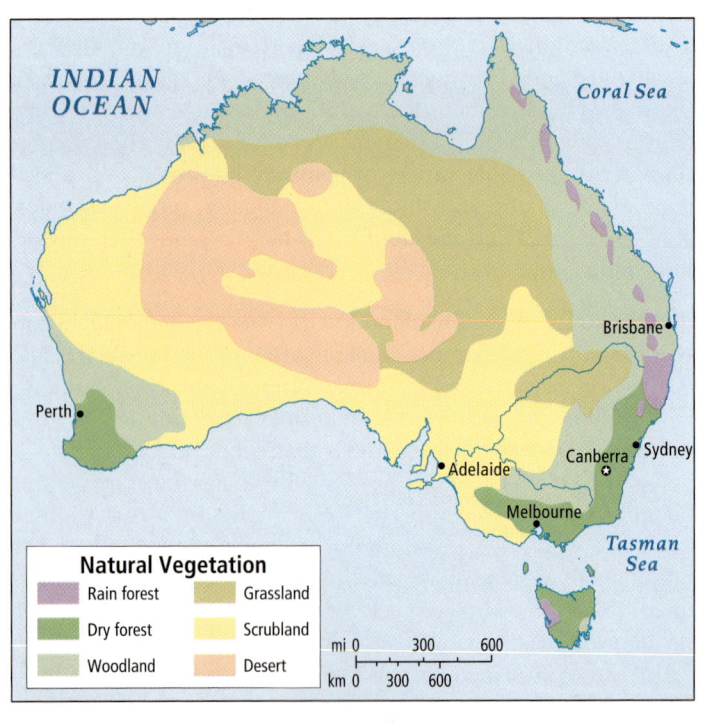

FIGURE 11.7 Australia's natural vegetation. Much of Australia is grassland and scrubland. A few forests can be found in the Eastern Highlands, the far southwest, and Tasmania.
[Source consulted: Tom L. McKnight, *Oceania* (Englewood Cliffs, NJ: Prentice Hall, 1995), p. 28]

up to 12 feet [3.7 meters] tall), were a major source of food for the Maori people. The moa was hunted to extinction before the Europeans arrived. Today, New Zealand may well be the country with the most nonnative species of mammals, fish, and fowl, nearly all brought in by European settlers.

THINGS TO REMEMBER

- This largest region of the world is primarily water and has a very small population of just 38.7 million people.
- The largest land area in Oceania is the continent of Australia.
- The thousands of islands of the Pacific were created either as a result of plate tectonics or volcanic activity.
- The climate of most of the region is warm and humid almost all the time.
- Due to their isolation from the life forms found on other major landmasses, the fauna and flora of the region are unique in many distinctive ways that have informed our knowledge about evolution. Many species, such as marsupials and monotremes, are endemic to the region.

ENVIRONMENTAL ISSUES

> **GEOGRAPHIC INSIGHT 1**
>
> **Environment:** Oceania faces a host of environmental problems and public awareness of environmental issues is keen. Global climate change, primarily warming, has brought rising sea levels and increasingly variable rainfall. Other major threats to the region's unique ecology have come from the introduction of many nonnative species and the expansion of herding, agriculture, fishing, and human settlements.

Global Climate Change

Oceania, with the exception of Australia, is a minor contributor of greenhouse gases. Australia has some of the world's highest greenhouse gas emissions on a per capita basis (see Figure 1.14 on page 22). Much of Australia's emissions, like those of the United States, result from the use of automobile-based transportation systems to connect a widely dispersed network of cities and towns. Additionally, Australia's heavy dependence on coal to generate electricity has led to large amounts of emissions, much as coal dependence has in the United States. Because Australia has a relatively small population (23.1 million people in 2013), it accounts for only slightly more than 1 percent of global emissions. However, despite the negligible contributions of greenhouse gases made by the islands of the Pacific, they, as well as Australia, are quite vulnerable to the effects of global climate change (Figure 11.8).

Sea Level Rise The best scientific research indicates that global warming is raising sea levels mainly through thermal expansion as rising temperatures cause the water in the ocean to expand in size, but also by melting glaciers and polar ice caps. Obviously, this issue is of great concern to residents of islands that already barely rise above the waves (see Figure 11.8A). If sea levels rise 4 inches (10 centimeters) per decade, as predicted by the International Panel on Climate Change, many of the lowest-lying Pacific atolls, such as Tuvalu, will disappear under water within 50 years. Other islands, some with already very crowded coastal zones, will see these zones shrink and become more vulnerable to storm surges and cyclones.

Other Water-Related Vulnerabilities Much of Oceania is particularly vulnerable to the droughts and floods that could result from global climate change. Parts of Australia and some low, dry Pacific islands are already undergoing prolonged droughts and freshwater shortages that are requiring changes in daily life and livelihoods. The fear is that the severe droughts are not the usual periodic dry spells but may represent permanent alterations in rainfall patterns that could also worsen wildfires. Such fires emerged as a major issue in Australia in February 2009, when 173 people died in a rural firestorm near Melbourne (see Figure 11.8B).

As fresh water becomes scarcer across the region, *virtual water* (see Chapter 1, page 26) becomes an issue. All export-related activities that permanently consume or degrade fresh water in their extraction or production processes (such as mineral mining; oil and gas extraction; meat, wool, and wheat production; and tourism-related construction and maintenance) are essentially extracting virtual water from places that are already under water stress. If the true costs of this freshwater depletion were counted and added to the price of the products, these exports might no longer be competitive on the world market—at least not until all global producers understood virtual water accountability to be in their best interests and raised their prices accordingly.

Another concern is the warming of the oceans, which can result in stronger tropical storms. Warmer ocean temperatures threaten coral reefs and the fisheries that depend on them by causing *coral bleaching* (see Figure 11.8C), a phenomenon that has affected all reefs in this region in recent years, especially the Great Barrier Reef. Because so many fish depend on reefs, coral bleaching also threatens many fishing communities. This is especially true on some of the Pacific islands that have few other local food resources.

Responses to Potential Climate Change Crises Oceania is pursuing a number of alternative energy strategies to reduce greenhouse gas emissions. Although Australia remains dependent on fossil fuel sales (primarily of coal, and also of crude oil and natural gas) to Asia, it is pursuing renewable energy strategies for domestic use. These include increasing power production from geothermal, solar, and biomass sources and, because of water shortages, decreasing the emphasis on hydropower. New Zealand has set a goal of obtaining 95 percent of its energy from renewable sources by 2025. Much of this energy will come from wind power, which has a great deal of potential in this region, especially in areas near the Roaring Forties. On low Pacific islands, solar energy is now the most widely used alternative to costly and polluting imported fuel.

FIGURE 11.8 PHOTO ESSAY: Vulnerability to Climate Change in Oceania

Oceania is vulnerable to a wide variety of hazards related to climate change, including sea level rise, stronger tropical storm intensity, and the availability of water becoming less certain. Fortunately, several countries in this region are already implementing solutions that are increasing resilience to climate hazards.

A A low-lying atoll in Tuvalu. With its highest point only 14.7 feet (4.5 meters) above sea level, Tuvalu is quite vulnerable to sea level rise. Combined with worse flooding during tropical storms, higher sea levels could make many low islands in this region uninhabitable. Tuvalu's government is already negotiating the future resettlement of parts of its population to nearby nations such as New Zealand.

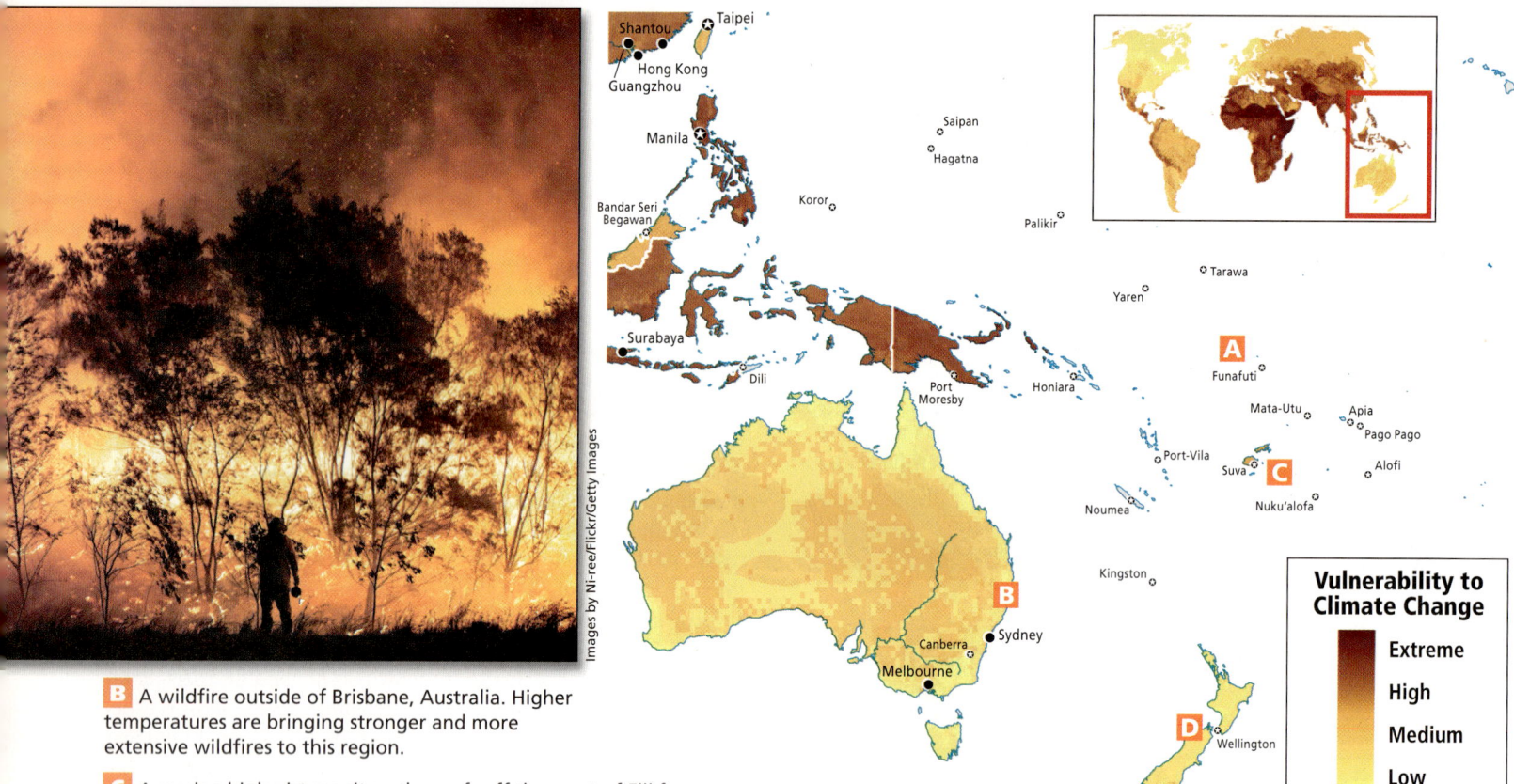

B A wildfire outside of Brisbane, Australia. Higher temperatures are bringing stronger and more extensive wildfires to this region.

C A marine biologist monitors the reefs off the coast of Fiji for signs of coral bleaching. Caused by warmer temperatures, coral bleaching threatens many fish species that live on coral reefs. In turn, human communities on Pacific islands are threatened by the loss of fisheries.

D A vineyard fitted with a drip irrigation system, visible at the base of the vines, in Marlborough, New Zealand. These systems provide resilience in the face of drought and use substantially less water than other irrigation methods.

Vulnerability to Climate Change
- Extreme
- High
- Medium
- Low

Environmental Issues **457**

Invasive Species and Food Production

European systems for producing food and fiber introduced into Oceania have had a profound effect on Oceania's environments. Many of the unique endemic plants and animals of Oceania have been displaced by **invasive species**, organisms that spread into regions outside of their native range, adversely affecting economies or environments. Many *exotic*, or *alien*, plants and animals were brought to Oceania by Europeans to support their food production systems. Ironically, many of these same species are now major threats to food production (Figure 11.9).

invasive species organisms that spread into regions outside their native range, adversely affecting economies or environments

ON THE BRIGHT SIDE

Water Conservation
Oceania is a world leader in implementing alternative water technologies. Some of these are simple but effective age-old methods, such as harvesting rainwater from roofs and the ground for household use. Most buildings in rural Australia, New Zealand, and many Pacific islands get at least part of their water this way, relieving surface and groundwater resources. Australia and New Zealand are now stretching water resources further by using very efficient drip irrigation technologies extensively in agriculture (see Figure 11.8D) and new low-cost water-filtration techniques.

Australia When Europeans first settled the continent, they brought many new animals and plants with them, sometimes intentionally, sometimes unintentionally. European rabbits are among the most destructive of these introduced species. Early British settlers who enjoyed eating rabbits brought them to Australia. Many were released for hunting, but with no natural predators, the rabbits multiplied quickly, consuming so much of the native vegetation that many indigenous animal species starved. Moreover, rabbits became a major source of agricultural crop loss and reduced the capacity of many grasslands to support herds of introduced sheep and cattle. Attempts to control the rabbit population by introducing European foxes and cats backfired as these animals became major invasive species themselves (Figure 11.10C). Foxes and cats have driven several native Australian predator species to extinction without having much effect on the rabbit population. Intentionally introduced diseases have proven more effective at controlling the rabbit population, though rabbits have repeatedly developed resistance to them.

Herding has also had a huge impact on Australian ecosystems. Because the climate is arid and soils in many areas are relatively infertile, the dominant land use in Australia is the grazing of introduced domesticated animals—primarily sheep, but also cattle. More than 15 percent of the land has been given over to grazing and Australia leads the world in exports of sheep and cattle products.

Dingoes, the indigenous wild dogs of Australia (see Figure 11.10B), prey on sheep and young cattle. To separate the wild dogs from the herds, the Dingo Fence—the world's longest fence—was built. It extends 3488 miles (5614 kilometers) and is unfortunately a major ecological barrier to other wild species. Meanwhile, kangaroos (the natural prey of dingoes) have learned to live on the sheep side of the fence, where their population has boomed beyond sustainable levels.

New Zealand New Zealand's environment has been transformed by introduced species and food production systems even more extensively than Australia's environment. No humans lived in New Zealand until about 700 years ago, when the Polynesian Maori people settled there. When they arrived, dense midlatitude rain forest covered 85 percent of the land. The Maori were cultivators who brought in yams and taro as well as other nonnative plants, rats, and birds. By the time of European contact (1642), forest clearing and overhunting by the Maori had already degraded many environments and driven several bird species to extinction.

European settlement in New Zealand dramatically intensified environmental degradation. Attempts to re-create European farming and herding systems in New Zealand resulted in environments that today are actually hostile to many native species, a growing number of which are becoming extinct. Only 23 percent of the country remains forested, with ranches, farms, roads, and urban areas claiming more than 90 percent of the lowland area (Figure 11.11).

Most of the cleared land is used for export-oriented farming and ranching. Grazing has become so widespread that today in New Zealand there are 15 times as many sheep as people, and 3 times as many cattle. Both farming and ranching have severely degraded the environment. Soils exposed by the clearing of forests proved infertile, forcing farmers and ranchers to augment them with agricultural chemicals. The chemicals, along with feces from sheep and cattle, have seriously polluted many waterways, causing the extinction of some aquatic species.

Pacific Islands In the Pacific islands, many unique species of plants and animals have been driven to extinction as the islands have been deforested and mined or converted to commercial agriculture. This has been the case in Hawaii, which is home to more threatened or endangered species than any other U.S. state, despite having less than 1 percent of the U.S. landmass. The extensive conversion of tropical Hawaiian forests to cropland—used for the export crops of sugar cane and pineapples—has caused the extinction of numerous plant, bird, and land species.

239. HAWAII CONSIDERED AMERICA'S ENDANGERED SPECIES CAPITAL

Globalization and the Environment in the Pacific Islands

As the Pacific islands have become more connected to the global economy over the years, flows of resources and pollutants have increased dramatically. Mining, nuclear pollution, and tourism are all examples of how globalization has transformed environments in the Pacific islands.

Mining in Papua New Guinea and Nauru Foreign-owned mining companies that took advantage of poorly enforced or nonexistent environmental laws are responsible for major environmental damage in Oceania. In the Ok Tedi Mine on Papua New Guinea,

FIGURE 11.9   PHOTO ESSAY: Human Impacts on the Biosphere in Oceania

Despite its vast size and relatively small population, Oceania has been severely impacted by human activity. Much of the damage has been done by people from distant countries, as well as by countries in Oceania that export resources outside the region.

A In 1946, on Bikini Atoll in the Marshall Islands (then a U.S. territory), the United States conducted one of the first underwater tests of a nuclear weapon and its effects on naval vessels. Since then, the United States and France have conducted over 300 nuclear tests in Oceania, some without sufficient attention to nuclear contamination.

B Once hunted to near extinction to protect sheep herds introduced to the area, the Tasmanian devil is now threatened by low genetic diversity, which leaves it vulnerable to disease.

C An endangered great white shark tangled in a fish net off the coast of New Zealand. Fleets from around the world come to Oceania to take advantage of its fisheries, many of which are now overexploited. Many Pacific islands still sell fishing rights to foreign fleets because they need the money.

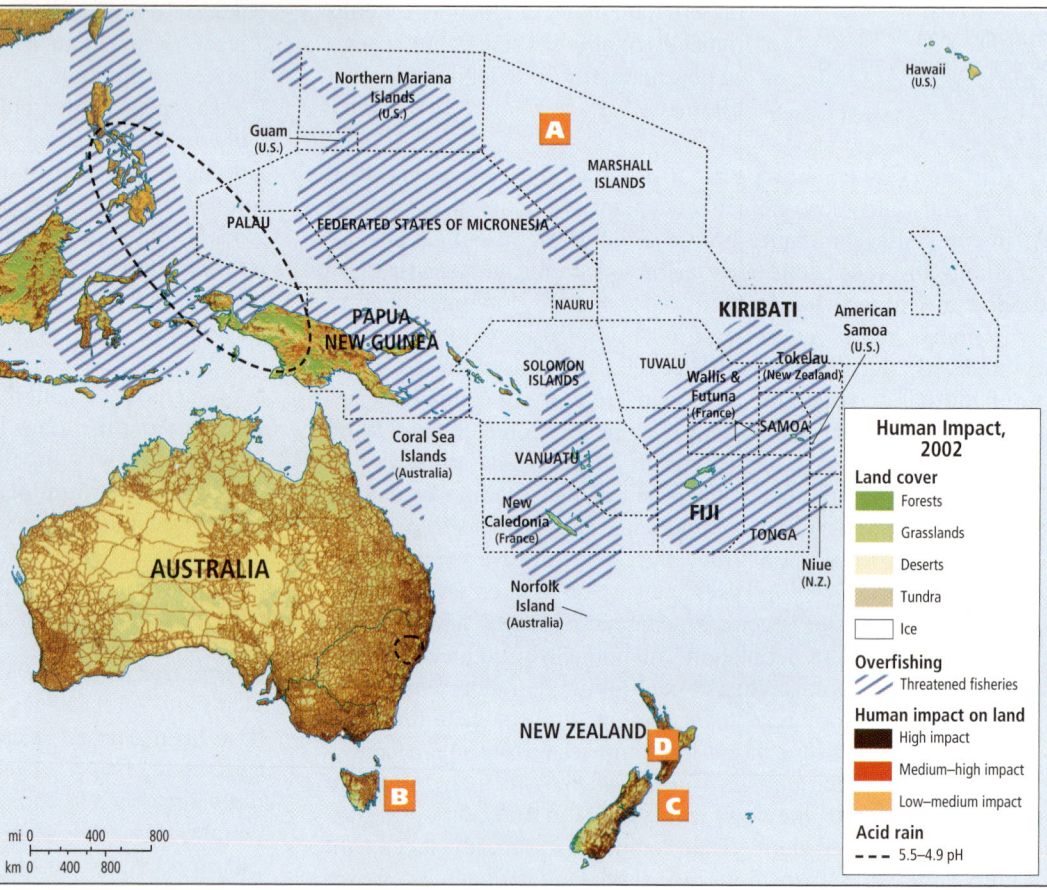

D A herd of sheep in New Zealand, where ranches, farms, roads, and urban areas cover 90 percent of the lowlands. Forests once covered 85 percent of New Zealand, but after two centuries of export-oriented agriculture and forestry, only 23 percent of the country remains forested. In recent decades, there have been increased efforts to conserve the remaining forests.

Human Impact, 2002

Land cover
- Forests
- Grasslands
- Deserts
- Tundra
- Ice

Overfishing
- Threatened fisheries

Human impact on land
- High impact
- Medium–high impact
- Low–medium impact

Acid rain
- - - 5.5–4.9 pH

huge amounts of mine waste devastated river systems (Figure 11.12). The environmental degradation forced tens of thousands of indigenous subsistence cultivators into new mining market towns where their horticulture skills were of little use and where they needed cash to buy food and pay rent.

ON THE BRIGHT SIDE

Environmental Awareness in New Zealand
New Zealand is a global environmental leader. It formed the world's first environmentally focused national political party in 1973 and spearheaded the world's first nuclear-free zone in 1984. New Zealand's government promotes a "clean and green" image internationally, but most New Zealanders acknowledge the severity of existing environmental problems and the need for further action.

In 2007, thirty thousand of these people sued the Australian parent mining company, which was then BHP Billiton, for U.S. $4 billion. Two villagers, Rex Dagi and Alex Maun, traveled to Europe and the United States to explain their cause and meet with international environmental groups. They and their supporters convinced U.S. and German partners in the Ok Tedi mine to divest their shares.

The most extreme case of environmental disaster caused by mining took place on the once densely forested Melanesian island of Nauru, which is one-third the size of Manhattan and located northeast of the Solomon Islands (see the Figure 11.1 map). Nauru's wealth was based on proceeds from the strip-mining of high-grade phosphates derived from eons of bird droppings (guano) that are used to manufacture fertilizer. The mining companies were owned first by Germany, then by Japan, and finally by Australia. For a time in the early 1970s, Nauru had the highest per capita income in the world. Today, the phosphate reserves are depleted, the proceeds ill spent, and the environment destroyed. Junked mining equipment sits on miles of bleached white sand where forest once stood.

FIGURE 11.10 LOCAL LIVES
People and Pets in Oceania

A A rainbow lorikeet, a small parrot that is native in much of Oceania. Rainbow lorikeets are popular pets because of their plumage and curious disposition. However, in parts of Australia and New Zealand, they are considered pests because they consume fruit in orchards, are noisy, and leave behind large droppings. [Vanessa Mylett/Flickr/Getty Images]

B An Australian dingo, a type of wild dog that lives mainly in the Australian outback. Dingoes were likely brought to Australia by Aborigines who used them as guard dogs and possibly as a food source. Shepherds consider dingoes to be pests, and in the 1880s they built a 3488-mile (5614-kilometer) fence to keep dingoes out of southeastern Australia. [Robin Smith/Photolibrary/Getty Images]

C A cat in New Zealand. Introduced throughout Oceania by Europeans who brought them in as pets and to control rodent populations, cats quickly became feral. This had disastrous impacts on native species of marsupials and birds. Cats have been completely removed from several of New Zealand's smaller islands, including those designated as native bird sanctuaries. [Kirsten Gamby/Flickr/Getty Images]

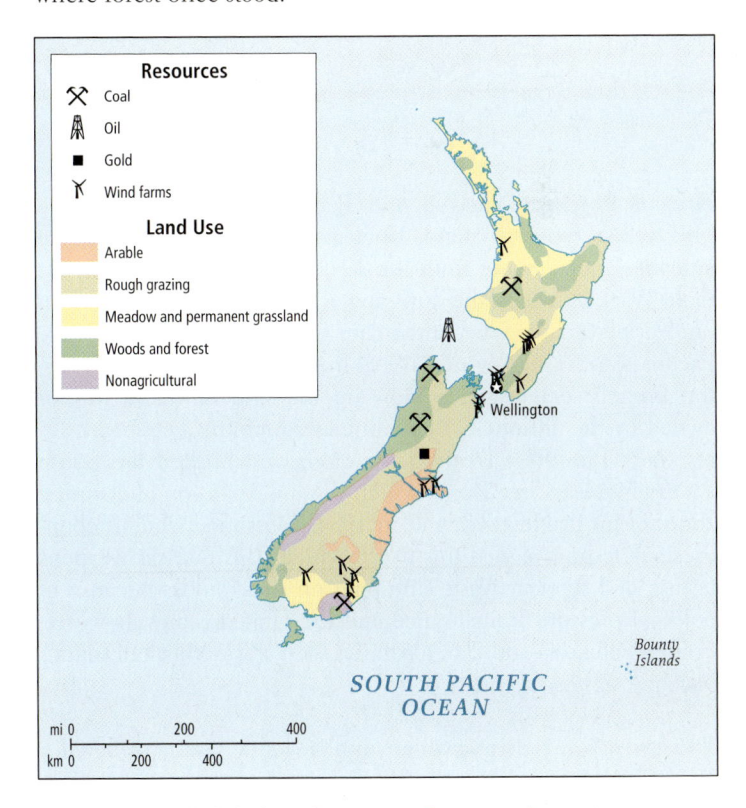

FIGURE 11.11 Land uses and natural resources of New Zealand. As a result of European settlement and the clearing of land for farming, only 23 percent of New Zealand remains forested. [Source consulted: Richard Nile and Christian Clerk, *Cultural Atlas of Australia, New Zealand, and the South Pacific* (New York: Facts on File, 1996), p. 194]

FIGURE 11.12 The Ok Tedi Mine.

(A) The Ok Tedi open-pit copper and gold mine. A large hole now exists where Mount Fubilan once stood.

(B) The Ok Tedi River as it flows downstream of the mine. Each year since its opening in 1984, the mine has discharged 80 million tons of contaminated mine tailings and eroded sediments, resulting in a once-deep river becoming clogged with poisonous runoff that kills many trees along its banks. Sediment from the mine river has killed innumerable fish in the river, contaminated 500 square miles (1300 square kilometers) of farmland, and adversely affected 50,000 people in 120 villages. Litigation against the mine owners is ongoing.

> **THINKING GEOGRAPHICALLY** (A) How many indigenous subsistence cultivators were forced off their land and into new mining market towns as a result of sediment and chemical pollution from the Ok Tedi Mine in Papua New Guinea? (B) What evidence shows that this part of the polluted river flows through lowlands where flooding has taken place?

Nuclear Pollution The geopolitical aspects of globalization have hit Oceania especially hard. Nuclear weapons testing by France and the United States from the 1940s to the 1960s (during the Cold War), as well as the dumping of nuclear waste from various nuclear powers, have become major environmental issues for the Pacific islands (see Figure 11.9A). In New Zealand in July 1985, the French secret service blew up the ship *Rainbow Warrior* owned by the antinuclear environmental group Greenpeace. In response, the 1986 Treaty of Rarotonga established the South Pacific Nuclear Free Zone, which was an expansion of a similar zone set up in New Zealand in 1984. Most independent countries in Oceania signed this treaty, which bans nuclear weapons testing and nuclear waste dumping on their lands. Because of political pressure from France and the United States, however, French Polynesia and U.S. territories such as the Marshall Islands have not signed the treaty.

Tourism Even tourism, which until recently was considered a "clean" industry, can create environmental problems. Foreign-owned tourism enterprises have often accelerated the loss of wetlands and worsened beach erosion by clearing coastal vegetation for hotel construction, golf courses, and waterfront-related entertainment.

Tourism has also strained island water resources because of showering, laundering, and other services that consume fresh water. Furthermore, inadequate methods of disposing of sewage and trash from resorts have polluted many once-pristine areas. Ecotourism (see Chapter 3, page 118) aimed at reducing these impacts is now a common element of development throughout the Pacific, but environmental impacts from ecotourism are still generally high.

The Great Pacific Garbage Island When oceanographer Charles Moore found himself surrounded by a massive floating island of plastic garbage in the north Pacific in 1997, he thought it was an anomaly. But his investigations revealed that the disposable, throwaway aspect of modern living was responsible. The island included plastic beverage bottles and caps along with Lego blocks, trash bags, toothbrushes, footballs, and kayaks—indeed, virtually every consumer product made. By 2008, several such garbage masses were floating in the Pacific and Atlantic oceans. Because modern plastics do not degrade, millions of seabirds and sea mammals die each year after ingesting bits of this trash. Whatever goes into the ocean eventually ends up on someone's dinner plate. Efforts to clean up the garbage island are currently being studied but have not yet been implemented.

The UN Convention on the Law of the Sea Based on the idea that all the problems of the world's oceans are interrelated and need to be addressed as a whole, the United Nations Convention on the Law of the Sea (UNCLOS) established rules governing all uses of the world's oceans and seas and has been ratified by 157 countries (although not the United States). The treaty allows islands to claim rights to ocean resources 200 miles (320 kilometers) out from their shores. Island countries can now make money by licensing privately owned fleets from Japan, South Korea, Russia, the United States, and elsewhere to fish within these offshore limits. As of yet, however, there is no overarching enforcement agency, and protecting the fisheries from overfishing by these rich and powerful licensees has turned out to be an enforcement nightmare for tiny island governments with few resources. Similarly, it has proven difficult to monitor and control the exploitation of seafloor mineral deposits by foreign mining companies.

241. ENDANGERED HAWAIIAN MONK SEAL POPULATION CONTINUES TO DECLINE

242. NEW SPECIES OF UNDERSEA LIFE FOUND NEAR INDONESIA

243. SCIENTISTS WARN OF DEPLETION OF OCEAN FISH IN 40 YEARS

THINGS TO REMEMBER

› GEOGRAPHIC INSIGHT 1

- **Environment** Oceania faces a host of environmental problems and public awareness of environmental issues is keen. Global climate change, primarily warming, has brought rising sea levels and increasingly variable rainfall. Other major threats to the region's unique ecology have come from the introduction of many nonnative species and the expansion of herding, agriculture, fishing, and human settlements.

- In an effort to combat global warming, renewable energy alternatives to fossil fuels are being pursued across the region. New Zealand has set a goal of obtaining 95 percent of its energy from renewable sources by 2025.

- Throughout Oceania, the introduction of food production systems from elsewhere has resulted in the spread of ecologically and economically damaging invasive plant and animal species.

- Globalization and the patterns of consumption by people who live far from the Pacific are seriously affecting human and animal life in this region.

HUMAN GEOGRAPHY

Oceania's past has been shaped by its ancient settlement and by the more recent populating of Australia and New Zealand by Europeans. By contrast, its present is increasingly being influenced by economic and geographic considerations, particularly its physical proximity to Asia.

HUMAN PATTERNS OVER TIME

The Peopling of Oceania

The longest-surviving inhabitants of Oceania are Australia's **Aborigines**, whose ancestors (the Australoids) migrated from Southeast Asia 50,000 to 70,000 years ago (Figure 11.13; see also Figure 11.14A), at a time when sea level was somewhat lower. It is possible that some memory of this ancient journey may be preserved in Aboriginal oral traditions, which recall mountains and other geographic features that are now submerged under water. At about the same time that the Aborigines were settling Australia and Tasmania, related groups were settling nearby areas.

Melanesians, so named for their relatively dark skin tones, a result of high levels of the protective pigment *melanin*, migrated throughout New Guinea and other nearby islands, giving this area its name, **Melanesia**. Archaeological evidence indicates

> **Aborigines** the longest-surviving inhabitants of Oceania, whose ancestors, the Australoids, migrated from Southeast Asia 50,000 to 70,000 years ago over the Sundaland landmass that was exposed during the ice ages
>
> **Melanesians** a group of Australoids named for their relatively dark skin tones, a result of high levels of the protective pigment melanin; they settled throughout New Guinea and other nearby islands
>
> **Melanesia** New Guinea and the islands south of the equator and west of Tonga (the Solomon Islands, New Caledonia, Fiji, and Vanuatu)

VIGNETTE

"With courage, you can travel anywhere in the world and never be lost. Because I have faith in the words of my ancestors, I'm a navigator."
—Mau Piailug

In 1976, Mau Piailug made history by sailing a traditional Pacific island voyaging canoe across the 2400 miles (3860 kilometers) of deep ocean between Hawaii and Tahiti. He did so without a compass, charts, or other modern instruments, using only methods passed down through his family. To find his way, he relied mainly on observations of the stars, the sun, and the moon. When clouds covered the sky, he used the patterns of ocean waves and swells as well as the presence of seabirds to tell him of distant islands over the horizon.

Piailug reached Tahiti 33 days after leaving Hawaii and made the return trip in 22 days. His voyage resolved a major scholarly debate over how people settled the many remote islands of the Pacific without navigational instruments thousands of years before the arrival of Europeans. Some thought that navigation without instruments was impossible and argued that would-be settlers simply drifted about on their canoes at the mercy of the winds, most of them starving to death on the seas, with a few happening upon new islands by chance. It was hard to refute this argument because local navigational methods had died out almost everywhere. However, in isolated Micronesia, where Piailug lives, indigenous navigational traditions still survive.

After the successful 1976 voyage, Piailug trained several students in traditional navigational techniques. His efforts have become a symbol of cultural rebirth and a source of pride throughout the Pacific. In 2007, the protégés of Mau Piailug sailed from Hawaii through the Marshall Islands to Yokohama, Japan, to celebrate peace and the human need to stay connected with nature. [Source: Facts on File, 2007. For detailed source information, see Text Sources and Credits.] ■

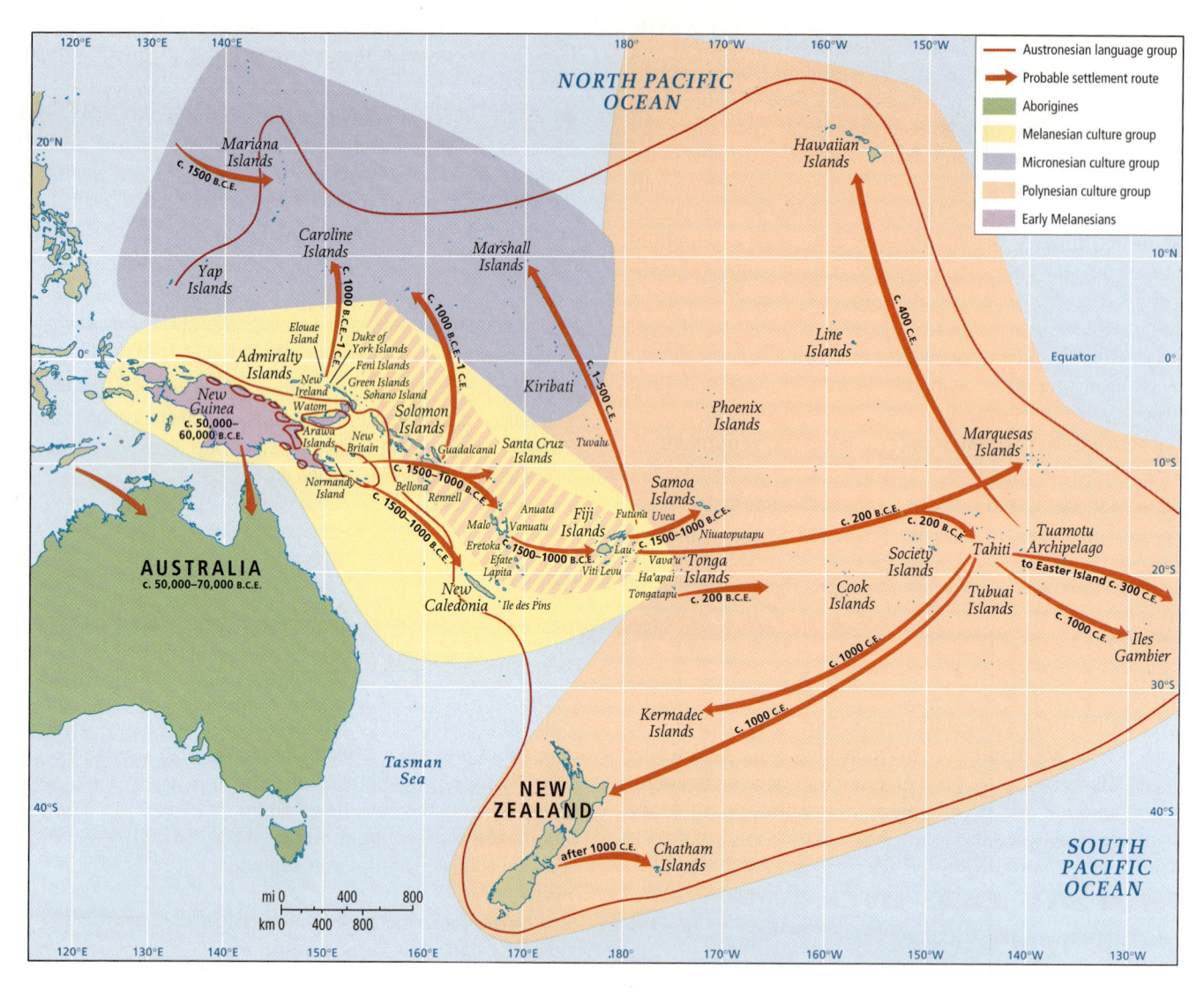

462 CHAPTER 11 Oceania: Australia, New Zealand, and the Pacific

FIGURE 11.13 Primary indigenous culture groups of Oceania. By 50,000 to 70,000 years ago, humans had come to New Guinea and Australia. About 25,000 years ago, people began moving across the ocean to nearby Pacific islands. Movement into the more distant islands began with the arrival of Austronesians, who went on to inhabit the farthest reaches of Oceania. [Source consulted: Richard Nile and Christian Clerk, *Cultural Atlas of Australia, New Zealand, and the South Pacific* (New York: Facts on File, 1996), pp. 58–59]

that they first arrived more than 50,000 to 60,000 years ago from Sundaland (see Figure 10.5 on page 410), a now-submerged shelf exposed during the Pleistocene epoch. They lived in isolated pockets, which resulted in the evolution of hundreds of distinct yet related languages. Like the Aborigines, the early Melanesians survived mostly by hunting, gathering, and fishing, although some groups—especially those inhabiting the New Guinea highlands—eventually practiced agriculture.

Much later, between 5000 to 6000 years ago and as recently as 1000 years ago, linguistically related *Austronesians* (a group of skilled farmers and seafarers from southern China who migrated to Southeast Asia; see Chapter 10, page 419) settled **Micronesia** and **Polynesia**, sometimes mixing with the Melanesian peoples they encountered. Micronesia consists of the small islands that lie east of the Philippines and north of the equator (see the

> **Micronesia** the small islands that lie east of the Philippines and north of the equator
>
> **Polynesia** the numerous islands situated inside an irregular triangle formed by New Zealand, Hawaii, and Easter Island

Figure 11.1 map). Polynesia is made up of numerous islands situated inside the large, irregular triangle formed by New Zealand, Hawaii, and Easter Island. (Easter Island, also called Rapa Nui, is a tiny speck of land in the far eastern Pacific, at 109° W 27° S, not shown in the figures in this chapter.) Recent experiments run by Polynesians (see the voyaging vignette on page 461) have provided evidence that ancient sailors could navigate over vast distances using seasonal winds, astronomic calculations, bird and aquatic life, and wave patterns to reach the most far-flung islands of the Pacific. The Polynesians were fishers, hunter-gatherers, and cultivators who developed complex cultures and maintained trading relationships among their widely spaced islands.

In the millennia that have passed since first settlement, humans have continued to circulate throughout Oceania. Some apparently set out because their own space was crowded and full of conflict or because food reserves were declining. It is also likely that Pacific peoples were enticed to new locales by the same lures that later attracted some of the more romantic explorers from Europe and elsewhere: sparkling beaches, magnificent blue skies, aromatic breezes, and lovely landscapes.

Arrival of the Europeans

The earliest recorded contact between Pacific peoples and Europeans took place in 1521, when the Portuguese navigator Ferdinand Magellan (exploring for Spain) landed on the island of Guam in Micronesia. The encounter ended badly. The islanders, intrigued by European vessels, tried to take a small skiff. For this crime, Magellan had his men kill the offenders and burn their village to the ground. A few months later, Magellan was himself killed by islanders in what later became the Philippines, which he had claimed for Spain. Nevertheless, by the 1560s, the Spanish had set up a lucrative Pacific trade route between Manila in the Philippines and Acapulco on the west coast of Mexico. Explorers from other European states followed, first taking an interest mainly in the region's valuable spices. The British and French explored Oceania extensively in the eighteenth century (**Figure 11.14C**).

The Pacific was not formally divided among the colonial powers until the nineteenth century. By that time, the United States, Germany, and Japan had joined France and Britain in taking control of various island groups. As in other regions, the European colonizers of Oceania emphasized extractive agriculture and mining. Because native people were often displaced from their lands or exposed to exotic diseases to which they had no immunity, their populations declined sharply.

The Colonization of Australia and New Zealand

Although all of Oceania has been under European or U.S. rule at some point, the most Westernized parts of the region are Australia and New Zealand. The colonization of these two countries by the British has resulted in Australia and New Zealand having many parallels with North America. In fact, the American Revolution was a major impetus for "settling" Australia because once the North American colonies were independent, the British wanted somewhere else to send their convicts. In early nineteenth-century Britain, a relatively minor theft—for example, of a piglet—might be punished with 7 years of hard labor in Australia (see Figure 11.14D).

A steady flow of English and Irish convicts arrived in Australia until 1868. Most of the convicts chose to stay in the colony after their sentences were served and are given credit for Australia's rustic self-image and egalitarian spirit. They were joined by a much larger group of voluntary immigrants from the British Isles who were attracted by the availability of inexpensive farmland. Waves of these immigrants arrived until World War II. New Zealand was settled in the mid-1800s, somewhat later than Australia. Although its population also derives primarily from British immigrants, New Zealand was never a penal colony.

Another similarity among Australia, New Zealand, and North America was the treatment of indigenous peoples by European settlers. In both Australia and New Zealand, native peoples were killed outright, annihilated by infectious diseases, or shifted to the margins of society. The few who lived on territory the Europeans thought undesirable were able to maintain their traditional ways of life. However, the vast majority of the survivors lived and worked in grinding poverty, either in urban slums or on cattle and sheep ranches. Today, native peoples still suffer from discrimination and maladies such as alcoholism and malnutrition. Even so, some progress is being made toward improving their lives (see pages 475–476). In 2008, the newly elected prime minister of Australia, Kevin Rudd, officially apologized to Aboriginal people for the treatment they have received since the land was first colonized.

Closely related to attitudes toward indigenous people were attitudes toward immigrants of any color other than white. By 1901, a whites-only policy (called the "White Australia policy") governed Australian immigration, with favored migrants coming from the British Isles and (after World War II) from southern Europe. This discrimination persisted until the mid-1970s, when the White Australia policy on immigration was ended. In New Zealand, where similar racist attitudes prevailed, there was never an official whites-only policy, and by the 1970s, students and immigrants were arriving from Asia and the Pacific islands. Controversy over immigration in Australia and New Zealand has recently centered on the arrival of refugees by boat from various parts of Asia (see Figure 11.17A on page 469).

Oceania's Shifting Global Relationships

During the twentieth century, Oceania's relationship with the rest of the world went through three phases: from a predominantly European focus, to identification with the United States and Canada, and finally to the currently emerging linkage with Asia.

Until roughly World War II, the colonial system gave the region a European orientation. In most places, the economy depended largely on the export of raw materials to Europe (see Figure 11.14E). Thus, even when a colony gained independence from Britain, as did Australia in 1901 and New Zealand in 1907, people remained strongly tied to their mother countries. Even today, the Queen of England remains the titular head of state in both countries. During World War II, however, the European powers provided only token resistance to Japan's invasion of much of the Pacific and its bombing of northern Australia. This European impotence began a change in the region's political and economic orientation. **244. VETERANS REMEMBER TRAGEDY OF WAR IN PACIFIC**

After the war, the United States, which already had a strong foothold in the Philippines, became the dominant power in the Pacific and U.S. investment became more important to the economies of Oceania. Australia and New Zealand joined the United States in a Cold War military alliance, and both fought alongside the United States in Korea and Vietnam, suffering considerable casualties and experiencing significant antiwar activity at home. U.S. cultural influences became strong, too, as North American

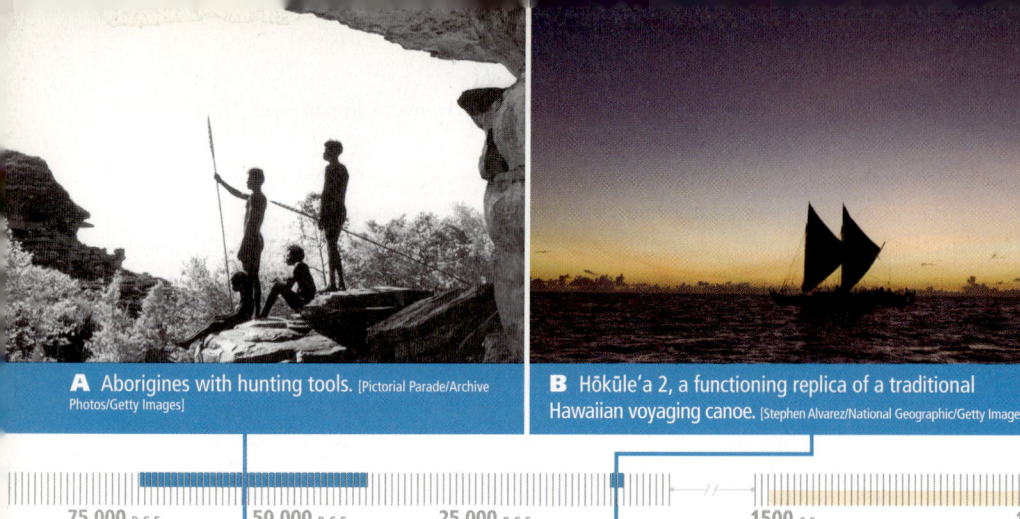

A Aborigines with hunting tools. [Pictorial Parade/Archive Photos/Getty Images]

B Hōkūle'a 2, a functioning replica of a traditional Hawaiian voyaging canoe. [Stephen Alvarez/National Geographic/Getty Images]

C An etching of early British contact with Pacific Islanders in 1783. [Hulton Archive/Getty Images]

75,000 B.C.E. | 50,000 B.C.E. | 25,000 B.C.E. | 1500 C.E. | 1550 C.E. | 1600 C.E. | 1650 C.E.

70,000–50,000 B.C.E. Australian Aborigines' ancestors migrate from Southeast Asia

4000–3000 B.C.E. Settlement of Micronesia and Polynesia

1500s C.E. Spanish explorations

FIGURE 11.14 VISUAL HISTORY OF OCEANIA

---

products, technologies, movies, and pop music penetrated much of Oceania.

By the 1970s, another shift was taking place as many of the island groups were granted self-rule by their European colonizers, and Oceania became steadily drawn into the growing economies of Asia. Since the 1960s, Australia's thriving mineral export sector has become increasingly geared toward supplying raw materials to Asian manufacturing industries (primarily Japan in the 1960s and China since the 1990s). Similarly, New Zealand's wool and dairy exports have gone mostly to Asian markets since the 1970s. Despite occasional backlashes against "Asianization," Australia, New Zealand, and the rest of Oceania are being transformed by Asian influences. Many Pacific islands have significant Chinese, Japanese, Filipino, and Indian minorities, and the small Asian minorities of Australia and New Zealand are growing in population (see Figure 11.14F). On some Pacific islands, such as Hawaii, Asians now constitute the largest portion (42 percent in Hawaii, for example) of the population.

THINGS TO REMEMBER

- Oceania can be divided into four distinct indigenous cultural regions: Australia and Tasmania, originally settled by Aborigines; Melanesia, settled by Melanesians; and Micronesia and Polynesia, settled by a variety of Austronesian peoples.
- Through colonization, Europeans were active in Oceania from the early sixteenth century until the end of World War II. During the 50 years after the war, the United States, Australia, and New Zealand were the principal powers in the Pacific.
- Since the 1970s, the influence of Asian countries has grown throughout Oceania.

GLOBALIZATION AND DEVELOPMENT

> **GEOGRAPHIC INSIGHT 2**
>
> **Globalization and Development:** Globalization, coupled with Oceania's stronger focus on neighboring Asia (rather than on its long-time connections with Europe), has transformed patterns of trade and economic development across Oceania. The transformation is being driven largely by Asia's growing affluence, its enormous demand for resources, and its similarly massive production of manufactured goods.

Globalization, Development, and Oceania's New Asian Orientation

One could say that globalization in Oceania began when the first European explorers came into the region, beginning the trend of influence by outsiders (primarily Europeans) on settlement, culture, and economics. More recently, the United States has exerted a powerful influence on trade and politics in the region. For the past several decades, however, globalization has reoriented this region toward Asia, which buys more than 76 percent of Australia's exports (mainly coal, iron ore, and other minerals). In 2011, China and India each purchased not only the output of mines, but also major shares of Australia's coal deposits. Both countries use coal to generate energy and are trying to secure future access to more coal. Asia also buys nearly 35 percent of New Zealand's exports (primarily meat, wool, and dairy products), as well as many other products and services from islands across Oceania (Figure 11.15).

Asia is also a major source of the region's imports. Because there is little manufacturing in Oceania, most manufactured goods are imported from China, Japan, South Korea, Singapore, and Thailand—Oceania's leading trading partners. Both Australia and New Zealand have free trade agreements either completed

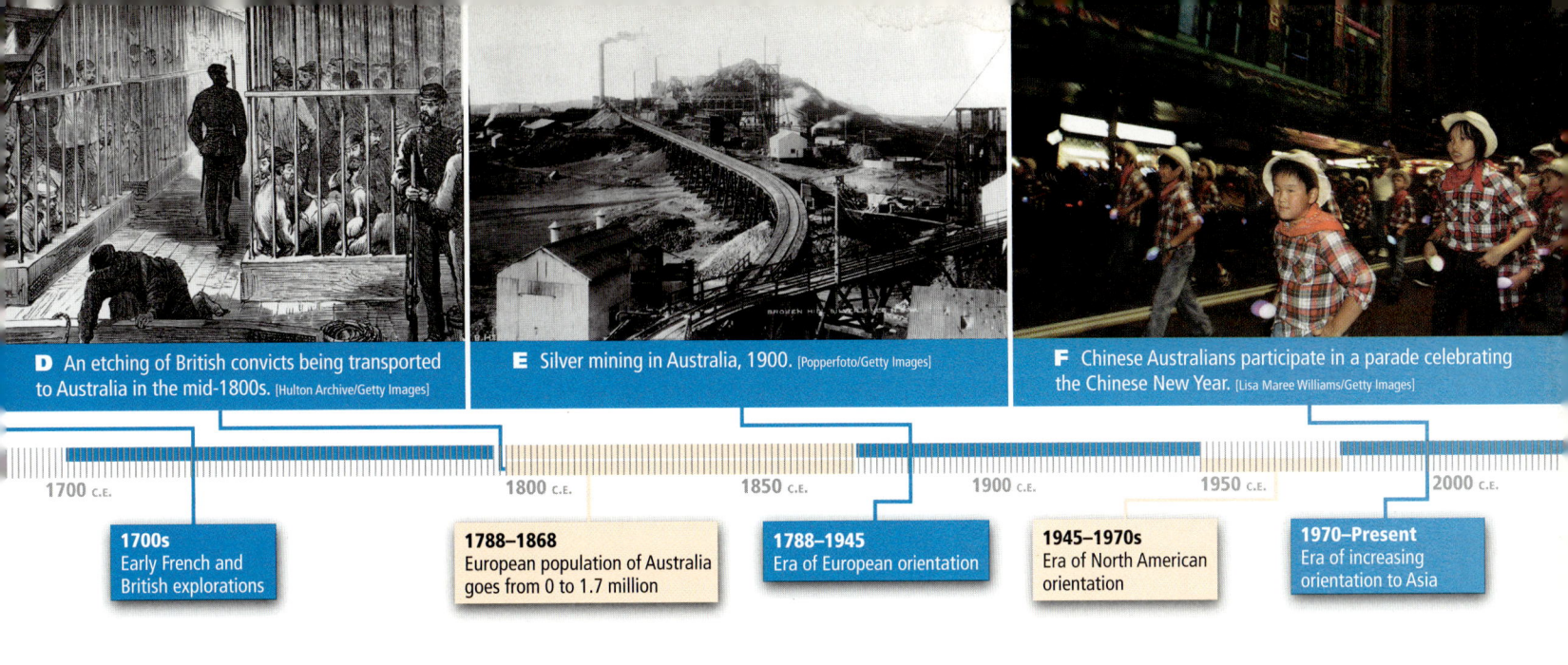

D An etching of British convicts being transported to Australia in the mid-1800s. [Hulton Archive/Getty Images]

E Silver mining in Australia, 1900. [Popperfoto/Getty Images]

F Chinese Australians participate in a parade celebrating the Chinese New Year. [Lisa Maree Williams/Getty Images]

1700s Early French and British explorations

1788–1868 European population of Australia goes from 0 to 1.7 million

1788–1945 Era of European orientation

1945–1970s Era of North American orientation

1970–Present Era of increasing orientation to Asia

---

or in continuous negotiation with Asia's two largest economies, China and Japan.

The Pacific islands are further along in their reorientation toward Asia than are Australia and New Zealand. Not only are coconut, forest, and fish products from the Pacific islands sold to Asian markets, but Asian companies own more and more of these industries on the various islands. Fleets from Asia regularly fish the offshore waters of Pacific island nations. Asians also dominate the Pacific island tourist trade, both as tourists and as investors in tourism infrastructure. And growing numbers of Asians are taking up residence in the Pacific islands, exerting widespread economic and social influence.

The Stresses of Asia's Economic Development "Miracle" on Australia and New Zealand

For Australia and New Zealand, Asia's global economic rise has meant both more trade and more competition with Asian economies in foreign markets. Throughout Oceania, local industries used to enjoy protected or preferential trade with Europe. They have lost that advantage because EU regulations stemming from the EU's membership in the World Trade Organization (WTO) prohibit such arrangements. No longer protected in their trade with Europe, local industries now face stiff competition from larger companies in Asia that benefit from much cheaper labor.

Australia and New Zealand are somewhat unusual in having achieved broad prosperity largely on the basis of exporting raw materials over a long period of time and to a wide range of customers (see Figures 11.14E and 11.15). Preferential trade with Europe allowed higher profits for many export industries. Strong labor movements in Australia and New Zealand meant that these industries' profits were more equitably distributed throughout society than the profits in other raw materials–based economies in Middle and South America and sub-Saharan Africa. Australian coal miners' unions successfully agitated not just for good wages, but also for the world's first 35-hour workweek. Other labor unions won a minimum wage, pensions, and aid to families with children long before such programs were enacted in many other industrialized countries. For decades, these arrangements were very successful. Both Australia and New Zealand had living standards comparable to those in North America but with a more egalitarian distribution of income. Since the 1970s, however, competition from Asian companies has meant that many workers in Australia and New Zealand have lost jobs and seen their hard-won benefits scaled back or eliminated.

Competition from Asian companies has also led to lower corporate profits in Oceania. As corporate profits have fallen, so have government tax revenues, which has resulted in cuts to previously high rates of social spending on welfare, health care, and education. The loss of social support, especially for those who have lost jobs, has contributed to rising poverty in recent years. Australia now has the second-highest poverty rate in the industrialized world. (The United States has the highest.)

Maintaining Raw Materials Exports as Service Economies Develop Although their monetary contribution to national economies remains high, industries that export raw materials are of decreasing prominence in the economies of Australia and New Zealand in that they now employ fewer people because of mechanization. This shift to replacing human labor with machinery has been essential for these industries to stay globally competitive with other countries where living standards are lower and workers are paid much less.

Today, the economies of both Australia and New Zealand are dominated by diverse and growing service sectors, which have links to the region's export sectors. Extracting minerals and managing herds and cropland have become technologically sophisticated enterprises that depend on many supporting services and an educated workforce. Australia is now a world leader in providing technical and other services to mining companies, sheep farmers, and winemakers, and New Zealand's well-educated workforce and

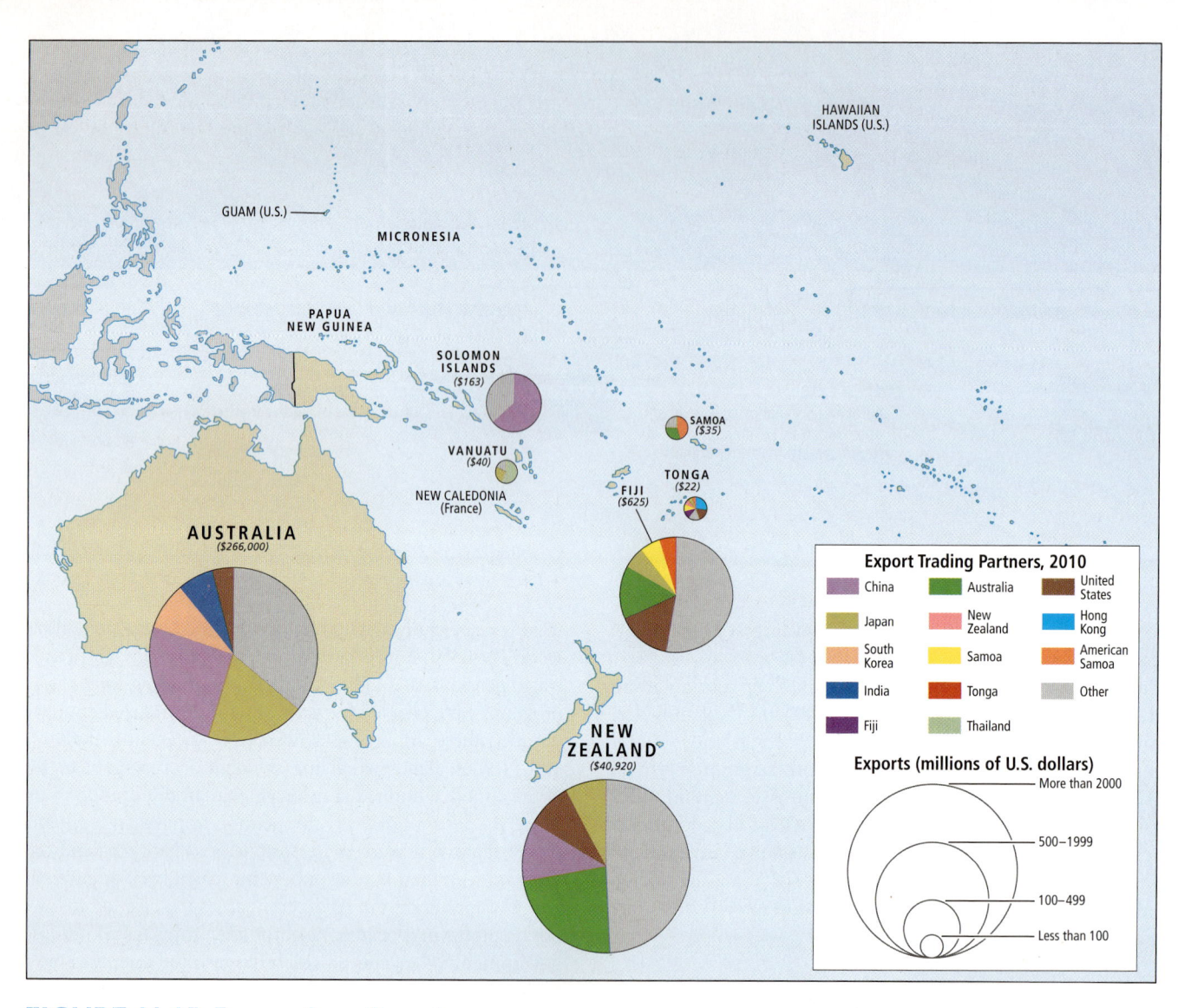

466 CHAPTER 11 Oceania: Australia, New Zealand, and the Pacific

FIGURE 11.15 Exports from Oceania. The colors of each pie chart indicate a country's export trading partners. The "other" sections can include trade with Canada, Mexico, the Caribbean, non-EU Europe, sub-Saharan Africa, and other locales, some of them new trading partners. (Figures for Hawaii do not include exports to other parts of the United States.) [Sources consulted: *The World Factbook,* Central Intelligence Agency, at https://www.cia.gov/library/publications/the-world-factbook/geos/xx.html; "East Asian and Pacific Affairs: Countries and Other Areas," U.S. Department of State, at http://www.state.gov/p/eap/ci/index.htm]

well-developed marketing infrastructure have helped it break into luxury markets for dairy products, meats, and fruits. Perhaps the most visible success has been New Zealand's global marketing of the indigenous kiwifruit. (In fact, *kiwi* is now a slang term for anyone from New Zealand.)

Economic Change in the Pacific Islands In general, the Pacific islands are also shifting away from extractive industries, such as mining and fishing, and toward service industries, such as tourism and government. On many islands, self-sufficiency and resources from abroad cushion the stress of economic change. Many households still construct their own homes and rely on fishing and subsistence cultivation for much of their food supply. On the islands of Fiji, for example, part-time subsistence agriculture engages more than 60 percent of the population, although it accounts for just under 17 percent of the economy. Remittances sent home from the thousands of Pacific Islanders working abroad are essential to many Pacific island economies and constitute more than half of all income on some islands. However, remittances are rarely part of official statistics.

In the relatively poor and undereducated nations (the Solomon Islands, Tuvalu, and parts of Papua New Guinea, for example), conditions typify what has been termed a **MIRAB economy**—one based on migration, remittances, aid, and bureaucracy. Foreign aid from former or present colonial powers supports government bureaucracies that supply employment for the educated and semiskilled. A MIRAB economy has little potential for growth. Nevertheless, Islanders who can be self-sufficient in food and shelter while saving extra cash for travel and occasional purchases of manufactured goods

> **MIRAB economy** an economy based on migration, remittances, aid, and bureaucracy

are sometimes said to have achieved **subsistence affluence**.

Where there is poverty, it is often related to geographic isolation, which means a lack of access to information and economic opportunity. Although computers and the new global communication networks are not yet widely available in the Pacific islands, they have the potential to significantly alleviate this isolation.

subsistence affluence a lifestyle whereby people are self-sufficient with regard to most necessities and have some opportunities to earn cash for travel and occasional purchases of manufactured goods

ON THE BRIGHT SIDE

Subsistence Affluence Practices Could Go Global
Many of the qualities of subsistence affluence practiced by Pacific Islanders, including local self-sufficiency and resource conservation strategies, have the potential for being expanded upon and adapted elsewhere in the region and across the world.

The Advantages and Stresses of Tourism

Tourism is a growing part of the economy throughout Oceania. Tourists come largely from Japan, Korea, Taiwan, Southeast Asia, the Americas, and Europe (Figure 11.16). In 2008 (the latest year for which complete figures are available), 17.7 million tourists arrived in Oceania. Of the tourists visiting Oceania, 23 percent came from Asia—16 percent from Japan alone; just 11 percent came from Europe, down from 17 percent in recent years; 33 percent came from North America; 19 percent were from within Oceania; and 12 percent came from other locations.

In some Pacific island groups, the number of tourists far exceeds the island population. Guam, for example, receives tourists annually in numbers equivalent to five times its population. Palau and the Northern Mariana Islands annually receive more than four times their populations. Such large numbers of visitors expecting to be entertained and graciously accommodated can place a special stress on local inhabitants. And although they bring money to the islands' economies, these visitors create problems for island ecologies, place extra burdens on water and sewer systems, and require a standard of living that may be far out of reach for local people. Perhaps nowhere else in the region are the issues raised by tourism as clear as they are in Hawaii.

CASE STUDY
Conflict over Tourism in Hawaii

Since the 1950s, travel and tourism have been the largest industries in Hawaii, producing nearly 18 percent of the gross state product in 2008. Tourism is related in one way or another to nearly 75 percent of all jobs in the state. (By comparison, travel and tourism account for 9 percent of GDP worldwide.) In 2008, tourism employed one out of every six Hawaiians and accounted for 25 percent of state tax revenues.

Dramatic fluctuations in tourist visits, often driven by forces far removed from Oceania, can wreak havoc on local economies. Decreases in tourism affect not just tourist facilities but supporting industries as well. For example, the construction industry thrives by building condominiums, hotels, resorts, and retirement facilities. The Asian recession of the late 1990s, the terrorist attacks of September 11, 2001, and the global recession of 2008–2009 all affected Hawaii's economy by creating dramatic slumps in tourist visits. In 2011, however, Hawaii's tourism industry rebounded: more than 7 million visitors arrived that year, up 11 percent over 2009.

Sometimes mass tourism can seem like an invading force to ordinary citizens. For example, an important segment of the Honolulu tourist infrastructure—hotels, golf courses, specialty shopping centers, import shops, and nightclubs—is geared to visitors from Japan, and many such facilities are owned by Japanese investors. Hawaiian citizens and other non-Japanese shoppers and vacationers can feel out of place.

Another example of the impositions of mass tourism is the demand by tourists for golf courses on many of the Hawaiian Islands, which has resulted in what Native (indigenous) Hawaiians view as desecration of sacred sites. Land that in precolonial times was communally owned, cultivated, and used for sacred rituals was confiscated by the colonial government and more recently sold to Asian golf course developers. Now the only people with access to the sacred sites are fee-paying tourist golfers.

As of 2012, there were more than 90 golf courses in Hawaii, and the golf industry alone contributed $1.6 billion to the state's economy—more than twice the amount from agriculture. Golf's total impact is $2.5 billion, which represents about 12.5 percent of the state's tourism sector income. Retired Americans who relocate to Hawaii, looking for a sunny spot—in what is often called *residential tourism*—have also had an effect on property values and the use of sacred lands by local citizens. [Sources: Hawaii Tourism Authority and a field report from Conrad M. Goodwin and Lydia Pulsipher. For detailed source information, see Text Sources and Credits.]

FIGURE 11.16 Tourism in Oceania. Tourism plays a major role in the economies of Oceania. In 2008, more than 17.7 million tourists visited the region, with just under one-third of these going to Australia. The origins of the tourists reflect changing trade patterns in the region, with more and more coming from Asia. Here you see hotels on Honolulu's Waikiki Beach, which provide lodging for a majority of Hawaii's 6 million yearly visitors.

The Future: Diverse Global Orientations?

Despite the powerful forces pushing Oceania toward Asia, important factors still favor strong ties with Europe and North America. In spite of the trade links and China's recent efforts to expand diplomatic and cultural relations with Australia, both Australia and New Zealand remain staunch military allies of the United States. Over the years, both have participated in U.S.-led wars in Korea, Vietnam, Afghanistan, and Iraq. In 2012, in an apparent effort to check the growing influence of the Chinese military in the South China Sea and the Indian Ocean, the Australian government gave the U.S. Marine Corps access to a large tract of land near Darwin (located in Australia's Northern Territory). The United States and Australia also opened discussions regarding the use of the Cocos Islands (Australian possessions in the Indian Ocean) for reconnaissance purposes.

In some of the Pacific islands, strong links to Europe and North America are also upheld by Europe and the U.S.'s continuing administrative control. In Micronesia, the United States governs Guam and the Northern Mariana Islands; in Polynesia, American Samoa is a U.S. territory. Just as the Hawaiian Islands are a U.S. state, the 120 islands of French Polynesia—including Tahiti and the rest of the Society Islands, the Marquesas Islands, and the Tuamotu Archipelago—are Overseas Lands of France. Any desire people in these possessions have for independence has not been sufficient to override the financial benefits of aid, subsidies, and investment money provided by France and the United States.

In 1989, Asia Pacific Economic Cooperation (APEC) formed as a group of 21 Pacific countries. Today, these states account for approximately 40 percent of the world's population, just over 50 percent of global production, and more than 40 percent of global trade. Its members include Oceania's most populous countries (Australia, Papua New Guinea, and New Zealand), as well as Brunei, Canada, Chile, China, Hong Kong, Indonesia, Japan, South Korea, Malaysia, Mexico, Peru, the Philippines, Russia, Singapore, Taipei, Thailand, the United States, and Vietnam. APEC was organized to enhance economic prosperity and strengthen the Asia–Pacific community, and while much is made of its potential, its inability to compel its membership to act in any way has led some to dismiss it as a pointless "talk shop." Nevertheless, should APEC ever evolve into a more potent force, the population and wealth of its membership would place it on par with the European Union (after which it is partially patterned), NAFTA, or the WTO. So far, APEC has tried to forge international cooperation on food security, climate change, and energy-efficient transportation.

ON THE BRIGHT SIDE

Sustainable Tourism

Some Pacific islands have attempted to deal with the pressures of tourism by adopting the principle of *sustainable tourism*, which is aimed at decreasing tourism's imprint and minimizing disparities between hosts and visitors. For example Samoa, with financial aid from New Zealand, now develops and monitors sustainable tourism components (beaches, wetlands, and forested island environments) and provides *knowledge-based tourism experiences* for visitors (information-rich explanations of political, social, and environmental issues).

---

growing affluence, its enormous demand for resources, and its similarly massive production of manufactured goods.

- Service industries are becoming the dominant income source for most of Oceania's economies, although extractive industries remain important.
- Tourism is a significant and growing part of the economies in Oceania, but it can produce stresses for the host countries.
- Oceania is at the center of a unique attempt (APEC) to forge international cooperation on food security, climate change, and energy-efficient transportation. ■

POWER AND POLITICS

> **GEOGRAPHIC INSIGHT 3**
>
> **Power and Politics:** In recent decades, stark divisions have emerged in Oceania over definitions of democracy—the system of government that dominates in New Zealand and Australia—versus the Pacific Way, a political and cultural philosophy based in the traditional cultures of the Pacific islands.

Democracy as it is practiced in Australia and New Zealand is a parliamentary system based on universal voting rights for adults, debate, and majority rule. The **Pacific Way** is based on traditional notions of power and problem solving and refers to a way of settling issues familiar to many Pacific Islanders. It favors consensus and mutual understanding over open confrontation, and respect for traditional leadership (especially the usually patriarchal leadership of families and villages) over free speech and other political freedoms. As such, the Pacific Way can embody very different definitions of fairness and corruption than parliamentary systems.

Pacific Way the regional identity and way of handling conflicts peacefully that grows out of Pacific Islanders' particular social experiences

As a political and cultural philosophy, the Pacific Way developed in Fiji around the time of Fiji's independence from the United Kingdom in 1970. It subsequently gained popularity in many Pacific islands, most of which gained independence in the 1970s and 1980s. The Pacific Way carries a flavor of resistance to Europeanization and has often been invoked to uphold the notion of a regional identity shared by Pacific islands that grows out of their unique history and social experience. It was particularly influential among educators given the task of writing new textbooks to replace those used by the former colonial masters. The new texts focused students' attention away from Britain, France, and the United States and toward their own cultures. Appeals to the Pacific

---

👉 THINGS TO **REMEMBER**

> **GEOGRAPHIC INSIGHT 2**
>
> - **Globalization and Development**
>
>   Globalization, coupled with Oceania's stronger focus on neighboring Asia (rather than on its long-time connections with Europe), has transformed patterns of trade and economic development across Oceania. The transformation is being driven largely by Asia's

FIGURE 11.17

PHOTO ESSAY: Power and Politics in Oceania

Politics in this region vary significantly between the more Europeanized areas of Australia, New Zealand, and Hawaii, and the more indigenous and traditional political cultures of New Guinea and many Pacific islands.

A A funeral at an immigration detention facility on Christmas Island, Australia, for refugees from Iraq and Iran who drowned while trying to reach Australia by boat. Immigration has been a political hot-button issue in Australia for many years. Controversy surrounds Australia's policy of sending refugees who arrive by boat to detention facilities on Pacific islands, some not governed by Australia. Here they may wait for years in prison-like conditions for their claims to be judged. Some Australians feel that because relatively few refugees come to their country, those who do come should be treated better and accepted more quickly. Others see refugees as a security risk or an economic drain and believe that the long detentions discourage refugees from coming to Australia.

Democratization and Conflict
Armed conflicts and genocides with high death tolls since 1990

※ 13,000 deaths

Democratization index
- Full democracy
- Flawed democracy
- Hybrid regime
- Authoritarian regime
- No data

B A Fijian soldier during the military coup of 2006, which overturned a fair election. Military takeovers also took place in Fiji in 1987 and 2000.

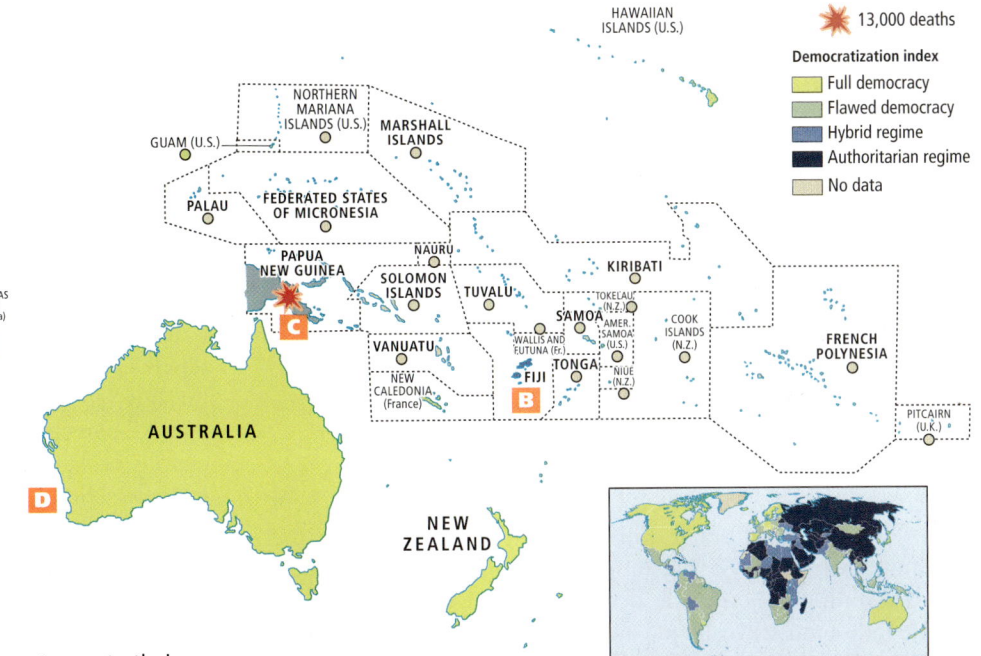

C Members of a violent criminal gang guard the entrance to their headquarters in Port Moresby, the capital of Papua New Guinea. Decades of rampant corruption have led to a breakdown of law and order in the capital and many other areas of Papua New Guinea.

D Corrie Bodney, an elder of the Ballaruk Aboriginal tribe, stages a sit-in at Perth International Airport, which is located on land the Ballaruk have occupied for thousands of years. In 2013, the government of Western Australia offered several Aboriginal tribes more than U.S.$1 billion to settle a larger claim, which covered the entire city of Perth.

Way have also been used to uphold attempts by Pacific island governments to control their own economic development and solve their own political and social problems.

In politics, the Pacific Way has occasionally been invoked as a philosophical basis for overriding democratic elections that challenge the power of indigenous Pacific Islanders. In 1987, 2000, and 2006, indigenous Fijians used the Pacific Way to justify coups d'état against legally elected governments (Figure 11.17B). All three of the overthrown governments were dominated by Indian Fijians, the descendants of people from India whom the British brought to Fiji more than a century ago to work on sugar plantations.

Fiji's population is now about evenly divided between indigenous Fijians and Indian Fijians. Indigenous Fijians are generally less prosperous and tend to live in rural areas where community affairs are still governed by traditional chiefs. Indian Fijians, in contrast, hold significant economic and political power, especially in the urban centers and in areas of tourism and sugar cultivation. In response to the coups, many Indian Fijians left the islands, resulting in a loss of badly needed skilled workers that has slowed economic development.

Political responses around Oceania to the Fiji coups have been divided. Australia, New Zealand, and the United States (via APEC and the state government of Hawaii) have demanded that the election results stand and the Indian Fijians be returned to office. But much of Oceania has referenced the Pacific Way in arguments supporting the coup leaders. As in Fiji, those who govern many of the Pacific islands are leaders of indigenous descent who have not always had the strongest respect for political freedoms, especially when their hold on power is threatened. Their decisions have at times upheld traditional Pacific values such as stability, respect for authority, and certain kinds of environmental awareness, while at other times contributing to corruption and civil disorder (see Figure 11.17C).

On the global stage, Fiji has been suspended from the Commonwealth of Nations (a union of former British colonies) for subverting majority-rule democracy. As a result, it is ineligible for Commonwealth aid and is not allowed to participate in Commonwealth sports events. Because sports play a central role in Pacific identity (see "Sports as a Unifying Force" on page 477), this latter sanction carries significant weight.

ON THE BRIGHT SIDE

The Value of the Pacific Way to Grassroots Sustainability

Regardless of its global political status, the Pacific Way is likely to endure, especially as a concept that upholds regional identity and traditional culture. Further, some organizations now use the Pacific Way as the basis of an integrated approach to economic development and environmental issues. For example, the South Pacific Regional Environmental Programme builds on traditional Pacific island economic activities—such as fishing and local traditions that require knowledge of and awareness about the environment—to promote grassroots economic development and environmental sustainability.

THINGS TO REMEMBER

> **GEOGRAPHIC INSIGHT 3**
- **Power and Politics** In recent decades, stark divisions have emerged in Oceania over definitions of democracy—the system of government that dominates in New Zealand and Australia—versus the Pacific Way, a political and cultural philosophy based in the traditional cultures of the Pacific islands.

- The Pacific Way, a political and cultural philosophy, was developed in Fiji around the time of Fiji's independence from the United Kingdom in 1970.

- In politics, the Pacific Way has occasionally been invoked as a philosophical basis for overriding democratic elections that challenge the power of indigenous Pacific Islanders.

- Governance in Oceania is for the most part based on democratic principles with regular elections; however, traditional power holders have staged coups that have negated or threatened political freedoms. ■

URBANIZATION

> **GEOGRAPHIC INSIGHT 4**

Urbanization: Oceania is only lightly populated but it is highly urbanized. The shift from agricultural and resource-based economies toward service economies is a major reason for the urbanization of the wealthiest parts of Oceania (Australia, New Zealand, Hawaii, Guam), where 80 to 100 percent of the population lives in cities. These trends are weakest in Papua New Guinea and many smaller Pacific islands.

The global trend of migration from the countryside to cities is quite visible in Oceania, where 66 percent of the overall population lives in urban areas. Australia and New Zealand have among the highest percentages of city dwellers outside Europe. More than 82 percent of Australians live in a string of cities along the country's relatively well-watered and fertile eastern and southeastern coasts. Similarly, 86 percent of New Zealanders live in urban areas. The vast majority of the people in these two countries live in modern comfort, work in a range of occupations typical of highly industrialized societies, and have access to tax-supported health-care and leisure facilities (Figure 11.18A). Vibrant, urban-based service economies employ about three-quarters of the population in both countries. Declining employment in mining and agriculture, where mechanization has dramatically reduced the number of workers needed, has also contributed to urbanization.

Throughout the Pacific, urban centers have transformed natural landscapes. In some small countries, such as Guam, Palau, and the Marshall Islands, they have become the dominant landscape. Although cities are places of opportunity, they can also be sites of cultural change, conflict, and environmental hazards (see Figure 11.18C).

Most Pacific island towns and all the capital cities are located in ecologically fragile coastal settings. Many of these waterfront towns were established during the colonial era as ports or docking facilities and were situated in places suitable for only limited numbers of people. Consequently, little land is available

FIGURE 11.18　　PHOTO ESSAY: Urbanization in Oceania

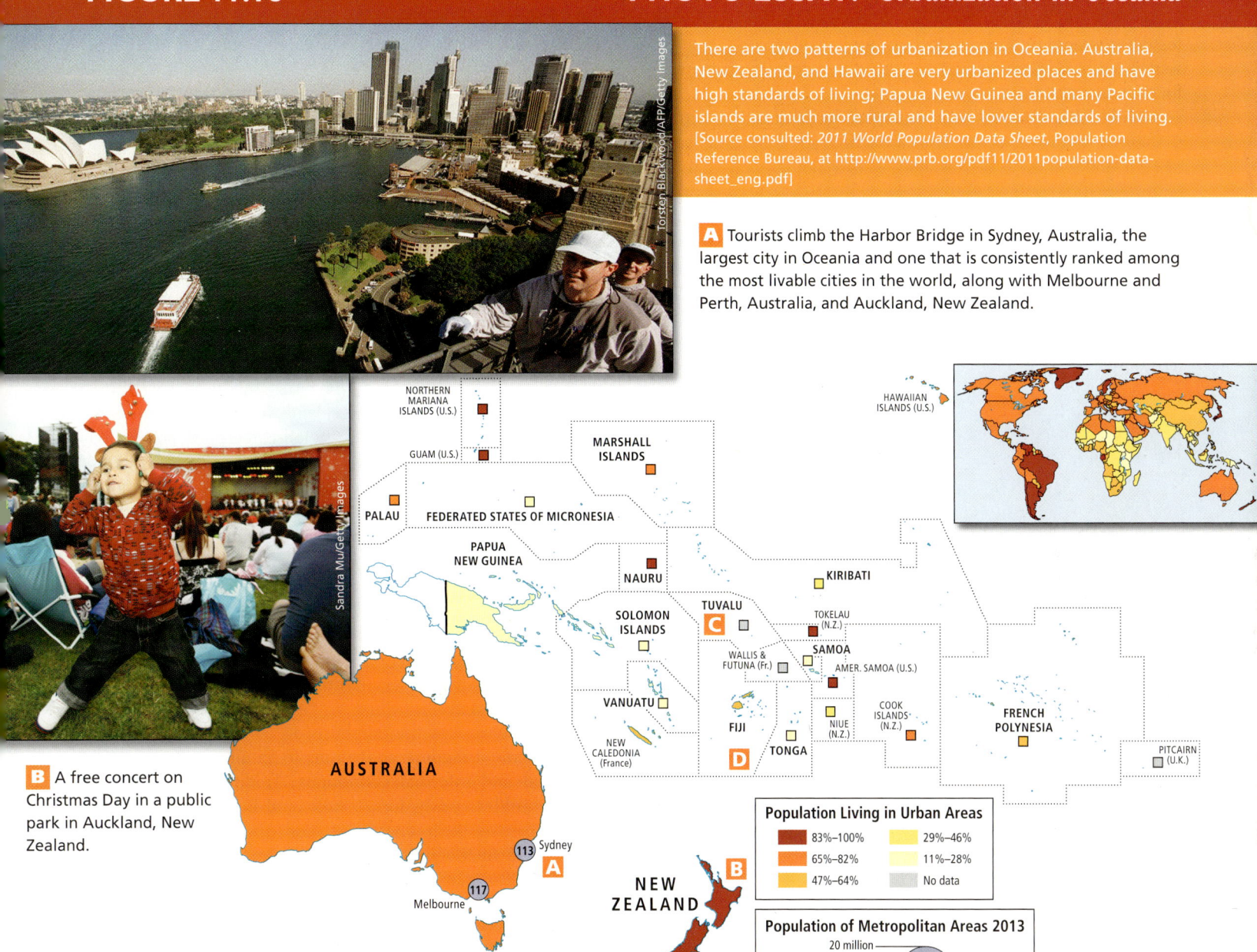

There are two patterns of urbanization in Oceania. Australia, New Zealand, and Hawaii are very urbanized places and have high standards of living; Papua New Guinea and many Pacific islands are much more rural and have lower standards of living. [Source consulted: *2011 World Population Data Sheet*, Population Reference Bureau, at http://www.prb.org/pdf11/2011population-data-sheet_eng.pdf]

A Tourists climb the Harbor Bridge in Sydney, Australia, the largest city in Oceania and one that is consistently ranked among the most livable cities in the world, along with Melbourne and Perth, Australia, and Auckland, New Zealand.

B A free concert on Christmas Day in a public park in Auckland, New Zealand.

Population Living in Urban Areas
- 83%–100%
- 65%–82%
- 47%–64%
- 29%–46%
- 11%–28%
- No data

Population of Metropolitan Areas 2013
- 20 million
- 10 million
- 5 million
- 3 million

Note: Symbols on map are sized proportionally to metro area population.

① **Global rank** (population 2013)

C Children wade through garbage during a high tide at Funafuti Atoll, capital of Tuvalu. A very densely populated country, with 4847 people per square mile (1871 per square kilometer), Tuvalu is one of the poorest nations in Oceania, with GDP (PPP) per capita of about U.S. $1600.

D A 4-year-old girl in a squatter settlement outside Suva, Fiji. Incomes are relatively low here, as is access to health care and education.

for development and access to housing is inadequate. Squatter settlements have been a visible feature of the region's urban areas for decades.

Multiculturalism has been enhanced by urbanization across the region. Many urban residents are letting go of the rural ways of their childhoods, as well as their ethnic identity and cultural commitments. As time goes on, more urban people are marrying across ethnic divisions, having only one or two children, and creating new patterns of social alliances and networks. Such cultural blending results in new social tensions, which change the very nature of social life in the island Pacific. Urban unemployment and unrest are on the rise, and low rates of economic growth restrict the revenue available to governments to manage urban development.

> **THINGS TO REMEMBER**

> **GEOGRAPHIC INSIGHT 4**
> • **Urbanization** Oceania is only lightly populated but it is highly urbanized. The shift from agricultural and resource-based economies toward service economies is a major reason for the urbanization of the wealthiest parts of Oceania (Australia, New Zealand, Hawaii, Guam), where 80 to 100 percent of the population lives in cities. These trends are weakest in Papua New Guinea and many smaller Pacific islands.

• The global trend of migration from the countryside to cities is quite visible in Oceania, where 66 percent of the overall population lives in urban areas.

• Most Pacific island towns and all the capital cities are located in ecologically fragile coastal settings.

• Many urban residents are letting go of the rural ways of their childhoods, as well as their ethnic identity and cultural commitments. ∎

POPULATION AND GENDER

> **GEOGRAPHIC INSIGHT 5**
>
> **Population and Gender:** In this largest but least populated world region, there are two main patterns relevant to population and gender. Australia, New Zealand, and Hawaii have older and more slowly growing populations, and relatively more opportunities for women. The Pacific islands and Papua New Guinea have much more rural, younger, and rapidly growing populations, with fewer opportunities for women.

A number of factors are influencing population growth in Oceania; among them are issues related to gender. While there is a trend toward equality across gender lines throughout Oceania, there is persistent gender inequality as well. A striking disparity is emerging between Australia and New Zealand (where women are gaining political and economic power) and Papua New Guinea and the Pacific islands (where change is much slower).

In Australia and New Zealand, women's access to jobs and policy-making positions in government has improved, particularly over the last few decades. New Zealand and the Australian province of South Australia were among the first places in the world to grant European women full voting rights (in 1893 and 1895, respectively). New Zealand has elected two female prime ministers, and in 2010 Australia elected a woman, Julia Gillard, to be prime minister. Moreover, according to the 2011 Inter-Parliamentary Union report on women in lower houses of parliament, in both countries, the proportion of women in national legislatures (32.2 percent in New Zealand, 24.7 percent in Australia) is well above the global average of 18 percent. In Papua New Guinea and the Pacific islands, women generally have far less political and economic power. No woman has yet been elected to a top-level national office, and women are a tiny minority in national legislatures when they are present at all.

In both Australia and New Zealand, young women are pursuing higher education and professional careers and postponing marriage and childbearing until their thirties. (This is also a trend in Hawaii, Guam, and the islands with French affiliation.) Nonetheless, both societies continue to reinforce the housewife role for women in a variety of ways. For example, the expectation is that women, not men, will interrupt their careers to stay home to care for young or elderly family members. Women in Australia receive, on average, only about 70 percent of the pay that men receive for equivalent work. This is, however, a smaller gender pay gap than in many other developed countries.

Throughout Papua New Guinea and the Pacific islands, gender roles and relationships vary greatly over the course of a lifetime. Because of the emphasis on community, male and female Pacific Islanders contribute to family assets through the formal and informal economies. Traditionally, men are the boat builders, navigators, fishers, and house builders. And they are usually the preparers of food, though women often supply some of the ingredients through their gathering and cultivating efforts (Figure 11.19). Most traders in marketplaces are women, and, aside from fish, the items they sell are also made and transported by women. Many young women today fulfill traditional roles as mates and mothers and practice a wide range of domestic crafts, such as weaving and basketry. In middle age, however, these same women may return to school and take up careers. With the aid of government scholarships, some Pacific Islander women pursue higher education or job training that takes them far from the villages where they raised their children. Thus, the expectation that Aurora (in this chapter's opening vignette) will study far from home and then support her family and elders is in line with evolving gender roles in Pacific ways of life. Aurora can expect that with age and accumulated experience, she will be boosted into a position of considerable power in her community.

Population Numbers and Distribution

Although Oceania occupies a huge portion of the planet, its total population is only 38.7 million people, close to that of the state of California (38 million) (Figure 11.20). The people of Oceania live on a total land area slightly larger than the contiguous United States but spread out in bits and pieces across an ocean larger

FIGURE 11.19 LOCAL LIVES
Foodways in Oceania

A A winemaker in Australia's Hunter Valley assesses the shiraz grapes at her winery. Australia is the fourth-largest exporter of wine in the world, after Italy, France, and Spain. [Peter Stoop/Fairfax Media/Getty Images]

B Taro is harvested in Hawaii. Grown in flooded fields, taro was originally brought to Oceania from Southeast Asia. It is now a major part of diets throughout the Pacific islands. While the leaves are also eaten, the root is particularly prized as a source of calories. [Robert Madden/National Geographic/Getty Images]

C Food is removed from a Maori earth oven, or *hangi*, in New Zealand. First a pit is dug, and then a fire is made to heat stones placed in the pit. Baskets of food are placed over the hot stones (which are covered with cloth and then earth) for several hours until the food is cooked. [Chris Jackson/Getty Images]

populated. For example, the Marshall Islands and Funafuti (see Figure 11.18C), the capital of Tuvalu, have 772 and 4847 people, respectively, per square mile (298 and 1871, respectively, per square kilometer).

THINGS TO REMEMBER

› GEOGRAPHIC INSIGHT 5
- **Population and Gender** In this largest but least populated world region, there are two main patterns relevant to population and gender. Australia, New Zealand, and Hawaii have older and more slowly growing populations, and relatively more opportunities for women. The Pacific islands and Papua New Guinea have much more rural, younger, and rapidly growing populations, with fewer opportunities for women.

- New Zealand and the Australian province of South Australia were among the first places in the world to grant European women full voting rights (in 1893 and 1895, respectively).

- In both Australia and New Zealand, young women are pursuing higher education and professional careers and postponing marriage and childbearing until their thirties.

- In some Pacific islands, women can inherit or personally accrue considerable power in their own communities over the course of a lifetime.

- Although Oceania occupies a huge portion of the planet, its total population is only 38.7 million people, close to that of the state of California (38 million). ∎

SOCIOCULTURAL ISSUES

The cultural sea change in Oceania, away from Europe and toward Asia and the Pacific, has been accompanied by new respect for indigenous peoples. Increasing economic interdependence with Asia has diminished historic discrimination against Asians in this region.

Ethnic Roots Reexamined

Until very recently, most people of European descent in Australia and New Zealand thought of themselves as Europeans in exile. Many considered their lives incomplete until they had made a pilgrimage to the British Isles or the European continent. In her book *An Australian Girl in London* (1902), Louise Mack wrote: "[We] Australians [are] packed away there at the other end of the world, shut off from all that is great in art and music, but born with a passionate craving to see, and hear and come close to these [European] great things and their home[land]s."

These longings for Europe were accompanied by racist attitudes toward both indigenous peoples and Asians. Most histories of Australia written in the early twentieth century failed to even mention the Aborigines, and later writings described them as amoral. At midcentury there were numerous projects to take Aboriginal children from their parents and acculturate them to European ways in boarding schools known for abuse and brutality. From the 1920s to the 1960s, whites-only immigration policies

than the Eurasian landmass. The Pacific islands have nearly 4.75 million people, including Hawaii's 1.74 million; Australia has 22.7 million; Papua New Guinea, 6.9 million; and New Zealand, 4.4 million.

Population densities remain low in Australia, at 7.8 people per square mile (3 per square kilometer) for the country as a whole and about 50 per square mile (20 per square kilometer) in Australia's settled southeast. New Zealand's density is also low, at 41.4 people per square mile (16 per square kilometer). In the Pacific islands, densities vary widely. Some are sparsely settled or uninhabited, while others—including some of the smallest, poorest, and lowest in elevation (which are thus some of the most exposed to rising sea levels)—are extremely densely

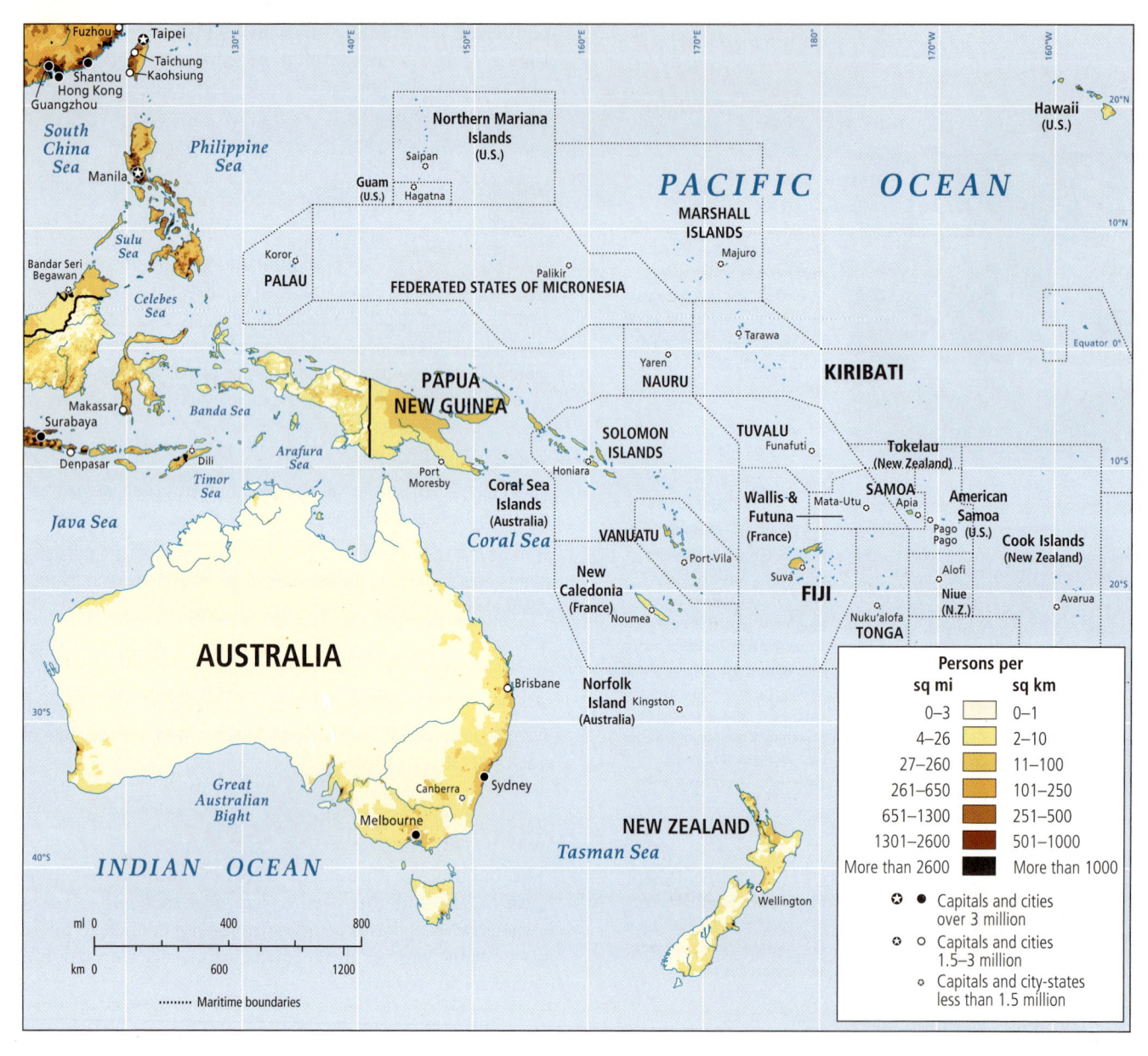

FIGURE 11.20 Population density in Oceania. Population growth is slowing throughout Oceania as more people move to the cities, where health care is better and women are more likely to pursue careers—factors that make large families less likely. Australia, New Zealand, and Hawaii are furthest along in this process and are quite urbanized, with smaller families and increasingly old populations.

barred Asians, Africans, and Pacific Islanders from migrating to Australia and discouraged them from entering New Zealand. As we have seen, trading patterns in that era further reinforced connections to Europe.

Weakening of the European Connection When migration from the British Isles slowed after World War II, both Australia and New Zealand began to lure immigrants from southern and eastern Europe, many of whom had been displaced by the war. Hundreds of thousands came from Greece, Italy, and what was then Yugoslavia. The arrival of these non-English-speaking people began a shift toward a more multicultural society. The whites-only immigration policy was abandoned and people began to arrive from many places. There was an influx of Vietnamese refugees in the early 1970s during the frantic exodus that followed the United States' withdrawal from Vietnam. More recently, skilled workers from India and elsewhere in Asia have been helping meet the growing demand for information technology (IT) specialists throughout the service sector.

As of 2010, more than one-fourth of the Australian population was foreign born. The fastest-growing group was from India. Nevertheless, while new immigration policies are increasing the numbers of immigrants from China, Vietnam, and India, people of Asian birth or ancestry remain a small percentage of the total population in both Australia and New Zealand. In 2006, the latest year for which complete statistics are available, 42 percent of Australia's foreign-born residents were from Europe and 15 percent

Australia's Foreign-Born Population, 2006
(percent by country of origin)

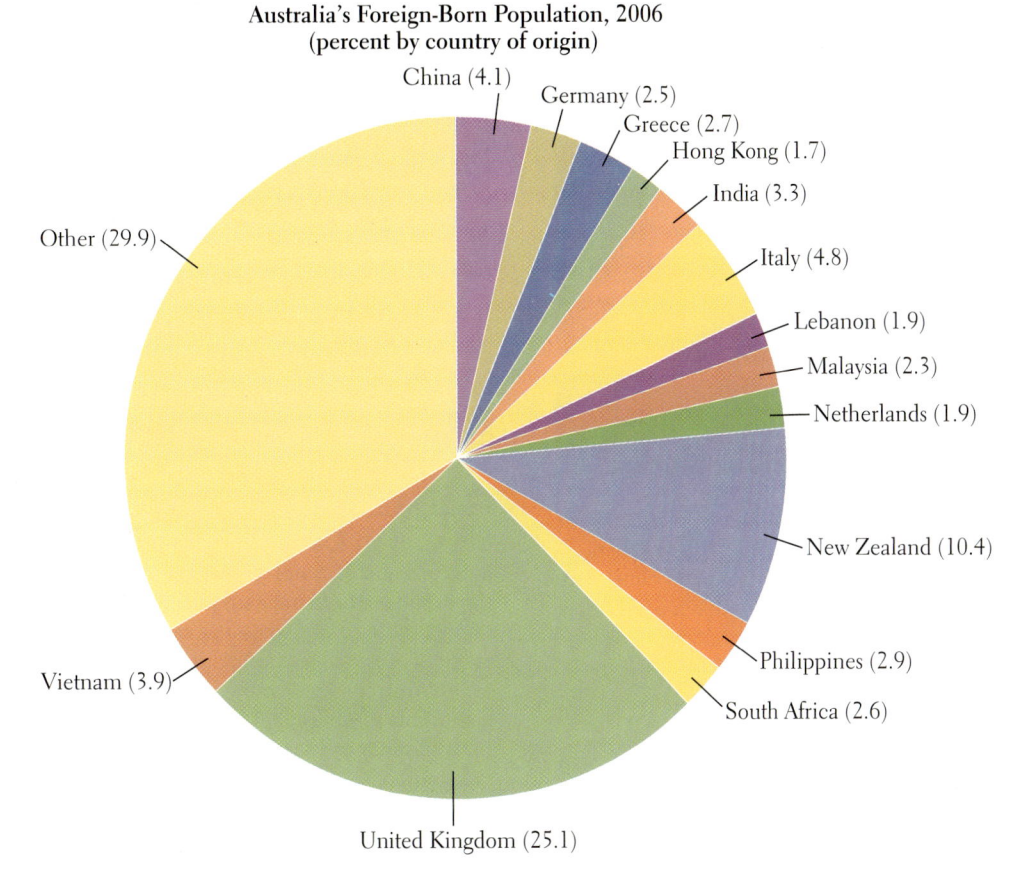

- China (4.1)
- Germany (2.5)
- Greece (2.7)
- Hong Kong (1.7)
- India (3.3)
- Italy (4.8)
- Lebanon (1.9)
- Malaysia (2.3)
- Netherlands (1.9)
- New Zealand (10.4)
- Philippines (2.9)
- South Africa (2.6)
- United Kingdom (25.1)
- Vietnam (3.9)
- Other (29.9)

FIGURE 11.21 Australia's cultural diversity in 2006. More than 24 percent (4.95 million) of Australia's people were born in other places, making Australia one of the world's most ethnically diverse nations. [Source consulted: Australian Bureau of Statistics, "Main Countries of Birth," *Year Book Australia*, 2008, Table 7.39, at http://www.ausstats.abs.gov.au/ausstats/subscriber.nsf/0/8D6ED0E197FE38A6CA2573E7000EC2AD/$File/13010_2008.pdf]

were from Asia (Figure 11.21). New Zealand has similar proportions among its foreign-born residents, and most immigrants continue to come from Europe. Although (because of low birth rates and high immigration rates) Europeans are decreasing as a percentage of the population in both Australia and New Zealand, they are projected to still constitute two-thirds or more of both countries' populations by 2021 (see http://tinyurl.com/7ejcmp3).

The Social Repositioning of Indigenous Peoples in Australia and New Zealand Perhaps the most interesting population change in Australia and New Zealand is one of identity. For the first time in 200 years, the number of people in both countries who claim indigenous origins is increasing. Between 1991 and 1996, the number of Australians claiming Aboriginal origins rose by 33 percent. By 2009, the Aboriginal population was estimated at 528,600. In New Zealand, the number claiming Maori background rose by 20 percent to 652,900 between 1991 and 2009.

These increases are due to changing identities, not to a population boom. More positive attitudes toward indigenous peoples have encouraged the open acknowledgment of Aborigine or Maori ancestry. Also, marriages between European and indigenous peoples are now more common. As a result, the number of people with a recognized mixed heritage is increasing.

As society has acknowledged that discrimination has been the main reason for the low social standing and impoverished state of indigenous peoples, respect for Aboriginal and Maori culture has also increased. The Australian Aborigines base their way of life on the idea that the spiritual and physical worlds are intricately related (Figure 11.22). The dead are present everywhere in spirit, and they guide the living in how to relate to the physical environment. Much Aboriginal spirituality refers to the *Dreamtime*, the time of creation when the human spiritual connections to rocks, rivers, deserts, plants, and animals were made clear. However, very few Aboriginal people continue to practice their own cultural traditions or live close to ancient homelands. Instead, many live in impoverished urban conditions. In New Zealand, where the Maori constitute about 15 percent of the country's population and Auckland has the largest Polynesian population (including Native Maori) of any city in the world, there are now many efforts to bring Maori culture more into the mainstream of national life.

Aboriginal Land Claims In 1988, during a bicentennial celebration of the founding of white Australia, a contingent of some 15,000 Aborigines protested that they had little reason to celebrate. During the same 200 years, they were assumed to have no

FIGURE 11.22 Aboriginal rock art. A Mimi spirit painted on a rock at Kakadu National Park, Australia. To the Aborigines, Mimi spirits are teachers who pass between this world and another dimension via crevices in rocks. They are responsible for many teachings on hunting, food preparation, use of fire, dance, and sexuality.

FIGURE 11.23 A performer at the Aboriginal Tent Embassy in Canberra, Australia. Intermittently since 1972, and continuously since 1992, Aboriginal activists have camped out on the grounds of Australia's parliament house in Canberra, Australia. Considered by many to be the most effective political action ever taken by Australian Aborigines, the first tent embassy was a response to the Australian government's denial of land ownership and other land rights to Aborigines in territories they had continuously occupied for thousands of years. As Aboriginal land rights have gained recognition, the tent embassy has championed other causes, including opposition to mining that threatens Aboriginal communities and cultural sites, as well as the plight of the Aboriginal urban poor, such as the community of Redfern in Sydney, Australia. The tent embassy remains controversial; it has been targeted by arsonists. The Australian government plans a more permanent structure but will then ban camping at the embassy.

prior claim to any land in Australia, they had lost basic civil rights, and they had effectively been erased from the Australian national consciousness. Into the 1960s, it was even illegal for Aborigines to drink alcohol.

British documents indicate that during colonial settlement, all Australian lands were deemed to be available for British use. The Aborigines were thought to be too primitive to have concepts of land ownership because their nomadic cultures had "no fixed abodes, fields or flocks, nor any internal hierarchical differentiation." The Australian High Court declared this position void in 1993. After that, Aboriginal groups began to win some land claims, mostly for land in the arid interior previously controlled by the Australian government. Figure 11.23 shows the Aboriginal Tent Embassy in 2012, versions of which have stood on the grounds of Parliament in Canberra for more than 40 years. The Aboriginal Tent Embassy was instrumental in raising public awareness of injustices and remains a national symbol of Aboriginal civil rights. Court cases and other efforts to restore Aboriginal rights and lands continue (see Figure 11.17D).

Maori Land Claims In New Zealand, relations between the majority European-derived population and the indigenous Maori have proceeded only somewhat more amicably than in Australia. In 1840, the Maori signed the Waitangi Treaty with the British, assuming they were granting only rights of land usage, not ownership (see Figure 11.24B). The Maori did not regard land as a tradable commodity, but rather as an asset of the people, used by families and larger kin groups to fulfill their needs. The geographer Eric Pawson writes: "To the Maori the land was sacred . . . [and] the features of land and water bodies were woven through with spiritual meaning and the Maori creation myth." The British assumed that the treaty had given them *exclusive* rights to settle the land with British migrants and to extract wealth through farming, mining, and forestry.

By 1950, the Maori had lost all but 6.6 percent of their former lands to European settlers and the government. Maori numbers had shrunk from a probable 120,000 in the early 1800s to 42,000 in 1900, and the Maori came to occupy the lowest and most impoverished rung of New Zealand society. In the 1990s, however, the Maori began to reclaim their culture, and they established a tribunal that forcefully advances Maori interests and land claims through the courts. Since then, nearly half a million acres of land and several major fisheries have been transferred back to Maori control. Nonetheless, the Maori still have notably higher unemployment, lower education levels, and poorer health than the New Zealand population as a whole.

Forging Unity in Oceania

Although wide ocean spaces and the great diversity of languages in the region sometimes make communication difficult, travel, sports, and festivals (Figure 11.24) are three forces that help bring the people of Oceania closer together.

Languages in Oceania The Pacific islands—most notably Melanesia—have a rich variety of languages. In some cases, the islands in a single chain have several different languages. A case in point is Vanuatu, a chain of 80 mostly high volcanic islands to the east of northern Australia. At least 108 languages are spoken

Sociocultural Issues **477**

FIGURE 11.24 LOCAL LIVES
Festivals in Oceania

A A young Aboriginal dancer at Garma Festival, which is held to encourage the practice of traditional dance, singing, visual art, and ceremony of the Yolngu people. The festival is held every year in Arnhem Land, which overlooks the Gulf of Carpentaria in Australia's Northern Territory. [Glenn Campbell/The Sydney Morning Herald/Fairfax Media/Getty Images]

B Waitangi Day in New Zealand, a national holiday that commemorates the signing of a treaty between the indigenous Maori of New Zealand and the British. The long boats shown here are Maori canoes, known as *waka*, and are part of a reenactment of the signing of the treaty. [Kenny Rodger/Getty Images]

C The aerial theater and comedy troop Dislocate performs at the Sydney Festival, a 3-week-long international arts festival that is held every January in Sydney, Australia. [Wendell Teodoro/WireImage/Getty Images]

by a population of just 180,000—an average of 1 language for every 1600 people.

Languages are both an important part of a community's cultural identity and a hindrance to cross-cultural understanding. In Melanesia and elsewhere in the Pacific, the need for communication with the wider world is served by a number of **pidgin** languages that are similar enough to be mutually understood. Pidgins are made up of words borrowed from several languages by people involved in trading relationships. Over time, pidgins can grow into fairly complete languages, capable of fine nuances of expression. When a particular pidgin is in such common use that mothers talk to their children in it, then it can literally be called a "mother tongue." In Papua New Guinea, a version of pidgin English is the official language.

pidgin a language used for trading; made up of words borrowed from the several languages of people involved in trading relationships

Interisland Travel One way in which unity is manifested in Oceania is interisland travel. Today, people travel in small planes from the outlying islands to hubs such as Fiji, where jumbo jets can be boarded for Auckland, Melbourne, and Honolulu. Cook Islanders call these little planes "the canoes of the modern age," and people travel for many reasons. Dancers from across the region attend the annual folk festival in Brisbane; businesspeople from Kiribati, Micronesia, can fly to Fiji to take a short course at the University of the South Pacific; a Cook Islands teacher can take graduate training in Hawaii; and sports fans can visit multiple locations over time.

Sports as a Unifying Force Sports and games are a major feature of daily life throughout Oceania. The region has shared sports traditions with and borrowed them from cultures around the world. Surfing evolved in Hawaii and, like outrigger sailing and canoeing, derives from ancient navigational customs that matched human wits against the power of the ocean. On hundreds of Pacific islands and in Australia and New Zealand, rugby, volleyball, soccer, and cricket are important community-building activities. Baseball is a favorite in the parts of Micronesia that were U.S. trust territories. Women compete in the popular sport of netball (similar to basketball but without a backboard).

Pan-Oceania sports competitions are the single most common and resilient link among the countries of the region. Attendance at regional sports events is so desirable that low-income islanders will hold yard sales and raffles to amass the cash necessary to make the trip. The centrality of such competitions in daily life encourages regional identity and provides opportunities for ordinary citizens to travel extensively around the region and to other parts of the world.

The *haka* (**Figure 11.25**) is an example of how, in the postcolonial modern era, indigenous culture in Oceania is being revived, celebrated, and appropriated in new places by those who wish to project a multicultural image. The haka is a highly emotional and physical dance traditionally performed by the Maori to motivate fellow warriors and intimidate opponents before entering battle. Dances like this have historically been a part of many cultures in the islands of Oceania, but the haka has

478 CHAPTER 11 Oceania: Australia, New Zealand, and the Pacific

team) have performed the haka: chanting, screaming, jumping, stomping their feet, poking out their tongues, widening their eyes to show the whites, and beating their thighs, arms, and chests.

Outside Oceania, those who perform the haka include the rugby teams at Jefferson High in Portland, Oregon, and Middlebury College in Vermont, and the football teams at Brigham Young University and the University of Hawaii. All these teams have players who are of Polynesian heritage. Most practitioners speak of the haka as filling them with the necessary exuberance, aggression, and spirituality to play a vigorous and successful game. To see videos of a haka, go to http://www.YouTube.com and type in *haka*.

THINGS TO REMEMBER

- Oceania's long-standing cultural and economic links to Europe are being challenged by reinvigorated native traditions and identities and by economic globalization, which is strengthening the region's links to Asia.

- The number of Asian immigrants into Australia and New Zealand has been increasing over the past two decades, while the number of Europeans has been declining. Asians, however, still make up only a small minority of the populations of both Australia and New Zealand.

- Sports and festivals are unifying forces for the region, inspiring fundraisers that allow ordinary citizens to travel to games, thus reinforcing regional and ethnic identity.

FIGURE 11.25 The haka, a Maori tradition. A haka performed by the New Zealand men's rugby team, the All Blacks, before a match against Australia.

now become an integral part of rugby, the region's most popular sport (Figure 11.26). Before almost every international match for the past century, the All Blacks (the New Zealand men's rugby

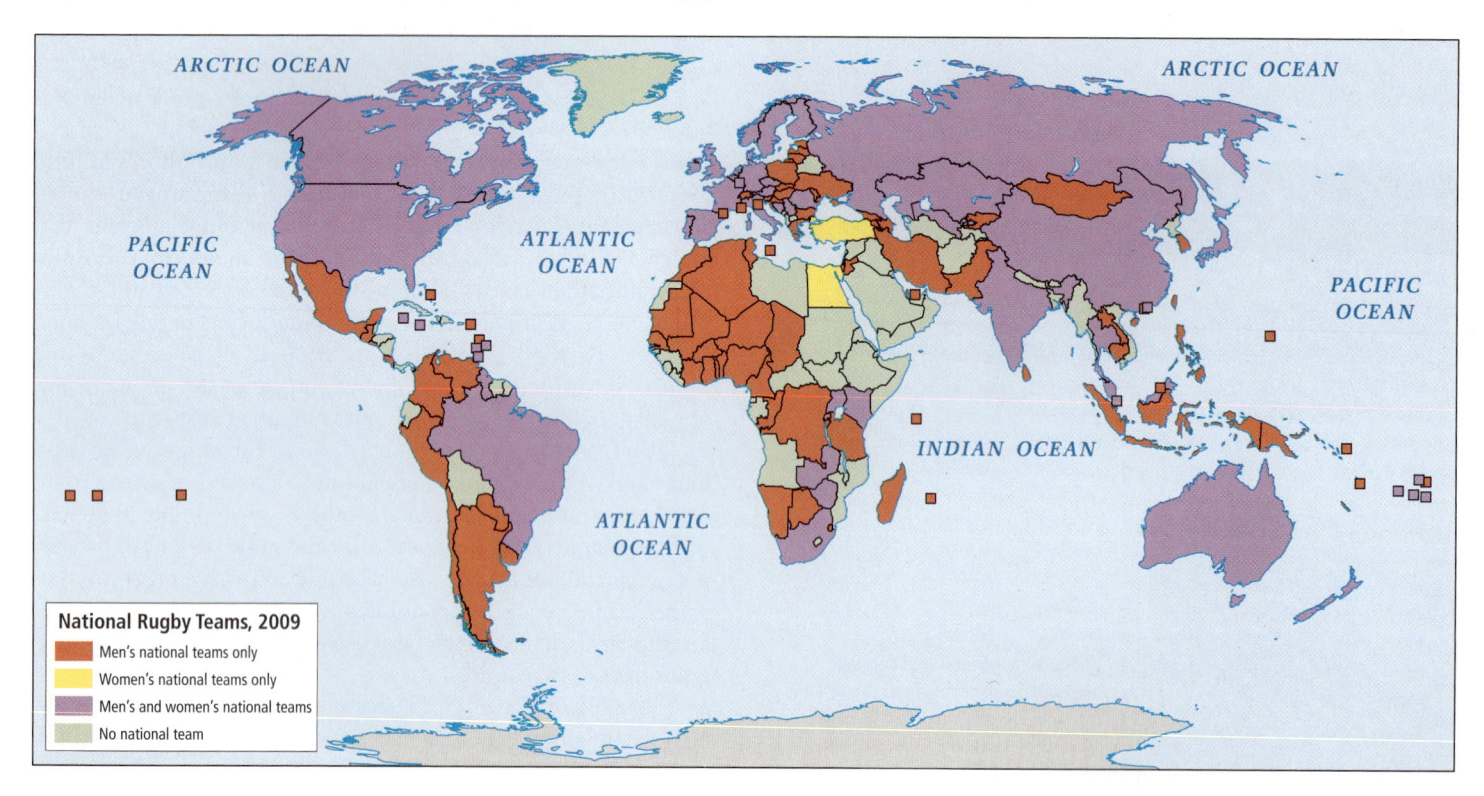

FIGURE 11.26 Rugby around the world. In more than 136 countries, women, men, boys, and girls play rugby. All of these countries have men's national rugby teams, and 58 of these countries have women's national teams. The men's World Cup rugby competition began in 1987 and the women's World Cup competition began in 1991. In April 2010, New Zealand had the top team for both men and women, but the ranking of the men's teams can change weekly. [Source consulted: http://www.irb.com/aboutirb/organisation/index.html]

GEOGRAPHIC INSIGHTS

Oceania: Review and Self-Test

1. Environment: Oceania faces a host of environmental problems and public awareness of environmental issues is keen. Global climate change, primarily warming, has brought rising sea levels and increasingly variable rainfall. Other major threats to the region's unique ecology have come from the introduction of many nonnative species and the expansion of herding, agriculture, fishing, and human settlements.

- Which parts of Oceania are most vulnerable to the impacts of climate change? Which parts of the region contribute the most and least to greenhouse gas emissions?
- Why are coral reefs important ecologically, and how are they threatened by climate change?
- How have European farm animals and crops introduced into Australia and New Zealand affected the economies of those countries? What is the purpose of the Dingo Fence? Has it worked?
- How do foreign patterns of consumption affect this region, and what are the threats to Oceania's food security?

2. Globalization and Development: Globalization, coupled with Oceania's stronger focus on neighboring Asia (rather than on its long-time connections with Europe), has transformed patterns of trade and economic development across Oceania. The transformation is being driven largely by Asia's growing affluence, its enormous demand for resources, and its similarly massive production of manufactured goods.

- What does Asia buy from the Pacific islands, and what are Asia's trade connections to Australia and New Zealand?
- How does the loss of preferential trading ties with Europe affect the region's workers and tax revenues?
- What sector is the basis of most economies in this region?
- Explain why many Pacific Islanders can be said to have subsistence affluence despite having rather low monetary incomes.

3. Power and Politics: In recent decades, stark divisions have emerged in Oceania over definitions of democracy—the system of government that dominates in New Zealand and Australia—versus the Pacific Way, a political and cultural philosophy based in the traditional cultures of the Pacific islands.

- How do ideas about the proper exercise of political power in the Pacific islands differ from those in Australia and New Zealand and some Europeanized islands?
- Where and when did the political philosophy known as the Pacific Way develop?
- In what situations has the Pacific Way been invoked?

4. Urbanization: Oceania is only lightly populated but it is highly urbanized. The shift from agricultural and resource-based economies toward service economies is a major reason for the urbanization of the wealthiest parts of Oceania (Australia, New Zealand, Hawaii, Guam), where 80 to 100 percent of the population lives in cities. These trends are weakest in Papua New Guinea and many smaller Pacific islands.

- Why, despite its low average population densities, is the region highly urbanized?
- Which parts of Oceania are both densely populated and threatened by climate change?
- What kind of cultural impact is urbanization having in Oceania?

5. Population and Gender: In this largest but least populated world region, there are two main patterns relevant to population and gender. Australia, New Zealand, and Hawaii have older and more slowly growing populations, and relatively more opportunities for women. The Pacific islands and Papua New Guinea have much more rural, younger, and rapidly growing populations, with fewer opportunities for women.

- Contrast the population growth rates in Australia and New Zealand with those in Oceania. In which parts of this region is the aging of the population of most concern?
- Which parts of this region were among the first in the world to grant European women the right to vote?
- Describe the possible changing gender roles of Pacific island women over the course of their lifetimes.
- What are some reasons for Oceania's relatively small overall population?

CRITICAL THINKING QUESTIONS

1. As Australia and New Zealand have moved away from their intense cultural and economic involvement with Europe, new policies and attitudes have evolved to facilitate their deeper involvement with Asia. If you were a college student in Australia or New Zealand, how might you experience these changes? Think about fellow students, career choices, language learning, and travel choices.

2. Discuss the emerging cultural identity of the Pacific islands, taking note of the extent to which Australia and New Zealand share or do not share in this identity. What factors are helping to forge a sense of unity across Polynesia and beyond? (First, review the spatial extent of Polynesia.)

3. Discuss the many ways in which Asia has historic, and now increasingly economic, ties to Oceania. In your discussion, include patterns of population distribution, mineral exports and imports, technological interactions, and tourism.

4. To what extent can the countries of Oceania exercise control over their future as the climate changes?

5. Australia and New Zealand differ from each other physically. Compare and contrast the two countries in relation to water, vegetation, and prehistoric and modern animal populations.

6. Indigenous peoples worldwide are taking action to safeguard their cultures, rights, and access to land and resources. Discuss how the indigenous peoples of Australia, New Zealand, and the Pacific islands are serving as leaders in this movement and what measures they are taking to reconstitute a sense of cultural heritage.

7. How is tourism both boosting economies and straining environments and societies throughout the Pacific islands? Describe the solutions that are being proposed to reduce the negative impacts of tourism.

8. Compare how women do and don't have political and economic power in Australia, New Zealand, Papua New Guinea, and the Pacific islands.

9. Compared with other regions, Australia and New Zealand are somewhat unusual in having become broadly prosperous on the basis of raw materials exports. How would you explain this achievement?

› THINKING GEOGRAPHICALLY

Now that you have read about Oceania, you should be able to answer the following questions about photos in this chapter. To answer these questions online, go to Geography LaunchPad.

Vulnerability to Climate Change (page 456)

A If sea levels rise 4 inches (10 centimeters) per decade, as predicted by the International Panel on Climate Change, how many years will it be before the lowest-lying Pacific atolls, such as Tuvalu, disappear under water?

C What kinds of actions could humans take to increase the resilience of coral reefs to climate change?

D Why would irrigation be needed in New Zealand, which has a wet climate?

Human Impacts on the Biosphere (page 458)

A Which nations have conducted the most nuclear tests in Oceania?

B Why were many plants and animals brought to Oceania from Europe?

C Why might islands in Oceania be in a quandary over the activities of foreign fishing fleets?

D What is the chief reason today for deforestation in New Zealand?

Visual History (page 464)

A From where did the ancestors of Aborigines migrate to Oceania?

B How did Polynesians use and navigate boats like the one in this painting?

C What were the early Europeans interested in when they came to Oceania in the sixteenth, seventeenth, and eighteenth centuries?

D By the time the flow of convicts to Australia from England and Ireland ended in 1868, what was Australia's total population?

E Until roughly World War II, on what did the economy in most parts of Oceania depend?

F Asians make up what percentage of Hawaii's population?

Power and Politics (page 469)

A When did Australia's whites-only immigration policy formally end?

B In Fiji, three coups d'état carried out by indigenous Fijians have removed legally elected governments headed by which ethnicity?

C How might corruption constrain political freedoms?

D Before 1993, why were Aborigines assumed to have no prior claim to land in Australia?

Urbanization (page 471)

A Approximately what percent of Australians live in cities?

C Notice the primary features of this photo. In what environmental zone is this settlement likely located?

CHAPTER KEY TERMS

Aborigines 461
atoll 451
endemic 454
Gondwana 450
Great Barrier Reef 450
hot spots 451
invasive species 457

Maori 451
marsupials 454
Melanesia 461
Melanesians 461
Micronesia 462
MIRAB economy 466
monotremes 454

Pacific Way 468
pidgin 477
Polynesia 462
Roaring Forties 451
subsistence affluence 467

EPILOGUE

PHOTO ESSAY: Antarctica

Despite its remote location and few human inhabitants, Antarctica is undergoing environmental change at an accelerating pace as temperatures rise and surrounding fisheries are exploited.

A Scientists measure sea level at Antarctica's Ross Sea. While most sea level rise predicted for the near future is caused by the thermal expansion of oceans, sea levels could rise 200 feet (60 meters) if Antarctica's massive glaciers were all to melt.

Territorial Claims of Antarctica According to the Antarctic Treaty
- New Zealand
- Australia
- France
- Norway
- United Kingdom
- Chile
- Multiple Claim: Argentina/U.K.
- Multiple Claim: Argentina/Chile/U.K.
- Multiple Claim: Chile/U.K.
- Unclaimed
- Ice Shelves

B A fishing vessel in the Southern Ocean, one of thousands that are placing increasing pressure on Antarctica's fisheries. Species like the Patagonian toothfish (marketed as Chilean sea bass) are severely overfished. Tiny shrimp known as *krill*, which form the basis of many oceanic ecosystems, are also increasingly being overfished.

C An emperor penguin dives beneath a hole in sea ice. Emperor penguins are found only in Antarctica and are proving extremely sensitive to climate change. In warmer years, declines in sea ice, the penguins' ideal hunting habitat, have resulted in widespread starvation among the penguins. Meanwhile, in colder years, fewer penguin chicks hatch.

D Amundsen–Scott South Pole Station, the dome-shaped structure, is a U.S. research station. Outside fly the flags of the first countries to sign the Antarctic Treaty of 1959, which bans all military and resource extraction–related activity, making the continent a scientific and nature reserve.

GLOSSARY

Aborigines (p. 461) the longest-surviving inhabitants of Oceania, whose ancestors, the Australoids, migrated from Southeast Asia 50,000 to 70,000 years ago over the Sundaland landmass that was exposed during the ice ages

acculturation (p. 145) adaptation of a minority culture to the host culture enough to function effectively and be self-supporting; cultural borrowing

acid rain (p. 64) precipitation that has formed through the interaction of rainwater or moisture in the air with sulfur dioxide and nitrogen oxides emitted during the burning of fossil fuels, making it acidic

agribusiness (p. 81) the business of farming conducted by large-scale operations that purchase, produce, finance, package, and distribute agricultural products

agriculture (p. 28) the practice of producing food through animal husbandry, or the raising of animals, and the cultivation of plants

agroecology (p. 341) the practice of traditional, nonchemical methods of crop fertilization and the use of natural predators to control pests

agroforestry (p. 282) growing economically useful crops of trees on farms, in conjunction with the usual plants and crops, to reduce dependence on trees from nonfarmed forests and to provide income to the farmer

Ainu (p. 400) an indigenous cultural minority group in Japan characterized by their light skin, heavy beards, and thick, wavy hair, who are thought to have migrated thousands of years ago from the northern Asian steppes

animism (p. 312) a belief system in which spirits, including those of the deceased, are thought to exist everywhere and to offer protection to those who pay their respects

apartheid (p. 291) a system of laws mandating racial segregation. South Africa was an apartheid state from 1948 until 1994

aquifers (p. 62) ancient natural underground reservoirs of water

archipelago (p. 409) a group, often a chain, of islands

assimilation (pp. 145, 184) the loss of old ways of life and the adoption of the lifestyle of another culture

Association of Southeast Asian Nations (ASEAN) (p. 425) an organization of Southeast Asian governments that was established to further economic growth and political cooperation

atoll (p. 451) a low-lying island or chain of islets, formed of coral reefs that have built up on the circular or oval rim of a submerged volcano

Australo-Melanesians (p. 419) a group of hunters and gatherers who moved approximately 40,000 to 60,000 years ago from what are now the northern Indian and Burman parts of southern Eurasia to the exposed landmass of Sundaland and present-day Australia

Austronesians (p. 419) a group of skilled farmers and seafarers from southern China who migrated south to various parts of Southeast Asia between 5000 and 10,000 years ago

authoritarianism (p. 38) a political system that subordinates individual freedom to the power of the state or of elite regional and local leaders

Aztecs (p. 122) indigenous people of high-central Mexico noted for their advanced civilization before the Spanish conquest

biodiversity (p. 114) the variety of life forms to be found in a given area

biosphere (p. 12) the entirety of the Earth's integrated physical spheres, with humans and other impacts included as part of nature

birth rate (p. 44) the number of births per 1000 people in a given population, per unit of time (usually per year)

Bolsheviks (p. 207) a faction of Communists who came to power during the Russian Revolution

brain drain (p. 139) the migration of educated and ambitious young adults to cities or foreign countries, depriving the communities from which the young people come of talented youth in whom they have invested years of nurturing and education

brownfields (p. 93) old industrial sites whose degraded conditions pose obstacles to redevelopment

Buddhism (p. 334) a religion of Asia that originated in India in the sixth century B.C.E. as a reinterpretation of Hinduism; it emphasizes modest living and peaceful self-reflection leading to enlightenment

capitalism (pp. 39, 169) an economic system based on the private ownership of the means of production and distribution of goods, driven by the profit motive and characterized by a competitive marketplace

capitalists (p. 207) usually a wealthy minority that owns the majority of factories, farms, businesses, and other means of production

carbon sequestration (p. 279) the removal and storage of carbon taken from the atmosphere

carrying capacity (p. 30) the maximum number of people that a given territory can support sustainably with food, water, and other essential resources

cartel (p. 249) a group of producers strong enough to control production and set prices for products

cartographers (p. 4) geographers who specialize in depicting geographic information on maps

cash economy (p. 48) an economic system that tends to be urban but may be rural, in which skilled workers, well-trained specialists, and even farm laborers are paid in money

caste system (p. 332) a complex, ancient Hindu system for dividing society into hereditary hierarchical classes

Caucasia (p. 223) the mountainous region between the Black Sea and the Caspian Sea

central planning (p. 169) a communist economic model in which a central bureaucracy dictates prices and output, with the stated aim of allocating goods equitably across society according to need

centrally planned, or socialist, economy (p. 208) an economic system in which the state owns all land and means of production, while government officials direct all economic activity, including the locating of factories, residences, and transportation infrastructure

Christianity (p. 245) a monotheistic religion based on the belief in the teachings of Jesus of Nazareth, a Jew, who described God's relationship to humans as primarily one of love and support, as exemplified by the Ten Commandments

civil disobedience (p. 335) protesting of laws or policies by peaceful direct action

G-1

civil society (p. 39) the social groups and traditions that function independently of the state and its institutions to foster a sense of unity and an informed common purpose among the general population

clear-cutting (p. 70) a method of logging that involves cutting down all trees on a given plot of land, regardless of age, health, or species

climate (p. 14) the long-term balance of temperature and precipitation that characteristically prevails in a particular region

climate change (p. 21) a slow shifting of climate patterns due to the general cooling or warming of the atmosphere

Cold War (pp. 169, 208) the contest that pitted the United States and western Europe, who were espousing free market capitalism and democracy, against the USSR and its allies, who were promoting a centrally planned economy and a socialist state

commercial agriculture (p. 283) farming in which crops are grown deliberately for cash rather than solely as food for the farm family

commodities (p. 293) raw materials that are traded, usually to other countries, for processing or manufacturing into more valuable goods

commodity dependence (p. 293) economic dependence on exports of raw materials

Common Agricultural Program (CAP) (p. 173) an EU program, meant to guarantee secure and safe food supplies at affordable prices, that places tariffs on imported agricultural goods and gives subsidies to EU farmers

communal conflict (p. 343) a euphemism for religiously based violence in South Asia

communism (pp. 39, 169, 207) an ideology, based largely on the writings of the German revolutionary Karl Marx, that calls on workers to unite to overthrow capitalism and establish an egalitarian society in which workers share what they produce; as practiced, communism was actually a socialized system of public services and a centralized government and economy in which citizens participated only indirectly through Communist Party representatives

Communist Party (p. 208) the political organization that ruled the USSR from 1917 to 1991; other communist countries, such as China, Mongolia, North Korea, and Cuba, also have ruling Communist parties

Confucianism (p. 376) a Chinese philosophy that teaches that the best organizational model for the state and society is a hierarchy based on the patriarchal family

contested space (p. 138) any area that two or more groups claim or want to use in different and often conflicting ways, such as the Amazon or Palestine

continental climate (p. 156) a midlatitude climate pattern in which summers are fairly hot and moist, and winters become longer and colder the deeper into the interior of the continent one goes

coral bleaching (p. 417) color loss that results when photosynthetic algae that live in corals are expelled

coup d'état (p. 134) a military- or civilian-led forceful takeover of a government

Creoles (p. 125) people mostly of European descent born in the Americas

crony capitalism (p. 424) a type of corruption in which politicians, bankers, and entrepreneurs, sometimes members of the same family, have close personal as well as business relationships

cultural homogenization (p. 154) the tendency toward uniformity of ideas, values, technologies, and institutions among associated culture groups

cultural pluralism (p. 438) the cultural identity characteristic of a region where groups of people from many different backgrounds have lived together for a long time but have remained distinct

Cultural Revolution (p. 379) a political movement launched in 1966 to force the entire population of China to support the continuing revolution

culture (p. 51) all the ideas, materials, and institutions that people have invented to use to live on Earth that are not directly part of our biological inheritance

czar (p. 206) the title of the ruler of the Russian empire

death rate (p. 44) the ratio of total deaths to total population in a specified community, usually expressed in numbers per 1000 or in percentages

delta (p. 14) the triangular-shaped plain of sediment that forms where a river meets the sea

democratization (p. 38) the transition toward political systems that are guided by competitive elections

demographic transition (p. 48) the change from high birth and death rates to low birth and death rates that usually accompanies a cluster of other changes, such as change from a subsistence to a cash economy, increased education rates, and urbanization

desertification (p. 239) a set of ecological changes that converts arid lands into deserts

detritus (p. 410) dead organic material (such as plants and insects) that collects on the ground

development (p. 35) a term usually used to describe economic changes such as the greater productivity of agriculture and industry that lead to better standards of living or simply to increased mass consumption

diaspora (p. 245) the dispersion of Jews around the globe after they were expelled from the eastern Mediterranean by the Roman Empire beginning in 73 C.E.; the term can now refer to other dispersed culture groups

dictator (p. 134) a ruler who claims absolute authority, governing with little respect for the law or the rights of citizens

digital divide (p. 76) the discrepancy in access to information technology between small, rural, and poor areas and large, wealthy cities that contain major governmental research laboratories and universities

divide and rule (p. 299) the deliberate intensification of divisions and conflicts by potential rulers; in the case of sub-Saharan Africa, by European colonial powers

domestication (p. 29) the process of developing plants and animals through selective breeding to live with and be of use to humans

domino theory (p. 421) a foreign policy theory that used the idea of the domino effect to suggest that if one country "fell" to communism, others in the neighboring region would also fall

double day (p. 187) the longer workday of women with jobs outside the home who also work as caretakers, housekeepers, and/or cooks for their families

dowry (p. 355) a price paid by the family of the bride to the groom (the opposite of *bride price*); formerly a custom practiced only by the rich

dual economy (p. 293) an economy in which the population is divided by economic disparities into two groups, one prosperous and the other near or below the poverty level

ecological footprint (p. 12) the amount of biologically productive land and sea area needed to sustain a person at the current average standard of living for a given population

economic core (p. 73) the dominant economic region within a larger region

economic diversification (p. 252) the expansion of an economy to include a wider array of activities

economies of scale (p. 172) reductions in the unit cost of production that occur when goods or services are efficiently mass produced, resulting in increased profits per unit

ecotourism (p. 118) nature-oriented vacations, often taken in endangered and remote areas, usually by travelers from affluent nations

El Niño (p. 116) periodic climate-altering changes, especially in the circulation of the Pacific Ocean, now understood to operate on a global scale

emigration (p. 47) out-migration from a place or country (see also *migration*)

endemic (p. 454) belonging or restricted to a particular place

erosion (p. 13) the process by which fragmented rock and soil are moved over a distance, primarily by wind and water

ethnic cleansing (p. 40) the deliberate removal of an ethnic group from a particular area by forced migration

ethnic group (p. 51) a group of people who share a common ancestry and sense of common history, a set of beliefs, a way of life, a technology, and usually a common geographic location of origin

ethnicity (p. 101) the quality of belonging to a particular culture group

euro (p. 172) the official (but not required) currency of the European Union as of January 1999

Euro zone (p. 35) those countries in the European Union that use the euro currency

European Union (EU) (p. 153) a supranational organization that unites most of the countries of West, South, North, and Central Europe

evangelical Protestantism (p. 147) a Christian movement that focuses on personal salvation and empowerment of the individual through miraculous healing and transformation; some practitioners preach to the poor the "gospel of success"—that a life dedicated to Christ will result in prosperity for the believer

Export Processing Zones (EPZs) (p. 130) specially created legal spaces or industrial parks within a country where, to attract foreign-owned factories, duties and taxes are not charged

export-led growth (p. 382) an economic development strategy that relies heavily on the production of manufactured goods destined for sale abroad

extended family (p. 146) a family that consists of related individuals beyond the nuclear family of parents and children

extraction (p. 35) mining, forestry, and agriculture

fair trade (p. 34) trade that values equity throughout the international trade system; now proposed as an alternative to free trade

favelas (p. 139) Brazilian urban slums and shantytowns built by the poor; called *colonias*, barrios, or *barriadas* in other countries

female genital mutilation (FGM) (p. 310) removing the labia and the clitoris and sometimes stitching the vulva nearly shut

female seclusion (p. 268) the requirement that women stay out of public view

feminization of labor (p. 423) the rising numbers of women in both the formal and the informal labor force

Fertile Crescent (p. 243) an arc of lush, fertile land formed by the uplands of the Tigris and Euphrates river systems and the Zagros Mountains, where nomadic peoples began the earliest known agricultural communities

floating population (p. 364) the Chinese term for people who live in a place other than their household registration location; many are jobless or underemployed people who have left economically depressed rural areas for the cities

floodplain (p. 14) the flat land along a river where sediment is deposited during flooding

food security (pp. 29, 369) the ability of a state to consistently supply a sufficient amount of basic food to the entire population

foreign direct investment (FDI) (p. 131) investment funds that come in to enterprises from outside the country

foreign exchange (p. 432) foreign currency that countries need to purchase imports

formal economy (p. 35) all aspects of the economy that take place in official channels

fossil fuel (p. 233) a source of energy formed from the remains of dead plants and animals

free trade (p. 33) the unrestricted international exchange of goods, services, and capital

Gazprom (p. 210) in Russia, the state-owned energy company; it is the largest gas entity in the world

gender (p. 49) the ways a particular social group defines the differences between the sexes

gender roles (p. 49) the socially assigned roles for males and females

genetic modification (GM) (p. 30) in agriculture, the practice of splicing together the genes from widely divergent species to achieve particular desirable characteristics

genocide (pp. 40, 300) the deliberate destruction of an ethnic, racial, or political group

gentrification (p. 93) the renovation of old urban districts by affluent investment, a process that often displaces poorer residents

Geographic Information Science (GISc) (p. 7) the body of science that supports multiple spatial analysis technologies and keeps them at the cutting edge

geopolitics (p. 39) the strategies that countries use to ensure that their own interests are served in relations with other countries

glasnost (p. 208) literally, "openness"; the policies instituted in the late 1980s under Mikhail Gorbachev that encouraged more transparency and openness in the workings of all levels of the Soviet government

global economy (p. 31) the worldwide system in which goods, services, and labor are exchanged

global scale (p. 11) the level of geography that encompasses the entire world as a single unified area

global warming (p. 21) the warming of the Earth's climate as atmospheric levels of greenhouse gases increase

globalization (p. 31) the growth of worldwide linkages and the changes these linkages are bringing about

Gondwana (p. 450) the great landmass that formed the southern part of the ancient supercontinent Pangaea

grassroots economic development (p. 297) economic development projects designed to provide sustainable livelihoods in rural and urban areas; these often use simple technology that requires minimal or no investment in imported materials

Great Barrier Reef (p. 450) the longest coral reef in the world, located off the northeastern coast of Australia

Great Leap Forward (p. 379) an economic reform program under Mao Zedong intended to quickly raise China to the industrial level of Britain and the United States

Green (p. 159) environmentally conscious

green revolution (p. 29) increases in food production brought about through the use of new seeds, fertilizers, mechanized equipment, irrigation, pesticides, and herbicides

greenhouse gases (GHG) (p. 21) gases, such as carbon dioxide and methane, released into the atmosphere by human activities, which become harmful when released in excessive amounts

Glossary

gross domestic product (GDP) per capita (p. 35) the total market value of all goods and services produced within a particular country's borders and within a given year, divided by the number of people in the country

gross national income (GNI) per capita (p. 48) the total production of goods and services in a country in a given year, divided by the mid-year population

groundwater (p. 284) water naturally stored in aquifers as many as 5000 years ago during wetter climate conditions

Group of Eight (G8) (p. 211) an organization of eight countries with large economies: France, the United States, Britain, Germany, Japan, Italy, Canada, and Russia

growth poles (p. 390) zones of development whose success draws more investment and migration to a region

guest workers (p. 184) legal workers from outside a country who help fulfill the need for temporary workers but who are expected to return home when they are no longer needed

Gulf states (p. 268) Saudi Arabia, Kuwait, Bahrain, Oman, Qatar, and the United Arab Emirates

hacienda (p. 128) a large agricultural estate in Middle or South America, more common in the past; usually not specialized by crop and not focused on market production

hajj (p. 245) the pilgrimage to the city of Makkah (Mecca) that all Muslims are encouraged to undertake at least once in a lifetime

Harappa culture (p. 331) *see* Indus Valley civilization

Hinduism (p. 331) a major world religion practiced by approximately 900 million people, 800 million of whom live in India; a complex belief system, with roots both in ancient literary texts (known as the *Great Tradition*) and in highly localized folk traditions (known as the *Little Tradition*)

Holocaust (p. 169) during World War II, a massive execution by the Nazis of 6 million Jews and 5 million gentiles (non-Jews), including ethnic Poles and other Slavs, Roma (Gypsies), disabled and mentally ill people, gays, lesbians, transgendered people, and political dissidents

Horn of Africa (p. 279) the triangular peninsula that juts out from northeastern Africa below the Red Sea and wraps around the Arabian Peninsula

hot spots (p. 451) individual sites of upwelling material (magma) that originate deep in Earth's mantle and surface in a tall plume; hot spots tend to remain fixed relative to migrating tectonic plates

***hukou* system** (p. 363) the system in China that ties people to their place of birth; each person's permanent residence is registered and any person who wants to migrate must obtain permission from authorities to do so

human geography (p. 3) the study of patterns and processes that have shaped human understanding, use, and alteration of the Earth's surface

human well-being (p. 36) various measures of the extent to which people are able to obtain a healthy and socially rewarding standard of living in an environment that is safe and sustainable

humanism (p. 165) a philosophy and value system that emphasizes the dignity and worth of the individual, regardless of wealth or social status

immigration (p. 47) in-migration to a place or country (see also *migration*)

import substitution industrialization (ISI) (p. 128) policies that encourage local production of machinery and other items that previously had been imported at great expense from abroad

Incas (p. 123) indigenous people who ruled the largest pre-Columbian state in the Americas, with a domain stretching from what is now southern Colombia to northern Chile and Argentina

income disparity (p. 127) the gap in income between rich and poor

indigenous (p. 111) native to a particular place or region

Indus Valley civilization (p. 331) the first substantial settled agricultural communities, which appeared about 4500 years ago along the Indus River in modern-day Pakistan and northwest India

industrial production (p. 35) processing, manufacturing, and construction

Industrial Revolution (p. 28) a series of innovations and ideas that occurred broadly between 1750 and 1850, which changed the way goods were manufactured

informal economy (p. 35) all aspects of the economy that take place outside official channels

infrastructure (p. 72) road, rail, and communication networks and other facilities necessary for economic activity

intertropical convergence zone (ITCZ) (p. 279) a band of atmospheric currents that circle the globe roughly at the equator; warm winds from both north and south converge at the ITCZ, pushing air upward and causing copious rainfall

intifada (p. 260) a prolonged Palestinian uprising against Israel

invasive species (p. 457) organisms that spread into regions outside their native range, adversely affecting economies or environments

Iron Curtain (p. 169) a long, fortified border zone that separated western Europe from (then) eastern Europe during the Cold War

Islam (p. 233) a monotheistic religion that emerged in the seventh century C.E. when, according to tradition, the archangel Gabriel revealed the tenets of the religion to the Prophet Muhammad

Islamism (p. 233) a grassroots religious revival in Islam that seeks political power to curb what are seen as dangerous secular influences; also seeks to replace secular governments and civil laws with governments and laws guided by Islamic principles

isthmus (p. 111) a narrow strip of land that joins two larger land areas

Jainism (p. 334) a religion of Asia that originated as a reformist movement within Hinduism more than 2000 years ago; Jains are found mainly in western India and in large urban centers throughout the region and are known for their educational achievements, nonviolence, and strict vegetarianism

jati (p. 332) in Hindu India, the subcaste into which a person is born, which traditionally defines the individual's experience for a lifetime

jihadists (p. 269) especially militant Islamists

Judaism (p. 245) a monotheistic religion characterized by the belief in one god (Yahweh), a strong ethical code summarized in the Ten Commandments, and an enduring ethnic identity

just-in-time system (p. 382) the system pioneered in Japanese manufacturing that clusters companies that are part of the same production system close together so that they can deliver parts to each other precisely when they are needed

***kaizen* system** (p. 382) a system of continuous manufacturing improvement, pioneered in Japan, in which production lines are constantly adjusted, improved, and surveyed for errors to save time and money and ensure that fewer defective parts are produced

Kyoto Protocol (p. 24) an amendment to a United Nations treaty on global warming, the Protocol is an international agreement, adopted in 1997 and in force in 2005, that sets binding targets for industrialized countries for reducing emissions of greenhouse gases

landforms (p. 13) physical features of the Earth's surface, such as mountain ranges, river valleys, basins, and cliffs

Latino (p. 61) a term used to refer to all Spanish-speaking people from Middle and South America, although their ancestors may have been European, African, Asian, or Native American

latitude (p. 4) the distance in degrees north or south of the equator; lines of latitude run parallel to the equator, and are also called parallels

legend (p. 4) a small box somewhere on a map that provides basic information about how to read the map, such as the meaning of the symbols and colors used

liberation theology (p. 146) a movement within the Roman Catholic Church that uses the teachings of Jesus to encourage the poor to organize to change their own lives and to encourage the rich to promote social and economic equity

lingua franca (p. 313) a language of trade

living wages (p. 34) minimum wages high enough to support a healthy life

local scale (p. 11) the level of geography that describes the space where an individual lives or works; a city, town, or rural area

longitude (p. 4) the distance in degrees east and west of Greenwich, England; lines of longitude, also called meridians, run from pole to pole (the line of longitude at Greenwich is 0° and is known as the prime meridian)

machismo (p. 146) a set of values that defines manliness in Middle and South America

Maori (p. 451) Polynesian people indigenous to New Zealand

map projections (p. 7) the various ways of showing the spherical Earth on a flat surface

maquiladoras (p. 131) foreign-owned, tax-exempt factories, often located in Mexican towns just across the border from U.S. towns, that hire workers at low wages to assemble manufactured goods which are then exported for sale

marianismo (p. 146) a set of values based on the life of the Virgin Mary, the mother of Jesus, that defines the proper social roles for women in Middle and South America

marketization (p. 129) the development of a free market economy in support of free trade

marsupials (p. 454) mammals whose babies at birth are still at a very immature stage; the marsupial then nurtures them in a pouch equipped with nipples

Mediterranean climate (p. 156) a climate pattern of warm, dry summers and mild, rainy winters

megalopolis (p. 91) an area formed when several cities expand so that their edges meet and coalesce

Melanesia (p. 461) New Guinea and the islands south of the equator and west of Tonga (the Solomon Islands, New Caledonia, Fiji, and Vanuatu)

Melanesians (p. 461) a group of Australoids named for their relatively dark skin tones, a result of high levels of the protective pigment melanin; they settled throughout New Guinea and other nearby islands

mercantilism (pp. 125, 166) a strategy for increasing a country's power and wealth by acquiring colonies and managing all aspects of their production, transport, and trade for the colonizer's benefit

Mercosur (p. 132) a free trade zone created in 1991 that links the economies of Argentina, Bolivia, Brazil, Paraguay, Uruguay, and Venezuela to create a common market

mestizos (p. 125) people of mixed European, African, and indigenous descent

metropolitan areas (p. 91) cities of 50,000 or more and their surrounding suburbs and towns

microcredit (p. 342) a program based on peer support that makes very small loans available to very low-income entrepreneurs

Micronesia (p. 462) the small islands that lie east of the Philippines and north of the equator

Middle America (p. 111) in this book, a region that includes Mexico, Central America, and the islands of the Caribbean

migration (p. 44) the movement of people from one place or country to another, often for safety or economic reasons

MIRAB economy (p. 466) an economy based on migration, remittances, aid, and bureaucracy

mixed agriculture (p. 283) farming that involves raising a diverse array of crops and animals on a single farm, often to take advantage of several environmental riches

Mongols (p. 205) a loose confederation of nomadic pastoral people centered in East and Central Asia, who by the thirteenth century had established by conquest an empire that stretched from Europe to the Pacific

monotheism (p. 245) the belief system based on the idea that there is only one god

monotremes (p. 454) egg-laying mammals, such as the duck-billed platypus and the spiny anteater

monsoon (p. 20) a wind pattern in which in summer months, warm, wet air coming from the ocean brings copious rainfall, and in winter, cool, dry air moves from the continental interior toward the ocean

Mughals (p. 331) a dynasty of Central Asian origin that ruled India from the sixteenth century to the nineteenth century

multicultural society (p. 52) a society in which many culture groups live in close association

multinational corporation (p. 32) a business organization that operates extraction, production, and/or distribution facilities in multiple countries

Muslims (p. 245) followers of Islam

nationalism (p. 168) devotion to the interests or culture of a particular country, nation, or cultural group; the idea that a group of people living in a specific territory and sharing cultural traits should be united in a single country to which they are loyal and obedient

nationalize (p. 131) to seize private property and place it under government ownership, with some compensation

nomadic pastoralists (p. 205) people whose way of life and economy are centered on tending grazing animals who are moved seasonally to gain access to the best pasture

nongovernmental organizations (NGOs) (p. 40) associations outside the formal institutions of government in which individuals, often from widely differing backgrounds and locations, share views and activism on political, social, economic, or environmental issues

nonpoint sources of pollution (p. 201) diffuse sources of environmental contamination, such as untreated automobile exhaust, raw sewage, and agricultural chemicals that drain from fields into water supplies

North American Free Trade Agreement (NAFTA) (pp. 75, 132) a free trade agreement made in 1994 that added Mexico to the 1989 economic arrangement between the United States and Canada

North Atlantic Drift (p. 156) the easternmost end of the Gulf Stream, a broad warm-water current that brings large amounts of warm water to the coasts of Europe

North Atlantic Treaty Organization (NATO) (p. 178) a military alliance between European and North American countries that was developed during the Cold War to counter the influence of the Soviet Union; since the breakup of the Soviet Union, NATO has expanded membership to include much of eastern Europe and Turkey, and is now focused mainly on providing the international security and cooperation needed to expand the European Union

nuclear family (p. 103) a family consisting of a married father and mother and their children

occupied Palestinian Territories (oPT) (p. 233) Palestinian lands occupied by Israel since 1967

offshore outsourcing (p. 340) the contracting of certain business functions or production functions to providers where labor and other costs are lower

oligarchs (p. 210) in Russia, those who acquired great wealth during the privatization of Russia's resources and who use that wealth to exercise power

OPEC (Organization of the Petroleum Exporting Countries) (p. 249) a cartel of oil-producing countries—currently, Algeria, Angola, Iran, Iraq, Kuwait, Libya, Nigeria, Qatar, Saudi Arabia, the United Arab Emirates, Ecuador, and Venezuela—that was established to regulate the production and price of oil and natural gas

organically grown (p. 82) products produced without chemical fertilizers and pesticides

orographic rainfall (p. 17) rainfall produced when a moving moist air mass encounters a mountain range, rises, cools, and releases condensed moisture that falls as rain

Ottoman Empire (p. 247) the most influential Islamic empire the world has ever known; begun in the 1200s when nomadic Turkic herders from Central Asia converged in western Anatolia (Turkey)

Pacific Rim (p. 97) a term that refers to all the countries that border the Pacific Ocean

Pacific Way (p. 468) the regional identity and way of handling conflicts peacefully that grows out of Pacific Islanders' particular social experiences

Partition (p. 336) the breakup following Indian independence that resulted in the establishment of Hindu India and Muslim Pakistan

pastoralism (p. 286) a way of life based on herding; practiced primarily on savannas, on desert margins, and in the mixture of grass and shrubs called *open bush*

patriarchal (p. 268) relating to a social organization in which the father is supreme in the clan or family

perestroika (p. 208) literally, "restructuring"; the restructuring of the Soviet economic system that was done in the late 1980s in an attempt to revitalize the economy

permafrost (p. 197) permanently frozen soil that lies just a few feet beneath the surface

physical geography (p. 3) the study of the Earth's physical processes: how they work and interact, how they affect humans, and how they are affected by humans

pidgin (p. 477) a language used for trading; made up of words borrowed from the several languages of people involved in trading relationships

plantation (p. 128) a large factory farm that grows and partially processes a single cash crop

plate tectonics (p. 13) the scientific theory that the Earth's surface is composed of large plates that float on top of an underlying layer of molten rock; the movement and interaction of the plates create many of the large features of the Earth's surface, particularly mountains

political ecologists (p. 36) geographers who study the interactions among development, politics, human well-being, and the environment

political freedoms (p. 38) the rights and capacities that support individual and collective liberty and public participation in political decision making

polygyny (p. 310) the practice of having multiple wives

Polynesia (p. 462) the numerous islands situated inside an irregular triangle formed by New Zealand, Hawaii, and Easter Island

population pyramid (p. 47) a graph that depicts the age and sex structures of a political unit, usually a country

populist movements (p. 146) popularly based efforts, often seeking relief for the poor

precipitation (p. 17) dew, rain, sleet, and snow

primary sector (p. 35) an economic sector of the economy that is based on extraction (see also *extraction*)

primate city (p. 139) a city, plus its suburbs, that is vastly larger than all others in a country and in which economic and political activity is centered

privatization (pp. 129, 210) the selling of formerly government-owned industries and firms to private companies or individuals

purchasing power parity (PPP) (p. 35) the amount that the local currency equivalent of U.S.$1 will purchase in a given country

purdah (p. 354) the practice of concealing women from the eyes of nonfamily men

push factors (p. 98) factors that get people to consider the drastic move of leaving family and friends and a familiar place to strike out into the unknown, with what are usually unknown resources

push/pull phenomenon of urbanization (p. 42) conditions, such as political instability or economic changes, that encourage (push) people to leave rural areas, and urban factors, such as job opportunities, that encourage (pull) people to move to the urban area

quaternary sector (p. 35) a sector of the economy that is based on intellectual pursuits such as education, research, and IT (information technology) development

Qur'an (or Koran) (p. 237) the holy book of Islam, believed by Muslims to contain the words Allah revealed to Muhammad through the archangel Gabriel

race (p. 53) a social or political construct that is based on apparent characteristics such as skin color, hair texture, and face and body shape, but that is of no biological significance

rate of natural increase (RNI) (p. 44) the rate of population growth measured as the excess of births over deaths per 1000 individuals per year without regard for the effects of migration

region (p. 9) a unit of the Earth's surface that contains distinct patterns of physical features and/or distinct patterns of human development

regional conflict (p. 344) a conflict created by the resistance of a regional ethnic or religious minority to the authority of a national or state government; currently these are the most intense armed conflicts in South Asia

regional self-sufficiency (p. 384) an economic policy in Communist China that encouraged each region to develop independently in the hope of evening out the wide disparities in the national distribution of production and income

regional specialization (p. 384) specialization (rather than self-sufficiency) in order to take advantage of regional variations in climate, natural resources, and location

religious nationalism (p. 344) the association of a particular religion with a particular territory or political unit to the exclusion of other religions

resettlement schemes (p. 431) government plans to move large numbers of people from one part of a country to another to relieve urban congestion, disperse political dissidents, or accomplish other social purposes; also called *transmigration*

responsibility system (p. 384) in the 1980s, a decentralization of economic decision making in China that returned agricultural decision making to the farm household level, subject to the approval of the commune

Ring of Fire (p. 13) the tectonic plate junctures around the edges of the Pacific Ocean; characterized by volcanoes and earthquakes

Roaring Forties (p. 451) powerful air and ocean currents at about 40° S latitude that speed around the far Southern Hemisphere virtually unimpeded by landmasses

Roma (p. 169) the now-preferred term in Europe for Gypsies

Russian Federation (p. 197) Russia and its political subunits, which include 21 internal republics

Russification (p. 216) the czarist and Soviet policy of encouraging ethnic Russians to settle in non-Russian areas as a way to assert political control

Sahel (p. 276) a band of arid grassland, where steppe and savanna grasses grow, that runs east-west along the southern edge of the Sahara

Salafism (p. 255) an extreme, purist Qur'an-based version of Islam that has little room for adaptation to modern times

salinization (p. 239) a process that occurs when large quantities of water are used to irrigate areas where evaporation rates are high, leaving behind dissolved salts and other minerals

scale (of a map) (p. 4) the proportion that relates the dimensions of the map to the dimensions of the area it represents; also, variable-sized units of geographical analysis from the local scale to the regional scale to the global scale

Schengen Accord (p. 184) an agreement signed in the 1990s by the European Union and many of its neighbors that allows for free movement across common borders

seawater desalination (p. 240) the removal of salt from seawater—usually accomplished through the use of expensive and energy-intensive technologies—to make the water suitable for drinking or irrigating

secondary sector (p. 35) an economic sector of the economy that is based on industrial production (see also *industrial production*)

secular states (p. 255) countries that have no state religion and in which religion has no direct influence on affairs of state or civil law

self-reliant development (p. 298) small-scale development in rural areas that is focused on developing local skills, creating local jobs, producing products or services for local consumption, and maintaining local control so that participants retain a sense of ownership over the process

services (p. 35) sales, entertainment, and financial services

sex (p. 49) the biological category of male or female; does not indicate how males or females may behave or identify themselves

sex tourism (p. 442) the sexual entertainment industry that serves primarily men who travel for the purpose of living out their fantasies during a few weeks of vacation

sex work (p. 350) the provision of sexual acts for a fee

shari'a (p. 246) literally, "the correct path"; Islamic religious law that guides daily life according to the interpretations of the Qur'an

Shi'ite (or Shi'a) (p. 246) the smaller of two major groups of Muslims who have different interpretations of shari'a; Shi'ites are found primarily in Iran and southern Iraq

shifting cultivation (pp. 122, 283) a productive system of agriculture in which small plots are cleared in forestlands, the dried brush is burned to release nutrients, and the clearings are planted with multiple species; each plot is used for only 2 or 3 years and then abandoned for many years of regrowth

Siberia (p. 197) a region of Russia that is located east of the Ural Mountains

Sikhism (p. 334) a religion of South Asia that combines beliefs of Islam and Hinduism

silt (p. 114) fine soil particles

Slavs (p. 205) a group of people who originated between the Dnieper and Vistula Rivers in modern-day Poland, Ukraine, and Belarus

slum (p. 42) densely populated area characterized by crowding, run-down housing, and inadequate access to food, clean water, education, and social services

smog (p. 64) a combination of industrial emissions, car exhaust, and water vapor that frequently hovers as a yellow-brown haze over many North American cities, causing a variety of health problems

social safety net (p. 89) the services provided by the government—such as welfare, unemployment benefits, and health care—that prevent people from falling into extreme poverty

social welfare (in the European Union, **social protection**) (p. 178) in Europe, tax-supported systems that provide citizens with benefits such as health care, affordable higher education and housing, pensions, and child care

South America (p. 111) the continent south of Central America

Soviet Union (p. 195) see *Union of Soviet Socialist Republics*

spatial distribution (p. 4) the arrangement of a phenomenon across the Earth's surface

spatial interaction (p. 4) the flow of goods, people, services, or information across space and among places

special economic zones (SEZs) (p. 390) free trade zones within China, which are commonly called export processing zones (EPZs) elsewhere

state-aided market economy (p. 382) an economic system based on market principles such as private enterprise, profit incentives, and supply and demand, but with strong government guidance; in contrast to the free market (limited government) economic system of the United States and, to a lesser degree, Europe

steppes (p. 197) semiarid, grass-covered plains

structural adjustment programs (SAPs) (p. 129) policies that require economic reorganization toward less government involvement in industry, agriculture, and social services; sometimes imposed by the World Bank and the International Monetary Fund as conditions for receiving loans

subcontinent (p. 320) a term often used to refer to the entire Indian peninsula, including Nepal, Bhutan, India, Pakistan, and Bangladesh

subduction zone (p. 112) a zone where one tectonic plate slides under another

subsidies (p. 173) monetary assistance granted by a government to an individual or group in support of an activity, such as farming, that is viewed as being in the public interest

Glossary

subsistence affluence (p. 467) a lifestyle whereby people are self-sufficient with regard to most necessities and have some opportunities to earn cash for travel and occasional purchases of manufactured goods

subsistence agriculture (p. 282) farming that provides food for only the farmer's family and is usually done on small farms

subsistence economy (p. 48) an economy in which families produce most of their own food, clothing, and shelter

suburbs (p. 91) populated areas along the peripheries of cities

summer monsoon (p. 322) rains that begin every June when the warm, moist ITCZ air first reaches the mountainous western Ghats

Sunni (p. 246) the larger of two major groups of Muslims who have different interpretations of shari'a

sustainable agriculture (p. 31) farming that meets human needs without poisoning the environment or using up water and soil resources

sustainable development (p. 36) the effort to improve current standards of living in ways that will not jeopardize those of future generations

taiga (p. 198) subarctic coniferous forests

Taliban (p. 357) an archconservative Islamist movement that gained control of the government of Afghanistan in the mid-1990s

temperate midlatitude climate (p. 156) as in south-central North America, China, and Europe, a climate that is moist all year with relatively mild winters and long, mild to hot summers

temperature-altitude zones (p. 114) regions of the same latitude that vary in climate according to altitude

tertiary sector (p. 35) an economic sector of the economy that is based on services (see also *services*)

theocratic states (p. 255) countries that require all government leaders to subscribe to a state religion and all citizens to follow rules decreed by that religion

total fertility rate (TFR) (p. 46) the average number of children that women in a particular population are likely to have at the present rate of natural increase

trade deficit (p. 77) the extent to which the money earned by exports is exceeded by the money spent on imports

trade winds (p. 114) winds that blow from the northeast and the southeast toward the equator

trafficking (p. 225) the recruiting, transporting, and harboring of people through coercion for the purpose of exploiting them

tsunami (p. 365) a large sea wave caused by an earthquake

tundra (p. 197) a treeless area, between the ice cap and the tree line of arctic regions, where the subsoil is permanently frozen

typhoon (p. 367) a tropical cyclone or hurricane in the western Pacific Ocean

UNASUR (p. 132) a union of South American nations that was organized in May of 2008; it supersedes Mercosur and the Andean Community of Nations, two previous customs unions

underemployment (p. 212) the condition in which people are working too few hours to make a decent living or are highly trained but working at menial jobs

Union of Soviet Socialist Republics (USSR) (p. 195) the multinational union formed from the Russian empire in 1922 and dissolved in 1991; commonly known as the *Soviet Union*

United Nations (UN) (p. 40) an assembly of 193 member states that sponsors programs and agencies that focus on economic development, general health and well-being, democratization, peacekeeping assistance in "hot spots" around the world, humanitarian aid, and scientific research

United Nations Gender Equality Index rank (GEI) (p. 36) a composite measure reflecting the degree to which there is inequality in achievements between women and men in three dimensions: reproductive health, empowerment, and the labor market. A high rank indicates that the genders are tending toward equality

United Nations Human Development Index (HDI) (p. 36) an index that calculates a country's level of well-being, based on a formula of factors that considers income adjusted to PPP, data on life expectancy at birth, and data on educational attainment

urban growth poles (p. 141) locations within cities that are attractive to investment, innovative immigrants, and trade, and thus attract economic development like a magnet

urban sprawl (p. 70) the encroachment of suburbs on agricultural land

urbanization (p. 42) the process whereby cities, towns, and suburbs grow as populations shift from rural to urban livelihoods

varna (p. 333) the four hierarchically ordered divisions of society in Hindu India underlying the caste system: *Brahmins* (priests), *Kshatriyas* (warriors/kings), *Vaishyas* (merchants/landowners), and *Sudras* (laborers/artisans)

veil (p. 269) the custom of covering the body with a loose dress and/or of covering the head—and in some places the face—with a scarf

virtual water (p. 26) the water used to produce a product, such as an apple or a pair of shoes

water footprint (p. 26) all the water a person consumes, including both virtual water and the water they consume directly

weather (p. 14) the short-term and spatially limited expression of climate that can change in a matter of minutes

weathering (p. 13) the physical or chemical decomposition of rocks by sun, rain, snow, ice, and the effects of life-forms

welfare state (p. 168) a government that accepts responsibility for the well-being of its people, guaranteeing basic necessities such as education, affordable food, employment, and health care for all citizens

West Bank barrier (p. 262) a 25-foot-high concrete wall in some places and a fence in others that now surrounds much of the West Bank and encompasses many of the Jewish settlements there

wet rice cultivation (p. 370) a prolific type of rice production that requires the plant roots to be submerged in water for part of the growing season

winter monsoon (p. 322) a weather pattern that begins by November, when the cooling Eurasian landmass sends the cooler, drier, heavier air over South Asia

World Trade Organization (WTO) (p. 34) a global institution made up of member countries whose stated mission is to lower trade barriers and to establish ground rules for international trade

Zionists (p. 260) those who have worked, and continue to work, to create a Jewish homeland (Zion) in Palestine

TEXT SOURCES AND CREDITS

Facts have been gathered from the following sources for text on the pages indicated.

CHAPTER 1
33 Adapted from Lydia Pulsipher's field notes, 1992–2000 (Olivia and Tanya), and Alex Pulsipher's field notes, 1999–2008 (Setiya).
52 Adapted from Lydia Pulsipher's field notes in Honolulu, 1995.

CHAPTER 2
61 Adapted from Jesse McKinley, "Drought Adds to Hardships in California," *New York Times*, February 21, 2009; David Leonhardt, "Job Losses Show Breadth of Recession," *New York Times*, March 3, 2009; and Catherine Rampell, "As Layoffs Surge, Women May Pass Men in Job Force," *New York Times*, February 5, 2009; Alissa Figueroa, *PBS Newshour*, "In Calif. Town, Prison May Fix One Problem, but Create Another," May 5, 2011, at http://www.pbs.org/newshour/bb/social_issues-jan-june11-mendota_05-05/.
82 Adapted from NPR Staff, *All Things Considered*, "What Recession? It's Boom Time for Nebraska Farms," February 25, 2011, at http://tinyurl.com/6jyljoa.

CHAPTER 3
111 Alex Pulsipher's field notes; Amazon Watch, 2006; Oxfam America, 2005; Juan Forero, "Rain Forest Residents, Texaco Face Off In Ecuador," *Morning Edition*, National Public Radio, April 30, 2009, at http://www.npr.org/templates/story/story.php?storyId=103233560; Gonzalo Solano, "Damages in Chevron Ecuador Suit Jump Billions," *San Francisco Chronicle*, September 18, 2010, at http://www.sfgate.com/default/article/Damages-in-Chevron-Ecuador-suit-jump-billions-3173946.php; Patrick Radden Keefe, "Reversal of Fortune," *New Yorker*, January 9, 2012, pp. 38–49; Reuters, "Ecuador: Villagers Try Again to Hold Chevron to $18bn Ruling," June 29, 2012, at http://www.independent.co.uk/news/world/americas/ecuador-villagers-try-again-to-hold-chevron-to-18bn-ruling-7897095.html.
129 NPR reports by John Idste, August 14, 2001, and August 16 and 25, 2003, and by Gary Hadden, August 27, 2003; David Bacon, "Anti-China Campaign Hides Maquiladora Wage Cuts," ZNet, February 3, 2003, at http://zcomm.org/znetarticle/anti-china-campaign-hides-maquiladora-wage-cuts-by-david-bacon; Maquila Portal, April 2006, at http://www.maquilaportal.com/cgi-bin/public/index.pl; "How Rising Wages Are Changing the Game in China," *Bloomberg Businessweek*, March 26, 2006, at http://www.businessweek.com/stories/2006-03-26/how-rising-wages-are-changing-the-game-in-china; Pete Engardio, "So Much for the Cheap 'China Price,'" *Bloomberg Businessweek*, June 4, 2009, at http://www.businessweek.com/magazine/content/09_24/b4135054963557.htm; "Avalanche of Chinese Investment in Mexico," Maquila Portal, December 1, 2011, at http://www.maquilaportal.com/index.php?blog/show/Avalanche-of-Chinese-investment-in-Mexico.html; Gustavo de Lima Palhares and Higor Uzzun Sales, "An Economic Analysis of the Role of Foreign Investments in Brazil," April 18, 2012, Council on Hemispheric Affairs, at http://www.coha.org/an-economic-analysis-of-the-role-of-foreign-investments-in-brazil.
130 "Foreign Direct Investment in Latin America and the Caribbean, 2011," UN Economic Commission for Latin America and the Caribbean, April 16, 2012, p. 37, at http://www.cepal.org/publicaciones/xml/2/46572/2012-182-liei-web.pdf; Alexei Barrionevo, "China's Interest in Farmland Makes Brazil Uneasy," *New York Times*, May 26, 2011, at http://www.nytimes.com/2011/05/27/world/americas/27brazil.html?pagewanted=all&_r=0; Juan Forero, "Latin America: Once A Risky Bet, Now EU's Hero?" *Morning Edition*, National Public Radio, December 5, 2011, at http://www.npr.org/2011/12/05/143131926/once-a-risky-bet-latin-america-tapped-to-aid-eurozone.
134 Adapted from Andrew Wheat, "Toxic Bananas," *Multinational Monitor* 17: 6–7, September 1996 (updated 2007); David Magney, "Costa Rican Bananas," David Magney Environmental Consulting, December 16, 2005 (updated July 9, 2007), at http://www.magney.org/photofiles/CostaRica-Bananas1.htm.
142 Lydia Pulsipher's field notes on Brazil, updated with the help of John Mueller, Fortaleza, Brazil, 2006.

CHAPTER 4
153 Conversations with geographer Margareta Lelea, a Romanian specialist, post-doctoral researcher, Departments of Entomology and Human and Community Development, University of California, Davis; Doreen Carvajal and Stephen Castle, "A U.S. Hog Giant Transforms Eastern Europe," *New York Times*, May 5, 2009, at http://www.nytimes.com/2009/05/06/business/global/06smithfield.html?pagewanted=all&module=Search&mabReward=relbias%3Ar.
159 Adapted from Rob Gifford, "Gardeners Brighten London Under Cover of Dark," *Morning Edition*, National Public Radio, May 15, 2006, at http://www.npr.org/templates/story/story.php?storyId=5404229.
175 Lydia Pulsipher's conversations with Vera Kuzmic and Dusan Kramberger, 1993–2012.

CHAPTER 5
195 Ellen Barry and Andrew E. Kramer, "In Biting Cold, Protesters Pack the Center of Moscow," *New York Times*, February 4, 2012, at http://www.nytimes.com/2012/02/05/world/europe/tens-of-thousands-protest-putin-in-moscow-russia.html?pagewanted=all; Simon Saradzhyan and Nabi Abdullaev, "Putin Election Victory Doesn't Pave an Easy Path Through His Third Presidential Term," *Christian Science Monitor*, March 5, 2012, at http://www.csmonitor.com/Commentary/Opinion/2012/0305/Putin-election-victory-doesn-t-pave-an-easy-path-through-his-third-presidential-term; Ellen Barry and Michael Schwirtz, "After Election, Putin Faces Challenges to Legitimacy," *New York Times*, March 5, 2012, at http://www.nytimes.com/2012/03/06/world/europe/observers-detail-flaws-in-russian-election.html?pagewanted=all; Michael Schwirtz, "Fear of Return to '90s Hardship Fuels Support for Putin," *New York Times*, March 3, 2012, at http://www.nytimes.com/2012/03/04/world/europe/in-russia-vote-fear-of-hardship-fuels-putin-support.html?pagewanted=all; Amanda Walker, "Last Anti-Putin Rally Before Russian Election," *Sky News*, February 26, 2012, at http://news.sky.com/story/929127/last-anti-putin-rally-before-russian-election.
202 "Shrinking Aral Sea," NASA Earth Observatory, August 25, 2000, at http://earthobservatory.nasa.gov/Features/WorldOfChange/aral_sea.php; "Central Asia: Uzbekistan," *World Factbook*, Central Intelligence Agency, at https://www.cia.gov/library/publications/the-world-factbook/geos/uz.html; "Cotton Fact Sheet: Uzbekistan," ICAC, at http://www.icac.org/econ_stats/country_fact_sheets/fact_sheet_uzbekistan_2011.pdf.
219 "2011 World Population Data Sheet," Population Reference Bureau, at http://www.prb.org/pdf11/2011population-data-sheet_eng.pdf; Timothy Heleniak, "Russia's Demographic Decline Continues," Population Reference Bureau, June 2002, at http://www.prb.org/Publications/Articles/2002/RussiasDemographicDeclineContinues.aspx; "Half of All Premature Deaths of Russian Adults Down to Alcohol," University of Oxford, June 26, 2009, at http://www.ox.ac.uk/media/news_stories/2009/090626.html.
226 "Overcoming Barriers: Human Mobility and Development," Human Development Report 2009, United Nations Development Program, at http://hdr.undp.org/en/content/human-development-report-2009.

CHAPTER 6
234 Mona Eltahawy, "Why Do They Hate Us?" *Foreign Policy*, April 23, 2012, at http://www.foreignpolicy.com/articles/2012/04/23/why_do_they_hate_us; Allison Good, "Debating the War on Women," *Foreign Policy*, April 24, 2012, at http://www.foreignpolicy.com/articles/2012/04/24/debating_the_war_on_women; "Talk to Al Jazeera: Why Arab Women Still 'Have No Voice,'" at http://www.aljazeera.com/programmes/talktojazeera/2012/04/201242111373249723.html.
239 A. M. MacDonald, H. C. Bonsor, B. E. O. Dochartaigh, and R. G. Taylor, "Quantitative Maps of Groundwater Resources in Africa," *Environmental Research Letters*, April 19, 2012, at http://iopscience.iop.org/1748-9326/7/2/024009/pdf/1748-9326_7_2_024009.pdf.

CHAPTER 7
277 Adapted from "Juliana Rotich—This Is What I See! Crowdsourced

TC-1

Text Sources and Credits

Crisis Mapping in Real Time," *99 Faces*, March 17, 2011, at http://99faces.tv/julianarotich/; Ushahidi Downloads, at http://download.ushahidi.com/; "Africa: The Next Chapter," *TED Radio Hour*, National Public Radio, at http://www.npr.org/2012/06/29/155904209/africa-the-next-chapter.

282 Adapted from "Silas Kpanan'Ayoung Siakor: A Voice for the Forest and Its People," Goldman Environmental Prize, at http://www.goldmanprize.org/node/442; Scott Simon, "Reflections of a Liberian Environmental Activist," *Weekend Edition Saturday*, National Public Radio, April 29, 2006, at http://www.npr.org/templates/story/story.php?storyId=5370987; author's personal communication with Siakor, September 2006.

310 James Fenske, "African Polygamy: Past and Present," February 15, 2012, at https://editorialexpress.com/cgi-bin/conference/download.cgi?db_name=CSAE2012&paper_id=115.

CHAPTER 8

319 Adapted from "Modi Goes on Fast over Narmada Dam," *India eNews*, April 16, 2006; Rahul Kumar, "Medha Patkar Ends Fast After Court Order on Rehabilitation," One World South Asia.

330 Lydia and Alex Pulsipher's field notes, Nilgiri Hills, June 2000; Government of Tamil Nadu, *Tamil Nadu Human Development Report* (Delhi: Social Sciences Press, 2003), at http://hdr.undp.org/sites/default/files/tn_nhdr_2003.pdf.

343 Beth Roy, *Some Trouble with Cows—Making Sense of Social Conflict* (Berkeley: University of California Press, 1994), pp. 18–19.

357 Internews Afghanistan, March 2008, at http://www.internews,org/bulletin/afghanistan/Afghan_200803.html; Maria Fantappie and Brittany Tanasa, "Fearless Women Use Radio to Make Ripples of Change in Herat, Afghanistan,"

September 7, 2011, at http://cima.ned.org/fearless-women-use-radio-make-ripples-change-herat-afghanistan.

CHAPTER 9

363 Adapted from Peter S. Goodman, "In China's Cities, a Turn from Factories," *Washington Post*, September 25, 2004, at http://www.washingtonpost.com/wp-dyn/articles/A48818-2004Sep24.html; Louisa Lim, "The End of Agriculture in China," *Reporter's Notebook*, National Public Radio, May 19, 2006, at http://www.npr.org/templates/story/story.php?storyId=5411325; background information from Kathy Chen, "Boom-Town Bound," *Wall Street Journal*, October 29, 1996, p. A6; "Life Lessons," *Wall Street Journal*, July 9, 1997.

369 Adapted from Maria Siow, "Desertification: One of the Challenges Faced by China," *Asia Pacific News*, October 29, 2009, at http://www.channelnewsasia.com/stories/eastasia/view/1014326/1/.html.

393 Paul Wiseman, "Chinese Factories Struggle to Hire," *USA Today*, April 11, 2005, at http://www.usatoday.com/money/world/2005-04-11-china-labor_x.htm; Louisa Lim, "The End of Agriculture in China," *Reporter's Notebook*, National Public Radio, May 17, 2006, at http://www.npr.org/templates/story/story.php?storyId=5411325; Mei Fong, "A Chinese Puzzle: Surprising Shortage of Workers Forces Factories to Add Perks; Pressures on Pay—and Prices," *Wall Street Journal*, August 16, 2004, p. B1; David Barboza, "Labor Shortage in China May Lead to Trade Shift," *New York Times*, April 3, 2006, Business Section, p. 1, at http://www.nytimes.com/2006/04/03/business/03labor.html?pagewanted=all; Qiu Quanlin, "Labor Shortage Hinders Guangdong Factories," *China Daily*, August 25, 2009, at http://www.chinadaily.com.cn/china/2009-08/25/content_8612599.htm; Nina Ying Sun, "Labor Shortage Returns to China," *Plastics News*, August 10, 2009, at http://www.pnchina.com/en/Detail.aspx?id=3831&cat=0; "Major Foreign Holders of Treasury Securities (in Billions of Dollars) Holdings 1/ At End of Period," Data and Charts Center, U.S. Department of the Treasury, at http://www.treasury.gov/ticdata/Publish/mfh.txt; "The Biggest Holders of US Government Debt," *CNBC Explains*, at http://www.cnbc.com/id/29880401/The_Biggest_Holders_of_US_Government_Debt.

CHAPTER 10

407 Adapted from the *Borneo Wire*, Jessica Lawrence, ed., Spring 2006, at http://borneoproject.org/article.php?id=623; "Community Stops Illegal Logging and Bulldozing Towards Protected Rainforests," Jessica Lawrence, ed., *Borneo Wire*, Fall 2006, at http://borneoproject.org/article.php?list=type&type=39; Mark Bujang, "A Community Initiative: Mapping Dayak's Customary Lands in Sarawak," presented at the Regional Community Mapping Network Workshop, November 8–10, 2004, Diliman, Quezon City, Philippines; Julia Zappei, "Malaysia Highest Court Affirms Tribes' Land Rights," Associated Press, May 10, 2009, at http://borneoproject.org/article.php?id=762; "US Announces Its Support of the UN Declaration of the Rights of Indigenous Peoples (UNDRIP), December 16, 2010," International Forum on Globalization, at http://ifg.org/programs/indig/USUNDRIP.html; video of deforestation in Indonesian Borneo, at http://news.mongabay.com/2012/0326-muaratae-video.html.

428 The fieldwork of Karl Russell Kirby, Department of Geography, University of Tennessee, 2010.

434 Adapted from notes by Kirsty Vincin, an Australian teacher in Hong Kong, November 17, 2009, and "The Filipina Sisterhood: An Anthropology of Happiness," *Economist*, December 20, 2001, at http://www.economist.com/node/883909.

437 Likhaan Center for Women's Health, 2010, at http://www.likhaan.org/; Carlos H. Conde, "Bill to Increase Access to Contraception Is Dividing Filipinos," *New York Times*, October 25, 2009, at http://www.nytimes.com/2009/10/26/world/asia/26iht-phils.html.

440 Adapted from personal communications with anthropologist Jennifer W. Nourse, University of Richmond, a specialist in Southeast Asia, 2010; Walter Williams, *Javanese Lives* (Piscataway, NJ: Rutgers University Press, 1991), pp. 128–134.

443 Adapted from the field notes of Alex Pulsipher and Debbi Hempel, 2000; coverage of the HIV-AIDS conference in Thailand, July 11–16, 2004, by the Kaiser Family Foundation; "Life as a Thai Sex Worker," *BBC News*, February 22, 2007, at http://news.bbc.co.uk/2/hi/asia-pacific/6360603.stm.

CHAPTER 11

448–449 Adapted from audio segments and articles written by Brian Reed for National Public Radio, "Preparing for Sea Level Rise, Islanders Leave Home," at http://www.npr.org/2011/02/17/133681251/preparing-for-sea-level-rise-islanders-leave-home; "Could People from Kiribati Be 'Climate Change Refugees?'" February 17, 2011, at http://www.npr.org/blogs/thetwo-way/2011/02/17/133848076/could-people-from-kiribati-be-climate-change-refugees; "Kiribati Mulls Fiji Land Purchase in Battle Against the Sea," *BBC News Asia*, March 8, 2012, at http://www.bbc.co.uk/news/world-asia-17295862.

INDEX

Abkhazia, 217
Aborigines. *See also* Oceania
    Ayer's Rock, *451*
    described, **461**–462, *462*, 464, *464*, 476, 480
    dingoes, *459*
    land claims, 475–476, 480
    racist attitudes toward, 473–474
Abortions, 103, 111, 395, 437
Abraham, 245
Abyssinia, 290
Accelerated globalization, Europe, 166–167
Acculturation, 74, **145**
Aceh province, tsunami (2004), 42, 321, *405*, 409, 427, 431
Acid rain, *64*, 66, 79, *119*, 161, *161*, 162, 200, 238, 281, 328, 330
Acquired immunodeficiency syndrome. *See* HIV/AIDS
Addis Ababa, 312
Adivasis, 333, 347
Affluence, subsistence, **467**, 479
Afghanistan
    Central Asia world region, 10–11, *11*
    climate change vulnerability, 325
    mujahedeen movement, 209, 346
    religious diversity, 333
    Soviet-Afghanistan war, 209, *215*, 228
    Taliban in, 227, 346, 356–**357**, 358
    U.S. involvement in, 85, 86–87, 106, 107, 178, 259
    war reconstruction in, 346
    women's status, 85, 90, 358
Africa (African continent). *See also* North Africa and Southwest Asia; South Africa; Sub-Saharan Africa
    Horn of Africa, **279**, 285, 300
    Pangaea and, 13
African Americans. *See also* Slavery
    ethnic composition in North America, *102*
    Obama and, 86, 87, 89, 90, 103, 341
    population, in states, *72*
    racial segregation in schools, 88
African grey parrots, 287
African National Congress, 292
African Plate, 13, *16*, 61, 155, 235, 409, 410
Afrikaners, 292
Afro-Asiatic language family, *55*
Afromusing, 277
Agent Orange, 422, *423*, 445
Agnostics, 103
Agra, Taj Mahal, 330, 331, 332, 359
Agribusiness, 38, 67, **80**–82, 99, 173, 190, 284, 338
Agriculture. *See also* Food production; Green revolution agriculture; *specific countries*
    commercial, **283**–284
    Common Agricultural Program, **173**–174, *175*

corporate, 174
defined, **28**
early human impact on, 28–29, 288, 331, 332
Fertile Crescent, *243*, **243**–244, 248, *248*
mixed, **283**, *285*
shifting cultivation, 15, 56, **122**, *126*, *135*, **283**
slash and burn, 415, *416*
subsidies, EU, **173**–174
subsistence, 28, **282**–283, 284, 466
sustainable, **31**, 118, 126, 417
wet rice cultivation, **370**–371, 415
Agroecology, **341**–342
Agroforestry, 118, **282**, 283, 287
Aguan River, *24*
Aguarico River, 110
Aguilar, Javier, 60–61
AIDS. *See* HIV/AIDS
Aila, Cyclone, *25*
Ainu people, *400*, **400**–401
Air pollution. *See also specific world regions*
    Beijing, 328
    Bulgaria, 162
    China, 374, *374*–375
    New Delhi, 328, 359
    smog, **64**, *66*, 374, 410
    Taiwan, 373, 374, 375
Air pressure, 16–17, 20
Ais kacang (cold dessert), *441*
Alaska
    Bering land bridge, 71, *72*, *122*, *126*
    continental climate, *19*
    Exxon Valdez disaster, 299
    pipeline, 68, *69*, 107
Alaskan Malamutes, 94
Albania, climate change vulnerability, *158*
Albers projection, 6, *7*
Alcohol abuse, 60, 75, 110, 194, 220, 222, 224, 228
Aleppo, Syria, 257
Alexander the Great, 331
Alexandria, 164, 238, *241*, 272
Alfombra, *145*
Algeria
    fossil water, 239, *239*
    Mediterranean climate, 236
    oil and gas reserves, 249
    women's rights, *234*
Alhambra, *166*
Alice Springs, desert, *452*
Allah, 52, 237, 245, 255
Allende, Salvador, 136
Alps, *151*, 155, 156
Altaic language family, *55*
Altitude-temperature zones, **114**, *116*, *117*
Altruism, 53
Amazon Basin, *109*, 110, 114, *117*, *119*, 122, 134, *145*, 148
Amazon River, 114, *120*

Amu Darya, 202
Amundsen-Scott South Pole Station, *481*
Ancient Greece/Rome, 164, *166*
Andes
    domesticated plants, 125
    formation, 112
    glacial melting, 120, *121*
    guinea pigs, *117*
    high-altitude climate zone, *18*
    Incas and, 123
    mining, 128
    photograph, *108*
    rain shadows, 114, *115*
Anglican Church, 312
Angola, 176, 296
Animal husbandry, 28–29, 31
Animals and plants. *See also* Domestication; Local lives; Vegetation; *specific animals*; *specific countries*
    endangered species, 282, 285, 286, 287, 295, 328, 329, 373, *414*, 457, 458, 461
    endemic species, **454**, 455, 457
    extinction, 287, *372*, 372, 455, 457, 458
    habitat loss, 15, 64, 68, 70, 91, 328, 411
    introduced species, 71, 457, 458, 459, 479
    invasive species, 282, 454, **457**, 461
    nonnative species, 70, 101, 282, 448, 455, 457, 461, 479
Animism, 299, 310, 311, **312**, 333, 334, 439, *439*, 440, 443
Antarctic Plate, *16*, *17*
Antarctic Treaty, *481*
Antarctica, photo essay, *481*
Anthropocene, 12
Apartheid, 289, **291**–292, *292*, 293, 302, 306
APEC (Asia Pacific Economic Cooperative), 468, 470
Appalachian Mountains, 59, 61, *62*, 64, 155, 197
Apple Corporation, 32, 34
Aqaba, Gulf of, *230*
Aquifers, Ogallala, 67, *67*, 82. *See also specific world regions*
Arab Spring movement
    commencement of, 39, 232
    described, 39, 254–256, *257*
    Egypt and, 39, 273, 277
    Libya and, 39, 273, 277
    NATO and, 178
    responses to, 271
    social networking and, 76, 258
    Syria and, 39, *41*, 262–263, 277
    Turkey and, 176
    Ushahidi and, 258, 277, *277*
    Uzbekistan and, 227
    women's status and, 233–234, *234*, 246, 258, 268, 269, 270
Arab world, 233

Arabian horses, *244*
Arabian Plate, 13, *16*, 235, 279
Arab–Islamic empire, 246–247, 248
Arab–Israeli conflict, 260
Aral Sea, 201–202, *202*, *203*, 204, 227, 228
Ararat, Mt., 231
Archipelagos, *405*, **409**, 419, 448, 449, 462, 468
Arctic climate, *19*, 63, 199
Arctic Ocean, 162, *163*, 197
Arctic Sea, 65, 196
Argentina
    asado, *136*
    Internet use, *133*
    ISI policies, 127, **128**–129, 134, 147
    pampas, *109*, 114, 134
    UNASUR, **132**, 148
    women in parliament, 89
Arid/semiarid climates, *18*, 63, *115*, 157, 199, 236, 239, 280, 323, 366, *412*, 452
Arizona–Mexico border, 100
Armed interventions, U.S., 136
Armenia, *11*, 178, 218, 226
Arranged marriages, 440, *440*
Arunachal Pradesh, 328
Asado, *136*
ASEAN (Association of Southeast Asian Nations), 33, **425**–426, *426*, 428, *428*
ASEAN Economic Community, 425
Asia. *See* Central Asia; East Asia; North Africa and Southwest Asia; South Asia; Southeast Asia
Asia Pacific Economic Cooperative. *See* APEC
Asian Highway, 426–427, *427*
Assimilation, 112, **145**, **184**–185, *186*
Association of Southeast Asian Nations. *See* ASEAN
Astana, Kazakhstan, 218–219, *220*, 228
Aswan dams, 242
Atacama Desert, *115*
Atheism, 103, 226, 227
Ati-Atihan Festival, *438*
Atlantic Coast, Hurricane Sandy, 23, *25*, *121*
Atlas Mountains, 230, 235, 236
Atolls, *461*
Aung San Suu Ki, 430, 442
AusAID program, 448
Australia. *See also* Aborigines; Oceania
    animals and plants, *454*, 454–455
    APEC and, 468
    climate change vulnerability, 455, *456*, 480
    climate zones, 451–454, *452*
    cultural diversity, 475
    development miracle, 465–467
    FGM and, 310–311, *311*
    future global orientations, 468
    GEI, 37
    gender roles, 472, 479

I-1

Australia. *See also* Aborigines; Oceania *(Continued)*
  HDI, 37
  invasive species, 457
  physical patterns, 450–451
  power and politics, 468–470, *469*
  refugees in, 463, 469, 474
  sociocultural issues, 473–478
  urbanization, 448, 470–472, *471*, 479, 480
  women in parliament, 89
An Australian Girl in London (Mack), 473
Australo-Melanesians, **419**, 422
Austria
  income disparities, 50
  population pyramid, 47, *48*
  women at work, *187*
Austro-Asiatic language family, 55
Austronesian language family, 55
Austronesians, 55, **419**, 422, *462*
Authoritarianism
  defined, **38**
  democratic systems compared to, 38–39, 42, 56
  European imperialism and, 176, 380
Automobiles, electric, 327, 339
Autonomous regions
  internal republics, 197, 216, 228
  Ningxia Huizu Autonomous Region, 367, 399
  Xinjiang Uygur Autonomous Region, 365, 378, 398
Average population density, 44, 46, 46–47
Ayers Rock, *451*
Azerbaijan, *11*, 178, 195, 217, 218, 226, 227
Aztecs, *117*, *122*, **122**–123, *123*, 145

Baby boomers, 95, 182
Backstreet abortions, 437
Badaga ethnic group, 330
Bagan, 422
Baghdad, 268
Bahrain
  greenhouse gas emissions, 22
  women in parliament, 258
  women's rights, 234
Baikonur Cosmodrome, 207
Balkans, 155
Ballaruk Aboriginal tribe, 469
Baltic states. *See* Estonia; Latvia; Lithuania
Banana haciendas, Costa Rica, 134
Bangalore, 340, 344
Bangkok, sex workers, 443
Bangladesh
  Bengal tigers, 329
  cows incident (vignette), 343, *343*
  Joypur, 354
  microcredit, 342
  religious diversity, 333
  sea level rise, 326
  subcontinent and, 320
  urbanization, 348
Banh mi sandwich, *441*
Banqiao Dam, 372
Baptists, 103, 104, 226
Barbados, 35, 50, *143*

Barrios, 43, 139, 142, 148
Basel, Rhine River, *151*
Bateria, 145
Bauxite, 128, *135*
Bazaars, 212, 295
Bechtel, 27, 32, 120
Beijing
  air pollution, 328
  Beijing–Tibet Railway, 399
  canals, 368
  dust storms, 367
  migration to, *391*
  Olympics, 374, 386
  pro-democracy movement, 387
  traditional Chinese medicine, 372
Belarus
  authoritarianism in, 214
  Chernobyl nuclear pollution, 201, 204
  population pyramid, 222
  unemployment, 212
  women in parliament, 226
Belém, tropical wet climate zone, *115*
Belgian town festival, *166*
Belgium, 170, 172, *187*
Belief systems, 52–53, *54*, 311–312
Belt, rain, 17, 20
Belt-tightening programs, 34, 129
Benares. *See* Varanasi
Bengal Famine, 337
Bengal tiger, 329
Benguela Current, 279
Bering land bridge, 71, 72, 122, 126
Berlin Conference, 290
Bhutan
  Bengal tigers, 329
  religious diversity, 333
  subcontinent and, 320
Bhutto, Benazir, 354
Bikini Atoll, 458
Bin Laden, Osama, 86, 346
Biodiversity, 14, **114**, 117. *See also* Amazon Basin
Biopiracy, 408
Biosphere, **12**–13. *See also* Human impact on biosphere
Birth control, Philippines, 437
Birth rates, 44. *See also* Demographic transition
Black Death, 43, *46*
Black market, 212, 217
Bleaching, coral reef, **417**
Bloodhound, *165*
Bodney, Corrie, 469
Boers, 292
Bolivia
  Amazon Basin, 109, 110, 114, 117, *119*, 122, 134, 145, 148
  cocaine trafficking, 138
  Cochabamba water utility, 27, 120
  glacial melting, 148
  highlands, *121*
  income disparities, 127, *127*
  population increase (1990–2011), *143*
  undernourishment, 30
Bollywood, 348
Bolsheviks, **207**–208, 209
Bombay. *See* Mumbai

Bombings. *See* Terrorism
Bopha, typhoon, *418*
Borneo Project, 406–407
Borscht, 225
Boryeong Mud Festival, 376
Bosnia civil war, *177*
Bosnia-Herzegovina, 155, *177*, *188*
Boston, 72, 91
Botswana
  demographic transition, 306
  groundwater discoveries, 284
  HIV-AIDS, 308
  income disparities, 50
  ITCZ and, 279
  Norilsk Nickel in, 201
  refugees in, 302
  women in parliament, 89
Bottled water, 27, 68
Brahmaputra River, 23, 321, 322, 324
Brahmin priests, 327, 333, 334, 357
Brain drain, **139**
Brazil
  Amazon Basin, 109, 110, 114, 117, *119*, 122, 134, 145, 148
  BRIC countries and, 87, 131, 134, 211–212
  Candomblé in, 146, 312
  favelas, 43, **139**, *141*, 141–142
  FDI in, *130*, **131**
  income disparities, 127, *127*
  landless movement, 138–139
  population increase (1990–2011), *143*
  Rondônia cattle, *119*
  Salvador, 140
  UNASUR, **132**, 148
Brazilian Amazon, 14
Brazzaville. *See* Congo
Breadbasket, 73
BRIC countries, 87, 131, 134, 211–212
Bride price, 355
British East India Company, 333, 334, 335
British Indian Empire, 32, 334, 336, 337, 338
British Isles, 32, 463, 473, 474
British Petroleum, 32, 68, 70
Brothels, 350, 442
Brownfields, **93**, 305
Bruegel, Pieter, *166*
Brundtland, Gro Harlem, 189
Brunei, 408, 430, 436, 438
Buddhism
  Falun Gong and, 386
  in North America, 103, *104*
  Ottoman Empire and, 247
  in Russia and post-Soviet states, 197, 226
  in South Asia, 331, 333, **334**
  Tibetan, 226, 399
  in world, *54*
Buffer state, 334
Bukhara, 206
Bulgaria
  air pollution, 162
  EU membership, 176
  population decline, 44
  women at work, *187*
  women in parliament, *188*
Bull riding, *101*

Burlaks, 207, 228
Burma (Myanmar)
  ethnic groups, 429
  Gorges, 404
  government transition, 408, 429, 430
  HIV-AIDS, 437
Burning Man, *101*
Bush, George H. W., 259
Bush, George W., 86, 87, 259
Bushmeat, 287, 295
Byaslag (cheese), 371

Cacao market, 124, 289
Cahokia, 71, 72, 107
Cairo, 264, 265, 268, 273
Calcutta. *See* Kolkata
Calgary Stampede, *101*
Cambodia
  fertility rates, 436
  killing fields, 422
  Tonlé Sap Lake, *418*
  women in parliament, 89
Camel race, *244*
Cameroon, 313
Campo de Dalía, greenhouses, *161*
Canada
  abroad, 86
  agribusiness, 67, **80**–82, 99
  agriculture, 80–83, *81*
  air pollution, 64, 66, 70, 79, 91
  Arctic tundra, *19*
  China–Canada economic relations, 78
  civil society and, 87
  climate change vulnerability, 64–68, 65, 107
  climate zones, 62–64, *63*
  cultural diversity, 61, 98, 98–100, *100*
  democratic system, 87–88
  ecological footprint, 12
  ethnicity in, 101–103, *102*
  families, 103–105, *105*
  FGM and, 310–311, *311*
  First Nations people, 74–75, 91, 101
  food production, 80–82
  G8 and, 211
  GEI, 37
  genetic modification, **30**–31, 82
  geopolitics, 76, 83, 86–87
  global economic recession and, 78–79, *79*
  globalization and, 75–83
  GNI per capita PPP, 37, 76, 78, 89–90
  green revolution agriculture, 81, 106
  greenhouse gas emissions, 64, 91, 93, 106
  HDI, 37
  human impact on biosphere, 68–70, *69*
  immigration and, 98, 98–100, *100*
  income disparities, 50, 101–102
  IT industries, 76, 78, 83
  oil/gas resources, 88
  political freedoms, 83–84
  political map, 89
  population patterns, 94–98, *97*
  population pyramid, 95
  power and politics, 83–87, *84–85*

regional map, 58–59
religions, 103, *104*
social safety net, 78, 89–90, *90*, 93
terms usage, 61
transportation networks, 83
urban sprawl, 43, 70, 87, 90, 91, 92, 93, *93*, 94, *104*, 106
urbanization, 90–94, *92*, 106, 107
U.S.–Canada relationships, 79, 79–80, 87–90
voting, 83–84, 90
water issues, 64, 66, 66–68
women at work, 80, *80*, *187*
women in parliament, 89, *188*
Canadian Rockies, 58
Candomblé, 146, 312
Canton. *See* Guangzhou
CAP. *See* Common Agricultural Program
Cape Town, 43, 275, 278, 289
Capitalism
crony, **424**–425, 428, 444
defined, 39, **169**, **207**
Carbon dioxide emissions, 4, 21–22, *22*, 24, 110, 117–118, 120, 147, 374, 411
Carbon footprint, 12
Carbon sequestration, 279
Caribbean
Caribbean Plate, *16*, 113
Dominican Republic, 123, 136, *143*
FDI in, *130*
Internet usage, *133*
plantations, 31, 32, *32*, 128
Carnaval, Rio de Janeiro, *145*
Carpathian Mountains, 155, 197
Carrying capacity, **30**, 306
Cartels, 135, **249**, 251
Cartography, **4**
Cash crops, 72, 128, 132, 135, 138, 213, 282, 295, 310, 417, 421
Cash economies, **48**
Caspian Sea, 159, 197, 201, 210, 212, 217, 223
Cassava root, 283
Caste system, **332**–334, 337, 344, 347, 354–355, 357
Castles, 164, 165
Castro, Fidel, 136, 138
Castro, Raúl, 136, 138
Çatalhöyük site, *243*, 244
Catalonia, 158
Catholicism. *See* Roman Catholicism
Cats, *244*, *411*, *457*, *459*
Cattle, in Rondônia, *119*
Caucasia
Central Asian world region, *11*
defined, **223**
Jews in, 226
political instability, *215*
population density, *223*
Caucasian language family, *55*
Caucasus Mountains, *192*, 198, 217
Ceiba tree, Ecuadorian Amazon, *120*
Celebration, Florida, *94*
Cendol (cold dessert), *441*
Central African Republic, 281, 298
Central America, *111*, 118, 122, *133*
Central Asia

contacts with world regions, *196*
5000 years ago, 205
Mughal Empire, 247, 331, 332, 334, *335*, 338, 421
world region, 10–11, *11*
Central Asian states, 195, *195*, 196, 201, 210, 226. *See also* Kazakhstan; Kyrgyzstan; Tajikistan; Turkmenistan; Uzbekistan
Central Europe, *154*, 155
Central lowlands, 58, 61, 62
Central planning, **169**, 208, **216**, 218, 220, 224, 228, 378, 379, 382
Central Siberian Plateau, *193*, 197, 205
Central Valley, 60–61, *61*
Chang Jiang (Yangtze River), 23, *361*, 364, 365, 367, 368, 369, 372, 396
Chapatis, 356
Charismatic evangelical Protestantism, **147**
Charleston, South Carolina, 33, *59*, *94*
Chávez, Hugo, 131, 132, 135
Chechnya, 197, 214, *215*, 216–217, 226, 227, 228
Cheese (*Byaslag*), *371*
Chernobyl nuclear power disaster, 160, *200*, 201, 204
Cherokee, 74
Chevron, 32, 110, 111
Chiang Mai, *411*, *441*
Chiao-Min Hsieh, 371
Chiapas, 138
Chicago River, 66
Children
female infanticide, 353, 396
labor, 289, 350, 352
landscape map, 2–3, *3*
North Africa and Southwest Asia, 271
one-child policy, 47, 182, 351, 362, 394, **395**–396, 397, 402
soldiers, 282, 302
South Asia, 342, 348, 350
two-child policy, 395, 437
Chile
Allende and, 136
Atacama Desert, *115*
copper mines, 128, 293
FDI in, *130*, **131**
income disparities, 127, *127*
Santiago, *140*, *144*
UNASUR, *132*, 148
Chimpanzees, 282, 287, *287*
China
agricultural zones, 370
air pollution, *374*, 374–375
BRIC countries and, 87, 131, 134, 211–212
cultural preference for boys, 362, 394, **395**–396, 397, 402
Cultural Revolution, 379, *379*
development, 363
earthquakes, 367, 389
ecological footprint, 12
emigration, 401
empires, *378*
ethnic groups, 398, **398**–400

export-led growth, 382
famines, 369, 372, 378, 402
FDI in, 384, 390, 425
feudalism, 375, 378, 379
floating population, **364**, 384, 392
food production, 369
GDP per capita, *385*, 390
glacial melting, 367
Great Leap Forward, **379**, 383
Great Wall, 376–377, *378*, 381, 402
greenhouse gas emissions, 22, 367, 373, *374*, 374
HIV/AIDS, 396
hukou system, **363**–364, 392, 393, 402
human patterns over time, 375–377
imperialism, 375
income disparities, 127, *127*
Islam in, *247*, 247
Japanese miracle, 382
Japan's expansions (1875–1942), *380*, 380
Li Xia vignette (rural-to-urban migration), 362–363, *364*, 390, 392
Mexico–China relations, 129
Muslims, 398–399
one-child policy, 351, 362, 394, **395**–396, 397, 402
Opium Wars (1839–1860), 377, 379, 393, 394, 402
population pyramid, 396
Qin empire, 375, 378, 379, 402
regional specialization, 370, **384**
responsibility system, **384**
RNI, 395
service economies, 390, 392, 394
SEZs and, **390**, 392
Sino-Tibetan language family, *55*
temperate midlatitude climate, 156
Three Gorges Dam, *361*, 372, *373*, 374, *375*, 402
twentieth century and, 377
urbanization, 363, 390–392
U.S.–China economic relations, 78
wet rice cultivation, 370, *371*
women in parliament, 89
China's Far Northeast (Manchuria), 364, 365, *366*, 378
Chinese Communist Party, 377–379, 381, 383, 385, 389, 398
Ching bo leung (cold dessert), *441*
Chipko (tree hugging), 329, *330*
Chivu, Grigore, 152–153
Cholera, 28, 305
Christ, 245
Christianity
Baptists, 103, *104*, 226
Evangelical Christianity, 103, 226, 312
Evangelical Protestantism, **147**
monotheism, 233, **245**, 248
Orthodox Christianity, 205, *206*, 217, 224, 226, 228, 312
Protestant Reformation, 166, *166*
Protestantism, 53, *54*, 103, 175, *177*
Roman Catholicism, *123*, 123–124, 146–147
Chukchi people, 220
Cienfuegos, Camillo, 137

Circular migration, 303, 305, 314
Cisco, 32
Cities
growth of, 42–43
livability, 91, *92*, 93–94, 107
megalopolis areas, **91**
primate, **139**, *140*, 142, 148, 218, *219*, 220, 303, 432, 434
suburbs and, **91**
world, 166, 401
Civil disobedience, **335**–336, 338
Civil rights, 88, 165, *195*, 292, 476
Civil society
Canada and, 87
defined, **39**, 42, 56
NGOs and, 39, 42
Pakistan and, 352
PRSFs and, 34
sub-Saharan Africa and, 291, 293, 303, 315
Civil War, U.S., 71, 72, 73, 74. *See also specific civil wars*
Clam chowder, 99
Cleansing, ethnic, **40**
Clear-cutting, 69, *70*, 279, 406
Climate, **14**, 20
Climate change
Antarctica, *481*
coral bleaching, **417**, *418*, 444, 455, *456*
described, **21**–23
desertification, **239**–240, 286, 367, 370
hurricanes and, 24, 107
Kyoto Protocol, **24**, 204, 370
Climate change vulnerability. *See also specific countries*
described, 23–28, *24*–*25*
exposure and, 23, 24, 25, 56, 64
resilience and, 23, 24, 25, 56, 64, 65, 70, 158, 159, 171, 203, 241, 272, 285, 325, 456
sensitivity and, 23, 24, 25, 56, 65, 158, 285, 325
Climate zones (climate regions), **14**, 16. *See also specific countries*
Clinton, Bill, 90
Clinton, Hillary, 90
Clouds, 17, 20, *20*, 21, 453
Clownfish, *411*
Coal ash spill, 70
Coal mining, *15*, 69, 70, *161*, 190, 373, 465
Coastal lowlands, *59*, 62, 143, 155, 278, 287, 316, 453
Coca-Cola, 32, 262, 327
Cocaine trafficking, 138
Cochabamba water utility, 27, 120
Cocos Islands, 468
Cocos Plate, *16*, 17, 113
Cold desserts, *441*
Cold War
alliances (pre-1989), 9
Cuba and, 136, 138
defined, **169**, **208**
geopolitics and, 39–40
Iron Curtain, **169**, *208*
Russia and post-Soviet states, 208–209, *209*

Collapse, of Soviet Union, 86, 169, 175, 194, 196, 198, 201, 204, 208, 220, 221, 223, 228
Colombia
Amazon Basin, 109, 110, 114, 117, 122, 134, 145, 148
civil war, 135
drug trade, 135–136
Incas and, 122, **123**, 123, 148
income disparities, 127, 127
Santos and, 40, 57
UNASUR, **132**, 148
Colonial regimes, in North Africa and Southwest Asia, 248, 250
Colonialism, European
beginnings, 166
decolonization and, 169
divide-and-rule tactics, **299**, 300, 303, 337
mercantilism, **125**, 127, **166**, 167, 293
Middle and South America, 124–126, 126
Oceania, 463
South Asia, 334–337, 335, 336, 337
Southeast Asia, 421
sub-Saharan Africa, 290–291, 291, 292
wealth transfers, colonies to Europe, 167
Colonias, 139, 148
Color revolutions, 214, 215, 216, 218, 227, 228
Columbia River, 72
Columbus, Christopher, 111, 123
Command economy, 208, 209, 210, 383
Commandments, Ten, 245, 312
Commercial agriculture, **283**–284
Commodities, **293**
Commodity dependence, **293**
Common Agricultural Program (CAP), **173**–174, 175
Commonwealth of Independent States, 195, 196
Commonwealth of Nations, 470
Communal conflict, **343**–344
Commune system, 383
Communism
capitalism compared to, 39, 169, 207
central planning, **169**, 208, 216, 218, 220, 224, 228, 378, 379, 382
defined, **39**, 207
domino theory, **421**
Europe and, 168
Marx and, 39, 168, 169, 207
Communist Party
Chinese Communist Party, 377–379, 381, 383, 385, 389, 398
Russia and post-Soviet states, 39, **208**, 226–227
Communist Revolution, 206–208
The Communist Manifesto (Marx), 168
Comoros, 298
Condom use, 350, 436, 437, 438, 443
Confit, of duck legs, 174
Conflicts and genocides. *See specific countries*

Confucianism
East Asia, 376–377, 378, 378, 395, 396, 402
Southeast Asia, 421, 428, 439, 439, 443
in world, 54
Congo (Brazzaville/Republic of Congo), 276, 278
Congo (Kinshasa/Democratic Republic of Congo), 41, 278, 279, 302, 304, 313
Conservative welfare systems, 179, 179
Container ship industry, 159
Contested space, **138**, 148, 337
Continental climates, 19, **156**, 236, 365
Continental plates, 13, 113
Cool humid climates, 18, 63, 115, 157, 199, 236, 323, 366, 412, 452
Coon cat, Maine, 94
Copenhagen Accord, 204
Copper mines, Chile, 128, 293
Coptic Christians, Egyptian, 240
Coral bleaching, **417**, 418, 444, 455, 456
Corn production, 82, 82, 136
Corporate agriculture, 174
Corruption, Russia and post-Soviet states, 194, 196, 212, 217
Corsica, summer dry climate zone, 157
Costa Rica, 89, 133, 134
Council of the European Union, 178
Coup d'état, 39, **134**, 135, 139, 258, 262, 291, 346, 408, 430, 469, 470, 480
Cows, Hinduism and, 329, 332, 343, 343
Creoles, **125**, 148
Croatia, 153, 155, 175, 187, 188
Crony capitalism, **424**–425, 428, 444
Crowdsourcing, 258, 277
Crowley's model of urban land use, 141
Cuba
population increase (1990–2011), 143
Santería, 146, 312
U.S.–Cuba relations, 136, 138
women in parliament, 89
Cultivation, shifting, 15, 56, **122**, 126, 135, **283**
Cultural diversity
Australia, 475
Canada, 61, 98, 98–100, 100
Middle and South America, 110, 145–146
Russia and post-Soviet states, 216
South Asia, 397–398
Southeast Asia, 419–421, 440–441
sub-Saharan Africa, 312–313
U.S., 61, 98, 98–100, 100
values and, 52
Cultural geography, 4, 51, 53–54
Cultural homogenization, **154**
Cultural pluralism, **438**, 438–439
Cultural preference, for boys, 47, 49
China, 362, 394, 395–396, 397, 402
India, 353
North Africa and Southwest Asia, 266
South Asia, 358
Southeast Asia, 406, 435, 438, 444
Cultural Revolution, **379**, 379

Culture groups, 51–52
Cultures
acculturation, 74, **145**
assimilation, 112, **145**, **184**–185, 186
described, **51**–52
material, 8, 8
Curitiba, 142, 148
Cuy, 117
Cyclones, 25, 42, 322, 325, 367, 455
Cyprus, 187, 188
Czars, **206**, **206**–207, 209, 216
Czech Republic, 155, 181, 187, 188, 190

Dadaab Refugee Camp, 285
Dalai Lama, 400
Dalal Street, 340
Dalits, 333, 334, 347
Dalmatians, 165
Dams
Aswan Dams, 242
Banqiao Dam, 372
global map, 319
Haditha Dam, 238
at Manantan, 281
North Africa and Southwest Asia, 242, 242
Sardar Sarovar Dam, 318, 319, 320, 336
Southeastern Anatolia Project, 242
Three Gorges Dam, 361, 372, 373, 374, 375, 402
on Tigris and Euphrates drainage basins, 242
Danube River, 151, 155, 156
Darfur region, 241
Dark Ages, 164
Death rates, 44
Debt crises, 84, 86, 131, 172–173, 175
Deccan Plateau, 317, 321, 334
Decolonization, 169, 175, 176
Deepwater Horizon oil spill, 68, 70
Deforestation
agroforestry and, 118, **282**, 283, 287
Amazon Basin, 109, 110, 114, 117, 122, 134, 145, 148
Brazilian Amazon, 14
carbon dioxide emissions, 4, 21–22, 22, 24, 118, 117–118, 120, 147, 374, 411
ecotourism and, 118, 120, 120, 122, 460
Indonesia, 411, 414, 415
Middle and South America, 14, 117–118, 120, 122, 147
oil palm trees and, 33, 38, 118, 119
shifting cultivation, 15, 56, **122**, 126, 135, **283**
South Asia, 329–330
Southeast Asia, 411, 414, 415
sub-Saharan Africa, 279–282
tree hugging and, 329, 330
Deindustrialization, 170, 334–335
Del Monte, 133
Delta Works, 158
Deltas, 14
Democratic Republic of Congo. *See* Congo
Democratization, **38**. *See also specific countries*

Demographic dividend, 352
Demographic transition, 48, 49. *See also specific world regions*
Deng Xiaoping, 379
Denmark, 187, 188
Derry, Northern Ireland, 177
Desalination, 161, **240**
Desert climates, 19, 236, 280, 452
Desertification, **239**–240, 286, 367, 370
Detritus, **410**
Development. *See also* Ecotourism; GEI; Globalization; GNI per capita; HDI; Human impact on biosphere; *specific world regions*
described, 34–38, 56
export-led growth, **382**, 423, 444
grassroots economic development, **297**–298, 299, 470
indigenous peoples and, 112, 408
measures of, 35–36
self-reliant, **298**, 315
service economies and, 35, 38
sustainable, 36, 38, 156, 283, 300
Dhaka
sea level rise, 326
slum, 349
street vendors, 45, 57, 356, 359
urbanization, 348
Dia de los Muertos, 145
Diaspora, **245**, 296, 299, 312
Dictators, **134**, 135, 136
Digital divide, **76**, 386, 388, 388
Dingoes, 457, 459, 479
Diseases. *See also* HIV/AIDS
Black Death, 43, 46
cholera, 28, 305
malaria, 28, 284, 304, 307–308, 309, 314
Native Americans, from Europeans, 71, 123, 124
schistosomiasis, 286, 307, 309
sub-Saharan Africa, 276, 277, 278, 303, 304, 305, 307
water pollution, 27, 28
Divide-and-rule tactics, **299**, 300, 303, 337
Dividend, demographic, 352
Djenné, 288, 289
Doctors Without Borders, 40
Dogrib people, 74–75
Dolphins, Amazon River, 120
Domestication
defined, **29**
Fertile Crescent, 243, 243–244, 248, 248
Middle and South America, 117, 124, 125
North America, 71
Dominican Republic, 123, 136, 143
Domino theory, **421**
Double day, 50, 187, 224
Douz, Tunisia, 238
Dowry, 353, **355**, 358
Dravidian language family, 55
Drip irrigation, 239, 327, 456, 457
Drug trade
Middle and South America, 110, 134, 135–136, 137, 138, 139, 147

Opium Wars (1839–1860), 377, 379, 393, 394, 402
Dual economies, 293–294
Duck legs, *174*
Duck-billed platypus, 454
Dumping, 174, 384
Dumping nuclear wastes, 460
Dust Bowl era, 73, *73*, 107, 367
Dysentery, 305
Dzerzhinsk, 201

Eagle, golden, *220*
Early extractive phase, Middle and South America, 127–128, *128*
Early human species, 28–29, 288, 331, 332
Early medieval period, 164
Earth oven, Maori, *473*
Earthquakes
  China, 367, 389
  East Asia, 365, 367, 369, 372, 389
  Haiti, 2010, 42, 113
  Japan, 365, 367, 369
  plate tectonics and, 13, 61, 113, 235
  Ring of Fire and, **13**, *17*, 365, 367, 409, 451
  South Asia, 321, 322
  Sumatra, *405*, 409, 431
  Three Gorges project, 372
  tsunamis and, 42, 321, **365**, 367, 369, *405*, 409, 427, 431
  Ushahidi program and, 258, 277, *277*
Earth's orientation to sun, angle of, *17*
East Asia (world region), 360–403. *See also* China; Japan; Mongolia; North Korea; South Korea; Taiwan
  air pollution, 374, 374–375
  animals and people, 371
  climate change vulnerability, *368*, 370, 402
  climate zones, 365–367, *366*
  conflicts and genocides, 387
  Confucianism, 376–377, 378, *378*, 395, 396, 402
  countries of, 364, *364*
  development, 381–384
  earthquakes, 365, 367, 369, 372, 389
  environmental issues, 367–375
  ethnic groups, 397–398
  FDI in, 392
  food production, 369–370
  GEI, 37
  gender roles, 363, 394
  global economy, 362, 369, 386, 393, 394, 402
  globalization and, 381–384
  GNI per capita PPP, 37, 393
  HDI, 37
  human impact on biosphere, *373*, 402
  human patterns over time, 375–382
  Internet usage, 388, *388*–389
  Japanese miracle, 382
  landforms, 365
  physical patterns, 364–367
  political map, *364*
  population patterns, 394–396, *396*
  power and politics, 384–390, *387*

regional map, *360*–361
RNI, 394
Russia (and post-Soviet states) interaction with, *196*
sociocultural issues, 397–401
terms usage, 364
urbanization, 362, 363, 390–394, *391*, 401, 402
visual history, 378–379
world region characteristics, 364
Easter Island (Rapa Nui), *108*, 451, 462
Eastern Europe, 39, 155, 169, 174, 474
Eastern Hemisphere, *6, 7*
Eastern Orthodox Christianity, 205, *206*, 217, 224, 226, 228, *312*
Ecology
  agroecology, 341–342
  ecological footprint, *12*–13
  political, 36, 38
Economic and technology development zones (ETDZs), 390, 392, 394, 398, 401
Economic Community of West African States, 298, *298*
Economic core, 72–73, *75*
Economic development. *See* Development
Economic diversification
  North Africa and Southwest Asia, 239, 252–253
  sub-Saharan Africa, **252**–253
  Uzbekistan, 204
Economic issues. *See specific countries*
Economies. *See also* Global economy; Informal economies; Service economies
  cash, *48*
  command, 208, 209, 210, 383
  dual, 293–294
  formal, **35**
  free market, 39, 129, 178, 208, 210, 382, 399, 424
  knowledge, 76
  MIRAB, **466**
  of scale, *172*
  sectors of, **35**, 76
  state-aided market economies, **382**, 406, 422, 428, 443
  subsistence, 29, *48*, 450
Ecotourism, 118, 120, *120*, 122, 287, 460
Ecuador
  Amazon Basin, *109*, 110, 114, 117, 122, 134, 145, 148
  ecotourism, 118, *120*
  oil development in, 110–111
  Secoya people, 110–111, *111*
  UNASUR, *132*, 148
Education expenditures, North Africa and Southwest Asia, *254*
Egypt
  Alexandria, 164, 238, *241*, 272
  Arab Spring and, 39, 273, 277
  Coptic Christians, *240*
  FGM in, 310–311, *311*
  fossil water, 239, *239*
  Islamist movements and, 255–256
  Mubarak and, 233, 256
  mummified cat, *244*

Muslim Brotherhood, 39, 233, 255, 256
  women's rights, *234*
El Niño phenomenon, 114–**116**, *115*
El Salvador, 147
Elderly, living arrangements in U.S., *96*
Electric cars, 327, 339
Elephants, 282, 287, *287*, 329, 409
Elite urban areas, 141
Eltahawy, Mona, 232, 233–234
Emperor penguins, *481*
Emigration, *47*, 401, 432, 434, 436
Empires. *See specific empires*
Endangered species, 282, 285, 286, 287, 295, 328, 329, 373, *414*, 457, *458*, 461
Endemic species, **454**, *455*, 457
Environmental issues. *See also* Air pollution; Climate change vulnerability; Deforestation; Globalization; Human impact on biosphere; Oil spills; Urbanization; Water pollution; *specific world regions*
  described, 12–31, *56*
  indigenous peoples and, *112*, 408
Erosion, *13*–14
Escarpments, 275, *278*
Estonia, 155, 169, 172, *187*, 195
ETDZs. *See* Economic and technology development zones
Ethanol, 26, 29, 82, 118, 282, 369
Ethiopia
  FGM in, 310–311, *311*
  Homo erectus in, 288
Ethnic cleansing, **40**
Ethnic groups
  Badaga, 330
  Burma, *429*
  China, 398, *398*–400
  culture groups, 51–52, *462*
  described, **51**–52, 55
  East Asia, 397–398
  Indonesia, 430
  Iraq, 259
  Kurds, 52, 233, 235, 257, 259, 273
  Nigeria, 300
  Oceania, 462
  Philippines, *422*
  Russia and post-Soviet states, *216*
  South Asia, 330, 331, 337
  sub-Saharan Africa, 291, 292, 299–300, 312–313
Ethnic xenophobia, 175
Ethnicity
  Canada and, 101–103, *102*
  culture compared to, 51–52
  sub-Saharan Africa and, 312–313
  U.S. and, 101–103, *102*
EU. *See* European Union
EU-28, 155, 159, 160, *179*
Euphrates River, 164, 235, 239, 242, 243, 244
Eurasian Plate, *16*, *17*, 155, 197, 235, 320, 365, 409, 450
Euro (€), *35*, *172*–173
Euro zone, **35**, 172, 173
Europe (world region), 150–191. *See also* Colonialism

aging population, 152, 180, 184, 190
air pollution, 162
assimilation, **145**, **184**–185, *186*
Central Europe, *154*, 155
climate change vulnerability, 156–159, *158*, 190
climate zones, 155–156, *157*
communism and, 168
conflicts and genocides, *177*
continental climate, **156**
countries of, *154*, 155
country alliances and relationships, *9*, 9–10, *10*
culture, sources, 164–165
debt crises, 172–173, *175*
deindustrialization and, 170
eastern, 39, 155, 169, 174, 474
energy resources, 159–160
environmental issues, 156–163
feudalism, 164, 165, 170
food security, 173–175
GEI, 37
gender roles, 184, 186–189
global economy, 152, 153, 166, 170, 175, 189
globalization and, 166
green policies, 159
HDI, 37
human impact on biosphere, 159–163, *161*, 190
human patterns over time, 164–170
immigration and, 184–186
imperialism, 176, 380
Industrial Revolution, **28**, 31, 32, 111, 166–168, 189, 190, 247, 289
informal economy, *187*
landforms, 155
medieval period, 164, 165, 170, 180, 246
Mediterranean climate, **156**
migration into, 184–186, *186*
multimodal transport, 159, 160, 163
Muslims in, 184–186, *186*
North Europe, *154*, 155, 162, 175, 176, 187
nuclear power, 159, 160, 163
physical patterns, 155–156
political map, *154*
population patterns, 180–184
power and politics, 175–180, *176–177*
regional map, *150–151*
RNI, 182, 394
Russia (and post-Soviet states) interaction with, *196*
seawater pollution, 162, *163*
service economies, 170–171
social welfare/protection systems, **178**–179, *179*
sociocultural issues, 184–189
South Europe, *154*, 155, *161*, 168, 179, 184, 187, *188*
southeastern, 40, 155
subregions, 155
temperate midlatitude climate, **156**
terms usage, 155
transportation, 159, *160*, 163
urbanization, 152, 165–169, 180, *181*, 182, 189, 190

Europe (world region) *(Continued)*
  vegetation, 155–156
  virtual water, 162
  visual history, 166–167
  water pollution, 162–163, *163*
  welfare states, 152, **168**–169, 175, 179–180, 189
  West Europe, *154*, 155, 156, 170, 179, 184, 187, *188*
  western, countries, 155
  women at work, 186–189, *187*
  women in parliaments, *188*
  world region characteristics, 152–155
European conquest, of Middle and South America, *123*, 123–124
European Economic Community, 170
European Russia, 197, *198*
European settlements, North America, 71–75
European Union (EU)
  agricultural subsidies and, **173**–174
  birth of, 169–172
  CAP, **173**–174, *175*
  cultural homogenization and, 154
  described, **153**–154
  economies of scale, 172
  EU-28, 155, 159, 160, *179*
  euro, 35, **172**–173
  euro zone, **35**, *172*, 173
  European Economic Community, 170
  expansion, 175–178
  food production and, 173–175
  Gazprom and, 210
  GDP per capita, *171*
  as global peacemaker, 178, 180, 189, 190
  GNI per capita PPP, *37*, *171*
  governing institutions, 178
  greenhouse gas emissions, 152, 156, 159, 162, 163, 189
  Internet use, 170
  members of, *171*
  NAFTA compared to, *171*
  Nobel Peace Prize to, 178
  population pyramid, *182*
  trading partners, *173*
  U.S. economy compared to, 171–172
Evangelical Christianity, 103, 226, 312
Evangelical Protestantism, **147**
Evapotranspiration, 26
Export Processing Zones, **130**–131, *390*, 423
Export trading partners, Oceania, 466
Export-led growth, **382**, 423, 444
Exposure, 23, 24, 25, 56, 64
Extended families, Middle and South America, 103, **146**, *147*
External processes, 13, 14, 20
Extinction
  animals, 287, 372, *372*, 455, 457, *458*
  languages, 53, 55
Extraction
  defined, **35**
  early extractive phase, Middle and South America, 127–128, *128*

extractive resources, 34–35
resource extraction, 201–204, 293, 314, *481*
Exxon Valdez disaster, 299

Facebook, 76, 225, 258, 262
Fair trade, **34**, 106
Falafels, *240*
Falun Gong, 386
Families
  Canada, 103–105, *105*
  extended families, Middle and South America, 103, **146**, *147*
  North Africa and Southwest Asia, 267–268
  North America, 103–105, *105*
  nuclear, **103**–104, 106, 146, 441, 442
  U.S., 103–105, *105*
Family farms, 75, 80, 82, 96, 174, 186, 213, 214, 306, 432
Famines
  agriculture, 29
  Bengal Famine, 337
  China, 369, 372, 378, 402
  Dutch forced-labor system in Southeast Asia, 421, *423*
  North Korea, 383
  Ofam International, 42
  Somalia, *42*
  sub-Saharan Africa, 283, 298, 315
Far East, Russian, 196, *197*
Farmland preservation, North America, 91
Fast-food industry, 33, 78, 118, 128
Favelas, 43, **139**, *141*, 141–142
FDI. *See* Foreign direct investment
Federal Security Service, 212
Female genital mutilation (FGM), 310–311, *311*
Female infanticide, 353, 396
Female literacy, in South Asia, 353
Female seclusion, **268**–269, *269*, 350, 357, 358, 440
Feminization of labor, **423**–424, *424*
Fertile Crescent, *243*, **243**–244, 248, *248*
Fertility rates. *See* Total fertility rate; *specific countries*
Festivals, *101*, 145, 172, 224, 246, *312*, 355, 376, 438, 477
Feudalism, 164, 165, 170, 375, 378, 379, 380
FGM. *See* Female genital mutilation
Fiji
  coral bleaching, *456*
  island formation, 451
  Kirabati and, 448–449
  Melanesia and, 461
  military coups, 469, 470, 480
  Pacific Way, 468, 470
  squatter settlement, *471*
Finland, 157, 170, *187*, 207
First Nations people, 74–75, 91, 101
Fisheries, threatened, 69, 119, *161*, 238, 281, 328, 373, *414*, 456, *481*
Five Pillars of Islamic Practice, 245–246, 248

Floating population, **364**, 384, 392
Floodplains, **14**, 321, 322
Florida, 65, *94*, 97, 103, 138
Food gardens, 68, 305
Food production, 28–31. *See also specific world regions*
Food security
  corn production, Hunnicutt case, 82, *82*
  described, **29**–31, **369**
  East Asia, 369–371
  Europe, 173–175
  global recession and, 29, 82
  Industrial Revolution and, 28
  Mali, *281*
  North Africa and Southwest Asia, 272
  Oceania, 468, 479
  sub-Saharan Africa, *281*, 282, 295, 305
Foodways, 99, *136*, *174*, 225, *240*, *283*, *356*, *371*, *441*, *473*
Footprints, 12–13, 26–27, *27*
Foreign direct investment (FDI)
  Canada–U.S. interaction, 79
  China, 384, 390, 425
  defined, **131**
  East Asia, 392
  Middle and South America, *130*, **131**
  Southeast Asia, 425
  sub-Saharan Africa, 295, 296
Foreign exchange, **432**
Foreign holders, of Treasury securities, 386
Foreign political involvement, in Middle and South America, 136, 138
Forests. *See* Deforestation; Logging; Rain forests; Tropical forests
Formal economies, **35**
Fortaleza, slums in, 142
Fossil fuels, 26, **233**
Fossil water, 67, 239, *239*, 240, 242, 286
Foundation for Integrated Education and Development, 118
France
  income disparities, 127, *127*
  women at work, *187*
  women in parliament, 89, *188*
Free market economies, 39, 129, 178, 208, 210, 382, 399, 424
Free trade. *See also* NAFTA
  described, 33–34
  fair trade, **34**, 106
  FTAA, 77–78, 132
  South Asia, 340
  WTO, **34**, 211, 296, 465, 468
Free Trade Area of the Americas. *See* FTAA
Free trade zones. *See* Export Processing Zones; Mercosur
French Polynesia, *447*, 460, 468
French Revolution, *167*, 168
Freshwater availability, 237
Frontal precipitation, 20
FTAA (Free Trade Area of the Americas), 77–78, 132
Fuji, Mt., *361*, 365, 379

Fujimori, Alberto, 145
Fukushima nuclear crisis, 160, 201, 369
Fuzziness, regional borders, 10

G8, **211**
Galápagos Islands, 454
Gandhi, Mohandas, 333, 335–336, 338, 343, 347
Ganga River (Ganges River), 23, 320, 321, 322, 326, 327, 328, 355
Ganga-Brahmaputra delta, *317*, 326, 351, *351*, 354
Ganges River. *See* Ganga River
Gansu Province, 368
Garbage island, Pacific, 460
Garma Festival, 477
Garzweiler open-pit coal mine, *161*, 190
Gaza Strip, 233, 240, 246, 259–262, *261*, 263, 265
Gazprom, **210**
GDP per capita, **35**–36. *See also specific countries*
GEI (Gender Equality Index), **36**, *37*
Gender, 49–51
Gender Equality Index. *See* GEI
Gender inequalities, 50–51. *See also specific world regions*
Gender roles, **49**–50, 51, 53, 56. *See also specific world regions*
Gendered spaces, 268–269
General Motors, 78, 93
Genetic modification (GM), **30**–31, 82
Genocides, **40**. *See also specific countries*
Gentrification, **93**
Geodesy, 7
Geographic Information Science (GISc), **7**
Geographic Information Systems (GIS), 7
Geography
  cultural, 4, 51, 53–54
  defined, 3–4
  human, **3**–4
  physical, **3**–4, 13–20
  regional, 4
  where/why questions, 2–3
Geoinformatics, 7
Geomatics, 7
Geopolitics. *See also specific world regions*
  defined, **39**
  global warming and, 21
  political freedoms and, 39–40, 42
  post-Cold War period, 39–40
Georgia (country)
  conflict in, 217
  Rose Revolution, 214
  women in parliament, 226
Germany
  G8 and, 211
  iron works in, *167*
  Nazis, 169, 175, 208, 260
  open pit coal mines, *15*
  population pyramid, *182*
  women at work, *187*
  women in parliament, 89, *188*
Gers (yurts), 205, 381
Ghettos, 43

Giant panda, 372
Gift of Givers, 40, 42
Ginza, iPhone purchases, 391
GIS, 7
GISc, 7
Glacial melting. *See also specific world regions*
　Andes, 120, *121*
　Antarctica, *481*
　Bering land bridge and, 71, 72, *122*, 126
　Bolivia, 148
　China, 367
　climate change and, 23
　ice ages and, 14, 62, 71, 75, 409, *410*, 419, 461
　Tajikistan, 203
Glasnost, **208**
Glastonbury Festival of Contemporary Performing Arts, *172*
Global economic recession (2007–2009)
　Arab Spring and, 254
　California produce and, 60, 61, *61*
　Canada and, 78–79, *79*
　China and, 379
　Chinese-Mexican relations and, 129
　euro and, 172–173
　food security and, 29, 82
　foreign workers expelled from Malaysia, 32–33
　globalization issues, 31
　Gulf states and, 251
　Hawaii and, 467
　Iceland and, 176
　informal economy and, 132
　Internet purchases and, 76
　IT jobs and, 78
　Las Vegas and, 92, 96–97
　Place of the Heart vignette and, 82, *82*
　remittances and, 29, 251
　Southeast Asia and, 425, 428, 442, 443
　undernourishment (global map), 30
　U.S. and, 78–79, *79*
Global economy, **31–34**. *See also specific world regions*
Global Footprint Network's calculator, 12
Global maps
　dams, 319
　domesticated plants, Middle and South America, 124, *125*
　freshwater availability, *237*
　GEI, 37
　GNI per capita PPP, 37
　greenhouse gas emissions, 22
　HDI rank maps, 37
　human impact on biosphere, *14–15*
　human well-being, 37
　income disparities, *127*
　indigenous groups, 112, 408
　languages, 55
　metropolitan areas, *44*
　national water footprints, 27
　Pangaea breakup, *16*
　population density, *46*
　religions, 54

rugby teams, 478
skin color, 55
undernourishment, 30
Ushahidi platform users, 277, *277*
Walmart, 77
world regions, *1*
Global patterns, local lives. *See also Vignettes*
　Central Valley unemployment, 60–61, *61*
　climate change issues, Kiribati, 448–449, *449*
　Li Xia (rural-to-urban migration), 362–363, 364, 390, 392
　pig farms, Romania, 152–153, *153*
　Putin protesters, 194–195
　Sarawak forest dwellers, 406–407, *407*
　Sardar Sarovar Dam, 318, *319*, 320, 336
　Secoya people, 110–111, *111*
　Ushahidi platform, 258, 277, *277*
Global peacemaker, EU and NATO, 178, 180, 189, 190
Global power challenges, U.S., 86–87
Global scale, **11**
Global warming
　Antarctica, *481*
　defined, **21**
　polar ice caps, 23, 71, 455
Globalization. *See also specific world regions*
　described, **31–34**, 56
　free trade compared to, 33–34
　maid trade and, **432**, 434, *434*
　self-sufficiency and, 2, 31, 34, 56
GNI (gross national income) per capita, 35–36, 37, **48**. *See also specific countries*
Gold Rush, 73, *73*
Golden Comanche, Chief, *101*
Golden eagle, *220*
Gondwana, *16*, **450**, 451, 454
Google, 32, 296, 388, 389
Gorbachev, Mikhail, 208–209, 225
Gorges, Burma, *404*
Gorillas, 287
Gospel of Success, 147, 312, 313
Grameen Bank, microcredit, 341, **342**, *342*, 359
Grand Coulee Dam, *319*
Grandmother hypothesis, 50, 51
Grassroots economic development, **297–298**, *299*, 470
Gravlax, 174
Great Barrier Reef, 447, **450**, 452, 455
Great Basin, 62, *63*, 73
Great Lakes, 62, 66, 73, *74*
Great Lakes Agreement, 66
Great Leap Forward, **379**, 383
Great Plains, 66–67, 73, 74, 75, 81, *81*
Great Rift Valley, 275, 279
Great Wall, 376–377, 378, 381, 402
Great Zimbabwe Empire, 288, *288*
Greece
　ancient Greece, 164
　debt crisis, 31, 153, 172, 173
　women at work, *187*

Green Belt Movement, 283
Green living, 68
Green policies, Europe, **159**
Green revolution agriculture
　described, **29–31**
　Middle and South America, 132–134, 138, 142
　North Africa and Southwest Asia, 263
　North America, 81, 106
　South Asia, 333, 337, 340–342
　Southeast Asia, 417
Greene, Mark, 91
Greenhouse gas emissions. *See also specific countries*
　carbon footprint, 12
　Kyoto Protocol, **24**, 204, 370
　reducing, incentive, 21
　urban sprawl, 43, **70**, 87, 90, 91, 92, 93, 93, 94, 104, 106
　world, 21–23, *22*
Greenhouse gases, **21**, *21*
Greenhouses, Campo de Dalía, *161*
Grey parrots, African, 287
Gross domestic product, **35–36**. *See also specific countries*
Gross national income. *See GNI per capita*
Groundwater
　Garzweiler open-pit coal mine, *161*, 190
　North Africa and Southwest Asia, 239, 239–240, 242
　Ogallala aquifer, 67, *67*, 82
　pumping, 240
　recharge, 319
　sub-Saharan Africa, **284**, 286, 305
Group of Eight (G8), **211**
Growth poles, **141**, 390, 394
Grozny, Chechnya, 215, 217
Guangzhou (Canton), 364, 379
Guatemala
　income disparities, 127, *127*
　liberation theology and, 147
　population increase (1990–2011), *143*
　Semana Santa in, *145*
Guerrilla Gardeners, 159
Guest workers, **184**, 265, 266, 267, 395
Guevara, Che, *137*
Guiana Highlands, 114
Guianas, *145*
Guilin, 366
Gulf of Aqaba, *230*
Gulf of Mexico, Mississippi River delta and, 62
Gulf states. *See also Bahrain; Kuwait; Oman; Qatar; Saudi Arabia; UAE*
　defined, **268**
　female seclusion, **268**–269, *269*, 350, 357, 358, 440
　OPEC and, 8, 86, **249**, 251–252, 253, *253*
Gulf War, 238, 254, 259
Guyana, 132
Gyanendra, King, 345, 347, 359
Gypsies (Roma), **169**, 184

Habitat loss, *15*, 64, 68, 70, 91, 328, 411
Haciendas, **128**, 132–133, *134*, 138
Haditha Dam, 238
Hainan, 364, 365, 366
Haiti
　Carnaval in, *145*
　climate change vulnerability, 148
　earthquake, 2010, 42, 113
　Hispaniola island and, 123, 128
　Hurricane Sandy, 23, 25, *121*
　income disparities, 123
　poverty, 127
　refugees, 84
　undernourishment, 30
　U.S. assistance and, 86, 107
　voodoo in, 312
Haiyan, typhoon, 417
hajj, **245**, 246
Half the Sky, 350
Halo-halo (cold dessert), *441*
hangi, 473
Harappa culture, **331**
Hawaii. *See also Oceania*
　endangered species, 457
　Hokule'a voyaging canoe, *464*
　Honolulu, 52, *467*, 477
　island formation, 451
　Mau Piailug sailing trip, 461
　norms and values, vignette, 52
　Oceania and, 450
　population patterns, 472–473, *474*, 479, 480
　power and politics, 469, 470
　tourism issues, 457
　tropical wet climate, *18*
　urbanization, 470, *471*, 472, 479
Hazare, Anna, 344
HDI (United Nations Human Development Index), **36**, 37, 50, 127, 259
Health expenditures, North Africa and Southwest Asia, 254
Hemispheres, 6, *7*
Herat, 357
Hezbollah, 260
High-altitude climates, *18*, 63, 115, 157, 199, 323, 366, 412
Highlands, Middle and South America, 112–113
Highways
　Asian Highway, 426–427, *427*
　Interstate Highway System, 78, 83, 88, 91
　Trans-Amazon Highway, 117
Hijras, 357
Hill rice, 15
Himalaya Mountains, *316*, 321, 322, 324, 329, 330, 358, 417, 438
Hindu Kush, 321, 351
Hinduism, **331**
　caste system, **332**–334, 337, 344, 347, 354–355, 357
　cows and, 329, 332, 343
　Hindu–Muslim relationship, 343–344
　Jainism and, 331, 333, **334**, 358
　purdah practice, 350, **354**, 355, 356, 358
Hindutva, 344

Hiroshima, 382
Hispaniola island, 123, 128
History. *See* Visual history; specific *world regions*
HIV/AIDS (human immunodeficiency virus/acquired immunodeficiency syndrome)
  China, 396
  FGM and, 310–311, *311*
  population patterns, 44
  sex workers, 225, 295, 308, **350**, *357*, 437, 442, 443, 444
  Southeast Asia, 437–438, 442, 443, 444
  sub-Saharan Africa, 44, 307–309, *309*, 314, 315
Ho Chi Minh, 421
Ho Chi Minh Trail, 427, *429*
Hodder, Ian, 244
Hog farm, *153*
Hokule'a 2, voyaging canoe, *464*
Holi festival, *355*
Holocaust, **169**, 248, 254
Homo erectus, 288, *288*
Homo sapiens, 53, 288, 291
Homogenization, cultural, **154**
Honduras, 24, 100, *121*, 127, 135, *143*, 148
Hong Kong
  described, 393–394
  nanny day (vignette), 434
  South China Sea and, *361*
Honolulu, 52, *467*, 477
Horn of Africa, **279**, 285, 300
Horses, *172*, 244, 376
Hot spots, *17*, 40, **451**
Households by type, U.S., 104–105, *105*
Huang He (Yellow River), 23, 364, 365, 367, 369
Hub-and-spoke network, 83
Hukou system, **363**–364, 392, 393, 402
Humans
  early, 28–29, 288, 331, 332
  well-being, **36**–38, *37*, *39*
Human Development Index. *See* HDI
Human geography, **3**–4
Human immunodeficiency virus. *See* HIV/AIDS
Human impact on biosphere. *See also* Animals and plants; Pollution; *specific world regions*
  Anthropocene and, 12
  described, **12**–13, *14*–*15*
  early human impacts, 28–29
  ecological footprint, **12**–13
  global map, *14–15*
Human patterns over time (history). *See specific world regions*
Human Rights Watch, 227, 259, 389, 432, 434
Human well-being, **36**–37, *37*, *39*
Humanism, **165**
Hungary
  air pollution, *162*
  EU membership, 9, 176
  population decline, 44
  women at work, *187*
  women in parliament, *188*

Hunnicutt case, corn production, 82, *82*
Hurricanes
  climate change and, 24, 107
  cyclones, 25, 42, 322, 325, 367, 455
  defined, **116**
  Florida and, 65
  frontal precipitation, 20
  Gulf of Mexico, 64
  Katrina, *65*
  Louisiana wetlands and, 62
  Middle and South America, 116, 117, *121*, 148
  Mitch, 24, *121*
  Sandy, 23, *25*, *121*
  typhoons, 367, 372, 417, *418*, *419*, 444
Husbandry, animal, 28–29, 31
Hutus–Tutsis conflict, 313

Iberian Peninsula, 123, 154, 246, 247
IBM, 32
Ice ages, 14, 62, 71, 75, 409, *410*, 419, 461
Ice caps, polar, 23, 71, 455
Iceland
  EU membership, 176
  women at work, *187*
  women in parliament, 89
Iftar, *246*
Ifugao tribal people, *481*
"Il Palio" horse race, *172*
IMF. *See* International Monetary Fund
Immigration, **47**
  Canada and, 98, *98*–100, *100*
  Europe and, 184–186
  European settlements, North America, 71–75
  international immigrants, 184, 189
  push/pull phenomenon and, 42, 98, 432
  U.S. and, 98, *98*–100, *100*
Imperialism
  China, 375
  divide-and-rule tactics, **299**, 300, 303, 337
  European, 176, 380
  Japan, 377, 379, 380, *380*
  Russian, 207
Import substitution industrialization (ISI), 127, **128**–129, 134, 147, 253, 338
Incas, 122, **123**, *123*, 148
Income disparities (wealth disparities), 50, **127**, *127*. *See also specific countries*
Independence struggles, Southeast Asia, 421–422, *423*
India
  BRIC countries and, 87, 131, 134, 211–212
  climate change vulnerability, *25*
  cultural preference for boys, 353
  FGM in, 310–311, *311*
  GDP income per capita PPP, 339
  greenhouse gas emissions, 22
  IT industries, 100, 337, 339, 340
  multiculturalism and, 344

  Pakistan–India Partition, 333, 335, **336**–337, *337*, 338, 343, 344
  privatization in, 339
  religious diversity, 333
  subcontinent and, 320
  Taj Mahal, 330, 331, 332, *359*
Indian Ocean tsunami (2004), 42, 321, *405*, 409, 427, 431
Indian-Australian Plate, 13, *16*, *17*, 320, 409, 451
Indigenous groups. *See also* Aborigines; Maori people; Native Americans
  Ainu, 400, **400**–401
  Austronesians, 55, **419**, 422, 462
  Aztecs, *117*, 122, **122**–123, *123*, 145
  defined, **111**
  environmental/development issues, *112*, 408
  First Nations people, 74–75, 91, 101
  global map, *112*, 408
  Incas, 122, **123**, *123*, 148
  languages, 55
  Mayans, *109*, *117*, 145
  Occidental Petroleum and, 110
  Oceania, 462
  peopling of Middle and South America, 122–123
  religions, 54
  Sarawak forest dwellers, 406–407, *407*
  Secoya people, 110–111, *111*
  skin color map, *55*
  Zapatistas and, 138
Indo-European language family, *55*
Indo-Gangetic Plain, *317*
Indonesia
  archipelagos, 409
  deforestation, 411, *414*, 415
  ethnic groups, 430
  FGM in, 310–311, *311*
  greenhouse gas emissions, 118, 279
  immigrants, in Malaysia, 33
  Islam in, 247, 423
  Japan's expansions (1875–1942), 380, *380*
  Kalimantan, 408, *414*, 441
  Pancasila philosophy, 430–431
  political freedoms, 430–431
  population pyramid, *437*
  refugees in, 345
  Yudhoyono and, 415
Indus river, 23, 321, 322, 331
Indus Valley civilization (Harappa culture), **331**
Industrial pollution, 162, 168, 201, 204, 227
Industrial production
  brownfields, **93**, 305
  defined, **35**
  deindustrialization, 170, 334
  ISI policies, 127, **128**–129, 134, 147, 253, 338
  Mid-Atlantic economic core, 72–73
  sub-Saharan Africa, early trade, 288–289
Industrial Revolution, **28**, 31, 32, 111, 166–168, 189, 190, 247, 289
Infant mortality rates

  Austria, 48
  Canada, 90, 101
  Cuba, 138
  EU, 178
  Israel, *259*
  Jordan, 48
  Mao Zedong's government and, 378
  Middle and South America, 143
  Palestine, *259*
  South Asia, 352
  Southeast Asia, 436, *437*
  sub-Saharan Africa, 306, 310
  U.S., 90, 101, 178
Infanticide, female, 353, 396
Informal economies
  described, **35**–36, 38
  Europe, *187*
  Middle and South America, 131–132, *141*, 142, 148
  Moscow, Natasha, 212
  North Africa and Southwest Asia, 265
  Pacific Islands, 472
  Russia and post-Soviet states, 212
  sub-Saharan Africa, 295, 299
Information technology (IT) industries
  India, 100, 337, 339, 340
  outsourcing, 78, 83, 340
  quaternary sector, **35**, 76
  South Asia, 338, 348, 353
  sub-Saharan Africa, 296
  U.S. and Canada, 76, 78, 83
Infrastructure, **72**
Injera, *283*
Inner-city decay, 93, 94
Interisland travel, Oceania, 477
Internal processes, *13*, 14, 20
Internal republics, 197, 216, 228
International immigrants, 184, 189
International Monetary Fund (IMF)
  Middle and South America help, 131
  PRSPs, 34, 131, 295
  SAPs, 129, 134, 294
  South Asia and, 338–339
  Southeast Asia and, 424–425
  sub-Saharan Africa help, 295
International Space Station, 207
Internet
  Arab Spring and, 76, 258
  crowdsourcing, 258, 277
  digital divide, **76**, 386, 388, *388*
  East Asia, 388, *388*–389
  EU, 170
  knowledge economy and, 76
  mail-order bride services, 225
  Middle and South America, 132, *133*, 144
  North Africa and Southwest Asia, 258, 262, 271
  North America, 76, 83, *133*
  secret mass surveillance programs, 38–39
  Ushahidi platform, 258, 277, *277*
  Zapatista rebellion, 138
Interstate Highway System, 78, 83, 88, 91
Intertropical convergence zone (ITCZ), **279**, 287, 321, 322, 366, 410, 412, 452

Intifada, **260**
Introduced species, 71, **457**, *458*, *459*, 479
Invasive species, 282, 454, **457**, 461
iPhone purchases, Ginza, *391*
Iran
    Central Asian world region, 10, *11*
    Kurds and, 52, 233, 235, 257, 259, 273
    Nauryz celebrations, *224*
    oil and gas reserves, 249
    population pyramid, 266
    women in parliament, 89
    women's rights, *234*
Iranian Spring, 256
Iran–Iraq War (1980–1988), 254, 259
Iraq, 10–11, *11*
    ethnic groups, 259
    FGM in, 310–311, *311*
    Kurds and, 52, 233, 235, 257, 259, 273
    oil and gas reserves, 249
    women in parliament, 258
    women's rights, *234*
Iraq War, 256
Iraq–Iran War (1980–1988), 254, 259
Ireland, Northern, 53, 105, 175, 177, 179, *187*
Iron Curtain, **169**, 208
Iron works, German, *167*
Irrigation, drip, 239, 327, **456**, 457
ISI. *See* Import substitution industrialization
Islam
    defined, **233**
    monotheism, 233, **245**, 248
    North African Islamic rule of Spain, *166*
    Pillars of Islamic Practice, 245–246, 248
    spread of, 246–247, *247*
Islamic extremists
    Islamism, **233**, 255–256
    jihadists, 255, 265, **269**
    mujahedeen movement, 209, 346
    Al Qaeda, 85, 86, 346
    Salafism, 255, 263
    Taliban, 227, 346, 356–**357**, 358
Islamic fundamentalism, 209, 226, 227, 233, 346, 399
Islamism, **233**, 255–256
Israel
    Arab–Israeli conflict, 260
    drip irrigation, 239
    infant mortality rates, 259
    oil and gas reserves, 249
    population pyramid, 266
    service economy, 252
    women in parliament, 258
Isthmus, Central America, **111**, 122
IT industries. *See* Information technology industries
Italy
    G8, 211
    women at work, *187*
    women in parliament, *188*
ITCZ. *See* Intertropical convergence zone

J curve, 43–44, 46
Jablonski, Nina, 53
Jainism, 331, 333, **334**, 358
Jaisalmer, desert, 323
Jalee, 356
Jamaica
    cultural diversity, 145
    greenhouse gas emissions, 22
    women in parliament, 89
Jamestown, 71
Jammu and Kashmir region, 337
Japan
    Ainu people, *400*, **400**–401
    air pollution, 373
    demographic transition, 394
    earthquakes, 365, 367, 369
    expansions (1875–1942), *380*, **380**
    feudal lords, 380
    food production, 369
    Fukushima nuclear crisis, 160, 201, 369
    G8 and, 211
    greenhouse gas emissions, 22, 367
    guest workers in, 395
    imperialism, 377, 379, *380*, **380**
    income disparities, 50
    Japanese miracle, 382
    just-in-time systems, **382**, 383, 384
    kaizen system, **382**, 384
    monsoon rainfall patterns, 369
    population aging, 394–395, 395, 402
    RNI, 394
    samurai festival, 376
    volcanoes and, 365
    wet rice cultivation, 370, *371*
    women at work, *187*
    women in parliament, 89, *188*
Jatis, 332–334, 347, 357
Jesus of Nazareth
    Ati-Atihan Festival, *438*
    in liberation theology, **146**–147
    Semana Santa procession, 145
Jews. *See also* Israel; Judaism
    Holocaust and, **169**, 248, 254
    yarmulkes, 186
Jihadists, 255, 265, **269**
Jilin, 366
Johnson-Sirleaf, Ellen, 282, *301*, 302, 303, 310, 315
Jordan
    camel race, *244*
    GNI per capita PPP, 48
    income disparities, 50, *127*, 127
    oil and gas reserves, 249
    population pyramid, 47, 48
    women's rights, *234*
Joypur, 354
Judaism
    converts, 52
    Middle and South America, 146
    monotheism, 233, **245**, 248
    North Africa and Southwest Asia, 243, 245, 248
    North America, 103
    South Asia, 331
    sub-Saharan Africa, *311*
    in world, 54
Just-in-time system, **382**, 383, 384

Kabaddi, 355
Kabul, 15
Kaizen system, **382**, 384
Kalahari Desert, 279, 280
Kalimantan, 408, *414*, 441
Kaliningrad, 195
Kamkwamba, William, 295, 298
Kang Kek Iew, 422
Karachi
    sea level rise, 326
    urbanization, 348
Karaoke singing, 392, *393*
Kasanka National Park, 287
Kashmir
    conflict in, 344, 345, 346
    high-altitude climate, 323
    Jammu and Kashmir region, 337
Kassem, Hisham, 256
Katrina, Hurricane, 65
Katz, Cindi, 271
Kawasaki, solar power station, 368
Kayford Mountain mine site, 69
Kazakhstan
    Aral Sea issues, 201–202, *202*, *203*, 204, 227, 228
    Astana, 218–219, 220, 228
    Central Asia world region, 10–11, *11*
    Central Asian states and, 195, *195*, 196, 201, 210, 226
    golden eagle, 220
    nuclear testing in, 201
    population pyramid, 222
    steppes, 199
    women in parliament, 226
KaZantip, 224
Kenya
    Anglicans in, 312
    FGM in, 310–311, *311*
    Homo erectus in, 288
    Maathai and, 283
    Ushahidi platform and, 258, 277, *277*
    women in parliament, 89
KGB, 212, 214
Khan, Genghis, 205, 377
Khartoum, 6, 7, 11
Khmer Rouge, 422
Khoisan language family, 55
Khubz, 240
Kilimanjaro, Mount, 274, 278
Killing fields, 422
Kimchi, *371*
Kinshasa. *See* Congo
Kiribati archipelago, 448–449, *449*
Knowledge economy, 76
Koli village, 348, *350*, *350*
Kolkata (Calcutta)
    Holi festival, 355
    sea level rise, 325, 326
    urbanization, 348
Köppen classification system, 14
Koran (Qur'an), **237**
Korea. *See* North Korea; South Korea
Korean Peninsula, 365, 375, 381, 396
Korean War, 381, 442
Kosovo Liberation Army, *177*
Krill, *481*
Kristof, Nicholas, 350
Kshatriyas, 333, 334
Kühtai, Alps in, *151*

Kumbh Mela, 355
Kurds, 52, 233, 235, 257, 259, 273
Kuwait
    Central Asian world region, *11*
    income disparities, 50
    oil and gas reserves, 249
    women's rights, 234
Kuzmic, Vera, *174*, 174–175
Kyoto Protocol, **24**, 204, 370
Kyrgyzstan
    Central Asia world region, 10–11, *11*
    Central Asian states and, 195, *195*, 196, 201, 210, 226
    Nauryz festival, 224
    population pyramid, 222
    Tulip Revolution, 214
    women in parliament, 89, 226

La Paz, *121*
La Rambla, *181*
La Tomatina, *172*
Labrador retrievers, *94*
Lagarde, Christine, 131
Land bridge, Bering, 71, 72, 122, 126
Land uses, New Zealand, *459*
Landforms. *See also specific world regions*
    described, **13**, 20
    external processes, 13, 14, 20
    internal processes, 13, 14, 20
Land-for-peace formula, 262
Landless movement, 138–139
Languages, of world, 55. *See also specific countries; specific languages*
Laos, 408, 409, 419, 420, *421*, 423, 430
Las Vegas, 92, 97
Latin America, 111, *130*
Latinos, 60, **61**, 81, 93, 96, *96*, 98, 101, 102, *102*, 103
Latitude, 4
Latitude lines (parallels), 4, 6, 7
Lattice screens, 356
Latvia, 169, 172, *187*, 188
Laurasia, 16
Lebanon
    military-health-education expenditures, 254
    oil and gas reserves, 249
    oil spill, 238
    Syria and, 262, 263
    working women, 270
Legends (maps), *4*, 5
Lenin, Vladimir, 208
Lettuce harvest, Central Valley, *61*
Li Xia vignette (rural-to-urban migration), 362–363, 364, *390*, 392
Liberation theology, **146**–147
Liberia
    civil war, 282
    Johnson-Sirleaf and, 282, *301*, 302, 303, 310, 315
    Taylor and, 282, 295, 302
Libya
    Arab Spring and, 39, 273, 277
    civil war, 256
    fossil water, 239, *239*
    oil and gas reserves, 249
    women in parliament, 258
    women's rights, *234*

Lima, squatter settlement, *140*
Lines of latitude (parallels), **4**, *6*, *7*
Lines of longitude (meridians), **4**, *6*, *7*
Lingua franca, **313**
Lions, 287
Lithuania, 155, 169, *187*, *188*, 195
Liu Xiaobo, 389
Livability, 91, *92*, 93–94, 107
Living, green, 68
Living wages, **34**, 56
Llamas, 114, *117*
Llanos, 114
Local lives. *See also* Global patterns, local lives
  animals, 94, *117*, 165, 220, 244, 287, 329, 372, *411*
  festivals, *101*, 145, *172*, 224, 246, 312, 355, 376, 438, *477*
  foodways, 99, *136*, *174*, 225, 240, 283, 356, 371, 441, *473*
Local scale, 3, **11**, 12
Location (photo interpretation), 8, *8*
Lodz, industrial pollution, *161*
Loess Plateau, 367, 369
Logging
  Brazilian Amazon, *14*, 56
  clear-cutting, *69*, **70**, 279, 406
  Middle and South America, 118, 122
  North America, 68–70, *69*
  Olympic National Park, *69*
  Pacific Northwest, 73
  South Asia, 329, 330
  Southeast Asia, 406, *407*, *408*, 411, *413*, 415
  sub-Saharan Africa, 276, 279, *281*, 282, 287, 302, 314
Lomo Saltado, *136*
Longitude, **4**
Longitude lines (meridians), **4**, *6*, *7*
Louisiana wetlands, *58*, 62
Lowlands
  central, *58*, *59*, 61, 62
  coastal, *59*, 62
  Middle and South America, *109*, 113–114, *125*, 143
Luther, Martin, 166
Luxembourg, *187*, *188*

Maathai, Wangari, 283
Macedonia, 155, 176, *187*, *188*
Machismo, **146**, 147
Machu Picchu, *122*, 148
Mack, Louise, 473
Mafia, Russian, 212, 225
Magellan, Ferdinand, 422, *463*
Maharashtra, 344
Maid trade, **432**, 434, *434*
Mail-order bride services, 225
Maine coon cat, *94*
Makkah (Mecca), 245, 289
Makoko slums, *304*
Malamutes, Alaskan, *94*
Malaria, 28, 284, 304, 307–308, 309, 314
Malaysia
  Islam in, 247, *247*
  Overseas Chinese in, 401, *440*, 441, *444*
  women in parliament, *89*
Maldives Islands, 326

Mali Empire, *288*, 289
Manantan, dam at, *281*
Manchuria. *See* China's Far Northeast
Mandela, Nelson, 292
Manufacturing employment decline, North America, 75–76
Mao Zedong, 378–379
Maori people
  described, **451**, 454–455, 457
  earth oven, *473*
  haka, 477–478, *478*
  identity changes, 475
  land claims, 476
  Waitangi Day, *477*
Map projections, *6*, *7*
Maps, 4–7. *See also* Global maps; Political maps; Regional maps; Scales
Maquiladoras, 120, 129, **131**, 132, 423
Marcos, Ferdinand, 442
Mardi Gras, *101*, 145, *224*
Marianismo, **146**, 147
Marketization
  phase of economic development, 127, **129**, 131
  water, 68, 120
  of women, Russia and post-Soviet states, 225
Marlborough, drip irrigation system, *456*
Marriages, arranged, 440, *440*
Marshall Islands, *458*, 460, 461, 470, 473
Marshall Plan, 169
Marsupials, **454**, 455, *459*
Marx, Karl, 39, 168, 169, 207
Maryland, temperate midlatitude climate, *63*
Mashrabiyas, *268*
Maslenitsa festival, *224*
Material culture, 8, *8*
Mato Grosso, soy fields, *119*
Mau Piailug sailing trip, 461
Mauritania, 89, *278*, 289, 298
Mauritius, 306
Mawazine festival, *246*
Mayan temple, *109*
Mayans, *109*, *117*, 145
Media
  civil society, *39*
  North Africa and Southwest Asia, 256–258
  political freedoms, *38*
  Putin and, 217, 227
Medieval period, Europe, 164, 165, 170, *180*, 246
Medina (Al Madinah), 245
Mediterranean climates, *18*, 74, **156**, 236
Megalopolis, **91**
Megawati Sukarnoputri, 442
Meghalaya, 323
Mekong River, 409, 427
Melanesia, **462**
Melanesians, 461–462, *462*
Mella, Juan Antonio, *137*
Mercantilism, **125**, 127, **166**, *167*, 293
Mercator projection, *6*, *7*
Mercedes-Benz, 340
Mercer Quality of Living Survey, 91

Mercosur (Southern Common Market), 33, **132**, 148
Meridians. *See* Lines of longitude
Messiah, 245
Mestizos, 111, **125**, 135, 145
Metropolitan areas
  East Asia, *390*, *391*
  Europe, *181*
  global map, *44*
  Middle and South America, *140*
  North Africa and Southwest Asia, 263, *264*
  North America, 23, *91*, *92*, 98, *105*
  Oceania, *471*
  Russia and post-Soviet states, *219*
  South Asia, 348, *349*
  Southeast Asia, *432*, *433*, *434*
  sub-Saharan Africa, *304*
Mexicali workers, Thompson Electronics, 129
Mexican hairless dog, *117*
Mexico. *See also* U.S.-Mexico border
  China-Mexico relations, 129
  drug war, *137*
  FDI in, *130*, **131**
  income disparities, 127, *127*
  ISI policies, 127, **128**–129, 134, 147
  maquiladoras, 120, 129, **131**, 132, 423
  population increase (1990–2011), *143*
  Sierra Madre, *108*, 112, 113, *115*
  Teotihuacan, *122*, 148
  Yucatán, *18*, *109*
  Zapatistas and, 138
Mexico City, *45*, 114, 120, 123, 136, 139, 143, 146
Microcredit, 341, **342**, *342*, 359
Mid-Atlantic economic core, 72–73
Middle America, **111**, 118, *133*
Middle and South America (world region), 108–149
  air pollution, 120, 122
  animals, *117*
  biodiversity, **114**, 117
  Catholicism and, *123*, 123–124
  climate change vulnerability, 117–120, *121*, 148
  climate zones, 114–117, *115*
  colonial heritage of, 124–126, *126*
  conflicts and genocides, *137*
  countries of, 111, *113*
  Crowley's model of urban land use, *141*
  cultural diversity, 110, 145–146
  deforestation, *14*, 117–118, 120, 122, *147*
  demographic transition, 143
  development, 118, 124–126
  domestication, *117*, 124, *125*
  drug trade, 110, 134, 135–136, *137*, *138*, 139, 147
  economic development phases, 127–131
  ecotourism, **118**, 120, *120*, 122
  environmental issues, 117–122, *147*, 148
  European conquest of, *123*, 123–124
  extended families, *103*, **146**, 147

  FDI in, *130*, **131**
  food production, 132–134, 147
  foreign political involvement in, *136*, 138
  GEI, *37*
  gender roles, 143, 144, 146, 147, 148
  glacial melting, 118, 120, *121*, 148
  global economy, 110, 111, 126, 132, 134, 147
  globalization issues, 110, 117, 126–134, 147
  GNI per capita PPP, *37*, *135*
  green revolution agriculture, 132–134, 138, 142
  greenhouse gas emissions, 110, 117, 118, 120, 147
  haciendas, **128**, 132–133, 134, 138
  HDI, *37*
  highlands, 112–113
  human impact on biosphere, *119*, 148
  human patterns over time, 122–126
  hurricanes, 116, 117, *121*, 148
  IMF help, 131
  income disparities, 125–134, *127*, 148
  informal economy, 131–132, *141*, 142, 148
  Internet usage, 132, *133*, 144
  ISI policies, 127, **128**–129, 134, 147
  landforms, 112–114
  Latin America, 111, *130*
  lowlands, *109*, 113–114, *125*, 143
  migration, 139–142
  mineral zones, *135*
  mining, *128*, 128
  North America compared to, 111
  peopling of, 122–123
  physical patterns, 112–117
  plantations, 114, 124, 128, 133–136, *135*, 145
  political freedoms, 110, 134, 136, *137*, 139, 147, 148
  political map, *113*
  population patterns, 143, *144*
  power and politics, 134–139, *137*
  primate cities in, **139**, *140*, 142, 148
  rain forests, 114, 117–118, 122, 128
  regional map, *108*–*109*
  regional trade blocs, 111, 132, 148
  religions, 146–147
  remittances, 131, *132*
  RNI, 143, *143*
  SAPs and, 127, **129**–135, 147
  sociocultural issues, 144–147
  temperature-altitude zones, **114**, *116*, 117
  terms usage, 111
  trade routes (circa 1600), Spanish-Portuguese, *124*, *124*
  transportation, 142
  underdevelopment, 124–126
  urbanization, 107, 120, 139–143, *140*, 147–148
  visual history, 122–123
  wars of independence from Spain and Portugal, *123*, 124–125
  water pollution, 118, 120, 122

world region characteristics, 110–112
Middle East, term usage, 232, 233. *See also* North Africa and Southwest Asia
Midlatitude climates, *18*, *63*, *156*, *452*
Migration. *See also* Rural-to-urban migration
   assimilation, 112, **145**, *184*–185, *186*
   brain drain, **139**
   circular, 303, *305*, *314*
   defined, **44**
   diaspora and, **245**, 296, 299, 312
   Europe and, 184–186, *185*
   maid trade and, *432*, 434, *434*
   Middle and South America, 139–142
   to primate cities, **139**, *140*, 142, 148, 218, *219*, 220, 303, 432, 434
   resettlement schemes, *413*, **431**
Milford Sound, *447*
Military expenditures, North Africa and Southwest Asia, *254*
Mindanao, *418*, *429*
Mineral zones, Middle and South America, *135*
Ming dynasty, 377
Mining. *See specific mines*
MIRAB economy, **466**–467
Mishra, Veer Bhadra, 327
Mississippi River, *62*, *66*, 73, *74*
Mississippi River delta, *62*
Mitch, Hurricane, *24*, *121*
Mixed agriculture, **283**, *285*
Moa bird, *454*–455
Mobile phone users
   Afghan women, *358*
   iPhone purchases, Ginza, *391*
   sub-Saharan Africa, *297*
Modest welfare systems, 179, *179*
Modi, Narendra, 319
Mohenjo-daro, 332
Moldova
   authoritarianism in, 214
   EU membership and, 178
   women in parliament, *226*
Monaco, *188*
Mongolia
   byaslag (cheese), *371*
   gers (yurts), *381*
   Naadam, *376*
   sheep shearing, *372*
   steppes in, *19*, *366*
   undernourishment, *30*
Mongols, **205**, 206, 226, 247, 377
Monotheism, 233, **245**, 248. *See also* Christianity; Islam; Judaism
Monotremes, **454**, 455
Monroe Doctrine, 136
Monsoons, **20**
   South Asia, *321*, 321–322, *324*, *325*, *327*, *331*
   summer, *17*, *20*, **322**, *327*, *365*, *366*, *367*, *368*, *369*, *412*
   winter, *321*, **322**, *324*, *365*, *367*, *452*
Montserrat, Soufrière Hills Volcano, *109*
Moore, Charles, 460
Mo'orea, *447*, *451*

Morales, Evo, 131
Mormon religion, *54*, *103*, *104*
Moro Islamic Liberation Front, *429*
Morocco
   Casablanca, *264*
   climate change vulnerability, *25*
   Islam in, *247*, *247*
   Mawazine festival, *246*
   oil and gas reserves, 249
   steppe, *236*
   tagines, *240*
   Wadi El Mellah, *231*
   women in parliament, 258
   women's rights, *234*
Moscow
   informal economy, Natasha, 212
   as primate city, 218, *219*, *220*
   Putin protests, *195*
   St. Basil's Cathedral, *206*, *206*
Moses, 245
Mountain West, Pacific Coast and, 73–74, *75*
MST movement (Movement of Landless Rural Workers), 138–139
Mubarak, Hosni, 233, 256
Mud Festival, Boryeong, *376*
Mudumali Wildlife Sanctuary, 330
Mugabe, Robert, 295, 302
Mughals, 247, **331**, 332, 334, 335, 338, 421
Muhammad, Prophet, 233, 237, 245–246, 247, 248, 269
Mujahedeen, 209, 346
Mullahs, 246
Mulley, Phillip, 330
Multiculturalism
   India, 344
   North Africa and Southwest Asia, 246, *247*
   Oceania, 430, 431, *472*, *474*, *477*
   U.S., *52*
Multimodal transport, Europe, 159, 160, *163*
Multinational corporations
   defined, **32**
   living wages and, 34, 56
Multiplier effect, 72
Mumbai (Bombay)
   Bangalore, 340, 344
   climate change vulnerability, *23*
   described, 349–350
   Koli village and, 348, 350, *350*
   sea level rise, *23*, *325*, *326*
   slums, *28*
   urban migrant success stories, *43*
   urbanization, 348, *349*
Mummified cat, *244*
Muslim Brotherhood, 39, 233, 255, 256
Muslims
   in China, 398–399
   defined, **245**
   in Europe, 184–186, *186*
   female seclusion, **268**–269, *269*, *350*, *357*, *358*, *440*
   Hindu–Muslim relationship, 343–344
   purdah practice, *350*, **354**, *355*, *356*, *358*

Shi'ite (Shi'a), 245, **246**, 259, 262, 263
Sunnis, *54*, 217, **246**, 259, 263
Myanmar. *See* Burma

Naadam, *376*
NAFTA (North American Free Trade Agreement)
   African Regional Economic Communities and, 296
   APEC and, 468, 470
   defined, **75**, **83**, **132**
   EU compared to, *171*
   FTAA and, 77–78, 132
   North America and, 98–99, 106
   pollution and, 120
   push factors and, 98–99
   as regional trade bloc, 33, 296
   Thompson Electronics and, 129
   trade deficits and, 77
   Zapatista rebellion and, 138
Nagarhole National Park, 329
Nagasaki, 382
Namibia, *19*, *280*, 298, 300
Nanny day, in Hong Kong, 434
Narmada River, Sardar Sarovar Dam, 318, *319*, 320, *336*
Natasha, informal economy and, 212
National water footprints, *27*
Nationalism, **168**, 344, 346, 430
Nationalize, **131**
Native Americans
   barbecues, 99
   disease and, 71, *123*, *124*
   equal opportunity, 101
   European arrival/settlements and, 71–75
   Mardi Gras Indians, *101*
   poverty, 102–103
   reservations, 74, *75*
   Trail of Tears, 73, *74*
   voting rights, 83
NATO (North Atlantic Treaty Organization)
   Afghanistan war, 87, 345
   Arab Spring and, 178
   Cold War alliances, pre-1989, *9*
   defined, **178**
   Georgia (country) and, 217
   as global peacemaker, 178, *180*, 189, 190
   Taliban and, 346
   Ukraine and, 214
Nauru, mining issues, 457, 459
Nauryz festivals, *224*
Nazarbayev, Nursultan, 218, 220
Nazca Plate, *16*, *17*, *113*
Nazis, 169, 175, 208, 260
Negative growth, *43*, *49*
Nehru, Jawaharlal, 343
Nenets people, 220
Nepal
   Bengal tigers, 329
   political issues, 347
   subcontinent and, 320
Netherlands
   FDI to Middle and South America, *130*
   sea level rise, *155*, 239

women at work, *187*
women in parliament, *89*
New Caledonia, 451, *452*, 461
New Delhi
   air pollution, 328, 359
   bicycle rickshaw driver, *349*
   Sardar Sarovar Dam project, 318, *319*, 320, *336*
   shopping mall, *333*
New England (1600s), 72, *75*
New Guinea. *See* Papua New Guinea
New Orleans, Katrina and, 65
New Urbanism projects, *94*, *94*
New York City, 23, *25*, 74, 86, *92*, *93*, 267
New Zealand. *See also* Maori people
   animals and plants, 454–455
   APEC and, 468
   climate change vulnerability, *456*, *480*
   climate zones, 451–454, *452*
   democratization, *41*
   development miracle, 465–467
   environmental awareness, 459
   FGM and, 310–311, *311*
   future global orientations, 468
   GEI, *37*
   gender roles, *472*, 479
   HDI, *37*
   invasive species, 457
   land uses, 459
   physical patterns, 450–451
   power and politics, 468–470, *469*
   refugees in, *463*, *474*
   sociocultural issues, 473–478
   urbanization, 448, 470–472, *471*, *479*, *480*
Newroz holiday, *246*
NGOs (nongovernmental organizations)
   civil society and, 39, 42
   color revolutions and, 214, *215*, 218
   described, **40**, 42
   Somalia, 42
   Ukraine crisis and, 214
Nicaragua
   poverty, 127
   primate cities, 139
   SAPs and, 131, 134
   U.S. armed interventions, 136
   women in parliament, *89*
Nicholas II, Czar, 207
Niger, slavery in, *290*
Niger River Delta, *275*, 299
Niger-Congo language family, *55*
Nigeria
   Anglicans in, 312
   conflict in, 299
   ethnic groups, *300*
   population pyramid, 306–307, *308*
Nike, 34, 424
Nile Delta, 235, 239, *241*, 242
Nile River, *231*, *242*, 272, 278
Nilgiri Hills, Mudumali Wildlife Sanctuary, 330
Nilo-Saharan language family, *55*
Ningxia Huizu Autonomous Region, 367, *369*, 399
Nishikigoi, *372*
Nitrogen dioxide, 66

Index

Nobel Prizes
  Aung San Suu Ki, 430, 442
  EU, 178
  Liu Xiaobo, 389
  Maathai, Wangari, 283
  Mandela, Nelson, 292
  Sen, Amartya, 36
  Yunus, Muhammad, 341, 342
Nobility, serfs and, 165
Nogales, Mexico, *121*
Nokia, 170
Nomadic pastoralists, **205**, 375, 383
Non bread, 225
Nongovernmental organizations. *See* NGOs
Nonnative species, 70, 101, 282, 448, 455, 457, 461, 479
Nonpoint sources of pollution, **201**
Nonviolence, civil disobedience, 335–336
Norilsk, *200*, 201, 228
Norms, cultural diversity and, 52
North Africa and Southwest Asia (world region), 230–273. *See also* Arab Spring movement; Islam
  agriculture, 243–246
  children and, 271
  climate change vulnerability, 239–240, *241*, 272
  climate zones, 236
  colonial regimes in, 248, *250*
  conflicts and genocides, 257
  countries of, 232, 233, *233*
  cultural preference for boys, 266
  dams, 242, *242*
  desertification, 239–240
  development, 249–253
  economic diversification, 239, 252–253, **252–253**
  education expenditures, *254*
  environmental issues, 237–242
  family values, 267–268
  Fertile Crescent, *243*, **243**–244, 248, *248*
  food production, 239
  GEI, *37*
  gender roles, 244, 268–271
  gendered spaces, 268–269
  geopolitical issues, 239, 248, 249, 258–263
  global economy, 232, 256, 263, 265, 271
  globalization issues, 249–253
  GNI per capita, *37*
  green revolution agriculture, 263
  groundwater, 239, 239–240, 242
  HDI, *37*
  health expenditures, *254*
  human impact on biosphere, *238*, 272
  human patterns over time, 243–248
  informal economy, 265
  Internet and, 258, 262, 271
  landforms, 235–237
  media and, 256–258
  Middle East compared to, 232, 233
  military expenditures, *254*
  multiculturalism, 246, 247
  oil/natural gas resources, 249–252, *251*
  OPEC, 8, 86, **249**, 251–252, 253, *253*
  patriarchal attitudes, 262, 267, 268, 270, 271
  physical patterns, 235–237
  political map, *233*
  population patterns, 265–267, *267*
  power and politics, 254–263
  regional map, *230–231*
  Russia (and post-Soviet states) interaction with, *196*
  sociocultural issues, 267–271
  terms usage, 233
  urbanization, 232, 263–265, *264*, 271, *273*
  vegetation, 235, 237
  virtual water, 240
  visual history, 248–249
  water issues, 237, *237*, 239
  women at work, 270
  women's status, 269–271
  world region characteristics, 232–234
North African Islamic rule, of Spain, *166*
North America (world region), 58–107
  agriculture, 80–83, *81*
  air pollution, 66
  climate change vulnerability, 64–68, *65*, 107
  climate zones, 62–64, *63*
  components of, 89
  demographic transition, 95
  development and, 75–83
  environmental issues, 64–70, 106, 107
  European settlements, 71–75
  families, 103–105, *105*
  food production systems, 80–82
  gender roles, 104, 110
  genetic modification, **30**–31, 82
  geopolitics, 76, 83, 86–87
  global economy, 76, 80, 82, 83, 106
  globalization and, 75–83
  green living, 68
  green revolution agriculture, 81, 106
  greenhouse gas emissions, 22, 64, 91, 93, 106
  habitat loss, 64, 68, 70, 91
  human impact on biosphere, 68–70, *69*
  human patterns over time, 70–75
  landforms, 61–62
  manufacturing employment decline, 75–76
  metropolitan areas, 23, **91**, 92, 98, 105
  Middle and South America compared to, 111
  peopling of, 71, 75
  physical patterns, 61–63
  political issues, 83–87
  political map, 89
  population patterns, 94–98, *97*
  power and politics, 83–87, *84–85*
  regional composition, 75
  regional map, *58–59*
  religions, 103, *104*
  RNI, 97
  service economies, 76
  service sector and, **35**
  sociocultural issues, 98–105
  temperate midlatitude climate, 156
  terms usage, 61
  transportation networks, 74, 83
  urban sprawl, 43, **70**, 87, 90, 91, 92, 93, 93, 94, 104, 106
  urbanization, 90–94, 92, 106, 107
  visual history, 72–73
  water pollution, 66
  world region characteristics, 60–61
North American Free Trade Agreement. *See* NAFTA
North American Plate, *16*, 61, 113
North Atlantic Drift, **156**, *157*
North Atlantic Treaty Organization. *See* NATO
North Europe, *154*, 155, 162, 175, 176, 187
North European Plain, *151*, 155, 156, 157, 162, 166, *192*, 197
North Korea
  air pollution, 373, 375
  famines, 383
  greenhouse gas emissions, 367
  Japan's expansions (1875–1942), 380, *380*
  Korean War, 381, 442
  monsoon rainfall patterns, 369
  nuclear weapons, 387
  undernourishment, *30*
  wet rice cultivation, 370, *371*
Northern Hemisphere, 6, *7*
Northern Ireland, 53, 105, 175, *177*, 179, *187*
Northern settlements, colonial U.S., 72–73
Norway
  EU and, *173*, 176
  Schengen Accord, 184
  women at work, *187*
  women in parliament, *188*
Nuclear families, **103**–104, 106, 146, 441, 442
Nuclear pollution, *112*, 201, 457, 460
Nuclear power
  Chernobyl disaster, 160, *200*, 201, 204
  Europe, 159, 160, 163
  Fukushima disaster, 160, 201, 369
  North America, 64
  renewable energy and, 26
Nuclear weapons
  on Hiroshima/Nagasaki, 382
  India, 346
  Kashmir conflict and, 346
  Kazakhstan, 201
  North Korea, 387
  Oceania, testing, *458*
  Pakistan, 346
  Southeast Asian Nuclear Weapons-Free Zone Treaty, 425
  terrorists and, 217
Nunavut territory, 74, 90

Obama, Barack, 86, 87, 89, 90, 103, 341
Obama administration, 87
Obamacare, 84, 89–90
Occidental Petroleum, 110
occupied Palestinian Territories (oPT), 232, **233**, 245, 260–262, *261*, 262, 271
Ocean currents
  Benguela Current, 279
  Peru Current, 114, *115*, 279
  Roaring Forties, **451**, *452*, 455
Oceania (world region), 446–480. *See also* Aborigines; Australia; Hawaii; Maori people; New Zealand; Pacific Islands; Papua New Guinea
  animals and plants, *454*, 454–455
  APEC and, 468
  climate change vulnerability, 455, *456*, 480
  climate zones, 451–454, *452*
  conflicts and genocides, *469*
  continent formation, 450
  countries of, *449*, 450
  culture groups, 462
  development, 464–468
  El Niño and, 114–**116**, *115*
  environmental issues, 455–461
  European colonization, 463
  export trading partners, *466*
  future global orientations, 468
  GEI, *37*
  gender roles, 472, 479
  geopolitics, 460
  global economy, 419, 457
  global relationships, 463–464
  globalization issues, 464–468
  GNI per capita, *37*
  HDI, *37*
  human impact on biosphere, *458*, 480
  human patterns over time, 461–464
  indigenous groups, *462*
  interisland travel, 477
  island formation, 451
  languages in, 476–477
  multiculturalism, 430, 431, 472, 474, 477
  peopling of, 461–463
  physical patterns, 450–455
  political map, *449*
  population patterns, 472–473, *474*
  power and politics, 468–470, *469*
  regional map, *446–447*
  sea level rise, *449*, 455, *456*
  service economies, 448, 465–466, 470, 472, 474, 479
  sociocultural issues, 473–478
  sports and, 477–478
  terms usage, 450
  tourism and, 460, 467–468
  urbanization, 448, 470–472, *471*, 479, 480
  virtual water, 455
  visual history, 464–465
  world region characteristics, 448–451
Oceanic plates, 13, 112–113
Offshore outsourcing, 78, 83, **340**, 424

Ogallala aquifer, 67, 67, 82
Ogoni people, 299
Oil palm plantations, 33, 38, 118, *119*
Oil spills
   Deepwater Horizon oil rig, 68, 70
   Exxon Valdez, 299
   Gulf War (1991), 238
   photo interpretation, 8, *8*
   Russian arctic, 200
   trans-Alaska pipeline, 68, 69, 107
Oil/natural gas resources. *See also*
    specific world regions
   Ecuador, 110–111
   OPEC, 8, 86, **249**, 251–252,
    253, *253*
   photo interpretation, 8, *8*
Ok Tedi Mine, 457, 459, *460*
Okonjo-Iweala, Ngozi, 302
Old urban core, *181*
Oligarchs, **210**, 212
Olympic National Park, 69
Olympics, Beijing, 374, 386
Oman
   oil and gas reserves, 249
   women's rights, 234
One-child family, 47, 182
One-child policy, China, 351, 362,
   394, 395–396, 397, 402
OPEC (Organization of Petroleum
   Exporting Countries), 8, 86,
   249, 251–252, 253, *253*
Opium Wars (1839–1860), 377, 379,
   393, 394, 402
oPT. *See* occupied Palestinian
   Territories
Orange Free State, 292
Orange Revolution, 214, 215
Orangutans, 406, 409, 411, *414*, 444
Organic agriculture, 31
Organically grown, **82**, 175
Organization of Petroleum Exporting
   Countries. *See* OPEC
Orographic rainfall, **17**, 20, 235,
   267, 453
Orthodox Christianity, 205, 206, 217,
   224, 226, 228, *312*
Ottoman Empire, 164, **247**–248, *249*,
   250, 260, 272
Outsourcing, 78, 83, **340**, 424
Overseas Chinese, 401, 440, 441, 444
Owens Valley, *20*
Ownership, of water, 27, 28, 329
Oxfam International, 40, 42, 111

Pacific Coast, Mountain West and,
   73–74, *75*
Pacific garbage island, 460
Pacific Islands. *See also* Oceania
   animals and plants, 454–455
   economic change in, 466–467
   El Niño and, 114–**116**, *115*
   globalization issues, 457, 459
   Great Barrier Reef and, *447*, 450,
    452, 455
   informal economy, 472
   Japan's expansions (1875–1942),
    380, *380*
   Pacific Way, 448, **468**–470, 472
   subsistence affluence and, **467**, 479
Pacific Mountain Zone, *193*

Pacific Ocean, Oceania and, 450
Pacific Plate, *16*, *17*, 61, 197, 409, 451
Pacific Rim, 97
Pacific Ring of Fire, 13, 17, 365, 367,
   409, 451
Pacific Way, 448, **468**–470, 472
Paddy rice, 415
Pahiyas Festival, *439*
Pakistan
   Central Asia world region,
    10–11, *11*
   civil society and, 352
   India-Pakistan Partition, 333, 335,
    336–337, *337*, 338, 343, 344
   Indus Valley civilization, 331
   population pyramid, 352, *353*
   religious diversity, 333
   subcontinent and, 320
Palace, of Peter the Great, 206
Palau, 450, 467, 470
Palestine, 234, 259, *259*–263, *261*
Palestinian Territories, 233. *See
   also* occupied Palestinian
   Territories
Palm Jumeirah, 251, *264*, 265
Pamirs, 203, *204*
Pampas, Argentina, *109*, 114, 134
Panama, 136
Panama Canal, 128
Pancasila philosophy, 430–431
Panda, giant, 372
Pangaea, 13, *16*, 61, 278, 450
Papua New Guinea. *See also* Oceania
   animals and plants, 454
   APEC and, 468
   crime, 469
   Melanesians, 461–462, *462*
   mining, 457, 459, *460*
   MIRAB economy, 466–467
   physical patterns, 450–453
   pidgin language, 477
   population patterns, 472–473, 479
   power and politics, 469
   refugees in, 345
   urbanization and, 470, *471*, 472, 479
Parachuting, *44*
Paraguay, 22, 127, 132, 136, 139
Parakeets, *165*
Parallels. *See* Lines of latitude
Paramilitaries, 40
Parilla, 136
Parliaments. *See* Women at work
Parrots, *165*, 287, 459
Partition
   colonizing of Africa, Berlin
    Conference, 290
   India–Pakistan, 333, 335, **336**–337,
    337, 338, 343, 344
   Plan of Partition with and
    Economic Union, 260, *261*
Pastoralism, **286**
Patriarchal attitudes, **268**. *See also*
   Confucianism
   Abraham, 245
   North Africa and Southwest Asia,
    262, 267, 268, 270, 271
   Pacific Way and, 468
   Tibet, 400
Peacemaker, EU and NATO, 178,
   180, 189, 190

Penguins, emperor, *481*
Peopling
   of Middle and South America,
    122–123
   of North America, 71, 75
   of Oceania, 461–463
   of sub-Saharan Africa, 288
Perestroika, **208**
Permafrost, **197**, 198
Persian Gulf
   fossil fuels, 233
   OPEC and, 8, 86, **249**, 251–252,
    253, *253*
   seawater desalination, 240
Peru
   Amazon Basin, *109*, 110, 114, 117,
    119, 122, 134, 145, 148
   cocaine trafficking, *138*
   earthquake, 2010, 113
   income disparities, 127, *127*
   Lima, squatter settlement, *140*
   Lomo Saltado, 136
   population increase (1990–2011),
    143
   UNASUR, *132*, 148
Peru Current, 114, *115*, 279
Peter the Great, Czar, 206
Philippine archipelagos, *405*, 419
Philippine Plate, *16*, *17*, 409, 451
Philippines
   birth control, 437
   ethnic groups, *422*
   income disparities, 127, *127*
   Magellan and, 422, 463
   maid trade and, 434
Phoenix, urban sprawl, 93
Photogrammetry, 7
Photograph interpretation, 8, *8*
Physical geography, **3**–4, 13–20. *See
   also* Geography
Physical patterns. *See specific
   world regions*
Pidgin language, **477**
Pig farms, Romania, 152–153, *153*
Pilgrimages, 245, 327
Pillars of Islamic Practice, 245–246, *248*
Pinochet, Augusto, 136
Piracy, 178, *408*
Pirozhki, 225
Pizarro, Francisco, 123
Place of the Heart (vignette), 82, *82*
Plains, Botswana, *274*
Plan of Partition with and Economic
   Union, 260, *261*
Plant cultivation, 28–29, 31. *See also*
   Animals and plants
Plantations
   Caribbean, 31, *32*, 32, 128
   defined, **128**
   haciendas, 128, 132–133, 134, 138
   Middle and South America,
    114, 124, 128, 133–136,
    135, 145
   multiplier effect and, 72
   oil palm, 33, 38, 118, *119*
   Southern colonies, U.S., 71–72,
    72, 81, 107
Plate tectonics
   African Plate, 13, *16*, 61, 155, 235,
    409, 410

   Antarctic Plate, *16*, 17
   Arabian Plate, 13, *16*, 235, 279
   Caribbean Plate, *16*, 113
   Cocos Plate, *16*, *17*, 113
   continental plates, 13, 113
   defined, **13**
   earthquakes and, 409
   Eurasian Plate, *16*, *17*, 155, 197,
    235, 320, 365, 409, 450
   Indian–Australian Plate, 13, *16*,
    17, 320, 409, 451
   Nazca Plate, *16*, *17*, 113
   North American Plate, *16*, 61, 113
   oceanic plates, 13, 112–113
   Pacific Plate, *16*, *17*, 61, 197,
    409, 451
   Pangaea and, 13, *16*, 61, 278, 450
   Philippine Plate, *16*, *17*, 409, 451
   Ring of Fire and, **13**, 17, 365, 367,
    409, 451
   South American Plate, *16*, *17*, 113
   subduction zones, **112**–113
Plateau continent, 13
Plateau of Tibet. *See* Tibet
Platypus, duck-billed, 454
Poaching, 287, *287*
Podrabinek, Alexandr, *195*
Poland
   EU membership, 9
   income disparities, 50
   Solidarity movement, 175–176
   women at work, 187
   women in parliament, 89, *188*
Polar ice caps, 23, 71, 455
Political ecology, **36**, 38
Political freedoms, **38**–42, *40–41*.
   See also Geopolitics; *specific
   countries*
Political maps
   East Asia, 364
   Europe, 154
   Middle and South America, 113
   North Africa and Southwest Asia,
    233
   North America, 89
   Oceania, 449
   Russia and post-Soviet states, *195*
   South Asia, 320
   Southeast Asia, 409
   sub-Saharan Africa, 278
Pollution. *See also* Air pollution;
   Water pollution
   industrial, 162, 168, 201, 204, 227
   nonpoint sources of, **201**
   nuclear, 112, 201, 457, 460
Polygyny, 270, **310**, *310*
Polynesia
   described, 462, **462**–463, *464*,
    464, 479, 480
   French Polynesia, *447*, 460, 468
Population patterns. *See also specific
   world regions*
   aging, 95, 152, 180, 184, 190, 352,
    394–395, *395*, 402
   average population density, 44, *46*,
    46–47
   described, 43–49, 56
   floating population, **364**, 384, 392
   global, 43–49
   J curve, 43–44, *46*

Population pyramids
  Austria, 47, 48
  Belarus, 222
  Canada, 95
  China, 396
  defined, 47
  EU, 182
  Germany, 182
  Indonesia, 437
  Iran, 266
  Israel, 266
  Jordan, 47, 48
  Kazakhstan, 222
  Kyrgyzstan, 222
  Nigeria, 306–307, 308
  Pakistan, 352, 353
  Qatar, 266
  Russia, 222
  South Africa, 306–307, 308
  Sri Lanka, 352, 353
  Sweden, 182
  U.S., 95
Populist movements, 135, **146**
Portugal
  Portuguese–Spanish trade routes (circa 1600), 124, *124*
  wars of independence from Spain and Portugal, 123, 124–125
  women at work, 187
  women in parliament, 188
Post-independence period, South Asia, 337
Potatoes, 123, *123*, 124, *125*
Potjiekos, 283
Potts, Deborah, 303
Poutine, 99
Poverty
  digital divide, **76**, 386, 388, *388*
  income disparities, 127, *127*
  microcredit and, 341, **342**, *342*, 359
  populist movements, 135, **146**
Poverty Reduction Strategy Papers. *See* PRSPs
Power and politics, 38–42, 40–41, 56. *See also specific countries*
PPP (purchasing power parity), **35**
Prague, Czech Republic, *181*, 190
Precipitation
  acid rain, **64**, 66, 79, 119, 161, *161*, 162, 200, 238, 281, 328, 330
  defined, **17**, 20
  frontal, 20
  Middle and South America, 114
  orographic rainfall, **17**, 20, 235, 267, 453
  rain shadows, 20, 114, *115*
Predjama Castle, *164*
Primary sector, **35**
Primate cities, **139**, 140, 142, 148, 218, *219*, 220, 303, 432, 434
Private spaces, 268, *268*, 269
Privatization
  defined, **129**, **210**
  IMF loans and, 34
  India and, 339
  Norilsk Nickel and, 201
  in Russia and post-Soviet states, 210, 212, 217

  SAPs and, 129, 131
  water, 27, 28, 329
Pro-democracy movement, 387
Projections, map, 6, *7*
Prostitution, 225, 289, 308, 377, 396, 443
Protestantism. *See* Christianity
PRSPs (Poverty Reduction Strategy Papers), 34, 131, 295
Public spaces, 268, *268*, 269, 271
Puerto Misahualli, 118
Puerto Rico, *123*, *126*
Pull/push phenomenon, **42**, 98, 432
Purchasing power parity, **35**
Purdah, 350, **354**, 355, 356, 358
Push/pull phenomenon, **42**, 98, 432
Pussy Riot band, 194, *195*
Putin, Vladimir, 194–195, 214, 217, 227
Pyroclastic flows, 113

Qaddafi, Muammar, 240
Al Qaeda, 85, 86, 346
Qatar
  oil and gas reserves, 249
  population pyramid, 266
  women's rights, 234
  working women, 270
Qin empire, 375, 378, 379, 402
Quaternary sector, **35**, 76
Québec, 88, 99, 103
Qur'an (Koran), **237**

Race
  cultural geography and, 3
  defined, **53**
  Middle and South America, 145–146
  skin color, 53, *55*, 101, 145–146
Racism, 88, 103, 473–474
Radio Sahar, 357
Railroad, Trans-Siberian, 218, *221*
Rain
  acid, **64**, 66, 79, 119, 161, *161*, 162, 200, 238, 281, 328, 330
  orographic rainfall, **17**, 20, 235, 267, 453
  shadows, 20, 114, *115*
Rain belt, 17, 20
Rain forests
  Middle and South America, 114, 117–118, 122, 128
  Oceania, 454, 457
  Southeast Asia, 411, 415
Rainwater harvesting, 120, 319
Ramadan, 245, 246, 262, 312, 356
Rapa Nui. *See* Easter Island
Rate of natural increase (RNI)
  China, 395
  described, 44, 46–47
  East Asia, 394
  Europe, 182, 394
  Japan, 394
  Middle and South America, 143, *143*
  North America, 97
  Southeast Asia, 435
  sub-Saharan Africa, 306
Rawalpindi, slum area, 328

Recession. *See* Global economic recession
Red Crescent, 40
Red Cross, 40
Red Sea, 230, 232, 235, 245, 279
Reed Dance, *312*
Reformation, Protestant, 166, *166*
Refugees
  in Australia, 463, *469*, 474
  Chechen, 217
  in Europe, 184
  Haitian, 84
  in Kenya, 285
  in New Zealand, 463, 474
  North Africa and Southwest Asia, 265
  Palestinian, 259–260, 265
  Rohingya, 429
  in Russia and post-Soviet states, 225
  from Sri Lanka, 330, 345, 347
  in Sub-Saharan Africa, 284, 285, 300, *301*, 302, 303
  Sudanese, *241*, 301
  Syrian, *41*, 57
  Tamil, 330
  in Uganda, 25
  in U.S., 99
  in Yemen, *241*, 272
  from Zimbabwe, 302
Regional conflicts, South Asia, **344**–347
Regional Economic Communities, sub-Saharan Africa, 296–297, 298
Regional geography, 4
Regional maps
  East Asia, 360–361
  Europe, *150–151*
  Middle and South America, *108–109*
  North Africa and Southwest Asia, *230–231*
  North America, *58–59*
  Oceania, *446–447*
  Russia and post-Soviet states, *192–193*
  South Asia, *316–317*
  Southeast Asia, *404–405*
  sub-Saharan Africa, *274–275*
Regional scales of analysis, 3, 4, **11**, 12
Regional self-sufficiency, **384**
Regional specialization, 370, **384**
Regional trade blocs. *See* Trade blocs
Regions, 9–12. *See also* Climate zones; World regions
Reindeer, 220
Religions. *See also specific countries; specific religions*
  belief systems, 52–53, *54*, 311–312
  FGM and, 310–311, *311*
  religious nationalism, **344**, 346
  of world, 52–53, *54*
Remittances
  China, 395
  defined, 35
  global recession and, 29, 251
  Middle and South America, 131, *132*
  MIRAB economy, **466**
  Pacific Islanders, 466

sub-Saharan Africa, 296, 299
Taiwan, 432
Renewable energy sources, 26, 68, 159, 162, 210, 252, 327, 369, 455, 461
Republic of Congo. *See* Congo
Republics, internal, 197, 216, 228
Reservations, Native Americans, 74, 75
Resettlement schemes, *413*, **431**
Resilience, 23, 24, 25, 56, 64, 65, 70, 158, 159, 171, 203, 241, 272, 285, 325, 456
Resource extraction, 201–204, 293, 314, *481*
Responsibility system, **384**
Revolutionary War, 73
Revolutions. *See specific revolutions*
Rhine River, *151*, 155
Rhodesian ridgeback, 287
Rice cultivation, wet, **370**–371, 415
Ring of Fire, **13**, *17*, 365, 367, 409, 451
Rio de Janeiro, 43, *109*, 141, *145*
RNI. *See* Rate of natural increase
Roaring Forties, **451**, *452*, 455
Robinson projection, 6, *7*
Rocky Mountains, 58, 61, 62, 73
Rodeo, bull riding, *101*
Rohingya refugees, 429
Roma (Gypsies), **169**, 184
Roman Catholicism, *123*, 123–124, 146–147
Romania
  EU membership, 152, 176
  pig farms, 152–153, *153*
  population decline, 44
  women at work, 187
  women in parliament, 188
Rome, ancient, 166
Rondônia, cattle, *119*
Rose Revolution, 214
Rose-ringed parakeet, *165*
Ross Sea, *481*
Rotich, Juliana, 277
Roy, Beth, 343
Rub'al Khali dunes, 230, 235
Rudimentary welfare systems, 179, *179*
Rugby, 478
RugMark, 350
Rural-to-urban migration
  brain drain, **139**
  demographic transition and, 48
  Li Xia and, 362–363, 364, 390, 392
  North Africa and Southwest Asia, 263, 268
  sub-Saharan Africa, 276, 303, 305, 314
  urbanization and, 42
Rus people, Slavs and, 205
Russia (Russian Federation)
  BRIC countries and, 87, 131, 134, 211–212
  continental climate, 19
  defined, **197**
  G8 and, 211
  greenhouse gas emissions, 22
  income disparities, 50
  population pyramid, 222
  Russian Empire, 195, 205–206, 207, 208, 209, 226
  women in parliament, 89, 226

Russia and post-Soviet states (world region), 192–229
  agriculture in, 212–213, *213*
  air pollution, 198, 201, 202
  Central Asian states, 195, *195*, 196, 201, 210, 226
  climate change vulnerability, 203, 228
  climate zones, *199*
  Cold War and, 208–209, *209*
  conflicts and genocides, *215*
  corruption, 194, 196, 212, 217
  countries of, *195*, 197
  cultural diversity, 216
  democratization/conflict issues, 214–217, *215*
  demographic transition, 220
  economic reforms, 194, 210, 213, 227
  environmental issues, 198–204
  ethnic groups, *216*
  European Russia, 197, 198
  food production, 212–213
  GDP per capita, 212
  GEI, 37
  gender inequalities, 221–223
  geopolitics, 195–196, *196*, 208, 209, 210
  glacial melting, 203, 204
  global economy, 211, 227, 228
  globalization issues, 209–212
  GNI per capita, 37
  HDI, 37
  human impact on biosphere, 200, 228
  human patterns over time, 205–209
  industrial areas and transport routes, *221*
  informal economy, 212
  landforms, 197
  nuclear pollution, 201
  oil/gas resources, 200, 209–212, *211*
  physical patterns, 197–198
  political map, *195*
  population patterns, 221–224
  power and politics, 214–218, 227
  primate cities in, 218, 223
  privatization in, 210, 212
  regional map, *192–193*
  religious revival, 226–227
  resource extraction, 201–204
  Russian Far East, 196, 197
  sociocultural issues, 224–227
  terms usage, 197
  transport routes and industrial areas, *221*
  urban pollution, 201
  urbanization, 194, 218–220, *219*, 227, 228
  vegetation, 198
  visual history, *206–207*
  women's status, 224–226
  world region characteristics, 194–197
  world regions' interaction with, *196*
  World War II and, 208–209
Russian arctic, oil spill, *200*
Russian Empire, 195, 205–206, *207*, 208, 209, 226

Russian Far East, 196, 197
Russian imperial expansion (1300–1945), *207*
Russian Mafia, 212, 225
Russification, **216**, 220

Saarschleife, uplands, *151*
Sahara Desert, 230, 276, 278
Sahel region, **276**, 279, 282, 286, 287, 288, 289, 312
Salafism, **255**, 263
Salinization, **239**, 240, 242, 341
Salvador, Brazil, *140*
Samurai festival, *376*
Sandy, Hurricane, 23, *25*, *121*
Santería, 146, 312
Santiago, Chile, *140*, 144
Santos, Juan Manuel, 40, 57
SAPs (structural adjustment policies)
  IMF and, 129, 134, 294
  Middle and South America, 127, **129**–135, 147
  PRSPs and, 34, 131, 295
  South Asia, 338–339
  sub-Saharan Africa, 294–295, 299
Sarajevo, *177*
Sarawak forest dwellers, 406–407, *407*
Sardar Sarovar Dam, 318, *319*, 320, 336
Satay (food dish), *441*
Saudi Arabia
  Arabian horses, *244*
  desert climate zone, *236*
  income disparities, *50*
  Makkah (Mecca), 245, 289
  oil and gas reserves, 249
  women in parliament, 89
  women's rights, *234*
Scale, economies of, **172**
Scale bar, 4, *5*
Scales (map scales), *4*, *5*
Scales of analysis, 3, 4, 11, 12
Schengen Accord, **184**
Schistosomiasis, 286, 307, 309
Schleswig-Holstein, North European Plain, *151*
Scotland, moist midlatitude climate zone, *157*
Scott-Amundsen South Pole Station, *481*
Sea level rise
  Antarctica and, *481*
  Bangladesh, 326
  described, 23–24
  Europe, 155, *158*
  India, *25*
  Mumbai, 23, *325*, 326
  Netherlands, 155, 239
  North Africa and Southwest Asia, *239*, *241*, 272
  North America, 64, *65*
  Oceania, *449*, *455*, *456*
  South Asia, 318, 322, 324, *325*, 326, 330, 358, 359
Seasons, 14, *17*, 20
Seawater desalination, **161**, *240*
Seawater pollution, Europe, 162, *163*
Seclusion, female, **268**–269, *269*, 350, 357, 358, 440
Secondary sector, **35**
Secoya people, 110–111, *111*

Secret mass surveillance programs, 38–39
Sectors, of economy, **35**, 76
Secular states, **255**, 344
Securocrats, 214
Segupova, Marina, 195
Self-reliant development, **298**, 315
Self-sufficiency
  China, 372, 375
  globalization and, 2, 31, 34, 56
  industrial, South Asia, 338
  Oceania, 450, 466, 467
  regional, **384**
  South Asia, 338
  Southeast Asia, 444
Semana Santa procession, *145*
Semiarid/arid climates, 18, 63, *115*, *157*, 199, 236, *239*, 280, *323*, 366, 412, 452
Semiautonomous regions, 197, 216
Sen, Amartya, 36
Sensitivity, 23, 24, *25*, 56, 65, 158, 285, 325
September 11th attacks, 40, 85, 86, 217, 259, 345, 346, 467
Sequestration, carbon, **279**
Serfs, 165
Service and Food Workers Union, *41*
Service economies
  China, 390, 392, 394
  development and, 35, 38
  Europe, 170–171
  Israel, 252
  North America, 76
  Oceania, 448, 465–466, 470, 472, 474, 479
  South Asia, 337, 338, 342
  Southeast Asia, 424, 433, 442
  urbanization and, 2, 42, 43, 56
Services, as economic sector, **35**
Sex, 49
Sex tourism, 427, 438, **442**–443
Sex trafficking, 100, 225, 438
Sex workers, 225, 295, 308, **350**, 357, 437, 442, 443, 444
Seychelles, 296, *298*, 306
Shadows, rain, 20, *114*, *115*
Shakira, *246*
Shamans, 226
Shanghai, *391*, 393–394
Shantytowns, 43, 139, 141, 142, 148, 305, 306, 348
Shari'a, **246**
Sheep, 372, *458*
Sheikh Hamad bin Khalifa, 258, 270
Shell Oil Company, 32, 299
Shifting cultivation, 15, 56, **122**, 126, 135, *283*
Shi'ite (Shi'a) Muslims, 245, **246**, 259, 262, 263
Shintoism, 54
Shiraz grapes, *473*
Shishmaref, 65
Siakor, Silas, 282
Siberia
  Central Siberian Plateau, *193*, 197, 205
  defined, **197**
  West Siberian Plain, 197, 198
  western, 206, *206*, 220

Siberian huskies, *220*
Sichuan Province, 362, *363*, 371
Sierra Madre, *108*, 112, 113, *115*
Sikhism, *54*, *333*, **334**, 337, 358
Silk Road, 29, 31, 205, *205*, *206*, 209, 228, 398, 399, 401
Siloviki, 214
Silt, 14, 62, **114**, 322, 324, 326
Singhalese nationalism, 346–347
Sino-Tibetan language family, *55*, 332
Skin color, 53, *55*, 101, 145–146
Skydiving, 44
Slash and burn agriculture, 415, *416*
Slavery
  European conquest of Middle and South America, *123*, 123–124
  in Niger, 290
  Southern plantations, 71–72, *72*
  sub-Saharan Africa slave trade, 289–290, *290*
  sugar production, 32, *123*, 148, 167–168
Slavs, **205**
Sleeping sickness, 307, 309
Slovakia, 155, 172, *187*, 188
Slovenia
  Balkans, 155
  EU membership, 172, 175
  green food production, *174*, 174–175
  Predjama Castle, *164*
  women at work, *187*
  women in parliament, 89, 187, *188*
Slums
  barrios, 43, 139, 142, 148
  colonias, 139, 148
  defined, **42**–43
  Dhaka, 349
  favelas, 43, **139**, *141*, 141–142
  ghettos, 43s
  Makoko, *304*
  Mumbai, 28
  Nogales, Mexico, *121*
  Rawalpindi, Pakistan, 328
  shantytowns, 43, 139, 141, 142, 148, 305, 306, 348
  squatter settlements, 139, *140*, 140–142, 148, 471, 472
  unplanned urbanization, 142
Smart growth, 91, 93–94, 106
Smog, **64**, 66, 374, 410
Soccer, 289, 291, 440, *477*
Social democratic welfare systems, 178–179
Social forestry movement, 329, 330
Social networking
  Arab Spring and, 76, 258
  Facebook, 76, 225, 258, 262
  Twitter, 76, 82, 258, 388
  YouTube, 76, 270, 357, 478
Social safety nets, 78, **89**–90, *90*, 93, 224
Social welfare/protection systems, **178**–179, *179*
Sociocultural issues, 51–55. *See also specific countries*
Soil salinization, 239, 240, 242, 341
Solar power, 26, 118, 162, 327, 368, 370

Solidarity movement, 175–176
Somalia
　NGOs in, 42
　refugees, in Kenya, 285
Songkran festival, 438
Soufrière, 32, 32
Soufrière Hills Volcano, 109
South Africa
　demographic transition, 306
　income disparities, 127, 127
　Mandela and, 292
　population pyramid, 306–307, 308
　subtropical climates, 18
South America, **111**, 128, 132–133, 133, 134, 138. *See also* Middle and South America
South American Plate, 16, 17, 113
South Asia (world region), 316–359
　agricultural zones, 322, 340, *341*
　air pollution, 329–330
　British colonial rule, 334–337, 335, 336, 337
　caste system, **332**–334, 337, 344, 347, 354–355, 357
　climate change vulnerability, 324–327, *325*, 359
　climate zones, 321–322, 323
　conflicts and genocides, 345
　countries of, 320
　cultural diversity, 397–398
　cultural preference for boys, 358
　deforestation, 329–330
　deindustrialization, 334–335
　demographic transition, 318, 330, 351, 352, 354, 358
　development, 338–342
　earthquakes, 321, 322
　environmental issues, 322–330
　ethnic groups, 330, 331, 337
　female literacy, 353
　food production, 340–342
　free trade, 340
　GDP per capita, 339, 340
　GEI, 37
　gender imbalance, 330
　geopolitics, 326
　glacial melting, 318, 321, 322, 324, 325, 330, 358
　global economy, 358
　globalization issues, 338–342
　GNI per capita, 37, 327, 337, 338, 342, 352
　green revolution agriculture, 333, 337, 340–342
　HDI, 37
　human impact on biosphere, 328, 359
　human patterns over time, 330–337
　IMF and, 338–339
　Indian Ocean tsunami (2004), 42, 321, *405*, 409, 427, 431
　invasions into, 331
　Islam in, 247, *247*
　IT industries, 338, 348, 353
　landforms, 320–321
　language groups, 332
　microcredit and, 341, **342**, *342*, 359
　monsoons, *321*, 321–322, 324, *325*, 327, 331

Mughal Empire, 247, 331, 332, 334, 335, 338, 421
physical patterns, 320–322
political map, 320
population patterns, 351, 351–354
post-independence period, 337
power and politics, 343–347, *345*
regional conflicts, **344**–347
regional map, *316*–*317*
religions, 331–334
religious nationalism, 344, 346
Russia (and post-Soviet states) interaction with, 196
sea level rise, 318, 322, 324, 325, 326, 330, 358, 359
service economies, 337, 338, 342
sociocultural issues, 354–358
subcontinent and, 320
terms usage, 320
total fertility rates, 352
undernourishment, 30, 342
urbanization, 318, 348–350, *349*, 352, 358, 359
vegetation, 321–322
virtual water, 326, *326*
visual history, 332–333
voting, 344
water issues, 324–329
world region characteristics, 318–321
South Carolina, Charleston, 33, *59*, 94
South China Sea, 361, 365, 430
South Europe, 154, 155, *161*, 168, 179, 184, 187, *188*
South Korea
　air pollution, 373, 375
　food security, 369, 371, 372
　greenhouse gas emissions, 367
　Japan's expansions (1875–1942), 380, *380*
　Korean War, 381, 442
　monsoon rainfall patterns, 369
　wet rice cultivation, 370, 371
South Pole Station, Amundsen-Scott, 481
South Sudan
　climate change vulnerability, 25
　independence, 11, 291
　Sudan–South Sudan hostilities, *301*
Southeast Asia (world region), 404–445. *See also* Indonesia
　ASEAN and, 33, **425**–426, *426*, 428, *428*
　Asian Highway, 426–427, *427*
　climate change vulnerability, 415–419, *418*, 444
　climate zones, 410, *412*
　conflicts and genocides, 429
　Confucianism, 421, 428, 439, *439*, 443
　countries of, 409
　crony capitalism, **424**–425, 428, 444
　cultural influences, 419–421
　cultural pluralism, 438, **438**–439
　cultural preference for boys, 406, 435, 438, 444
　deforestation, 411, *414*, 415
　demographic transition, 436
　emigration, 432, 434, 436

environmental issues, 411–419
European colonization, 421
FDI in, 425
food production, 415–417
GEI, 37
gender patterns, 441–443
gender roles, 406, 433, 435, 436, 438, 441, 442, 443, 444
glacial melting, 417, *418*, 419
globalization issues, 424–428
GNI per capita, 37
green revolution agriculture, 417
HDI, 37
HIV/AIDS in, 437–438, 442, 443, 444
human impact on biosphere, 414, 444
human patterns over time, 419–422
IMF and, 424–425
independence struggles (1896–1975), 421–422, 423
Islam in, 247, *247*
Japan's expansions (1875–1942), 380, *380*
landforms, 409–410
logging, 406, 407, 408, 411, 413, 415
maid trade and, **432**, 434, *434*
Overseas Chinese in, 401, 440, 441, 444
physical patterns, 409–410
political map, 409
population patterns, 435, 435–438
power and politics, 428–431, *429*
primate cities, 432, 434
regional map, *404*–*405*
religions, 438–440, *439*
RNI, 435
service economies, 424, 433, 442
sociocultural issues, 438–443
state-aided market economies, 406, 422, 428, 443
terms usage, 408
tourism, 417, 426–428, *427*, 442–443
typhoons, 417, *418*, 419, 444
urbanization, 406, 431–434, *433*, 442, 444, 445
vegetation, 410
visual history, 422–423
volcanoes, 419
world region characteristics, 407–408
Southeast Asian Nuclear Weapons–Free Zone Treaty, 425
Southeastern Anatolia Project, 242
Southeastern Europe, 40, 155
Southern Common Market. *See* Mercosur
Southern Hemisphere, 6, 7
Southern settlements, colonial U.S., 71–72
Southwest Asia. *See* North Africa and Southwest Asia
Soviet Union (Union of Soviet Socialist Republics/USSR)
　collapse, 86, 169, 175, 194, 196, 198, 201, 204, 208, 220, 221, 223, 228

defined, **195**
Gorbachev and, 208–209, 225
perestroika, **208**
Russification process, **216**, 220
Soviet-Afghanistan war, 209, *215*, 228
Soy fields, Mato Grosso, 119
Spain
　climate change vulnerability, 25
　North African Islamic rule, 166
　Spanish-Portuguese trade routes (circa 1600), 124, *124*
　wars of independence from Spain and Portugal, 123, 124–125
　women at work, 187
　women in parliament, 188
Spanish Inquisition, 146
Spatial analysis, 4, 7
Spatial distribution, **4**
Spatial interaction, **4**
Special economic zones (SEZs), **390**, 392
Species, introduced. *See* Animals and plants
Spoke-and-hub network, 83
Sports, Oceania and, 477–478
Squatter settlements, 139, *140*, 140–142, 148, 471, 472
Sri Lanka
　civil war, 346–347
　population pyramid, 352, 353
　refugees from, 330, *345*, 347
　religious diversity, 333
St. Basil's Cathedral, 206, *206*
Stalin, Joseph, 207, 208, 209, 217
Stalin Era, 207, 208, 209, 217
Starvation. *See also* Famines; Undernourishment
　Chechen population, Stalin, 217
　emperor penguins, *481*
State of Palestine, 233. *See also* occupied Palestinian Territories
State-aided market economies, **382**, 406, 422, 428, 443
Steppes, 18, *19*, **197**, *199*, 366
Street vendors, Dhaka, 45, 57, 356, 359
Strip mining, 70, 459
Structural adjustment policies. *See* SAPs
Subcontinent, **320**
Subduction, 13
Subduction zone, **112**–113
Subregions, of Europe, 155
Sub-Saharan Africa (world region), 274–315
　agriculture, 282–286
　apartheid and, 289, **291**–292, 292, 293, 302, 306
　civil society and, 291, 293, 303, 315
　climate change vulnerability, 282–286, *285*, 315
　climate zones, 280
　conflicts and genocides, *301*
　countries of, 278
　cultural diversity, 312–313
　deforestation, 279–282
　democratization/conflict issues, 302
　demographic transition, 276, 303, 305, 306, *307*, 309, 314
　desertification, 286

diaspora, 296, 299, 312
early agriculture, industry, trade, 288–289
economic issues, 293–299, *294*
environmental issues, 279–287
ethnic groups, 291, 292, 299–300, 312–313
ethnicity and, 312–313
European colonialism, 290–291, *291*, 292
European slave trade, 289–290, *290*
FDI in, 295, 296
FGM in, 310–311, *311*
gender issues, 302, 310–311, 314
geopolitics, 300
global economy, 293, 314
globalization issues, 293–299
GNI per capita, 37
groundwater, 284, 286, 305
HDI, 37
HIV/AIDS in, 44, 307–309, *309*, 314, 315
human impact on biosphere, *281*, 315
human patterns over time, 288–291
IMF help, 295
informal economies, 295, 299
IT industries, 296
landforms, 278–279
language groups, 313, *313*
local economic development, 296–299
logging, 276, 279, *281*, 282, 287, 302, 314
mobile phone users, 297
peopling of, 288
physical patterns, 278–287
political map of, *278*
population patterns, 288, 305–309
power and politics, 291, 299–303, *302*–303
primate cities in, 303
refugees, 284, 285, 300, *301*, 302, 303
Regional Economic Communities, 296–297, *298*
regional economic development, 296–299
regional map, *274–275*
religions, *311*, 311–312
RNI, 306
SAPs, 294–295, 299
sociocultural issues, 309–313
terms usage, 278
trade organizations, *298*
undernourishment, *30*
urbanization, 276, 303–305, *304*, 314, 315
vegetation, 279
visual history, 288–289
water issues, 284–286
wildlife issues, 286–287
world region characteristics, 276–277
Subsidence, 62, 68, 242
Subsidies, EU and, **173**–174
Subsistence affluence, **467**, 479
Subsistence agriculture, 28, **282–283**, 284, 466

Subsistence economies, 29, **48**, 450
Subtropical climates, *18*, 280
Suburbanization, 91
Suburbs, **91**
Success, Gospel of, 147, 312, 313
Sudan
climate change vulnerability, *241*
fossil water, *239*, 239
guest workers in Europe, 184
oil and gas reserves, *249*
refugees, *241*, *301*
women in parliament, 258
Sudras, 333, 334, 347
Sugar production, 32, *123*, 148, 167–168
Suicide bombings, 260, 262
Sumatra, earthquake, 405, 409, 431
Summer monsoons, *17*, 20, **322**, 327, 365, 366, 367, 368, 369, *412*
Sun, angle of Earth's orientation to, *14*, *17*, 20
Sunni Muslims, 54, 217, **246**, 259, 263
Supercontinents. See also Plate tectonics
Gondwana, *16*, **450**, 451, 454
Laurasia, *16*
Pangaea, 13, *16*, 61, 278, 450
plate tectonics and, 13
Suriname, *119*
Sustainable agriculture, **31**, 118, 126, 417
Sustainable development, **36**, 38, 156, 283, 300
Swamp taro, *449*
Sweden
income disparities, *50*
population pyramid, *182*
women at work, *187*
women in parliament, *89*, *188*
Swidden cultivation, 415, *416*
Switzerland, *89*, *187*, *188*
Syria
Arab Spring and, *39*, *41*, 262–263, 277
civil war, *39*, *41*, 257, 262, 277
oil and gas reserves, *249*
women's rights, *234*
working women, *270*

Tai language family, 55
Taiga, **198**
Taiwan
air pollution, *373*, 374, 375
development, 362
earthquakes, 367
export-led growth, 382
food production, 369
globalization, 362
greenhouse gas emissions, 367
Japan's expansions (1875–1942), *380*, 380
monsoon rainfall patterns, 369
political map of East Asia, *364*
power and politics, 362
uncertain status, 381
wet rice cultivation, *370*, 371
Taj Mahal, 330, 331, 332, 359
Tajikistan
Central Asian states and, 195, *195*, 196, 201, 210, 226

civil war, 227
glacial melting, 203
women in parliament, 226
Taliban, 227, 346, 356–**357**, 358
Tamil Nadu, 327, 330
Tan Phuc vignette, 427–428
Tannur, 240
Tapirs, 111
Tariffs, 33. See also ASEAN
Taro (plant), 473
Tasmanian devil, *458*
Taylor, Charles, 282, 295, 302
Tea crops, for South Africa, *285*
Temperate climates, *18*, 63, *115*, 157, 199, 236, 280, 323, 366, *412*, 452
Temperate midlatitude climate, 63, **156**
Temperature, 16–17
Temperature-altitude zones, **114**, *116*, 117
Temporary workers, 184, 251
Ten Commandments, 245, 312
Tenochtitlán, *122*, *123*, *124*
Tent villages, 43
Teotihuacan, *122*, 148
Territorial claims, Antarctica, *481*
Terrorism. See also Islamic extremists
bin Laden, 86, 346
Islamic fundamentalists, 209, 226, 227, 233, 346, 399
nuclear weapons and, 217
Al Qaeda, 85, 86, 346
September 11th attacks, 40, 85, 86, 217, 259, 345, 346, 467
suicide bombings, 260, 262
War on Terror, 86, 87, 255, 346
Tertiary sector, **35**
Tet celebration, *438*
Texaco, in Ecuador, 110–111
Thailand
flooding, *418*
income disparities, *127*, 127
satay (food dish), *441*
slash-and-burn agriculture, 416
Southeast Asia political map, *409*
Thali, 356
Thematic concepts, 12–57. See also Development; Environmental issues; Gender; Globalization; Population patterns; Power and politics; Urbanization
Theocratic states, 209, **255**–256
Thomas, Saint, 334
Thompson Electronics, Mexicali workers, 129
Three Gorges Dam, *361*, 372, 373, 374, 375, 402
Tiananmen Square, 386, 387, 402
Tibet (Xizang)
Beijing–Tibet Railway, 399
Buddhism in, 54
in ethnic groups of China, 398
high altitude climate, *19*
occupation, by Mao's government, 378
Plateau of Tibet, 321, 360, 364, 367, 368, 369
Sino-Tibetan language family, 55, 332
term usage, 364

Tibetans, 399–400
Tibetic-Burmic languages, 55, 332
Tibetan Buddhism, 226, 399
Tierra caliente, 114, *116*, 143
Tierra del Fuego, *108*, 112, 114, *115*, 117
Tierra fria, 114, *116*, 143
Tierra helada, 114, *116*, 143
Tierra templada, 114, *116*, 143
Tigers, 329, 330, 347, 409, 411
Tigris River, 164, 235, 239, 242, 243, 244
Tijuana, *145*
Timbuktu (Tombouctou), 247, 288, 289
Timgad, *166*
Timkat festival, *312*
TivaWater jug, *306*, 306
Tombouctou. See Timbuktu
Tong, Anote, 448
Tonlé Sap Lake, *418*
Tordesillas, Treaty of, 123
Tornadoes, 23, 62, 454
Tortellini, *174*
Torture, 135, 146, 147, 227, 346, 422
Total fertility rate (TFR), **46–47**, 352, 436
Tourism
Antarctica, *481*
Asian Highway, 426–427, *427*
ecotourism, **118**, 120, *120*, 122, 287, 460
Hawaii and, 467
indigenous peoples and, 112, 408
Las Vegas, 92, 97
Oceania, 460, 467–468
Pacific Northwest, 70
Palm Jumeirah, 251, *264*, 265
sex, 427, 438, **442**–443
Southeast Asia, 417, 426–428, *427*, 442–443
U.S.–Canada, 79, 87
Toyota, 32, 370, 382, 383
Trade. See also Free trade; specific countries
BRIC and, 87, 131, 134, 211–212
EU's trading partners, *173*
fair trade principles, **34**, 106
maid trade and, **432**, 434, *434*
SEZs and, **390**, 392
Silk Road, 29, 31, 205, *205*, *206*, 209, 228, 398, 399, 401
Spanish-Portuguese trade routes (circa 1600), *124*, *124*
World Trade Organization, **34**, 211, 296, 465, 468
Trade blocs, regional
ASEAN, 33, **425**–426, *426*, 428, *428*
ASEAN Economic Community, 425
formation, 33
FTAA, 77–78, 132
Mercosur (Southern Common Market), 33, **132**, 148
Middle and South America, 111, 132, 148
Regional Economic Communities, sub-Saharan Africa, 296–297, *298*
UNASUR, **132**, 148
Trade deficit, 77, 171, 172

Trade winds, **114**, *115*
Traditional Chinese medicine, 372
Trafficking, 100, *138*, **225**, 438
Trail of Tears, 73, 74
Tram, *181*
Trans-Alaska pipeline, 68, 69, 107
Trans-Amazon Highway, 117
Transportation. *See specific countries*
Trans-Siberian Railroad, 218, *221*
Trayvon Martin shooting, 103
Treasury securities, foreign holders of, *386*
Treaty of Tordesillas, 123
Tree cropping, 118
Tree hugging (*Chipko*), 329, 330
Tropical cyclones, 322, 367
Tropical forests, 38, 114, 117–118, 282, 330
Tropical humid climates, *18*, *63*, *115*, *236*, *280*, *323*, *366*, *412*, *452*
the Troubles, 177
Trudeau, Pierre, 87
Tsawwassen First Nation, 91
Tsunamis, 42, 321, **365**, 367, 369, *405*, 409, 427, 431
Tuamotu Archipelago, *462*, 468
Tulip Revolution, 214
Tuna, large, *371*
Tundra, *19*, **197**
Tunisia
  Arab Spring and, 39
  Douz, 238
  oil and gas reserves, 249
  women's rights, *234*
Tupai Island, *447*
Turkey
  Arab Spring and, 176
  Çatalhöyük site, *243*, 244
  Central Asia world region, 10–11, *11*
  continental climate zone, 236
  income disparities, 127, *127*
  Kurds and, 52, 233, 235, 257, *259*, 273
  oil and gas reserves, 249
  Ottoman Empire, 164, **247**–248, *249*, *250*, 260, 272
  Southeastern Anatolia Project, 242
  women in parliament, 89
Turkmenistan
  Central Asia world region, 10–11, *11*
  Central Asian states and, 195, *195*, 196, 201, 210, 226
  women in parliament, 226
Tutsis–Hutus conflict, 313
Tuvalu, 455, *456*, 466, *471*, 473, 480
Twitter, 76, 82, 258, 388
Two child policy, 395, 437
Two-state solution, 260
Typhoid, 305
Typhoons, 367, 372, 417, *418*, 419, 444

UAE (United Arab Emirates), *11*
  FGM in, 310–311, *311*
  greenhouse gas emissions, 22
  oil and gas reserves, 249
  women's rights, *234*
Uganda
  agriculture system, *285*
  Anglicans in, 312
  climate change vulnerability, *25*, *285*
  deforestation, 279
  European colonies, 1914, *291*
  HIV/AIDS education, 309
  Homo erectus in, 288
  primate cities, 303
  TivaWater jugs, 306, *306*
UK. *See* United Kingdom
Ukraine
  authoritarianism in, 214
  Chernobyl nuclear power disaster, 160, *200*, 201, 204
  crisis in, 214, *216*
  NATO and, 214
  Orange Revolution, 214, *215*
  women in parliament, 226
Umbanda, 146, 147
UN. *See* United Nations
UNASUR, **132**, 148
Underdevelopment, Middle and South America, 124–126
Underemployment, **212**, 265, 342, 364
Undernourishment
  Congo (Kinshasa/Democratic Republic of Congo), *41*
  global map, *30*
  indigenous peoples (Australia and New Zealand), 463
  microcredit (South Asia) and, 341, **342**, *342*, 359
  Middle and South America migrant families, 142
  South Asia, *30*, 342
  Zapatista rebellion and, 138
Unemployment
  Central Valley, 60–61, *61*
  deindustrialization and, 170, 334
  Export Processing Zones and, **130**–131, 390, 423
  Sardar Sarovar controversy and, 318, *319*, 320, 336
  underemployment, **212**, 265, 342, 364
UNHDI. *See* HDI
Union of Soviet Socialist Republics. *See* Soviet Union
United Arab Emirates. *See* UAE
United Kingdom (UK)
  G8 and, 211
  income disparities, *50*
  women at work, *187*
United Nations (UN), **40**
United Nations Convention on the Law of the Sea, 460–461
United Nations Food and Agricultural Organization, 31, 305
United Nations Framework Convention on Climate Change, 24
United Nations Gender Equality Index. *See* GEI
United Nations Human Development Index. *See* HDI
United States (U.S.). *See also* NAFTA; U.S.–Mexico border
  abroad, 86
  acid rain and, **64**, 66, 79
  Afghanistan and, 85, 86–87, 106, 107, 178, 259
  agribusiness, 67, **80**–82, 99
  agriculture, 80–83, *81*
  air pollution, 64, 66, 70, 79, 91
  armed interventions, Middle and South America, 136
  Canada–U.S. relationships, 79, 79–80, 87–90
  China–U.S. economic relations, 78
  climate change vulnerability, *25*, 64–68, *65*, 107
  climate zones, 62–64, *63*
  colonial period, 145, 147
  Cuba-U.S. relations, 136, 138
  cultural diversity, 61, *98*, 98–100, *100*
  debt issues, 84, 86
  democratic system, 87–88
  ecological footprint, 12
  elderly, living arrangements, *96*
  ethnicity in, 101–103, *102*
  EU economy compared to, 171–172
  families, 103–105, *105*
  FGM and, 310–311, *311*
  food production, 80–82
  G8 and, 211
  GEI, 37
  genetic modification, **30**–31, 82
  geopolitics, 76, 83, 86–87
  global economic recession and, 78–79, *79*
  global power, challenges, 86–87
  globalization and, 75–83
  GNI per capita PPP, 37, 76, 78, 89–90, 171
  green revolution agriculture, 81, 106
  greenhouse gas emissions, 22, 64, 91, *93*, 106
  Haiti and, 86, 107
  HDI, 37
  households by type, 104–105, *105*
  human impact on biosphere, 68–70, *69*
  immigration and, *98*, 98–100, *100*
  income disparities, *50*, 78–79, 101–102, *102*, 127, *127*
  IT industries, 76, 78, 83
  Latinos, 60, *61*, 81, 93, 96, *96*, 98, 101, 102, *102*, 103
  multiculturalism, 52
  oil/gas resources, 88
  political freedoms, 83–84
  political map, 89
  population patterns, 94–98, *97*
  population pyramid, *95*
  power and politics, 83–87, 84–85
  regional map, 58–59
  religions, 103, *104*
  social safety net, 78, **89**–90, *90*, 93
  terms usage, 61
  transportation networks, 74, 83
  urban sprawl, 43, **70**, 87, 90, 91, 92, 93, *93*, 94, 104, 106
  urbanization, 90–94, *92*, 106, 107
  virtual water, 67
  voting, 83–84, 90
  water issues, 64, *66*, 66–68
  women at work, 80, *80*, 187
  women in parliament, 89, *188*, 226
Unplanned urbanization, 142
Uplands, Saarschleife, *151*
Ural Mountains, *192*
Uralic language family, 55
Urban core, old, *181*
Urban food gardens, 68, 305
Urban growth poles, **141**
Urban pollution, Russia and post-Soviet states, 201
Urban sprawl, 43, **70**, 87, 90, 91, 92, 93, *93*, 94, 104, 106
Urban squatters, 139, *140*, 140–142, 148, *471*, 472
Urbanization. *See also specific world regions*
  described, **42**–43, *44*–*45*, 56
  habitat loss and, *15*, 64, 68, 70, 91, *328*, 411
  push/pull phenomenon, **42**, 98, 432
  service economies and, 2, 42, 43, 56
  smart growth, 91, 93–94, 106
  suburbanization and, 91
  unplanned, 142
Urdu, 331, 332
Uruguay, 131, 132, *143*, 146
U.S. *See* United States
Ushahidi platform, 258, 277, *277*
U.S.–Mexico border
  Arizona border detention, 100
  climate change vulnerability, *65*
  maquiladoras, 120, 129, **131**, *132*, 423
  Nogales, Mexico, *121*
USSR. *See* Soviet Union
Utah, desert, *63*
Utah Valley, 81, 96
Uzbekistan
  Arab Spring and, 227
  Aral Sea issues, 201–202, *202*, *203*, 204, 227, 228
  Central Asia world region, 10–11, *11*
  Central Asian states and, 195, *195*, 196, 201, 210, 226
  economic diversification, 204

Values, cultural diversity and, 52
Vancouver, livable city, 92
Varanasi (Benares), 320, 327, *328*, 329, 357
Varna system, **333**–334, 347
Vedas, 331–332
Vegetation. *See also* Animals and plants
  Europe, 155–156
  North Africa and Southwest Asia, 235, *237*
  Oceania, *454*, 454–455
  photo interpretation and, *8*, 8
  Russia and post-Soviet states, 198
  South Asia, 321–322
  Southeast Asia, 410
  sub-Saharan Africa, 279
Veils, **269**, *269*, 271, 354
Venediktova, Galina, 194
Venezuela
  Chávez and, 131, 132, 135
  income disparities, 127, *127*

Vietnam
  Agent Orange, 422, 423, 445
  feminization of labor, 424
  fertility rates, 436
  Tan Phuc vignette, 427–428
  war, 421–422
Vignettes
  Arizona–Mexico border problems, 100
  arranged marriages, 440, 440
  banana hacienda, Costa Rica, 134
  birth control, Philippines, 437
  Central Valley unemployment, 60–61, 61
  corn production, Hunnicutt case, 82, 82
  cows incident, Bangladesh, 343, 343
  ecotourism, Puerto Misahualli, 118
  favelas, in Fortaleza, 142
  green food production, Slovenia, 174, 174–175
  Koli village, 348, 350, 350
  Mexicali workers, Thompson Electronics, 129
  nanny day in Hong Kong, 434
  norms and values, Honolulu, 52
  Radio Sahar, 357
  Secoya people, 110–111, 111
  sex workers, Bangkok, 443
  Siakor and Liberian logging, 282
  Soufrière, 32, 32
  Tsawwassen First Nation, 91
  Wang Youde, 367, 369
  workers in global economy, 32, 32–33
Virtual water, 26, 26–27, 28, 67, 162, 240, 326, 326, 455
Visual history
  East Asia, 378–379
  Europe, 166–167
  Middle and South America, 122–123
  North Africa and Southwest Asia, 248–249
  North America, 72–73
  Oceania, 464–465
  Russia and post-Soviet states, 206–207
  South Asia, 332–333
  Southeast Asia, 422–423
  sub-Saharan Africa, 288–289
Volcanoes
  Japan and, 365
  Mt. Fuji, 361, 365, 379
  Ring of Fire and, 13, 17, 365, 367, 409, 451
  Soufrière Hills Volcano, 109
  Southeast Asia, 409
  tectonic plates and, 13, 17, 113, 197
Volga River, 192
Vologada region, 199
Voodoo, 146, 312

Voting
  Afghanistan, women, 85
  South Asia, 344
  U.S. and Canada, 83–84, 90

Wadi El Mellah, 231
Wadi Mur, 241
wadis, 235
Wages, living, **34**, 56
Waitangi Day, 477
Waitangi Treaty, 476, 477
Walmart, 34, 76, 77, 77
Wang Youde, 367, 369
War on Drugs, 136
War on Terror, 86, 87, 255, 346
Wars. See specific wars
Water
  bottled water, 27, 68
  fossil, 67, 239, 239, 240, 242, 286
  marketization, 68, 120
  ownership, 27, 28, 329
  privatization, 27, 28, 329
  rainwater harvesting, 120, 319
  virtual water, 26, **26**–27, 28, 67, 162, 240, 326, 326, 455
Water footprints, 26–27, 27
Water pollution and scarcity, 26–28. See also specific world regions
Wealth. See also Income disparities
  mercantilism and, **125**, 127, 166, 167, 293
  oligarchs and, **210**, 212
Weather, 14
Weathering, **13**, 14, 20
Wegener, Alfred, 13
Welfare states, 152, **168**–169, 175, 179–180, 189
West Bank, 233, 259–262, 261, 263, 265
West Bank barrier, **262**, 262
West Europe, 154, 155, 156, 170, 179, 184, 187, 188
West Germany, 169, 170, 175
West Papua, 54, 418, 429, 431
West Siberian Plain, 197, 198
West Virginia, Kayford Mountain, 69
Western Europe, countries, 155
Western Hemisphere, 6, 7
Western Siberia, 206, 206, 220
Wet rice cultivation, **370**–371, 415
Wetland loss, Louisiana, 58, 62
Where/why questions, geography, 2–3
"Why Do They Hate Us?" (Eltahawy), 233–234
Why/where questions, geography, 2–3
Wildebeests, 285
Wildlife issues, sub-Saharan Africa, 286–287
Winds, trade, **114**, 115
Winter monsoons, 321, **322**, 324, 365, 367, 452

Women
  dowry customs, 353, 355, 358
  female infanticide, 353, 396
  female seclusion, **268**–269, 269, 350, 357, 358, 440
  FGM and, 310–311, 311
  grandmother hypothesis, 50, 51
  maid trade and, **432**, 434, 434
  maquiladoras and, 120, 129, **131**, 132, 423
  marianismo and, **146**, 147
  purdah practice, 350, **354**, 355, 356, 358
  sex trafficking, 100, 225, 438
  Taliban and, 356–357
  veiling practice, **269**, 269, 271, 354
Women at work
  Canada, 80, 80
  double day, 50, **187**, 224
  Europe, 186–189, 187
  feminization of labor, **423**–424, 424
  North Africa and Southwest Asia, 270
  in parliaments, 89, 188, 226, 258
  prostitution, 225, 289, 308, 377, 396, 443
  sex workers, 225, 295, 308, **350**, 357, 437, 442, 443, 444
  U.S., 80, 80
Women's status
  Afghanistan, 85, 90, 358
  Arab Spring and, 233–234, 234, 246, 258, 268, 269, 270
  human rights, 234
  North Africa and Southwest Asia, 269–271
  Russia and post-Soviet states, 224–226
Work. See also Unemployment
  child labor, 289, 350, 352
  in global economy (vignette), 32, 32–33
  guest workers, **184**, 265, 266, 267, 395
  living wages, **34**, 56
  outsourcing, 78, 83, **340**, 424
  temporary workers, 184, 251
World Bank
  China and, 384
  Okonjo-Iweala and, 302
  PRSPs, 34, 131, 295
  SAPs, 129, 294–295
  Sardar Sarovar Dam and, 319
  Taiwan and, 381
  Three Gorges Dam and, 372, 374
World cities, 166, 401
World Cup, 289, 291, 478
World regional approach, 4

World regional scale, 11
World regions, 1, 4, 11–12. See also East Asia; Europe; Middle and South America; North Africa and Southwest Asia; North America; Oceania; Russia and post-Soviet states
World Trade Organization (WTO), **34**, 211, 296, 465, 468
World War I, 103, 168, 169, 207, 248, 250, 290
World War II
  Holocaust and, **169**, 248, 254
  Nazis, 169, 175, 208, 260
  North Africa and Southwest Asia and, 248
  Russia and post-Soviet states and, 208–209
World Wildlife Fund, 40, 411, 413
WTO. See World Trade Organization
WuDunn, Sheryl, 350

Xian, 51, 399
Xinjiang Uygur Autonomous Region, 365, 378, 398
Xizang. See Tibet
Xolo, Mexican hairless dog, 117

Yachana Lodge, 118
Yangtze River. See Chang Jiang
Yanukovych, Viktor, 214
Yarmulkes, 186
Yasukuni Shrine, 387
Yellow River. See Huang He
Yemen, 11
  FGM in, 310–311, 311
  refugees in, 241, 272
  Wadi Mur, 241
  Women's rights, 234
  working women, 270
Yokohama, 397, 402, 461
Yom Kippur War, 249, 260
YouTube, 76, 270, 357, 478
Yucatán, Mexico, 18, 109
Yudhoyono, Susilo Bambang, 415
Yunus, Muhammad, 341, 342
Yurts (gers), 205, 381

Zambia, 287, 294, 296
Zapata, Emiliano, 138
Zapatistas, 138
Zharkeshov, Yernar, 218, 220
Zimbabwe
  Great Zimbabwe Empire, 288, 288
  Mugabe and, 295, 302
Zimmerman, George, 103
Zionists, **260**
Zones. See Climate zones; specific zones
Zoroastrians, 224, 246, 247

150°W
120°W
90°W
60°W
30°W
0°
ARCTIC OCEAN
Ellesmere Island
Prime Meridian
Greenland
Beaufort Sea
Baffin Bay
Baffin Island
Greenland Sea
Mt. McKinley (Denali) 20,320 ft.
Yukon
Mackenzie
Great Bear Lake
Davis Strait
Iceland
60°N
Bering Sea
Aleutian Islands
Gulf of Alaska
Peace
Great Slave Lake
ROCKY MOUNTAINS
CANADIAN SHIELD
Hudson Bay
Labrador Sea
British Isles
Ireland
Saskatchewan
Lake Winnipeg
Newfoundland
Columbia
Missouri
Great Lakes
St. Lawrence
Great Salt Lake
GREAT PLAINS
Appalachian Mts.
Iberian Peninsula
Ohio
Azores
Death Valley -282 ft.
Colorado
Arkansas
Mississippi
NORTH PACIFIC OCEAN
NORTH ATLANTIC OCEAN
Atlas Mts.
30°N
Rio Grande
Baja California
Canary Islands
Gulf of Mexico
Greater Antilles
S A H
Hawaiian Islands
Cuba
Lesser Antilles
Hispaniola
Caribbean Sea
S A
Niger
Orinoco
0° Equator
Galápagos Is.
Gulf of Guinea
Amazon
Marañón
Solimões
AMAZON BASIN
Madeira
São Francisco
Polynesia
ANDES
BRAZILIAN HIGHLANDS
Paraná
SOUTH ATLANTIC OCEAN
Easter Island
30°S
Pampas
Aconcagua 22,834 ft.
Land Elevations
meters feet
4877 16,000
3353 11,000
2134 7000
914 3000
305 1000
152 500
0 0
Ocean Depths
meters feet
0 0
300 984
3500 11,483
5000 16,404
Río de la Plata
SOUTH PACIFIC OCEAN
Valdes Peninsula -131 ft.
Patagonia
Falkland Islands
Tierra del Fuego
60°S
Scale at Equator
mi 0 500 1000 1500 2000
km 0 500 1000 1500 2000
Weddell Sea
Vinson Massif 16,066 ft.